Annotated Bibliography of Quaternary Vertebrates of Northern North America

with Radiocarbon Dates

Annotated Bibliography of Quaternary Vertebrates of Northern North America

with Radiocarbon Dates

EDITED BY C.R. HARINGTON

Annotated by D. Naughton, C.R. Harington, A. Dalby, M. Rose, and J. Dawson

Radiocarbon-Date Table compiled by G. Harington

Published by the University of Toronto Press in collaboration with the Canadian Museum of Nature

Toronto Buffalo London
Printed in Canada

ISBN 0-8020-4817-X

Printed on acid-free paper

Published in collaboration with the Canadian Museum of Nature

Canadian Museum of NATURE
Musée canadien de la NATURE

National Library of Canada Cataloguing in Publication

Main entry under title:
Annotated bibliography of Quaternary vertebrates of northern North America : with radiocarbon dates / edited by C.R. Harington ; annotated by D. Naughton ... [et al.].

ISBN 0-8020-4817-X

1. Vertebrates, Fossil – Canada – Bibliography. 2. Vertebrates, Fossil – Alaska – Bibliography. 3. Vertebrates, Fossil – Greenland – Bibliography. 4. Paleontology – Quaternary – Bibliography.
I. Harington, C.R., 1933–

Z6033.V45A55 2003 016.566′0971 C2002-903092-7

University of Toronto Press acknowledges the financial assistance to its publishing program of the Canada Council for the Arts and the Ontario Arts Council.

University of Toronto Press acknowledges the financial support for its publishing activities of the Government of Canada through the Book Publishing Industry Development Program (BPIDP).

To friends and colleagues, past and present,
of the Geological Survey of Canada and
Archaeological Survey of Canada

CONTENTS

PREFACE

The aim of this book is to provide basic information (e.g., concerning species, locality, geological age, stratigraphy) on more than 1000 sources (ranging from books, monographs, and theses to shorter scientific and popular publications, as well as newspaper articles) produced between 1260 and 2000.* These sources deal with remains of fishes, amphibians, reptiles, birds, and (mainly) mammals that lived from about 2 million to 5000 years ago in Canada, Alaska, and Greenland (Kalaallit Nunaat, Grønland). Ole Bennike's (1997) masterful review of Quaternary vertebrates from Greenland has been particularly useful where that country is concerned. In some cases annotations of publications that focus on taxonomic (e.g., morphological, serological, DNA) and zoogeographic evidence relating to Quaternary vertebrates of the region are included.

In addition to using square brackets to enclose information in parentheses, I have sometimes employed them to interject recent comments about the validity of statements made in the annotations and to note other references that may be pertinent to the subject at hand. Where taxa are mentioned in the annotations, I usually include common name followed in parentheses by my view of the most appropriate scientific (Latin) name – where necessary indicating its synonymy with terms used in the original reference [e.g., steppe bison (*Bison priscus* = *Bison crassicornis*)]. Five indexes (Scientific Names; Common Names; Localities and Stratigraphic Terms; Personal Names and Institutions; and Subjects) are provided to make the contents of the book more accessible. To obtain maximum value, the reader should use these indexes and their components in conjunction – cross-checking items sought where possible.

A significant feature of the book is the extensive radiocarbon date table, which includes more than 2500 direct dates, mainly on bone, of Quaternary vertebrates listed by genus, species, date, laboratory number, material dated, location, references, and remarks. The main reason for including this table is to try to establish, within the range of radiocarbon dating, 'who was where, when.' Among other things, it will allow a much clearer view of the nature and timing of extinctions occurring near the end of the Pleistocene in northern North America.

This is the first attempt to produce such an annotated bibliography. It is hoped that it will be useful to North American Quaternary scientists specializing in paleobiology (especially vertebrate paleontology), paleoecology, archaeology, and geochronology, as well as to zoologists and taxonomists. It may also serve as a useful introduction to the literature for those entering this field of study and as a handy perspective for professionals, scientific writers, journalists, and the interested public. Consequently, oversights are inevitable, and I would greatly appreciate receiving comments that would help to improve this work. Although the data are computerized so that they may be readily corrected and updated, such a bibliography is never complete. A complementary database – FAUNMAP (1994) – documents Quaternary mammal distribution in the 48 contiguous states of the United States for the last 40,000 years. Further, it is worth noting that the *Bibliography of Fossil Vertebrates* (published by the Society of Vertebrate Paleontology) provides world-

* Reports in 'The Kings Mirror' (Anonymous, about 1260) indicating that wolves formerly occupied West Greenland are the earliest known for Greenland and northern North America. The earliest published record for Canada seems to be Kalm's (1749) concerning the skeleton of a large whale from Quebec. Otto van Kotzebue's (1821) report of a mammoth cranium from Eschscholtz Bay, Alaska, is one of the earliest from Alaska.

wide coverage of references on vertebrates, beginning with the earliest Paleozoic and extending to the Pleistocene. Valuable background information on most species and some faunas mentioned in the text is given in Kurtén and Anderson, *Pleistocene Mammals of North America* (1980).

This work had its origin in an annotated bibliography of 139 items appended to the paper 'Pleistocene Mammal Research in Canada' (Harington 1963) presented at the first Canadian Society of Zoologists Meeting in Ottawa. This particular project ('Canadian Pleistocene Vertebrates: Bibliographical Study') was initiated in 1969, when a student assistant, Lila Jeletzky, began work on a straight bibliography. The focus shifted from Canada to northern North America in order to provide a broader, more natural coverage, and the first annotations were produced by another student assistant, Jennifer Dawson, in 1984 and, on contract, by Maureen Rose in 1985. The source of the annotations was my extensive reprint file on the topic. The project was brought to preliminary fruition in 1993 with the help of Andrew Dalby, a third-year geology student from the University of Waterloo, who produced a 94-page computerized 'Annotated Bibliography of Pleistocene and Early Holocene Vertebrates of Canada and Alaska' with 516 annotations. This book has resulted from: the earlier annotations; those contributed and checked by my research assistant, Donna Naughton, over the past five years; my annotations and revisions contributed since 1998, with five locality maps; the compilation of the radiocarbon-date table; and word processing of the whole by my wife, Gail.

ACKNOWLEDGMENTS

I am particularly grateful to: (1) those who helped me compile the material, especially Maureen Rose, Andrew Dalby, Donna Naughton, and Gail Harington; (2) Drs A.S. Dyke and Roger McNeely (Geological Survey of Canada, Ottawa), as well as Dr Richard E. Morlan (Archaeological Survey of Canada, Gatineau, formerly Hull), for providing access to their extensive radiocarbon databases and for much other pertinent advice that has helped to improve the coverage; (3) Dr Ole Bennike (Geological Survey of Denmark and Greenland, Copenhagen) for advising me of current, pertinent publications on Greenland; (4) Dr T.H. Heaton (University of South Dakota at Vermillion) for providing an extensive list of his radiocarbon dates on Quaternary vertebrates from caves in southeastern Alaska, and Dr Alwynne Beaudoin (Provincial Museum of Alberta, Edmonton) for refining and augmenting her radiocarbon data for use in the radiocarbon-date table; (5) Drs Holmes Semken (University of Iowa, Iowa City) and Jim I. Mead (Northern Arizona University, Flagstaff) for their constructive comments on a rough draft of this book; (6) the Canadian Museum of Nature for its support, particularly to Dory Cameron (Canadian Museum of Nature publishing) for help in arranging for the publication, as well as providing funds for computer expertise regarding the radiocarbon-date table, to Susan Hopkinson, who greatly improved the appearance of that table, and to CMN Library staff, particularly Arch Stewart and Mike Wayne; (7) Barbara Porter and Suzanne Rancourt of University of Toronto Press, as well as James Leahy, freelance editor for University of Toronto Press.

ABBREVIATIONS*

AB	Alberta
AK	Alaska
AMS	Accelerator Mass Spectrometry method of radiocarbon dating
asl	above sea level
BC	British Columbia
Bldg.	Building
BP	radiocarbon years before present (taken to be AD1950)
Bp	base pairs
CAA	Canadian Arctic Archipelago
CAN	Canada
cf.	comparable to
Ch.	Channel
Ck.	Creek(s)
cm	centimetre(s)
CMC	Canadian Museum of Civilization (Hull, Quebec)
CMN	Canadian Museum of Nature (Ottawa, Ontario)
DNA	deoxyribonucleic acid (genetic analysis of fossil material)
E	east
et al.	and others (authors)
ft	foot or feet (measurement)
^{14}C	radioactive isotope of carbon
g	gram(s)
GIS	Geographic Information System
GRE	Greenland
GSC	Geological Survey of Canada
Hwy.	Highway
I.	Island(s)
in	inch(es)
kg	kilogram(s)
L.	Lake
LGM	Last Glacial Maximum
Lb(s)	pound(s)
Loc	Locality
Lsd	Legal subdivision
m	metre(s)
Ma	million years ago
MB	Manitoba
mer.	meridian
mi	mile(s)
mm	millimetre(s)
mtDNA	mitochondrial DNA
N	north
NB	New Brunswick
NF	Newfoundland and Labrador
nov. gen.	new genus
NMC	National Museums of Canada – see CMN
NS	Nova Scotia
NT	Northwest Territories
NU	Nunavut
ON	Ontario
oz	ounce(s)
p./pp.	page/pages
PE	Prince Edward Island
Pen.	Peninsula
PMA	Provincial Museum of Alberta (Edmonton)
Pk.	Park
Pt.	Point
QC	Quebec
R.	River
rge.	range
S	south
SD	standard deviation
sec.	section
SK	Saskatchewan
sp./spp.	species (singular and plural)
twp.	township
UK	United Kingdom
USA	United States of America
W	west
yd	yard(s)
YT	Yukon Territory

*See page 331 for a list of acronyms for radiocarbon-date laboratories.

Figure 1 Partial carcass of an extinct steppe bison (*Bison priscus*) discovered in August 1951 during placer gold-mining operations on Dome Creek, near Fairbanks, Alaska, and radiocarbon dated at 31,400 +2040/ -1815 (St-1721). The greatest of Alaskan Pleistocene mammal collectors, Otto Geist, examines a fragment of hide. Photograph by T.L. Péwé.

Figure 2 Right mandible with teeth - note long, strongly ribbed cutting tooth or incisor - of a giant beaver (*Castoroides ohioensis* CMN-16587) collected by C.R. Harington in June 1968 about 6 miles downstream from Old Crow, Yukon (Porcupine Loc. 100). Late Pleistocene. Photograph by Harry Foster.

Figure 3 Bones [e.g., pelvic region (top)] of a partial skeleton of a ringed seal (*Phoca hispida*) collected by L.A. Simonarson from last (Sangamonian) interglacial clay (Pátorfik beds, Nugssuak) near Thule, Greenland. Photograph by Geert Brovad and courtesy of Dr Kim Aaris-Sørensen, Zoologisk Museum, Copenhagen, Denmark.

Locality Maps

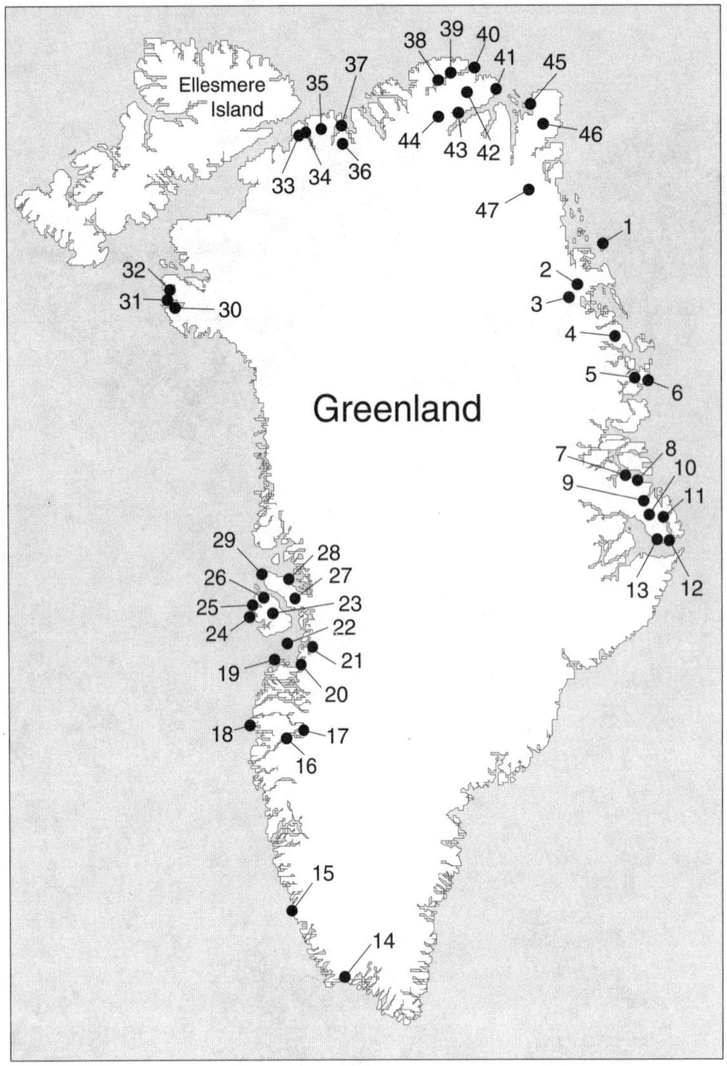

Map 1. Greenland *(Kalaallit Nunaat – showing mainly radiocarbon-dated Quaternary vertebrate localities and place names used by Bennike 1999, Fig. 1):*

1 Île de France site of large whale (T-9367)
2 Whale site [*Balaenoptera physalus* (K-6096)] near Storstrømmen
3 Storstrømmen
4 Hochstetter Forland
5 Wollaston Forland site of *Salix herbacea* (AAR-1217)
6 Wollaston Forland
7 Skeldal site of *Rangifer tarandus* (Lu-1096)
8 Jameson Land site of whale remains
9 Jameson Land site of *Canis lupus* (AAR-829)
10 Jameson Land
11 Jameson Land site of *Canis lupus* (AAR-831)
12 Hesteelv
13 Fynselv
14 Maniitsup tunua site of *Sebastes* cf. *S. marinus*
15 Frederikshåb Isblink site of *Odobenus rosmarus* (K-5979)
16 Søndre Strømfjord
17 Søndre Strømfjord site of *Rangifer tarandus* (ETH-4583)
18 Holsteinborg site of *Balaena mysticetus* (I-6235)
19 Egedesminde site of *Gadus morhua*
20 Orpigsuup tasia site of *Phoca vitulina* and *Phoca* sp.
21 Jakobshavn Isbrae site of *Odobenus rosmarus* (Ua-2350)
22 Disko Bugt
23 Disko (island)
24 Disko site of large whales (K-5969, K-3161)
25 Disko site of ?*Balaena mysticetus* (I-16356, I-16366)
26 Disko site of ?*Balaena mysticetus* (Ua-1785)
27 Nuussuaq
28 Pattorfik
29 Nuussuaq site of large whale (I-16414)
30 Thule site of *Balaena mysticetus* (W-48)
31 Thule (Saunders Ø) site of large whale (GSC-2257)
32 Thule (Iterdlak) site of *Balaena mysticetus* (Lu-3661)
33 Hall Land
34 Hall Land site of *Phoca hispida* (K-4687)
35 Nyboe Land
36 Warming Land
37 Warming Land site of *Salix* sp. (Ua-948)
38 Peary Land site of *Canis lupus* (Ua-1346)
39 Peary Land site of large whale (Lu-3658)
40 Peary Land site of *Salix arctica* (AAR-1122)
41 Peary Land (Kap København) site of *Rangifer tarandus* (Ua-1112)
42 Peary Land
43 Peary Land (Jørgen Brønlund Fjord) sites of *Canis lupus* (AAR-827), *Ursus maritimus* (AAR-1357), *Odobenus rosmarus* (K-5980, Lu-3660), *Rangifer tarandus* (K-3865), large whales (Lu-3658, 3659)
44 Peary Land site of *Canis lupus* (AAR-830)
45 Kronprins Christian Land (Station Nord) site of ?*Balaena mysticetus* (K-5977)
46 Kronprins Christian Land
47 Kronprins Christian Land site of *Canis lupus* (Ua-950)

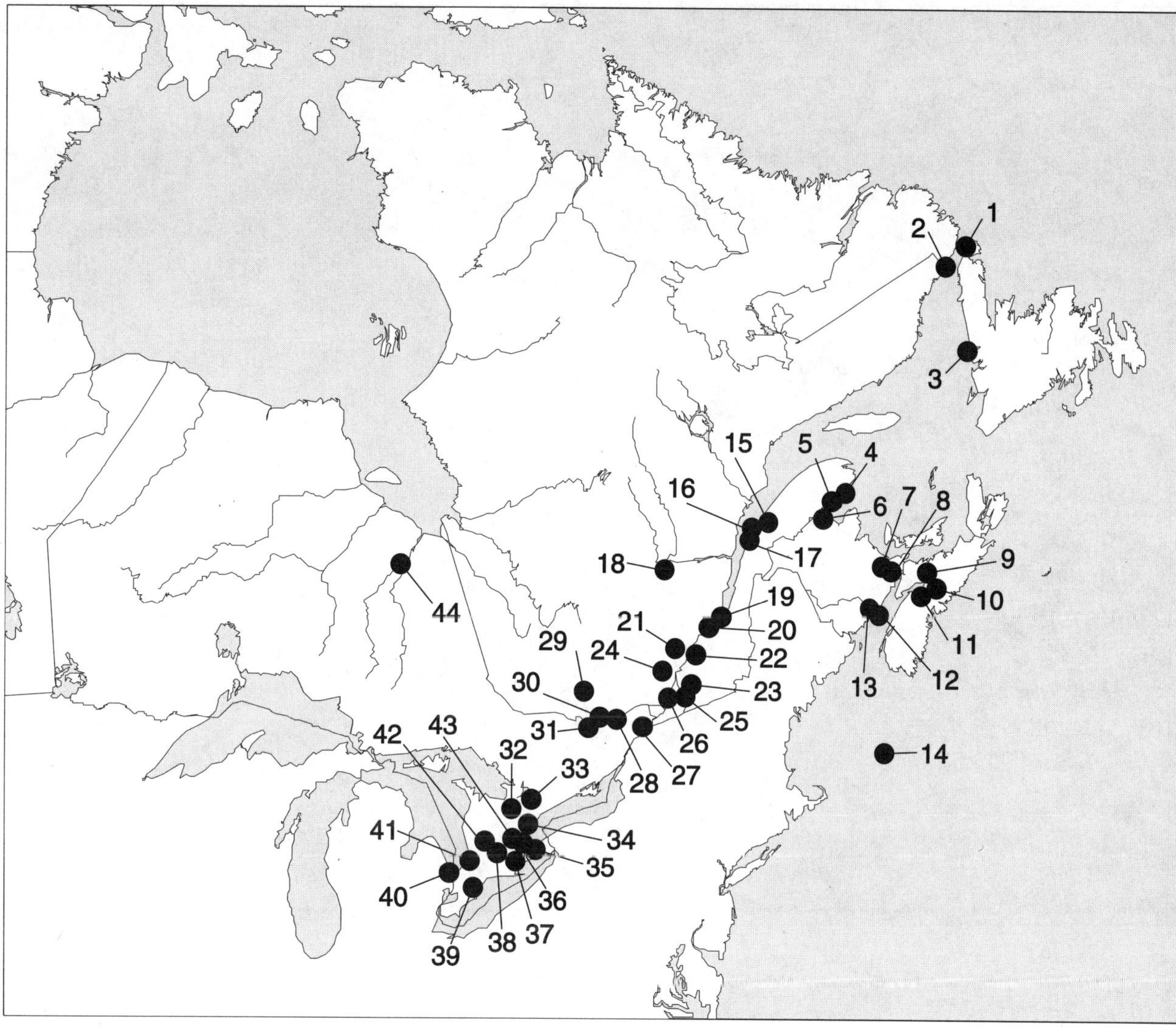

Map 2. Eastern Canada

1 Cook's Harbour
2 Forteau
3 Trout River
4 Anse-aux-Gascons
5 St-Elzéar-de-Bonaventure Cave
6 Jacquet River
7 Moncton
8 Hillsborough
9 Debert site
10 East Milford
11 Miller Creek
12 Black River
13 Fairville
14 Georges Bank
15 La Redemption Caves (Trou Otis, Spéos-de-la-Fée)
16 Rimouski (Plano site)
17 Bic
18 Pointe-de-Chambord (Lac St-Jean)
19 La Durantaye
20 St-Nicolas
21 Shawinigan
22 Daveluyville
23 Mont-St-Hilaire
24 St-Félix-de-Valois
25 St-Césaire
26 Montreal area (Montreal East, etc.)
27 Cornwall
28 Ottawa area (Green Creek, Besserer's Wharf, Besserer's Grove, Ottawa East, Jock River, Nepean, Foster's Sandpit, etc.)
29 Laflèche Cave
30 Mine Cave, Breckenridge and Eardley
31 White Lake, Pakenham
32 Alliston site
33 Udora site
34 Toronto area (Don Brickyard, Scarborough Bluffs, Woodbridge, etc.)
35 Wellandport
36 Hamilton area (Hamilton Bay, Coote's Paradise, Burlington Bar, Burlington Heights, etc.)
37 Marburg
38 Innerkip
39 Muirkirk
40 Sarnia
41 Parkhill site
42 Rostock
43 Kelso Cave
44 Moose River Crossing

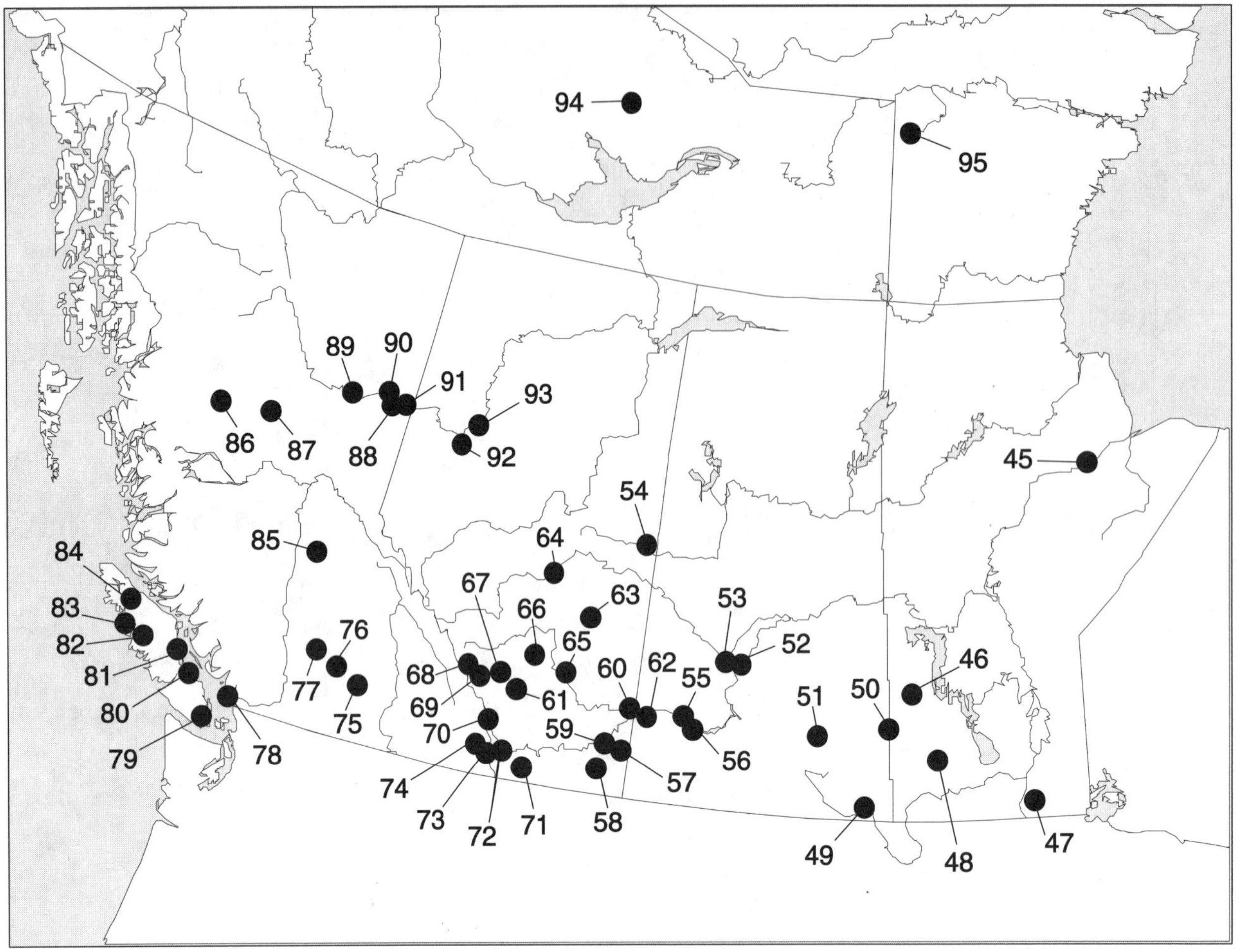

Map 3. Western Canada

45 Bird
46 Swan River
47 Grunthal Quarry
48 Minnedosa
49 Oxbow Dam site
50 Shell River
51 Fort Qu'Appelle (Bliss Gravel Pit)
52 Saskatoon area (Riddell, Saskatoon, Gowen and Duh sites, etc.)
53 Harder site
54 Cold Lake
55 Kyle
56 Wellsch Valley
57 Maser-Frisch site
58 Fletcher site
59 Medicine Hat area (Surprise, Mitchell and Island bluffs, etc.)
60 Empress and Bindloss area
61 Calgary area (Galleli Pit, Galvin Pit, Mona Lisa site, Tuscany site, etc.)
62 Heron Eden site
63 Killam
64 Edmonton area [Fort Saskatchewan, Beverly, Clover Bar Pit, Standard General Riverview Pit, Consolidated Concrete Pit 48, Consolidated Concrete Pit 45 (near Villeneuve) etc.]
65 Hand Hills
66 Three Hills (Milan site)
67 Cochrane (Griffin South, Griffin North and Clarke pits)
68 Vermilion Lakes site
69 Exshaw (Rat's Nest Cave)
70 January Cave
71 St. Mary Reservoir (Wally's Beach site)
72 J-Crossing site
73 Maple Leaf site
74 Eagle Cave
75 Lumby
76 Westwold
77 Kamloops area (Cherry Bluff, etc.)
78 Vancouver area
79 Saanich area
80 Qualicum Beach
81 Courtenay area
82 Windy Link Pot Cave
83 Port Eliza Cave
84 Resonance and Pellucidar caves
85 Quesnel Forks
86 Babine Lake
87 Chuchi Lake
88 Taylor (Ostero Pit)
89 Portage Pass
90 Charlie Lake Cave site
91 Clayhurst Crossing
92 Watino area
93 Peace River
94 Lower Carp Lake
95 Grant Lake site

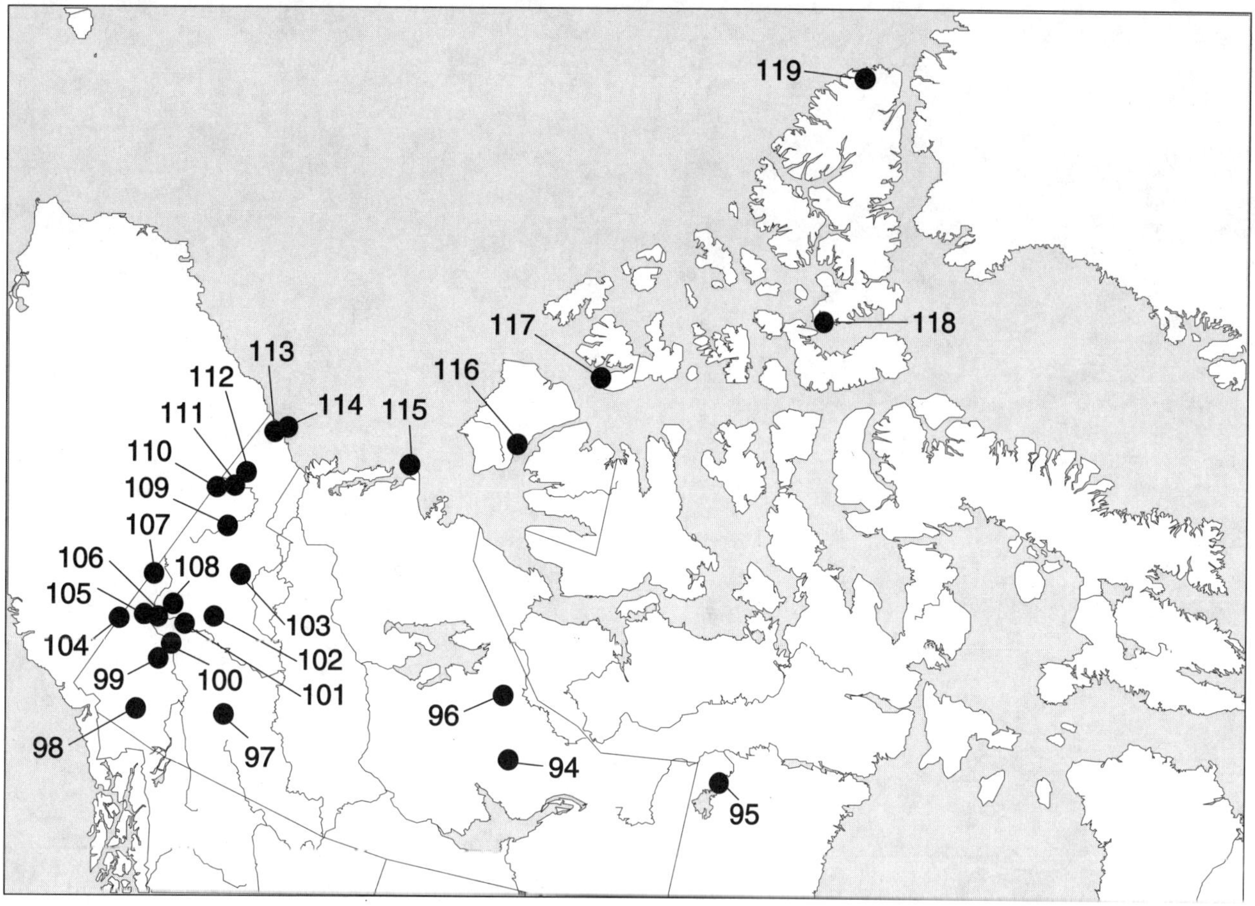

Map 4. Northern Canada

94 Lower Carp Lake
95 Grant Lake site
96 Acasta Lake
97 Ketza River
98 Canyon site
99 Big Creek area (Revenue Creek, Boliden Creek, etc.)
100 Fort Selkirk area
101 Ash Bend
102 Dublin Gulch
103 Hungry Creek
104 Scottie Creek
105 Thistle Creek
106 Brewer Creek
107 Sixtymile area
108 Dawson City area (Hunker Creek, Last Chance Creek, Dominion Creek, Sulphur Creek, Gold Run Creek, Eldorado Creek, Quartz Creek, etc.)
109 Whitestone River site
110 Bluefish Caves
111 Porcupine Locality 100
112 Old Crow Basin
113 Engigstciak
114 Herschel Island
115 Baillie Islands
116 Jesse Bay (Morgan Bluffs), Banks Island
117 Cape James Ross, Melville Island
118 Cape Storm, Ellesmere Island
119 Clements Markham Inlet, Ellesmere Island

*Note that many Holocene bowhead whale, walrus, and other marine mammal localities too numerous to map here are detailed by Dyke and others [e.g. Dyke, Hooper, Harington and Savelle (1999), Dyke and Morris (1990)]

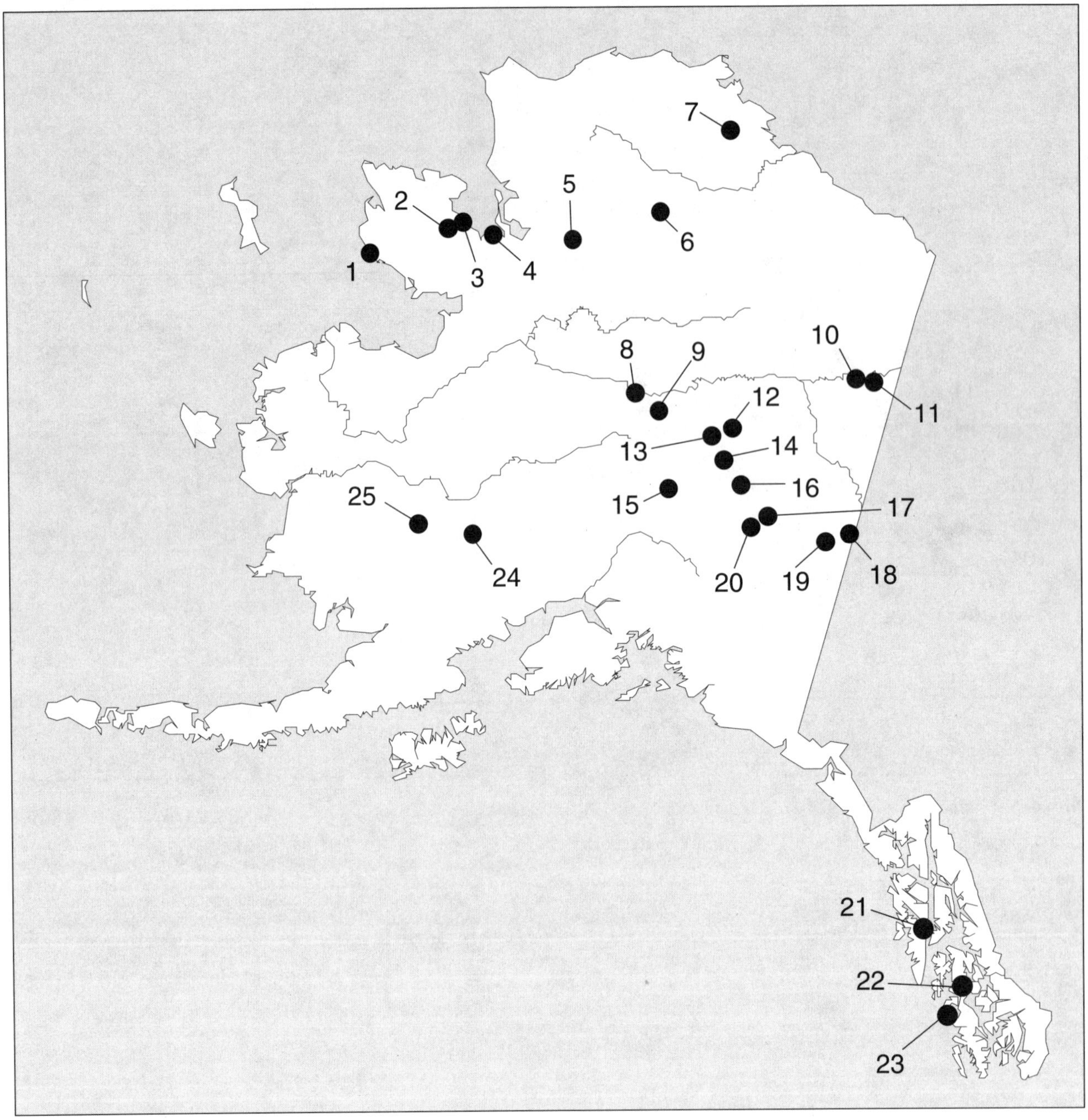

Map 5. Alaska

1 Nome
2 Trail Creek Caves
3 Cape Deceit
4 Eschscholtz Bay (Elephant Point, etc.)
5 Epiguruk Bluff
6 Mesa site
7 Teshekpuk Lake
8 Palisades
9 Tofty Mining District and Dalton Gulch
10 Porcupine River Cave 1 and Lower Rampart Caves
11 Arch Cave
12 Fairbanks area (Engineer Creek, Gold Hill, Upper Cleary Creek, Pearl Creek, Fox Creek, Cripple Creek, Eva Creek, Dawson Cut, Fairbanks Creek, Dome Creek, Ester Creek, Ready Boullion Bench, Sheep Creek, Chena Ridge, West Dawson Cut, etc.)
13 Joy Creek Ridge
14 Chugwater site
15 Dry Creek and Carlo Creek sites
16 Tanana River and Broken Mammoth sites, and Canyon Creek
17 Healy Lake
18 Jack Wade Creek
19 Lost Chicken Creek
20 Gerstle River
21 Hidden Falls site
22 Northern Prince of Wales Island cave sites (On Your Knees, El Capitan, Kushtaka, Devil's Canopy, Bumper, Blowing in the Wind caves, etc.)
23 Chuck Lake site

Annotated Bibliography of Quaternary Vertebrates of Northern North America

ANNOTATIONS

1. Ackerman, R.E. (1996a): Ilnuk site. *In*: American Beginnings: The Prehistory and Palaeoecology of Beringia (F.H. West, ed.). University of Chicago Press, Chicago and London. pp. 464–70.
This site (SLT 067) on the downslope side of USGS triangulation station ILNUK overlooks a bend in the Holitna River in the Kuskokwim drainage of southwestern Alaska (Figure 10-4). It may be as old as 4390 ± 100 BP (AA-1721, based on a radiocarbon date on calcined bone). Several dispersed calcined bone fragments apparently come from a cervid (Cervidae), and tooth enamel fragments may represent either bison (*Bison*) or wapiti (*Cervus*) according to R.D. Guthrie. The remains of large herbivores support the use of the site as a hunting camp for large game. A midsection fragment of a bone arrowhead (*Homo sapiens*) slotted to receive inset microblades (Figure 10-5:h) was recovered with other pieces of broken, burned bone from the centre of the excavation.

2. Ackerman, R.E. (1996b): Cave 1, Lime Hills. *In*: American Beginnings: The Prehistory and Palaeoecology of Beringia (F.H. West, ed.). University of Chicago Press, Chicago and London. pp. 470–7.
This cave, at the eastern extremity of the Lime Hills at an estimated 527 m asl in the Kuskokwim drainage of southwestern Alaska, extends back from the entrance as a narrow, curving corridor for 17.7 m. Dr C. Gustafson identified the following bones from the site: fish (Pisces), bird (Aves), caribou (*Rangifer tarandus*), hare (*Lepus americanus* or *Lepus othus*), Dall sheep (*Ovis dalli*), porcupine (*Erethizon dorsatum*), possible grizzly bear (cf. *Ursus arctos*), fox (*Alopex lagopus* or *Vulpes vulpes* = *Vulpes fulva*), possible moose (?*Alces alces*), arctic ground squirrel (*Spermophilus parryii* = *Citellus parryii*), and bones of at least three different species of microtine rodent. All of the species noted were part of a Late Pleistocene fauna and, except for bison (*Bison*), can be found in the region today. The microtine rodent bones may have been brought into the cave mouth by raptors and may help delineate the nature of the paleoenvironment. Three bone samples were selected for radiocarbon dating to determine the age of the cave component: (1) a caribou metapodial (a possible fleshing tool) yielded a date of 13,130 ± 180 BP (Beta-67671); (2) a caribou humerus with probable cut-marks gave a date of 15,690 ± 140 BP (Beta-67669), indicating that the cave was occupied (*Homo sapiens*) during the Late Pleistocene; and (3) a bison astragalus yielded a Mid-Wisconsinan date [27,950 ± 560 BP (Beta-67670)]. Since no stone artifacts have yet been recovered from the lower component, the question whether humans (*Homo sapiens*) or animals were responsible for the faunal remains mentioned cannot be resolved. Two artifacts from the upper component were made of bone or antler.

3. Ackerman, R.E. (ed.) (1996c): Bluefish Caves. *In*: American Beginnings: The Prehistory and Palaeoecology of Beringia (F.H. West, ed.). University of Chicago Press, Chicago and London. pp. 511–13.
Bluefish Caves (I-III; 67°09′N, 140°45′W) are 54 km S of Old Crow village, Yukon. The caves are small cavities in a Devonian limestone ridge in the Keele Mountains S of Bluefish Basin. Within Unit B (three loess layers overlying the frost-spalled and lag-covered bedrock of Unit A and overlain by the humus-rich Unit C formed after 10,000 BP and containing only Holocene fauna) are bones of: mammoth (*Mammuthus*), bison (*Bison* cf. *Bison priscus*), horse (*Equus lambei*), Dall sheep (*Ovis dalli*), caribou (*Rangifer tarandus*), moose (cf. *Alces*), wapiti (*Cervus elaphus*), saiga (*Saiga tatarica*), tundra muskox (*Ovibos moschatus*), American lion (*Panthera leo atrox*), cougar (*Felis concolor*), brown bear (*Ursus

arctos*), wolf (*Canis lupus*) and many smaller mammals, birds (Aves), and fishes (Pisces). Split long bones showing traces of whittling and with areas of high polish may be tool forms. One, a split caribou tibia that could have been a broken fleshing tool, has a date of about 24,800 BP. A mammoth bone core and reattached flake from the bottom of Unit B in Cave II indicate a well-developed bone technology (*Homo sapiens*). Separate dates for the core and flake yield an average age of 23,500 BP. Unit B formed during about 15,000 years. The pollen record indicates a shift from dry, herbaceous steppe-tundra (Unit B) to moist shrub tundra (Unit C). [See Cinq-Mars (1979, 1982, 1990) and Cinq-Mars and Morlan (1999).]

4. Ackerman, R.E., Georgina, D., Mastrogiuseppe, J., and Ruter, A. (1997): The archaeology of Lime Hills Cave, southwestern Alaska: A multidisciplinary approach. Beringian Paleoenvironments Workshop (September 20–3, Florissant, Colorado). Program and Abstracts. p. 11.

Excavation of 17 1 × 1 m units at Lime Hills Cave, southwestern Alaska, in 1993 and 1995 provided an extensive sample of organic and lithic artifacts, cave sediments, pollen, plant macrofossil, and animal remains. The early cave occupation is reflected by a date-cluster for horse (*Equus*), mammoth (*Mammuthus*), and bison (*Bison*) ranging from 38,000 to 27,000 BP, and a second date-cluster for caribou (*Rangifer tarandus*) ranging from 15,000 to 13,000 BP. The third date-cluster of 9500 to 8400 BP is associated with clear evidence for human (*Homo sapiens*) presence in the cave. Following this occupation, there is one isolated date at 3780 BP and a late cluster of dates from 500 to 200 BP. Artifacts from the 9500 to 8400 BP interval include side-slotted antler arrowheads and a side-slotted antler spearhead. The potential for mixing is evident by the presence of rodent (Rodentia) bones throughout all levels of the cave, and bear (*Ursus*) teeth at the cave entrance. Plant macrofossil and pollen data suggest a grassland transition from a Late Pleistocene steppe/herb zone, through a shrub zone, followed by the appearance of spruce. The faunal data are less continuous, with a sharp break between Pleistocene and Holocene faunas. Onset of forest conditions is reflected by the appearance of beaver (*Castor canadensis*), American marten (*Martes americana*), and red fox (*Vulpes vulpes*).

5. Ackerman, R.E., Reid, K.C., Gallison, J.D., and Roe, M.E. (1985): Archaeology of Heceta Island: A survey of 16 timber harvest units in the Tongass National Forest, southeastern Alaska. Center for Northwest Anthropology, Washington State University, Pullman. 178 pp. (pp. 110–45).

The Chuck Lake site (CRG-237), containing six localities, is situated on the SW side of Chuck Lake (Figures 3, 29) on the eastern half of Heceta Island, Alaska. Radiocarbon dates on charcoal and shell from localities 1 to 3 (which yielded bone; pp. 121–5) varied between 8220 ± 125 BP (WSU-3241) and 5140 ± 90 BP (WSU-3245). Some of the mammal bones were quite fragmentary. Table 18 provides a summary of the taxa recovered, number of specimens, minimum number of individuals, and the percentage of the total of each taxon. The inventory of bone in the middens of the Chuck Lake site indicates that fishes (Pisces), including nearly 3% Salmonidae, constitute 95% of the total sample. Birds [Aves, only the Cormorant (*Phalacrocorax auritus*) is specifically identified] were 2.6% of the sample. Clearly land mammals (Mammalia) were minor items in the human (*Homo sapiens*) diet when the site was occupied. Sea mammals [seal (*Phoca*) and northern sea lion (*Eumetopias jubatus*)] were the next most important food items. The low frequency of land mammals [e.g., wolf or dog (*Canis*), mink (*Mustela vison*), beaver (*Castor canadensis*), and deer (or caribou)] suggests that terrestrial resources were of minor importance. Subsistence emphasized exploitation of marine resources.

6. Agadzhanyan, A.K. (1984): The history of collared lemmings in the Pleistocene. *In*: Beringia in the Cenozoic Era (V.L. Kontrimavichus, ed.). Amerind Publishing Co. Pvt. Ltd., New Delhi. pp. 379–88.

Based on the morphology of the M^{1-2}, the author agrees with Zazhigin (1984) that each evolutionary stage of *Dicrostonyx* can be established and that transformation of one species into another is linear and occurs by way of replacement of the dominant morphotype. The author further concludes that *Dicrostonyx simplicior* (the direct-line progenitor to both *Dicrostonyx torquatus* and *Dicrostonyx hudsonius*) has teeth that more closely resemble those of *Dicrostonyx hudsonius*. Hence, *Dicrostonyx hudsonius* is probably a relict of the Middle Pleistocene. The evolution of collared lemmings from the Pleistocene to the Holocene shows that close affinities existed between the populations of North America and northern Asia in the Early, Middle, and Late Pleistocene.

7. Agassiz, L. (1833–43): Recherches sur les poissons fossiles 5. Neuchatel.
'In general, I may say,' the author remarks, when discussing the closeness with which Tertiary fishes agreed with recent ones, 'that I have not yet found a single species which was perfectly identical with any marine existing fish, except the little species (*Mallotus*), which is found in nodules of clay, of unknown age, in Greenland.' [See footnote 2, in Brown 1870, p. 691.]

8. Agenbroad, L.D. (1984): New World mammoth distribution. *In*: Quaternary Extinctions (P.S. Martin and R.G. Klein, eds.). University of Arizona Press, Tucson, Arizona. pp. 90–108.
Following a discussion of mammoth (*Mammuthus*) taxonomy, the locality of known fossils is reviewed. Species-by-species distribution maps are provided. Deposition and chronology are discussed and all radiocarbon dates are listed in table form. All included radiocarbon dates have been published previously. There have been mammoths in the New World since 1.7 Ma. All species became extinct approximately 10,000 years ago. About 30% of the fossil sites younger than 15,000 years ago show evidence of human (*Homo sapiens*) activity. The author concludes that it would be an extraordinary coincidence if cultural impacts had little or nothing to do with the timing of New World mammoth extinction.

9. Agenbroad, L.D. (1985): The distribution and chronology of mammoth in the New World. Acta Zoologica Fennica 170:221–4.
Published data on the discovery of mammoth remains provides a fossil population of approximately 3100 individuals from 1427 localities. Using the species identification provided by the investigators, maps of species frequency and geographic distributions were constructed for the southern mammoth (*Mammuthus meridionalis*), imperial mammoth (*Mammuthus imperator*), Columbian mammoth (*Mammuthus columbi*), and woolly mammoth (*Mammuthus primigenius*) – representing a transition from the primitive to the more progressive. Absolute dates from 75 of the sites provide a time depth of nearly 1.8 million years for mammoth in the Americas. An isochronal map of mammoth dated younger than 15,000 BP provides evidence for an extinction model.

10. Agenbroad, L.D., and Barton, B.R. (1991): North American mammoths: An annotated bibliography, 1940–1990. Scientific Papers of the Mammoth Site of Hot Springs, South Dakota, Volume 2:1–118.
Entries in this publication (450 references) on mammoth (Mammuthus) are summarized by author's surname, date, title, publication, location, number of specimens, species, context (type of deposit), date, and comment. The publication is indexed according to: author, country, province/state (Canada, Alberta, British Columbia, Manitoba, Northwest Territories, Ontario, Saskatchewan, and Yukon are listed on p. 97 and Alaska on pp. 97–8), genera/species, geological context, and age/date.

11. Ager, T.A. (1980): Late Quaternary environmental history of the Tanana Valley, Alaska. Institute of Polar Studies Report No. 54:1–117.
The paleoenvironmental framework presented in this paper and others provides a context to understand the human ecology of early man (*Homo sapiens*) in the region. Two archaeological sites are located in the Tanana River drainage basin. Both are among the oldest reliably dated sites yet discovered. The Village site at Healy Lake has occupancy dates going back as far as 10,500 or 11,000 years ago. The initial dates from the Dry Creek site suggest occupancy at least as early as 10,690 ± 250 BP. Bone fragments of extinct bison (*Bison*) and horse (*Equus*) were found in association with evidence of human occupancy at Dry Creek. The Healy region was probably shrub tundra at that time, based on the palynological evidence. The Dry Creek site, while occupied around the same time, was probably sparsely vegetated with grasses and herbs, but this conclusion is tentative and is still under examination. The present study investigates the vegetational history of the area in sufficient detail to speculate about the possible inter-relationships between vegetation change, early people, and the Late Pleistocene large mammals in interior Alaska. It suggests that there was a dramatic shift from steppe-tundra to shrub tundra about 14,000 years ago. Although this would have drastically reduced the potential food and caused the withdrawal of the large herds of grazing animals, there were pockets of south-facing slopes and glacial outwash in the Yukon-Tanana Upland in which the grassland vegetation persisted and bison and horse

could have remained to feed early humans 11,000 years ago. The Dry Creek site was likely a seasonal hunting camp where the occupants took advantage of large grazing animals concentrated by the rigorous habitats. Human impact upon the grazers was probably less critical than the ultimate loss of even those remaining marginal habitats as the warmer conditions at the end of Wisconsinan time permitted the rapid spread of forest vegetation over much of the region.

12. Ahnert, G.T. (1986): Fossil bones: Clues to locating heavy minerals in alluvial gravels. California Mining Journal, February:3–5.

Pleistocene placer deposits and mammal bones are often found together in the Yukon and Alaska, so the presence of one will often indicate the presence of the other. Discussed are remains of woolly mammoth (*Mammuthus primigenius*), Yukon horse (*Equus lambei*), steppe bison (*Bison priscus*), moose (*Alces alces*), caribou (*Rangifer tarandus*), and some rare occurrences of mastodon (*Mammut americanum*). Photographs are included.

13. Aigner, J.S., and Lively, R. (1986): Excavations at the Chugwater Site, Alaska. Archaeology 39(6):58–9, 76.

In the course of reviewing known Alaskan sites that had been occupied by early people (*Homo sapiens*), 'who lived in a land of tundra grasslands at the warming of the last glacial period more than 8400 years ago,' the authors mention: (1) Onion Portage on the Kobuk River in northwestern Alaska, where people camped at a strategic caribou (*Rangifer tarandus*) portage to interrupt the animals during their spring and fall migrations; (2) Healy Lake in central Alaska, where a related band fished (Pisces), captured fowl (Aves), and hunted the ever-dwindling herds of bison (*Bison*) in summer; (3) Carlo Creek vicinity in the Nenana Valley near Mt Denali, where a family struggled to survive a lean, late winter 8500 years ago by catching hibernating ground squirrels (*Spermophilus parryii*) in their nests. In the Tanana Valley near Fairbanks, two promontory sites are known – Campus and Chugwater – where for at least 7000 years people left discarded food, broken tools, and stone flakes. A bison- and sheep- (*Ovis dalli*) hunting camp at Dry Creek near Denali Park produced evidence of visitations between 10,000 and 11,000 years ago. Also fossilized nests of ground squirrels, digestive stones (gastroliths) from grouse (Tetraonidae), and traces of bison and sheep teeth in windblown loess were found at Dry Creek.

14. Akersten, W.A., and McDonald, H.G. (1991): *Nothrotheriops* from the Pleistocene of Oklahoma and paleogeography of the genus. Southwestern Naturalist 36(2):178–85.

In their discussion of the distribution of the small Pleistocene ground sloth (*Nothrotheriops*), the authors mention the Stalker et al. (1982) report of the genus from the Irvingtonian of Medicine Hat, Alberta, based on a second phalanx (LOB-53). However, they do not believe the specimen can be assigned to either *Nothrotheriops* or *Megalonyx*, another ground sloth of similar size during the Irvingtonian. The reported Rancholabrean occurrence of *Nothrotheriops* in the Medicine Hat area of Alberta (Stalker et al. 1982) is based on part of a humerus shaft (GIB-69-41) and a partial glenoid region of a scapula (MB-68-431). The authors believe that instead of *Nothrotheriops*, the humerus fragment represents a juvenile proboscidean (Proboscidea), and the scapula fragment is best referred to a slightly atypical equid (Equidae). Thus, there is no definite record of *Nothrotheriops* from E of the North American Cordillera during Rancholabrean time.

15. Alberta, Research Council of (1959): [Photographs of modern plains bison and fossilized Late Pleistocene bison skull]. 40th Annual Report, p. 22.

This report includes a photograph of two Alberta bison skulls: a modern plains bison (*Bison bison bison*), and a Late Pleistocene bison (*Bison*) skull.

16. Alberta, Research Council of (1962): [Mammoth tusk from near Camrose and horse bones from near Edmonton]. 43rd Annual Report, pp. 30–1.

This report includes several photographs of Pleistocene mammal remains retrieved from Saskatchewan gravels and sands in Alberta: a mammoth (*Mammuthus*) tusk from near Camrose; and horse (*Equus*) bones from near Edmonton. A stone tool (*Homo sapiens*) was found with the latter.

17. Alison, B. (1988): Early Yukoners. [Abridged version of article in Up Here: 'Life in Canada's North']. Reader's Digest 133(795):124.

Based on the work of W. Irving and his associates from

the University of Toronto, the author explains that some human (*Homo sapiens*) artifacts were found in the Old Crow Basin, Yukon, which could be as old as 150,000 BP. The earliest previous evidence of humans in North America was generally accepted to be around 12,000 BP.

18. Allen, J.A. (1903): Report on the mammals collected in northeastern Siberia by the Jesup North Pacific Expedition, with itinerary and field notes, by N.G. Buxton. Bulletin of the American Museum of Natural History 19:101–84.

The author (pp. 182–4) comments on American affinities of some eastern Siberian mammals. He mentions the close relationship between boreal mammals of the two continents – 'a relationship so intimate that it could only have been brought about by a former land bridge connecting the two areas, the existence of which in comparatively recent time, geologically speaking, is generally conceded, if not practically demonstrated.' Probably most of the more northern mammals on the two continents are slightly modified descendants of types that formerly had a circumarctic distribution, but which have slowly differentiated since the disruption of the former land connection at Bering Strait – e.g., carnivores such as bears (Ursidae), canids (Canidae), cats (Felidae), and some weasels (Mustelidae); cervids such as wapiti (*Cervus*), caribou (*Rangifer*), moose (*Alces* = *Paralces*), mountain sheep (*Ovis*), tundra muskoxen (*Ovibos*), and possibly bison (*Bison*); shrews (*Sorex*), rodents such as collared lemming (*Dicrostonyx*), brown lemming (*Lemmus*), ground squirrel (*Spermophilus* = *Citellus*); plus the hare (*Lepus*) and pika (*Ochotona*). He mentions that the 'Kamchatkan Bighorn' (*Ovis nivicola*) is more closely related to American mountain sheep than to any Asiatic species. He notes that eastern Siberia has derived some of its living mammals from boreal America relatively recently.

19. Allen, J.A. (1913): Ontogenetic and other variations in muskoxen, with a systematic review of the muskox group. Memoir of the American Museum of Natural History, New Series 1(4):103–226.

This monograph deals mainly with the taxonomy of tundra muskoxen (*Ovibos moschatus*), but it includes pertinent sections on: *Ovibos yukonensis* [now synonymized with *Ovibos moschatus*] based on an old male skull from Pleistocene silts along the Palisades of the Yukon River, some 35 mi downstream from Tanana, Alaska (pp. 201–2); and *Bootherium*, *Symbos*, and *Liops* [all now considered as helmeted muskox (*Bootherium bombifrons*) – see McDonald and Ray 1989]. *Symbos tyrrelli* from 70 ft below the surface at Lovett Gulch, Bonanza Creek near Dawson City, Yukon, is described. Two other fragmentary helmeted muskox skulls are mentioned from the Dawson City area, Yukon – one donated by J.B. Tyrrell to the National Museum (Smithsonian Institution) in Washington, DC, and the other seen by L.S. Quackenbush at Magnet Gulch near Dawson City.

20. Ami, H.M. (1884): List of fossils from Ottawa and vicinity. Ottawa Field-Naturalist 2:54–62.

The purpose of this paper is to list the species represented by the fossils in the Ottawa area. The vertebrates of Quaternary age found in Leda Clay and Saxicava Sand are: lump sucker (*Cyclopterus lumpus*), capelin (*Mallotus villosus*), and harp seal (*Phoca groenlandica*).

21. Ami, H.M. (1887): The great ice age and subsequent formations at Ottawa, Ontario. Ottawa Naturalist 1:65–74, 81–8.

This paper discusses theories of Quaternary glaciation and how it affected sediment deposition in the Ottawa area. The fossils referred to are: (1) a bone from Odell's Brickyard, Ottawa, Ontario, sent to Professor Cope for identification and presented to the writer by A.P. Low of the Geological Survey of Canada; (2) seal (*Phoca*), capelin (*Mallotus villosus*), lump sucker (*Cyclopterus lumpus*), sculpin (*Cottus ricei*), and some feather (Aves) impressions. Ami notes that the lump sucker and the sculpin are still extant in the St Lawrence River (p. 83). Nodules containing such fossils are found in marine sediments (Leda Clay) in several localities in the area, in particular, Green Creek, Ontario, and the Lièvre River, Quebec.

22. Ami, H.M. (1892): Additional notes on the geology and palaeontology of Ottawa and its environs. Ottawa Naturalist 6:73–8.

These notes outline work completed by Ami in 1890 for the Ottawa Field-Naturalist's Club in and around the Ottawa area. The Pleistocene vertebrates are: eastern chipmunk (*Tamias striatus*), found at Moose Creek, Ontario, in stratified gravels N of the C.A.R. track [see Ray 1965]; and some harp seal bones (*Phoca groenlandica*), found at

Odell's Brickyard just SE of Ottawa in Leda Clay. Ami's stated purpose is to present 'some of the best fossiliferous localities in Ottawa and Hull for the use of collectors and students in geology.'

23. Ami, H.M. (1897): Contribution to the palaeontology of the post-Pliocene deposits of the Ottawa Valley. Ottawa Naturalist 11:20–6.
This paper lists the fossils collected in the Ottawa area, Ontario and Quebec, until 1897. It includes as much detail about individual specimens as possible. Overall, it is a good indication of paleontological knowledge of the Ottawa area at the time. The vertebrates of Quaternary age include: (1) the Stewart collection (1893) of fossils from Green Creek, Russell, Ontario, with capelin (*Mallotus villosus*) and sculpin (*Cottus ricei*); (2) the Ells collection (1893) including fossils from Besserer's Grove, such as capelin (*Mallotus villosus*) and, from near Cornwall, bones of white whale (*Delphinapterus leucas*); (3) the Geological Survey of Canada collection: unspecified bird (Aves) feather impressions (1881) and the bones of a seal (*Phoca*) thought to have been a young harbour seal (*Phoca vitulina*) from near Tétreauville, Quebec (1889) [but see Harington and Sergeant 1972]; (4) a portion of the lower jaw of the harp seal (*Phoca groenlandica*) acquired by the Redpath Museum from the Pleistocene of the Ottawa Valley [see Grant 1883]; (5) at Green Creek, Geological Survey staff collected lumpfish (*Cyclopterus lumpus*), sculpin (*Cottus*), and capelin (*Mallotus villosus*); a stickleback (*Gasterosteus*) is also mentioned for this site.

24. Ami, H.M. (1898): The mastodon in western Ontario. Science 7(160):80.
This paper describes the recovery of two Ontario mastodon (*Mammut americanum*) specimens: one from Essex county, buried in sand and gravel, and another from Norfolk county, buried in clay and shell marl.

25. Ami, H.M. (1902): Lists of fossils to accompany report by Dr R.W. Ells on the city of Ottawa map. Geological Survey of Canada Annual Report (new series) 12:51–6.
This appendix lists Pleistocene fossils of Ottawa up to 1899, the majority of which are invertebrates and plants. The vertebrates referred to include: capelin (*Mallotus villosus*), collected (1) at Green Creek, Ontario, by John Stewart in 1893, (2) from a creek near Cyrville Bridge by John Stewart in 1893, and (3) from Green Creek by J.W. Dawson and members of the Geological Survey of Canada; sculpin (*Cottus ricei* = *Cottus uncinatus*), collected at Green Creek by John Stewart in 1893, and at Green Creek by J.W. Dawson and Geological Survey of Canada staff; lump sucker (*Cyclopterus lumpus*), collected by J.W. Dawson and members of the Geological Survey of Canada from Green Creek; stickleback (*Gasterosteus*), collected by J.W. Dawson and members of the Geological Survey of Canada; and seal (*Phoca*) from Odell's Brickyard, Ottawa East, collected by W. and M. Odell, R.L. Burland, H. Nelson, and H.M. Ami, 1889–96, and from Wright's Brickyard N of Tétreauville, Quebec [ringed seal *Phoca hispida* = 'probably young of harbour seal *Phoca vitulina*' – see Harington and Sergeant 1972].

26. Anderson, E. (1970): Quaternary evolution of the genus *Martes* (Carnivora, Mustelidae). Acta Zoologica Fennica 130:1–132.
This monograph on *Martes* includes morphological, environmental, and zoogeographic data bearing on the dispersal history of the group (true martens of the boreal regions, the North American fishers, and the Oriental yellow-throated martens). Pertinent here are data on the American marten (*Martes americana*), noble marten (*Martes nobilis*), diluvial marten (*Martes diluviana*), and the fisher (*Martes pennanti*). The American marten first appears in the East in the Late Pleistocene. It is thought to have crossed the Bering Isthmus in Early Wisconsinan time and migrated across Canada to the E coast of North America. Then it was isolated by the Laurentide ice sheets, evolving into the '*americana*' group. In postglacial times, it moved westward to Alaska. The '*caurina*' group seems to be a later invader, showing closer relationship to the sable (*Martes zibellina*) than to the '*americana*' group. It spread southward along the coast into the Sierra Nevada and central Rocky Mountains. The noble marten is known only from the Late Pleistocene of western United States [but see Harington 1977a and Youngman and Schueler 1991]. It probably reached America in Early Wisconsinan times and migrated southward to California, Idaho, and Wyoming, where it became isolated and died out in early postglacial times. The fisher – the largest living member of the genus – is related to the early Chinese marten (*Martes paleosinensis*) – diluvial marten (*Martes diluviana*) group. Its earliest-

known occurrence is Late Wisconsinan, but it may have come across from China in the Late Illinoian or Early Wisconsinan. Since the diluvial marten first appears in Illinoian faunas of eastern United States, it is postulated that it reached North America via the Bering Isthmus during Late Yarmouthian or Early Illinoian times from eastern Asia.

27. Anderson, E. (1973): Ferret from the Pleistocene of central Alaska. Journal of Mammalogy 54(3): 778–9.

A ferret (*Mustela eversmanni* cf. *Mustela michnoi*) specimen in the Frick Collection, American Museum of Natural History, was reported to have been unearthed 16 km W of Fairbanks, Alaska (64°50′N, 148°00′W). Pleistocene ferrets are known from the Medicine Hat area, Alberta, and the Old Crow Basin (Loc. 65), Yukon. It was compared with *Mustela nigripes*, *Mustela putorius*, *Mustela eversmanni beringiae*, and *Mustela vison*.

28. Anderson, E. (1977): Pleistocene Mustelidae (Mammalia, Carnivora) from Fairbanks, Alaska. Bulletin of the Museum of Comparative Zoology 148(1): 1–21.

Five Late Wisconsinan mustelids – ermine (*Mustela* cf. *Mustela erminea*), ferrets (*Mustela vison, Mustela eversmanni beringiae*), wolverine (*Gulo gulo*), and badger [(*Taxidea taxus*) the northernmost known example] – were found in deposits near Fairbanks, Alaska (between 64°45′N and 65°N, on the N side of Tanana River). They coexisted with: giant short-faced bear (*Arctodus simus*), grizzly bear (*Ursus arctos*), American lion (*Panthera leo atrox*), scimitar cat (*Homotherium serum*), and wolf (*Canis lupus*). Systematic descriptions of the mustelids are given, as well as extensive bone measurement tables and figures of specimens.

29. Anderson, E. (1984): Review of the small carnivores of North America during the last 3.5 million years. *In*: Contributions in Quaternary Vertebrate Paleontology: A volume in memorial to John E. Guilday (H.H. Genoways and M.R. Dawson, eds.). Carnegie Museum of Natural History Special Publication No. 8:257–66.

This publication reviews the sites that have provided fossils, age of first appearances, known radiocarbon dates, habitat, geographic distributions, known size, and food preferences. Most of this information is presented in handy table format. Pertinent to this study are the following small carnivores of Rancholabrean Land Mammal Age (Table 3): fisher (*Martes pennanti*) from Yukon; American marten (*Martes americana*) from Yukon; long-tail weasel (*Mustela frenata*) from southern Canada; ermine (*Mustela erminea*) from Alaska and Yukon; black-footed ferret (*Mustela nigripes*) from Canada; wolverine (*Gulo gulo*) from Alaska and Canada; badger (*Taxidea taxus*) from Alaska; spotted skunk (*Spilogale putorius*) from southern Canada; coyote (*Canis latrans*) from southern Canada; dhole (*Cuon alpinus*) from Beringia; red fox (*Vulpes vulpes*) from North America; arctic fox (*Alopex lagopus*) from Yukon; and lynx (*Felis lynx* = *Lynx canadensis*) from Alaska and southern Canada.

30. Anonymous (about 1260): Konúngs Skuggsjo ['The Kings Mirror']. Grønlands Historiske Mindesmaerker 3:264–392 [printed in 1845].

According to Bennike (1997, p. 904), reports in this ancient work indicate that the wolf (*Canis lupus*) formerly inhabited West Greenland.

31. Anonymous [F.C. Raben] (1826): Udtag af en dagbog holdet paa en rejse i Grönland i sommeren 1823. Tiddskrift for Naturvidenskaberne 3:271–89.

According to Bennike (1997, p. 901), the author reports capelin (*Mallotus villosus*) remains in calcareous clay nodules from West Greenland.

32. Anonymous (1900): More fossil remains: More musk ox found buried on American Gulch. Dawson Daily News, January 11.

A skeleton, thought to be a muskox (Ovibovini) or bison (*Bison*) was found in good condition on American Gulch near Dawson City, Yukon. [A check showed that this article was not found under either January 11 or November 1].

33. Anonymous (1901a): Strange find on Dominion: Animal perfectly preserved. Dawson Daily News, March 29.

A carcass, with flesh, hide, hair, and antlers ['horns'] intact, thought to be a very large moose (*Alces*), was found by prospectors William Gorbracht and a Mr Kendall while sinking a mining shaft on Veronica Creek (a tributary of Dominion Creek near Dawson City, Yukon) that entered Dominion at Claim 231 below Lower Discovery. The carcass, having hair that 'was long for that of a moose,' lay on a huge layer of ice beneath 5 ft of muck and 4.5 ft of gravel.

34. Anonymous (1901b): Huge prehistoric jawbone found. Dawson Daily News, March 29.

The mandible of a large, unknown animal was found at a depth of 40 ft on Claim 1 above Radford's Discovery on Quartz Creek near Dawson City, Yukon, by A.A. Johnson. The description indicates that a huge tooth (10 in long by 3 in wide) was rooted in each side of the jaw and that it measured 2 ft across the posterior ascending rami. [Probably a woolly mammoth (*Mammuthus primigenius*) is represented, although the article states: 'It evidently is no part of a mammoth or mastodon.']

35. Anonymous (1903a): Skull found on Gold Run. Dawson Daily News, August 12.

In April 1902, J.B. Bannerman found the skull of a small mare [clearly a Pleistocene *Equus lambei* in nearly pristine condition according to an excellent photograph showing a right-side view of articulating cranium and mandible] in frozen gravel 35 ft below the surface on Claim 34, Gold Run Creek near Dawson City, Yukon. Another horse jaw lay nearby. Bannerman planned to take the specimen to show David Starr Jordan, President of Stanford University. The skull is 26 in long, 10½ in deep, 9 in wide, and 'is not fossilized.' It was suspected to be a domesticated horse. T. Obalski, a French scientist, estimated the age of the skull at 500 years. A veterinarian, Dr Strong, estimated the individual age of the horse to be 15 years.

36. Anonymous (1903b): Mammoth bones found in Klondike gravels. Dawson Daily News, September 11.

The article mentions mammoth (*Mammuthus*) bones found in the Klondike goldfields. It is illustrated by a picture of a woolly mammoth (*Mammuthus primigenius*) skull with tusks [in the wrong sockets], a pelvic bone, humerus, molar, and other bones.

37. Anonymous (1904a): That new mastodon. Dawson Daily News, March 11.

William Foster, having visited Claim No. 5 below Amack's Discovery on Quartz Creek near Dawson City, Yukon, reported the discovery of a 5-ft-long cranium (with two teeth and lacking the mandible) with tusks about 6 ft long and 7 in across. Apparently several vertebrae and ribs were associated with the specimen, which was embedded in frozen ground some 38 ft below the surface. A graphic description is given of the release of the specimen [probably a mammoth (*Mammuthus*)] from frozen ground using steam points.

38. Anonymous (1904b): Complete mastodon. Dawson Daily News, March 14.

A complete specimen of a mastodon [probably a mammoth (*Mammuthus*)] was found at Quartz Creek near Dawson City, Yukon. Apparently the skin was attached, but the body was badly decomposed. The only other complete specimen found is from Siberia. [A check did not reveal the article under this date.]

39. Anonymous (1904c): Examine mastodon: Details are given. Dawson Daily News, March 28.

A relatively well-preserved 'mastodon' [probably a young mammoth (*Mammuthus*)] cranium with solid tusks, two molar teeth, and two or three ribs was found in place in the pay-streak of gold-bearing gravels 2 ft above bedrock on Quartz Creek near Dawson City, Yukon. [See reports of March 11 and 14 above.] The inner tips of the tusks converged.

40. Anonymous (1904d): Mastodon now en route. Yukon World [Dawson City], August 14.

R.E. Voshell and John Pearson of Quartz Creek, Yukon, were on their way to show a 'monster mastodon skeleton' [a good skull with tusks, teeth, and some postcranial bones probably of a woolly mammoth (*Mammuthus primigenius*)] at the St Louis World's Fair. One loose tooth measures 12 in in length and weighs about 12 lbs. They are under bond to return it to the Canadian Government within two years, as the specimen was exhumed on property leased, for mining privileges only, from the Dominion Government. Apparently the specimen was exhibited in Vancouver, and a Professor Odlum took accurate measurements of the different parts.

41. Anonymous (1905): [Extinct ruminant related to the muskox *Ovibos* sp. in Klondike gravels near Dawson, Yukon Territory]. Science 21(542):788.

This article refers to a paper written by W.H. Osgood which compares an extinct animal with the modern muskox (*Ovibos*) and the American bison (*Bison bison bison*). The specimen was recovered from Klondike gravel near Dawson City, Yukon.

42. Anonymous (1906): [Fossil bison and mammoth from Yukon in Carnegie Museum]. Science 24:711.
An excellently preserved steppe bison (*Bison priscus* = *Bison crassicornis*) skull and associated remains of a mammoth (*Mammuthus*) were found in gold-bearing gravel at the bottom of a mine located on Sulphur Creek, 65 km E of Dawson City, Yukon. The bison was the most perfect specimen of its kind so far discovered.

43. Anonymous (1907): Uncovered mastodon. Dawson Daily News, March 2.
Perhaps the largest mastodon [probably mammoth (*Mammuthus*)] ever uncovered in Alaska was excavated at a depth of 80 ft during the last six months near Circle City (more precisely on Alice Creek, a tributary of Mineral Creek which flows into Woodchopper Creek) by Max Lohbrunner. A tusk from the skeleton is over 9 ft 8 in long and has a circumference of 18 in. The jaws contain four teeth, each of which weighs 50 lbs. The other parts of the skeleton are poorly preserved compared with the tusk. Part of the specimen is stored in Circle City, where it will be held until it is sent to Seattle. There, it will be mounted for display at the Alaska-Yukon-Pacific Exposition, if not secured by the Smithsonian Institution. Lohbrunner stated that at Alice Creek skeletons of muskox (*Ovibos moschatus*), caribou (*Rangifer tarandus*), bear (Ursidae), and other animals have been found.

44. Anonymous (1913): Camel's bones found in this Territory. Dawson Daily News, July 18.
The Smithsonian Institution of Washington, DC, announced the discovery of camel (Camelidae) bones about 50 mi from the mouth of Old Crow River, Yukon, in the summer of 1912 by Copely Amory, Jr, a museum worker. It was believed that such finds supported the hypothesis of a land connection between Alaska and Siberia and mild climate at earlier times in the region. [See Harington 1997b.]

45. Anonymous (1915): Dawson prospector says he has found ancient mammoth. Dawson Daily News, July 22.
Two geologists, Dr Alfred Brooks and Dr Cairnes [presumably D.D. Cairnes] were to examine a sample of dried mammoth (*Mammuthus*) flesh found with a tusk about 30 ft above water level on the Yukon River by a Dawson City prospector named R.A. Fox.

46. Anonymous (1960): [Mammoth tusk from Bowness, Alberta]. Bow-Mont Banner 1(50), January 28.
A mammoth (*Mammuthus*) tusk was excavated behind Frank's Lumber Company yard 5 ft below the surface by Peter Nielson and Ed Kersey in December 1959. [The tusk was sent to Dr C. Brant, Department of Archaeology, University of Alberta, Edmonton, and seems to have been lost there.]

47. Anonymous (1964): Fossil pit yields best treasure yet. Calgary Herald. September 10.
A gravel pit [the Beverly pit], 2 mi E of Edmonton, Alberta, which has proved an unparalleled source of fossil remains of prehistoric animals, may have yielded its last treasure. Dr L.A. Bayrock, a geologist with the Research Council of Alberta said the fossilized skull of a sabretooth tiger (*Smilodon fatalis*) is the most interesting specimen recovered from the pit. Bayrock said, 'For about five years we have been receiving fossils from this particular pit which show that several species of animal existed several thousand years after they had generally been thought extinct. The pit is just about worked out now.' The skull mentioned was found several years ago by C. Morrison of Edmonton when he was working at the pit. He turned it over to the Alberta Research Council last week. [Presumably this fossil represented another mammal than the sabretooth, since it was never reported in fossils from Beverly since turned over to the Provincial Museum of Alberta.]

48. Anonymous (1967): The centennial exhibit of the Eva Brook Donly Museum of Simcoe, Ontario. Pamphlet issued on May 3, pp. 1–4.
On page 2 there is a note on the Marburg Mastodon – an exhibit loaned by the National Museum of Canada: 'The mastodon [*Mammut americanum*] tusks and skull … were found in June, 1897, on the farm of Charles Challand in Woodhouse township near Marburg, Ontario. Museum experts believe this mastodon was one of the largest ever found in North America. Dr H. Ami of the National Museum of Canada [actually the Geological Survey of Canada], who collected the remains, estimated that it

would have been five times [?] the size of an ordinary elephant. The longer tusk measures 9 ft, 5 in.'

49. Anonymous (1971a): It's a mammoth? Ottawa Citizen, June 25.

Workmen excavating an open-pit copper mine near Burns Lake, in central interior British Columbia have discovered the remains of what appears to be a mammoth (*Mammuthus*). A Noranda Mines Ltd geologist said the remains could be more than 10,000 years old. [See Harington et al. 1974.]

50. Anonymous (1971b): What about the dog that buried it? Vancouver Province. July 3, Third Section. p. 19.

Geological Survey of Canada crews carefully unearthed a huge bone, believed to be either a mammoth (*Mammuthus*) or a mastodon (*Mammut americanum*), found recently at an open-pit copper mine near Burns Lake, British Columbia. It is estimated that the bone is between 15,000 and 30,000 years old. Included in the find are ribs, tusks, teeth, and a complete thigh bone which will be sent to the National Museum in Ottawa for detailed analysis. Workers of Noranda Mines Ltd discovered the bones in the mine on Newman Peninsula at Babine Lake. This caption accompanies a photograph of an excavated limb bone with the excavating crew behind it. [See Harington et al. 1974.]

51. Anonymous (1971c): Mastodon bones arrive – 12 barrels. Ottawa Journal, July 9.

A dozen barrels of 20,000-year-old bones have arrived at the National Museum of Natural Sciences in Ottawa this week. They are the remains of an ice-age elephant, thought to be a mastodon (*Mammut americanum*). Museum scientists will not be able to say for sure until the remains are uncrated and studied. Dr H.W. Tipper of Smithers, British Columbia, reported the find to Dr C.R. Harington of Ottawa, curator of Quaternary zoology. [See Harington et al. 1974.]

52. Anonymous (1971d): A mammoth-sized jigsaw puzzle. Ottawa Citizen, July 21, p. 1.

The bones of a 'mastodon' [actually a Columbian mammoth (*Mammuthus columbi*) from near Babine Lake, British Columbia] are shown in two black and white photographs. National Museum technician Gilles Danis holds a leg bone, which comes up to his waist. [See Harington et al. 1974.]

53. Anonymous (1972a): Tusk uncovered. Victoria Times, May 18, p. 8.

A 2 ft tusk [probably belonging to a mammoth (*Mammuthus*)] was found beneath 80 ft of gravel in Central Saanich Municipal gravel pit just N of Victoria, British Columbia, on Wednesday afternoon. Workers noticed the ivory-coloured tusk as they excavated gravel with a loader. They left their find at the municipal hall where it will be examined by provincial government experts.

54. Anonymous (1972b): The mammoth of Liverpool Bay. Arctic Digest 4(6):7–8.

A mammoth (*Mammuthus*) tusk was discovered and then exhumed by a float plane pilot. It was found 160 km W of Tuktoyaktuk, Northwest Territories (69°30′N, 129°00′W). The tusk was radiocarbon dated at approximately 12,000 BP at the request of Dr W.O. Kupsch at the University of Saskatchewan. The area in which the tusk was found was covered by continental glaciers about 15,000 BP.

55. Anonymous (1974a): Babine Lake skeleton find pre-historic mammoth. Lake District News – Houston Today, April 24.

During June 1971, as stripping was in progress at Noranda's Bell copper mine on Babine Lake, the bones and tusks of a mammoth (*Mammuthus*) were uncovered. Dr H.W. Tipper of the Geological Survey of Canada was notified and spent two days at the site excavating fossils. The reporter then reviews some of the scientific findings in a recently published reprint of the *Canadian Journal of Earth Sciences*. The mammoth, which roamed the Babine Lake area 40,000 years ago, fed on grasses, shrubs, birch, and willows. It became bogged in a muddy pond and in an effort to get out, worked itself in deeper. It may have stood 11 ft 4 in and had large, curved tusks (too far gone to be sent to Ottawa). Some of the earliest exposed material at the site was taken by souvenir hunters. The article reviews some of the paleoenvironmental questions posed by the find. Two illustrations are included: a sketch of the Babine Lake Mammoth as it may have appeared in its natural shrub-tundra environment during the Olympia Interglaciation and a photograph of the mammoth scapula being excavated. [See Harington et al. 1974.]

56. Anonymous (1974b): Bison facts unearthed. Lethbridge Herald, July 7.

The skull of a large buffalo (*Bison*), estimated to be three to four years old at death, was unearthed by workers digging trenches for the new sewer system in Bellevue, Alberta. Skull measurements are: 85 cm from tip to tip of the horncores (a modern buffalo would not be more than 60 cm), 64 cm from tip of the nose to the top of the skull; the horncores were 32 cm long with a circumference of 35 cm at the base, and 10 cm across the eye sockets. The dew claws on the animal were twice the size of a modern buffalo (*Bison bison bison*). A knife (*Homo sapiens*), flaked from quartzite, recovered nearby, was said by Dr Brian Reeves of the Department of Archaeology at University of Calgary to be 'the earliest evidence yet found of man in the Crowsnest Pass.' [See Anonymous 1974c.]

57. Anonymous (1974c): Ancient bison skull. Lethbridge Herald, July 31.

The horncores of this ancient bison (*Bison*) skull found recently at Bellevue, Alberta, measures 35 in from tip to tip. A skull from a buffalo killed today might measure 24 in. A quartzite knife (*Homo sapiens*) found nearby has been termed 'the earliest evidence yet found of man in the Crowsnest Pass.' There is a black and white photograph of the front of the cranium with horncores. The nasal region is badly damaged. [See Anonymous 1974b.]

58. Anonymous (1976a): [Mammuthus primigenius or Mammut americanum: The fossil from Fort McMurray]. Great Canadian Oil Sands News 9(6).

The pelvic bones of either a woolly mammoth (*Mammuthus primigenius*) or mastodon (*Mammut americanum*) were found in gravels in a mine shaft in Fort McMurray, Alberta, in July 1976. The bones were identified by Dr P. Currie of the Provincial Museum of Alberta. The specimen included the right half of the pelvis and a fragment of the left half.

59. Anonymous (1976b): U.S. scientists to seek origins of first migrants. Calgary Herald, September 29.

The article outlines a three-year project focusing on eight sites in Alaska and one in Yukon sponsored jointly by the National Geographic Society and U.S. National Park Service. Dr William Irving of the University of Toronto said he made a rare find that summer – the jaw and tooth of a human (*Homo sapiens*). He said the jaw [from Old Crow Basin, Yukon] has not been dated but is probably more than 20,000 years old. Some of the bone tools found in the area date back more than 25,000 years. The United States part of the project is expected to concentrate at an archaeological dig known as Dry Creek, 75 mi S of Fairbanks, Alaska. The site had been discovered by Charles Holmes in 1973. Evidently human hunters were there about 11,000 years ago, and flaked stone tools and mammal (Mammalia) bone middens have been found. Perhaps earlier traces of people will be found in this narrow, ice-free corridor that existed along a natural migratory route during glacial times.

60. Anonymous (1977): Save our mammoth, Wallace demands. Vancouver Province, June 24.

Scott Wallace, provincial Conservative leader, urged that mammoth (*Mammuthus*) bones recently discovered at the B.C. Hydro Site One dam be properly protected from souvenir hunters. Bjorn Simonson, the Provincial Archaeologist, said: 'One of the interesting things about this find is that the bones are well preserved.' Accompanying photographs show: Christopher Boatman, Site One project manager, with a partial tusk and bones; and the spot where the mammoth bones were unearthed in a gravel bank.

61. Anonymous (1978a): Early man in the Yukon. North Renfrew Times, January 11.

The writer reports on a talk on the above subject by Dr Richard E. Morlan (Archaeological Survey of Canada) presented to the Deep River Science Association on January 6. Dr Morlan outlined the earlier evidence for Paleolithic man (*Homo sapiens*) in North America, which led to his unique discoveries in Old Crow, Yukon. There, fresh mammoth (*Mammuthus*) bones have been worked to produce flakes and cores, as well as polished facets, and butchering marks. Only a handful of stone tools has been found and these are impossible to date. In the summer of 1977 Dr Morlan for the first time identified evidence of bones, altered by humans, in place high up the frozen banks of Old Crow River. The new material has been dated at 35,000 BP, pushing the antiquity of people in North America back in time considerably.

62. Anonymous (1978b): Whale bone discoveries spark interest in area. Pembroke Observer, January 27.

In a talk presented to the Ottawa Valley Historical Society in Pembroke, Ontario, Clyde Kennedy mentioned several finds of whale bones at the Hanson Sand and Gravel Pit near White Lake, Ontario (about 8 mi SW of Arnprior): (1) In the summer of 1975 Kennedy examined the Hanson Pit and decided it was being excavated in a Champlain Sea shore feature. On October 10, 1975, Allan Jones found bones of the right forefin of a bowhead whale (*Balaena mysticetus*) while loading sand; (2) On September 23, 1977, Terry Bandy, while loading sand at the same pit, found pieces of a whale (*Balaena mysticetus*) rib; (3) On November 5, 1977, Clyde Kennedy found scraps of bone and the end of a bone about 1 ft wide. Finally, he and assistants completed the excavation, with the aid of trowels and brushes, of a 9-ft-long bowhead whale jaw bone. All of the bones were donated to the National Museum of Natural Sciences [now Canadian Museum of Nature]. The bones were identified by a paleontologist C.R. Harington.

63. Anonymous (1978c): Mammoth tusk major find. Ottawa Citizen, August 23.

Photograph shows Owen Beattie and Rick Percy (Department of Archeology, Simon Fraser University) examining a section of a 3 m ivory tusk uncovered in a gravel pit near Sardis in the Fraser Valley, British Columbia. The tusk is thought to be from a woolly mammoth (*Mammuthus primigenius*). The piece was displayed in the provincial museum at Victoria.

64. Anonymous (1980a): Edmonton workers unearth mammoth or mastadon [sic] tusk. Calgary Herald, April 25.

The tip of a mammoth (*Mammuthus*) or mastodon (*Mammut*) tusk, more than .5 m long, was discovered by workers excavating an underground parkade at Jasper Avenue and 103 Street in downtown Edmonton, Alberta. It was first spotted by labourer Pat McGinnis on December 7 after it fell from a backhoe scoop. A sample was sent to the University of Alberta for radiocarbon dating. Jeffrey Doran of the Provincial Museum of Alberta said the tusk was found in till identified as between 12,000 and 25,000 years old, and that the tusk could be older than the till because it was likely carried to the site by glacial action. A photograph of Doran with the tusk accompanies the article.

65. Anonymous (1980b): Mammoth tusks. Calgary Herald, June 30.

Photograph shows Jeff Doran of the Provincial Museum of Alberta examining mammoth (*Mammuthus*) tusks that were unearthed in a gravel pit E of Edmonton, Alberta. The tusks, at least 8,000 years old, are the largest found in Alberta.

66. Anonymous (1980c): Une vertèbre de baleine trouvée à Sherbrooke. La Tribune, Sherbrooke. 21 juillet. p. A-3.

Did whales visit this area 10,000 years ago? That is the question posed by the discovery of a vertebra of a large whale [like a bowhead whale (cf. *Balaena mysticetus*)] found in December 1979 near Wilson Street in Sherbrooke, Quebec, and housed in Le Musée du Séminaire de Sherbrooke. It measured 34 in long and 22 in high. The curator of the museum, Jean-Paul Mercier, confirmed that the specimen was a whale vertebra, and suggested it represented a Champlain Sea whale about 10,000 years old. The vertebra was found by two students, Isabelle Thériault and Nathalie Brousseau. The Cookshire, Quebec, specimen was mentioned [see O'Neil 1951], but was said to have been found in 1815 rather than 1951. [This lumbar vertebra yielded a radiocarbon date of 750 ± 60 BP (Beta-70094), so was definitely not of Champlain Sea age, but was more likely transported inland by people (*Homo sapiens*) – see Harington 1988].

67. Anonymous (1984a): Bone find a treasure trove. Calgary Herald, December 15.

Bear (*Ursus*) bones, from a cave on the NW coast of Vancouver Island discovered by speleologists, have yielded a Simon Fraser University radiocarbon date of 9380 ± 140 BP. Archaeologist Grant Keddie said there were complete skeletons preserved at the site – perhaps the deepest cave complex in Canada. Deer (*Odocoileus*) and bear are the only species of mammal reported from the site so far. A black and white photograph shows Keddie holding what appears to be a black bear (*Ursus americanus*) skull. [See Nagorsen et al. 1995.]

68. Anonymous (1984b): Through the ages in the Swan River valley. Historic Resources Branch. Manitoba Culture, Heritage and Recreation. 12 pp.

By roughly 9500 BC (11,500 BP), the southwestern corner of Manitoba was free of ice. The Swan River valley appears to have been ice-free by this time also, although probably it was partly inundated by glacial meltwater for the next few centuries. The oldest artifact found in the area is part of a Clovis point from near Benito. This discovery attests to the presence of Clovis people (*Homo sapiens*) in the district, but the general impression is that population density throughout southwestern Manitoba was low during this cultural period. Although Clovis people probably subsisted on a variety of plants and animals, the mammoth (*Mammuthus*) was the most favoured game species. As southwestern Manitoba was largely covered by coniferous forests then, there would have been little inducement for mammoths and their hunters to visit the region. Understandably, therefore, mammoth remains are rare in Manitoba, 'and it seems probable that those which have been found date prior to the last Ice Age.' By about 9000 BC (11,000 BP), mammoths were extinct and hunters living S of the ice sheet were turning their attention to other species – notably the now extinct varieties of bison (*Bison*), which, like their modern counterparts, occupied grassland environments. Like earlier Clovis people, Lindenmeier populations were uncommon in Manitoba. Lands above the level of Glacial Lake Agassiz remained under spruce-dominated forest, which failed to attract significant numbers of grassland-adapted bison. A warming trend, perhaps aided by forest fires, was responsible for the rapid transition from spruce forest to grassland over much of southern Manitoba, including the Swan River valley. Into this district came bison that were being driven out of their original range to the S by increasing drought. Bison bones from the Swan River valley have been dated to about 8000 BC (10,000 BP). Glacial Lake Agassiz then stood at high levels, and the bones were recovered from a beach which was being formed at this time (p. 4).

69. Anonymous (1984c): Bones bare big beaver's past. Outdoor Canada 12(8):18.

Bones of a giant beaver (*Castoroides ohioensis*), which possibly weighed as much as 218 kg, have been excavated by Dr Richard Harington of the National Museum of Natural Sciences in the Old Crow Basin, Yukon. Harington found the giant beaver remains in the same layer of sediments as those of the modern beaver (*Castor canadensis*), indicating that the two species lived at the same time. They probably thrived when what is now the Yukon was moist, with scattered lakes and tall forests. Then, about 25,000 years ago, the area became drier and cooler, giving way to grassy tracts with sagebrush. This radical change in environment probably stressed the giant beaver, which may have had difficulty moving to new and suitable habitat. There is no evidence that it shared its small relative's ability to shape its own world by building dams. [An accompanying illustration shows a restoration of the giant beaver, but the photograph of the giant beaver mandible below it is upside down!]

70. Anonymous (1989a): [Ice age mammal bones at Keno City Mining Museum]. Whitehorse Star, August 9, p. 35.

Bernie Walter, a former employee of United Keno Hill silver mines in Elsa, Yukon, has been collecting fossil bones from the Sulphur Creek area near Dawson City since 1980. He has donated his collection to the Keno City Mining Museum, where they are on public display. The specimens include skulls, leg bones, shoulder blades, lower jaws, and molars from such animals as the woolly mammoth (*Mammuthus primigenius*), the American mastodon (*Mammut americanum*), steppe bison (*Bison priscus*), Dall sheep (*Ovis dalli*), helmeted muskox (*Bootherium bombifrons*), moose (*Alces*), and caribou (*Rangifer tarandus*). A photograph of the display in the Keno City Mining Museum shows many of these bones.

71. Anonymous (1989b): Tusks dated 70,000 years. Calgary Herald, December 15, p. C7.

Pieces of mastodon (*Mammut americanum*) tusks found this week near Windsor, Nova Scotia, are thought to be 70,000 years old – the oldest remains of this elephant-like animal found in the province. The find occurred within 7 m of the spot where a 1-m-long piece of tusk was discovered in October. Both were in a block of peat-like sediment and are thought to be from the same animal. The Nova Scotia Museum in Halifax plans to resume its search for other parts of the mastodon.

72. Anonymous (1990): Yukon goldfields mined for ivory. Canadian Geographic, August/September, p. 10.

Yukon goldfields are yielding increasing quantities of ivory. An unbroken mammoth (*Mammuthus*) tusk encountered during gold mining can fetch up to $1,000 for lucky miners. Whole tusks, when dried and polished, are often sold to collectors or museums for $12,000 or more. Tusk fragments may be turned into rings, bracelets, pendants, and other jewellery by sawing them into pieces, carving them, and setting them with semi-precious stones. A Calgary company is producing 15,000 pieces of mammoth ivory jewellery a year and selling each piece for between $80 and $400. Although some tusks have been radiocarbon dated at 22,000 years old, they are deemed of little interest to paleontologists, who find other mammoth teeth and bones provide more clues to life in the Pleistocene.

73. Anonymous (1991): Sproule Associates collection. Calgary Herald, September 31.
Among ancient and modern mammal bones on display in the offices of Sproule Associates (a Calgary geological engineering firm) are mammoth (*Mammuthus*) and narwhal (*Monodon monocerus*) tusks, as well as muskox (*Ovibos moschatus*) and walrus (*Odobenus rosmarus*) skulls. [There is a colour photograph of the firm's Noel Cleland with what appear to be a modern bull muskox skull and an older-looking walrus cranium with very long tusks, like that of a male Pacific walrus.]

74. Anonymous (1992a): Mammoth bone unearthed. Calgary Herald, March 14.
A piece of mammoth (*Mammuthus*) pelvis was recovered from gravel near the Fraser River just N of Prince George, British Columbia. Brian Kulchiski of Prince George found the partly buried bone last September while hiking. He took the bone to Judith Johnson, a College of New Caledonia zoologist, who sent it to archaeologists at the Royal British Columbia Museum in Victoria. They confirmed it belonged to a mammoth.

75. Anonymous (1992b): Miners' hunt for gold yields mammoth find. Calgary Herald, September 4.
Mammoth (*Mammuthus*) bones were unearthed on a claim of Norm Ross on Gold Run Creek in the Klondike gold fields near Dawson City, Yukon, in early August while a pit was being stripped in preparation for mining. The article mentions a feeling prevalent at the time 'that miners would bulldoze bones and artifacts rather than stop work, enter the tangle of paperwork and turn over the find.' However, Ross said he saw preserving the fossils as a responsibility he was willing to accept. The recovered bones comprise a little over half of the mammoth's skeleton – the pelvis, some vertebrae, and the ribs are missing. The single molar tooth is as big as a person's head [suggesting that it is a third molar of an adult and not a 'young' individual, as stated near the beginning of the article]. The femur is 'nearly as tall as a man and weighs about 22 kg.' An accompanying photograph shows Jeff Hunston, Director of the Yukon Heritage Branch, with the femur. Hunston remarked that miners on Upper Gold Run Creek turned over a complete lower jaw of a mammoth [presumably a woolly mammoth (*Mammuthus primigenius*)] that was found earlier that summer.

76. Anonymous (1993): Egg find one of the best. Calgary Herald, August 20.
The bones of a six-year-old mammoth (*Mammuthus*) were recovered from a gypsum mine near Milford, Nova Scotia. About 5% of the skeleton was found including the left lower mandible, a humerus, an ulna, a scapula, several ribs, some vertebrae, and a number of foot bones. The specimen was discovered about 100 m from the skeleton of an adult male discovered two years earlier. [The article's title refers to a photograph published alongside the 'mammoth' text that illustrates a 70 million-year-old nest of sauropod eggs being offered for auction in London. Presumably the mammoth bones referred to are, in fact, mastodon (*Mammut americanum*) remains – see Anonymous 1994a.]

77. Anonymous (1994a): Two Nova Scotia mastodons may be 70,000 years old. Mammoth Trumpet 9(2):8.
Remains of two American mastodons (*Mammut americanum*) recently recovered from a gypsum quarry at East Milford, Nova Scotia, are of late Sangamonian (about 70,000 BP) age. About 60–70% of the bones of an adult and less than 10% of a juvenile (about seven years old) were recovered according to Dr Derek S. Davis, collections manager at the Nova Scotia Museum of Natural History in Halifax. Bone, tusk, and tooth material was obtained from both animals. Good collections of molluscs, insects, as well as amphibian (Amphibia), reptiles (Reptilia), possibly bird (Aves) and other mammal (Mammalia) bones were recovered also. The excavation from sink holes in the gypsum (National Gypsum Company) quarry continued from October 1991 to August 1993 under the direction of Robert Grantham, curator of

geology of the museum. Conservation of some specimens may take up to three years. Dr Daniel Fisher of the University of Michigan is examining the mastodon remains. Davis is analysing the molluscs.

78. Anonymous (1994b): Une baleine en Estrie. La Presse. 19 decembre.

A very small vertebra of a Champlain Sea (about 12,000–10,000 BP) whale (Cetacea) was found at Saint-Césaire, Quebec, near Granby in the Eastern Townships [=Estrie]. The discovery was made by amateur paleontologist Jean-Marc Morin, who lives in Saint-Césaire. Also on September 18, 1993, Dave Sangster, a member of the Ottawa Paleontological Association who visited the Saint-Césaire sandpit, found a whale vertebra 50 cm long. The bones were found in sediments containing marine mollusc shells.

79. Anonymous (1996): Magnificent mastodon. St John Times Globe, February 16. p. 1 (Local News).

Fossilized bones [actually casts of the original bones] of an eight-tonne American mastodon (*Mammut americanum*) that became mired in the mud near the Petitcodiac River 80,000 years ago have been assembled for display at the New Brunswick Museum, Market Square, by Research Casting International. A photograph shows Dr Randall Miller, the museum's curator of geology and paleontology, dwarfed by the mounted mastodon skeleton. This species lived in the forests of eastern Canada during the last ice age. Visitors may view the skeleton beginning April 27, when the new facility opens. [See Harington, Grant, and Mott 1993.]

80. Anonymous (1997): No bones about it – Construction has scientific value. Ottawa Citizen, May 9. p. D9.

Excavation for the foundation of the Canadian Museum of Nature's new National Heritage Building in Aylmer, Quebec, has allowed an intriguing glimpse into the world 10,000 years ago, when the entire region lay at the bottom of a frigid sea. This glimpse was provided by a single, small upper-arm bone of a ringed seal (*Phoca hispida*). Museum vertebrate paleontologist Dr Richard Harington said that the fossil indicated that the Champlain Sea seal probably lived on tiny shrimp-like animals and fishes, where large sheets of sea ice would have extended along the shores of estuaries – now the south-facing valleys of the Gatineau Hills. Called landfast ice, this is the same environment in which ringed seals, the smallest and best ice-adapted seals, raise their young in today's Arctic. An almost complete skeleton found in Hull nearly a century ago supports the conclusion that a population of ringed seals probably inhabited the region near the close of the last glaciation. Other fossils found near Ottawa reveal that marine mammals, birds (Aves), fishes (Pisces), and invertebrates such as starfish, barnacles, and clams lived in and around this part of the Champlain Sea. Bowhead whales (*Balaena mysticetus*) more than 15 m long might have been seen along these shores, as well as humpback (*Megaptera novaeangliae*), finback (*Balaenoptera physalus*), and white whales (*Delphinapterus leucas*) and harbour porpoises (*Phocoena phocoena*). Other marine mammals included harp (*Phoca groenlandica*), bearded (*Erignathus barbatus*) and harbour (*Phoca vitulina*) seals, as well as walrus (*Odobenus rosmarus*). Harington said he owed the discovery to the watchful eyes of the museum's environmental auditor, Dr Erich Haber.

81. Anonymous (No Date a): Old buffalo kill found at Taber. Calgary Herald.

At a meeting of the Archaeology Society of Alberta, Brian Reeves of the University of Alberta at Calgary said that a driver with his earth-moving machine discovered a buffalo kill site more than 7000 years old while working at a water hole SW of Taber last March. Preliminary investigations at the site were carried out by Dr R.G. Forbis, head of archaeology at the University of Alberta at Calgary. Reeves and his field party recovered tools (*Homo sapiens*) used for slaughtering and butchering the bison.

82. Anonymous (No Date b): Saskatchewan to display mammoth. Calgary Herald.

A replica of the site of the bones of a 10,000-year-old mammoth (*Mammuthus*) discovered near Kyle, Saskatchewan, will be reconstructed at the Saskatchewan Museum of Natural History said F.G. Bard, the museum's director. Samples of the soil enclosing the bones, a floor model of the site exposed on three sides, with the bones placed as they were found, will be used in the reconstruction. Within the next eight to ten days, the lower jaw containing a tooth will be placed with mammoth bones already on display. There were no signs of early humans (*Homo sapiens*) at the site.

83. Anonymous (No Date c): Picnic 'find' proves to be ancient tusk. Calgary Herald.

The remains of a giant elephant tusk were found on the shore of the Gouverneur Dam reservoir just E of Cadillac, Saskatchewan, and about 100 mi SW of Moose Jaw by W. Baxter of Cadillac and I. Isbister of nearby McCord while on a picnic with their families. Bruce McCorquodale, provincial paleontologist, says the chunk of ivory (of 7 in diameter and 4 ft long) is the basal section of a tusk that was more than 8 ft long. He says it probably belonged to a grazing mammoth (*Mammuthus*), although mastodons (*Mammut americanum*) also roamed the area at one time. The tusk is estimated to be at least 30,000 years old.

84. Anonymous (No Date d): The Hillsborough mastodon. The New Brunswick Museum (Pamphlet). 3pp.

This brochure highlights a specimen of mastodon (*Mammut americanum*) that was found in 1936 at Hillsborough, Albert County, New Brunswick. It came from blue clay 3 m below the bottom of a recently drained lake. This was the first mastodon from the maritime provinces that was nearly complete and found *in situ*. The animal lived 10,000–15,000 BP. [The specimen is now considered to be of last (Sangamonian) interglacial age – see Harington, Grant, and Mott, 1993.]

85. Antevs, E. (1928): The last glaciation: With special reference to the ice retreat in northeastern North America. American Geographical Society Research Series No. 17:1–292.

The ice sheet in eastern North America apparently was bordered by a narrow belt with plants and animals [e.g., tundra muskoxen (*Ovibos moschatus*), and reindeer/caribou (*Rangifer tarandus*)] characteristic of the modern tundra of the North. The author mentions the presence of reindeer, probably originating during the last glaciation, at Toronto, Ontario (p. 29). As the ice sheet retreated, parts of the arctic flora and fauna withdrew to higher ground, in some cases surviving to the present day as relicts. While arctic biota largely migrated in front of the growing ice sheet and then retreated northward as the ice melted back, the same flora and fauna to some extent survived at least the last glaciation in the Gulf of St. Lawrence, Torngat Mountains of NE Labrador and in the Canadian Arctic Islands. In Quaternary beds of the Kuzitrin Lowland (65°N, 165°W) on Seward Peninsula, Alaska, in regions now treeless, there have been found together remains of mammoth (*Mammuthus*), horse (*Equus*), and other mammals, as well as large tree trunks (p. 31). The mammoth and horse probably died out during one of the glaciations when temperature was considerably lower than at present. Probably Alaskan stratigraphic studies will reveal evidence of retreats and advances of the biota as the ice cap waxed and waned (p. 32).

86. Apland, B., and Harington, C.R. (1994): Pleistocene bison skeleton (*Bison bison* cf. *occidentalis*) from Clayhurst Crossing, British Columbia. Géographie physique et Quaternaire 48(2):213–23.

A large, old, male bison skeleton from a gravel pit at 56°07′40″N, 120°05′00″W was tentatively referred to the extinct western bison (*Bison bison* cf. *Bison bison occidentalis*). It is the most complete Late Pleistocene 'medium-horned' bison so far reported from Canada. Four (AMS) radiocarbon dates on the left tibia provide a weighted average age of 10,500 ± 90 BP [10,750 ± 180 BP (RIDDL-220); 10,600 ± 160 BP (AA-1219); 10,580 ± 210 BP (CAMS-398); and 10,340 ± 150 BP (CAMS-150)]. A *Bison* scapula collected at the same site was dated at 10,230 ± 140 BP (AECV-1558c) (Peter Bobrowsky, pers comm.). These dates support the earlier suggestion that river terraces in the mid-Peace area began developing before 10,000 years ago and that the region was habitable for Late Pleistocene flora and fauna, including people (*Homo sapiens*), to travel between the Beringian refugium and the southern interior Plains of North America prior to 10,500 BP.

87. Appel, S. (1968): A summer with vertebrates. Ottawa Citizen, July 11.

The article lists activities of Lila Jeletzky, a summer student assistant of Richard Harington, a vertebrate paleontologist with the National Museum of Canada in Ottawa. Lila's assignment is to catalogue all vertebrate specimens of Champlain Sea age (about 12,000 to 11,000 BP), finding out where they are kept, measuring them, and illustrating or photographing them for use in a scientific publication. She is working on fossils in the museum's George Street building and will travel to Montreal and Quebec City in search of specimens at McGill and Laval universities. She hopes to visit Green Creek in Gloucester to actively collect some Champlain Sea fossils. There is a black and white photograph of Jeletzky with a Champlain Sea white whale (*Delphinapterus leucas*) cranium.

88. Appelt, M., Gullov, H.C., and Kapel, H. (1998): The gateway to Greenland: Report on the field season 1996. *In*: Man, Culture and Environment in Ancient Greenland. Report on a Research Program. (J. Arneborg and H.C. Gullov, eds.). Danish Polar Center Publication 4:136–96.

The authors provide a normalized radiocarbon date of 1175 ± 40 BP (AAR-3222) on an arctic fox (*Alopex lagopus*) mandible from Inglefield Land, Greenland.

89. Archibald, J.D., Gingrich, P.D., Lindsay, E.H., Clemens, W.A., Krause, D.W., and Rose, K.D. (1987): First North American land mammal ages of the Cenozoic era. *In*: Cenozoic Mammals of North America, Geochronology and Biostratigraphy (M.O. Woodburne, ed.). University of California Press, Los Angeles. pp. 24–76.

This chapter provides the most current biochronostratigraphic information available for mammals in North America. The key species are listed for each age and region.

90. Ardley, E. (1907): [Cornwall specimen of *Delphinapterus leucas*]. Report of the State Geologist of Vermont, pp. 81–2.

This report refers to a white whale (*Delphinapterus leucas*) skeleton found in Pleistocene clay, 5 m below the ground surface, in 1906 in Cornwall, Ontario. The lower jaw was well preserved, unlike the skull and upper jaw that were damaged. Also included were 35 vertebrae, both shoulder blades, most of the ribs, the hyoid, and one stylohyal.

91. Ardley, E. (1916): Note on the discovery of a skeleton of *Beluga catodon* (white whale) in the Pleistocene (Leda clay) at the town of Montreal East, Quebec. Canadian Record of Science 9(8):490–3.

A near-perfect white whale (*Delphinapterus leucas* = *Balaena catodon*) was found 4.6 m above the St Lawrence River and 4 m below the surface in Leda clay near Montreal (5 m above the present St Lawrence River). The bones retrieved were: cranium and lower jaw, three teeth, hyoid, 42 vertebrae, a chevron, sternum, several ribs, scapulae, humeri, and ulnae. The skeleton was associated with the following marine mollusc shells: *Natica clausa*; *Portlandia arctica* (= *Leda glacialis*); *Macoma calcarea* (= *Macoma proxoma*); *Macoma balthica* (= *Macoma greenlandica*); and *Tethya logani* (= *Tethea logani*). This was the twelfth such whale found in North America up to this point. The dates and locations of previous North American Quaternary whale (Cetacea) finds (1849–1906) are listed, without reference to species. The Canadian specimens mentioned are: (1) 20 vertebrae (mainly caudal) collected by J.W. Dawson in 1858 for the Geological Survey of Canada Museum, Ottawa [no locality is given]; (2) a few detached bones from Rivière-du-Loup, Quebec, collected by J.W. Dawson in 1864 for the Redpath Museum, Montreal; (3) a nearly complete skeleton from Cornwall, Ontario, collected by E. Billings in 1870 for the Geological Survey Museum; (4) an imperfect skeleton from Jacquet River, New Brunswick, collected in 1874 for the Provincial Museum, Halifax, Nova Scotia; (5) a few vertebrae and a rib fragment from Smiths Falls, Ontario, collected by J.W. Dawson in 1883 for the Redpath Museum; (6) part of a lower jaw (11 ft long) of a large whale from Little Metis, Quebec, collected by J.W. Dawson in 1891 for the Redpath Museum; (7) a nearly complete skeleton from Smith's Brickyard, Papineau Road, Montreal, collected by J.W. Dawson in 1895 for the Redpath Museum; (8) ten vertebrae, ribs, and part of a cranium from Smith's Brickyard, Montreal, collected in 1901 for the Redpath Museum; (9) a few bones, hyoid, etc. from Williamstown, Ontario, collected by E. Ardley for the Redpath Museum; (10) most of a skull and several vertebrae from Pakenham, Ontario, reported by J.F. Whiteaves [now in Canadian Museum of Nature collections – see Harington and Fitzgerald 1973]; (11) a nearly complete skeleton [reported above] from Queen's City Oil Company, Montreal East, Quebec, collected in 1916 by E. Ardley for the Redpath Museum.

92. Arneborg, J., Heinemeier, N., Rud, N., and Sveinbjörnsdóttir, A. (1998): AMS dates from the hall (XVII). *In*: Man, Culture and Environment in Ancient Greenland (J. Arneborg and H.C. Gulløv, eds.). Danish National Museum and Danish Polar Center, Copenhagen, Danish Polar Center Publication No. 4:27–30.

Several AMS radiocarbon dates are provided from a Viking-age hall at the 'Farm beneath the Sand' near Nuuk in SW Greenland (64°06′N, 50°05′W). Among them is a date of 890 ± 55 BP [AAR-3395 (GUS x3702) calibrated age AD1170 (1040–1220)] on caribou (*Rangifer tarandus*) bone from the floor beneath the burnt destruction layer.

93. Ashley, G.M., and Hamilton, T.D. (1983): Epiguruk Bluff – A Late Quaternary environment record from northwestern Alaska. Abstracts with Programs, 96th Annual Meeting of the Geological Society of America, p. 518.
Mammal remains have been found in Epiguruk Bluff (12–36 m high; 3.5 km long) on the Kobuk River, Alaska. It was formed by Late Quaternary alluviation, eolian activity, and soil formation. Mammoth (*Mammuthus*) and horse (*Equus*) bones were found. Radiocarbon dates given for the former are approximately 37,000, 23,000, 19,000, and 14,500 BP; for the latter 35,000 BP.

94. Baker, F.C. (1920): The Life of the Pleistocene or Glacial Period as Recorded in the Deposits Laid Down by the Great Ice Sheets. University of Illinois. 448 pp.
This book lists Canadian Pleistocene fossils. The vertebrates include: unidentified fish (Pisces), capelin (*Mallotus villosus*), beaver (*Castor canadensis*), wapiti (*Cervus elaphus*), Scott's moose (*Alces scotti* = *Cervalces borealis*), caribou (*Rangifer tarandus*), bison (*Bison*), mastodon (*Mammut americanum*), Columbian mammoth (*Mammuthus columbi* = *Elephas*), and white whale (*Delphinapterus leucas*). It includes detailed stratigraphic information of the fossil sites and descriptions of the fossils. [See pp. 163–4, 178, 182, 329–30, 333, 357.]

95. Barendregt, R.W., Churcher, C.S., and Stalker, A.M. (1988): Stratigraphy, paleomagnetism, and vertebrate paleontology of Quaternary preglacial sediments at the Maser-Frisch site, southeastern Alberta. Geological Society of America Bulletin 100:1824–32.
Vertebrate remains were found at the Maser-Frisch site, 5 km NE of Irvine, Alberta, 4 km N of the Trans-Canada Highway (49°59′15″N, 110°02′10″W). A map of the area and paleomagnetic graphs are incorporated. Systematic descriptions are included of: rabbit (*Hypolagus* or *Sylvilagus*), ground squirrel (cf. *Spermophilus*), prairie dog (*Cynomys*), bog lemming (*Synaptomys*), bone-eating dog (cf. *Borophagus diversidens*), coyote (*Canis latrans*), unidentified cat (Felidae), cheetah or puma(?) (*Felis* cf. *Felis concolor* or *Felis trumani*), badger (*Taxidea taxus*), elephant (cf. *Mammuthus*), Pacific horse (*Equus* cf. *Equus pacificus*), complex-toothed horse (*Equus* cf. *Equus complicatus*), camel (cf. *Camelops*), unidentified deer (Cervidae), and an unidentified bovine (Ovibovini).

96. Barendregt, R.W., Thomas, F.F., Irving, E., Baker, J., Stalker, A.M., and Churcher, C.S. (1991): Stratigraphy and paleomagnetism of the Jaw Face section, Wellsch Valley site, Saskatchewan. Canadian Journal of Earth Sciences 28:1353–64.
The basal portion of the Jaw Face section at the Wellsch Valley site, some 50 km NNW of Swift Current, Saskatchewan (59°39′55″N, 107°52′30″W), contains mammalian fossils that are considered to be Late Blancan to Early Irvingtonian in age. It has been sampled for paleomagnetic studies through a thickness of 11 m. The section above 8.8 m has normal polarity; that below has reversed polarity. A tephra, which has yielded a minimum fission-track age of 0.69 ± 0.11 Ma, lies within the reversely magnetized part, near the top of the fossiliferous zone and just below the reversal at the 8.8 m level. Therefore, the change from normal to reversed polarity is probably the Brunhes-Matuyama reversal at 0.73 Ma. The Jaw Face section could be interpreted as extending from the upper part of the Mayuyama Zone into the lower Brunhes Zone; that is, the mammalian faunas are entirely Irvingtonian in age. A stratigraphic column is illustrated. Mammalian fossils discovered include: ground sloth (*Megalonyx*), Minidoka camel (*Camelops minidokae*), Cope's peccary (*Platygonus bicalcaratus*), Pacific horse (*Equus pacificus*), complex-toothed horse (*Equus complicatus*), imperial mammoth? (*Mammuthus imperator*?), southern mammoth? (*Mammuthus meridionalis*?), bone-eating dog (*Borophagus diversidens*), bog lemming (*Synaptomys*), bog lemming (*Mictomys*), vole (*Pliophenacomys*), vole (*Allophaiomys*), Hibbard's tundra vole (*Microtus paroperarius*), pocket gopher (*Thomomys*), Richardson's ground squirrel? (*Spermophilus* cf. *Spermophilus richardsonii*? = *Citellus* cf. *Citellus richardsonii*?), Rexroad ground squirrel? (*Spermophilus* cf. *Spermophilus rexroadensis*? = *Citellus* cf. *Citellus rexroadensis*?), black-tailed prairie dog (*Cynomys* cf. *Cynomys meadensis*), and ancient rabbit (*Hypolagus vetus*).

97. Barnett, V. (1987): Fossil tusk sales a dicey business. Calgary Herald, March 24.
The writer questions the sale of jewellery incorporating pieces of mammoth (*Mammuthus*) tusk in the Glenbow

Museum [Calgary, Alberta] shop. Although the Glenbow's jewellery is described as 'mastodon' tusk, it is probably from mammoths, which were much more common in northern Canada and Alaska than the American mastodon (*Mammut americanum*). The writer notes that 'part of this country's heritage is being trivialized for adornment' and mentions that the Yukon government is becoming concerned about the problem because considerable quantities of tusk fragments and other fossils are being unearthed in placer mining operations. According to the article, heritage legislation is being drafted in the Yukon, but in the meantime nobody knows what is being lost. Jim Burns, curator at the Provincial Museum of Alberta is quoted as saying he would be pleased to take responsibility for the conservation of donated tusks. One man in Los Angeles was offered $200 for a 110-lb piece of tusk. The writer suggests that somewhere in northern Alberta or in the Yukon it would be appropriate to commemorate in an exhibit the fact that mammoths once lived there.

98. Barr, W. (1971): Postglacial isostatic movement in northeastern Devon Island: A reappraisal. Arctic 24(4):249–68.
Amended emergence and uplift curves are presented for the Truelove Inlet area of Devon Island, Nunavut. The new curves are based mainly on radiocarbon dates from whale bone, driftwood, and basal peat. Whalebone material was collected from bowhead whale (*Balaena mysticetus*) skulls embedded in sand and gravel of the raised beaches. The upper surface of the whalebone is usually level with the ground surface, and can be detected from a distance by lusher vegetation. It is believed that the whale skulls attained their present positions as a result of natural strandings of whales on the gently shelving beaches. Mainly dense, solid ear bones were collected for dating. Of 18 radiocarbon dates listed (Table 1), three are from whales: 8270 ± 150 BP (GSC-991) from a skull embedded in silt 3.6 km ESE of Base Camp at 42.4 m asl; 6100 ± 125 BP (S-432) from bone embedded in limestone beach rubble 0.4 km W of Base Camp at 11 m asl; and 2900 ± 85 BP (S-433) from a skull embedded in the lowest raised beach 2.4 km NW of Base Camp at 3 m asl.

99. Bayrock, L.A. (1964): Fossil *Scaphiopus* and *Bufo* in Alberta. Journal of Paleontology 38(6):1111–12.
In the summer of 1956 the author collected a large number of skeletons of the spadefoot toad (*Scaphiopus bombifrons*) and Great Plains toad (*Bufo cognatus*) from a postglacial fine- to medium-grained sand deposit near the town of Killam, Alberta (52°48′N, 111°51′W). Most specimens lay between 3.0 and 1.5 m below the surface, and were found 160 km N of their present range.

100. Bayrock, L.A. (1966): Bison in Alberta. Archaeological Society of Alberta Newsletter 8 (Spring):1–4.
The author describes the history, evolution, and extinction of various bison species in Alberta. Horncore-spread measurements were given for the following species: steppe bison (*Bison priscus* = *Bison crassicornis*), western bison (*Bison bison occidentalis* = *Bison occidentalis*), wood buffalo (*Bison bison athabascae*), and plains buffalo (*Bison bison bison*). Other species mentioned in the article are: giant bison (*Bison latifrons*) and a southern variant of western bison (*Bison bison antiquus* = *Bison antiquus*). The history covers 14,000 BP (Late Wisconsinan) to the present. The article deals with species as a whole and does not refer to any particular specimens.

101. Bayrock, L.A. (1969): Incomplete continental glacial record of Alberta. *In*: Quaternary Geology and Climate. United States National Academy of Science, Publication 1701:99–103.
The Saskatchewan Gravels are deposits (valley fills of rivers) of sand and gravel in Alberta and parts of Saskatchewan that underlie clearly recognizable glacial deposits and overlie Cretaceous and Paleocene shale and sandstone. Since the gravels do not contain Canadian Shield material, the gravels antedate continental glaciation of the province. Recent fossil evidence suggests that these deposits are of Late Pleistocene age – perhaps between Early Wisconsinan and before the Wisconsinan ice spread into the region – and that during all previous glaciations the terrain from Montana to the Arctic Ocean along the E side of the Rocky Mountains was ice-free. The following Pleistocene vertebrate remains from Saskatchewan Gravels in Alberta, with exact locality information, are reported (pp. 101–2): (1) woolly mammoth (*Mammuthus primigenius*) and horse (*Equus*) bones from W and E banks of the South Saskatchewan River; (2) elephant [presumably mammoth] and horse bones from three gravel pits at Camrose; (3) two incomplete lower molars of elephant [evidently a Columbian mammoth (*Mammuthus columbi*)] from near Edmonton. The first (Bi-M-2) has a lamellar frequency of 6 and has a maximum width of 82 mm and height of 120 mm, as well as a length

of 160 mm. The second (Bi-M-1) has a lamellar frequency of 7 to 8 and a maximum width of 90 mm; (4) horse bones were found at two more exposures a few mi W of Edmonton (Bayrock and Hughes 1962). Evidently during Nebraskan, Kansan, and Illinoian glaciations the terrain from Montana through Alberta, Northwest Territories, northern Alaska, and Bering Strait was open for migration of people (*Homo sapiens*) and other animals from Asia to North America. During the Wisconsinan glaciation this passage was repeatedly closed by the advances of the Laurentide ice sheet. [See Reimchen 1968.]

102. Bayrock, L.A., and Hughes, G.M. (1962): Surficial geology of the Edmonton District, Alberta. Research Council of Alberta Preliminary Report 62(6):16–17.
Bones of a species of *Equus* were found in sands and gravels in the Edmonton district (Lsd 9, sec 14, twp 52, rge 25). The bones have not been identified positively, but are similar to the following species: Scott's horse [(*Equus scotti* = *Equus midlandensis*) Kansan and Late Sangamonian or Wisconsinan], and Pacific horse [(*Equus pacificus*) Kansan]. It has been established that the fossils in these sands are Early Pleistocene in age based on other fossils in the adjacent overlying sands and gravels [but see Burns and Young (1994) for data indicating that such bones are Late Pleistocene].

103. Beaudoin, A.B. (1987a): Collection and submission of samples for radiocarbon dating. *In*: Archaeology in Alberta, 1986 (M. Magne, compiler). Archaeological Survey of Alberta Occasional Paper No. 31:188–96.
The author outlines criteria that should be considered when selecting samples for radiocarbon dating. A sample submission sheet for the now-extinct Alberta Environmental Centre radiocarbon laboratory is included (p. 195).

104. Beaudoin, A.B. (1987b): Alberta radiocarbon dates 1982–1986. *In*: Archaeology in Alberta, 1986 (M. Magne, compiler). Archaeological Survey of Alberta Occasional Paper No. 31:197–213.
This list includes all dates produced for Archaeological Survey of Alberta staff since the compilation of Brumley and Rushworth (1983), and includes some dates generated earlier that did not appear on that list and a few dates from other sources. Most dates in Table 20 are uncorrected dates. Species represented by bone samples are not given.

105. Beaudoin, A.B. (1988): Alberta radiocarbon dates 1986–1987. *In*: Archaeology in Alberta, 1987 (M. Magne, compiler). Archaeological Survey of Alberta Occasional Paper No. 32:159–67.
This paper contains a list of all radiocarbon dates produced for the Archaeological Survey of Alberta between 1986 and 1987 and follows the previous list by Beaudoin (1987). In 1987, the Radiocarbon and Tritium Laboratory at the Alberta Environmental Centre, Vegreville, began producing dates that were ^{13}C-corrected. Most dates presented in Table 13 include this correction. Species represented by bone samples are not given.

106. Beaudoin, A.B. (1991): Alberta radiocarbon dates 1988–1989. *In*: Archaeology in Alberta, 1988 and 1989 (M. Magne, ed.). Archaeological Survey, Provincial Museum of Alberta Occasional Paper No. 33:239–53.
This paper contains a list (Table 23) of all radiocarbon dates produced for the Archaeological Survey of Alberta between 1988 and 1989, subsequent to the previous listing by Beaudoin (1988). Significantly, when bone has been dated, the species (mainly *Bison*) from which the bone was derived, if known, is included.

107. Beaudoin, A.B. (1998): Bison, birds, and bulrushes: Early Holocene macroremains at the Fletcher site (DjOw-1), Alberta, and implications for plains landscape and climate. Geological Society of America, Abstracts with Programs 30(7):A168.
Screening of matrix from the Cody-complex level at the Fletcher site in southern Alberta is yielding a detailed picture of the flora and fauna in and around a prairie lake about 9000 BP. Abundant bison (*Bison*) remains are found in clay at a depth of 220–270 cm overlain by mainly sandy sediment. Other vertebrate remains include: beaver (*Castor canadensis*) and deer (*Odocoileus*), indicating woody or shrubby vegetation in the vicinity, as well as vole (*Microtus*), mouse (*Peromyscus*), and tiger salamander (*Ambystoma tigrinum*). Morphological characters of bird-shell fragments suggest that several bird (Aves) species are represented. Evidently the plains had more permanent water sources in the Early Holocene than at any time since.

108. Beaudoin, A.B., Wright, M., and Ronaghan, B. (1996): Late Quaternary landscape history and archaeology in the 'Ice-Free Corridor': Some recent results from Alberta. Quaternary International 32:113–26.
Multidisciplinary research programs conducted within the Alberta portion of the 'Ice-Free Corridor' since 1986 as part of the Archaeological Survey's 'First Albertans' project have used studies of paleogeography and paleoenvironments to formulate archaeological search strategies. This approach has resulted in the discovery of two new important sites – Saskatoon Mountain (GhQt-4) and the James Pass Meadow Complex (EkPu-3 to 9) – that are yielding information on terminal Late Wisconsinan/Early Holocene human (*Homo sapiens*) occupation of the corridor. Both sites are dated by AMS or conventional radiocarbon dates to the millennium 10,000–9,000 BP. In the fifteen years since the 1978 AMQUA Conference on the Corridor, other researchers have identified four additional significant Paleoindian sites in this region. So far, none confirms human occupation of the 'Ice-Free Corridor' region before about 11,000 BP.

109. Beaupré, M., and Caron, D. (1986): Découvrez le Québec souterrain. Québec Science Éditeur, Presses de l'Université du Québec, Sillery. 254 pp.
Information is provided on various Quebec caves containing Quaternary vertebrate remains: (1) a map of Spéos de la Fée near La Redemption is provided on p. 77, and the cave and its Quaternary vertebrates are mentioned on pp. 115 and 245; (2) a photograph of the inside of Saint-Elzéar, with a map and description, are given on pp. 114 and 115, and a photograph of two black bear (*Ursus americanus*) skulls [not just one as mentioned in the caption] and a pelvis are presented on p. 243, and another photograph of a calcified wolverine (*Gulo gulo*) skeleton is shown on p. 244; (3) Laflèche Cave is described on p. 128. Of interest to cave paleontologists are four caves near the mouth of the Saguenay River (see map on p. 240 and an inventory of specimens from each of the caves on p. 241): Cave I has three human (*Homo sapiens*) specimens and two other mammal (Mammalia) bones; Cave II has 191 human specimens, 35 other mammal bones, 39 bird (Aves) bones, and six fish (Pisces) bones; Cave III has five human specimens, 20 other mammal bones, and 28 bird bones; Cave IV has only two specimens of other mammals.

110. Bednarski, J. (1990): An Early Holocene bowhead whale (*Balaena mysticetus*) in Nansen Sound, Canadian Arctic Archipelago. Arctic 43(1):50–4.
About 10,000 BP, northern Axel Heiberg and Ellesmere islands underwent a climatic amelioration that ended the last glaciation. Generally, by 8000 BP, accelerated retreat left extensive coastal areas ice-free. The occurrence of an Early Holocene bowhead whale (*Balaena mysticetus*) skeleton in glaciomarine silts adjacent to Nansen Sound on the NE coast of Axel Heiberg Island, Nunavut (80°42′N, 90°43′W, about 6 km SE of the mouth of Lightfoot River, approximately 2.5 km inland and at least 40 m asl), several hundred kilometres N of its present range, is in accord with other biological and glaciological evidence indicating that the Early Holocene climate in the High Arctic was less severe than at present. The skeleton formed the core of a large turf-covered mound from which only a few ribs and mandible fragments protruded. The skeleton, although not articulated, was roughly in order with vertebrae scattered over a distance of about 19 m. The cranium (approximately 1.5 m wide – see Figure 4) was found 26 m downslope of the rest of the skeleton on a solifluction surface. An ear bone yielded a radiocarbon date of 7475 ± 220 BP (S-3035), indicating that the whale dates from the postglacial climatic optimum. The author postulates that the paleomigration route is from Jones Sound to Norwegian Bay and then northward through Eureka Sound and is marked by at least two undated bowhead skulls found near Eureka (80°00′N, 85°57′W) on Slidre Fiord, and a whale (Cetacea) skull of unknown species radiocarbon dated to 1380 ± 130 BP (GSC-452; Lowdon and Blake 1968) from the junction of Eureka Sound and Bay Fiord (78°54′N, 85°10′W). The author also mentions a narwhal (*Monodon monoceras*) tusk from Wootton Peninsula (82°N), northern Ellesmere Island, that was radiocarbon dated to 6830 ± 50 BP (Evans 1989).

111. Beebe, B.F. (1978): Two new Pleistocene mammal species from Beringia. American Quaternary Association, Abstracts of the Fifth Biennial Meeting (September 2–4, Edmonton, Alberta). p. 159.
Two new species have been recorded in the Old Crow Basin, Yukon, from Loc. 11A: a domestic dog (*Canis familiaris*) – complete right dentary; and a peccary (cf. *Platygonus*) – proximal part of left radio-ulna. The latter is 2400 km N of the previously northernmost peccary

from Wellsch Valley, Saskatchewan. Both have a minimum radiocarbon age of 20,000 BP. [The dog specimen has since yielded a Late Holocene age by AMS radiocarbon dating.]

112. Beebe, B.F. (1980a): Pleistocene peccary, *Platygonus compressus*, Le Conte, from Yukon Territory, Canada. Canadian Journal of Earth Sciences 17(9):1204–9.
This paper describes the discovery of flat-headed peccary (*Platygonus compressus*) remains on Old Crow River in the northern Yukon. A portion of the left radius was found roughly 1 km from the mouth of King Edward Creek and is believed to be Mid-Wisconsinan in age. This specimen was found 3000 km N of its known Pleistocene range.

113. Beebe, B.F. (1980b): A domestic dog (*Canis familiaris L.*) of probable Pleistocene age from Old Crow, Yukon Territory, Canada. Canadian Journal of Archaeology 4:161–8.
A complete right dentary of one of the earliest-known domesticated dogs (*Canis familiaris*) was recovered from Old Crow Basin, Yukon. The specimen has not been radiocarbon dated but is believed to be Pleistocene in age based on relative stratigraphic position. [The specimen has since yielded a Late Holocene age by AMS radiocarbon dating.]

114. Beebe, B.F. (1983): Evidence of carnivore activity in a Late Pleistocene/Early Holocene archaeological site (Bluefish Cave I), Yukon Territory, Canada. *In*: Carnivores, Human Scavengers and Predators: A Question of Bone Technology (G.M. LeMoine and A.S. MacEachern, eds.). Proceedings of the 15th Annual Conference of the Archaeological Association of the University of Calgary. University of Calgary. pp. 1–14.
Bluefish Cave I, a small rock shelter, has yielded the only recorded faunal assemblage in Yukon to encompass the Wisconsinan–Holocene transition. The fauna contains more than 40 species of fish (Pisces), birds (Aves), and mammals (Mammalia), the oldest dating from about 15,000 BP. Human activity, indicated by the occurrence of lithic artifacts, may be responsible for the accumulation of the bone. However, carnivore activity is indicated by the extensive damage to the bone. A series of photographs of carnivore-gnawed bone fragments is used to illustrate and discuss various effects of gnawing and chewing. Fracture type and tooth marks suggest that a wolf (*Canis lupus*) or large dog (*Canis familiaris*) [wolverine (*Gulo gulo*) activity is also a possibility] is the likely carnivore candidate. Although humans (*Homo sapiens*) could have opened the marrow cavity initially, no evidence of human complicity could be clearly proven.

115. Beebe, B.F., and Hulland, T.J. (1988): Mandibular and dental abnormalities of two Pleistocene American lions (*Panthera leo atrox*) from Yukon Territory. Canadian Journal of Veterinary Research 52:468–72.
This report deals with two partial American lion (*Panthera leo atrox*) mandibles from the Yukon, one (NYRP Porcupine River-77-1) was collected from a gravel bar on Porcupine River between Driftwood River and Rat Indian Creek, the other (NMC 35316) from Dawson Locality 37 on Hunker Creek. Both specimens were X-rayed. The first showed evidence of chronic periodontitis, an extensive sclerosing osteomylitis, and congenital absence of left lower incisor teeth 1 and 2. The second showed longstanding loss of the left lower canine tooth with subsequent obliteration of the alveolus by lamellar and spicular bone and the congenital absence of two incisors in each half of the mandible. The fact that two of seven American lion mandibles known from the Yukon and Alaska show major mandibular bone disease and that they share a dental anomaly suggests an unusually high incidence of pathological change.

116. Bélanger, N. (1988): The Whitestone Mammoth. BIOME 8(3):3.
In July 1967, during his second year of field work near Old Crow in the northern Yukon (the richest area for collecting ice age vertebrate remains in Canada), Dick Harington heard a strange story handed down to some of the natives of Old Crow. The tale involved a 'monster' that had broken out of a lake bed near the upper Porcupine River, trudged up the river, and died under a bank on Whitestone River, a tributary of Porcupine River. Bearing in mind that such legends may have a core of truth to them that could relate to large, extinct ice age mammals, Dick decided to investigate. With his field assistant, Peter Lord, he left Old Crow in late July. Several days later, they stopped briefly at Johnson Village, an abandoned

Kutchin settlement, checking the base of a peat-capped bluff on the next bend with little success. Finally, there was no use proceeding farther because the stream was too shallow and they were up on the tundra level with no more bluffs in sight. Having turned back on August 5, Dick recalled that they had not examined the upstream part of the bluff near Johnson Village. While Dick was securing the boat, Peter, who was walking downstream, let out a cry and pointed toward the river. While running toward him, Dick noticed a complete woolly mammoth (*Mammuthus primigenius*) jaw with teeth embedded in sand at the water's edge and then saw the upturned part of a mammoth skull farther out in the water. The rest of the day was spent in freezing water excavating the skeleton. The tusks were excellently preserved, but most of the right one had been broken off before the animal died. A radiocarbon date on a rib indicates that this relatively small adult female died about 30,000 years ago. The 'monster legend' was, indeed, partly true.

117. Belcher, E. (1855): The Last of the Arctic Voyages; Being a narrative of the Expedition in H.M.S. Assistance, under the Command of Captain Sir Edward Belcher, C.B., in Search of Sir John Franklin, during the years 1852–53–54 ..., Vol. II. Lovell Reeve, London. 419 pp.
After descending from Mount Parker (near the eastern coast of Grinnell Peninsula, 76°50′N, 93°45′W, p. 261) on Devon Island, Nunavut, a Royal Navy party on May 11, 1853 'unexpectedly stumbled upon some bones of great size projecting from the earth, which on closer examination, proved to be the head and probably the entire skeleton of a very large whale [presumably a bowhead (*Balaena mysticetus*)], imbedded at a height of not less than five hundred feet above the sea.' After they failed to dislodge a rib, part of the skull was broken away, charged with oil, and ultimately deposited at the Royal College of Surgeons in London, England. On the party's descent from Cape Disraeli (76°44′N, 93°17′W, about 840 ft high, p. 266) on May 14, 1853, another imbedded whale [also presumably a bowhead (*Balaena mysticetus*)] was found 'at about the same elevation' [approximately 500 ft asl].

118. Bell, R. (1879): [Mastodon jaw from Moose River]. Geological Survey of Canada, Report of Progress for 1877–8:7c.
Bell recorded an American mastodon (*Mammut americanum*) jaw with a tooth from the bed of Moose River, Ontario, near the N side below the forks, 46 mi from Moose Factory. An Indian broke the tooth out of the jaw with his axe and carried it to Moose Factory.

119. Bell, R. (1898a): On the occurrence of mammoth and mastodon remains around Hudson Bay. Bulletin of the Geological Society of America 9:369–90.
This review summarizes a paper given by Robert Bell before the Geological Society of America. The important points are the discovery of some mastodon (*Mammut americanum*) bones near the Mattagami and Missinaibi rivers (James Bay) in 1877, and a small mammoth (*Mammuthus*) tooth on Long Island, Nunavut, in Hudson Bay (54°50′N, 79°25′W).

120. Bell, R. (1898b): Mastodon and mammoth remains found near Hudson Bay. Science 7(160):80.
This paper refers to the 1877 discovery of the following specimens near and around Hudson Bay that are mentioned in the previous annotation: mastodon (*Mammut americanum*) and mammoth (*Mammuthus*). Bell attempts to reconstruct the geology and paleoclimatology of this region based on the presence of these species. He also describes the biology and subsequent extinction of these proboscideans.

121. Bell, R. (1903): Summary Report of the Geological Survey of Canada for 1902. 472 pp.
Parts of a mammoth (*Mammuthus*) skeleton were found by J.H. Peck of Stanley, New York, on the Haffey lot, W half of lot 14, concession 6, Adjala, Ontario (44°05′N, 79°55′W – near Alliston) in 1887 (p. 292A). This is the only instance known by the author of mammoth bones having been found in the Georgian Bay Basin. One of the molars is in Elmira College, New York; the other bones are in the Geological Museum of Lafayette College, New York.

122. Bell, T. (1996): The last glaciation and sea level history of Fosheim Peninsula, Ellesmere Island, Canadian High Arctic. Canadian Journal of Earth Sciences 33:1075–86.
In this paper the author mentions a radiocarbon date of 7970 ± 90 BP (TO-1966) (date normalized using $\delta^{13}C$ = –25‰ and corrected for a reservoir effect of 410 years) on a narwhal (*Monodon monoceros*) tusk from diamicton 5

m asl at Site 8 about 7 km W of Cape Lockwood on northern Fosheim Peninsula, Nunavut (80°17′N, 84°18′W; Table 1). Deglaciation of the peninsula was underway by 9500 BP; however, local ice caps may have persisted through the warmest period of the Holocene until 6000 to 5000 BP. This was likely a function of reduced sea-ice conditions and increased moisture availability.

123. Bendix, B. (1998): Appendix 2: The gateway to Greenland, Quaternary zoology. Preliminary results from analysis of bones from Qeqertaaraq House 1, midden area. *In*: Man, Culture and Environment in Ancient Greenland (J. Arneborg and H.C. Gulløv, eds.). Danish National Museum and Danish Polar Center, Danish Polar Center Publication No. 4:188–90.

Of 5,811 bone fragments from the House 1 midden, 86% are mammal (Mammalia) and 14% bird (Aves – not yet identified). Of 2,763 identified mammal bone fragments, approximately 40% are arctic fox (*Alopex lagopus*); 28% seal [ringed seal (*Phoca hispida*), bearded seal (*Erignathus barbatus*), and hooded seal (*Cystophora cristata*)]; 23% walrus (*Odobenus rosmarus*); 8% tundra muskox (*Ovibos moschatus*); and 1% arctic hare (*Lepus arcticus*). A few specimens of polar bear (*Ursus maritimus*) and caribou (*Rangifer tarandus*) are also reported. There is no indication of the age of the midden.

124. Bendix-Almgreen, S.E. (1976): Palaeovertebrate faunas of Greenland. *In*: Geology of Greenland (A. Escher and W.S. Watt, eds.). Greenland Geological Survey, Copenhagen. pp. 536–73.

While this chapter deals with all time periods, the known Quaternary fossils of Greenland are outlined in the last two pages. Species mentioned are: capelin (*Mallotus villosus*), Atlantic cod (*Gadus*), rosefish (*Sebastes* cf. *Sebastes marinus*), bowhead whale (*Balaena mysticetus*), walrus (*Odobenus*), harbour seal (*Phoca vitulina*), ringed seal (*Phoca hispida = Pusa hispida*), small seal (*Phoca*), and caribou (*Rangifer tarandus*). Most of these finds derive from raised marine deposits formed during postglacial time, and range in age from 9000 BP to 400 BP. The *Phoca hispida* specimen, an articulated but incomplete skeleton, may belong to an interglacial fauna. [See Bennike et al. 1989 for the radiocarbon date of this specimen.]

125. Bennike, O. (1983): Palaeoecological investigations of a Holocene peat deposit from Vølvedal, Peary Land, North Greenland. Rapport Grønlands Geologiske Undersogelse 115:15–20.

This paper reports on the collection of peat sections at 83°02′N, 34°21′W during field work in 1979 and 1980. Remains of plants and invertebrates are noted along with a Lapland Bunting (*Calcarius lapponicus*) feather. [See Bennike and Dyck 1986 for further details.]

126. Bennike, O. (1986): Nordgrønlands kvartærgeologi. *In*: Geologi i Grønland No. 3 (N. Henriksen, ed.). Grønlands Geologiske Undersøgelse, Copenhagen, Denmark. pp. 36–41.

This booklet in Danish provides an illustrated perspective on the geology of northern Greenland. Mention is made of a skeleton of a bowhead whale (*Balaena mysticetus*) found 28 m asl at Station Nord and dated at 7000 BP. A colour photograph (Figure 1) is included. There is also a photograph of finely bedded sand about 2 Ma at Kap København that produced Early Pleistocene mammal (Mammalia) and plant remains (Figure 2).

127. Bennike, O. (1987): Quaternary geology and biology of the Jørgen Brønlund Fjord area, North Greenland. Meddelelser om Grønland, Geoscience 18:3–23.

Jørgen Brønlund Fjord area was deglaciated 8000–7600 BP. The Early Holocene marine limit is about 65m asl in the fiord, and about 80m E of the fiord. The slow initial rate of emergence was soon followed by a period of rapid emergence. Emergence continued until about 1000 BP. A table of radiocarbon dates includes caribou (*Rangifer tarandus*) antlers from Kajakelv, 7980 ± 115 BP, (K-3865); Jørgen Brønlund Fjord, 6240 ± 100 BP (K-4348); Kap Harald Moltke, 5470 ± 95 BP (K-3866); near Brønlundhus, 3710 ± 80 BP (K-4059); Paralleldal, 2080 ± 75 BP (K-3867); and tundra muskox (*Ovibos moschatus*) bones from Midternaes, 3830 ± 85 BP (K-3364); and Kap Mylius-Erichsen, 2430 ± 75 BP (K-3363). A large number of bones considered to be of Holocene age were found scattered on the ground surface around two archaeological ruins, and the fauna includes many species no longer found in the area. A table listing these animals is included. The partial skeleton of a large whale, tentatively identified as a bowhead whale (*Balaena mysticetus*), found about 46m asl near Uglegylphøj, could be as old as 6000 BP based on the emergence curve. The whale bones could have been deposited during the 'Hypsithermal.'

128. Bennike, O. (1990): The Kap København Formation: Stratigraphy and palaeobotany of a Plio-Pleistocene sequence in Peary Land, North Greenland. Meddelelser om Grønland, Geoscience 23:1–85.
The Kap København Formation in eastern Peary Land of North Greenland comprises essentially two facies: silt and sand. The former was deposited in sublittoral environments below wave base, while the latter was deposited in the littoral zone and was much disturbed by wave action. Part of this last facies was deposited near river mouths and contains allochthonous remains of freshwater and terrestrial organisms as well as the expected marine species. This paper deals mainly with the vascular plant remains. The mammal remains discussed are from Repenning et al. 1987. They include a tooth of *Hypolagus*, a large extinct rabbit, and a maxillary fragment of *Lepus*, an extant genus. The simultaneous occurrence of these lagomorphs – *Hypolagus* is known from 2.0 Ma [since *Hypolagus* arose as early as Lower Miocene, presumably this should read 20 Ma] to 0.8 Ma, *Lepus* from 2.3 Ma to the present – suggests an age of 2.0 to 2.3 Ma for the Kap København Formation. This Late Pliocene age is supported by the formanifera, ostracod, mollusc, and paleomagnetic data. The forest-tundra vegetation of the time implies a warmer climate than today, with summer temperatures about 7–8°C higher. This temperature regime was not consistent with other northern climates at the time and could have been a localized North Atlantic phenomenon.

129. Bennike, O. (1991): Marine mammals in Peary Land, North Greenland. Polar Record 27:357–9.
This note records the marine mammal sightings and material collected in North Greenland during 1982, 1983, and 1986. Fossil and subfossil specimens collected include a sub-recent skeleton of an adult female hooded seal (*Cystophora cristata*) – a northern extralimital record for the species – and a polar bear (*Ursus maritimus*) jaw from an archaeological site. The small right mandible of a polar bear dated to 1440 ± 45 AD (K-352 calibrated) and came from the Kølnaes archaeological site. The hooded seal remains were discovered at 82°24′N, 21°00′W near Kap København and are presumed to have been deposited within the previous 75 years.

130. Bennike, O. (1995): Palaeoecology of two lake basins from Disko, West Greenland. Journal of Quaternary Science 10(2):149–55.
Macrofossils, lithology, and chronology of two lake basins on Disko in central West Greenland were investigated. Both lakes were isolated from the sea in Middle Holocene time as a result of relative land uplift. A marine – brackish – limnic sequence was recovered from Qivittut Lake (69°26.5′N, 53°40′W). In Zones 1 and 2 of the core from this lake, marine fishes (Pisces) are represented, being most common in Zone 1. A characteristic of the upper marine part of Zone 3 are the bones of sticklebacks (*Gasterosteus aculeatus*). This small predacious fish lives in lakes, streams, brackish water, and in the sea, but is most common in brackish water. Since few other animal remains are found in the same core interval as the sticklebacks, the author suggests that these fish fed on copepods, because several copepod specimens are common in brackish water, and because the chitinous skeletons of these crustaceans are seldom preserved.

131. Bennike, O. (1997): Quaternary vertebrates from Greenland: A review. Quaternary Science Reviews 16:899–909.
Remains of fishes, birds, and mammals are rarely reported from Quaternary deposits in Greenland. The oldest archaeological sites in Greenland date from between 4500 and 4000 years ago and analyses of bone material from the sites indicate that probably all common species were present by then. The oldest Quaternary vertebrate remains come from Late Pleistocene deposits and comprise Atlantic cod (*Gadus morhua*), hare (*Lepus*), rabbit (*Hypolagus*), and ringed seal (*Phoca hispida*). Interglacial and interstadial deposits have yielded remains of cod (*Gadus morhua*), Little Auk (*Alle alle*), collared lemming (*Dicrostonyx torquatus*), ringed seal (*Phoca hispida*), reindeer (*Rangifer tarandus*), and bowhead whale (*Balaena mysticetus*). Early and Mid-Holocene finds include capelin (*Mallotus villosus*), polar cod (*Boreogadus saida*), red fish (*Sebastes* cf. *Sebastes marinus*), sculpin (Cottidae/Cottunculidae – species unidentified), three-spined stickleback (*Gasterosteus aculeatus*), Lapland Longspur (*Calcarius lapponicus*), Arctic hare (*Lepus arcticus*), collared lemming (*Dicrostonyx torquatus*),

wolf (*Canis lupus*), ringed seal (*Phoca hispida*), reindeer (*Rangifer tarandus*), and bowhead whale (*Balaena mysticetus*). It is considered unlikely that any vertebrate species could survive in Greenland during the peak of the last glaciation, but many species had probably already immigrated by the Early Holocene. Pollen and macrofossil analyses of Early Holocene lake sediments show that a fairly rich vegetation was established very soon after the last deglaciation. The author goes on to propose possible immigration routes for various species including people (*Homo sapiens*). Some previously unpublished radiocarbon dates are presented in this paper: a walrus (*Odobenus rosmarus*) from Peary Land (K-5980) dated at 1490 ± 75 BP (1340 ± 75 BP corrected); a walrus from Peary Land (Lu-3660) dated at 1870 ± 60 BP (1320 ± 60 BP corrected); a walrus from Frederikshåb Isblink (K-5979) dated at 2630 ± 60 BP; a walrus from Holsteinsborg (K-5978) dated at 3870 ± 90 BP; a bowhead (*Balaena mysticetus*) from Thule (Lu-3661) dated at 8290 ± 100 BP (8140 ± 100 BP corrected); a whale (cf. *Balaena mysticetus*) from Kronprins Christian Land (K-5977) dated at 6050 ± 105 BP (5900 ± 105 BP corrected); an unidentified large whale from Peary Land (Lu-3658) dated at 4740 ± 70 BP (4240 ± 70 BP corrected); and another unidentified large whale from Peary Land (Lu-3659) dated at 7550 ± 90 BP (7000 ± 90 BP corrected). Some previously published dates include: a walrus (*Odobenus rosmarus*) from Jakobshaven Isbrae dated at 3890 ± 100 BP (Ua-2350); a caribou (*Rangifer tarandus*) from Peary Land dated at >40,000 BP (Ua-1112); a caribou from Peary Land dated at 7980 ± 115 (K-3865); a caribou from Skeldal dated at 6200 ± 70 BP (Lu-1096); a caribou from Søndre Strømfjord dated at 5800 ± 80 BP (ETH-4583); a fin whale (*Balaenoptera physalus*) from Storstrømmen dated at 3480 ± 90 BP (K-6096); a bowhead whale (*Balaena mysticetus*) from Thule dated at 8300 ± 200 BP (W-48); a bowhead from Holsteinsborg dated at 5645 ± 115 BP (I-6235); a ?bowhead (cf. *Balaena mysticetus*) from Scoresby Land dated at 8580 ± 85 BP (Lu-1095); a large whale from Thule dated at (39,000 BP (GSC-2257); a large whale from Île de France dated at 6955 ± 40 BP (T-9367); a large whale from Kong Oscars Fjord dated at 7000 ± 190 BP (I(UM)-429); and a large whale from Disko dated at 2750/80 ± 80 BP (K-3161).

132. Bennike, O. (1999): Colonisation of Greenland by plants and animals after the last ice age: A review. Polar Record 35(195):323–36.

For more than 100 years an intense debate has taken place among biologists and geologists concerning the history of the plants and animals of Greenland. The debate concerns whether the existing species survived the last ice age in ice-free areas within Greenland or whether they immigrated after deglaciation. The history of this debate is outlined. In the light of data derived from analyses of cores through the Greenland Ice Sheet covering ice age climate, and paleoecological studies of interglacial and Early Holocene deposits, the idea that a large portion of Greenland's plants and animals may have survived during the ice ages is evaluated. While ice-free areas [refugia] existed, the author concludes that only hardy, cold-adapted species could have survived, which also explains why so few clearly endemic species are present in Greenland. Most members of the present biota are considered to be postglacial immigrants. Some species reached Greenland by walking or flying, but most arrived by passive, long-distance, chance dispersal carried by wind, ocean currents, and in particular birds. Considering the great size of Greenland, the existing flora and fauna are relatively small (three non-marine fishes, 40 non-marine birds, and seven non-marine mammals among the vertebrates). This is mainly due to the northern geographical position and low temperatures, but also partly due to Greenland's isolated situation. At the beginning of the Quaternary, forest tundra existed in northern Greenland. From the last interglacial, non-marine flora and fauna records are now established for three sites in northern Greenland, and from many localities in Jameson Land. In Jameson Land, the summer temperature during the last interglacial was about 5°C higher than at present, and many species of plants and animals lived much farther north than today. No plant or animal remains have so far been recovered from the colder parts of the last glacial stage. Two significant revelations of the ice-core studies are that: (1) the transition from the last glacial to the present interglacial was amazingly rapid and took place over only a few decades; (2) the climate during the last temperature minimum was much colder than previously believed – the mean annual temperature was as much as 25°C below the present. The Early Pleistocene biota of Greenland, as well as the last interglacial fauna, were very different from those of the Holocene. Among vertebrates, obviously the tundra muskox (*Ovibos moschatus*), collared lemming (*Dicrostonyx torquatus*), and ermine (*Mustela erminea*) entered Greenland across the sea ice of 25-km-wide Nares Strait. Paleoeskimos (*Homo sapiens*) reached Greenland by the same route over 4400 years ago. The oldest wolves (*Canis lupus*) are about 7600 years old, but they may have arrived earlier considering that caribou (*Rangifer tarandus*), their main prey, had reached Greenland some 8900 years ago.

133. Bennike, O., and Dyck, J. (1986): A subfossil Lapland Bunting *Calcarius lapponicus* feather from Vølvedal, North Greenland. Ornis Scandinavica 17(1):75–7.

Subfossil feathers are rarely recorded, though they have a high preservation potential in peat deposits. In the lower part of a peat deposit at 83°02′N, 34°21′W, a well-preserved feather (complete with calamus, shaft, and barbules) was found and identified as originating from a Lapland Bunting (*Calcarius lapponicus*) on the basis of the structure of the downy barbules and the pigmentation pattern. The age of the feather is estimated at 6000 to 8000 BP. It represents the oldest-known bird from Greenland. This circumpolar extant species is presently not found in the extreme northern parts of Greenland and Canada. Its presence at Vølvedal indicates that the peat was formed in a warmer climate than the present one – presumably during the Holocene climatic optimum, between 8000 and 4000 BP.

134. Bennike, O., and Kelly, M. (1987): Radiocarbon dating of samples collected during the 1984 expedition to North Greenland. Rapport Grønlands Geologiske Undersogelse 135:8–10.

The results of 18 radiocarbon age determinations on shells (14), plants (2), and mammals (2) are summarized. Ringed seal [*Phoca hispida* (GGU 313897)] collected at 81°15′N, 59°36′W in eastern Hall Land, dated to 8680 ±120 BP (K-4687), is one of the mammals. The other mammal date is from a metacarpal of a caribou [*Rangifer tarandus* (GGU 313935)] found on the surface at 81°27′N, 54°38′W (mouth of Alma Glück Dal), dated to 2155 ± 90 BP (K-4574). [See also Bennike et al. 1989.]

135. Bennike, O., and Weidick, A. (1999): Observations on the Quaternary geology around Nioghalvfjerdsfjorden, eastern North Greenland. Geology of Greenland Survey Bulletin 183:57–60.

Samples collected along a transect between Blåsø in the inland area and the outer coast included radiocarbon-dated bones. After deglaciation, marine molluscs and marine mammals were able to live in Nioghalvfjerdsfjorden. Five seal, including probably the ringed seal (*Phoca hispida = Pusa hispida*), and whale bones, collected on raised beaches, on raised marine deposits and along the shore of Blåsø, have been dated between about 7700 and 4500 BP (Figure 2). The distribution of dates shows that the fiord was not glaciated during this period in the Middle Holocene. Molluscs and ringed seals may have lived in the fiord during period of permanent sea ice, whereas driftwood and whales [probably bowheads (*Balaena mysticetus*)] could only enter the fiord during periods of open water. The driftwood and whale dates are restricted to the period 7000–5400 BP, and perhaps this period was preceded and followed by a period when the fiord was covered by sea ice throughout the year. Figure 3A is a colour photo of two whale (probably bowhead whale) ribs found on a raised beach, and Figure 3B shows a seal (probably ringed seal) right femur and *Mya truncata* shells from raised marine deposits from near Blåsø in the inner part of Nioghalvfjerdsfjorden.

136. Bennike, O., Higgins, A.K., and Kelly, M. (1989): Mammals of central Greenland. Polar Record 25(152):43–9.

Although this paper deals mainly with distribution of living species, some fossils were collected during the surveys. These include a hare (*Lepus*) mandible – also described in Bennike (1990), which was estimated to be about two million years old; some droppings very similar to those of modern collared lemming (*Dicrostonyx*), which were considered to be 130,000–115,000 BP; a subfossil skeleton of a ringed seal (*Phoca hispida*) from eastern Hall Land radiocarbon dated to 8680 ± 120 BP (K-4687) – hence the oldest Holocene mammal recorded in Greenland (see also Bennike and Kelly 1987); and two separate vertebrae of the same narwhal (*Monodon monoceros*) aged at about 1000–2000 BP based on the glacial history of the area.

137. Bennike, O., Meldgaard, M., Heinemeier, J., and Rud, N. (1994): Radiocarbon AMS dating of Holocene wolf (*Canis lupus*) remains from Greenland. The Holocene 4(1):84–8.

Radiocarbon AMS dating of six polar wolf (*Canis lupus*) remains from E and N Greenland and one from Ellesmere

Island document the presence of this large carnivore in Greenland at around 6800, 1600, 800, and 500 BP. The actual data are: a left humerus (MGUH VP 3332) found near Vølvedal, Peary Land, dated at 6785 ± 100 BP (Ua-1346); a left mandible (ZMK 112a (1950)) found at Midsommersøen, Peary Land, dated at 1570 ± 100 BP (AAR-830); limb bones from some adults found near Blåsø dated at 790 ± 100 BP (Ua-950); and a left mandible (CN 3961) in Heilprin Land dated at 520 ± 70 BP (AAR-827). Three of the remains yielded modern ages. These results demonstrate the presence of the wolf in Greenland since the Early Holocene, and it is argued that it was probably present during most of the Holocene. In the early 1900s the wolf was trapped almost to extinction in Greenland. Since then the populations have recovered in the N and E. West Greenland is the only Arctic region besides Svalbard with a population of caribou (*Rangifer tarandus*) but no wolves. The authors support the hypothesis that the Greenland wolf is synonymous with *Canis lupus arctos* and that it reinvaded Greenland from the Queen Elizabeth Islands, having survived the last full glacial period in the Beringian Refugium. The age of Greenland's wolves is supported by the oldest-known caribou from northern Greenland which also dates to the Early Holocene [7980 ± 115 BP (K-3865)].

138. Bennike, O., Hansen, K.B., Knudson, K.L., Penney, D.N., and Rasmussen, K.L. (1994): Quaternary marine stratigraphy and geochronology in central West Greenland. Boreas 23:194–215.

A new stratigraphic framework is proposed for the Quaternary portion of central West Greenland based primarily on faunal and geochronologic studies of shallow shelf deposits. Data from 24 known sites and 18 new sites are presented. Four pre-Holocene marine events are described. The interglacial Ivnaarssuit and Nordre Laksebugt marine events are considered to be Middle Pleistocene in age. The interstadial Laksebugt marine event is considered to be Late Middle Pleistocene, whereas the Svartenhuk marine event is correlated with the last interglacial. For the last glacial period an extensive ice shelf is proposed W of Disko. The oldest postglacial deposits are dated at 10,470 ± 130 BP. Tables of macrofossils, bryozoans, foraminifera, ostracods, and radiocarbon dates are presented along with stratigraphic illustrations, maps, and many specimen photographs. The macrofossils include fishes – capelin (*Mallotus villosus*), sculpin (Cottidae/Cottunculidae) [including the Arctic staghorn sculpin (*Gymnocanthus tricuspis*)], sticklebacks (Stichaeidae/Lumpeniidae), Greenland cod (*Gadus ogac*), and Atlantic cod (*Gadus morhua*); birds – Dovekie (*Alle alle*); and mammals – large whales. None of the fish was radiocarbon dated, but the whale material includes: a bone of a large whale from Nuussuaq Peninsula dated at 380 ± 80 (I-16414); a bone of ?bowhead (cf. *Balaena mysticetus*) from Disko dated at 6510 ± 80 (Ua-1785); a ?bowhead (cf. *Balaena mysticetus*) rib dated at 7190 ± 150 (I-16366); a possible bowhead (cf. *Balaena mysticetus*) bone from Disko dated at 7350 ± 120 (I-16356); and the inner part of a jaw of a large whale (Cetacea) from Disko dated at 8400 ± 90 (K-5969).

139. Bennike, O., Björck, S., Böcher, J., Hansen, L., Heinemeier, J., and Wohlfarth, B. (1999): Early Holocene plant and animal remains from North-east Greenland. Journal of Biogeography 26:667–77.

Early Holocene vertebrates of NE Greenland (Figure 2 shows the localities mentioned in the text) are occasionally mentioned in this paper: (1) the Zackenberg, Wollaston Forland delta deposits yielded the oldest Holocene arctic hare (*Lepus arcticus*) record from Greenland – Table 2 lists this species as common in sample 9201 dating to about 7900 BP; (2) a macrofossil diagram (Figure 3) from Peters Bugt Sø indicates that sticklebacks (*Gasterosteus aculeatus*) were present in layers 3–4 (about 6800 BP) and the more recent layers 1–2. Evidently the sticklebacks are N of their present northern geographical range limits (now about 100 km S of Peters Bugt Sø at approximately 74°N), so their previous occurrence on Hochstetter Forland implies slightly higher summer temperatures than at present (p. 669). The authors also mention that geese (Anserinae) may have migrated to NE Greenland in the Early Holocene, and that these large migratory birds constitute a main agent for the dispersal of plants and invertebrates to Greenland. Among vertebrates, the E Greenland reindeer/caribou (*Rangifer tarandus eogroenlandicus*) is probably a synonym of the Peary caribou (*Rangifer tarandus pearyi*) from northernmost Canada (Meldgaard 1986), and the Greenland wolf (*Canis lupus orion*) is considered a synonym of the arctic wolf (*Canis lupus arctos*) because the former was described using a NW Greenland specimen that was undoubtedly a straggler from Canada (Bennike et al. 1994). Both caribou and wolf remains from N Greenland have been dated to the Early or Mid-Holocene (Bennike 1997), and these species were presumably also present in NE Greenland this early.

140. Bensley, B.A. (1913): A *Cervalces* antler from the Toronto interglacial. University of Toronto Studies, Geological Series 8:1–3.
The author reports the discovery of a right antler of Scott's moose (*Alces scotti* = *Cervalces borealis* = *Cervalces scotti* = *Cervalces roosevelti* = *Cervus americanus*) found in a gravel pit in the Pleistocene deposits of Toronto, Ontario, by A.P. Coleman in 1909. The antler is an isolated fragment and is believed to have come from elsewhere. It is compared to a similar *Cervalces roosevelti* antler described in 1909, and to the moose (*Alces alces*). Bensley speculates (p. 2) that the Toronto specimen may be of Aftonian age, and also that the greater length of the antler beam may be a more primitive condition in this group.

141. Bensley, B.A. (1923): A muskox skull from Iroquois Beach deposits at Toronto: *Ovibos proximus sp. nov*. University of Toronto Studies, Geological Series 23:3–11.
The author describes a tundra muskox (*Ovibos moschatus* = *Ovibos proximus*) skull found in a gravel pit of the York Sand and Gravel Company that is part of a gravel bar formation of the Iroquois Beach deposit E of Toronto, Ontario. The pit was 8 m deep, but the depth from which the skull was recovered is unknown. The skull itself, which is almost complete, includes the braincase, the upper portion of the face, and the horn cores. It is from the youngest of three Toronto Pleistocene beds (Don, Scarborough, and Iroquois). This specimen is compared to *Ovibos moschatus* and *Ovibos yukonensis* (= *Ovibos moschatus*) and skull measurements are provided. Measurements are provided (pp. 9–10), as are four illustrations of the specimens (Figures 1–4).

142. Berezowsky, M. (1997): Paleontologists dig up ice-age remains around Medicine Hat. Calgary Herald, August 23. p. A4.
During the last three weeks, Jim Burns, an ice-age fossils curator at the Provincial Museum in Edmonton was part of a five-man team looking for fossils near Medicine Hat, Alberta. Scouring the banks of the South Saskatchewan River, the group found remains of camels (Camelidae), horses (*Equus*), mammoths (*Mammuthus*), prairie dogs (*Cynomys*), antelope (*Antilocapra*), and a giant ground sloth (Xenarthra) dating between 1.6 Ma and 500,000 BP. Burns said the high point of the fieldwork was finding fossilized camel remains. In a 4 m by 5 m section, the team found 68 fossils, including two camel shoulder blades. A photograph shows the five party members holding two large mammoth teeth.

143. Bethune, B. (2001): Mystery of the first North Americans. Maclean's, March 19:24–9.
This magazine article, prompted by Elaine Dewar's book *Bones: Discovering the First Americans*, virtually mirrors the article 'Hunt for the First Americans' (Parfit 2000), in that three main hypotheses for human (*Homo sapiens*) immigration to America are discussed: (1) the traditional 'Clovis Theory,' whereby people crossed the Bering Isthmus from Siberia into Alaska and Yukon, eventually working their way S into the heartland of North America (however, the author notes that studies by Geological Survey of Canada geologists support the idea that the way was blocked by ice 'until long after Clovis culture was flourishing far to the south'); (2) the 'Coastal Theory,' whereby people adapted to living on coastal resources, moved S of the Bering Isthmus and S along the Pacific coasts of North, Central, and South America (this hypothesis holds much sway at present); and (3) the 'Atlantic Crossing Theory' supported by Dennis Stanford of the Smithsonian Institution, who suggests Clovis's origins may be in the Solutrean of Europe. Arguments by geneticists and linguists suggest that humans may have entered North America 50,000 to 36,000 years ago. The article is accompanied by a map showing the main archaeological sites and the migratory routes mentioned.

144. Bigelow, N., and Ager, T.A. (2000): Interstadial vegetation at the Eva Creek site, central Alaska. Troy Péwé Memorial Workshop: Paleoclimates and Paleoenvironments in Eastern Beringia and the Bering Land Bridge (August 21–4, Chena Hot Springs, Alaska). Abstracts, p. 41.
Preliminary analysis of plant macrofossil remains collected from about 40,000–36,000 BP (Mid-Wisconsinan) interstadial deposits that also include a nearly complete short-faced bear (*Arctodus simus*) skeleton, at Eva Creek, near Fairbanks, Alaska, indicates a dense willow and herb vegetation – at least at this locality. [So it appears that the bear died in a dense willow thicket with marshy and disturbed soil covered by herbaceous vegetation nearby.]

145. Bigsby, J.L. (1825): Notice of a cave containing bones, in Lanark, Upper Canada. American Journal of Science 9:354–5.

The author discovered bones of a large, unidentified animal (Mammalia?) in the mouth and on the floor of a cave, located 40 km N of Perth, Ontario, on the Mississippi River. The age of the animal is also undetermined.

146. Billings, E. (1856): On the mammoth and mastodon. Canadian Naturalist and Geologist 1(54):379–90.
This paper discusses mammoths (*Mammuthus*) and mastodons (*Mammut americanum*) in general, and specifically tusks and part of a lower jaw of a mammoth discovered in 1851 at Burlington Heights, 30 ft beneath the surface in from Hamilton, Ontario (pp. 380–1). By this time, it was known that mammoth bones could be found in North America from 40°N northward to the Arctic Circle. The author also mentions (p. 383) abundant mammoth remains from cliffs of frozen mud on the E side of Bering Strait at Eschscholtz Bay in 'Russian America' (now Alaska). [See Kotzebue 1821.]

147. Billings, E. (1863): On the remains of the fossil elephant found in Canada. Canadian Naturalist and Geologist 8(11):135–47.
The remains of a woolly mammoth (*Mammuthus primigenius = Euelephas Jacksoni*) were found in 1852 at Burlington Heights, near Hamilton, Ontario, at the western extremity of Lake Ontario, 13 m below the surface and 20 m above the present lake level. The bones were recovered from a deposit of coarse sand overlain by 10 m of stratified gravel (with pebbles of limestone). The age of the specimen has not been determined. Also included are descriptions of the more important bones: jaw, tooth, tusk, and portions of scapulae.

148. Billings, E. (1869): Note on the bones of a mastodon found near Dunnville, October, 1868. Geological Magazine of London 6:38–9.
The author gives an account of the 1868 find of a mastodon (*Mammut americanum*) from Dunnville, Ontario, situated at the E end of Lake Erie. The fossil was not complete but included the following: seven teeth, a few vertebrae, a few rib fragments, and part of the right ramus of the lower jaw. The remains were buried in a swamp, partly embedded in a layer of fine sand 0.75 m thick overlying boulder clay. Above the sand was 0.45 m of black vegetable mould. According to the author, the 'animal lived long after the glacial period.'

149. Billings, E. (1870): [Bones of a whale lately discovered near Cornwall, Ontario]. Proceedings of the Natural History Society of Montreal. Canadian Naturalist and Quarterly Journal of Science 5:438–9.
Billings refers to a specimen found 0.5 mi from the railway station at Cornwall, Ontario, which is believed to be closely related to the white whale (*Delphinapterus leucas = Beluga leucas*). The site is located 20 m above the St Lawrence River and more than 65 m asl. The lower jaw is nearly perfect, but the skull and upper jaws were damaged. Also included were 35 vertebrae, the scapulae, most of the ribs, the hyoid, and one stylohyal. These bones were discovered in Quaternary clay. The author mentions another specimen discovered near Montreal several years before that consists of only half a back bone. [See also Ardley 1907.]

150. Bird, J. (1939): Artifacts in Canadian river terraces. Science 89(2311):340–1.
Eighty fragmentary bones of a large and small species of horse (*Equus*), an unidentified deer (Cervidae) antler, as well as several artifacts (*Homo sapiens*) were found in a gravel pit near the North Saskatchewan River, 5 km upstream from Edmonton, Alberta. The bones were found unarticulated and randomly scattered throughout the gravel. Artifacts found *in situ* in the pit walls consist of a large quartzite core and two large quartzite flakes – one seemingly retouched into a rough sidescraper. Professor P.S. Warren of the Department of Geology, University of Alberta, agreed that they were not intrusive. The paper also refers to a metatarsal resembling that of a wapiti (*Cervus elaphus*) discovered in about 6 m of gravel at Peace River, Alberta, in a large pit excavated in an old terrace 100 ft above present water level at the W end of the bridge crossing the river from the town. Some roughly flaked, waterworn quartzite cobbles were found in loose gravel here too.

151. Björck, S., Wohlfarth, B., Bennike, O., Hjort, C., and Persson, T. (1994): Revision of the Early Holocene lake sediment based chronology and event stratigraphy on Hochstetter Forland, NE Greenland. Boreas 23:513–23.
In Peters Bugt Sø (Figures 1A, B) on Hochstetter Forland, remains of three-spined sticklebacks (*Gasterosteus aculeatus*, Table 4, layer 5 at depths of 172–168 cm) were found in sediments deposited just after the lake was isolated about 6000 BP. [See Bennike 1997, p. 902.]

152. Black, R.F. (1966): Late Pleistocene to Recent history of Bering Sea – Alaska coast and man. Arctic Anthropology 3(2):7–22.

Despite the fact that many scientists feel this is a likely region to find the earliest human (*Homo sapiens*) sites, the combination of fluctuating sea levels, crustal instability, volcanism, damage by succeeding glacial ice sheets, and the fact that the earliest sites are likely to be currently underwater along with the usual events such as loess deposition, weathering, isostatic and eustatic changes, all contribute to the improbability of finding an early site and properly aging it. Some of the areas where these disturbances are less significant are indicated, and it is suggested that more research be done in these regions. The author reviews the stratigraphy and glacial history of the area and attempts to make some generalities about chronology.

153. Blake, W., Jr (1974): Studies of glacial history in Arctic Canada. II. Interglacial peat deposits on Bathurst Island. Canadian Journal of Earth Sciences 11(8):1025–42.

The author mentions (p. 1036) a tusk [presumably of a woolly mammoth (*Mammuthus primigenius*)], collected in southern Melville Island, Northwest Territories, in 1916 by the Northern Party of the Canadian Arctic Expedition 1913–18, that dated to 21,900 ± 320 BP (GSC-1760). A search of V. Stefansson's diary for 1916 failed to reveal any mention of the tusk being collected, but on April 5, 1917, while at the NE corner of Borden Island, Stefansson wrote a letter to Captain H. Gonzales, who was then visiting the base camp near Liddon Gulf, Melville Island. In it he requests that 'the scientific specimens, including the mammoth tusk' should all be taken to the ship *Polar Bear*, which was then at winter quarters in Victoria Island. Later Stefansson (1918) reported that 'We found a tusk, but no other mammoth remains, in Melville Island,' but no details of the collection site are given. [See Harington 1990b.]

154. Blake, W., Jr (1975a): Pattern of postglacial emergence, Cape Storm and South Cape Fiord, southern Ellesmere Island, N.W.T. Geological Survey of Canada Paper 75-1C:69–77.

Eighteen radiocarbon dates were done on six whale [probably bowhead whale (cf. *Balaena mysticetus*)] bone samples found embedded in raised beaches at Andersrag Beach near Cape Storm, southern Ellesmere Island, Nunavut. One specimen at an elevation of 118 m yielded a date of 9410 ± 90 BP (GSC-1496C); a specimen from a whole skeleton at an elevation of 80.5 m on the SW side of Cape Fiord (Fig. 13.7) provided a date of 9070 ± 90 BP (GSC-1748). About 24 km E of Cape Storm another whale bone from 82 m was dated at 8940 ± 90 BP (GSC-964-2). Many cross-check age determinations between driftwood and whale bones have shown that in this type of arctic environment using the collagen fraction of whale bones gives reliable results, whereas in most cases determinations on the bone apatite fraction give ages that are too young. The NW coast of Jones Sound first became open to the sea between 9500 and 9000 years ago, during postglacial time. [See Lowdon and Blake 1980 for a list of dates.]

155. Blake, W., Jr (1975b): Radiocarbon age determinations and postglacial sequence at Cape Storm, southern Ellesmere Island, Arctic Canada. Geografiska Annaler 57, Series A, Nr A62:1–71.

The author provides a section on whale bones (pp. 38–42). Because of the abundance of large whale [presumably mainly the bowhead whale (*Balaena mysticetus*)] bones at Cape Storm, Ellesmere Island, Nunavut, many specimens were collected – especially those that appeared to be well preserved. Reasons for the collection were: (1) radiocarbon dates on whale bones would supplement data provided by driftwood, and at the sites where the two types of material occurred together, comparative ages would be of interest; (2) it was an opportunity to date various fractions of bones in order to determine which fraction gives the most reliable results. Many cross-checks have shown that, in such an Arctic environment, the organic (collagen) fraction of whale bones gives reliable results, as opposed to the bone apatite fraction, which commonly yields ages that are too young. Not only is the taphonomy of the whale remains discussed, but so too are collecting techniques and dating procedures. Of particular interest are: (1) Fig. 10, which shows seven locations (circles) from near the present shoreline to near the top of Andersrag Beach N of Cape Storm; (2) Fig. 13, a colour air photo showing the position of a 5+ m long whale mandible dated to 4490 ± 60 BP (GSC-1021) lying on the surface; (3) Fig. 14 showing the posterior cranial fragment of a bowhead whale dated to 1610 ± 50 BP (GSC-1227) lying on a raised beach at an elevation of 3 m; (4) Fig. 17 showing several articulated bowhead vertebra centra and ribs exposed at the edge of a gully cut through raised beaches W of South Cape Fiord at an elevation of 80.5 m. A rib was dated at 9070 ± 90 BP (GSC-1748). The position of the specimen is circled on an air photo (Fig. 24). Radiocarbon age determinations on marine mollusc shells and whale bones indicate that during postglacial time the northwestern coast of Jones Sound first became open to the sea between 9500 and 9000 BP. [See also Dyke and Morris 1990, and Dyke, Hooper, and Savelle 1996.]

156. Blake, W., Jr (1980): Mid-Wisconsinan interstadial deposits beneath Holocene beaches, Cape Storm, Ellesmere Island, Arctic Canada. American Quaternary Association, Abstracts of the Sixth Biennial Meeting (August 8–20, Orono, Maine). pp. 26–7.

At Cape Storm, southern Ellesmere Island, Nunavut, a series of raised beaches extends 130 m asl. Whale bones embedded in these beaches at 118 m asl are radiocarbon dated at 9340 ± 80 BP (GSC-1496-3). Older deposits at 63 m asl contained an ulna of a Dovekie (*Alle alle*). This bone represents the first evidence of a Pleistocene bird from the High Arctic of Canada. Ice-free enclaves existed in this part of the Arctic during Mid-Wisconsinan time.

157. Blake, W., Jr (1982): Geological Survey of Canada radiocarbon dates XXII. Geological Survey of Canada Paper 82-7:1–22.

This report includes a radiocarbon-dated mammoth (*Mammuthus*) bone from the Trio Ready Mix Gravel Pit, Saanich Peninsula, Vancouver Island, British Columbia (48°31′52″N, 123°22′45″W), at an elevation of 70 m. Dated at 17,000 ± 240 BP (GSC-2829), the specimen was 16.5 m below the base of a 4.5 m-thick till deposited near the end of the Fraser Glaciation. Also reported is an age of 48,100 ± 1100 BP (GSC-3032-2) for willow (*Salix*) wood, which came from a permanently frozen peat bed exposed by placer mining along Revenue Creek, Yukon (62°21′N, 137°12′30″W). The peat bed also contained *Mammuthus* bones. There is some evidence suggesting the bones from this site may be artifacts.

158. Blake, W., Jr (1983): Geological Survey of Canada radiocarbon dates XXIII. Geological Survey of Canada Paper 83-7:1–34.

This paper reports the radiocarbon age of the Hillsborough mastodon (*Mammut americanum*). This animal died after becoming mired in a swamp near Hillsborough, New Brunswick (45°54.6′N, 64°39.8′W). A radiocarbon date of 13,600 ± 220 BP (GSC-1222) was provided by a fragment of treated bone. Other related dates taken from surrounding peat [>43,000 BP (GSC-1680)] and organic detritus taken from supposed coprolites [37,200 ± 1310 (GSC-2469)] suggest a much older age. [See Harington, Grant, and Mott 1993.]

159. Blake, W., Jr (1984): Geological Survey of Canada radiocarbon dates XXIV. Geological Survey of Canada Paper 84-7:1–35.

This paper reports several Pleistocene bone dates. A mastodon (*Mammut americanum*) femur turned up by a plow in a field near the settlement of Lower Middle River, Cape Breton, Nova Scotia (46°8.1′N, 60°55.2′W) was radiocarbon dated at 31,300 ± 500 BP (GSC-1220-2). The age and sedimentary context of this specimen suggest that the fossil has been reworked by fluvial or perhaps glacial action. A vertebra of a balaenopterid whale (cf. *Balaenoptera acutorostrata*) – NMC-36170 from the base of a 2-4 m-thick clay unit in a fresh ditch excavated near the S bank of the Kouchibouguacis River, New Brunswick (46°43.4′N, 64°59.6′W) – was dated at 12,600 ± 140 BP (GSC-3284). This same specimen had previously been aged at 18,570 ± 500 BP (S-1969) and 11,830 ± 950 BP (S-1969[B]), by the Saskatchewan Research Council Radiocarbon Dating Laboratory. The Saskatchewan age as reported by S-1969 is questioned. Ages from GSC-3284 and S-1969(B) together suggest that the area was submerged by over 8 m of water around 12,500 BP. Stellar sea lion (*Eumetopias jubatus*) remains associated with aged serpulid worm tubes [dated to 12,400 ± 160 BP (GSC-3726)] were uncovered on Bowen Island, British Columbia (49°23.8′N, 123°21.8′W) during excavations in 1981. The pit is 85 m asl and was dug about 3 m deep. These specimens relate to the retreatal stages of the Late Wisconsinan (Fraser) glaciation. Organic components of an arctic ground squirrel (*Spermophilus parryii*) nest (NMC 21094) from Dominion Creek, Yukon (63°48′N, 138°41′W), were dated at 12,200 ± 100 BP. The nest contained a nearly complete ground squirrel skeleton. This is the first date on a ground squirrel nest from the Yukon Territory. A right horn core of a saiga antelope (*Saiga tatarica*; USGS M1422) from a sand bar of the Usuktuk (also spelled Esuktuk) River, Alaska (69°58.4′N, 156°33′W), was dated to 37,000 ± 990 BP (GSC-3050). This specimen is the ninth known from North America, and the second reported from the Arctic Slope of Alaska; it is not far from the locality of a metacarpal (USNM 'RSFVL 75-30') from the bank of the Kuk River [see Harington 1980d]. These two specimens, along with the one from Baillie Islands, Northwest Territories, suggest a Late Pleistocene migration of saigas eastward across Bering Isthmus to northwestern Canada along the Arctic Slope.

160. Blake, W., Jr (1986): Geological Survey of Canada radiocarbon dates XXV. Geological Survey of Canada Paper 85-7:1–32.
A bison (*Bison*) cranium came from an excavation at 4th and 5th streets, Calgary, Alberta, on a terrace 5 m above the Bow River [elevation 1045 m (51°02′55″N, 114°04′25″W)]. The bone, recovered from crossbedded gravels referred to the Late Pleistocene Bighill Creek Formation, was assigned a radiocarbon age of 10,200 ± 280 BP (GSC-3065).

161. Blake, W., Jr (1987): Geological Survey of Canada radiocarbon dates XXVI. Geological Survey of Canada Paper 86-7:1–60.
This paper reports a number of radiocarbon dates of mammals from northern North America. Packrat (*Neotoma cinerea*) faeces from the Grotte Valerie Cave on the South Nahanni River, Northwest Territories (61°17′N, 124°05′W), elevation 760 m, were dated at 110 ± 60 BP (GSC-2035). A tusk fragment of a mammoth (cf. *Mammuthus*) was dated at 21,600 ± 2300 BP (GSC-1760-2). This specimen was probably collected near Cape James Ross, Melville Island, Northwest Territories, in 1916 by the Northern Party of the Canadian Arctic Expedition (V. Stefansson – leader) [Kindle 1924 and Harington 1990b]. A walrus (*Odobenus rosmarus*) cranium with tusk was collected from the upper part of a 15 m sea cliff on the W side of Coburg Island, Nunavut (75°52.5′N, 79°02′W). The tusk's date of 8690 ± 100 BP (GSC-2899) agrees closely with other nearby marine pelecypods and marine algae fossils. A whale rib, of either *Balaena*, *Balaenoptera*, or *Megaptera* (78°30.8′N, 75°32.5′W) at an elevation of 80 m, was the highest sample of marine origin found in the area. It was dated at 6500 ± 260 BP (GSC-3600) but is considered too young for the elevation. Other specimens, mainly marine pelecypods from lower elevations, were aged at 8150 ± 80 BP (GSC-3089) and 8190 ± 110 BP (GSC-2913). A narwhal (*Monodon monoceras*) tusk from the E side of Erik Harbour, Ellesmere Island (78°37.5′N, 74°44′W) at about 3 m asl, provided a date of 1180 ± 50 BP (GSC-1918). A probable bowhead whale (cf. *Balaena mysticetus*) rib was found imbedded in beach gravel and sand on top of a moraine ridge (elevation of 54 m) on the SW side of Rosse Bay, Ellesmere Island, Nunavut (78°38′N, 74°48′W). This specimen supplied a date of 6920 ± 90 BP (GSC-3055). A walrus tusk found attached to a partial cranium imbedded in sandy silt in an upland (non-beach) site near Read Bay, Cornwallis Island (75°01.2′N, 93°34′W), provided a date of 3510 ± 50 BP (GSC-2951). This age is unexpectedly young, suggesting that deposition so far from the sea may have been influenced by external factors. A few of the dates reported in this paper refer to Greenland vertebrate specimens. These are: whale (Cetacea) bone, perhaps a mandible (sample BS-74-20), >39,000 BP (GSC-2257), from Uvdlisaitut W at 9 m asl; marine pelecypod shells in association with whale (Cetacea) bones [12,200 ± 160 BP (GSC-2346] from Uvdlisaitut W at the highest raised beach section. A series of dates were done on marine pelecypod shells and a whale bone from beach ridges near Uvdlisaitut, northern Greenland (76°36′N, 69°45′W). Barnacle fragments associated with a narwhal (*Monodon monoceros*) tusk fragment (NMC 43811) were collected from an exposure along the E coast of Nordvestø, Greenland (76°42.8′N, 73°07′W), about 1.5 m above the high-tide level. The barnacles dated to 3460 ± 130 BP (GSC-2102). The present date list, GSC XXVI, is the fifteenth to be published directly in the Geological Survey of Canada's Paper series. Lists prior to GSC XXII were published first in the journal *Radiocarbon* and were reprinted as GSC Papers.

162. Blake, W., Jr (1992): Holocene emergence at Cape Herschel, east-central Ellesmere Island, Arctic Canada: Implications for ice sheet configuration. Canadian Journal of Earth Sciences 29(9):1958–80.
Twenty-five radiocarbon age determinations on marine mollusc shells, basal organic pond sediments, charred remains in archaeological sites, and other materials [two dates on a whale (Cetacea) rib from the region at an elevation of 54 m yielded 6920 ± 90 BP (GSC-3055) and 7120 ± 25 (TO-33)] have allowed the construction of an emergence curve for Cape Herschel, Ellesmere Island, Nunavut (78°35′N, 74°40′W). The author notes that marine mammal bones and tusks are scarce in raised marine deposits in the field area compared with the abundance of seal (*Phoca*), walrus (*Odobenus rosmarus*), and narwhal (*Monodon monoceras*) there today. Archaeological sites

on Cape Herschel contained charred bone fragments [charred bone (Mammalia) from an elevation of 10.5 m yielded a date of 2800 ± 60 BP (GSC-3064), whereas a narwhal tusk from about 3 m yielded a date of 1180 ± 50 BP (GSC-1918)] and other charred residues, both probably mainly derived from marine mammals; and the scarcity of bones on and in the raised beaches may also reflect their use by Paleoeskimos (*Homo sapiens*).

163. Blake, W., Jr. (1996): Interstadial deposits from Ellesmere and Coburg islands, Arctic Canada. Program and Abstracts of the 14th Biennial Meeting of the American Quaternary Association (May 20–2, Flagstaff, Arizona):151.
Near Cape Storm, southern Ellesmere Island, Nunavut, where an older generation of marine deposits underlies the Holocene raised beaches, a series of finite age determinations in the 35,000–42,000-year range have been reported. New AMS radiocarbon dates on an Alcidae sternum [presumably from a Dovekie (*Alle alle*)] of 35,560 ± 1550 BP (Ua-2212) and on a whale (Cetacea) rib of 39,130 ± 550 BP (TO-5241) have reinforced the results on marine pelecypod shells and marine algae. In addition, new amino acid determinations on *Mya truncata* shells from the main organic detritus level at 63 m asl have yielded ratios averaging 0.017 (AAL-7700) for the hydrolysate fraction, a value even lower than the 0.020 to 0.022 ratios (AAL-4389 to 4390) previously obtained on the same species from the same horizon. These data, with organic material dated between 43,000 and 20,000 BP from near Makinson Inlet and Coburg Island, support the concept of a high-latitude Cape Storm Nonglacial Interval.

164. Bliss, L.C., and Richards, J.H. (1982): Present day arctic vegetation and ecosystems as a predictive tool for the arctic-steppe mammoth biome. *In*: Paleoecology of Beringia (D.M. Hopkins, J.V. Matthews, Jr, C.E. Schweger, and S.B. Young, eds.). Academic Press, New York. pp. 241–57.
An ecosystem model is presented for Beringia 25,000 to 11,000 BP based on the dominance of four major plant communities and six large herbivores [woolly mammoth (*Mammuthus primigenius*), horse (*Equus*), steppe bison (*Bison priscus*), caribou (*Rangifer tarandus*), muskox (*Ovibos moschatus*), and moose (*Alces*) – see Figure 2 showing the energy flow within a theoretical ecosystem in Beringia about 25,000 BP]. Assuming daily forage requirements of the herbivores, net annual plant production, and forage harvest rates for a sustained yield, an estimated herbivore biomass and annual production are calculated, as well as human (*Homo sapiens*) population density (.02–.03 km^{-2}). The authors believe that only a diverse vegetation assemblage could have supported a diversity of herbivores and thus a human population. They postulate that the earlier concept of an 'upland sedge–grass steppe–mammoth–man' ecosystem could not have maintained a diversity of herbivores and carnivores without a diversity of plant communities.

165. 'B.M.P.' (1967): Remains of mastodon found on Woodhouse farm in 1897. Simcoe Reformer, January 18, p. 9.
The first article gives information on the loan of a display (Centennial exhibit) of the Marburg Mastodon (*Mammut americanum*) to the Eva Brook Donly Museum in Simcoe by the National Museums of Canada. The second article, from *The Simcoe Reformer* of June 17, 1897, provides details of the find on June 11 on the farm of Charles Challand on Concession 5, Lot 15, in Woodhouse Twp. A third article by Prudence M. Austin of Woodhouse Township provides a human interest story about the discovery of the mastodon by a farm worker while digging to install a tile drain. Finally, there is an official summary of the find as outlined in a report of the Geological Survey of Canada Vol. X, 1897. Evidently Dr H.M. Ami spent six weeks collecting in the area and noting the geological context. Accompanying illustrations include a reconstruction of an American mastodon and a map of Woodhouse Twp showing the spot where the mastodon remains were found.

166. Bobrowsky, P., and Rutter, N.W. (1992): The Quaternary geologic history of the Canadian Rocky Mountains. Géographie physique et Quaternaire 46(1):5–50.
This is a review of the Quaternary geology of the Canadian Rocky Mountains, especially the stratigraphy and sedimentology associated with past glacial and nonglacial events that have affected the region during the Pleistocene. The main focus is on the last glaciation in the area and its glacial geology, although paleoecological (botanical and paleontological) and paleoenvironmental evidence is mentioned. The region is arbitrarily divided into northern, central, and southern sections (Figure 1). Table 1, a list of Pleistocene radiocarbon dates for the Canadian Rocky Mountains and surrounding areas, in-

cludes 17 dates on bone, tusk, or tooth of various ice age mammals. A section on vertebrate paleontology (pp. 31–2) mentions: horse (*Equus*) and mammoth (*Mammuthus*) from preglacial and glacial sediments near Dawson Creek, British Columbia; ground squirrel (*Spermophilus*), Niobrara horse (*Equus niobrarensis*), bison (*Bison*), and a proboscidean (Proboscidea) described from interglacial or older sediments at Watino, Alberta; pre–Late Wisconsinan mammoth (*Mammuthus*) from Taylor and Portage Mountain moraine, British Columbia; Columbian mammoth (*Mammuthus* cf. *Mammuthus columbi*) from Babine Lake, British Columbia, 34 species of Mid-Wisconsinan vertebrates from Eagle and January caves in southwestern Alberta; mountain goat (*Oreamnos*) and groundsloth (*Megalonyx*) from Sangamonian deposits near Quesnel and Quesnel Forks, British Columbia; a vertebrate faunal assemblage (beginning about 10,800 BP) from stratified deposits at Charlie Lake Cave, British Columbia; isolated bison (*Bison*) remains from near Fort St John and Taylor, British Columbia; to the E, several postglacial large mammal taxa [woolly mammoth (*Mammuthus primigenius*), Mexican horse (*Equus* cf. *Equus conversidens*), Niobrara horse (*Equus niobarensis*), steppe bison (*Bison priscus*), western bison (*Bison bison occidentalis*), wood bison (*Bison bison athabascaae*), muskox (*Ovibos* cf. *Ovibos moschatus*), and camel (*Camelops*)]; wapiti (*Cervus elaphus*) skeleton about 9000 BP from near Watino, Alberta; bighorn sheep (*Ovis canadensis*) about 9300 BP from the northern Rocky Mountain Trench, British Columbia. Assemblages from Medicine Hat and Cochrane, Alberta, are mentioned. There is also a section on archaeology (p. 34).

167. Bobrowsky, P.T., Catto, N., and Levson, V. (1991): Reconnaissance Quaternary geological investigations in Peace River District, British Columbia (93P, 94A). British Columbia Geological Survey Branch Geological Fieldwork 1990, Paper 1991-1:345–58.

A bison (*Bison*) tibia, recovered from stratified pebble-gravel at 56°06.8′N, 121°47.9′W, was radiocarbon dated at 10,240 ± 160 BP (AECV-1206C). It was found in the Ostero Gravel Pit near Taylor, British Columbia, 5.5 m above the base of the gravel pit and 12.8 m below the upper ground surface [see Apland and Harington 1994]. A previous find of mammoth bone (*Mammuthus*) aged at 27,400 ± 580 BP (GSC-2034) from the same section led Mathews (1978a) to correlate the gravel with other gravel of Mid-Wisconsinan age. The new date suggests that the gravels may have been reworked postglacially. Detailed stratigraphy is discussed and illustrated.

168. Bocherens, H., Fizet, M., Mariotti, A., Gangloff, R.A., and Burns, J.A. (1994): Contribution of isotopic biogeochemistry (^{13}C, ^{15}N, ^{18}O) to the paleoecology of mammoths (*Mammuthus primigenius*). Historical Biology 7:187–202.

Carbon isotopic abundances in Alaskan, Albertan, and Russian mammoths indicate a diet of C_3-plants. The relatively high nitrogen isotopic abundances suggest arid conditions in Alberta and Alaska during the last ice age [presumably Late Pleistocene], and similar conditions in Russia. Nitrogen isotopic abundances are higher in mammoths relative to coeval herbivores, which may be due to differences in protein content of their diet. Oxygen isotopic abundances are similar in mammoths and in modern mammals from the same areas. The characteristic isotopic signature of mammoth ivory allows distinction from elephant or marine mammal ivory. Six specimens from Alaska include molars from Lost Chicken Creek, Chatanika, Koyukuk River, Eielson Air Force Base and tusks from Fairbanks and Alaska [unspecified]. Alberta bones sampled for δ^{13}C and δ^{15}N from Pit 48 (53°39′N, 113°17′W) near Edmonton include: horse (*Equus*), mammoth (*Mammuthus*), and ground sloth (*Megalonyx*), and span the period 40,000–25,000 BP.

169. Bocherens, H., Pacaud, G., Lazarev, P.A., and Mariotti, A. (1996): Stable isotope abundances (^{13}C, ^{15}N) in collagen and soft tissues from Pleistocene mammals from Yakutia: Implications for the paleobiology of the Mammoth Steppe. Palaeogeography, Palaeoclimatology, Palaeoecology 126:31–44.

A suite of skeletons and frozen carcasses of Upper Pleistocene mammals from Yakutia (Sakha Republic, Russia) has been analysed for stable isotopic abundances of carbon and nitrogen. The isotopic enrichment between herbivorous and carnivorous species is similar in modern and Pleistocene specimens, except for mammoths (*Mammuthus primigenius*), which show more negative δ^{13}C values and higher δ^{15}N values of collagen relative to other herbivorous species. A similar isotopic pattern can be seen in Upper Pleistocene mammoths from Alaska, and this pattern suggests paleobiological significance. The authors discuss the similarity of nitrogen values of mammoths from Siberia, Alaska, and Alberta on p. 40 and provide a chart (Figure 3) of $\delta^{13}C$ and $\delta^{15}N$ values of bone collagen from Upper Pleistocene samples from Siberia and Alaska (mammoths, herbivores, and carnivores).

170. Bombin, M., and Muehlenbachs, K. (1985): $^{13}C/^{12}C$ ratios of Pleistocene mummified remains from Beringia. Quaternary Research 23:123–9.
The purpose of this study is to use the $\delta^{13}C$ of mummified fossil tissues from Late Pleistocene frozen sediments of Yukon, Alaska, and Siberia as indicators of relative abundances of C_3/C_4 plants in extinct ecosystems of Beringia. The samples are ideal for study by the stable isotope technique because diagenesis has been minimal and the animals were all herbivores. Carbon stable isotope ratios of Beringian Late Pleistocene mummified remains [dried flesh, tendon, hide, hair, and hornsheaths of bison (*Bison*), horses (*Equus*), mammoth (*Mammuthus*), caribou (*Rangifer tarandus*), muskox (Ovibovini), moose (*Alces*), woolly rhinoceros (*Coelodonta antiquitatis*), and other undetermined species] found preserved in permafrost indicate that these large herbivores fed exclusively on C_3 plants (Table 1). In paleoclimatic terms, the results indicate that, even if radiation and the xeric nature of the Beringian environments were compatible, the growing season was probably too short and/or cool for C_4 graminoids to compete with C_3 species. The higher productivity of the Beringian steppes was probably based on temperate C_3 graminoids, which are adapted to maximize their growth during the spring and early summer (cool-season species), including wheat grasses (*Agropyron*), blue grasses (*Poa*), fescues (*Festuca*), needle grasses (*Stipa*), and xeric sedges.

171. Bonnichsen, R. (1978): Critical arguments for Pleistocene artifacts from the Old Crow Basin, Yukon: A preliminary statement. *In*: Early Man in America. From a Circum-Pacific Perspective (A.L. Bryan, ed.). Occasional Papers No. 1, Department of Anthropology, University of Alberta, Published by Archaeological Researchers International, Edmonton. pp. 102–18.
The fossil sites in the Old Crow area, Yukon, which offer evidence of altered bones, are reviewed and mapped, and arguments that support a Pleistocene age for these redeposited artifacts (*Homo sapiens*) are presented. Careful examination of the partially mineralized Pleistocene specimens indicate that it would not have been feasible for the users to have originally created the tools from partially fossilized material but rather that the bones were green when the artifacts were made. An illustrated section explains how the tools were created. A table of apatite radiocarbon dates is provided with ages for 22 altered specimens. The earliest uncorrected radiocarbon age included is [>39,900 BP (I-4428) corrected date and lab number: 33,800 ± 2000 BP] on a mammoth (*Mammuthus*) bone. Illustrations and photographs of many of the artifacts support the text.

172. Bonnichsen, R. (1979): Pleistocene bone technology in the Beringian Refugium. National Museum of Man Mercury Series, Archaeological Survey of Canada, Paper No. 89:1–297.
This study, undertaken between 1973 and 1978, focuses on the alteration of Pleistocene mammal bone from sites in Eastern Beringia (particularly Old Crow Basin, Yukon) by geological, biological, and cultural [human (*Homo sapiens*)] agencies. Bones of the mammoth steppe fauna – e.g., woolly mammoth (*Mammuthus primigenius*), horse (*Equus*), steppe bison (*Bison priscus*), and caribou (*Rangifer tarandus*) – are mainly considered in the study. As a control, a reference collection of bones modified by known processes was assembled and bone-breaking experiments were conducted for the purpose of separating fossil bones altered by the three processes mentioned. Several hundred bones show spiral fractures that are often associated with marrow extraction, and more than 90 artifacts were identified (e.g., bone and/or ivory bifaces; small blade cores; bone-edge spalls like microburins; flaked tools modified by whittling, grinding, planing, and polishing). Many of the bone artifacts were probably employed for butchering and hide-working. These simple tools may have been discarded at butchering stations – new artifacts could be rapidly produced at each new butchering site. This study suggests that emphasis was placed on bone-working in Mid-Wisconsinan times in the Yukon. As extinction of the Pleistocene megafauna occurred, humans probably responded by developing new technology emphasizing the use of lithic artifacts. Many examples of modified bones are illustrated, and Table 29 lists 14 radiocarbon dates on bison (*Bison*), helmeted muskox (*Bootherium bombifrons* = *Symbos*), and caribou (*Rangifer tarandus*) from Alaskan and Yukon specimens.

173. Bonnichsen, R., and Sorg, M.H. (eds.) (1989):

Bone Modification. Published by the Center for the Study of the First Americans, University of Maine, Orono, Maine. 535 pp.
Papers are cited individually in this volume. [See Irving et al. 1989.]

174. Borden, C.E. (1960): An early site in the Fraser Canyon, British Columbia. National Museum of Canada Bulletin 162:101–18.
Paleoindians (*Homo sapiens*), presumably adapted to life along rivers, and particularly to salmon (*Oncorhynchus*) fishing, occupied a site near Yale, British Columbia, on the Fraser River.

175. Bornhold, B. (ed.) (1997): Paleoenvironmental studies in Saanich Inlet, British Columbia. Delta 8(1):17–19.
Report of a 1996 international research team that drilled for two days in Saanich Inlet, Vancouver Island, British Columbia. Their intent was to obtain an ultra-high-resolution record of climate, oceanography, terrestrial vegetation, and oceanic productivity since deglaciation, 12,000 to 14,000 BP. Many of the cores contained abundant wood and charcoal as well as fish (Pisces) bones and scales. Evidence of earthquakes, ash falls (Mazama), and floods were identified.

176. Bostock, H.S. (1961): Physiography and resources of the northern Yukon. Canadian Geographical Journal 63(4):12–119.
The author notes (p. 118) that placer mines in Yukon and Alaska have yielded skeletons and bones of mammoth (*Mammuthus*), mastodons (*Mammut americanum*), camels (Camelidae), horses (*Equus*), and bison (*Bison*) – some even preserved in the flesh by permafrost. Northern Yukon explorers from the early days have remarked on the tusks seen there, and the native people are reported as saying that the course of the Old Crow River was the best locality in Alaska and Yukon to find them. Prophetically, Bostock states: 'Certainly the geological setting of the Old Crow Plain with its recent subsidence and unglaciated character suggests that it should be.' [See Harington 1977a, 1989c.]

177. Bouchard, M.A., Harington, C.R., and Guilbault, J.-P. (1993): First evidence of walrus (*Odobenus rosmarus* L.) in Late Pleistocene Champlain Sea sediments, Quebec. Canadian Journal of Earth Sciences 30:1715–19.
The almost complete skull (MCQ 91-117) of a walrus was discovered in Champlain Sea sediments at Sainte-Julienne-de-Montcalm (45°57.9′N, 73°41.9′W), altitude 94 m, 60 km N of Montreal, Quebec, on August 22, 1990. It was the first report of this species in glacial marine deposits W of Quebec City. The skull has been dated by accelerator mass spectrometry at 10,090 ± 60 BP (TO-2224) and the shells in the sediment adhering to it at 10,880 ± 60 BP (TO-2225). The age of the skull implies that walruses were part of the Champlain Sea marine mammal fauna consisting of three species of seals and five whale species. As the sea drained, walruses withdrew toward the Gulf of St Lawrence, where they have been hunted to extinction in historic times. A photograph of the specimen, a schematic of the stratigraphy, and a location map are provided. In addition, a table of known Quebec Quaternary walrus remains is supplied.

178. Bowers, P.M. (1978): Geology and archaeology of the Carlo Creek site, an Early Holocene campsite in the central Alaska Range. American Quaternary Association, Abstracts of the Fifth Biennial Meeting (September 2–4, Edmonton, Alberta). p. 188.
Early Holocene vertebrate remains were found at a former human (*Homo sapiens*) occupation at the Carlo Creek site, a former sandbar on the upper Nenana valley, central Alaska Range. The remains included butchered ground squirrel (*Spermophilus*), Dall sheep (*Ovis dalli*), and caribou (*Rangifer tarandus*). Radiocarbon dates on hearth-charcoal yield ages of 8400 ± 200 BP (WSU-1700), 8690 ± 330 BP (GX-5132), and 10,040 ± 440 BP (GX-5131).

179. Bowyer, V., Kuzyk, G.W., Russell, D.E., Farnell, R.S., Gotthardt, R.M., Hare, P.G., and Blake, E. (1999): Caribou remains at Thandlät: Archaeology and paleoecology of some well-preserved sites on ice patches in the southwestern Yukon. Programme and Abstracts, Canadian Archaeological Association, 32nd Annual Conference (April 28–May 2, Whitehorse, Yukon). pp. 22–3.
In 1997, well-preserved prehistoric organic artifacts and faunal material were discovered melting out of a permanent snowpatch in the Kusawa Lake area of the southwestern Yukon. Caribou (*Rangifer tarandus*) faecal remains from stratified deposits within the snowpatch have

been radiocarbon dated to about 4300 BP. Caribou were present in the area at least 2500 years ago, although they are absent in the region today. A wooden dart/arrow shaft (*Homo sapiens*) recovered from the edge of the snow-patch is a rare example of Mid-Holocene organic technology in North America. The site offers an opportunity to explore the ecology of prehistoric caribou as well as implications of climatic change for caribou.

180. Brassard, G.R., and Blake, W., Jr (1978): An extensive subfossil deposit of the arctic moss *Aplodon wormskioldii*. Canadian Journal of Botany 56:1852–9.
A 2.6-m-thick subfossil peat deposit, composed entirely of the arctic moss (*Aplodon wormskioldii*), from Nordvestø (76°44.2′N, 73°13′W), in the Carey Islands of northwestern Greenland, yielded tiny barbs of bird (Aves) feathers and one nearly-intact feather embedded in the peat at a depth of 122 to 121 cm (p. 1856). Presumably the feathers would date to about 5000 BP, since radiocarbon dates on peat at depths of 253–258 cm and 15–18 cm at the site are 6300 ± 80 BP (GSC-2368) and 4390 ± 140 BP (GSC-2415), respectively (Table 1). On p. 1858, the authors mention that occasional fragments of bird feathers along the core suggest that gulls (Laridae) or other cliff-nesting birds, or birds preying on them, frequented the site during the peat-forming period. [See Bennike 1997, p. 902.]

181. Brend, Y. (1997): Ancient fish fossils excite scientists. Whitehorse Star, November 25:1-2.
Bernard Lauriol (University of Ottawa) found remains of two ancient fishes (Pisces) along Porcupine River downstream from Old Crow, Yukon last July. One is estimated to be 700,000 to one million years old. Soft mud perfectly preserved it down to each ray of its fins and crease of its gills. Steve Cumbaa, a paleontologist who specializes in fossil fish at the Canadian Museum of Nature in Ottawa, says the specimen looks more like a grayling (*Thymallus*) than anything else. Nobody is sure of the exact species yet, but the specimen is about 40 cm long. Accompanying the article are photographs of archaeologist Ruth Gotthardt holding a siltstone rock with a clear outline of the skeleton (p. 1), and Bernard Lauriol holding the fish fossil in the field (p. 2).

182. Brennin, R., Murray, B.W., Friesen, M.K., Maiers, L.D., Clayton, J.W., and White, B.N. (1997): Population genetic structure of beluga whales (*Delphinapterus leucas*): Mitochondrial DNA sequence variation within and among North American populations. Canadian Journal of Zoology 75:795–802.
Belugas (*Delphinapterus leucas*) are migratory over much of their range, congregating in small groups around shallow river estuaries in summer and overwintering in areas of broken pack ice. To investigate the potential genetic structuring within North American belugas, the authors analysed variation in mitochondrial DNA. Using 10 restriction enzymes, eight haplotypes were identified among 95 belugas from 12 sampling locations. The haplotypes formed two lineages, one occurring primarily in whales from the St Lawrence estuary and eastern Hudson Bay, and the other primarily in belugas sampled in the waters of western Hudson Bay, southern Baffin Island, western Greenland, the Canadian High Arctic, and eastern Beaufort Sea. These lineages may represent the original Pacific and Atlantic refugial stocks that colonized the Arctic after deglaciation.

183. Brigham-Grette, J., and Hopkins, D.M. (1991): The first paleomagnetic evidence from Cape Deceit: A critical Late Pliocene/Early Pleistocene mammal, plant macrofossil, and beetle locality in Central Beringia, Alaska. Geological Society of America Annual Meeting (October 21–4, San Diego), Abstracts with Programs 23(5):99.
Coastal exposures of unconsolidated sediments on the S shore of Kotzebue Sound near Deering, NW Alaska, contain a complex sequence of interbedded organic-rich gravel, sand, and silt subdivided in ascending order into the Cape Deceit Formation, Inmachuk Formation, and Deering Formation (Guthrie and Matthews 1971). The Cape Deceit Formation is important because of its enclosed fauna, which includes the earliest-known occurrences of the collared lemming lineage (*Predicrostonyx hopkinsi*) and of caribou (*Rangifer*). A peat bed in the Cape Deceit Formation contains remains of larch-forest tundra. Matthews (1974) suggested an age between 0.4 and 0.9 Ma for the Cape Deceit fauna. More recent estimates are 1.2–1.8 Ma (Sher 1986) and about 2.4 Ma (Repenning et al. 1987) based on comparison and telecorrelations with other mammal localities in Eurasia, particularly the Olyor fauna of NE Siberia. Paleomagnetic inclination data through silt-rich Unit 2 of the Cape Deceit Formation indicates the entire unit is normally magnetized. The Cape Deceit Formation may have been de-

posited during the Olduvai subchron, but given the associated fauna, the authors think it more likely that it was deposited during the Jaramillo subchron [about 0.9–0.96 Ma].

184. Brown, R. (1870): On the physics of arctic ice, as explanatory of the glacial remains in Scotland. Quarterly Journal of the Geological Society of London 26:671–701.
The author mentions finding 'fossiliferous clay' up to a height of more than 500 ft asl on the banks overlooking glaciers in central West Greenland. He continues (p. 691): 'I found this clay everywhere along the coast and in Leer Bay, south-west of Claushavn; in knots [concretions?] of this clay are found impressions of the Angmaksaett (*Mallotus arcticus*, O. Fabr.), a fish still quite abundant in Davis Strait.' [This fossil is of capelin (*Mallotus villosus = Mallotus arcticus*.]

185. Brumley, J.H., and Rushworth, C.A. (1983): A summary and appraisal of Alberta radiocarbon dates. *In*: Archaeology in Alberta, 1982 (D. Burley, ed.). Archaeological Survey of Alberta Occasional Paper No. 21:142–222.
The authors have attempted to collect all presently available radiocarbon dates from Alberta (Table 4). The list is intended as a reference tool and not as a primary database. Researchers interested in dates relating to a specific area of the province, time interval, or cultural complex can use it to quickly identify primary data sources, where such information is available. The date list is estimated to comprise 80 to 90% of all radiocarbon dates available for the province. Table 3 summarizes abbreviations and conventions used in Table 4, which contains 341 dates. Table 5 summarizes the number of radiocarbon-dated sites and the total number of dates from the various parts of the province. Table 6 summarizes the cultural association relative to the number of dates (e.g., AV 21 means 21 dates are associated with the Avonlea culture). An extensive list of references follows. Where bone is dated in Table 4, unfortunately for those interested in taxa represented, there is no indication of the species or skeletal part sampled.

186. Bryan, A.L. (1994): Early human occupations in the southern part of the 'ice-free corridor.' Current Research in the Pleistocene 11:121–3.
Recent research in Alberta leads to the conclusion that conditions in the 'ice-free corridor' between about 11,500 BP and 21,500 BP were unsatisfactory for occupation by large mammals (Mammalia), including humans (*Homo sapiens*). The presence of the Laurentide ice and severe climatic conditions were the principle mitigating factors. The author reviews the history of fluted points in Alberta and cautions future researchers to search early sites for tools other than fluted points, as there is as yet no clear proof that the early (before the Laurentide ice) inhabitants of Alberta, and hence western North America, used fluted points.

187. Bryant, H.N. (1987): Wolverine from the Pleistocene of the Yukon: Evolutionary trends and taxonomy of *Gulo* (Carnivora: Mustelidae). Canadian Journal of Earth Sciences 24:654–63.
The author describes wolverine (*Gulo gulo*) specimens of possible Sangamonian to Wisconsinan age collected by the Northern Yukon Research Programme between 1975 and 1982 in the Old Crow Basin, Yukon. This location is approximately 160 km N of the Arctic Circle with a total area of about 8300 km^2. Dental and mandibular measurements were made on the 28 specimens and compared with extant *Gulo gulo* and other fossil species. The Old Crow Basin specimens average 2.7% smaller than comparable measurements of the living wolverine, and the width of P_4 is statistically significant. These results do not differ from Harington's (1977a) previous analysis of a similar sample from Old Crow Basin.

188. Buckland, W. (1831): On the occurrence of the remains of elephants, and other quadrupeds, in the cliffs of frozen mud, in Eschscholtz Bay, within Bering's Strait, and in other distant parts of the shores of Arctic seas. *In*: Narrative of a Voyage to the Pacific and Bering's Strait (F.W. Beechey, ed.). Colburn and Bentley, London. pp. 593–612.
Remains of mammoth (*Mammuthus*) and other Pleistocene mammals were collected from the coastal bluffs at Eschscholtz Bay, Alaska, in 1828 by Captain Beechey, Royal Navy, and Mr Collie. They were presented to the British Museum and were first reported on by the author.

189. Buckley, A. (1994): It's the granddaddy of all bones. Yukon News, February 4:1–2.
A caribou (*Rangifer tarandus*) leg bone found in the summer of 1989 in silty sand under a layer of volcanic rock near Fort Selkirk, Yukon, is the oldest known, according to Richard Harington, curator of ice age zoology at Ot-

tawa's Canadian Museum of Nature. The specimen was dated to more than one million years old according to radioisotopic and paleomagnetic methods of dating. Vancouver-based geologist Lionel Jackson was doing fieldwork in southern Yukon when he and his assistants found the 3–4-in-long bone fragment. 'The specimen doesn't look any different than that of the same bone of a modern caribou,' Harington said. 'Also, we originally thought the age of this volcanic rock was 1.2 to 1.3 million years. Recent discussions indicate it may be even older than that.' Jackson also found a remnant of hare (*Lepus*), as well as the bones of a medium-sized rodent (Rodentia) and teeth from an extinct vole-like creature [since identified as *Lasiopodomys* by C.A. Repenning]. 'It's a unique and interesting situation that should be probed further,' says Harington, who will be in the Yukon on February 19 to talk about the find.

190. Burke, A., and Cinq-Mars, J. (1996): Dental characteristics of Late Pleistocene *Equus lambei* from the Bluefish Caves, Yukon Territory, and their comparison with Eurasian horses. Géographie physique et Quaternaire 50(1):81–93.
Bluefish Caves I, II, and III of northern Yukon have yielded the earliest *in situ* evidence of human (*Homo sapiens*) occupation of Eastern Beringia, associated with one of the largest and most diverse Late Pleistocene faunal assemblages recovered in the region. This assemblage contains a large sample of teeth (524 cheekteeth so far) of the Yukon horse (*Equus lambei*). Remains of *Equus lambei* from the caves have been radiocarbon dated between 12,900 ± 100 BP (GSC 2881, femur) and 17,440 ± 220 BP (RIDDL 278, metatarsal). A smallish, broad-skulled horse, *Equus lambei* strongly resembles Late Pleistocene equids from Siberia. The broad, wide muzzle is attributed to an adaptation to cold conditions during the Late Pleistocene. Overall size, estimated using tooth dimensions only, is approximately the same as contemporaneous European horses and many modern caballoids. Ordinarily in horses, the females display significantly weaker or nonexistent canines. This trait is used to identify the sex ratios. A much-higher-than-usual proportion, almost 85%, of the Bluefish *Equus lambei* adult mandibles possess well developed canine teeth. Either more bachelor herds are represented in the sample or many of the females retain the canine. If the latter is the case, this characteristic could be diagnostic for the species.

191. Burke, A., and Cinq-Mars, J. (1998): Paleoethological reconstruction and taphonomy of *Equus lambei* from the Bluefish Caves, Yukon Territory, Canada. Arctic 51(2):105–15.
Bluefish Caves, northern Yukon, have yielded evidence of pre-Holocene human (*Homo sapiens*) occupation of Eastern Beringia. The three Bluefish Caves contain a large Late Pleistocene fauna [including remains of predators such as wolf (*Canis lupus*), bears (*Ursus arctos*), and lions (*Panthera leo*)] *in situ*. Radiocarbon dates on *Equus lambei* bones from Bluefish Caves are: 12,900 ± 100 BP (GSC-2881) femur, Cave I; 12,290 ± 440 BP (NMC-1236) bone fragment, Cave III; 17,440 ± 220 BP (RIDDL-278) metatarsal, Cave I; and 22,680 ± 530 BP (NMC-1237) bone fragment, Cave II. Research on mortality patterns and paleoethology of the Yukon horse (*Equus lambei*) – a dominant component of the Bluefish fauna – was based on dental remains. Mortality profiles for *Equus lambei* indicate that predators were likely primary agents of bone accumulation at Cave I, whereas Caves II and III bones apparently accumulated through accidental or natural deaths, probably regularly monitored by humans and other predators/scavengers. Paleoethological reconstruction for *Equus lambei* supports the suggestion that Bluefish Basin was not a polar desert during the Late Pleistocene. The use of tooth height/age tables to establish age profiles of fossil equid populations is demonstrated to be limited to establishing broad, relative-age categories.

192. Burns, J.A. (1980): The brown lemming, *Lemmus sibiricus* (Rodentia, Arvicolidae), in the Late Pleistocene of Alberta and its postglacial dispersal. Canadian Journal of Zoology 58(8):1507–11.
Fossil remains of brown lemming (*Lemmus sibiricus*) were found in January Cave, Alberta (50°11′N, 114°31′W) at an elevation of 2040 m asl. In all, 45 specimens were found, consisting of 43 teeth, a frontal bone, and a toothless mandible. A radiocarbon date on bone collagen from the four lowest levels from the NE extension of the cave where three of the specimens were found was 23,100 ± 860 BP (GaK-5438). Thermoluminescence dating on specimens of ground squirrel (*Spermophilus*) and marmot (*Marmota*) from the second-lowest level of Area SE yielded a date of 22,355 ± 1105 BP (MATL-79-4-1). The author argues for the existence of a southern relict population of the brown lemming prior to a north-

ward postglacial dispersal to British Columbia and the Yukon Territory.

193. Burns, J.A. (1982): Water vole *Microtus richardsoni* (Mammalia, Rodentia) from the Late Pleistocene of Alberta. Canadian Journal of Earth Sciences 19(3):628–31.
A Late Pleistocene water vole (*Microtus richardsoni*) specimen (left mandible fragment with M_1) was recovered from Eagle Cave in the Rocky Mountains of southern Alberta (49°38′N, 114°38′W). It was exhumed from a bed that was radiocarbon dated at 22,700 ± 1000 BP (GaK-2336). This find is significant as it represents the oldest-known example of this species; Pleistocene finds are rare. The presence of water voles in southwestern Alberta during the Late Wisconsinan suggests that local conditions along the southern part of the mountain front were not climatically extreme.

194. Burns, J.A. (1986): A 9000-year-old wapiti (*Cervus elaphus*) skeleton from northern Alberta, and its implications for the Early Holocene environment. Géographie physique et Quaternaire 40(1):105–8.
A complete skeleton in good condition of a male wapiti (*Cervus elaphus*) was discovered near Watino, Alberta, in 1984 in the Wakaluk Quarry. The first radiocarbon date of collagen from its antler gave 9075 ± 305 BP (S-2614), and a second date on ribs was 9920 ± 220 BP (AECV-272C). The paper refers to other wapiti finds.

195. Burns, J.A. (1987): Late Quaternary zoogeography of the northern pocket gopher, *Thomomys talpoides*, in southwestern Alberta. Canadian Field-Naturalist 101(3):419–22.
Specimens of northern pocket gophers (*Thomomys talpoides*) were found in two locations. The first one is January Cave in northern Alberta (50°11′N, 114°31′W) at an elevation of 2040 m asl. Three radiocarbon dates have established that the sedimentation took place in the Mid-Wisconsinan Interstadial: 23,100 ± 860 BP (GaK-5438), 31,900 ± 1400 BP (QL-1738), and 33,500 ± 1100 BP (QL-1737). The second location is Eagle Cave in Crowsnest Pass in southwestern Alberta, 10 km W of Coleman, at an elevation of 1377 m asl. Two radiocarbon dates for the unit indicate a Mid-Wisconsinan age: 22,700 ± 1000 BP (GaK-2336) and >33,000 BP (S-1963). A map (Figure 1) shows location of fossil and recent occurrences in Alberta.

196. Burns, J.A. (1989): Fossil vertebrates from Rats Nest Cave, Alberta. Canadian Caver 89(Spring):41–3.
The author lists vertebrate taxa found in 1986 and 1988 digs in Rat's Nest Cave on Grotto Mountain near Exshaw, Alberta. Among the remains are several snake (Serpentes) vertebrae, frog and toad (Salientia) bones, and a wide range of avian (Aves) and fish (Pisces) bones. Radiocarbon dating on bone protein gave ages ranging from 2480 ± 80 BP (AECV-275C) to 7060 ± BP (AECV-295C).

197. Burns, J.A. (1990): Paleontological perspectives on the ice-free corridor. *In*: Megafauna and Man: Discovery of America's Heartland (L.D. Agenbroad, J.I. Mead, and L.W. Nelson, eds.). Mammoth Site of Hot Springs, South Dakota, Inc. Scientific Papers 1:61–6.
From about 1975 to 1990, vertebrate paleontological research in Alberta has confirmed the existence of an ice-free, cool, dry steppe tundra from 40,000–20,000 BP. This includes work on fossiliferous mountain caves, extinct prairie dog (*Cynomys*) towns, and buried valley fill in the Edmonton region. However, no evidence of human (*Homo sapiens*) occupation has been identified for this interval. From 20,000 to 14,000 BP, no archaeological or paleontological sites are known for the area in question. The first human presence in the corridor is about 10,770 BP, associated with a bone date from Charlie Lake Cave in the Peace River District of British Columbia. The oldest date for human occupation in the Alberta part of the corridor is associated with a radiocarbon date of 10,400 BP at the Vermilion Lakes site near Banff, and the earliest postglacial vertebrate remains are about 11,200 BP from Cochrane, west of Calgary. Geological evidence of a human migration route to the heartland of North America suggests that passage may have been possible by 14,000 BP. The author summarizes many of his discoveries and those of others concerning Quaternary vertebrates and paleoenvironments in Alberta during the last 40,000 years. He gives a previously unpublished species list of Mid-Wisconsinan mammals from gravel pits in the Edmonton area: horse (*Equus*), bison (*Bison*), wapiti (*Cervus elaphus*), barren ground caribou (*Rangifer tarandus*), mammoth (*Mammuthus*), American mastodon (*Mammut americanum*), wolf (*Canis lupus*), camel (*Camelops*), muskox (*Ovibos*), American lion (*Panthera*

leo atrox), and giant short-faced bear (*Arctodus simus*). The dates on bone collagen range from 39,960 to 22,750 BP. A date on a mammoth tusk from a locality near High Level, Alberta (Figure 1), of 22,020 BP, with similar dates from the Hand Hills and Empress, suggests a synchronous ice advance into Alberta along a front from NW to SE and from a dispersal centre to the NE.

198. Burns, J.A. (1991): Mid-Wisconsinan vertebrates and their environment from January Cave, Alberta, Canada. Quaternary Research 35:130-143.
Fossil remains of vertebrates were found in January Cave, Alberta (50°11′N, 114°31′W), at an elevation of 2040 m asl. The six most common taxa in January Cave are: pika (*Ochotona princeps*), hoary marmot (*Marmota caligata*), Columbian ground squirrel (*Spermophilus columbianus*), wood rat (*Neotoma cinerea*), long-tailed vole (*Microtus* cf. *Microtus longicaudus*), deer mouse (*Peromyscus maniculatus*), and arctic grayling (*Thymallus arcticus*). Also found were many fish (Pisces) scales and vertebrae. Roughly 90% of all identified fossils were rodents (Rodentia). Radiocarbon dates from collagen of mixed species provided dates of 33,500 ± 1100 BP (QL-1737), 31,900 ± 1400 BP (QL-1738), and 23,100 ± 860 BP (GaK-5438).

199. Burns, J.A. (1996a): Review of the Pleistocene zoogeography of prairie dogs (genus *Cynomys*) in western Canada with notes on their burrow architecture. *In*: Palaeoecology and Palaeoenvironments of Late Cenozoic Mammals: Tributes to the career of C.S. (Rufus) Churcher (K.M. Stewart and K.L. Seymour, eds.). University of Toronto Press, Toronto. pp. 34–53.
The fossil history, zoogeography, and paleoenvironment of prairie dogs (*Cynomys*) in western Canada are described. The temporal span may be from as early as 'Nebraskan glacial' to as young as the late nineteenth century, although the majority of the fossils fall into the Late Wisconsinan interstadial. The early finds were misidentified as a black-tailed species, but a comprehensive review of all available diagnostic material shows them to be white-tailed prairie dogs (subgenus *Leucocrossuromys*). A map illustrates the 24 known fossil localities. A table accompanies the map to provide the latitude, longitude, name, and age of each site. Some of the other vertebrates from the sites are identified. An extensive paleoburrow excavation at the Winter site of Hand Hills, Alberta, is also illustrated. Table 1 includes estimates of the ages of sites containing fossil prairie dogs (*Cynomys*) in Alberta and Saskatchewan, the sites being mapped in Figure 2.

200. Burns, J.A. (1996b): Vertebrate paleontology and the alleged ice-free corridor: The meat of the matter. Quaternary International 32:107–12.
Based on over 30 radiocarbon dates from across Alberta which show a gap of chronology during the 20,000 BP to 11,600 BP time period, the author strongly suggests that the use of the term 'ice-free corridor' is inaccurate. The author suggests the area be called the 'western interior route' with a pre- or postglacial modifier. He contends that there is no evidence for the corridor concept but rather that the province prior to 21,000 BP was an unglaciated steppe landscape followed by about 10,000 years of complete ice coverage. By 11,600 BP, southwestern Alberta was able to support a viable and fairly complex ecosystem, but the archaeological evidence suggests that humans (*Homo sapiens*) did not make their way into the area for another thousand years or so. By the time humans arrived, the broad expanse was hardly the essence of a spatially restricted 'corridor.' The ice had receded and left migrants a broad plains landscape, newly vegetated and replete with game. Table 1 includes nine radiocarbon dates on mammoth (*Mammuthus*), four dates on horse (*Equus*), one date on caribou (*Rangifer tarandus*), four dates on prairie dogs (*Cynomys*), and three dates on bison (*Bison*). All are bone collagen dates.

201. Burns, J.A., and McGillivray, W.B. (1989): A new prairie dog, *Cynomys churcherii*, from the Late Pleistocene of southern Alberta. Canadian Journal of Zoology 67:2633–9.
This paper describes a new taxon of Churcher's prairie dog (*Cynomys churcherii*), recovered at Hand Hills, Alberta, 28 km NE of Drumheller (51°34′N, 112°20′W) at an elevation of 1090 m asl. Radiocarbon dates on bone collagen set the time of occupation at about 22,000 BP (RIDDL-681). The holotype is an almost complete skeleton of a young adult. The animal is larger than *Cynomys ludovicianus*, *Cynomys leucurus*, and *Cynomys gunnisoni*. [See Goodwin 1993.]

202. Burns, J.A., and Young, R.R. (1993): The Late Pleistocene paleontology and stratigraphy of the Hand Hills, Alberta: *In*: The Palliser Triangle; A Re-

gion in Space and Time [a festschrift honouring Dr Archibald MacSween Stalker]. University of Lethbridge. pp. 207–14.

This is a preliminary paleontological/geological/stratigraphic study of bones and burrows of prairie dogs (*Cynomys*) first discovered in the Hand Hills (40 km NE of Drumheller), Alberta, in 1983. The Winter site (Figure 1), located in the NW corner of the Hand Hills, 6 km S of Delia, contains the most extensive network of *Cynomys* burrows of the nine sites recognized in the Hand Hills. Although prairie dogs are unknown in the Holocene of Alberta, they have been recovered from 18 Middle to Late Pleistocene sites in southern Alberta. All diagnostic specimens of *Cynomys* from the 18 sites are 'white-tailed' prairie dogs referable to the subgenus *Leucocrossuromys* – perhaps a new species. They are larger than any extant prairie dog. AMS radiocarbon dates on bones from the Winter and Courtney sites in the Hand Hills, and the Schowalter site in the Wintering Hills, 40 km S across the Red Deer River valley, range from about 33,460 to 22,200 BP [no laboratory numbers are given]. Near the end of this Mid-Wisconsinan interstadial, characterized by prairie dog occupation, climatic conditions more favourable to preservation of organic soil prevailed. Prairie dogs were subsequently extirpated by changing conditions during the last glaciation of Alberta. Many tens of metres of infilled tunnels at the Winter site were mapped, excavated, and screened for bone. At least 10 vertebrate species (including several ground squirrel species, pocket gopher (*Thomomys*), hare (*Lepus*), mice, voles (*Microtus*), an amphibian (Amphibia), and two bird (Aves) species were associated in the contemporary fauna.

203. Burns, J.A., and Young, R.R. (1994): Pleistocene mammals of the Edmonton area, Alberta. Part I. The carnivores. Canadian Journal of Earth Sciences 31(2):393–400.

Four major Pleistocene fossil sites are described from the Saskatchewan gravels near Edmonton: A. Consolidated Concrete Pit 48 – 53°38′30″N, 113°17′0″W; B. Clover Bar Sand and Gravel Ltd. Pit – 53°35′50″N, 113°21′30″W; C. Standard General, Riverview Pit – 53°36′30N, 113°18′20″W; D. Consolidated Concrete Pit 45 – 53°48′30″N, 113°40′30″W. This report provides descriptions and measurements of wolf (*Canis* cf. *Canis lupus*), short-faced bear (*Arctodus simus*) – first record for Alberta – and American lion (*Panthera leo atrox*) bones found at these sites. A faunal list of 15 Mid-Wisconsinan taxa identified from the Edmonton area is given in Table 1. The herbivores (at least 13 taxa) are discussed in Young et al. (1994). A woolly mammoth (*Mammuthus primigenius*) humerus from Pit 48(A) yielded a radiocarbon date of 26,750 ± 790 BP (AECV-1102C). A mammoth (*Mammuthus*) pelvis from the Clover Pit (B) has provided a date of 22,820 ± 520 BP (AECV-538c). Also from (B), a bison (*Bison*) metacarpal returned a date of 11,620 ± 170 BP (AECV-1203c). The Riverview Pit (C) is dated by two samples – an unidentified mammal bone gave a date of 31,290 ± 1960 BP (AECV-941c) and a bison horncore provided a date of >41,260 BP (AECV-942c).

204. Burns, J.A., Young, R.R., and Arnold, L.D. (1993): Don't look fossil gift horses in the mouth; they, too, have tongues of ice. Program and Abstracts of the Joint Annual Meeting of the Geological Association of Canada and the Mineralogical Association of Canada (May 17–19, Edmonton). p. A-14.

This abstract mentions over 1000 Pleistocene fossils donated to the Provincial Museum of Alberta in Edmonton over the past four years from 10 local gravel pits. In descending order of numbers these consist of horses (*Equus*), bison (*Bison*), mammoth (*Mammuthus*), caribou (*Rangifer tarandus*), muskoxen (*Ovibos*), ground sloth (*Megalonyx*), mastodon (*Mammut americanum*), camel (*Camelops*), beaver chewed wood (*Castor canadensis*), wolf (*Canis lupus*), lion (*Panthera leo atrox*), and short-faced bear (*Arctodus simus*). Radiocarbon dates on this material range from 42,900 BP to 21,300 BP (40 dates) with four 'greater-thans.' The evidence suggests the existence of a significant cool-steppe environment over much of Alberta during this period, suitable for abundant grazers and attendant predators. It is only after 11,600 BP that vertebrates reappear in Edmonton, strongly suggesting that ice precluded all vertebrates and higher plants from the region between 21,300 and 11,600 BP. There is thus no vertebrate evidence for an ice-free corridor as it was originally defined.

205. Butler, D. (1971): Oh, dem bones ... Ottawa Citizen, June 5, p. 43.

This article, on the opening of a new paleontology hall at the Victoria Memorial Museum Building of the National Museum in Ottawa, Ontario, mentions exhibits of whales (Cetacea) and seals (Phocidae) of Champlain Sea age that occupied the Ottawa Valley about 10,000 years ago, when sea water covered Parliament Hill to a depth of 60 ft. Fossils of marine and freshwater fish (Pisces) of similar age from Green Creek E of Ottawa were to be displayed. Fossils of American lion (*Panthera leo atrox*), short-faced bear (*Arctodus simus*), giant moose (*Alces latifrons*), and giant beaver (*Castoroides ohioensis*) were to be featured too.

206. Byun, A.S., Koop, B.F., and Reimchen, T.E. (1997): North American black bear mt DNA phylogeography: Implications for morphology and the Haida Gwaii refugium controversy. Evolution 51:1647–53.
The controversial role of Haida Gwaii (Queen Charlotte Islands, British Columbia) as a biological refugium on the NW coast of North America has been discussed for more than 50 years. The presence of morphologically divergent subspecies on Haida Gwaii is one of the major lines of evidence suggesting the islands' role as a refugium during the Wisconsinan glaciation. However, since morphological distinction can be derived postglacially as well as in extended isolation, such evidence is ambiguous. To examine this question, the authors carried out a phylogenetic analysis of cytochrome b sequences (719 bp) of black bear (*Ursus americanus*), one of the distinctive endemics of Haida Gwaii, and compared these with conspecifics from across North America – focusing mainly on the NW coast. They found that Haida Gwaii bears are indistinguishable from coastal bears of British Columbia, but are highly distinct from continental bears. The coastal mitochondrial lineage occurs in each of the three recognized coastal subspecies, suggesting that the morphological characteristics differentiating these taxa may be postglacially derived. The data are consistent with recent suggestions that a glacial refugium existed on the now-submerged continental shelf connecting Haida Gwaii, Vancouver Island, and the coastal fringe of mainland British Columbia. This refugium would have been an additional source for postglacial recolonization of northwestern North America.

207. Cameron, A.W. (1951): Greenland right whale recorded in Gaspé County, Quebec. *In*: Annual Report of the National Museum of Canada for the Fiscal Year 1949–50. National Museum of Canada Bulletin 123:116–19.
The author discusses a specimen of Greenland right or bowhead whale (*Balaena mysticetus*) that was exhumed near Sainte-Anne-des-Monts in Gaspé County, Quebec. The whale was buried under 2 m of sand and beach gravel, and extended 2.5 m beneath that; it was 5 m above the present high-tide level. The whale remains were found nearly 2000 km S of the species' present range. The skeleton is incomplete, as the smaller bones of the specimen are damaged, and rostrum and mandibles are missing. The bone measurements are included in the paper; its age was unspecified.

208. Cameron, A.W. (1958): Mammals of the islands in the Gulf of St Lawrence. National Museum of Canada Bulletin 154:1–165.
This study, conducted from 1949 to 1955, covers the mammalian zoogeography of Cape Breton Island, Prince Edward Island, Anticosti, Magdalen Islands, and Newfoundland – with comparisons from adjacent mainland areas. Most islands located on the continental shelf have had rather recent land connections, and therefore their faunas are usually identical with those of the nearby mainland. That such a situation does not exist in three of the five larger islands in the Gulf is due to the fact that they probably have had no postglacial land connections. They might be designated 'postglacial islands.' The three northern islands – Newfoundland, Anticosti, and the Magdalens – differ from the neighbouring mainland in that more than 50% of mainland mammal species are absent, and at least one order and three families are unrepresented. On the other two, Cape Breton Island and Prince Edward Island, which probably have had land connections in the last 12,000 years, no less than four species are absent. Mammals that colonized the three northern islands were admitted by a sweepstakes route, while a filter bridge existed in the case of the southern islands. Geological evidence indicates that Newfoundland, Anticosti, and the Magdalens were cut off from the mainland before the ice retreat; therefore the mammals now occurring on them were obliged to cross a water barrier. The types of barriers that have isolated these populations are discussed. The author suggests that there were very few colonizations by any one species and that each colonization consisted of a very few individuals. Therefore, inbreeding has probably been partially responsible for the rapid rate of subspeciation.

209. Campbell, J.D. (1952): Paleobotany and stratigraphic sequence of the Pleistocene Klondike 'muck deposits.' PhD thesis, McGill University, Montreal. 131 pp.

This study presents the first pollen diagram worked out for the central Yukon, and, as a result of field observation in the summers of 1950 and 1951 at a mining camp on Hunker Creek near Dawson City and laboratory examination, a newly formulated chronology of Pleistocene events in the Klondike. The author notes that at Hunker Creek, the bones of large extinct mammals are found only at middle and lower levels of valley-bottom deposits. Remains of certain moose-like (cf. *Alces*) animals are found in the lowermost silt beds a little above 'grade,' while Pleistocene horses (*Equus*) and some forms of large bison [probably the steppe bison (*Bison priscus*)] and muskoxen (Ovibovini) appear in the region of interspersed coarse beds. At grade level a few mammoth (*Mammuthus*) specimens are found, and in levels dug by the dredge below grade, considerable amounts of ivory are recovered, often in good condition. Mastodon (*Mammut americanum*) remains are very rare in the Klondike, having been recovered only once or twice and then from muck of doubtful position. Campbell thought that the mastodon became extinct in this region before the mammoth (p. 58). Campbell gives a first-hand account of a steppe bison (*Bison priscus*) that Otto Geist had just excavated from one of the creeks near Fairbanks, Alaska: 'It had the skin of head, forefeet and most of the body preserved like tanned leather, and about half a bushel of loose hair was found around it.' He notes that no remains of flesh and hide are reported from the Klondike, but bones are plentiful in the muck of valley bottoms, especially in the coarse lower deposits and a little above 'grade' level on Hunker Creek. He found, or was present at the finding of, remains of: grizzly bear (*Ursus arctos*); horse (*Equus*); wapiti (*Cervus elaphus*); moose (*Alces*); extinct moose-like cervids [giant moose? (*Alces latifrons*?)]; several kinds of bison (*Bison*); muskox-like bovids (Ovibovini); and mammoth (*Mammuthus*) (p. 101).

210. Canby, T. (1979): Search for the first Americans. National Geographic 156(3):330–63.
This article deals with the first humans (*Homo sapiens*) in Beringia and their subsequent dispersal into the Americas. As well, it refers to certain other mammals (Mammalia) that also entered the Americas about the same time as humans, who were probably hunting them: mammoth (*Mammuthus*), bison (*Bison*), and horse (*Equus*). The article includes maps of Siberia, Beringia, and America showing details of Quaternary human and other mammal sites. [See Parfit 2000.]

211. Carl, G.C. (1949): Provincial Museum of Natural History and Anthropology. Report for the year 1949. Province of British Columbia, Department of Education. 23 pp.
This report mentions that a mammoth (*Mammuthus*) from Victoria, British Columbia, and mastodon (*Mammut americanum*) tooth from Shawnigan Lake, British Columbia, were donated to the Provincial Museum of Natural History and Anthropology [now the Royal British Columbia Museum].

212. Carlson, A.K., and Carlson, R.L. (1998): Dating the Likely mammoth. The Midden 30(4):4–5.
In May 1991 the authors visited Likely, British Columbia, where Paul Leward and Michael Mattice had found a mammoth (*Mammuthus*) tooth 33 cm long × 9 cm wide in a gravel pit at the NW end of Little Lake (52°37′N, 121°44′W), 7 mi W of the bridge over the Quesnel River at Likely. Some tooth fragments were still *in situ* in glaciofluvial gravels and sand at an elevation of about 2850 ft asl. Traces of soil in the interstices of the tooth indicated that it originated from the surface of the gravels, then slipped downslope. A sample of the tooth yielded a radiocarbon date of 20,190 ± 190 BP (Beta-50901). Other mammoth (*Mammuthus*) bones in the lower Fraser Valley (Hicock et al. 1982) range from 22,700 ± 320 BP to 21,400 ± 240 BP, indicating that a suitable mammoth habitat existed then on the Fraser Plateau and the lower Fraser River.

213. Carlson, C.C., and Klein, K. (1996): Late Pleistocene salmon of Kamloops Lake. *In*: Life in Stone: A Natural History of British Columbia's Fossils (R. Ludvigsen, ed.). University of British Columbia Press, Vancouver. pp. 274–80.
Some concretions eroding from glacial-lake sediments along the S shore of Kamloops Lake contain skeletons of salmonids (family Salmonidae). The late geologist and paleontologist Richard Hughes identified these fossils as sockeye salmon (*Oncorhynchus nerka*). Hughes originally collected at this site in the 1970s, acquiring specimens for the Caribou College geology section. His suggested Late Pleistocene age for the salmon was undoubtedly based on the age of the host sediment – a layer of fine-grained, silty, grey clay extending along the Kamloops Lake shore from above the high-water line to below the low-water level. The site was relocated in the fall of 1991, and the authors made several collecting trips be-

tween 1992 and 1994, recovering more than 175 specimens (none complete). Because of their age (older than those from archaeological sites), the fossils are important for understanding the zoogeography of salmon and their potential for use by early humans (*Homo sapiens*). One specimen measured 42 cm from the snout to what appears to be the base of the tail. The presence of a prominent hooked jaw on many of the specimens suggests that they were spawning fish that died near the lake shore – thus accounting for the concentration of fish at one locality. The authors agree with Richard Hughes that the fish is *Oncorhynchus*, but are uncertain of the species, suspecting them to be Kokanee, the landlocked and stunted forms of sockeye (the heads are small and ^{13}C isotope evidence from bone collagen in the fossils show that they did not eat protein from marine sources). A bone sample collected from a fossil near the surface yielded a date of 15,480 ± 60 BP (Beta-70750 CAMS-11912). Samples from buried specimens yielded a date of 18,110 ± 90 BP (Beta-76622 CAMS-16369), indicating that the Thompson River valley was not covered by glacial ice near the peak of the last glaciation as previously thought. Also, it casts doubt on Fladmarks's (1975) hypothesis that salmon were not available for human consumption in the Interior until some 5000 BP.

214. Carlson, R.L. (1994): Early cultural traditions in British Columbia west of the Rocky Mountains. Current Research in the Pleistocene 11:123–5.

The author provides a review of the five cultural traditions known in British Columbia before 8000 BP. These are in order of antiquity: Fluted Point Tradition, Stemmed Point Tradition, Plano Tradition, Pebble Tool of Foliate Biface Tradition, and the Microblade Tradition. All are named for the lithic tool type used. Information on Tradition age, archaeological sites, tool descriptions, and Tradition derivation is included. The Fluted Point, Stemmed Point, and Plano traditions are interior hunting traditions that reached British Columbia from the continental interior. The Pebble Tool and Microblade traditions are probably coastal derivatives of pre-Denali and Denali Beringian traditions.

215. Carlton, S. (1982): Ancient bones near Sarnia excite experts. London Free Press, August 23. p. A5.

A construction crew clearing land for a diversion channel unearthed a 2000-year-old collection of animal bones at a site 8 km E of Sarnia, Ontario. Dr John H. McAndrews of the Royal Ontario Museum in Toronto, who visited the site with a team of specialists, termed it 'the biggest collection of well-preserved bones in Ontario' outside of an archaeological site. The team identified skeletal remains of: wolf (*Canis lupus*); deer (*Odocoileus*); moose (*Alces alces*); wapiti (*Cervus elaphus*); rabbit (*Sylvilagus*); muskrat (*Ondatra zibethicus*); beaver (*Castor canadensis*); frog (*Rana*); three kinds of turtle (Chelonia); ducks and geese (Anatidae); and fish (Pisces). No signs of domesticated animals such as cows were discovered, nor were any artifacts, indicating that people were not involved with the animals' deaths. The animals may have become mired in part of a lake that once covered the region – the bones were 1 m below the surface between a layer of peat and marl. McAndrews said that it was rare to find so many wapiti of various ages in one place. Samples of fossil pollen from the mud around the bones were taken for analysis at the Royal Ontario Museum in order to find out more about the environment when the animals lived.

216. Carrier, L. (1989): Le karst de Kingsmere; étude de ses remplissages, parc de la Gatineau, Quebec. MA thesis, University of Ottawa. 120 pp.

Mine Cave is located in Gatineau Park, Quebec, about 20 km NW of Ottawa and near the Kingsmere Estate (Figures 1.1; 6.1). Among local karst features, Mine Cave possesses the best archaeological potential. A paleontological study of Mine Cave's filling has permitted an inventory of the Holocene fauna living in Gatineau Park. The fauna consists of 15 species: raccoon (*Procyon lotor* 13%); woodchuck (*Marmota monax*); muskrat (*Ondatra zibethicus*); black bear (*Ursus americanus*); big brown bat (*Eptesicus fuscus* 18.1%); white-footed mouse (*Peromyscus leucopus*); porcupine (*Erethizon dorsatum*); meadow vole (*Microtus pennsylvanicus*), white-tailed deer (*Odocoileus virginianus*); grey squirrel (*Sciurus carolinensis*); eastern chipmunk (*Tamias striatus*); masked shrew (*Sorex cinereus*); star-nosed vole (*Condylura cristata*); red fox (*Vulpes vulpes*); and American toad (*Bufo americanus* 62.6%). Humans (*Homo sapiens*) can be added according to incisions on some bones (p. 57, Table 4). Raccoon, big brown bat, and American toad remains were found in all 10 of the 10-cm-deep stratigraphic levels at the site (Table 6; Annexes).

217. Carter, L.D. (1982): Late Wisconsin desertification in northern Alaska. Geological Society of America, 95th Annual Meeting. Abstracts with Programs 14(7):461.
Remains of mammoth (*Mammuthus*), horse (*Equus*), and bison (*Bison*) were found in sand-dune deposits on the Alaskan Arctic Coastal Plain between the Colville and Ikpikpuk rivers. Radiocarbon dates of organic material above the bones range from 30,000 to 12,000 BP; the mammal remains are estimated to be older than 28,000 BP.

218. Carter, L.D., Brigham-Grette, J., and Marincovich, L., Jr (1986): Late Cenozoic Arctic Ocean sea ice and terrestrial paleoclimate. Geology 14:675–8.
Sea otter (*Enhydra lutris*) remains found in deposits of two marine transgressions (Bigbendian and Fishcreekian) of the Alaskan Arctic Coastal Plain that occurred between 2.4 and 3 Ma suggest that during these two events the southern limit of seasonal sea ice was at least 1600 km farther N than at present in Alaskan waters. Perennial sea ice must have been severely restricted or absent, and winters were warmer than at present during these two sea-level highstands.

219. Casteel, R.W. (1975): An early post-glacial record of the Pacific sardine, *Sardinops sagax*, from Saanich Inlet, Vancouver Island, British Columbia. Copeia 3:570–9.
At McCurdy Point, Saanich Inlet, off the coast of Vancouver Island, British Columbia (48°33.9′N, 123°30.2′W), 11 cores of 6-m-long overlapping sections were taken, of which one contained a left operculum of a sardine (*Sardinops sagax*). Its corrected radiocarbon age lies between 8850 and 9840 BP. This is significant because the oldest known *S. sagax* is no older than 2000 BP.

220. Catto, N., and Mandryk, C. (1990): Geology of the postglacial ice-free corridor. *In*: Megafauna and Man: Discovery of America's Heartland (L.D. Agenbroad, J.I. Mead, and L.W. Nelson, eds.). Mammoth Site of Hot Springs, South Dakota, Inc. Scientific Papers 1:80–5.
The postulated ice-free corridor extends from the Richardson and Mackenzie mountains southward along the foothills of the Rocky Mountains to the Canada/United States border. The primary control on the existence, location, and timing of the corridor was the eastward extent of the Cordilleran and montane glaciers, rather than the extent of the Laurentide ice. Although the Laurentide ice approached or reached its maximum westward extent during the Late Wisconsinan, the restricted Cordilleran and montane glaciers allowed an unglaciated corridor to exist between the eastern and western glaciers. The corridor area remained available as a geographically suitable migration route for flora and fauna throughout most of Wisconsinan time. There is, however, no reliable archaeological evidence indicating occupation of the Canadian prairies by people (*Homo sapiens*) prior to the warming phase of the Late Wisconsinan.

221. Champagne, D.E., Harington, C.R., and McAllister, D.E. (1979): Deepwater sculpin, *Myoxocephalus thompsoni* (Girard) from a Pleistocene nodule, Green Creek, Ontario, Canada. Canadian Journal of Earth Sciences 16(8):1621–8.
A nodule containing the cranium of a deepwater sculpin (*Myoxocephalus thompsoni*) was discovered in Pleistocene Champlain Sea deposits at Green Creek, 10 km E of Ottawa, Ontario (45°26′12″N, 75°35′05″W). It represents the first fossil reported for this species. The specimen is deposited in the Paleontology Collection of the Canadian Museum of Nature (formerly the National Museum of Natural Sciences) and is catalogued as NMC 34968. There is an interesting comparison of photographs which highlights the usefulness of ultraviolet light to improve contrast between bones and matrix. The discovery of this specimen (estimated age – 10,000 BP) provides an additional basis for refuting derivation of the species from a post-Wisconsinan marine submergence and suggests an origin at the beginning of the Wisconsinan or earlier. It also suggests that the Champlain Sea, or a nearby lake basin in that area, was an oligotrophic lacustrine environment with bottom summer temperatures below 8°C.

222. Chandler, G. (2000): Where ancient hunters gathered: Receding waters and erosive winds expose a lode of tracks, bones, and artifacts. Canadian Geographic 120(3):61–4.
In the fall of 1998, construction of a new spillway for the dam at St Mary Reservoir, 200 km S of Calgary, Alberta, forced the province to drop the water level 10 m below normal. Then powerful southwesterly winds whipped away exposed layers of fine sand and other sediments be-

fore Shayne Tolman, a high-school teacher from nearby Cardston, stumbled across several stone spearpoints in the reservoir bed while on a family outing. Tolman's find attracted the attention of University of Calgary researchers, including geologist Len Hills, who visited the site several times in the fall of 1998 and winter of 1999. In late February, the afternoon sun revealed hundreds of well-preserved footprints. Hills immediately surmised that the largest were mammoth (*Mammuthus*) prints and the others from long-extinct megafauna. Tolman had found North America's only archaeological site containing evidence of Late Pleistocene mammoths, camels (Camelidae), bovids (Bovidae), horses (*Equus*), and possibly wolves (*Canis lupus*), coupled with human (*Homo sapiens*) artifacts (spearpoints, other stone tools, and hearths with fire-cracked rock). The evidence suggests that the area was frequented by animals and hunters for 11,000 years, right up to the time of European contact. Hills and Brian Kooyman, a University of Calgary archaeologist, quickly assembled a team of 12 mostly volunteer archaeology and geology students to work on the project and secured emergency funding. Work began on April 19, 1999, at the 2-square-km 'Wally's Beach' site. 'One of our biggest heartaches was the loss of a set of human footprints,' said Hills. 'We found them just before they were obliterated on a particularly stormy day, before we could record and photograph them.' Some mammoth prints were so well preserved that doctoral candidate Paul McNeil could deduce that the average stride length for an adult was 2.6 m. Juvenile tracks, about 20 cm across, and a mammoth with a limp were also noted. What may be the hoofprints of a Late Pleistocene horse were discovered also. As well, complete skulls and articulated backbones of horses (*Equus conversidens*) were recovered. Examples of every projectile point style that ever existed in Alberta are represented at the site. Some of the earlier (Clovis) spear points, up to 13 cm long and designed for use against large animals like mammoths, are more than 11,000 years old. Over 200 intact projectile points and at least 1000 partial or complete stone knives, hide scrapers, and drills uncovered suggest that hunters slaughtered animals and prepared carcasses at the site. Some 1000 bones – mainly of muskox (Ovibovini) and horse (but none of mammoth) have been recovered. Radiocarbon dates of extinct muskox and horse bones place them in the same period as the Clovis points. The team was permitted at least one more study season – the reservoir was kept low again until the end of June 2000. Illustrations include: colour photographs of the site, bovid tracks, horse (*Equus conversidens*) bones, and Paul McNeil preserving and collecting tracks; as well as an excellent map of the area and a restoration of the ice-age megafauna as it may have appeared at the site.

223. Chaney, R.W. (1937): Notes on the finding of mammals and plants in frozen Pleistocene deposits near Fairbanks, Alaska. Geological Survey of America Proceedings 1936:399 (Abstract).

Chaney refers to the Fairbanks Exploration Company, and the fact that many mammals (Mammalia) and plants have been dug up during their regular dredging activities. The Pleistocene muck containing the fossils is said to have accumulated due to low-temperature slumping rather than more normal deposition processes. The plant samples indicate a similarity of environmental conditions at the time of deposition to the current conditions. No list of specimens or species is presented.

224. Chapdelaine, C., and LaSalle, P. (1995): Physical environments and cultural systems in the Saint Lawrence valley, 8,000 to 3,000 B.P.: A multidisciplinary framework. *In*: Archaeological Geology of the Archaic Period in North America (E.A. Bettis III, ed.). Geological Society of America Special Paper 297:115–29.

This paper reviews the state of knowledge of the changing paleoenvironment of the St Lawrence Lowlands and relates this to the archaeological site distribution. A detailed outline of the historic periods and archaeological evidence for each is presented. Few sites are known for the period before 6000 BP. The present configuration of the St Lawrence River was probably attained between 8000 and 6000 BP. After 7000 BP, only minor fluctuations of river level are recorded. The earliest evidence for human (*Homo sapiens*) occupation of the St Lawrence valley is at the Plano site in Rimouski, excavated in 1991, radiocarbon dated at 8150 ± 130 BP (Beta-47978). The Plano people were hence the first to occupy the St Lawrence Lowlands as they emerged from the waters of the Champlain and Goldthwait seas.

225. Chapman, E.J. (1858): Mastodon remains, Morpeth, Canada West. Canadian Journal of Industry, Science and Art 3:356–7.

A mastodon (*Mammut americanum* = *Mastodon ohioticus*) molar was found 2 m below the surface in drift sand above a limestone ridge in Morpeth, Ontario. In the drawings sent to the author, five distinct crown ridges were evident.

226. Chartier, M., Cournoyer, M., Daoust, N., Harington, C.R., Bouchard, M.A., and Guilbault, J.-P. (1997a): A new vertebrate assemblage from Latest Pleistocene marine sediments near St-Nicolas, Quebec. Canadian Quaternary Association, 8th Biennial Meeting, Program with Abstracts:10.
Approximately 50 vertebrate specimens have been recovered from Champlain Sea sediments exposed in sand and gravel pits near St-Nicolas, Quebec. The fossils belong to at least six taxa, including two species of fish [Atlantic wrymouth (*Cryptacanthodes maculatus*) and eelpout (*Lycodes*)], two species of aquatic birds [Thick-billed Murre (*Uria lomvia*) and Oldsquaw duck (*Clangula hyemalis*)], and two species of marine mammals [seal (*Phoca*) and white whale (*Delphinapterus leucas*)]. The fish and bird records are new for the Champlain Sea deposits. The material consists of well-preserved but disarticulated and isolated (mostly postcranial) elements. The stratigraphy of the site is summarized. [See Chartier et al. 1997c, and Occhietti et al. in press.]

227. Chartier, M., Cournoyer, M., Daoust, N., Harington, C.R., Bouchard, M.A., and Guilbault, J.-P. (1997b): A new assemblage of vertebrates from Champlain Sea sediments (Latest Pleistocene) near St-Nicolas, Quebec. Geological Association of Canada/Mineralogical Association of Canada Annual Meeting, Abstracts 22:A-25.
Next to Ottawa-area assemblages (e.g., Green Creek, Eardley, Breckenridge), the St-Nicolas locality is, in both sample size and taxonomic diversity, the richest known deposit of Champlain Sea vertebrates. [See Chartier et al. 1997a for a list of vertebrate fossils, and Occhietti et al. in press.]

228. Chartier, M., Dubreuil, M., Cournoyer, M., Daoust, N., Harington, C.R., Cumbaa, S.L., Bouchard, M.A., and Guilbault, J.-P. (1997c): New records of vertebrates from Champlain Sea deposits (Latest Pleistocene) near Quebec City, Canada. Fifty-seventh Annual Meeting, Society of Vertebrate Paleontology (October 8–11, Chicago, Illinois), Abstracts and Papers. Journal of Vertebrate Paleontology 17 (Supplement to Number 3):37.
Champlain Sea deposits are the result of a brief marine transgression-regression cycle following the northward retreat of the Laurentide ice from the St Lawrence valley in Late Wisconsinan time. Some 50 vertebrate specimens have been recovered from a regressive Champlain Sea sequence exposed in sand and gravel pits near St-Nicolas, S of Quebec City, Quebec. Fossils belong to at least six species, including two bottom-dwelling fishes [Atlantic wrymouth (*Cryptacanthodes maculatus*) and eelpout (*Lycodes*)], two aquatic birds [Thick-billed Murre (*Uria lomvia*) and Oldsquaw duck (*Clangula hyemalis*)], and two marine mammals [seal (*Phoca*) and white whale (*Delphinapterus leucas*)]. The fish and bird records are new for Champlain Sea deposits. Equally noteworthy is the unusually high concentration of avian remains in the sample (>20%). The material consists of well-preserved but disarticulated and isolated (mostly postcranial) elements, indicating transport and/or redeposition. The bones, together with a rich invertebrate macrofauna, occur in the upper part of a 5-m-thick unit of cross-bedded coarse to fine sand overlying marine clays and underlying shallow marine silts and clays. This unit is interpreted as a subaqueous delta-fan formation, based on its sedimentary character and fossil content. In both sample-size and taxonomic diversity, the St-Nicolas assemblages rank among the most important Champlain Sea vertebrate localities, being surpassed only by some assemblages from the Ottawa area of Ontario and Quebec. [See Occhietti et al. in press.]

229. Chernyavskii, F.B. (1984): Systematic relationship of some Old and New World land mammals in connection with the problem of Beringia. *In*: Beringia in the Cenozoic Era (V.L. Kontrimavichus, ed.). Amerind Publishing Co. Pvt. Ltd., New Delhi. pp. 501–10.
The author analyses the taxonomic relationships of certain species of Eurasian and North American mammals to distinguish the following Holarctic species complex: masked shrew (*Sorex cinereus*); arctic ground squirrel (*Spermophilus parryii* = *Citellus parryi*); red-backed vole (*Clethrionomys rutilus*); brown lemming (*Lemmus sibiricus*); tundra vole (*Microtus oeconomus*); wolf (*Canis lupus*); red fox (*Vulpes vulpes*); arctic fox (*Alopex lago-*

pus); brown bear (*Ursus arctos*); ermine (*Mustela erminea*); least weasel (*Mustela nivalis*); wolverine (*Gulo gulo*); lynx (*Felis lynx = Lynx lynx*); caribou (*Rangifer tarandus*); and moose (*Alces alces*). The following species occupy NE Siberia, and in NW ['northeastern' (sic)] America are replaced by vicars, e.g., hare (*Lepus timidus – Lepus othus*); pika (*Ochotona hyperborea – Ochotona collaris*); vole (*Microtus gregalis – Microtus miurus*); mountain sheep (*Ovis nivicola – Ovis dalli*). A disjunction is assumed to have taken place in the Middle Pleistocene. The following species-pairs are less closely related than those in the preceding list: squirrel (*Sciurus vulgaris – Sciurus carolinensis*); chipmunk (*Eutamias sibiricus – Eutamias minimus*); marten (*Martes zibellina – Martes americana*); mountain sheep (*Ovis nivicola – Ovis canadensis*); etc., and indicate an even earlier separation. These data suggest that the Bering Land Bridge formed at least three times in the Pleistocene, and that intensive faunal exchange took place between Asia and America – migration from Asia being predominant. An ecological analysis of Quaternary mammals of Beringia shows that tundra, lowland and mountain 'tundra-steppe,' forest-tundra and flood-forest landscapes were prevalent on the Bering Land Bridge in the second half of the Pleistocene.

230. Chesemore, D.L. (1968): Muskox skull from Teshekpuk Lake, northern Alaska. Canadian Field-Naturalist 82(2):152.
Muskoxen (*Ovibos moschatus*), once common on the coastal plain of northern Alaska, have not occupied the area since the late nineteenth century. During the summer of 1962, the author found the light-coloured upper part of a muskox skull with hornbases embedded in the surface of a low ice-core mound at the E end of Teshekpuk Lake (70°33′29″N, 152°37′40″W).

231. Chin, G. ed. (2000): Population genetics of frozen bears. Science 287:1365.
Leonard et al. (2000) have analysed mitochondrial DNA sequences from brown bear (*Ursus arctos*) samples preserved for more than 15,000 years in Alaska [and Yukon] permafrost. Surprisingly, the sequences within a single population of ice age bears showed a similar range of variation to modern North American bears. If this variation existed before the populations became geographically subdivided, then the current divergence is not the result of post-isolation genetic drift, but must have arisen by a 'sampling effect,' as the original population became fragmented and dispersed. The analysis of ancestral populations is a rare bonus for the population geneticist, and the exceptional preservation of ice age brown bears is unlikely to be repeated in many organisms. But reconstructions of past population history by phylogenetic inference to extant populations now have an added twist.

232. Chlachula, J. (1994): A Paleo-American (Pre-Clovis) settlement in Alberta. Current Research in the Pleistocene 11:21–3.
Investigations in the Bow River valley of southwestern Alberta provide evidence for a prehistoric occupation pre-dating the earliest Paleoindian ('projectile point') record known from western Canada. The two sites are in the city of Calgary. Site 1, found 24 m below the present soil surface, contained 22 artifacts including three finished tools, two cores, and many derived flakes, some of which could be related to the cores. Site 2, probably originally situated near Site 1, was discovered later and had been displaced by the valley ice advance. A number of the lithic specimens exhibit glacial striations superimposed on artificially flaked faces. In general the lithic industry displays analogous technological and typological traits of the Late Pleistocene Palaeolithic of NE Eurasia. An early Late-Wisconsinan age (ca. >21,000 BP) is supported by palynological evidence for Site 1. Site 2 predates the Bow Valley Advance and is assumed to be Mid- to early Late-Wisconsinan (ca. 25,000–21,000 BP). The geoarchaeological data from these sites suggest that the opening of the 'ice-free corridor' along the eastern flanks of the Rocky Mountains during the final Late Wisconsinan should no longer be regarded as a major precondition for the initial peopling of North America S of the continental ice, as people evidently occupied this area before the time of glacial coalescence in NW Alberta.

233. Chlachula, J. (1996): Geology and Quaternary environments of the first preglacial Palaeolithic sites found in Alberta, Canada. Quaternary Science Reviews 15(4):285–313.
A pebble-tool industry, including two chronologically different stone artifact assemblages reminiscent of the Eurasian Paleolithic, has been recorded at two locations [Site 1 (Varsity Estates) and Site 2 (Silver Springs) are located 2.5 km apart on the N side of the Bow River near

the NW margin of Calgary (Figure 1)] in the Bow River valley, southwestern Alberta. Geological context suggests a fluctuating braided river environment during the early occupation (Silver Springs). Following deglaciation of the valley, a later occupation is shown by a flaked-stone assemblage excavated in place on top of the till at nearby Varsity Estates. This occupation surface was buried under 24 m of sediments after submergence of the river valley by Glacial Lake Calgary, which was dammed by the Laurentide ice advance into the eastern Calgary area, implying a Late Wisconsinan age of >21,000 BP for the lithic industry. Palynological data associated with the later assemblage do not support the idea that an extremely cold, inhospitable glacial environment characterized the eastern slopes of the Canadian Rocky Mountains during the Late Wisconsinan. The author summarizes (p. 308) vertebrate fossil evidence indicating steppe-parkland communities throughout western Alberta during the Mid-Wisconsinan Non-Glacial Interval.

234. Chlachula, J. (1998): Reply to comment of R.R. Young, R.B. Rains, and G. Osborn on the paper by Chlachula (1996). Geology and Quaternary environments of the first preglacial Paleolithic sites found in Alberta, Canada. Quaternary Science Reviews, 16, 285–313. Quaternary Science Reviews, 17:455–7.
The author sees no reason to withdraw his interpretation that lithic artifacts on top of Bow Valley till and below Glacial Lake Calgary clays at Site 1 in Calgary, Alberta, are in primary position, although undated. Consequently, the site is evidence for a hiatus between Cordilleran and Laurentide ice sheets in the western Calgary area during the Late Wisconsinan. Clearly some kind of biota must have been present, although the available evidence is limited. It is unlikely that prehistoric people (*Homo sapiens*) lived on the floor of the forming proglacial lake near the Laurentide ice front as would be implied by the scenario of Young et al. The author hopes that further investigation will reconcile the model of Young et al. with the archaeological evidence he has reported. [See Young et al. 1998.]

235. Choquette, L.P.E., Harington, C.R., and Archibald, J. (1975): Paleopathology: Exostoses of the third metacarpal in Pleistocene horses from the Yukon Territory. Canadian Journal of Earth Sciences 12(6):1053–8.
Two third metacarpals from extinct Yukon Pleistocene horses display pathological exostoses. Both bones are stored at the Canadian Museum of Nature (formerly the National Museum of Natural Sciences). One specimen, a Yukon horse (*Equus (Asinus) lambei*), NMC 13473, from Gold Run Creek (63°43.5′N, 138°41′W), is of probable Late Pleistocene age. The other, from a larger horse (such as *Equus (Plesippus) verae*), NMC 14355, was collected from Old Crow River Loc. 15 (67°51.7′N, 139°48.5′W) and may be of Middle Pleistocene age or somewhat later.

236. Christiansen, E.A. (1968): Geology of the Saskatoon site. Napao 1(2):28–31.
This paper provides a stratigraphic context for the Saskatoon Pleistocene mammal site with comments on the age and origin of the layers. This site is a sandpit overlain and underlain by tills. Features of the sand layers, pebble wedges, and unconformities suggest that the sand was deposited in close proximity to a glacier. The sand unit lies between two tills of the Floral Formation and is estimated, based on radiocarbon-dated wood, to be older than 34,000 years. [For information on the fauna and archaeological aspects of the site see Lammers 1968 or Pohorecky and Wilson 1968.]

237. Christiansen, H.H., and Humlum, O. (1993): Glacial history and periglacial landforms of the Zackenberg area, Northeast Greenland: Preliminary results. Geografisk Tidsskrift 93:19–29.
About 8000 BP a delta some 20 m thick was built out into Young Sound in the southern part of Zackenberg lowland (Figure 2) after sea level had dropped to about 15–20 m above present sea level. Willow (*Salix herbacea*) leaves from the upper part of the fluvial forests of the delta have been dated to 7920 ± 140 BP (AAR-1217). This plant-rich debris contained droppings assigned to arctic hare (*Lepus arcticus*) on the basis of size and shape (MGUH VP3345), and the date has been corrected to 7140 ± 140 BP by Bennike (1997, p. 902, Table 1).

238. Christie, W.J. (1854–6): [Extracts from a letter regarding the finding of mastodon bones near Shell River, British America]. Boston Society of Natural History Proceedings 5:265–6.
Sir John Richardson sent a copy of this communication (to Sir George Simpson from Christie, dated Fort Pelly, Swan River District, January 28, 1855) to the president of

the society to be read to members. Christie states regarding remains of an American mastodon (*Mammut americanum* = *Mastodon*): 'I procured in October last an Indian guide, and proceeded to Shell River [Manitoba], where the bones of that animal were found. The shoulder bones, referred to by Sir John Richardson [see Richardson 1854 and 1861], together with other bones and the *teeth*, were found in the river by the Indians several years since, and were taken out and laid on the bank. Owing to some superstitious fear, the Indians, at the time of their discovery, would not bring them here; but two years afterwards they were sent for, and the shoulder bones, together with a tooth were brought here [Fort Pelly]; but whether or not the tooth was sent to Red River with the shoulder bones, I cannot ascertain. On reaching Shell River last fall, I carefully examined the spot where the Indians laid the bones, and found them; but having been buried, they had decayed so much that they fell to pieces on being touched. A portion of a large bone, and some fragments of what is supposed to have been a tooth, I brought here; these will be carried to Norway house [Norway House, Manitoba] next spring and disposed of as you may direct.'

239. Churcher, C.S. (1966): A report on some Pleistocene localities in southern Alberta and Saskatchewan and their vertebrate fossil faunas. Preliminary report of progress to the Geological Survey of Canada. 96 pp. [Copy in Quaternary Vertebrates of Northern North America file at Canadian Museum of Nature.]

The purpose of this field work was to investigate Pleistocene localities that Dr A.M. Stalker knew had produced vertebrate fossils and to discover other Pleistocene localities. A summary (pp. 89–90) indicates: (1) the presence of horse (*Equus conversidens*), wapiti (*Cervus elaphus* = *Cervus canadensis*), mountain sheep (*Ovis canadensis*), and plains bison (*Bison bison*) at Cochrane Pits, Alberta, that date to about 8000 years ago; (2) the presence of Spruce Grouse (*Canachites canadensis*), black-tailed prairie dog (*Cynomys* cf. *Cynomys ludovicianus*), Richardson's ground squirrel (*Spermophilus ?richardsonii* = *Citellus ?richardsonii*), fox (*Vulpes*), mammoth? (*Mammuthus*?), wapiti (*Cervus elaphus* = *Cervus canadensis*), mountain sheep (*Ovis* cf. *Ovis canadensis*), horse (*Equus*), and possibly humans (*Homo sapiens*) from Island Bluff near Medicine Hat, Alberta. The Stewart Valley fauna from Saskatchewan comprises: extinct peccary (*Platygonus*), mammoth (*Mammuthus*), extinct camel (?*Camelops*), and horse (*Equus*). It suggests an Aftonian or Kansan age – the strong mineralization supports a 'lower Pleistocene age.' Faunal identifications are summarized in Table 16.

240. Churcher, C.S. (1967): A second report on some Pleistocene localities in southern Alberta and Saskatchewan and their vertebrate fossil faunas. Preliminary report of progress to the Geological Survey of Canada. 181 pp. [Copy in Quaternary Vertebrates of Northern North America file at Canadian Museum of Nature.]

In 1966 the Cochrane Pits, Alberta, were visited and more fossils were collected as a basis for a scientific paper. [See Churcher 1968a, 1975.] Some 150 specimens were obtained from seven of 10 bluffs in the Medicine Hat area of Alberta. Apparently specimens were derived from the following levels: post-Wisconsinan, early classical Wisconsinan interstadial, pre-classical Wisconsinan (Altonian), and earlier pre-Wisconsinan (possibly Sangamonian or Illinoian or very early Wisconsinan). Artifacts (*Homo sapiens*) were obtained from the Altonian level at Island Bluff and Mitchell Bluff. More fossils were recovered from the Wellsch Valley site in Stewart Valley, Saskatchewan. Brief visits were made to localities near Fort Macleod and Lethbridge, Alberta; along the South Saskatchewan River between Stewart Valley, Saskatchewan, and Medicine Hat, Alberta; and Fort Qu'Appelle, Saskatchewan. Figures 1–10 are locality maps of sites visited.

241. Churcher, C.S. (1968a): Pleistocene ungulates from the Bow River gravels at Cochrane, Alberta. Canadian Journal of Earth Sciences 5(6):1467–88.

The following vertebrates were found in the gravels constituting the second major terrace above the Bow River's N bank at Cochrane, Alberta: wapiti (*Cervus elaphus* = *Cervus canadensis*); caribou (*Rangifer tarandus*); bighorn sheep (*Ovis canadensis*); extinct western bison (*Bison bison occidentalis* = *Bison occidentalis*); and extinct Mexican horse (*Equus conversidens*). The previously known range of *Equus conversidens* was from the Late Pleistocene of the southern United States and Mexico. This fauna represents the first extensive kiang-type equine from the Canadian prairies. Radiocarbon dates on the '*Bison occidentalis* bones' range from 10,760 ± 160 BP (GSC-612) to 11,370 ± 170 BP (GSC-613). This paper includes measurements and photographs of the specimens, which now form part of the collection of the Canadian Museum of Nature (NMC). [The wapiti dentary fragment has been reidentified as *Bison*.]

242. Churcher, C.S. (1968b): A third report on some Pleistocene localities in southern Alberta and Saskatchewan and their vertebrate fossil faunas. Preliminary report of progress to the Geological Survey of Canada. 142 pp. [Copy in Quaternary Vertebrates of Northern North America file at Canadian Museum of Nature.]
During 1967 visits were made to 14 bluffs in the Medicine Hat, Alberta, area and vertebrate fossils were obtained from all but one, and from perhaps seven distinct horizons. New specimens include: a sloth (Xenarthra) phalanx, diagnostic horse (Equidae) teeth from the lowest gravels, camel (*Camelops*), caribou (*Rangifer tarandus*), and a mammoth (*Mammuthus*) tooth and artifacts (*Homo sapiens*) from gravels just below the tills. Of 250 specimens collected, about 150 were identifiable. Drs Stalker and Churcher focused their efforts on Wellsch Valley, N of Swift Current, Saskatchewan, where some 95 specimens were collected including remains of mammoth, horse, camel, and possibly prongbuck (Antilocapridae). Visits were also made to the Taber Early Man site, Alberta [but see Wilson et al. 1983], and Fort Qu'Appelle, Saskatchewan, gravel pits. New sites, briefly seen, included Rattlesnake Cut and Wolf Island Cut W of Medicine Hat, Alberta, and Dundurn Sand Hills S of Saskatoon, Saskatchewan (see maps Figures 1 to 3 for positions of localities).

243. Churcher, C.S. (1968c): Mammoth from the Middle Wisconsin of Woodbridge, Ontario. Canadian Journal of Zoology 46:219–21.
Fragments of a tusk and a partial lower molar of a woolly mammoth (*Mammuthus primigenius*) were recovered in glacial till from above the Early Wisconsinan Sunnybrook Till at Woodbridge, Ontario (43°45′50″N, 79°35′30″W). Although the specimens were not radiocarbon dated, their age has been estimated as 40,000–50,000 BP (Port Talbot Interstade). Radiocarbon-date ranges are given for other mastodon (*Mammut americanum*) and mammoth species similar to the one that was found. The specimens were separated by 65 m and have been compared to Jefferson's mammoth (*Mammuthus jeffersoni*) and Columbian mammoth (*Mammuthus columbi*).

244. Churcher, C.S. (1969a): The vertebrate faunas of Surprise, Mitchell, and Island Bluffs near Medicine Hat, Alberta. Notes to accompany Dr A.M. Stalker's paper entitled 'Quaternary stratigraphy in Southern Alberta: II Sections near Medicine Hat.' From the 19th Field Conference of the Mid-Western Friends of the Pleistocene.
This paper provides a faunal list (45 species in nine orders) indicating the positions of Quaternary vertebrate finds in the bluffs near Medicine Hat, Alberta. [See Churcher 1970a.]

245. Churcher, C.S. (1969b): A fourth report on some Pleistocene localities in southern Alberta and Saskatchewan and their vertebrate fossil faunas. Preliminary report of progress to the Geological Survey of Canada. 411 pp. [Copy in Quaternary Vertebrates of Northern North America file at Canadian Museum of Nature.]
The author describes new specimens collected in 1968 from the Cochrane Pits near Calgary, Alberta, and the Empress Gravel Pits at Empress, Alberta. More details of the Kansan mammalian faunas from Stewart Valley, Saskatchewan (the Wellsch Valley Local Fauna), and the Medicine Hat Buried Valley are provided.

246. Churcher, C.S. (1970a): The vertebrate faunas of Surprise, Mitchell and Island Bluffs, near Medicine Hat, Alberta (72L). *In*: Quaternary Geochronology and Paleontology. Geological Survey of Canada, Report on Activities, Paper 70-1, (Part A):158–60.
This paper lists all vertebrates identified to the end of 1968 from the following locations near Medicine Hat, Alberta: Surprise Bluff (S¼ sec. 34, twp. 12, rge. 6, W 4th mer.); Mitchell Bluff (NE¼ sec. 32, twp. 13, rge. 5, W 4th mer.); and Island Bluff (SW¼ sec. 4, twp. 14, W 4th mer.) The fauna listed by the units in which they were found include 45 species in nine orders.

247. Churcher, C.S. (1970b): A fifth report on some Pleistocene localities in southern Alberta and Saskatchewan and their vertebrate fossil faunas. Preliminary report of progress to the Geological Survey

of Canada. 133 pp. [Copy in Quaternary Vertebrates of Northern North America file at Canadian Museum of Nature.]

During 1969, field work in the Medicine Hat area, Alberta, included visits to all but two of the bluffs. Collecting was concentrated on the Kansan Basal Gravels and the ?Yarmouthian and ?Illinoian Lower Sands, as well as postglacial and inter-till levels. The map (frontispiece) shows all sites visited in 1969. Drs Stalker and Churcher had the assistance of an 'Archaeological Team from Calgary' to excavate at Mitchell Bluff. Western bison (*Bison bison occidentalis = Bison occidentalis*) was identified from the Cochrane Pits near Calgary, Alberta. Gravel pits near Empress, Alberta, yielded: mammoth (*Mammuthus* cf. *Mammuthus columbi = Mammuthus* cf. *Mammuthus jeffersoni*), and horse (*Equus* cf. *Equus conversidens*) from the South Pit; and woolly mammoth (*Mammuthus primigenius*), horse (*Equus conversidens*), caribou (*Rangifer*), camel? (*Camelops*?), and western bison (*Bison bison* cf. *occidentalis = Bison* cf. *Bison occidentalis*) from the North Pit. A partial outline for a publication 'A Kansan Mammalian Fauna from Stewart Valley, Saskatchewan; the Wellsch Valley Local Fauna' indicates the presence of 10 mammals: ground sloth (?*Megalonyx*); unidentified lagomorphs (Lagomorpha) and canids (Canidae); bobcat (Felis cf. *Felis rufus = Lynx* cf. *Lynx rufus*); southern mammoth (*Mammuthus* cf. *Mammuthus meridionalis = Mammuthus* cf. *Mammuthus haroldcooki*); horses (*Equus pacificus*, and *Equus complicatus*); peccary (*Platygonus*); camel (*Camelops*); and an antilocaprid (Antilocapridae). These fossils were collected from lag-gravel and minor-gravel lenses beneath tills and they overlay Oldman and Bearpaw [Cretaceous] deposits on the heights above the S bank of the South Saskatchewan River at the W end of Diefenbaker Lake, NW of Stewart Valley, Saskatchewan. A list of taxa represented in the Medicine Hat Buried Valley deposits include: a toad (*Bufo*); at least two birds (Aves); and nearly 40 species of mammals (pp. 52–121).

248. Churcher, C.S. (1972a): Imperial mammoth and Mexican half-ass from near Bindloss, Alberta. Canadian Journal of Earth Sciences 9(11):1562–7.

The mammals recovered on the Red Deer River between Bindloss and Empress, Alberta (50°57′N, 110°08′W), are: imperial mammoth (*Mammuthus imperator*); extinct Mexican horse (*Equus conversidens*); and American lion (*Panthera leo atrox*). These specimens were found in stratified gravels 30 m above the level of the Red Deer River, at 650 m asl. A post-Wisconsinan age of >8000 BP is suggested for the deposit.

249. Churcher, C.S. (1972b): A sixth report on some Pleistocene localities in southern Alberta and Saskatchewan and their vertebrate fossil faunas. Preliminary report of progress to the Geological Survey of Canada. 75 pp. [Copy in Quaternary Vertebrates of Northern North America file at Canadian Museum of Nature.]

The author reports on Quaternary mammal remains collected during 1970 and 1971 field work. Identifiable rodent (Rodentia) and rabbit (Lagomorpha) specimens are represented for the first time from Wellsch Valley, Saskatchewan. The Medicine Hat Buried Valley, Alberta, produced some 300 new specimens – the total taxa are now 53 [new species are: spotted skunk (*Spilogale* cf. *Spilogale putorius*); black-footed ferret (*Mustela nigripes*); beaver (*Castor canadensis*); Harlan's ground sloth (*Glossotherium harlani = Paramylodon harlani*); and possibly shrubox (?*Euceratherium*). Fossils from the Empress area, Alberta, now comprise nine species of which seven are known from one site. In this report only the new taxa or taxa new to a site or level are reported on in detail.

250. Churcher, C.S. (1975): Additional evidence of Pleistocene ungulates from the Bow River Gravels at Cochrane, Alberta. Canadian Journal of Earth Sciences 12:68–76.

Mammal specimens found since 1968 at Cochrane, Alberta, in Bow River gravels are: extinct western bison (*Bison bison occidentalis = Bison occidentalis*) – four horn cores; bighorn sheep (*Ovis canadensis*) – four lower cheek teeth, radius; and extinct Mexican horse (*Equus conversidens*) – phalanx. Radiocarbon dates for the Cochrane gravels pits are 10,760 ± 160 BP (GSC-612) and 11,370 ± 170 BP (GSC-613). [See also Churcher 1968a.]

251. Churcher, C.S. (1980): Did the North American mammoth migrate? Canadian Journal of Anthropology 1(1):103-105.

This paper examines the habits of the Columbian mammoth (*Mammuthus columbi*) and the woolly mammoth

(*Mammuthus primigenius*) based on circumstantial evidence of taphonomic observations, geographical distribution of remains (e.g., Old Crow Basin, Yukon), geological settings and faunal associations, and extrapolations from extant elephants. The author concludes that mammoths were migratory and that their human hunters followed them. Archaeological examinations of fluted spear-points from 14,000 to 10,000 BP support this hypothesis.

252. Churcher, C.S. (1983): Faunal correlation of Pleistocene deposits in western Canada. *In*: Correlation of Quaternary Chronologies (W.C. Mahaney, ed.). Geo Books, J.W. Arrowsmith Ltd., Bristol, England. pp. 145–58.
Before 1960, Quaternary mammal finds were virtually unknown in the Canadian prairies, and most remains that were found prior to this date were not positively identified. This paper lists all identified Quaternary mammals on the prairies with their locations and radiocarbon dates.

253. Churcher, C.S. (1985): Equids (genus *Amerhippus*) from the Sangamon at Medicine Hat, Alberta. Fourth International Theriological Congress (August 13–20, Edmonton, Alberta). No. 92.
A find of a horse [resembling the extinct *Equus (Amerhippus) santae elenae* from Peru] near Medicine Hat, Alberta, extends the known range of *Amerhippus*, which was believed only to inhabit the Pleistocene of South America. The Medicine Hat find includes 17 specimens (nine incisor teeth, a lower molar, and seven postcranial bones).

254. Churcher, C.S. (1986): A mammoth measure of time: Molar compression in *Mammuthus* from the Old Crow Basin, Yukon Territory, Canada. Current Research in the Pleistocene 3:61–4.
Seventy Proboscidean molars from frozen unconsolidated sediments along the Old Crow River, Yukon, include: woolly mammoth (*Mammuthus primigenius*); Columbian mammoth (*Mammuthus columbi*); and imperial mammoth (*Mammuthus imperator, Mammuthus trogontherii, Mammuthus armeniacus*) or southern mammoth (*Mammuthus meridionalis*). A graph shows average ridge-plate frequency vs. average enamel thickness. This paper also refers to other mammalian fauna found in the Old Crow Basin. The author speculates that the mammoth teeth indicate faunal ages between 3 and 2 Ma and about 10,000 BP.

255. Churcher, C.S. (1989): *Mammuthus* as an indicator of the age of deposits in the Old Crow Basin, Yukon Territory, Canada. Fifth International Theriological Congress 1:143–4.
The extent of molar compression in mammoths in the Old Crow Basin, Yukon, was used to indicate the age of the deposits. These specimens were derived from mammoths that lived during the last two to three million years, in other words from the Late Pliocene to the Pleistocene-Holocene boundary. The species mentioned are: southern mammoth (*Mammuthus meridionalis, Mammuthus trogontherii*); Columbian mammoth (*Mammuthus columbi*); steppe mammoth (*Mammuthus armeniacus*); imperial mammoth (*Mammuthus imperator*); and woolly mammoth (*Mammuthus primigenius*).

256. Churcher, C.S., and Dods, R.R. (1979): *Ochotona* and other vertebrates of possible Illinoian age from Kelso Cave, Halton County, Ontario. Canadian Journal of Earth Sciences 16(8):1613–20.
Remains of vertebrates were found in a travertine-cemented dolomite breccia from a demolished cave in the Lockport Formation of the Niagara Escarpment W of Milton, Ontario. These are: American toad (*Bufo americanus*); Spruce Grouse or Ruffed Grouse (*Canachites canadensis* or *Bonasa umbellus*); varying hare (*Lepus americanus*); cottontail (*Sylvilagus floridanus*); large pika (*Ochotona*); woodland deer mouse (*Peromyscus maniculatus*); muskrat (*Ondatra zibethicus*); little brown bat (*Myotis lucifugus*); and striped skunk (*Mephitis mephitis*). Radiocarbon dates of coexisting plant remains range from 10,500 BP to 9500 BP. However, the presence of *Ochotona* would indicate an earlier Illinoian age for the assemblage. [See Mead 1987; Mead and Grady 1996.]

257. Churcher, C.S., and Fenton, M.B. (1968): Vertebrate remains from the Dickson Limestone Quarry, Halton County, Ontario, Canada. National Speleological Society Bulletin 30(1):11–16.
Remains of vertebrates were found in a crevice cave in the Niagara Escarpment SW of Milton, Ontario. These are: painted turtle (*Chrysemys picta*); short-tailed shrew (*Blarina brevicauda*); smoky shrew (*Sorex fumeus*); big

brown bat (*Eptesicus fuscus*); long-eared bat (*Myotis keenii*); black-backed vole (*Clethrionomys gapperi*); meadow vole (*Microtus pennsylvanicus*); red squirrel (*Tamiasciurus hudsonicus*); deer mouse (*Peromyscus maniculatus*); muskrat (*Ondatra zibethicus*); and wapiti (*Cervus elaphus* = *Cervus canadensis*). Although no bones were radiocarbon dated, the authors suggest that the assemblage has a minimum age of 215 BP due to the presence of wapiti, which have been extinct in the area since 1750 AD.

258. Churcher, C.S., and Karrow, P.F. (1963): Mammals of Lake Iroquois age. Canadian Journal of Zoology 41:153–8.
Remains of mammals were found in a buried soil horizon at Scarborough Bluffs, Toronto, Ontario. These are: grey fox (*Urocyon cinereoargenteus*); eastern chipmunk (*Tamias striatus*); and meadow vole (*Microtus pennsylvanicus*); they have been radiocarbon dated from wood charcoal at 5550 ± 70 BP (S-115A) and 5240 ± 100 BP (S-115B). Other bones from the site are: vertebrae from a colubrid snake (Colubridae); a snake (Serpentes) rib; and a juvenile fox (Canidae) frontal. The authors believe that this assemblage is contemporaneous with the Hamilton Bay fauna, including pine vole (*Microtus pinetorum* = *Pitymys pinetorum*) and muskrat (*Ondatra zibethicus*). [See Wetmore 1958.]

259. Churcher, C.S., and Karrow, P.F. (1977): Late Pleistocene muskox (*Ovibos*) from the Early Wisconsin at Scarborough Bluffs, Ontario, Canada. Canadian Journal of Earth Sciences 14(2):326–31.
A cannon bone of a muskox (*Ovibos moschatus*) was found on the lower slope in a mudflow apparently originating in seepage from stream sediments about 1 to 2 m above the present beach at Scarborough Bluffs, Ontario. This may be the first Early Wisconsinan muskox specimen from the site. [The specimen is too long and slender to represent a muskox and apparently represents a right metatarsal (not metacarpal) of modern cattle (*Bos taurus*). The bone was AMS radiocarbon dated at 290 ± 50 BP (TO-3651) – see Morris et al. 1993.]

260. Churcher, C.S., and Morgan, A.V. (1976): A grizzly bear from the Middle Wisconsin of Woodbridge, Ontario. Canadian Journal of Earth Sciences 13:341–7.
A fragment of the humerus of a grizzly bear (*Ursus arctos*) was found above the Early Wisconsinan Sunnybrook Till at Woodbridge, Ontario, from the same horizon that yielded remains of a woolly mammoth (*Mammuthus primigenius*). Radiocarbon ages for this find are from organic beds dated to 49,700 BP (GSC-203) and from a peat ball in channel fill dated to 45,000 ± 900 BP (GSC-1181). This is the second such grizzly bear specimen found in eastern Canada. The first was from a gravel sequence at Barrie, near Lake Simcoe. Bone from that specimen yielded a radiocarbon date of 11,700 ± 250 BP (Peterson 1965b).

261. Churcher, C.S., and Peterson, R.L. (1982): Chronologic and environmental implications of a new genus of fossil deer from Late Wisconsin deposits at Toronto, Canada. Quaternary Research 18:184–95.
A fragmentary cranium including basal portions of antlers of new cervine deer (*Torontoceros hypogaeus*) [now called Toronto deer] was found in Early Lake Ontario-age deposits on the exposed bench of Glacial Lake Iroquois at Toronto, Ontario. The radiocarbon date derived from a sample of the antler is 11,315 ± 325 BP (GX-4914). The deer was about the size of a caribou. The specimen is deposited at the Royal Ontario Museum in Toronto and is catalogued as ROM 75974.

262. Churcher, C.S., and Pinsof, J.D. (1987): Variation in the antlers of North American *Cervalces* (Mammalia; Cervidae): Review of new and previously recorded specimens. Journal of Vertebrate Paleontology 7(4):373–97.
The authors review most of the known specimens of moose with long antler beams from North America. In their view, the genus is *Cervalces* and apparently comprises two species: the North American Scott's moose (*Alces scotti* = *Cervalces scotti*) with branching antlers whose variations are explicable through age, maturational, geographical, or possibly stratigraphic differentiation; and the Eurasian species of giant moose (*Alces latifrons* = *Cervalces latifrons*) with flat palmate antlers. Pertinent here are the mapped specimens (Figure 1, localities 38–41): *Alces scotti* = *Cervalces roosevelti* from Fort Qu'Appelle, Saskatchewan; *Alces latifrons* from the Dawson area, Yukon; *Alces latifrons* from Old Crow Basin, Yukon; and *Alces latifrons* = *Cervalces alaskensis* from Fairbanks area, Alaska.

263. Churcher, C.S., and Stalker, A.M. (1970): A late, postglacial horse from Pashley, Alberta. Canadian Journal of Earth Sciences 7(3):1020–6.
A few fragments of the left hind foot of an extinct Mexican horse (*Equus conversidens*) were found in the Pashley pit 12.8 km ESE of Medicine Hat, and 5.6 km W of Pashley, Alberta. The bones were found in unsorted gravel overlain by 1.8 m of sand and soil. The bones of the horse were too small for radiocarbon dating, but are believed to have been 2000 years younger than the youngest known indigenous horse (8000 BP based on comparative stratigraphy).

264. Churcher, C.S., and Wilson, M. (1979): Quaternary mammals from the eastern Peace River District, Alberta. Journal of Paleontology 53(1):71–6.
Six genera and 10 species of mammals are known from postglacial gravels of the region. These gravels exposed in pits or river bluffs, including three of four terraces in the Smoky River and Peace River valleys, include fossils of: woolly mammoth [*Mammuthus primigenius* – molar dated to 27,400 ± 580 BP (GSC-2034)]; small horse (*Equus* cf. *Equus conversidens*); Niobrara horse (*Equus* cf. *Equus niobrarensis*); steppe bison (*Bison priscus*); western bison (*Bison bison occidentalis*); camelid (possibly *Hemiauchenia* or *Camelops*); and several extant taxa such as wapiti (*Cervus elaphus* = *Cervus canadensis*), wood bison (*Bison bison athabascae*), plains bison (*Bison bison bison*), and tundra muskox (*Ovibos* cf. *Ovibos moschatus*). Remains of four genera and five species were derived from Sangamonian or earlier deposits near Watino on the Smoky River: ground squirrel (*Spermophilus*), a proboscidean (*Mammuthus* or *Mammut*), small horse (*Equus*), Niobrara horse (*Equus* cf. *Equus niobrarensis*), and bison (*Bison*). Both faunas may have existed in either a plains environment or aspen-poplar parkland.

265. Churcher, C.S., and Wilson, M. (1990): Methods in Quaternary Ecology #12. Vertebrates. Geoscience Canada 17(2):59–78.
The study of Quaternary vertebrate fossils continues to be very important in contributing to our knowledge of Quaternary ecology. This review paper discusses: taphonomy, deposition characteristics, collecting techniques, laboratory and preparation techniques, paleoenvironmental reconstruction applications, zooarchaeology, and biostratigraphy. Several examples, including illustrations, from Canada are mentioned, including: (1) a summary of marine mammal finds in Champlain Sea deposits (p. 60); (2) a photograph (Figure 1) of Quaternary vertebrate fossils [Minidoka camel (*Camelops minidokae*), western camel (*Camelops hesternus*), Shasta ground sloth (*Nothrotheriops shastense*), and caribou (*Rangifer tarandus*)] from the South Saskatchewan River valley of Alberta and Saskatchewan; (3) part of an 11,000 BP bison (*Bison bison*) cranial fragment from the Bow River valley, Alberta (Figure 2); (4) a collection of Quaternary vertebrate bones from Old Crow Locality 11A, Yukon (Figure 7); (5) a partial molar of a woolly mammoth (*Mammuthus primigenius*) from Late Pleistocene gravel at West Peace River, Alberta (Figure 10); (6) a giant bison (*Bison latifrons*) skull in sands and gravels at Mitchell Bluff, South Saskatchewan River near Medicine Hat, Alberta (Figure 13); (7) a western camel metacarpal in sandy silt at Lindoe Bluff, South Saskatchewan River, Medicine Hat, Alberta (Figure 14); (8) metacarpals of western camel (*Camelops hesternus*), ancient bison (*Bison bison antiquus*), Mexican horse (*Equus conversidens*), and large mountain sheep (*Ovis canadensis catclawensis*) about 11,000 BP from Cochrane gravel pits, Alberta (Figure 15); (9) a photograph of excavation at Mitchell Bluff, Medicine Hat, Alberta, showing a Sangamonian or Mid-Wisconsinan fossil exposure (Figure 16); (10) a diagram showing chronoclinal variation in bison horncore-spread on the northern Great Plains from 11,000 BP to present (Figure 18); and (11) geographic clinal variation in giant moose (*Alces latifrons* = *Cervalces*) antler beams – note particularly the degree of overlap between polygons for Yukon and Alaska (Figure 19).

266. Churcher, C.S., Morgan, A.V., and Carter, L.D. (1993): *Arctodus simus* from the Alaskan Arctic slope. Canadian Journal of Earth Sciences 30:1007–13.
A left humerus of the Yukon short-faced bear (*Arctodus simus yukonensis*) was recovered from a point bar on the Ikpikpuk River, Alaska (69°41′N, 154°54′W). The bone is lacking its unfused proximal epiphysis and hence represents an immature individual. This is the northernmost record for this species. The specimen is radiocarbon dated at 27,190 ± 280 BP (TO-2539), which lies within the observed age distribution of *Arctodus*. The individual is larger than average for the species, despite its immaturity, and may represent a male. A series of photographs of

the bone are included. Also presented is a table including all the dated *Arctodus* specimens known to the authors.

267. Churcher, C.S., Nielsen, E., and Lammers, G.E. (1988): A woolly mammoth (Proboscidea, *Mammuthus primigenius*) molar from the Hudson Bay Lowland of Manitoba. Canadian Journal of Earth Sciences 25:933–8.
A woolly mammoth (*Mammuthus primigenius*) molar was found in a point bar on the Limestone River, 1.5 km N of Bird, Manitoba. The specimen is believed to have lived any time from the Late Sangamonian to the Early Wisconsinan.

268. Churcher, C.S., Pilny, J.J., and Morgan, A.V. (1990): Late Pleistocene vertebrate, plant and insect remains from the Innerkip site, southwestern Ontario. Géographie physique et Quaternaire 44(3):299–308.
A peat deposit in southern Ontario, 50 km NE of London (43°10′N, 80°43′W) along the banks of the Thames River, has produced the following Quaternary vertebrate fossils: white-tailed deer (*Odocoileus virginianus*), meadow vole (*Microtus pennsylvanicus*), muskrat (*Ondatra zibethicus*), and Blanding's turtle (*Emydoidea blandingi*). In addition, pollen, wood fragments, and seeds of 21 plant genera were identified along with remains from 19 families of Coleoptera. The deposit is aged at >50,000 BP based on radiocarbon dating of the peat. Five samples were dated: the youngest was 33,230 ± 610 BP (MGS-134B), the oldest >50,000 BP (GSC-2010-2). The deposit was formed in a small, relatively acidic, marshy pond during the Sangamonian Interglaciation or one of the Early Wisconsinan interstadials. The fossils recovered represent species that were deposited near where they lived, and represent a local flora and fauna which is a better climatic indicator than an assemblage of the larger and more mobile hardy megafaunal mammals. A stratigraphic column is graphically represented, along with a map and photograph of the site and photographs of the fossils.

269. Cinq-Mars, J. (1979): Bluefish Cave 1: A Late Pleistocene Eastern Beringian cave deposit in the northern Yukon. Canadian Journal of Archaeology 3:1–32.
Many Pleistocene vertebrates were excavated from the Bluefish Caves 54 km SW of Old Crow in the northern Yukon. The fish are: northern pike (*Esox lucius*) and a fish like an inconnu (cf. *Stenodus leucichthys*). The birds are: ptarmigan or grouse (Tetraonidae) and other unidentified birds [see McCuaig-Balkwill and Cinq-Mars 1998]. The mammals are: unidentified shrew (Soricidae), snowshoe hare (*Lepus americanus*), arctic hare (*Lepus arcticus*), arctic ground squirrel (*Spermophilus parryii*), mouse (cf. *Peromyscus*), brown lemming (*Lemmus sibiricus*), collared lemming (*Dicrostonyx torquatus*), three species of voles (cf. *Microtus*), red-backed vole (*Clethrionomys* cf. *Clethrionomys rutilus*), undetermined Canidae, arctic fox (*Alopex lagopus*), red fox (*Vulpes vulpes*), weasel (*Mustela*), Yukon horse (*Equus* cf. *Equus lambei*), caribou (*Rangifer tarandus*), wapiti (*Cervus elaphus* = *Cervus canadensis*), moose (cf. *Alces alces*), Dall sheep (*Ovis dalli*), and bison (*Bison*). The age of the deposit ranges from 13,000 to 10,000 BP; many radiocarbon dates are given.

270. Cinq-Mars, J. (1982): Les grottes du Poisson-Bleu. Geos 11(1):19–21.
This article deals with the archaeology and fauna in Eastern Beringia from 17,000 to 12,000 BP; it examines the area as a periglacial isolated paleoenvironment. [See Cinq-Mars 1979 for a faunal list.]

271. Cinq-Mars, J. (1990): La place des grottes du Poisson-Bleu dans la préhistoire Béringienne. Révista de Arqueologia Americana 1:9–32.
The Bluefish Caves contain the first *in situ* remains that may clarify certain aspects of the initial peopling (*Homo sapiens*) of Beringia in the last 40,000 years. The paper deals primarily with anthropology. [See Cinq-Mars 1979 for a faunal list and Morlan 1987 and Morlan and Cinq-Mars 1982.]

272. Cinq-Mars, J., and Morlan, R.E. (1999): Bluefish Caves and Old Crow Basin: A new rapport. *In*: Ice-Age People of North America: Environment, Origins, and Adaptations (R. Bonnichsen and K.L. Turnmire, eds.). Oregon State University Press for the Center for the Study of the First Americans, Corvallis. pp. 200–12.
Bluefish Caves, Yukon, located 54 km SW of the village of Old Crow at an altitude of 250 m in northern Yukon (Figure 1), have yielded evidence of episodic human

(*Homo sapiens*) activity from about 25,000 to 10,000 BP. An important specimen from Cave II is a mammoth (*Mammuthus*) limb bone flake and its parent core (Figure 5) AMS radiocarbon dated to 23,500 BP (both flake and core were dated, the dates overlapping at one sigma, providing the average date listed). Since the content of Bluefish Caves constrains the range of taphonomic processes that could account for the breaking and flaking of a large mammoth bone, the authors conclude that the flake and core were produced by humans. [Also, a split caribou (*Rangifer tarandus*) tibia, reminiscent of a broken fleshing tool, has been likewise dated by Erle Nelson to 24,800 BP – see Morlan and Cinq-Mars 1982, Figure 9.] The Pleistocene fauna of Bluefish Caves includes: mammoth (*Mammuthus*), bison (*Bison* cf. *Bison priscus*), horse (*Equus lambei*), sheep (*Ovis dalli*), caribou (*Rangifer tarandus*), moose (cf. *Alces*), wapiti (*Cervus elaphus*), saiga (*Saiga tatarica*), tundra muskox (*Ovibos moschatus*), American lion (*Panthera leo atrox*), cougar (*Felis concolor*), brown bear (*Ursus arctos*), wolf (*Canis lupus*), and many smaller mammals, birds (Aves), and fishes (Pisces). Lithic artifacts, made of exotic, high-quality cherts, were mainly found in Unit B loess that contained remains of the Late Pleistocene vertebrate fauna – microflakes in the unit apparently span much of the 25,000 to 10,000 BP interval. The bone core and flake from Bluefish are reminiscent of some fresh-fractured mammoth bones from several Old Crow Basin localities occurring some 60 km farther NE in Yukon. Lack of relevant site context at the Old Crow localities renders interpretation more difficult, but many of the mammoth bones have been AMS-dated between 40,000 and 25,000 BP. The authors argue that at least some of the bones comprise evidence of human presence in this part of Eastern Beringia during the dated interval. Evidently human bone-flaking technology represented part of a cultural adaptation to northern habitats during the Late Pleistocene – a view having precedent in other such examples from sites scattered from Europe to Siberia.

273. Cinq-Mars, J., Morlan, R.E., and Rolland, N. (1999): Filtering the bathwater: A re-examination of Eastern Beringian Late Pleistocene bone technology. Programme and Abstracts, Canadian Archaeological Society 32nd Annual Conference (April 28–May 2, Whitehorse, Yukon). pp. 26–7.

News in the mid-1980s that a caribou (*Rangifer tarandus*) fleshing tool once dated at about 27,000 BP was Late Holocene in age according to AMS radiocarbon dating led to uncritical dismissal of much other evidence interpreted as indicating human (*Homo sapiens*) presence in Eastern Beringia prior to the peak of the last glaciation, and perhaps even earlier. In the light of recent and continuing studies at Bluefish Caves, Yukon, and various areas of Eurasia and North America, it is argued that the 'dismissal' was premature, and that there were indeed people living at the easternmost edge of the Mammoth Steppe biome as early as 40,000 years ago. This evidence must be considered in order to better appreciate complex human dispersal processes that led to the colonization of the New World and that can be traced back to the Middle-to-Upper Paleolithic transition of Eurasia.

274. Cinq-Mars, J., Harington, C.R., Nelson, D.E., and MacNeish, R.S. (1991): Engigstciak revisited: A note on Early Holocene AMS dates from the 'Buffalo Pit.' *In*: NOGAP Archaeology Project: An Integrated Archaeological Research and Management Approach (J. Cinq-Mars and J.-L. Pilon, eds.). Canadian Archaeological Association Occasional Paper No. 1:33–44.

Three AMS radiocarbon dates on butchered bison [closer to the steppe bison (*Bison priscus*) than the western bison (*Bison bison occidentalis*)] bones, together with other available lines of evidence from the lower stratigraphic units of the 'buffalo Pit' at Engigstciak (NiVk-1, see map Figure 1) on the Firth River, northern Yukon, support the notion that a form of bison procurement was being implemented by hunters (*Homo sapiens*) along parts of the Yukon Coastal Plain in the Early Holocene (between about 9800 and 9400 BP). Clear photographs of the butchered bison bones are given (Figures 2–4). Radiocarbon dates on the bones are: (1) distal end of a right tibia of bison [specimen NiVk-1 (S475 W270-1)] 9870 ± 180 BP (RIDDL-362); (2) distal end of a right metacarpal (specimen NiVk-1:135) 9770 ± 180 BP (RIDDL-281); (3) distal end of a left metatarsal (specimen NiVk-1:459) 9400 ± 230 BP (RIDDL-319).

275. Clague, J.J. (1980): Late Quaternary geology and geochronology of British Columbia: Part I: Radiocarbon dates. Geological Survey of Canada Paper 80(13):1–28.

This paper provides an update on Fulton (1971). Tables

of radiocarbon dates are listed as follows: (1) dates beyond the radiocarbon-dating range; (2) finite dates on sediments of the Olympia nonglacial interval; (3) and (4) dates pertaining to the advance and recessional phases of the Fraser Glaciation, respectively; (5) dates bearing on postglacial sea levels; (6) dates relating to volcanic flows and tephras; (7) dates bearing on postglacial climates; and (8) miscellaneous dates. Some significant Quaternary mammal radiocarbon dates mentioned include: a mammoth (*Mammuthus*) tusk from Rocky Mountain Portage formerly dated to >11,600 BP (I-2244A) redated at 25,800 ± 320 BP (GSC-2859); mammoth bone from Babine Lake dated at 34,000 ± 800 BP(GSC-1754); mammoth bone from the Saanich Peninsula dated at 17,000 ± 240 BP (GSC-2829); a mammoth tusk from Chilliwack dated at 22,700 ±320 BP (GSC-2232); a mammoth tooth from Taylor dated at 27,400 ± 580 BP (GSC-2034); a bighorn sheep (*Ovis canadensis*) horncore from Finlay Forks dated at 9280 ± 200 BP (GSC-1497); and a series of elk (*Cervus elaphus*) antlers – from Maze Lake dated at 110 ± 55 BP (S-537), from Pennask Lake dated at 184 ± 75 BP (S-493), from Upper Loon Lake dated at 367 ± 75 BP (S-497), from Upper Loon Lake dated at 390 ± 75 BP (S-496), from Moose Meadow dated at 615 ± 70 BP (S-539), from Chezacut dated at 1105 ± 65 BP (S-541), from Squawk Lake dated at 1200 ± 65 BP (S-540), from Dominic Lake dated at 1430 ± 85 BP (S-495), from McGlashan Lake dated at 1740 ± 75 BP (S-454), from Kelowna dated at 2225 ± 90 BP (I-9993), and from Dog Creek dated at 3625 ± 75 BP (S-538).

276. Clague, J.J. (1981): Late Quaternary geology and geochronology of British Columbia: Part 2: Summary and discussion of radiocarbon-dated Quaternary history. Geological Survey of Canada Paper 80(35):1–41.
This summary interprets, by region, the paleogeography, paleoecology, and paleoclimatology based on the radiocarbon dating of specimens presented in Clague (1980).

277. Clark, T.H. (1927): A fossil bison skull from Yukon Territory. Canadian Field-Naturalist 41(3): 45–7.
An incomplete skull of an extinct steppe bison (*Bison priscus = Bison crassicornis*) was found 20 m below the surface in gold-bearing gravel near Dawson City, Yukon. The article includes photographs.

278. Clarkson, M. (1988): Mystery monsters. St Catharines-Niagara Standard, May 7. p. 15.
This newspaper article with photographs describes two mastodon (*Mammut americanum*) specimens that were unearthed in Wainfleet Twp, Niagara, Ontario. The first, an incomplete skeleton (five teeth, the skull, tusks 1.3 m long, vertebrae, and ribs), was found in 1963 on Gordon Stayzer's dairy farm. The second, called the Wellandport mastodon, was found in 1912. The latter was virtually complete, but with only one tusk.

279. Cockfield, W.E. (1921): Sixtymile and Ladue rivers area, Yukon. Geological Survey of Canada Memoir 123:1–60.
At Sixtymile, Yukon, stumps and trunks of trees and Pleistocene mammal bones were found near the top of bench gravels of Claims 11 to 19 on the N side of the river and 50 ft above its bed. The gravels were evidently deposited by a stream. L.M. Lambe, GSC vertebrate paleontologist, identified the bones as: (1) the posterior part of a bison (*Bison*) [probably the steppe bison (*Bison priscus*)] cranium; (2) part of the shaft of a mammoth (*Mammuthus*) limb bone and a tusk; and (3) a left femur, left tibia, right radius with part of an ulna shaft of a horse (*Equus*) smaller and lighter in stature than Scott's horse (*Equus scotti*) [probably the Yukon horse (*Equus lambei*)]. The section at this point is as follows (from bottom up): 5 ft frozen ground; 5 ft of muck, moss, tree remains and a volcanic ash layer; 15 ft of gravel; 65 ft of muck. Presumably the bones are from the top of the gravels and the base of the muck.

280. Cockfield, W.E., and Walker, J.F. (1933): Geology and placer deposits of Quesnel Forks area, Cariboo District, British Columbia. Summary Report of the Geological Survey of Canada for 1932, Part AI:76A–143A.
Under the heading Interglacial Deposits (p. 89A), the authors mention the raising of a fossil mountain goat (*Oreamnos americanus = Oreamus* [sic] *montanus*) skull from the sluice-tunnel at Bullion Mine (see pp. 118A–122A) in the Quesnel Forks area of British Columbia – probably placing these lower gravels as Pleistocene. C.M. Sternberg stated of this specimen: 'The portion of a fossil cranium is a Rocky Mountain goat, erroneously named since it is really an antelope (*Oreamus montanus*) belonging to the chamois group of the true antelope family (Antilopi-

dae). This genus was evidently a late immigrant from the Old World, arriving on this continent during Pleistocene time.' The authors mention that, consequently, there is some evidence that the gravel may be interglacial and that there were several periods of ice retreat in the region (p. 89A) [See Harington 1971b]. Other fossils collected from Pleistocene deposits at various hydraulic pits in the area are: (1) a fragment of a woolly mammoth (*Mammuthus primigenius* = *Elephas primigenius*) from the Morehead Pit (on Morehead Creek and Little Lake Creek); (2) a left humerus of a bison (*Bison bison*) from the Bullion Pit (on the South Fork of the Quesnel River about 4 mi SE of Quesnel Forks); (3) a fragment of a Columbian mammoth (*Mammuthus columbi* = *Elephas columbi*; pp. 91A, 107A) said to have come from 'bedrock gravels' at the Matthias Gold Mining Company Pit (N side of the North Fork of Quesnel River near Wolverine Creek). C.M. Sternberg says of the bison that it is not earlier than Late Pleistocene and is more likely Recent. Unfortunately, there is no clear indication of the source of such fossils in the sections that they have been washed from (p. 91A).

281. Cohen, L. (1982): Taber theory dashed. Equinox 1(4):17–18.
The remains of what archaeologists call the Taber Child (*Homo sapiens*) consist of a handful of bone shards, weighing only 1.5 oz and forming a small part of the skull and the right tibia of a four- to nine-month-old infant (photograph on p. 17). Many experts believed that these fragments were from a child who died near Taber, Alberta, more than 37,000 years ago. In the spring of 1982, however, advanced radiocarbon data were released indicating the remains were only about 3680 years old. Further, Ian Wainwright's study at Canadian Conservation Institute, using an infrared spectrophotometer to determine the amount of protein, concluded that the specimen's protein content was similar to that of human bones from the 300 to 5100 BP Gray site in Saskatchewan. Dr Archibald Stalker, who was responsible for collecting the specimen in 1961 and for estimating its geological age based on stratigraphic position, objects to the date because: (1) he doubts that all the preservative on the bone could have been removed; (2) the valley it was found in 'could not have been formed in 3680 years'; and (3) there are no cracks in the glacial deposits through which the bones could drop into lower, older deposits. [Normalized AMS radiocarbon dates on Taber Child bone are: 3550 ± 500 (CRNL-1229), and 3390 ± 90 BP (Oxa-773).]

282. Cole, G. (1979): Of land bridges, ice-free corridors, and early man in the Americas. Field Museum of Natural History Bulletin 50(1):15–27.
Pertinent is the mention that bones of various Pleistocene mammals have been found in abundance at many Yukon sites along the Porcupine and Old Crow rivers. Along with these bones were found several hundred bone artifacts [bones altered by humans (*Homo sapiens*) through butchering activities, breaking to extract marrow, or as raw material for tool making]. Several bone tools have yielded radiocarbon dates in the range of 29,000 to 25,000 BP. The presence of people in Beringia 30,000 to 25,000 years ago, and perhaps to more than 50,000 years ago, does not necessarily mean that these people figure in the peopling of the Americas. Various Asiatic animals are known from Eastern Beringia that either never established themselves elsewhere in North America or did so at a time long after their appearance there. This may also have been true of some early human inhabitants of the area. A few archaeological sites in Alaska have yielded material for radiocarbon dates in the range of 12,000 to 10,000 years ago. A long gap separates these dates from the 25,000 BP and older dates from the Yukon. To help explain this, T.D. Hamilton of the U.S. Geological Survey suggested that 'the distribution of Early Man in northwestern North America may not have been continuous in either space or time.' There would be nothing unusual in early human inhabitants of Eastern Beringia dying out or withdrawing during the deteriorating climate of the Late Wisconsinan glaciation (e.g., the failed Norse settlements in Greenland and vicinity). The possibility of early people entering the heartland of North America by the 'ice-free corridor' is also mentioned. There is no reason to think that early people arrived either in Eastern Beringia or in the central part of North America as a result of purposeful migration. Traversing the corridor would not necessarily have taken a great deal of time, but people could not have done it until sources of food, clothing, and shelter were available there.

283. Coleman, A.P. (1894): Interglacial fossils from the Don Valley, Toronto. American Geologist 13: 85–95.
Pleistocene vertebrates mentioned are: deer (Cervidae),

beaver (Castoridae), and mastodon (*Mammut americanum*) from central and western Ontario.

284. Coleman, A.P. (1895): Glacial and interglacial deposits near Toronto. Journal of Geology 3:622–45.
This article mentions previous finds of proboscidean (Proboscidea) teeth from the Toronto area, Ontario, and some unidentified teeth found with leaves of oak, beech, and willow at a rock-cut in the Don Valley locally called the Convict Cutting. A mammoth (*Mammuthus*) tooth was found in stratified sand near the Don River, 12 km N of the city centre. A mastodon (*Mammut americanum*) tooth was found near the Don River, while another was found in a sandpit some 5 km E of the city centre. Coleman notes that these fossils may be postglacial rather than interglacial.

285. Coleman, A.P. (1899a): The Iroquois Beach. Transactions of the Canadian Institute 6:29–44.
Most of the information on Quaternary vertebrates is given in the section 'Fossils of the Iroquois Beach Deposits' (pp. 35–9). Generally the Iroquois Beach is barren of fossils. However, at relatively low levels, or where the gravel has been cemented by lime, fossils are likely to be preserved. Following are a series of reports of fossil bones: (1) Excavation of the Desjardins Canal at Burlington Heights, Ontario, revealed mammal bones including: woolly mammoth (*Mammuthus primigenius* = *Euelephas jacksoni*), wapiti (*Cervus elaphus* = *Cervus canadensis*) antlers, and a beaver (*Castor canadensis*) jaw. The mammoth bones occurred at 70 ft above the present lake level, and the other remains 7 ft higher, or from 30 to 40 ft below the Glacial Lake Iroquois Beach level; (2) A Toronto, Hamilton and Buffalo Railway cutting at Hunter Street in Hamilton revealed a number of bones (Mammalia?) probably from a layer of brown clay – perhaps an old soil containing larch and spruce remains. Several partly decayed elephant (Proboscidea) bones were found buried in blue Erie Clay at the base of the Hunter Street cutting – 'some of them being now in the museum at Hamilton, others apparently carried off by onlookers.' Also, the scapula of a moose (*Alces alces*) or large deer (Cervidae) was obtained from a sandy beach bed about 1 yd or so above the clay according to the workmen. Colonel Grant got a decayed elephant (Proboscidea) rib in a Slabtown gravel pit previously. Grant also mentioned that bison (*Bison*) horncores, etc. were carried away from the site, in addition to animal remains secured by Sir William Logan for the Geological Survey of Canada. It is somewhat doubtful if the elephant remains from the Hunter Street tunnel came from Iroquois Beach deposits or from an earlier land surface. A.E. Walker obtained the lower jaw of a carnivore (Carnivora), probably from sand overlying the till at Hunter Street; (3) A mammoth (*Mammuthus*) tooth is recorded from the York gravel pits; (4) In the Carlton sand and gravel pit, at a depth of 20 ft, an 'Indian [*Homo sapiens*] flint arrowhead' was found, along with several massive bones and antlers of large deer (Cervidae) from the same pit. In the summer of 1897, three fragile caribou (*Rangifer tarandus* – probably barren ground caribou) antlers were found just above a clay layer at a depth of between 12 and 20 ft in a gravel pit on the N side of this pit. Details are recorded on p. 38. Many other deer bones and antlers are reported from gravel pits in the region. Coleman later (p. 43) states: The caribou and elephants of the bars near Toronto suggest a distinctly colder climate than the present though not necessarily a glacial one. [See Coleman 1899b.]

286. Coleman, A.P. (1899b): The Iroquois Beach at Toronto and its fossils. Ottawa Naturalist 12:195.
Mammalian remains found in Glacial Lake Iroquois Beach deposits in the Toronto, Ontario, area are: caribou (*Rangifer tarandus*) antlers common in the Carlton bar, W of downtown; and mammoth (*Mammuthus*) teeth in the bar near York, to the E.

287. Coleman, A.P. (1900): On the Pleistocene near Toronto. Report of the Seventieth Meeting of the British Association for the Advancement of Science, pp. 328–34.
A few mammalian remains were found in sandpits of probable Sangamonian age near Christie and Shaw streets in Toronto, Ontario, W of downtown: (1) a tusk of a mammoth (*Mammuthus*) or mastodon (*Mammut americanum*), and (2) an atlas vertebra of a plains bison (*Bison bison bison* = *Bison americanus*). The author mentioned that this is the most easterly locality in Canada where bison remains have been found. He later lists 'possibly mammoth or mastodon and bison, and an undetermined fish (Pisces) as members of the "Warm Climate Beds," Don Valley.'

288. Coleman, A.P. (1901a): Marine and freshwater beaches of Ontario. Bulletin of the Geological Society of America 12:129–46.
The author mentions (p. 131) the great numbers of concretions in the Ottawa area, Ontario (Green Creek and Besserer's Wharf), containing capelin (*Mallotus villosus*) and 'two or three other fish.' In addition to these marine forms, land and freshwater plants, several birds (Aves) and mammals, including a seal (*Phoca*) and chipmunk [*Tamiasciurus* – see Ray 1965], have been obtained from the concretions, showing that the Leda clay was formed not far from a shore where clay, with drift materials, was brought down by rivers. Welch's Gravel Pit (see map Figure 1) N of Smiths Falls, Ontario, is famous for the bones of a humpback whale (*Megaptera novaeangliae* = *Megaptera longimana*) found there in a railway ballast pit at a height of 440 ft asl. Probably the bones were deposited in a beach cut on one side of the ridge and not in the ridge itself (p. 133). Further mention is made of the chipmunk (common in the same region now), as well as 'feathers and bone of a bird' not yet identified (p. 137). He mentions that 'seals, dolphins [none has been reported], and whales' (Cetacea) found in the Champlain Sea deposits are still living in the Gulf of Saint Lawrence.

289. Coleman, A.P. (1901b): Sea beaches of eastern Ontario. Report of the Ontario Bureau of Mines 10:215–27.
The author notes that many concretions from marine (Leda) clay near Ottawa, Ontario (Green Creek and Besserer's Wharf), contain well-preserved capelin (*Mallotus villosus*). Feather impressions and bird (Aves) and seal (*Phoca*) bones are rarely found. Concretions containing fish (Pisces) remains are reported at other points down the Ottawa River, toward Montreal, and as far upstream as Lake Coulonge, Quebec (about 97 km NW of Ottawa). In 1882, two vertebrae and a rib of a humpback whale (*Megaptera novaeangliae* = *Megaptera longimana*) were found in Welch's Gravel Pit about 5 km N of Smiths Falls at 136 m asl. Bones of a white whale (*Delphinapterus leucas* = *Beluga catodon*) were found in Saxicava sand near Cornwall, Ontario. At D.J. McMillan's gravel pit some 2.5 km SE of Finch, decayed bones were found in the gravel but were lost. The list of vertebrates on p. 224 includes: harp seal (*Phoca groenlandica*) – Green Creek; white whale (*Delphinapterus leucas* = *Beluga catodon*) – Cornwall; humpback whale – Smiths Falls; chipmunk (*Tamias striatus*) – Ottawa; bird (Aves) feathers and a limb bone – Green Creek; capelin (*Mallotus villosus*); smelt (*Osmerus mordax*); spoonhead sculpin (*Cottus ricei* = *Cottus uncinatus*); lumpfish (*Cyclopterus lumpus*) and three-spined stickleback (*Gasterosteus aculeatus*?) – all from Green Creek. [See McAllister et al. 1981.]

290. Coleman, A.P. (1901c): Glacial and interglacial beds near Toronto. Journal of Geology 9:285–310.
Mammalian remains found in 'Warm Climate Beds' at Taylor's Brickyard (Don Valley, Toronto, Ontario) in clay and blue peaty clay include a caribou (*Rangifer tarandus*) antler and an unidentified fish (Pisces). A mammoth (*Mammuthus*) or a mastodon (*Mammut americanum*) ulna, smoothed and scratched by glacial action, was found in a sewer cut near the corner of Dupont and Bathurst streets. Proboscidean (Proboscidea) tusk fragments and a large atlas of a plains bison (*Bison bison bison* = *B. americanus*) were found at Adare's Sandpit.

291. Coleman, A.P. (1904): The Iroquois Beach in Ontario. Bulletin of the Geological Society of America 15:347–68.
Bones of a mammoth (*Mammuthus*), wapiti (*Cervus elaphus*), and beaver (*Castor canadensis*) came from a cut through the Dundas Bay bar at about 22 to 25 m above the lake. In a 1.6 m layer of brown stratified clay (a paleosol?) at the Hunter Street tunnel near Hamilton, bones of a mammoth and other animals were found. The author later refers to unworn mammoth tusks and bones about 26 m below the Iroquois gravel bar at Hamilton, Ontario, and mentions that mammoths lived at Hamilton and at York, near Toronto, and that caribou (*Rangifer tarandus*) were common at Toronto Junction.

292. Coleman, A.P. (1913): Excursion B2-Toronto and vicinity. *In*: Excursions in vicinity of Toronto and to Muskoka and Madoc. (Excursions B2, B5, B6, B8, and B10). Ontario Bureau of Mines, Guide Book No. 6:1–29.
The Don Valley Brickyard produced: a bone of a large bear (Ursidae), bones or horncores of bison (*Bison*), as well as white-tailed deer ('Virginia red deer') (*Odocoileus virginianus*) and (?)caribou (cf. *Rangifer*

tarandus) remains. Mammalian remains found in interglacial beds near Christie and Shaw streets in the western part of Toronto include: an antler of a Scott's moose (*Alces scotti* = *Cervalces borealis*), an atlas vertebra of a bison (*Bison*), part of a lower jaw of a bear (*Ursus*), and tusk fragments of a mammoth (*Mammuthus*) or a mastodon (*Mammut americanum*).

293. Coleman, A.P. (1922): Glacial and Post-Glacial lakes in Ontario. University of Toronto Studies. Publications of the Ontario Fisheries Research Laboratory No. 10:15–76.
Vertebrate remains were found all over Ontario: a catfish (*Ictalurus punctatus*) spine from the Don beds (Sangamonian) near Toronto; bison (*Bison*) bones from freshwater beds in the Rainy River district; bones of a small fish and a bony plate of a sturgeon (*Acipenser*) from stratified clay of Glacial Lake Agassiz at Fort Frances; mammoth (*Mammuthus*) bones in Queen Victoria Park, Toronto; remains of mammoth, wapiti (*Cervus elaphus*), bison (*Bison*), and beaver (*Castor canadensis*) in a gravel bar extending N from Hamilton toward Burlington as well as a mammoth tusk and bones from gravel pits and brickyards W of the bar; caribou (*Rangifer tarandus*) antlers and proboscidean bones (including a mammoth tooth) from gravel bars at Toronto; some duck (Anatidae) bones, a few feathers of an unidentified bird, and bones of a chipmunk (*Tamias striatus*) in marine clay at Green Creek near Ottawa.

294. Coleman, A.P. (1933): The Pleistocene of the Toronto region (including the Toronto Interglacial Formations). Forty First Annual Report of the Ontario Department of Mines 41(7) (1932):1–69.
Vertebrate remains found in the Toronto area up to 1932 include: part of a lower jaw of a bear – 'as large as a grizzly' (*Ursus*); giant beaver (*Castoroides ohioensis*) cutting tooth or incisor; bison (*Bison*); caribou (*Rangifer tarandus*); catfish (*Ictalurus punctatus*); skull of a young groundhog (*Marmota monax* = *Arctomis monax*); mammoth (*Mammuthus*) or mastodon (*Mammut americanum*); muskox (*Ovibos moschatus*); and an antler fragment of a Scott's moose (*Alces scotti* = *Cervalces borealis*). The author also mentions 'remains of mammoth, wapiti (*Cervus elaphus*), beaver (*Castor canadensis*), and probably buffalo (*Bison*)' from the bar at Burlington, Ontario.

295. Coleman, A.P. (1936): Lake Iroquois. Forty Fifth Annual Report of the Ontario Department of Mines 45(7):1-36.
Vertebrate remains found in Glacial Lake Iroquois beach deposits in southern Ontario include: bones of a mammoth (*Mammuthus*); antlers of a wapiti (*Cervus elaphus*); bison (*Bison*) horncores; and the jaw of a beaver (*Castor canadensis*) in the Burlington Heights area of Hamilton; a mammoth tooth and other remains from gravel pits in the Homer bar at St Catharines; bison (*Bison*) horncores; a moose (*Alces alces*) shoulder blade from the Hamilton bar; a mammoth or a mastodon (*Mammut americanum*) tusk; mammoth tusk and bones; bones and antler of caribou (*Rangifer tarandus*); bones and teeth of a mammoth; and a skull of a muskox (*Ovibos moschatus*) from Toronto.

296. Coleman, A.P. (1941): The Last Million Years: A History of the Pleistocene in North America. University of Toronto Press. 216 pp.
This book is the product of wide observations extending over more than half a century, and of extensive study of the literature of glacial geology. Following introductory chapters on the principles of glacial geology and on the preglacial aspect of North America, the core of the book is arranged regionally rather than stratigraphically. The Mississippi region is discussed first, followed by the Lower Great Lakes Region with emphasis on the Toronto area, then the St Lawrence Valley, Eastern, Central, Cordilleran and Far North regions. The author lists the known Pleistocene deposits and the fossil faunas, and interprets their significance. Comments on Quaternary vertebrates of Canada and Alaska include the following: Don Beds (Sangamonian) of Toronto, Ontario – catfish (*Ictalurus*), bison (*Bison*), large bear (*Ursus*), giant beaver (*Castoroides*), and two kinds of deer (Cervidae) (p. 76); Sandpit 87, Toronto, Ontario (Late Sangamonian?) – Scott's moose (*Alces scotti* = *Cervalces*), bison (*Bison*), bear (*Ursus*), and mammoth or mastodon (Proboscidea) (p. 81); Toronto, Ontario – striated mammoth (*Mammuthus*) bone just below upper boulder clay of a final cold stage when mammoths lived near the front of an advancing glacier (p. 83); Champlain Sea (Ottawa area, Ontario) – capelin (*Mallotus villosus*), seal (Phocidae); Smiths Falls – humpback whale (*Megaptera*); Finch, Ontario – seal (Phocidae); Montreal – white whale (*Delphinapterus leucas*); general – chipmunk (*Eutamias*)

and birds (Aves) (pp. 94–5); Moose River Basin, Ontario – mammoth and mastodon (*Mammut americanum*) (p. 129); Klondike, Yukon – mentions 15 species reported by Tyrrell (1912, pp. 147–8); Alaska – Von Kotzebue's finds of mammoth in ice (p. 148); Arctic – caribou (*Rangifer tarandus*), muskox (*Ovibos moschatus*), arctic hare (*Lepus arcticus*), and probably many other animals were finding plenty of pasture and living N of the Arctic Circle when more southerly areas were deeply buried under ice for thousands of years (p. 150); Hamilton Bar, Ontario – remains of mammoth, bison, wapiti (*Cervus elaphus*), and beaver (*Castor canadensis*) were found 83 ft below the bar surface, and animals that had lived on an old shore of Glacial Lake Iroquois and some elephant (*Mammuthus*?) bones were found in a railway-cut with trunks of cool-climate spruce (*Picea*) and larch (*Larix*) (p. 165); Fort Frances on Rainy River, Ontario – parts of a small fish (Pisces) and a large bony plate of a sturgeon (*Acipenser*) were found with six genera of freshwater molluscs in a Glacial Lake Agassiz beach (p. 171); Champlain Sea deposits, Quebec and Ontario – if white whale (*Delphinapterus leucas*), harbour seal (*Phoca vitulina*), and walrus (*Odobenus rosmarus*) remains from Bic represent living taxa, 'we may suppose that subarctic marine mammals' pushed far inland.

297. Collier, A.J. (1908): Geography and geology. *In*: The Gold Placers of Part of Seward Peninsula, Alaska. United States Geological Survey Bulletin 328 (see pp. 89–91).

Fragments of mammoth (*Mammuthus*) bones were found at Council and White Mountain near Golofnin Sound, Alaska. At Council, several feet of sandy silt are overlain by several feet of peat. Also, mammoth remains have been found on gravel bars of streams in the Ophir Creek area of the Fish River Basin. The gravels, sand, and silt are considered to be 'ancient,' whereas the overlying peat is inferred to be Holocene. 'The presence of mammoth and other mammalian [Mammalia] remains indicates that the apparent conformity of the beds throughout the section exposed suggests continuous deposition from Pleistocene to recent time.'

298. Conference on Vertebrate Paleontology in Alberta. (1965): Vertebrate Paleontology in Alberta; Report of a conference held at the University of Alberta (August 20–September 3, Edmonton, Alberta). 76pp.

The paper in this volume of interest to Pleistocene researchers is cited under Fuller and Bayrock, 1965.

299. Conroy, C.J., and Cook, J.A. (2000): Molecular systematics of a Holarctic rodent (*Microtus*: Muridae). Journal of Mammalogy 81(2):344–59.

Beringia was the intermittent connection that allowed exchange of mammals between Asia and North America. Because some mammalian genera are widely distributed on both continents, recovery of phylogenetic histories of species within these genera may help reconstruct the sequence of intercontinental exchanges. The authors tested phylogenetic and biogeographic hypotheses in the widespread voles (*Microtus*) through parsimony and likelihood analysis of mt DNA sequence data. Living specimens of *Microtus* in North America are thought to be derived from multiple invasions from Asia or, alternatively, as a single invasion followed by autochthonous speciation. Mitochondrial cytochrome-b gene sequences were obtained for 78 individuals representing 24 specimens of *Microtus*. Data supported: (1) a clade of taiga voles (*pennsylvanicus, montanus, townsendii* and *canicaudus*); (2) a clade of Asian species (*kikuchii, fortis, montebelli* and *middendorffi*), plus the Holocene tundra vole (*Microtus oeconomus*) and several other previously unidentified clades (see maps, Figures 1, 2). The narrow-skulled vole (*Microtus gregalis*) also was found to be distant from *Microtus abbreviatus* and the singing vole (*Microtus miurus*), thus contradicting monophyly of the subgenus *Stenocranius*. Monophyly of the North American species was supported, but weakly. Basal relationships reflect a single pulse of diversification about 1.3 Ma (Figure 5). Because the authors' data address only the history of living species, there may have been other invasions of North America whose descendants became extinct. For example, Hibbard's tundra vole (*Microtus paroperarius*) and the Cape Deceit vole (*Microtus deceitensis*), now extinct but present in North America in the Early Pleistocene, share the 4-triangle lower first molar with the tundra vole, a Holarctic species thought to be a Late Pleistocene colonizer of North America (Lance and Cook 1998). Because of the apparent monophyly of endemic North American species, only two invasions may have occurred: the first resulting in species restricted to North America, the second in the tundra vole.

300. Cook, J.P. (1996): Healy Lake. *In*: American Be-

ginnings: The Prehistory and Palaeoecology of Beringia (F. Hadleigh West, ed.). University of Chicago Press, Chicago and London. pp. 323–7.
The Healy Lake Village site (64°00′N, 144°45′W) is on a ridge projecting into Healy Lake, Alaska. The commonest features were areas of burned earth, charcoal, and calcined bone [mostly bird (Aves), small mammal (rabbit/squirrel), and some large mammal, e.g., caribou (*Rangifer tarandus*) and sheep (*Ovis*)]. There are 42 radiocarbon dates on material (mainly charcoal) from the site, only two of which are on unidentified bone: 8960 ± 150 BP (GX-1340) from Level 4 and 11,090 ± 170 (GX-1341) from Level 8. These dates are within the range of the Chindadn complex.

301. Cooke, H.B.S., Harington, C.R., and Sollows, J.D. (1993): Undescribed mammoth (*Mammuthus*) teeth from Georges Bank and Nova Scotia. Proceedings of the Nova Scotia Institute of Science 40:19–28.
Ten molar teeth of mammoths are described, nine of them dredged by fishermen from the NE corner of Georges Bank, while the tenth was in gravel from near James River, Nova Scotia. Measurements, morphological features, and photographs of the specimens are presented and discussed and the molars are referred to woolly mammoth (*Mammuthus primigenius*), although a few specimens are lower-crowned than is usual and may represent a local variant. [One of the specimens from Georges Bank, a left upper third molar YCM NS 92.6 has since yielded a radiocarbon date of 12,270 ± 60 BP (Beta-115203)].

302. Cooper, A., and Shapiro, B. (2000): Population genetics in Late Pleistocene mammals. Troy Péwé Memorial Workshop. Paleoclimates and Paleoenvironments in Eastern Beringia and the Bering Land Bridge (August 21–4, Chena Hot Springs, Alaska). Abstracts, p. 45.
Permafrost-preserved mammalian (Mammalia) bones contain a genetic record of past events, recording such factors as: population size; local extinction and replacement; and genetic diversity. Radiocarbon dating of pertinent bones allows these genetic changes to be related to paleoenvironmental events. The authors have analysed many brown bear (*Ursus arctos*) remains, as well as American lion (*Panthera leo atrox* = *Leo atrox*) and steppe bison (*Bison priscus*) from the greater Fairbanks area, Alaska, and Sixtymile, Yukon. The data reveal large changes in brown bear population structure prior to 20,000 BP, and surprising genetic continuity thereafter.

303. Cope, E.D. (1871): Fossil whale in the drift. American Naturalist 5:125.
Bones of a whale were found in Cornwall, Ontario. This whale seems to be the same as the white whale (*Delphinapterus leucas* = *Beluga Vermontana* = *Beluga leucas*).

304. Cope, E.D. (1890): The Cetacea. American Naturalist 24:600–16.
This paper deals mainly with Tertiary whales, but mentions Pleistocene white whale (*Delphinapterus leucas* = *Delphinapterus vermontanus*) remains in the Champlain Sea clays of the St Lawrence Basin (p. 610). White whale remains from Canada are mentioned in a list of Extinct Cetacea of North America (p. 616).

305. Corbeil, M.R. (1995): The archaeology and taphonomy of the Heron Eden site, southwestern Saskatchewan. MA thesis, Department of Anthropology and Archaeology, University of Saskatchewan, Saskatoon. 179 pp.
The Heron Eden (EeOi-11) archaeological site lies 13 km S and 1.6 km E of Prelate near the NW edge of the Great Sand Hills, Saskatchewan. The site was discovered when Ruth and Fulton Heron found projectile points of the Eden-Scottsbluff complex eroding out on the surface. Two complete points were excavated with bone of a large form of bison (*Bison*) from a dark brown paleosol below the plough zone. Bison bone collagen from the cultural layer yielded a normalized age of 9010 ± 120 BP (S-3114). Associated taxa with bison are: pronghorn antelope (*Antilocapra*), wolf (*Canis lupus*), northern pocket gopher (*Thomomys talpoides*), and Richardson's ground squirrel (*Spermophilus richardsonii*). [See Morlan 1999.]

306. Cottle, T. (1852): Fossil Pachydermata in Canada. Annals and Magazine of Natural History, Series 2(10):395–6.
The right ramus of a lower jaw and a tusk from a Columbian mammoth (*Mammuthus columbi* = *Elephas primigenius*?) at Burlington Heights, near Hamilton, Ontario, were found 13 m below the surface in Glacial Lake Iroquois beach sand, 20 m above the present level of Lake Ontario [The lower jaw with molar tooth has since

been identified as that of a Columbian mammoth.] Near this site, where a railway was planned to cross the Desjardins Canal, were a scapula and bone fragments of a herbivorous animal about the size of a fallow deer buried in silt. The author also refers to 'the Mastodon's remains mentioned by Lyell' found on the right bank of the Niagara River.

307. Cottle, T. (1853): On fossil Pachydermata in Canada. American Journal of Science, Series 2, 15:282–3.

The author records 'the first discovery of the remains of one of the large, extinct Pachydermata in Canada' near the end of January in excavating a railway cutting through a narrow spit of land at the head of Lake Ontario known as Burlington Heights, Ontario. This feature seems to have been a bar formed at the mouth of a large estuary, which entered Lake Ontario. To the W are the Dundas marshes, which exit into Burlington round the point of Burlington Heights, and through which the Desjardins Canal is carried. Two mammoth (*Mammuthus columbi* = *Elephas primigenius*?) specimens were discovered: a right ramus of a lower jaw to beyond the symphysis, and a tusk (6 ft 8 in long on the outer curve). The jaw measured: 19 in from the angle to the symphysis; 2 ft 2 in from the condyloid process to the symphysis; and 1 ft from the base of the angle to the tip of the coracoid [coronoid?] process. The jaw contained only one molar measuring 3¼ in wide × 13 in long (of which 4¾ in had been worn). The remains were found 40 ft below the surface and 60 ft above the level of Lake Ontario in sand covered by successive layers of cemented gravel and sand. In sinking a coffer dam near this spot for the foundation of a bridge where the railroad was planned to cross the Desjardins Canal, were found, deep in the silt, a scapula and bone fragments of a herbivore the size of a fallow deer (Cervidae). [See Cottle 1852.]

308. Cowan, I.M. (1941): Fossil and subfossil mammals from the Quaternary of British Columbia. Transactions of the Royal Society of Canada 35, Section 4:39–49.

This review is mainly based on Quaternary vertebrate fossils from collections of the British Columbia Provincial Museum [now the Royal British Columbia Museum] and the University of British Columbia Department of Geology. Following are basic data on the specimens: a radius of an immature Steller's sea lion (*Eumetopias* cf. *Eumetopias jubatus*) was discovered in sewer excavations in Vancouver (54th Avenue near south Granville Street) – probably from Late Pleistocene raised delta and marine deposits; Columbian mammoth (*Mammuthus columbi* = *Parelephas* (*Elephas*) *columbi*) remains are abundant – (1) tooth fragment from the Matthias Gold Mining Company near Quesnel Forks; (2) limb bone fragments, including the head of a humerus (No. 492) from the E side of Chum Creek; (3) tooth fragment (No. 296) weathered from a cliff on James Island, Gulf of Georgia; (4) basal 30 in of a small tusk (No. 484) found on Island View Beach, Saanich; (5) part of a tooth (No. 249) from the beach at Cadboro Bay, Victoria; (6) tooth fragment (No. 485) from a sandpit at Mount Tolmie, Victoria; (7) a tooth fragment with 18 plates (No. 297) from Shuswap Lake; (8) also – anterior seven plates of a premolar from cliffs at Port Townsend, Washington [USA]; remains of woolly mammoth (*Mammuthus primigenius* = *Mammonteus* (*Elaphas*) *primigenius*) have been reported from gold-bearing gravels at Quesnel Forks (Cockfield and Walker 1932) – also two molars from Ester Creek near Fairbanks, Alaska, two tusks from near Dawson City, Yukon, and a tooth and part of a tusk from Chicken Creek, near Forty-Mile, Yukon; part of a lower molar (No. 377) from Vancouver Island (James Island?) of an imperial mammoth (*Mammuthus imperator* = *Archidiskodon* (*Elephas*) *imperator*) collected prior to 1895; a second lower premolar of a Pacific horse (*Equus pacificus*, identified by W.D. Matthews) from within 6 mi of Horsefly; a left third premolar (No. 675) of a horse (*Equus*) from a gravel pit near Langford, Saanich, area was found at a depth of 60 ft; Holocene? wapiti (*Cervus elaphus* = *Cervus canadensis*) specimens – (1) a cranium with antler pedicels (No. 682) from a depth of about 10 ft in a river bank 15 mi S of Kamloops; (2) a cranium (No. 702) of a large male with antlers removed by cutting (*Homo sapiens*) from 18 in below the surface in a stream-bed at Australian Ranch, Australian; (3) a large antler from Hanging Valley Creek; a Holocene moose (*Alces*) partial cranium with antler bases from muskeg near McBride: a cranium and left antler of a mule deer (*Odocoileus hemionus hemionus*) from a depth of 8 ft at Chilancoh Ranch, 18 mi W of Alexis Creek, overlain by a 2 ft bed of glacial till; a right antler of a female caribou (*Rangifer tarandus*) was recovered from presumably interglacial gravel 3 ft above bedrock and 165 ft below river level at a placer mine at

Wingdam; a partial cranium with horncores (No. 678) of a bison (*Bison bison*) from 4 ft below the surface in muskeg at Wilson Creek near Atlin; a broken thoracic vertebra and a fragment from the base of a horncore from gold-bearing gravels at Germansen Creek may belong to bison (*Bison*); two steppe bison (*Bison priscus = Superbison* cf. *Superbison crassicornis*) specimens: (1) a fifth cervical vertebra (No. 604) from McCulloch Station, Kettle Valley Railroad; (2) a horncore base with adjacent frontal bone (No. 493) from a sewer excavation at Bay and Cook streets in Victoria – also the posterior of a cranium from near Dawson City, Yukon; [a tooth from Midway is probably referable to a recent sheep (*Ovis*) rather than to the tundra muskox (*Ovibos moschatus*) (Lambe 1907); and another posterior cranial fragment (No. 680) from Dease Lake referred to *Ovibos* by Cowan represents a helmeted muskox (*Bootherium bombifrons*) instead (Harington 1968)]; mountain goat (*Oreamnos montanus*) has been reported from gold-bearing gravels of Bullion Mine at Quesnel Forks by Cockfield and Walker (1933) [see Harington 1971b]; several Pleistocene or Early Holocene partial crania of mountain sheep (*Ovis canadensis*) are from Tranquille Creek and adjacent Hanging Valley Creek; a whale (Balaenidae?) vertebral centrum from a depth of 20 ft in Pleistocene gravels of the Colwood Delta at Langford; a periotic of a killer whale (*Orcinus orca = Orca*) may come from relatively recent deposits near Cape Beale. Most specimens mentioned have been recovered from glacial till and reworked gravels in two widely separate areas of the province, southern Vancouver Island and parts of Puget Sound, and from the Cariboo District.

309. Cowan, I.M. (1954): The occurrence of the Pleistocene wolf *Canis dirus* in the Rocky Mountains of central Alberta. Canadian Field-Naturalist 68:44.
A right lower canine tooth of a dire wolf (*Canis dirus*) was found in Banff National Park, Alberta. This was the northernmost record of the animal in North America. [Recent examination indicates that both the identification and Pleistocene age of the specimen are doubtful.]

310. Craig, B.G., and Fyles, J.G. (1960): Pleistocene geology of Arctic Canada. Geological Survey of Canada Paper 60-10:1–21.
On western Banks Island, Northwest Territories, outside the inferred boundary of the Wisconsinan Laurentide ice sheet, till deposited by an older(?) Laurentide ice sheet rests upon Beaufort Formation gravels. Overlying the till are pockets of pond silt (three localities are marked on the W coast of Banks Island, Figure 2) that contained peat yielding a small amount of pollen, and that are associated in one place [Worth Point? – the Worth Point Formation is now considered to be 'preglacial – Late Pliocene?' (Vincent 1989, Table 2.1)] with small trees and beaver-cut (Castoridae) sticks. The pollen flora of spruce, pine, birch, alder, and tundra plants record climatic conditions considerably warmer than those of the area today 'and hence probably dates from an interglacial.' Wood from one of the trees has a radiocarbon age of >35,000 BP [I(GSC)-19] and a date of >38,000 BP [I(GSC)-26] has been assigned to willow twigs from peat overlying this succession.

311. Cranz, D. (1765): Historie von Grönland. 2 vols., Leipzig.
According to Bennike (1997, p. 901), the author reports capelin (*Mallotus villosus*) in calcareous clay nodules from West Greenland.

312. Crossman, E.J., and Harington, C.R. (1970): Pleistocene pike, *Esox lucius*, and *Esox* sp., from the Yukon Territory and Ontario. Canadian Journal of Earth Sciences 7(4):1130–8.
Late Pleistocene northern pike (*Esox lucius*) dentary fragments were found in the Old Crow Basin, Yukon, at Old Crow Loc. 69 (67°51′N, 139°48′W) with remains of hare (*Lepus*), woolly mammoth (*Mammuthus primigenius*), horse (*Equus*), caribou (*Rangifer tarandus*), and bison (*Bison*). Adjacent bivalves were radiocarbon dated to 10,850 ± 160 BP (I-4224), but the pike remains are probably of earlier Late Wisconsinan age. A tooth fragment from the Don beds (Sangamonian) at Toronto, Ontario, is referred to pike (*Esox*). Comparisons with extant and pre-Quaternary pikes were made.

313. Cumbaa, S.L., McAllister, D.E., and Morlan, R.E. (1981): Late Pleistocene fish fossils of *Coregonus*, *Stenodus*, *Thymallus*, *Catostomus*, *Lota*, and *Cottus* from the Old Crow Basin, northern Yukon, Canada. Canadian Journal of Earth Sciences 18(11):1740–54.
Mid-Wisconsinan fish remains were found at Old Crow

Loc. 15, Yukon Territory (67°51′30″N, 139°48′40″W) including: broad whitefish (*Coregonus nasus*), inconnu (*Stenodus leucichthys*), longnose sucker (*Catostomus catostomus*), burbot (*Lota lota*), and freshwater sculpin (*Cottus*). All except the sculpin are the first fossils of their kind in North America. The presence of the Arctic grayling (*Thymallus arcticus*) is now extended from 32,000 to about 60,000 BP. The fossils were found 16.5 m below the surface under Holocene peats, silts, glaciolacustrine clays, then silts and sands [radiocarbon dates from near the top of the last unit: 38,800 ± 2000 BP (GSC-2756) and 41,100 ± 1650 BP (GSC-2574)].

314. Cwynar, L.C. (1982): A late Quaternary vegetation history from Hanging Lake, northern Yukon. Ecological Monographs 52(1):1–24.

Pertinent are comments on the 'Arctic-steppe biome' (pp. 18–19). The author notes that paleontologists have agreed that the Late Wisconsinan vegetation of Eastern Beringia must have been a grassland because the megafauna was dominated by grazers, mostly bison (*Bison*), mammoth (*Mammuthus*), and horse (*Equus*). This view is strengthened by the fossil occurrence of saiga (*Saiga tatarica*), which is restricted to Siberian steppes today. The growing recognition that the herb zone is a consistent feature of Eastern Beringian pollen diagrams led to the idea that a unique, productive, now-extinct grassland (Arctic-Steppe Biome, Matthews 1976) existed during the herb zone from about 30,000 to 14,000 BP. The author notes that 'there are no extensive deposits of loess' in the northern Yukon which could support large herds of grazers during a full-glacial climate, despite many Late Pleistocene mammal sites there. Since Cwynar's pollen sequence from Hanging Lake 'does not support the concept of a glacial-age grassland on the uplands of northern Yukon,' how were large herds able to survive? He makes the point that many of the Pleistocene mammal bones could have been reworked and redeposited several times and may not represent a community. Probably extinct horse, bison, and mammoth roamed the full-glacial landscape in herds; these herds may have been small with a low density of animals. A low-density, patchy distribution of animals in alluvial, wet-meadow habitats would account for the widespread distribution of these fossils in fluvial deposits. [See Guthrie 1985.]

315. Dacks, B. (1999): Amazing discovery. Legacy, Alberta's Cultural Heritage Magazine 4(4):10–13.

Tracks of woolly mammoth (*Mammuthus primigenius*) and bones of pony-sized horses (*Equus*), camels (Camelidae), helmeted muskox (*Bootherium bombifrons*), and bison (*Bison*), as well as human (*Homo sapiens*) artifacts, indicate that they shared an environment in southern Alberta over 11,000 years ago, near the end of the ice age. The proof of their existence is spread over 2 sq km at the bottom of St Mary Reservoir, about 15 km NE of Cardston, Alberta. For two years, Shayne Tolman, a Cardston elementary school teacher, had been exploring Wally's Beach (the provincial recreation area at the reservoir) when he found: (1) Clovis spear points in great concentration; (2) clear ungulate tracks pressed firmly into the mud; (3) a skull later identified as a muskox; and (4) mammoth tracks. Tolman alerted archaeologist Barry Wood, then geologist/paleontologist Len Hills, and archaeologist Bryan Kooyman at the University of Calgary. Hills' graduate student Paul McNeil joined the group. From the number of tracks they surmised there were healthy-sized herds. Kooyman stated that it looked like a place where people were watching for game – scrapers for taking flesh from hides were found in the area. The landscape, as they envisioned it, consisted of low shrubs in a broad, grassy valley with gentle slopes near a river where animals came to drink. In addition to the mammal remains mentioned previously, badgers (*Taxidea taxus*), wolves (*Canis lupus*), muskrats (*Ondatra zibethicus*), beavers (*Castor canadensis*), rabbits (Leporidae), and ground squirrels (*Spermophilus*) were found. Some of the artifacts were made from stone (chert) found in quarries about 400 km away in Montana. One tool was found only 75 cm from the muskox bones. With camel tracks the party found abundant evidence of extinct North American horses – not only tracks but also five fairly complete skeletons in one area and two complete skulls. Because much of the field evidence, tracks especially, was so transitory, everything was measured and photographed in minute detail. Some of the actual tracks were removed in plaster jackets to the laboratory. The site automatically became a protected historical resource under Alberta's historical resources legislation, and several provincial government departments are searching for funds to provide for long-term study of the artifacts from the reservoir.

316. Dall, W.H. (1869): [Bones of muskox, bison and elephant near Yukon River, Alaska]. Proceedings of the Boston Society of Natural History 13:136–7.

Dall stated that in Alaska near the great bend of the Yukon River, bones of muskox (*Ovibos moschatus*), 'buffalo,' or bison (*Bison*) were frequently found on the ground surface, still having 'an animal odor about them.' Fossil elephant [probably woolly mammoth (*Mammuthus primigenius*)] bones occurred with them but were black and fragile.

317. Dall, W.H. (1921): Pliocene and Pleistocene fossils from the arctic coast of Alaska and the auriferous beaches of Nome, Norton Sound, Alaska. United States Geological Survey Professional Paper 1250:23–34.

This paper deals almost exclusively with invertebrate fossils – particularly marine molluscs. The author (p. 23) alludes to early discoveries of vertebrate fossils from arctic and subarctic shores of Alaska during the voyages of Kotzebue in 1816 and of Beechey in 1826. The author, while in the Coast Survey service in 1880, examined the Pleistocene deposits in Kotzebue Sound and near Bering Strait, collecting many vertebrate remains.

318. Dall, W.H., and Harris, G.D. (1892): Correlation papers. Neocene. United States Geological Survey Bulletin 84:1–339.

Pertinent to this publication is a section Distribution of Fossil Vertebrates under Alaska (p. 266). The following fossil localities are mentioned: (1) on the arctic coast at about 71°N at Skull Cliff, Beechey's party obtained remains of an elephant (Proboscidea) in clay overlying a layer of ice; (2) tusks, teeth, and bones of woolly mammoth (*Mammuthus primigenius = Elephas primigenius*) and Columbian mammoth (*Mammuthus columbi = Elephas columbi*) were collected near Topanica Creek, Norton Sound, by Wossnessenski; (3) other remains of the same sort have been collected on the coast between Bristol Bay and Norton Sound; (4) elephant (Proboscidea) teeth, bones of bison (*Bison = Bison antiquus*) and especially of muskox (*Ovibos moschatus*) 'are not rare on the tundra of the Yukon valley, whence specimens were brought by Dall in 1868'; (5) Kotlo and Inglutalik rivers 'have the reputation of affording these bones in extraordinary numbers'; (6) along the arctic coast E of Point Barrow, where the bones and ivory occur frozen in the clays, they are so common as to serve Eskimo (*Homo sapiens*) carvers for economic purposes – e.g., in 1880 Dall obtained a deep ladle carved from a mammoth tusk that was said to have come from the mouth of Colville River; (7) ice age animal bones are also associated with clays in the Ulukak River valley, and a lake near Nushagak (on Russian authority), and are reported from the upper part of the Knik or Fire River which enters Cook Inlet; (8) a mammoth (*Mammuthus*) tooth was found on St George Island in the Pribilof Islands in 1836 according to Veniaminoff; and (9) mammoth tusks and teeth were discovered on Unalaska Island in the Aleutians in 1801 according to a report of Dr Stein published in St Petersburg, Russia, in 1830.

319. Dallimore, S.R., Wolfe, S.A., Matthews, J.V., Jr, and Vincent, J.-S. (1997): Mid-Wisconsinan eolian deposits of the Kittigazuit Formation, Tuktoyaktuk Coastlands, Northwest Territories, Canada. Canadian Journal of Earth Sciences 34(11):1421–41.

The Kittigazuit Formation is a Late Quaternary sand unit commonly observed throughout the Tuktoyaktuk Coastlands and in subbottom sediments of the southern Beaufort Shelf. It is eolian in nature, and plant and insect fossils suggest that, although the summer climate during deposition was as warm or slightly warmer than today, conditions were likely more arid. The authors mention that there are several clues [e.g., Pleistocene horse (*Equus*) evidence from the Alaskan Coastal Plains (Guthrie and Stoker 1990)] that this sea sand area was dry and cold, with little snow during Late Wisconsinan time. AMS radiocarbon dates indicate that the Kittigazuit Formation was deposited between 37,000 and 33,000 BP (Mid-Wisconsinan). A date on postglacial mammoth (*Mammuthus*) bone from Tununuk (see map site 11, Figure 1), Northwest Territories, is given as 19,440 ± 290 BP (I-8578), with a note that it may establish a minimum age for deglaciation of Richards Island. It is also noted that an age of 14,920 ± 160 BP (Beta-25119, ETH-3898) from a saiga antelope (*Saiga tatarica*) from Baillie Islands indicates that regional climate was quite arid, at least during winter during this time. [See Harington and Cinq-Mars 1995.]

320. Dalquest, W. (1988): *Astrohippus* and the origin of Blancan and Pleistocene horses. Occasional Papers of the Museum Texas Tech University 116:1–21.
In this review of North American Late Cenozoic horses, the author discusses opposing views of European and American paleontologists, referring, for example, to whether the Yukon wild ass (*Equus lambei*) is a subgenus (*Asinus*) (Harington and Clulow 1973) or a caballine horse (Forsten 1986). [The species is now considered a Yukon horse (*Equus lambei*).]

321. Dalzell, K.E. (1968): The Queen Charlotte Islands 1774–1966. C.M. Adam, SS1, Terrace, British Columbia. 340 pp.
Abe Johnson and Ed Hewitt, former prospectors from Rossland and Nelson, British Columbia, had learned of the existence of very large bones near Sandspit. They carefully excavated the bones, packed an assortment of them in crates, and with public notices in the *Vancouver Province* (July 12, 1907 – 'Rancher at Skidegate discovered mammoth [*Mammuthus*] rib of Proboscides Ungulata') and the *Victoria Times* (July 13, 1907) had them shipped on the *Princess Beatrice* in August 1907. According to a newspaper account, 'one tusk measures eighteen feet, six inches and is twenty-five inches in diameter at the large end. It is nearly straight having two feet curvature. The bones so far recovered show that the mastodon [?*Mammut americanum* – perhaps the 'tusk' is a mandible of a large whale (Cetacea)] from which it came was sixteen to seventeen feet across the rump and seven feet between the eyes' [leading to the headline, 'Seven Feet between the Eyes – This Mastodon Must Have Been Forty Feet High – And the Bones Are in a Good State of Preservation,' in the Vancouver paper which heralded the arrival of the bones. Hewitt and Johnson put their find on display on Hastings Street for 30 cents a look and drew large crowds with their sign 'Prehistoric Mastodon Bones From the Queen Charlotte Islands.' The publicity sent a rash of bone hunters to the region to dig for more specimens. Agnes Mathers stated that she was only able to save part of a rib and vertebrae before all the bones were carted away. They lay in the shade of a tree in her vegetable garden 'today' [presumably about the time when the book was written in the 1960s].

322. Danish/Greenlandic Network (1999) Archaeological Research in Greenland. http://www.natmus.dk/arg.
The aim of this homepage is to serve as a link between Danish and Greenlandic researchers working on Greenlandic prehistory to stimulate dialogue. It is intended to be of interest to the public in providing current views of Greenland's prehistoric cultures. The site was created by researchers connected to the Danish/Greenlandic Network, sponsored by the Danish Research Councils. Radiocarbon dates pertaining to each of seven prehistoric cultures are given under the heading 'C14 Datings.' Most dates are on caribou (*Rangifer tarandus*), muskox (*Ovibos moschatus*), and arctic hare (*Lepus arcticus*), besides turf and wood, but dates on caribou antler and driftwood are excluded. Dates on marine mammal bones are presented too. Most dates are δ^{13}C-corrected and calibrated according to Stuiver and Pearson's (1993) curve.

323. Darwent, C., and Darwent, J. (1998): GIS modeling of changes in mammalian resources in Eastern Arctic prehistory. 8th International Congress of the International Council for Archaeozoology (August 23–9, Victoria, British Columbia). Final Program and Abstracts. p. 94.
The prehistory of the North American Eastern Arctic is generally viewed as including a shift from terrestrial resources during the early periods to marine-based resources. A marine or maritime adaptation is often characterized by the procurement of large, open-water marine mammals such as walrus (*Odobenus rosmarus*) and whale (Cetacea). However, the hypothesized resource shift has not been tested on a regional scale using fossil-mammalian faunas from the Canadian Arctic and Western Greenland. Variation in mammalian communities over the last 4000 years was caused by climatic shifts, including fluctuations in ice leads and polynias. The High Arctic islands, in particular, were strongly affected by these shifts in resource availability. Changes in technology, in part as a response to what was available, enabled the exploitation of different mammalian resources, including for example the introduction of toggling harpoons, large whaling harpoons, and umiaks. Using a Geographical Information System (GIS), the authors model spatial-temporal changes in the locations and diversity of mammalian species in the Eastern Arctic from early In-

dependence I through Pre-Dorset, Dorset, Thule, and Historical stages. Data are primarily generated from archaeological collections, but paleontological specimens are also included. This approach is used to address problems of preservation, taphonomy, demography, and local and regional sampling of zoological material.

324. Davies, J.L. (1958): Pleistocene geography and the distribution of the northern pinnipeds. Ecology 39:97–113.
Present distributions of northern pinnipeds were affected by Pleistocene glaciations. The author points out that expansions of glacial and marine ice pushed seal populations southward, tending to split Atlantic and Pacific populations, and contractions of the ice led to general northward movement. Often previously isolated stocks would tend to regain contact and mix. Species discussed are: northern fur seal (*Callorhinus ursinus*), northern sea lion (*Eumetopias jubatus*), walrus (*Odobenus rosmarus*), harbour seal (*Phoca vitulina*), ringed seal (*Phoca hispida*), harp seal (*Phoca groenlandica*), ribbon seal (*Phoca fasciata*), bearded seal (*Erignathus barbatus*), grey seal (*Halichoerus grypus*), and hooded seal (*Cystophora cristata*). This paper includes distribution maps and tables of phylogenetic relationships.

325. Davies, W.E., Krinsley, D.B., and Nicol, A.H. (1963): Geology of the North Star Bugt area, Northwest Greenland. Meddelelser om Grønland 162 (12):1–68.
In a section of this publication titled Glacial Geology, Daniel Krinsley mentions (p. 58 and Figure 22) that many large bones and a small amount of baleen of a bowhead whale (*Balaena mysticetus*) were excavated from a raised marine delta of Siorqap kûa during August 1953. The bone assemblage, their conspicuous lack of erosion, and their relative position indicate that the whale was probably intact when rapidly buried by fluviomarine sand. Edible mussel (*Mytilus edulis*) shells found with the bones indicate that burial occurred within the littoral zone. Baleen found near one of the jaw bones and at 13 m asl, 3.6 m beneath a prominent 17 m marine terrace was radiocarbon dated to 8500 ± 200 BP (W-48, see Suess 1954. United States Geological Survey radiocarbon dates I. Science 120(3117):467–73). According to Bennike (1997, p. 906) this specimen represents the oldest dated bowhead from Greenland.

326. Davis, S.D. (1990): Prehistory of southeastern Alaska. *In*: The Handbook of North American Indians (W. Sturtevant, ed.). W. Suttles, Northwest Coast – Smithsonian Institution, Washington. pp. 197–202.
Vertebrate remains are mentioned from the following sites (see map, Figure 1): (1) At Hidden Falls, a stratified multicomponent site situated about 8 m above low tide dating to approximately 7500 BC, fish (Pisces) bone and marine shells suggest a coastal marine subsistence for the people (*Homo sapiens*). Early Phase (Component II 2670 ± 110 BC to 1265 ± 80 BC) occupation is associated with: dog (*Canis*), deer (Cervidae), sea mammals (Pinnipedia), anadromous and marine fish (Pisces); Middle Phase (Component III 1000 BC to 650 AD) occupation fauna includes: marine fish – Pacific gray cod (*Gadus macrocephalus*), salmon (*Oncorhynchus*), rockfish (*Sebastes*), herring (Clupeidae), and halibut (*Hippoglossus stenolepis*) – deer (Cervidae), dog (*Canis familiaris*), sea otter (*Enhydra lutris*), harbour seal (*Phoca vitulina*), and bird (Aves) remains. (2) At the Chuck Lake site, Heceta Island, dated at about 6300 BC were seal, sea lion, waterfowl (Anatidae), beaver (*Castor canadensis*), a canid (Canidae), and possible deer.

327. Dawson, G.M. (1894a): Geological notes on some of the coasts and islands of Bering Sea and vicinity. Geological Society of America Bulletin 5:117–46.
This paper is of general interest regarding early ideas on a land connection between Siberia and Alaska.

328. Dawson, G.M. (1894b): Notes on the occurrence of mammoth remains in the Yukon district of Canada and in Alaska. Quarterly Journal of the Geological Society 1:1–9.
The author states that Robert Campbell of the Hudson's Bay Company first reported mammoth remains in the Yukon in *The Discovery and Exploration of the Yukon (Pelly) River* printed at Winnipeg in 1885. He wrote: 'I saw the bones, heads and horns of Buffaloes [probably the large-horned bison (*Bison priscus*)]; but this animal had become extinct before our visit, as had some species of elephant [probably the woolly mammoth (*Mammuthus primigenius*)], whose remains were found in various swamps. I forwarded an elephant's thigh bone [actually a tibia] to the British Museum, where it may still be seen.' Evidently these bones were found near the former site of Fort Selkirk before 1852. Dawson also notes that gold

miners had frequently encountered mammoth bones in the Yukon River valley, particularly near Forty-Mile Creek. Alaskan mammoth remains reported include: tusk fragments from a lake near Nushergak, Inglutalik River, and Kotlo River; bones, tusks, and molars collected by F. Mercier [some now in the Canadian Museum of Nature Collection] near Tanana, Kotzebue Sound (where bones were first collected by Kotzebue in 1816). Species listed by W.H. Dall include: woolly mammoth (*Mammuthus primigenius* = *Elephas primigenius*); Columbian mammoth (*Mammuthus columbi* = *Elephas columbi*); horse (*Equus* = *Equus major*); moose (*Alces alces* = *Alces americanus*); caribou (*Rangifer tarandus* = *Rangifer caribou*); tundra muskox (*Ovibos moschatus*); helmeted muskox (*Bootherium bombifrons* = *Symbos cavifrons* = *Ovibos maximus*?); and large-horned bison (*Bison priscus* = *Bison crassicornis*); mammoth (*Mammuthus*). Mammoth remains were found at St George and St Paul islands (Pribilof Islands) and at Unalaska Island (Aleutian Islands). The author also reports a mammoth tooth from southern Vancouver Island [there are several records from Saanich Peninsula near Victoria] and a large mammoth(?) bone from preglacial gold-bearing gravels at Cherry Creek [Okanagan District, British Columbia]. The author provides several reasons for concluding that a land connection existed between NE Siberia and NW North America 'at the time of the existence of this mammoth, or for some portion of that time.' In the discussion, Henry Woodward mentions that in 1850 Captain Kellett [Royal Navy] bought remains of muskox and mammoth to the British Museum from Kotzebue Sound, Alaska; and in 1873 the Reverend R. McDonald (one of the Hudson's Bay Company's chaplains) from Fort McPherson gave mammoth, muskox, large-horned bison (*Bison priscus*), and horse remains to the National Collection. Woodward also remarks that mastodon (*Mammut americanum* = *Mastodon*) 'has lately been found in Kent County, Ontario.'

329. Dawson, G.M. (1896): Summary report on the operations of the Geological Survey for the year 1895. Geological Survey of Canada Annual Report (New Series) 8, Part A.

The report on paleontology and zoology (pp. 130A–149A) was submitted by J.F. Whiteaves and includes: (1) a humerus, ulna, and part of a radius, probably of a harp seal (*Phoca groenlandica*), from Champlain Sea clay at Graham's Brickyard near Ottawa, Ontario, submitted by Alex Graham; (2) part of a mammoth (*Mammuthus*) molar found about 6 mi upstream from Edmonton, Alberta, submitted by James Gibbons; and (3) N.J. Giroux mentions a white whale (*Delphinapterus leucas* = *Delphinapterus catodon*) skeleton from near Coteau du Lac church, Quebec (22 ft below the surface in an excavation for the Soulange Canal). Unfortunately only one of the vertebrae could be preserved, as the rest of the bones fell to pieces (p. 73A).

330. Dawson, G.M. (1901a): Summary report of the Geological Survey of Canada for 1898, Report A. 208 pp.

In 1895 the author obtained from J. Gibbons a woolly mammoth (*Mammuthus primigenius* = *Elephas primigenius* or *Elephas americanus*) tooth from workings in Saskatchewan gravels some 6 mi upstream from Edmonton, Alberta. The author also saw a small, well-preserved mammoth tusk at Edmonton that was picked up on a bar of the North Saskatchewan River near Goose Encampment [presumably near Edmonton] (p. 19A). J.F. Whiteaves submitted the following report on paleontology and zoology (pp. 173A–194A). In late September, a visit was made to Muirkirk, Ontario, where mammoth (*Mammuthus*) remains had been discovered. The bones were purchased and notes were made on the circumstances attending their discovery, their position relative to the deposits in which they were found, and the physical features of their surroundings (pp.189A, 190A). The purchase is outlined thus (p. 193A): mammoth remains found on the farm of Charles Fletcher about 1.5 mi NE of Muirkirk in 1895 – lower jaw with teeth in place; upper molars with parts of the cranium; parts of the tusks; a few vertebrae and ribs; part of a scapula; two humeri; ulna and radius; and all bones of the hind limbs, except some of the smaller foot bones. The skull of a muskox (*Ovibos moschatus*) 'from the roof of a [mining] tunnel run in the north bank of the Saskatchewan [River] about one mile below [downstream from] Edmonton'. Dawson (p. 19A) thought that this specimen, representing an old individual, was derived from Saskatchewan gravels. [This specimen has been radiocarbon dated at 43,100±490BP (Beta-115201).]

331. Dawson, G.M. (1901b): [Skull of extinct bison]. Summary Report of the Geological Survey of Canada for the year 1901:185A.

Part of a skull of an extinct bison [probably the steppe bison (*Bison priscus* = *Bison laticornis*)] found in a layer of 'muck' about 15 ft below the ground surface on claim 17, Gold Run Creek, near Dawson City, Yukon, was collected by, or received from, R.G. McConnell in 1900. [See Harington and Clulow 1973.]

332. Dawson, J.W. (1857): On the newer Pliocene and post-Pliocene deposits of the vicinity on Montreal, with notices of fossils recently discovered in them. Canadian Naturalist and Geologist 2(6):401–26.
Several whale (Cetacea) caudal vertebrae and a seal (*Phoca*) pelvis were found at the Montreal brickyards, apparently in Leda clay, by Sir William Logan. Bones of the posterior extremities of a seal (*Phoca*) were collected at Green Creek, Ontario. In collections of the Geological Survey of Canada were: part of a seal pelvis; a whale vertebra obtained by Sir William Logan from clay near Montreal; nodules containing capelin (*Mallotus villosus*) and lumpfish (*Cyclopterus lumpus* = *Lumpus anglorum*) from Greek Creek, Ontario. Further, this paper includes the first published notice of a sculpin (*Cottus*) from Green Creek.

333. Dawson, J.W. (1871): Notes on the post-Pliocene geology of Canada. Canadian Naturalist 6:116–87, 247–59, 365–416.
At Green Creek, E of Ottawa, Ontario, capelin (*Mallotus villosus*), lumpfish (*Cyclopterus lumpus*), stickleback (*Gasterosteus*), unidentified fish (Pisces) vertebrae and other bones, and bird (Aves) feathers were found. In Leda clay of Montreal, harp seal (*Phoca groenlandica*) and white whale (*Delphinapterus leucas* = *Beluga vermontana*) were found. A white whale skeleton was also found in Saxicava sand near Cornwall, Ontario. Remains of a woolly mammoth (*Mammuthus primigenius* = *Euelephas jacksonii*) were found in superficial gravels [from Burlington Heights near Hamilton, Ontario – see Billings 1863].

334. Dawson, J.W. (1878): Note on the fossil seal from the Leda clay of the Ottawa Valley. Canadian Naturalist and Quarterly Journal of Science 8:340–1.
Fossils found in clay nodules at Green Creek, E of Ottawa, Ontario, are: harp seal (*Phoca groenlandica*), capelin (*Mallotus villosus*), lumpfish (*Cyclopterus lumpus*), stickleback (*Gasterosteus*), and sculpin (*Cottus*). The seal remains, obtained by Dr Grant, included the left ramus of the lower jaw, containing a canine and four molars. The author alludes to an earlier (1856) report of the bones of the 'hinder extremity of a young seal' enclosed in a nodule from Green Creek obtained by Mr Billings, and mentions skeletons of larger individuals 'of this species' found in post-Pliocene clays of Montreal, parts of which are in the collections of the Geological Survey of Canada and McGill University.

335. Dawson, J.W. (1883a): On portions of the skeleton of a whale from gravel on the line of the Canadian Pacific Railway, near Smith's Falls, Ontario. Canadian Naturalist and Quarterly Journal of Science 10(7):385–7.
Bones of a white whale (*Delphinapterus leucas* = *Beluga vermontana*) were found: (1) in clay from Rivière du Loup, Quebec; (2) at an Intercolonial Railway excavation on the southern shore of Baie des Chaleurs; (3) in brick clays near Montreal; (4) in Saxicava sand near Cornwall, Ontario; and (5) from clay at an elevation of 46.5 m in Charlotte Township, Vermont, U.S.A. Two vertebrae (dorsal and lumbar) and part of a rib of a humpback whale (*Megaptera novaeangliae* = *Megaptera longimana*) were found in Welch's Gravel Pit on the Canadian Pacific Railway line 5 km N of Smiths Falls, Ontario, at about 136 m asl.

336. Dawson, J.W. (1883b): On portions of the skeleton of a whale from gravel on the line of the Canadian Pacific Railway, near Smith's Falls, Ontario. American Journal of Science, Series 3, 25:200–2.
[See Dawson 1883a.]

337. Dawson, J.W. (1883c): Canadian Pleistocene. Geological Magazine 10:111–13.
Saxicava sand and gravel of the St Lawrence River valley may contain remains of whales (Cetacea). Mastodon (*Mammut americanum* = *Mastodon*) and mammoth (*Mammuthus* = *Elephas*) remains may be found in postglacial deposits (alluvial, peaty, and lacustrine). The author refers particularly to bone of a large humpback whale (*Megaptera novaeangliae* = *Megaptera longimana*) from gravel N of the outlet of Lake Ontario [near Smiths Falls, Ontario], and mentions that the Upper Leda

Clay on the Ottawa River [near Green Creek, Ontario] contains feathers of birds (Aves) and skeletons of capelin (*Mallotus villosus*).

338. Dawson, J.W. (1891): Note on a fossil fish and marine worm found in the Pleistocene nodules of Green's Creek on the Ottawa. Canadian Record of Science 4:86–8.
Vertebrate fossils found in clay nodules at Green Creek, Ontario, include sculpin (*Cottus ricei* = *Cottus uncinatus*), capelin (*Mallotus villosus*), lumpfish (*Cyclopterus lumpus*), three-spined stickleback (*Gasterosteus aculeatus*), and American smelt (*Osmerus mordax*). Three sculpin specimens are mentioned.

339. Dawson, J.W. (1893): The Canadian Ice Age. William V. Dawson, Montreal. 301 pp.
Dawson summarizes his previous papers on this subject. A list of Canadian Pleistocene vertebrates (pp. 265–9) includes specimens from Green Creek, Ottawa, Ontario: harp seal (*Phoca groenlandica*), capelin (*Mallotus villosus*), smelt (*Osmerus mordax*), sculpin (*Cottus ricei* = *Cottus* (*Centrodermichthys*) *uncinatus*), lumpfish (*Cyclopterus lumpus*), three-spined stickleback (*Gasterosteus aculeatus*), and unidentified bird (Aves) feather impressions. Found at the Goose River mouth on the N shore of the St Lawrence River was cod (*Gadus* = '*Salmo salar*'); from Montreal, Quebec, Leda clay: harp seal (*Phoca groenlandica*); from Rivière du Loup and Montreal, Quebec, and Cornwall, Ontario: white whale (*Delphinapterus leucas* = *Beluga catodon*); from 5 km N of Smiths Falls, Ontario: humpback whale (*Megaptera novaeangliae* = *Megaptera longimana*). The author states 'we can scarcely include … the Mastodon [*Mammut americanus*] or Mammoth [*Mammuthus*], and their contemporaries, as their remains, so far as is known in Canada, are rather Postglacial or Modern.'

340. Dawson, J.W. (1895): Note on a specimen of *Beluga catodon*, from Leda Clay, Montreal. Canadian Record of Science 6:351–4.
A nearly complete skeleton of white whale (*Delphinapterus leucas* = *Beluga catodon*) was found at a depth of 6.8 m in brick clay near Papineau Road, Smith's Brickyard, Montreal, Quebec, about 31 m above the St Lawrence River. The specimen was associated with marine shells and a fragment of black spruce [probably *Picea nigra*] wood. 'The best specimens heretofore found in Canada are one discovered in Peel's Brickyard, Montreal, one found near Cornwall, and another discovered at Bathurst, N.B., and described by Gilpin and Honeyman.' [See Gilpin 1874 and Honeyman 1874a, b.]

341. Deagle, B.E., Reimchen, T.E., and Levin, D.B. (1996): Origins of endemic stickleback from the Queen Charlotte Islands: Mitochondrial and morphological evidence. Canadian Journal of Zoology 74:1045–56.
Uncertainties remain as to whether the more divergent forms of the threespine stickleback (*Gasterosteus aculeatus*) from the Queen Charlotte Islands have been recently derived in postglacial periods or are relicts of an ancient lineage that may have survived in ice-free refugia postulated for the area. The authors' data indicate that the 'relict' lineage was largely limited (18 of 20 populations) to adjacent watersheds in the NE corner of the Queen Charlotte Islands close to a suspected glacial refugium, but they also found the lineage in two remote lakes on the W coast of the islands, distant from any known refugia. They obtained a sample of 33 sticklebacks from the mid-Pacific and found both mitochondrial DNA lineages, strongly suggesting continuing dispersal of these fishes across the Pacific and inconsistent with previous suggestions of relict status for one of the lineages.

342. Debicki, R.L. (1983): Placer deposits: Their formation, evolution, and exploitation. *In*: Yukon Placer Mining Industry 1978–1982. Indian and Northern Affairs Canada, Exploration and Geological Services, Northern Affairs Program, Whitehorse. pp. 18–34.
Pertinent are remarks on p. 24. Deposits of wind-blown silt and decayed organic material, commonly called black muck, blanket gravel deposits in many stream valleys in Yukon. Such deposits may be up to 20 m thick. Fossils such as the steppe bison (*Bison priscus*) cranial fragment with horncores and sheaths from Sulphur Creek near Dawson City (Figure 23) are found in black muck. Other fossils found in the Klondike area, including remains of Yukon horses (*Equus lambei*), sabre-tooth tigers [probably scimitar cats (*Homotherium serum*)], mastodons (*Mammut americanum*), and mammoths (*Mammuthus*) lived between 39,900 and 22,200 BP (Harington and Clu-

low 1973). 'Eight of the thirteen species are now extinct, while two of the others no longer live in Yukon.'

343. Deller, D.B., and Ellis, C.J. (1988): Early Palaeo-Indian complexes in southwestern Ontario. *In*: Late Pleistocene and Early Holocene Paleoecology and Archaeology of the Eastern Great Lakes Region (R.S. Laub, N.G. Miller, and D.W. Steadman, eds.). Bulletin of the Buffalo Society of Natural Sciences 33:251–63.
The authors indicate the presence of three time-sequential Paleoindian complexes in southwestern Ontario. In Gainey times, large, extensive sites occur throughout the NE. By Parkhill time, these seem to occur only in the most northern locations. In Crowfield times, they seem to have disappeared. Perhaps large sites are primarily associated with spruce parklands. Large sites are often interpreted as a product of the communal hunting of migratory herbivores, specifically caribou (*Rangifer tarandus*) (e.g., Parkhill). Given that sites in the NE are related to communal hunting, the authors suggest their northward trend and eventual disappearance are tied into the development of environments in which this procurement pattern became less viable. After 10,600 BP, the environments became more closed (especially with the appearance of pine forests in Ontario), and range suitable for migratory caribou may have become reduced and limited to more northerly regions.

344. Demers, D., and Locat, J. (1985): Quaternary stratigraphy and fossil seal, La Durantaye area, Quebec. Géographie physique et Quaternaire 39(1):25-34.
The authors report a nearly complete seal (Phocidae) skeleton [including the trunk, tibiae, and fibulae, but lacking the shoulder blades and ends of both fore and hind limbs – the frontals were the only part of the cranium that was found (Figure 6 and 8)]. Discovered by L. Carriere on the E bank of Rivière Boyer, Quebec [2.5 km W of the town of La Durantaye (Figure 1)], the fossil came from black, silty marine clay (Unit 4 about 1.2 m deep, Figure 3). Other fossils found in the same layer as the skeleton were: many shells of the edible mussel (*Mytilus edulis*); common sea urchins (*Strongylocentratus drobuchiensis*); and flat clusters of sponge spicules [Porifera, probably (*Tethya logani*)]. A preliminary identification has revealed it to be a grey seal (*Halichoerus grypus*) or hooded seal (*Cystophoca crystata*). [A precise identification is required, as well as a radiocarbon date on the bone.]

345. Dempster, L. (1997): Ice age fossil sheds new light on Alberta's past. Calgary Herald, August 15.
An ice age camel (Camelidae) has led two Alberta researchers to knowledge about the landscapes of Alberta a million years ago. The discovery of camel shoulder blades, along with bits of mammoth (*Mammuthus*), hare (*Lepus*), and giant sloth (Xenarthra) fossils will help scientists obtain a clearer picture of rich grazing plains and turbulent rivers. Robert Young of the University of Alberta said how rare it was to find ice age vertebrate remains in glaciated parts of Canada. Expedition co-leader Jim Burns of the Provincial Museum of Alberta said two well-preserved camel scapulae were excavated from their original burial site 120 m below the cutbanks of the South Saskatchewan River about 30 km E of Medicine Hat, Alberta. Some of the bones could be as old as 1.6 Ma, and the camel was probably 6–7 ft at the shoulder.

346. Deschamps, E.M., and Lauriol, B. (1989): An Early Holocene cave deposit, Caverne de la Mine (Québec, Canada). Canadian Archaeological Association 32nd Annual Conference (April 28–May 2, Whitehorse, Yukon). Programme and Abstracts. pp. 30–1.
Caverne de la Mine is located some 20 km NE of Ottawa and contains significant fossils. The cave, partly a natural trap, accumulated many bones and teeth which are useful for paleoecological reconstruction. The upper 100 cm, containing a relatively recent fauna dated at 5000 BP, is characterized by black bear (*Ursus americanus*), white-tailed deer (*Odocoileus virginianus*), raccoon (*Procyon lotor*), mouse (*Peromyscus*) and big brown bat (*Eptesicus fuscus*). The bottom 70 cm of infill dating between 8230 ± 80 BP and 5020 ± 70 BP is being studied. Based on cranial elements recovered, 22 mammal species have been identified from this unit. The assemblage contains many and varied small mammals and no black bear or deer remains. Two species, the woodland vole (*Microtus pinetorum*) and Ungava lemming (*Dicrostonyx hudsonius*), no longer occupy the area. The former is found mainly in eastern United States, while the latter is found in northern Quebec and Labrador, suggesting Holocene range shifts to the S and N respectively.

347. Dillon, L.S. (1956): Wisconsin climate and life zones in North America. Science 123(3188):167–76.

This paper deals mainly with paleoclimatic evidence, including past plant distributions (based on pollen analyses) to reconstruct Wisconsinan climate and life zones. Muskoxen are used as examples as well. The muskox (*Ovibos moschatus*) is taken as a fairly reliable indicator of tundra conditions. That such conditions prevailed just S of the Wisconsinan ice sheets is indicated by the past distribution of *Ovibos* remains. Many other extinct muskox genera [*Symbos, Bootherium*, and *Gidleya* – now all assigned to *Bootherium bombifrons*], because their remains have been found throughout the grass belt of middle North America, seem to be 'prairie' not tundra indicators. Hence, there is no good evidence that severe polar conditions existed within the United States, except close to ice sheets. Since no ecological factor is known that would explain the absence of muskoxen (*Ovibos*) from the E side of Hudson Bay, the author suggests that the central part of the Laurentide ice disappeared before the ice S and E of the bay – a point in accord with precipitation patterns postulated.

348. Dixon, E.J. (1984): Context and environment in taphonomic analysis: Examples from Alaska's Porcupine River Caves. Quaternary Research 22(2):201–15.
Alterations of vertebrate bones from Alaska's Porcupine River caves were examined. Most of the work was conducted in Cave 1, 68 m above the river (approximately 67°N, 143°W) and bones included: arctic hare (*Lepus othus*); snowshoe hare (*Lepus americanus*); ground squirrel (*Spermophilus parryii*); hoary or Alaska marmot (*Marmota* cf. *Marmota caligata* or *Marmota broweri*); collared lemming (*Dicrostonyx torquatus*); brown lemming (*Lemmus sibiricus*), red-backed vole (*Clethrionomys*); yellow-cheeked vole (*Microtus xanthognathus*); muskrat (*Ondatra zibethicus*); red fox (*Vulpes vulpes*); black bear (*Ursus americanus*); brown bear (*Ursus arctos*); mink or ermine (*Mustela* cf. *Mustela vison* or *Mustela erminea*); horse (*Equus*); caribou (*Rangifer tarandus*); Dall sheep (*Ovis dalli*); bison (*Bison*); a proboscidean (Proboscidea); a bird (Aves); and a fish (Pisces). Radiocarbon analyses of a bison metacarpal and a horse ulna yielded dates of 21,050 +320/–340 BP (DIC-1333) and 21,780 ± 310 BP (DIC-1334) respectively.

349. Dixon, E.J. (1985): The origins of the first Americans. Archaeology 28(2):22–7.
It is generally agreed that humans (*Homo sapiens*) came originally to the Americas from Asia. The author argues that they might not have needed the Bering land bridge for they could have crossed in skin-hulled boats or over pack ice. Other mammals mentioned are: wolverine (*Gulo gulo)*, bear (*Ursus*), wolf (*Canis lupus*), horse (*Equus*), and mammoth (*Mammuthus*).

350. Dixon, E.J. (1993): Quest for the Origin of the First Americans. University of New Mexico Press, Albuquerque. 154 pp.
The book has three major themes: (1) summary and interpretation of the early prehistory of northwestern North America, where North America and Eurasia were repeatedly connected by land, allowing much biotic interchange; (2) the process of scientific inquiry, involving discovery of new data and techniques and collaboration between scientists; and (3) the history of archaeology in Alaska. Chapter 3 will be of interest to paleobiologists, as it deals with collecting of important horse (*Equus*), bison (*Bison*), moose (*Alces*), and mammoth (*Mammuthus*) carcasses by Otto Geist in Alaska for the Frick Collection, American Museum of Natural History. Tables 3.6 and 3.7 list 10 radiocarbon dates for mammoths and 16 for bison in Eastern Beringia. Chapter 4 mentions artifacts and pseudoartifacts, dealing with evidence from Yukon and Alaska for human working of supposed Pleistocene bone. Chapter 7 deals with the Nenana Complex involving 10 early human (*Homo sapiens*) sites in central Alaska (Figure 7.5) [however, Jay Creek Ridge, Carlo Creek (component I), Broadway, and Owl Ridge (component II) sites range between 9500 and 8500 BP] dating between about 11,500 and 10,500 BP. Although poorly preserved faunal remains of sheep (*Ovis*), elk (*Cervus elaphus*), and grouse and ptarmigan (*Lagopus*) were found at Dry Creek, the Broken Mammoth and Mead sites are the only ones of this period containing a wide variety of well-preserved remains documenting the animals hunted by these people. Preliminary analysis of the bones by Yesner and Holmes indicate that cranes, ducks, swans, and geese were consumed. Bones of hare (*Lepus*), ground squirrel (*Spermophilus*), beaver (*Castor canadensis*), caribou (*Rangifer tarandus*), and possibly elk (*Cervus elaphus*) or moose (*Alces*) have been identified also, as well as a salmonid fish (Pisces) scale, and mammoth (*Mammuthus*) or mastodon (*Mammut*) tusk fragments. A small stone chip imbedded in a worked tusk fragment from the Broken Mammoth site suggests that mammoths were

hunted or scavenged by people. Analysis of residues preserved on stone artifacts from Jay Creek Ridge indicates that they had come in contact with mammoth, bison, and sheep. Woolly mammoth (*Mammuthus primigenius*) blood residue on fluted projectile points from Eastern Beringia indicated that people had hunted them and that mammoths had persisted as late as 9500 BP at Jay Creek Ridge. Table 9.1 lists the results of residue analysis of fluted projectile points from 11 sites in Eastern Beringia.

351. Dixon, E.J. (2001): Human colonization of the Americas: Timing, technology and process. Quaternary Science Reviews 20(2001):277–99.
Geological and archaeological research indicates that people (*Homo sapiens*) first colonized the Americas by watercraft along the southern coast of the Bering Isthmus and the western coast of the Americas. Early dates from some archaeological sites indicate that human colonization began before 13,000 BP. Several distinctive cultural traditions had developed in Eastern Beringia by 11,000–10,000 BP, and geological, biological, and linguistic evidence, as well as dated human skeletal remains, suggest that people occupied the Americas before about 11,500 BP. Glacial geology indicates colonization could have begun about 14,000–13,000 BP along the western coasts of the Americas and ended about 5000 BP with the deglaciation of the Canadian Eastern Arctic and coastal Greenland. A model for early coastal and subsequent inland colonization of the Americas along large ecological zones best fits current geological and archaeological data.

352. Dixon, E.J., and Smith, G.S. (1986): Broken canines from Alaskan cave deposits: Re-evaluating evidence for domesticated dog and early humans in Alaska. American Antiquity 5(2):341–51.
Broken canines from two caves (Cave 1 and Bear Cave about 5 km apart and situated nearly 100 km downstream from the Alaska/Yukon border) on the Porcupine River, Alaska, are analysed with respect to size, development, morphology, breakage pattern, and stratigraphic position. These analyses indicate that they are deciduous bear teeth similar to canines recovered from various stratigraphic levels at Trail Creek Caves 2 and 9 on the Seward Peninsula, Alaska (Larsen 1968). So the Trail Creek specimens (including specimens from Late Pleistocene levels) are also naturally exfoliated bear (*Ursus*) teeth and not dog (*Canis familiaris*) canines broken out of jaws by people (*Homo sapiens*), as originally suggested.

353. Dixon, E.J., and Thorson, R.M (1984): Taphonomic analysis and interpretation in North American Pleistocene archaeology. Quaternary Research 22:155–9.
The authors state that there is no hard evidence that there were humans (*Homo sapiens*) in North America prior to 12,000 BP because reworked mammal (Mammalia) bones may have been altered by geological processes, not necessarily by man.

354. Dixon, E.J., Plaskett, D.C., and Thorson, R.M. (1979): Report of the 1979 archaeological and geological reconnaissance of cave deposits, Porcupine River, Alaska. University of Alaska. 47 pp.
During the summers of 1978 and 1979, the University of Alaska Museum investigated the Alaskan portion of the Porcupine River in search of caves that could have been used by humans (*Homo sapiens*) prior to 14,000 years ago. Eleven caves out of many found were considered to have been appropriate for human use and four were tested in 1979. Of these, Arch Cave showed inconclusive results. Large mammal bones (including fractured bones and a complete *Alces* tibia) were recovered from the sand and silt stratum (7b) and estimated to be of Pleistocene age. Bear Cave, about 60 m above the Porcupine River near the centre of the Lower Ramparts, proved to be the most culturally significant of the caves surveyed. Seven (and possibly more) periods of occupation were identified. The oldest was aged at about 5485 BC (approximately 7500 BP). A butchered and burned bone of a black bear (*Ursus americanus*) was discovered and aged at about 2300 BC (approx. 4300 BP). This cave has up to 8 m of deposits, of which less than two have been excavated. During the initial survey of the area, two mammoth (*Mammuthus*) tusks and a molar fragment were found near Canyon Village in a thick cap of alluvial sediments. A fragment of one of the tusks was radiocarbon dated at 27,490 +670 / –740 BC (DIC-1570) [29,420 +670 / –740 BP]. Radiocarbon dates of 24,620 +680 / –750 BC (DIC-1571) [26,570 +680 / –750 BP] and 28,640 +730 / –810 BC (DIC-1573) [30,590 +730/ –810 BP] for wood indicate that the highest terrace formed during aggradation of the middle Porcupine valley between 25,000 and 30,000 years ago, or Mid-Wisconsinan. The presence of at least

10 m of sand and fossil-bearing alluvium along the Middle Porcupine River was noted.

355. Dixon, E.J., Thorson, R.M., and Plaskett, D.C. (1985): Cave deposits, Porcupine River, Alaska. National Geographic Research Reports 20:129–53.
Archaeological investigation of the Alaskan part of the Porcupine River drainage was undertaken by the University of Alaska Museum during the summers of 1978–80. The primary focus of the study was to document the presence of people (*Homo sapiens*) in Alaska before 14,000 BP, when North America and Asia were connected. Caves were selected for study because of their common association with human activity and because they provide depositional environments favouring organic preservation. Two mammoth (*Mammuthus*) tusks, one *in situ*, were located near Canyon River in a thick cap of alluvial sediments being eroded by the river. Radiocarbon dates of 26,570 +680/ –750 BP (DIC-1571) from wood fragments stratified 20 m above the *in situ* tusk, and 29,440 +670/ –740 BP (DIC-1570) from a fragment of the tusk were obtained. A rib of a large mammal was also found *in situ*. A mammoth (*Mammuthus*) molar fragment and several unidentified large mammal bone fragments that had eroded from the sediments were collected. Artifacts were obtained in several layers at a probable Holocene site near the mouth of Coleen River, some of which contained bone fragments, ash and fire-cracked rock. At Arch Cave, deep levels (Pleistocene?) contain well-preserved faunal remains. Table 1 lists 13 radiocarbon dates on woolly mammoth (*Mammuthus primigenius*), caribou (*Rangifer tarandus*), moose (*Alces*), and bison (*Bison*) or moose bone, most of which are between 35,000 and 22,000 BP (exact dates and laboratory numbers are not given). At a site called The Dunes, part of a moose (*Alces*) calcaneum, and both mandibles and a cranial fragment of an ermine (*Mustela erminea*) were recovered from Unit 5a. A Holocene deposit in Bear Cave (LRC 79-2) yielded butchered and burned foot bones of a black bear (*Ursus americanus*) and an Arctic Small Tool tradition stone projectile point.

356. Dixon, E.J., Heaton, T.H., Fifield, T.E., Hamilton, T.D., Putnam, D.E., and Grady, F. (1997): Late Quaternary regional geoarchaeology of southeast Alaska karst: A progress report. Geoarchaeology: An International Journal 12(6):689–712.
Karst-systems, sea-caves, and rock-shelters within the coastal temperate rainforest of Alaska's Alexander Archipelago preserve important records of regional archaeology, sea-level history, glacial and climatic history, as well as vertebrate paleontology. Two AMS radiocarbon dates on human bone [the mandible of an adult male (*Homo sapiens*) containing all teeth but four incisors, three vertebrae, and a partial right innominate – apparently from the same individual] discovered in On Your Knees Cave (49-PET-408) on Prince of Wales Island document the oldest reliably dated human in Alaska to about 9800 BP. A sample extracted from the mandible dated to 9730 ± 60 BP (CAMS-1991), and another sample from the pelvis (innominate) was dated to 9880 ± 50 BP (CAMS-3238). In addition to the human remains, three artifacts were discovered in 1996: a piece of mammal bone modified to make what appears to have been a flint flaker; a fragment of a large, barbed bone projectile point apparently made from marine mammal rib; and a grey-black chert biface. Locality 1 at the Chuck Lake archaeological site on Heceta Island, just W of Prince of Wales Island, has been dated to 8200 BP. It is important because faunal remains are preserved there: shellfish and bottom and rock fish (the most abundant subsistence resources), terrestrial and marine mammals, and waterfowl (Anseriformes). A series of AMS radiocarbon dates from cave deposits in the region span the past 40,000 years and provide the first evidence of Pleistocene faunas from the NW coast of North America. Vertebrate fossil deposits from caves on Prince of Wales Island and smaller surrounding islands date from 6400 BP to 12,300 BP (postglacial) and from 35,400 BP to more than 44,500 BP (prior to the last glacial). Recent discoveries include well preserved vertebrate fossils spanning the last glacial interval. Most significant has been the discovery of 15 brown bears (*Ursus arctos*) ranging in age from 7205 ± 65 BP (AA-15224) to 12,295 ± 120 BP (AA-10445). The earliest radiocarbon date reported for black bear (*Ursus americanus*) on Prince of Wales Island is 11,565 ± 115 BP (AA-10448), suggesting that black and grizzly bears coexisted there for at least 1800 years. The most significant cave discovered (On Your Knees Cave) has produced an older vertebrate fauna providing the first firm evidence of a glacial refugium in the Alexander Archipelago. Two radiocarbon dates, one on a brown bear [35,365 ± 800 BP (AA-15227)] and another on a black bear [41,600 ± 1500 BP (AA-16831)] suggest that these two species cohabited

Prince of Wales Island, and even the same caves, before and after the last glacial maximum. Two ringed seal (*Phoca hispida*) specimens from On Your Knees Cave have yielded the following dates: 17,565 ± 160 BP (AA-18450) and 20,670 ± 80 BP (CAMS-33980) from the same seal ulna; and 13,690 ± 240 BP (AA-21564), suggesting that the cave was ice-free during the entire glacial interval. Other extirpated species found in that cave include: marmot (*Marmota*) dated to 32,900 ± 2400 BP (AA-21565); heather vole (*Phenacomys intermedius*); lemming (*Lemmus*); red fox (*Vulpes vulpes*); and an unidentified bovid (Bovidae). Large numbers of bird (Aves) and fish (Pisces) remains have been recovered also. In 1995, an AMS radiocarbon date of 8600 ± 60 BP (CAMS-24378) was obtained from a black bear rib and another date of 8725 ± 70 BP (AA-17451) was obtained on its femur. These bones were thought to be associated with a bone projectile-point fragment found by Heaton in Kushtaka Cave on Prince of Wales Island, but a subsequent AMS date of 2820 ± 60 BP (CAMS-27263) on that bone point shows that the human (*Homo sapiens*) artifact and the bear remains were not contemporaneous. Fossils and genetic evidence suggest that brown and black bears have been continuous, long-term residents of the region.

357. Dodge, W.P., Korff, S.A., and Vose, C.R. (1953): Exploration fund of the Explorer's Club: Otto Geist brings back specimens from Yukon. *In*: Exploration expeditions contemplated and in the field. Explorer's Journal 31(1-2):11.
This is an announcement of a completed expedition on the Old Crow River, Yukon. Geist returned with one-third of a tonne of specimens which included: horse (*Equus*), bison (*Bison*), tundra muskox (*Ovibos moschatus*), moose (*Alces alces*), and caribou (*Rangifer tarandus*). [See Geist 1955 and 1956?.]

358. Donnelly, J., and Donnelly, R. (1972): Remains of white whale kept in a crate at Pakenham. Perth Courier, June 29.
Members of Project Echo (an Opportunities for Youth project) went to Pakenham to see white whale (*Delphinapterus leucas*) remains from Champlain Sea deposits dating between 12,000 and 9800 BP. The article tells about the chance fossil discovery on the farm of Patrick Cannon in September 1906. Instead of being housed in a museum, the bones were kept in an old wooden crate, tucked away in the corner of a barn belonging to Alphonse Herrick – Patrick Cannon's nephew. The Herrick family became keepers of the whale about 1959, when Cannon sold his farm to R.E. Nanne. Only part of the skeleton was excavated – the rest is presumed to be buried on the Nanne farm. [See Harington and Fitzgerald 1973 – the specimen was donated by Alphonse Herrick to the Canadian Museum of Nature.]

359. Dorr, J.A., and Eschman, D.E. (1970): Geology of Michigan. University of Michigan Press, Ann Arbor. 476 pp.
The authors discuss finds of remains of whales (Cetacea) and walrus (*Odobenus rosmarus*) from Michigan sand, gravel, and swamp deposits (pp. 384, 408–11), and speculate on routes these marine mammals may have used to end up where they did (see map, Figure XIV-22). They state: 'There is some evidence, for example, of a direct sea level connection from the Atlantic down the St Lawrence valley into the lakes at one stage, which would have given access much the same as into Hudson Bay today' (pp. 384). [See Harington 1988.]

360. Dreimanis, A. (1962): Postglacial mastodon remains at Tupperville, Ontario. *In*: Abstracts for 1961. Geological Society of America, Special Paper 68:167.
Bones and teeth of mastodons (*Mammut americanum*) were found in two localities NE of Lake St Clair in southwestern Ontario. The first is 0.5 km NE of Tupperville, in calcareous sands of a beach bar of ancient Lake St Clair, 180 m asl; and the second is 3 km SW of Tupperville, below the continuation of the same sand bar, 179 m asl. The minimum age of the second specimen is 6230 ± 240 BP (S-16, gyttja inside skull). [See also Dreimanis 1967.]

361. Dreimanis, A. (1966): Lake Arkona-Whittlesey and post-Warren radiocarbon dates from 'Ridgetown Island' in southwestern Ontario. Ohio Journal of Science 66(6):582–6.
An ancient sandbar (Ridgetown 'Island'), now a few kilometres from the N shore of Lake Erie between 81°45′N and 82°00′W, contained mastodon (*Mammut americanum*) remains. This specimen was found at the NE end of Ridgetown Island, 19 km NE of Wade's Gravel Pit. Two radiocarbon dates are given: on the muck surrounding the bones 12,000 ± 500 BP (S-30), and on some wood with the bones 11,400 ± 450 BP (S-29).

362. Dreimanis, A. (1967): Mastodons, their geologic age and extinction in Ontario, Canada. Canadian Journal of Earth Sciences 4(4):663–75.
Most Canadian occurrences of mastodon (*Mammut americanum*) have been found below the Glacial Lake Warren shore, in southern Ontario. This would indicate an age of less than 12,400 BP. The oldest mastodon found in this area was radiocarbon dated at 12,000 ± 450 BP (I-586). The Tupperville mastodon was dated using bone collagen to 8910 ± 150 BP (GSC-614) and the Thamesville mastodon dated, also using bone collagen, to 11,380 ± 170 BP (GSC-611). A mastodon found in the James Bay Lowlands was radiocarbon dated at >53,000 BP (GRO-1435, lignite). The author concludes that these animals became extinct around 9000 BP. Accompanying the paper are a map (Figure 1) showing the location of mastodons (54) in southern Ontario, and a list (Table 1) of radiocarbon dates on or associated with mastodon bones from Ontario. To explain the extinction of mastodons in Ontario, Dreimanis proposes that they did not find their way to the northern boreal spruce forests, being separated from them by a wide belt of pine and hardwood forests, which had developed on the better-drained morainic areas of the region. Weakened by poorer forage, mastodons became more sensitive to diseases and an easier prey for Paleoindians (*Homo sapiens*).

363. Dreimanis, A. (1968): Extinctions of mastodons in eastern North America: testing a new climatic environmental hypothesis. Ohio Journal of Science 68(6):257-271.
More than 600 Late Wisconsinan mastodon (*Mammut americanus*) occurrences are known from the glaciated and periglacial parts of eastern North America (including southern Ontario). Most are from poorly drained lowlands, swamps, and valleys, and on the continental shelf. Of the 28 radiocarbon dates on mastodon bones, or associated wood, from eastern North America, 80% are 12,000–9000 BP. The author gives evidence suggesting the mastodons of eastern North America were associated with open spruce woodlands or spruce forests. He speculates that their extinction was initiated by increasing dryness 11,000–10,000 BP, which caused retreat of spruce forests into moister lowlands before their disappearance from mastodon habitat. Mastodon migration from relict spruce enclaves toward the more abundant northerly spruce forests was hampered because the two areas were probably separated by a rapidly expanding, intervening belt of pine and hardwood forests on better-drained areas of north-central Michigan, central Ontario, and central New York state.

364. Drew, M. (1999): Mammoth undertaking. Calgary Sun, May 7:44–5.
Shayne Tolman, a Grade 5 teacher from Cardston, Alberta, found an 11,000-year-old stone projectile point (*Homo sapiens*) on the exposed bottom of the St Mary Reservoir when the water was drawn down so work could be done on the spillway. There were artifacts everywhere – tools, flakes from tool-making, and fire-cracked rocks. Bones lay scattered on the surface – nearly all from extinct animals, but some from rabbits (*Lepus*), beaver (*Castor canadensis*), badgers (*Taxidea taxus*), and ground squirrels (*Spermophilus*). Even ground squirrel burrows were preserved. But when the archaeologists arrived they made the greatest finds – tracks of extinct animals still preserved in the ancient mud. When camels (*Camelops hesternus*) and mammoths (*Mammuthus*) walked here along the ancient banks of the St Mary River, the great continental ice sheets of the Wisconsinan glaciation had receded far to the N and the land was covered with grass and shrubs. However, with the protecting layers of overlying soil gone, the incessant southern Alberta wind is slowly scouring the tracks away. Runoff is coming and the reservoir will fill, flooding the site. When the spillway work is done, water in the reservoir will rise even higher. The article includes excellent colour photographs of the site, mammoth and camel tracks, a ground squirrel skull, and a Folsom projectile point.

365. Driver, J.C. (1982): Early Prehistoric killing of bighorn sheep in the southeastern Canadian Rockies. Plains Anthropologist 27(98-1):265–72.
Remains of three bighorn sheep (*Ovis canadensis*) were found in Holocene pond deposits in the Crowsnest Pass, Alberta, N of the Crowsnest River, SE of Bellevue. The sheep bones were found at the bottom of a blue-black clay layer overlying outwash gravels. At another section at a stratigraphically similar position, bison (*Bison*) remains were radiocarbon dated at 8550 ± 270 BP (RL-873). Above the clays is a layer of marl with more bison bones dated at 7200 ± 230 BP (RL-876). The author concludes that the sheep remains are about 8550 BP. Other sheep bones found in outwash gravels W of Calgary yield

ages of 10,760 ± 160 BP, 11,370 ± 170 BP (GSC-612, 613) (Churcher 1968) [the original source – Churcher 1968, indicates that these bones were of western bison (*Bison bison occidentalis* = *Bison occidentalis*) not *Ovis*]; and in the Peace River region, British Columbia, 9280 ± 200 BP (GSC-1497) (Rutter et al. 1972). A detailed map of the sites, a stratigraphic section, and a description of the bones are included.

366. Driver, J.C. (1988): Late Pleistocene and Holocene vertebrates and palaeoenvironments from Charlie Lake Cave, northeast British Columbia. Canadian Journal of Earth Sciences 25:1545–53.
Excavations during the summer of 1983 by Fladmark outside Charlie Lake Cave (HbRf-39, 56°16′35″N, 120°56′15″W), Peace River District, British Columbia, revealed deposits dating from 10,700 BP to the present. The earliest fauna (10,700–10,000 BP) was deposited when the newly deglaciated landscape was largely unforested and included bison (*Bison*), ground squirrel (*Spermophilus*), a large hare (*Lepus*), snowshoe hare (*Lepus americanus*), and a variety of small mammals and birds (Aves), including Cliff Swallow (*Hirundo pyrrhonota*). By 10,000 BP snowshoe hare was the most abundant mammal, indicating the development of forested conditions. By 9000 BP, the fauna resembled the modern Peace River fauna prior to European settlement. A radiocarbon table is provided [see also annotation for Driver et al. 1996], along with a stratigraphic diagram and a full faunal listing. The author reports on three radiocarbon dates on bison from the Peace River district of Alberta reported in Churcher and Wilson 1979. These are: 10,200 ± 100 BP (GSC-2895); 10,200 ± 100 BP (GSC-2902); and 9880 ± 130 BP (GSC-2865).

367. Driver, J.C. (1998a): Human exploitation of pioneering post-glacial communities in northeastern British Columbia, Canada. 8th International Congress of the International Council for Archaeozoology (August 23–9, Victoria, British Columbia). Final Program and Abstracts. p. 107.
Charlie Lake Cave in northeastern British Columbia is the only archaeological site in the central region of the 'ice-free corridor' (on the eastern side of the Rocky Mountains) where dated archaeological and faunal assemblages from the immediate postglacial period occur. Human (*Homo sapiens*) occupation begins at 10,500 BP, a date that is the same as the earliest dates on postglacial fauna from the region. Faunal remains, although sparse, suggest an association of species not found in the region today, and confirm palynological evidence of an open landscape. The main human prey was bison (*Bison*). By about 10,000 BP there is evidence for the establishment of productive aquatic habitats, and by 9500 BP the faunal assemblages were essentially modern and dominated by southern boreal forest species. So people entered postglacial landscapes as soon as it was possible to hunt large game. These landscapes changed rapidly, and the establishment of widespread coniferous forest by 9500 BP must have resulted in rapid changes in the ways people subsisted. This can be seen in the later periods at Charlie Lake Cave, where snowshoe hare (*Lepus americanus*) and fish (Pisces) are important.

368. Driver, J.C. (1998b): Late Pleistocene collared lemming (*Dicrostonyx torquatus*) from northeastern British Columbia, Canada. Journal of Vertebrate Paleontology 18(4):816–18.
The author reports on several specimens of the collared lemming (*Dicrostonyx torquatus*) from Charlie Lake Cave (56°16′35″N, 120°56′15″W) in the Peace River District of northeastern British Columbia: two left-lower first molars from different individuals, an isolated right-lower first molar (almost certainly redeposited), and a nearly complete right mandible with a complete tooth row (Figure 1 in text). Layer 105, from which the mandible and a molar were recovered, is the main layer in subzone IIb [AMS radiocarbon dates from this layer are: 10,290 ± 100 BP (CAMS-2317) on a Raven (*Corvus corax*) scapula, and 10,560 ± 80 BP (CAMS-2134) on a bison (*Bison*) phalanx]. Besides the Raven and bison, other associated vertebrates from subzones IIa and IIb are: duck (Anatinae), woodpecker (Picidae), Cliff Swallow (*Hirundo pyrrhonota*), large hare (*Lepus*), snowshoe hare (*Lepus americanus*), deer mouse (*Peromyscus*), a marmot or woodchuck (*Marmota*), ground squirrel (*Spermophilus*), and a wolf-sized canid (*Canis*). The Charlie Lake Cave specimens are firmly dated to about 10,500 BP (average of six dates from subhorizons IIa and IIb), and are the youngest securely dated specimens beyond the modern range of *Dicrostonyx torquatus*. Of three hypotheses, the author prefers the idea that collared lemmings moved N from the central Rockies as biotic communities were re-established in Alberta and British

Columbia at the end of the last glaciation. After 10,000 BP the appearance of spruce forests signalled the arrival of essentially modern vertebrate communities. One of the oldest artifacts from Charlie Lake Cave suggests initial migration of human (*Homo sapiens*) populations from the south to this area.

369. Driver, J.C., Handly, M., Fladmark, K.R., Nelson, D.E., Sullivan, G.M., and Preston, R. (1996): Stratigraphy, radiocarbon dating, and culture history of Charlie Lake Cave, British Columbia. Arctic 49(3):265–77.
The Charlie Lake Cave, above the N side of Stoddart Creek, British Columbia, has after three seasons of excavation revealed a sequence of stratified deposits that spans the Late Pleistocene and entire Holocene. Analyses of sediments, radiocarbon dates, faunal remains, and artifacts show that the site was first occupied by humans at about 10,500 BP, when local environments were more open than today. By 9500 BP, boreal forest had moved into the area, and human use of the site was minimal until about 7000 BP, when a brief occupation of the site probably included a human burial. Use of the site intensified after about 4500 BP, possibly because the cave became more accessible. The site was used both as a residential base and as a more temporary hunting station or lookout. Radiocarbon dates (13 from animal bone) are provided in a table. Species radiocarbon dated include: Raven (*Corvus corax*) – 10,290 ± 100 BP (CAMS-2137) and 9490 ± 140 BP (CAMS-2138); bison (*Bison*) – 10,380 ± 160 BP (SFU-378); 10,450 ±150 BP (SFU-300); 10,560 ± 80 BP (CAMS-2134); 10,770 ± 120 BP (SFU-454); 9670 ± 150 BP (CAMS-2136); 9760 ± 160 BP (SFU-355); 9990 ± 150 BP (RIDDL-393); 4270 ± 160 BP (SFU-382); and sciurid (Sciuridae) – 10,100 ± 210 BP (RIDDL-392). Some evidence suggests that the Raven skeletons were deposited by humans (*Homo sapiens*) during ritual activity. Work on the faunal assemblage is still in progress. Many of the specimens probably reached the site as a result of natural processes. Faunal preservation is good, particularly in the zone dating from about 10,800 BP to about 9600 BP, where rapid deposition preserved many quite fragile bones. The results of the last two field seasons, 1990 and 1991, appear to be quite similar to the already-reported results of the first field season in 1983. [See Driver 1988.] The most obvious faunal change occurs at the Pleistocene-Holocene boundary, where the dominant small mammal changes from ground squirrel (*Spermophilus*) to snowshoe hare (*Lepus americanus*). This change appears to mark a transition to a boreal forest from a predominantly open landscape at around 9500 BP. Further evidence for the open-landscape theory is provided by the presence of a single specimen of collared lemming (*Dicrostonyx torquatus*) recovered at the Zone IIb level (~10,300–10,700 BP). This species is today confined to tundra, although during the Late Pleistocene it lived S of the ice sheets. [See also Driver 1998b.]

370. Dubrovo, I. (1990): The Pleistocene elephants of Siberia. *In*: Megafauna and Man: Discovery of America's Heartland (L.D. Agenbroad, J.I. Mead, and L.W. Nelson, eds.). Mammoth Site of Hot Springs, South Dakota, Inc. Scientific Papers 1:1–8.
Siberia has proboscidean fossil remains of *Paleoloxodon*, the southern mammoth (*Mammuthus meridionalis* = *Archidiskodon meridionalis*), as well as steppe (*Mammuthus trogontherii*) and woolly (*Mammuthus primigenius*) mammoths. Only the southern and woolly mammoths migrated to North America. Evidence from Yakutia, Russia and western North America indicate that the southern mammoth was the ancestral mammoth of North America, arriving not earlier than 1.7 to 2.0 million years ago. The woolly mammoth (*Mammuthus primigenius*) is known throughout Siberia and northern North America. It is the late form that migrated to North America. Lands near the Bering Isthmus (= Bering Land Bridge) are considered to be the primary areas of research for the chronology and character of the spread of elephants from Asia to North America.

371. Dufresne, F. (1966): My Way Was North: An Alaskan Autobiography. Holt, Rinehart and Winston, New York. 274 pp.
The chapter 'Old Bones' (pp. 46–53) describes the author's experiences at Elephant Point, Alaska. Local Inuit told him of this place they called 'Old Bones' just across the bay from the reindeer corral where Dufresne was working. Napook, an aged native, recalled that when he was a boy the site had projected much farther into the sea – now the waves and sun had eroded it back until it looked like the broken end of a bridge. The natives told the author that the icy façade was studded at times with bony remains of beasts of another age, and as the exposed front melted away and crumbled, many strange objects

toppled into the surf below. For years the natives had been visiting the place to gather fossil ivory for carving into tools and artifacts, and of late for making into strings of beads and bracelets to sell commercially. He recreates the scene of his visit to the site vividly: 'In the swill of the surf, Napook picked up a mammoth [*Mammuthus*] molar big as a cobblestone. I pried loose a massive bison skull half buried in the wet sand, then began struggling with a thighbone ... There were bones sticking up all around, and I was hurrying from one to another when Napook pointed upward and yelled, "No more diggum! Here comes whole one!" ... It looked like an emaciated elephant in the very act of breaking through the wall of its icy crypt where it had been in deep-freeze for hundreds of centuries. Two enormous tusks of stove-pipe diameter, curling back against themselves like fish-hooks, hung suspended in the air. Part of an incredibly huge skull was exposed with ice water trickling from its open jaws. A poised foreleg the size of a tree trunk gave the impression of probing cautiously for a foothold which wasn't there any more because it had melted away.' Napook was astounded, and said such a beast lived in the legends of his forefathers. Since they could not wait for the gigantic skeleton to melt out and plunge into the sea, they 'waded into the chilly breakers under the awesome spectacle, gathering handfuls of dark hair long as those of a horse's tail and picked up bones with fragments of hide still attached.' Dufresne (p. 50) also mentions that gold miners on Seward Peninsula 'did not always appreciate all the prehistoric big game.' 'Long strands of hair and rust-colored wool sometimes clogged the riffles and sluice boxes. Skulls bigger than the camp stove, too bulky to be carted off in wheelbarrows, had to be rolled out of the way. Ivory tusks up to 15 feet long required the strength of two men to carry them from the pits, and occasionally the water streams would loose nauseous odors and have everybody holding their noses.' When his vessel *Hazel* headed back to Nome, her forward lockers were crammed with king-size bones, and many mammoth tusks and 'super-bison' [steppe bison (*Bison priscus*)] horns were lashed down on the open deck. He sent several huge wooden crates of the bones to Washington, D.C.

372. Dumais, P., Poirier, J., and Rousseau, G. (1993): Squatec (ClEe-9), a Late Pleistocene/Early Holocene site in southeastern Québec, Canada. Current Research in the Pleistocene 10:14–17.

The site was discovered in the Témiscouata region of southeastern Quebec (about 50 km inland from the S shore of the St Lawrence River between Témiscouata and Des Aigles lakes), and is thought to be part of a glacial delta that formed between 11,500 and 10,500 BP. Test pits yielded a collection composed mainly of debitage and resharpening flakes that are considered to be derived from a chert outcrop about 17 km away. A biface fragment and scraper have been collected, but no diagnostic tools were found. A probable mineralized mammalian (Mammalia) bone fragment was unearthed also, but its small size prevents species identification. The deepest artifacts were found in the topset bed of the oldest unit (60 cm and lower) – a glacial delta. Artifacts were found also in the second unit – a beach deposit of Glacial Lake Madawaska, which drained about 10,000 BP. The authors propose that the oldest human (*Homo sapiens*) occupation of the site is Paleoindian. It occurred in an environment characterized by tundra vegetation where local ice sheets were still receding. If this interpretation is correct, Squatec would be the first early-Paleoindian site recognized in Quebec and would be the most northerly site (48°N) of this age in the NE. The location of this site on a major trend offering direct access to the S shore of the St Lawrence River points to a possible presence of early Paleoindians on the coast of the Goldthwait Sea.

373. Dyck, W. (1967): The Geological Survey of Canada radiocarbon dating laboratory. Geological Survey of Canada Paper 66(45):1–45.

This paper outlines the concepts and techniques used by the Geological Survey of Canada, Radiocarbon Dating Laboratory.

374. Dyck, W., and Fyles, J.G. (1963): Geological Survey of Canada radiocarbon dates I and II. Geological Survey of Canada Paper 63-21:1–31.

This paper reports on a significant age of a piece of wood from a layer underlain by silty gravel containing mammal (Mammalia) bones. The aged wood came from the base of a woody peat unit at a depth of 6 m, at Hunker Creek, Yukon Territory, and provided a radiocarbon date of 9520 ± 130 BP (GSC-73). The silty gravel layer immediately below the dated sample has supplied an almost identical age: 9510 ± 220 BP (I(GSC)-196, Isotopes II). Wood from the layer below the bone-bearing gravel has supplied an age of >35,000 BP (I(GSC)-181, Isotopes II).

375. Dyck, W., and Fyles, J.G. (1964): Geological Survey of Canada radiocarbon dates III. Geological Survey of Canada Paper 64-40:1–15.
A plant sample from Hunker Creek, Yukon Territory [see Dyck and Fyles 1963], has provided a radiocarbon date of 30,800 +1600, −1400 BP (GSC-88). It came from about a metre below the top of a frozen silt layer beneath 6 m of woody, silty peat at the mouth of Last Chance Creek (64°01′N, 139°06′W). This date, along with the dates reported in Dyck and Fyles (1963), confirms a major stratigraphic break between silt and overlying peat, probably related to erosion during the interglacial.

376. Dyck, W., Fyles, J.G., and Blake, W., Jr (1965): Geological Survey of Canada radiocarbon dates IV. Geological Survey of Canada Paper 65-4:1–23.
This paper reports a radiocarbon date for wood from clayey alluvium exposed in a diversion channel cut in the Assiniboine River floodplain, 9.6 km W of Russell, Manitoba (50°46′N, 101°26′W). The sample came from an organic-clay zone containing twigs and bone including a skull of steppe bison? (*Bison priscus*? = *Bison preoccidentalis*?). The wood was dated at 6320 ± 140 BP (GSC-280). The organic-clay zone is overlain by 4.5 m of silty clay containing beds of pelecypod-bearing silt, sand, and gravel, and is overlain in turn by 18 m of clay, in part probably lacustrine.

377. Dyck, W., Lowdon, J.A., Fyles, J.G., and Blake, W., Jr (1966): Geological Survey of Canada radiocarbon dates V. Geological Survey of Canada Paper 66-48:1–32.
Several radiocarbon dates are pertinent to this bibliography. A date of 10,420 ± 150 BP (GSC-454) was supplied by bones of a white whale (*Delphinapterus leucas*) discovered at a depth of about 6 m (altitude approximately 90 m) in the Foster Sand Pit S of the centre of Ottawa, Ontario (45°20′N, 75°42′W). The sand was deposited in a spit formed by shore currents of the Champlain Sea. Grass from inter-till silt exposed by a 25-m-high road cut near Minnedosa, Manitoba (50°22′N, 99°54′W), was dated at >31,300 BP. Bones of a large ground squirrel (*Spermophilus* cf. *Spermophilus parryii* = *Citellus* cf. *Citellus undulatus*) and a large vole (*Microtus* cf. *Microtus oeconomus*) associated with the grass suggest the silt is an interstadial deposit. A wedge of elk horn (*Cervus elaphus*) was associated with charcoal from a depth of 180 cm exposed in a natural cut on Pemberton Creek, British Columbia (50°41′25″N, 119°50′50″W). The charcoal, taken from the lower portion of a 5 cm band of volcanic ash was dated to 3410 ± 130 BP (GSC-345) and was above the elk antler which was incorporated in massive sandy silt 25 cm below the volcanic ash. This places a minimum age on the antler artifact (*Homo sapiens*). A bowhead whale (*Balaena mysticetus*) vertebra, collected at 67 m asl, 11 km inland from the mouth of Cunningham River, Somerset Island, Nunavut (73°59′N, 93°40′W), is reported to be 8990 ± 140 BP (GSC-450). This sample is from the same locality and 1.5 m higher than GSC-150, a marine pelecypod sample dated to 9180 ± 170 BP. Both samples were redone, and dates agree within limits of error. Organic material from a 5–20-cm-thick layer whose top is at the 15–30 cm depth, overlying till, and underlying alluvium (or possibly slope deposit) along a small creek on Bathurst Island, Nunavut (75°45′N, 98°32′W), 52 m asl, has an age of 4070 ± 140 BP (GSC-401). Organic material is believed to be in place, and the presence of lemming (*Dicrostonyx torquatus*) droppings suggests that climate was similar to that existing today.

378. Dyke, A.S. (1979a): Radiocarbon-dated Holocene emergence of Somerset Island, central Canadian Arctic. Geological Survey of Canada Paper 79-1B:307–18.
This paper includes tables with localities and radiocarbon dates of whale (unspecified Cetacea) and walrus (*Odobenus rosmarus*) specimens from Somerset Island, Nunavut. The purpose of this is to establish the Holocene emergence of the island; these dates seem to be far too young with respect to plotted curves based on emergence rates. The next paper provides radiocarbon corrections that subsequently agree with these calculated rates. A map of localities of finds and photographs of specimens are included. [See Dyke 1980 for corrected dates.]

379. Dyke, A.S. (1979b): Glacial and sea-level history of southwestern Cumberland Peninsula, Baffin Island, N.W.T., Canada. Arctic and Alpine Research 11(2):179–202.
Southwestern Cumberland Peninsula, Nunavut (Fig. 1), has been glacierized at various times by eastward-flowing Laurentide ice, by an expanded Penny Ice Cap, and by fiord and valley glaciers fed from cirques. Radiocarbon dates and amino acid age estimates on fossils from

raised marine deltas and correlation with the stratigraphic record of northern Cumberland Peninsula allow establishment of a tentative 'absolute' chronology. Of particular interest here is a date of 1560±120 BP (QU-240) on whalebone [presumably bowhead whale (*Balaena mysticetus*)] from Millut Bay (66°38′N, 67°34′W) at an elevation of 1.5 m, and another on whalebone dated to 770±80 BP (QU-241) from Shark Fiord (66°36′N, 66°41′W) at an elevation of 0 m (Table 1).

380. Dyke, A.S. (1980): Redated Holocene whale bones from Somerset Island, District of Franklin. Geological Survey of Canada Paper 80-1B:269–70.
This paper provides corrected radiocarbon dates for Dyke (1979). The whalebone (*Balaena mysticetus*) dates were far younger than predicted by island-emergence curves because lichen and moss growing on the bones contaminated the samples. The incorrect dates were: 6140 ± 165 BP (S-1381); 4465 ± 85 BP (S-1386); and 5205 ± 70 BP (S-1388). The revised dates (and predicted dates based on the emergence curves) are respectively: 9590 ± 115 BP (S-1381) for specimen DCA-77-B21; 7105 ± 90 BP (S-1386) for specimen DCA-77-B12; and 8805 ± 95 BP (S-1388) for specimen DCA-77-B1. Graphs are provided.

381. Dyke, A.S. (1990): Sudden changes in range of the bowhead whale and in frequency of driftwood penetration during the Holocene in the Canadian Arctic. Canadian Quaternary Association, American Quaternary Association, Programme and Abstracts of the First Joint Meeting (June 4-6, Waterloo, Ontario). p. 17.
Remains of bowhead whales (*Balaena mysticetus*) and boreal driftwood logs have been collected from hundreds of raised-beach sites in the central and eastern Canadian Arctic Islands, Nunavut, and many have been radiocarbon dated. The frequency distribution of whale bone through time and with elevation is strongly bimodal. Bowheads were abundant throughout the region during the Early Holocene, when Bering Sea and Davis Strait stocks were able to mix. At 8500 BP a perennial sea-ice barrier formed in the central channels of the Arctic Islands, separating the two whale stocks. The barrier retreated, perhaps disappearing between 5000 and 3000 BP, likely due to increased ablation of sea ice.

382. Dyke, A.S. (1993): Glacial and sea level history of Lowther and Griffith islands, Northwest Territories: A hint of tectonics. Géographie physique et Quaternaire 47(2):133–45.
Lowther and Griffith islands (Figure 1), in the centre of Parry Channel, Nunavut, were overrun by the Laurentide ice sheets early in the last glaciation. The author constructs emergence curves (from about 10,000 BP to the present) for both islands, partly based on radiocarbon-dated bowhead whale (*Balaena mysticetus*) bones (Figure 7). Five bone dates (S-3134, S-3310, P-2141, S-3288, and S-3289) from the islands are given with their elevations, radiocarbon ages (uncorrected), and coordinates (Table I). The dates are discussed on p. 142. Three were based on ear bone samples. The highest sample at 98.5 m gave an uncorrected age of 9110±110 BP (S-3310) – only about a century younger than expected based on the ages from Lowther Island and Resolute Bay.

383. Dyke, A.S. (1998): Holocene delevelling of Devon Island, Arctic Canada: Implications for ice sheet geometry and crustal response. Canadian Journal of Earth Sciences 35:885–904.
The raised beaches of Devon Island, Nunavut (see map, Fig. 2 for localities mentioned), contain abundant dateable materials. A large set of radiocarbon dates (228) include 73 dates on bowhead whale (*Balaena mysticetus*) bones from raised beaches, 6 on walrus (*Odobenus rosmarus*), 5 on seal (Phocidae) bones from archaeological sites, and 2 from caribou (*Rangifer tarandus*) from archaeological sites that are listed in Table A1, along with laboratory numbers, relative sea level, latitude and longitude, and $\delta^{13}C$ (‰) readings. The Devon Island evidence supports Blake's (1970) Innuitian Ice Sheet hypothesis.

384. Dyke, A.S. (1999): Last Glacial Maximum and deglaciation of Devon Island, Arctic Canada: Support for an Innuitian Ice Sheet. Quaternary Science Reviews 18:393–420.
The glacial geology of Devon Island, Nunavut, strongly supports the Late Wisconsinan Innuitian Ice Sheet hypothesis. The author mentions seven radiocarbon-dated bowhead whale (*Balaena mysticetus*) specimens from the region (Tables 2, 3): a skull dated to >33,000 BP (S-3565) from Sheills Peninsula is the oldest and came from gravel underlying Holocene beach gravel that caps the

section; bone (94DCA245) dated to 9920 ± 180 BP (S-3595) from an elevation of 13 m at Cape Home (74°35′39″N, 83°31′45″W) – the oldest-dated postglacial marine fossil pertaining to the deglaciation of the island; bone (94DCA203) dated to 9540 ± 200 BP (S-3580) from an elevation of 32.5 m at Cape Home, Croker Bay (74°34′27″N, 83°39′27″W); bone (94DCA173) dated to 9420 ± 200 BP (S-3572) from an elevation of 64.5 m in the Cape Sparbo-Hardy Lowland (75°46′24″N, 83°51′17″W) – from a beach well below the marine limit; bone (94DCA261) dated to 9120 ± 200 BP (S-3597) from an elevation of 76.5 m at Lovell Point (74°57′38″N, 91°59′55″W).

385. Dyke, A.S. (2000a): Surficial geology, Phillips Creek, Baffin Island. Geological Survey of Canada Map 1961A.
This 1:250,000 scale map focuses on part of Baffin Island, Nunavut. It includes localities of radiocarbon-dated remains of bowhead whales (*Balaena mysticetus*).

386. Dyke, A.S. (2000b): Surficial geology, Milne Inlet, Baffin Island. Geological Survey of Canada Map 1962A.
This 1:250,000 scale map of part of Baffin Island, Nunavut, includes localities of radiocarbon-dated remains of: white whale (*Delphinapterus leucas*), bowhead whale (*Balaena mysticetus*), and walrus (*Odobenus rosmarus*).

387. Dyke, A.S. (2000c): Surficial geology, Moffet Inlet and Fitzgerald Bay, Baffin Island. Geological Survey of Canada Map 1963A.
This 1:250,000 scale map of part of Baffin Island, Nunavut, includes localities of radiocarbon-dated remains of: narwhal (*Monodon monoceras*), bowhead whale (*Balaena mysticetus*), and walrus (*Odobenus rosmarus*).

388. Dyke, A.S. (2000d): Surficial geology, Arctic Bay, Baffin Island. Geological Survey of Canada Map 1964A.
This 1:250,000 scale map of part of Baffin Island, Nunavut, includes localities of radiocarbon-dated remains of: narwhal (*Monodon monoceras*), bowhead whale (*Balaena mysticetus*), and walrus (*Odobenus rosmarus*).

389. Dyke, A.S. (2000e): Surficial geology, Navy Board Inlet, Baffin Island. Geological Survey of Canada Map 1965A.
This 1:250,000 scale map of part of Baffin Island, Nunavut, includes localities of radiocarbon-dated remains of: bowhead whale (*Balaena mysticetus*) and walrus (*Odobenus rosmarus*).

390. Dyke, A.S., and Hooper, J. (2000): Surficial geology, Berlinguet Inlet and Bourassa Bay, Baffin Island. Geological Survey of Canada Map 1960A.
This 1:250,000 scale map focuses on part of Baffin Island, Nunavut. It includes localities of radiocarbon-dated remains of the following marine mammals: narwhal (*Monodon monoceras*), bowhead whale (*Balaena mysticetus*), and walrus (*Odobenus rosmarus*).

391. Dyke, A.S., and Morris, T.F. (1990): Postglacial history of the bowhead whale and of driftwood penetration: Implications for paleoclimate, central Canadian Arctic. Geological Survey of Canada Paper 89-24:1–17.
This report is the first to document changes in the summer range of the bowhead whale (*Balaena mysticetus*) throughout postglacial time in the central Canadian Arctic. The current and former range of the whale is governed by sea-ice conditions. Four postglacial phases include: (1) 11,000–8500 BP – a large bowhead population extended in summer from Beaufort Sea to Baffin Bay with the first ones arriving in the central Arctic from Beaufort Sea, while access to Baffin Bay was likely blocked by glacier ice; (2) 8500–5000 BP – bowheads were excluded from the central Arctic channels; (3) 5000–3000 BP – bowheads from Baffin Bay reoccupied the central Arctic channels either in lower numbers or less frequently than during the Early Holocene; (4) 3000 BP–present – bowheads were excluded from the central Arctic channels during most summers. Appendix 1 lists 53 radiocarbon dates on bowhead whales from the central Canadian Arctic with information on material dated, elevation, and location (see Figure 8). Undated bowhead samples (59) with specimen numbers, elevation, location, and minimum ages are listed in Table 2.

392. Dyke, A.S., and Savelle, J.M. (2001): Holocene

history of the Bering Sea bowhead whale (*Balaena mysticetus*) in its Beaufort Sea summer grounds off southwestern Victoria Island. Quaternary Research 55:371–9.

Fossil remains of 43 bowhead whales (*Balaena mysticetus*) were located on raised beaches of western Wollaston Peninsula, Victoria Island, Northwest Territories, near the historic summer range of the Bering Sea stock in the Beaufort Sea. Radiocarbon dates were obtained on 16 of the specimens (Table 2 lists the skeletal part dated, radiocarbon date, laboratory number, elevation, and coordinates), indicating that both an Early and a Late Holocene group are present (Figure 4). Radiocarbon samples in the Early Holocene group (56 to 28 m elevation) range from 10,210 ± 70 BP (TO-7755) to 8830 ± 80 BP (TO-7760). During this interval bowheads ranged widely from the Beaufort Sea to Baffin Bay. The ages of samples between 22 and 3 m elevation are from 1660 ± 60 BP (TO-8011) to 1040 ± 60 BP (TO-8010). Evidently no whales reached Wollaston Peninsula between these periods, indicating relatively extensive sea-ice cover. Although the Late Holocene recurrence may relate to the expansion of pioneering Thule Culture whalers (*Homo sapiens*) eastward from Alaska, there are few Thule sites and limited evidence of Thule whaling in the area to support this suggestion.

393. Dyke, A.S., Hooper, J., and Savelle, J.M. (1996): A history of sea ice in the Canadian Arctic Archipelago based on postglacial remains of the bowhead whale (*Balaena mysticetus*). Arctic 49(3):235–55.

The distribution and radiocarbon ages of bowhead whale subfossils show that the range of the whale has expanded and contracted abruptly several times over the last 10,000 years. Each expansion or contraction was followed by nearly stable conditions that persisted for millennia. More than 400 radiocarbon dates were used to reconstruct paleo-ranges. Since bowheads are adapted to live in the loose edges of the polar sea ice, their range can be used to reconstruct summer sea-ice minima. Postglacial time is divided into four intervals: (1) 10,000–8500 BP – a large bowhead population extended in the summer all the way to retreating glacier margins and ultimately from the Beaufort Sea to Baffin Bay; (2) 8500–5000 BP – bowheads were excluded from most of the archipelago because most of the channels failed to clear of sea ice; summer conditions for most of this time were more severe than during historical times; (3) 5000–3000 BP – bowheads reoccupied the central channels of the Arctic Islands, and their range extended beyond historical limits; (4) 3000–0 BP – sea ice excluded whales from the central channels, as it does today.

394. Dyke, A.S., McNeely, R.N., and Hooper, J. (1996): Marine reservoir corrections for bowhead whale radiocarbon age determinations. Canadian Journal of Earth Sciences 33:1628–37.

Twenty-two pairs of radiocarbon dates on driftwood and bowhead whale (*Balaena mysticetus*) bones from raised beaches, dates on whale bone and terrestrial plant detritus from a stratigraphic section, and 25 additional dates on whale bones from the lowest (≤1000 BP) raised beaches in the eastern Canadian Arctic suggest that a marine-reservoir correction of about –200 years is appropriate for normalized age determinations on bone collagen from the bowhead whales in this region. This is considerably less than the correction (–400 years) normally applied to carbonate shells of marine molluscs from this region. The carbon in bowhead collagen appears to be derived from the whales' zooplankton food rather than from marine bicarbonate. Two appendices provide the full details of the radiocarbon dates and localities.

395. Dyke, A.S., Morris, T.F., and Green, D.E.C. (1991): Postglacial tectonic and sea level history of the central Canadian Arctic. Geological Survey of Canada Bulletin 397:1–56.

More than 130 new radiocarbon dates on driftwood, whale bone, marine shells, and organic detritus from Prince of Wales and adjacent smaller islands, Nunavut, form the basis for 14 local emergence curves representing small sample areas. Of particular interest here are radiocarbon-dated vertebrate remains from various shorelines in the area, including: 63 bowhead whale (*Balaena mysticetus*); 4 seal (Phocidae); 3 narwhal (*Monodon monoceras*), and 2 fox [presumably arctic fox (*Alopex lagopus*)] listed according to date, laboratory number, elevation, latitude and longitude, and comments in Table 1. There is an excellent photograph on the cover of a field worker looking at remains of a bowhead whale skeleton dated at 9335 ± 145 BP (S-2913) on a raised beach at an elevation of 58 m on the W side of Prescott Island. Figure 6 shows the base of a bowhead cranial fragment and a rib on raised beach gravel covered by peat, before and after excavation. Figure 7 shows an otic capsule and periotic bone of a bowhead whale. Figure 8 shows a bowhead whale cranial base with earbone.

396. Dyke, A.S., Hooper, J., Harington, C.R., and Savelle, J.M. (1999): The Late Wisconsinan and Holocene record of walrus (*Odobenus rosmarus*) from North America: A review with new data from Arctic and Atlantic Canada. Arctic 52(2):160–81.
The Late Wisconsinan and Holocene record of the Atlantic walrus is known from many collections of bones and tusks from Arctic Canada S to North Carolina, as well as from many archaeological sites in the Arctic and Subarctic. In contrast, the Pacific walrus has no dated Late Wisconsinan or Early Holocene record in North America, and it may have been displaced into the northwestern Pacific at Last Glacial Maximum (LGM). The Atlantic walrus rapidly exploited newly deglaciated territory, moving northward from its LGM refugium and reaching the Bay of Fundy by 12,800 BP, the Grand Banks by 12,500 BP, southern Labrador by 11,500 BP, and the central Canadian Arctic Archipelago (CAA) by 9700 BP. Its southern range limit may have retracted to the Bay of Fundy by about 7500 BP. Within the CAA, walrus remain clustered in two main periods: 9700 to 8500 BP and 5000 to 4/3000 BP. This pattern strongly resembles the distribution of the bowhead whale (*Balaena mysticetus*) radiocarbon ages from the same area, which suggests a common control by sea-ice conditions. Walrus remains from archaeological sites are also discussed. There is a map of walrus distribution in northern North America showing present range, extralimital sites, and walrus remains in archaeological sites, as well as extensive lists of radiocarbon dates (Tables 1 and 3).

397. Egginton, P.A., and Andrews, J.T. (1989): Sea levels are changing; current and future relative sea levels in Canada. Geos 18(2):15–22.
Geologists and geomorphologists, working in emergent coastal areas, try to reconstruct sea-level histories by dating (among other items) the preserved bones of sea mammals, such as whales (Cetacea) and walrus (*Odobenus rosmarus*). This paper mentions a find of whale bones from an ancient beach 58 m asl, on Prescott Island, Nunavut. The bones were radiocarbon dated at about 9300 BP. A photograph of the specimen is included (Figure 3 and cover).

398. Eisely, L.C. (1943): Did the Folsom bison survive in Canada? Scientific Monthly 56:468–72.
The author states his belief that the wood bison (*Bison bison athabascae*) is the slightly modified remainder of the Pleistocene western bison (*Bison bison occidentalis* = *Bison occidentalis*).

399. Eisenmann, V. (1979): Les métapodes d'*Equus sensu lato* (Mammalia, Perissodactyla). Geobios 12:863–86.
The author notes that metapodials of the Yukon horse (*Equus lambei*) are slender compared with those of Przewalskii's horse (*Equus przewalskii*), which Forstén (1986) regards as *Equus caballus przewalskii*. [See Forstén 1986.]

400. Eisner, W.R., Törnqvist, T.E., Koster, E.A., Bennike, O., and Leeuwen, van, J.F.N. (1995): Paleoecological studies of a Holocene lacustrine record from the Kangerlussuaq (Søndre Strømfjord) Region of West Greenland. Quaternary Research 43:55–66.
Kangerlussuaq is a 175-km-long, WSW-ENE-oriented fiord. The airport lies at the head of the fiord, some 25 km W of the margin of the Greenland ice sheet. The authors show, in a macrofossil concentration diagram from Lake 31 (67°03′N, 50°28′W, Figures 1, 2), E of the airport [a core is radiocarbon dated to 5030 ± 60 BP (UtC-1586) at a depth of 91–2 cm and to 3950 ± 50 BP (UtC-1585) at a depth of 65–6 cm] the presence of a fish (Pisces). A single bone of a three-spined stickleback (*Gasterosteus aculeatus*) was reported in gyttja at a depth of about 70 cm in the core – perhaps dating to about 4000 BP (Figure 10).

401. Elias, S. (1995): The Ice-Age History of Alaskan National Parks. Smithsonian Institution Press, Washington. 150 pp.
Focusing on more than 30,000 years of Alaskan prehistory, this book describes the geology, climate, ancient flora and fauna, and human presence in four of Alaska's national parks and preserves: Denali, Kenai Fjords, Glacier Bay, and Bering Land Bridge. There is a section on vertebrate fossils, their nature and significance (p. 10), as well as the nature of the 'Mammoth Steppe' and its mammalian fauna, dominated by woolly mammoths (*Mammuthus primigenius*), horses (*Equus*), and steppe bison (*Bison priscus*). Causes of megafaunal extinction about

11,000 BP and the arrival of early humans (*Homo sapiens*) in the New World are mentioned in chapter 5 (pp. 70–7). A useful pyramid diagram (Figure 5.13) of the ancient steppe-tundra ecosystem shows the principle predators ['Saber-Tooth Cat' should be replaced by 'scimitar cat' (*Homotherium serum*), however] and herbivores, dominant vegetation types, and soil conditions.

402. Elias, S. (1997): Bridge to the past. Earth, April(1997): 51–5.
New data from sediment cores suggest that the Bering Land Bridge sank not 14,500 years ago but, rather, around 11,000 years ago, a date more in keeping with the archaeological evidence. The Bridge was not a great place for large grazing animals. It probably produced only enough grass to keep migrating herds alive on their way to grassy upland plains.

403. Elias, S.A. (2001): Beringian paleoecology: Results from the 1997 workshop. Quaternary Science Reviews 20(2001):7–13.
In this summary of the workshop, Elias mentions Quaternary vertebrates twice: (1) the late-glacial period (14,000–10,000 BP) was an interval of rapid environmental change throughout Beringia. Climatic fluctuations brought about wholesale changes in the distribution of Beringian plants and animals, and may have played the most important role in the regional extinction of many megafaunal species; (2) the colonization of interior Eastern Beringia began by 12,000 BP according to evidence from reliably dated sites (Yesner 2001). Three possibly older cave sites – Bluefish Caves, Lime Hills Caves, and Trail Creek Caves – have stratigraphic and taphonomic problems that are not easily resolved [What are they?]. The Pleistocene–Holocene transition was a period of rapid environmental change that saw the extinction of many megafaunal mammals in Beringia – the chief prey of early hunters (*Homo sapiens*) – so their extinction affected humans, forcing development of new hunting strategies and lifestyle changes. The extinction of obligate grazers, such as mammoths (*Mammuthus*) and horses (*Equus*), may have been forced by climatic change, but bison (*Bison*) and wapiti (*Cervus elaphus*) seem to have been favoured, at least until 9000 BP. Faunal data from the central Tanana valley shows that people used many different animals for food, including small game, waterfowl (Anatidae), and fishes (Pisces). We know more about the diet of people at Broken Mammoth, Alaska, than we do for most other Beringian sites, for organic materials are so well preserved there.

404. Elias, S.A., and Brigham-Grette, J. (eds.) (2001): Beringian Paleoenvironments Festschrift in Honour of D.M. Hopkins. Quaternary Science Reviews 20(1-3):1–574.
Much progress has been made in Beringian studies since the publication of *Paleoecology of Beringia* in 1982. The 1977 Beringian Paleoenvironments workshop attracted Russian, Canadian, and U.S. scientists, giving rise to the 20 paleoenvironmental and archaeological papers included here. The subjects of the papers focus on: (1) timing and environments associated with the Old Crow tephra in Eastern Beringia; (2) last (Sangamonian) interglacial environments of Eastern Beringia; (3) interstadial environments from the middle of the last (Wisconsinan) glaciation in Beringia; (4) full glacial environments of Beringia; (5) late-glacial environments in Beringia; (6) Early Holocene environments of Beringia; (7) Late Pleistocene archaeology of Siberia; and (8) the timing, adaptations, and possible migration routes of people (*Homo sapiens*) entering the New World.

405. Elias, S.A., Hamilton, T.D., Edwards, M.E., Begét, J.E., Krumhardt, A.P., and Lavoie, C. (1999): Late Pleistocene environments of the western Noatak Basin, northwestern Alaska. Geological Society of America Bulletin 111(5):769–89.
The authors discuss the stratigraphy and paleoecology of two bluff sections that include organic deposits considered assignable to part of substage 5e (Sangamonian Interglacial) and to part of the transition between stage 6 and stage 5. One section includes the Old Crow tephra (about 130,000–140,000 BP), whereas both sections include organic sediments that probably represent one or more older interglacials. A large fragment of caribou (*Rangifer tarandus*) antler from a buried organic soil near the NE end of bluff Nk-29a (Figure 7) has a radiocarbon age of at least 52,500 +2700/–2000 BP (USGS-1843). An antler fragment from the same unit has a uranium-series age of 152,000 ± 5000 BP (B.J. Szabo, U.S. Geological Survey 1995, personal communication). It lay horizontally along a bedding plane within flood-plain deposits at

the top of Unit 3. The sample was analysed by isotope dilution alpha spectrometry, using a combined ^{229}Th-^{236}U spike. The ^{230}Th/^{232}Th value is greater than 20, indicating little in the way of contaminating detrital materials that might contain U or Th; thus no correction for 'inherited' ^{230}Th is required. Insect remains from peat near the top of Unit 3 suggest a boreal–arctic environment.

406. Ells, R.W. (1898): Sand and clays of the Ottawa Basin. Bulletin of the Geological Society of America 9:211–22.

The author comments (p. 216) on the presence of 'fish-bearing nodules ... obtained similar to those so abundant in the vicinity of Greens creek [Green Creek] a short distance below Ottawa city' in clays of the Ottawa River expansion known as Lac Coulonge, Quebec, at about 370 ft asl. He also mentions that the discovery of whale (Cetacea) bones in gravels of the great kame-like ridge near Smiths Falls, Ontario, SW of Ottawa at an elevation of 440 ft shows that 'the seas must at that time have been at a higher level than the present elevation of Lake Ontario.'

407. Ells, R.W. (1907): Report on the geology and natural resources of the area included in the northwest Quarter-sheet, Number 122, of the Ontario and Quebec series, comprising portions of the counties of Pontiac, Carleton and Renfrew. Geological Survey of Canada Publication No. 977:1–71.

Billings (1856) under the subheading 'The Lawrencian formation' records the capelin (*Mallotus villosus*) from the clays of Flat Rapids, Madawaska River, Renfrew County (p. 52). This record is again mentioned (p. 71) under 'The Pleistocene Deposits. The Marine Sediments. Leda Clay or "Green Creek Formation."'

408. Elson, J. (1962): History of Glacial Lake Agassiz. Problems of the Pleistocene and Arctic, Vol. 2. McGill University Museum Publication. p. 8.

The author reports bison (*Bison*) bones occurring in alluvial fill in small gullies tributary to the Assiniboine River near Treesbank, Manitoba. [These remains may belong to the western bison (*Bison bison occidentalis* = *Bison occidentalis*) since bones attributed to that species were found in the vicinity between 1914 and 1916 (Hay 1924, p. 200).]

409. Engstrom, M.D., Baker, A.J., Eger, J.L., Boonstra, R., and Brooks, R.J. (1993): Chromosomal and mitochondrial DNA variation in four laboratory populations of collared lemmings (*Dicrostonyx*). Canadian Journal of Zoology 71:42–8.

Genetic differentiation among populations and speciation in *Dicrostonyx* is hypothesized to have resulted from either allopatric divergence in glacial refugia during the Wisconsinan or sympatric processes uncorrelated with refugial isolation. The authors examined chromosomal and mitochondrial DNA variation in four laboratory colonies, representing three species, in a preliminary evaluation of these hypotheses. Sex chromosome morphology is geographically structured and correlated with proposed southern and northern refugia. Restriction fragment analysis of mitochondrial DNA revealed two ancient, divergent genotypic assemblages, corresponding to geographic distributions of sex chromosomes. Autosomnal variation, and any resulting reproductive isolation, probably is recent and uncorrelated with refugial history, whereas divergence of sex chromosomes and disparate mitochondrial assemblages likely predate the Wisconsinan glaciation. So the evidence strongly suggests that at the level of mt DNA, the genetic architecture of collared lemmings has been shaped by historical zoogeographic events in the Pleistocene, albeit at an earlier time than envisioned by Macpherson (1965).

410. Estabrook, B. (1982): Bone age man. Equinox 1(2):84–96.

This article notes the many finds of ice age mammal bones and artifacts in Old Crow Basin, northern Yukon, since the discovery of a 27,000-year-old caribou (*Rangifer tarandus*) flesher in 1966 [AMS radiocarbon dating now indicates the tool is only about 1300 years old]. The article describes the roles of various archaeologists and paleontologists (e.g., W.N. Irving, Richard Morlan, Robson Bonnichsen, and Richard Harington) and their interpretations of the bones they found. It emphasizes the conflicting views of adherents to the accepted hypotheses (Clovis Theory) that humans (*Homo sapiens*) only reached the New World about 12,000 years ago (e.g., C. Vance Haynes and Paul S. Martin) and those of the Yukon researchers who believed that people entered North America much earlier. [See Cinq-Mars and Morlan 1999, and Harington 1989c.]

411. Evans, D.J.A. (1989): An Early Holocene narwhal tusk from the Canadian High Arctic. Boreas 18:43–50.

A narwhal (*Monodon monoceras*) tusk was found on the NW coast of Ellesmere Island, 24 km ENE of Cape Woods on the Wootton Peninsula (82°16′N, 85°20′W). It was 34 m asl and radiocarbon dated at 6830 ± 50 BP (TO-476). The tusk is now in the collection of the Canadian Museum of Nature. Maps and a photograph are included.

412. Eyles, N., and Williams, N.E. (1992): The sedimentary and biological record of the last interglacial–glacial transition at Toronto, Canada. Geological Society of America Special Paper 270:119–37.

A substantial Sangamonian interglacial and subsequent Wisconsinan glacial sedimentary record is preserved at Toronto, Ontario. An interglacial sequence records climatic deterioration from warm temperate to subarctic. The interglacial Don Beds, resting on presumed Illinoian till, were deposited on a storm-influenced shoreface of an ancestral Lake Ontario in water depths that increased over the recorded time interval from about 2 to 20 m. Pollen and faunal (e.g., mainly using remains of ostracods, caddisflies, chironomids, and diatoms) analyses indicate a climatic deterioration in the upper Don Beds from warm-temperate (mean annual temperatures some 2°C warmer than present) to cool temperate conditions (temperatures dropped by about 3°C). Continual cooling is recorded in overlying deeper-water deltaic sediments of the Scarborough Formation, but later climatic amelioration and the return of mixed forests are suggested by the pollen record and caddisfly fauna. The youngest deltaic sediments lying below a Wisconsinan glacial complex were deposited in a subarctic setting in an ice-dammed lake (mean annual temperature dropped by at least 7°C). The likely continuity of the Don Beds and Scarborough Formation is emphasized, and they are placed in an expanded 'Sangamon' interglacial possibly equivalent to the whole of isotope stage 5 (130,000 to 75,000 BP). Schematic paleogeographic diagrams (Figure 12) illustrate the evolution of the Ontario Basin from about 130,000 BP to the Late Wisconsinan maximum (after 25,000 BP). Vertebrate remains recovered from the Don Beds [pike (*Esox*), catfish (*Ictalurus punctatus*), woodchuck (*Marmota monax = Marmota arctos*), mammoth or mastodon (Proboscidea), white-tailed deer (*Odocoileus virginianus*), bison (*Bison*), and giant beaver (*Castoroides ohioensis* – the pictured skeleton (Figure 5) is misleading in that only an incisor tooth fragment was collected)] are mentioned briefly (p. 130) without a supporting reference.

413. Fagan, B.M. (1987): The Great Journey: The Peopling of Ancient America. Thames and Hudson, London and New York. 288 pp.

This book, about the search for the first Americans, contains most pertinent data in Chapter 5 on Beringia and Chapter 6 on Alaska and Yukon. The author notes that between 75,000 and 45,000 BP and from 25,000 to 14,000 BP, the Bering Isthmus was exposed for crossing by Upper Paleolithic people (*Homo sapiens*) and the animals they may have hunted. On the other hand, crossing by watercraft at other times was possible. There is little direct evidence of the nature of vertebrate life on the land connection between Eurasia and North America, but some fossil mammoth (*Mammuthus*) bones have been brought up by seabottom grabs from the Bering Strait. Varying views of the nature of the Beringian landscape are presented, with the conclusion that: (1) Beringia was largely treeless; vegetation was dominated by herbaceous plants, with alder (*Alnus*), dwarf birch (*Betula*), and heath shrubs found in rare, more sheltered places; (2) judging from the mammalian population, vegetation may have been more productive than modern tundra or polar desert. Pages 111–17 are devoted to a description of the 'mammoth' fauna, which contained nearly twice the number of species as the modern tundra: woolly mammoth (*Mammuthus primigenius*), muskox (*Ovibos moschatus*), saiga antelope (*Saiga tatarica*), caribou (*Rangifer tarandus*), steppe bison (*Bison priscus*), and wild horse (*Equus*), etc. Probably these animals were not scattered over the landscape, but concentrated in lowland meadows and near streams. The author speculates that people, relying heavily on terrestrial hunting (like the Dyukhtai), crossed between the continents as early as 25,000 BP. The author dwells (pp. 122–6) on archaeological evidence from Old Crow Basin, Yukon [beginning with the caribou (*Rangifer tarandus*) 'flesher' first dated at about 27,000 BP and later dated at about 1300 BP], because 'it is an excellent illustration of the difficulties of interpreting very early American sites.' He mentions that if radiocarbon dates on bone from Bluefish Caves, Yukon, are correct, then people (*Homo sapiens*) wielding Dyukhtai-like tools were living in Eastern Beringia about 12,000 BP. Many other Alaskan Late Wisconsinan–Early Holocene archaeological sites are mentioned, some producing vertebrate bones – e.g., Trail Creek [caribou (*Rangifer tarandus*), horse (*Equus*) and bison (*Bison*)]. It is likely that as the glaciers retreated, hunters pursued big-game animals along the ever-widening corridor between Cordilleran and Laurentide ice sheets to reach the North American plains between 14,000 and 12,000 years ago. Chapter 10 deals with bison hunting, and pp. 199–201 outline the author's views on bison evolution and adaptation. Most archaeological sites mentioned here are in the United States, an exception being Head-Smashed-In Buffalo Jump in western Alberta.

414. Farrand, W.R. (1961): Frozen mammoths and modern geology. Science 133(3455):729–34.
The author summarizes what research has been done with the few frozen (bones, flesh, and skin) woolly mammoth (*Mammuthus primigenius*) carcasses that have been found in Siberia and Alaska. Skin and flesh of an infant woolly mammoth found in frozen muck in Alaska has been radiocarbon dated at 21,300 ± 1300 BP (L-601), and a steppe bison (*Bison priscus* = *Bison crassicornis*) horn-sheath, also found in muck, at 16,000 ± 2000 BP (M-38). [See Guthrie 1990a.] [The statement that even bones of the woolly mammoth are lacking in the Canadian Arctic Islands (but see footnote 18) must be modified in the light of specimens from Banks and Melville islands, Northwest Territories – see Harington 1990b.]

415. FAUNMAP (1994): FAUNMAP: A database documenting Late Quaternary distributions of mammal species in the United States. Illinois State Museum Scientific Papers 25(1 and 2):1–690.
This work (co-directors and principal authors: R.W. Graham and E.L. Lundelius, Jr; compilers: M.A. Graham, R.F. Stearley, and E.K. Schroeder; with 14 regional collaborators) is a synthetic electronic database created to document the Late Quaternary distribution of mammal (Mammalia) species in the 48 contiguous states of the United States for the last 40,000 years. Criteria for selection of sites for FAUNMAP were: (1) a known geographic location (at least to county level); (2) geochronological controls (e.g., radiometric dates, stratigraphy, biostratigraphy, cultural affiliation) for sites younger than 40,000 years and older than AD1500; and (3) voucher specimens (actual specimen, cast, or photograph) in a public institution. More than 2500 references (pp. 162–265) have been recorded in an electronic database, each linked with the locality or localities for which data were derived. FAUNMAP is composed of two databases: (1) the archival database contains data as they were originally reported in the literature; and (2) the research database has been filtered to standardize taphonomy, establish time periods, accept and reject 'absolute' dates, and eliminate mixed faunal assemblages. The main objective of FAUNMAP was to create a database that could be used to investigate specific questions about the evolution of mammalian communities in fluctuating Late Quaternary environments. Maps are provided of selected species [marked in bold in the Systematic List (pp. 293–302)] in relation to continental ice sheets, and, where appropriate, to the species' modern range (pp. 314–609).

416. Fay, L.P. (1984): Mid-Wisconsin and Mid-Holocene herpetofaunas of eastern North America: A study in minimal contrast. *In*: Contributions in Quaternary Vertebrate Palaeontology: A Volume in Memorial to John E. Guilday (H.H. Genoways and M.R. Dawson, eds.). Carnegie Museum of Natural History, Special Publication 8:15–19.
Two herpetofaunas (Strait Canyon, Virginia – Mid-Wisconsinan; and St Elzear, Quebec – Mid-Holocene) provide the first view of nonglacial Quaternary environments in eastern North America. Nine amphibians and 16 reptiles were recovered at Strait Canyon, six amphibians [blue-spotted salamander (*Ambystoma laterale*), spotted salamander (*Ambystoma maculatum*), American toad (*Bufo americanus*), spring peeper (*Hyla crucifer*), northern leopard frog (*Rana pipiens*), and wood frog (*Rana sylvatica*)] and one snake [garter snake (*Thamnophis sirtalis*)] at St Elzear. Of the last seven species, the American toad (*Bufo americanus*) is by far the most common (about 879) herptile. All herpetological members of these local faunas presently occur at or near the fossil localities, indicating conditions nearly identical to the current climates for western Virginia 30,000 years ago and about 5,000 years ago in eastern Quebec. No Altonian Stadial (Strait Canyon) or Hypsithermal (St Elzear) effects are evident in the herpetofaunas, and no apparent similarity to the steepened mammalian faunal gradient of eastern North America is revealed. *Ambystoma laterale* is reported from the fossil record for the first time, possibly due to the fact that St Elzear is the first locality within range of this species.

417. Fay, M. (1980): Stones and bones. Calgary Magazine 3(1):42–5.
This paper refers to the finding of the Taber (Alberta) child, a fossil human being (*Homo sapiens*), believed to have been between 30,000 and 70,000 years old. This date was based on relative stratigraphy [see Stalker 1969a, 1977; Cohen 1982; and Wilson et al. 1983].

418. Fedje, D.W., White, J.M., Wilson, M.C., Nelson, D.E., Vogel, J.S., and Southon, J.R. (1995): Vermilion Lakes site: Adaptations and environments in the Canadian Rockies during the Latest Pleistocene and Early Holocene. American Antiquity 60(1):81–108.
This site in Banff National Park, Alberta, contains evidence for 12 cultural components spanning the last 11 millennia. It has yielded cultural remains suggesting a possible mountain sheep-hunting focus during the Paleoindian (*Homo sapiens*) period. Colluvial and aeolian deposition has resulted in a well-stratified record for the period from 10,800 to about 9000 BP. This paper focuses on the early period of occupation, representing the earliest securely dated site found thus far in the southern ice-free corridor area. The lowest sediments observed were lacustrine, overlain by terrestrial sediments containing the cultural record, covered and protected by colluvium, followed by aeolian sediments and finally by the Mazama eruption, which added 10–20 cm of tephra at about 6850 BP. Thirty-four radiocarbon dates were obtained, of which 23 were done by the RIDDL AMS group. Ten of these are bone collagen; the remainder are charred wood. Faunal remains from the oldest layer are dominated by mountain sheep (*Ovis canadensis* cf. *Ovis canadensis catclawensis*), which show frequent spiral fracturing and occasional negative cones of percussion. Other mammal remains include a cervid (cf. *Rangifer tarandus*), another cervid (cf. *Odocoileus*), an unidentified ground-squirrel–sized rodent (Rodentia), and an unidentified mouse-sized rodent. Seasonality could not be determined. Faunal remains from the next oldest layer (around 9600 BP) are also dominated by mountain sheep, but include snowshoe hare (*Lepus americanus*), probable beaver (cf. *Castor canadensis*), and bison (*Bison*). Many of the broken elements exhibit spiral fractures and negative cones of percussion. Limited evidence suggests an autumn occupation. Several stratigraphic diagrams are provided along with tables of radiocarbon dates, although the dated bones are not identified.

419. Fenton, M.M. (1982): Quaternary stratigraphy of the Canadian Prairies. Joint Annual Meeting of the Geological Association of Canada and Mineralogical Association of Canada (May 17–19, Winnipeg). Program with Abstracts 7:48.
The author refers to vertebrate fossil sites in southern Alberta and adjacent Saskatchewan that provide evidence of at least three glaciations – the oldest of possibly Nebraskan and definitely of Illinoian vertebrate age. Sites of Sangamonian age have been found in southern and west-central Alberta, southern and central Saskatchewan, and west-central Manitoba.

420. Ferns, P.N. (1978): General observations of mammals. *In*: Joint Biological Expedition to North East Greenland 1974 (G.H. Green and J.J.D. Greenwood, eds.). pp. 164–7.
The author records finding large whale [Cetacea, presumably the bowhead whale (*Balaena mysticetus*)] vertebrae along the beach at Kap Seaforth and Holms Bugt, Greenland. A whale skeleton was also found at Holms Bugt, partially buried in a raised beach about 1 km inland and approximately 30 m asl (p. 167).

421. Fielden, H.W. (1877): The Post-Tertiary beds of Grinnell Land and North Greenland. Annals and Magazine of Natural History, Series 4, 20:488.
The author mentions ringed seals (*Phoca hispida*), caribou (*Rangifer tarandus* = *Cervus tarandus*), tundra muskoxen (*Ovibos moschatus*), and collared lemming (*Dicrostonyx torquatus* = *Myodes torquatus*) from raised beaches from 1000 ft asl to present sea level in this region. [Grinnell Land is now part of northern Ellesmere Island, Nunavut.]

422. Fielden, H.W., and de Rance, C.E. (1878): Geology of the coasts of the Arctic lands visited by the late British expedition under Captain Sir George Nares, R.N., K.C.B., F.R.S. Quarterly Journal of the Geological Society of London 34:556–67.
Remains of the following were found on Grinnell Land (Ellesmere Island, Nunavut) and northern Greenland: collared lemming (*Dicrostonyx torquatus* = *Myodes torquatus*), ringed seal (*Phoca hispida*), caribou (*Rangifer tarandus* = *Cervus tarandus*), and tundra muskox (*Ovibos moschatus*). They were found from sea level to 308 m asl, and they are all extant on the island. A

map is included showing the shell stations from which the bones came.

423. Fitzgerald, G.R. (1978): Pleistocene grebes from the Old Crow Basin, Yukon Territory. Canadian Journal of Earth Sciences 15(12):1887–92.
Grebe remains were found at two sites on the Old Crow River, Yukon, that include: an unidentified grebe (*Podiceps*), a Red-necked Grebe (*Podiceps grisegena*), and two Horned Grebes (*Podiceps auritus*). A detailed map of the sites is included. The relative stratigraphy exhibits an Early Pleistocene to Recent age, and the condition of the fossil bones indicates a Late Pleistocene age.

424. Fitzgerald, G.R. (1980): Pleistocene loons of the Old Crow Basin, Yukon Territory, Canada. Canadian Journal of Earth Sciences 17(11):1593–8.
Ten loon fossils [Red-throated Loon (*Gavia stellata*), Arctic Loon (*Gavia arctica*), Common Loon (*Gavia immer*), and Yellow-billed Loon (*Gavia* cf. *Gavia adamsii*)] were found in the Old Crow Basin, Yukon, which is presently outside the range of *Gavia adamsii*. A Yellow-billed Loon specimen is from beds that correlate with beds dating >54,000 BP and are presumed to be of Sangamonian interglacial age. Two specimens of Red-throated Loon and one of the Yellow-billed Loon have minimum dates of about 10,700 BP [based on a date on a freshwater mussel (*Anodonta beringiana*) commonly found in organic sandy gravel containing some bird bones]. It is worth noting that the shaft of part of a right humerus of a Common Loon has been modified by humans (*Homo sapiens*) (p. 1594).

425. Fitzgerald, G.R. (1991): Pleistocene ducks of the Old Crow Basin, Yukon Territory, Canada. Canadian Journal of Earth Sciences 28(10):1561–71.
Thirteen species of Pleistocene ducks were found in the Old Crow Basin, Yukon. These are: Mallard (*Anas platyrhynchos*), Northern Pintail (*Anas acuta*), American Wigeon (*Anas americana*), Northern Shoveler (*Anas clypeata*), Blue-winged Teal (*Anas discors*), Green-winged Teal (*Anas crecca*), Lesser Scaup (*Anthya affinis*), Ring-necked Duck (*Anthya collaris*), Oldsquaw (*Clangula hyemalis*), Common Eider (*Somateria mollissima*), Black Scoter (*Melanitta nigra*), White-winged Scoter (*Melanitta fusca*), and the Surf Scoter (*Melanitta perspicillata*). These range in age from latest Illinoian to Recent. The oldest specimens occur at a 140,000 BP paleomagnetic event. Others were beyond the radiocarbon-date range >54,000 BP (GSC-2066) based on bones and wood; this assemblage also included a specimen of a short-faced skunk (*Brachyprotoma obtusata*), which would indicate a warmer climate than at present. The author concludes that this area, during the Sangamonian, was warmer than today, and as a result, the treeline extended at least to this area. [See Harington 1990a.]

426. Fitzhugh, W. (1973): Environmental approaches to the prehistory of the North. Journal of the Washington Academy of Science 63(2):39–53.
One of the most intriguing aspects of the geological investigation of Hamilton Inlet, Newfoundland (Labrador), results from the reported existence of a few fossil whale (Cetacea) skeletons N of the Groswater Bay and near Cape Harrison, where they are claimed to lie nearly 1000 ft asl. If the elevation of these whale fossils is correct, they are far above the most recent postglacial marine limit, and they pose a unique challenge to geological interpretation. A substantial revision of current thought (pp. 42–5) may be required.

427. Fladmark, K.R. (1975): A paleoecological model for Northwest Coast prehistory. National Museum of Man, Mercury Series, Ottawa. No. 43.
The author hypothesizes that salmon (*Oncorhynchus*) were not available in large numbers for human (*Homo sapiens*) consumption in the interior of British Columbia until Mid-Holocene time (about 5000 BP). [See Carlson and Klein 1996.]

428. Fladmark, K.R. (1979): Routes: Alternate migration corridors for early man in North America. American Antiquity 44(1):55–69.
The author reviews the relative feasibility of interior and coastal routes for early people (*Homo sapiens*) entering southern North America from Beringia during the Late Pleistocene. Paleoenvironmental and archaeological data suggest that a chain of sea-level refugia around the North Pacific coast could have provided a real alternative to the interior 'ice-free' corridor, and that maritime cultural adaptations may have been among the first to arrive S of Canada. Faunal remains in the corridor area are mentioned: e.g., a bison skull with an associated pebble tool at the Bayrock site in southern Alberta and charred

bones from the Cochrane terraces near Calgary, both estimated to be about 11,000 years old; as well as a tusk from the Peace River area of British Columbia indicating that mammoths had reached that area by about 11,600 BP. The author also mentions that many biologists have argued for the existence of a significant biotic refugium on the Queen Charlotte Islands to account for the highly disjunct and endemic insular flora and fauna. The endemic species of caribou (*Rangifer dawsoni*) suggests the survival of an early cold-adapted fauna on the islands. Further, the author indicates that the arrival of the Dawson caribou and the black bear (*Ursus americanus*) of the islands may have corresponded with a lowered sea level during and immediately following the last glaciation. Mammoth (*Mammuthus*), mastodon (*Mammut americanum*), and bison (*Bison*) are reported for both Olympia Interstadial and early postglacial deposits of the Olympic Peninsula and Puget Sound area, and a muskox (*Bootherium bombifrons* = *Symbos cavifrons*) was found in early Fraser Glaciation outwash on southern Vancouver Island. Fragmentary sea lion (Otariinae) and whale (Cetacea) remains have been reported for late glacial and early postglacial deposits of the Fraser Lowlands and southern Vancouver Island respectively. The author speculates that migratory waterfowl (Anseriformes) may have used the chain of North Pacific coastal refugia as a flyway, that cold-adapted land mammals such as caribou (*Rangifer tarandus*) probably existed in larger refugia, and that most marine fishes (Pisces) and sea mammals were probably at least as abundant as at present.

429. Fladmark, K.R. (1983): Times and places: Environmental correlates of Mid- to Late Wisconsinan human population expansion in North America. *In*: Early Man in the New World (R. Shutler, Jr, ed.). Sage Publications, Beverly Hills. pp. 13–41.
Human (*Homo sapiens*) occupation of North America cannot be proven environmentally 'impossible' for any time in the last 60,000 years, although the climate of the last glaciation (particularly 18,000 to 15,000 BP) would seem least favourable. Before or after glacial maxima, human dispersals through North America were much more feasible. Mid-continental cultures prior to 14,000 to 12,000 BP must have been well adapted to stringent Arctic-type environments. However, on the Pacific Coast, more mesic environments may have permitted greater cultural flexibility within limitations of seacoast adaptations. After 14,000 to 13,000 BP, warm temperatures, retreating glaciers, and brief biotic continuity probably presented the best conditions for human expansion S of Beringia at any time during the last 60,000 years. During the Late Wisconsinan only three routes existed for people moving out of Beringia: (1) an Arctic Atlantic route seems least likely because of environmental harshness; (2) a midcontinental corridor seems increasingly supported by geological data, but an ice-free strip adjacent to the maximum Laurentide ice sheet would have been marginal for human survival; (3) the Pacific Coast refugia, although milder and with higher biotic productivity than the other areas during the glacial climax, cannot be proven to have been accessible from Beringia. If occupied, the last route would have entailed fishing, shellfish collecting, and possibly marine-mammal hunting; and boats seem mandatory.

430. Flerov, K.K. (1967): On the origin of the mammalian fauna of Canada. *In*: The Bering Land Bridge (D.M. Hopkins, ed.). Stanford University Press, Stanford, California. pp. 271–80.
Zoogeography of northern Nearctic and Palaearctic mammals in the Pleistocene is discussed. The role of the Bering Land Bridge, glaciation, habitat and climatic change, dispersal abilities, and ecology are some of the factors covered. The author supports mammals as reliable indicators of ecological change since they exhibit recognizable bony changes related to changes in food procurement or mode of movement. Much of the discussion involves conjectural relationships between mammals of northern North America and their Eurasian source-species. Bison (*Bison*) and mountain sheep (*Ovis*) dispersal history and evolution are emphasized.

431. Forbis, R.G. (1968): Fletcher: A Paleo-Indian site in Alberta. American Antiquity 33(1):1–10.
Paleoindians (*Homo sapiens*) evidently killed and butchered bison (*Bison*) at the Fletcher site (DjOw-1) on the N edge of Chin Coulee, 50 km SSE of Taber, Alberta. Alberta and Scottsbluff projectile points were derived from the bone layer, which was deposited between 11,000 and 7000 BP according to stratigraphic evidence. [This Cody complex kill site was discovered in 1963 and includes deer (*Odocoileus*) as well as bison. The site has been difficult to radiocarbon date. Normalized dates on

bison bone collagen are: 7735 ± 110 BP (S-1084) and 6040 ± 170 BP (RL-560) – see Morlan 1999.]

432. Forstén, A. (1986): *Equus lambei* Hay, the Yukon wild horse, not ass. Journal of Mammalogy 67(2):422–3.

The author asserts that the Yukon horse (*Equus lambei*) was a caballine horse rather than an ass, based on skull measurements. These measurements are taken from the research of Hay (1917), Quinn (1957), Harington and Clulow (1973), Eisenmann (1979), and Savage (1951). *Equus lambei* seems to be most closely related to the Eurasian caballine horse (*Equus caballus przewalskii*). [See Burke and Cinq-Mars 1996, 1998.]

433. Foster, J.B. (1965): The evolution of the mammals of the Queen Charlotte Islands, British Columbia. Occasional Paper of the British Columbia Provincial Museum No. 14:1–130.

The Queen Charlotte Islands are the most isolated islands in British Columbia. After a review of the glacial history of the region, the author concludes that glacial refugia probably existed on the islands and that a land bridge extended to the mainland at the close of the last glaciation. Two summers were spent collecting mammal specimens from 45 of the islands. All native species of mammals, except the caribou (*Rangifer tarandus*), could have found the glacial refugia of the islands suitable for habitation during the last glaciation – allowing at least 100,000 years for the evolution of characteristics found only in the insular races. Analysis of specimens suggests that the ancestors of black bear (*Ursus americanus carlottae*), marten (*Martes americana nesophila*), ermine (*Mustela erminia haidarum*), and one shrew (*Sorex obscurus elassodon*) lived and became distinct in the refugia during the last glaciation.

434. Foster, J.H., and Stalker, A.M. (1976): Palaeomagnetic stratigraphy of the Wellsch Valley site, Saskatchewan. Geological Survey of Canada Paper 76(1C):191–3.

Many mammalian remains were found at the Wellsch Valley site near Swift Current, Saskatchewan, from an exposure of sediments called 'Jaw Face' (50°40′N, 107°52′25″W). This is on the E side of the valley trending N toward the South Saskatchewan River. Most remains are from a sand layer 6.5 m deep up to an ash layer at 2.5 m. The fauna includes: unidentified large ground sloth (Xenarthra), Gazin's marsh rabbit (*Hypolagus* cf. *Hypolagus limnetus*), ground squirrel (*Spermophilus* = *Citellus*), Meade prairie dog (*Cynomys* cf. *Cynomys meadensis*), southern pocket gopher (*Thomomys bistylis* n. sp.), northern pocket gopher (*Thomomys* cf. *Thomomys talpoides*), Kansan bog lemming (*Synaptomys (Myctomys) kansasensis*), Cape Deceit vole (*Microtus deceitensis*), Deering vole (*Pliomys* cf. *Pliomys deeringensis*), rice rat (?*Oryzomys*), unidentified rodent remains, bone-eating dog (*Borophagus diversidens*), bobcat? (*Felis* cf. *Felis rufus* = *Lynx* cf. *Lynx rufus*), southern mammoth (*Mammuthus meridionalis* = *Mammuthus imperator haroldcooki*), Pacific horse (*Equus pacificus*), eastern horse (*Equus complicatus*), Cope's peccary (*Platygonus* cf. *Platygonus bicalcaratus*), camel (*Camelops*), pronghorn buck (?*Tetrameryx*), and shrubox (Ovibovini cf. ?*Preptoceras*). This faunal assemblage is consistent with the beginning of the Irvington Land Mammal age. If this faunal assignment is correct, the reversal pattern found at 'Jaw Face' should be the Olduvai paleomagnetic event [about 1.8 Ma].

435. Fox, W. (1998): Muskox and wapiti in the northern Yukon. IsoTrace Laboratory, Canadian Centre for Accelerator Mass Spectrometry at the University of Toronto, 1996 Annual Report:15–16.

Accelerator mass spectrometry dates on muskox (*Ovibos moschatus*) and wapiti (*Cervus elaphus*) bone from Engigstciak, Yukon, were dated at 870 to 3330 BP (TO-5678 and 5680) and 3010 BP (TO-5681) respectively [radiocarbon errors not given]. These dates confirm the former presence of muskoxen in northern Yukon until their recorded extirpation [near Barrow, Alaska] toward the close of the nineteenth century. The wapiti date confirms its presence in the archaeological deposits of this multi-component site.

436. Fraser, T.A., and Burn, C.R. (1997): On the nature and origin of 'muck' deposits in the Klondike area, Yukon Territory. Canadian Journal of Earth Sciences 34:1333–44.

Organic-rich 'muck' deposits, which blanket auriferous gravels in the Klondike area, comprise two principal stratigraphic units: (1) a silty Late Pleistocene deposit, and (2) Holocene organics lying unconformably on the silt. The silt is both primary (massive) and redeposited (bedded) loess. The deposits are found predominantly in valley bottoms and, if undisturbed, are normally perennially frozen. Within the silt unit, paleosols were identified in three stratigraphic positions. The mineral-rich Late Pleistocene silt layer frequently contains mammal bones usually within a few decimetres of the underlying gravel in the lowest paleosol. Radiocarbon dates indicate that the loess was deposited during Late Wisconsinan McConnell glaciation beginning after 27,000 years BP. The loess was likely derived from the floodplain of the Yukon River during periods of low flow. Turf in growth position suggests a grassland environment for the area during the McConnell glaciation. A mummified Yukon horse (*Equus lambei*) carcass (Harington and Eggleston-Stott 1996) in the silts indicates that some of these sediments have been frozen since shortly after deposition. Radiocarbon dates show that peat growth began at the start of the Holocene when the climate became abruptly wetter. Stratigraphy is illustrated both photographically (Figures 1, 4, 7), and a composite section of the Klondike muck deposits, compiled from an examination of 23 sections is shown in Figure 5 – the paleosol contain Pleistocene mammal bones lying at the base of the silt and above the basal gravel. Figure 2 is a map of the Klondike placer-mining area showing the sites studied.

437. Fredskild, B. (1983): The Holocene vegetational development of the Godthåbsfjord area, West Greenland. Meddelelser om Grønland, Geoscience 10:1–28.
Analysis of sediments dating to about 9000 BP from Johannes Iversen Sø (64°24′N, 50°12′W, approximately 100 m asl), Greenland, yielded abundant three-spined stickleback (*Gasterosteus aculeatus*) bones (Figure 15). The author notes that this species is presently known as far N as 74°.

438. Fredskild, B. (1985): The Holocene vegetational development of Tugtuligssuaq and Qeqertat, Northwest Greenland. Meddelelser om Grønland, Geoscience 14:1–20.
In both West and East Greenland, remains of the three-spined stickleback (*Gasterosteus aculeatus*) have been recovered in Holocene lake sediments N of the modern range limit of the species [see Bennike 1997]. In Melville Bugt [Langesø (75°22′N, 58°36′W), Tugtuligssuaq, Figures 1, 2] a bone from the sternal armature of this species was found in sediments dating to about 6800 BP (transition between Paleovegetation zones B and C). This, with other evidence, indicates lake conditions warmer than today, undoubtedly connected with a longer ice-free period (p. 18).

439. Fredskild, B., and Røen, U. (1982): Macrofossils in an interglacial peat deposit at Kap København, North Greenland. Boreas 11:181–5.
In 1979, a 90-m-thick till-covered sequence with bands of organic debris was found 4 km SW of Kap København, eastern Peary Land (about 82°27′N, 20°45′W; 40–130 m asl), Greenland. The interglacial age of the peat is confirmed by an infinite radiocarbon age, and by amino acid analysis of shells indicating pre-Eemian age. A few scales and a broken spine of the three-spined stickleback (*Gasterosteus aculeatus*) are reported. The size of the spine makes the identification clear. [Apparently this 'interglacial' was later determined to be Early Pleistocene, about 2 Ma – see Bennike 1991, p. 902, which states that the spine is questioned because no more remains of this fish species have been recovered in spite of an intensive search.]

440. Freeman, E.B. (1976): Toronto's geological past – An introduction. Ontario Division of Mines Miscellaneous Publication [unpaginated pamphlet].
This attractive, profusely illustrated pamphlet provides excellent maps and stratigraphic diagrams showing the glacial features of the Toronto area. One fold-out page is devoted to 'Remains of Life.' Pertinent to this publication is a list of vertebrate remains from the Don Formation (Sangamonian interglacial layered sands, silts, and clays formed under warm-temperate climatic conditions 'similar to southern Pennsylvania and Ohio'): young groundhog (*Marmota monax*), giant beaver (*Castoroides ohioensis*), two species of deer (Cervidae), bison (*Bison*), large bear (*Ursus* cf. *Ursus arctos*), and catfish (*Ictalurus*). An interesting interpretation of a Don Formation scene is quoted from Coleman (1941, pp. 76–7), which reads in part: 'one can see the ancient forest of maples and oaks and many other trees on the river shore with deer coming down to drink, bears tearing open a rotten log for its small inhabitants; and at some creek mouth the giant beaver fells a tree with a slash to feed on its branches; while openings in the forest show buffalo grazing.'

441. Freuchen, P. (1915): General observations as to natural conditions in the country traversed by the expedition [The First Thule Expedition 1912]. Meddelelser om Grønland 51(9):341–425.
On June 12, 1912, on a raised beach about 10 m asl near Cape Harald Moltke, Greenland (82°09′N, 29°45′W), the author 'found the remains of a whale [Cetacea – probably bowhead whale (*Balaena mysticetus*)] skeleton lying loose on the slope' (p. 359).

442. Frick, C. (1930): Alaska's frozen fauna. Natural History 30:71–80.
The author tells of the cooperative effort of Alaska College [now University of Alaska-Fairbanks] and the American Museum of Natural History in the summer of 1930 'to take advantage of the unusual opportunity afforded by the present vast stripping operations of the mining industry' (e.g., United States Smelting and Mining Company; Fairbanks Exploration Company). The field work was in the charge of Peter Kaisen (pictured on p. 75 with remains of steppe bison, American lion, and mammoth) of the museum. The cooperative field party collected 28 large cases of skulls, jaws, and other bones of ice age mammals – rare and important evidence of the prehistory of Alaska, which otherwise would have been lost to science. At one site, the 'newly washed surface of former Alaskan accumulations' exposed: an 11.5 ft mammoth (*Mammuthus*) tusk; a horse (*Equus*) limb bone; broken stumps of an ancient forest; a moose (*Alces*) hind limb with hoof and some hair attached; skulls of wolverine (*Gulo gulo*) and badger (*Taxidea taxus*); the widespread horns of a huge bison (probably *Bison priscus*); and the skull of 'a very African-like lion' (*Panthera leo atrox*). Frick mentions: (1) the rareness of American mastodon remains compared to woolly mammoths; (2) collecting the skull of an extremely old mammoth – without teeth or alveoli, 'the best pair of grinders having been worn through and shed'; (3) the great American lion 'now for the first time shown to have been present in the Pleistocene of Alaska'; he also notes that the skulls of the American lion (*Panthera leo atrox* = *Felis atrox alaskensis*) and the European cave lion (*Panthera leo spelaea* = *Felis spelaea*) 'are distinctively lionine rather than tigrine'; and (4) that Alaskan ice age mammals may, at times, have been hunted (*Homo sapiens*) according to the find of a deeply buried arrowhead. A preliminary list (p. 79) of extinct Pleistocene mammals ('Pre-tundra Fauna of Alaska-Yukon') includes: giant short-faced bear (*Arctodus simus yukonensis* = *Arctodus yukonensis*); dire wolf (*Canis dirus* = *Aenocyon dirus alaskensis*, new subspecies) [but see Kurtén and Anderson 1980, p. 171 regarding paleorange]; American lion (*Panthera leo atrox* = *Felis atrox*, new subspecies); western camel (*Camelops hesternus* = *Camelops*); steppe bison (*Bison priscus* = *Bison crassicornis*); helmeted muskox (*Bootherium bombifrons* = *Symbos tyrrelli* [sic] = *Bootherium sargenti*); horse (*Equus* = *Equus alaskae*); woolly mammoth (*Mammuthus primigenius* = *Mammonteus* (*Elephas*) *primigenius*); and American mastodon (*Mammut americanum* = *Mastodon americanus*).

443. Friesen, T.M. (1986): Mammoth bone breakage patterns from Locality 11A, Old Crow, northern Yukon Territory. *In*: Student Research in Canada's North (W.P Adams and P.G. Johnson, eds.). Proceedings of the National Student Conference on Northern Studies (November 18–19). pp. 479–85.
The Old Crow Basin, Yukon, is rich in remains of woolly mammoths (*Mammuthus primigenius*), Columbian mammoths (*Mammuthus columbi*), and mastodons (*Mammut americanum*). Broken bones of such are examined, as there appears to be evidence of bone breakage before desiccation and fossilization. This may indicate the presence of man (*Homo sapiens*) in the area anywhere from 120,000 to 25,000 BP, for these breakages would not have occurred from reworking of sediments. The author explains that carnivores (Carnivora) may have broken the bones at the time of death, but teeth marks are absent. A table of broken bones is given.

444. Friesen, T.M. (1989): Analysis of fracture patterns on proboscidean bones from Locality 11A, Old Crow, northern Yukon Territory. MA thesis, University of Toronto. 147 pp.
Fresh-fractured woolly mammoth (*Mammuthus primigenius*) bones (variously dated from 25,000 to 150,000 BP) from the Old Crow Basin, Yukon, have been advanced as evidence of human (*Homo sapiens*) presence. The author outlines the nature of the breaks (including photographs and illustrations) and reviews the possible causes of the bone breakage. Experimental data on controlled breakage of moose (*Alces alces*) bone are presented (also including

photographs). The author concludes that some of the bones were broken by carnivores (Carnivora), some possibly by trampling by other proboscideans, and because most were deposited in river valleys they may have been affected by fluvial action or ice break-up. The role of freezing temperatures, both to prolong the 'fresh' stage of the bone and to reduce the force required to break the bone, is discussed. The author concludes that the Old Crow 'cores' were produced by natural agencies. The possible role of early humans is unlikely but cannot be completely eliminated. Table 5.1 includes a list of Quaternary mammals identified from deposits at Old Crow Locality 11A.

445. Friesen, T.M. (1997): Effects of collector bias on Pleistocene bone breakage frequencies from Old Crow, northern Yukon Territory. Current Research in the Pleistocene 14:113–14.
This report presents quantified data that reveal biases in frequencies of bone-breakage types introduced by selective collection at Locality 11A in the Old Crow Basin, northern Yukon. Only proboscidean bones were studied (no teeth or tusks were used because they break differently from bones), more than 95% of which are from mammoths (*Mammuthus*), the rest representing American mastodon (*Mammut americanum*). Paleontological collections were made with a bias toward complete bones, whereas 'archaeological-paleontological' collecting led to the preservation of more spirally fractured bones – thought to have been evidence for human (*Homo sapiens*) activity. The 1975 collection (n = 1322) of 'all' bones and bone fragments from designated areas on the surface of this point bar acted as an unbiased datum. Breakage categories used were: (1) bones broken while fresh (spiral fractures); (2) broken bones lacking spiral fractures; and (3) virtually complete bones. Complete bones represented only 2% of the 'unbiased' 1975 sample, whereas 'archaeological-paleontological' collecting contained 6.5% and 11.6% (data for two separate years) complete bones, and paleontological collecting contained 33.3% and 54.5% of intact bones. These results are significant because they illustrate biases produced by different collecting methods at one site, allowing different interpretation of the site's taphonomic history.

446. Frison, G.C., and Walker, D.N. (1990): New World paleoecology at the Last Glacial Maximum and the implications for New World prehistory. *In*: The World at 18 000 BP, Volume 1, High Latitudes (O. Soffer and C. Gamble, eds.). Unwin Hyman, London. pp. 312–30.
In discussing the problems of Clovis (*Homo sapiens*) origins, the authors mention the claims of the Old Crow region, Yukon, based on a caribou (*Rangifer tarandus*) tibia fleshing tool that was originally dated at about 27,000 BP (bone apatite fraction) and re-dated by accelerator mass spectrometry to less than 3000 BP. Evidence of pre-Clovis occupation of the Old Crow Basin is considered controversial. The authors note that the large mammal species [including mammoth (*Mammuthus*), bison (*Bison*), horse (*Equus*), caribou (*Rangifer tarandus*), and muskox (*Ovibos moschatus*)] living together in Eastern Beringia were many times more diverse than the extant tundra faunas. This implies that the Late Wisconsinan vegetation was able to support more herbivores than does today's vegetation.

447. Fuller, W.A., and Bayrock, L.A. (1965): Late Pleistocene mammals from central Alberta, Canada. *In*: Vertebrate Paleontology in Alberta. University of Alberta, Edmonton. pp. 53–63.
Pleistocene mammals were found in two distinct deposits from near Edmonton, Alberta. They are postglacial alluvial deposits of the terraces of many streams, and Saskatchewan gravels and sands. The fossils found include: horse (*Equus*), mammoth (*Mammuthus*), giant bison (*Bison latifrons*), steppe bison (*Bison priscus* = *Bison crassicornis*), western bison (*Bison bison occidentalis*), which may be the same as wood bison (*Bison bison athabascae*), and camel (cf. *Camelops*). Radiocarbon dating of the specimens had not been complete by the time this paper was published. Tables of bone measurements and photographs of bison horncores are included.

448. Fulton, R.J. (1971): Radiocarbon geochronology of southern British Columbia. Geological Survey of Canada Paper 71-37:1–28.
The radiocarbon-dated Quaternary history of southern British Columbia extends over the past 52,000 years. This interval has been subdivided into three major geologic-climate units: Olympia Interglaciation (52,000–19,000 BP), Fraser Glaciation (in the lowlands from 19,000 to 10,000 BP) and Postglacial (10,000–present). At least part of the Olympia Interglaciation was warm

and humid enough to support populations of large herbivores such as bison (*Bison*) and Mexican horse (*Equus* cf. *Equus conversidens*). All of the radiocarbon dates upon which the three geologic-climate units are based were performed on organic mucks or organic material besides bone. [See Clague 1980 for further dates from this region.]

449. Fulton, R.J. (1975): Quaternary geology and geomorphology, Nicola-Vernon area, British Columbia. Geological Survey of Canada Memoir 380:11–18.
Vertebrate remains were found in a sand lens in the silt from beneath the Okanagan Centre drift at Westwold, British Columbia. The nature of the sediments indicates nonglacial conditions within the Wisconsinan. The fauna includes: unidentified fish (Pisces) and rodents (Rodentia), and bison (*Bison*). Twelve kilometres E of Lumby, British Columbia, mammals were found in Bessette sediments. The climate of deposition is believed to be similar to the present for this location. The fauna includes: horse (*Equus*), Mexican horse (*Equus conversidens*), bison (*Bison*), unidentified mammoth (*Mammuthus*), and Columbian mammoth (*Mammuthus* cf. *Mammuthus columbi*).

450. Fulton, R.J. (1995): Proboscidean tusk of middle Wisconsinan age from sub-till gravel, near Turtle Mountain, southwestern Manitoba. Geological Survey of Canada, Current Research 1995-E:91–6.
An AMS radiocarbon age of 33,860 ± 330 BP (TO-4639) was obtained for a partial tusk discovered in a gravel pit about 22 km SW of Boissevaine, Manitoba. As this age is about 10,000 years older than the advance of the last ice sheet, either it was incorporated from older gravels, or the tusk comes from an unrecognized, Middle Wisconsinan–aged gravel which underlies the gravel exposed immediately below the till. This find adds to earlier findings at Medicine Hat (see Stalker 1977), which indicate that large mammals were present in the southern part of the Canadian Prairie Provinces during Middle Wisconsinan. The specimen currently resides in the Moncut Museum in Boissevaine, Manitoba.

451. Fulton, R.J., and Smith, G.W. (1978): Late Pleistocene stratigraphy of south-central British Columbia. Canadian Journal of Earth Sciences 15(6): 971–80.
This paper deals primarily with Late Pleistocene sedimentation, but does mention vertebrate remains in two locations. First are the Westwold, British Columbia, sediments of Sangamonian age, and second are the Bessette sediments >43,800 to 19,000 BP. [See Fulton 1975.]

452. Fulton, R.J., Anderson, T.W., Gadd, N.R., Harington, C.R., Kettles, I.M., Richard, S.H., Rodriques, C.G., Rust, B.R., and Shilts, W.W. (1987): Summary of the Quaternary of the Ottawa Region. *In*: INQUA 87 (R.J. Fulton, ed.). Quaternary of the Ottawa Region and Guides for Day Excursions. pp. 7–20.
This article provides the context for INQUA Day Excursions in the Ottawa region of Ontario and Quebec and an introduction and summary of the Quaternary of the region for XII INQUA Congress delegates. Maps showing physiographic and bedrock, ice-flow, marine limits, surficial geology, and radiocarbon localities are provided. Tables of radiocarbon data, pollen stratigraphy, paleosalinity, macrofaunal associations, and animal and plant species contained in nodules of Champlain Sea age support the text and maps. Illustrations of the emergence curve of the area and a correlation diagram showing the relationship between events of the western basin of the Champlain Sea and those in the southern Glacial Lake Agassiz, Lake Huron, and Lake Ontario basins over time complete the picture.

453. Funder, S., Abrahamsen, N., Bennike, O., and Feyling-Hanssen, R.W. (1985): Forested Arctic: Evidence from North Greenland. Geology 13:542–6.
Shallow-water marine sediments of the Kap København Formation at 82°30′N, Peary Land, North Greenland, contain an abundance of well-preserved remains of terrestrial vegetation and invertebrate fauna indicative of a mosaic of forest-tundra environments having similarities to present conditions in Labrador. The presence of foraminifera faunas, marine molluscs, and one mammal (*Hypolagus*) suggests an age of 2 Ma, at the Pliocene-Pleistocene transition. At this time, the arctic treeline was located 2500 km to the N of its present position in the NW Atlantic region, and forest-tundra vegetation existed in lowland areas bordering an Arctic Ocean without perennial sea-ice cover.

454. Gadd, N.R. (1961): Surficial geology of the Ottawa area. Department of Mines and Technical Sur-

veys Canada, Report of Progress Paper 61-19:5–11.
Two distinct bodies of Pleistocene clay exist at Green Creek, near Ottawa, Ontario; one is marine, containing invertebrate marine fossils and capelin (*Mallotus villosus*), and the other, freshwater, contains land mammals (Mammalia), bird (Aves) feathers, and plant matter.

455. Gadd, N.R. (1980): Maximum age for a concretion at Green Creek, Ontario. Géographie physique et Quaternaire 34(2):229–38.
Radiocarbon dating has been done on plant matter and mollusc shells found in clay nodules at Green Creek, near Ottawa, Ontario. These dates range from 9860 to 14,400 BP. Mentioned are bird (Aves) feathers, capelin (*Mallotus villosus*), deepwater sculpin (*Myoxocephalus thompsoni*), and lake trout (*Cristivomer namaycush*). The first two are marine fish, and the last one is a freshwater species, indicating the paleoenvironmental change that took place during the Late Pleistocene.

456. Gaffin, J. (1976): Digging down to bare bones. Calgary Herald, September 17.
After 11 years of digging in Yukon, Dr William Irving, professor of anthropology at University of Toronto finally made a major discovery – the right half of a human (*Homo sapiens*) jawbone. The mandible, though yet to be radiocarbon dated, is thought to be 20,000 years old. The previous year, working with six PhD students and eight Old Crow Indians, Dr Irving collected two tons of fossil bones that were shipped to the university for study during the winter.

457. Galusha, T. (1975): Childs Frick and the Frick Collection of fossil mammals. Curator 18(1):9–12.
This paper, mainly a biography of Childs Frick, mentions his enormous Alaskan fossil vertebrate collection and some of his techniques for achieving such success. His total collection is estimated between one-quarter and half a million specimens.

458. Gard, L.M., Louis, G.E., and Whitmore, F.C. (1972): Steller's sea cow in Pleistocene interglacial beach deposits on Amchitka, Aleutian Islands. Geological Society of America Bulletin 83(3):867–9.
The first Pleistocene Steller's sea cow (*Hydrodamalis gigas*) was found in deposits of interglacial beach sand and gravel 35 m asl at the head of South Bight, Amchitka, Aleutian Islands, Alaska. Uranium/Thorium dating of the bone yields dates of 135,000 ± 12,000 and >122,000 BP.

459. Geist, O.W. (1953): Collecting Pleistocene fossils in Alaska, Proceedings of the Second Alaskan Science Conference. Science in Alaska, pp. 171–2.
Fossil vertebrates found in many localities in Alaska include: American lion (*Panthera leo atrox*), sabretooth tiger (*Smilodon floridanus*), unidentified cat, lynx (*Lynx canadensis*), bear (*Ursus*), wolf (*Canis lupus*), coyote (*Canis latrans*), wolverine (*Gulo gulo*), mink (*Mustela vison*), marten (*Martes americana*), badger (*Taxidea taxus*), fox (*Alopex lagopus*), ground sloth (*Megalonyx*), camel (*Camelops*), bighorn sheep (*Ovis canadensis*), prongbuck (*Antilocapra americana*), Saiga antelope (*Saiga tatarica*), mountain goat (*Oreamnos*), wapiti (*Cervus elaphus*), giant elk(?) (*Cervalces*), moose (*Alces alces*), caribou (*Rangifer tarandus*), bison (*Bison*), tundra muskox (*Ovibos moschatus*), helmeted muskox (*Bootherium bombifrons* = *Bootherium sargenti*), horse (*Equus*), mastodon (*Mammut americanum*), mammoth (*Mammuthus*), porcupine (*Erethizon dorsatum*), ground squirrel (*Citellus*), vole (*Microtus*), beaver (*Castor canadensis*), giant beaver (*Castoroides ohioensis*), rabbits, and birds. [See Guthrie 1968a.]

460. Geist, O.W. (1955): Vertebrate paleontological reconnaissance of the Old Crow River area, Yukon Territory, Canada. Abstract. Geological Society of America Bulletin 66:1702.
Many Pleistocene vertebrate fossils were found in sand and gravel bars in the Old Crow Flats, Yukon. They include: moose (*Alces alces*), caribou (*Rangifer tarandus*), superbison (*Bison*), tundra muskox (*Ovibos moschatus*), beaver (*Castor canadensis*), giant beaver (*Castoroides ohioensis*), and wolf (*Canis lupus*).

461. Geist, O.W. (1956): Biogeography – The fossil record. Proceedings of the Fourth Alaskan Science Conference. pp. 200–1.
This abstract deals briefly with Pleistocene geographic ranges of the following: bison and superbison (*Bison*); camel (*Camelops*); mammoth (*Mammuthus*); mastodon (*Mammut americanum*); horse (*Equus*); and sabretooth cat (*Smilodon floridanus*).

462. Geist, O.W. (1956?): Scientific investigations in the Old Crow and Porcupine River regions of Alaska and Yukon Territory, 1952–1953. Unpublished report. 108 pp.

Sponsored by the University of Alaska and financed by the Explorer's Club Exploration Fund, Otto Geist began his boat trip up the Old Crow River on August 3, 1952. This 1952 reconnaissance of Old Crow Basin comprises Part I (pp. 11–65) of the report, and is the only section relevant to Quaternary vertebrates. Peter Lord and Charlie Linklater from the settlement of Old Crow acted as his guides. By August 5, after passing Schaeffer and Johnson creeks, Geist had begun to collect a few fossils on some of the exposed gravel bars. On August 8, some were found 'in place in the hard bases of some of the high cliffs.' Horse (*Equus*) and mammoth (*Mammuthus*) specimens were most common along this stretch of the river, but some beaver (Castoridae) fossils were collected. Above 'Last Cutbank' [C.R. Harington's Locality 51], Geist stated 'We could see the fossils on the river bottom as we travelled along. We collected only the valuable ones, mostly mammoth teeth.' Evidently fossils were scarce or lacking beyond this point. On August 21 the party climbed Ammerman Mountain, which is just W of the Yukon–Alaska border. On their return, Geist walked up Thomas Creek a few miles and collected some fossils. Above Timber Creek fossils were collected on bars and at the foot of bluffs. 'We also dug out a cottonwood (poplar) stump which evidently had been cut down by beavers, and from the width of the tooth marks, it could have been cut by a giant beaver (*Castoroides ohioensis*). This stump we found still buried very near the line which separated the harder lower part of the high bluff from the soft overlaying mass [probably an interglacial deposit at C.R. Harington's Locality 44].' Fossils were collected near the mouth of Black Fox Creek on August 28, and many specimens, mainly mammoth (*Mammuthus*), were found on rocky bars farther south. They returned to Old Crow on August 30, where the fossils were cleaned, shellacked and packed for shipment. Of 380 specimens collected by Geist during this trip up Old Crow River in 1952, mammoth (*Mammuthus*), horse (*Equus*) and bison (*Bison*) were most common [see Harington 1977a, Table 2]. At least nine species were represented in his collection, five of which (*Mammuthus, Bison, Equus*, Ovibovini, and *Castoroides*) no longer occur naturally in Yukon.

463. Geist, O.W. (1961): Collecting Pleistocene fossils and natural history material in arctic Alaska river basins – 1959, 1960, 1961. Office of Naval Research, Arctic Institute of North America. Report for Projects 247 and 254.

Wood was found associated with Pleistocene animal fossils along the Ikpikpuk River system on the North Slope of Alaska.

464. Geist, V. (1989): Did large predators keep humans out of North America? *In*: The Walking Larder. Patterns of Domestication, Pastoralism, and Predation (J. Clutton-Brock, ed.). Unwin Hyman, London. pp. 282–94.

Had people (*Homo sapiens*) reached North America when it had an intact Rancholabrean fauna, they would have met, among several large carnivores, the giant short-faced bear (*Arctodus simus*) – a bear larger than the grizzly (*Ursus arctos*). The author presumes that the short-faced bear populations would have been large enough, fierce enough, and sufficiently aggressive in approaching humans that they would have prevented people from entering Alaska until after about 12,000 BP, when *Arctodus* was on the way to extinction. He also presumes that early humans entering North America from Siberia were unable to defend themselves adequately against *Arctodus*.

465. Genoways, H.H., and Dawson, M.R., eds. (1984): Contributions in Vertebrate Paleontology: A volume in Memorial to John E. Guilday, Carnegie Museum of Natural History Special Publication No. 8, Pittsburgh. 538 pp.

The papers presented in this book which are relevant to the bibliography are annotated individually. See 1984 papers by Fay, Morlan, LaSalle, Guthrie, and Harington.

466. Gidley, J.W. (1908): Descriptions of two new species of Pleistocene ruminants of the genera *Ovibos* and *Bootherium*, with notes on the latter genus. Proceedings of the U.S. National Museum 34(1627): 681–4.

Specimens of Pleistocene muskoxen were found in the Palisades of the Yukon River, Alaska, 56 km below the Tanana River. They include: tundra muskox (*Ovibos moschatus*), helmeted muskox (*Bootherium* bombifrons = *Bootherium sargenti* = *Symbos tyrrelli*). A table of bone measurements is included.

467. Gidley, J.W. (1913): Notice of the occurrence of a Pleistocene camel north of the Arctic Circle. Smithsonian Miscellaneous Collections 60(26):1–2.
Pleistocene mammals were found in the Yukon near the Alaskan boundary, 80 km from the mouth of the Old Crow River. The specimens included a phalanx of a camel (cf. *Camelops* or cf. *Camelus*), woolly mammoth (*Mammuthus primigenius*), horse (*Equus*), and bison (*Bison*).

468. Giesecké, C. (1861): Catalogue of geological and geographical collection of minerals from the arctic regions, from Cape Farewell to Baffin's Bay. Lat. 59°14′N, to 76°32′N. Journal of the Royal Dublin Society 3:198–215.
The author mentions collecting 'Impressions of the Salmo arcticus, in Clay' at his locality 211 Auppadlartok in southwestern Greenland (apparently not far from the present Nuuk, which is his locality 194). Bennike (1997, p. 901) considers that such fossils most likely represent capelin (*Mallotus villosus*) found in calcareous nodules or concretions, some of which were brought to Europe by whalers more than 200 years ago (Bendix-Almgreen 1976). J.W. Tayler in a letter appended to this paper (p. 215), dated February 28, 1861, states that among several 'interesting new places he collected at is locality 120, where The Greenlanders say that large bones [possibly bowhead whale (*Balaena mysticetus*) bones from raised beaches?] are to be found high up on this mountain.'

469. Gillespie, J.M. (1970): Mammoth hair: Stability of alpha-keratin structure and constituent proteins. Science 170(3962):1100–2.
Hair from a woolly mammoth (*Mammuthus primigenius*) was found at Dome Creek, Alaska, and was radiocarbon dated at 32,700 ± 980 BP (ST-1632). The alpha-keratins are compared with those of extant elephants.

470. Gilmore, C.W. (1908): Smithsonian exploration in Alaska in 1907 in search of Pleistocene fossil vertebrates. Smithsonian Miscellaneous Collections 51(1807):1–38.
This is a systematic description of the Pleistocene vertebrate fauna of the Yukon and Alaska. The Yukon site, Fox Gulch on Bonanza Creek 19 km from Dawson, contains: mammoth (*Mammuthus*), bison (*Bison*), horse (*Equus*), and moose (*Alces alces*). The Alaskan fauna, from many tributaries of the Yukon River, includes: woolly mammoth (*Mammuthus primigenius*), Columbian mammoth (*Mammuthus columbi*), American mastodon (*Mammut americanum*), horse (*Equus*), steppe bison (*Bison priscus* = *Bison crassicornis* = *Bison alleni*), western bison (*Bison bison occidentalis*), helmeted muskox (*Bootherium bombifrons* = *Symbos cavifrons*), tundra muskox (*Ovibos moschatus* = ?*Ovibos maximus*), moose (*Alces alces*), caribou (*Rangifer tarandus*), bear (*Ursus*), and beaver (*Castor canadensis*). This paper includes photographs and an excellent map of the Dawson area.

471. Gilmore, C.W. (1941): A history of the Division of Vertebrate Paleontology in the United States National Museum. Proceedings of the United States National Museum 90(3109):305–77.
This history of the Division is based on early records of the Smithsonian Institution as well as an examination of its published papers and annual reports supplemented by information resulting from the author's 37 years of service in the Division. All phases of its development, organization, personnel, explorations, field trips, and records covering 90 years (from the establishment of the Institution to 1940) are included in the history. Pertinent items mentioned include: (1) expeditions led by A.G. Maddren to collect Pleistocene mammals in Alaska, Old Crow, and Yukon river valleys in 1904, and by C.W. Gilmore to Alaska and Yukon river valley in 1912; (2) Pleistocene fossils from Canada presented by J.W. Dawson in 1861; (3) skull of a steppe bison (*Bison priscus* = *Bison alleni*) from near Rampart, Alaska, presented by D. McLean in 1904; (4) collection of Pleistocene mammals from Alaska made by Maddren and transferred to the Smithsonian Institution in 1905; (5) helmeted muskox (*Bootherium bombifrons* = *Symbos tyrrelli*) skull from Yukon Pleistocene presented by J.B. Tyrrell through W.H. Osgood in 1906; (6) collection of Pleistocene mammals made by Gilmore in Alaska in 1907 and received by the Museum in 1908; (7) steppe bison skulls (*Bison priscus* = *Bison crassicornis*) and a muskox skull (*Ovibos moschatus* = *Ovibos yukonensis*, holotype) received in 1908; (8) a collection of Pleistocene fossils [containing the foot bone of a camel (Camelidae)] made by Copely Amory, Jr, along Old Crow River, Yukon, transferred in 1913, and a skull of the Alaskan horse (*Equus niobrarensis alaskae*) and a

mastodon (*Mammut americanum*) tooth from the Pleistocene of Alaska presented by C.P. Snyder; (9) a well-preserved skull and five articulated vertebrae of a steppe bison (*Bison priscus – Bison alleni*) from the Pleistocene of Alaska purchased in 1914; (10) the type skull with lower jaws of the Yukon horse (*Equus lambei*) purchased in 1917; and (11) a well-preserved skull of the Alaskan horse (*Equus niobrarensis alaskae = Equus alaskae*) from the Alaskan Pleistocene collected for the Museum by James Ford in 1933.

472. Gilpin, J.B. (1874): Observations on some fossil bones found in New Brunswick, Dominion of Canada. Proceedings and Transactions of the Nova Scotian Institute of Natural Sciences 3:400–4.
Remains of a small whale were found on the Jacquet River, Baie des Chaleurs, New Brunswick. The author believed that it was a white whale (*Delphinapterus leucas*), but Perkins (1910) suggested that it was a narwhal (*Monodon monoceras*). Fragments of every part of the skeleton were present from the surface down to 3 m below the sand and gravel.

473. Godfrey-Smith, D.I., Grist, A., and Scallion, P. (1997): ESR and OSL geochronology of the East Milford, Nova Scotia, Canada, mastodon locality, and the timing of the Illinoian-Sangamon transition. Atlantic Geology 33(1):61.
The unconsolidated sediments around the mastodon (*Mammut americanum*) fossil were dated at an average of 135,000 ± 15,000 years ago using optical dating (OSL), while the fossil dental enamel was dated at 74,500 ± 6900 years ago using electron spin resonance (ESR). The sediments in which the mastodon foundered were probably deposited during Late Illinoian deglaciation (end of oxygen isotope stage 6), or at the beginning of the Sangamonian interglacial (oxygen isotope stage 5). Pollen data confirm an interglacial climate. The fossil-enamel date indicates death at the end of oxygen isotope stage 5a, when the climatic cooling leading to the Wisconsinan glaciation had already begun. The specimen became mired in a cavity that was already infilled with much older, probably waterlogged sediments. The dosimetric ages are in agreement with independent, nonfinite radiocarbon ages on the mastodon and fossil wood from an overlaying organic horizon. The results confirm that pre-Sangamonian glacial and interglacial sediments are preserved in Nova Scotia. [See also Harington, Grant, and Mott (1993) for a discussion of the similarly aged Hillsborough, New Brunswick, mastodon.]

474. Goetcheus, V.G., and Birks, H.H. (2001): Full-glacial upland tundra vegetation preserved under tephra in the Beringia National Park, Seward Peninsula, Alaska. Quaternary Science Reviews 20(2001): 135–47.
A closed, dry, herb-rich [dominated by *Kobresia myosuroides*, other sedges (*Carex*) and grasses] tundra-grassland with a continuous moss layer, growing on calcareous soil that was continuously supplied with loess, was exposed beneath a blanket of more than 1 m of volcanic ash on Seward Peninsula, Alaska (Figs. 1, 2) in 1968. It represents full-glacial (about 21,500 years old) vegetation of Beringia. This vegetation would have provided range for grazing mammals (Mammalia), whose presence has been confirmed by radiocarbon-dated bones. The ages of large mammal fossils range throughout the full-glacial (LGM), and probably open steppe-like vegetation provided food for them and for smaller mammals. Although the relative roles of small and large mammals in maintaining the *Kobresia* communities cannot be estimated, direct evidence of abundant small-mammal droppings on the soil surface and the discovery of a vole (*Microtus*) nest show that there was a long-term balance between plant growth and grazing in this ecosystem.

475. Goetcheus, V.G., and Hopkins, D.M. (1997): A moment in time: The landscape of the full-glacial Bering Land Bridge at 18,000 years B.P. Beringian Paleoenvironments Workshop (September 20–3, Florissant, Colorado). Program and Abstracts. p. 64.
A vegetated land surface on Seward Peninsula, Alaska, that had been buried by a tephra was dated to 18,000 BP – near the peak cold period of the last (Wisconsinan) glaciation. Abundant macrofossils, pollen, and insect remains indicate the environment was cold and dry. The vegetation was dominated by grasses, sedges, and herbs. The only shrubs found were two species of willow (*Salix*). So the landscape, occupied by caribou (*Rangifer tarandus*), many small animals, and probably some of the megafauna, was covered by a 'dry, meadowy tundra with some steppe-tundra affinities' during the full-glacial.

476. Goldring, W. (1922): The Champlain Sea. New York State Museum Bulletin No. 239-240:153–87.
The author reports lumpfish (*Cyclopterus lumpus*) and stickleback (*Gasterosteus aculeatus*) from the shore of the Ottawa River between Green Creek and Besserer's Wharf near Ottawa, Ontario.

477. Goodwin, H.T. (1993): Patterns of dental variation and evolution in prairie dogs, genus *Cynomys*. *In*: Morphological Change in Quaternary Mammals of North America (R.A. Martin and A.D. Barnosky, eds.). Cambridge University Press. pp. 107–33.
Prairie dogs are large, distinctive ground squirrels of the genus *Cynomys* that occur today on the Great Plains and high plateaus and basins of the Rocky Mountains from southern Saskatchewan to northern Mexico. The author refers Churcher's prairie dog (*Cynomys churcherii*), a new species described by Burns and McGillivray (1989), to the Niobrara prairie dog (*Cynomys niobrarius churcherii*) (p. 109), suggesting that the southern subspecies (*Cynomys niobrarius niobrarius*) probably gave rise to this northern subspecies. A map of fossil sites (Figure 7.2) shows the Hand Hills, Alberta, site as well as a site near Saskatoon, Saskatchewan.

478. Gordon, B.C. (1970): *Bison antiquus* from the Northwest Territories. Arctic 23(2):132–3.
A skull of extinct bison (*Bison bison antiquus* = *Bison antiquus*) was found in alluvial sand and silt at the base of an 18 m bank of the Blackstone River, Northwest Territories (61°5′25″N, 123°0′00″W). Included is a table of bison skull measurements comparing different species.

479. Gordon, B.C. (1976): Migod – 8,000 years of barrenland prehistory. National Museum of Man Mercury Series, Archaeological Survey of Canada Paper No. 56:1–298.
This publication describes the fluctuation of cultures (*Homo sapiens*), caribou (*Rangifer tarandus*), and climate at the Migod archaeological site at Grant Lake on the lower Dubawnt River, Northwest Territories. It was occupied by caribou hunters and fishing peoples for the past 8000 years. Situated at a prominent water crossing and fishing spot, it was first occupied following drainage of Glacial Lake Hyper-Dubawnt by Northern Plano Agate Basin peoples about 8000 BP. Following a 1700-year hiatus, Shield Archaic hunters occupied the site until 3700 BP, moving S to Manitoba after 3500 BP, perhaps in response to a prolonged cold period. The later human history is also discussed.

480. Gordon, B.C. (1996): People of sunlight: People of starlight: Barrenland archaeology in the Northwest Territories of Canada. Canadian Museum of Civilization Mercury Series 154:1–259.
Hunters (*Homo sapiens*) of the Beverly caribou (*Rangifer tarandus*) range of the Northwest Territories have been dependent on and influenced by seasonal migrations for 8000 years. Archaeological sites in this range extend from the northern calving ground, S to the boreal forest. Sites are small near the calving ground, large and stratified at major water crossings near treeline where the herd massed, and small in the forest where the herd dispersed in winter. Analyses of more than 13,000 artifacts from 1002 hunting camps of four major archaeological traditions are analysed and combined with 131 radiocarbon estimates to indicate that Northern Plano people (8000 to 7000 BP) arrived soon after deglaciation and evolved into Shield Archaic (6500 to 3500 BP), which represents forest dwellers during a warm period. Pre-Dorset (3450 to 2650 BP), the only non-Indian culture, came from the N in response to a world-wide climatic deterioration. Taltheilei (2600 to present) appeared when the climate warmed. Late Taltheilei developed into the modern Dene.

481. Gotfredsen, A.B. (1992): Nyt fra Saqqaq-kulturen. Forskning i Grønland/tussat 92:41–5.
According to Bennike (1997, p. 904), the second-oldest polar bear (*Ursus maritimus*) bones in Greenland come from a Saqqaq archaeological site in West Greenland dated to about 2900 BP.

482. Gotfredsen, A.B. (1998): The faunal material of the Saqqaq site Nipisat I, Sisimiut district, West Greenland. *In*: Man, Culture and Environment in Ancient Greenland (J. Arneborg and H.C. Gulløv, eds.). Danish National Museum and Danish Polar Center, Copenhagen, Danish Polar Center Publication No. 4:124–32.
For centuries Nipisat Island (15 km S of Sisimiut in West Greenland) was visited by Saqqaq people in summer and autumn. There they took advantage of the great abundance of seabirds – gulls (Laridae) in particular, which bred on and near the island. Harbour seal (*Phoca vituli-*

na) was the dominant seal. Caribou (*Rangifer tarandus*) played a surprisingly important role in the diet of the inhabitants, and is the dominant game in the younger parts of the midden. The wide variety of species [Table 1 lists at least five species of fishes (Pisces), 24 species of birds (Aves), and 13 species of mammals (Mammalia) based on 65,800 bone fragments] shows that the Saqqaq people had a broad spectrum of resources.

483. Graham, M.A., Wilson, M.C., and Graham, R.W. (1987): Paleoenvironments and mammalian faunas of Montana, southern Alberta and southern Saskatchewan. *In*: Late Quaternary Mammalian Biogeography and Environments of the Great Plains and Prairies (R.W. Graham, H.A. Semken Jr, and M.A. Graham, eds.). Illinois State Museum Scientific Papers 22. pp. 410–59.

This paper reviews, in detail, the fossil record from the area and mentions many published and a few previously unpublished radiocarbon dates. These include: a mammoth (*Mammuthus*) tibia from a terrace deposit along James River, Alberta, dated to 10,240 ± 325 BP (GX-2918D); bones of prongbuck (*Antilocapra americana*) from the Metke site near High River, Alberta, dated to 6930 ± 260 BP (RL-903); and plains bison (*Bison bison bison*) ribs from the Harder site near Saskatoon, Saskatchewan, which dated to 3360 ± 120 BP and 3425 ± 105 BP [no lab numbers provided]. The authors place the fossil discoveries in not only chronological sequence, but also paleoenvironmental context and use them to show evolving climatic changes. Tables list the modern taxa from the two areas (Montana and southern Alberta/southern Saskatchewan) and the Pleistocene taxa by site. Another table lists the radiocarbon data.

484. Graham, R.W. (1990): Evolution of new ecosystems at the end of the Pleistocene. *In*: Megafauna and Man: Discovery of America's Heartland (L.D. Agenbroad, J.I. Mead, and L.W. Nelson, eds.). Mammoth Site of Hot Springs, Inc. Scientific Papers 1:54–60.

Environmental fluctuations of the terminal Pleistocene produced new ecosystems. Changes in plant distributions and subsequent plant–animal interactions resulted from climatic changes. These new plant communities may have been lower in effective nutritional value, thereby lowering the carrying capacity for large herbivores. Reduction of growing season plus greater snow cover for longer periods of time may have created stress in the large herbivores. Pleistocene herbivore communities were altered, shifted geographically, and, in many cases, became extinct. It is difficult to isolate a single cause for the extinction event at the end of the Pleistocene (about 10,000 BP). In North America alone, more than 32 genera of mammals died out. Proponents of both human (*Homo sapiens*) predation (Overkill hypothesis) and climatic schools of thought tend to agree that the extinction of the large carnivores is a result of the demise of the herbivores.

485. Grant, D. (1982): Mammoth skeleton turns up in field. Toronto Globe and Mail, May 10.

In 1981, a farmhand, while ploughing on Terrence Poole's farm (near Rostock, Ontario – about 40 km W of Kitchener), turned up in 1 ft of peat and mud an upper left molar tooth, neck vertebra, the upper part of a front leg, and part of a shoulder blade, as well as a piece of tusk of a mammoth (*Mammuthus*). John McAndrews of the Royal Ontario Museum analysed fossil pollen from the skull, concluding, 'there was abundant spruce pollen ... indicating Ontario was woodland, much like today.' The bones were estimated to be about 11,000 years old. The size and degree of wear on the tooth suggest that it is from an adolescent, about five years old.

486. Grant, D.R. (1991): Surficial geology, Stephenville, Newfoundland. Geological Survey of Canada Map 1737A.

This 1:250,000 scale map of part of Newfoundland includes localities of radiocarbon-dated remains of bowhead whale (*Balaena mysticetus*).

487. Grant, J.A. (1864): On the geology of the Ottawa. Canadian Naturalist 1:424–5.

This paper mentions that the most common fossil found in clay nodules at Green Creek (36 m asl), E of Ottawa, Ontario, is the capelin (*Mallotus villosus*). It has also been found in clay nodules on Chaudière Lake, Quebec (56 m asl), on the Madawaska River (60 m asl), and at Fort Coulonge Lake (101 m asl).

488. Grant, J.A. (1883): On the inferior maxilla of *Phoca groenlandica*, from Green's Creek, Gloucester, Russell Co. Ontario. Royal Society of Canada, Proceedings and Transactions 1:286.

This abstract reports that the lower jaw of a harp seal (*Phoca groenlandica*) was found in a calcareous clay nodule at Green Creek (36 m asl).

489. Grantham, R.G. (1989): Mastodon tusk rare find. Toronto Sunday Star, October 15.
This newspaper article reported the find of a mastodon (*Mammut americanum*) tusk found in Fundy Gypsum Company's quarry near Windsor, Nova Scotia. R.G. Grantham, a curator at the Nova Scotia Museum, Halifax, was interviewed.

490. Grantham, R.G., and Kozera-Gillis, K.A. (1992): The East Milford mastodon dig progress report to May 1, 1992. Nova Scotia Museum, Halifax. 39 pp.
Fragments of bone and four molars of a mastodon (*Mammut americanum*) were found in Milford's Gypsum quarry, between Carroll's Corner, Halifax County, and East Milford, Nova Scotia. The specimens were found in sediments in a gypsum sinkhole. Wood from a peat layer was radiocarbon dated beyond radiocarbon range at >50,000 BP (GSC-1642), so two uranium-thorium dates were taken; they were 84,200 BP (UQT-186) and 84,900 BP (UQT-185). The authors estimate that the mastodon itself was 70,000 to 80,000 years old. Many maps and extensive descriptions of the specimens are included. The authors also list previous finds of mastodons from Nova Scotia: (1) the first one was found on Lower Middle River, Victoria County, Cape Breton Island, about 1883; (2) a tooth was dredged up from George's Bank in 1965; (3) another tooth was found in glacial gravels near St Francis Xavier University in Antigonish; and (4) a complete tusk, a fragment of a second, and bone fragments were found in Fundy Gypsum Company's quarry near Windsor, in 1989.

491. Gravlund, P., Meldgaard, M., Pääbo, S., and Arctander, P. (1998): Polyphyletic origin of the small-bodied High-Arctic subspecies of tundra reindeer (*Rangifer tarandus*). Molecular Phylogenetics and Evolution 10(2):151–9.
Sequence analysis of mtDNA of the three small-bodied High-Arctic subspecies of reindeer or caribou [Peary caribou (*Rangifer tarandus pearyi*) from the Canadian Arctic Islands; East Greenland caribou (*Rangifer tarandus eogroenlandicus*) from East Greenland, extinct since 1900; and the Svalbard reindeer (*Rangifer tarandus platyrhynchus*)] suggests Peary caribou and East Greenland caribou are most closely related, and probably evolved in the High Arctic of North America. Svalbard reindeer probably evolved from Eurasian large-bodied reindeer. Thus the two groups are ecotypes of relatively recent origin that evolved convergently, probably as a result of similar climatic and nutritional conditions.

492. Grayson, D.K. (1984a): Archaeological associations with extinct Pleistocene mammals in North America. Journal of Archaeological Science 11:213–21.
It has been argued that there are insufficient associations between extinct Pleistocene mammalian (Mammalia) genera and human (*Homo sapiens*) activity to support the argument that human predation played a role in causing the extinctions. This argument assumes that there are many records for most or all of the extinct mammals that date to the very end of the Pleistocene. An analysis of a sample of radiocarbon-dated sites (younger than 12,000 BP) with associated extinct mammal fossils suggests that this criticism is weakly based. There are not enough sites (younger than 12,000 BP) containing both human and extinct Pleistocene mammal evidence to support the argument that this lack is due to human overkill.

493. Grayson, D.K. (1984b): Time of extinction and nature of adaptation of the noble marten, *Martes nobilis*. Carnegie Museum of Natural History Special Publication No. 8:233–40.
The noble marten (*Martes nobilis*) is known from at least 12 sites in western United States and Yukon. It is generally thought to have become extinct at the end of the Pleistocene, and to have been adapted to cool, if not boreal, conditions. However, the author mentions two sites (Hidden Cave, Nevada, and Dry Creek Rockshelter in Idaho) that date to 3500–3000 BP, and argues that the noble marten may not have become extinct until the Late Holocene. Furthermore, associated vertebrates suggest that this species was broadly adapted and could not have suffered from competition with the American marten (*Martes americana*) throughout its range. [See Harington (1977a) and Youngman and Schueler (1991).] Why did the noble marten survive the major environmental changes associated with the end of the Pleistocene at about 10,000 BP only to become extinct during Late Holocene time?

494. Griffin, J.B. (1965): Late Quaternary prehistory in the northeastern woodlands. *In*: The Quaternary of the United States (H.E. Wright, Jr, and D.G. Frey, eds.). Princeton University Press, Princeton. pp. 655–67.
The author notes (p. 657) that one of the most interesting facets of the late-glacial fauna in the Great Lakes and St Lawrence River is the presence of marine mammals. Whales (Cetacea), walrus (*Odobenus rosmarus*), and seals (Phocidae) have been found in beach and other deposits in Michigan, Ontario, and Quebec. None has been adequately dated, but the most likely time for their appearance would be during the intrusion of the early Champlain Sea. From there, the marine mammals could have moved into the Great Lakes by the Ottawa–Mattawa Channel during the Algonquin discharge about 11,000 BP. [See Harington 1988.]

495. Grønnow, B., and Meldgaard, M. (1988): Boplads i dybfrost. Naturens Verden 1988:409–40.
According to Bennike (1997, p. 902), bones of arctic hare (*Lepus arcticus*) and ringed seals (*Phoca hispida*) from archaeological sites show that the first people (*Homo sapiens*) arrived in Greenland about 4500 years ago and hunted these species.

496. Gruchy, C.G. (1968): Two Late Quaternary Salmonidae (Pisces) from the Ottawa region, Canada. Acta Societatis Zoologicae Bohemoslovacae 32(4):337–41.
Two Late Quaternary salmonids were found in clay nodules at Besserer's Springs, 8 km E of Ottawa, Ontario. The first was identified only as Salmonidae, and the second was a lake trout (*Cristivomer namaycush*). Also found were: capelin (*Mallotus villosus*), smelt (*Osmerus mordax*), and (?)Atlantic salmon (cf. *Salmo salar*). Photographs of the specimens are included.

497. Gruchy, C.G. (1971): Reidentification of a Pleistocene *'Salmo'* as the first fossil *Gadus* from Canada. Canadian Journal of Zoology 49(4):427–30.
The first Canadian example of a Pleistocene cod (*Gadus*) was found at Rivière des Outardes [formerly Goose River], Canton de Tremblay, Quebec (49°04′N, 68°28′W). Skull bones and vertebrae were present and they were identified as Atlantic cod (*Gadus* cf. *Gadus morhua*).

498. Guilday, J.E. (1963): Pleistocene zoogeography of the lemming, *Dicrostonyx*. Evolution 17:194–7.
The collared lemmings, genus *Dicrostonyx* Gloger, are currently divided into two subgenera. *Misothermus* Hensel contains a single species, the Ungava lemming (*Dicrostonyx hudsonius* Pallas), isolated on the tundra of northern and coastal Ungava from all other *Dicrostonyx*. The subgenus *Dicrostonyx* Gloger contains the remaining species of the genus, the collared lemming (*Dicrostonyx torquatus* Pallas) of the Palaearctic, *Dicrostonyx groenlandicus* (Traill) of the Nearctic, and *Dicrostonyx exsul* G.M. Allen confined to St Lawrence Island in the Bering Strait. The *torquatus-groenlandicus-exsul* group may be conspecific. Using fossil evidence and modern *Dicrostonyx* ranges in North America, the author concludes that the Ungava lemming (*Dicrostonyx hudsonius*) is a relict of a former (pre-Wisconsinan) Holarctic distribution.

499. Guthrie, M.L. (1988): Blue Babe: The Story of a Steppe Bison Mummy from Ice Age Alaska. White Mammoth, Fairbanks. 31 pp.
This well-illustrated popular booklet summarizes much of the information in Dale Guthrie's [Mary Lee's husband] book *Frozen Fauna of the Mammoth Steppe. The Story of Blue Babe* (1990). The story begins in the summer of 1979, with Walter Roman's (a placer miner in the Fairbanks area of Alaska) report of a steppe bison (*Bison priscus*) carcass, and continues with details of collecting, preserving, describing, and casting of a body form for display of the specimen. The author tells of visits to European Paleolithic cave sites where bison depictions were common. She summarizes the history of steppe bison, and remarks on their American lion (*Panthera leo atrox*) predators and on the dissection of the carcass of the 36,000 BP steppe bison male.

500. Guthrie, R.D. (1966a): The extinct wapiti of Alaska and Yukon Territory. Canadian Journal of Zoology 44:47–57.
Pleistocene wapiti (*Cervus elaphus*) are discussed, listing 13 fossil localities in Alaska and the Yukon. A population (large-antlered forms with a comparatively high frequency of supernumerary proximal tines) which existed in the northern part of North America during the Wisconsinan glaciation became extinct at its close. This paper deals with their competition with other ungulates, their geo-

graphic distribution, their chronology, and their historical zoogeography. Other species mentioned include: bison (*Bison bison*), caribou (*Rangifer tarandus*), and moose (*Alces alces*).

501. Guthrie, R.D. (1966b): Pelage of fossil bison – A new osteological index. Journal of Mammalogy 47(4):725–7.
The theory presented here is that the extent of orbital protrusion is directly correlated to the depth of facial display hair. Although there is an area around each eye in which the hair is very short, the facial hair reduces peripheral vison, a problem alleviated by extending the eye out away from the skull. Northern forms with more facial hair for cold protection would be expected to have more protruding orbits. This turns out to be the case. This character also appears to be independent of horn shape and size. The ancient bison (*Bison bison antiquus* = *Bison antiquus*) and the giant bison (*Bison latifrons*), both southern forms, exhibit far less orbital protrusion than the more northern steppe bison (*Bison priscus* = *Bison crassicornis*). The modern bison, both the plains (*Bison bison bison*) and wood (*Bison bison athabascae*) forms, have orbital protrusions very similar to the northern bison rather than to the southern species, supporting the hypothesis that *Bison bison* is a recent invader of the southern part of the continent.

502. Guthrie, R.D. (1966c): Paleontological notes: Bison horn cores – Character choice and systematics. Journal of Paleontology 40(3):738–40.
The relative value of using horncores for taxonomic purposes is discussed. Considering the magnitude of the variation, caution is suggested when basing a phylogeny or identification solely on horncore measurements.

503. Guthrie, R.D. (1967): Differential preservation and recovery of Pleistocene large mammal remains in Alaska. Journal of Paleontology 41(1):243–6.
The author discusses differential preservation of fossil elements. Examples are provided using horse (*Equus*) limb bones and cranial elements of large mammals. Preservation is not random. The size of the element and its composition/durability are the main factors that determine the probability of recovery.

504. Guthrie, R.D. (1968a): Paleoecology of the large-mammal community in interior Alaska during the Late Pleistocene. American Midland Naturalist 79(2):346–63.
This study of four fossil assemblages (Fairbanks Creek 65°04′N, 147°10′W, Engineer Creek 64°57′N, 147°38′W, Cripple Creek 64°49′N, 148°01′W, and Gold Hill 64°51′N, 147°59′W) of large mammals from Late Pleistocene (sediments are mainly Wisconsinan, except for those at Gold Hill, which are mainly pre-Wisconsinan) sediments near Fairbanks, Alaska, emphasizes the structure, composition, habitat, and pattern of subsequent extinctions of the community. All four faunas were composed predominantly of grazers: steppe bison (*Bison priscus*), horse (*Equus*), and woolly mammoth (*Mammuthus primigenius*). Other species mentioned are: brown bear (*Ursus arctos*), moose (*Alces alces),* caribou *(Rangifer tarandus*), muskox (*Ovibos moschatus*), wapiti (*Cervus elaphus*), Dall sheep (*Ovis dalli*), saiga antelope (*Saiga tatarica*), extinct moose-like animal (*Alces latifrons* = *Cervalces alaskensis*), wolf (*Canis lupus*), coyote (*Canis latrans*), American lion (*Panthera leo atrox* = *Felis*), helmeted muskox (*Bootherium bombifrons* = *Symbos* and *Bootherium*), camel (*Camelops*), yak [*Bos* – now considered modern cattle rather than yak], American mastodon (*Mammut americanum* = *Mastodon americanum*), and sabretooth cat [probably the scimitar cat (*Homotherium serum* = *Smilodon*)]. It should be noted that only limb bones were used to make calculations. The high percentage of grazers in the fossil community suggests that interior Alaska was a grassland environment during the Late Pleistocene.

505. Guthrie, R.D. (1968b): Paleoecology of a Late Pleistocene small mammal community from interior Alaska. Arctic 21(4):223–44.
A number of sites in the Fairbanks area were surveyed during the summer of 1965. Additional sites in the Yukon-Tanana upland were sampled but no small mammal fossils were found. The narrow-skulled vole (*Microtus gregalis* = *Microtus miurus)* was found in every major stratigraphic unit ranging in age from Illinoian to Late Wisconsinan. Three other microtine species were present in the assemblage: brown lemming (*Lemmus sibiricus*), collared lemming (*Dicrostonyx torquatus*), and yellow-cheeked vole (*Microtus xanthognathus*). None of the *Lemmus* occurs in the middle zones thought to have been deposited during a full glacial stage. Five of the six *Di-*

crostonyx specimens occur in beds representing Late or Middle-Late Wisconsinan time. In addition to microtines, bones of arctic ground squirrel (*Spermophilus parryii = Citellus undulatus)* were also found. Three species found in the area as fossils (*Spermophilus parryii, Microtus gregalis* and *Dicrostonyx torquatus*) are now absent from the region. It is suggested that after the close of the Wisconsinan, the environment became wetter, and plant communities tended towards woodland. If these changes resulted in a habitat with even more restricted grassy alpine tundra vegetation than at present, it may account for the elimination of those three species of small mammals.

506. Guthrie, R.D. (1970): Bison evolution and zoogeography in North America during the Pleistocene. Quarterly Review of Biology 45(1):1–15.
The fossil record and information about contemporary forms provide evidence that the evolutionary pattern of bison (*Bison*) cannot be interpreted as either a unidirectional decrease in horn size or a series of successive invasions to the New World from the Old. Bison arose in Eurasia and have had a much longer history there than in North America. In spite of this longer history in the Old World, bison have undergone greater evolutionary changes in North America. This can be explained by a different mode and intensity of competition in the New World. The major points presented are: (1) The giant bison (*Bison latifrons*) was a New World product. (2) *Bison priscus* (= *Bison crassicornis*), the steppe bison, appeared early as a Holarctic northern species and remained in that niche until the Late Wisconsinan. (3) Most of the other bison species in the Late Pleistocene were derived indirectly or directly from this widespread northern species. (4) Middle and Late Pleistocene bison can be placed in four species: *Bison priscus*, which can be dated at least as far back as Early Mid-Pleistocene; *Bison latifrons*, which extends back at least to Late Illinoian time [perhaps *Bison latifrons* gave rise to the ancient bison (*Bison bison antiquus*); if so the species *Bison alleni* should be maintained]; *Bison bison antiquus*, which originated during the early to middle part of the Wisconsinan glaciation; and the plains bison (*Bison bison bison*), which was a Late Wisconsinan product. (5) *Bison latifrons* became extinct, at least over most of its range, in pre-Wisconsinan time. *Bison priscus* and *Bison bison antiquus* became extinct in the Late Wisconsinan, and *Bison bison* still exists in relict populations. (6) Two or more species of bison have not occurred sympatrically for extended periods of time. (7) Neither the 'orthogenetic' nor the 'wave' theory adequately accounts for the evolution of bison in North America; rather, the fossils can only be explained by a combination of invasions from Siberia and evolutionary changes that occurred in the new environment.

507. Guthrie, R.D. (1972): Re-creating a vanished world. National Geographic 141(3):295–302.
This popular article gives an idea of the nature of mammalian (Mammalia) life on the steppe-like grasslands of unglaciated Alaska about 12,000 years ago. Of 31 species listed (and shown in a reproduction of a mural by Jay Matternes for the Smithsonian Institution in Washington, D.C.), 11 are now extinct. Most of the species came from Eurasia via a broad land connection (central Beringia) joining it to North America during the last glaciation. The ancient landscape favoured large grazers – especially woolly mammoths (*Mammuthus primigenius*), steppe bison (*Bison priscus*), and horses (*Equus*). Carnivore remains are much rarer – e.g., only one of several hundred bones is from the American lion (*Panthera leo atrox = Panthera atrox*). Some of the most striking finds from ice-age Alaska are carcasses, and two excellent examples are photographs of: a 31,000-year-old mineral-stained partial carcass of a steppe bison collected near Dome Creek; and a 17,000-year-old, hornless, helmeted muskox (*Bootherium bombifrons = Bootherium nivicolens*) carcass. A highlight of the article consists of a series of four illustrations by Jay Matternes showing the steps he takes in restoring flesh and fur to a skeleton of an extinct short-faced bear (*Arctodus simus*) and then fitting the animal into its natural surroundings. [A problem with the often-used mural (wallchart) showing most of the different species together in a small piece of terrain is that, although successful diagrammatically, the landscape is highly overstaffed!]

508. Guthrie, R.D. (1973a): Hot spots for North American Pleistocene fossil mammals, and how they got that way. Plaster Jacket 20:1–13.
This illustrated popular article explains how tar pits [e.g., Rancho La Brea], frozen muck [e.g., Alaska] and melted ice wedges, and sinkholes become foci for fossil accumulation. The value of these preserved assemblages is discussed.

509. Guthrie, R.D. (1973b): Mummified pika (*Ochotona*) carcass and dung pellets from Pleistocene deposits in interior Alaska. Journal of Mammalogy 54(4):970–1.
A mummified pika (*Ochotona princeps collaris*) was collected in 1933 at Chatanika, approximately 30 mi N of Fairbanks (65°06′N, 147°29′W). The specimen, now at the American Museum of Natural History, has some hair remaining around the digits and is flecked with blue vivianite. The occurrence of this alpine species corroborates the theory that the mountain environments extended into the lowlands during glacial times, resulting in a drier, more steppe-like environment throughout the interior basins and the Arctic Plains. The existence of one subspecies of *Ochotona princeps* in Alaska during the Wisconsinan glaciation supports the biogeographic inference that subspeciation was probably due to isolation during glaciation.

510. Guthrie, R.D. (1980): Bison and man in North America. Canadian Journal of Anthropology 1(1): 55–73.
Rapid evolutionary changes in bison (*Bison*) morphology and the frequent occurrence of bison in archaeological sites make them particularly important in North American Quaternary studies. Major morphological trends in bison evolution are discussed and a conceptual model is proposed to account for the changes in bison evolution. Central to this picture is the steppe bison (*Bison priscus*), which dominated the Holarctic throughout the last half of the Pleistocene. It was used as a food item by humans in Eurasia and Beringia but was not a major prey species. In North America, however, bison became the staple meat resource at the end of the last glacial. The increasing density of bison resulted from vegetational change and the extinction of competitor species. Seasonality of available resources was critical in bison adaptations and numbers, which in turn affected the modes of human bison hunting and human (*Homo sapiens*) density in North America. The author is a proponent of the school of taxonomy that believes *Bison preoccidentalis*, *Bison occidentalis*, *Bison latifrons*, *Bison crassicornis*, and the living subspecies *Bison bison athabascae* and *Bison bison bison* are all just forms [subspecies?] of the steppe bison (*Bison priscus*). Some of the morphological data are discussed and related to possible ethology.

511. Guthrie, R.D. (1982): Mammals of the mammoth steppe as paleoenvironmental indicators. *In*: Paleoecology of Beringia (D.M. Hopkins, J.V. Matthews, Jr, C.E. Schweger, and S.B. Young, eds.). Academic Press, Toronto. pp. 307–26.
The reasons for the diversity, dietary specialization, and gigantisms of mammals of the mammoth steppe are examined in an attempt to provide clues to the general paleoenvironment. Mammalian growth patterns suggest a long growing season, a seasonally rich vegetation with very different antiherbivory defences from the present northern vegetation, cold and windy but variable winters with little snow, and warm, dry summer soils which have a deep summer thaw. [See also Vereshchagin and Baryshnikov 1982.] The Mammoth Steppe had a more finely grained 'mosaic' distribution of flora communities than exists today. The habitat was mainly open plains with a wide variety of grasses, and a scattering of riparian and upland short-shrub growth. The different dietary requirements of the mammalian species would allow for a high diversity under such conditions. The general lack of winter snowcover would also allow the spring rains and warmth to penetrate the soils and bring on an early onset of spring vegetation, hence effectively extending the growing season and allowing for greater individual size of the herbivores. The grassland environment and the lack of winter snow cover would have greatly increased the potential carrying capacity of the mammoth steppe. Some time around 14,000–10,000 BP, the climate began to change. The Holocene featured wetter summers and more winter snows. This produced cooler soils, a shorter growing season, an increase in cushion and evergreen plants, limited nutrient availability favouring plants which produce more antiherbivory chemicals, less windswept grasses accessible during the winter, and limited species whose mobility depended on lack of snow cover. The high plateaus of central Asia come close to what the Mammoth Steppe must have been like. This habitat appears to have escaped the dramatic ecological change that occurred in the North and has been relatively stable in the Late Quaternary. It could be considered a Late Pleistocene relict that existed S of the Mammoth Steppe during the last glaciation.

512. Guthrie, R.D. (1983): Paleoecology of the site and its implications for early hunters. *In*: Dry Creek: Ar-

chaeology and Paleoecology of a Late Pleistocene Alaskan Hunting Camp (W.R. Powers, R.D. Guthrie, and J.F. Hoffecker, eds.). National Park Service, Washington, D.C. pp. 211–87.
[See Powers et al. 1983.]

513. Guthrie, R.D. (1984a): Alaskan megabucks, megabulls, and megarams: The issue of Pleistocene gigantism. *In*: Contributions in Quaternary Vertebrate Palaeontology: A Volume in Memorial to John E. Guilday (H.H. Genoways and M.R. Dawson, eds.). Carnegie Museum of Natural History, Special Publication 8:482–510.
Pleistocene large mammals from Alaska appear to be larger than their living counterparts. This seems to be the case for cervids [caribou (*Rangifer*), wapiti (*Cervus*), moose (*Alces*)], bovids [mountain sheep (*Ovis*), muskoxen (*Ovibos*), and bison (*Bison*)]. It is concluded that the primary reason behind such large-bodied, large-horned or -antlered individuals was twofold: (1) they belonged to populations which were kept understocked by high winter mortality and (2) they experienced a long peak in nutrient availability during the growth season. The length of this seasonal peak in nutrient availability is seen as critical to studies of body-size changes. Wild sheep are used as an example and are examined in greater detail than other species. Large nonruminant ungulates seem to respond differently than ruminants to the changing conditions from Mid-Pleistocene to Late Pleistocene through the Holocene. This may be due to their more conservative life histories. Pleistocene ground squirrels (*Spermophilus*) in Alaska seem to have responded differently than ungulates in their body-size changes. Their special life-history features, particularly hibernation and reproduction, may account for this difference.

514. Guthrie, R.D. (1984b): Environmental influences on body size, social organs, population parameters and extinction of Beringian mammals. *In*: Beringia in the Cenozoic Era (V.L. Kontrimavichus, ed.). Amerind Publishing Co. Pvt. Ltd., New Delhi. pp. 389–421.
The author proposes that the variations in 'seasonality' pattern influence mammalian body size, the degree of development of social organs, reproductive rate, population density, and several other aspects of life history. The seasonal bottleneck largely controls density or quantity, and the luxuriance of the growth season largely controls individual condition or quality. [See also Guthrie 1982.] Evidence suggests that the Pleistocene tundra-steppes represented an unparalleled extreme in high-quality mammalian populations. Most species were larger in body size and had better-developed social organs than their counterparts elsewhere. It is suggested that the disappearance of the tundra-steppe environment (and the accompanying extinction of the megafauna) resulted from a decline in wind and a rise in summer temperatures leading to warm humid summers that would encourage the growth of trees, mosses, and marshland. These phenomena are more marked in the early stages of the post-Wisconsinan (post-Würm) than in the preceding interstadial and interglacial periods.

515. Guthrie, R.D. (1984c): Mosaics, allelochemics and nutrients: An ecological theory of Late Pleistocene megafaunal extinctions. *In*: Quaternary Extinctions: A Prehistoric Revolution (P.S. Martin and R.G. Klein, eds.). University of Arizona Press, Tucson. pp. 259–98.
The author discusses his ideas about extinctions of Alaskan Pleistocene mammals. During the Pleistocene a large, complex grassland existed in northwestern North America that stretched across most of northern and central Asia and eastward to cover much of Europe – the Mammoth Steppe. A curious aspect of this Mammoth Steppe is its peculiar mixture of biota without counterpart among modern biomes. Many forms appeared in Alaska that are now associated with southern grasslands: badger (*Taxidea*), ferrets (*Mustela*), and bison (*Bison*). Eurasian forms like saiga antelope (*Saiga*), horses (*Equus*), and lion (*Panthera*) were mixed with North American short-faced bears (*Arctodus*) and camels (*Camelops*). Then there are scimitar cats (*Homotherium*) and helmeted muskoxen (*Bootherium* = *Symbos*). All of these appeared in a heterogeneous mix with large mammals in Alaska today: moose (*Alces*), caribou (*Rangifer*), tundra muskoxen (*Ovibos*), Dall sheep (*Ovis*), and others. At the end of the last glaciation many of these animals became extinct in Alaska, and the complex northern grassland that had existed throughout most of the Pleistocene disappeared. Range for saigas, horses, and mammoths, etc. contracted and those species died out, while bison, wapi-

ti (*Cervus*), badgers, etc. contracted southward. If climatically induced ecological changes are responsible for such a phenomenon, a central biotic explanation is likely. In addition to a new hypothesis accounting for Pleistocene extinctions, the author presents a model for the network of ecological events near the end of the Pleistocene – events of which megafaunal extinctions were but a part. The model follows: (1) Climatic changes in seasonal regimes decreased diversity, increased plant zonation, and caused a shift in net antiherbivory defence strategies. (2) This change in plant community resulted in a shorter, less diverse growing season for ungulates, decreasing net annual quality and quantity of resources available for many large mammals. (3) Restriction in available resources decreased local faunal diversity, body size, ranges, and frequently resulted in extinction. The author argues that the Pleistocene/Holocene change was an unparalleled jolt across a threshold toward more constricted growth seasons that sharply reduced resources available for ungulate growth. The results: (1) breakdown of many complex plant associations ('plaids') into homogenous zones ('stripes'); (2) a resulting decline in the large-mammal biomass; (3) a general dwarfing of large mammals in the Holocene with related decrease in size of antlers, tusks, horns, etc.; (4) a separation of species that formerly lived together; (5) many large and small mammal populations were reduced (regional extinctions); (6) due to earlier adaptations to plant diversity and antiherbivory defences, some ungulates were affected by all-out changes more than others. Some members of the Holarctic megafauna expanded to dominate the new plant zones: caribou (*Rangifer*) in northern tundra; moose (*Alces*) in the boreal forest; deer (*Odocoileus* and *Capreolus*) in the deciduous forest; bison (*Bison*) in the parkland and plains; wapiti (*Cervus*) as a filler between these species; and brown bear (*Ursus arctos*) as a predator-scavenger. Above all, the shift at the beginning of the Holocene occurred with rapidity. The author concludes that the present can be used to understand the past only with sensitive discretion. Much of the past may have no modern analogue.

516. Guthrie, R.D. (1985): Woolly arguments against the mammoth steppe – A new look at the palynological data. Quarterly Review of Archaeology 6(3):9–16.
The author argues against images of a relatively barren landscape for Beringia during the last glaciation developed by Colinvaux and West (1984), Colinvaux (1980), Ritchie (1984), Ritchie and Cwynar (1982), and Cwynar and Ritchie (1980). He makes the following points: (1) Herb pollen influx analysis indicates a grassy peak during the last glacial maximum. (2) Radiocarbon dates show that dominant members of the mammoth fauna [bison (*Bison*), horse (*Equus*) and mammoth (*Mammuthus*)] were present continuously in the far north during the Late Pleistocene (i.e., the last interstadial, full glacial peak and late glacial). (3) There are, in the Beringian fossil record, flora and fauna associated with more xeric environments farther S. (4) Pleistocene large-mammal communities of Eastern Beringia were the easternmost part of similar mammalian communities that extended across Eurasia to the Atlantic coast, and evidence from that faunal complex suggests a similarly cold, dry environment dominated by grasses. (5) The mammals (particularly large ones) were grazing specialists. (6) Foot morphology indicates that they were adapted to a firm substrate, unlike the moister softer terrain of the area today. (7) Body size indicates that animals were not on a marginal range, but could feed on much high-quality forage during the summer. (8) Fossils of most members of this large-mammal fauna are widespread in Beringia. (9) Most of these large grazing mammals would have been unable to tolerate present snow depths, and had local access to adequate winter range, making possible greater diversity and density of large mammals than exists today. Three tables of radiocarbon dates on bison [(*Bison*), 34 dates], horse [(*Equus*), 26 dates] and mammoth [(*Mammuthus*) 14 dates] from Alaska and Yukon are provided.

517. Guthrie, R.D. (1988): Bone litter from an Alaskan Pleistocene carnivore den. Current Research in the Pleistocene 5:69–71.
The accumulation of bones within a Pleistocene-aged den in a loess cliff overlooking the Tanana River (64°18′N, 146°15′W), central Alaska, has been attributed primarily to the short-faced bear (*Arctodus simus*), although probably several other carnivores used the site over many years. Species recovered from the partly destroyed cave are all members of the Mammoth Steppe fauna: snails (*Succinea*), steppe bison (*Bison priscus*), horse (*Equus*), woolly mammoth (*Mammuthus primigenius*), caribou (*Rangifer tarandus*), and wolverine (*Gulo gulo*).

518. Guthrie, R.D. (1990a): Frozen Fauna of the

Mammoth Steppe. The Story of Blue Babe. University of Chicago Press, Chicago. 323 pp.
This book uses the story of 'Blue Babe,' a mummified steppe bison (*Bison priscus*) evidently killed by American lions (*Panthera leo atrox*) about 36,000 BP and discovered by placer miners near Pearl Creek, Alaska, to highlight knowledge about other such long-frozen ice age mammals from Alaska and Siberia. This subject is also used as an introduction to current information on the paleoenvironment and biota of the now-defunct 'Mammoth Steppe' [cool, dry grassy plains that supported the Pleistocene Holarctic megafauna, including horses (*Equus*), woolly mammoths (*Mammuthus primigenius*), and steppe bison (*Bison priscus*)] as well as the evolution of bison through time. Four tables of radiocarbon dates [Table 9.1, Chronological Distribution of Alaskan Pleistocene Bison (34 dates); Table 9.2, Chronological Distribution of the Horse (26 dates from Alaska and Yukon); Table 9.3, Chronological Distribution of the Mammoth (13 dates from Alaska, Yukon, and Northwest Territories); and Table 9.4, Chronological Distribution of the Ground Squirrel (13 dates from Alaska and Yukon)] show that the species concerned lived in Eastern Beringia throughout the cold peak of the last glaciation. [See Guthrie 1985.]

519. Guthrie, R.D. (1990b): Late Pleistocene faunal revolution: A new perspective on the extinction debate. *In*: Megafauna and Man: Discovery of America's Heartland (L.D. Agenbroad, J.I. Mead, and L.W. Nelson, eds.). Mammoth Site of Hot Springs, South Dakota, Inc. Scientific Papers 1:42–53.
This paper focuses on extraordinary faunal events – specifically megafaunal extinctions and their possible causes beginning about 12,000 years ago, near the end of the Birch Period. This period was marked by rapid evolutionary change, fractionation of biotic communities, and great range reductions – changes unparalleled in the last 400,000 years. The author argues that these changes were driven by climatic and vegetational change. He examines the impact on Late Pleistocene mammals of the Eurasian Mammoth Steppe and their cousins on the American Great Plains. The Mammoth Steppe (Figure 2) was dominated by large grazers – e.g., saigas (*Saiga tatarica*), horses (*Equus*), and woolly mammoths (*Mammuthus primigenius*) – and arid, steppe-like vegetation, but included more mesic-adapted species like caribou (*Rangifer tarandus*) and tundra muskoxen (*Ovibos moschatus*). These complex mammalian communities became even more closely related when a large 'tail' of the mammoth steppe biome was pinched off during the last glaciation and drawn into the northern Great Plains (Figures 3 and 4). Species now associated with northwestern North America reaching the Great Plains included: arctic ground squirrel (*Spermophilus parryii*), tundra muskox (*Ovibos moschatus*), caribou (*Rangifer tarandus*), collared lemming (*Dicrostonyx torquatus*), brown lemming (*Lemmus sibiricus*), arctic shrew (*Sorex arcticus*), yellow-cheeked vole (*Microtus xanthognathus*), singing vole (*Microtus miurus*), northern bog lemming (*Synaptomys borealis*), least weasel (*Mustela rixosa*), ermine (*Mustela erminia*), and some extinct mammals like woolly mammoths (*Mammuthus primigenius*). Other Mammoth Steppe species had reached the Great Plains earlier in the Pleistocene – e.g., lion (*Panthera leo*), sheep (*Ovis canadensis*), wapiti (*Cervus elaphus*), wolf (*Canis lupus*), red fox (*Vulpes vulpes*), black-footed ferret (*Mustela nigripes*), Columbian mammoth (*Mammuthus columbi*), and ancient bison (*Bison bison antiquus* = *Bison antiquus*). Some species from the Great Plains moved N, including: short-faced bears (*Arctodus simus*), helmeted muskox (*Bootherium bombifrons*), western camel (*Camelops hesternus*), and badger (*Taxidea taxus*). The author argues that the extinction list is inflated. There are no extinct Pleistocene bison. Other types of inflation relate to: mountain sheep (*Ovis canadensis catclawensis* = *Ovis catclawensis*), armadillo (*Dasypus bellus*), American lion (*Panthera leo atrox* = *Panthera atrox*), and Harrington's mountain goat (*Oreamnos harringtoni*). By setting extinctions aside, the extent of these ecologically driven faunal events during the Late Pleistocene becomes strikingly apparent. [See Guthrie 1984c.]

520. Guthrie, R.D. (1990c): New dates on Alaskan Quaternary dogs and wolves. Current Research in the Pleistocene 7:109–10.
Wolves (*Canis lupus*) were domesticated in Eurasia before 12,000 BP. Clearly, dogs (*Canis familiaris*) accompanied people (*Homo sapiens*) to the New World rather early, as dog fossils occur in Paleoindian sites. In an attempt to discover the age of dog skulls collected for the American Museum of Natural History from gold-mining exposures in interior Alaska, several were AMS radiocarbon dated: F:AM-688 70932 from Fairbanks Creek, 105

± 50 BP (AA-3900); F:AM-672 8171 from the Fairbanks region, 355 ± 60 BP (AA-3901); F:AM-82 1006 from Ester Pit, 140 ± 80 BP (AA-3903); F:AM-199 8698 from Ester Creek, 490 ± 50 BP (AA-3904 [erroneously listed as 3903]); F:AM-30434 from Cleary Creek, 600 ± 70 BP (AA-3905); F:AM-30473 from the Fairbanks area, 330 ± 70 BP (AA-3907); F:AM-200 4345 from Little Eldorado Creek, 130 ± 50 BP (AA-3908); F:AM-197 8697 from Ester Creek (New Cut), 230 ± 55 BP (AA-3909); F:AM-340 1038 from Little Eldorado Creek, 140 ± 80 BP (AA-3910); F:AM-482 2548 from Cripple Creek, 370 ± 105 BP (AA-3911). These dates show that recent dog skulls were mixed with Pleistocene bones. A wolf skull, identically stained and from the same collection (F:AM-295 5219) from Fairbanks Creek dated by the same technique yielded an age near the LGM, 18,610 ± 165 BP (AA-3912). Although there are many undated wolf fossils, this is the first wolf skull dated from Alaska.

521. Guthrie, R.D. (1990d): New dates on Alaskan Quaternary moose *Cervalces – Alces* – Archaeological, evolutionary, and ecological implications. Current Research in the Pleistocene 7:111–12.
Moose-like animals in the form of *Alces gallicus* (=*Cervalces gallicus*) first appeared in the Eurasian fossil record during the Early Pleistocene. The New World Scott's moose (*Alces scotti* = *Cervalces scotti*), distinguished by its bizarre, convoluted antler palms, diverged from the simpler-palmed Old World giant moose (*Alces latifrons* = *Cervalces latifrons*) while retaining other early *Alces* features: e.g., broad occiput, long antler beam, and long nasal bones. During the last glaciation, Eurasian giant moose underwent rapid evolutionary changes toward modern moose (*Alces alces*). Scott's moose became extinct at the end of the Pleistocene. Perhaps the giant moose entered the New World via Alaska during isotope stage 3, but Scott's moose, which evolved in the heart of North America, never spread to Alaska. The author provides radiocarbon dates on some of the giant moose and moose specimens found near Fairbanks, Alaska. Three *Alces latifrons* carcasses date from the Boutellier Interval (isotope stage 3): F:AM-274 4002, yearling bull from Little Eldorado Creek, 33,000 ± 750 BP (AA-3897 on skin); F:AM-274 4001, young cow from Little Eldorado Creek, 25,330 ± 200 BP (AA-3896 on skin); UA V64, young bull from Livengood, 32,040 +870/–980 BP (DIC-3090 on skin). However, three 'fossil' antlers of (*Alces alces*) are of Holocene age: F:AM-2039 (Box 76) from the Fairbanks region, 8740 ± 70 BP (DIC-2414 on antler collagen) – the oldest moose yet recorded from Alaska, also showing human (*Homo sapiens*) use in the form of cut-marks; F:AM-539 from Goldstream Creek, 740 ± 40 BP (DIC-2415 on antler collagen) – with human cut-marks; and uncatalogued from Pearl Creek, 5380 ± 55 BP (DIC-2416 on collagen from a large antler collected by the author). It is notable, so far, that: (1) there are no dates from the LGM (isotope stage 2); (2) transitional giant moose date from the Boutellier Interval (isotope stage 3); and (3) modern moose are confined to the Holocene. Evidently the modern moose evolved rapidly in mid-latitude Asia and accompanied Asiatic peoples moving into Beringia at the end of the last glaciation and thence S into the New World.

522. Guthrie, R.D. (1991): New paleoecological and paleoethological information on the extinct helmeted muskoxen from Alaska. Annales Zoologici Fennici 28(3-4):175–86.
A newly discovered left hornsheath of an adult female helmeted muskox (*Bootherium bombifrons*) recovered from a placer gold mine on Lilian Creek, Alaska (65°30'N, 148°30'W), is described and illustrated. This is the first complete female hornsheath described for that species. The horn shape is compared with *Ovibos* and male *Bootherium*. The different aspect of the horn compared with *Ovibos* suggests a different behavioural emphasis in its use. New information on *Bootherium* diet is also presented, based on remains of ingested food and possible faecal pellets, which indicates that this species is an eclectic feeder, capable of feeding in woodlands but also able to utilize the more xeric grassland vegetation. Comparisons are made between *Ovibos* and *Bootherium* dietary niches.

523. Guthrie, R.D. (1994): The fossil record of habitat alteration and intercontinental migration. American Quaternary Association, 13th Biennial Meeting (June 19–22, Minneapolis, Minnesota). Program and Abstracts. p. 12.
The Mammoth Steppe expanded and contracted, more or less in rhythm with Milankovitch cycles. During cold/arid extremes the outliers of the Mammoth Steppe reached westward to the Atlantic and eastward to Alaska and Yukon. The author hypothesizes that northern mam-

mals of the Mammoth Steppe, because of intense seasons of food abundance and scarcity, produced high-fat milk for their young, and that windy, cold conditions forced most of the species to conserve energy (e.g., by thick pelage, reduced ears). Because of a need to eat more food faster, fossil molar teeth are a proxy record of dietary pressures – thus the trend to increasing tooth complexity and durability. Most evolution on the Mammoth Steppe occurred within species. The northern Pleistocene faunas are markedly similar: woolly mammoth (*Mammuthus primigenius*), steppe bison (*Bison priscus*), caballine horse (*Equus*), reindeer/caribou (*Rangifer tarandus*), wolverine (*Gulo gulo*), lemmings (*Dicrostonyx* and *Lemmus*), voles (*Microtus*), and ground squirrels (*Spermophilus*).

524. Guthrie, R.D. (1995a): Landscape change and the human colonization of Alaska. American Association for the Advancement of Science, Proceedings of the 46th Arctic Division Science Conference (September 19–22, Fairbanks, Alaska). p. 6.
The author hypothesizes that the treeless Mammoth Steppe of northern Asia prevented Late Pleistocene human (*Homo sapiens*) populations from reaching Alaska. These northern limits were broached at the warming phases of the last glaciation by Mongoloids that had adapted to living in harsh landscapes during the extremes of that period. Like other mammals that experienced the harsh Late Pleistocene climates, they had evolved rapidly, aiding their northern spread once climates began to ameliorate. Transformations in northern landscapes during the Pleistocene/Holocene shift were important to the whole biota, including people. Extinctions, cloud cover, tree line, tussock tundra, moose, and mammoths are all part of the story.

525. Guthrie, R.D. (1995b): Mammalian evolution in response to the Pleistocene-Holocene transition and the break-up of the Mammoth Steppe: Two case studies. Acta Zoologica Cracovia 38(1):139-54.
The Pleistocene-Holocene boundary (13,000–9000 BP, centred on 12,000 BP) seems to be one of the most informative periods for understanding rapid evolution in response to climatic change. This was a time of great faunal revolution for large mammals (Mammalia) in particular, with great distributional changes, rapid evolution, extinctions, and radical changes in community structure. When the Mammoth Steppe (the arid steppe dominant from Alaska to England) broke up, some of the northern megafauna, like the woolly mammoth (*Mammuthus primigenius*), suffered from the reduction of solid footing and disappearing grasses and forbs. Other species, like moose (*Alces alces*), were favoured by the climatic amelioration, resulting in growth of forests and wetlands. These two species are emblematic of the varying results of climatic change. People (*Homo sapiens*) moving into the N during this period complicate our assessment of such interactions between climate and mammals, but new radiocarbon dates on moose and mammoth suggest that interaction between people and large mammals, while significant, was not the chief cause of the faunal revolution at this critical period. Guthrie thinks that the moose was the only wild mammal that accompanied people as they dispersed into the New World (p. 143). A theoretical model of moose, giant moose (*Alces latifrons = Cervalces latifrons*), and Scott's moose (*Alces scotti = Cervalces scotti*) distribution, extinction, and evolution relative to Late Pleistocene (35,000–11,000 BP) climatic change is presented (Fig. 1). The blitzkrieg model of woolly mammoth decline and extinction as caused by people from Eurasia (Fig. 2) and the climatic-ecological model of woolly mammoth decline and extinction due to changes in plant communities (Fig. 3) during the 15,000–5000 BP period are also presented.

526. Guthrie, R.D. (1996): Four Late-Pleistocene large-mammal localities in interior Alaska. *In*: American Beginnings: The Prehistory and Palaeoecology of Beringia (F.H. West, ed.). University of Chicago Press, Chicago and London. pp. 119–29.
The author discusses four collections of large Pleistocene mammal remains in Alaska's interior made between the 1930s to the 1950s by O.W. Geist for the Frick Laboratory of the American Museum of Natural History in New York. This material is abridged by the editors from 'Paleoecology of the large-mammal community in Interior Alaska during the Late Pleistocene' [see Guthrie 1968a].

527. Guthrie, R.D. (1997a): Gold Hill: Fossil mammals in loess deposits in Beringia. *In*: Terrestrial Paleoenvironmental Studies in Beringia (M.E. Edwards, A.V. Sher, and R.D. Guthrie, eds.). Alaska Quaternary Center, University of Alaska, Fairbanks. pp.79–81.
This paper reports an attempt to extract mammalian fos-

sils from a primary loess deposit at Gold Hill (64°51′N, 147°59′W) near Fairbanks, Alaska. It offers explanations for why fossils are usually rare in these sediments and describes the effort that must be expended to unearth them. Although the techniques are similar to those used for valley-bottom sediments, much larger volumes of silt must be processed. The fossils at this location differ from most other Alaskan fossils in that they come from primary deposits and not reworked sediments and they include species not encountered at any other Alaskan locality [e.g., the scimitar cat (*Homotherium*) and ground sloth (*Megalonyx*)]. Using stratigraphy, paleomagnetic, and tephra-dating techniques, there is now a dated tephra sequence going back to Pliocene times. A highly fossiliferous bed that lies along a major unconformity at the uppermost paleomagnetic reversal appears to date from about 200,000 BP, or even later, judging from the relatively complex collared lemming (*Dicrostonyx*) molars.

528. Guthrie, R.D. (1997b): Origin and causes of the Mammoth Steppe: A story of cloud cover, woolly mammals, and the little buckle. Beringian Paleoenvironments Workshop (September 20–3, Florissant, Colorado). Program and Abstracts. p. 67.

The eastward extent of the Mammoth Steppe was partially interrupted at Bering Strait. Even during glacial maxima a load of intermittent maritime cloud cover, created by the narrow bite on either side of Bering Strait, produced an ecological bottleneck or 'buckle' in the extensive steppe belt. This special habitat became a minor refuge for some mesic plants and animals. While this mesic bottleneck did not act as an ecological barrier to some steppe-adapted species like woolly mammoths (*Mammuthus primigenius*), ferrets [e.g., Beringian ferret (*Mustela eversmanni*)], saiga antelopes (*Saiga tatarica*), steppe bison (*Bison priscus*), and horses (*Equus*), it seems to have checked the spread of woolly rhinoceroses (*Coelodonta antquitatis*), camels (*Camelops hesternus*), short-faced bears (*Arctodus simus*), American badgers (*Taxidea taxus*), and so on. On either side of this narrow mesic band during full-glacial time was the broad belt of complex and heterogeneous, arid-adapted communities of the Mammoth Steppe, which virtually circled the globe.

529. Guthrie, R.D. (2001): Origin and causes of the mammoth steppe: A story of cloud cover, woolly mammoth tooth pits, buckles, and inside-out Beringia. Quaternary Science Reviews 20(2001): 549–74.

The author proposes that frequent clear skies was the most important factor causing the vastness of northern arid steppes during Pleistocene glacial episodes. Conversely, low maritime cloud cover best accounts for today's tundra, and the cloudy Polar Front for the taiga zone. Even during glacial maxima, the nearness of the sea to the Bering Isthmus created intermittent maritime cloud cover. This regional cloud cover interrupted (or caused a 'buckle' in) the arid steppe belt (Fig. 8). While this Beringian mesic buckle was not an intercontinental paleoenvironmental barrier to most steppe-adapted species [e.g., saiga antelope (*Saiga tatarica*), woolly mammoth (*Mammuthus primigenius*), and caballoid horses], it seems to have limited the spread of woolly rhinoceroses (*Coelodonta antiquitatis*), western camels (*Camelops hesternus*), American long-legged kiang-like horses (*Equus*), giant short-faced bears (*Arctodus simus*), American badgers (*Taxidea taxus* = *Taxus*), and helmeted muskoxen (*Bootherium bombifrons*). By studying plant remnants preserved in small enamel pits in the teeth of steppe bison (*Bison priscus*), helmeted muskoxen, caribou (*Rangifer tarandus*), Pleistocene horses, woolly rhinoceroses, and moose (*Alces alces* – of Holocene age only), as well as recording stomach contents of frozen woolly mammoths, the author has created pie diagrams suggesting the nature of their diets on the Mammoth Steppe. This study confirms the view that the 'grazers' were indeed using grasses as a staple diet (Fig. 12). Relatively high specimen diversity, large individual body size, and wide distribution all suggest that the 'mammoth fauna' was not uncommon. [See Guthrie 1997b.]

530. Guthrie, R.D., and Greenwalt, A. (1988): Saw marks on Pleistocene megafaunal remains from Alaska. Current Research in the Pleistocene 5:109–11.

The authors provide several examples of cut marks and saw marks that could easily be considered proof of Pleistocene humans (*Homo sapiens*) without closer scrutiny. All the examples are stained over the saw marks, and hence the alteration of the bones appears to be made while the bone was fresh. In one example the bones are of modern cattle (*Bos*); in another, the marks on a moose (*Alces alces*) antler radiocarbon dated at 749 ± 40 BP may represent alteration prior to European colonization;

in yet another the authors propose that a mammoth (*Mammuthus primigenius*) tooth radiocarbon dated to 32,340 + 1070/–1250 BP, from the Ikpikpuk River in northern Alaska, had its roots sawn off by modern natives probably for use as an ivory substitute.

531. Guthrie, R.D., and Guthrie, M.L. (1990): On the mammoth's dusty trail. Natural History 99(7):34–41.
This popular article features several colour restorations of the Mammoth Steppe fauna and its environment. A generalized discussion follows which highlights the appearance, behaviour, and diet of some of the more spectacular members of the fauna such as woolly mammoth (*Mammuthus primigenius*), steppe bison (*Bison priscus*), horses (*Equus*), and lions (*Panthera leo atrox*). The overall paleoecology of the region and how it affected the preservation of the fossil remains is explained using some of the more famous mummified specimens such as 'Blue Babe' (a Mid-Wisconsinan steppe bison) as examples.

532. Guthrie, R.D., and Matthews, J.V., Jr (1971): The Cape Deceit fauna – Early Pleistocene mammalian assemblage from the Alaskan Arctic. Quaternary Research 1:474–510.
Pleistocene exposures near Cape Deceit, Alaska (66°05′N, 162°50′W) contain a complex sequence of organic sediments which have yielded fossil vertebrates of taxonomic, evolutionary, and zoogeographic significance. Species found in the Cape Deceit Formation – lowermost of the three major stratigraphic units recognized at the site – include: wolf (*Canis*), caribou (*Rangifer*), wapiti (*Cervus* cf. *Cervus elaphus*), giant pika (*Ochotona whartoni* sp. nov.), brown lemming (*Lemmus* cf. *Lemmus sibiricus*), Hopkins' lemming (*Predicrostonyx hopkinsi* gen. sp. nov.), Cape Deceit vole (*Microtus deceitensis* sp. nov.), Deering vole (*Pliomys deeringensis* sp. nov.), marmot (*Marmota*), ground squirrel (*Spermophilus = Citellus*), shrew (*Sorex*), and horse (*Equus*). The stratigraphy of the sediments and evolutionary stage of the fossils indicates the age of the site to be at least pre-Cromerian. This is then the earliest North American record of several of the included genera in addition to the only North American record of *Pliomys*. This site records the oldest-recognized Eastern Beringian land mammal fauna to date. Other stratigraphic features reveal that the area was treeless tundra at that time, showing that the contemporary tundra mammal fauna has a lengthy arctic-adapted evolutionary history.

533. Guthrie, R.D., and Stoker, S. (1990): Paleoecological significance of mummified remains of Pleistocene horses from the North Slope of the Brooks Range, Alaska. Arctic 43:267–74.
During the summer of 1983, a skull, radius, and forefoot with hoof, all with tissues still attached, of a horse (*Equus*) was recovered from the frozen Pleistocene silt along the Titaluk River at 69°35′N, 156°35′W. This specimen was radiocarbon dated to 17,190 ± 240 BP (DIC-2418) from hoof keratin. Other horse fossils from the general area (Ikpikpuk River) provided dates of >40,000 BP (I-9320); >40,000 BP (I-9319); 32,270 ± 1500 BP (I-9275); 23,910 ± 470 BP (I-9318); 20,810 ± 410 BP (I-9274); and 19,250 ± 360 BP (I-9371). These radiocarbon dates from horse fossils found on the North Slope of Alaska show that horses lived there during the last peak glacial (Duvanny Yar Interval, Marine Isotope Stage 2). Some previous paleoecological studies have assumed the region's climate during that period was too extreme for large mammals. Hundreds of Pleistocene horse bones have been found on the North Slope, where horse is more commonly found than bison, the opposite of ratios in interior Alaska. The structure of the Titaluk hoof suggests that the animal survived on winter-range characterized by low snowfall and/or snow removal by wind and low forage quality but high quantity. The rate of hoof growth and wear suggests a relatively sedentary existence and little digging through snow for food. Bones with mummified soft tissue may have been buried and preserved by wind-drifted eolian silt.

534. Guthrie, R.D., Sher, A.V., and Harington, C.R. (2001): New radiocarbon dates on saiga antelopes (*Saiga tatarica*) from Alaska, Canada, and Siberia: Their paleoecological significance. *In*: Zooarchaeology in the North: People and Animals in Alaska, the Arctic and Northwest North America; A Festschrift in Honor of R. Dale Guthrie (S.C. Gerlach and M. Murray, eds.). British Archaeological Reports. International Series 944:50–7.
Saiga antelopes, now native to central Asia, were able to spread westward into Europe as far as southwestern England and eastward to the Northwest Territories of Canada during the Pleistocene. Seventeen dates on Beringian saigas reveal two peaks within radiocarbon-dating range,

and these mainly lie outside the maximum cold period of the last glaciation, marine isotope stage 2 (about 25,000 to 15,000 BP). The first cluster of dates corresponds closely to isotope stage 3 (about 40,000 to 25,000 BP), and the second to the juncture between isotope stages 1 and 2 (15,000 to 12,000 BP). Habitat specialization of saigas, reviewed in connection with this new chronology, provides climatic insights as to why saigas were able to colonize Beringia. Also these can account for the species' rarity or absence during peak glacial times and during the Holocene, as well as why, unlike other Beringian ungulates, saigas were never able to colonize southward into the mid-continent of North America.

535. Hall, D.A. (1993): Coprolites give evidence mastodon died accidentally: New Brunswick find dates to before Wisconsin glaciation. Mammoth Trumpet 8(4):6.
One of the most complete mastodon (*Mammut americanum*) specimens in Canada was discovered in 1936 when a sinkhole near Hillsborough, New Brunswick, was being excavated to form a fish pond. Researchers conclude that the mastodon was a young adult that weighed about 8.3 metric tons and dates to the end of the Sangamonian Interglacial. They are confident that spheroidal masses found with the skeleton are coprolites. The condition of the bone indicates that the animal sank into the mud on its right side. Three radiocarbon dates (13,600 BP on bone collagen, >43,000 BP on associated peat, and 51,500 ± 1270 BP on calcite cement in a coprolite) are less valuable geochronologically than pollen analysis. The pollen assemblage is virtually identical to those beds in the vicinity that are thorium/uranium-dated to the end of the Sangamonian Interglacial. Forest similar to that near the northern part of today's boreal forest surrounded the ancient bog where the mastodon became mired. Nine other finds of fragmentary mastodon fossils in the region are mentioned. [See Harington, Grant, and Mott 1993.]

536. Hall, D.A. (1995): Tanana sites connect Alaska with Eurasia. Mammoth Trumpet 10(4):1, 12–13.
Approximately 95 km SE of Fairbanks, Alaska, along the Tanana River, the Broken Mammoth site yielded thousands of animal bones during extensive excavations in 1990–3. Following an interview with D.R. Yesner, team leader of the site study, the author reports that in addition to the bones, lithic tools (microblades), discovered in association with woolly mammoth (*Mammuthus primigenius*) bones, were aged by stratigraphy at 11,700 BP. If this dating is valid, then the Nenana cultural complex had microblade technology despite current theories to the contrary. A number of radiocarbon-dated (AMS) samples verify that the earliest human (*Homo sapiens*) occupation of the site is between 11,000 and 11,800 BP. AMS dating of the mammoth ivory places it at 12,050 BP, placing it at only 300 years older than the earliest human occupation date. This causes some debate that mammoths and humans could have coexisted in the area. Bison (*Bison*) and wapiti (*Cervus elaphus*) are the main species in the faunal record, with caribou (*Rangifer tarandus*), bear (Ursidae), voles (*Microtus*) and lemmings (*Dicrostonyx* and/or *Lemmus*), shrews (Soricidae), ground squirrels (*Spermophilus*), pikas (Ochotonidae), hares (Leporidae), dire wolf [(*Canis dirus*) – identification needs checking], and arctic fox (*Alopex lagopus*) also found. Some bone tools have also been discovered at this site. The analysis of the bone remains is as yet incomplete. Look for papers by Yesner et al. in the future.

537. Hall, D.A. (1999a): Charting the way into the Americas: Following the Pacific Coast to America. Mammoth Trumpet 14(1):1–11.
This article includes results of a Peopling of the Americas Symposium held in Portland, Oregon, in August 1998 – sponsored by the Center for the Study of the First Americans and the Oregon Museum of Sciences and Industry. The Pacific Rim route for humans (*Homo sapiens*) has advanced to a predominant hypothesis that has passed significant tests. Archaeologists described revelations including: (1) retrieving a stone tool (a basalt flake about 10 cm long) from a 10,200-year-old shoreline at Gwaii Haanas identified as a likely archaeological site more than 50 m under the sea (Daryl Fedje of Parks Canada and Heiner Josenhans of the Geological Survey of Canada using Canadian Coast Guard vessels and Canadian Hydrographic Service instruments); (2) finding bones of a man who lived in SE Alaska (Prince of Wales Island) more than 9000 years ago and determining that his diet was almost exclusively marine food (Timothy Heaton, University of South Dakota, and James Dixon, Denver Museum of Natural History); (3) discovering that before 12,000 years ago there was a great expanse of open tundra (a herb zone with dwarf shrubs such as willows) off the coast of British Columbia (Hecate Strait) that is now

under at least a 100 m of sea water (paleoecologist Rolf Mathewes of Simon Fraser University). Knut Fladmark, an archaeologist from Simon Fraser University, stated that while his name 'seems to be fairly commonly associated with the coastal-route hypothesis, I certainly cannot take credit or blame for its invention.' He said the idea may go back as far as Aleš Hrdlicka in 1932, but certainly to Calvin Heusser, a palynologist, who presented evidence for a series of glacial refugia along the northwest coast in his doctoral thesis published in 1960. Fladmark said there is a simple reason for so few coastal sites in the world being older than 8000 or 9000 years: rise in sea levels simply drowned or washed away evidence of nearly all earlier coastal adaptations everywhere – including the northern Pacific coast of North America. He also suggested that drowned coastal areas may have been focal points for cultural innovation and evolution.

538. Hall, D.A. (1999b): Alberta scientists track mammoths across recently exposed landscape. Mammoth Trumpet 14(4):1, 6–10.

The draining of the large St Mary irrigation reservoir near Cardston in southern Alberta, and subsequent wind erosion, have chanced to expose sensational paleontological and archaeological findings. Shayne Tolman, a Cardston school teacher, was the first to collect some of the ice age bones and artifacts. An ancient bison (*Bison bison antiquus* = *Bison antiquus*) skull and helmeted muskox (*Bootherium bombifrons* – a photograph of the back of a cranium is figured) were the first good fossils found, along with the skeleton of an extinct horse (*Equus conversidens*) – portions of five have been found. They also found stone artifacts and tracks – mammoth (*Mammuthus*), horse (*Equus*), and ungulate tracks (there are clear photographs of a mammoth track and trackway and the prints of an unknown ungulate's track and trackway), as well as a camel – probably western camel (*Camelops hesternus*) track. In all, Dr L.V. Hills, of the University of Calgary, states that remains of between 20 and 25 species of vertebrates have been recorded [including two canids (Canidae), badger (*Taxidea taxus*), rabbit (Leporidae), ground squirrels (*Spermophilus*), and birds (Aves)]. Some small stone flakes made by people (*Homo sapiens*) seem to be associated with horse. AMS radiocarbon dates on *Bootherium bombifrons*, *Bison bison antiquus*, and *Equus* yielded the following ages, respectively: 10,980 ± 90 BP, 11,130 ± 70 BP, and 11,330 ± 80 BP from upper to lower layers. The sequence of cultural material extends from Clovis artifacts to those from historic time: three Clovis points, a Folsom point, through other Paleoindian points to a historic metal trade point. Some artifacts were tested for blood residues, to help discover which animals were being hunted by humans.

539. Hamilton, T.D., and Bischoff, J.L. (1984): Uranium-series dating of fossil bones from the Canyon Creek vertebrate locality in central Alaska. *In*: The United States Geological Survey in Alaska: Accomplishments during 1982 (K.M. Reed and S. Bartsch-Winkler, eds.). United States Geological Survey Circular 9:26–9.

A mammoth (*Mammuthus*) cranial fragment and the humerus, metacarpal, and radius of some horses (*Equus*) were found in a road cut near Canyon Creek, 85 km SE of Fairbanks, Alaska, on the Tanana River (map provided). The bones were radiocarbon dated by Weber et al. (1981) at 39,360 ± 1740 BP, but the authors doubted this figure, as tephra found with the assemblages exceeded 56,600 BP (USGS-1253). The bones were aged again using uranium-series dating. The mammoth fragment and the horse metacarpal and humerus yielded an age of 80,000 BP; the latter's radius yielded an age of 36,000 BP, and was believed to have been contaminated. A stratigraphic cross-section and two tables of radiometric dates are included.

540. Hamilton, T.D., Ashley, G.M., Reed, K.M., and Schweger, C.E. (1993): Late Pleistocene vertebrates and other fossils from Epiguruk, northwestern Alaska. Quaternary Research 39:381–9.

Sediments exposed at Epiguruk Bluff, a large cutbank on the Kobuk River about 170 km inland from Kotzebue Sound (about 67°04′N, 158°08′W) record multiple episodes of glacial-age alluviation followed by interstadial down-cutting and formation of paleosols. The identified vertebrates include: woolly mammoth [(*Mammuthus primigenius*); a jaw found at the base of the unit was radiocarbon dated at 23,620 ± 110 BP (USGS-1438)]; bison (*Bison*); horse (*Equus*); caribou (*Rangifer*); an unidentifiable canid (Canidae); arctic ground squirrel (*Spermophilus parryii*); collared lemming (*Dicrostonyx* cf. *Dicrostonyx torquatus*); brown lemming (*Lemmus sibiricus*); and vole (*Microtus*). Based on bone, wood, and peat, the deposit represents material laid down between 37,000 and 14,000 BP. The pollen record indicates that the area supported tundra vegetation with abundant riparian willows during Middle and Late Wisconsinan time. This area would have been a favourable refugium for plants, animals, and possibly humans (*Homo sapiens*) throughout the last glaciation.

541. Hamilton, T.D., Ashley, G.M., Reed, K.M., and Van Etten, D.P. (1984): Stratigraphy and sedimentology of Epiguruk Bluff – A preliminary account. *In*: The United States Geological Survey in Alaska: Accomplishments during 1982 (K.M. Reed and S. Bartsch-Winkler, eds.). United States Geological Survey Circular 939:12–15.
Mammal remains were found *in situ* at Epiguruk Bluff on the Kobuk River 14 km W of Ambler, Alaska (map provided). The species include: woolly mammoth (*Mammuthus primigenius*), mammoth (*Mammuthus*), horse (*Equus* cf. *Equus lambei*), brown lemming (*Lemmus sibiricus*), arctic ground squirrel (*Spermophilus parryii*), collared lemming (*Dicrostonyx*), and vole (*Microtus*). There appear to be no direct dates on these bones and publishing information is lacking as to their stratigraphic context. Molluscs were found beneath the lower paleosol and in thaw-lake sediments. The fauna is consistent with a cool, wet-tundra or riparian environment. A stratigraphic diagram (Figure 8) is included. Radiocarbon dates suggest that the upper paleosol began forming before 35,000 BP and was buried beneath loess and alluvium some 24,000 BP. The lower paleosol was not dated.

542. Hansen, L.A., Jorgensen, M.E., Houmark-Nielsen, M., and Kronborg, C. (1994): Late Pleistocene stratigraphy and depositional environments of the Fynselv area, Jameson Land, East Greenland. Boreas 23:385–97.
The Fynselv area is located on the S coast of Jameson Land (Figure 1). Fragments of marine shells and whale bones [probably from skull fragments of a bowhead whale (*Balaena mysticetus*) according to Jeppe Møhl] were found in Subunit Ic (p. 389, Figures 3, 4, 5), up to 10 m thick consisting of beds of sand interbedded with a few metre-thick heterolithic beds of fine to medium sand and mud. Sedimentary structures indicate a general movement of sediment towards the S (Figure 6). The marine environment is indicated by the presence of marine fossils. Wave ripples indicate that deposition took place above wave-base, and outsized clasts are interpreted as dropstones. Evidently, the whale bones were deposited in a marine delta. Relative sea level was at least 20 m above the present, and Unit I is correlated with the Langelandselv [Sangamonian/Eemian] Interglaciation – a thermoluminescence age on redeposited marine shells and whale bones is given as 121,000 ± 10,000 BP (Figure 4).

543. Harder, P., Jensen, A.S., and Laursen, D. (1949): The marine Quaternary sediments in Disko Bugt. Meddelelser om Grønland 149:85.
Near Zirphaea Pynt (Plate 1), West Greenland, at a section along Ilulialik River, a number of concretions, including one containing a fish (Pisces) vertebra (p. 30), were found on the beach. Evidently this specimen is of Holocene age (Bennike 1997, p. 902). At Sandpynten, in the upper layer of clay, a caribou (*Rangifer tarandus*) atlas was found along with concretions carrying impressions of fishes (Pisces). Near 'Engell's Profil' [Engell's Section] there is a deposit laid down in shallow water modified by a considerable current where a lake near Orpigsûp tasia had a narrow connection to the sea. There, surface finds on the slope include: a fish (Pisces) otolith and vertebra, and an otic capsule (ear bone) of a harbour seal (*Phoca vitulina*) (p. 42). A sample just less than 1 m below the top of the section in the same area yielded two fish vertebrae (p. 43). With loose marine mollusc shells high up on the slope on the north shore of the river and the lake, an axis vertebra of a seal (*Phoca* sp.) and the bones of a cod [*Gadus morhua*?] skull were collected (p. 48). W of Niaqornârusuaq (Figure 16) in Portlandia clay overlying moraine sediment a concretion containing a fish (Pisces) vertebra fragment was collected (p. 70). Along the cliff in the bay E of Niaqornârusuaq, many clay concretions with skeletal parts of fishes (Pisces) and starfish were collected (p. 75). Fish bones were also collected in the *Mytilus* layer in the southern bay at Claushavn (p. 79).

544. Hare, G. (1994): Field report on the nature and extent of Pleistocene bone deposits at Scottie Creek, Yukon. Report prepared for Heritage Branch, Government of Yukon. 22 pp.
In October 1993, road construction crews working on the Shakwak Highway Reconstruction Project encountered many Pleistocene mammal (Mammalia) bones, including those of woolly mammoth (*Mammuthus primigenius*), steppe bison (*Bison priscus*), and Yukon horse (*Equus lambei*) in buried context from road-cuts between Kilo-

metres 1958 and 1961 of the Alaska Highway (see map), near Scottie Creek in western Yukon. According to C.R. Harington, a paleontologist with the Canadian Museum of Nature, this is the most southwesterly ice age mammal site in the Yukon. Yukon Heritage Branch staff visited the site at the time but were unable to document the stratigraphy due to recent snowfall and covering of the road-cut with geotextile cloth. Work at the site resumed in mid-April 1994, and the author revisited the site between April 18 and 22. He reports small quantities of Pleistocene bones from several localities. Most bones are heavily fragmented and appear to have been redeposited, probably through glacial stream action. Where recovered *in situ*, bone deposits are concentrated in the lowest buried organic horizon above fractured bedrock. The depth of this horizon varies from 1.5 to 5 m below the surface depending on the thickness of silt deposits. Most of the work in the area was completed by April 24. The author notes that Pleistocene bone 'will probably be present and available for collecting in waste pits alongside this new portion of the Alaska Highway.' He reports an AMS radiocarbon date of 20,660 ± 100 BP (laboratory number not included) on an *Equus lambei* mandible recovered at Kilometre 1961. Colour photographs show sites at Kilometre 1961 and 1959 where bones were recovered, Plate 6.3 shows a bison bone *in situ*, and Plate 6.9 shows part of a Yukon horse cranium on the surface at Kilometre 1958.5. Plate 6.11 shows 37 bones or bone fragments recovered during the 1994 field visit. [Plate 6.15 shows two horse teeth rather than 'teeth of a probable bison.'] Five selected drill-core logs from near Scottie Creek are appended.

545. Harington, C.R. (1963): Pleistocene mammal research in Canada. Typescript of paper presented at the First Canadian Society of Zoologists Meeting (January 8, Ottawa, Ontario). 56 pp. [Copy in Quaternary Vertebrates of Northern North America Publication File.]

Increased evidence concerning botany, geology, climatology, glaciology, and oceanography of the past million years is leading to a new era of paleoecology where Pleistocene mammals may be thought of as existing in close to their natural environments rather than as bizarre, isolated heaps of bone. Historical activity in Canadian Pleistocene mammal research from 1851 to 1961 is indicated by analyses of major contributions over 20-year intervals. In productivity, the period from 1891 to 1910 was highest, 28 papers having been published. Well over half were written by officers of the Geological Survey of Canada, who made a significant contribution to the study of Pleistocene mammals between 1863 and 1937. Finds of mammoths (*Mammuthus*), mastodons (*Mammut americanum*), and marine mammals have been relatively common in Canada. Fossil bison (*Bison*), horses (*Equus*), and muskoxen (Ovibovini) were evidently discovered more frequently with increased exploration of western and northern regions. So far, about 40 genera and 53 to 60 species of ice age mammals have been reported for Canada. The present status of Pleistocene mammal research in Canada is reviewed and some problems are discussed. Satisfactory answers to our questions about life in this country during the ice age will only result from co-operation of the paleontologist with the geologist, paleobotanist, and archaeologist. Discoveries concerning Pleistocene mammals may provide a valuable perspective on the study of living mammals. Likewise, the mammalian paleontologist can learn much from the mammalogist and ecologist. There are two appendices: Appendix I is a tentative list of Canadian Pleistocene mammals by provinces and territories; Appendix II is a tentative list of Canadian Pleistocene mammals. These are followed by an annotated bibliography of 139 items.

546. Harington, C.R. (1968): A Pleistocene muskox (*Symbos*) from Dease Lake, British Columbia. Canadian Journal of Earth Sciences 5(5):1161–5.

A skull of *Symbos* [first identified by Cowan (1941) as *Ovibos*] was found exposed by the receding floodwaters of a small stream near Dease Lake, British Columbia. There was grey mudstone on the bones, and the eroded fragment is believed to have been transported. The age of the specimen is unknown, but is believed to be between 25,000 and 6000 BP. It is compared to helmeted muskoxen (*Symbos tyrrelli* and *Symbos cavifrons*). Photographs of the specimen are included. Analysis of fecal pellets associated with a well-preserved *Symbos* skeleton (F:AM 4254 from Little Eldorado Creek near Fairbanks, Alaska) suggests that the muskox was feeding on a dry-site type of vegetation with grass and sedge under winter conditions. [See Guthrie 1991.] [All muskoxen described in this annotation are now considered helmeted muskoxen (*Bootherium bombifrons*).]

547. Harington, C.R. (1969): Pleistocene remains of the lion-like cat (*Panthera atrox*) from the Yukon Territory and northern Alaska. Canadian Journal of Earth Sciences 6(5):1277–88.

American lion (*Panthera leo atrox* = *Panthera atrox*) skull and forelimb fragments found at a muck and gravel interface near Dawson City, Yukon Territory, are the first record of the species for Canada. The partial skull of *Panthera leo atrox* from Quartz Creek, Yukon (63°47′N, 139°06′W), is currently in the collections of the Canadian Museum of Nature (CMN) and is numbered 12230. A horncore of steppe bison (*Bison priscus* = *Bison crassicornis*) from this site has provided a radiocarbon date of 30,300 ± 1850 BP (I-3571). A right humerus and an ulna of a lion (cf. *Panthera leo atrox*) were found at Dominion Creek (63°48′N, 130°41′W) and a radiocarbon date of a Yukon horse (*Equus*) with this assemblage yielded a date of 14,870 ± 260 (I-3569). A right maxillary fragment of another lion (cf. *Panthera leo atrox*) was found at the Kaolak River, Alaska (70°N, 160°W). This specimen significantly extends the known northward range of the species. The species has been reported from 26 localities from Peru [but see Seymour 1983, who has removed the lion from the faunal list at Talara, Peru] to Alaska, and may have migrated from Eurasia just prior to the Sangamon interglacial. Photographs of the specimens are included. The author suggests that *Panthera atrox* is conspecific with the Eurasian cave-lion (*Panthera leo spelaea*) and the Chinese lion-like cat (*Panthera youngi*). Other animals were found in the assemblage at Quartz Creek: woolly mammoth (*Mammuthus primigenius*); small horse (*Equus lambei*); caribou (*Rangifer tarandus*); and wolf (*Canis* cf. *Canis lupus*). The assemblage at Dominion Creek included: arctic ground squirrel (*Spermophilus parryii* = *Spermophilus undulatus*); woolly mammoth (*Mammuthus primigenius*); moose (*Alces alces*); caribou (*Rangifer tarandus*), and steppe bison.

548. Harington, C.R. (1970a): A Pleistocene muskox (*Ovibos moschatus*) from gravels of Illinoian age near Nome, Alaska. Canadian Journal of Earth Sciences 7(5):1326–31.

A tundra muskox (*Ovibos moschatus*) cranial fragment was dredged from Illinoian age outwash gravels near Third Beach, Nome, Alaska (64°32′50″N, 165°22′30″W). This is one of the earliest specimens of *Ovibos* from North America, and the author thinks that the specimen is evidence that the tundra muskox first entered North America via Bering Isthmus during the Illinoian glaciation. Photographs and a map are provided.

549. Harington, C.R. (1970b): A postglacial muskox (*Ovibos moschatus*) from Grandview, Manitoba, and comments on zoogeography of *Ovibos*. National Museum of Natural Science. Publications in Paleontology 2:1–13.

A tundra muskox (*Ovibos moschatus*) cranial fragment was found buried 3 m in an alluvial sand and gravel deposit from the S side of the Valley River, 3 mi E of Grandview Manitoba (NE¼ Lsd 10, sec 28, twp 25, rge 23). The fossil provided a radiocarbon date of 8620 ± 190 BP (I-1623). Photographs are included. This specimen is placed in paleogeographic and paleoenvironmental context in a section describing the dispersal history of *Ovibos* (pp. 4–6).

550. Harington, C.R. (1970c): Ice Age mammal research in the Yukon Territory and Alaska. *In*: Early Man and Environments in Northwest North America (R.A. Smith and J.W. Smith, eds.). Produced by the University of Calgary Archaeological Association, published by The Students' Press. pp. 35–51. [Reprinted in 1971 in University of British Columbia Law Review 6(1):59–73.]

The author summarizes chronologically the history of Quaternary mammal research in the areas most productive of fossils in Eastern Beringia – Alaska and Yukon. The areas chosen, with highlights, are: (1) Eschscholtz Bay, Alaska, where the Russian Admiral Kotzebue (1821) was the first to publish an account mentioning ice age mammals from Eastern Beringia. Richardson (1861) later listed remains of six mammalian genera collected by the Admiral. Quackenbush (1909) led an American Museum of Natural History expedition in search of Pleistocene fossils – mainly near Eschscholtz Bay – from which he reported 10 genera and 14 species of Pleistocene mammals. (2) Old Crow area, Yukon, where the first recorded ice age mammal remains were collected along Porcupine River prior to 1873 by the Rev. R. McDonald. Otto Geist (1955) of the University of Alaska collected many vertebrate fossils on a trip up Old Crow River, listing nine species. The author began fieldwork in the area in 1966. Horse (*Equus*), mammoth (*Mammuthus*), and bison (*Bison*) specimens were common, whereas moose (*Alces alces*), giant moose (*Alces latifrons*), caribou (*Rangifer tarandus*), helmeted muskox (*Bootherium*), mastodon (*Mammut americanum*), fox (*Alopex* and *Vulpes*), ground squirrel (*Spermophilus*), fish (Pisces), and bird (Aves) were rarer. Other specimens of interest were scimitar cat (*Homotherium*) and giant beaver (*Castoroides*). A list of 19 Pleistocene mammals from Old Crow Loc. 14N are given in Table I. (3) Dawson City area, Yukon, where G.M. Dawson (1894b and 1901b) mentioned Pleistocene mammal remains from Yukon. In 1904 Obalski visited this area, and mentioned nine Pleistocene mammal species he observed there. L.M. Lambe (1912b) mentioned a large short-faced bear cranium from the area, and listed 14 species of Pleistocene mammals from the Yukon. Among the more interesting specimens collected by the author were: American lion (*Panthera leo atrox* = *Panthera atrox*); wolverine (*Gulo gulo*); wolf (*Canis* cf. *Canis lupus*); a ground squirrel (*Spermophilus parryii* = *Spermophilus undulatus*) nest with bones of an individual. Ten species from Gold Run Creek are listed in Table II. (4) Fairbanks area, Alaska, where large-scale gold mining in 1928 began to show that it was a promising site. A program financed by Childs Frick of the American Museum of Natural History in cooperation with the University of Alaska and mining companies allowed the collecting of thousands of important fossils in the Fairbanks area since then (Péwé 1957). R.D. Guthrie (1968a, b) of the University of Alaska published on both the large mammals and rodents from deposits in that area. It should be noted that the coastal bluff fossil sites near Eschscholtz Bay and along the deeply cut banks of Old Crow River are a result of natural exposure, whereas the Dawson City and Fairbanks sites are exposed in the course of placer mining for gold. Pleistocene mammalogists should work in conjunction with other paleoenvironmental indicators such as plant, shell, beetle, fish, and bird remains.

551. Harington, C.R. (1971a): A Pleistocene lion-like cat (*Panthera atrox*) from Alberta. Canadian Journal of Earth Sciences 8(1):170–4.
A tooth-bearing facial fragment of a lion-like cat (*Panthera leo atrox* = *Panthera atrox*) was found in a gravel pit 61 m above the level of the Red Deer River near Bindloss, Alberta (50°57′N, 110°08′W). The specimen was 1 m below the surface; photographs of it are included. Other Sangamonian-age fossils found here are: Mexican horse (cf. *Equus conversidens*) and an elephant (Proboscidea cf. *Mammuthus*). [See Churcher 1972a.]

552. Harington, C.R. (1971b): A Pleistocene mountain goat from British Columbia and comments on the dispersal history of *Oreamnos*. Canadian Journal of Earth Sciences 8(9):1081–93.
A mountain goat (*Oreamnos*) cranial fragment was found 914 m asl in gravels underlying glacial tills from the Bullion Mine, 6.4 km SW of Quesnel Forks, British Columbia (52°38′N, 121°36′W). The animal is considered to be of Late Sangamonian age, or older. The author discusses the specimen in relation to the dispersal history of *Oreamnos*, and states that despite McConnell's (1905) report of mountain goat from the Klondike, there is no solid evidence of the species from Eastern Beringia. Photographs of the specimen are included. An assemblage of fossils of different ages found in the Cariboo District of British Columbia are represented by: moose (*Alces alces*); bison (*Bison*); woolly mammoth (*Mammuthus primigenius*); Columbian mammoth (*Mammuthus columbi*); caribou (*Rangifer tarandus*); mule-deer (*Odocoileus hemionus*); small horse [*Equus* = *Equus (Asinus)*]; ground sloth (*Megalonyx*); and unidentified rodent (Rodentia) and fish (Pisces) remains.

553. Harington, C.R. (1971c): A postglacial freshwater drum (*Aplodinotus grunniens*) from Ontario, and comments on the zoogeography of the species. Canadian Journal of Earth Sciences 8(9):1137–44.
A fossil freshwater drum (*Aplodinotus grunniens*) was found N of New Liskeard, in northern Ontario in the Clay Belt (former lake bed of Glacial Lake Barlow-Ojibway). This paper includes a map of Canadian occurrences of *Aplodinotus grunniens* and discusses the dispersal history of the group.

554. Harington, C.R. (1971d): Ice age mammals in Canada. Arctic Circular 22(2):66–89.
The author reviews the known Canadian Pleistocene faunal localities and some significant isolated fossils by region. Each regional discussion includes a description of the site and the species found and provides some insight into the possible age and paleoenvironment. Environmental warming and increase in the number of human (*Homo sapiens*) hunters are suggested as possible factors in the extinction of large ice age mammals between 10,000 and 5,000 years ago. [See Harington 1978b and Kurtén and Anderson 1980.]

555. Harington, C.R. (1971e): Part I: The history and environment of the Champlain Sea. *In*: The Champlain Sea and Its Vertebrate Fauna. Trail and Landscape 5(5):137–41.
This paper is a general description of the Champlain Sea Pleistocene fauna and environment. It refers to the following vertebrates: harp seal (*Phoca groenlandica*), white whale (*Delphinapterus leucas*), capelin (*Mallotus villosus*), and sculpin (*Artediellus uncinatus*). [See Harington 1971b.]

556. Harington, C.R. (1972): Part II: Vertebrates of the Champlain Sea. *In*: The Champlain Sea and Its Vertebrate Fauna. Trail and Landscape 6(1):33–9.
This paper refers to the following vertebrates found in clay nodules in Green Creek, near Ottawa, Ontario: lake trout (*Salvelinus namaycush*), smelt (*Osmerus mordax*), three-spined stickleback (*Gasterosteus aculeatus*), sculpin (*Artediellus uncinatus*), lumpfish (*Cyclopterus lumpus*), and capelin (*Mallotus villosus*). Unidentified bird (Aves) bones were also found in Montreal and sands near Uplands Airport (Ottawa International), Ottawa, Ontario.

557. Harington, C.R. (1973a): A short-faced bear from ice age deposits at Lebret, Saskatchewan. Blue Jay 31(1):11–14.
A fragment of the mandible of a giant short-faced bear (*Arctodus* cf. *Arctodus simus*) was found in sandy gravels at Lebret, Saskatchewan. Its age is anywhere from Sangamonian interglacial to Late Wisconsinan. Photographs of the specimen are included. Comparisons are made to the giant short-faced bear (*Arctodus simus*), eastern short faced-bear (*Arctodus pristinus*), North American spectacled bear (*Tremarctos floridanus*), and spectacled bear (*Tremarctos ornatus*).

558. Harington, C.R. (1973b): The ice age mountain goat. The Beaver 304(1):11–13.
A mountain goat (*Oreamnos americanus*) skull fragment collected in 1932 near Quesnel Forks, British Columbia, was found in a gold-bearing gravel layer below some 275 ft of bedded boulder clay and gravel. The stratigraphy related to its burial suggests an age of about 70,000 BP. This age corresponds to an interglacial period when the climate was similar to today's. A discussion follows which details the routes mountain goats may have followed to reach North America and how the species may have survived the glaciations.

559. Harington, C.R. (1974): Animal life in the ice age. Canadian Geographical Journal 88(4):38–43.
This popular article on Canadian ice age mammals focuses on paleoenvironmental evidence derived from the study of several interesting specimens: (1) a nearly complete ringed seal (*Phoca hispida*) skeleton from Champlain Sea deposits near Hull, Quebec, which suggests that fast ice existed near the western margin of the sea during winter and spring some 10,000 years ago; (2) the Columbian mammoth (*Mammuthus* cf. *Mammuthus columbi*) skeleton from the Babine Lake area of central British Columbia evidently occupied a dwarf birch shrub tundra in Mid-Wisconsinan time; (3) a giant beaver (*Castoroides ohioensis*) specimen that lived during relatively warm interglacial conditions in what is now Old Crow Basin, Yukon; (4) a saiga antelope (*Saiga tatarica*) fossil from Baillie Islands, Northwest Territories, which is a good indicator that steppe-like grasslands extended as far E as this toward the close of the last glaciation. Each of the specimens is illustrated and mapped.

560. Harington, C.R. (1975a): Evidence of early man in the Old Crow Basin, Yukon Territory. Arctic Circular 22(3):118–28.
The author reviews the results of his 1966, 1971, and 1973 field seasons in the Old Crow Basin, Yukon. Evidence of human (*Homo sapiens*) presence is indicated by the discovery of several lithic and bone tools. A caribou (*Rangifer tarandus*) tibia modified into a hide scraper has been radiocarbon dated to 27,000 (+3000/-2000) BP. A mammoth (*Mammuthus*) long bone fragment and a radius, both of which had flakes removed by heavy blows when fresh, were 25,750 (+1800/–1500) BP and 29,100 (+3000/–2000) BP respectively. All of the tools are figured. [See Nelson et al. (1986) for a revised AMS date on the caribou artifact.]

561. Harington, C.R. (1975b): A bone tool found with ice age mammal remains near Dawson City, Yukon Territory. Arctic Circular 23(1):2–5.
During the summer of 1973 at Hunker Creek (63°55′N, 138°52′W) near Dawson City, a large bullet-shaped piece of caribou antler was found in association with fossil mammal bones. Specimens of steppe bison (*Bison

priscus = *Bison crassicornis*), woolly mammoth (*Mammuthus primigenius*), Yukon horse (*Equus lambei*), caribou (*Rangifer tarandus*), American lion (*Panthera leo atrox*), and wolf (*Canis lupus*) are known from the same site. The antler tool is estimated to be over 15,000 years old. Artifacts of Late Pleistocene age have been reported from the lower, more northerly Old Crow Basin, but this is the first evidence suggesting the presence of early man (*Homo sapiens*) in this higher region of the Yukon. [See Harington and Morlan 1992.]

562. Harington, C.R. (1975c): A postglacial walrus (*Odobenus rosmarus*) from Bathurst Island, Northwest Territories. Canadian Field-Naturalist 89(3): 249–61.
A cranial fragment with a tusk of a walrus (*Odobenus rosmarus*) was found on central Bathurst Island, Nunavut, 16 km W of Goodsir Inlet (75°43′N, 98°25′W). Photographs are included of the specimen, which was buried in raised-beach deposits 53 m asl. Additional Quaternary walrus records for Canada are listed. The paper also mentions a bowhead whale (*Balaena mysticetus*) bone, found 50 m asl, near Resolute, Cornwallis Island, Nunavut (74°42′N, 94°59′W). The radiocarbon dates for the walrus and whale bones, respectively, are 7320 ± 120 BP [I-7795 – corrected from 'I-7769'] and 7380 ± 140 BP (GSC-1193), indicating that they occupied the area about the same time.

563. Harington, C.R. (1975d): Pleistocene muskoxen (*Symbos*) from Alberta and British Columbia. Canadian Journal of Earth Sciences 12(6):903–19.
Two cranial fragments of helmeted muskox (*Bootherium bombifrons* = *Symbos cavifrons*) were found in Late Pleistocene gravels in Alberta and British Columbia. The first was from 4.8 km SW of Fort Saskatchewan, Alberta, buried 6.4 m deep (53°39′40″N, 113°17′00″W). The other was found on the Saanich Peninsula on Vancouver Island, British Columbia, buried 15.2 m deep (48°33′54″N, 123°25′05″W). A restoration of the helmeted muskox, a map showing these fossil localities in relation to other North American *Bootherium* localities, and photographs of the specimens are included. A list of land mammals and localities on Vancouver Island is given (Table 1a) and includes: American mastodon (*Mammut americanum*), imperial mammoth (*Mammuthus imperator*), Columbian mammoth (*Mammuthus columbi*), horse (*Equus*), bison (*Bison*), and wapiti (*Cervus elaphus*). [But see Kool 1979.] Table 1b lists Pleistocene land mammals of the islands between the mainland and Vancouver Island, including imperial and Columbian mammoths and bison, whereas Table 1c lists Pleistocene land mammals from the adjacent mainland coast including mammoths, wapiti, and bison. The discovery of a fossil helmeted muskox (and remains of other ice age land mammals) on southeastern Vancouver Island suggests that land connected the island with the mainland during the Late Pleistocene. Causes for such a land connection are considered. [See Harington 1976.]

564. Harington, C.R. (1976): Mammoths on Vancouver Island. Daily Colonist ('The Islander'), Victoria, British Columbia, May 16:6-7.
During the last ice age, herds of mammoths (*Mammuthus*) lived on what is now southeastern Vancouver Island, British Columbia. Not only did mammoths live there, but also American mastodons (*Mammut americanum*), horses (*Equus*), helmeted muskoxen (*Bootherium bombifrons* = *Symbos cavifrons*), and bison (*Bison bison*). Most of the 18 fossils of ice age mammals known from Vancouver Island have come from gravel pits in the Saanich Peninsula. Exceptions are mastodon teeth from Shawnigan Lake and Courtenay area, and mammoth specimens from near Courtenay and Sooke. [But see Kool 1979.] Geological evidence gathered from the sequence of sediments at gravel pits on the Saanich Peninsula indicate that the animals died before the peak of the last (Wisconsinan) glaciation some 20,000 years ago. Likely some of the species such as the imperial mammoth (*Mammuthus imperator*) and American mastodon could have crossed large floodplains that filled the Strait of Georgia during the Olympia Interglaciation (>50,000 BP to <20,000 BP). The article is illustrated by restorations of the helmeted muskox, imperial mammoth, American mastodon, a lower molar tooth of the imperial mammoth, and a photograph of the Vancouver Island marmot (*Marmota vancouverensis*), whose ancestor is considered by the author to have reached the island during the second-last (Illinoian) glaciation. [See Harington 1996a.]

565. Harington, C.R. (1977a): Pleistocene mammals of the Yukon Territory. PhD thesis, Department of Zoology, University of Alberta, Edmonton. 1060 pp.
This study is based on a collection of about 14,000 Early,

Middle, and Late Pleistocene fossils from Yukon, mainly from the Dawson City and Old Crow Basin areas. Sixty-four species are represented in the fossil record to date; 40% are now extinct and about 60% no longer occur in Yukon. Paleoenvironmental implications of detailed studies, largely on fossil pollen, plant macrofossils, molluscs, ostracods, and vertebrates of ?Sangamon (>54,000 BP) and Mid-Wisconsinan (32,400 ± 700 BP) ages provide interesting glimpses into Pleistocene Yukon paleoenvironments. An estimated 75% of the species (mainly cold-adapted) were derived from Eurasia or Beringia, while 25% (mainly dry, scrub-grassland species) seem to have been derived from southern North America. Fifty-six of the fossils were radiocarbon dated using bone collagen whenever possible. These dates are included in Appendix II, a list of criteria used for suggesting the geological age of Yukon Pleistocene mammal fossils.

566. Harington, C.R. (1977b): Marine mammals in the Champlain Sea and the Great Lakes. Annals of the New York Academy of Sciences 288:508–37.
This is a systematic list of marine mammals found in the Champlain Sea and the Great Lakes region from 12,500 BP to about 8000 BP. It includes maps and detailed descriptions of all the specimens. They include: capelin (*Mallotus villosus*), sculpin (*Artediellus uncinatus*), three-spined stickleback (*Gasterosteus aculeatus*), lake trout (*Salvelinus namaycush*), smelt (*Osmerus mordax*), lumpfish (*Cyclopterus lumpus*), unidentified birds (Aves), snowshoe hare (*Lepus americanus*), eastern chipmunk (*Tamias striatus*), American marten (*Martes americanus*), sperm whale (*Physeter catodon*), white whale (*Delphinapterus leucas*), narwhal (*Monodon monoceras*), harbour porpoise (*Phocoena phocoena*), finback whale (*Balaenoptera physalus*), humpback whale (*Megaptera novaeangliae*), bowhead whale (*Balaena mysticetus*), unidentified whales (Cetacea), walrus (*Odobenus rosmarus*), ringed seal (*Phoca hispida*), harp seal (*Phoca groenlandica*), bearded seal (*Erignathus barbatus*), hooded seal (*Cystophora cristata*), and unidentified seals (*Phoca*). [See Harington 1988, Harington and Ochietti 1988, and McAllister et al. 1988.]

567. Harington, C.R. (1977c): Wildlife in B.C. during the ice age. B.C. Outdoors 33(6):28–32.
A picture is presented of British Columbia from pre- to post-Wisconsinan times. Imperial mammoths (*Mammuthus imperator*), Columbian mammoths (*Mammuthus columbi*), mountain goat (*Oreamnos*), Pacific salmon (Salmonidae), helmeted muskox (*Bootherium bombifrons = Symbos cavifrons*), and Dall sheep (*Ovis dalli*) are some of the fossils which date back to the Early Wisconsinan (about 80,000–65,000 BP). Mastodons (*Mammut americanum*), ground sloths (*Megalonyx*), woolly mammoths (*Mammuthus primigenius*), wild horses (*Equus*), bighorn sheep (*Ovis*), bison (*Bison*), mule deer (*Odocoileus*), moose (*Alces alces*), and caribou (*Rangifer tarandus*) fossils date to Late Wisconsinan times (about 25,000–13,000 BP). A discussion follows suggesting the climatic conditions at those times and the method and time of arrival of the various species.

568. Harington, C.R. (1978a): Faunal exchanges between Siberia and North America – Evidence from Quaternary land mammal remains in Siberia, Alaska and the Yukon Territory. American Quaternary Association, Abstracts of the Fifth Biennial Meeting (September 2–4, Edmonton, Alberta). pp. 56–77.
On the basis of fossils collected in Yukon alone, the movement of mammals from Eurasia to North America has been extensive. About 72% of North American Quaternary mammals are derived from Eurasia, the remainder having come from southern North America. There is some evidence that Beringia may have not only served as a 'bridge' but also as an important speciation centre. The earliest known mammoth (*Mammuthus*) and bison (*Bison*) fossils are discussed, as are possible earliest dates for American lion (*Panthera leo atrox*), muskoxen (*Soergelia, Praeovibos* and *Ovibos*), large horse (*Equus verae*), giant moose (*Alces latifrons*), saiga (*Saiga tatarica*), dhole (*Cuon*), moose (*Alces alces*), mountain sheep (*Ovis*), and grizzly bear (*Ursus arctos*). Use of mammoth and bison as indicator species is proposed. [See Harington 1980b.]

569. Harington, C.R. (1978b): Quaternary vertebrate faunas of Canada and Alaska and their suggested chronological sequence. National Museum of Natural Sciences (Canada) Syllogeus No. 15:1–105.
This is a comprehensive detailed work dealing with Quaternary vertebrate faunas of northern North America. Thirty-one Canadian and Alaskan faunas are reviewed ranging in age from 1.8 Ma to 5000 BP. Each fauna is placed in its paleoenvironmental, geological, stratigraph-

ic, and chronological context. Known radiocarbon dates from bone or associated organic materials are tabled. The faunas are compared to those of similar age from the Palaearctic side of Beringia and also to those S of the North American ice sheets. The publication is illustrated by photographs of some of the best specimens and ink sketches of some of the extinct species as they may have appeared in life. A map (Figure 2) shows localities of Quaternary vertebrates of Canada and Alaska described in the text. There is an extensive reference list (pp. 87–99) and an Appendix providing details on nine of the most important Quaternary vertebrate faunas of north-eastern Siberia.

570. Harington, C.R. (1979): Wildlife and man in B.C. during the ice age. *In*: Pioneer Days in British Columbia (A. Downs, ed.). Volume 4, Heritage House Publishing Company Ltd., pp. 142–7.
This popular article reviews vertebrate faunas of British Columbia during the last 2 million years. The prolific fossil sites are named along with the species they have produced. The evidence for human (*Homo sapiens*) occupation is also reviewed. [See Harington 1996a.]

571. Harington, C.R. (1980a): A preliminary list of faunal remains from two caves (Trou Otis and Spéos de la Fée) in Gaspé, Québec. *In*: Le karst de plate-forme de Boischâtel et le karst barré de La Rédemption, état des connaissances (J. Schroeder, ed.). Société Québécoise de Spéléologie, Montréal. pp. 93–105.
Preliminary collections of faunal remains from sediments from the two caves near La Rédemption, Quebec, have yielded seven species of mammals: eastern long-eared bat (*Myotis keenii* = *Myotis septentrionalis*), Ungava lemming (*Dicrostonyx* cf. *Dicrostonyx hudsonius*), American porcupine (*Erethizon dorsatum*), red fox (*Vulpes vulpes*), grizzly bear (*Ursus arctos*), caribou (*Rangifer tarandus*), and moose (*Alces alces*). Bats and porcupines, which commonly spend part of their lives in caves or subterranean shelters, are most frequently represented. The lemming and grizzly bear are of great interest because of their rarity in the faunal record of this part of eastern North America. It is postulated that they occupied Gaspé when patches of tundra like habitat existed there, perhaps in early postglacial time. Species other than the lemming and the grizzly bear still occupy the region and tend to be common in forested areas.

572. Harington, C.R. (1980b): Faunal exchanges between Siberia and North America: Evidence from Quaternary land mammal remains in Siberia, Alaska and Yukon Territory. Canadian Journal of Anthropology 1(1):45–9.
This paper compares Beringian species in order to establish faunal-exchange patterns over Bering land connections during glacial maxima. Specimens mentioned include: steppe bison (*Bison priscus* = *Bison crassicornis*) (>39,000 BP); horse (*Equus*); moose (*Alces*); giant moose [*Alces latifrons* antler (33,800 ± 2000 BP)]; Staudinger's muskox (*Praeovibos priscus*); tundra muskox (*Ovibos moschatus*); Soergel's muskox (*Soergelia* cf. *Soergelia elisabethae*); helmeted muskoxen (*Bootherium* = *Symbos*); pika (*Ochotona*); collared lemming (*Dicrostonyx*); brown lemming (*Lemmus*); vole (*Microtus*) and red-backed vole (*Clethrionomys*); caribou (*Rangifer tarandus*); lion-like cat (*Panthera leo atrox*); Saiga antelope (*Saiga tatarica*); mountain sheep (*Ovis*); brown bear (*Ursus arctos*); short-faced bear (*Arctodus*); shrew (*Sorex*); ground sloth (*Megalonyx*); beaver (*Castor*); porcupine (*Erethizon*); badger (*Taxidea*); lynx (*Felis lynx* = *Lynx canadensis*); and mastodon (*Mammut*), mammoth (*Mammuthus*), and camelids (*Camelops*).

573. Harington, C.R. (1980c): Pleistocene mammals from Lost Chicken Creek, Alaska. Canadian Journal of Earth Sciences 17:168–98.
Ten species of Pleistocene mammals are represented in a collection of fossils from a placer mining site at Lost Chicken Creek (64°03.2′N, 141°52.6′W), E of Chicken, Alaska. Among these are four extinct species: woolly mammoth (*Mammuthus primigenius*); Yukon horse (*Equus lambei* = *Equus* (*Asinus*) *lambei*); helmeted muskox (*Bootherium bombifrons* = *Symbos cavifrons*); and steppe bison (*Bison priscus* = *Bison crassicornis*); an extinct subspecies American lion (*Panthera leo atrox*); two species now extinct in the region – kiang (*Equus* cf. *Equus kiang*) and wapiti (*Cervus elaphus*); and three species that still occupy the area – caribou (*Rangifer tarandus*); humans (*Homo sapiens*), and Dall sheep (*Ovis dalli* = *Ovis ?dalli*). The kiang is recorded for the first time from Alaska. All of the Lost Chicken Creek specimens are now part of the Canadian Museum of Nature (CMN) collections. An American lion mandible was collected near the same location at an earlier date [see Whitmore and Foster 1967]. A steppe bison tibia (CMN-

25845) with a distinct impact mark and concentric and radial cracks (indicating breakage by humans) provided a radiocarbon date of 10,370 ± 160 BP (I-8582). Other bison fossils and a mammoth limb bone also showed evidence of human breakage. A horse bone collected in 1965 by H.L. Foster has yielded a date of 26,760 ± 300 BP (SI-355, Péwé 1975b). A horncore of the helmeted muskox (CMN-25892) provided a date of 20,500 ± 390 BP (I-10649). A shed left antler of wapiti (CMN-25873) gave a date of 10,050 ± 150 BP (I-9998). The relatively great spread in dates could be explained by massive downslope movement of remains from higher, former steppe-like areas. Other dated Alaskan mammoth remains include: (1) dried flesh from a partial carcass found on Fairbanks Creek yielded a date of 21,300 ± 1300 BP (L-601); and (2) hair from a skull collected at Dome Creek gave a date of 32,700 ± 980 BP (SI-1632). In nearby Yukon, a mammoth from Whitestone River has provided a radiocarbon date of 30,300 ± 2000 BP (I-3576). The species is considered an indicator of tundra-like habitat. An *Equus lambei* metacarpal collected *in situ* near Dawson City, Yukon, gave a date of 14,870 ± 260 BP (I-3569) and a caribou antler also from near Dawson City yielded a date of 23,900 ± 470 BP (I-8580), showing that both of these species were present near the peak of the Wisconsinan glaciation. A radio-ulna of a caribou from Dawson Loc. 8 with a date of 5010 ± 100 BP (I-8642) and an antler from Old Crow dated to 6450 ± 135 BP (I-4221) show that caribou were in the NW Yukon also during the warmest part of the postglacial. Other helmeted muskox remains from the region (winter fecal pellets from Little Eldorado Creek, Alaska) gave a date of (40,000 BP (SI-291); hornsheaths from two specimens from Upper Cleary Creek and Dome Creek, Alaska, yielded dates of 25,090 ± 1070 BP (SI-850) and 17,695 ± 445 BP (SI-851), supporting the Lost Chicken Creek specimen date, which suggests that *Bootherium* occupied Eastern Beringia during the peak of the Wisconsinan glaciation. A Dall sheep horncore from Dawson Loc. 2, Yukon, gave a radiocarbon date of 23,000 ± 600 BP (I-4225). The Lost Chicken Creek fauna suggests that extensive tracts of cool, steppe-like grassland existed in the region during the Late Wisconsinan. Some of the younger dates on the fauna suggest that there was a late-surviving large-mammal assemblage in Eastern Beringia, possibly providing a source of food for early humans.

574. Harington, C.R. (1980d): Radiocarbon dates on some Quaternary mammals and artifacts from northern North America. Arctic 3(4):815–32.
Ten radiocarbon dates on five genera of Quaternary mammals from northern North America are discussed. Of particular interest are: (1) a 29,000-year-old artifact from the Yukon; (2) the first evidence that steppe mammoths (*Mammuthus columbi* or *Mammuthus armeniacus*) occupied Eastern Beringia during the peak of the Wisconsinan glaciation; (3) dates indicating that saiga antelopes (*Saiga tatarica*) and Yukon short-faced bears (*Arctodus simus yukonensis*) occupied the Yukon–Alaska region in Mid-Wisconsinan time; (4) dates indicating that bison (*Bison* sp.) lived near the arctic coast of the Northwest Territories, and tundra muskoxen (*Ovibos moschatus*) lived in the western Yukon in late postglacial time; and (5) dates suggesting that tundra muskoxen have occupied the central Canadian Arctic Islands for the last 7000 years. Radiocarbon dates mentioned in the text include: (1) a mammoth (Proboscidea cf. *Mammuthus*) limb-bone fragment from Old Crow Basin, Yukon, modified by humans [29,300 ± 1200 BP (I-11051; the correct lab number is I-11050)]; (2) a possible wood bison (*Bison* = *Bison bison athabascae*?) tibia shaft from Hunker Creek near Dawson City, Yukon, possibly broken by humans [1465 ± 85 BP (I-11051)]; (3) a giant short-faced bear (*Arctodus simus*) humerus from Hunker Creek near Dawson City, Yukon [29,600 ± 1200 BP (I-11037)]; (4) a steppe or Columbian mammoth (*Mammuthus columbi* or *Mammuthus armeniacus*) partial mandible with tooth [29,190 ± 400 BP (I-10971)]; (5) a bison (*Bison*) hornsheath from Baillie Islands, Northwest Territories [1810 ± 90 BP (I-5407)]; (6) saiga antelope (*Saiga tatarica*) horncore fragment from Usuktuk River, Alaska [37,000 ± 990 BP (GSC-3050)]; (7) muskox (*Ovibos moschatus*) cranial fragment from Brewer Creek, Yukon [2830 ± 100 BP (I-3568)]; (8) muskox cranium with left hornsheath from Miller Creek, Yukon [3280 ± 90 BP (I-10985)]; (9) muskox horncore from Goodsir Inlet, Bathurst Island, Nunavut [6725 ± 130 BP (I-10919)]; (10) muskox mandible from Bathurst Island, Nunavut [2950 ± 90 BP (I-9996)]. Associated dates are also mentioned, and photographs of eight specimens are included, as is a locality map (Figure 1).

575. Harington, C.R. (1981a): Pleistocene saiga antelopes in North America and their paleoenvironmen-

tal implications. *In*: Quaternary Paleoclimate (W.C. Mahaney, ed.). Geo Books, University of East Anglia, Norwich, England. pp. 193–225.
Eight Nearctic saiga antelope (*Saiga tatarica*) fossils are discussed, including one from the Baillie Islands, E of the Mackenzie Delta, Northwest Territories. This species seems to have dispersed from Eurasia during the Late Pleistocene, then became extinct in Europe and North America near the end of the Wisconsinan glaciation, but survives in central Asia. Saiga fossils, based on the species' current adaptations, appear to be indicators of somewhat arid conditions, with shallow snow cover, steppe-like vegetation, and generally flat terrain. Other species mentioned are: caribou (*Rangifer tarandus*), bear (*Ursus*), and bison (*Bison*). Photographs and measurements of the fossils are provided, as well as a map of the species' Pleistocene and Recent distributions (Figure 11) in the Holarctic.

576. Harington, C.R. (1981b): Whales and seals of the Champlain Sea. Trail and Landscape 15(1):32–47.
A description of the history of this 12,000–10,000 BP inland sea precedes an impression of its appearance about 11,500 BP. Then brief descriptions of the appearance, habits, and Pleistocene remains of the following marine mammals are given: bowhead whale (*Balaena mysticetus*), finback whale (*Balaenoptera physalus*), humpback whale (*Megaptera novaeangliae*), white whale (*Delphinapterus leucas*), unidentified seals (*Phoca*), bearded seal (*Erignathus barbatus*), harbour seal (*Phoca vitulina*), harp seal (*Phoca groenlandica*), ringed seal (*Phoca hispida*), harbour porpoise (*Phocoena phocoena*), and walrus (*Odobenus rosmarus*). The paleoenvironmental implication of the fossils, their radiocarbon dates, and the value of collectors' contributions are mentioned in the Conclusion. The various specimens are illustrated and there are maps of whale (Figure 2) and seal (Figure 3) remains.

577. Harington, C.R. (1981c): Ice age animals and early man in the Old Crow region, Yukon Territory. Program and Abstracts of the Third International Symposium on Arctic Geology (June 28–July 1, Calgary, Alberta). p. 61.
A comprehensive program of collecting ice age vertebrate fossils in the Old Crow Basin, Yukon, was initiated by the National Museum of Natural Sciences [now Canadian Museum of Nature] in 1966. So far, approximately 61 species of mammals (Mammalia), 32 species of birds (Aves), and seven species of fishes (Pisces) are known from the basin's Pleistocene deposits. Most of the fossils are of Late Pleistocene age. About 75% of the mammalian species were derived from Eurasia, and the remainder entered the Yukon from southern North America. Evidently, people (*Homo sapiens*) reached the basin 30,000 years ago or earlier. While Paleolithic Europeans were developing stone technology and art to a brilliant crescendo, the earliest Yukoners were not only exploring a new continent, but had successfully adapted to survival in the Arctic with the aid of a highly advanced bone technology.

578. Harington, C.R. (1983a): Ice age animals and early man in the Old Crow region, Yukon Territory. Talk presented at Geological Society of America Meeting. (November 1, Indianapolis, Indiana). Abstracts with Programs. 15(6):591.
A comprehensive program of collecting ice age vertebrate fossils in the Old Crow Basin, initiated in 1966 by the National Museum of Natural Sciences, Ottawa, led to the establishment of two multidisciplinary groups: the Yukon Refugium Project and the Northern Yukon Research Programme. As a result of these projects, much has been learned about the Quaternary history of the region. So far, approximately 61 species of mammals (Mammalia), 32 species of birds (Aves), and 7 species of fish (Pisces) are known from the basin's ice age deposits. Most of the fossils are of Late Pleistocene age. About 75% of the mammalian species were derived from Eurasia, and the remainder entered the Yukon from southern North America. Mammoths (*Mammuthus*), giant beavers (*Castoroides ohioensis*), ground sloths (*Megalonyx*), and large camels (Camelini) are among the more interesting mammals represented. Evidently, people (*Homo sapiens*) reached the basin 30,000 years ago, or earlier. However, apart from a fragment of a human jaw, possibly associated with ice age faunal remains, only indirect evidence of early humans is available. While Paleolithic Europeans were developing stone technology and art to a brilliant crescendo, the earliest known Yukoners were not only exploring a new continent, but had successfully adapted to survival in the Arctic with the aid of a highly advanced bone technology. [See Harington 1981c.]

579. Harington, C.R. (1983b): Significance of the fos-

sil locality at Green Creek, Ontario. Trail and Landscape 17(3):164–78.
This popular article deals with the importance of the fossils of Green Creek, E of Ottawa, Ontario. Vertebrate remains, generally preserved in calcareous clay nodules, found here include: cisco (*Coregonus artedii* or *Coregonus zenithicus*); lake trout (*Salvelinus namaycush*); capelin (*Mallotus villosus*); smelt (*Osmerus mordax*); sucker (*Catostomus*); cod (*Gadus*); Atlantic tomcod (*Microgadus tomcod*); hookear sculpin (*Artediellus*); deepwater sculpin (*Myoxocephalus thompsoni*); blenny-like fish (Blennoidea); lumpfish (*Cyclopterus lumpus*); three-spined stickleback (*Gasterosteus aculeatus*); unidentified bird (Aves) feather impressions; unidentified seal (*Phoca*); harp seal (*Phoca groenlandica*); and American marten (*Martes americana*). Nodules found in the vicinity of Green Creek contain a broad, often excellently preserved sample of plants and animals that lived in the Ottawa area about 10,000 years ago. The author makes a specific proposal for preserving land along Green Creek and building a display room readily accessible from the highway. [A Green Creek Conservation Area has since been established and is part of the National Capital Commission's Greenbelt Management Plan.] The paper includes photographs of some of the best specimens from the locality: (1) a nodule containing three skeletons of capelin; (2) a marchfly; (3) a feather impression; (4) a nodule containing an American marten skeleton; and (5) a sucker skeleton – the largest complete fish specimen from a Green Creek nodule.

580. Harington, C.R. (1983c): The woolly mammoth. Neotoma 17, National Museum of Natural Sciences.
This museum pamphlet contains a general description of the morphology, biology, and geographic distribution of the woolly mammoth (*Mammuthus primigenius*). [See Harington 1996b.]

581. Harington, C.R. (1983d): *Soergelia* – An indicator of Holarctic Middle Pleistocene deposits? Correlation of Quaternary Chronologies Symposium (May 26–30, York University, Toronto). Abstracts with Program and Field Guide. pp. 66–7.
This primitive muskox probably looked like a cross between a tundra muskox (*Ovibos moschatus*) and a mountain sheep (*Ovis*). It preferred cool, tundra-like or cool, dry parkland habitat and was widely distributed from Europe through Siberia to Texas and Kansas. *Soergelia* fossils appear to provide support for cross-continental correlation of deposits of Kansan age in North America with those of Mindel or Elster age in Eurasia. So far the only northern North American fossils have come from undated sediments in the Old Crow Basin, Yukon. About two dozen specimens have been collected from several localities. The redeposited sediments in Old Crow Basin have not provided a satisfactory age for these bones, but relatively complete crania have been collected from deposits of Late Kansan age at Rock Creek, Texas, and Courtland Canal, Kansas. [See Harington 1989a.]

582. Harington, C.R. (1983e): History and future prospects for collecting Quaternary vertebrates and artifacts in the Dawson Area, Y.T. Solicited Report to Yukon Placer Mining Guidelines Public Review Committee. [Typescript in Quaternary Vertebrates of Northern North America Publications File at CMN]. 10 pp.
This report includes sections on: (1) the history of Quaternary vertebrate collecting in the Dawson City area previous to 1966; (2) collecting by representatives of the National Museum of Natural Sciences [now Canadian Museum of Nature] from 1966 to 1980. Some highlights are: (1) The 1966 collection of many excellently preserved ice age mammal bones from Schmidt's placer operation on Dominion Creek, including the first American lion (*Panthera leo atrox*) bones from Canada. The fine preservation of fossils at this site was attributed to use of a massive sprinkler system which 'melted away' the surrounding frozen organic silt. (2) The 1967 collection of several good Yukon horse (*Equus lambei*) skulls and rare specimens of American badger (*Taxidea taxus*) from Gold Run and Dominion creeks. At Schmidt's locality (see above) bones of an individual arctic ground squirrel (*Spermophilus parryii*), with droppings, seeds from a cache, and nesting grasses were collected. (3) Wood bison (*Bison bison athabascae*) that lived in the Dawson City area some 1350 years ago according to a radiocarbon date on a skull from Quartz Creek. (4) Organizing and monitoring a 1972 Dawson City display of the best ice age mammals collected in the area for the public and 40 geologists from an International Geological Congress field trip. (5) The 1973 collection and identification of the oldest-known human (*Homo sapiens*) artifact from the area, a caribou (*Rangifer tarandus*) antler punch,

from Ericson's and Leidtke's operation on Hunker Creek [see Harington and Morlan 1992]. (6) In 1975, the beginning of a collection of very rich and important Pleistocene mammal fauna from Sixtymile, Yukon. (7) The 1980 discovery of new ice age mammal localities at Big and Mattson creeks. The author, in Review and Future Prospects, emphasizes the great contribution that Yukon placer miners have made to our knowledge of the Quaternary history of the Dawson City area and the fact that the best return in fossils and artifacts resulted from the use of ground-sluicing techniques (e.g., sprinklers, automatic gates and monitors) as opposed to water restriction and use of ripping of the overburden using tractor blades – highly destructive of bones and artifacts. It is ironic that fish preservation is being promoted by restricting water use for placer mining when none of the fishes under consideration is endangered. Indeed, such a high percentage of Yukon is occupied by fish, whereas such a small percentage of Yukon is highly productive of gold – a significant contribution to Canada's gross domestic product. There is a three-page appendix by the author addressed to the president of the Klondike Placer Miners' Association dated January 4, 1980.

583. Harington, C.R. (1984a): Whales and seals of the Champlain Sea. Ottawa Archaeologist 11(2):3–18. Reprinted from Trail and Landscape.

[See Harington 1981b.] Some interesting photographs have been included: (1) John Hanson holding a humerus of the White Lake, Ontario, bowhead whale (*Balaena mysticetus*); (2) excavation of the 2.4-m-long jaw of the White Lake bowhead; and (3) completion of excavation of the jaw mentioned at the Hanson Pit.

584. Harington, C.R. (1984b): Quaternary marine and land mammals and their paleoenvironmental implications – Some examples from northern North America. *In*: Contributions in Quaternary Vertebrate Paleontology: A volume in memorial to John E. Guilday (H.H. Genoways and M.R. Dawson, eds.). Carnegie Museum of Natural History Special Publication No. 8:511–25.

This paper issues an appeal to vertebrate paleontologists to ensure they collect as much information as possible from fossil evidence. Examples are provided illustrating how important it is to properly identify specimens, how important the stratigraphic and sedimentological evidence can be, how much information about the paleoenvironment can be gained from careful study of the associated plant and animals, and how intestinal contents and droppings can reveal food habits. Taken one step further, this information must be presented to the public to help them understand the magnitude of changes that can occur in landscapes and their biota over even relatively short periods of geological time. Specific examples include radiocarbon dates for land mammals and Champlain Sea marine mammals. All are discussed in more detail in specific papers. The terrestrial mammals mentioned are: Columbian mammoth (*Mammuthus* cf. *Mammuthus columbi*) from Babine Lake, British Columbia, (from wood) 42,900 ± 1860 (GSC-1657), (from a silty layer) 43,800 ± 1830 (GSC-1687), and (from a bone) 34,000 ± 690 (GSC-1754); a saiga antelope (*Saiga tatarica*) from the Baillie Islands, Northwest Territories (an Alaskan example is 37,000 ± BP (GSC-3050); and an arctic ground squirrel (*Spermophilus parryii*) from Dawson City, Yukon, (from nesting grasses) 12,200 ± 100 (GSC-2641). The marine mammals include: a bowhead whale (*Balaena mysticetus*), ringed seal (*Phoca hispida*), harbour seal (*Phoca vitulina*), bearded seal (*Erignathus barbatus*), which date between 12,000 and 10,000 BP; and a walrus (*Odobenus rosmarus*) from Qualicum Beach, Vancouver Island, British Columbia >40,000 BP (I-11617).

585. Harington, C.R. (1984c): Mammoths, bison and time in North America. *In*: Quaternary Dating Methods (W.C. Mahoney, ed.). Elsevier Science Publishers, Amsterdam. pp. 299–309.

Mammoths and bison are important indicator species to approximate geological time in the Quaternary, especially where more precise dating methods are lacking. The earliest mammoths (*Mammuthus*) mark the beginning of the Irvingtonian Land Mammal Age (about 1.8 Ma), and the earliest bison (*Bison*) remains mark the beginning of the Rancholabrean Land Mammal Age (about 1.2 Ma). Over time, mammoths display an increase in number and compression of enamel plates, and a thinning of enamel in molar teeth, and bison show (at least over the last 12,000 years) a progressive reduction in horncore size. Southern mammoths (*Mammuthus meridionalis*) seem to be useful indicators of Early (Nebraskan to Kansan) Pleistocene deposits, whereas woolly mammoths (*Mammuthus primigenius*) indicate Late (mainly Wisconsinan) Pleistocene deposits. A diagram (Figure 1) shows the suggested chronology and regional distribution of mammoths and bison in North America during the Quaternary. The Appendix includes the description of a Late Holocene [3240 ± 90 BP (I-11638)] western bison (*Bison bison occidentalis*) from the Banff Archives site, Alberta.

586. Harington, C.R. (1985): Comments on Canadian Pleistocene mammals. Acta Zoologica Fennica 170:193–7.
The most productive Pleistocene mammal localities occur around the margin of the relatively barren, formerly heavily glaciated Precambrian Shield. The most significant land-mammal sites occur in the Prairies and unglaciated parts of the Yukon. In Saskatchewan, the Wellsch Valley site has yielded our best evidence of an Early Pleistocene fauna. An impressive sequence of mammalian remains, extending from Middle Pleistocene to the present, is exposed at Medicine Hat, Alberta. The richest known sites in Canada are the Dawson and Old Crow areas of the Yukon. Dawson localities have produced unusually well-preserved specimens of Late Wisconsinan age, which are generally exposed in the course of placer-mining for gold. In contrast, bones exposed by the winding Old Crow River tend to be more poorly preserved, but represent a greater variety of species. Canada is bounded by three oceans – Atlantic, Pacific, and Arctic – a fact that is obvious when marine mammal remains are considered. Of particular interest in this regard are recently discovered walrus (*Odobenus rosmarus*) skeletons found near Qualicum Beach, Vancouver Island, and Moncton, New Brunswick. They indicate the more southerly distribution of this species during Early and Late Wisconsinan time, respectively. In future, the author suggests that a systematic paleontological testing of Canadian caves would be valuable. Other important Pleistocene mammal sites may be found in the western Canadian Arctic Islands, particularly Banks Island.

587. Harington, C.R. (1986a): Ice Age animals and early man in the Old Crow region, Yukon Territory. *In*: Living Explorers of the Canadian Arctic (S. Milligan and W. Kupsch, eds.). Outcrop, The Northern Publishers, Yellowknife. pp. 27–39.
This paper refers to many species of Pleistocene vertebrates found in the Old Crow Basin, Yukon Territory. The species are arranged in tabular form by age. Human (*Homo sapiens*) artifacts (both stone and bone) from the area are discussed and the evolution of the topography of the Old Crow Basin is described.

588. Harington, C.R. (1986b): The impact of changing climate on some vertebrates in the Canadian Arctic. *In*: Proceedings of a Canadian Climate Program Workshop on Impact of Climatic Change on the Canadian Arctic (H.M. French, ed.). (March 3–5, Orillia). pp. 100–13.
Great changes in climate and vertebrate life have occurred in the Canadian Arctic, based on our knowledge of the geological record. This paper briefly reviews some finds from the past as they relate to prevailing climate of the period. Generally, climatic implications derived from fossils demonstrate a marked cooling from tropical in Siluro-Devonian time to arctic in the Quaternary. In dealing with the present, the author considers the known distribution of some prominent arctic vertebrates with emphasis on core areas of range, and some of their climatic controls. Finally, possible future effects on selected arctic vertebrates are considered based on two scenarios: (1) extreme warming of climate; and (2) extreme cooling of climate. The Tertiary-Quaternary transition includes comments on Kap København, Greenland, and Ocean Point, Alaska faunas. The remains of lemming (*Dicrostonyx torquatus*) and ptarmigan? (*Lagopus*?) from interglacial deposits (perhaps 700,000 years old) from near Jesse Bay on southwestern Banks Island, and Dovekie (*Alle alle*) from Mid-Wisconsinan (about 40,000 years ago) deposits at Cape Storm on southern Ellesmere Island are mentioned. Late Wisconsinan (about 25,000 to 10,000 years ago) mammoth [presumably tundra-adapted woolly mammoths (*Mammuthus primigenius*)] remains are known from Cape James Ross on southwestern Melville Island and northern Banks Island, Northwest Territories, and other localities along the Beaufort Sea coast of the Northwest Territories and near the Mackenzie Delta. Possibly tundra muskoxen (*Ovibos moschatus*) survived Wisconsinan glaciation on a western Banks Island refugium, for remains from Bernard and Masik river areas have been radiocarbon dated at >34,000 BP and about 11,000 BP, respectively. Presumably a saiga antelope (*Saiga tatarica*) skull fragment from Baillie Islands, Northwest Territories, is indicative of steppe-like grasslands along the arctic coast of northwestern North America during the last glaciation. [See Harington and Cinq-Mars 1995, and Guthrie et al. in press.] Beginning about

12,000 years ago, the Pleistocene megafauna that occupied refuges in northwestern North America became extinct, and examples and possible causes are mentioned. Following the extinction some 10,000 years ago, the modern Canadian Arctic vertebrate fauna was left. Holocene marine mammals from raised beaches, as well as land mammals and birds (Aves), are considered (pp. 102–3).

589. Harington, C.R. (1986c): The American mastodon. Neotoma 19, National Museum of Natural Sciences.
This museum pamphlet contains a general description of the morphology, biology, and geographic distribution of the American mastodon (*Mammut americanum*). [See Harington 1996c.]

590. Harington, C.R. (1986d): The giant beaver. Neotoma 20, National Museum of Natural Sciences.
This museum pamphlet contains a general description of the morphology, biology, and geographic distribution of the giant beaver (*Castoroides ohioensis*). [See Harington 1996f.]

591. Harington, C.R. (1987a): Stop 25: 80 Pup site, Hunker Creek. *In*: INQUA 87 Excursion Guide Book A-20(a) and (b), Quaternary Research in the Yukon (S.R. Morison and C.A.S. Smith, eds.). pp. 65–8.
This paper was provided to participants at the 1987 INQUA Conference as an overview of this Hunker Creek site. The site has been a source of Pleistocene fossil vertebrates since 1973. During this time more than 260 mammal bones have been collected. The finds consist of: steppe bison (*Bison priscus* = *Bison crassicornis*, 46.4%), small horse (*Equus lambei*, 19.2%), woolly mammoth (*Mammuthus primigenius*, 11.1%), Dall sheep (*Ovis dalli*, 11.1%), caribou (*Rangifer tarandus*, 2.7%), arctic ground squirrel (*Spermophilus parryii*, 1.2%), vole (*Microtus*, 0.4%), wolf (*Canis lupus*, 1.9%), dog-like canid (*Canis* cf. *Canis familiaris*, 0.4%), arctic fox (*Alopex lagopus*, 0.8%), short-faced bear (*Arctodus simus*, 0.4%), American lion (*Panthera leo atrox*, 0.8%), large horse (*Equus*, 0.8%), Scott's? horse (*Equus* cf. *Equus scotti*, 1.2%), moose (*Alces alces*, 1.2%), and Alaskan bison (*Bison alaskensis*, 0.8%). The high percentage of grazers suggests the presence of productive tracts of steppe-like grassland during the last glaciation. Some of the specimens exhibit pathological signs. Site stratigraphy is illustrated, and a locality map is provided.

592. Harington, C.R. (1987b): Stop 27: Pleistocene vertebrates of Old Crow Locality 11A. *In*: INQUA 87 Excursion Guide Book A-20(a) and (b), Quaternary Research in the Yukon (S.R. Morison and C.A.S. Smith, eds.). pp. 71–5.
This paper was provided to participants at the 1987 INQUA Conference as an overview of the Old Crow site. Site stratigraphy is illustrated, and a locality map and an extensive list of fossils are included. [See Holland and Johnson 1974, and Friesen 1989.]

593. Harington, C.R. (1987c): A Pleistocene pig with wet feet. BIOME 7(1):4.
In February 1986, Bob Grantham, Curator of Geology of the Nova Scotia Museum, submitted a right lower jaw with molar teeth of a pig-like mammal for study. It had been pulled up from the western margin of Canada's portion of Georges Bank (off the New England coast) by a scallop dredger – encrusting marine invertebrates like barnacles and bryozoans attest to the fossil's source on the sea floor. The author compared the fossil with extinct peccaries and found that it was closest to the long-nosed peccary (*Mylohyus nasutus*). Size and lack of heavy wear on the teeth indicated that the specimen represented a large individual that died in early adulthood. This large, long-legged peccary probably lived in open areas near cold-temperate forest margins in what is now eastern and central United States during the ice age. Fossils of other large ice age mammals such as mammoths (*Mammuthus*), mastodons (*Mammut americanum*), Scott's moose (*Alces scotti* = *Cervalces scotti*), and helmeted muskoxen (*Bootherium bombifrons* = *Symbos cavifrons*) have also been dredged up from the Atlantic coastal shelf during the twentieth century. Evidently these extinct animals occupied the region when world sea level was much lower during the peak of the last glaciation about 25,000 to 15,000 years ago. The author attempts to reconstruct a scene with the peccary in its natural environment about that time, and an ink sketch illustrates it.

594. Harington, C.R. (1988): Marine mammals of the Champlain Sea, and the problems of whales in Michigan. *In*: The Late Quaternary Development of the

Champlain Sea Basin (N.R. Gadd, ed.). Geological Association of Canada, Special Paper 35. pp. 225–40.
Five species of whales particularly adapted to cool inshore conditions have been reported from Champlain Sea deposits – mainly white whale (*Delphinapterus leucas*), but also harbour porpoise (*Phocoena phocoena*), humpback whale (*Megaptera novaeangliae*), common finback whale (*Balaenoptera physalus*), and bowhead whale (*Balaena mysticetus*). A new record of white whale [from a sandpit on the S side of Mont St-Hilaire, Quebec – radiocarbon dated at 9470 ± 170 BP (Beta-27511)] is added. Seals such as harp (*Phoca groenlandica*), bearded (*Erignathus barbatus*) and ringed (*Phoca hispida*), all adapted to breeding on pack ice or landfast ice, have also been discovered. The author adds a ringed seal record (a mandible – CMN 36247) found in a sandpit near Ottawa, Ontario (45°19′55″N, 75°41′40″W) estimated at 10,600 BP; a small phocid bone (CMN 43801), heavily eroded, from a gravel pit near Bearbrook, Ontario (45°23′30″N, 75°20′15″W) with an approximate age of 10,200 ± 100 BP; a left tibia (New York State Museum 892-D/8569; plaster cast CMN 37543) found at Plattsburgh and previously tentatively identified as hooded seal (*Cystophora cristata*), now referred to harbour seal (*Phoca vitulina*) with an estimated age of 11,000–10,000 BP; and a fragmentary bearded seal lumbar vertebra (CMN 37586) from a sandpit on the southeastern flank of Mont St-Hilaire, Quebec (45°31.9′N, 73°08.7′W). A complete list (Table 1) of all known marine mammal specimens from the Champlain Sea deposits with localities and radiocarbon dates is given. As a follow-up, the author summarizes the other known findings of marine mammal material from inland sites (e.g., Michigan whale remains) in eastern North America and concludes that they are human (*Homo sapiens*) imports, all less than 750 BP.

595. Harington, C.R. (1988/89): Ice age animals of the Yukon Territory. Dansk Naturhistorisk Forening Årsskrift 1988/89:38.
The author summarizes the importance of the Yukon in the study of Canadian Pleistocene vertebrates. Forty thousand fossils have been collected from the Old Crow area alone, including: ground sloth (*Megalonyx*), giant pika (*Ochotona whartoni*), giant beaver (*Castoroides ohioensis*), short-faced bear (*Arctodus simus*), hyena (*Chasmaporthetes*?), scimitar cat (*Homotherium serum*), gigantic camels (Camelini), and Soergel's muskox (*Soergelia*). At Bluefish Caves, there is evidence that as early as 15,000 BP humans (*Homo sapiens*) hunted steppe bison (*Bison priscus*), small horse (*Equus lambei*), woolly mammoth (*Mammuthus primigenius*), and caribou (*Rangifer tarandus*). Farther south, in the Dawson area, fossils of western camel (*Camelops hesternus*), helmeted muskox (*Bootherium bombifrons*), and tundra muskox (*Ovibos moschatus*) have been unearthed.

596. Harington, C.R. (1989a): *Soergelia*: An indicator of Holarctic Middle Pleistocene deposits? Second Annual Muskox Symposium (October 1–4, 1987, University of Saskatchewan). Published by the National Research Council, Ottawa. (ISBN 0-660-54820-8). pp. A1–A9.
Soergelia is argued to be a primitive muskox rather than the previously identified 'caprine' or steppe goat, due to the structure of the horncores, metapodials, and dentition. It was lighter and more slender-limbed than the modern tundra muskox (*Ovibos moschatus*) and preferred cool tundra-like or parkland habitat. It may be most closely related to the Lower Pleistocene Chinese *Boopsis*, and *Preptoceras* and *Euceratherium* from North America. *Soergelia* evidently dispersed from Asia westward to Europe and eastward to North America following the spread of cool tundra-like or parkland habitat during the Middle Pleistocene (Mindel-Elster in Europe and Kansan in North America). The author proposes that *Soergelia* has significant value biostratigraphically. *Soergelia* fossils may allow correlation of Kansan deposits of Texas and Kansas with Olyor deposits and their equivalents in Russia, and Mindel and Elster sediments in Europe. Also, it may be inferred that there are (or were) unrecognized fossil-bearing sediments of Kansan age exposed within the Old Crow Basin of the Yukon due to the presence of *Soergelia* specimens there, and that although there are no fossils reported from Alaska, undoubtedly it lived there also.

597. Harington, C.R. (1989b): Ice age fossils and vanished vertebrates. *In*: The Natural History of Ontario (J. Theberge, ed.). McClelland and Stewart, Toronto. pp. 156–64.
Ontario's vertebrate fossil record is virtually confined to the last few hundred thousand years of the Quaternary. Either the older fossil-bearing rocks were not deposited or if they were, they were eroded away, perhaps during

glaciations. Kelso Cave, W of Milton, apparently contains remains of 200,000-year-(Illinoian)-old vertebrates [but see Mead and Grady 1996]. The most remarkable species found there is a large pika (*Ochotona*). Toronto's Don Valley Brickyard contains fossils of Sangamonian interglacial age (about 120,000 years ago). The most interesting mammal fossil from this location is the giant beaver (*Casteroides ohioensis*), represented by a single incisor. Deposits of similar age in the Moose River Basin of northern Ontario are also showing signs of being important fossil-bearing Sangamonian sites. Early Wisconsinan deposits are well represented near Toronto, where bison (*Bison*), bear (*Ursus*), Scott's moose (*Alces scotti = Cervalces*), tundra muskox (*Ovibos moschatus*), and mammoth or possibly mastodon (*Mammut americanum*) have been found. Middle Wisconsinan deposits (Innerkip and Woodbridge) have revealed fossils of mammoth (*Mammuthus*), grizzly bear (*Ursus arctos*), Blanding's turtle (*Emydoidea blandingi*), rodents (Rodentia), and deer (Cervidae). More abundant Late Wisconsinan deposits have provided bones of grizzly bear, tundra muskox, caribou, a caribou-sized deer (*Torontoceros hypogaeus*), and mammoths and mastodons. Many modern species have been recovered from clay nodules formed about 12,000 to 10,000 years ago (Late Wisconsinan) along the shores of the former Champlain Sea in the Ottawa area. Bones of whales and seals such as white whale (*Delphinapterus leucas*), bowhead whale (*Balaena mysticetus*), ringed seal (*Phoca hispida*), harp seal (*Phoca groenlandica*), bearded seal (*Erignathus barbatus*), and walrus (*Odobenus rosmarus*), also found in Champlain Sea deposits, support the theory that it was a cold, salt sea.

598. Harington, C.R. (1989c): Pleistocene vertebrate localities in the Yukon. *In*: Late Cenozoic History of the Interior Basins of Alaska and the Yukon (L.D. Carter, T.D. Hamilton, and J.P. Galloway, eds.). U.S. Geological Survey Circular 1026:93–8.

This paper identifies eight productive Pleistocene fossil sites (see map, Figure 1) in the Yukon and provides stratigraphic and location data, as well as listings of the species present. The localities are: Old Crow Basin, Bluefish Caves, Sixtymile, Dawson area, Dublin Gulch, Big Creek, Ketza River, and Herschel Island. In addition to the extensive fossil listings, a number of important radiocarbon dates are given: Old Crow – a woolly mammoth (*Mammuthus primigenius*) 13,820 ± 840 (Beta-13867); and domestic dog (*Canis familiaris*) 2,110 ± 40 (TO-276); Bluefish Caves – small horse (*Equus lambei*) 12,900 ± 100 (GSC-2881), and mammoth (*Mammuthus* cf. *Mammuthus primigenius*) 15,500 ± 130 (GSC-3053); Sixtymile area – western camel (*Camelops hesternus*) 23,320 ± 640 (Beta-8864), tundra muskox (*Ovibos moschatus*) 21,160 ± 280 (Beta-13869), black-footed ferret (*Mustela nigripes*) 39,560 ± 490 (TO-214), arctic ground squirrel (*Spermophilus parryii*) 47,500 ± 1900 (Beta-16157), brown bear (*Ursus arctos*) 36,500 ± 1150 (Beta-16162), mastodon (*Mammut americanum*) 24,980 ± 1300 (Beta-16163), and spruce wood, 26,080 ± 300 (Beta-13870) [mentioned because open spruce forest seems to be an important habitat requirement for American mastodons]; Dawson area – caribou (*Rangifer tarandus*) 23,900 ± 470 (I-8580), brown bear 41,000 ± 1050 (Beta-16159), short-faced bear (*Arctodus simus yukonensis*) 29,600 ± 1200 (I-11037), grass (ground squirrel nest) 12,200 ± 100 (GSC-2641), Alaskan bison (*Bison alaskensis*) >39,900 (I-5405), steppe bison (*Bison priscus*) 22,200 ± 1400 (I-3570), and woolly mammoth 32,350 ± 1750 (I-4226); Dublin Gulch – horse (*Equus*) 31,450 ± 1300 (I-10935); Big Creek – willow wood (associated with American lion (*Panthera leo atrox*), steppe bison, and woolly mammoth) 48,100 ± 1100 (GSC-3032-2); Ketza River – bison (*Bison*) 26,350 ± 280 (TO-393); Herschel Island – small horse (*Equus lambei*) 16,200 ± 150 (SFU:RIDDL-765).

599. Harington, C.R. (1990a): Vertebrates of the last interglaciation in Canada: A review, with new data. Géographie physique et Quaternaire 44:375–87.

Vertebrate fossils and faunas that are reasonably inferred to be of last (Sangamonian) interglacial age are considered in geographic order from E to W to N in Canada. Data on localities, vertebrate taxa, stratigraphy, geochronology, paleoenvironment, and paleoclimate are considered. Information on key faunas from Toronto, Ontario, Fort Qu'Appelle and Saskatoon, Saskatchewan, Medicine Hat and Old Crow River, Yukon, are supplemented by data on smaller faunas and, in some cases, individual specimens. New data are included for several localities. Fishes (Pisces), such as whitefish (*Coregonus* sp.) and pike (*Esox* sp.), had broad distributions from eastern to northwestern Canada. Except for a turtle (*Emydoidea blandingi*) from Innerkip, Ontario, amphibians and reptiles have not yet been reported from Canada during the Sangamonian interglaciation. Several species of grouse-like birds (Tetraonidae) are known from western Canada. Among the mammals, American mastodons (*Mammut americanum*), mammoths (*Mammuthus*), beavers (*Castor canadensis*), giant beavers (*Castoroides ohioensis*), muskrats (*Ondatra zibethicus*), voles (*Microtus*), white-tailed deer (*Odocoileus virginianus*), Scott's moose (*Alces scotti* = *Cervalces*), bison [perhaps mainly giant bison (*Bison latifrons*)], and muskoxen (Ovibovini) were evidently most widespread during the last interglacial interval. The western plains had a characteristic large-mammal fauna that included Columbian mammoths (*Mammuthus columbi*), Scott's horses (*Equus scotti*), small horses (*Equus conversidens*), western camels (*Camelops hesternus*), pronghorns (Antilocapridae), giant bison (*Bison latifrons*), and helmeted muskoxen (*Bootherium bombifrons* = *Symbos cavifrons*).

600. Harington, C.R. (1990b): Ice age vertebrates in the Canadian Arctic Islands. *In*: Canada's Missing Dimension: Science and History in the Canadian Arctic Islands (C.R. Harington, ed.). Vol. 1. Canadian Museum of Nature, Ottawa. pp. 140–60.

Remains of Pleistocene vertebrates are rarely found in the Canadian Arctic. From earliest to latest they include: a few fossils of Middle Pleistocene interglacial age (perhaps 700,000 years ago) – tundra-adapted animals from Jesse Bay, southern Banks Island, Northwest Territories – collared lemming (*Dicrostonyx torquatus*) and ptarmigan (*Lagopus*); small seal (*Phoca*), large whale [perhaps a bowhead (cf. *Balaena mysticetus*)], small horse (*Equus lambei*), and a bison (*Bison*) from last interglacial (or earlier) deposits near the Beaufort Sea coast, Yukon and Northwest Territories; mammoth (*Mammuthus*) and tundra muskox (cf. *Ovibos moschatus*) of possible Early Wisconsinan age from Garry Island, Northwest Territories; a Middle Wisconsinan Dovekie (*Alle alle*) from Cape Storm, southern Ellesmere Island, Nunavut, and a tundra muskox (*Ovibos moschatus*) of the same age (or older) from Banks Island; and several species [mammoth (*Mammuthus*); Yukon horse (*Equus lambei*); saiga antelope (*Saiga tatarica*); steppe bison (*Bison priscus*); helmeted muskox (*Bootherium bombifrons* = cf. *Symbos cavifrons*)]; and tundra muskox from the Late Wisconsinan deposits of the western islands (Melville, Banks, Baillie, and Herschel). Early Holocene (about 10,000 to 5000 BP) vertebrates are more commonly found, and include Oldsquaw duck (*Clangula hyemalis*), ringed seal (*Phoca hispida*), walrus (*Odobenus rosmarus*), narwhal (*Monodon monoceros*), bowhead whale, Peary caribou (*Rangifer tarandus*), tundra muskox, and collared lemming. Radiocarbon dates on these specimens indicate relatively warm (open water) conditions about 9000 and 4000 BP.

601. Harington, C.R. (1990c): Arctic bison. BIOME (a publication of the Canadian Museum of Nature) 10(2):4.

Following a synopsis of North American fossil bison systematics, new evidence is presented which suggests that wood bison (*Bison bison athabascae*) lived on the Arctic coast of Canada, possibly up until the 1500s. The most northerly historic wood bison record, near Lac la Martre, is over 900 km SE. Three specimens discussed are the following: (1) a hornsheath from the Baillie Islands, Northwest Territories, just E of the Mackenzie Delta, dated to about 1800 BP; (2) a partial metacarpal from Harrowby Bay (about 40 km SE of Baillie Islands) dated to about 5200 BP; and (3) a partial skull of a small male wood bison from the Old Horton River Channel (48 km SE of Baillie Islands) dated to about 420 BP. This latter specimen has a number of cut marks on the upper surface of the skull showing that people used it as a cutting board. An accompanying map shows the Baillie Islands and Lac La Martre, and there is a photograph of the wood bison cranial fragment from Old Horton River Channel.

602. Harington, C.R. (1991): Walrus near Montreal! BIOME 11(3):4.

Having studied remains of Champlain Sea marine mammals for many years, the author wondered as early as 1977 why walrus (*Odobenus rosmarus*) bones had never been reported. After all, they had been excavated from 14 localities in the eastern approaches to the Champlain Sea – almost to Quebec City [considered as the entrance to this inland sea that existed between about 12,000 and 10,000 BP]. The solution to this apparent problem was more a matter of time and luck than paleoenvironmental factors [barriers] controlling walrus distribution, for in August 1990, farmer Jacques Daviau pulled some bones out of an excavation near his home at Ste-Julienne, Quebec, 50 km N of Montreal. Michel Bouchard, a geologist at the University of Montreal, sent the bones to the author a few weeks later. They consisted of four pieces of an adult male walrus skull with fragments of sea shells plastered to the roof of the cranium. In September, with the help of a backhoe to expose the marine sediments, fieldwork at the site suggested that walruses occupied the area after the sea had reached its maximum extent, and had probably fed on several of the species of shellfish exposed in the excavation. In March, Bouchard received a radiocarbon date indicating that the Ste-Julienne walrus had died a little more than 10,000 years ago. Comparison of the fossil with modern known-age adult male walruses, and sectioning its cementum layer around the root of a premolar tooth suggested that the Ste-Julienne walrus was about 14 years old at death. It is not only the first record of a walrus from the Champlain Sea, but it fills an important gap in the species' postglacial range, helping us to understand better its northward shift since the Laurentide ice sheet began its retreat. [Bouchard et al. 1993; Dyke et al. 1999.]

603. Harington, C.R. (1995a): Field trip guide; vertebrate palaeontology of the Western Basin of the Champlain Sea. Proceedings of the 28th Annual Meeting of the American Association of Stratigraphic Palynologists (October 14, Ottawa, Ontario). 7 pp.
This is a discussion of locations and species of Champlain Sea fossils focusing mainly on Foster's Sandpit near the Ottawa International Airport, Ontario, and an Eardley, Quebec, fossil site. A table listing animal and plant species contained in nodules of Champlain Sea-age is provided.

604. Harington, C.R. (1995b): Pleistocene mammals of Dublin Gulch and the Mayo District, Yukon Territory. Palaeoecology and Palaeoenvironments of Late Cenozoic Mammals. Tributes to the Career of C.S. (Rufus) Churcher (K.M. Stewart and K.L. Seymour, eds.). University of Toronto Press, Toronto. pp. 346–74.
Pleistocene mammal bones from near the base of an organic silt/colluvium unit overlying glacial till at Dublin Gulch, Yukon, represent the following taxa (in order of abundance of specimens): small horse (*Equus lambei*), bison [probably steppe bison (*Bison priscus*)], Dall sheep (*Ovis dalli*), caribou (*Rangifer tarandus*), moose (*Alces* cf. *Alces alces*), American lion (*Panthera leo atrox*), and possibly mammoth (cf. *Mammuthus*). A radiocarbon date of 31,450 ± 1,750 BP on a horse bone indicates that this fauna is of Middle Wisconsinan age. Species represented in the Dublin Gulch fauna evidently lived in a nearly treeless environment like the present low arctic tundra, but drier.

605. Harington, C.R. (1995c): Jefferson's ground sloth. Beringian Research Notes, No. 1:1–4.
This popular series, published by the Heritage Branch of Yukon Tourism in Whitehorse, provides a general description of Jefferson's ground sloth (*Megalonyx jeffersonii*) with background on its appearance, evolution, dispersal, and postulated habits and extinction. This long-haired, ox-sized animal was one of several ice age mammals, including the American mastodon (*Mammut americanum*) and flat-headed peccary (*Platygonus compressus*), to enter Yukon and Alaska from southern North America. This ground sloth was the only one that roamed as far N as Yukon (Old Crow Basin), Alaska (Fairbanks area), and Northwest Territories (Lower Carp Lake). *Megalonyx* has also been reported from Saskatchewan, Alberta (Medicine Hat, Edmonton), and British Columbia (Quesnel Forks). Illustrations include a sketch of the species in its postulated natural habitat, a left-side view of a cranium, and a side view of an upper molar from Old Crow Basin, Yukon. [See McDonald et al. 2000.]

606. Harington, C.R. (1996a): Quaternary animals: Vertebrates of the ice age. *In*: Life in Stone: A Natural History of British Columbia's Fossils (R. Ludvigsen, ed.). UBC Press, Vancouver. pp. 259–73.
The Quaternary vertebrates of British Columbia are discussed by district. The fossils are placed in geological, stratigraphic, paleoenvironmental, geochronological, and biogeographical context. Species mentioned include: Columbian mammoth (*Mammuthus columbi*), American mastodon (*Mammut americanum*), brown bear (*Ursus arctos*), black bear (*Ursus americanus*), helmeted muskox (*Bootherium bombifrons = Symbos cavifrons*), Vancouver Island marmot (*Marmota vancouverensis*), walrus (*Odobenus rosmarus*), killer whale (*Orcinus orca*), Steller's sea lion (*Eumetopias jubatus*), Pacific sardine (*Sardinops sagax*), Pacific salmon (*Oncorhyncus*), mountain goat (*Oreamnos americanus*), bighorn sheep (*Ovis canadensis*), ground sloth (*Megalonyx*), western bison (*Bison bison* cf. *Bison bison occidentalis*), giant bison (*Bison latifrons*), Dawson caribou (*Rangifer daw-*

soni), and three-spined stickleback (*Gasterosteus aculeatus*).

607. Harington, C.R. (1996b): Woolly mammoth. Beringian Research Notes, No. 2:1–4.
This popular publication provides a general description of the woolly mammoth (*Mammuthus primigenius*) with background on its appearance, evolution, dispersal, and postulated habits, as well as its relationship to people and extinction. During the last glaciation, when most of Canada was covered by ice, the species was isolated in refuges N and S of the ice sheets. In the northern area (Eastern Beringia), one of the best-preserved specimens consists of much of the front part of the body of a young woolly mammoth from Fairbanks Creek, Alaska, tentatively radiocarbon dated at about 21,000 BP. In life this baby, called 'Effie,' would have weighed nearly 100 kg. One of the best Canadian specimens from this area is a skeleton from Whitestone River, Yukon [see Bélanger 1988]. The mammoth died there about 30,000 BP, according to a radiocarbon date. In the southern refuge, remains of woolly mammoths have been found in southern British Columbia, Alberta, Saskatchewan, Manitoba, and Ontario (in addition to North Dakota, South Dakota, Minnesota, New York, Virginia, and the Atlantic continental shelf off Virginia). In Yukon, mammoth limb bones show signs of breakage by humans (*Homo sapiens*), and in Alaska, a fluted point made by an early hunter was found with bones of a young elephant – probably those of a woolly mammoth. The paper is illustrated by a restoration of a woolly mammoth, a top view of a lower jaw with teeth of the Whitestone mammoth from northern Yukon, and a left-side view of a male, showing its skeletal structure. [See Cooke et al. 1993.]

608. Harington, C.R. (1996c): American mastodon. Beringian Research Notes, No. 3:1–4.
This popular publication provides a general description of the American mastodon (*Mammut americanum*) with background on its appearance, evolution, dispersal, habits, and extinction. In 1739 an expedition from Montreal under the leadership of Charles Le Moyne, Baron de Longueuil, found bones and teeth on a journey down the Ohio River to the Mississippi. But it was not until 1799 that the great French anatomist Georges, Baron Cuvier, first recognized that they belonged to an elephant-like animal he called *mastodon*. Toward the end of the last glaciation (10,000 years ago), American mastodons ranged from Alaska and Yukon to central Mexico, and from Pacific to Atlantic coasts [see Harington, Grant, and Mott 1993]. Indeed, teeth have been dredged up by fishermen off the Atlantic coast. In Canada, most mastodon remains (more than 60 specimens) have been found in postglacial sediments of southern Ontario. Fossils are known from the Northwest Territories, but not Nunavut, and every province except Prince Edward Island and Newfoundland. One of the most remarkable specimens is the skeleton of a young adult from Hillsborough, New Brunswick, that had become mired in a swamp during the last (Sangamonian) interglacial about 100,000 years ago. Mastodon remains of similar age are known from sinkholes at Miller Creek and East Milford, Nova Scotia. At the last site, remains of an adult skeleton and a 7-year-old were found with frog (Amphibia) and turtle (Reptilia) bones. Illustrations show: a restoration of a mastodon in its open spruce forest habitat, and side and top views of a mastodon lower-third molar from Old Crow Basin, Yukon.

609. Harington, C.R. (1996d): North American short-faced bear. Beringian Research Notes, No. 4:1–4.
This popular publication provides a general description of North American short-faced bears (*Arctodus simus* and *Arctodus pristinus*) with background on their appearance, evolution, dispersal, habits, and extinction. These unusually tall and highly carnivorous bears ranged from Alaska and Yukon to Mexico and from Pacific to Atlantic coasts. The giant short-faced bear (*Arctodus simus*) is important here since *Arctodus pristinus* is not represented in northern North America. A striking feature of the largest known skull of *Arctodus simus*, from Gold Run Creek, Yukon, is its great width relative to length [nearly 80% compared with less than 70% for recent brown and black bears (*Ursus arctos* and *Ursus americanus*)]. The giant short-faced bear had reached its northernmost range (as well as its maximum size) in Yukon and Alaska by the Mid-Wisconsinan interstadial. Radiocarbon dates of five Yukon specimens are approximately: 44,000 BP, 29,600 BP, 26,000 BP, 25,000 BP, and 20,000 BP (on an excellently preserved cranium of an adult female from Ophir Creek – see Figure 3). Other illustrations are a restoration of *Arctodus simus* attacking a steppe bison (*Bison priscus*) and a diagram showing the relative size of brown and giant short-faced bears (*Ursus arctos* vs *Arctodus*

simus). Apart from Yukon, the only other recorded Canadian specimens are from Mid-Wisconsinan deposits at Edmonton, Alberta, and possibly last (Sangamonian) interglacial deposits at Lebret, Saskatchewan.

610. Harington, C.R. (1996e): American lion. Beringian Research Notes, No. 5:1–4.
This popular publication provides a general description of American lions (*Panthera leo atrox*) with background on their appearance, evolution, dispersal, habits, and extinction. American lions were among the largest flesh-eating land animals of the ice age. They ranged from Alaska and Yukon as far S as Mexico [the report from Talara, Peru, applies to a large jaguar, not a lion – see Seymour 1983, p. 1]. Two items suggest that these lions may have had faintly striped pelts (Figure 1 shows this species attacking an extinct Yukon horse): a detailed wall engraving from Les Combarelles Cave (France) and a mammoth-tusk figurine of a person wearing the striped pelt of a large, long-tailed cat from Mal'ta (Russia). Two other depictions from French caves, the author suggests, may be among the first recorded attempts at motion pictures! One is an engraved, painted lioness (with a cub's head below her belly) on the wall of Les Trois Frères showing her head and tail in at least three different positions (Figure 2) suggesting that her head wagged as her tail thrashed up and down. A side view of a lion from Grotte Marie with too many front and hind legs suggests walking. Ice age lions were able to pursue bison (*Bison*), horse (*Equus*), and mammoth (*Mammuthus*) herds as they slowly spread across the Bering Isthmus, which existed during Late Pleistocene glaciations. Probably Eurasian cave lions (*Panthera leo spelaea*) gave rise to American lions after they entered Alaska during the Illinoian glaciation. During the last (Wisconsinan) glaciation, lions were isolated in refuges in northwestern North America (Eastern Beringia) and S of the ice sheets. In Canada, most American lion remains have been found in Yukon deposits. The best is a virtually complete skull (side view in Figure 3) from the Dawson City area. Other Yukon fossils are from Old Crow Basin and Bluefish Caves (N of the Arctic Circle), Sixtymile, Dublin Gulch, and Big Creek. Other Canadian fossils are from Edmonton, Bindloss, and Medicine Hat in Alberta. The frozen carcass of a steppe bison (*Bison priscus*) found near Fairbanks, Alaska, in 1979 showed signs of being killed by lions in early winter some 36,000 years ago [see Guthrie 1990a]. Obviously American lions had their problems, for two Yukon fossils show severe pathological damage to the front of the lower jaws – it is not difficult to imagine them being kicked when attacking horses.

611. Harington, C.R. (1996f): Giant beaver. Beringian Research Notes, No. 6:1–4.
This popular publication provides a general description of giant beavers (*Castoroides ohioensis*) with background on their appearance, evolution, dispersal, habits, and extinction [Figure 1 shows a skeleton in side view with lower jaws and teeth of a giant beaver from Porcupine River, Yukon, and a modern beaver (*Castor canadensis*)]. The giant beaver was the largest ice age North American rodent. *Castoroides* ranged from Florida to Yukon and Alaska (where the only evidence is a small fragment of a cutting tooth from Cripple Creek near Fairbanks). The most northerly records are from Old Crow Basin, Yukon (which lies about 150 km N of the Arctic Circle), where many fossils have been found in deposits varying in age from the last (Sangamonian) interglacial to the Early Wisconsinan glaciation. Fossils show that both giant and modern beavers coexisted in Old Crow Basin. The only giant beaver fossil found elsewhere in Canada is a cutting tooth from last interglacial deposits in the Don Valley, Toronto. [Another cutting tooth is reported from Indian Island, New Brunswick – see Young 2000 and Miller et al. 2000.] The author speculates that giant beavers spread rapidly northwestward into Yukon from the S via chains of lakes which tend to form along the southern margins of the Canadian Shield (e.g., Lakes Superior, Manitoba, Athabasca, and Great Slave and Great Bear lakes). A likely time for this northerly shift would have been during the last interglacial. Figure 2 shows a photograph of a life-like model of a giant beaver 2.5 m tall, and Figure 3 is a restoration (ink sketch) of the front part of the body of a giant beaver.

612. Harington, C.R. (1996g): American scimitar cat. Beringian Research Notes, No. 7:1–4.
This popular publication provides a general description of scimitar cats (*Homotherium serum*) with background on their appearance, evolution, dispersal, habits, and extinction. No large ice age North American cat is more interesting than the scimitar cat. Although much rarer than the sabretooth cat (*Smilodon fatalis*) and the American lion (*Panthera leo atrox*), it seems to have been the only

sabretooth-like cat from Eastern Beringia (unglaciated Alaska, Yukon, and adjacent Northwest Territories). It ranged broadly in North America from Yukon and Alaska to Texas and Florida. The first-known record for Yukon and Canada was a mandible with teeth from Old Crow River, Yukon. An apparent clue to the appearance of *Homotherium* may come from a 16-cm-long Paleolithic stone carving of a cat, lacking feet, from the cave of Isturitz in southwestern France. Although the carving has often been referred to the cave lion (*Panthera leo spelaea*), the short tail and deep-set lower jaw indicate that it is best identified as *Homotherium latidens*, a Late Pleistocene European scimitar cat. Detailed surface markings on the carving indicate that *Homotherium* may have had fine spots on the body with paler underparts and sheathed 'scimitars' (canine teeth). Evidence from Friesenhahn Cave, Texas, indicates that scimitar cats lived in caves and preyed heavily on young mammoths (Figure 1). The author attempts to reconstruct a typical hunt by a scimitar cat. Figure 2 is a sketch of a side view of a scimitar cat skull.

613. Harington, C.R. (1996h): Steppe bison. Beringian Research Notes, No. 8:1–4.

This popular publication provides a general description of steppe bison (*Bison priscus*) with background on their appearance (Figure 1 shows two bulls jousting), evolution, dispersal, habits, and extinction. This large-horned species was well adapted to cool, steppe-like grasslands that existed throughout the Holarctic (northern Eurasia and North America) from England eastward to the Northwest Territories of Canada during the ice age. Images from Paleolithic caves provide many insights into the appearance and behaviour of these bison (Figure 3, a mural on the roof of Altamira Cave, Spain, shows a charging bull). Perhaps the most striking scene in Paleolithic art is on the wall of Lascaux Cave, France. It consists of a man falling back with arms outstretched before a charging bison (with head down and tail raised) that has been wounded and is spilling its entrails. 'Blue Babe,' a nearly complete 8- to 9-year-old male carcass found in 1979 at Pearl Creek, a placer mining site near Fairbanks, Alaska, was radiocarbon dated to about 36,000 BP. Its weight was estimated to be 700 to 800 kg and its body was mainly rich dark brown with blackish peripheral regions (e.g., front of face, beard, anterior and posterior humps, tail, and legs). Partial carcasses of other steppe bison have been found in Yukon, Alaska, and Siberia. In addition to Blue Babe, a large male carcass radiocarbon dated to about 31,000 BP was found in 1952 at Fairbanks Creek, and two lower legs were found in 1936 at Cleary and Goldstream creeks, Alaska. The Yukon specimens are two lower forelegs – one with abundant blackish-brown fur – from the Dawson City area. Among the most significant and best-preserved steppe bison specimens is a series of bones from Old Crow Basin Locality 11(1). Skeletal parts include excellent skulls with hornsheaths preserved, representing more than seven individuals of both sexes [Figure 2 is a photograph of the right side of a female cranium with hornsheaths from Old Crow Locality 11(1)] and various ages. Perhaps they are from a herd that broke through thin or rotting ice in early winter or late spring. Their high stratigraphic position and several overlapping radiocarbon dates indicate that they died about 12,000 BP – the latest solid evidence of Yukon steppe bison. Other Canadian specimens are from Edmonton, Alberta, and Roaring River, Manitoba. Probably steppe bison reached their maximum distribution and numbers during the last (Wisconsinan) glaciation – it's the typical bison of Yukon and Alaska during that period. Analyses of 'tooth jam' (plant fragments caught in pits in Alaskan steppe-bison cheek teeth) suggest that grasses were predominant in the diet – the two most common in Blue Babe are *Agropyron* and *Danthonia*. About 3.5% of more than 3000 Yukon steppe bison bones examined showed signs of pathological problems: twisted horncores, periodontal disease, suppurative swellings of the lower jaw, bony swellings on vertebrae (including a series of thoracic vertebrae with spines broken, offset and healed, probably resulting from a fall), broken and healed ribs, fractured and healed leg and upper foot bones, and apparent osteoarthritic swellings near the joints of a thigh bone. Claw and canine puncture marks on the carcass of Blue Babe indicates that this adult male steppe bison was killed in winter by American lions (*Panthera leo atrox*). Then the carcass was scavenged by smaller mammals (Mammalia) and birds (Aves). Early people (*Homo sapiens*) hunted and butchered steppe bison because, at the Engigstciak, Yukon, archaeological site near the Beaufort Sea coast, three bones dating between 10,400 and 9,400 BP show well-preserved, high-impact ring fractures attributable to human butchering. The author suspects that steppe bison were hunted by people in the Dawson City area (Nugget Gulch) even earlier, for a similar ring fracture was detected on a bison upper foreleg dated to approximately 31,000 BP. The author suggests a three-step hypothesis for extinction of this species: (1) rapidly changing climate and environment toward the close of the last glaciation resulted in replacement of steppe-like grassland range by boreal forest and tundra; (2) these bison withdrew to 'islands' of higher, better-drained country where their habitat would survive longer; (3) in such places, human hunters could have accelerated their extinction (e.g., Engigstciak?). The author thinks that smaller-horned bison such as Early Holocene western bison (*Bison bison occidentalis*) and wood bison (*Bison bison athabascae*) are descendants of the steppe bison.

614. Harington, C.R. (1997a): Pleistocene vertebrates of Sixtymile, Yukon Territory: A preliminary discussion. *In*: Terrestrial Paleoenvironmental Studies in Beringia (M.E. Edwards, A.V. Sher, and R.D. Guthrie, eds.). Alaska Quaternary Center, University of Alaska, Fairbanks. pp. 83–90.

This paper focuses on the long-term excavations (1975–86) at Sixtymile Loc. 3, a prolific site located in an unglaciated area of Yukon. Approximately 1300 specimens represent 20 species ranging from birds (Aves) to bears (*Ursus*), ferrets (*Mustela*), muskox (*Ovibos*), mastodon (*Mammut americanum*), wolverine (*Gulo gulo*), lion (*Panthera leo atrox*), and camels (*Camelops hesternus*). Apart from the variety and numbers of bones, the quality of the fossils (including well-preserved carcasses) and the number of radiocarbon-dated samples, make this site significant. The site stratigraphy, paleoenvironment, geochronology, and Pleistocene fauna, as well as a discussion of some important finds and their implications, are presented. The fauna and radiocarbon data are provided in table form. Radiocarbon dates from Loc. 3 range from 40,000 to 20,000 BP. Dates include: western camel 23,320 ± 640 BP (Beta-8864); brown bear (*Ursus arctos*) 36,500 ± 1150 BP (Beta-16162); and black-footed ferret (*Mustela nigripes*) 39,560 ± 490 BP (TO-214). Six more specimens have been submitted for AMS radiocarbon dating, including snow goose (*Chen caerulescens*) and wolverine, which have never been dated from Yukon. Radiocarbon dates reported in the paper also include those known from other Sixtymile sites nearby. [See Harington 1989c, for these dates.]

615. Harington, C.R. (1997b): Ice age Yukon and Alaskan camels. Beringian Research Notes, No. 10:1–4.

This popular publication provides a general description of ice age camels (Family Camelidae) with background on their evolution, dispersal, habits, and extinction. Camelid bones recovered from Old Crow Basin in northern Yukon are chiefly from camels much larger than either of the living species (*Camelus bactrianus* and *Camelus dromedarius*). The fossils are closest in shape and size to a very large member of the true camels (Camelini) like *Titanotylopus* (see restoration, Figure 1). Could these Yukon *Titanotylopus*-like camels be relics of an earlier migration to Eurasia (Odessa and Cherkassy, Russia) about 5 million years ago? Bones of another smaller camel [western camel (*Camelops hesternus*) – see restoration Figure 2] have been recovered on the banks of Sixtymile River on the Yukon side of the Alaska-Yukon border. Oddly, all the bones are from one locality (Sixtymile Loc. 3) that has yielded remains of many other ice age vertebrate species. One of the *Camelops* bones was radiocarbon dated to about 23,000 BP – near the cold peak of the last glaciation. Thirty-six western camel remains have been reported from mining sites near Fairbanks, Alaska (about two-thirds are from Cripple Creek and Gold Hill; others are from Engineer, Fairbanks and Ester creeks). Dates on the bones range between 40,000 and 25,000 BP. So, perhaps western camels did not enter Yukon and Alaska from the S until the relatively warm Mid-Wisconsinan interstadial (about 50,000 to 25,000 BP), dying out there toward the peak of the last glaciation. Western camels were most abundant in the western United States and southwestern Canada (Alberta and Saskatchewan) and central Mexico during the last part of the ice age. They probably reached unglaciated Yukon and Alaska by migrating northward via dry terrain on the eastern flanks of the Rocky Mountains during a relatively warm period.

616. Harington, C.R. (1998a): Walrus tusks, whale bone date back 9,000 years ago. Northern Pen [St Anthony's, Newfoundland] 19(16):A-12.

An adult male Atlantic walrus (*Odobenus rosmarus*) skull and partial skeleton discovered in 1994 from a sandpit near Cooke's Harbour, Newfoundland, and a rib fragment of a large whale (cf. *Balaena mysticetus*) from the same raised-beach deposit site have yielded radiocarbon dates of about 8700 BP (neck vertebra) and 9000 BP respectively. These dates indicate that the two species rapidly moved into the shallow coastal waters of Newfoundland shortly after the end of the last glaciation some 10,000 years ago. A female walrus partial cranium dredged up from the Grand Banks (one of only two known specimens from the area) has provided a radiocarbon date of 12,500 BP. This date is the second-oldest (oldest about 12,800 BP from the Bay of Fundy) dated walrus from Atlantic Canada. These early dates represent herds that were spreading northward as the sea ice withdrew. By about 9500 BP, walruses had penetrated the Canadian Arctic Islands and become common there by 8500 to 9000 BP and again from 3000 to 5000 BP. [See Dyke et al. 1999.]

617. Harington, C.R. (1998b): North American saiga. Beringian Research Notes, No. 11:1–4.

This popular publication provides a general description of saiga antelopes (*Saiga tatarica*) with background on their evolution, dispersal, habits, and extinction in North America. Saigas (Figure 1 is an illustration of a male and a female running), presently confined to Central Asia, spread westward to England and eastward to the Northwest Territories of Canada during the ice age. The species is a valuable paleoenvironmental indicator of dry, steppe-like grasslands. During the last glaciation these fertile, highly mobile saigas moved rapidly with the expansion of fresh, steppe-like range, reaching as far westward as Somerset in southwestern England via the 'English Isthmus' (now English Channel) and as far eastward as Baillie Islands, E of the Mackenzie Delta in Canada via the 'Bering Isthmus' (now Bering Strait). The oldest radiocarbon-dated saiga known is about 37,000 BP from Usuktuk River, northern Alaska, whereas the latest survivor, also from Alaska, dates to about 12,000 BP. Only three Canadian specimens are known: a left horncore with adjacent cranial bone dated to about 15,000 BP from Baillie Islands, Northwest Territories; a similar specimen dated to about 13,400 BP from Bluefish Caves, Yukon (Figure 2); and a lower foreleg bone from Old Crow Basin, Yukon, dated to about 13,200 BP. These dates, when combined with many other radiocarbon dates on specimens from Alaska and Siberia [Guthrie et al. (in press)], create two spikes on the graph, suggesting that saigas may have been most numerous before (about 40,000 to 20,000 BP) and after (about 15,000 to 12,000 BP) the peak of the last glaciation. Northern saigas may have been more adaptable than previously thought because: (1) their putative steppe-like range in northern Siberia, Alaska, Yukon, and Northwest Territories was peppered with unfamiliar arctic plant species; (2) in reaching Lost Chicken Creek, Alaska, they probably had to travel over unusually rough ground reaching altitudes close to 1500 m asl; (3) they endured fewer frost-free days and longer periods of winter darkness. Nevertheless, North American saigas probably died out about 12,000 BP (as they seem to have at the opposite extremity of their range in western Europe) because of rapid changes in climate and plantscapes occurring then, as former steppe-like terrain was replaced by spruce forest and tundra. The author speculates that what might have been most damaging to the survival of northern saigas were periods of sudden winter warming, creating tough ice-crusts that severely checked herd movement and made foraging difficult [see Harington and Cinq-Mars 1995]. The specimen from Bluefish Caves, Yukon, although evidently not butchered, comes from the earliest-known archaeological site in North America.

618. Harington, C.R. (1999a): Yukon Pleistocene vertebrates: A perspective. Canadian Archaeological Association, 32nd Annual Conference (April 28–May 2, Whitehorse, Yukon). Programme and Abstracts. pp. 40–1.

This presentation summarizes the history of collecting and research, highlighting the contributions of such people as Robert McDonald, A.G. Maddren, Copely Amory, Jr, O.J. Murie, and Otto Geist, as well as a few long-term projects such as those of the Canadian Museum of Nature, Northern Yukon Research Programme (University of Toronto), and the Yukon Refugium Project. Important recent developments are the appointment of a Yukon Archaeologist and establishment of the Yukon Beringian Interpretive Centre, Whitehorse (including its popular series Beringian Research Notes). Studies of Sixtymile and Dublin Gulch faunas have been published, and the author lists other faunal, paleopathological, and DNA projects that are in progress.

619. Harington, C.R. (1999b): Ancient caribou. Beringian Research Notes, No. 12:1–4.

This popular publication provides a general description of early caribou/reindeer (*Rangifer tarandus*) with background on their evolution, dispersal, and habits. Caribou (see painting of a bull caribou with steppe bison in background, Figure 1) is the only member of the deer family (Cervidae) adapted to the harsh arctic and subarctic environments of North America and Eurasia. With the horse (*Equus*), steppe bison (*Bison priscus*), and woolly mammoth (*Mammuthus primigenius*), it is one of the commonest species known from ice age Beringia. However, unlike those species, caribou survived the period of megafaunal extinction that occurred toward the close of the last glaciation about 10,000 BP. The oldest-known caribou are from the Early Pleistocene of Eastern Beringia [e.g., Fort Selkirk, Yukon, about 1.6 Ma, and Cape Deceit, Alaska, probably slightly younger (perhaps 0.9 Ma according to Brigham-Grette and Hopkins 1991)] and presumably originated there. Apart from the Fort Selkirk specimen, some of the oldest Yukon caribou finds are from Old Crow Basin. Fossils have been excavated from the bottom of the high bluffs that may be of Illinoian age or older, as well as from higher *in situ* deposits probably of last (Sangamonian) interglacial age (about 130,000 years old). Three other Yukon specimens are of interest: (1) a tibia (lower leg bone) fleshing tool from Old Crow Basin originally radiocarbon dated to about 27,000 BP (now considered to be only about 1300 BP) that was largely responsible for the 'great bone rush' in that region following 1966; (2) a partial beam of a caribou antler from Old Crow Basin with four polished facets on the base (interpreted as a pestle) radiocarbon dated to about 25,000 BP; and (3) a caribou antler 'punch,' used to flake stone tools (*Homo sapiens*), from the Dawson City area radiocarbon dated to about 11,300 BP. It is remarkable that North American caribou occupied three major ice-free refugia of likely significance for American biological and linguistic (*Homo sapiens*) differentiation. Beringian caribou produced the barren-ground race; the now-extinct Dawson caribou (*Rangifer dawsoni*) occupied the Northwest Coast refugium, dying out about 1910 on the Queen Charlotte Islands; and the southern refugium produced the woodland caribou. In addition, the small, pale Peary caribou (*Rangifer tarandus pearyi*) of the Canadian Arctic Islands may have originated in a western Banks Island refugium. Caribou had reached northernmost Greenland more than 40,000 BP, but there is no evidence that they survived there during the last glaciation. Genetic studies indicate that of five major lineages of North American caribou, the most ancient include woodland caribou from southern Canada. Northern populations tend to be concentrated in another lineage. Among the oldest-known examples of cave art, going back about 40,000 BP, are several lively depictions of reindeer from Chauvet Cave in southern France (see the mural of a resting bull in Figure 2).

620. Harington, C.R. (in press). Quaternary vertebrates of Quebec: A summary. Géographie physique et Quaternaire.

Marine mammals have played a prominent part in the Quaternary vertebrate history of Quebec since Per Kalm's (1749) mention of a whale (Cetacea) skeleton found inland from the St Lawrence River near Quebec City. Many other records of Champlain Sea (about 12,000–10,000 years ago) marine mammals have come from Montreal and Ottawa areas [where finds in calcareous clay nodules of fish (Pisces) and bird (Aves) remains, besides those of marine mammals, have come from sites near Eardley and Breckenridge, Quebec, as well as Green Creek, Ontario]. Recent collecting near Saint-Nicolas and Saint-Césaire has yielded interesting fish, bird, seal (*Phoca*), walrus (*Odobenus rosmarus*), and white whale (*Delphinapterus leucas*) remains. Narwhal (*Monodon monocerus*) bones have been reported from the N and S coasts of Baie des Chaleurs. Table 1 lists vertebrates of the Champlain Sea and vicinity. Among the most remarkable specimens of Champlain Sea age from Quebec are: (1) most of a common finback whale (*Balaenoptera physalus*) found near Daveluyville and dated to about 11,400 BP; (2) a nearly complete skeleton of a ringed seal (*Phoca hispida*) from Hull, which suggests the presence of land-fast sea ice near the western margin of the sea during winter and spring; (3) the near-perfect skeleton of a leopard frog (*Rana pipiens*) found in a clay nodule at Eardley – perhaps the best-preserved ice age frog in North America; (4) a complete 4-m-long walrus skeleton found near Bic in 1869 that probably was destroyed by fire at Rimouski; and (5) a nearly complete bearded seal (*Erignathus barbatus*) skeleton from near La Durantaye. Perhaps cave faunas will become increasingly important in sorting out the Quaternary faunal sequence in Quebec. The finds from caves near Saint-Elzéar [three amphibians, two reptiles, four birds, 34 mammals including several specimens that indicate colder climate, e.g., arctic shrew (*Sorex arcticus*), arctic hare (*Lepus arcticus*), heather vole (*Phenacomys intermedius*), Ungava lemming (*Dicrostonyx hudsonius*), and least weasel (*Mustela nivalis*)] and La Rédemption in

Gaspé [seven mammal species, including Ungava lemming and grizzly bear (*Ursus arctos*)] are significant. In the Gatineau area, Mine Cave is characterized by black bear (*Ursus americanus*); white-tailed deer (*Odocoileus virginianus*); raccoon (*Procyon lotor*); mouse (*Peromyscus*); and big brown bat (*Eptesicus fuscus*) in the upper 100 cm, and by fishes, amphibians (Amphibia), reptiles (Reptilia), and 23 mammals in the lower 70 cm – including woodland vole (*Microtus pinetorum*) and Ungava lemming that are no longer found in the area. Laflèche Cave, about 27 km N of Hull, has produced four birds and 21 species of mammals (Table 2), including several arctic-adapted species: Snowy Owl (*Nyctea scandiaca*), arctic hare, arctic fox (*Alopex lagopus* – a jaw was dated at about 10,000 BP), and Ungava lemming. Evidently such species retreated northward with the melting Laurentide ice sheet, until they reached northernmost Quebec and Labrador, where they presently survive. Perhaps the evidence for Ungava grizzlies fits this hypothesis too. Caves with faunal remains have also been reported from near the mouth of Saguenay River. Table 3 lists 32 radiocarbon dates on Quaternary vertebrates from Quebec. An American mastodon (*Mammut americanum*) molar tooth from the Lac-St-Jean area is the only representative of the extinct North American megafauna known from the province. When radiocarbon dated, it may be of key importance in testing a hypothesis concerning the dispersal of mastodons in eastern North America during late-glacial time. Where large mammals that survived the late-glacial extinctions are concerned, caribou (*Rangifer tarandus*) are known to have occupied the Rivière-du-Loup area as early as Mid-Wisconsinan time (about 40,600 BP).

621. Harington, C.R., and Ashworth, A.C. (1986): A mammoth (*Mammuthus primigenius*) tooth from Late Wisconsin deposits near Embden, North Dakota, and comments on the distribution of woolly mammoths south of the Wisconsin ice sheets. Canadian Journal of Earth Sciences 23:909–18.

A well-preserved third molar of a rather old woolly mammoth (*Mammuthus primigenius*) was recovered from sand and gravel forming the highest (Herman) prominent strandline of Glacial Lake Agassiz near Embden in western Cass County, North Dakota. The Herman strandline is estimated to have formed about 11,500 BP, and presumably the tooth is of similar age. Lack of wear of the fossil and its position in the high-energy strandline deposits suggests that the animal lived nearby along the western shore of Glacial Lake Agassiz. This tooth is placed in context with other mammoth fossils from North America that indicate that woolly mammoths occupied discontinuous tundra-like range S of the ice sheets, extending from southern British Columbia to the Atlantic coast (Fig. 3).

622. Harington, C.R., and Beard, G. (1992): The Qualicum walrus: A Late Pleistocene walrus (*Odobenus rosmarus*) skeleton from Vancouver Island, British Columbia, Canada. Annales Zoologica Fennici 28:311–19.

An adult female walrus (*Odobenus rosmarus*) skeleton (CMN 38490) from Early Wisconsinan glaciomarine clay near Qualicum Beach, Vancouver Island, British Columbia (49°23′00″N, 124°34′50″W), is the best-preserved Pleistocene walrus specimen recorded from the W coast of North America. The radiocarbon age of the fossil is >40,000 BP (I-11617), which concurs with a stratigraphic assessment of the site, both suggesting an Early Wisconsinan age. The skeleton consists of most of the cranium and mandible; some of the ribs and vertebrae; most of the forelimbs except the right scapula and part of the left, and most of the carpals, metacarpals, and phalanges; most of the hindlimbs (although some are damaged) and most of the tarsals, metatarsals and phalanges. The chronological age of the specimen has been determined through tooth-cementum counts to be 12 years. Paleoenvironmental, stratigraphic, and geochronological data on the Qualicum walrus are all in accord with known habits and habitat of modern walruses, as well as with the knowledge of the species' past distribution. Other southerly fossil walruses are listed. The conclusion is drawn that Pacific walruses made southerly advances during Pleistocene glaciations and northward withdrawals during interglaciations.

623. Harington, C.R., and Cinq-Mars, J. (1995): Radiocarbon dates on saiga antelopes (*Saiga tatarica*) fossils from Yukon and the Northwest Territories. Arctic 48(1):1–7.

Saiga antelopes (*Saiga tatarica*), presently confined to Central Asia, spread westward to England and eastward to the Northwest Territories of Canada during the Late Pleistocene. Two adult male saiga cranial fragments from the Yukon Territory and the Northwest Territories have yielded radiocarbon dates of 13,390 ± 180 BP (RIDDL-

279) and 14,920 ± 160 BP (Beta-25119, ETH-3898) respectively. Thus, saigas occupied the easternmost part of their known Pleistocene range toward the close of the Wisconsinan glaciation. Saigas probably died out between 13,000 and 10,000 years ago in North America because of rapid changes in climate and plantscapes occurring about that time, as former steppe-like terrain was replaced by spruce forest and tundra. [See Harington 1998b].

624. Harington, C.R., and Clulow, F.V. (1973): Pleistocene mammals from Gold Run Creek, Yukon Territory. Canadian Journal of Earth Sciences 10(5):697–759.

Remains of 13 species of mammals are reported from Pleistocene deposits at Gold Run Creek (63°43.5′N, 138°41′W) near Dawson City, Yukon Territory. Eight species are extinct – giant short-faced bear (*Arctodus simus yukonensis*), American lion (*Panthera leo atrox*), American mastodon (*Mammut americanum*), woolly mammoth (*Mammuthus primigenius*), Yukon horse (*Equus lambei* = *Equus (Asinus) lambei*), helmeted muskox (*?Bootherium*), Alaskan bison (*Bison alaskensis*), and steppe bison (*Bison priscus* = *Bison crassicornis*); two are no longer living in the Yukon – American badger (*Taxidea taxus*) and kiang (*Equus kiang* = *Equus* cf. *Equus (Asinus) kiang*); and three are still extant in the area – wolf (*Canis* cf. *Canis lupus*), moose (*Alces alces*), and caribou (*Rangifer tarandus*). The most common elements of the fauna are the Yukon horse and steppe bison. The badger and Alaskan bison are reported for the first time from the Yukon Pleistocene. All species are photographically illustrated. These mammals may have inhabited a cool grassland or open parkland during Late Wisconsinan time. Steppe bison and mammoth bone from Gold Run Creek have yielded radiocarbon dates of 22,200 ± 1400 BP (I-3570) [this date is incorrectly quoted as 22,000 ± 1400 BP on p. 742 of the text] and 32,250 ± 1750 BP (I-4226) respectively. The Alaskan bison is evidently older than the remainder of the fauna, with a radiocarbon date of over 39,900 BP (I-5405). This specimen was found separately in a deep sink in the bed of Gold Run Creek and could very well have been deposited earlier than the other specimens. A map of the fossil locality (Figure 1), a stratigraphic diagram (Figure 2), and 26 tables of measurements are included.

625. Harington, C.R., and Eggleston-Stott, M. (1996): Carcass of a small Pleistocene horse from Last Chance Creek near Dawson City, Yukon. Current Research in the Pleistocene 13:105–7.

A partial carcass of a Yukon horse (*Equus lambei*) was found by placer miners at 15 Pup on Last Chance Creek near Dawson. It consists of most of the right foreleg, with dried flesh, skin, and some lower-leg hair; the hide from ear to tail with tail, mane, and some body hair; and the gut and its contents. A bone sample yielded an AMS radiocarbon date of 26,280 ± 210 BP (Beta-67407) indicating that the horse died during a relatively warm interval (Mid-Wisconsinan interstadial) before the peak of the last (McConnell/Wisconsinan) glaciation about 20,000 to 18,000 years ago. Study has shown that little microbial or geochemical deterioration has occurred since the horse died. Genomic DNA studies are currently underway, as are detailed investigations of the hair, hide, and searches for endoparasites and ectoparasites. This well-preserved specimen has provided an opportunity to assess the evolutionary relationship between *Equus lambei* and modern horses.

626. Harington, C.R., and Fitzgerald, G.R. (1973): The Pakenham Whale. Ottawa Journal, April, 28. p. 37.

This article describes relocating the skeleton of a white whale [*Delphinapterus leucas* – dated at 10,400 ± 80 BP (Harington 1988)] found near Pakenham, Ontario, in 1906 and initially described by Whiteaves (1907). The article is accompanied by a photograph of all of the bones preserved, a close-up of the skull, a sketch of its appearance in the Champlain Sea, and a map of the locality.

627. Harington, C.R., and Fitzgerald, G.R. (1987): Excursion M: Vertebrate paleontology of the Western Basin of the Champlain Sea. *In*: INQUA 87 (R.J. Fulton, ed.). Quaternary of the Ottawa Region and Guides for Day Excursions. pp. 18, 115–17.

A history of collecting in the region is provided along with a map of the excursion sites and a description of the geology, paleoecology, and fossil species from the two sites (Foster's Sandpit near the Ottawa International Airport, Ontario, and the Eardley sites, Quebec).

628. Harington, C.R., and Irving, W.N. (1967): Some Upper Pleistocene middens near Old Crow, Yukon

Territory. National Museums of Canada. 10 pp. [Mimeographed preliminary report in Quaternary Vertebrates of Northern North America Publication file at Canadian Museum of Nature].
This report was stimulated by the discovery of what appeared to be artifacts (*Homo sapiens*) at Loc. 14N during Harington's 1966 fieldwork in Old Crow Basin, Yukon. After giving background information on the geography and stratigraphy of the basin, the authors discuss the Pleistocene mammals and the artifacts. [But see Cinq-Mars and Morlan 1999, Cinq-Mars et al. 1999, and Nelson et al. 1986.] Some Pleistocene vertebrate fossils mentioned are: beaver (*Castor canadensis*), giant beaver (*Castoroides*), canid (*Canis*), arctic fox (*Alopex lagopus*), woolly mammoth (*Mammuthus primigenius*), horse (*Equus*), caribou (*Rangifer tarandus*), bison (*Bison*), tundra muskox (*Ovibos moschatus*), helmeted muskox (*Bootherium*), ground sloth (*Megalonyx*), hare (*Lepus*), camelid (*Camelops*), and moose (*Alces alces*). Radiocarbon dates are given for the peat at the top of the section – 6430 ± 140 BP (GSC-121) – and a log at the bottom – >41,300 (GSC-199). The authors have identified artifacts made from mammoth and other bones that they believed are from middens dating from before the peak of the Wisconsinan glaciation.

629. Harington, C.R., and Morlan, R.E. (1992): A Late Pleistocene antler artifact from the Klondike District, Yukon Territory, Canada. Arctic 45:269–72.
A modified caribou (*Rangifer tarandus*) antler, interpreted as a flintknapper's (*Homo sapiens*) punch, was collected with hundreds of other Pleistocene mammal bones at Hunker Creek (63°55′N, 138°52′W) near Dawson City, Yukon. It has yielded an AMS radiocarbon date of 11,350 ± 110 BP (Beta-27512, ETH-4582), which fits within the span of the Nenana complex, characterized both by core and blade and by core and flake technology. Although the specimen was not found in stratigraphic context, its probable burial history is inferred from the radiocarbon age and surface alteration. Remains of the following Pleistocene mammals were found at the same locality about the same time: steppe bison (*Bison priscus*), woolly mammoth (*Mammuthus primigenius*), Yukon horse (*Equus lambei*), caribou (*Rangifer tarandus*), helmeted muskox (*Bootherium bombifrons* = *Symbos cavifrons*), American lion (*Panthera leo atrox*), and wolf (*Canis lupus*). A date of 23,900 ± 470 BP (I-8580) for a caribou antler from the same collection is given.

630. Harington, C.R., and Occhietti, S. (1980): Pleistocene eider duck (*Somateria* cf. *mollissima*) from Champlain Sea deposits near Shawinigan, Québec. Géographie Physique et Quaternaire 34(2):239–45.
Most of a pelvic bone of a large Common Eider (*Somateria* cf. *Somateria mollissima*) was found in a unit of stratified clay and sandy silt near Shawinigan Quebec (46°34′35″N, 72°43′40″W). A radiocarbon date of bivalves on marine mollusc shells just above the fossil yielded a date of 10,300 ± 100 BP (GSC-2101). The fossil is important because it is the first bird (Aves) specimen from Champlain Sea deposits to be identified to the generic level. Evidently the duck died and was deposited near the northern shore of the sea a few hundred years before it drained.

631. Harington, C.R., and Occhietti, S. (1988): Inventaire systématique et paléoécologie des mammifères marins de la Mer de Champlain (fin du Wisconsinien) et de ses voies d'accès. Géographie physique et Quaternaire 42:45–64.
The Champlain Sea covered a large area between Quebec City and Lake Ontario from approximately 12,400 years ago to 9300 years ago. Several species of whales, particularly those adapted to cool inshore conditions, lived in this sea. Approximately 80% of whale species recorded from Champlain Sea deposits are white whales (*Delphinapterus leucas*). Other whale species represented are: humpback (*Megaptera novaeangliae*); bowhead (*Balaena mysticetus*); finback (*Balaenoptera physalus*); and harbour porpoise (*Phocoena phocoena*). Two narwhal (*Monodon monoceros*) specimens are recorded from the eastern approaches to the Champlain Sea, but their deep water preference evidently prevented them from entering the sea. Seals, particularly those adapted to breeding on pack ice, such as harp (*Phoca groenlandica*) and bearded (*Erignathus barbatus*), and those adapted to breeding on land-fast ice, such as ringed (*Phoca hispida*), also lived in the Champlain Sea. An open coastal water species, the harbour seal (*Phoca vitulina*) has also been found near the southern margin of the sea. Although walrus (*Odobenus rosmarus*) specimens have been recorded from 14 localities in the eastern approaches to the sea, the species has not yet been recorded from Champlain Sea deposits. This is somewhat surprising given the abundance of prey

species, both fish and mollusc, which were available [but see Bouchard et al. 1993]. Many paleoenvironmental aspects of the Champlain Sea are discussed, including the invertebrate components. New reports include: a white whale (*Delphinapterus leucas*) from Montreal, a ringed seal (*Phoca hispida*) from Ottawa, and a bearded seal (*Erignathus barbatus*) from Saint-Hillaire, Quebec. Maps of Champlain Sea whales (Figure 1) and seals (Figure 2) are provided, as well as a table of vertebrates, algae, echinoderms, and sponge fossils from the sea and its eastern approaches (Table I); the biostratigraphy of the sea (Table II); and radiocarbon dates on vertebrates, marine molluscs, and plants from the sea (Table III). Six of the marine mammal specimens are illustrated.

632. Harington, C.R., and Sergeant, D.E. (1972): Pleistocene ringed seal skeleton from Champlain Sea deposits near Hull, Quebec – A reidentification. Canadian Journal of Earth Sciences 9(8):1039–51.

A seal skeleton from Champlain Sea deposits near Hull, Quebec, formerly identified as probably belonging to a young harp (*Phoca groenlandica*) or harbour seal (*Phoca vitulina*), actually represents a seven-year-old ringed seal (*Phoca hispida*). The skeleton was deposited *in situ*, probably during an early cold phase of the Champlain Sea about 12,000–11,000 BP. The specimen suggests that fast ice existed near the western margin of the Champlain Sea during winter and spring. Two photographs of the mounted skeleton (Figures 3A, B) and a photomicrograph of a canine-tooth section, to show how the age of this individual is estimated (Figure 4), are given, as are pictures of the mandibles of the fossil compared with those of modern ringed, harp, and harbour seals (Figures 5A, B) to show how they can be identified. Table 3 lists additional seal records from Champlain Sea deposits near Ottawa, Ontario.

633. Harington, C.R., and Shackleton, D.M. (1978): A tooth of *Mammuthus primigenius* from Chestermere Lake near Calgary, Alberta, and the distribution of mammoths in southwestern Canada. Canadian Journal of Earth Sciences 15(8):1272–83.

A molar of a woolly mammoth (*Mammuthus primigenius*) was found 100 m from the western shore of Chestermere Lake (51°02′N, 113°50′W). It is believed to be Late Wisconsinan in age. Included are photographs of this specimen, which is one of many mammoth remains found in the Calgary area. The Chestermere Lake fossil is considered in relation to 94 records of mammoth cheek teeth from Western Canada (Manitoba 5, Saskatchewan 35, Alberta 17, British Columbia 17 – Figure 2, Table 2). Some problems concerning the relationships of North American and Eurasian mammoths (*Mammuthus*) are mentioned.

634. Harington, C.R., Anderson, T.W., and Rodrigues, C.G. (1993): Pleistocene walrus (*Odobenus rosmarus*) from Forteau, Labrador. Géographie Physique et Quaternaire 47(1):111–18.

A partial skeleton of a walrus (*Odobenus rosmarus*) was found in raised-beach deposits, 4 m below the surface, 2.5 km NW of Town Wharf, Forteau, Newfoundland (Labrador) (51°29.3′N, 56°58.6′W). A radiocarbon date on bone collagen from two rib fragments was 11,490 ± 160 BP (Beta-16518). The article includes a map, photographs, and a photomicrograph of a section of a premolar of the specimen. Other walrus finds mentioned are: a complete skeleton from a gravel pit, near Moncton, New Brunswick, 9700 ± 130 BP (Beta-16161); a deteriorated anterior skull fragment dredged from the bottom of the Bay of Fundy, 9360 ± 90 BP (TO-1554); a cranium also from the Bay of Fundy, 12,760 ± 90 BP (TO-1927); a cranial fragment from W of Ste-Julienne, Quebec, 10,090 ± 60 BP (TO-2224); and one from Bathurst Island, Nunavut, 7320 ± 120 BP (I-7796) [correct lab number is I-7795]. Data from associated pollen and marine mollusc shells suggest that the Forteau specimen was reworked and deposited by a receding sea some 10,300 to 10,400 years ago, when sparse tundra vegetation evidently bordered on a glaciomarine environment.

635. Harington, C.R., Bonnichsen, R., and Morlan, R.E. (1975): Bones say man lived in the Yukon 27,000 years ago. Canadian Geographic Journal 91(1 and 2): 42–8.

This article is a synthesis of the prevailing arguments for human (*Homo sapiens*) presence in the Old Crow Basin, Yukon, before the peak of the last glaciation, over 20,000 BP. The senior author, during a field trip in 1966, discovered the first evidence of such ancient human presence in the area. A map of fossil and artifact localities, and photographs of bone and stone tools, support the arguments. [See Cinq-Mars and Morlan 1999, Morlan et al. 1990, and Nelson et al. 1986.]

636. Harington, C.R., Grant, D.R., and Mott, R.J. (1993): The Hillsborough, New Brunswick, mastodon and comments on other Pleistocene mastodon fossils from Nova Scotia. Canadian Journal of Earth Sciences 30:1242–53.
This paper presents a detailed description of the remains of a young adult (15–18 years old) mastodon (*Mammut americanum*) found in 1936 in peaty silt and clay underlying till in gypsum karst near Hillsborough, New Brunswick. Three hundred and twelve bones and bone fragments and some associated spheroids believed to be coprolites are documented. Radiocarbon ages of the site include peat aged at >43,000 BP (GSC-1680) and wood fragments from a coprolite aged at 37,200 ± 1310 BP (GSC-2467). Other attempts at aging yielded unreliable dates. Extensive measurements of the fossils are provided along with photographs of both the Hillsborough specimen and some of the Nova Scotia specimens. A pollen analysis was undertaken of the spheroids and of clay and peat sediments. The data suggest that the mastodon died during the Sangamonian interglaciation about 100,000 years ago, when the habitat was similar to that of postglacial mastodons found elsewhere in North America. The authors then present a paleogeographic hypothesis to explain the apparent absence of postglacial mastodons in the Maritime provinces.

637. Harington, C.R., Plouffe, A., and Jetté, H. (1996): A partial bison (*Bison* cf. *B. latifrons*) skeleton from Chuchi Lake, and its implications for the Middle Wisconsinan environment of central British Columbia. Géographie physique et Quaternaire 50(1):73–80.
Fragmentary but massive left and right horncores, found with eight postcranial bones (in the collection of the University of British Columbia Museum of Anthropology), from a clay unit underlying a diamicton of the last (Fraser) glaciation at Chuchi Lake, British Columbia, probably represent an individual giant bison (*Bison* cf. *Bison latifrons*). A radiocarbon (AMS) date of 30,740 ± 220 BP (TO-3653) from one of the horncores contrasts with overlapping dates on an associated humerus of 34,800 ± 420 BP (Beta-78573, CAMS-17566) and 35,480 ± 1080 BP (Beta-78574). Despite this discrepancy, these dates are in accord with the suspected stratigraphic age of the clay unit. Pollen evidence from the site suggests the massive-horned bison occupied an open forest in the vicinity. The Chuchi Lake specimen is important because it is the first indication of giant bison from British Columbia, and it appears to be one of the latest-known survivors of this species.

638. Harington, C.R., Tipper, H.W., and Mott, R.J. (1974): Mammoth from Babine Lake, British Columbia. Canadian Journal of Earth Sciences 11(2):285–303.
A partially articulated Columbian mammoth (*Mammuthus* cf. *Mammuthus columbi*) skeleton was found in silty pond deposits overlain by a thin layer of gravel and a thick layer of glacial till on Babine Lake, British Columbia (55°00′N, 126°14′W). No molars were found, so the identification was based on limb proportions. Given were the following radiocarbon dates: two on wood from the silty layer 42,900 ± 1860 BP (GSC-1657), 43,800 ± 1830 (GSC-1687); and 34,000 ± 690 BP (GSC-1754) on a mammoth bone. The animal probably was trapped in a silty pond and died there. Paleobotanical evidence indicates that during this phase of the Olympia Interglaciation, the vegetation near Babine Lake was similar to present shrub-tundra just beyond the treeline in northern Canada.

639. Harris, S.A. (1985): Evidence for the nature of the Early Holocene climate and paleogeography, High Plains, Alberta, Canada. Arctic and Alpine Research 17(1):49–67.
Mammals of the Bow River Valley, Cochrane, and Calgary, Alberta, are discussed relative to climate and paleogeography (pp. 63–4). This paper includes a chart (Figure 12) with radiocarbon-age ranges (12,500–6000 BP) for the following species: bison (*Bison*), western camel (*Camelops* cf. *Camelops hesternus*), wapiti (*Cervus elaphus*), Mexican horse (*Equus conversidens*), mammoth or mastodon (*Mammuthus* or *Mammut*), bighorn sheep (*Ovis canadensis*), caribou (*Rangifer tarandus*), pronghorn antelope (*Antilocapra americana*), wolf (*Canis lupus*), coyote (*Canis latrans*), beaver (*Castor canadensis*), painted turtle (*Chrysemys picta*), white-tailed jackrabbit (*Lepus townsendii campanius*), deer (*Odocoileus*), muskrat (*Ondatra zibethicus*), raccoon (*Procyon lotor hirtus*), cottontail (*Sylvilagus nuttallii*), and fox (*Vulpes*).

640. Harrison, J.E. (1973): Dated paleosol from below Mazama (?) tephra. Geological Survey of Canada Paper 73-1 (Part B):127–8.
Mammal bone fragments were found 1.6 km W of Cal-

gary, 90 m N of Highway 1. They were in a pit with 5 m of massive sands and silts containing a tephra bed and buried paleosol 61 m above the Bow River, 1128 m asl. Associated twig fragments gave a radiocarbon age of 8400 ± 150 BP (GSC-1819) for the paleosol, so the small mammal [Rodentia?] bones just below are older than that.

641. Hay, O.P. (1902): Bibliography and catalogue of the fossil vertebrata of North America. Bulletin of the United States Geological Survey No. 179, 868 pp.
A bibliography of authors, titles and a catalogue of species. [See Hay 1929.]

642. Hay, O.P. (1912): The recognition of Pleistocene faunas. Smithsonian Miscellaneous Collections 59(20):1–16.
This paper includes Pleistocene faunal distribution maps for North America (but not N of 60°N). They are: mastodon (*Mammut americanum*), woolly mammoth (*Mammuthus primigenius*), Columbian mammoth (*Mammuthus columbi*), giant beaver (*Castoroides ohioensis*), peccary (*Platygonus compressus*), muskox (*Ovibos*), horses (*Equus*), tapirs (*Tapirus*) and giant ground sloths (*Megatherium*, *Mylodon*, *Paramylodon*, and *Megalonyx*). Regarding the giant beaver, the author specifically mentions that 'it is now known to have lived at one time on the Old Crow River, Yukon Territory' (p. 8).

643. Hay, O.P. (1913a): Description of the skull of an extinct horse found in central Alaska. Smithsonian Miscellaneous Collections 61(2):1–18.
A fossil skull of a Niobrara horse (*Equus niobrarensis alaskae*) was found at Tofty, Alaska, a small mining town on Sullivan Creek, 38 km SW of Rampart and 11 km NW of Hot Springs (map provided). Photographs and a table of bone measurements are included. Species from the Old Crow River, Yukon, 80 km from its mouth include mammoth (*Mammuthus*), bison (*Bison*), and Niobrara horse (*Equus* cf. *Equus niobrarensis alaskae*). Other horses have been found throughout Alaska and Yukon. A fragmentary pelvis of a horse from mine tailings at Fox Gulch near Dawson City, Yukon, and the right ramus of the lower jaw of a young horse – part of a metatarsal and a first phalanx collected on the Porcupine River, probably not far from Rampart House – are mentioned (p. 14).

644. Hay, O.P. (1913b): The extinct bisons of North America; with descriptions of one new species, *Bison regius*. Proceedings of the United States National Museum 46(2021):161–200.
The author gives a taxonomic review of the bison fossils known at the time. He discusses and illustrates ancient bison (*Bison bison antiquus* = *Bison antiquus*), western bison (*Bison bison occidentalis* = *Bison occidentalis*), steppe bison (*Bison priscus* = *Bison alleni* = *Bison crassicornis*), giant bison (*Bison latifrons*), and describes and illustrates a new species (*Bison regius*) [now considered *Bison latifrons* – see Kurtén and Anderson 1980].

645. Hay, O.P. (1914): The Pleistocene mammals of Iowa. Iowa Geological Survey 23(4403):263.
Regarding northern North American specimens, the author mentions: (1) A fragment of a caribou (*Rangifer tarandus*) antler from Toronto – probably from Sangamonian deposits (pp. 34, 274). (2) The shaft of an antler of Scott's moose (*Alces scotti* = *Cervalces borealis*) from probable Sangamonian deposits at Toronto. A humerus of the type was collected near Brantford, Ontario – apparently the entire skeleton was present, but was discarded. The age of the specimen was estimated to be post-Wisconsinan (p. 263). (3) The report of a camel (*Camelops*) phalanx and remains of woolly mammoth (*Mammuthus primigenius*), horse (*Equus*), and extinct bison (*Bison*) (p. 230) found in the summer of 1912 on Old Crow River, Yukon. (4) Photographs of: helmeted muskox (*Bootherium bombifrons* = *Symbos tyrrelli*) type specimen from the Dawson City area, Yukon, Plate 37); steppe bison (*Bison priscus* = *bison alleni*) from Alaska, Plate 41; woolly mammoth (*Mammuthus primigenius*) tusk from Alaska, Plate 52; part of an upper right side of a jaw with a third milk molar and first molar from Alaska, Plate 55; an upper right third molar from Alaska, Plate 58; and a Columbian mammoth (*Mammuthus columbi*) lower third milk molar from Old Crow River, Yukon.

646. Hay, O.P. (1915): Contributions to the knowledge of the mammals of the Pleistocene of North America. Proceedings of the United States National Museum 48(2086):515–75.
The only northern North American specimens mentioned in this paper are: helmeted muskox (*Bootherium bombifrons* = *Bootherium nivicolens*) from Elephant Point, Eschscholtz Bay, Alaska (p. 523), and a Niobrara horse (*Equus niobrarensis*) from an unspecified locality in Alaska.

647. Hay, O.P. (1917): Description of a new species of extinct horse, *Equus lambei*, from the Pleistocene of Yukon Territory. Proceedings of the United States National Museum 53:435–43.
A complete skull of a Yukon horse [(*Equus lambei*) was found at Gold Run Creek, Yukon, buried 10 m below the surface. It is described as a new species, and is distinct from a horse skull described by J. Tscherski (1893) from Liakhof Island in northeastern Siberia, Russia (p. 443). Excellent photographs of the specimen are given (Plates 56–8).

648. Hay, O.P. (1923a): Characteristics of sundry fossil vertebrates. I: Additional species of *Equus lambei* from the Yukon. Pan American Geologist 39:101–5.
A nearly complete skull of a Yukon horse (*Equus lambei*), and a maxilla of another were found by M. Moldestrad at a gold-mine near Dawson City, Yukon. Included are photographs and tables of comparative measurements (p. 102) of the specimens.

649. Hay, O.P. (1923b): The Pleistocene of North America and its vertebrated animals from states east of the Mississippi and from Canadian Provinces east of longitude 95°. Carnegie Institute of Washington Publication No. 322. 499 pp.
Included is a detailed list of vertebrate specimens found in North America. Champlain Sea fossils of capelin (*Mallotus villosus*), lump sucker (*Cyclopterus lumpus*), smelt (*Osmerus mordax*), sculpin (*Cottus ricei* = *Cottus uncinatus* = *Artediellus atlanticus*), and stickleback (*Gasterosteus aculatus*) (p. 287) are mentioned, and this paper also contains distribution maps for the following groups or species from eastern Canadian fossil localities: cetaceans (unidentified whales), white whale (*Delphinapterus leucas*), walrus (*Odobenus rosmarus*), harp seal (*Phoca groenlandica*), harbour seal (*Phoca vitulina*), mastodon (*Mammut americanum*), woolly mammoth (*Mammuthus primigenius*), Columbian mammoth (*Mammuthus columbi*), deer (*Odocoileus*), wapiti (*Cervus elaphus*), caribou (*Rangifer tarandus*), tundra muskox (*Ovibos moschatus*), and bison (*Bison*).

650. Hay, O.P. (1924): The Pleistocene of the middle region of North America and its vertebrated animals. Carnegie Institute of Washington Publication No. 322a. 385 pp.
The Canadian discoveries mentioned in this book are: a Columbian mammoth (*Mammuthus columbi*) or woolly mammoth (*Mammuthus primigenius*) tooth, in surficial deposits on a bank of the North Saskatchewan River, 10 km upstream from Edmonton, Alberta (p. 84); Columbian mammoth (*Mammuthus* cf. *Mammuthus columbi*) teeth found near Moose Jaw, Saskatchewan (p. 121); a pair of mastodon (*Mammut americanum*) shoulder blades found on Swan River near Fort Pelly, Saskatchewan [actually from Manitoba – see Richardson 1854]; an unidentified proboscidean tooth found in Calgary, Alberta, and another found in the bank of Loon River, Alberta (57°41′N, 115°00′W); a woolly mammoth (*Mammuthus* cf. *Mammuthus primigenius*) tusk from Melville Island, Northwest Territories, at the head of Liddon Gulf (75°N); a horse (*Equus*) molar found in Empress, Alberta (p. 154), and a horse (*Equus*) skull found in York Factory, Manitoba, at the mouth of the Nelson River; a muskox (*Ovibos* cf. *Ovibos moschatus*) skull found on the bank of the North Saskatchewan River, about 1.5 km downstream from Edmonton, Alberta (p. 185); a bison (*Bison* cf. *Bison bison*) skull found on the Red Deer River, near Red Deer, Alberta (p. 209); and bison (*Bison*) remains found about 1.5 km W of Coleman, Alberta.

651. Hay, O.P. (1927): The Pleistocene of the western region of North America. Carnegie Institute of Washington Publication No. 322b. 346 pp.
A horse (*Equus*) molar was found at Empress, Alberta, and another one was found just N of Edmonton, Alberta (p. 373). Capelin (*Mallotus villosus*) fossils were found at Renfrew and Flat Rapids, Ontario (p. 273). Pronghorn (*Antilocapra americana*), bison (*Bison*), caribou (*Rangifer tarandus*), badger (*Taxidea taxus*), wolf (*Canis lupus*), and bear (*Ursus*) specimens were found N of Dundurn Saskatchewan (p. 284).

652. Hay, O.P. (1929): Second bibliography and catalogue of the fossil vertebrata of North America. Carnegie Institute of Washington Publication No. 390, 916 pp.
With the 1902 publication, this book lists most papers on fossil vertebrates in North America to 1928.

653. Haynes, G. (1980): Evidence of carnivore gnawing on Pleistocene and Recent mammalian bones. Paleobiology 6(3):341–51.
Based on inspection of gnawing damage to bones of modern prey animals, sets of typical damage types or patterns are recognized for certain skeletal elements. Observations are made of these damage types on bones of Pleistocene and Recent North American mammals, including bison (*Bison*), proboscideans (Proboscidea), moose (*Alces*), horse (*Equus*), wapiti (*Cervus elaphus*), and caribou (*Rangifer tarandus*). Pleistocene bones used in the study are from Alaskan and Yukon mammalian remains collected by O.W. Geist for the University of Alaska Museum in Fairbanks, and from Department of Paleobiology collections at the Smithsonian Institution, Washington, D.C. The intent of the paper is to note the presence of certain types of damage in the Pleistocene bones, and to describe some of the types of damage that are not readily apparent to researchers who have no experience in examining modern carnivore assemblages. Predictable *sequences* of gnawing were noted in some species and for some elements. Examples of gnawing damage to Pleistocene bones are documented in the following photographs: (1) a caribou antler from central Alaska showing probable gnawing damage (furrows) (Fig. 2); (2) humeri of horse, bison, and moose from Alaska with gnawed greater tuberosities; (3) bison femora from central Alaska showing gnawing damage at distal ends (Fig. 7); (4) tibiae of four horses and a bison from central Alaska showing gnawing damage at the proximal ends (Figs. 10, 11); (5) moose metapodial distal ends showing gnawing damage (Fig. 13). Of particular interest is a partial set of thoracic vertebrae from a single proboscidean [most likely a mammoth (*Mammuthus*)] from central Alaska. These bones show probable heavy gnawing only on transverse processes on the right side, suggesting that the carcass was lying on the left side throughout its early decay period. 'Some cultural modification of animal bones, resulting from dismemberment, marrow extraction, or use of bone as raw material for tool manufacture, cannot be reliably distinguished from natural modification until more detailed attribute analyses have been performed for each potential agency.'

654. Heard, D.C. (1977): The behaviour of Vancouver Island marmots, *Marmota vancouverensis*. MSc thesis, University of British Columbia, Vancouver. 129 pp.
The author hypothesizes (following McCabe and Cowan 1945 and Harington 1976) that the Vancouver Island marmot (*Marmota vancouverensis*) survived the last glaciation in refugia on Vancouver Island.

655. Heaton, T.H. (1994): Variation in fossil and modern *Ursus arctos* from Alaska. Abstracts of Papers, Fifty-fourth Annual Meeting, Society of Vertebrate Paleontology (October 19–22, Seattle, Washington), Journal of Vertebrate Paleontology 14 (Supplement to Number 3):28A.
Fossil grizzly or brown bear (*Ursus arctos*) skeletons have been found in three caves on Prince of Wales Island, Alaska, dating back to 12,295 ± 120 BP. Grizzly bears do not currently occupy the island. Late Pleistocene-Early Holocene bears of Prince of Wales Island are much larger than their modern counterparts of the southern Alaskan coast: some are as large as or larger than any living *Ursus arctos*. The author notes that *Ursus arctos* is highly variable in skull shape and size as well as in tooth morphology.

656. Heaton, T.H. (1995a): The 1994 excavation of a Quaternary vertebrate fossil deposit from Bumper Cave, Prince of Wales Island, Alaska. Geological Society of America, 29th Annual North-Central and South-Central Sections Meeting (April 27–8, Lincoln, Nebraska). Abstracts with Program 27(3):57.
Bumper Cave, 5 mi E of Mount Calder on northern Prince of Wales Island, was discovered and mapped in 1993 by Kevin Allred and David Love, who found an extensive bone deposit along the entire length of the passage. Remains of at least nine bears were recovered, all probably of grizzly bears (*Ursus arctos*) – not a current inhabitant of the island. A nearly complete skeleton of a female was found in the back of the cave, and parts of two juveniles were found nearby. Remains closer to the entrance were more disarticulated, less complete, and more weathered. An exceptionally large lower molar, probably of a male grizzly, was found in the entrance room. The faunal diversity of the cave is low, as screenwashing of nearly a ton of sediment produced only bear and vole (*Microtus*).

657. Heaton, T.H. (1995b): Colonization of southeast Alaska by *Ursus arctos* prior to the peak of Wisconsin glaciation. Journal of Vertebrate Paleontology, Abstracts of Papers 15(Supplement to Number 3):34A.
Until recently brown bears (*Ursus arctos*) were thought not to have reached the Alexander Archipelago until after the Wisconsinan glaciation. This was surmised from their current distribution and from the late appearance S of the ice sheets. Evidence on Prince of Wales Island, SE Alaska, does not support this hypothesis. Remains of at least 13 postglacial brown bears have been recovered from three caves dating from 12,295 ± 120 BP to 7,205 ± 65 BP showing that this species once inhabited the island. Remains of *Ursus arctos* from a fourth cave dated to 35,365 ± 800 BP are associated with remains of black bear (*Ursus americanus*). This last date is Middle Wisconsinan and predates the Late Wisconsinan glacial maximum.

658. Heaton, T.H. (1995c): Middle Wisconsin bear and rodent remains discovered on Prince of Wales Island, Alaska. Current Research in the Pleistocene 12:92–4.
The extensive karst lands of Prince of Wales Island in the southern Alexander Archipelago have yielded early postglacial vertebrate deposits in several different habitats. The dominant species are grizzly bear (*Ursus arctos*) and black bear (*Ursus americanus*) dating to at least 12,300 BP. However, two caves containing vertebrate remains that predate the Late Wisconsin glacial maximum (21,000-16,000 BP) provide the first glimpses of the Archipelago's more distant biotic past. On Your Knees Cave yielded a partial femur of a large bear [apparently a grizzly nearly as large as that reported from El Capitan Cave] dated to 35,365 ± 800 BP (AA-15227). Remains of a smaller bear – probably a black bear, a bird (Aves) femur, and an abundance of ground fish (Pisces) bone found in a surface layer may be of Holocene age. A tibia, apparently of a small black bear, was radiocarbon dated to 41,600 ± 1500 BP (AA-16831). The association of grizzly and black bears, river otters (*Lutra canadensis*), and fish is identical to the El Capitan Cave fauna. Apparently both black and grizzly bears either colonized the island twice or survived on it through the peak of the Wisconsinan glaciation in a refugium (like that found on the Queen Charlotte Islands farther S). A vertebrate microfauna of even great age found in Devil's Canopy Cave, located 35 km S of On Your Knees Cave, yielded marmot (*Marmota*) remains [an incisor being dated at >44,500 BP (AA-8871A)], a deer mouse (*Peromyscus*) mandible, and skeletal elements of small rodents. The marmot molar is closer in size to the yellow-bellied marmot (*Marmota flaviventris*) rather than the hoary marmot (*Marmota caligata*). These two sites, as well as refugia on the Queen Charlotte Islands, suggest that low-elevation ridges are the best places to search for fossil deposits predating the peak of the last glaciation.

659. Heaton, T.H. (1995d): Interpretation of $\delta^{13}C$ values from vertebrate remains of the Alexander Archipelago, S.E. Alaska. Current Research in the Pleistocene 12:95–7.
Late Pleistocene vertebrate remains have been found and radiocarbon dated from seven caves on Prince of Wales Island and several smaller islands nearby. Twenty-one dated samples have ^{13}C corrections, and these stable isotope values offer insights into the ecology and diet of the island's prehistoric vertebrates. The purpose of this analysis is to determine the diet of prehistoric island mammals and establish which species is responsible for the extensive ground fish remains in coastal caves. [At El Capitan Cave a large volume of ground fish (Pisces) bone – flatfish (Pleuronectiformes), sculpin (Cottidae), and at least 20 other marine species – was found associated with black bear (*Ursus americanus*), grizzly bear (*Ursus arctos*), river otter (*Lutra canadensis*), shrews (Soricidae), bats (Chiroptera), and several species of rodent (Rodentia)]. Low $\delta^{13}C$ values (mean of –20.5 ‰) in four postglacial and one Mid-Wisconsinan, or older, black bears suggests that they obtained food from terrestrial sources. Seven early postglacial grizzly bears dated from Prince of Wales Island [El Capitan Cave (coast) 12; Bumper Cave (subalpine) 4; Blowing in the Wind Cave (alpine) 1)] have a mean $\delta^{13}C$ value of –18.1‰. These values are all higher than the highest value for black bears (–18.7‰) and they suggest that grizzly bears had a marine component to their diet. A ninth grizzly bear from On Your Knees Cave dates to the Mid-Wisconsin period and has a $\delta^{13}C$ value of –15.9‰, the highest of any bear in the study. Plant food may have been scarce, a predominantly fish diet accounting for this high value. The value for river otter bone from El Capitan Cave is –10‰, the highest of the 21 samples. This suggests a predominantly marine diet, and that the otters are responsible for the extensive fish deposits in On Your Knees and El Capitan caves.

This conclusion is consistent with studies of river otter diets and den sites in the Alexander Archipelago. The value of –23.7‰ for a marmot (*Marmota*) tooth indicates a terrestrial plant diet, as would be expected. A deer (Cervidae) humerus from Nautilus Cave, Heceta Island, radiocarbon dated at 8180 ± 70 BP (AA-10574), has a $\delta^{13}C$ value of –25.2‰, by far the lowest of the 21 samples. The deer likely obtained its plant food from a closed-canopy forest, suggesting that dense rainforest was established in the Alexander Archipelago before 8000 BP.

660. Heaton, T.H. (1996a): The Late Wisconsin vertebrate fauna of On Your Knees Cave, northern Prince of Wales Island, Southeast Alaska. Abstracts of Papers, Fifty-sixth Annual Meeting, Society of Vertebrate Paleontology (October 16–19, New York, New York), Journal of Vertebrate Paleontology 16 (Supplement to Number 3): 40A–41A.

This small coastal cave in the Alexander Archipelago appears to contain a continuous fossil record through the Wisconsinan glaciation. The cave, situated 500 m from Sumner Strait and at an elevation of 85 m, consists of 68 m of crawlway passage. Mammal and fish remains have been found throughout the cave. A nearly complete skeleton of the otter (*Lutra canadensis*) was found on the surface and is clearly responsible for thick deposits of ground fish bone nearby. Buried deeper in the same room was a partial femur of the brown bear (*Ursus arctos*) dated at 35,365 ± 800 BP (AA-15227), found in association with remains of black bear (*Ursus americanus*) and deer (*Odocoileus hemionus*). A partial black bear tibia dated at 41,600 ± 1500 BP (AA-16831) was associated with other black bear bones and fish (Pisces) and rodent (Rodentia) remains. Parts of a small seal, possibly the ringed seal (*Phoca hispida*), found on the surface dated to 17,565 ± 160 BP (AA-18450). Recent genetic studies of modern *Ursus arctos*, together with remains from this cave, suggest that coastal refugia existed in the archipelago throughout the last glaciation and were inhabited by these bears.

661. Heaton, T.H. (1996b): The fossil gold mine. National Speleological Society News, July:172–5.

This is a popular account, covering the period 1991–5, of the highlights of a project to examine caves on Prince of Wales Island in the Alexander Archipelago of southeastern Alaska. The author mentions radiocarbon dates of 10,745 BP and 9760 BP on a black bear (*Ursus americanus*) skeleton and giant brown bear (*Ursus arctos*) remains, respectively from El Capitan Cave. The oldest brown bear was dated at 12,295 BP. Remains of two juvenile brown bears were recovered from alpine Blowing in the Wind Cave in 1992. The 1994 excavation of Bumper Cave yielded remains of at least nine brown bears (including three juveniles), a caribou (*Rangifer tarandus*), and several voles (*Microtus*). Radiocarbon dates ranged from 11,640 BP to 7205 BP. On Your Knees Cave yielded brown bear bone dating to 35,365 BP, and a black bear tibia from the site was dated at least to 41,600 BP. This added weight to the theory of ice age coastal refugia, but did not prove it. In 1995, the ulna of a small seal (*Phoca*) from a different part of On Your Knees Cave was dated to 17,565 BP – near the peak of the last glaciation – suggesting ice-free coastal habitat then. In another coastal cave, the first human (*Homo sapiens*) artifact – the tip of a bone spear associated with two black bear skeletons – was recovered.

662. Heaton, T.H., and Grady, F. (1992a): Two species of bear found in Late Pleistocene/Early Holocene den in El Capitan Cave, Prince of Wales Island, southern Alaska coast. Abstracts of Papers, Fifty-second Annual Meeting, Society of Vertebrate Paleontology (October 28–31, Toronto, Ontario), Journal of Vertebrate Paleontology 12 (Supplement to Number 3):32A.

A complete skeleton and two partial black bear (*Ursus americanus*) skeletons were found in a chamber in El Capitan Cave. The bear represented by the first skeleton, AMS dated at 10,745 ± 75 BP, presumably produced associated piles of dung containing ground (Pisces) fish bone. The other two skeletons are probably older. Fragments of a large bear, tentatively identified as a very large brown bear (*Ursus arctos*) were dated at 9760 ± 75 BP. These are the first Pleistocene vertebrate fossils recovered from the Alexander Archipelago and suggest early postglacial colonization followed by some extirpation, since *Ursus americanus* is the only bear now inhabiting Prince of Wales Island.

663. Heaton, T.H., and Grady, F. (1992b): Preliminary report on the fossil bears of El Capitan Cave, Prince of Wales Island, Alaska. Current Research in the Pleistocene 9:92–9.

In 1990, Kevin Allred found a complete skeleton of a black bear (*Ursus americanus*), parts of at least two other black bears, and fragments of a larger species of bear, as well as piles of small ground fish (Pisces) bones – apparently remnants of the bear's dung. Presumably the passage was used as a den. Work in 1991 revealed more detail on the bear remains, as well as the skull of an ermine (*Mustela erminea*) and bones of several bats (Chiroptera). Two of the black bears are young adults, spanning the upper size range for modern black bears. The largest bear apparently represents a grizzly (*Ursus arctos*). Samples of the complete black bear and the large bear yielded radiocarbon dates of 10,745 ± 75 BP (AA-7793) and 9760 ± 75 BP (AA-7794) respectively. Thus the age of the deposit falls near the end of the Wisconsinan glacial, and the two bear species must have been early postglacial inhabitants of Prince of Wales Island.

664. Heaton, T.H., and Grady, F. (1993): Fossil grizzly bears (*Ursus arctos*) from Prince of Wales Island, Alaska, offer new insights into animal dispersal, interspecific competition, and age of deglaciation. Current Research in the Pleistocene 10:98–100.
In 1992, the authors excavated parts of at least four black bears (*Ursus americanus*) and three grizzly bears (*Ursus arctos*) in El Capitan Cave on northern Prince of Wales Island. Skulls of two juvenile grizzlies [the larger dated at 9,995 ± 95 BP (AA-10451)] were recovered from deep shafts in Blowing in the Wind Cave in the alpine karst on El Capitan Peak. These fossils show that grizzlies colonized Prince of Wales Island and occupied diverse habitats there before disappearing. The large bear previously reported from El Capitan (Heaton and Grady 1992 a, b) has now been identified as an enormous, aged grizzly radiocarbon dated at 9760 ± 75 BP (AA-7794). A second, smaller individual was later identified from the same site. A third grizzly, smaller still, was radiocarbon dated at 12,295 ± 120 BP (AA-10445). The oldest black bear from El Capitan Cave was radiocarbon dated at 11,565 ± 115 BP (AA-10448), so black and grizzly bears coexisted on the island for at least 1800 years. Since El Capitan Cave is located in a glacial valley, the date of 12,295 ± 120 BP on a grizzly bear specimen provides a new minimum age for substantiated deglaciation on northern Prince of Wales Island, which is near the centre of the Alexander Archipelago.

665. Heaton, T.H., and Grady, F. (1997): The preliminary Late Wisconsin mammalian biochronology of Prince of Wales Island, southeastern Alaska. Abstracts of Papers, Fifty-seventh Annual Meeting, Society of Vertebrate Paleontology, Journal of Vertebrate Paleontology 17 (Supplement to Number 3): 52A.
Extensive excavation of three caves on northern Prince of Wales Island provides a preliminary biochronology for island mammals across the last glacial maximum. Most cave deposits contain extinct species such as black bear (*Ursus americanus*), river otter (*Lutra canadensis*), mule deer (*Odocoileus hemionus*), and several micromammals. Extirpated mammals include 20 postglacial individuals of brown bear (*Ursus arctos*) radiocarbon dated from 12,295 ± 120 BP (AA-10445) to 7205 ± 65 BP (AA-15224). Specimens of red fox (*Vulpes vulpes*) were recovered, one dating to 11,275 ± 90 BP (AA-21567) and the other postglacial. Two caves contain fossils that predate the peak of the last glaciation. One brown bear femur dates to 35,365 ± 800 BP (AA-15227) and three black bear specimens date to 41,600 ± 1500 BP (AA-16831), 29,820 ± 400 BP (AA-21570) and 28,700 ± 360 BP [correct date is 28,695 ± 360 BP] (AA-21569). Specimens known only from before the last glacial maximum include: marmot (*Marmota*) dating to (44,500 BP (AA-8871A), 32,900 ± 2400 BP (AA-21565), and 23,560 ± 770 BP [correct date is 23,660 ± 770 BP] (AA-21566); a bovid (Bovidae) horncore dated to 32,000 ± 2200 BP (AA-22883); as well as undated specimens of heather vole (*Phenacomys*) and brown lemming (*Lemmus*) that have not been found in postglacial deposits and do not live on the island today. Ringed seal (*Phoca hispida*) lived in the region during the peak of the last glaciation according to a bone sample, given minimal pretreatment, that was dated to 17,565 ± 160 BP (AA-18450), and later re-dated to 20,670 ± 80 BP (CAMS-33980). Another bone believed to be from the same animal was dated at 20,060 ± 500 BP (AA-22884), and a bone from a different individual was dated at 13,690 ± 240 BP (AA-21564). Since these seals appear scavenged, probably terrestrial mammals survived on the island as well.

666. Heaton, T.H., and Grady, F. (1998): Quaternary artiodactyls of the Alexander Archipelago, southeast Alaska. Fifty-eighth Annual Meeting, Society of Vertebrate Paleontology (September 30–October 3,

Snowbird, Utah), Journal of Vertebrate Paleontology 18 (Supplement to Number 3): 49A–50A.
Postglacial fossils of deer (*Odocoileus hemionus*) have been found in Nautilus Cave, Heceta Island [8180 ± 70 BP (AA-10574)] and On Your Knees Cave, Prince of Wales Island [5210 ± 60 BP (CAMS-31069)]. Low $\delta^{13}C$ values indicate a dense forest canopy then. A caribou (*Rangifer tarandus*) metacarpal from Bumper Cave, Prince of Wales Island [10,555 ± 110 BP (AA-18449R)] with a $\delta^{13}C$ value of –19.7 demonstrates that caribou ranged through much of the archipelago prior to extensive forest development. A horncore tip from On Your Knees Cave [32,000 ± 2200 BP (AA-22883)] closely matches saiga (*Saiga tatarica*). The complete skeleton of an unusual cervid (Cervidae) with Asian affinities was found in Zina Cave, Prince of Wales Island. These artiodactyls, together with brown bear (*Ursus arctos*) dating to 35,365 ± 800 BP (AA-15527) from On Your Knees Cave, suggest an Asian connection as far S as the Alexander Archipelago during Middle Wisconsinan time.

667. Heaton, T.H., and Grady, F. (2000): Vertebrate biogeography, climate change, and an ice age coastal refugium in southeastern Alaska. Journal of Vertebrate Paleontology, Abstracts of Papers 20 (Supplement to Number 3):48A.
Caves on Prince of Wales Island, and nearby islands in southeastern Alaska, have produced a vertebrate record spanning nearly the entire available range of radiocarbon dating. Both den and natural trap caves have been excavated. On Your Knees Cave, the site with the most complete record, has been used as a den by bears, foxes, and otters (*Lutra canadensis*). Some species were only present before the Last Glacial Maximum (LGM), such as: hoary marmot (*Marmota caligata*, six dates from >40,000–23,560 BP); brown lemming (*Lemmus sibiricus*), and heather vole (*Phenacomys ?intermedius*). Others were present only after the LGM such as mule deer (*Odocoileus hemionus*, eight dates 8180–260 BP and still present). Mammals dated to before and after, but not yet during, the LGM are: brown bear (*Ursus arctos*, 13 dates 35,365–26,820 BP and 12,295–7205 BP); black bear (*Ursus americanus*, 28 dates >40,000–28,695 BP and 11,565–3960 BP and still present), and caribou (*Rangifer tarandus*, five dates 37,990–26,670 BP and 11,560–10,555 BP). Ringed seal (*Phoca hispida*), a climatic indicator of land-fast sea ice, dates mainly to the LGM (10 dates, 23,560–13,690 BP). Fox [arctic fox (*Alopex lagopus*) and possibly red fox (*Vulpes vulpes*) four dates 19,480–10,050 BP] have been found from the LGM to the end of the Wisconsinan glaciation. The mammalian evidence indicates: tundra conditions in coastal Alaska preceding the LGM; cold during the LGM (leading to several island extinctions); tundra following the LGM; then a warming trend and development of the modern rainforest in the Early Holocene. This ice-free coastal refugium was inhabited by a maritime arctic fauna during the Late Wisconsinan.

668. Heaton, T.H., and Love, D. (1995): The 1994 excavation of a Quaternary vertebrate fossil deposit from Bumper Cave, Prince of Wales Island, Alaska. Geological Society of America, Abstracts with Programs 27(3):57.
Bumper Cave (SE1/4, sec 22, Twp 65S, rge 77E on the Petersburg A-5 NE quadrangle at 1733 ft asl) is situated in a subalpine, forested, metamorphic karst area on northern Prince of Wales Island, southeastern Alaska. Remains of at least nine bears, probably grizzlies (*Ursus arctos*) – not a current inhabitant of the island – were excavated there in 1994. A nearly complete skeleton of a female was found in the back of the cave, with parts of two juveniles nearby. Among the remains in the entrance room was the lower molar of an exceptionally large bear – probably a male brown bear. Screen washing of nearly a ton of cave sediment produced remains of only two genera: *Ursus* (bears) and *Microtus* (voles).

669. Heaton, T.H., Talbot, S.L., and Shields, G.F. (1996): An ice age refugium for large mammals in the Alexander Archipelago, southeastern Alaska. Quaternary Research 46:186–92.
Prior to this study, brown (*Ursus arctos*) and black (*Ursus americanus*) bears of the Alexander Archipelago were considered postglacial immigrants. Vertebrate fossils from Prince of Wales Island caves now demonstrate that those bears coexisted there before and after the last glaciation. Differences in mitochondrial DNA sequences suggest that living brown bears of the Alexander Archipelago comprise a distinct clade and are more closely related to polar bears (*Ursus maritimus*) than to their mainland conspecifics. Brown bears and perhaps other large mammals

have continuously occupied the archipelago for at least 40,000 years, so habitable refugia were available throughout the last glaciation. Seal (*Phoca* cf. *Phoca hispida*) remains from On Your Knees Cave, Prince of Wales Island, lived in the area near the peak of the last (Fraser) glaciation, suggesting that the area was ice-free. Remains of brown and black bears from that cave date to the preceding nonglacial (Olympia) interval. Stable carbon isotope studies of bears from that cave indicate: (1) an unusually large marine component in the brown bears' diet, and (2) a primarily terrestrial plant diet for black bears. Table 2 gives 25 AMS radiocarbon dates based on bone collagen and associated carbon isotope values from El Capitan Cave [10 dates on brown bear (*Ursus arctos*), black bear (*Ursus americanus*), fish (Pisces) and river otter (*Lutra canadensis*)], Bumper Cave [five dates on brown bear and caribou (*Rangifer tarandus*)], Blowing in the Wind Cave [one date on brown bear], Kushtaka Cave [three dates on black bear and a human (*Homo sapiens*) artifact], Enigma Cave, Dall Island [one date on black bear], Nautilus Cave, Heceta Island [one date on mule deer (*Odocoileus hemionus*)], On your Knees Cave [three dates on brown bear, black bear and ringed seal (*Phoca hispida*)], and Devil's Canopy Cave [one date on marmot (*Marmota*)].

670. Heinemeier, J., and Rasmussen, K.L. (1998): The gateway to Greenland, Appendix 4: Radiocarbon datings. *In*: Man, Culture and Environment in Ancient Greenland (J. Arneborg and H.C. Gulløv, eds.). Danish National Museum and Danish Polar Center, Copenhagen. Danish Polar Center Publication No. 4:192–3.

The following radiocarbon dates from Dorset and Thule site bones are from the Smith Sound area of Greenland: Conventional dates, Sample 3 from Structure 1, muskox (*Ovibos moschatus*) 985 ± 46 BP (K-6703), calibrated age [1025 (1015–1155)]; Sample 8 from Structure 4, muskox jawbone 925 ± 38 BP (K-6706), calibrated age AD1060–1155 (AD 1035–1170): AMS dates, Sample 9 from Structure 4, arctic hare (*Lepus arcticus*) 770 ± 40 BP (AAR-3219), calibrated age AD1280 (AD1240–1285); Sample 12 from Structure 161, arctic fox (*Alopex lagopus*) jawbone 1175 ± 40 BP (AAR-3222) reservoir corrected age 930 ± 40 BP, calibrated age AD1050–1155, (AD1030–1170); Sample 15 from Structure 294, caribou (*Rangifer tarandus*) rib 565 ± 65 BP (AAR-3223), calibrated age AD1400 (AD1310–1430).

671. Herman, W. (1981): Has scientist found giant step for man? Toronto Star (Sunday Star), April 19:A16.

The article states that Bill Irving, a University of Toronto archaeologist, found evidence from Old Crow Basin in northern Yukon that humans (*Homo sapiens*) entered that part of North America before the Wisconsinan glaciation – earlier than 110,000 years ago. The evidence consists of mammoth (*Mammuthus*) bones apparently broken by people (*Homo sapiens*) in stratigraphic position below a paleomagnetic reversal that occurred about 110,000 years ago. Irving said that Brenda Beebe, a University of Toronto vertebrate paleontologist, supplied critical evidence by noting that with the tools were remains of a lemming (*Dicrostonyx*) common about 150,000 years ago, but extinct for the past 60,000 years. All the fossil vertebrate evidence she found supports Irving's conclusions that the site is not a mixture of new and old. [See Jopling et al. 1981.]

672. Herrington, S.J. (1986): Phylogenetic relationships of the wild cats of the world. PhD dissertation, University of Kansas. 415 pp.

Relevant information on the lions and 'tigers' of Eastern Beringia are dealt with in a section 'Pleistocene *Panthera* of the New World' (pp. 325–47). This study shows that several 'tigers' (*Panthera tigris* subsp. nov.) were inadvertently included among North American Pleistocene 'lions': 11 American Museum of Natural History (Frick Collection) Pleistocene dentaries from Alaska: 69004, 69005, 69006 (Figure 4.14), 69009, 69016 (Figure 3.14), 69020, 69047, 30465, 30761 and 30763, as well as United States National Museum (Smithsonian Institution) dentary 23619 are assigned to this new subspecies of tiger. Further analysis of the Beringian material is necessary before other specimens can be identified positively. The discordant morphological variation patterns between American lions (*Panthera leo atrox* = *Panthera atrox*) and the Beringian lions and their Eurasian counterpart (*Panthera leo spelaea*) are considered to be strong evidence that the two populations were evolving independently. The variation is not explained by an ancestral cline, since the American lion does not follow the same pattern as the Beringian–Asian lions. The author recommends recognizing the southern lions as a distinct species, *Panthera atrox*.

673. Herrington, S.J. (1987): Subspecies and the conservation of *Panthera tigris*: Preserving genetic heterogeneity. *In*: Tigers of the World (R.L. Tilson and U.S. Seal, eds.). Noyes Publications, Park Ridge, New Jersey. pp. 51–61.

The possibility that both lions (*Panthera leo*) and tigers (*Panthera tigris*) were included in identifications of the American lion (*Panthera leo atrox*) seems never to have been considered. In a recently completed study of these fossils, the author developed a set of morphometric characters (Figure 5) that distinguished between lions and tigers with 100% accuracy in discriminant function analyses. In addition, there were several qualitative cranial and dental characters that could distinguish lions from tigers with a high degree of accuracy (Figure 4). Modern lions and tigers were compared with the fossil North American material. All specimens from the contiguous United States evidently represented lions. However, the material from Eastern Beringia (Fairbanks area and Lost Chicken Creek, Alaska) included both lions and tigers. Apparently lions invaded Eastern Beringia during the Illinoian glaciation, and arrived in the area of the contiguous United States during the Sangamonian interglacial. The author hypothesizes that tigers and a different population of lions penetrated as far as Eastern Beringia during the Wisconsinan glaciation in a separate dispersal event. [See Herrington 1986.]

674. Hibbard, C.W. (1949): Pleistocene vertebrate paleontology in North America. Bulletin of the Geological Society of America 60:1417–28.

This is a critical review of past trends and present requirements in North American Pleistocene vertebrate paleontology.

675. Hibbard, C.W. (1958): Summary of North American Pleistocene mammalian local faunas. Papers of the Michigan Academy of Sciences, Arts and Letters 43:1–32.

This publication is complementary to this work in that it provides data on Pleistocene mammal faunas of the conterminous 48 states of the United States and Mexico, i.e., southern North America.

676. Hibbard, C.W., Ray, C.E., Savage, D.E., Taylor, D.W., and Guilday, J.E. (1965): Quaternary mammals of North America. *In*: The Quaternary of the United States (H.E. Wright, Jr, and D.G. Frey, eds.). Princeton University Press. pp. 509–25.

This paper deals with all North American Late Pliocene and Pleistocene mammalian genera, and provides a complete listing of these. It also features a current listing of all the known faunal sites (S of the glacial margin) that feature mammalian assemblages. A general discussion follows, outlining some of the zoogeographical and evolutionary implications, and highlighting some problem areas that need more research.

677. Hibben, F.C. (1943): Evidences of early man in Alaska. American Antiquity 8:254–9.

A review of the state of information of early humans (*Homo sapiens*) in Alaska. The majority of the evidence presented is lithic. Photographs of the stone tools and locality maps are included. The author (p. 256) states that most mammal (Mammalia) remains are dismembered and disarticulated, even though some fragments retain parts of big animals, skin, hair, and flesh. He mentions that Otto Geist suggested that Alaskan Pleistocene mammals were completely destroyed in one, or a series of, volcanic eruptions, and the loess with the dismembered animal and plant remains washed into creek valleys from the surrounding hills, where they were buried or later died.

678. Hicock, S.R., Hobson, K., and Armstrong, J.E. (1982): Late Pleistocene proboscideans and early Fraser glacial sedimentation in eastern Fraser Lowland, British Columbia. Canadian Journal of Earth Sciences 19(5):899–906.

Three proboscidean (Proboscidea) tusk fragments were found in the Chilliwack River Gravel Pit near Chilliwack, British Columbia. Their locations and radiocarbon ages are: (1) 49°5.6′N, 121°57.0′W – 21,400 ± 240 BP (SFU-65); (2) (same location) – 21,600 ± 240 BP (SFU-66); (3) 49°6.6′N, 121°55.5′W (Bailey Sanitary Landfill Pit, 2.5 km NE of the Chilliwack River Gravel Pit) – 22,700 ± 320 BP (GSC-2232). The authors suggest (p. 905) that the tusks are probably from mammoths that fed on grassy early Fraser outwash deposits in the eastern Fraser Lowland between 23,000 and 21,000 BP. They also mention that ice age mammals could have migrated southward and westward (away from the early Fraser ice advance) into the Fraser Lowland, across the ancestral Strait of Georgia via the Quadra sandur, and onto southwestern

Vancouver Island where the earliest Fraser ice advance did not occur until after 17,000 BP. [See Harington 1975d, 1976.]

679. Hill, C.G. (1984): Journey through time: An introduction to the archaeology and cultural history of the MSTW [Morden-Stanley-Thompson-Winkler] Planning District. Papers in Manitoba Archaeology, Popular Series No. 6:1–100.

Humans (*Homo sapiens*) apparently entered postglacial Manitoba between 10,000 and 8000 BC (12,000 and 10,000 BP), based on scanty and sparsely distributed finds of stone tools. Very little information has yet been recovered as to the kinds of animals that were hunted in Manitoba at that time (p. 34). In dealing with climatic and environmental changes in this region since the retreat of the Wisconsinan ice [which began about 13,000 BC (15,000 BP)], the author mentions the extinction of western bison (*Bison bison occidentalis = Bison occidentalis*) perhaps as recently as 4200 BC (6200 BP), marking the emergence of the modern southern Manitoba mammalian fauna (p. 37). There are line drawings showing two views of 'an extinct bison' cranial fragment with horncore found in the bank of the Assiniboine River near Millwood, Manitoba (p. 38). Bison became common in southwestern Manitoba about 8000 BC (10,000 BP), when the region became an extensive grassland. A handy diagram correlates the history of Glacial Lake Agassiz with the Paleoindian history of southern Manitoba based on finds of fluted and Plano points (pp. 48, 49). The paper includes excellent black and white sketches of Paleoindian hunters attacking bison (e.g., pp. 46, 47).

680. Hill, J.E. (1942): *Citellus parryii* from the Pleistocene of Alaska. Geological Society of America Bulletin 53(12):1842.

A large series of ground squirrel skulls and skeletons in the Childs Frick collection in the American Museum of Natural History agree closely with Recent arctic ground squirrels (*Spermophilus parryii ablusus = Citellus parryii ablusus*) from the same region. They were associated with mammoth (*Mammuthus*), giant bison [presumably steppe bison (*Bison priscus*)], and horse (*Equus*) remains. Studies by Dr A.C. Whitford showed that hairs of mummified specimens of these ground squirrels, also in the Frick collection, agreed better with hairs of *ablusus* than do hairs of other Recent races of the same species. Cranially the Pleistocene ground squirrels show a slightly wider zygomatic spread and relatively greater mastoid breadth. The latter measurement is the only one statistically significant.

681. Hillaire-Marcel, C. (1980): Les faunes des mers post-glaciaires du Québec: Quelques considérations paléoécologiques. Géographie physique et Quaternaire 34(2):3–55.

This paleoecological study of the fossil fauna of the postglacial seas of Quebec focuses on molluscs. Plate III (#19) illustrates a beluga (*Delphinapterus leucas*) vertebra from Champlain Sea sediments near Saint-Nicolas, Quebec. The average age of molluscs in the same layer of sediment as the beluga is about 10,000 BP. [See Chartier et al. 1997a.]

682. Hillerud, J.M. (1980): Bison as indicators of geologic age. Canadian Journal of Anthropology 1(1):77–80.

A phylogeny of fossil bison is presented that conforms to the existing knowledge of fossil age based on stratigraphy. A proposal is presented that: (1) Giant bison (*Bison latifrons*) may have been an early immigrant to North America and may have been ancestral to other North American bison species. (2) Steppe bison (*Bison priscus*) is an artificial agglomeration of samples of many bison populations over an undefinable and overly extended period of time – the 'species' need reevaluation. (3) *Bison crassicornis* [considered here steppe bison (*Bison priscus*)], while probably not a valid paleospecies, could be retained to describe the North American high-latitude, boreal group. (4) Bison arrived in North America in pre-Illinoian times and have survived since then as a continuous lineage, with a continuing 'genetic plexus' to the present, evolving by 'phylogenesis,' diminishing in size, accepting alien contributions from Eurasian bison, and producing short-lived aberrant forms in geographically isolated areas. (5) In Late Pleistocene and Holocene time, two 'chronoclines' developed: a northern one including a sequence of steppe bison (*Bison priscus = Bison crassicornis*) to western bison (*Bison bison occidentalis = Bison occidentalis*) to wood bison (*Bison bison athabascae*); and a southern one of giant bison (*Bison latifrons*) to the *Bison alleni-cheneyi* group [considered here as giant bison (*Bison latifrons*)] to ancient bison (*Bison bison antiquus = Bison antiquus barbouri*) to another ancient bison (*Bison bison antiquus* = *Bison antiquus taylori*) to at least a part of the plains bison (*Bison bison bison*). Once a workable phylogeny has been accepted, the fossils can be used as indicator species to age the human (*Homo sapiens*) artifacts with which they frequently are associated.

683. Hills, L.V., McNeil, P.E., and Tolman, S. (1999): First reported Canadian proboscidean trackway from the St. Mary Reservoir, Alberta, Canada. Fifty-ninth Annual Meeting of the Society of Vertebrate Paleontology (October 20–3, Denver), Abstracts of Papers. Journal of Vertebrate Paleontology 19(3):51A.
Construction of a new spillway resulted in a 10-m water drop of the St Mary Reservoir, Alberta. Wind erosion of the reservoir bottom sediments has revealed a rich assortment of human (*Homo sapiens*) artifacts, and skeletal remains of many mammals (Mammalia), birds (Aves), and fishes (Pisces). Extensive trace fossils include many rodent (Rodentia) burrows, and tracks of camels (Camelidae), horses (*Equus*), and bovids (Bovidae). In addition, the authors describe a well-preserved trackway (a series of nine excellent tracks stretching over 11 m) of (probably) a woolly mammoth (*Mammuthus primigenius*). The tracks are circular to slightly ellipsoidal in shape with a diameter of 55 to 70 cm. They consist of concentric rings of sediment pushed up while the animal walked on mud. The innermost ring of the tracks represents the true foot impression and has a diameter of 45 cm, suggesting a mature female or immature male. The average stride length of 2.5 m is consistent with a modern elephant (Elephantidae) with a shoulder height of 2.5 m walking at slow speed. Only a proboscidean, and likely a woolly mammoth (*Mammuthus primigenius*), is considered the maker of these tracks.

684. Hills, L.V., Kooyman, B., Tolman, S., and McNeil, P. (1999): The St. Mary Reservoir Late Pleistocene paleontological and archaeological site. Canadian Quaternary Association – Canadian Geomorphology Research Group Meeting (August 23–7, Calgary), Program and Abstracts: 28.
The Wally's Beach site (DhPg-8) is a Late Pleistocene to Recent archaeological and paleontological site in the St Mary Reservoir of southern Alberta. Faunal material recovered includes: extinct Mexican horse (*Equus* cf. *Equus conversidens*), helmeted muskox (*Bootherium bombifrons*), ancient bison (*Bison bison antiquus* = *Bison antiquus*), badger (*Taxidea taxus*), several rodents (Rodentia), a lagomorph (Leporidae), at least two canids (Canidae), and a number of other mammals and birds (Aves). Radiocarbon dates for bison and horse are 11,130 ± 90 BP and 11,300 ± 70 BP, respectively. A series of Pleistocene horse, camel (probably the western camel (*Camelops hesternus*), proboscidean (probably the woolly mammoth (*Mammuthus primigenius*), and bovid (Bovidae) tracks were recovered. The trackways are spread over several hundred metres and show dominant patterns of movement into and out of the Pleistocene river valley. The proboscidean trackways have allowed reconstruction of animal size, foot morphology, and gait. Archaeological flakes and simple tools were associated with muskox and horse remains, suggesting contemporaneity and perhaps human (*Homo sapiens*) use of these species. Archaeological material, including a roasting pit, is present in the same paleosol as the tracks, and its position on the edge of a river valley suggests a hunting lookout. Ground squirrel (*Spermophilus*) remains in hibernacula, including many that died during the flooding of the reservoir, suggest it is possible to partially reconstruct the ancient topography and estimate the amount of erosion that has taken place since the reservoir was flooded.

685. Hinchley, H. (1975): Whale bones found near White Lake pit. Renfrew Mercury, December 3:1.
This article describes unidentified whale bones [later identified as a bowhead whale (*Balaena mysticetus*)] found in a sandpit near White Lake, Renfrew County, Ontario. The bones from the right forefin were each over 1 m long and are believed to be older than 10,000 BP. [See Harington 1988.]

686. Hind, H.Y. (1860): Narrative of the Canadian Red River Exploring Expeditions of 1857 and of the Assiniboine and Saskatchewan Exploring Expedition of 1858. Longman, Green, Longman and Roberts, London.
In volume I (pp. 312–13) this explorer received an account of 'Great Bones' from Mr McKay at Fort Ellice in 1857. He mentions that they were seen many years ago protruding from the bank of Shell Creek [Shell River, Manitoba], 20 ft below the prairie level. McKay instructed the hunters at the post to bring them to him, but no Indians would touch them, and the Métis only brought 'a

tooth and collar bone [scapula?], which were stated by a medical gentleman to whom they were shown to have formerly belonged to a mammoth.' Mr Christie of Fort Pelly visited Shell River in order to collect more specimens, but obtained only some crumbling ribs that were sent to the settlement at Red River. 'The Indians had long regarded these ancient relics as the bones of Manitou and worthy objects of veneration.' [See Richardson 1854, and Tyrrell 1892 – the bones were later identified as mastodon (*Mammut americanum*).] Hind (volume II, p. 60) mentions other similar skeletal remains described to him by an old Ojibwa hunter who lived near Dauphin Lake, Manitoba. Ta-wa-pit, while pointing out a spot near where he obtained salt from the edges of a spring, 'described at great length the appearance and virtues of some gigantic bones exposed in the bank of Valley River near where it cuts through the old Lake Ridge.' Hind goes on to say how Ta-wa-pit crushed bone fragments to powder, using them as a medical preparation. 'From his descriptions I infer that the bones are those of a mammoth [*Mammuthus*]; his rough drawing of the ribs and teeth in the sand correspond, in point of dimensions, with those of that gigantic animal.'

687. Hinde, G.J. (1878): The glacial and interglacial strata of Scarboro' Heights, and other localities near Toronto, Ontario. Canadian Journal of Science, Literature and History, New Series 15:388–411.
Although vertebrate fossil remains are not mentioned from this locality, it is well to remember this early classic study of climatic change in the Toronto area through various Pleistocene cold and warm phases as discerned from the sequence of sedimentary deposits. Hinde's research is a forerunner to Coleman's [e.g., Coleman 1932] later contributions, which do mention vertebrates, and he clearly recognizes the Sangamonian interglacial (Don Beds: 'Complete disappearance of the glacier; the land covered with vegetation; Lake Ontario filled with fresh water to a higher level by at least 150 feet than at present; in the lake extensive beds of clay and sand were formed, in which plants and other remains eroded off the land were imbedded. Climate temperate, similar to the present' (pp. 412–13).

688. Hoefs, M. (2001): Prevalence of mandibular oligodonty in wild sheep: Possible evolutionary relevance. Journal of Mammalogy 82(2):401–6.
The author agrees with Korobitsyna et al. (1974) that early pachyceriform sheep evolved their distinctive characteristics in the Beringian refugium, probably during the Riss-Illinoian glacial period. Pachyceriform (thick-horned) sheep then migrated S when glaciers blocking the route melted, and where they differentiated into modern bighorn sheep (*Ovis canadensis*) after isolation during the Würm-Wisconsinan glaciation. Isolation of pachyceriform sheep in eastern Siberia from those in Alaska, probably in the Eemian-Sangamonian interglacial, but possibly as late as the Holocene (beginning 10,000 years ago), led to the differentiation of Siberian snow sheep (*Ovis nivicola*) from Dall sheep (*Ovis dalli*). Mandibular oligodonty [lack of development of the first cheek tooth (P_2)] was widespread in New World sheep (e.g., snow sheep 60.3% – also having the most advanced karyotype in *Ovis*, Dall sheep 23%, bighorn sheep 10.1%) but was much lower on the average in Old World sheep [e.g., argalis (*Ovis ammon*) and mouflons (*Ovis gmelini*) 1.7% and urials (*Ovis vignei*) 1%]. Generally more recently evolved taxa showed the anomaly more than did primitive ones.

689. Hoffecker, J.F., Powers, W.R., and Bigelow, N.H. (1996): Dry Creek. *In*: American Beginnings: The Prehistory and Palaeoecology of Beringia (F.H. West, ed.). University of Chicago Press, Chicago and London. pp. 343–52.
The Dry Creek site (HEA-005, 63°53′N, 149°02′W, about 470 m asl) is located in the north-central foothills of the Alaska Range, some 125 km SW of Fairbanks, Alaska. Faunal remains include: bird (Aves) gastroliths in the size range of ptarmigan (cf. *Lagopus*), and a small number of poorly preserved mammal remains. Dall sheep (*Ovis dalli*) and wapiti (*Cervus elaphus* = *Cervus canadensis*) remains were excavated from component I assigned to the Nenana complex (*Homo sapiens*). Remains from component II assigned to the Denali complex are steppe bison (*Bison priscus*) and Dall sheep. The Dall sheep and wapiti exhibit larger-than-modern body size.

690. Hoffecker, J.F., Powers, W.R., and Goebel, T. (1993): The colonization of Beringia and the peopling of the New World. Science 259:46–53.
The discovery of a Paleoindian complex in central Alaska, combined with recent redating of the Bering Land Bridge and key archaeological sites, suggests that Beringia was

settled during the final Pleistocene interstadial (12,000 to 11,000 BP). Probably Beringia was colonized in response to changes in climate and vegetation that occurred then. The 'Beringian Environmental Debate' is considered carefully, the authors concluding that Last Glacial environments probably contained a complex mosaic of vegetation that included some steppe-like communities, and that lack of adequate mammalian prey is unlikely to have been the barrier to human (*Homo sapiens*) settlement before 14,000 to 12,000 BP. In stating that 'Central Alaska contains the most important early sites in Beringia,' the authors downplay the significance of Bluefish Caves, Yukon. They note that a small quantity of faunal remains [bison (*Bison*), sheep (*Ovis*), and wapiti (Cervus)] was recovered from Dry Creek, Alaska. The Broken Mammoth site at the mouth of Shaw Creek, Alaska, in addition to artifacts, produced a large quantity of faunal remains [including bison, wapiti, caribou (*Rangifer*), and various small vertebrates]. Trail Creek Cave 9 on Seward Peninsula contained a bison bone that appeared to have been broken by humans about 13,070 BP. So members of the Nenana Complex hunted large mammals such as bison and wapiti in the Tanana Valley and sheep in the northern foothills of the Alaska Range; also they may have exploited smaller mammals, birds (Aves), and fish (Pisces).

691. Hoffecker, J.F., Waythomas, C.F., and Powers, W.R. (1988): Late glacial loess stratigraphy and archaeology in the Nenana Valley, central Alaska. Current Research in the Pleistocene 5:83–6.

Four multi-component archaeological sites in the Nenana Valley, about 100 km SW of Fairbanks, Alaska, contain one or more occupation levels dated to 12,000–10,000 BP. Two distinct archaeological complexes (*Homo sapiens*) appear to be present: an industry lacking microblade technology dated to about 11,800–11,000 BP and a microblade industry dated to about 10,600–7000 BP. Late-glacial faunal remains from both the Nenana Valley and the neighbouring Teklanika Valley include woolly mammoth (*Mammuthus primigenius*), steppe bison (*Bison priscus*), and wapiti (*Cervus elaphus* = *Cervus canadensis*). Associated faunal remains from Dry Creek comprise wapiti and Dall sheep (*Ovis dalli*) and date between about 11,800 and 11,000 BP. Steppe bison and Dall sheep remains associated with the Denali complex, overlying the Nenana complex horizon at Dry Creek, date to 10,690 ± 250 BP (SI-1561 – based on charcoal).

692. Hoffmann, R.S. (1984): An ecological and zoogeographical analysis of animal migration across the Bering Land Bridge during the Quaternary period. *In*: Beringia in the Cenozoic Era (V.L. Kontrimavichus, ed.). Amerind Publishing Co. Pvt. Ltd., New Delhi. pp. 464–81.

Recent North American mammals may be grouped according to major ecological communities. The proportion of species and superspecies having Holarctic distribution is very high among tundra mammals (Mammalia) and birds (Aves), somewhat lower among taiga and alpine species, and very small for the rest (deciduous forest, steppe, and desert). Further, the degree of taxonomic affinity is highest for tundra species and less for taiga and alpine forms. Ecological and zoogeographical data indicate: (1) the environment of the Bering Isthmus was tundra or cold-steppe in the Late Pleistocene, taiga (in part) in the Middle or Early Pleistocene, and steppe (in part) in the Early Pleistocene or Late Pliocene; (2) isolating barriers between Nearctic and Palearctic were competitive as well as ecological and physical; (3) there is a connection between time of isolation in 'Beringia,' and degree of phenetic difference in related Holarctic populations, which may permit estimates of rates of evolution in various mammals and bird lineages.

693. Hoffmann, R.S., and Peterson, R.S. (1967): Systematics and zoogeography of *Sorex* in the Bering Strait area. Systematic Zoology 16(2):127–36.

Shrew populations on islands in the Bering Strait region (*Sorex pribilofensis, Sorex jacksoni, Sorex hydrodromus*) were regarded as related to the arctic shrew (*Sorex arcticus*) of the Alaskan mainland. The authors' data indicate that *Sorex pribilofensis* and *Sorex jacksoni* are members of the *Sorex* (*Otiosorex*) *cinereus* group, which occurs both in Alaska and eastern Siberia. *Sorex hydrodromus* apparently does not exist on Unalaska Island, and is a synonym of *Sorex pribilofensis*. The zoogeographic pattern of Beringian shrews (*Sorex*), determined mainly by morphometric analysis, suggests the following pattern of dispersal and isolation. In the Early to Middle Pleistocene, a forest-adapted shrew ancestral to the *araneus-arcticus* group occurred in Eurasia. In the Mindel/Kansan glacial, the Bering land connection probably supported taiga (in part), and these shrews colonized North America. There, it was subsequently isolated, first by Bering Strait, and then S of the continental Illinoian ice, where it evolved into typical *Sorex arcticus*. In Eurasia, the ancestral shrew evolved into modern *Sorex araneus* during the later (Riss-Würm) Pleistocene. During the Riss/Illinoian, *Sorex arcticus* moved into Beringia, giving rise to a Siberian population. In the Würm/Wisconsinan, some of these *arcticus* shrews were isolated in the Beringian refugium, where tundra prevailed. This population diverged sufficiently to produce the tundra-adapted *Sorex arcticus tundrensis* and *Sorex arcticus borealis*. Divergence has apparently reached the species level in *Sorex araneus*, where the two forms are sympatric. In North America, *tundrensis* is contiguously allopatric with other *Sorex arcticus*.

694. Hoffmann, R.S., Koeppel, J.W., and Nadler, C.F. (1979): The relationships of the amphiberingian marmots (Mammalia: Sciuridae). Occasional Papers of the Museum of Natural History, University of Kansas, 83:1–56.

The authors suggest that the Vancouver Island marmot (*Marmota vancouverensis*) is a Pleistocene isolate derived from mainland populations of the hoary marmot (*Marmota caligata*).

695. Holland, W.J. (1915): A skull of *Bison crassicornis*. Annals of the Carnegie Museum (Pittsburgh) 9:225.

In 1907 the Carnegie Museum purchased from Frank Caldwell of Indianapolis the cranium (posterior to the orbits) with horncores and partial sheaths (Plate 43) of a steppe bison (*Bison priscus* = *Bison crassicornis*), which came from an excavation in gold-bearing gravel near Dawson City, Yukon. It was found at a depth of 44 ft below the surface and was associated with 'mastodon' [probably mammoth (*Mammuthus*)] remains. The dentition of upper and lower jaw were missing. Fourteen measurements on the specimen are provided. The greatest distance between tips of the horns is 1155 mm.

696. Holland, E.A., and Johnson, R. (1974): A report on the faunal analysis of Locality 11A, Old Crow River, Yukon Territory. [A preliminary report carried out at University of Toronto and deposited in Quaternary Vertebrates of Northern North America file at Canadian Museum of Nature].

The aims of this typescript report are: (1) to determine the identity of the animals represented in the 11A collection, and (2) to determine if the bones of these animals exhibited modifications that could be attributed to early human (*Homo sapiens*) activity. Old Crow Locality 11A (67°48′N, 130°56′W) is on a sharp bend on the left limit of Old Crow River (Map 5). A list of Pleistocene mammal specimens from the site (Table 2) follows: woolly mammoth (*Mammuthus primigenius*); horse, large and small (*Equus*); moose (*Alces* cf. *Alces alces*); caribou (*Rangifer* cf. *Rangifer tarandus*); western camel (*Camelops* cf. *Camelops hesternus*); saiga antelope (*Saiga tatarica* = *Saiga* cf. *Saiga tatarus* [sic]); muskox (*Ovibos* cf. *Ovibos moschatus*); steppe bison (*Bison priscus* = *Bison* cf. *Bison crassicornis*); smaller bison than steppe bison (*Bison*); American scimitar cat (*Homotherium* cf. *Homotherium serum*); brown bear (*Ursus* cf. *Ursus arctos*); giant beaver (*Castoroides* cf. *Castoroides ohioensis*); and beaver (*Castor* cf. *Castor canadensis*). Age distribution of the various species studied are given in percentages of: immature, mature, and indeterminate (Table 3). Frequency distribution of horse elements and mammoth elements are provided (Tables A-1 and A-5); and percentage composition of the various genera identified are listed in Table A-9 – highest are horse (41%), mammoth (37%), bison (12%), and caribou (2.9%). The first three animals also show signs of having been modified by people, although the techniques used on each kind were different. [See Friesen 1989, 1997.]

697. Holman, J.A. (1995): Pleistocene Amphibians and Reptiles in North America. Oxford University Press, New York, Clarendon Press, Oxford. 243 pp.

Three of the 19 important sites mentioned in this book are in Canada: (1) Kelso Cave, Halton County, Ontario, which provided a fossil of the American toad (*Bufo* cf. *Bufo americanus*) – the presence of a large pika (*Ochotona*) dates the site to probable Illinoian age; (2) Innerkip site, southwestern Ontario, which has been radiocarbon dated to > 50,000 BP and has yielded a Blanding's turtle (*Emydoidea blandingi*) fossil; and (3) the Glacial Lake Iroquois site, near Toronto, Ontario, has been radiocarbon dated at 11,315 ± 325 BP based on the antler of a Toronto deer (*Torontoceros*), and has yielded a frog (*Rana*) fossil.

698. Holman, J.A. (1996): The leap not taken. Natural Sciences (College of Natural Sciences, Michigan State University) 10(2):8–9.

A 10,000-year-old, two-thirds-grown leopard frog (*Rana pipiens*) was identified from a skeleton in a clay nodule of Champlain Sea age collected from Eardley, Quebec, near Ottawa, Ontario. The specimen, which may be the best-preserved Pleistocene frog in North America, was passed to Professor J.A. Holman for identification to the species level by Dr C.R. Harington of the Canadian Museum of Nature in Ottawa. Illustrations for the article include: (1) a black and white picture (with an excellent full-colour picture of the skeleton on the journal's cover); (2) a map of the Champlain Sea near its maximum extent, with the Eardley site marked; and (3) a photograph of a modern leopard frog in nearly the same position as the skeleton in the nodule.

699. Holman, J.A., and Clouthier, S.G. (1995): Pleistocene herpetofaunal remains from the East Milford mastodon site (ca. 70 000–80 000 BP), Halifax County, Nova Scotia. Canadian Journal of Earth Sciences 32(2):210–15.

The East Milford site (45°00′30″N, 63°25′00″W), consists of a gypsum sinkhole infilled with Pleistocene sediments. The discovery in 1991 of the skeleton of a mastodon (*Mammut americanum*) [see Grantham and Kozera-Gillis 1992, and Harington, Grant, and Mott 1993] prompted an excavation by the Nova Scotia Museum of Natural History. This excavation led to the discovery of the remains of three species of reptiles (Reptilia) and amphibians (Amphibia); a leopard frog (*Rana pipiens*), NSM 92.108.01; painted turtles (*Chrysemys picta*), NSM 92.70.03 [not 97.70.03 as in Figure 4], 92.108.01, and 92.150.01; and wood turtles (*Clemmys insculpta*), NSM 92.82.01, 92.80.01, 92.212.07, 92.236.01, 92.96.01, and 99.82.01. Preliminary stratigraphic studies and radiocarbon dating of the wood and peat matrix suggests an approximate age of 70,000–80,000 BP for the site. All of the herpetological species found are cold-tolerant forms that inhabit Nova Scotia today. The presence, especially of the turtles, indicates that the area had Early Wisconsinan summer temperatures at least as warm as today's. Subsequent glaciations obliterated their habitat, so the present analogues of these herpetological species must have reinvaded Nova Scotia from S of Canada after the retreat of the Laurentide ice.

700. Holman, J.A., Harington, C.R., and Mott, R.J. (1997): Skeleton of a leopard frog (*Rana pipiens*) from Champlain Sea deposits (ca. 10 000 BP) near Eardley Quebec. Canadian Journal of Earth Sciences 34(8): 1150–5.

A well-preserved and unusually complete skeleton of a subadult leopard frog (*Rana pipiens*), CMN 46835, was recovered from a calcareous nodule collected near Eardley, Quebec (45°33′N, 76°06′W). Such nodules from the Ottawa area evidently date to the Ottawa-Delta phase of the Champlain Sea, ca. 10,000 BP, and may contain the remains of fishes, birds, mammals, marine mollusc shells and other invertebrates, and plants. This is the first record of an amphibian from Champlain Sea deposits. Palynological analysis of matrix from the specimen suggests either: (1) that forest communities were prominent in the area where frogs lived, or (2) that, presuming the specimen was from a large body of water, the environment supported herb and shrub-tundra communities with only a few trees. The specimen indicates the presence of grassy meadows near the seacoast.

701. Holmes, C.E. (1996): Broken Mammoth. *In*: American Beginnings: The Prehistory and Palaeoecology of Beringia (F.H. West, ed.). University of Chicago Press, Chicago and London. pp. 312–18.

The Broken Mammoth site (XBD-131; 64°16′N, 146°07′W) is located near the confluence of Shaw Creek and Tanana River approximately 22 km NW of Delta Junction, Alaska. The name derives from the recovery of mammoth (*Mammuthus*) tusk fragments associated with other faunal material and lithic tools (*Homo sapiens*) from an eroding bluff face. Approximately 20,000 items (about half lithic and half faunal) were recovered from 200 sq m. A distinctive characteristic of the site are the well-preserved faunal remains associated with Cultural Zones II, III, and IV. CZ I (about 4500–2000 BP) yielded only a few unidentified burned bone fragments. CZ II (about 7700–7200 BP) yielded a substantial fauna including: small rodents (Rodentia), ground squirrel (*Spermophilus*), hare (*Lepus*), beaver (*Castor canadensis*), caribou (*Rangifer tarandus*), moose (*Alces alces*), and bison (*Bison*). CZ III (centring on 10,300 BP) and IV (11,800–11,200 BP) are similar in faunal composition: bison, wapiti (*Cervus elaphus*), caribou, small rodents, ground squirrel, snowshoe hare (*Lepus americanus*), possible otter (*Lutra canadensis*?), swans [probably (*Olor columbianus*)], geese (Anserinae), ducks (Anatidae), other birds (Aves), fish (Salmonidae), and burned-bone fragments. Proboscidea [probably mammoth (cf. *Mammuthus*)] tusk fragments were associated with either CZ III or IV. A cache of three ivory artifacts associated with CZ IV yielded an AMS date of 15,830 ± 70 (CAMS-9898, Beta-67690), indicating that 'old' ivory was collected for use in artifact manufacture. Waterfowl are common in CZ IV. Flotation samples have yielded: mammal hairs, plant macrofossils, and insect remains. Other radiocarbon dates (Table 6-4) on vertebrates are AMS dates of 11,060 ± 90 BP (CAMS-7203) on bone of a large mammal (Mammalia) from CZ-3/4A?; 11,060 ± 90 BP (CAMS-7204) on wapiti (*Cervus elaphus*) bone from CZ-3/4A; and 11,500 ± 80 BP (CAMS-8261) on a swan [probably (*Olor columbianus*)] from CZ-4B.

702. Holmes, C.E. (1998): Archaeological testing and evaluation of the Gerstle River Quarry, East-Central Alaska, 1996. Division of Parks and Outdoor Recreation, Alaska Department of Natural Resources, Office of History and Archaeology Report Number 65:1–18.

The Gerstle River Quarry site (XMX-246) is 1 km SE of Gerstle River Bridge and about 1 km N of the Alaska Highway between Delta Junction and Tok Junction, Alaska. Remains of birds [probably grouse or ptarmigan (gastroliths) and a duck (?*Anas*)], wapiti (*Cervus elaphus*), bison (*Bison*), black bear (*Ursus americanus*), horse (*Equus*), and a distal humerus fragment probably of a saiga antelope (*Saiga tatarica*, Figure 11) are reported from the site (Table 2). Collagen from a horse radius yielded a date of 15,090 ± 70 BP (Beta-109267). Ages of other remains can be estimated by other radiocarbon dates at the site. Gastroliths and bones (including bison) with human (*Homo sapiens*) microblades and flakes date between about 10,000 and 8000 BP, and a wapiti pelvic bone was found in place and dates between about 7600 and 6300 BP. A black bear fibula and wapiti humerus were found in loess dating between about 6300 and 3400 BP. A concentration of 'small mammal bones' was recovered 2.54 m below datum, and therefore is likely older than 10,000 BP.

703. Holmes, C.E., and Bacon, G. (1982): Holocene bison in central Alaska: A possible explanation for technological conservatism. 9th Annual Meeting of the Alaska Anthropological Association (April 2–3, Fairbanks). pp. 1–11.

The authors report recovery of a proximal tibia fragment of a bison (*Bison*) from archaeological site XMH-297, located approximately 33 km S of Delta Junction along Delta River, Alaska. It was found in place in a loess unit directly above a layer of volcanic ash thought to be Jarvis Creek Ash. Charcoal from a hearth stain associated with artifacts and calcined bones some 35 cm above the bone yielded a date of 2285 ± 145 BP (GX-6750), whereas a date of 3980 ± 150 BP (GX-6752) was obtained on charcoal from a paleosol about 35 cm below the bone. Therefore, bison were present in the Delta River valley between approximately 2030 BC and 330 BC, just after the fall of Jarvis Ash about 1550 BC. Also noteworthy is a 'fossil animal [ground squirrel (*Spermophilus*)?] burrow' in the lowest loess horizon, and therefore >8555 ± 380 BP. A few other bison remains from Alaska and Yukon have been radiocarbon dated: (1) a hornsheath collected near Chester Creek Bridge along the Campbell Airstrip road in the Municipality of Anchorage in 1969 by Frederick Hadleigh West and referred to western bison (*Bison bison occidentalis* = *Bison occidentalis*) by Guthrie yielded a date of 470 ± 90 BP (SI-852); (2) a hornsheath collected in 1939 from the Goldstream area near Fairbanks and tentatively identified as steppe bison (*Bison priscus* = *Bison preoccidentalis*) [if the date is correct perhaps the hornsheath is more likely to be from a western bison (*Bison bison occidentalis*)] yielded a date of 5340 ± 110 BP (SI-815) [should read SI-845]; Basal Canyon archaeological site in the SW Yukon where bones dated to about 7200 BP (Workman 1974); and Baillie Islands, Northwest Territories, where bison (*Bison*) has been radiocarbon dated to about 2100 BP (Harington 1980d). 'There is no direct reference to bison in the recorded ethnographic present.' Human (*Homo sapiens*) predation, in conjunction with environmental constraints, particularly adverse snow conditions, contributed directly to the extinction of bison in Alaska. [See Stephenson et al. in press.]

704. Holmes, C.E., Vander Hoek, R., and Dilley, T.E. (1996): Swan Point. *In*: American Beginnings: The Prehistory and Palaeoecology of Beringia (F.H. West, ed.). University of Chicago Press, Chicago and London. pp. 319–23.

The Swan Point site (XBD-156; 63°18′N, 146°02′W; 322 m asl) is located on the northern edge of Shaw Creek Flats in the central Tanana River valley, Alaska. Although faunal remains are poorly preserved compared with the nearby Broken Mammoth site, goose (*Branta*) and large cervid (Cervidae) remains have been identified from the terminal Pleistocene component. Mammoth (*Mammuthus*) ivory from SP-7 (containing the oldest cultural material) has yielded a date of 12,060 ± 70 BP (NSRL-2001, CAMS-17045) (Table 6-5). Moose (*Alces alces*) remains have been recovered from Mid-Holocene and historic components (*Homo sapiens*).

705. Hone, E. (1934): The present status of the muskox. Special Publication of the American Committee for International Wildlife Protection, Number 5:1–87.
The author lists (pp. 5–8) ancient remains of the muskox (*Ovibos*) from its circumpolar range. She mentions Captain Beechey's 1828 finds at Eschscholtz Bay, Alaska, as well as Richardson's (1861) mention of them at the same site. Seton (1929) lists two Pleistocene records from Alaska and one each from British Columbia and Alberta. J.A. Allen (1913) mapped fossil muskox and helmeted muskox (*Bootherium bombifrons* = *Symbos* = *Liops*). Allen (1913) also provides a note given him by R.M. Anderson about finding an old muskox skull in August 1908 on a sandspit on southwestern Herschel Island, Yukon. On June 2, 1909, his party found remains along Colville River, Alaska, and on June 24, part of an old skull in beach muds about 2 mi W of Colville River mouth. Anderson (Allen 1913) reported that Leffingwell found a skull fragment in a cut bank 10 mi E of Collinson Point, Alaska. Hone provides a map showing fossil remains of *Ovibos*: one in Yukon and eight in Alaska.

706. Honeyman, D. (1874a): Skeleton of a whale in the Quaternary of New Brunswick. American Journal of Science, Series 3, 7:597.
A narwhal (*Monodon monoceras*) skeleton was found in clay at Jacquet River, Dalhousie, New Brunswick. It was found in a cutting of the International Railway [actually the Intercolonial Railway – see Perkins 1910] at a depth of 12 ft 6 in, 400 yds from the sea and 25 ft asl. Clay attached to the vertebrae contained barnacle fragments. [See Perkins 1910, who considered it probably represented a narwhal rather than a white whale (*Delphinapterus leucas* = *Beluga vermontana*?).]

707. Honeyman, D. (1874b): On the Quaternary containing the New Brunswick fossil cetacean. American Journal of Science, Series 3, 8:219–20.
This paper repeats the information in the previous annotation and adds that the narwhal (*Monodon monoceras*) was buried in clay overlain by 3 to 5 m of sand and gravel. Twelve species of marine mollusc shells were associated with the whale remains. It also mentions an unidentified mammal (Mammalia) vertebra found on the Tittigouche River, New Brunswick.

708. Hooper, C.L. (1884): Report of the cruise of the U.S. Revenue Steamer 'Corwin' in the Arctic Ocean, 1881. Senate Document, 1st Session, 48th Congress 1883–1884, 8(204):1–147.
Regarding his visit to Elephant Point, Eschscholtz Bay, Alaska in 1880, the author (p. 24) mentions not seeing fossils in the ice of the cliff or the soil above. However, on the beach below, after the tide fell, he found 'a large number of mammoth [*Mammuthus*] bones and tusks and some smaller bones belonging, probably, to the 'aurock' [steppe bison? (*Bison priscus*?)] and musk ox [helmeted muskox (*Bootherium bombifrons* = *Bootherium nivicolens*)].'

709. Hooper, J., and Dyke, A.S. (2000): Surficial geology, Agu Bay and Easter Cape, Baffin Island. Geological Survey of Canada Map 1959A.
This 1:250,000 scale map focuses on part of Baffin Island, Nunavut. It includes localities of radiocarbon-dated remains of the following marine mammals: narwhal (*Monodon monoceras*), bowhead whale (*Balaena mysticetus*), and walrus (*Odobenus rosmarus*).

710. Hopkins, D.M. (1959): Cenozoic history of the Bering land bridge. Science 129:1519–28.
This is a carefully documented account of the making and breaking of the Asian–North American land connection and its past environments, which have an important bearing on Pleistocene-mammal exchange between the continents.

711. Hopkins, D.M. (ed.) (1967a): The Bering Land Bridge. Stanford University Press, Stanford, California. 495 pp.
The papers presented in this book were first given at the VII Congress of the International Association for Quaternary Research, held at Boulder, Colorado, August 30–September 5, 1965. Works on geology, paleogeogra-

phy, radiocarbon chronology, vegetation, fossil mammals – both terrestrial and marine – paleobiogeography, stratigraphy, marine invertebrate fossils, and human migrations are included. [Salient articles are cited individually – see particularly Péwé and Hopkins 1967.]

712. Hopkins, D.M. (1967b): The Cenozoic history of Beringia – A synthesis. *In*: The Bering Land Bridge (D.M. Hopkins, ed.). Stanford University Press, Stanford, California. pp. 451–84.
The author provides a synthesis and review of the facts surrounding the timing, climate, and extent of the Bering Land Bridge. Glaciations, vegetation, sea-level fluctuations, animal and plant dispersal, and human (*Homo sapiens*) migration are discussed.

713. Hopkins, D.M. (1972): The paleogeography and climatic history of Beringia during Late Cenozoic time. Internord, International Journal of Arctic and Nordic Studies 12:121–50.
Along with discussions of changing climate and landforms and related flora and fauna of Beringia, the author mentions and maps the known distribution (fossil sites) of the saiga antelope (*Saiga tatarica*). This distribution suggests that the ground was smooth and also that the winter snowcover was thin and discontinuous. Further, if Bering Strait froze smooth during winter in the period from 10,000 to 8000 years ago, when it was not continuous but the shrubby vegetation reached almost the limits of the land, then forest animals such as moose (*Alces alces*), porcupine (*Erethizon dorsatum*), mountain sheep (*Ovis*), and even beaver (*Castor canadensis*) could easily have crossed. [Note that the caption with Figure 2 belongs with Figure 5, the caption with Figure 5 goes with Figure 3, and the caption with Figure 3 goes with Figure 2.]

714. Hopkins, D.M. (1973): Sea level changes in Beringia during the past 250,000 years. Quaternary Research 3(4):520–40.
This article is a comparison of sea-level changes in the Caribbean and the Bering Strait/Isthmus regions during the last 250,000 years, including a detailed discussion of the tectonic and isostatic history, ocean circulation, and glacial melting as they affected Beringia. [Although it does not mention vertebrates, this outline of Beringian sea-level history is important regarding dispersal of vertebrates through the region.]

715. Hopkins, D.M. (1979): Landscape and climate of Beringia during Late Pleistocene and Holocene time. *In*: The First Americans: Origins, Affinities and Adaptations (W.S. Laughlin and A.B. Harper, eds.). Gustav Fischer, New York. pp. 15–41.
This chapter summarizes current knowledge of the changing landscape of Beringia during times when the earliest aboriginal Americans (*Homo sapiens*) were probably present, followed by a discussion of some of the resources and environmental constraints that would have affected the early human populations. The chapter is broken into three sections: Beringia during the last glaciation (20,000–14,000 years ago); Beringia during deglaciation (14,000–10,000 years ago), and Beringia during Holocene time (10,000 years ago to present). During the 20,000–14,000 BP period, the land-mammal fauna was dominated by several species of large, gregarious herbivores (p. 26) including: bison (*Bison*), horse (*Equus*), mammoth (*Mammuthus*), and caribou/reindeer (*Rangifer tarandus*). Present in smaller numbers were: muskox (*Ovibos moschatus*), wapiti (*Cervus elaphus*), and mountain sheep (*Ovis*), which were preyed on by more diverse carnivores than present. Grazing animals were predominant. The presence of western camel (*Camelops hesternus*) and saiga antelope (*Saiga tatarica*) emphasizes the steppe-like character of the megafauna. The standing crop of game animals was far larger than present. During deglaciation (14,000–10,000 BP), the range of large herbivores deteriorated greatly (pp. 31–32). The once-interconnected arctic steppe eventually broke down into isolated relict grasslands on the few remaining active glacial outwash plains, patches of isolated steppe vegetation on S-facing valley slopes, and river bluffs scattered throughout central Chukotka, central and eastern Alaska, and Yukon. Large herbivores became increasingly vulnerable to human predators. Mammoth and horse were quickly exterminated, and bison persisted only in Eastern Beringia and in small numbers. By 10,000 BP, the rich arctic-steppe fauna had been replaced by the impoverished modern ungulate fauna: caribou and a few muskoxen in the tundra, moose (*Alces alces*) in the taiga, and a few bison in relict steppe patches. The author mentions that the youngest mammoth remains in Alaska are 15,380 ± 300 BP (SI-456). The youngest directly dated horse remains reported so far from Beringia are 13,640 ± 410 BP (I-9422). Horse, bison, and probable proboscidean (Proboscidea) bones occur in component II of the Dry Creek

archaeological site – about 10,600 years old. Bison remains as young as 5340 ± 110 BP (SI-845) have been found in the Fairbanks area of central Alaska, and two dates on a bison bone from near Anchorage, Alaska, are 470 ± 90 BP (SI-851) and <180 BP (I-9277) [see Stephenson et al. in press]. The author notes that the dispersal of humans along the S coast of Beringia was accompanied by a wave of extinctions comparable to Late Pleistocene extinctions in which people seem to be implicated. Steller's sea cow (*Hydrodamalis gigas*) and the flightless Spectacled Cormorant (*Phalacorcorax perspicillatus*) were historically confined to unoccupied and undiscovered northern islands, but once ranged around the northern Pacific. Steller's sea cow was exterminated within a couple of decades and the Spectacled Cormorant within little more than a century after the discovery of relict populations on the Commander Islands by the Bering Expedition in 1741. Consideration should be given to the possibility that sea-mammal hunting originated somewhere in southern Beringia before 10,000 BP (p. 35).

716. Hopkins, D.M. (1982): Aspects of the paleogeography of Beringia during the Late Pleistocene. *In*: Paleoecology of Beringia (D.M. Hopkins, J.V. Matthews, Jr, C.E. Schweger, S.B. Young, V. Stanley, eds.). Academic Press, Toronto. pp. 3–28.

This paper reviews knowledge of Beringian glacial chronology, shoreline history, paleoceanography, and paleoclimatology. In the synthesis the author makes the point that most large-mammal carcasses found in Alaska and northeastern Siberia date either from 45,000–30,000 BP or from 14,000–10,000 BP, suggesting that similar conditions favourable for burial and preservation prevailed during both the time of the birch zone and the latter part of the Boutellier interval. The birch-zone climate featured snowy winters, heavy spring runoff, and very dry summers. Despite evidence of great aridity, there are enough well-dated fossil remains of Duvanny Yar age [this corresponds to the herb zone in long Beringian pollen sequences and lasted from about 30,000 to 14,000 BP] to establish firmly the presence of mammoth (*Mammuthus*), horse (*Equus*), bison (*Bison*), caribou (*Rangifer tarandus*), and saiga (*Saiga tatarica*). 'Whether or not human hunters (*Homo sapiens*) were also present remains an open question for all parts except the western area of Beringia' (p. 27). An interesting characteristic of Beringian large-mammal faunas of 13,000–11,000 BP is the appearance of wapiti (*Cervus elaphus*) – perhaps reflecting the proliferation of more shrubby but still open and generally treeless vegetation.

717. Hopkins, D.M., and Marincovich, L., Jr. (1984): Whale biogeography and the history of the Arctic Basin. Proceedings of the International Symposium on Arctic Whaling, University of Groningen, The Netherlands (February, 1983), University of Groningen, Works of the Arctic Centre 8:8–24.

The oldest-known record of whales (Cetacea) from the shores of the Arctic Ocean consists of vertebrae of unidentified species from the Early and Middle Pleistocene of the Gubik Formation of Alaska. Other whale remains from Upper Pleistocene and Holocene deposits include: (1) probably the complete skeleton of a grey whale (*Eschrichtius robustus*); (2) scattered remains of probably white whales (*Delphinapterus*?) found by L.D. Carter in the youngest part of the Gubik Formation near Teshekpuk Lake, also from northern Alaska; and (3) many remains of bowhead whales (*Balaena mysticetus*) in raised beaches of Early Holocene age on southern Ellesmere Island (Blake 1975). The lack of a lengthy fossil record could indicate that whales were absent from the Arctic Basin until relatively recently. A review of cetacean fossils is included, followed by a discussion of the importance of the Arctic Ocean to modern whale distribution and evolution.

718. Hopkins, D.M., Giterman, R.E., and Matthews, J.V. Jr. (1976): Interstadial mammoth remains and associated pollen and insect fossils, Kotzebue Sound area, northwestern Alaska. Geology 4(3):169–72.

A proboscidean, probably a woolly mammoth (*Mammuthus primigenius*) but possibly mastodon (*Mammut americanum*), skeleton from the Baldwin Peninsula, is associated with plant remains radiocarbon dated at 26,900 (+2,400, –3,400) BP (AU-90). Pollen from the same level records shrub-tundra dominated by dwarf birch, sedges, and grasses; spruce pollen is lacking, and alder is represented by a few stray grains, indicating that forest vegetation lay hundreds of kilometres away. The radiocarbon age and the pollen spectrum suggest that the animal lived during a Late Wisconsinan interstadial interval equivalent to the Plum Point Interstade of the Great Lakes region of North America and the final mild episode of the Kargin Interstade of northwestern Siberia. A stratigraphic column is illustrated and the site is mapped (Figure 1). Tables present the insect fauna data and the pollen data.

719. Hopkins, D.M., MacNeil, F.S., and Leopold, E.G. (1960): The coastal plain at Nome, Alaska: A Late Cenozoic type section for the Bering Strait region. Report of the 21st International Geological Congress 1960, Part 4:46–57.
Fish (Pisces) remains recovered from estuarine silt of Sangamonian interglacial age include most of a postcranial skeleton of a starry flounder (*Platichthys stellatus*), identified by W.I. Follet, and two premaxillae and a mandible of a rock sole (*Lepidopsetta bilinsata*) identified by J.E. Fitch. This material was found with a pair of walrus (*Odobenus rosmarus*) tusks and a mollusc fauna of distinctive Pelukian (Sangamonian) marine deposits in northwestern Alaska. [See Harington 1978, p. 73.]

720. Hopkins, D.M., Smith, P.A., and Matthews, J.V. Jr. (1981): Dated wood from Alaska and the Yukon: Implications for forest refugia in Beringia. Quaternary Research 15:217–49.
Information on radiocarbon-dated and identified fossil wood and macrofossils from Alaska and NW Canada are synthesized and augmented by results of palynological studies in an effort to show the persistence of some, and total extinction of other, tree and large-shrub species. Possible dispersal routes taken by species that reinvaded Beringia in postglacial times are also reconstructed from the fossil record. [This paper is pertinent in that the distribution of trees and shrubs at various times in the Late Pleistocene obviously would influence distribution and numbers of vertebrates.]

721. Hopkins, D.M., Matthews, J.V. Jr, Schweger, C.E., Young, S.B., and Stanley, V. (eds.) (1982): Paleoecology of Beringia. Academic Press, Toronto. 489 pp.
These papers were presented at the Wenner-Gren Foundation for Anthropological Research Symposium #81, Burg Wartenstein, Austria, June 8–17, 1979. The book includes chapters on paleogeography, vegetation, steppe-tundra, paleoclimate, Pleistocene ungulates (Ungulata), and humans (*Homo sapiens*) in ancient Beringia. Some papers are cited separately.

722. Hoskin, C.M., Guthrie, R.D., and Hoffman, L.P. (1970): Pleistocene, Holocene and Recent bird gastroliths from interior Alaska. Arctic 23(1):14–23.
Polished and rounded grains of quartz, secondary quartz, and chert have been found in loess and peat in interior Alaska. The intermediate diameter of these grains is between 35 and 0.07 mm, with grains of 0.07 mm, 2–4 mm, and 10–12 mm being most abundant. Polish on these grains has been produced by abrasion in bird (Aves) gizzards. Gizzard-bearing birds are living in interior Alaska today and produce glossy polish on their gastroliths comparable to the polish on gastroliths from the Holocene and Pleistocene deposits. The polish and roundness of Recent bird gastroliths vary, being more pronounced in those living birds in late winter. Some of the Holocene and Pleistocene polished gastroliths are believed to have been produced by species of birds not now living in interior Alaska, because modal size classes of these ancient gastroliths do not coincide with modal size classes of polished gastroliths from birds living there today. Comparative diagrams include measurements and weights of gizzard stones for Rock Ptarmigan (*Lagopus mutus*), White-Tailed Ptarmigan (*Lagopus leucurus*), Spruce Grouse (*Canachites canadensis*), Blue Grouse (*Dendragapus obscurus*), Willow Ptarmigan (*Lagopus lagopus*), and Ruffed Grouse (*Bonasa umbellus*) (Figures 5 and 6).

723. Huard, V.A. (1908): Capture d'un morse. Le Naturaliste Canadien 35(4):49–51.
A complete 13-ft-long walrus (*Odobenus rosmarus*) skeleton was found near Bic, Quebec, by Intercolonial Railway construction workers in 1869, buried under 4.5 m of clay and more than 100 ft asl. The specimen was presented to the Séminaire de Rimouski, but probably was destroyed by fire many years later (p. 51).

724. Hubbard, O.P. (1887): Skeleton of a whale found over 130 years since in the St. Lawrence River Valley, near Quebec. American Journal of Science 33:242.
A skeleton of an unidentified species of whale was found near Quebec City, Quebec. It was located some French mi from Quebec City and 1 French mi from the St Lawrence River. The skeleton was of considerable size. [See Kalm 1749.]

725. Hughes, O.L., Harington, C.R., Janssens, J.A., Matthews, J.V. Jr, Morlan, R.E., Rutter, N.W., and

Schweger, C.E. (1981): Upper Pleistocene stratigraphy, paleoecology, and archaeology of the northern Yukon interior, Eastern Beringia: I. Bonnet Plume Basin. Arctic 34(4):329–65.
Stratigraphic and chronometric data show that Bonnet Plume Basin in northeastern Yukon was glaciated in Late Wisconsinan time, rather than during an earlier advance of Laurentide ice. The helmeted muskox (*Bootherium bombifrons*) and giant moose (*Alces latifrons*) are mentioned as having been collected from Old Crow Basin, Yukon, by V. Rampton in 1962 and identified by C.R. Harington. The giant moose had been previously reported from Eurasian Pleistocene deposits only. Mid-Wisconsinan mammal remains from the Hungry Creek, Yukon, site in Bonnet Plume Basin include: (1) ground squirrel (*Spermophilus* cf. *Spermophilus parryii*) and collared lemming (*Dicrostonyx*) from organic silt lying within the basal nonglacial gravel of Unit 1 at Station 1 – suggesting the occurrence of well-drained areas nearby before the deposition of the overlying varved-glacial lake sediments; (2) a few postcranial elements tentatively referred to brown lemming (?*Lemmus*) and vole (?*Microtus*) also came from this unit, as did four lemming-like scats; (3) a partial mandible including all incisors and LP_2 (NMC 35851) of a Yukon horse (*Equus* cf. *Equus lambei* = *Equus* cf. *Equus* (*Asinus*) *lambei*) found in place in the gravel of Unit 1. Yukon horse localities in northwestern North America are discussed and mapped (Figure 5), and the mandible fragment is illustrated (Figure 6).

726. Hughes, O.L., Harington, C.R., Schweger, C., and Matthews, J.V., Jr (1987): Stop 15: Ash Bend section. *In*: INQUA 87 Excursion Guide Book A-20(a) and (b) (S.R. Morison and C.A.S. Smith, eds.). Quaternary research in the Yukon. pp. 50–3.
This paper was provided to participants of the 1987 INQUA Conference as a perspective on the Old Crow, Yukon, sites. Site stratigraphy is illustrated and an extensive list of fossils is included. The fossils [mainly bison (*Bison*) with a few mammoth (*Mammuthus*) and moose (*Alces*)] come from just below an ash layer aged at approximately 80,000 BP. Some of the bones exhibit spiral fractures.

727. Hume, M. (1977): Mammoth skeleton found on Hydro site. Vancouver Province, June 23:1, 10.
Mammoth (*Mammuthus*) bones were unearthed by construction workers from a gravel pit near BC Hydro's Site One dam (14 mi downstream from Bennet Dam). The bones were discovered by a front-load operator on June 9, while digging into the gravel face of a 25-ft-high shelf. Only the front section of the skeleton was recovered at the time and the bones were in good condition. Archaeologist Ross Brand said: 'There was a tusk, the lower part of a jawbone and some other bones about 2½ feet long and four inches thick. We've found parts of tusks here before, but never anything like this.'

728. Hunston, J. (1993): A mammoth task. Dawson International Gold Show Program. pp. 7, 14–15.
In August 1992, Miles Johnson of Ross Mining Services uncovered woolly mammoth (*Mammuthus primigenius*) bones as he worked on Lisle Gatenby's property on Gold Run Creek near Dawson City, Yukon. In consultation with Norm Ross, a recovery team led by Dr Ruth Gotthardt investigated the site. Although some parts of the mammoth (estimated to be 15,000 to 12,000 years old) were articulated (the complete skeleton was not there), the silty layer also contained remains of bison (*Bison*), horse (*Equus*), and caribou (*Rangifer tarandus*). Clearly, the bones had been moved by an ancient stream. Several of the large bones had unfused articular ends, indicating that a juvenile or young adult was represented. However, C.R. Harington's examination of the lower jaw with heavily worn teeth indicated an old individual was present too. Sediment from the jaw will be analyzed for fossil pollen by Dr Les Cwynar in order to characterize the Late Pleistocene environment of the area. Three photographs of the excavation are included.

729. Hunter, H.A., and Langston, W., Jr (1965): Odontoma in a northern mammoth. Palaeontology 7(4):674–81.
A fossil odontoma probably from a woolly mammoth (*Mammuthus primigenius*) was found on Hunker Creek between Too Much Gold and Gold Bottom creeks, Yukon (65°58′N, 138°58′). Radiocarbon dates are given for adjacent peat 9510 ± 220 BP (GSC-196) and spruce wood >35,000 BP (GSC-181). Photographs of the specimen are included. So far, the majority of such cases among these animals are from northwestern North America and northeastern Asia [Beringia] – a region that experienced some degree of faunal interchange when mammoths were liv-

ing. The Yukon horse (*Equus lambei*), is also mentioned (Figure 1).

730. Irving, W.N. (1971): Recent early man research in the north. Arctic Anthropology 8(2):68–82.
This article reviews the field information concerning the arrival of humans (*Homo sapiens*) in North America. Recent evidence supports the hypothesis that people arrived before the maximum Wisconsinan glaciation and apparently before 30,000 BP. Sites mentioned in Alaska are: Trail Creek, Batza Tena, Sagavanirktok, Fairbanks, Healey Lake, Groundhog Bay, and Anangula; in Canada: Fisherman's Lake, Acasta Lake, Grant Lake, Taber, Medicine Hat, and Sheguiandah (Figure 1). A number of significant artifacts and radiocarbon dates are mentioned and a sizeable reference list is included.

731. Irving, W.N. (1975): Pleistocene archaeology in Eastern Beringia. Northern Yukon Research Programme, Contribution No. 4:1–12. [Typescript in Quaternary Vertebrates of Northern North America Publications File, Canadian Museum of Nature].
This paper was presented at the Symposium for the 'Correlation of the Ancient Cultures of Siberia and Adjoining Territories of the Pacific Coast' October 20–2, 1975, at the Siberian Division of the Academy of Sciences of the USSR, Novosibirsk. The purpose of the paper is to discuss the evidence for human (*Homo sapiens*) activity in the Yukon Territory at a time before the Classical Wisconsinan, and to suggest how this evidence can be understood in relation to the Paleolithic of Siberia. The main study area is Old Crow Flats, Yukon Territory. This region contains many exposures of Pleistocene sediments and has yielded numerous archaeological and fossil sites. Many of the fossils exhibit fractures and markings attributed to human activity. The author apparently concludes (p. 8) 'that man was breaking mammoth (*Mammuthus*), horse (*Equus*) and bison (*Bison*) bones while they were fresh, in the Canadian part of Beringia.' Current radiocarbon dates indicate a Paleolithic culture as old as 25,000 to 29,000 years, but the possibility is strong that the culture was even older yet. [See Irving 1978.]

732. Irving, W.N. (1978): Pleistocene archaeology in Eastern Beringia. *In*: Early Man in America. From a Circum-Pacific Perspective (A.L. Bryan, ed.). Occasional Papers No. 1, Department of Anthropology, University of Alberta, Published by Archaeological Researchers International, Edmonton. pp. 96–101.
The author presents evidence for human (*Homo sapiens*) activity in Yukon at a time before the Classical Wisconsinan glacial stage and suggests a tentative relationship to the Paleolithic of Siberia. Study of bone remains suggests some of the hunting techniques. Bone tools made from fresh bones were created by combinations of striking with a hafted hammer, flaking with an antler striker, grooving, whittling, and grinding. The author mentions that near Old Crow Locality 11, 'both upstream and downstream, we have recovered more than 20,000 well-preserved bones.' Most bones are from secondary alluvial deposits (e.g., Localities 14N and 11A). [See Irving 1975.]

733. Irving, W.N. (1985): Context and chronology of early man in the Americas. Annual Reviews of Anthropology 14:529–55.
The author discusses three theories about when modern humans (*Homo sapiens*) reached the Americas via Beringia. The first, called 'Clovis was First,' claims that hunters, who used the Clovis point, arrived at the end of the Wisconsinan 12,000–10,000 BP. Position Two, the most popular view, is based on bone tools dated from 70,000–20,000 BP, which is seen as circumstantial evidence of human arrival, and the fact that the Bering land bridge existed. Position Three asserts that people were there much earlier, 80,000–150,000 BP, or more, based on bone-breakage patterns attributed to humans. The author adheres to the third view and explains that some bones, which had seemingly unnatural alterations, were found in an assemblage with the lemming (cf. *Predicrostonyx*) and an absence of bison (*Bison*) remains, indicating a Sangamonian age. Pertinent is the section on Old Crow River (pp. 547–9), which gives a good perspective on archaeological work there. Irving notes that experimental analogy does not prove that any of the Old Crow specimens are really artifacts, but it shows that a large number of inferences drawn about the way Old Crow bones had been modified could be replicated today in ways that natural forces could not emulate. He criticizes Thorson and Guthrie's (1984) attempt to explain the early Old Crow artifacts by river-ice forces by noting that their experiments cannot even approximate the systematic nature of flaking observed in Old Crow bone and in the Ginsberg experiment. He concludes that 'there are many

tens of thousands of years of human activity in the Old Crow Basin represented almost exclusively by bone artifacts.' He accepts membership in Position III, while recognizing that few archaeologists 'are prepared by training to recognize a New World Early Palaeolithic or describe it for properly skeptical colleagues.'

734. Irving, W.N. (1987): New dates from old bones. Natural History 1987(2):8–13.
A caribou (*Rangifer tarandus*) flesher (bone tool) with a serrated edge was found at Loc. 14N, Old Crow Flats, Yukon (69°51′N, 139°46′W), indicating human (*Homo sapiens*) presence. This was found in an assemblage with two mammoth (*Mammuthus*) bones and bison (*Bison*) remains. Radiocarbon dating of the apatite in the bones yielded an average date of 27,000 BP for each of the flesher [27,000 +3000 –2000 (GX-1640)] and the mammoth bones [see Irving and Harington 1973]. At that time, this seemed to be conclusive evidence that man had been there then. However, it was re-dated, based on the bone collagen of the flesher, at 1350 ± 150 BP (RIDDL-145). [See Nelson et al. 1986.]

735. Irving, W.N., and Beebe, B.F. (1984): Northern Yukon Research Programme, director's report 1975–80. University of Toronto Anthropological Series Number 18:1–181.
This publication outlines the multidisciplinary approach of the program and describes six years of research in the biogeography and human (*Homo sapiens*) cultures of the Old Crow Basin, Yukon. The discovery, dating, and interpretation of evidence for humans of Pleistocene age were primary objectives. Approximately 34,000 Pleistocene vertebrate fossils were collected and catalogued during the period, representing an estimated 25 to 30% of the total number of specimens examined in the field. In the autumn of 1976 Dr B.F. Beebe was appointed Curator of Pleistocene Collections, with the additional responsibility of conducting field and laboratory research as a zoologist specializing in ice age vertebrates. Professor C.S. Churcher was originally appointed as a Principal Investigator in Pleistocene vertebrates, but only spent the summer of 1976 in the field. Curation of the fossils is described (pp. 51–60). A section on Pleistocene Zoology (pp. 93–101) provides most of the relevant information on vertebrate fossils. Significant finds include a flat-headed peccary (*Platygonus compressus*) well N of its known paleorange; two specimens of a hyena (tentatively termed *Adcrocuta eximia borealis* n. subsp. by Beebe); and a domestic dog (*Canis familiaris*) jaw thought to be of Pleistocene age [but see more recent radiocarbon date in Irving et al. 1989, p. 359].

736. Irving, W.N., and Cinq-Mars, J. (1974): A tentative archaeological sequence for Old Crow Flats, Yukon. Arctic Anthropology 11:65–81.
Recent archaeological surveys in the middle Porcupine Basin and especially the Old Crow Flats area have provided a cultural sequence (*Homo sapiens*) that appears to span most of the postglacial period and segments of Wisconsinan times. Assemblages are, for the most part, comparable to those found in neighbouring interior Alaska, although complexes such as Kayuk and Denbigh Flint, and later early Eskimo manifestations, appear to be missing or poorly represented. The paper provides a locality map (Figures 6.26, 6.27) identifying the various sites (there are many good photographs of the most important fossil localities), numerous photographs of the artifacts found, and a site-by-site written description. A significant conclusion drawn is that sometime during the postglacial period, a major change appears to have occurred in the pattern of land utilization, from an early intensive use of lookout-site areas to a more riverine-adapted hunting pattern. Three different hypotheses are presented that might explain such a shift.

737. Irving, W.N., and Harington, C.R. (1973): Upper Pleistocene radiocarbon-dated artefacts from the northern Yukon. Science 179(4071):335–40.
A caribou flesher with a serrated edge was found at Loc. 14N, Old Crow River, Yukon (69°51′N, 139°46′W). This was radiocarbon dated (based on bone apatite) at 27,000 +3000 –2000 BP (GX-1640) [this has since been re-dated as a Late Holocene artifact at 1350 ± 150 BP (RIDDL-145); see Nelson et al. 1986]; two mammoth (*Mammuthus*) bones found with the flesher were dated in the same manner: (radius) 29,100 +3000 –2000 BP (GX-1567) and (long bone) 25,750 +1800 -1500 BP (GX-1568). [Since the caribou flesher date was found in error, it is likely that the mammoth bone dates are also in error; see Irving 1987 and Nelson et al. 1986]. This would put humans (*Homo sapiens*) in the region earlier than the Clovis culture [see Irving 1985]. Other radiocarbon dates include: a mammoth (*Mammuthus*) femur 22,600 ± 600

BP (I-3573), and a large bison (*Bison*) humerus 33,800 ± 2000 BP (I-4227). The following animals were associated with the 'human' artifacts: giant pika (*Ochotona*), hare (*Lepus*), beaver (*Castor canadensis*), giant beaver (*Castoroides ohioensis*), muskrat (*Ondatra zibethicus*), singing vole (*Microtus miurus*), dhole (*Cuon*), arctic fox (*Alopex lagopus*), wolverine (*Gulo gulo*), American lion (*Panthera leo atrox*), scimitar cat (*Homotherium*), mastodon (*Mammut americanum*), woolly mammoth (*Mammuthus primigenius*), large horse (*Equus*), western camel (*Camelops hesternus*) [but see Harington 1997b], moose (*Alces alces*), caribou (*Rangifer tarandus*), and steppe bison (*Bison priscus* = *Bison crassicornis*). The paper includes a map and photographs.

738. Irving, W.N., Jopling, A.V., and Beebe, B.F. (1986): Indications of pre-Sangamon humans near Old Crow, Yukon, Canada. *In*: New Evidence for the Pleistocene Peopling of the Americas (A.L. Bryan, ed.). University of Maine, Orono, Maine. pp. 49–63.
Polished and cut bones of mammoths (*Mammuthus*) [presumably by humans (*Homo sapiens*)] were found in an assemblage with other (unspecified) mammal remains including extinct Hensel's lemming (*Dicrostonyx henseli*) at Old Crow River Loc. 12 (67°49.5′N, 139°52.5′W). The lemming is believed to have become extinct at the end of the Illinoian, and the deposit is stratigraphically lower than the Old Crow Tephra, both events tentatively dated at 80,000 BP. The authors believe that this puts humans in the Yukon at this time. The paper includes a map, an air photograph showing Locs. 300, 11A, 11, 12, 13 and 14N, specimen photographs, and detailed stratigraphic sections.

739. Irving, W.N., Jopling, A.V. and Kritsch-Armstrong, I. (1989): Studies of bone technology and taphonomy, Old Crow Basin, Yukon Territory. *In*: Bone Modification (R. Bonnichsen and M.H. Sorg, eds.). Published by the Center for the Study of the First Americans. University of Maine, Orono, Maine. pp. 347–79.
Mammal remains found at Old Crow River Loc. 12 seem to indicate a terminal Illinoian age for the assemblage (67°49.5′N, 139°52.5′W). Many of the larger bones seem to have been altered by humans (*Homo sapiens*). The associated remains of extinct lemming (*Dicrostonyx henseli*), collared lemming (*Dicrostonyx torquatus*), Ungava lemming (*Dicrostonyx hudsonius*), and possibly Hopkins' lemming (*Predicrostonyx*) seem to indicate a Middle Pleistocene age, but these species also represent a large evolutionary range and do not seem to have lived at the same time. This may be explained by either rapid evolution or long-term deposition at the site. Their presence and the absence of any bison (*Bison*) indicate a pre-Rancholabrean mammal age. Other fauna (Table 1 lists the fauna from Old Crow Loc. 12) include: whitefish (*Coregonus*), sucker (*Catostomus*), longnose sucker (*Catostomus catostomus*), unidentified bird (Aves) remains, pika (*Ochotona princeps*), giant pika (*Ochotona* cf. *Ochotona whartoni*), wolf (*Canis lupus*), ermine (*Mustela erminea*), wolverine (*Gulo gulo*), mastodon (*Mammut americanum*), mammoth (*Mammuthus*), horse (*Equus*), caribou (*Rangifer tarandus*), muskox (Ovibovini), hare (*Lepus*), varying hare (*Lepus americanus*), arctic hare (*Lepus arcticus*), beaver (*Castor canadensis*), giant beaver (*Castoroides ohioensis*), arctic ground squirrel (*Spermophilus parryii*), red-backed vole (*Clethrionomys rutilus*), brown lemming (*Lemmus sibiricus*), muskrat (*Ondatra zibethicus*), yellow-cheeked vole (*Microtus* cf. *Microtus xanthognathus*), and northern bog lemming (*Synaptomys borealis*). A steppe bison (*Bison* cf. *Bison priscus*) also found near the vicinity, was estimated at 10,000–15,000 BP. The paper shows detailed bone studies, which include graphs, charts, tables, a map, and photographs.

740. Irving, W.N., Mayhall, J.T., Melbye, F.J., and Beebe, B.F. (1977): A human mandible in probable association with a Pleistocene faunal assemblage in Eastern Beringia: A preliminary report. Canadian Journal of Archaeology 1:81–93.
A fragmentary mandible of a human child (*Homo sapiens*) was found in point-bar deposits at Old Crow Loc. 11A, Yukon (67°45′N, 139°48′W). Unfortunately, the human bone fragment is too small for conventional radiocarbon dating. Some of the other vertebrate remains were radiocarbon dated and fell into two distinct groups: 11,000–12,000 BP and >20,000 BP. Based on the bone colour differences between the two ages, the human falls into the >20,000 BP group. Other species found at the site included: unidentified fish (Pisces) and bird (Aves) remains, giant beaver (*Castoroides ohioensis*), vole (*Microtus*), bog lemming (cf. *Synaptomys*), American lion (*Panthera leo atrox*), canids (cf. Canidae), woolly mammoth (*Mammuthus primigenius*), Columbian mammoth

(*Mammuthus* cf. *Mammuthus columbi*), Yukon horse (*Equus* cf. *Equus lambei*), horse (cf. *Equus*), western camel (*Camelops hesternus*), wapiti (*Cervus elaphus*), caribou (*Rangifer* cf. *Rangifer tarandus*), bison (*Bison*), helmeted muskox (*Bootherium bombifrons* = *Symbos cavifrons*), and other unidentified ungulates. This paper includes: bone and teeth measurements of the mandible; photographs and an x-ray of it; and a map.

741. Jackson, L.E., Jr, and C.R. Harington (1991): Middle Wisconsinan mammals, stratigraphy, and sedimentology at the Ketza River site, Yukon Territory. Géographie physique et Quaternaire 45(1):69–77.
Fragmentary remains of Middle Wisconsinan mammoth (*Mammuthus*), bison [(*Bison*) – probably *Bison priscus* – an AMS radiocarbon analysis of a left radius fragment (NMC 43794) yielded a date of 26,350 ± 280 BP (TO-393)], Yukon horse (*Equus lambei*), moose (*Alces*), hare (*Lepus*), ground squirrel (*Spermophilus* cf. *S. parryii*), and lemming (*Dicrostonyx*, cf. *Dicrostonyx torquatus*) were recovered from sediments that stratigraphically underlie till of the Late Wisconsinan McConnell Glaciation. The mammalian fossils reflect either coexisting steppe-like and wetland environments or the cumulative record of succeeding environments. The fossil locality is about 40 m above the Ketza River and 30 km SE of the settlement of Ross River. The bone-bearing sediments are a complex of colluvium, fan, and outwash sediments. Similar fossil sites are likely to occur where glaciofluvial watercourses abut hillside deposits laid down prior to the onset of the McConnell Glaciation.

742. Jackson, L.E., Jr, and Pawson, M. (1984): Alberta radiocarbon dates. Geological Survey of Canada Paper 83(25):1–27.
Included are listings of radiocarbon dates with regional significance to the Late Quaternary geochronology and geomorphology of Alberta. One of the tables deals with dates pertaining to past mammalian (Mammalia) fauna. A related map locates each of the collection sites. The fauna reported includes: bison (*Bison*); steppe bison (*Bison* cf. *Bison priscus*); western bison (*Bison bison occidentalis* = *Bison occidentalis*); medium to small-horned bison (*Bison bison*); western camel (*Camelops* cf. *Camelops hesternus*); woolly mammoth (*Mammuthus primigenius*); Mexican horse (*Equus conversidens*); bighorn sheep (*Ovis canadensis*); and wapiti (*Cervus elaphus* = *Cervus canadensis*). All radiocarbon dates included have been previously published: a bison bone (*Bison*) from Calgary dated at 10,200 ± 280 BP (GSC-3065); a cervical vertebra and tibia of steppe bison (*Bison priscus*) from Watino dated at 10,200 ± 100 BP (GSC-2895) and 10,200 ± 100 BP (GSC-2902) respectively; two steppe bison tibias from Peace River dated at 9880 ± 130 BP (GSC-2865); and a bone from a plains bison (*Bison bison bison*) from Fort Macleod dated at 5410 ± 300 BP (GSC-803), which is the oldest of its kind recorded to date.

743. Jackson, L.E., Jr, MacDonald, G.M., and Wilson, M.C. (1982): Periglacial origin for terraced river sediments in Bow Valley, Alberta. Canadian Journal of Earth Sciences 19(12):2219–31.
Fluvial terraces flank the Bow River for 100 km from the eastern margin of the front ranges of the Rocky Mountains to Calgary, Alberta, and beyond. The terraces are cut predominantly in gravel fill, which ranges in thickness from about 10 m near Calgary to 30 m near the mountain front. The gravels have been deposited in a braided-stream environment. Radiocarbon dates on ungulate remains from the gravels indicate the main period of fill occurred about 11,500 to 10,000 BP. The mammalian fossil sites at Cochrane and Calgary are mapped (Figure 1). The Cochrane pits have yielded bones of: Mexican horse (*Equus conversidens*); wapiti (*Cervus elaphus*), caribou (*Rangifer tarandus*), large mountain sheep (*Ovis* cf. *Ovis canadensis*), and bison (*Bison*) [see Churcher 1968a, 1975]. The fauna from Calgary pits includes: mammoth (*Mammuthus*), Mexican horse, western camel (*Camelops* cf. *Camelops hesternus*), and bison [see Wilson 1981 and Wilson and Churcher 1978]. Camel and bison have been recorded from pits downstream from Calgary as well [see Wilson 1981]. Radiocarbon dates from two of the Calgary sites are in close agreement with those from Cochrane. Table 2 lists dates on 'bone' from Cochrane [10,760 ± 160 BP (GSC-612), 11,370 ± 170 BP (GSC-613), 11,100 ± 160 BP (GSC-989)] and Calgary [11,300 ± 290 BP (RL-757) and 10,200 ± 280 BP (GSC-3065)]. The faunas as a whole indicate 'a well established steppe vegetation' (p. 2221).

744. Jackson, L.E., Jr, Telka, A.M., and Jetté, H. (1996): A pre-Wisconsinan glacial/interglacial record, Revenue Creek, central Yukon. Poster, published by the Geological Survey of Canada.

This poster provides important paleoenvironmental background information for the Pleistocene mammals reported from the area [See Harington 1978b, 1989c]. Revenue Creek (62°20′N, 137°15′W), in the Big Creek area, lay beyond the limits of ice cover during the penultimate (Reid) glaciation in central Yukon and within terrain last glaciated before the beginning of the Brunhes Chron (0.78 Ma). Placer mining has created a 10 m exposure of peaty gravels and peat from surface to bedrock in the centre of the valley. Pollen with plant and insect macrofossils from these sediments record an interglacial/glacial/interglacial environmental succession. The basal zone is dominated by partly open-canopy spruce forest. The middle zone ranges from shrub to dry herb tundra. Insect remains outnumber plant fossils. The upper zone is a closed-canopy forest assemblage, which is tentatively correlated with the last (Sangamonian) interglacial. Abundant wood has yielded ages of >38,000 and >40,000 BP. Four tables indicate correlations of the three zones of bryophytes and vascular plant macrofossils with insect remains and pollen diagrams at the upstream and downstream ends of the section at Revenue Creek. There are photographs of the section, and maps of the site.

745. Jackson, L.J. (1979): Late Wisconsin environments and Palaeo-Indian occupation in the northeastern United States and Ontario. MSc thesis, Trent University, Peterborough. University Microfilms International, Ann Arbor, Michigan. 423 pp.

The thesis deals with Late Wisconsinan humans (*Homo sapiens*). There is hard evidence of people living in North America from 13,000 BP to the present, and circumstantial evidence earlier than that. Pages 49–51, titled *Eastern subsistence associations*, deals generally with human relationship with the vertebrate fauna. Chapter Seven, *Late Wisconsinan vertebrate fauna of Southern Ontario* (249–324), gives a detailed account of the fauna. A distribution map and a table of specimen localities accompanies the following list of Ontario Late Wisconsinan species: eastern chipmunk (*Tamias striatus*), beaver (*Castor canadensis*), white whale (*Delphinapterus leucas*), humpback whale (*Megaptera novaeangliae*), bowhead whale (*Balaena mysticetus*), bear (*Ursus*), grizzly bear (*Ursus arctos*), American marten (*Martes americanus*), mastodon (*Mammut americanum*), Columbian mammoth (*Mammuthus columbi*), woolly mammoth (*Mammuthus primigenius*), bearded seal (*Erignathus barbatus*), ringed seal (*Phoca hispida*), harp seal (*Phoca groenlandica*), caribou (*Rangifer tarandus*), white-tailed deer (*Odocoileus virginianus*), moose-like animal (*Alces*), wapiti (*Cervus elaphus*), bison (*Bison bison*), tundra muskox (*Ovibos moschatus*), lake trout (*Salvelinus namaycush*), smelt (*Osmerus mordax*), lumpfish (*Cyclopterus lumpus*), three-spined stickleback (*Gasterosteus aculeatus*), sculpin (*Artediellus uncinatus*), and capelin (*Mallotus villosus*). If Late Wisconsinan parkland environments were the preferred habitat of North American Paleoindians, the regional dominance of large, gregarious, and migratory herbivores of like adaptation in these parklands may have provided the basis for homogeneous regional Paleoindian economies (p. 383).

746. Jackson, L.J. (1983): Geochronology and settlement distribution in the early Palaeo-Indian occupation of southern Ontario, Canada. Quaternary Research 19(3):388–99.

Examination of a sample of 150 fluted-point localities from southern Ontario, in relation to datable features of Late Wisconsinan ice retreat, discloses maximum possible ages for early Paleoindian (*Homo sapiens*) occupation and reveals selection of specific physiographic situations. General relationship to maximum ice-advance positions suggests occupation during the Two Creeks Interstade after the Port Huron ice retreat about 12,300 BP. Specific relation to radiocarbon-dated preglacial Great Lakes strands favours occupation during the North Bay Interstade after Greatlakean ice retreat about 11,500 BP. Locality frequency on Glacial Lake Algonquin strands suggests contemporaneity with the main stage of this lake about 11,500 BP, with a small number of lake-plain localities indicating minor post-Algonquin persistence. Radiocarbon dates for fluted-point sites elsewhere in the glaciated Northeast place occupation coeval with the southwestern Folsom fluted-point transition of about 10,850 to 10,200 BP. Locality situation in regions dominated by preglacial sediments, on lake-edge features adjacent to strand-dissecting tributaries, and within major river valleys, implies selectivity reflecting primary adaptations. Fluted-point associations with caribou (*Rangifer tarandus*) and elk (*Cervus elaphus*) remains suggest that ‘focal’ adaptation to cervids, comparable to southwestern Folsom bison (*Bison*) exploitation, might underlie the homogeneous nature and distribution of early Paleoindian localities throughout the Northeast. Charts (Table 3) provide the locality number, elevation, strand situation, landform, soil type, local unit, and physiographic unit along with the nearest drainage. Selected radiocarbon dates are provided in Table 3.

747. Jackson, L.J. (1987): Ontario paleoindians and proboscideans: A review. Current Research in the Pleistocene 4:109–12.

In Ontario, 110 proboscidean (Proboscidea) fossil specimens have been found since 1800. The author describes those that are associated with evidence of human (*Homo sapiens*) activity, since many of the previous paleontological investigations of these sites have ignored the human remains or artifacts in those assemblages. Fluted points with mastodon (*Mammut americanum*) remains were found at an unspecified farm in the southwestern part of the province and at another site in Norfolk County (Dreimanis 1967), and a possible skinning tool was recorded from the Ferguson farm mastodon site near Tupperville (Dreimanis 1968). Deer (*Odocoileus*) bones and fluted points were found in peat above mastodon remains at Marburg, Norfolk County (Ami 1898). McAndrews and Jackson (1986) document some 75 mastodons and about 28 mammoth (*Mammuthus*) sites, as well as seven other unidentified proboscideans in Ontario. Human artifacts were only found with the mastodons. The author indicates that even though human over-hunting is believed to have been a cause of North American Proboscidean extinction, other hunted species are still living, such as deer (*Odocoileus*), wapiti (*Cervus elaphus*), and caribou (*Rangifer tarandus*). The author suspects that earliest Clovis analogues in Ontario may have hunted mastodon and/or mammoth. Communal hunting of other species by later Paleoindians could have been accelerated by dwindling proboscidean populations as spruce landscape disappeared after 11,000 BP.

748. Jackson, L.J. (1988): Fossil cervids and fluted point hunters: A review for southern Ontario. Ontario Archaeology 48:27–41.

This paper examines the distribution of deer (Cervidae) fossils in southern Ontario in relation to 160 Early Paleoindian (*Homo sapiens*) fluted point localities (Figs. 1, 2). The frequency of deer in the Ontario fossil record is surpassed only by that of mammoth (*Mammuthus*) and American mastodon (*Mammut americanum*). Caribou (*Rangifer tarandus*), Toronto deer (*Torontoceras hypogaeus*), wapiti (*Cervus elaphus*), Scott's moose (*Alces scotti = Cervalces scotti*), white-tailed deer (*Odocoileus virginianus*), and other species were present during interstadials after the Port Huron ice retreat about 12,300 BP. Radiocarbon and other geochronological evidence shows that the ages of these fossils overlap those of Early Paleoindian occupation during main Glacial Lake Algonquin time (11,500 to 10,400 BP). Elsewhere in the glaciated Northeast, deer are the only large herbivores found at Early or Late Paleoindian sites. Sixty per cent of Ontario cervid fossils are from physiographic regions with many fluted-point localities. Deller and Ellis (1988) have identified a time-sequence of Paleoindian groups (Gainey–Barnes–Crowfield) characterized in Ontario by different tool kits. Perhaps evidence will be uncovered of the earliest (Gainey) fluted points and tool kits with remains of American mastodon, so abundant in southwestern Ontario. Species such as Scott's moose, Toronto deer, and caribou also seem to have been available to Ontario Gainey people. Later Barnes-type, fluted points on Parkill complex sites may be more likely associated with migratory caribou. Both American mastodon and mammoth were disappearing at this time, and the persistence of species such as Scott's moose and Toronto deer is questionable. The latest, Crowfield-type, fluted points and tool kits might occur with remains of migratory caribou and newly appearing, more stationary species such as wapiti and white-tailed deer. Tentative dating of Crowfield about 10,300 BP places these people during major environmental changes, including the draining of Glacial Lake Algonquin and replacement of spruce parkland with pine forest. Such changes must have had major effects on faunal ranges and assemblages.

749. Jackson, L.J. (1989): Late Pleistocene caribou from northern Ontario. Current Research in the Pleistocene 6:72–4.

A large, shed caribou (*Rangifer tarandus*) antler (Fig. 1 shows an outline of the antler with measurements as well as a locality map) was recovered in 1956 during dredging to reach iron ore at the bottom of Steep Rock Lake near Atikokan, Ontario. It was embedded in about 20 m of silt and clay, and was sent to the Royal Ontario Museum for conservation. It was returned to Atikokan, where it was displayed for many years – ultimately in the Centennial Museum. A tine of the antler was cored, yielding an AMS radiocarbon date of 9940 ± 120 BP (AA-3285) – the first late-glacial date on caribou from Ontario. The author considers that the antler was shed by a bull returning from its summer range north of Atikokan, indicating that barren-ground caribou occupied range near the edge of the retreating Laurentide ice sheet. The author discusses possible caribou hunting by Paleoindians (*Homo sapiens*) occupying sites in the region, such as the Cummins site near Thunder Bay.

750. Jackson, L.J. and McKillop, H. (1988): Ontario fossil cervids: Neglected data and dating. Current Research in the Pleistocene 5:73–5.
Human (*Homo sapiens*) remains have been found with cervids, as well as proboscideans. The Late Pleistocene (12,000–10,000 BP) Ontario species that are commonly found with human artifacts and remains are: mammoth (*Mammuthus*), mastodon (*Mammut americanum*), unspecified Rodentia and Carnivora, caribou (*Rangifer tarandus*), wapiti (*Cervus elaphus*), Scott's moose (*Alces scotti* = *Cervalces scotti*), deer (*Odocoileus*), and other unidentified Cervidae.

751. Jackson, L.J. and McKillop, H. (1991): Approaches to Palaeo-Indian economy: An Ontario and Great Lakes perspective. Midcontinental Journal of Archaeology 16(1):34–68.
This article examines the historical development of Great Lakes Paleoindian (*Homo sapiens*) studies, and the effect of research biases on our ability to develop cultural explanations. Large Paleoindian strandline sites occupied briefly in spring and/or fall might coincide with caribou (*Rangifer tarandus*) migration and major interception events. Consequently, this could easily represent unusual activity in terms of the overall Paleoindian schedule. Inland hunting during periods of summer and/or winter herd dispersal may have occupied a much greater portion of total exploitation time, and account for the frequency of small interior sites. Much greater effort is needed to establish fossil animal ranges. Shed male caribou antlers from Glacial Lake Iroquois deposits provide important new range and seasonality information for Late Pleistocene herds in southern Ontario. A radiocarbon date of 9440 ± 120 BP on caribou antler from northern Ontario also documents persistence of this species after Glacial Lake Algonquin ended, suggesting a possible range shift following the Early Paleoindian period. Significantly, the Rice Lake sites are found in an area with caribou and other cervid (Cervidae) fossils. The Sandy Ridge site has a hearth feature with inclusive calcined cervid bone that may be caribou. However, there are too few radiocarbon-dated fossil finds to determine much about Pleistocene caribou ranges or direction of migrations.

752. Janossy, D. (1984): Some data on faunal exchange across the Bering Land Bridge. *In*: Beringia in the Cenozoic Era (V.L. Kontrimavichus, ed.). Amerind Publishing Co. Pvt. Ltd., New Delhi. pp. 359–66.
Comparisons are made between some Eurasian and North American gallinaceous birds (Tetraonidae) and small mammals based on osteological and paleontological criteria. Present zoogeography and paleontology of the tundra ptarmigan (*Lagopus*) suggests that these forms reached the New World the same way as people (according to Müller-Beck's 1966 hypothesis) not later than during the Early Wisconsinan. This form was subsequently isolated by later glacial cover resulting in the origin of White-Tailed Ptarmigan (*Lagopus leucurus*), now widely distributed in the Rocky Mountains and northern New Mexico. Secondary colonization of recent Eurasian species in North America took place toward the end of the Rancholabrean: Willow Ptarmigan (*Lagopus lagopus*) and Rock Ptarmigan (*Lagopus mutus*) took part in this colonization (p.362). The author hypothesizes that the Spruce Grouse (*Canachites canadensis*) reached North America when forest conditions prevailed on the Bering Isthmus before the Irvingtonian. Of Eurasian origin, the Ruffed Grouse (*Bonasa umbellus*) may have had a similar dispersal history. The origin of the lemmings (*Lemmus* and *Dicrostonyx*) is associated with Beringia. Other rodents are discussed.

753. Janusitis, A. (1947): Fossil monster. Montreal Standard, November 22, p. 2.
The article relates how Marcel Soucy, the son of a Daveluyville, Quebec, farmer discovered the skeleton of a large whale [now identified as a common finback whale (*Balaenoptera physalus*) radiocarbon dated at 11,400 ± 90 BP (GSC-2871) – see Harington 1988] while digging a ditch on the farm 4 mi outside the village. As he dug, he found more bones including some 10 ft long and weighing 200 lbs. Herman Fournier, a furniture dealer, bought

the skeletal remains for $300 and 50% of any future profits and arranged for excavation of any other bones left at the site. Reverend J.W. Laverdière, a paleontologist at Laval University, stated: 'To say that it is 10,000 years old would be nearer the truth. However, no definite age can yet be set.' Laverdière's theory was that the whale died and sank into the soft bottom of the Champlain Sea, and as the sea receded, the earth hardened and preserved the skeleton. Photographs include: a series of vertebrae; a portrait of Laverdière; and what appears to be parts of the skull.

754. Jenness, D. (editor) (1933): The American Aborigines. Their Origin and Antiquity. University of Toronto Press.
Published for presentation at the 5th Pacific Science Congress, Canada. A collection of papers by ten authors. [See Johnston 1933, and Romer 1933.]

755. Jensen, D.E. (2000): The population genetics of ice age brown bears. MSc thesis, University of Oxford. 46 pp.
Perhaps the world's largest preserved collection of vertebrate genetic material from the Late Pleistocene are bones from frozen ground in northern regions. In this case, the ancient DNA was extracted from 25 out of 50 ice age brown bear (*Ursus arctos*) samples, as well as five black bear (*Ursus americanus*) samples and one sample tentatively identified as a giant short-faced bear (*Arctodus simus*) from fossil localities in Alaska and Yukon (Table 1). Preliminary studies reveal that ancient DNA techniques and the polymerase chain reaction can recover, amplify, and sequence mitochondrial DNA of up to 200 base-pairs in length from such specimens. Leonard et al. (2000) demonstrated that ancient DNA could be extracted and sequenced from brown bear bones in frozen gravel that were radiocarbon dated from 43,000 BP to 14,000 BP. Reliable sequence dates were obtained for 18 samples and radiocarbon dates for 12 samples. After results were combined with those of Leonard et al. (2000), phylogenetic trees were constructed to compare changes in genetic diversity and distribution between ancient and modern brown bear populations, and paleoecological interpretations were developed. It is hypothesized that the modern distribution of brown bears into four distinct geographic clades is a consequence of population fragmentation caused by climatic change and interspecific competition with short-faced bears during the Late Pleistocene. Understanding genetic relationships between past populations and their recent counterparts may provide insights for management-planning and species translocations as well as establishment of Evolutionary Significant Units for conservation. Also, this approach may revolutionize population genetics and systematics. Figure 8 comprises four maps of western North America showing distribution and relationships of brown bears at: present; 10,000 to 20,000 BP; 20,000 to 35,000 BP; and 35,000 to 50,000 BP.

756. Johnson, P.H., Olson, C.B., and Goodman, M. (1985): Prospects for the molecular biological reconstruction of the woolly mammoth's evolutionary history: Isolation and characterization of deoxyribonucleic acid from the tissue of *Mammuthus primigenius*. Acta Zoologica Fennica 170:225–31.
Although the tissue used in this study came from Siberian specimens, the results have an application that is broader than just the Palaearctic. The authors managed to isolate DNA from frozen tissues radiocarbon dated between 53,000 to 10,000 years old. This purified DNA was compared to that of the Asian elephant (*Elephus maximus*). Only a small fraction of DNA isolated from mammoth tissue was homologous to Asian elephant DNA – the remaining DNA may be the result of macrobial contamination from the distant past.

757. Johnston, W.A. (1917): Pleistocene and Recent deposits in the vicinity of Ottawa, with a description of the soils. Geological Survey of Canada Memoir 101:1–69.
On page 25, the author mentions that Quaternary deposits of calcareous clay nodules at Green Creek and Besserer's Wharf, near Ottawa, Ontario, are rich in the following animals: sculpin (*Cottus*), capelin (*Mallotus*), seal (Phocidae), 'chipmunk' [this may refer to an excellent specimen of the American marten (*Martes americana*), but see Ray (1965)], and birds (Aves). The author makes other pertinent comments: (1) that calcareous concretions from which many plant and animal remains have been collected occur in the upper as well as lower clays, and that they occur most commonly in the upper part of the lower silty clays in the Ottawa River valley between Green Creek

and Besserer's Wharf, as well as along Green Creek; (2) that probably the concretions were formed after the withdrawal of the marine waters and largely since the establishment of the present drainage; (3) that most concretions found along river banks were originally in place in the clays; (4) that perhaps some concretions may have been found in clays distributed by landslides, and that they include recent organic remains that became buried in the clay. The author thinks that the occurrence of chipmunk [probably American marten (*Martes americana*?)] bones, bird (Aves) feather impressions, and plant remains are of doubtful significance in implying abundant vegetation and climatic conditions similar to the present climatic conditions.

758. Johnston, W.A. (1933): Quaternary geology of North America in relation to the migration of man. *In*: The American Aborigines: Their Origin and Antiquity (D. Jenness, ed.). University of Toronto Press, Toronto. pp. 11–45.
In his extensive review of this subject, the author mentions the following: (1) Seven mammals from fossiliferous interglacial deposits were found at Toronto, Ontario (p. 25). (2) Silt deposits in the unglaciated Yukon valley contained remains of Quaternary mammals, some of which probably are interglacial in age. Conditions probably were as favourable for migration of Quaternary mammals as they were for early people (*Homo sapiens*) during at least one interglacial as in postglacial time (p. 28). (3) Some of the larger Quaternary mammals such as mammoth (*Mammuthus*), bison (*Bison*), and horse (*Equus*) may have crossed from Siberia to Alaska on the ice (which even now forms each winter across Bering Strait), but it seems improbable that smaller animals would have followed this route (p. 28). (4) The main Pleistocene migratory route of mammals invading North America appears to have been by Bering Strait or across a land bridge in that region (p. 32). (5) Coastal sediments of northwestern Alaska contain, in places, many mammoth (*Mammuthus*), horse (*Equus*), bison (*Bison*), and muskox (*Ovibos*) remains (p. 30). (6) Probably elephants [e.g., *Mammuthus*] and possibly other large Quaternary mammals did not live in Alaska in late-glacial and postglacial time, but were present in large numbers in interglacial time and were killed off by severe climatic conditions during the last glaciation that affected the mountainous parts of Alaska and Yukon. The interglacial age of Alaskan mammoths has also been suggested by Antevs (1928) (p. 34). (7) Human migration between Siberia and Alaska probably did not occur until the area had become populated with the existing animals in late-glacial or early postglacial times (p. 34). (8) The occurrence of bison (*Bison*) remains in one of the Glacial Lake Agassiz beaches in Manitoba (Hay 1924) shows that these animals lived in the region when the lake (dammed by the retreating ice sheet) existed (p. 37).

759. Johnston, W.A., and Uglow, W.L. (1933): Placer and vein gold deposits of Barkerville, Cariboo District, British Columbia. Summary Report of the Geological Survey of Canada for 1932, Part AI:1A–75A.
Fossils in deposits overlying bedrock in the Barkerville area, British Columbia (see map, Figure 4), are not abundant. A fossil tooth (p. 16A) was found in tailings from the New Waverly hydraulic pit on Grouse Creek. Evidently it had come from glacial gravels exposed in the pit, and was identified by Charles Sternberg (Geological Survey of Canada) as that of a woolly mammoth (*Mammuthus primigenius* = *Elephas primigenius*). A similar tooth was found about 1900 by James Craig of Quesnel in bedrock gravels of the deeply buried channel of the old Bonanza Mine on lower Lightning Creek. These fossils indicate that their source deposits are Pleistocene in age.

760. Jopling, A.V., Irving, W.N., and Beebe, B.F. (1981): Stratigraphic, sedimentological, and faunal evidence for the occurrence of pre-Sangamon artifacts in northern Yukon. Arctic 34(1):3–33.
The authors report preliminary results of geological and paleontological studies pertaining to archaeological specimens of human (*Homo sapiens*) workmanship excavated at Pleistocene exposures along the lower middle course of Old Crow River, Yukon (see map, Figure 1; air photograph showing the locations of the sites studied, Figure 2; as well as photographs of excavations at site OCR12, Figure 12). They state that their interdisciplinary studies show that artifacts (see descriptions pp. 26–9) made by humans occur in deposits of Glacial Lake Old Crow laid down before Sangamonian time, probably during a phase of the Illinoian glaciation. Geological events surrounding and following deposition of Glacial Lake Old Crow were complicated by changing lake levels, localized flowage

of soft sediments, pingo formation and dissolution, and colluvial transport of vertebrate fossils and artifacts. Following deepwater stages of the lake, an environment similar to present is suggested by the excavated vertebrate fauna and by permafrost features, although Sangamonian warming is likely. Most important to vertebrate paleontologists is a list of Pleistocene vertebrate taxa excavated from Old Crow Basin Units 1b ('latest Illinoian'), 2 ('Sangamonian'), and the upper part of 2 ('Early Wisconsinan'). A list of species identified follows: whitefish (*Coregonus*), inconnu (*Stenodus leucichthys*), arctic grayling (*Thymallus arcticus*), longnose sucker (*Catostomus catostomus*), burbot (*Lota lota*), ducks and geese (Anatidae), goose (*Chen*), grouse and ptarmigan (Tetraonidae), perching birds (Passeriformes), giant pika (*Ochotona* cf. *Ochotona whartoni*), varying hare (*Lepus americanus*), arctic hare (*Lepus arcticus*), beaver (*Castor canadensis*), giant beaver (*Castoroides ohioensis*), arctic ground squirrel (*Spermophilus parryii*), red-backed vole (*Clethrionomys rutilus*), brown lemming (*Lemmus sibiricus*), collared lemming (*Dicrostonyx torquatus*), muskrat (*Ondatra zibethicus*), singing vole (*Microtus miurus*), meadow vole (*Microtus pennsylvanicus*), tundra vole (*Microtus oeconomus*), chestnut-cheeked vole (*Microtus xanthognathus*), wolf (*Canis lupus*), arctic fox (*Alopex lagopus*), fisher (*Martes pennanti*), wolverine (*Gulo gulo*), mammoth (*Mammuthus*), Vera's horse (*Equus verae*), wapiti (*Cervus elaphus*), caribou (*Rangifer tarandus*), bison (*Bison*), muskox (Ovibovini), and Soergel's muskox (*Soergelia*).

761. Josenhans, H.W., Fedje, D.W., Conway, K.W., and Barrie, J.V. (1995): Post glacial sea levels on the Western Canadian continental shelf: Evidence for rapid change, extensive subaerial exposure, and early human habitation. Marine Geology 125:73–94.

Grounded piedmont-type glaciers inundated and isostatically loaded the deep troughs that indent the Western Canadian continental shelf, as far W as the shelf edge. Glaciers do not appear to have covered the offshore lands E of the Queen Charlotte Islands (Haida Gwaii). Ice retreated from the shelf about 13,500 BP. A rapid emergence of the crust followed the ice retreat and resulted in a relative sea-level fall. At 10,400 BP relative sea level on the continental shelf was more than 100 m below that of today, and large areas of the Queen Charlotte Basin were subaerially exposed. Eustatic sea-level rise, coupled with subsidence of a glacioisostatic forebulge, allowed sea levels to rise very rapidly and reach the present shoreline of the Queen Charlotte Islands by about 9100 BP. Dated shoreline deposits (shells), submerged wood, and barnacles on stone tools at elevations between −110 m and +14 m suggest a sea-level rise of 6.3 cm per year between 12,200 and 11,000 years ago. Reconstructions of the paleogeography and paleoenvironments suggest that a hospitable environment for human (*Homo sapiens*) habitation existed in areas that are now submerged. Stone tools excavated from intertidal deposits support this interpretation. Close examination of former shorelines in Gwaii Haanas (Southern Moresby Island) led to the discovery of several archaeological sites dating from about 9000 (or more) to 5000 years ago. A flaked tool, to which a barnacle was attached, was excavated from raised estuarine deposits at Arrow Creek on Matheson Inlet. A date of 9810 ± 190 BP (CAMS 9984) (corrected to 9210 ± 190 BP) was derived from the barnacles showing that people occupied Gwaii Haanas at least by this time (see illustrations of stone tools on p. 86). At Echo Bay, some 30 km N, a large assemblage of stone tools (including bifacially flaked basalt knives) was excavated from sediments in the modern intertidal zone. At this site a sea otter (*Enhydra lutris*) bone dated to 9270 ± 100 BP (no lab number given) with stone tool cut marks on it was recovered in close association. From a study of pollen assemblages, the regional climate in the area shifted from warm and moist to warm and dry at 9000 BP. This was followed by a warm moist period at about 7500 BP and a cool moist period, similar to that of today, beginning about 5500 BP.

762. Josenhans, H., Fedje, D., Pienitz, R., and Southon, J. (1997): Early humans and rapidly changing Holocene sea levels in the Queen Charlotte Islands-Hecate Strait, British Columbia, Canada. Science 277:71–4.

Marine cores from the continental shelf-edge of British Columbia demonstrate that sea level at the shelf-edge was 153 m below present 14,000 calendar years ago and more than 30 m lower than the maximum eustatic low of 120 m. Dated artifacts, including stone tools, indicate that humans (*Homo sapiens*) occupied this region by at least 10,200 calendar years before present. Local sea level rose rapidly (5 cm per year) during the period of early human occupation as a result of the eustatic sea-level rise and glacio-isostatic forebulge movement. This shelf-edge site was first elevated and then subsided. The exposed shelf-edge was available for human occupation and may have served as a migration route during times of lowered sea levels between 13,500 and 9500 BP. The recent discovery of human remains (9880 ± 50 BP) in a cave on Prince of Wales Island [see Dixon et al. 1997], immediately N of the Queen Charlottes, highlights the potential for eventual discovery of early postglacial human occupation of the Gwaii Islands and the adjacent submerged areas of Hecate Strait. Gwaii Haanas Indian oral history abounds in legends of rapidly rising seas. The lack of evidence for human use of the early postglacial landscape may be largely attributable to the drastic and rapid changes in past sea levels. The opportunity to locate evidence of human activity dating to the period of lowered sea levels may be better some 150 km to the E, along the British Columbia coast, where isostatic dynamics resulted in coeval raised marine deposits now situated >200 m above modern sea levels.

763. Julien, P. (1968): Un poisson inattendu. Photo-Journal (Montréal), July 10–17:19.
This article in French describes Clovis Duchesneau's find, apparently most of a skeleton of a narwhal (*Monodon monoceros*) preserved in the freshly opened face of a gravel pit at Anse-aux-Gascons, Gaspé, Quebec. His attention was drawn to a relatively straight tusk located in front of the skeleton that was rusty in colour and measured about 6.5 ft long, and he collected it. 'The fossilized skeleton of what was perhaps a narwhal is now scattered for many miles under a few inches of beautiful black asphalt.' A photograph shows a view of the tusk from the root.

764. Julig, P., Jopling, A., Beebe, B., Alcock, J., D'Andrea, C., and Irving, W. (1983): Excavation report on an *in situ* bone assemblage from Locality 12, Old Crow River, northern Yukon. *In*: Carnivores, Human Scavengers and Predators: A Question of Bone Technology (G.M. LeMoine and A.S. MacEachern, eds.). Proceedings of the 15th Annual Conference of the Archaeological Association of the University of Calgary. Calgary. pp. 15–37.
This paper reports on 1982 field excavations at Loc. 12, Old Crow River (67°50′N, 139°51′W), in stratigraphic Unit 1b, commonly referred to as the 'Reworked Lower Lake deposits.' The controlled excavations provided a substantial quantity of flaked and broken large-mammal bones and ivory in association with numerous pebbles and cobbles. Differential weathering on flaked proboscidean (Proboscidea) bone was apparent within the assemblage. Two shape indices relating to sphericity and flatness were used to examine differences between two *in situ* flaked proboscidean bone assemblages at Loc. 12 and a secondary point-bar deposit (Loc. 11A) to test inferences regarding differences in assemblage form due to transport versus *in situ* modification. Since taphonomic considerations are the focus of this paper, the taxa are incidental. A total of 1011 bone and ivory specimens were recorded from controlled excavations in Unit 1b at Loc. 12. Many horse (*Equus*) bones were included, but proboscidean bone and ivory elements comprised 37% of the total. Nearly 11% of the 259 proboscidean bone specimens showed curvilinear fracture, suggesting that they were broken by percussion while fresh. Several such specimens the authors interpret to be caused by people (*Homo sapiens*). A flattened, rounded ivory point, bifacially ground and subsequently broken by torsion, particularly suggests human modification. The presence of several vertebrate fossil specimens tentatively assigned to pre-Illinoian times [e.g., peccary (*Tayassu*) and Soergel's muskox (*Soergelia*)] suggests allochthonous input to these Unit 1b assemblages (p. 24). [The identification of *Tayassu* is questionable. The only peccary specimen known from the Old Crow Basin is part of a radius of the flat-headed peccary (*Platygonus compressus*) from Locality 74 (Beebe 1980).]

765. Kalm, P. (1749): Canada-Fort Saint-Frederic, June 28. *In*: Voyage de Pehr Kalm au Canada en 1749. Annotated French translation of the travel journal by J. Rousseau and G. Béthume. Pierre Tisseyre, Montreal. pp. 112–13.
Kalm was told by the commander of the fort about a complete skeleton of a large whale (Cetacea) that was found somewhere along the St Lawrence Valley, on dry land (p. 112). [Kalm, a Swedish botanist, was in America from 1748 to 1751, and learned of the whale specimen in 1750, which puts this publication date in doubt – see Hubbard 1887.]

766. Karrow, P.F. (1984): Quaternary stratigraphy and history, Great Lakes–St. Lawrence region. Geological Survey of Canada Paper 84-10:138–53.
Vertebrate remains were found in two localities in the Toronto area. The Don Formation contains warm-climate fossils of Sangamonian age which include: groundhog (*Marmota*), deer (*Odocoileus*), bear (*Ursus*), bison (*Bison*), giant beaver (*Castoroides ohioensis*), catfish (*Ictalurus punctatus*), pike (*Esox*), and other unidentified fish (p. 141). The Pottery Road Formation occupying valleys cut in the Scarborough Formation at Toronto contains cool-climate fossils of Early Wisconsinan age (vertebrates are rare here) which include: deer (*Odocoileus*), bear (*Ursus*), bison (*Bison*), and muskox (*Ovibos*) (p. 143). Middle Wisconsinan or older fossil sites (excluding the Toronto interglacial sequence) in southwestern Ontario are listed with their radiocarbon ages (Table 1) and mapped (Figure 6). Vertebrate fossils are noted to occur at: Port Talbot II (13 dates average 44,000 BP); Woodbridge (45,000 BP); and Innerkip (50,000 BP). [See Harington 1990a.]

767. Karrow, P.F. (1987a): Comments on 'Geology of the Sheguiandah early man site: Key concepts and issues' by Robert E. Lee. Géographie physique et Quaternaire 41(3):403–6.
The author agrees with Lee's supposition that more study is needed to determine if humans (*Homo sapiens*) were living on Manitoulin Island since 12,000 BP, or if any lived there 30,000 BP.

768. Karrow, P.F. (1987b): Quaternary geology of the Hamilton-Cambridge area, southern Ontario. Ontario Geological Survey Report 255:1–94.
Walker (1896) and Coleman (1937) reported the discovery of mammoth (*Mammuthus*) bones under bar gravels on top of till in the Hunter Street tunnel in Hamilton, Ontario. Logan (1863) reported the finding of mammoth and beaver (*Castor canadensis*) bones, and a wapiti (*Cervus elaphus*) antler in the Desjardins Canal when it was cut through the Hamilton Bar. Grant (1895) described the discoveries and mentioned finding bison (*Bison*) horns there. Part of a left calcaneum of a bison was also found at a building site in Iroquois sand and gravel at Jackson and Bay streets, Hamilton, in 1974. In the Brant Museum in Burlington, there is part of a tusk from the Aldershot Bar (Photo 16). The label states that it represents a mammoth and that it was found by James Syvret on December 7, 1953, at a depth of 28 ft in the J. Cooke and Son Cement Block Company Gravel Pit at Aldershot. The author states that the occurrence of the tusk in Iroquois gravels indicates its age to be about 12,000 BP according to dates on wood from Lake Iroquois sediments. Vertebrate fossils found on the Westdale Plain behind the Hamilton (Glacial Lake Iroquois) Bar is estimated as Iroquois age. The following list (p. 51) was compiled from the work of Weber (1955), Wetmore (1958), and C.S. Churcher: northern pike (*Esox lucius*), white sucker (*Catostomus commersoni*), brown bullhead catfish (*Ictalurus nebulosus*), unidentified sunfishes (Centrarchidae), toad (*Bufo*), bullfrog (*Rana catesbiana*), blue racer (*Coluber constrictor*), pilot black snake (*Elaphe obsoleta*), timber rattlesnake (*Crotalus horridus*), rattlesnake (*Sistrurus*), Wood Duck (*Aix sponsa*), Oldsquaw (*Clangula hyemalis*), Barred Owl (*Strix varia*), Red-Winged Blackbird (*Agelaius phoeniceus*), Common Grackle (*Quiscalus quiscula*), short-tailed shrew (*Blarina brevicauda*), eastern grey squirrel (*Sciurus carolinensis*), eastern chipmunk (*Tamias striatus*), red squirrel (*Tamiasciurus hudsonicus*), northern flying squirrel (*Glaucomys sabrinus*), woodchuck (*Marmota monax*), deer mouse or white-footed mouse (*Peromyscus*), red-backed vole (*Clethrionomys gapperi*), meadow vole (*Microtus pennsylvanicus*), pine vole (*Microtus pinetorum = Pitymys pinetorum*), muskrat (*Ondatra zibethicus*), and red fox (*Vulpes vulpes*). Grant (1891) reported similar specimens from Burlington Heights Ridge, near Hamilton.

769. Karrow, P.F. (1990): Interglacial beds at Toronto, Ontario. Géographie Physique et Quaternaire 44(3):289–97.
This paper includes an extensive list of all fossils found in the Sangamonian Don Formation at Toronto. A radiocarbon date on wood in the assemblage yielded >46,000 BP (L-409). The vertebrates (Table 11, p. 295) include: bison (*Bison*), giant beaver (*Castoroides ohioensis*), pike (*Esox*), catfish (*Ictalurus punctatus*), groundhog (*Marmota monax*), two species of deer (Cervidae), large bear (*Ursus*), and many unidentified small vertebrate remains [mainly fish scales and bones – see Harington 1990a, p. 380, for addition of fish species identified by S.L. Cumbaa in 1987]. At the top of the York Till at Woodbridge, Ontario (Figure 1), rodent (Cricetidae) and wolverine? (*Gulo gulo*?) remains were found. The climate of this

time was anywhere from present temperatures to 3°C higher. Included are a map, stratigraphic section, and photographs.

770. Karrow, P.F., and Warner, B.G. (1988): Ice, lakes, and plants, 13,000 to 10,000 years B.P.: The Erie–Ontario lobe in Ontario. *In*: Late Pleistocene and Early Holocene Paleoecology and Archeology of the Eastern Great Lakes Region. Bulletin of the Buffalo Society of Natural Sciences 33:39–92.

Mammoth (*Mammuthus*), caribou (*Rangifer tarandus*), muskox (*Ovibos*), and bison (*Bison*) remains have been found in several localities in Glacial Lake Iroquois beach deposits. These are the earliest postglacial vertebrates so far known in Ontario. A new deer species, now called Toronto deer (*Torontoceros hypogaeus*), has been discovered in the Toronto area. Mastodon (*Mammut americanum*) remains have been found in scattered locations on former lake plains of Lake Erie, but very few are known from the western Lake Ontario basin. A diverse bird (Aves), reptile (Reptilia), and mammal (Mammalia) assemblage from Hamilton, Ontario, originally attributed to Lake Iroquois time (Wetmore 1958), was later reinterpreted as intrusive and of much younger age (p. 45).

771. Karrow, P.F., and Warner, B.G. (1990): The geological and biological environment for human occupation in southern Ontario. *In*: The Archaeology of Southern Ontario to A.D. 1650 (C. Ellis and N. Ferris, eds.). Occasional Publications of the London Chapter, Ontario Archaeological Society, No. 5:5–35 (with an appended 15 pp. of references).

This paper summarizes the geological and biological history of southern Ontario as a background for consideration of the history of human (*Homo sapiens*) migration, survival, and settlement. Controversy and a lack of data leave open the question of when human occupation began. Suggestions of arrival before the last (Wisconsinan) glaciation (which peaked about 20,000 BP) are well emplaced in the literature, but the authors' account begins at the glacial maximum. A series of maps indicates phases of glaciations of the region until about 10,000 BP. Of particular interest is a map (Figure 2.10) showing sites of postglacial mastodons (*Mammut americanum*) and mammoths (*Mammuthus*) in Ontario.

772. Karrow, P.F., Churcher, C.S., and Morgan, A.V. (1979): Quaternary paleontology of the Toronto Region [Chart exhibited at the 1979 meeting of the International Geological Correlation Programme in Czechoslovakia]. Published by the University of Waterloo Department of Earth Sciences, Waterloo, Ontario.

This is a chart of a stratigraphic section of Wisconsinan and Sangamonian deposits, graphically showing what fossils are present in each formation. The Late Wisconsinan Glacial Lake Iroquois deposits (12,000 BP) include: Columbian mammoth (*Mammuthus columbi*), woolly mammoth (*Mammuthus primigenius*), and bison (*Bison*). The Middle Wisconsinan Port Talbot deposits (<45,000 BP) include: woolly mammoth (*Mammuthus primigenius*) and grizzly bear (*Ursus arctos*). The Early Wisconsinan Pottery Road Formation (80,000 BP) includes: Columbian mammoth (*Mammuthus columbi*), bison (*Bison*), grizzly bear (*Ursus arctos*), Scott's moose (*Alces scotti* = *Cervalces borealis*), white-tailed deer (*Odocoileus virginianus*), and muskox (*Ovibos*). The Sangamonian Don Formation includes: bison (*Bison*), black bear (*Ursus americanus*), Scott's moose (*Alces scotti* = *Cervalces borealis*), white-tailed deer (*Odocoileus virginianus*), giant beaver (*Castoroides ohioensis*), groundhog (*Marmota monax*), catfish (*Ictalurus punctatus*), and pike (*Esox*).

773. Keddie, G. (1979): The late ice age of southern Vancouver Island. The Midden 11(4):16–22.

In 1976 a mammoth (*Mammuthus*) humerus was found in Saanichton Gravels N of Victoria, British Columbia (map provided). A radiocarbon date on the bone yields an age of 17,000 ± 240 BP (GSC-2829). The remains (in 1975 part of a large mammoth tusk had been found at the same level and site – the Trio Ready Mix Ltd. Gravel Pit on Cordova Bay Road in Saanich) were found about 20 m below the surface, within 1 m of the erosional surface between the Saanichton Gravels and the Quadra Sands below. This find alters the previously postulated timing of the glaciation of southern Vancouver Island.

774. Keddie, G. (1995): List of all mammoth and mastodon fossils in the collections of the Royal British Columbia Museum as of October 12, 1995, and list of fossils once in the collection that cannot be located. 15

pp. [Typescript in Quaternary Vertebrates of Northern North America Publications File, Canadian Museum of Nature, Ottawa.]
In 1994 new accession numbers were assigned to all proboscidean (Proboscidea) remains in the paleobiology collection. Old catalogue numbers follow the new ones. Important sources for background information on the specimens were found in: (1) Government Records (GR 111) including museum correspondence between about 1916 and mid-1960, much of which is not duplicated in Museum Archive records; (2) Annual Reports of the Provincial Museum; (3) C.F. Newcombe Catalogue, 'Glacial Period in B.C. Mammoth and Marine fossils,' pp. 21–9; (4) C.F. Newcombe Catalogue, 'Fossils. Register of Localities where collected' (there is one relevant reference on p. 36); (5) Large Old Catalogue, 'Earth History Catalogues, Fossils and Minerals Accession No. 1 to 1222.' The purpose of this list is to compile existing information on the proboscidean [mammoth (*Mammuthus*) and American mastodon (*Mammut americanum*)] remains.

775. Keele, J., and Johnston, W.A. (1913): The superficial deposits near Ottawa. *In*: Excursions in the Neighbourhood of Montreal and Ottawa (Excursions A6, A7, A8, A10, A11). Geological Survey of Canada. Guide Book No. 3:126–35.
Unspecified whale bones were found in a sand and gravel deposit 134 m asl near Smiths Falls, in eastern Ontario [this probably refers to a humpback whale (*Megaptera novaeangliae*) first reported by J.W. Dawson 1883a]. At Green Creek and Besserer's Wharf, E of Ottawa, several fossils were found in calcareous nodules: seal (Phocidae), capelin (*Mallotus villosus*), unspecified fish (Pisces) and birds (Aves) – both feathers and bones.

776. Keenan, T.J., and Cwynar, L.C. (1992): Late Quaternary history of black spruce and grasslands in southwest Yukon Territory. Canadian Journal of Botany 70:1336–45.
A wood bison (*Bison bison athabascae*) specimen was found at the Canyon Archaeological site between Long Last Lake and Two Horsemen Pond, near Dawson City, Yukon, radiocarbon dated at 7195 ± 130 BP (SI-1117). Another wood bison fossil, found at the Kusawa Bluff site at Kusawa Lake, S of Two Horsemen Pond, was dated at <5380 ± 100 BP (Beta-14402). Another was found at the Chimi site at Aishihik Lake and was dated (from peat) at 6420 ± 110 BP (S-1281). This species represents a varied forest to semi-open prairie animal, not a plains animal. This area at the hypsithermal (about 6000 BP) was believed to have been open prairie, but these finds change that view. The climate was warmer than the present, the warmest of the Holocene. See Table 2 for a list of nine radiocarbon dates pertaining to the wood bison in Yukon.

777. Kehoe, T. (1964): The Kyle mammoth. Saskatchewan Archaeology Newsletter 8:1–4.
A mammoth (*Mammuthus*) was found near Kyle, southern Saskatchewan. The bones were uncontaminated and were submitted for radiocarbon dating; the results were not available when the paper was published. The Kyle specimen is one of approximately 70 mammoths discovered in the province. More of the skeleton was recovered from the Kyle animal than from any of the earlier finds: a shoulder blade; an upper molar tooth; a near-complete lower jaw with molar in place; a nearly complete tusk; eight vertebrae; ribs; a femur; and three feet were exposed and taken to the Saskatchewan Museum of Natural History in Regina. There was no evidence of human (*Homo sapiens*) hunters associated with this skeleton.

778. Kelly, M. (1975): A note on the implications of two radiocarbon dated samples from Qaleragdlit imâ, South Greenland. Bulletin of the Geological Society of Denmark 24:21–6.
Fish (Pisces) remains are recorded from fossiliferous carbonate concretions that occur fairly abundantly in glacial tills of the moraines near Qaleragdlit imâ. Shells of several species of marine molluscs are found in these concretions too – especially *Hiatella arctica* and *Mytilus edulis*. A whole concretion from this sample yielded a radiocarbon date of 4540 ± 130 BP (Birm-454).

779. Kelly, M., and Bennike, O. (1992): Quaternary geology of western and central North Greenland. Grønlands Geologiske Undersøgelse Rapport 153: 1–34.
The authors mention a single otolith of the polar cod (*Boreogadus saida*) from Nyeboe Land, North Greenland (MGUH VP 3343). From its altitude at 29 m asl, a

Middle Holocene age seems likely (p. 25; Table 4). A ringed seal (*Phoca hispida*) skeleton from Hall Land was dated to about 8700 BP, which is the oldest-known Holocene mammal from Greenland (Bennike et al. 1989). Much of the present mammalian fauna was present in the Middle Holocene, and Peary's reindeer/caribou (*Rangifer tarandus pearyi*) at least were present in the Early Holocene.

780. Kemp, B. (1995): Mastodon makeover a mammoth job – 80,000-year-old giant found in N.B. will be centrepiece of museum exhibit. St. John Evening Times Globe, September 15. pp. A1–A2.

This article describes the Hillsborough Mastodon (*Mammut americanum*) and how it may have died after becoming mired in pond deposits. The skeleton was found by workmen digging a dam near Hillsborough, New Brunswick, and the bones were stored in the New Brunswick Museum. Research Casting International of Toronto was given a $60,000 contract by the museum to mount a cast of the skeleton for display as the centrepiece of a new outlet in Market Square. A photograph shows Matt Fair and Peter May of Research Casting with Dr Randall Miller of the New Brunswick Museum wrapping the mastodon bones for shipment to Toronto.

781. Kennedy, C.C. (1977): Whale bones found. Renfrew Mercury, November, 30. p. A12.

Bowhead whale (*Balaena mysticetus*) right flipper bones, jawbone, and other bone fragments were found in the Hanson Pit, at White Lake, 13 km SW of Arnprior, Ontario. A radiocarbon date on bone yielded an age of 11,400 BP. The article provides several photographs of the specimens. Photographs of two bones of the forelimb of the specimen, excavation of the jaw bone, and a whale rib from the site are included.

782. Kenyon, A., and Churcher, C.S. (1965): A flake tool and a worked antler fragment from Lake Agassiz. Canadian Journal of Earth Sciences 2:237–46.

A stone tool [indicating the presence of humans (*Homo sapiens*)] and a scraper made from a ?moose (cf. *Alces alces*) antler were found buried 3.7 m in Glacial Lake Agassiz beds near Morson, Rainy River District, Ontario. The antler was radiocarbon dated at 7861 ± 423 BP (SM-696-1). A map (Figure 4) of Recent and Late Pleistocene records of moose and Scott's moose (*Alces scotti* = *Cervalces scotti*) or giant moose (*Alces latifrons* = *Cervalces alaskensis*) is included.

783. Kermode, F. (1916): Palaeontology. *In*: Province of British Columbia, Report of the Provincial Museum of Natural History for the year 1915. p. N21 and Plate XI.

The following Pleistocene mammal remains were accessioned or commented on: (1) A mammoth (*Mammuthus*) tooth from Mount Tolmie gravel near Victoria, Vancouver Island (48°28′N, 123°21′W) found early in 1915. Photographs of three such teeth from Vancouver Island and two from the Yukon were sent to Dr O.P. Hay of Washington, who stated that the northern specimens were of the woolly mammoth (*Mammuthus primigenius* = *Elephas primigenius*), while those from the S (Vancouver Island), which had fewer and wider plates, belonged to the Columbian mammoth (*Mammuthus columbi* = *Elephas columbi*). (2) A worn vertebra of a species of whale (Cetacea) from Pleistocene gravels between Parson's Bridge and Langford Lake, Vancouver Island, which has been in the museum for many years, was identified by O.P. Hay. It was found while taking ballast for the Esquimalt and Nanaimo Railway. (3) In 1913 a fine skull (occiput with horncores) of a bison (*Bison*) was found in the Atlin district of British Columbia by Mr E.M. Anderson. L.M. Lambe identified it as *Bison bison*. (4) Also in 1913 a fragment of bison (*Bison*) horn was found by Dr C.F. Newcombe in low-lying ground between George Jay Public School and the Protestant Orphanage, E of Cook Street, Victoria. This specimen was identified by L.M. Lambe of the Geological Survey of Canada as *Bison bison*. (5) A mastodon (*Mammut americanum* = *Mastodon americanus*) tooth from Yukon (presumably the one shown in Figure 2 in Plate XI that was found in 1903 at Dago Gulch, Hunker Creek near Dawson City) was identified as a left lower third molar by L.M. Lambe. Other specimens figured in Plate XI are: Figure 3, a mammoth (*Mammuthus*) tooth from Cordova Bay, near Victoria (about eight enamel plates are present, but the roots are poorly preserved and the rest appears to be heavily eroded); and Figure 1, a mammoth (*Mammuthus*) tooth from Shuswap Lake (about 16 enamel plates are preserved).

784. Khan, E. (1970): Biostratigraphy and paleontology of a Sangamon deposit at Fort Qu'Appelle,

Saskatchewan. National Museums of Canada Publications in Palaeontology 5:1–82.
Mammal remains were found in the Bliss Pit containing Echo Lake gravels, which are 52 m below overlying till; this is 0.4 km SW of Fort Qu'Appelle, Saskatchewan. The fauna included: canid (*Canis*), badger (*Taxidea*), Columbian mammoth (*Mammuthus* cf. *Mammuthus columbi*), Scott's horse (*Equus scotti*), camel (*Camelops* cf. *Camelops hesternus*), Scott's moose (*Alces scotti* = *Cervalces roosevelti*), giant bison (*Bison latifrons*), and helmeted muskox (*Bootherium bombifrons* = *Symbos cavifrons*). The assemblage was radiocarbon dated (from mollusc shells) at >30,000 BP (GSC-987). The gravels are estimated to range from Kansan to Early Holocene. The biostratigraphic evidence strongly suggests a Sangamonian interglacial age for the faunal assemblage. Photographs and measurements of the best specimens are included.

785. Khan, E. (1971): Suggestive evidence for the presence of early man. Napao 3(1):32–3.
Some broken horse (*Equus*) bones were found in channel B of the third gravel bed at the Bliss Pit, Fort Qu'Appelle, Saskatchewan. The breaks may have been made by humans (*Homo sapiens*), because they resemble the same type of reworking of confirmed human artifacts found in Siberia; the breaks cannot be accounted for by natural geological processes. The bones of this assemblage were 90% horse (*Equus*), an animal easily hunted by man. The gravel pit is of Sangamonian age. There is an excellent photograph of the site (Figure 1) and a specimen [CMN 11924 – unidentified] from it.

786. Kindle, E.M. (1923): Ottawa Valley concretions. *In*: Canadian Pleistocene Concretions. Bulletin of the Geological Society of America 34(3):620–47.
The remains of capelin (*Mallotus villosus*) and American marten (*Martes americana*) were found in clay nodules at Green Creek, E of Ottawa. Bones of small fish are often found within the nodules. The age of the nodules is Late Wisconsinan. Plate 8 shows photographs of: the opposite sides of a concretion from the Ottawa River downstream from Green Creek showing the skull and other parts of an American marten – probably the first published illustration of this excellent specimen; and both sides of a concretion from Green Creek showing parts of a capelin skeleton.

787. Kindle, E.M. (1924): Note on a mammoth tusk from the Arctic Archipelago (Article 12). American Journal of Science 7(207):183–5.
A well-preserved mammoth (*Mammuthus* cf. *Mammuthus primigenius*) tusk fragment about 3 ft long and 5¾ in in diameter was found on Melville Island, Northwest Territories at 75°N (erroneously reported as Banks Island, 73°N, and corrected in the Geological Survey of Canada's library copy). The specimen was collected in 1916 by V. Stefansson's northern party of the Canadian Arctic Expedition near Cape James Ross on the S side of the entrance to Liddon Gulf. This extends the northern limit of the species. Another mammoth (*Mammuthus* cf. *Mammuthus primigenius*) tusk was found at Port Hope, Alaska, along with two mastodon (*Mammut americanum*) teeth. [The specimen has since been radiocarbon dated at 21,900 ± 320 BP (GSC-1760) – see Harington 1990b, p. 158.]

788. King, K. Jr. (1978): Gamma-carboxyglutamic acid in fossil bones and its significance for amino acid dating. Nature 273:41–3.
The author discusses amino acid racemization reactions as a possible new dating technique; one problem with this would be leaching, thus contamination, of *in situ* specimens. A whale [probably bowhead whale (*Balaena mysticetus*)] rib was found in permafrost of a raised marine delta at a fiord on Baffin Island, Nunavut, and its minimum age is estimated at 48,000–44,000 BP, according to G.H. Miller.

789. King, R.H. (1991): Paleolimnology of a polar oasis, Truelove Lowland, Devon Island, N.W.T., Canada. Hydrobiologia 214:317–25.
A marine transgression covered the Truelove Lowland of Devon Island, Nunavut, up to an elevation of 65 m by about 9700 BP. Ages and elevations of bowhead whale (*Balaena mysticetus*) remains and driftwood mark the emergence from the sea (Table 1). Radiocarbon dates on whalebone are: 8920 ± 140 BP (Beta-14819) at 68.6 m asl; 8270 ± 150 BP (GSC-991) at 42.4 m asl; 6790 ± 90 BP (Beta-19201) at 21.8 m asl; 6100 ± 125 BP (S-433) [should read 6110 ± 125 BP (S-432)] at 9.7 m asl; 2900 ± 85 BP (S-433) at 4.7 m asl.

790. Kinniburgh, C. (1999): Discovery may reveal ancient secrets. Calgary Herald, June 6:A1–A2.

Calgary archaeologists excavated a Plainview point estimated to be 11,000 years old in soil beside a prehistoric campfire on the sandy shore of Lake Minnewanka near Banff, Alberta. The 700-m-long site is subject to flooding as part of a reservoir. An Agate Basin point was also discovered with parts of an extinct bison (*Bison*) skull. Three of Alberta's oldest stone-tool discoveries, including this most recent find, have been located in the Bow Valley of Banff National Park – within a few kilometres of each other. A photograph shows University of Calgary archaeologist Alison Landals holding the Plainview point.

791. Kishchinskii, A.A. (1984): Basic elements of mountain faunas of north-east Siberia and north-west America and stage of formation thereof (based on data of biogeographic analysis). *In*: Beringia in the Cenozoic Era (V.L. Kontrimavichus, ed.). Amerind Publishing Co. Pvt. Ltd., New Delhi. pp. 482–91.
Basic zoogeographic elements of mountain faunas in northeastern Siberia and northwestern North America are distinguished. A general picture of the evolution of the faunas has been hypothesized on the basis of the probable development of each species. In the Early Quaternary ancient mountain sheep [*Ovis* (*Pachyceros*)], pika (*Ochotona*), marmot (*Marmota*), etc. invaded Beringia, moving from inner Asia along steppe and mountain-steppe habitats. In the Late Pleistocene a vast common area of alpine evolution developed in northeastern Asia and Alaska. Recent species of sheep, marmots, pikas, arctic ground squirrels (*Spermophilus parryii*), etc. evolved there. In North America, S of the Cordilleran ice sheet, mountain faunas developed independently. Subalpine faunas in northeastern Siberia began to form by the end of the Pleistocene and in northwestern America perhaps only in the Holocene, when present ranges formed and recent species and subspecies evolved.

792. Kitts, D.B. (1953): A Pleistocene muskox from New York and the distribution of the musk-oxen. American Museum Novitates No. 1607:1–8.
While describing a tundra muskox specimen from Vestal, New York, the author mentions specimens collected by Beechey (1831) in Alaska, and a tundra muskox (*Ovibos moschatus = Ovibos yukonensis*) partial skull from the Palisades of the Yukon River in Alaska. The article includes a map (Figure 2) of Pleistocene tundra muskox and helmeted muskox (*Bootherium bombifrons = Symbos cavifrons*) sites for North America.

793. Klassen, R.W., and Delorme, L.D. (1967): Geology and paleontology of Pleistocene deposits in southwestern Manitoba. Canadian Journal of Earth Sciences 4:433–47.
Mammal remains found in a Pleistocene intertill (see Minnedosa River stratigraphic section on p. 437) silt on Riding Mountain, southwestern Manitoba (50°23.2′N, 99°54.5′W) include: arctic ground squirrel (*Spermophilus* cf. *Spermophilus parryii = Citellus* cf. *Citellus undulatus*), meadow vole (*Microtus* cf. *Microtus pennsylvanicus*), and tundra vole (*Microtus* cf. *Microtus oeconomus*). A radiocarbon date on associated grass yielded an age of >31,300 BP (GSC-297). The specimens seem to correlate well with Port Talbot interstade sediments in the Lake Erie region, though it is inferred that they are older and correlative either with an Early Wisconsinan interstade or with the Sangamonian interglacial. [See Fenton 1982.]

794. Klein, D.R. (1965): Postglacial distribution patterns of mammals in the southern coastal regions of Alaska. Arctic 18(1):7–20.
During the Wisconsinan glaciation the present land areas of the coastal regions of Alaska bordering the Gulf of Alaska were nearly completely overridden by ice. The now-existing flora and fauna of the region have presumably become established in the 10,000 years since the recession of the ice. The present distribution and affinities of small mammals in the Alaskan southern coastal area can be explained by either or both of the following hypotheses: (1) refugia existed in some offshore areas now submerged at least during the Wisconsinan glaciation, as well as land connections between islands and the mainland; (2) the affinities of some species of small mammals occurring on widely separated islands are the result of parallel morphological changes occurring under similar environmental stimuli in postglacial times. Various species of mouse (*Peromyscus*), vole (*Microtus*), shrew (*Sorex*), and marmot (*Marmota*) are discussed in this connection. All large land mammals of the region apparently are postglacial arrivals from refugia either in Beringia or from the southern refugium (S of the continental ice sheets). Taxonomic relationships, dispersal requirements, and source refugia for the following species are discussed: brown bear (*Ursus arctos*), black bear (*Ursus*

americanus), wolf (*Canis lupus*), black-tailed deer (*Odocoileus hemionus sitkensis*), moose (*Alces alces*), caribou (*Rangifer tarandus*), wapiti (*Cervus elaphus* = *Cervus canadensis*), mountain goat (*Oreamnos americanus*), and Dall sheep (*Ovis dalli*).

795. Klein, R.G. (1975): The relevance of Old World archaeology to the first entry of man into the New World. Quaternary Research 5:391–4.
Archaeological evidence from the USSR suggests that cultural adaptations to the most rigorous environments of northern Eurasia were not achieved until 35,000–40,000 BP. This presumably sets an absolute basement date for the entry of man (*Homo sapiens*) into Alaska through the region of Beringia. The absence of evidence for pre-14,000-yr-old man in the 48 adjacent United States, comparable in any sense to the evidence that has been developed for man prior to 14,000 BP in the Old World, suggests that movement S out of Alaska only occurred after 14,000 BP.

796. Knuth, E. (1965): Pearlyland's arkaeologi 1. Naturens Verden 1965:170–84.
According to Bennike (1997, p. 905), a small artifact made of walrus (*Odobenus rosmarus*) tusk from the Neoeskimo site Kølnaes in eastern Peary Land of Independence II age (3100–2400 BP) may have been imported, so walruses were not necessarily hunted locally then.

797. Knuth, E. (1967): Archaeology of the musk-ox way. École Pratique des Hautes Études. Sixième section. Contributions du Centre d'Études Arctiques et Finno-Scandinaves 5. Sorbonne, Paris. 78 pp.
Section XII deals with animal bones from the 'Muskox Way' – Steensby's route for Paleoeskimos (*Homo sapiens*) from northern Ellesmere Island across Nares Strait via Robeson Channel and Smith Sound to Greenland [see Section III and Plate 1]. The analysis of the bones was undertaken by Ulrik Møhl, and the distribution of vertebrates at the most important Peary Land settlements may be seen from the table on p. 32a. The overall picture shows clearly that tundra muskoxen (*Ovibos moschatus*) were the most important game both for Independence I and II peoples. Few caribou (*Rangifer tarandus*) remains were uncovered. Other species listed from archaeological sites of Independence I age (approximately 4500–3600 BP) are: seal (*Phoca*), arctic hare (*Lepus arcticus*) [found at almost the same number of sites as muskoxen], arctic fox (*Alopex lagopus*), trout [arctic charr (*Salvelinus alpinus*)], Brant Goose (*Branta bernicla*), Rock Ptarmigan (*Lagopus mutus*), Glaucous Gulls (*Larus hyperboreus*), Ivory Gull (*Pagophila eburnea*), Oldsquaw or Longtailed Duck (*Clangula hyemalis*), King Eider (*Somateria spectabilis*), Barnacle Goose (*Branta leucopsis*), and Red Knot (*Calidris canutus*). Apparently a phalanx of a domesticated dog (*Canis familiaris*) was found in 1965 at the Burin Delta site (Independence I; 81°03′N, 70°05′W) in Ella Bay, northern Ellesmere Island, Nunavut. This was the only evidence of a dog associated with either of the Independence cultures along the Muskox Way. A list of vertebrate bones (p. 32) found in archaeological sites of the Independence II culture (about 3200–2500 BP) includes: walrus (*Odobenus rosmarus*), seal, tundra muskox, arctic hare, arctic fox, arctic charr, Brant Goose, ptarmigan, Glaucous Gull, Oldsquaw Duck, and caribou antlers.

798. Knuth, E. (1968): The Independence II bone artifacts and the Dorset-evidence in North Greenland. Folk 10:61–80.
Of 439 Independence II artifacts (*Homo sapiens*) found along the 'Muskox Way,' 170 (about 39%) comprise bone implements, which are mainly of muskox (*Ovibos moschatus*) bone, but nine are of walrus (*Odobenus rosmarus*) tusk and two of walrus rib. Bird (Aves) and arctic hare (*Lepus arcticus*) bones have been used for needles. A few small artifacts made of walrus tusk have been recovered from Independence II (about 3100–2400 BP) ruins at Jørgen Brønlund Fiord (Figure 1), Peary Land.

799. Knuth, E. (1978): The 'Old Nûgdlît Culture' site at Nûgdlît Peninsula, Thule District, and the 'Mesoeskimo' site below it. Folk 19–20:15–47.
The author mentions H.P. Steensby's hypothesis that: (1) the first people (*Homo sapiens*) in Greenland (Independence I) followed the track of muskox (*Ovibos moschatus*) and crossed Nares Strait near its narrowest point (Robeson Channel), whence the 'Muskox Way' led them N of Greenland, across Peary Land and down the E coast of Greenland; (2) the later caribou (*Rangifer tarandus*) hunters crossed Smith Sound to Inglefield Land and moved down the W coast of Greenland via the Thule District. Radiocarbon dates for three marine-adapted mammals from Old Nûgdlît in Northwest Greenland (approximately 76°48′15″N, 70°35′20″W; Figure 4) are (p. 39):

(1) polar bear (*Ursus maritimus*) bone 5850 BP; (2) large whale [bowhead whale? (*Balaena mysticetus*?)] bone 4920 BP; and (3) walrus (*Odobenus rosmarus*) tusk 4060 BP [no degrees of error or laboratory numbers are listed]. [Bennike (1997) notes that the polar bear date is the oldest for that species from Greenland, and that a bone from the site was dated to 5060 ± 95 BP (K-2560).]

800. Knuth, E. (1981): Greenland news from between 81° and 83° North. Folk 23:91-111.
A few small artifacts (*Homo sapiens*) made of walrus (*Odobenus rosmarus*) tusk were recovered from Independence II (3100–2400 BP) ruins at Jørgen Brønlund Fjord (Knuth 1968), but they were probably imported, since the lack of bone remains in the ruins indicates that the species was not hunted locally. This may also apply to a small artifact made of walrus tusk from the Neoeskimo site Kølnæs (82°35′25″N, 19°40′00″W) in eastern Peary Land dated at about 1450 AD (calibrated). Further, the oldest tundra muskox (*Ovibos moschatus*) records come from Independence I sites in western and eastern North Greenland dated to about 4500 BP (Bennike 1997). Thirteen dates on muskox bone are listed from about 4365 BP to 1450 BP (Figures 8, 9). [See Radiocarbon-date Table.]

801. Knuth, E. (1983): The northernmost ruins of the globe. Folk 25:5–21.
Bennike (1997) remarks that two vertebrae and a rib (ZMK51/1980) of a large whale (?*Balaena mysticetus*) were collected at Hvalterrasserne ('Whale Terraces') at Frigg Fjord, North Greenland (83°6.7′N, 32°58′W; Figures 3, 5). The whalebone yielded a radiocarbon age of 4740 ± 70 (Lu-3658) (Bennike 1997, Table 1). Knuth (p. 17) gives radiocarbon dates ['presumably from muskoxen' (*Ovibos moschatus*)] from the first and second floors of the Main Ruin (*Homo sapiens*) as: 3670 ± 80 BP and 3630 ± 80 BP [no laboratory numbers are provided].

802. Koch, I.P. (1916): Survey of Northeast Greenland. Meddelelser om Grønland 46(2):1–387.
According to observations of the elevation of strand lines above the sea, Germania Land has risen about 400 m asl (Figure 20). Great stretches of Germania Land have been below the sea surface, and traces left by the sea consist of eroded strand lines and sea shells. At Hvalsletten, an old beach surface 20 km long and about 10 km wide between Moskusoksefjaeldene and Saelsøen lies between 40 and 45 m asl. The name of this plain owes its origin to a stranded whale, probably a bowhead whale (*Balaena mysticetus*), the skeleton of which was found a few kilometres from the shore.

803. Kontrimavichus, V.L. (editor) (1984): Beringia in the Cenozoic Era. Originally published in Russian in 1976 by the Academy of Sciences of the USSR. Translation by R. Chakravarty, Amerind Publishing Co. Pvt. Ltd., New Delhi. 724 pp.
This book results from papers presented at a meeting 'The Bering land bridge and its role in the history of Holarctic floras and faunas in the Late Cenozoic' held in Khabarovsk, May 10–15, 1973. Chapter headings include Geological History, Vegetative Cover and Palaeogeography, Beringia and Fauna of the Holarctic, Beringia and Recent Fauna of the Holarctic, and Beringia and the Problem of Human Migration to America. Individual papers of interest are annotated separately.

804. Kool, R. (1979): Clarification of two *Mammuthus* molars from the British Columbia Provincial Museum. Canadian Journal of Earth Sciences 16(10): 2058–9.
Two molars found in British Columbia that were thought to have belonged to mammoth (*Mammuthus*) are now identified as extant elephant teeth. One found at Butler's Dump, a log dump in Sooke, British Columbia, was reidentified as an African elephant (*Loxodonta africana*) tooth. It was referred to in Harington (1975d) and Harington and Shackleton (1978), but they did not have the tooth for examination. The other molar was found in Cobble Hill, N of Victoria, British Columbia, and was reidentified as Indian elephant (*Elephas indicus*). Upon further examination, none of these was found *in situ*.

805. Korobitsyna, K.V., Nadler, C.F., Vorontsov, N.N., and Hoffmann, R.S. (1974): Chromosomes of the Siberian snow sheep, *Ovis nivicola*, and implications concerning the origin of amphiberingian wild sheep (subgenus *Pachyceros*). Quaternary Research 4:235–45.
The chromosomes of the snow sheep (*Ovis nivicola*), described for the first time, exhibit 2n = 52, the lowest diploid number reported for wild sheep and goats. The new chromosomal data, together with a review of the fossil history of *Ovis*, lead the authors to conclude that thick-horned wild sheep (subgenus *Pachyceros*) evolved their

distinctive characteristics while isolated in the Beringian refugium and then migrated southward into western North America when glacial barriers melted. A seeming paradox is that the snow sheep with more primitive morphology and behaviour has a more recently evolved karyotype. [See Hoefs 2001.]

806. Korotkevich, E.L., and Topachevskii, V.A. (1984): Related elements among the Neogene mammalian fauna of Eastern Europe, Asia and North America. *In*: Beringia in the Cenozoic Era (V.L. Kontrimavichus, ed.). Amerind Publishing Co. Pvt. Ltd., New Delhi. pp. 289–95

The paper describes finds of Neogene mammals on the northern coasts of the Black Sea and compares them with North American, Asian, and other Eastern European faunas. Genera discussed are *Blarina*, *Pratilepus*, *Microtoscoptes*, *Promephitis*, *Plesiogulo*, *Parataxidea*, *Hyaenarctos*, *Procapreolagus*, *Eostylocerus*, *Muntiacus*, and *Gigantocamelus*. The find of *Gigantocamelus* is especially interesting as it is the first one in western Europe and confirms their eastern entry to Europe from North America.

807. Kotzebue, O. von (1821): A voyage of discovery into the South Sea and Behring's Straits, for the purpose of exploring a northeast passage, undertaken in the years 1815–1818. Longman, Hurst, Rees, Orme, and Brown, London. Vol. 1 (pp. 218–20), Vol. 3 (p. 299).

A remarkable discovery was made on August 8 when the ship was anchored at a sandy point on the E side of Eschscholtz Bay, Kotzebue Sound. The crew came upon 'masses of the purest ice, of the height of an hundred feet, which are under a cover of moss and grass.' At a part of the coastal bluffs that had fallen in and was melting, a 'quantity of mammoths' [*Mammuthus*] teeth and bones were exposed to view by the melting, and among which I myself found a very fine tooth. We could not assign any reason, for a strong smell, like that of burnt horn, which we perceived in this place.' 'By a good observation, we found the latitude of the tongue of land 66°15′36″, north.' The author praises the ship's physician, Dr Eschscholtz, for the discovery (Vol. 1, p. 220). In Vol. 3 (p. 299), the author remarks about the same place: 'We found, near the ice ground of the point of land where we bivouacked, and where the natives had stayed before us, some grinders, which perfectly resembled those of mammoth, and also a tusk which, by its greater thickness at the root, and its simple curvature considerably differed from the well known mammoth's horns, and seemed to have much more resemblance to the teeth of the present race of elephants. During the night our watch-fire was partly kept up with such ivory.'

808. Kowalski, K. (1980): Origin of mammals of the arctic tundra. Folia Quaternaria 51:3–16.

This paper presents a generalized discussion on arctic mammals along with an assessment of the vegetation and climate of the region over time. A cooling trend in northern latitudes developing from Miocene times gradually brought on a changing vegetation and winter temperatures of –50°C in arctic North America. For mammals the limiting factors in arctic living are not necessarily temperature but rather year-round food availability and amount of snowcover. Mammalian evolution in a large part of the territory occupied today by arctic tundra ecosystems (Siberia and Beringia) was continuous. Lemmings (*Lemmus* and *Dicrostonyx*) and reindeer (*Rangifer tarandus*) began to develop as tundra forms from their forest ancestors beginning in the Pliocene. They were joined later by the muskox (*Ovibos moschatus*), which is of steppe origin. The typical carnivores of the tundra, arctic fox (*Alopex lagopus*), and polar bear (*Ursus maritimus*) developed much later. The mammalian fauna of the arctic tundra has an entirely different history from that of alpine mammals. All arctic tundra mammals originated in the Old World (including Beringia). The postglacial disappearance of the steppe-tundra ecosystem and the development of a continuous taiga belt has no analogy in the interglacials and was the main factor of extinction of large mammalian species in Eurasia.

809. Krause, P. (1971): Mammoths and mastodons. Zoology 12(4):22–5.

Proboscidean (Proboscidea) remains were found in many localities in Manitoba. Mammoth (*Mammuthus*) teeth have been found in: Benito, Bird's Hill, Dufresne, Rathwell, Rivers, Snake Creek, Springfield, and Transcona. Mastodon (*Mammut americanum*) remains have been found in Moosenose and Swan River. A figure shows clearly the differences between lower molars of mammoth and mastodon with a human molar for scale.

810. Krohne, D.T. (1982): The karyotype of *Dicrostonyx hudsonius*. Journal of Mammalogy 63(1): 174–6.
This report deals with the karyotype of the Ungava lemming (*Dicrostonyx hudsonius*). Diploid and fundamental numbers of *Dicrostonyx hudsonius* most closely resemble those of *Dicrostonyx richardsoni* (Table 1) that probably was isolated in the same periglacial refugium during the Pleistocene. Despite the karyotypic similarity of these taxa, the fossil record indicates that they diverged before the arrival of *Dicrostonyx* in the New World. According to Macpherson (1965), *Dicrostonyx* entered North America before the Pleistocene and subsequently was isolated in different glacial refugia. The karyotypic variation noted by the author could have arisen in allopatry during this period. Those taxa isolated in periglacial tundra have karyotypes different (higher 2N and FN) from those isolated in Beringia.

811. Krueger, H.W., and Weeks, C.F. (1966): Geochron Laboratories, Inc. radiocarbon measurements II. Radiocarbon 8:142–60.
Two items are pertinent: (1) An arctic ground squirrel (*Spermophilus parryii* = *Citellus undulatus*) nest from the bank of the Chatanika River, ¼ mi N of the Elliott Highway crossing, Alaska (65°05′N, 147°45′W) from 8 m below the surface in Wisconsinan organic silt yielded a radiocarbon date of 14,760 ± 850 BP (GX-250). According to T.L. Péwé, this dates a ground squirrel living in a thawed layer above frozen ground (p. 145). (2) A horse (*Equus*) bone, unaltered in appearance, from a gravel pit near Hemlock Park Farm, Ontario (44°18′40″N, 76°28′20″W), apparently from glacial outwash, yielded a collagen date of 815 ± 110 BP (GX-201), and a carbonate date of 310 ± 115 BP (p. 146). W.A. Gorman of Queen's University, Kingston, who collected the bone in 1963, states that 'the structure of the teeth suggested a fossil horse.' [It is difficult to credit either date, especially the collagen date, which should be the more reliable of the two. An AMS check date on bone collagen from a third carpal of this horse, best considered *Equus caballus*, is recent 110 ± 40 BP (Beta-151996).]

812. Kupsch, W.O. (1959): Annotated bibliography of Saskatchewan geology (1823–1958). Saskatchewan Department of Mineral Resources Report No. 9:1–198.
References to Pleistocene vertebrates are on p. 194.

813. Kupsch, W.O. (1961): Annotated bibliography of Saskatchewan geology, 1959 and 1960. Saskatchewan Department of Mineral Resoures Report No. 9 (Supplement):1–54.
References to Pleistocene vertebrates are on p. 51.

814. Kurtén, B. (1963): Notes on some Pleistocene mammal migrations from the Palaearctic to the Nearctic. Eiszeitalter und Gegenwart 14:96–103.
Relatively little has been published on the correlation between the Pleistocene mammalian (Mammalia) faunas of Eurasia and North America. This study is based on case histories of intercontinental migration of several carnivores (Order Carnivora) by means of taxonomic and evolutionary analyses. The author assumes that mammalian migration between the Old and New World occurred exclusively across Beringian land connections during glacial phases. If immigrants first appear in an interglacial fauna, they are assumed to have migrated in the preceding cold phase, and it is a prerequisite that a probable ancestor is known to have lived at the given time in the assumed area of origin. Accordingly, the following times for mammalian migration from Palaearctic to Nearctic are suggested: (1) sabretooth cats of the genera *Megantereon* and *Smilodon*, Elster (Kansan) (Table 1); black bears of the genus *Ursus*, Elster (Kansan) (Table 2); brown bears (*Ursus arctos*), Würm (Wisconsinan); wolverines of the genus *Gulo*, Saale (Illinoian). Correlation between the mammalian faunas in the Palaearctic and Nearctic should be based on compilation of many other cases of migration.

815. Kurtén, B. (1964): The evolution of the polar bear: *Ursus maritimus* Phipps. Acta Zoologica Fennica 108:1–30.
The known fossil and subfossil material was examined, including some new finds. The Late Pleistocene polar bear (from early Würm deposits), herein described as a new subspecies (*Ursus maritimus tyrannus*), was considerably larger than the living form. The author proposes that modern polar bears originated from early Middle Pleistocene European brown bears (*Ursus arctos*). As late as the European Yoldia Period, polar bears still retained many brown-bear traits. This close relationship hypothesis is further supported by the fact that brown bear/polar

bear crosses sometimes produce fertile hybrids.

816. Kurtén, B. (1967): Pleistocene bears of North America: 2. genus *Arctodus*, short-faced bears. Acta Zoologica Fennica 117:1-60.

This paper deals with short-faced bears (*Arctodus*), with detailed descriptions of each species. Remains of giant short-faced bear (*Arctodus simus*) have been found at Upper Cleary Creek, Alaska; Ester Creek, Alaska; and Gold Run Creek, Yukon; all are estimated as Late Pleistocene in age. For the large form, the earliest available name is *Arctodus simus yukonensis* (Lambe) (p. 57).

817. Kurtén, B. (1973): Transberingian relationships of *Ursus arctos* Linné (brown and grizzly bears). Societas Scientarum Fennica, Commentationes Biologicae 65:1–10.

Size and skull-width allometry in brown bears (*Ursus arctos*) on both sides of Bering Strait was studied. Size isophenes are repeated in mirror fashion in NE Asia and NW North America. Kamchatka and Kodiak bears are relicts of a broad-skulled population along the southern coast of Beringia.

818. Kurtén, B. (1984): Trans-Beringian carnivore populations in the Pleistocene. *In*: Beringia in the Cenozoic Era (V.L. Kontrimavichus, ed.). Amerind Publishing Co. Pvt. Ltd., New Delhi. pp. 339–45.

In four carnivore families (Canidae, Ursidae, Hyaenidae and Felidae), up to 15 instances of possibly circumpolar populations are found in the Pleistocene and Holocene: Early Blancan, Late Blancan, and the Irvingtonian-Rancholabrean transition. [See the table of trans-Beringian carnivores on p. 343 that groups them under the following geological categories: Recent, Late Pleistocene, and Early Pleistocene and Pliocene.] It seems that especially intensive migration of northern mammals occurred some 4 to 3.5 Ma. The Late Villafranchian mutual exchange led to the appearance of circumpolar populations of jaguars and coyotes as well as southern mammoths (*Mammuthus meridionalis*). The beginning of the Rancholabrean age in North America once again brought large-scale migrations [e.g., lions (*Panthera leo atrox*)].

819. Kurtén, B. (1985): The Pleistocene lion of Beringia. Annales Zoologici Fennici 22:117–21.

The Late Pleistocene lion of Beringia was a distinctly smaller animal than the great American lion (*Panthera leo atrox*) S of the Wisconsinan ice. It probably was on a cline with the slightly larger European cave lion (*Panthera leo spelaea*) and is referred to the same taxon. Morphometric data are presented in table form for size of canines, size of carnassials, and size of mandibles. The author concludes that the Beringian lions are most properly referred to the cave lion (*Panthera leo spelaea*). There is an intriguing discussion of the nature and relationships of the Wisconsinan large-mammal fauna of Beringia (p. 120).

820. Kurtén, B., and Anderson, E. (1980): Pleistocene Mammals of North America. Columbia University Press, New York. 442 pp.

This is a much cited, general compendium on North American Pleistocene mammals, including those from Canada and Alaska. It includes, among other things, information on: geographical ranges, sites, faunal lists, museum collections, radiometric dates, and many references and figures (including restorations).

821. Kuzyk, G.W., Russell, D.E., Farnell, R.S., Gotthardt, R.M., Hare, P.G., and Blake, E. (1999): In pursuit of prehistoric caribou on Thandlät, southern Yukon. Arctic 52(2):214–19.

In 1997, Gerald Kuzyk noted a large concentration of caribou (*Rangifer tarandus*) fecal pellets and a caribou antler on a permanent snowpatch near Kusawa Lake, southern Yukon (Figure 1, map). Caribou are absent from the area today. Coring of the snowpatch revealed continuous deposits of droppings [containing sedge (36%), lichen (23%), moss (11%), and grass (10%)] to depths of at least 160 cm. Part of a wooden dart or arrow shaft found on the edge of the site represents one of the few organic examples of Mid-Holocene hunting technology found in Canada. An age of 2450 ± 50 BP (TO-6871) was obtained for the fecal matter from about 1.6 m below the surface of the snowpatch, and the shaft fragment yielded a date of 4360 ± 50 BP (TO-6870). These dates indicate that aboriginal Yukoners (*Homo sapiens*) hunted caribou at this site for at least 4000 years.

822. Lambe, L.M. (1898): On the remains of mammoths in the museum of the Geological Survey Department. Ottawa Naturalist 14:136–7.

The author lists all the mammoth (*Mammuthus*) specimens in the Geological Survey of Canada museum prior to 1898. An almost complete skeleton (parts are listed) of a woolly mammoth (*Mammuthus* cf. *Mammuthus primigenius = Elephas primigenius*? = *Elephas Jacksoni*) was found 2.4 km NE of Muirkirk, Ontario. Bone fragments (right ramus and symphysis of a lower jaw, part of a large tusk, parts of two scapulae, etc.) of a mammoth (*Mammuthus* cf. *Mammuthus primigenius*) were found at Burlington Heights, near Hamilton, Ontario. [See Cottle 1852, 1853.] Other mammoth remains were also found at the following localities: three teeth from Norton Sound, Alaska; several fragments from the Yukon River 64 km downstream from the mouth of the Tanana River, Alaska; a molar under an opera house, on Queen Street, St Catharines, Ontario; a molar near Nulato, on the Yukon River, Alaska; a tooth fragment in drift 10 km upstream from of Edmonton, Alberta; and a cast of a molar (original in the British Columbia Provincial Museum, Victoria) from the shore of Shuswap Lake, British Columbia.

823. Lambe, L.M. (1905): Progress of vertebrate paleontology in Canada. Transactions of the Royal Society of Canada 2(10), Section 4:13–56.
The author mentions nine known Pleistocene mammal species for Canada.

824. Lambe, L.M. (1907): On a tooth of *Ovibos*, from Pleistocene gravels near Midway, B.C. Ottawa Naturalist 21:15–18.
An ungulate (cf. *Ovibos*) tooth was found on a rock surface beneath 60 m of unconsolidated gravel at a placer mining operation in Rock Creek, British Columbia, 13 km above its entry into the Kettle River, 6.4 km N of the Canada–U.S. border. This specimen was compared with a tundra muskox (*Ovibos moschatus*) tooth from Fort Rae, Great Slave Lake, Northwest Territories, and a bighorn sheep (*Ovis canadensis*) tooth; the specimen was smaller than the former, and larger than the latter, so it could be either. [The specimen looks like a modern sheep tooth]. The author reviews records of Pleistocene *Ovibos* remains in Canada, including a helmeted muskox (*Bootherium bombifrons = Symbos cavifrons, = Symbos tyrrelli*) skull found in gravels at Bonanza Creek, Yukon.

825. Lambe, L.M. (1910): Palaeontology and zoology. Summary Report of the Geological Survey of Canada for 1909:269–73.
Pleistocene vertebrates mentioned include: a white whale (*Delphinapterus leucas*) caudal vertebra buried 8 m deep in Ottawa East (Vanier), Ontario; a woolly mammoth (*Mammuthus primigenius*) molar found at Eldorado Creek, near Dawson City, Yukon; and a Yukon horse (*Equus lambei*) skull buried 10 m in placer gravel at Gold Run Creek, Yukon.

826. Lambe, L.M. (1911a): Paleontological Division. Summary Report of the Geological Survey of Canada for 1910:269–73.
In his annual progress report, Lambe notes that the following Pleistocene mammal specimens have been added to the Geological Survey of Canada paleontological collection: (1) part of a molar tooth of a mammoth (*Mammuthus*) from a clay bank on Loon River (40 mi from its mouth), a tributary of the Peace River, Alberta, was donated by H.A. Conroy; (2) a fine skull of a horse [the holotype of *Equus lambei*] from Claim 34, Gold Run Creek, Yukon, was presented to the Geological Survey Museum by J.M. Morrison and J.B. Bannerman of Carcross, Yukon – not purchased as stated in the previous year's Summary Report; (3) a well-preserved skull without mandible of the giant short-faced bear (*Arctodus simus = Arctotherium* cf. *A. simum* Cope) from Pleistocene deposits on Gold Run Creek (from frozen gravel at a depth of 40 ft) was purchased for the Geological Survey collection.

827. Lambe, L.M. (1911b): On *Arctotherium* from the Pleistocene of Yukon. Ottawa Naturalist 25(2):21–6.
An almost complete bear (*Arctotherium*) skull was found at Gold Run Creek, Yukon. It was compared to the giant short-faced bear (*Arctodus simus*), and described as the holotype of the Yukon short-faced bear (*Arctodus simus yukonensis = Arctotherium yukonense*). Bone measurements and photographs are provided.

828. Lambe, L.M. (1912a): Paleontological Division. Summary Report of the Geological Survey of Canada for 1911:346–57.
The author provides information on bison (*Bison*) remains from Alberta and mammoth (*Mammuthus*), horse (*Equus*), caribou (*Rangifer tarandus*), bison, and mountain sheep (*Ovis*) from Yukon.

829. Lambe, L.M. (1912b): The past vertebrate life in

Canada. Transactions of the Royal Society of Canada 3(5), Section 4:3–15.
The author lists 18 species of Canadian Pleistocene mammals, 10 of which were found only in Yukon.

830. Lambe, L.M. (1914a): Report of the Vertebrate Paleontologist. Summary Report of the Geological Survey of Canada for 1912:397–403.
The author mentions: mammoth (*Mammuthus*) remains from Yukon, bison (*Bison*) from Alberta, and wolf (*Canis lupus*) from Manitoba.

831. Lambe, L.M. (1914b): Report of the Vertebrate Palaeontologist. Summary Report of the Geological Survey of Canada for 1913:293–9.
This list includes: two capelin (*Mallotus villosus*) in nodules from Leda Clay at Besserer's Grove 8 mi E of Ottawa, Ontario; white whale (*Delphinapterus leucas*) caudal vertebra and bison (*Bison*?) femur fragment from Pleistocene gravel on Lot 15, Concession 5 of Nepean Township near Jock River, Ontario; unidentified 'dog' (Canidae) partial mandible with teeth and deer (Cervidae) leg bones between the Canadian Pacific Railway and the Thompson River, 2.4 km W of the bridge at Savona, British Columbia; a mammoth (*Mammuthus*) molar from Teller, Alaska; steppe bison (*Bison priscus = Bison crassicornis*) partial cranium and horns buried in 3 m of muck, 615 m above Upper Dominion Creek near the head of 18 Pup, Yukon; the backs of two steppe bison skulls with horncores from Upper Dominion Creek, Yukon; a tusk socket of a mammoth (*Mammuthus*?), also from Upper Dominion Creek; and a mastodon (*Mammut americanum*) molar from McQuesten Creek, Stewart River, Yukon [there is a detailed description of a hard precipitate, calculus or tartar, occurring in two of the 'valleys' of the tooth].

832. Lammers, G.E. (1968): A note on the Saskatoon site, Saskatoon, Saskatchewan. Napao 1(2):32–4.
The remains of several Late Rancholabrean mammals were found at North Saskatoon Station, Saskatchewan (52°10′N, 106°35′W). The specimens (radiocarbon dated at >34,000 BP) are: mammoth [*Mammuthus* (*Parelephas*)], Niobrara? horse (*Equus* cf. *Equus niobrarensis*), bison (*Bison*), a camel, tentatively referred to the western camel (*Camelops*), and an unidentified deer (Cervidae). [See Christiansen (1968) for geology and Pohorecky and Wilson (1968) for a preliminary archaeological report; see also SkwaraWoolf 1981 and Skwara 1988; identifications are upgraded in Harington 1990a.]

833. Lammers, G.E., and Wrigley, R.E. (1984): The history of Manitoba's animals. *In*: Natural Heritage of Manitoba: Legacy of the Ice Age (J.T. Teller, ed.). pp. 127–59.
The sections 'Animal life during the ice age' and 'Animal life after the ice age' (pp. 140–1) are pertinent to this bibliography. Manitoba was covered by glaciers and therefore was uninhabitable through much of the Pleistocene. During many glacial advances, plants and animals were forced slowly southward into ice-free areas of the United States and Canada. Bones and teeth of animals that lived during the interglacial periods have been found in many localities in Manitoba and nearby areas to the S and W. The authors refer to grazers such as mammoths (*Mammuthus*), pronghorns (*Antilocapra*), horses (*Equus*), ancient bison (*Bison bison antiquus = Bison antiquus antiquus*), western bison (*Bison bison occidentalis = Bison antiquus occidentalis*), western camels (*Camelops hesternus = Camelops*), and llama-like mammals (*Hemiauchenia*). As an example, the authors refer to the Fort Qu'Appelle fauna (>30,000 BP) from southeastern Saskatchewan including a turtle (Reptilia), badger (*Taxidea*), western camel (*Camelops hesternus = Camelops*), horse (*Equus*), mammoth (*Mammuthus*), and helmeted muskox (*Bootherium bombifrons = Symbos*). Fossil mammal remains are common in southern Manitoba, although they have not been collected systematically. Their distribution was determined largely by the extent of Glacial Lake Agassiz. Few fossil vertebrates have been found within the great area of this lake. Ice age rodents, including a large ground squirrel (*Spermophilus*) and a vole (*Microtus*), have been recovered from a silt deposit on Riding Mountain. Grass associated with these bones was more than 30,000 years old [see Klassen and Delorme 1967]. Mammoth teeth and ivory tusks have been discovered in gravel pits at Souris, Transcona, Lac Du Bonnet, and Birds Hill (Figure 7 shows a mammoth molar from Rathwell and a tusk fragment from Transcona). Horse teeth have been collected from deposits at Shellmouth Dam and Souris (Figure 8 shows a molar fragment from Souris). Remains of tundra muskox (*Ovibos moschatus*) have been located in southern Manitoba (a partial cranium from Grandview is shown in

Figure 9), indicating an earlier, cool grassland or tundra region. Excavations in southern Manitoba frequently turn up bison (perhaps of three subspecies of *Bison bison*) remains – bones and horncores are often exposed along river banks. Remains of Holocene deer (*Odocoileus*), wapiti (*Cervus elaphus*), moose (*Alces alces*), and bison are frequently exposed during clearing and ploughing land, and in archaeological surveys. Most are no older than 3000 years, yet some records are from areas outside the animals' historic ranges (e.g., bison skulls have been unearthed 62 km N of Gypsumville and along the banks of the Winnipeg River E of Pinawa). Thus species ranges have shifted N and S during the present interglacial. Following glaciations, exposed areas of southern Manitoba were invaded first by plants and animals of the tundra and boreal coniferous forest biomes (12,000 to 10,000 years ago), then grassland and southern transition species (10,000 to 3000 years ago).

834. Lance, E.W., and Cook, J.A. (1998): Biogeography of tundra voles (*Microtus oeconomus*) of Beringia and the southern coast of Alaska. Journal of Mammalogy 79(1):53–65.

Allozymic, chromosomal, and morphological variation among six subspecies of tundra vole (*Microtus oeconomus*) was examined and related to the biogeographical history of Beringia with a particular focus on the Prince William Sound region. Chromosomal and allozymic similarities between Alaskan and Russian tundra voles corroborate previous hypotheses of a relatively recent invasion across the Bering land connection from Russia. In populations from the southern coast of Alaska (Figure 1), unique alleles were detected in *Microtus oeconomus operarius* from Hinchinbrook Island (which may have served as a refugium for voles during the last glacial maximum, p. 62), and *Microtus oeconomus elymocetes* from Montague Island in Prince William Sound. *Microtus oeconomus operarius* from the Copper River Delta, and *Microtus oeconomus macfarlani* from interior Alaska exhibited higher allozymic variation, indicating a larger effective population size and longer history than populations from the southern coast of Alaska. Allozymic evidence of gene flow between *macfarlani* and *operarius* from the Copper River Delta suggests that the southern coast of Alaska was colonized from interior Alaska since the last glacial advance. The larger body size of *elymocetes* from Montague Island suggests that this population may have an older age structure compared with other populations.

835. Landals, A.J. (1986): The Maple Leaf site: An interpretation of prehistoric hunting and butchering strategies in the southern Alberta Rockies. MA thesis, Department of Archaeology, University of Calgary. 244 pp.

The Maple Leaf site (DjPo-47) is located on a bedrock terrace along the bank of Crowsnest River, within the Livingstone Range, Crowsnest Pass, Alberta. Landals describes three components in her zooarchaeological study: (1) Late Paleoindian (*Homo sapiens*) based on a projectile point that may be a variant of the Pryor stemmed-type, as well as a normalized radiocarbon date on bison (*Bison*) bone collagen of 8630((270 BP (RL-873) – these items came from a clay layer above cobbles and below marl; (2) an Elko Eared point in marl just below Mazama tephra; (3) Bitteroot/Salmon River points, two Oxbow points and another (unclassified) from clay above the Mazama tephra. Associated with the bison remains were canid (*Canis*) and bighorn sheep (*Ovis canadensis*). [See Morlan 1999.]

836. Langston, W., and Oschinsky, L. (1963): Notes on Taber 'Early Man' site. Anthropologica 5(2):147–50.

The remains of a human (*Homo sapiens*) child were found 7.5 km N of Taber, Alberta, on the E bank of the Oldman River in cross-bedded sand. The authors did not estimate the age of the specimen. [See Cohen 1982 and Stalker 1969a.]

837. Larsen, H. (1968): Trail Creek, final report on the excavation of two caves at Seward Peninsula, Alaska. Acta Arctica 15:7–79. København.

[See West (1996) for summary and update.]

838. LaSalle, P. (1984): Geological setting and preliminary faunal report for the St-Elzéar Cave, Québec, Canada. *In*: Contributions in Quaternary Vertebrate Paleontology: A volume in memorial to John E. Guilday (H.H. Genoways and M.R. Dawson, eds.). Carnegie Museum of Natural History Special Publication No. 8, Pittsburgh. pp. 332–46.

This paper provides a radiocarbon date (about 5000 BP) for the specimens listed in LaSalle and Guilday (1980). A map of major Pleistocene mammal assemblages of North

America is provided. Two of the rodents – Ungava lemming (*Dicrostonyx hudsonius*) and chestnut-cheeked vole (*Microtus xanthognathus*) – no longer live in this area.

839. LaSalle, P., and Guilday, J.E. (1980): Caverne de Saint-Elzéar-de-Bonaventure, Rapport préliminaire sur les fouilles de 1977 et 1978. Direction de la Géologie DPV-750:1–31. [An English translation is in the Quaternary Vertebrates of Northern North America Publications File at the Canadian Museum of Nature.] After two seasons of excavation, Saint-Elzéar-de-Bonaventure Cave, which is on a plateau N of Baie des Chaleurs, Quebec (65°21′30″N, 48°14′20″W), has produced remains of: one beetle, 531 snails, 181 amphibians (three species), two reptiles (one species), four birds (four species), and 4503 mammals (34 species). Nearly 80% of the fossils are parts of small mammals (including hares). Amphibians (Amphibia) include: salamander (*Ambystoma laterale*), American toad (*Bufo americanus*), and pickerel frog (*Rana palustris*). Reptiles (Reptilia) include: common garter snake (*Thamnophis sirtalis*). Birds (Aves) include: Red-breasted Nuthatch (*Sitta canadensis*), Blue Jay (*Cyanocitta cristata*), White-winged Crossbill (*Loxia* cf. *Loxia leucoptera*), and an undetermined species from the family Fringillidae. Mammals (Mammalia) include: star-nosed mole (*Condylura cristata*), arctic shrew (*Sorex arcticus*), masked shrew (*Sorex cinereus*), smoky shrew (*Sorex fumeus*), Gaspé shrew (*Sorex gaspensis*), American water shrew (*Sorex palustris*), pygmy shrew (*Sorex hoyi*), short-tailed shrew (*Blarina brevicauda*), little brown bat (*Myotis lucifugus*), snowshoe hare (*Lepus americanus*), arctic hare (*Lepus arcticus*), woodchuck (*Marmota monax*), eastern chipmunk (*Tamias striatus*), American red squirrel (*Tamiasciurus hudsonicus*), northern flying squirrel (*Glaucomys sabrinus*), deer mouse (*Peromyscus maniculatus*), Gapper's red-backed vole (*Clethrionomys gapperi*), southern bog lemming (*Synaptomys cooperi*), northern bog lemming (*Synaptomys borealis*), heather vole (*Phenacomys intermedius*), Ungava lemming (*Dicrostonyx hudsonius*), meadow vole (*Microtus pennsylvanicus*), rock vole (*Microtus chrotorrhinus*), chestnut-cheeked vole (*Microtus xanthognathus*), woodland jumping mouse (*Napaeozapus insignis*), meadow jumping mouse (*Zapus hudsonius*), beaver (*Castor canadensis*), porcupine (*Erethizon dorsatum*), black bear (*Ursus americanus*), least weasel (*Mustela nivalis*), ermine (*Mustela erminea*), wolverine (*Gulo gulo*), moose (*Alces alces*), and caribou (*Rangifer tarandus*). Several of these species are no longer extant in the region and their presence seems to indicate a period of colder climate. These species include: arctic shrew (*Sorex arcticus*), arctic hare (*Lepus arcticus*), heather vole (*Phenacomys intermedius*), Ungava lemming (*Dicrostonyx hudsonius*), and least weasel (*Mustela nivalis*). The sediments containing the fossils indicate a very slow and gradual accumulation, and due to the chemistry of the cave, the bone is unsuitable for radiocarbon dating. A radiocarbon date of 940 ± 120 BP (QU-978) was provided by a piece of surface wood. Currently the mineralogy, palynology, and sedimentology of the cave is under study, and it is hoped that a more reliable minimum-maximum age of the deposits may be achieved through this work.

840. Laucius, J. (1994): Yukon miners' discovery of ice age horse is the equivalent of striking ... prehistoric gold. Ottawa Citizen, February 6:D4. Cooperation between Yukon gold miners and scientists has yielded the world's best-preserved example of the extinct Yukon horse (*Equus lambei*). The carcass, from Last Chance Creek near Dawson City, Yukon, includes the pelt and a well-preserved foreleg. Some miners have been collecting and donating ice age mammal [e.g., mammoth (*Mammuthus*), camel (*Camelops*), and bison (*Bison*)] bones to the Canadian Museum of Nature in Ottawa for about 25 years. This 26,000-year-old specimen is one of the most spectacular finds. The horse's shoulder bone was discovered by children the previous year. Later, their fathers, gold miners Lee Olynyk and Ron Toews, found the 'freeze-dried' foreleg and alerted Yukon archaeologist Ruth Gotthardt, who later investigated and shipped the specimen to Dick Harington at the Canadian Museum of Nature. He sent a small sample of bone to a radiocarbon laboratory for dating, and the result confirmed that the specimen was from the extinct horse, *Equus lambei*. The pelt was sent to Ottawa's Canadian Conservation Institute for preservation. DNA studies were to probe *Equus lambei*'s relationship to current domestic and other horses. Scientists were to examine the carcass for tooth and claw marks to see if the horse had fallen victim to predators, and also to see if there were traces of surface parasites. The pelt alters ideas about the appearance of the Yukon horse: it had a long, flowing blondish mane, and lacked the dorsal stripe often shown in restorations. Gotthardt was interested in the horse because it may provide information on ancient environments to which early humans (*Homo sapiens*) in the Yukon may have adapted. An accompanying illustration shows Harington holding a skull of *Equus lambei* (not the Last Chance Creek specimen), with the foreleg, showing dried skin and hair, in the foreground.

841. Laughlin, W.S. (1967): Human migration and permanent occupation in the Bering Sea area. *In*: The Bering Land Bridge (D.M. Hopkins, ed.). Stanford University Press, Stanford, California. pp. 409–50.
Taking into account genetics, language, morphology, and the archaeological evidence, the author concludes that in the Old World, the ancestral Aleuts and Eskimos were the permanent coastal residents and the ancestral American Indians were the interior inhabitants. Both of these groups took advantage of the Bering Land Bridge to move into the New World. Conditions inland on the land bridge were severe and likely only a few of its inhabitants found their way to North America. The coastal conditions were considerably more congenial. The author estimates that 15,000 years ago the peoples of the Old and New worlds became genetically separated and that the Aleuts and Eskimos are more similar to Asiatic Mongoloids than to American Indians.

842. Lauriol, B., Carrier, L., and Thibaudeau, P. (1988): Topo climatic zones and ice dynamics in the caves of the northern Yukon, Canada. Arctic 41(3):215–20.
The authors describe an ice-free zone in Bear and Tsitché-han caves near the Arctic Circle in Yukon. They note that drying here affects the whole cave environment, including pieces of wood introduced by porcupines (*Erethizon dorsatum*), as well as the remains of other mammals (Mammalia), some of which are as old as 4270 [no SD given] BP (UL-257). The skeletons of small rodents (Rodentia) are very well-preserved in this zone (p. 218).

843. Lauriol, B., Prévost, C., Deschamps, E., Cinq-Mars, J., and Labreque, S. (2001): Faunal and archaeological remains as evidence of climate change in freezing caverns, Yukon Territory, Canada. Arctic 54(2):135–41.
Animal and plant remains, some associated with prehistoric artifacts, were collected in freezing caverns of northern Yukon. Radiocarbon dates show that the oldest remains are Middle Wisconsinan. The absence of material of Late Wisconsinan age suggests that the caves were infilled by ice during this cold period. Climatic warming and ice melting during the Holocene allowed animals and prehistoric hunters (*Homo sapiens*) to visit these caves regularly. Large mammal remains are rare in Tsi-tche-Han Cave (66°49′N, 139°20′W, 800 m asl – just N of a point where Porcupine River crosses the Arctic Circle in Yukon – Figs. 1, 2a, 3). Two femora of caribou (*Rangifer tarandus*) lay on the silts in the outer room, but were not radiocarbon dated. Fecal remains of porcupine (*Erethizon dorsatum*) are common, with many droppings in the second room – a sample collected at random yielded a date of 4270 ± 70 BP (UL-257, Table 1). Freeze-dried bear [like those of the short-faced bear? (cf. *Arctodus*)] droppings dated at 37,940 ± 460 BP (TO-2211) were found beyond an ice plug that starts 35 m inside the Grotte des Méandres in the same vicinity as Tsi-tche-Han Cave. Pollen and macrofossil analyses of the bear dung revealed a diet composed primarily of juniper (*Juniperus*), which modern bears do not feed on. Remains of marmots (*Marmota*) are common in northern Yukon caves, and an entire carcass collected from Grand Caverne Glacée yielded a modern age of 70 ± 50 BP (TO-2210, Table 1). Nests that the authors seem to attribute to marmots were found in Caverne Glacée 85 and dated at 1300 ± 100 BP (UQ-1282, Table 1), and in a small, unnamed cave in the same area and dated at 860 ± 100 BP (UQ-1584, Table 1).

844. Laursen, D. (1944): Contributions to the Quaternary geology of northern West Greenland and especially the raised marine deposits. Meddelelser om Grønland 135(8):1–125.
An Atlantic cod (*Gadus morhua* = *Gadus callarias*) otolith (ear bone) is reported from the Pátorfik beds on the N coast of Nuussuaq, West Greenland (see Figure 26 for a map of the localities mentioned). These beds are dated by foraminiferal biostratigraphy to the Early Pleistocene and may be about 2 Ma (Bennike 1997). This otolith from Mellemkløft in sand about 60 m asl is illustrated (Figure 16), and evidently represents a fish that was 16–17 years old at death and between 90 and 120 cm long (p. 80). Another otolith of this species was found in a Holocene deposit at Iginiarfik near Egedesminde (p. 81; Bennike 1997, p. 901). At Mellemkløft (p. 25) the cliff is 64 m high. At 60 m asl lies a consolidated sand bed nearly 1 m thick, and directly above are sands containing re-

mains of caribou (*Rangifer tarandus*), harp seal (*Phoca groenlandica*), and Thick-Billed Murre (*Uria lomvia* = *Uria uria*). The murre bone was identified by Dr M. Degerbøl. The caribou remains ('some antlers and bones') were partly embedded and partly on the surface, and the author suggests that they may be relatively recent. The harp seal remains consist of some lower jaws and loose teeth. At a distance of 1900 m E of Pátorfik in a sand bed 40 m asl some ringed seal (*Phoca hispida* = *Phoca foetida*) jaws and harp seal remains were recovered, indicating, with associated marine mollusc shells, an arctic climate (pp. 27, 81). In a sand bed 10.5 m asl at Quaersuarssuk kitdleq, about 8.5 km W of Quaersut, two fragments of narwhal (*Monodon monoceras*) bones were found. All of the mammals mentioned live today off, or in, Greenland, and thus yield no information of a climate different from the present.

845. Laursen, D. (1950): The stratigraphy of the marine Quaternary deposits in West Greenland. Meddelelser om Grønland 151:1–142.
The author reports the following vertebrate remains of probable Holocene age between 64° and 72°N: (1) a cranial bone of a fish (Pisces) with marine mollusc shells reported from a site about 25 m inland on a terrace near 'Engell's Profil' on the W side of Orpiqsûp tasia kitdleq (p. 44); (2) two pieces of fish vertebrae from the Zirphaea layer at the same locality (p. 45, Figure 16); (3) a fish otolith from Hunde Ejland on the terrace below the church at 16 m asl – probably a surf beaten shore deposit made under arctic conditions (p. 52); (4) part of a fish vertebra with marine mollusc shells from a podzolic layer 27.5 m asl at a lake near Skalvigen (p. 65, Figure 20); (5) fish otoliths and vertebrae from Horizon F, deposited in shallow water (p. 109) in the area investigated (Plate 1) – a suggested correlation of Horizon F, West Greenland with the Baltic indicates it is about the age of the Littorina Sea in Denmark (p. 133).

846. Laverdière, J.W. (1950): Baleine fossile de Daveluyville, Québec. Naturaliste Canadien 77:271–82.
A large whale thought to be a little piked whale (cf. *Balaenoptera acutorstrata*) [but best considered a common finback whale (*Balaenoptera physalus*) – see Sternberg 1951] was found on the Soucy farm, Canton de Bulstrode, Compté d'Arthabaska, 3.6 km SW of Daveluyville, Quebec. Other whale sites are also mentioned: (1) a bowhead whale (*Balaena mysticetus*) found at Ste-Anne-des-Monts, Quebec; (2) Ontario white whale (*Delphinapterus leucas*) remains found in Nepean; Ottawa East (Vanier); Pakenham, Lanark County; Cornwall; Williamstown, Glengarry County; (3) Quebec white whale (*Delphinapterus leucas*) specimens found in Montreal and Rivière-du-Loup; (4) humpback whale (*Megaptera novaeangliae* = *Megaptera longimana*) remains found near Smiths Falls, Ontario, and Métis, Rimouski County, Quebec; (5) unidentified Quebec whale remains found at Bic, Rimouski County; Sept-Îles; Longue Pointe de Mingan; Baie des Anglais, Anticosti; as well as from Jacquet River, near Bathurst, New Brunswick. Other mammal sites mentioned are: mastodon (*Mammut americanum*) from Cape Breton Island, Nova Scotia; walrus (*Odobenus rosmarus*) from Bic, Rimouski County, Quebec; harp seals (*Phoca groenlandica*) from Gloucester, Ontario, and Ste-Marie, a suburb of Montreal, Quebec; and a ringed seal (*Phoca hispida*) from Tétreauville, Ottawa County, Quebec [this ringed seal was incorrectly identified as a harbour seal (*Phoca vitulina*) – see Harington and Sergeant 1972]. [See Harington 1988.]

847. Lea, P.D. (1996): Vertebrate tracks in Pleistocene eolian sand-sheet deposits of Alaska. Quaternary Research 45:226–40.
Deformation structures interpreted as vertebrate hoof and foot tracks occur within upper-Pleistocene eolian sand-sheet deposits in the stabilized Kantishna sand sea of central Alaska and in the Nushagak lowland of southwestern Alaska. Exposures of tracks are generally limited to cross-sections, which reveal concave deformation structures in which pre-existing strata diminish downward. The amount of detail provided in the tracks is extremely variable, depending not only on the conditions present when the track was made but also on the consequent conditions of preservation. Deposits in both areas contain tracks that are 6 to 16 cm in diameter and are divided by a central ridge, reflecting formation by artiodactyl (Artiodactyla) ungulates. Larger (21–34 cm) tracks without a central ridge, observed only in the Nushagak lowland, were formed by proboscideans (Proboscidea), probably woolly mammoths (*Mammuthus primigenius*). Numerous photographs illustrate the tracks. The presence of tracks at multiple stratigraphic levels and preservation of roots and rhizocretions within the eolian deposits suggest that vegetated sand sheets may have formed a locally

important grazing habitat for large herbivores during at least part of the last (Wisconsinan) glaciation.

848. LeBlanc, R.J. (1994): Early cultures of the Mackenzie drainage to the Keewatin barrenlands area of northern Canada. Current Research in the Pleistocene 11:144–6.
The evidence for early (about 8000 BP and older) human (*Homo sapiens*) occupation of areas of northern Alberta, British Columbia E of the Rockies, and the districts of Mackenzie and Keewatin, Northwest Territories, is scant. This paper reviews the evidence and discusses the cultural and environmental significance of the known sites.

849. Leechman, D. (1950): Facts and comments: An implement of elephant bone from Manitoba. American Antiquity 16:157–60.
A bone tool made from a mammoth (*Mammuthus*) or a mastodon (*Mammut americanum*) fibula was found near Lac du Bonnet, Manitoba, about 80 km NE of Winnipeg. The locality was at the bottom of Glacial Lake Agassiz, so the specimen is older than the lake (>12,000 BP). Photographs are provided.

850. Lehman, D. (1984): Bear bones bonanza. Victoria Times-Colonist. November 29.
Speleologists exploring the deepest cave complex in Canada (it runs at least 2.6 km underground and may be as long as 10 km), on the NW coast of Vancouver Island, British Columbia, found remains of black bear (*Ursus americanus*) and deer (*Odocoileus*). The bones were sent to Grant Keddie of the British Columbia Provincial Museum for examination and study. A radiocarbon date carried out at Simon Fraser University indicates the black bear was 9380 ± 140 BP [a laboratory number is not given]. The bones are well preserved, and fractures in some evidently resulted when the animals fell over steep drops in the cave floor. Complete skeletons were preserved with the surrounding sediments, which contain pollen that can yield important paleoenvironmental information. A photograph with the article shows Grant Keddie holding a nearly complete black bear skull.

851. Leidy, J. (1852): [Fossils in gold deposits of the St. Lawrence Valley, Québec]. Geological Survey of Canada Report of Progress for 1850–1851. p. 8.
The fossils found in this area include: whales (Cetacea), seals (Phocidae), capelin (*Mallotus villosus*), and lumpfish (*Cyclopterus lumpus*).

852. Leidy, J. (1856a): Notice of the remains of a species of extinct seal, from the Post-Pliocene deposit of the Ottawa River. Canadian Naturalist and Geologist 1:238–9.
E. Billings of Ottawa, Ontario, sent the author a specimen consisting of part of a large concretion of indurated clay containing embedded bones. The specimen was found by the writer Peter McArthur in a bed of blue [Champlain Sea] clay containing boulders and marine shells and fishes (Pisces) in Gloucester, Ontario. It came from a 30-ft-high bank of the Ottawa River. The author identifies the bone in the nodule as parts of the articulated hindlimbs of a young seal (Phocidae). Tibia and fibula are crushed together and bones of foot and ankle are well preserved, but the epiphyses of the foot bones are separated and partly developed. The surrounding matrix is marked by impressions of hair and skin (Plate III). Capelin (*Mallotus villosus*) and lumpfish (*Cyclopterus lumpus)* were also found there in similar nodules.

853. Leidy, J. (1856b): Notice of the remains of a species of extinct seal, from the Post-Pliocene deposit of the Ottawa River. Proceedings of the Philadelphia Academy of Natural Sciences 8:90–1.
[See Leidy 1856a.]

854. Leith, E.I. (1949): Fossil elephants of Manitoba. Canadian Field-Naturalist 63:135–7.
Remains of mastodon (*Mammut americanum = Elephas rupertianus = Mastodon giganteus*) were found at Shell River, S of Swan River, Manitoba, at 51°43′N (erroneously reported as Swan River and Pelly, Saskatchewan). A tooth of another mastodon (cf. *Mammut americanum*) was found in a pit at Moosenose. Mammoth (*Mammuthus* cf. *Mammuthus primigenius*) remains were found at the following localities: tusk fragments on the S bank of Swan River 5.6 km N of Benito, and buried 10 m in gravels at a pit 6.5 km E of Transcona (see map, Figure 1 and photographs of mammoth remains in Plate 1).

855. Lent, P.C. (1999): Muskoxen and Their Hunters: A History. University of Oklahoma Press, Norman. 324 pp.

The evolution and dispersal history of muskoxen (Ovibovini) is considered in Chapter 2 of this book. Pertinent are sections on *Ovibos* in North America and muskoxen of the Mammoth Steppe. Figure 2.4 shows a map of North America with sites of helmeted muskox (*Bootherium bombifrons*) and tundra muskox (*Ovibos moschatus*) fossils.

856. Leonard, J., Wayne, R.K., and Cooper, A. (2000): Population genetics of ice age brown bears. Proceedings of the National Academy of Sciences, USA 97(4):1651–4.
The Pleistocene was a dynamic period for Holarctic mammals, complicated by episodes of glaciation, local extinctions, and intercontinental migration. The genetic consequences of these events are difficult to resolve from the study of modern populations. To provide a direct view of population genetics in the Late Pleistocene, the authors measured mitochondrial DNA sequence variation in seven Pleistocene brown bear (*Ursus arctos*) specimens (from Sixtymile, Yukon, and near the Fairbanks area, Alaska) radiocarbon dated from 42,000 to 14,000 BP. About 36,000 years ago, the Eastern Beringian brown bear population had a higher genetic diversity than any extant North American population, but by 15,000 years ago, genetic diversity appears similar to that of the present. The older, genetically diverse Eastern Beringian population contained sequences from three clades now restricted to local regions of North America, indicating that current phylogeographic patterns may provide misleading data for evolutionary studies and conservation management. The Late Pleistocene phylogeographic data also indicate possible colonization routes to areas S of the Cordilleran ice sheet. Figure 2 shows: (1) sequence clades II, III, and IV in the Sixtymile–Fairbanks region of Eastern Beringia at 35,000 to 45,000 BP with a brown bear dated to >40,000 BP on Prince of Wales Island; (2) sequence clade IIIb in the same region at 15,000 BP (with no brown bears in the Pacific Northwest in both periods); and (3) the present situation with sequence clade IIIb mainly in northwestern Canada, IIIa throughout most of Alaska, II on the Admiralty, Baranof and Chichagof islands of southeastern Alaska, and IV in the Rocky Mountain region of southern Canada and northern United States. Corrected radiocarbon dates on Quaternary brown bear specimens (1) from the Fairbanks area, Alaska are: right ulna (AMNH 95679) from Lower Goldstream 15,370 ± 60 BP (UCR-3742, CAMS-51806); right humerus (AMNH 95630) from Lower Goldstream 14,980 ± 60 BP (UCR-3741, CAMS-51805); right radius (AMNH 95665) from Gold Hill 13,760 ± 50 BP (UCR-3743, CAMS-54128); left humerus (AMNH 95629) from Fairbanks Creek 15,680 ± 50 BP (UCR-3745, CAMS-54129); radius (AMNH 95664) from Cripple Creek 42,850 ± 850 BP (UCR-3746, CAMS-54130); (2) from Sixtymile, Yukon are: right ulna (CMN 42381) 35,970 ± 660 BP (UCR-3744, CAMS-51808); right humerus (CMN 38279) 36,500 ± 1150 BP (Beta-16162).

857. Lincoln, A. (1982): Beechey in Alaska. Pacific Discovery 35(1):13–22.
Within this biographical account of Captain F.W. Beechey's voyage aboard *HMS Blossom* in 1825–6 are mentions of the specimens collected by the expedition scientists. Illustrated are some of the mammoth (*Mammuthus*) bones from Elephant Point in Eschscholtz Bay, Alaska, which were collected along with big long-horned oxen [probably steppe-bison (*Bison priscus*)] and large deer. [See Buckland (1831) for a more detailed description of the fossils collected on the voyage.]

858. Lindsey, C.C. (1978a): Aquatic zoogeography and the Ice-Free Corridor. American Quaternary Association 5th Biennial Meeting, Abstracts. pp. 31–4.
Unlike botanical evidence, the paleontology of fish (Pisces) has contributed little to reconstruction of the Quaternary climate and events in the Ice-Free Corridor. The presence or absence of certain aquatic species in lakes in or near the Ice-Free Corridor in the Northwest Territories and Yukon Territory indicate that some drainage patterns have been drastically altered by ice during glacial maxima. To examine the possibility that waters (southerly western tributaries to Mackenzie River) drained northward and then westward along the Laurentide ice front and eventually entered Peel River Basin, fish and zooplankton were sampled from some lakes in the Mackenzie Mountains. The northern samples support the possibility of a much more extensive temporary Yukon River drainage than has been suspected. That drainage could have had its source at or even S of the northern Alberta border.

859. Lindsey, C.C. (1978b): Prairie fish distribution. Manitoba Nature 19(4):8–15.

Ground whitefish (*Prosopium cylindraceum*), lake whitefish (*Coregonus clupeaformis*), northern pike (*Esox lucius*), lake trout (*Salvelinus namaycush*), fourhorn sculpin (*Myoxocephalus quadricornis quadricornis*), deepwater sculpin (*Myoxocephalus quadricornis thompsoni*), and brassy minnow (*Hybognathus hankinsoni*) are discussed in relation to the origin of prairie fishes from such ice age refugia as the Mississippi drainage system S of the Laurentide ice and the Beringian refugium.

860. Linnamae, U., Walker, E.G., and Kelly, D.L. (1988): A summary of the archaeology of the Saskatoon area. *In*: Out of the Past: Sites, Digs, and Artifacts in the Saskatoon Area (U. Linnamae and T.E.H. Jones, eds.). Saskatoon Archaeological Society. pp. 155–62.

The authors state that the Saskatoon site located in the N end of Saskatoon, near the sewage treatment plant might represent a Pre-Clovis occupation (p. 156). Clovis-culture humans [(*Homo sapiens*) 11,500–10,500 BP] who hunted mammoth (*Mammuthus*), camels (*Camelops*), giant bison (*Bison latifrons*), ground sloth[?], and a variety of horses (*Equus*) are represented in the Saskatoon area by a single Clovis-like projectile point (Figure 12.2a) found at the Saskatoon cemetery.

861. Lively, R.A. (1996): Chugwater. *In*: American Beginnings: The Prehistory and Palaeoecology of Beringia (F.H. West, ed.). University of Chicago Press, Chicago and London. pp. 308–11.

The Chugwater site (FAI-35; 64°41′30″N, 147°13′05″W; 224 m asl) is located on Moose Creek Bluff, 35 km SE of Fairbanks, Alaska. A small amount of fragmentary bone and several bird (Aves) gastroliths have been recovered from test excavations.

862. Livingstone, D. (1951): A new record for the mastodon *Mastodon americanus* Kerr, from Nova Scotia. Proceedings of the Nova Scotian Institute of Science 22(4):15–16.

A mastodon (*Mammut americanum* = *Mastodon americanus*) right lower third molar tooth was found in glacial till under the Naval Magazine at Burnside, Halifax County. This is believed to be the first evidence of mastodon on the Nova Scotia mainland. The tooth originally came from a gravel pit near Nine-Mile River, near Emsdale, Nova Scotia.

863. Logan, W.E. (1847): On the topography and geology of the Ottawa River and some of its tributaries. Geological Survey of Canada, Report of Progress, 1845–46.

The author reports capelin (*Mallotus villosus*) remains from the mouth of Gatineau River, Quebec.

864. Logan, W.E. (1863): Superficial geology. Geology of Canada (Geological Survey of Canada Report of Progress from Its Commencement to 1863). Dawson Brothers, Montreal. pp. 886–930.

The Quaternary section in this book, drawn from J.W. Dawson's work, lists several vertebrate finds and their localities: a deer (*Odocoileus*) bone found in gravel beds at the 13th and 14th lot in Enniskillen, Ontario; woolly mammoth (*Mammuthus primigenius*), wapiti (*Cervus elaphus*), and beaver (*Castor canadensis*) found 22 m above Lake Ontario at the Desjardins Canal, Burlington Heights, near Hamilton, Ontario; capelin (*Mallotus villosus*) found at Green Creek, E of Ottawa, Ontario, and in neighbouring Quebec (Chaudière Lake, Madawaska, and Fort Coulonge Lake); lumpfish (*Cyclopterus lumpus*), sculpin (*Cottus*), and seal (Phocidae) from Green Creek; white whale (*Delphinapterus leucas*) from Mile End Quarries in stratified sand and gravel deposits, Montreal, Quebec; harp seal (*Phoca groenlandica*) found in the Peel and Comté Claypit in Montreal; a young seal (Phocidae) and an unidentified bird (Aves) at Bulmer and Shepherd's Brickyard, both localities in Ste-Marie, a Montreal suburb; and a whale (Cetacea) and a walrus (*Odobenus rosmarus*) found between Bic and Matane, Rimouski County, Quebec. Illustrations of a partial harp seal skull [locality not mentioned] and capelin are included (p. 965).

865. Lowdon, J.A. (1985): Geological Survey of Canada Radiocarbon Dating Laboratory. Geological Survey of Canada Paper 84-24:1–19.

This paper explains the procedures and techniques practised by the Radiocarbon Dating Laboratory of the Geological Survey of Canada. It is an update from Dyck 1967.

866. Lowdon, J.A., and Blake, W., Jr (1968): Geological Survey of Canada Radiocarbon Dates VII. Geological Survey of Canada Paper 68-2B:207–45.
In a gravel pit in Kent County, 2.5 km NE of Thamesville, Ontario (42°34′N, 81°57.4′W), altitude 189 m, mastodon (*Mammut*) bones were exposed at the 1.8 m level. They were dated to 11,380 ± 170 BP (GSC-611). Mastodon (*Mammut americanum*) bones from Chatham Twp., 2 km SSW of Tupperville, Ontario (42°33.9′N, 82°17.3′W) at altitude 178 m, are dated to 8910 ± 150 BP (GSC-614). The bones were exposed during excavation of a drainage ditch in a slight depression of lacustrine deposits. Wood collected during excavation for bones at NW end of 'Mitchell Bluff,' S bank of South Saskatchewan River, 10 km N of Medicine Hat, Alberta (50°07′45″N, 110°38′40″W) at altitude 670 m, was dated to >30,000 BP (GSC-780). The wood was about 1 m below surface in sand and gravel with extensive mammal (Mammalia) fauna and chert fragments possibly worked by man (*Homo sapiens*). Bone sample of bison (*Bison*) and horse (*Equus*) from 'Lindoe Bluff' on the W bank of the South Saskatchewan River 3 km N of Medicine Hat (50°04′50″N, 110°39′20″W), altitude 663 m, was dated to 11,200 ± 200 BP (GSC-805). Bones were in gravel laid down when the river bed was 7–15 m above its present level. The date ages the extensive mammal fauna of the site and indicates that the South Saskatchewan River incised rapidly immediately following the last (Wisconsinan) glaciation.

867. Lowdon, J.A., and Blake, W., Jr (1970): Geological Survey of Canada Radiocarbon Dates IX. Geological Survey of Canada Paper 70-2(B):46–86.
Mollusc shells from the Bliss Gravel Pit, Fort Qu'Appelle, Saskatchewan (50°46′N, 103°48′W) at altitude 492 m, were dated to >30,000 BP (GSC-987). These shells were from gravel and sand beds, 15 m thick, containing abundant vertebrate fossils, overlain by thick drift including two tills and underlain by one or more tills. The intertill deposit is thought to be of Sangamonian age. Bone from the middle terrace of three postglacial terraces of the Griffin Gravel Pit on the Bow River near Cochrane, Alberta, 23 m above the river level, (51°10′40″N, 114°27′10″W) was dated to 5670 ± 150 BP (GSC-988). The fauna of the site includes specimens of western bison (*Bison bison occidentalis*), Mexican horse (*Equus conversidens*), bighorn sheep (*Ovis canadensis*), and wapiti (*Cervus elaphus = Cervus canadensis*). [For an additional date from this site see Lowdon and Blake 1975.]

868. Lowdon, J.A., and Blake, W., Jr (1973): Geological Survey of Canada Radiocarbon Dates XIII. Geological Survey of Canada Paper 73-7:1–61.
A wood sample from a sinkhole near Milford, Nova Scotia (45°00′20″N, 63°25′10″W), altitude 23–30 m, has been dated at >50,000 BP (GSC-1642). This site includes unspecified mammal (Mammalia) remains and beaver-cut (*Castor canadensis*) sticks [Harington 1978b, p. 14]. The deposit is stratified, with gypsum as base, overlain by a dark grey silty clay, a dark grey till, red silty till, buff-red sand, pebbly sand, and a red clayey till. These deposits may represent an Early Wisconsinan interstadial deposit or, more probably, the cooler part of an interglacial interval.

869. Lowdon, J.A., and Blake, W., Jr (1975): Geological Survey of Canada Radiocarbon Dates XV. Geological Survey of Canada Paper 75-7:1–32.
Two unidentified bone fragments from the Canadian Pacific Gravel Pit near Empress, Alberta (50°57′50″N, 100°00′50″W), were dated to 14,200 ± 1120 BP (GSC-1199) and 20,400 ± 320 BP (GSC-1387). The site has yielded specimens of woolly mammoth (*Mammuthus primigenius*), imperial mammoth (*Mammuthus imperator*), Mexican horse (*Equus conversidens*), Niobrara horse (*Equus* cf. *Equus niobrarensis*), western camel (*Camelops* cf. *Camelops hesternus*), caribou (*Rangifer tarandus*), and western bison (*Bison* cf. *Bison bison occidentalis*). Although the younger date must be questioned, both dates are substantially older than expected, and neither allows sufficient time for post-Wisconsinan incision of the large Red Deer valley. The possibility that the valley was cut earlier must now be considered. In addition, the presence of a large, well-developed fauna so close to the time of the Classical Wisconsinan Glaciation seems improbable. If the 20,400 BP date is correct, the fauna may have lived in a prolonged interval between two advances of the Wisconsinan glacier. An additional date of unidentified bone from the Griffin Gravel Pit at Cochrane, Alberta (see Lowdon and Blake 1970), is presented – 7220 ± 480 BP (GSC-612-2). The fauna of the

site includes specimens of western bison, Mexican horse, bighorn sheep, and wapiti (*Cervus elaphus* = *Cervus canadensis*) [but this identification is questionable].

870. Lowdon, J.A., and Blake, W., Jr (1976): Geological Survey of Canada Radiocarbon Dates XVI. Geological Survey of Canada Paper 76-7:1–21.
This paper reports on a buried paleosol with unidentified bone and carbonized wood fragments. The 30–46-cm-thick paleosol was exposed in the wall of a pit near Calgary, Alberta (51°04′30″N, 114°12′40″W), at an altitude of 1135 m. The exposure is in a high terrace of the Bow River. Carbonized twig fragments were dated at 8400 ± 150 BP (GSC-1819). The date provides a minimum for the drainage of Glacial Lake Calgary. [For other papers in this series see also: Blake 1982, 1983, 1984, 1986, 1987; Dyck and Fyles 1963, 1964; Dyck et al. 1965, 1966; Lowdon 1985; Lowdon and Blake 1968, 1970, 1973, 1975, 1979, 1980, 1981; Lowdon et al. 1967, 1969, 1970, 1971, 1972, 1974, 1977; McNeely 1989; McNeely and McCuaig 1991].

871. Lowdon, J.A., and Blake, W., Jr (1979): Geological Survey of Canada Radiocarbon Dates XIX. Geological Survey of Canada Paper 79-7:1–58.
This paper reports the radiocarbon dates of a number of sediment samples. A humerus of bowhead whale [(*Balaena mysticetus*) CMN CR-75-55] from Hanson's Gravel Pit near White Lake, Ontario (45°22′20″N, 76°31′40″W), at an elevation of about 170 m, has a radiocarbon date of 11,500 ± 90 BP (GSC-2269). This sample marks the early, deep cold phase of the Champlain Sea. [See also Lowdon and Blake 1981:GSC XXI]. Bison (*Bison*) bones (sacrum, ribs, and vertebrae) were recovered at about 1.5 m depth in a sandy silt alluvium stream cutbank along the North Saskatchewan River (54°03′N, 112°14′W), at an elevation of about 565 m. The sample was dated to 2800 ± 140 BP (GSC-1195-2). The bones date a well-developed terrace level (the lower of two) which was deposited when the river was about 12 m above the present level. The date corresponds well with both GSC-1205 and GSC-1207 (see below). Bison bones (2 tibias) recovered in a gravel pit near Athabasca, Alberta (54°43′N, 113°17′W), at an elevation of about 550 m, provided a date of 10,200 ± 160 BP (GSC-1205). The stratigraphy of this sample is in question; however it is considered that the bones came from gravels about 4–5 m below the surface. This sample provides a date for the beginning of postglacial river valley development in the area. Bison bones (tibia and rib) were recovered from 1.8 m depth in a sandy silt alluvium stream cutbank along the North Saskatchewan River (54°02′N, 112°20′W), at an elevation of 570 m. This sample dated to 6040 ± 140 BP (GSC-1207). This dates a well-developed terrace (the middle of three) which was deposited when the river floodplain was about 20 m higher than the present level. This date marks the initial period of postglacial sedimentation by the Athabasca River following the draining of Glacial Lake Athabasca. Bison bones were recovered about 1.7 m below surface from an 18-m-high exposure (altitude about 550 m) in a silt and clay deposit of Glacial Lake Athabasca. Freshwater gastropod shells associated with the bones provided a radiocarbon date of 10,200 ± 280 BP (GSC-1380). The sample dates the next event in the glacial history of the Athabasca area – i.e., the beginning of the development of the Athabasca Valley following the draining of Glacial Lake Athabasca. Western bison (*Bison bison occidentalis* = *Bison occidentalis*) ribs recovered from sand, clay, and diamicton exposed in a specially dug excavation near Three Hills, Alberta (51°41′30″N, 113°04′30″W), at an elevation of between 808 and 814 m, provided a radiocarbon date of 9720 ± 300 BP (GSC-1894). A skeleton of wapiti (*Cervus elaphus nelsoni*) accompanied bison remains. Bones (horncores, tibias, and smashed skulls) of bison (*Bison* cf. *Bison priscus*) from the Vera Lane Gravel Pit in Peace River, Alberta (56°13′40″N, 117°18′10″W), elevation about 35 m above the current level of Peace River and nearly 200 m below prairie level have been dated at 9880 ± 130 BP (GSC-2865). This date correlates nicely with GSC-2895 and GSC-2902 (see below), which are from a similar stratigraphic position at another site. The date indicates that the river had already incised its valley to near the present depth by 9880 BP. This is a surprisingly fast rate of downcutting if the area had been covered by Classical Wisconsinan ice only a few thousand years before. Other species of mammal known from this site are wapiti (*Cervus elaphus* = *Cervus canadensis*) and woolly mammoth (*Mammuthus primigenius*). A large bison (*Bison*) cervical vertebra from the Northern Alberta Railway Ballast Pit, near Watino, Alberta (55°43′00″N, 117°37′30″W), elevation about 410 m, has provided a radiocarbon date of 10,200 ± 100 BP (GSC-2895). The bones came from near the base of the alluvial sand and

gravel forming the second terrace above the river. The site has since been destroyed. A left tibia of a large bison (*Bison*), from the same site as GSC-2895 (above) has provided a radiocarbon date also of 10,200 ± 100 BP (GSC-2902). A skull of Alaskan bison (*Bison* cf. *Bison alaskensis*), now at the Provincial Museum of Alberta (P71.7), came from this pit in 1970. The dated material (2895 and 2902) probably derived from this species. A tusk fragment, possibly mammoth (*Mammuthus*), from sandy gravel exposed in a fresh face of the Bailey Gravel Pit, near Vedder Crossing, British Columbia (49°06′20″N, 121°55′20″W), at elevation of about 60 m, has provided a radiocarbon date of 22,700 ± 320 BP (GSC-2232). This seems to be the earliest record of Quaternary elephant remains from the Fraser Lowland. The date of the tusk suggests that such elephants occupied the Fraser Lowland prior to the peak of the Late Wisconsinan glaciation. Several species of large Quaternary mammals, including mastodons (*Mammut americanum*) and mammoths, probably reached southeastern Vancouver Island from the mainland about this time (30,000–17,000 BP) by crossing large, vegetated floodplains that filled the Strait of Georgia. An anterior rib fragment (CMN 17915) of a Columbian mammoth (*Mammuthus* cf. *Mammuthus columbi*) was recovered from a 6-m-thick silt layer at the Noranda Open Pit Mine, Babine Lake, British Columbia (55°00′00″N, 126°14′00″W), at elevation of about 703 m. The silt unit was underlain by a clay layer 6 m thick and overlain by 0.3–1.5 m of gravel with a light grey till of up to 24 m in thickness over the gravel. The bone provided a radiocarbon date of 34,000 ± 690 BP (GSC-1754). The sample indicates that mammoths occupied central British Columbia during the Olympia nonglacial interval. A skull of bighorn sheep (*Ovis canadensis*) exposed during operation of a gravel pit on the W side of the Finlay-Parsnip Access Road, British Columbia (55°48′00″N, 123°38′30″W), at an elevation of about 730 m, provided a radiocarbon date of 9280 ± 200 BP (GSC-1497). The specimen was found about 3 m below the ground surface in dry ice-contact fluvial gravels. This date provides a minimum age for deglaciation in Parsnip Valley. [See Rutter et al. 1972].] A mammoth (*Mammuthus*) tusk from a gravel pit about 12.5 km from Hudson Hope, British Columbia, in the Portage Mountain moraine (approximately 56°0.6′N, 122°0.7′W), at an elevation of about 760 m, provided a radiocarbon date of 25,800 ± 320 BP (GSC-2859). If this is an uncontaminated date it raises the possibility that the tusk is either redeposited or that the Cordilleran climax at Portage Mountain was early in Classical Wisconsinan time and that Cordilleran glacial deposits still further to the E are significantly older. A fragment of woolly mammoth (*Mammuthus primigenius*) upper molar (CMN 17679), recovered from the Ostero Gravel Pit near Taylor, British Columbia (56°08.5′N, 120°40.5′W), at an elevation of about 455 m, provided a radiocarbon date of 27,400 ± 580 BP (GSC-2034). The tooth was found about 15–18 m below surface. A moose (*Alces alces*) metatarsal was also found at the same site. Beaver-gnawed (*Castor canadensis*) sticks, collected from a road-cut along the Dempster Highway near Frog Creek, Northwest Territories (67°25′N, 134°28′W), has provided a radiocarbon date of 9490 ± 90 BP (GSC-1814). Part of an ancient beaver dam, this sample suggests that beaver occupied a habitat only 2° S of their present interglacial limit of 69°N in this region by the close of the Wisconsinan glaciation. Beaver-gnawed (*Castor canadensis*) sticks were recovered from the headwall of a retrogressive thaw-flow slide beside an unnamed lake in the Grandview Hills, Northwest Territories (67°06′N, 131°13′W). The sample provided a radiocarbon date of 9560 ± 120 BP (GSC-2298), which suggests a minimum date for initiation of the thermokarst pond which was the organic silt's interpreted deposition source. The date is also a minimum for the postglacial occurrence of poplar and beaver in the region. Fragments of a single Yukon horse (*Equus* cf. *Equus lambei*) femur obtained from a bone concentration in the lower stratigraphic unit (VII) of Bluefish Cave I, Yukon, has provided a radiocarbon date of 12,900 ± 100 BP (GSC-2881). Palynological and stratigraphic data support this age of Late Pleistocene, post–Late Wisconsinan maximum. This date for a Yukon horse is the youngest on record from Eastern Beringia, and is made even more interesting by the presence of lithic artifacts (*Homo sapiens*) in the same unit. The date suggests that the demise of the herb-rich tundra may have occurred later in some of the upland environments. It also dates the complex faunal accumulation at Bluefish Caves that includes proboscideans (Proboscidea) and appears to be, in part, culturally induced.

872. Lowdon, J.A., and Blake, W., Jr (1980): Geological Survey of Canada Radiocarbon Dates XX. Geological Survey of Canada Paper 80-7:1–28.

This paper reports the radiocarbon dates of a number of

pertinent samples. A series of whalebone samples from raised beaches N of Cape Storm, Ellesmere Island, Nunavut, were radiocarbon dated in order to provide an idea of the age of the ridges and the rate of emergence. Each sample was dated at least twice. (1) whale (Cetacea) vertebra dated to 980 ± 60 BP (GSC-980) from 76°22.5′N, 87°30′W and 940 ± 70 BP (GSC-980-2); (2) bowhead whale (*Balaena mysticetus*) cranial fragment dated to 1610 ± 50 BP (GSC-1227) from 76°24′N, 87°30′W and 1870 ± 90 BP (GSC-1227-2); (3) rib associated with bowhead mandibles dated to 4490 ± 60 BP (GSC-1021) from 76°24′N, 87°30′W and 4580 ± 60 BP (GSC-1021-2) and 4450 ± 70 BP (GSC-1021-3); (4) bowhead whale rib dated to 5600 ± 60 BP (GSC-979) from 76°24′N, 87°30′W and 5420 ± 70 BP (GSC-979-2) and 5050 ± 180 BP (GSC-979-3); (5) whale skull fragment dated to 7260 ± 80 BP (GSC-1498) from 76°24.5′N, 98°30′W and 6560 ± 170 BP (GSC-1498-2) and 7240 ± 80 BP (GSC-1498-3).

873. Lowdon, J.A., and Blake, W., Jr (1981): Geological Survey of Canada Radiocarbon Dates XXI. Geological Survey of Canada Paper 81-7:1–22.
This paper reports the radiocarbon dates of a number of pertinent samples. A core sample from the vertebra of a common finback whale (*Balaenoptera physalus*) excavated from a trench in marine clay near Daveluyville, Quebec (46°10.5′N, 72°10′W) at an elevation of about 85 m has provided a radiocarbon date of 11,400 ± 90 BP (GSC-2871). The almost complete skeleton resides at the Archaeological Museum at Université du Québec at Trois-Rivières. The radiocarbon date of this specimen is in accord with the only other date on a large Champlain Sea whale – a bowhead [*Balaena mysticetus* (GSC-2269); Lowdon and Blake 1979] which dated at 11,500 ± 90 BP.

874. Lowdon, J.A., Fyles, J.G., and Blake, W., Jr (1967): Geological Survey of Canada Radiocarbon Dates VI. Geological Survey of Canada Paper 67-2(B):1–42.
This paper reports on three pertinent radiocarbon dates. A bison (*Bison*) jaw bone from the Castle River, 6 mi W of Pincher Creek, Alberta (49°29′N, 114°03′30″W) dated to 6150 ± 140 BP (GSC-447). Associated teeth provided a confirming age of 6100 ± 180 BP (GSC-490). Western bison (*Bison bison occidentalis*) bone [see Churcher 1968a] from the Griffin Gravel Pit near Cochrane, Alberta (51°10′40″N, 114°27′10″W) provided a date of 10,760 ± 160 (GSC-612). Western bison bone from the Clarke Gravel Pit near Cochrane, Alberta (51°10′40″N, 114°27′00″W), provided a date of 11,370 ± 170 BP (GSC-613).

875. Lowdon, J.A., Robertson, I.M., and Blake, W., Jr (1971): Geological Survey of Canada Radiocarbon Dates XI. Geological Survey of Canada Paper 61-7:255–324.
This paper reports the radiocarbon dates of a number of pertinent samples. Bison (*Bison*) bones recovered at 2.7 m depth in lacustrine clayey silt at the Mona Lisa site, downtown Calgary, Alberta (51°02′N, 114°05′W), provided a radiocarbon date of 8080 ± 150 BP (GSC-1209). A 7.5 cm volcanic ash layer (Mazama?) was about 0.3 m above the bone horizon. The deposit probably was deposited in a lake in the former flood plain of the Bow River. Organic sediments collected near Woodbridge, Ontario (43°45′39″N, 79°35′30″W), about 6 m below ground surface, have produced a radiocarbon date of 45,000 ± 900 BP (GSC-1181). The dated sample was collected within 2 m of a partial molar of a woolly mammoth (*Mammuthus primigenius*) and is believed to roughly date the enclosed tooth. The date confirms assignment of the organic layer to the Port Talbot interstade.

876. Lowdon, J.A., Robertson, I.M., and Blake, W., Jr (1977): Geological Survey of Canada Radiocarbon Dates XVII. Geological Survey of Canada Paper 77-7:1–25.
This paper reports the radiocarbon dates of a number of pertinent samples. Vertebrae and ribs of snakes (*Coluber constrictor*, *Elaphe obsoleta*, and *Crotalus horridus*) from consolidated and unconsolidated bands of $CaCO_3$-cemented shore debris in Westdale Ravine, Hamilton, Ontario (43°16′N, 79°54′W) at an elevation of 88 m, have provided a radiocarbon date of 4330 ± 210 BP (GSC-1681). A sample of marine shell fragments from the Bailey Gravel Pit, near Chilliwack, British Columbia (49°06′20″N, 121°55′20″W), at an altitude of 46 m, has provided a radiocarbon date of >34,000 BP (GSC-2230). The dated sample is in a layer of glaciomarine clayey silt with a mammoth? (*Mammuthus*?) tusk found 15 m away. A rib, probably from a bowhead whale (cf. *Balaena mysticetus*), collected from a gravel beach ridge at South

Cape Fiord, Ellesmere Island, Nunavut (76°25.6′N, 85°03′W), at an altitude of 80.5 m, has produced a radiocarbon date of 9070 ± 90 BP (GSC-1748). A second determination on the same rib produced a radiocarbon date of 9070 ± 90 BP (GSC-1748-2).

877. Lowdon, J.A., Wilmeth, R., and Blake, W., Jr (1969): Geological Survey of Canada Radiocarbon Dates VIII. Geological Survey of Canada Paper 69-2(B):22–42.
Charcoal (NMC-84) from the Drynoch Slide site (EcRi-1), British Columbia (50°20′N, 121°24′W), has been dated to 7530 ± 270 BP (GSC-530). The sample was selected from a midden of fish (Pisces) remains and artifacts (*Homo sapiens*). The material is overlain by ash, possibly Mazama Ash dated about 6600 BP. A microblade from the site is so far the oldest found in British Columbia.

878. Lowdon, J.A., Wilmeth, R. and Blake, W., Jr (1972): Geological Survey of Canada Radiocarbon Dates XII. Geological Survey of Canada Paper 72-7:1–26.
An horizon of buried soil from a depth of 2.7 m, from the DlPo-20 site about 30 m above the Oldman River, Alberta (49°51.5′N, 114°22.5′W), has provided a radiocarbon date of 8000 ± 150 BP (GSC-1158). This sample is taken from a living-floor (*Homo sapiens*) and is closely associated with charred bison (*Bison*) bones dated at 9520 ± 240 BP (GX-956).

879. Lowey, M. (1992): Whale prints. Calgary Herald, April 18, p. A18.
This article deals with the fieldwork of Arthur Dyke of the Geological Survey of Canada. By finding and collecting specimens and radiocarbon dating remains of bowhead whales (*Balaena mysticetus*) from raised beaches in the Canadian Arctic Islands (mainly Nunavut), the scientist has been able to construct climate over the past 10,000 years. On beaches older than 8000 years, researchers had no trouble excavating more than 1000 bowhead skulls over eight field seasons. Dyke said that they usually only packed out the large, dense, well-preserved ear bones from the whales – ideal samples for radiocarbon dating. Bone data reveal periods during the past 10,000 years when warmer climate melted sea ice now separating western and eastern Arctic bowheads (e.g., there is evidence of warmer temperatures between about 11,000 and 8500 BP). About 5000 to 3000 BP, warmer summers again allowed bowheads to expand their summer range. A photograph of most of a bowhead skeleton on a high raised beach in the Canadian Arctic is included. [See Dyke and Morris 1990.]

880. Lowey, M. (1993): Early arrivals. Calgary Herald, April 24.
Bones of fossilized ice age animals show that humans (*Homo sapiens*) could have reached the Great Plains some 23,000 years ago. Rob Young, a University of Calgary geographer noted that this view contradicts that of the scientific establishment. The idea that Alberta was covered by an ice sheet about 23,000 BP is incorrect since ice age mammal bones from gravel pits near Edmonton indicate that wildlife was abundant then. Young, with colleagues from the University of Calgary and the Provincial Museum of Alberta, have collected bones from the Edmonton pits including: Alberta's first record of a giant short-faced bear (*Arctodus simus*); a skull of an American lion (*Panthera leo atrox*); a rare mastodon (*Mammut americanum*) tooth; wolves (*Canis lupus*); mammoths (*Mammuthus*); small caribou (*Rangifer tarandus*); bison (*Bison*); ground sloths (*Megalonyx*); camels (*Camelops*); and many horses (*Equus*). Young stated that only one glaciation affected Alberta during the Late Pleistocene, and that the ice did not enter Alberta until after 23,000 to 22,000 BP, based on the ages of the bones. Jim Burns, curator of Quaternary paleontology at the Provincial Museum of Alberta in Edmonton, considers the case pretty solid, stating that more dates on the fossils have been obtained in the last four years than had been gathered previously (since the 1950s). He said that the glaciation had demolished most ice age bones except those protected in gravel pits. A few years ago researchers reached an agreement with several gravel pit operators including Standard General, Consolidated Concrete, and Lafarge. In one field season alone, a single gravel pit near Edmonton yielded 900 fossils – most plucked from conveyor belts during the mining operation. Burns says, next to Yukon, which remained largely unglaciated, Alberta has the best collections of ice age mammal remains in terms of diversity (the gravel pits have yielded bones of 16 species) and numbers of specimens. As far as early humans (*Homo sapiens*) are concerned, there is evidence at Bluefish Caves, Yukon, as early as 25,000 BP. In Alberta an ancient hearth beside Vermilion Lakes near Banff revealed

a nomadic band of hunters camped there 10,500 years ago. Young and colleagues believe that a cool, dry climate, with less seasonal variation than now, prevailed when these ice age mammals were present, and that the landscape was mainly broad grasslands, like the steppe of southern Asia. Young's and Burns's work in the Hand Hills of Alberta turned up bones and burrows of large prairie dogs (*Cynomys*) that lived 22,000 years ago. Illustrations include: a photo of a life-size model of the ground sloth (*Megalonyx*), a photo of Jim Burns and Gene Seal holding a mammoth leg bone, and a diagram showing the relative sizes of a giant short-faced bear and a grizzly bear (*Ursus arctos*).

881. Lowey, M. (1998): Hikers find well-preserved buffalo bones – 9,000 feet up. Calgary Herald, March 30. pp. A1–A2.
Jim Fosti and Gordon Scruggs, Calgary climbers, found the sun-bleached bones of a young male wood buffalo (*Bison bison athabascae*) near a 9000-ft mountain ridge on the eastern border of the Siffleur Wilderness, SW of Rocky Mountain House, Alberta. Radiocarbon dating of bone collagen has confirmed that the skeleton is about 700 years old. The bones, recovered by helicopter, include a skull with horncores and sheaths, leg bones, ribs, and a lower jaw with teeth. A colour photograph shows the bones, discoverers, and University of Calgary staff, and there is a map showing the locality.

882. Lowey, M. (1999): Spring runoff threatens fossil find. Calgary Herald, April 29.
A team of University of Calgary archaeologists and geologists unearthed the bones or fossilized footprints of several extinct ice age mammals including horse (*Equus*), bison (*Bison*), muskox (*Bootherium*), wolf (*Canis lupus*), mammoth (*Mammuthus*), and camel (probably *Camelops hesternus*). Along with the ice age mammal fossils were ancient bones of badger (*Taxidea taxus*), rabbit (*Lepus*), beaver (*Castor canadensis*), and muskrat (*Ondatra zibethicus*), as well as some 2000 ground squirrel (*Spermophilus*) specimens. Also scattered on the exposed floor of the St Mary Reservoir near Cardston, Alberta, were many spear and arrow points, including 11,000- to 11,500-year-old Clovis spear points, which were used to hunt mammoths. Preliminary radiocarbon dating has shown the bison and horse specimens to be about 11,000 years old. Scientists speculate that the site, covering about 2 square km, was the shore of a prehistoric river or lake where animals came to drink, scavenge, and hunt. Shayne Tolman, a Cardston elementary school teacher, found an ancient point when his sons were playing on mudflats exposed by dropping water in the reservoir. The fossils were discovered by the Tolmans in the summer of 1996, but the site's significance only became apparent in the fall of 1998, when Tolman called in University of Calgary scientists.

883. Lowey, M. (1999). Emptied reservoir like treasure trove for archaeologists. Calgary Herald, May 8:B-12 (see also A1–A2).
Water levels in the St Mary Reservoir near Cardston, Alberta, are at their lowest since the artificial lake was filled in 1949 to accommodate construction of a new spillway at the dam. The sandy bottom, exposed to wind and wave erosion, has yielded evidence of bones, ice age mammal tracks and burrows, as well as artifacts and fire pits to a University of Calgary team of archaeologists, geologists, and student volunteers. This group has so far unearthed the skull of a prehistoric bison (*Bison*) and remains of at least six Pleistocene horses (*Equus*). Barely 50 m from some ancient ungulate tracks, the sand yielded a partial skull and several vertebrae of an extinct muskox (*Bootherium*) – a species never found before in southern Alberta. A short distance away are pie-plate-sized footprints of an 11,000-year-old woolly mammoth (*Mammuthus primigenius*). Dozens of stone spear and arrow points (including Clovis points estimated at 11,000–11,500 BP) have been sifted from the reservoir's sand. Excavation also produced stone scrapers, slicers, and a rock-lined fire pit. Illustrations show: an ancient camel track, a 'muskox' track, and a tibia of a Pleistocene horse (*Equus*).

884. Loy, T., and Dixon, E.J. (1998): Blood residues on fluted points from Eastern Beringia. American Antiquity 63(1):21–46.
Blood residues have been microscopically and chemically detected on 21 of 34 projectile points (presumably mainly dating to between 11,500 and 10,000 BP) from Eastern Beringia. From these residues a variety of large mammal species were identified as prey, including: woolly mammoth (*Mammuthus primigenius*) from Girls Hill, Kugururuk River, Lisburne, and Point site [all sites are in Alaska unless otherwise indicated – see map, Figure 1]; bison (*Bison bison*) from Girls Hill and Utukok River;

sheep (*Ovis dalli*) from Lisburne; bear (*Ursus arctos*) from Kugururuk River and Lisburne; caribou (*Rangifer tarandus*) from Batza Tena, Driftwood Creek, Girls Hill, Old Crow area, Yukon, Putu, and Utukok River; and muskox (*Ovibos moschatus*) from Hanks Hill. Because there are no radiocarbon dates on mammoths, the authors reject the hypothesis that fluted projectile points were invented later and independently in Eastern Beringia and bear no historical or cultural relationship to the Llano complex.

885. Lucas, F.A. (1899): The fossil bisons of North America. Proceedings of the U.S. National Museum 21 (1172):755–71.
Lucas undertook this first revision of North American bison since Allen (1876) in an attempt to assign definite characters to the various species of bison occurring in a fossil condition in North America and to disentangle the complicated synonymy in which they have been involved (p. 755). He recognized seven species of bison (*Bison*) in his effort to stabilize bison taxonomy. At least two of the species he mentioned [the western bison (*Bison bison occidentalis* = *Bison occidentalis*; Plate 65 shows a specimen from Fort Yukon, Alaska) and steppe bison (*Bison priscus* = *Bison crassicornis* – he considers Alaskan bison (*Bison alaskensis*) to be a synonym of *Bison crassicornis* (p. 756); Plates 73–76 show specimens from Point Barrow and Eschscholtz Bay, Alaska)] had fossils in northern North America (e.g., Alaska and Yukon). [See Skinner and Kaisen 1947 and McDonald 1981.]

886. Lundelius, E.L. Jr, Downs, T., Lindsay, E.H., Semken, H.A., Zakrzewski, R.J., Churcher, C.S., Harington, C.R., Schultz, G.E., and Webb, S.D. (1987): The North American Quaternary sequence. *In*: Cenozoic Mammals of North America (M.O. Woodburne, ed.). University of California Press, Berkeley. pp. 211–35.
This chapter examines the last three land-mammal ages of the Cenozoic: the Blancan, the Irvingtonian, and the Rancholabrean. Although the Blancan age is largely Pliocene, the faunal break between the Blancan and the succeeding Irvingtonian age is less pronounced than that between the Blancan and the preceding Hemphillian. The Blancan mammalian fauna is therefore more closely related to that of the Pleistocene than to that of preceding land-mammal ages. Key species for each age are identified, along with a discussion of the palaeoecology of the age. A map showing the location of the faunas mentioned supports the text. An extensive bibliography is provided. Pertinent to this bibliography are comments on: (1) Wellsch Valley fauna (about 1.88 to 1.72 Ma) near Swift Current, Saskatchewan, that contains the Blancan survivor, a bone-eating dog (*Borophagus*), and one of the earliest southern mammoth (*Mammuthus meridionalis*) records in North America; (2) *Soergelia*, a primitive muskox of possible Irvingtonian age from Old Crow River Loc. 11A, Yukon; (3) Cape Deceit fauna from western Alaska containing early occurrences of the lemming (*Dicrostonyx*), caribou (*Rangifer tarandus*), and wapiti (*Cervus*); (4) Medicine Hat fauna in southern Alberta ranging in age from Irvingtonian ('Mid-Kansan') to Early Holocene – the richest known vertebrate assemblage from the northern plains; (5) presence of bison (*Bison*) in Alaska and Yukon since Illinoian time or earlier; (6) the large mammal fauna of Eastern Beringia during Wisconsinan time, including the steppe bison (*Bison priscus*), horse (*Equus*), woolly mammoth (*Mammuthus primigenius*), steppe mammoth (*Mammuthus columbi/armeniacus*), western camel (*Camelops hesternus*), helmeted muskox (*Bootherium bombifrons*), Dall sheep (*Ovis dalli*), and American lion (*Panthera leo atrox*); (7) the fact that Alaska is likely to differ faunally from the continental interior not only because of climatic differences but also because Alaska is on the dispersal route between Asia and central North America, and there seem to have been barriers to dispersal e.g., confinement of the saiga antelope (*Saiga tatarica*) and the early muskox (*Praeovibos*) to the Eastern Beringian region.

887. Lydekker, R. (1885): [Northern North American examples]. Catalogue of the fossil Mammalia in the British Museum of Natural History, Part 2:24–39.
The Canadian and Alaskan species in this compilation are: steppe bison (*Bison priscus*) found along the Upper Porcupine River, Yukon (67°00′N), and at Eschscholtz Bay, Kotzebue Sound, Alaska (66°30′N); a muskox (*Praeovibos priscus* = *Ovibos*) found at the Upper Porcupine River site, Yukon. Specimens from Porcupine River, Yukon, were presented to the museum by the Reverend Robert McDonald in 1873, whereas those from Kotzebue Sound, Alaska, were presented by Captain F.W. Beechey, R.N., in 1831 and by Captain Kellett and Lieutenant Wood, R.N., in 1850.

888. Lydekker, R. (1886): [Northern North American examples]. Catalogue of the Fossil Mammalia in the British Museum of Natural History, Part 4:204–5.
This is similar to Lydekker (1885), repeating some of the Porcupine River, Yukon, and Eschscholtz Bay (Kotzebue Sound), Alaska, examples.

889. Lyell, C. (1845): Travels in North America in the Years 1841–1842, with Geological Observations in the United States, Canada, and Nova Scotia. Volume 2. Wiley and Putnam.
Capelin (*Mallotus villosus*) in concretions obtained by W.E. Logan is included in a faunal list of specimens from the Ottawa River near Bytown [Ottawa, Ontario].

890. Lysek, C.A. (1997): Ancient Alaskan bones may help to prove coast migration theory. Mammoth Trumpet 12(4):8–11, 20.
Excavation of a cave designated 49-PET-408 on northern Prince of Wales Island in the Alexander Archipelago of Alaska yielded the mandible (with all of the teeth except four incisors) of an adult human male in his early 20s. Three human vertebrae and a partial right pelvis found in the same chamber, about 10 m from the cave entrance, appear to be from the same individual. Samples from the mandible and pelvis yielded AMS radiocarbon dates of 9730 ± 60 BP (CAMS-29873) and 9880 ± 50 BP (CAMS-3238) respectively, making them the oldest reliably dated human (*Homo sapiens*) remains found so far in Alaska and Canada. These bones could lend support to the hypothesis that early people migrated to North America along the Pacific Coast. Radiocarbon dates from the cave span the past 40,000 years, indicating the presence of refugia there during the Late Pleistocene. Bones of a brown bear (*Ursus arctos*) yielded a radiocarbon date of 35,365 ± 800 BP (AA-15227) and of a black bear (*Ursus americanus*) gave 41,600 ± 1500 BP (AA-16831) – apparently both species lived there through the last glacial maximum, probably living on fish (Pisces). Pleistocene marmot (*Marmota*) and seal (*Phoca*) bones have been found in caves in the area also.

891. MacDonald, G.F. (1968): Debert: A Palaeo-Indian site in central Nova Scotia. National Museums of Canada, Anthropology Papers No. 16:1–207.
One of the most important Paleoindian sites in eastern North America is Debert, Nova Scotia. Evidently a fairly large population (*Homo sapiens*) lived there about 10,600 BP. Cooperative hunting at the site may have concentrated on the woodland caribou (*Rangifer tarandus*). Unfortunately, high soil acidity destroyed all traces of faunal remains at the site.

892. MacDonald, J. (1988): Gatineau labyrinth: The Laflèche Cave. *In*: Up the Gatineau! (Historical Society of the Gatineau, Chelsea, Quebec) 14:25–30.
This article written in 1978 gives interesting background information on Laflèche Cave, one of the important sites for Quaternary vertebrates near Hull, Quebec. Apparently the site was discovered by a hunter named Dubois in 1865. The first scientist to visit the cave was Dr James A. Grant in 1869, who published an account of it in *Transactions of the Ottawa Natural History Society*. The author reports that 'no trace of primitive man [*Homo sapiens*] was ever found – or is ever likely to be found – in the cave. The climatic events of the ice age made this part of the world uninhabitable until less than 10,000 years ago. The only bones discovered in the cave were those of foxes [presumably the red fox (*Vulpes vulpes*)], beavers [*Castor canadensis*], and otters [*Lutra canadensis*].' Don Hogarth, an Ottawa geologist, stated that Laflèche Cave probably formed in the Pleistocene or earlier as a result of rainwater, made acidic by dissolved carbon dioxide, percolating throughout the limestone and enlarging some of the openings. [See Harington (in press).]

893. MacEwan, G. (1964): Monsters in these parts. Calgary Herald, after October 19.
This article in 'Our Natural Heritage' – a weekly series by MacEwan – notes the recent discovery of mammoth (*Mammuthus*) bones near Kyle, Saskatchewan, some 60 mi N of Swift Current. The bones were discovered on October 19, 1964, by William MacEvoy while working on the road 1.5 mi W of Kyle. The scraper blade cutting into clay exposed large bones. Fortunately MacEvoy recognized the potential significance of the bones, and the news brought scientists, collectors, and spectators (approximately 20,000 on a weekend) to excavate and view the bones. A tusk was found 3 ft below the surface and bones were located up to 8 ft deep. Presumably the dead body of the mammoth settled in the mud and sand of a glacial pond about 10,000 years ago [radiocarbon analysis of bone yielded a date of 12,000 ± 200 BP (S-246) – see Harington 1978b, p. 26], when the area was still in

the grip of a warming ice age. The bones were treated by removing the clay and coating them with shellac and alcohol to prevent further deterioration, then encasing them in plaster jackets and shipping them to their new home in the Saskatchewan Museum of Natural History, Regina. Museum director F.G. Bard said it was the most extensive discovery of mammoth bones made so far in Saskatchewan.

894. MacIntosh, G.D. (1997): Palaeoecology of Scottie Creek District, Beaver Creek, Yukon Territory: Life on the edge. MS thesis, University of Alaska, Fairbanks. 114 pp.

The region around Beaver Creek, Yukon Territory (approx. 62°30′N, 141°00′W) represents the far [south] eastern extent of Eastern Beringia. During the Late Wisconsinan-Holocene transition, the juxtaposition of an arid, unglaciated interior and large piedmont glacier of the Macauley-Kluane Glaciation strongly influenced local climate and vegetation. Two pollen records from Daylight Coming Out Lake and Island Lake, plus a reinterpretation of the 1971 Antifreeze Pond record, provide new information on Late Pleistocene to Holocene climatic changes. The Island Lake records also provide insight on thaw-lake evolution during the Holocene. In the Late Wisconsinan, the region experienced extreme polar conditions that delayed the birch rise until 11,000 BP, 1000 years later than in neighbouring regions. Recent archaeological finds have placed human (*Homo sapiens*) occupation of the region as early as 10,130 BP. It is likely that the extremely harsh climate and lack of woody vegetation would have forestalled human occupation prior to the birch rise. Several associated fragments of a Yukon horse (*Equus lambei*) dated to 20,660 ± 100 BP (Beta-70102) were recovered from along the Beaver Creek portion of the Alaska Highway in 1993 during a realignment and upgrade. The highway work in the area destroyed many of the fossil localities. All of these sites were located in loess or glaciolacustrine deposits of Wisconsinan age. The preservation potential of the sediments was rather poor and only large bones were recovered, mainly horse and bison (*Bison*) with only one example each of mammoth (*Mammuthus*) and caribou (*Rangifer tarandus*) [see Hare 1994]. Also four archaeological sites in the area yielded hundreds of bone fragments, most unidentifiable due to extensive burning [see Walde 1994]. The oldest of these sites dates to older than 7810 ± 80 BP. In general the fossils indicate that the region probably had the usual Late Pleistocene megafauna community and that the megafauna were present throughout the Late Wisconsinan. The extended duration of the Herb Zone due to the cold climate may have made this area a refugium for the last of the Late Pleistocene megafauna.

895. Mackay, J.R. (1958): The Anderson River map-area, N.W.T. Department of Mines and Technical Surveys, Geographical Branch Memoir 5:24–6.

A steppe bison (*Bison* cf. *Bison priscus*) skull was found roughly 6.5 km SW of Tuktoyaktuk, Northwest Territories. Mammoth (*Mammuthus*) remains were found in several localities: Nicholson Peninsula (tusk fragments); on the beaches of the wave-cut bluffs of Bathurst Peninsula (tusks and teeth); mouth of Holmes Creek (tooth); and tusks from a beach 10 km SW of Cape Bathurst, 1.6 km S of Stanton and the NE side of the MacKenzie River Delta near Kittigazuit. These fossils, scattered from Richards Island to Darnley Bay, suggest that they are from interglacial to late-glacial formations.

896. Mackay, J.R. (1963): The Mackenzie Delta area, N.W.T. Department of Mines and Technical Surveys, Geographical Branch Memoir 8:15–21.

This paper repeats the mammoth (*Mammuthus*) remains and steppe bison (*Bison* cf. *Bison priscus* = *Bison* cf. *Bison crassicornis*) finds mentioned in Mackay (1958), in addition to other mammoth finds: central portion of Eskimo Lakes (skull); between Cape Bathurst and Herschel Island (bones); Anderson River; the mouth of the Kugaluk River; and the East Channel N of Tununuk to Kittigazuit (all tusks and teeth). A Late Pleistocene age is reckoned based on the mammoth teeth and the bison horns.

897. Mackay, J.R., Mathews, W.H., and MacNeish, R.S. (1961): Geology of the Engigstciak archaeological site, Yukon Territory. Arctic 14(1):25–52.

The authors report two subfossil assemblages from Engigstciak, about 16 km upstream from the mouth of Firth River in northern Yukon (see map, Figure 1). The upper layer of Quaternary sediment contained remains of: caribou (*Rangifer tarandus*), tundra muskox (*Ovibos moschatus*), Dall sheep (*Ovis dalli*), arctic or coloured fox (*Alopex lagopus* or *Vulpes vulpes*), grizzly bear (*Ursus arctos*), moose (*Alces alces*), rodents (Rodentia), fish (Pisces), birds (Aves), and sea mammals (Pinnipedia). The lower layer contained remains of: caribou, modern bison (*Bison bison*), wapiti (*Cervus elaphus*), Dall sheep [I have since reidentified the mountain goat (*Oreamnos americanus*) as Dall sheep], small mammals, fish, birds, and seal (*Phoca*). A caribou antler from this layer has been radiocarbon dated at 3250 ± 800 BP (P-228). Vertebrate remains occurring in a discontinuous organic layer on the upper SW slope of the plateau include: large bison, caribou, and 'a jaw, probably of a horse (*Equus*)' [if so, it would probably be older than Holocene].

898. MacNeish, R.S. (1955): Two archaeological sites on Great Bear Lake, Northwest Territories, Canada. National Museums of Canada Bulletin 136:54–84.
At the E end of Great Bear Lake, Northwest Territories (Plate I), a stratigraphic sequence of three cultural complexes has been found. The earliest (Franklin Tanks component) seems to have been contemporaneous with mammoth (*Mammuthus*) in that area, and is definitely contemporaneous with a beach and lake level of Great Bear Lake 35 ft higher than the present lake level. This earliest culture has affiliations with various sites farther S that have Plainview points (*Homo sapiens*), some of which have been dated. Radiocarbon dates for these sites range from about 8000 to 10,500 BP. Evidently, fragments of Plainview-like points and mammoth or mastodon (*Mammut americanum*) tusk fragments (Plate III, 2) have been picked up at the Franklin Tanks Site (N.W.T. 54) during a reconnaissance trip in 1951. The author notes (p. 59) that about 2600 BC, just as forest and tundra invaded the area, people hunted and killed the 'diminutive Polar Caribou' [Peary caribou (*Rangifer tarandus pearyi*)], an animal that at present is not in this area, and that is adapted to arctic and glacial conditions.

899. MacNeish, R.S. (1959): Men out of Asia; As seen from the northwest Yukon. Anthropological Papers of the University of Alaska 7(2):41–70.
The author suggests that the movements of humans (*Homo sapiens*) across the Bering Land Bridge was in both directions, for cultural advances occurred concurrently in eastern and in western Beringia. Archaeological studies of the Firth River site (Engigstciak) in northern Yukon seem to indicate this. Fossil pollen studies indicate that the area around Engigstciak was a grassy plain with trees in the valley flanks – drier and perhaps warmer than at present. Faunal remains (p. 48) support this interpretation, for besides the ever-present caribou (*Rangifer tarandus*) bones there were those of bison (*Bison*), wapiti (*Cervus elaphus*), and mountain goat (*Oreamnos americanus*) [reidentified as Dall sheep (*Ovis dalli*) – see Mackay et al. 1961]. Vertebrate remains from Flint Creek (p. 46) show that these people killed mainly bison, caribou, hare (*Lepus*), and a few Arctic birds (Aves).

900. Macpherson, A.H. (1965): The origin of diversity in mammals of the Canadian Arctic tundra. Systematic Zoology 14(3):153–73.
Seventeen modern Canadian Arctic tundra mammals share only six broad patterns of taxonomic variation and geographic distribution (see maps, Figures 3–18). The species discussed include polar bear (*Ursus maritimus*), arctic fox (*Alopex lagopus*), arctic hare (*Lepus arcticus*), collared lemming (*Dicrostonyx torquatus*), muskox (*Ovibos moschatus*), caribou (*Rangifer tarandus*), gray wolf (*Canis lupus*), ermine (*Mustela erminea*), masked shrew (*Sorex cinereus*), red-backed vole (*Clethrionomys rutilus*), brown lemming (*Lemmus sibiricus*), tundra vole (*Microtus oeconomus*), arctic ground squirrel (*Spermophilus parryii*), wolverine (*Gulo gulo*), red fox (*Vulpes vulpes*), and humans (*Homo sapiens*). Tundra species were able to maintain populations in the northern refugia (Beringia and Pearyland [but see Bennike 1999]) if the species was widespread (e.g., collared lemming), and only in the former if confined to western North America (e.g., brown lemming). Tree-line species (e.g., caribou) gave rise to woodland and tundra forms, the former from populations isolated S of the Laurentide ice, and the latter from populations trapped in the cooling northern refugia. Some species maintained themselves only in Beringia (e.g., humans), and others only S of the Laurentide ice sheets (e.g., wolverine). Two species, polar bear and arctic fox, find sea ice no barrier to distribution, and their present ranges are of little zoogeographical relevance.

901. Maddren, A.G. (1905): Smithsonian exploration in Alaska in 1904 in search of mammoth and other fossil remains. Smithsonian Miscellaneous Collections 49 (Publication 1584):1–117.
This report covers the 1904 expedition but also includes

observations on travels in 1899 along the Yukon River, in 1900 along the Bering Sea and Arctic Ocean, and in 1902–3 in Alaska. Although the author was primarily interested in proboscidean (Proboscidea) bones, he also collected bison (*Bison*), horse (*Equus*), bear (*Ursus*), and other large, unidentified mammal fossil remains. Part of his explorations included the Old Crow Basin, Yukon, where he collected many fossils (some of which were later lost when his canoe tipped over). The expedition was looking for the source of the many scattered bones, where they hoped to discover entire skeletons of Pleistocene mammals. This objective was unsuccessful. Proboscidean remains from Alberta and British Columbia are also mentioned (pp. 24–5).

902. Maher, W.J. (1968): Muskox bone of possible Wisconsin age from Banks Island, Northwest Territories. Arctic 21(4):260–6.
Tundra muskox (*Ovibos moschatus*) remains are: a metacarpal fragment from 72 km inland near the Bernard River at the confluence with a large unnamed tributary (73°23′N, 121°54′W) radiocarbon dated at >34,000 BP (S-288); a bone from Thesiger Bay, southern Banks Island indirectly dated at 10,660 ± 170 BP (GSC-240, on peat). Dispersal of this species as well as caribou (*Rangifer tarandus*), arctic fox (*Alopex lagopus*), and polar bear (*Ursus maritimus*) is suggested to be via sea ice. The possible existence of a Wisconsinan refugium on Banks Island is discussed. Evidence suggests that a refugium existed there, isolated from Beringia. In addition to plant species, two small mammal species [brown lemming (*Lemmus sibiricus*) and arctic hare (*Lepus arcticus*)] and possibly muskoxen survived in a Banks Island refugium. [See Harington 1990b.]

903. Mandrak, N.E., and Crossman, E.J. (1992): Postglacial dispersal of freshwater fishes into Ontario. Canadian Journal of Zoology 70:2247–59.
The present distributions of 117 native freshwater fishes (Pisces) of Ontario have been shaped by processes active following the Wisconsinan glaciation. During that period, these species survived in unglaciated refugia. Refugial origins of 91 species were resolved using a 'refugial index' [see map, Figure 3, summarizing their refugial origins and dispersal routes]. Seventy-two species resided in the Mississippian refugium, 13 species in the Atlantic Coastal refugium, four species in dual Atlantic Coastal–Mississippian refugia, one species in a Missourian refugium, and one species in all three refugia. Most species with similar distributions in Ontario shared the same refugia and dispersal routes in eastern North America, so the authors hypothesize that historical processes were important in shaping present ranges of Ontario freshwater fishes.

904. Mandryk, C.A. (1990): Could humans survive the ice-free corridor?: Late-glacial vegetation and climate in west central Alberta. *In*: Megafauna and Man: Discovery of America's Heartland (L.D. Agenbroad, J.I. Mead, and L.W. Nelson, eds.). Mammoth Site of Hot Springs, South Dakota Inc. Scientific Papers 1:67–79.
A radiocarbon-dated sediment core from Mitchell Lake located 18 km SW of Rocky Mountain House, Alberta, provides paleoenvironmental data from at least 17,960 BP. It is unlikely that the initial cold, semi-arid *Artemisia* steppe vegetation could have supported a fauna capable of sustaining human (*Homo sapiens*) populations. During the birch zone, however, particularly by 14,000 BP, during deglaciation and widening of the corridor, primary productivity would have increased greatly and could have supported increasingly viable flora and fauna.

905. Mandryk, C.A.S., Josenhans, H., Fedje, D.W., and Mathewes, R.W. (2001): Late Quaternary paleoenvironments of northwestern North America: Implications for inland versus coastal migration routes. Quaternary Science Reviews 20(2001):301–4.
Results of recent studies of glacial geology and ancient sea levels support the possibility of coastal migration from Beringia to the Pacific Northwest between about 14,000 and 10,000 BP, and preliminary paleoecological data suggest that the coastal landscape was in part vegetated, and probably able to support a terrestrial fauna, including people (*Homo sapiens*). The same cannot be said about the 'ice-free corridor.' Major portions of the corridor were blocked from 30,000 to at least 11,500 BP, and even when land was exposed by about 18,000 to 13,000 BP, the environment was nonviable (see graph of estimated nutritional viability of the corridor, Fig. 3).

906. Mann, D., Peteet, D., Reanier, R., Kunz, M., and Durand, S. (1997): Late-glacial changes in an arctic landscape: The northern flank of the Brooks Range

during Paleoindian occupation. Beringian Paleoenvironments Workshop (September 20–3, Florissant, Colorado). Program and Abstracts. pp. 101–2.
Paleoindian people (*Homo sapiens*) occupied the Mesa site in the northern foothills of the Brooks Range, Alaska, between 10,300 and 9700 BP. An earlier occupation may have occurred about 11,700 BP. As part of the interdisciplinary Mesa Project, the authors are reconstructing the history of ecosystem changes in the region for the 14,000–8000 BP interval. They note that the rapid spread of tussock tundra after 10,000 BP 'may have been the proximate cause for megafauna and Paleoindians leaving the region.' They suspect that Paleoindians used the North Slope of Alaska during warm intervals during the late glacial when firewood had become available yet modern tussock tundra was of limited extent.

907. Martin, L.D. (1979): The biostratigraphy of arvicoline rodents in North America. Transactions of the Nebraska Academy of Sciences 7:91–100.
Arvicoline rodents are of considerable biostratigraphic use in North America for the correlation of continental sediments of Blancan age or younger. The Sappa Formation (1.2 Ma) contains the earliest-dated record of a *Microtus*-like microtine (*Allophaiomys*) from North America. This arvicoline may be an immigrant from Eurasia that entered North America in the Early Pleistocene. If so, this species could extend the use of arvicoline fossils for international correlations as well. The earliest known lemmings (*Lemmus* and *Dicrostonyx*) in North America are mentioned.

908. Martin, P.J. (1982): Digestive and grazing strategies of animals in the arctic steppe. *In*: Paleoecology of Beringia (D.M. Hopkins, J.V. Matthews, Jr, C.E. Schweger, and S.B. Young, eds.). Academic Press, New York. pp. 259–66.
The author speculates on the physiological and digestive strategies of large herbivores of the arctic steppe by examining the strategies used by species that appear to be modern analogues of these animals. Horses (*Equus*) and elephants (Proboscidea) are monogastrics with large caecums that are able to exist on low-protein, high-fibre forages by consuming large quantities per unit of body weight, and by using microbial protein from caecal fermentation. Ruminants, such as bison (*Bison*), can consume less feed per unit of body weight than horses or elephants, and hence require forage somewhat higher in protein and lower in fibre. Herbivores use physiological and behavioural adaptations to survive in very cold weather. Selection would likely favour large animals that could grow rapidly and attain large body size.

909. Martin, P.S. (1973): The discovery of America. Science 179:969–74.
This paper proposes that people (*Homo sapiens*) invaded the New World toward the end of the last (Wisconsinan) ice age and spread southward in such explosive numbers to attain sufficient density to wipe out much of their prey. The author supports the hypothesis of the massive – to the point of extinction – exploitation of unwary and hence vulnerable species unused to human hunting strategies.

910. Martin, P.S. (1982): Pattern and meaning of Holarctic mammoth extinctions. *In*: Paleoecology of Beringia (D.M. Hopkins, J.V. Matthews, Jr, C.E. Schweger, and S.B. Young, eds.). Academic Press, New York. pp. 399–408.
Climatic change at the time of the last (Wisconsinan) glaciation is unlikely to have been the primary cause of megafaunal [including mammoth (*Mammuthus*)] extinctions. The mammoth fauna had survived similar periods of climatic stress. In western Europe, mammoth extinction apparently preceded the disappearance of steppe tundra, the mammoth's supposed habitat. The global pattern of all extinctions of large land mammals appears to follow Paleolithic people's footsteps. The loss of many large herbivores in North and South America, apparently suddenly, between 12,000 and 10,000 BP is consistent with this pattern. The presence of a hunting culture in America during that period is accepted by archaeologists. Computer-modelling of the needs of a hunting population growing at only 1% per year and advancing southward from the Bering Isthmus suggests that mammoth extinction from human (*Homo sapiens*) predation alone could have occurred in less than 500 years.

911. Martin, P.S., and Klein, R.G. (eds.) (1984): Quaternary Extinctions: A Prehistoric Revolution. University of Arizona Press, Tucson. 892 pp.
An important and extensive treatment of Quaternary extinctions on a global basis. [For an Alaskan view, for example, see Guthrie 1984c.]

912. Mason, O.K., Bowers, P.M., and Hopkins, D.M. (2001): The Early Holocene Milankovitch thermal maximum and humans: Adverse conditions for the Denali complex of Eastern Beringia. Quaternary Science Reviews 20(2001):525–48.
Calibration of radiocarbon ages has forced the revision of paleoecological relationships with the Milankovitch thermal maximum of 10,000–9000 cal [calibrated radiocarbon years] BP. In Eastern Beringia, paleogeographic factors had dramatic consequences for people (*Homo sapiens*) – many represented by the Late Pleistocene–Early Holocene (13,000–7000 cal BP) Denali complex. Calibration of 71 radiocarbon ages associated with Denali occupations [see map, Fig. 1] indicates that the earliest precede both the Milankovitch Maximum and the spread of spruce in interior Alaska by more than 2000 years. Denali occupations peaked during the worldwide temperature decline between 8500 and 8000 cal BP – equivalent to the 'Mesoglacial' or 'younger Younger Dryas' [see histogram of calibrated radiocarbon ages for the Denali complex in Eastern Beringia, Fig. 2]. Fewer sites date to the Milankovitch Maximum, suggesting an inverse relationship between warmer climate and human population. Considerable speculation accompanies discussions of the hunting preferences and capabilities of Denali people. Microblades are an ideal weapon for wounding caribou (*Rangifer tarandus*), especially if used with an atlatl and inset in burin-grooved caribou antler. Caribou were widespread in interior Alaska during the Little Ice Age, but the postglacial appearance of moose (*Alces alces*) in Beringia occurs with the first evidence of people (*Homo sapiens*) – which could involve a culinary interest. Perhaps the disappearance of bison (*Bison*) and Denali microblades before 2759–1837 cal BP implies a causal linkage. Also, fluted-point-using, but not Denali, people evidently preyed on mammoths (*Mammuthus*). Of the many radiocarbon dates listed for Denali sites in Eastern Beringia (Table 2), two on caribou bone from Tingmeakpuk (KIP-273) in the Brooks Range of Alaska are provided: 8435 ± 85 BP (Beta-49163) and 7060 ± 70 BP (Beta-49166).

913. Mason, R.J. (1960): Early man and the age of the Champlain Sea. Journal of Geology 68:366–76.
The author reinterprets the geological sequence in the Ontario Basin and St Lawrence valley of Quebec and Ontario that seems to be supported by radiocarbon and geological data. He suggests that available Champlain Sea radiocarbon dates are substantially correct, that Glacial Lake Iroquois is older than previously thought, and that this reinterpretation is more compatible with Paleoindian (*Homo sapiens*) archaeological sequences outside this region. This view allows Paleoindians entry into the Champlain Sea and Glacial Lake Iroquois areas by 11,000 to 10,500 BP.

914. Matheus, P.E. (1994): Pleistocene mammals. Alaska Geographic 21(4):54–69.
This popular article notes how physical and biological processes shaped the ecology of Alaska's mammal community through the Late Pleistocene and Early Holocene, setting the stage for the modern ecosystems. A table on p. 56 lists (using common and Latin names and designating their supposed habitat requirements) Alaska's ice age mammals (separating herbivores and carnivores) under the headings Early and Middle Pleistocene Mammals (Pre-Wisconsinan) and Late Pleistocene Mammals (Wisconsinan). Based on their abundance as fossils, mammoth (*Mammuthus*), bison (*Bison*), and horse (*Equus*) dominated the Mammoth Steppe. Pages 59, 60, and 61 provide a graphic restoration of the mammals of Alaska's Mammoth Steppe during the Late Pleistocene, with a key to their identification. [It should be noted that the giant ground sloth (*Megatherium*), depicted in the restoration on p. 65, has never been reported from Eastern Beringia: the smaller Jefferson's ground sloth (*Megalonyx jeffersonii*) is the only ground sloth so far known from the region.] [See McDonald et al. 2000.]

915. Matheus, P.E. (1995): Diet and co-ecology of Pleistocene short-faced bears and brown bears in Eastern Beringia. Quaternary Research 44:447–53.
Carbon and nitrogen stable-isotope analysis of fossil bone collagen reveals that Pleistocene short-faced bears (*Arctodus simus*) of Beringia were highly carnivorous, while contemporaneous brown bears (*Ursus arctos*) had highly variable diets that included varying amounts of terrestrial vegetation, salmon, and small amounts of terrestrial meat. A reconsideration of the short-faced bear's highly derived morphology indicates that they foraged as scavengers of widely dispersed large mammal carcasses and were simultaneously designed both for highly efficient locomotion and for intimidating other large carnivores. This allowed *Arctodus* to forage economically over a large home range and seek out, procure, and de-

fend carcasses. The isotope data and this reconstruction of *Arctodus*'s foraging behaviour refute the hypothesis that competition from brown bears was a significant factor in the extinction of short-faced bears. Sixteen short-faced bear and 14 brown bear fossils were analysed along with 13 recent brown bear bones from two modern populations. Radiocarbon dates are available for two of the *Arctodus* samples [CMN 7438, cranium and teeth from Gold Run Creek, Yukon, 26,040 ± 270 BP (TO-2696), and CMN 37577, humerus from Hunker Creek, Yukon, 29,600 ± 1200 BP (I-11037)].

916. Matheus, P.E. (1997): Paleoecology and ecomorphology of the giant short-faced bear in Eastern Beringia. PhD thesis, University of Alaska Fairbanks. 282 pp. (UMI Dissertation Services, Ann Arbor, Michigan, 1999).

The short-faced bear (*Arctodus simus*) was a widespread Tremarctine bear indigenous to North America until its extinction around 11,500 BP. *Arctodus* inhabited Pleistocene ice-free refugia in Eastern Beringia (the northwestern limit of its range) until at least 20,000 BP. Because of its gracile, long-legged build and extremely large size, most paleontologists believe this bear was a high-speed-pursuit predator that had preyed on the largest herbivores of Pleistocene North America. Alternatively, energetic arguments have been used to suggest that *Arctodus* was too large to be carnivorous and evolved its large size within a herbivorous or omnivorous niche. To test these competing hypotheses, the author reconstructed aspects of *Arctodus*' trophic position and paleodiet by analysing stable isotope ratios (δ^{13}C and δ^{15}N) in bone collagen extracted from Eastern Beringian fossils. Other bears (*Ursus*) and carnivores (Carnivora) from Beringia were analysed to help interpret the results. Isotopes reveal that *Arctodus* was highly carnivorous: it fed on herbivores that consumed C_3 vegetation, and it did not eat salmon. The herbivore/omnivore hypothesis is thus rejected. Predatory hypotheses predict that we should find certain morphological features in a predatory bear that would enhance one or more of the following skills: top running speed, acceleration, or manoeuvrability at high speeds. Postcranial morphology of *Arctodus* was reanalysed and data on running speed and bone strength in other large mammals were used to show that a bear the size of *Arctodus* with long, gracile limbs would not have been able to endure the extreme dynamic forces incurred during predatory activities. Instead, *Arctodus*' morphology and body size indicate that it had evolved to maximize locomotor efficiency using a pacing gait. It is suggested that *Arctodus* evolved as a specialized scavenger adapted to cover an extremely large home range in order to seek out, procure, and defend large-mammal carcasses from other carnivores. By modelling herbivore populations and their mortality, the author shows that enough carcass biomass was being produced in Pleistocene Beringia to make this scavenging niche energetically feasible. The model helps show that *Arctodus*' extinction probably is best tied to a reduction of year-round carcasses on the landscape, a condition that arose in the Holocene when the herbivore fauna became less diverse and began to experience more seasonal mortality.

917. Matheus, P., Payton, W., and Kunz, M. (1999): Isotope ecology of Late Quaternary mammals in Eastern Beringia. Canadian Archaeological Association, 32nd Annual Conference (April 28–May 2, Whitehorse, Yukon). Programme and Abstracts. pp. 47–8.

These are initial findings of a long-term study of paleodiets, paleoecology, paleoclimate, and chronology of Late Quaternary mammals in Eastern Beringia. Stable-isotope levels (δ^{13}C and δ^{15}N) were measured in purified collagen extracted from bones of 12 species of Late Quaternary mammals – mainly from central and northern Alaska. The difference between the two isotopic levels can help to show difference in the ecology, physiology, and diets of each species, as well as for reconstructed predator–prey relationships. Pleistocene caribou (*Rangifer tarandus*) and muskox (*Ovibos moschatus*) display values indicative of typical tundra diet, like their modern counterparts. Bison (*Bison*) data suggest a homogeneous diet consisting of mostly C_3 grasses. High levels of δ^{15}N in mammoths (*Mammuthus*) indicate that Late Pleistocene ecosystems were arid. Preliminary data suggest that: scimitar cats (*Homotherium serum*) preyed mainly on mammoths; American lions (*Panthera leo atrox*) preyed mainly on steppe bison (*Bison priscus*) and perhaps muskoxen and caribou, but little on horses (*Equus*) or mammoths; wolves (*Canis lupus*) probably preyed on caribou and muskoxen; and short-faced bears (*Arctodus simus*) and wolverines (*Gulo gulo*) fed eclectically on a variety of herbivores. Fifty new AMS radiocarbon dates from steppe bison bone from Alaska's Arctic Slope indicate continuous presence between 40,000 and 10,000 BP with a possible hiatus between 16,000 and 13,000 BP. The first Holocene bison (6400 BP) from the Arctic Slope was detected.

918. Mathews, W.H. (1978a): The Hudson Hope mammoth of 1977. Discovery 7(2):40–3.
Remains of a mammoth (*Mammuthus*) (cranium – the only specimen found in place; tusk fragment; jaw and two molars; a vertebra; parts of each leg had been collected in the spoil at the foot of the face a few feet farther W) were found in a gravel pit for British Columbia Hydro's 'Site One' dam, a few kilometres upstream from Hudson Hope, British Columbia, on the Peace River. Presumably the whole mammoth had been buried within the gravel, but excavation had removed all but the cranium. This is an interglacial deposit as wood from the gravel was radiocarbon dated at >40,000 BP [no lab number]. A leg bone had a small fragment of dried flesh and dark brown hair attached. Another mammoth tooth from the same deposit, but at Taylor, British Columbia, 80 km downstream from Site One, was radiocarbon dated at 27,400 BP [see Mathews 1978b].

919. Mathews, W.H. (1978b): Quaternary stratigraphy and geomorphology of Charlie Lake (94A) map-area, British Columbia. Geological Survey of Canada Paper 76-20:1–24.
The author mentions the following Pleistocene mammal remains: a woolly mammoth (*Mammuthus primigenius*) tooth was found buried 15–18 m deep in a postglacial terrace of the Ostero Gravel Pit at Taylor, British Columbia (56°09′00″N, 120°42′30″W). A radiocarbon date on it was 27,400 ± 580 BP (GSC-2034, apatite). From this same locality were found: a bison (*Bison* cf. *Bison priscus*) maxilla with broken teeth buried at the same level as the mammoth; and moose (*Alces alces*) remains, buried 3 m deep in gravel deposits. A mammoth (*Mammuthus*) tusk was found in a kame moraine in Rocky Mountain Portage, 13 km W of Hudson Hope, British Columbia (56°02′00″N, 121°54′30″W) and was radiocarbon dated at 11,600 BP (I-2244A) [correction of an earlier date of 7670 ± 170 BP (I-2244)]. The second sample was limited in size; hence there is no estimate of counting provision. Sheep (*Ovis*) remains were found in fluvial gravels near the Finlay Forks area 112 km W of Hudson Hope. [See Clague 1980.]

920. Matthew, G.F. (1879): Report on the superficial geology of southern New Brunswick, 1878. Geological Survey of Canada Report of Progress 1877–1878: 1–36EE.
The author reports (p. 23EE) that a 'right' whale (*Beluga*) [probably meant *white* whale (*Delphinapterus leucas*) – see Harington 1977b] worn lower jaw was found in Leda Clay at the mouth of the Pocologan River, New Brunswick [actually near the mouth of the Pocologan River, Charlotte County 45°07′N, 66°35′W] and mentions that it had 'lost a great part of the gelatine once contained in its pores.' A harp seal (*Phoca groenlandica* = *Phoca Gröenlandica*) partial skeleton was found in organic black sand buried 8 m at Hopkins Brickyard in St John (= Fairville), New Brunswick [45°20′N, 65°50′W]; its skull and limbs were virtually complete, but the specimen was later lost in a fire in June 1877. [See Miller 1990.]

921. Matthews, J.V., Jr (1968): A paleoenvironmental analysis of three Late Pleistocene coleopterous assemblages from Fairbanks, Alaska. Quaestiones Entomologicae 4:202–24.
This paper deals mainly with Pleistocene insect fossils from the Eva Creek exposure near Fairbanks Alaska. Some of these are compared with associated mammal remains (p. 214) for biostratigraphic and paleoecological implications, which include: mammoth (*Mammuthus*), horse (*Equus*), camel (*Camelops*), steppe bison (*Bison priscus*), caribou (*Rangifer tarandus*), [probably Dall sheep (*Ovis dalli*) rather than snow sheep (*Ovis nivicola*)], tundra muskox (*Ovibos moschatus*), lemming (*Dicrostonyx*), and singing vole (*Microtus gregalis*). The beetle, pollen, and mammal fossils lead to the conclusion that at the time of their deposition, the paleoenvironment of lowland interior Alaska was similar to alpine tundra, a higher elevation or coastal tundra in other parts of Alaska. The environmental differences of the three main insect assemblages seem to indicate local differences rather than changes in the macroclimate of interior Alaska.

922. Matthews, J.V., Jr (1970): Quaternary environmental history of Interior Alaska: Pollen samples from organic colluvium and peats. Arctic and Alpine Research 2(4):241–51.
This paper suggests the importance of pollen samples from the thick Pleistocene frozen silts of interior Alaska

in reconstructing the paleoenvironment of the Late Pleistocene. This evidence, along with similar evidence from fossil mammals (p. 248) and insects, shows that an arctic environment once prevailed in large areas of lowland interior Alaska, unlike the spruce woodlands of today.

923. Matthews, J.V., Jr (1974): Wisconsin environment of interior Alaska: Pollen and macrofossil analysis of a 27 meter core from the Isabella Basin (Fairbanks, Alaska). Canadian Journal of Earth Sciences 11(6):828–41.

Information on the Wisconsinan environment of interior Alaska has been obtained from study of pollen as well as the plant and animal macrofossils contained in the sediments of a 27-m core taken at Isabella Creek, N of Fairbanks, Alaska (see map, Figure 1). Sediments from the core range in age from 34,900 BP to 4510 BP. The lower portion of the sequence contains the first botanical and zoological evidence from interior Alaska of a Mid-Wisconsinan interstadial. Following this interstadial, in Late Wisconsinan time, a cold, treeless steppe-tundra environment existed in the lower portions of the Isabella Basin. The author (p. 835) states that, based on fossil insects and vertebrates, lowland areas near Fairbanks were essentially treeless during cold periods of the Late Pleistocene. Vertebrate fossils particularly imply that such lowland treeless environments were more steppe-like than present Alaskan tundra. Around 8500 years ago spruce-deciduous woodland once again occupied the area. Based on data obtained from plant macrofossils, the present Alaska flora, like the present mammalian fauna, is impoverished compared with that of the Pleistocene.

924. Matthews, J.V., Jr (1976): Arctic-steppe – An extinct biome. American Quaternary Association, Abstracts of the Fourth Biennial Meeting (October 9–10, Tempe, Arizona). pp. 73–7.

This is an eclectic review of evidence indicating that the Late Pleistocene environment of Eastern Beringia was highly productive – the eastern end of a huge grassland biome, unequalled today in size or character. It has been called periglacial steppe, tundra steppe, and steppe tundra; however, the term arctic steppe seems more appropriate since it reflects both the regional extent of the environment as well as its dissimilarity to contemporary tundra. Only two ungulates, caribou (*Rangifer tarandus*) and muskox (*Ovibos moschatus*), live on the tundra, and their standing crop is relatively low. In contrast, arctic steppe was characterized by at least seven ungulates, most known or presumed to have been grazers. Steppe bison (*Bison priscus*) and woolly mammoth (*Mammuthus primigenius*) dominated. Caribou and muskox were less frequent and shared the ungulate niche with such nontundra species as wapiti (*Cervus elephas*), Dall sheep (*Ovis dalli*), yak (*Bos*), saiga antelope (*Saiga tatarica*), and possibly camel (Camelidae). The author estimates that the productivity of this biome was more than 50 times that of contemporary tundra [but see Matthews 1982: 'comparisons with the fauna of the African plains are inappropriate']. Because of the high nutritive quality of arctic plants, arctic steppe may have supported ungulates at a much lower level of primary productivity than exists in regions with an analogous fauna. The presence of large and small mammals etc. that could not live in boggy terrain suggests that arctic steppe also differed from tundra by its better-drained substrate. Arctic steppe climate was 'arctic' in terms of mean annual temperature, but because of its continentality, summers may have been relatively warm and dry. Only during the Late Pleistocene did northern treeless communities take on attributes of arctic steppe. In Alaska-Yukon, the decline of the arctic steppe is signalled by an abrupt increase of shrub birch 14,000 to 12,000 BP. Steppe conditions persisted in some regions, possibly due to the feedback effect of the large-mammal community – some members apparently surviving until 10,000 BP. By 8500 BP the arctic steppe biome had disappeared. Humans (*Homo sapiens*) were part of the Wisconsinan arctic steppe, and its demise may be a function of their contribution to extinction of the large ungulates as their range (and gene pools) waned as forests and tundra expanded. Behaviour and distribution of living mammals related to arctic steppe species is informative. For example, the social organization of saiga antelopes suggests that the few Alaskan and Canadian saiga fossils greatly underrepresent its former abundance. [See Matthews (1982) for an updated review.]

925. Matthews, J.V., Jr (1982): East Beringia during Late Wisconsin time: A review of the biotic evidence. *In*: Paleoecology of Beringia (D.M. Hopkins, J.V. Matthews, Jr, C.E. Schweger, S.B. Young, and V. Stanley, eds.). Academic Press, Toronto. pp. 127–50.

This comprehensive review provides pollen, plant, insect, and vertebrate fossil evidence that has been used to determine the nature of the Late Pleistocene environment of Beringia. Several lines of evidence show that the number of large-mammal species existing together in Eastern Beringia during the Late Pleistocene was several times larger than that of the existing tundra fauna. Prominent were: woolly mammoth (*Mammuthus primigenius*), steppe bison (*Bison priscus*), horse (*Equus*), several types of muskox [the tundra muskox (*Ovibos moschatus*) and helmeted muskox (*Bootherium bombifrons* = *Symbos*)], and caribou (*Rangifer tarandus*). 'In spite of low pollen production in Late Pleistocene plant communities, sufficient forage must have existed to support this unique ungulate community.' The most probable reconstruction is of a treeless region composed of a mosaic of communities, among which were large tracts that can only be termed steppe-like. The Late Wisconsinan is viewed as the last time when presently disjunct steppe species in Russia and Eastern Beringia had continuous ranges. Figure 2 (finite radiocarbon dates, mainly on bone collagen, of Quaternary mammals from Alaska and Yukon) was of critical importance in demonstrating which species lived in Eastern Beringia during the cold peak of the last (Wisconsinan) glaciation.

926. Matthews, J.V., Jr, Harington, C.R., Hughes, O.L., Morlan, R.E., Rutter, N.W., Schweger, C.E., and Tarnocai, C. (1987): Schaeffer Mountain Lookout and Old Crow Basin stratigraphy/paleontology. *In*: INQUA Excursion Guide Book A-20(a) and (b), Quaternary research in the Yukon (S.R. Morison and C.A.S. Smith, eds.). pp. 75–83.
This paper, specially produced for the INQUA Conference, is provided as an overview of Old Crow Basin, Yukon. It includes a map, illustrations of the stratigraphy and geology, a list of fossils found and a brief discussion of the archaeological aspects of the region.

927. McAllister, D.E. (1963): Fish remains from a 600-yr-old Yukon archaeological site. Canadian Field-Naturalist 77(4):232.
Remains attributable to northern pike (*Esox lucius*), lake trout (*Salvelinus namaycush*), and burbot (*Lota lota*) were found in Bennett Lake cultural phase deposits at the Little Arm site at Kluane Lake, Yukon.

928. McAllister, D.E., and Harington, C.R. (1969): Pleistocene grayling *Thymallus*, from Yukon, Canada. Canadian Journal of Earth Sciences 6(5):1185–90.
Eleven arctic grayling (*Thymallus arcticus*) scales were found in the southern bank of the Porcupine River, 9.7 km SSW of Old Crow (67°28′N, 139°54′W) (Figure 1). The scales were from the same horizon as freshwater shells dated at 32,400 ± 770 BP (GSC-952) and wood dated >37,000 BP (GSC-958) and provide the first fossil record of *Thymallus* for North America. Furthermore, the scales confirm previous suggestions that arctic grayling survived the Wisconsinan glaciation in the Beringian refugium. Associated faunal and floral evidence suggests that the graylings lived in a cool, shallow lake with wet meadow habitat and coniferous trees nearby. Photographs of the specimen scales and modern analogues are included.

929. McAllister, D.E., and St-Onge, D. (1981): Postglacial fossil fishes from Coppermine River, Northwest Territories, Canada. Canadian Field-Naturalist 95:203–5.
Fish remains were found in freshwater sandy silt deposits in the valley of the Coppermine River (66°00′N, 115°55′W). The fish, which were found at the uppermost level of the ancient lake, were: arctic charr (*Salvelinus alpinus*), arctic grayling (*Thymallus arcticus*), northern pike (*Esox* cf. *E. lucius*), and slimy sculpin (*Cottus cognatus*). At another major section at Quicksand Creek (66°49′00″N, 116°21′00″W), which may correlate with the first section, a piece of willow from the base of the channel and some peat from the top of the channel were dated at 8400 ± 80 BP (GSC-2959) and 3210 ± 60 BP (GSC-2998), respectively. Therefore, the estimated date of the fossil-bearing deposit was 8400–9000 BP based on the maximum level of Glacial Lake Coppermine.

930. McAllister, D.E., Cumbaa, S.L., and Harington, C.R. (1981): Pleistocene fishes (*Coregonus, Osmerus, Microgadus, Gasterosteus*) from Green Creek, Ontario, Canada. Canadian Journal of Earth Sciences 18(8):1356–64.
Specimens of cisco (*Coregonus* cf. *Coregonus artedii*) (CMN 36127) and Atlantic tomcod (*Microgadus tomcod*) (CMN 35800) are reported for the first time from calcareous nodules at Green Creek. The occurrence of rainbow smelt (*Osmerus mordax*) (CMN 35801 and CMN 36097) and fully plated threespine stickleback (*Gasterosteus aculeatus*) (CMN 36610) is confirmed. Analysis of

the 10 species of fishes now known of Late Wisconsinan or early postglacial age from this region suggests the following not necessarily contemporaneous paleoenvironments: oligotrophic lakes; a cool, deep marine bay with gravel beaches; and a brackish estuary fed by creeks. Photographs of cisco (CMN 36127) and rainbow smelt (CMN 35801) are included.

931. McAllister, D.E., Harington, C.R., Cumbaa, S.L., and Renaud, C.B. (1988): Paleoenvironmental and biogeographic analyses of fossil fishes in peri-Champlain Sea deposits in eastern Canada. *In*: The Late Quaternary Development of the Champlain Sea Basin (N.R. Gadd, ed.). Geological Association of Canada Special Paper 35:241–58.
The occurrence of 11 species of fishes is documented for peri-Champlain Sea post-Pleistocene deposits. The spoonhead sculpin (*Cottus ricei* – previously misidentified as *Artediellus uncinatus*) and longnose sucker (*Catostomus catostomus*) are reported from Champlain Sea deposits for the first time with additional records of lake cisco (*Coregonus artedii*), Atlantic cod (*Gadus morhua*), and Atlantic tomcod (*Microgadus tomcod*). The biogeographic and environmental significance of the ichthyofauna is discussed. The mean latitude of the ranges of the peri-Champlain Sea species is in southern Labrador, more than 8° farther N than during Champlain Sea times. This implies that the Champlain Sea environment was cooler than it is at present. Habitat analysis of peri-Champlain Sea fishes suggests four main ecological groupings: (1) those that occupy cool lakes or streams throughout their lives (five freshwater species, only one of which never enters seawater); (2) anadromous fishes (two species); (3) predominantly euryhaline fishes (one species); and (4) more strictly marine fishes (three species). From a biogeological viewpoint, the fossils suggest that the lake cisco, lake charr (*Salvelinus namaycush*), longnose sucker, spoonhead sculpin, and deepwater sculpin (*Myoxocephalus thompsoni*) might have survived in a freshwater refugium on the Atlantic Coastal Plain E of the Appalachians. The rainbow smelt (*Osmerus mordax*), Atlantic cod, Atlantic tomcod, lumpfish (*Cyclopterus lumpus*), and threespine stickleback (*Gasterosteus aculeatus*) survived on the western N Atlantic coast – either S of the ice sheet (most likely for the anadromous species, smelt and stickleback) or either offshore of or S of the ice sheet. The article contains excellent photographs of most of the fossils.

932. McAndrews, J.H. (1982): Holocene environment of a fossil bison from Kenora, Ontario. Ontario Archaeology 37:41–51.
A bison skull (Manitoba Museum of Man and Nature V-780) found in a bog near Hayes Lake, Ontario, not far from Kenora, is tentatively identified as steppe bison (*Bison priscus = Bison crassicornis*). Studies were made of the sediment associated with the bone to detail its composition, radiocarbon age, fossil pollen, and plant macrofossils. The sediment found inside the skull was radiocarbon dated at 4850 ± 60 BP. The skull is therefore identified as Mid-Holocene in age. This suggests that bison were present in Ontario during the interval of the Hypsithermal woodland (9200 to 3600 years ago) and could have been available as prey to late Paleoindian and Archaic hunters. The identification of the skull is acknowledged as questionable. [Certainly this skull needs reexamination, and a radiocarbon date based on the bone would test that based on the sediment.]

933. McAndrews, J.H. (1986): Jumbo and other Ontario elephants. Royal Ontario Museum Archaeological Newsletter 2(16):1–4.
Mammoth (*Mammuthus*) bones with a tusk were found at Poole Farm, Rostock, Ontario, and are estimated at 10,790 BP. Other mammoth and mastodon (*Mammut americanum*) finds in Ontario ranging from 125,000 to 10,000 BP are discussed.

934. McAndrews, J.H. (1994): Pollen diagrams for southern Ontario applied to archaeology. *In*: Proceedings of Great Lakes Archaeology and Paleoecology (R.I. MacDonald, ed.). Symposium presented by the Quaternary Sciences Institute at University of Waterloo, 21–3 September, 1991. pp. 179–95.
Pollen diagrams from the Late Pleistocene (Rostock Mammoth site) and from the Holocene (Hams Lake site) were selected as regional postglacial standards because these sites have continuous sedimentation, good pollen preservation, and abundant radiocarbon dates. Also the sites are centrally located along the boundary between the deciduous and mixed-forest regions. The pollen diagrams reflect the vegetational changes and hence climatic changes. The Rostock Mammoth site is a swale in ground moraine containing pond sediment overlain by 30

cm of peat. The bones were found near the surface of the pond layer with no associated artifacts [see McAndrews and Jackson 1988]. The pollen diagram indicates that the Holocene (zone 2) is contained in the peat layer, and the pond layer contains the spruce zone (zone 1). Three subzones are distinguished in zone 1: 1p is dominated by jack/red pine pollen; 1a has high sedge pollen counts; and 1b contains predominantly spruce pollen. Zone 1p represents periglacial desert. Zone 1a indicates a warming trend beginning around 13,000 years ago that allowed tundra sedges, willows, birch, and poplar to invade the periglacial desert. Continued warming caused spruce to spread over the landscape to form a boreal woodland suitable for browsing mastodon (*Mammut americanum*) and grazing mammoth (*Mammuthus*) and caribou (*Rangifer tarandus*) (zone 1b). Other southern Ontario pollen analyses of mud surrounding mammoth and mastodon bones, including a recent examination of the Highgate Mastodon site; all confirm the period of their lives to be within zones 1a or 1b.

935. McAndrews, J.H., and Jackson, L.J. (1988): Age and environment of Late Pleistocene mastodon and mammoth in southern Ontario. *In*: Late Pleistocene and Early Holocene Paleoecology and Archaeology of the Eastern Great Lakes Region (R.S. Laub, N.G. Miller, and D.W. Steadman, eds.). Bulletin of the Buffalo Society of Natural Sciences 33:161–72.
This paper lists 27 postglacial [15,000 BP to present] mammoth (*Mammuthus*) and 61 mastodon (*Mammut americanum*) finds in southern Ontario. Some radiocarbon dates and a map of the sites are given. The mammoths include: the Rostock 4290 ± 120 BP (WAT-945, bone collagen) [questionable date] and 10,790 ± 150 BP (WAT-999, tusk collagen); the Muirkirk 8310 ± 200 BP (Beta-17869, bone collagen). The mastodons include: the Rodney 11,400 ± 450 BP (S-29, muck) and 12,000 ± 500 BP (S-30, muck); the Ferguson 6250 ± 240 BP (S-16, gyttja in skull) and 8910 ± 150 BP (GSC-614, bone collagen) [questionable dates]; the Perry 12,000 ± 200 BP (S-172, plant matter) and 11,860 ± 170 BP (GSC-211, plant matter); the Verbeke 10,000 ± 140 BP (BGS-219, wood) and 9830 ± 140 BP (BGS-228, muck); and the Thamesville 11,380 ± 170 BP (GSC-611, bone collagen). Other undated Ontario mastodon sites include: Ingersol, Harwich, Blenheim, Wellandport, Welland, Bothwell, Dutton, and Point Edward. The authors theorize that the young dates (<9000 BP) of some specimens represent Holocene persistence of some small populations before they became extinct.

936. McAtee, C. (1979): Late-Quaternary chronology, vegetation, and depositional history of Klondike 'muck' deposits, Yukon Territory, Canada. Northern Affairs Program, Indian Affairs and Northern Development, Contract #Y8MG-11. Final Report. 58 pp.
The author has obtained information on Late Quaternary environments of the Klondike region by study of pollen and spores contained in 'muck' deposits of Dominion and Hunker creeks near Dawson City, Yukon. The muck sequences were probably deposited during a Mid-Wisconsinan interstadial that had a climate warmer than that of the present (or less likely, during Sangamonian interglacial times). McAtee mentions bones of more than 10 species of Pleistocene mammals that are commonly found at the base of the muck (Harington and Clulow 1973), but notes that 'they are also deposited in the silts or with the gravel layers, as seen in Ken Tatlow's 1978 stripping operation on Hunker Creek' (Figure 1). In the section at Tony Kosuta's placer operation on the left limit of lower Hunker Creek (Figure 1), the author notes 'two fossil mammal bones' in place in sandy silt at a depth of 330 cm, and 'fossil mammal leg bones' in the top part of a sand and gravel unit about 650 cm deep (p. 20). Of particular interest is the pollen analysis (Table 9) of an arctic ground squirrel (*Spermophilus parryii*) nest and droppings from the Miben Section at the confluence of Last Chance Creek and Hunker Creek (64°01′N, 139°06′W, 425 m asl) from the base of Unit 1 (clayey silt – see p. 24 for section): Gramineae 73.5%; *Artemisia* 14.6%; Chenopodiineae 7.4%; *Salix* 2.6%; Compositae 1.0%; unknown 0.6%; *Betula* 0.3%; *Botrychium* 0.3%; *Lycopodium* present; *Sphagnum* present; indeterminate 4.9%; charcoal present; fungal spores very abundant. [So indications are that the landscape was very grassy with a few shrubs.]

937. McCabe, T.T., and Cowan, I.M. (1945): *Peromyscus maniculatus macrorhinus* and the problem of insularity. Transactions of the Royal Canadian Institute 25:117–215.
In order to explain the origin and spread of the *rubiventer* subspecies of the deer mouse (*Peromyscus maniculatus*) on Hunter and small adjacent islands off Bella Coola,

British Columbia, the authors suggest a tendency to mutation toward intensification of pigmented areas (dark body, black stripe and reddish belly), which the present population exemplifies. This probably took place during or after the withdrawal of ice sheets from the area and before the subsequent elevation of the coast. The authors postulate that the subspecies originated on Hunter Mountain (which rises from the N shore of Hunter Island) and spread over the local island group, which was previously submerged. The authors also hypothesized that the Vancouver Island marmot (*Marmota vancouverensis*) survived the last (Wisconsinan) glaciation in refugia on Vancouver Island.

938. McConnell, R.G. (1905): Report on the Klondike Gold Fields. Summary Report of the Geological Survey of Canada for 1901:29B.

While discussing the gravels of the Klondike Gold Fields, Yukon, the author separates the gold-bearing gravels into three classes from oldest to youngest: (1) High-level gravels ('White channel') consisting of the Yellow gravels below and the White gravels above; (2) gravels at intermediate levels (Terrace gravels); and (3) Low-level gravels (lowest to highest – River gravels, Creek gravels, and Gulch gravels). Of these, the Low-level gravels are most important and floor the valley bottoms from a depth of 4 to 10 ft, resting on bedrock usually consisting of decomposed and broken schists and overlaid by black, frozen muck from 2 to 30+ ft thick. The Creek gravels frequently enclose remains of leaves, roots, and other vegetation, as well as bones (p. 29B) of various extinct and extant northern mammals such as mammoth (*Mammuthus*), bison (*Bison*), bear (*Ursus*), muskox (Ovibovini), mountain sheep (*Ovis dalli*), and [mountain] goat (*Oreamnos americanum*). [The mountain goat remains have since been considered to represent Dall sheep – see Harington 1971b.]

939. McCorquodale, B.A. (1957): Giant bison found. Blue Jay 15(1):42.

A portion of a skull of a giant bison (*Bison latifrons*) was found in 1956 in stratified interglacial beds at a gravel pit near Fort Qu'Appelle, Saskatchewan. The fossil assemblage included: muskox-like bovid (cf. *Ovibos*), horse (*Equus*), and mammoth (*Mammuthus*). There is a comparative photograph showing the giant bison specimen with the cranium of a modern bison (*Bison bison bison*). [See Khan 1970, and Harington 1990a.]

940. McCuaig-Balkwill, D., and Cinq-Mars, J. (1998): Migratory birds from Bluefish Caves, Eastern Beringia. 8th International Congress of the International Council for Archaeozoology (August 23–9, Victoria, British Columbia). Final Program and Abstracts. p. 194.

The northern Yukon Bluefish Caves (I, II, and III) site has yielded the oldest-known archaeological evidence in Canada. A mammoth (*Mammuthus*)-bone tool from Cave II has been AMS radiocarbon dated to 24,800 BP, making it the earliest-dated artifact from a documented stratigraphic context in Eastern Beringia. The three small caves were also found to contain a rich and diverse array of Late Pleistocene vertebrate remains ranging from >33,000 BP to 10,000 BP. These are thought to have accumulated as a result of carnivore, raptor, and human (*Homo sapiens*) activity, as well as *in situ* mortality. Remains of migratory birds were found in all three caves, although they are sparse in the shallower deposits of Caves II and III. The following birds are represented from Cave II: Lesser Golden Plover (*Pluvialis dominica*), Black-bellied Plover (*Pluvialis squatarola*), swallows (Hirundinidae), American Robin (*Turdus migratorius*), and Tree/Chipping Sparrow (*Spizella arborea/passerina*). Migratory species from Cave III include swallows and the Red-tailed Hawk (*Buteo jamaicensis*). The larger Cave I assemblage includes bones of almost 500 swallows, 21 shorebirds [including Golden Plover, Black-bellied Plover, Spotted Sandpiper (*Actitis macularia*)], and a few passerines representing flycatchers (*Sayornis* cf. *S. saya*), thrushes (*Catharus*), and probably waxwing (*Bombycilla*). The Bluefish Caves loess spans much of the full- and late-glacial sequence, so the presence of shorebird and breeding swallow remains in the lower loess unit reinforces the idea that the environment around the caves about 25,000 years ago (approaching full-glacial time) showed – as suggested by archaeological and associated faunal evidence – a relatively high level of productivity.

941. McCulloch, D., and Hopkins, D. (1966): Evidence for early Recent warm interval in northwestern Alaska. Geological Society of America Bulletin 77: 1089–1108.

A warm interval that began at least 10,000 years ago and lasted until 8300 years ago is recorded in the coastal tundra-covered area of northwestern Alaska by the presence of fossil wood of tree size or tree species, fossil beaver-gnawed wood found beyond the modern range of beaver (*Castor canadensis*), evidence of ice-wedge melting, buried soils, and soils that extend below the top of modern permafrost. Dating of the warm interval is based on a series of radiocarbon dates: a beaver-gnawed wood fragment from a dam has a radiocarbon age of 8550 ± 400 years (W-1249); wood fragment aged 9020 ± 400 years (W-1255); wood from beaver dam aged 9480 ± 160 years (Y-1351) and 7270 ± 350 years (W-1250); twigs aged 11,340 ± 400 years (W-1254); wood aged 10,200 ± 800 years (L-137G); poplar log aged 8800 ± 1000 years (L-117F); wood from beaver dam aged 3600 ± 500 (L-117E); wood fragment from beaver dam aged 9400 ± 750 (L-137N); and twigs aged 9690 ± 400 years (W-485). Prior to about 10,000 years ago, the Kotzebue Sound–Seward Peninsula was clothed in treeless tundra. This severe climate ended possibly about 11,000 years ago, but surely by 10,000 years ago. In the period that followed, warmer summers resulted in lowering of the permafrost table, melting of ice wedges, local development of relatively deep soils, and advance of trees and beavers into the area. The warm interval in this area came to a close about 8300 years ago, and with the return of cooler conditions, the depth of seasonal ground thawing was reduced, ice wedges began to grow again, and trees and beaver retreated eastward.

942. McCulloch, D., Taylor, D.W., and Rubin, M. (1965): Stratigraphy, non-marine mollusks, and radiometric dates from Quaternary deposits in the Kotzebue Sound area, western Alaska. Journal of Geology 73(3):442–53.

A dated sequence of interbedded marine and nonmarine Quaternary sediments in the Kotzebue Sound area records marine transgressions of Yarmouthian and Sangamonian age, glaciations of Illinoian and Wisconsinan age, and two periods (Sangamonian and Late Wisconsinan to Recent) when the climate was warmer than at present. Nonmarine molluscan faunas from beds of Late Illinoian to Recent age show little significant change. $Thorium_{230}$-$Uranium_{238}$ ages of 170,000 ± 17,000 and 175,000 ± 16,000 years were determined on shells from the marine sediments of Yarmouthian age. A chart of radiometric ages is provided, including 12 dated samples, as is a locality map and a stratigraphic illustration.

943. McDonald, H.G., Harington, C.R., and DeIuliis, G. (2000): The ground sloth *Megalonyx* from Pleistocene deposits of the Old Crow Basin, Yukon, Canada. Arctic 53(3):213–20.

The bear-sized ground sloth (*Megalonyx*), endemic to North America, was widespread, reaching as far N as Alaska, Yukon and Northwest Territories during the Pleistocene. Twenty-two specimens collected from 10 localities in the Old Crow Basin, northern Yukon, include several bones that can be referred to Jefferson's ground sloth (*Megalonyx jeffersonii*) based on distinctive morphology of the upper and lower caniniform teeth. All of the Yukon specimens are relatively small, suggesting a geological age earlier than Wisconsinan – probably Sangamonian. The Alaskan, Yukon, and Northwest Territories records imply that the species occupied a broad east-west range in northwestern North America during a warm phase of the Late Pleistocene.

944. McDonald, J.N. (1981): North American Bison. Their Classification and Evolution. University of California Press, Berkeley. 316 pp.

Includes chapters on bison (*Bison*) classification, evolution, morphology, adaptation, possible hybridization, as well as an appendix of radiocarbon dates. Many Canadian and Alaskan specimens are mentioned.

945. McDonald, J.N. (1984): An extinct muskox mummy from near Fairbanks, Alaska: A progress report. Proceedings of the First International Muskox Symposium. (D.R. Klein, R.G. White, and S. Keller, eds.). Biological Papers of the University of Alaska, Special Report No. 4:148–52.

The carcass of a young helmeted muskox (*Bootherium bombifrons*) was collected from frozen muck at Fairbanks Creek (64°58′N, 147°10′W), Alaska, in 1940, and is part of the Frick Collection at the American Museum of Natural History (F:AM A-293-5268). The horncores and some soft tissues are missing and the right side of the trunk has been extensively damaged. The limbs are longer and relatively more slender, and the tail is longer than the same parts of the tundra muskox (*Ovibos moschatus*). Two radiocarbon dates apply to this specimen: on hair from the hind limb [17,210 ± 500 BP (SI-454)] and on muscle from under the scalp [24,140 ± 2200 BP (SI-455)]. They indicate that the animal died about 18,000 years ago – near the peak of the last glaciation.

Hair is best preserved on the lower limbs and varies from blackish-brown to orange and yellow-brown – generally darker than most tundra muskox (*Ovis moschatus*). Much of it is from 12 to 28 mm long, but the dark guard hairs reach about 55 mm. Plant matter is preserved in the gastrointestinal tract. A table lists external measurements of the specimen, and a map of helmeted muskox localities is included.

946. McDonald, J.N., and Ray, C.E. (1989): The autochthonous North American musk oxen *Bootherium*, *Symbos*, and *Gidleya* (Mammalia: Artiodactyla: Bovidae). Smithsonian Contributions to Paleobiology No. 66:1–77.
The authors examined over 170 specimens before coming to the conclusion that *Bootherium*, *Symbos*, and *Gidleya* are synonymous and in fact represent one species, the helmeted muskox (*Bootherium bombifrons*). They based this on measurements from horncores and cranial characters such as occipital height, mastoid breadth, frontal breadth, postorbital width, and angle of basioccipital-basisphenoid flexion. Histograms of the results are provided. Maps showing the localities of all the specimens used in the study and high-quality photographs of many of the important specimens are included. Radiocarbon dates mentioned include: (1) *Bootherium* from the Fairbanks area dated to 17,210 ± 500 BP (A-293-5268[hair], SI-454); 24,140 ± 2200 BP (A-293-5268[tissue], SI-455); and 22,540 ± 900 BP (AM-Frick Coll. 30508[hornsheath], SI-292); (2) '*Symbos*' from Little Eldorado Creek, Alaska, dated to >40,000 BP (A-204-4254[fecal material], SI-291); (3) from Upper Cleary Creek, Alaska, dated to 25,090 ± 1070 BP (AM-Frick Coll. 33124[hornsheath], SI-850); (4) from Dome Creek, Alaska, dated to 17,695 ± 445 BP (AM-Frick Coll. A-651-3006[hornsheath], SI-851); and (5) from Lost Chicken Creek, Alaska, dated to 20,500 ± 390 BP (CMN 25892[bone], I-10649).

947. McDonald, J.N., Ray, C.E., and Harington, C.R. (1991): Taxonomy and zoogeography of the musk ox genus *Praeovibos* Staudinger, 1908. *In*: Beamers, Bobwhites, and Blue-Points (J.R. Purdue, W.E. Klippel, and B.W. Styles, eds.). Tributes to the Career of Paul W. Parmalee, Illinois State Museum Scientific Papers 23:285–314
The authors examined 32 of the 51 known *Praeovibos* crania (all partial) and concluded that the genus contains three recognized species: *Praeovibos priscus* Staudinger, 1908; *Praeovibos beringiensis* Sher, 1971; and *Praeovibos recticornis* (Ryziewicz), 1933. Most of the specimens have been recovered from Middle Pleistocene sites in Eurasia and North America. The genus is not known to have survived into the Late Pleistocene. *Praeovibos*, ecologically eurytopic, occupied habitats ranging from Mediterranean woodlands and interglacial mesophytic forests to boreal woodlands and tundra. *Praeovibos* probably gave rise to the helmeted muskox (*Bootherium bombifrons*), its North American vicar, and the tundra muskox (*Ovibos moschatus*), the physiologically and ecologically specialized muskox of the Holarctic tundra. Many significant specimens are illustrated, and maps of the collection sites are provided.

948. McDowell, P.F., and Edwards, M.E. (2001): Evidence of Quaternary climatic variations in a sequence of loess and related deposits at Birch Creek, Alaska: Implications for the Stage 5 climatic chronology. Quaternary Science Reviews 20(2001):63–76.
Of interest in a description of stratigraphic units at Birch Creek (Fig. 1, Table 1), Alaska are: (1) spruce, poplar, and/or willow logs with beaver (Castoridae)-cut ends in Unit F-1 – deposits in a lake possibly initiated by [beaver-?] damming of a small tributary stream; and (2) a 'mastodon' (*Mammut americanum*?) tusk found about 20 m above the river in Unit D-1 – a primary loess deposit.

949. McNeely, R. (1989): Geological Survey of Canada Radiocarbon Dates XXVIII. Geological Survey of Canada Paper 88-7:1–93.
This paper reports a pertinent radiocarbon date. Beaver-chewed (*Castor canadensis*) wood from the upper part of Unit 2, Hungry Creek, Yukon (65°34.5′N, 135°30′W) provided an uncorrected radiocarbon date of 36,900 ± 300 BP (GSC-2422). This provides a minimum for the advance of the Laurentide ice sheet over Hungry Creek.

950. McNeely, R. (2000): The Geological Survey of Canada 'Date Locator File.' http://sts.gsc.nrcan.gc.ca/radiocarbon (or e-mail radiocarbon_dating@gsc.nrcan.gc.ca).
The Geological Survey of Canada Radiocarbon Dating Laboratory has developed a computerized database ('Date Locator File') of all GSC radiocarbon dates. This

database provides the research community immediate access to the relevant information produced by the GSC Laboratory. The database allows dates to be interactively selected on key parameters such as: Locality, Latitude/Longitude, Submitter, Type of Material, and Laboratory Codes. A series of Open Files are also available on the GSC Web site that includes not only GSC dates but all published dates within a region [e.g., British Columbia – 866 selected dates; Saskatchewan – 665 (all) dates; Manitoba – 890 (all) dates; Gaspésie – 200 dates (GSC Map 1804A, Veillette and Cloutier 1993)]. Databases of the GSC Radiocarbon Laboratory include: 'Date Locator File' (6100 GSC dates); 'Other dates' (3095 non-GSC dates – basic information only); GEOTOP I (45 University of Quebec at Montreal radiocarbon dates I); Newfoundland (560 unverified dates); AMS dates (800 mainly IsoTrace, but many Beta and CAMS dates); Alaska (2250 dates from the United States Geological Survey Open File); INSTARR (815 dates published in Institute of Arctic and Alpine Research, Boulder, Colorado, Date Lists I-VII); 'Bone Dates' [330 radiocarbon dates on bone, mainly of marine mammals, done at the Saskatchewan Laboratory (S)] for a total of 14,295 dates as of August 15, 1998. The onus is on all users to provide information to the database and thereby enhance its usefulness to the scientific community.

951. McNeely, R., and Jorgensen, P.K. (1992): Geological Survey of Canada Radiocarbon Dates XXX. Geological Survey of Canada Paper 90-7:1–84.

This paper reports pertinent radiocarbon dates. Mammal (Mammalia) fecal pellets from peaty tundra-pond deposits near Komakuk Dew Line Station, 12 km E of the mouth of the Backhouse River, Yukon (69°35.5′N, 140°14′W) were dated to 10,580 ± 370 BP (TO-651). This age and others indicate that vegetation was well established and that tundra ponds were evolving through thermokarst activity at the end of the Late Wisconsinan and during the earliest Holocene. Conditions at the time of the tundra pond deposits were probably similar to those at present.

952. McNeely, R., and Jorgensen, P.K. (1993): Geological Survey of Canada Radiocarbon Dates XXXI. Geological Survey of Canada Paper 91-7:1–85.

This paper reports a pertinent radiocarbon date. A bone of bowhead whale (*Balaena mysticetus*) (probably the 14th or 15th lumbar vertebra, CMN 45042) embedded in silt about 0.7 km W of the mouth of the Romaines River near the village of Romaines, Newfoundland (48°33.22′N, 58°41.02′W), provided a radiocarbon age of 13,345 ± 130 BP (S-3047). This bone is from the marginal zone of the Robinsons Head Readvance and relates to an associated deglacial relative sea level of about 21 m. A bowhead whale bone collected (89-DCA-28) 4.2 km SW of Cape York (30 m elevation), N coast of Brodeur Peninsula, NW Baffin Island, Nunavut (73°47.5′N, 87°08′W) was dated at 8820 ± 130 BP (S-3138). This correlated strongly with other dated bowhead remains from this site, which collectively indicate that relative sea level fell to present levels at about 6000 BP, then fell somewhat lower before recovering to current levels. A bowhead earbone (89-DCA-10) from Lowther Island, Nunavut, at an elevation of 4.5 m (74°35.8′N, 97°30.2′W), provided an uncorrected radiocarbon date of 1510 ± 80 BP (S-3134).

953. McNeil, P., Hills, L., Kooyman, B., and Tolman, S. (1999a): First reported Canadian proboscidean trackway, probably mammoth, from the St. Mary Reservoir, Alberta, Canada. Third British Columbia Paleontological Symposium (University of Victoria, May 14–17). Abstracts. pp. 51–2.

Construction of a new spillway resulted in a 10-m lowering of the St Mary Reservoir near Cardston, Alberta, which exposed fine aeolian sediment and paleosols of late-glacial or Early Holocene age that have been severely eroded by wind since. Erosion has revealed a rich assortment of human (*Homo sapiens*) artifacts as well as bones of 11 mammals. Trace fossils include many rodent burrows and a well-preserved trackway of a proboscidean [probably a mammoth (*Mammuthus*)]. A series of nine tracks (figured on p. 52) evenly spaced over a distance of 11 m were discovered by Shayne Tolman and Ed Boyce in the winter of 1999. The tracks are circular to slightly ellipsoidal with a diameter of 55 to 65 cm. They consist of concentric rings of sediment pushed up while the animal walked on soft muddy ground. The innermost ring represents the true track made by the foot and has a diameter of 35 to 40 cm. In four of the tracks there is a clear indication of hind feet overstepping fore feet – showing stride pattern (average length of 2.4 m is consistent with a hip height of about 2 m, walking slowly). Thus, the authors consider that a mammoth was the most likely track maker.

954. McNeil, P.E., Hills, L.V., Kooyman, B. and Tolman, S. (1999b): Late Pleistocene fauna from the St. Mary Reservoir, southwestern Alberta. Ninth Canadian Paleontology Conference (August 20–1, Calgary), Program and Abstracts No. 9:18.
Construction of a new spillway resulted in a 10-m water-level drop of the St Mary Reservoir in southern Alberta. The reservoir floor was exposed to intense wind erosion, in places removing 1.5 to 2 m of sediment and exposing a diverse assemblage of vertebrate fossils, ancient trackways, and Paleoindian artifacts. This provides a unique look at the Late Pleistocene environment of southern Alberta. AMS radiocarbon analyses of helmeted muskox (*Bootherium bombifrons*), ancient bison (*Bison bison antiquus*), and an excellent specimen of the Mexican horse (*Equus conversidens*) have yielded dates of 10,980 ± 80 BP (TO-7691), 11,130 ± 90 BP (TO-7693), and 11,330 ± 70 BP (TO-7696), respectively. Other fossils include complete specimens of American badger (*Taxidea taxus*), two other mustelids (Mustelidae), at least two canids (Canidae – one probably a wolf *Canis lupus*), a cat (Felidae), four types of ground squirrel (*Spermophilus*), a rabbit (Leporidae), beaver (*Castor canadensis*), muskrat (*Ondatra zibethicus*), seven genera of birds (Aves), and a fish (Pisces) – some are not in stratigraphic context. Trackways and trampled areas of the megafauna have been found: woolly mammoth (probably *Mammuthus primigenius*), western camel (*Camelops hesternus*), Mexican horse, and an unidentified bovid (Bovidae). The tracks are found in association with each other, with fossil bones, and human (*Homo sapiens*) artifacts indicating the coexistence of the fauna with Paleoindians. The location of the site on an old river channel or island with gentle banks, allowing access to water, explains the abundance and diversity of the fauna.

955. Mead, E.M., and Mead, J.I. (1989): Quaternary zoogeography of the Nearctic *Dicrostonyx* lemmings. Boreas 18:323–32.
This review lists all the known lemming (*Dicrostonyx*) fossils from North America in table form including: collared lemming (*Dicrostonyx torquatus*) from Bluefish Cave I, Yukon, dated at 12,900 ± 100 BP (GSC-2881); collared lemming from Eagle Cave, Alberta, dated at 22,700 ± 1000 BP (GaK-2336); collared lemming from January Cave, Alberta, dated at 23,100 ± 860 BP (GaK-5438); other lemmings (cf. *Dicrostonyx*, *Dicrostonyx* cf. *Dicrostonyx simplicior*, and *Dicrostonyx gulielmitorquatus*) all from Old Crow, Yukon, dated at between >54,000 BP (GSC-2066) and >51,000 BP (GSC-2559); Ungava lemming (*Dicrostonyx hudsonius*) from St-Elzéar Cave, Quebec, dated at about 5000 BP or earlier; and Ungava lemming from Trou Otis Cave, Quebec, dated at 'Late Pleistocene.' All of the above dates are based on material accessory to the fossil itself or on the stratigraphy of the locality; they are not direct dates on *Dicrostonyx*.

956. Mead, J.I. (1987): Quaternary records of pika, *Ochotona*, in North America. Boreas 16:165–71.
This review lists all the known pika fossils in table form including: giant pika (*Ochotona whartoni*) from Eagle Cave, Alberta, dated at >33,000 BP; pika (*Ochotona princeps*) from January Cave, Alberta, dated at >32,000 BP; pika (*Ochotona*) from Kelso Cave, Ontario, undated but estimated to be Illinoian; and giant pika and pika from Old Crow Basin, Yukon, undated but estimated to be Late Illinoian-Sangamonian. Pika have lived in North America since the Hemphillian land-mammal age (Late Pliocene). At least three species – *Ochotona whartoni* (extinct giant pika), *Ochotona princeps* (the extant species), and unidentified large and small forms – lived in North America during the Quaternary. Pika lived in northeastern portions of the continent during the Illinoian glacial and possibly earlier. During the Wisconsinan glacial period, pika stayed in the mountainous W, with one exception in northern Nebraska. *Ochotona* have not always been found to live only in rocky or talus habitats. Giant pika is believed to have lived in a cold grassland, as do many of the extant Asian species. Cooler summers may have been all that was needed for pika to migrate to new territories, areas without talus. [See also Mead and Grady 1996.]

957. Mead, J.I., and Grady, F. (1996): *Ochotona* (Lagomorpha) from Late Quaternary cave deposits in eastern North America. Quaternary Research 45:93–101.
Pikas had a wider distribution during the Pleistocene than they do today. Five hundred and twenty-six Pleistocene fossil specimens from five caves, not currently within pika distribution, were examined. These bones represent two forms: a large extinct form, the giant pika (*Ochotona whartoni*) and a small form [probably the extant pika

(*Ochotona princeps*)]. The giant pika is known from the Cape Deceit fauna in Alaska (~1.9 to 0.9 BP), Gold Hill in Alaska (>~125,000 BP), and at least 10 locations in Old Crow Basin, Yukon (~125,000–80,000 BP). Two of the five cave localities in eastern North America provided bones of the large form [Elba (44°01′N 80°10′W), Ontario, 8670 ± 220 BP (TO-2566); and Kelso (43°30′N, 70°55′W), Ontario, Late Wisconsinan]. The Elba cave femur, recovered loose on an exposed ledge about 25 m below the cave entrance, is the youngest-dated pika specimen from eastern North America. Of the 20 taxa found at Elba, only two – the pika and pine marten (*Martes americana*) – were found not to be of the local modern fauna. The marten was extirpated in the early 1900s, leaving the pika as the only extinct species recovered from the cave. Pollen data from Kelso Cave suggest an age of about 10,500 BP for the matrix around the *Ochotona* specimen. As is the case at Elba, all the faunal elements, with the exception of *Ochotona*, are part of the modern fauna. It is possible that *Ochotona* survived as a relict population in the rocky areas of Ontario along the Niagara Escarpment, making the species, at best, suspect as an indicator of Illinoian age for eastern North America [e.g., Churcher and Dods 1979].

958. Meldgaard, M. (1986): The Greenland caribou – Zoogeography, taxonomy, and population dynamics. Meddelelser om Grønland Bioscience 20:1–88.

A comprehensive survey of all available geological, archaeological, historical, ethnohistorical, statistical, and game-biological data concerning the past and present status and distribution of the Greenland caribou about the size of Peary caribou is given. The main conclusions are: (1) Prior to 6000 BC (about 8000 BP), small caribou (*Rangifer tarandus pearyi*) migrated from Ellesmere Island. (2) Prior to 2000 BC (about 4000 BP), large caribou the size of American tundra caribou (*Rangifer tarandus groenlandicus*) migrated across Nares or Davis Strait to the northern part of West Greenland. (3) By 1500 BC (about 3500 BP),West Greenland was inhabited by both large and small caribou separated by Sukkertoppen Iskappe. (4) Between 1500 BC (about 3500 BP) and 1000 AD (about 1000 BP), the small West Greenland caribou disappeared. (5) By 1900 AD (about 100 BP), the small Northeast caribou became extinct. (6) Through the last 1000 years, the range of the Greenland caribou has been reduced through population extinctions beginning in Southeast Greenland and progressing N and S along the coast. (7) Greenland caribou are currently, and possibly always have been, subject to dramatic population fluctuations caused by climatic changes, and secondarily by overgrazing and predation. In addition to *Rangifer* dates [reported also in Bennike 1987], this paper lists dates of antler from: Kap Morris Jesup, 1830 ± 70 BP (K-3868); Independence Fjord, 740 ± 65 BP (K-4349); Vahl Fjord, 1680 ± 70 BP (K-4138); Angmagssalik Fjord, 1250 ± 70 BP (K-4343); Sermilik Fjord, 1110 ± 70 BP (K-4139); Kulusuk, 990 ± 70 BP (K-4140); Kulusuk, 970 ± 70 BP (K-4345); Angmagssalik Ø, 940 ± BP (K-4344); Angmagssalik Ø, 870 ± 65 BP (K-4346); and Tingmiarmiut Fjord,1260 ± 70 BP (K-4350).

959. Meldgaard, M. (1988): The Great Auk, *Pinguinus impennis* (L.) in Greenland. Historical Biology 1:145–78.

This paper presents all available archaeozoological, ethnohistorical, and historical evidence on the Great Auk (*Pinguinus impennis*) in Greenland. Data from 132 Great Auk bones from seven archaeological sites are presented along with the first published English translation of Otto Fabricius' 1808 handwritten notes on the Great Auk. Great Auks probably occurred farther N in Greenland around 2000 BP than in historical times. It also was a prized prey of Inuit hunters (*Homo sapiens*) through 4500 years. The last Great Auk recorded in Greenland was killed by an Inuk hunter in 1815.

960. Meldgaard, M. (1991): New perspectives on the zoogeography of the Greenlandic caribou (*Rangifer tarandus*). *In:* 4th North American Caribou Workshop Proceedings (C. Butler and S.P. Mahoney, eds.), (October 31–November 3, 1989, St John's). pp. 37–63.

An antler dated to >40,000 BP documents the presence of *Rangifer* in North Greenland in pre-Holocene time, probably during the Sangamonian/Eemian interglacial. The existence of caribou in this High Arctic environment during the last interglacial indicates that the evolution of polar caribou ancestral to Peary caribou (*Rangifer tarandus pearyi*), East Greenland caribou (*Rangifer tarandus eogroenlandicus*), and possibly Spitsbergen caribou (*Rangifer tarandus platyrhynchus*) took place before about 100,000 BP. It is not likely that caribou survived the last glaciation in a North Greenland–Ellesmere Island refugium. Climatic conditions were presumably too severe. In the Early Holocene, about 9000 BP, small polar caribou immigrated to North Greenland probably from High Arctic Canada. They spread by island hopping to all parts of Greenland, formed distinct populations, and followed different evolutionary paths. In North Greenland they disappeared after 500 BP; in Northeast Greenland they died out about 90 BP, following severe winters; in Southeast Greenland they were probably exterminated by Inuit (*Homo sapiens*) about 900 BP; and in West Greenland they were most likely out-competed by American tundra caribou (*Rangifer tarandus groenlandicus*) that migrated to the area from Baffin Island, Nunavut, about 4000 BP. The paper includes a list (Table 1) of zoogeographically significant caribou remains (30 bone dates) from Greenland. [See Radiocarbon-date Table.]

961. Meldgaard, M., and Bennike, O. (1989): Interglacial remains of caribou (*Rangifer tarandus*) and lemming (*Dicrostonyx torquatus*(?) from North Greenland. Boreas 18:359–66.

A male caribou (*Rangifer tarandus*) antler (Figures 2, 3) from Kap København in Peary Land, North Greenland (82°29′N, 21°01′W, 53 m asl), found directly above the Kap København Formation, has been radiocarbon dated at >40,000 BP. A conventional date on the antler gave an age of 31,680 +1600 / –1380 BP (K-4918), whereas an AMS date yielded an age of >40,000 BP (Ua-1112). Both analyses were on the collagen fraction of the antler. The conventional date is probably finite because of contamination by microorganisms in the natural environment. Five lemming (*Dicrostonyx torquatus* (?)) droppings from Skalhøjene in Warming Land (81°57′N, 53°38′W, 115 m asl, Figure 6) were found with allochthonous plant material dated to 32,270 +1120 / –1200 BP (K-4574), whereas a willow (*Salix*) twig from the same sample yielded an age of >40,000 BP (Ua-948). Because the caribou antler and remains from Skalhøjene point to biotic and climatic conditions no more severe than modern conditions, the authors propose that the fossils in question belong to the Sangamonian interglacial. It is unlikely that caribou and lemming survived in North Greenland during the last (Wisconsinan) full glacial period. They probably reentered Greenland in the Early Holocene.

962. Mertie, J.B., Jr (1937): The Yukon-Tanana region, Alaska. U.S. Geological Survey Bulletin 872:1–276.

A partial list of Pleistocene mammals from the Fairbanks area of Alaska is given on p. 191.

963. Miller, R.F. (1990): New records of postglacial walrus and a review of Quaternary marine mammals in New Brunswick. Atlantic Geology 26:97–107.

New Brunswick Quaternary walrus (*Odobenus rosmarus*) records include: a nearly complete skeleton in a gravel ballast pit near Moncton, found in 1871; a tusk from a gravel pit roughly 100 m from Chaleur Bay, Clifton; fragments of cranium, teeth, and tusk from an offshore bar, E of Tracadie; deteriorated anterior skull from 80 m depth in the Bay of Fundy 6.4 km SSE of Cape Spencer, radiocarbon dated at 9360 ± 90 BP (TO-1554); a cranium, possibly from a gravel pit on Portage Island, Miramichi Bay, Northumberland County; a cranium from a gravel pit at Village St Paul, Gloucester County; a cranium from a beach near Beresford, NW of Bathurst; a cranium from a scallop drag 17.7 km off Black River, radiocarbon dated at 12,760 ± 90 BP (TO-1927); and a lower jaw from beach sand, Dune de Buctouche (46°31′N, 64°40′W). Other Atlantic Canada walrus tusks include: two from scallop drags off Pictou Island, Northumberland Strait, Nova Scotia; from a scallop drag near Granville Ferry, Nova Scotia; from a 31-m-deep scallop drag 9.7 km NW of Miminegash, Prince Edward Island; and two specimens from scallop drags from Georges Bank (41°32′N, 66°11′W) and another from 41°41′00″N, 66°20′30″W. Previous records of New Brunswick marine mammals include: humpback whale (*Megaptera novaeangliae*) axis vertebra (incorrectly identified as a walrus) from a gravel pit near Grande Anse on the N shore of the province; finback(?) whale (cf. *Balaenoptera physalus*) vertebra from gravel dumped in Fredericton (source of gravel unknown); ice age narwhal (*Monodon monoceras*) from the Jacquet River on the Intercolonial Railway, on the north shore, reported by Gilpin (1874), Honeyman (1874), and Perkins (1908); white whale (*Delphinapterus leucas*) worn lower jaw ramus from Leda clay at the mouth of the Pocologan River, Charlotte Co. (45°7′N, 66°35′W); harp seal (*Phoca* cf. *Phoca groenlandica*) partial skeleton from organic black sand buried 8 m at Hopkins Brickyard in St John (45°20′N, 65°50′W) reported by Harington (1977b), Harington and Occhietti (1988), and Matthew (1879); and minke whale (cf. *Balaenoptera acutorostrata*) vertebra from a clay unit on the S bank of Kouchibouguacis River, 1.5 km W of St-Louis-de-Kent (46°3.4′N, 64°59.6′W) reported by Rampton et al. (1984) and Rampton and Paradis (1981), radiocarbon dated at 12,600 ± 140 BP (GSC-3284) and 11,830 ± 950 BP (S-1969B). This paper includes photographs of some of the walrus specimens and a map of the New Brunswick sites.

964. Miller, R.F. (1997): New records and AMS radiocarbon dates on Quaternary walrus (*Odobenus rosmarus*) from New Brunswick. Géographie physique et Quaternaire 51(1):107–11.

This paper updates Miller (1990), including six new fossils and four AMS radiocarbon dates, of late-glacial to postglacial walrus (*Odobenus rosmarus*) in New Brunswick. Two of the radiocarbon dates refer to specimens reported in the 1990 paper and two from this paper. The dates are: 2890 ± 40 BP (Beta-71157) on a cranium (NBMG 4584) from Portage Island, Miramichi Bay; 9980 ± 60 BP (Beta-69386) on a mandible (NBMG 8621) from the Bay of Fundy off Quaco Head; 10,270 ± 70 BP (Beta-89281) on a mandible (NBMG 9105) from the Bay of Fundy S of St Martins; and 12,760 ± 90 BP (TO-1927) on bone from the cranium of an almost-complete skeleton in the Huntsman Collection from the Bay of Fundy 18 km off Black River. The oldest-dated walrus fossil (12,760 ± 90 BP) indicates that these marine mammals inhabited the Bay of Fundy no more than a thousand years after it became ice free.

965. Miller, R.F., Harington, C.R., and Welch, R. (2000): A giant beaver (*Castoroides ohioensis* Foster) fossil from New Brunswick, Canada. Atlantic Geology 36:1–5.

An isolated, large, upper-right incisor tooth (NBMG 10368) of a giant beaver (*Castoroides ohioensis*) recovered from Indian Island, New Brunswick (44°56′N, 66°58′W; Figure 1) near the mouth of the Bay of Fundy represents the easternmost record of this species in Canada. Other giant beaver specimens are from Old Crow Basin, northern Yukon, and Toronto, Ontario (Don Beds) (Figure 1). Unfortunately, no stratigraphic context is available for the specimen, which was recovered loose near a beach also known to yield archaeological materials much younger than the giant beaver tooth. The specimen may have been deposited on a storm beach from near-shore deposits, but it is also possible that the tooth was carried to the site by people (*Homo sapiens*). Other Quaternary terrestrial and marine mammal remains of New Brunswick are mentioned, and a table (Table 1) of comparative measurements of giant beaver upper incisor teeth from Canada and two photographs (Figure 2) of the tooth are provided.

966. Miller, R.R. (1965): Quaternary freshwater fishes of North America. *In*: The Quaternary of the United States (H.E. Wright, Jr, and D.G. Frey, eds.). Princeton University Press, Princeton, New Jersey. pp. 569–81.

In his section 'Pleistocene Fishes,' the author mentions the following Canadian taxa (pp. 571–3): (1) Pacific salmon (*Oncorhynchus*) from the Thompson River of British Columbia, giving the date as Late Wisconsinan and acknowledging a personal communication from C.C. Lindsey, 1964 – based on salmon remains in the British Columbia Provincial Museum, catalogue number 470-474, 526; (2) minnows (Cyprinidae) from Lillestrom, Saskatchewan, estimated to be 10,000 years old; (3) sticklebacks (*Gasterosteus*) from Pleistocene [Champlain Sea] beds in the Ottawa River valley, Ontario [see Dawson 1871 and McAllister et al. 1988]; (4) presumably catfish (*Ictalurus punctatus*) from the Sangamonian interglacial deposits (Don Beds) at Toronto, Ontario [see Coleman 1932]; and unspecified fish from Goose River, Quebec [see Lambe 1905, and Uyeno and Miller 1963].

967. Moffit, F.H. (1905): The Fairhaven gold placers, Seward Peninsula, Alaska. United States Geological Survey Bulletin 247:1–85.

The section 'Mammalian Remains' (pp. 41–2) is pertinent. These observations arise from a geological and topographical survey made during the summer of 1903 of an important placer-gold district in NE Seward Peninsula of Alaska (see map, Plate I). The author briefly mentions Elephant Point, Eschscholtz Bay, where he says there appear to be two ice beds overlain by clay, containing remains of Pleistocene mammals, and covered by a good growth of dwarf birches and herbage. Mammoth (*Mammuthus*) teeth and parts of tusks are found in valley deposits and on gravel bars of streams. Bones are known from all the larger streams, and are usually found in muck overlying the gravels, being exposed by stream erosion. 'In some cases it may be that the animals were mired in bogs where now found, but it is also possible that the remains have been washed to their present location in the muck from some other point.' A skeleton was exposed on the surface – two large tusks, badly decayed and partly concealed by mosses, were noted by the survey party near Goodhope Bay. 'Teeth and a considerable portion of a skeleton were found by prospectors on Inmachuk River, and portions of tusks were picked up by the writer on Kugruk River. Prospectors report numerous remains on the Buckland [River] of both mammoth and musk ox [*Ovibos moschatus*].' Two Laplanders prospecting on Quartz Creek, a tributary to Kievalik River, uncovered a muskox skull, which lay near bedrock beneath 8 ft of gravel. A similar skull was found on Bear Creek, a tributary of Buckland River.

968. Møbjerg, T. (1998): The Saqqaq culture in the Sisimiut municipality elucidated by the two sites Nipisat and Asummiut. *In*: Man, Culture and Environment in Ancient Greenland (J. Arneborg and H.C. Gulløv, eds.). Danish National Museum and Danish Polar Center, Copenhagen, Danish Polar Center Publication No. 4:98–118.

The natural boundaries of the Sisimiut district are Nordre Strømfjord to the N, Sukkertoppen Icecap to the S, Davis Strait to the W, and the Inland Icecap to the E (Figure 1). The Paleoeskimo (*Homo sapiens*) site of Asummiut (SIK 901) yielded a small collection of burned bones, but the amount of collagen was too small for reliable dating at the Århus AMS laboratory. However, 19 radiocarbon dates on caribou (*Rangifer tarandus*) bones (10 conventional dates and the rest AMS dates, Figure 21) are listed from the Nipisat site, covering the period 3670 ± 90 BP (K-6192) – 2860 ± 80 BP (K-5864). [See radiocarbon-date table for details.]

969. Møhl, J. (1986): Dog remains from a Paleoeskimo settlement in West Greenland. Arctic Anthropology 23 (1, 2): 81–9.

A sample of 15,000 bones (Table 1) from Qajâ, Jakobshavn Icefjord, West Greenland, of Sarqaq age is from 1925 BC to 925 BC in age according to radiocarbon dates. Apparently the people in the settlement depended most heavily on seals [especially harp seals (*Phoca groenlandica* = *Pagophilus groenlandicus*) and ringed seals (*Phoca hispida*)]. Among birds (Aves), eiders (*Somateria*), gulls [Glaucous Gulls (*Larus hyperboreus*) and Iceland Gulls (*Larus glaucoides*)], and Thick-Billed Murres (*Uria lomvia*) were most common. Dogs (*Canis familiaris*), arctic foxes (*Alopex lagopus*), and caribou/reindeer (*Rangifer tarandus*) were the most common mammals. Evidently too, the first people (*Homo sapiens*) who arrived in Greenland about 4500 years ago hunted hares (*Lepus arcticus*). The 18 dog bones are the first proof of this animal's existence in Paleoeskimo times.

970. Moodie, G.E.E., and Reimchen, T.E. (1973): Endemism and conservation of sticklebacks in the Queen Charlotte Islands. Canadian Field-Naturalist 87(2): 173–5.

Populations of the threespine stickleback (*Gasterosteus aculeatus*) in the Queen Charlotte Islands, British Columbia, show degrees of melanism, gigantism, spine loss, and lateral plate reduction, which are seldom or never exceeded elsewhere in this widely distributed coastal species. Although the study is incomplete and causes of the peculiarities of the sticklebacks are far from understood, there are already some obvious correlations between their morphology and environmental variables. Most of the populations showing lateral plate reductions or spine loss occur in lakes with no outlets and apparently no other species of fish (Pisces). The absence of both competitors and predators has seemingly allowed these fishes to evolve novel adaptations to their ecologically unusual environments.

971. Moodie, G.E.E., and Reimchen, T.E. (1976): Glacial refugia, endemism, and stickleback populations of the Queen Charlotte Islands, British Columbia. Canadian Field-Naturalist 90(4):471–4.

The biota of the Queen Charlotte Islands, British Columbia, has been recognized since the 1880s as being distinct from that of the mainland. The authors discuss previously described populations of the threespine stickleback (*Gasterosteus aculeatus*) in terms of the controversy regarding postulated refugia and the origin of other endemic groups on the islands. Despite the temptation to expect a common basis for all endemism in the Queen Charlotte Islands, the authors believe sticklebacks in these islands have undergone rapid postglacial evolution, perhaps accelerated by the limited diversity of the fish (Pisces) fauna. They emphasize that the evolutionary history for different taxa must be determined independently on the basis of evidence relevant to that group.

972. Morgan, A.V., McAndrews, J.M., Pilny, J.J., Goodwin, A.G., and Morgan, A. (1983): Paleoecological reconstruction at the Rostock Mammoth site, Ontario. Geological Association of Canada, Mineralogical Association of Canada, Joint Annual Meeting. p. A48.
This abstract reports a find of mammoth (*Mammuthus*) remains at the surface of a cultivated wetland near Rostock, Ontario. The remains include: portions of skull with molar; tusk remnants; vertebra; scapula; and limb bone. An anomalous radiocarbon date yielded a Late Holocene age, but further palynological studies show the mammoth to be between 10,000 and 10,500 BP.

973. Morlan, R.E. (1977): Fluted point makers and the extinction of the Arctic-Steppe biome in Eastern Beringia. Canadian Journal of Archaeology 1:95–108.
A speculative outline of possible events and relationships between the extinction of the arctic steppe environment of Late Pleistocene Beringia and the appearance of fluted-point makers in the interior of North America S of Wisconsinan limits is presented. Data and inferences concerning the paleoenvironmental circumstances, the status of the potential corridors of emigration, and the possible connection of the expansion of human (*Homo sapiens*) fluted-point culture and the Late Pleistocene extinction phenomena are discussed.

974. Morlan, R.E. (1978): Early man in northern Yukon Territory: Perspectives as of 1977. *In*: Early Man in America from a Circum-Pacific Perspective (A.L. Bryan, ed.). Department of Anthropology, University of Alberta, Occasional Papers No. 1. Published by Archaeological Researchers International, Edmonton. pp. 78–95.
An introduction to the Yukon Refugium Project launched in 1975 is provided along with arguments to support a Pleistocene origin of the altered bone tools found during the fieldwork. Photographs of some of the tools and discussions of how they may have been created and used support the argument. The tools are made from horse (*Equus*), caribou (*Rangifer tarandus*), bison (*Bison*), and proboscidean [probably mammoth (*Mammuthus*)] bones. The complexities of the redeposition and stratigraphy, and implications of radiocarbon dates of 29,100 BP to 25,700 BP, are discussed. The data suggest evidence of a circum-Pacific distribution for the peopling (*Homo sapiens*) of the New World rather than the circumpolar distribution of Mid-Wisconsinan age as is currently favoured. A radiocarbon date of 22,600 ± 600 BP (I-3573) on a mammoth (*Mammuthus*) femur from Old Crow Loc. 14N, Yukon, is mentioned.

975. Morlan, R.E. (1979): A stratigraphic framework for Pleistocene artifacts from Old Crow River, northern Yukon Territory. *In*: Pre-Llano Cultures of the Americas: Paradoxes and Possibilities (R.L. Humphrey and D. Stanford, eds.). Anthropological Society of Washington. pp. 125–45.
After more than a decade of paleontological and archaeological research in Old Crow Basin, Yukon, the author outlines a framework for Mid-Wisconsinan human (*Homo sapiens*) occupation of the region. An extensive part on stratigraphy provides a composite section (Figure 2) on which positions where bones and artifacts were found are marked. In all, eight units are given detailed description. Units 2–4 and 7 are the main bone-bearing units. Figure 3 is important, graphing radiocarbon dates from the Old Crow region with two-sigma errors for Units 3a, 4–5, 7b, and 8 (terraces), as well a separate column for 20 bone collagen dates. Information on 25 other dates is also provided in the figure caption. The bone dates show three clusters. Intermediate between the Holocene and Mid-Wisconsinan clusters is a series of steppe bison (*Bison priscus* = *Bison crassicornis*) bone dates ranging between 11,900 and 12,500 BP (I-3574, I-7764, I-7765). The third and largest cluster extends from 28,000 BP to beyond the range of the radiocarbon method. If apatite dates are excluded, there is, as yet, no direct chronometric evidence for the antiquity of Pleistocene artifacts from the region. Although recovery of spirally fractured bones and a polished splinter from Unconformities A and B at Loc. 15 point the way to a solution of the archaeological problem in Old Crow Basin, it cannot be assumed that these unconformities are the only, or even the major, source of artifacts found in the reworked deposits throughout much of the basin. A remarkable feature of collecting in the area is the rarity of stone artifacts. In a closing section, the author mentions: (1) a caribou (*Rangifer tarandus*) tibia flesher; (2) a mammoth (*Mammuthus*) or mastodon (*Mammut americanum*) bone core used as an anvil or cutting-board, (3) an Oldsquaw Duck (*Clangula hyemalis*) humerus and White-fronted Goose (*Anser albifrons*) tibiotarsus with cuts probably made by a stone tool during butchering of the birds; (4) a caribou antler billet cut on two facets on

the bez tine known to be at least 10,700 years old. A shorter version of this paper was presented at the 11th Annual Meeting of the Canadian Archaeological Association in Quebec City, April 1978.

976. Morlan, R.E. (1980a): Taphonomy and archaeology in the Upper Pleistocene of the northern Yukon Territory: A glimpse of the peopling of the New World. National Museum of Man Mercury Series, Archaeological Survey of Canada Paper No. 94. 380 pp.
This paper offers evidence that northeastern Beringia was populated by humans (*Homo sapiens*) much earlier than has been shown farther S in the Western Hemisphere – perhaps as long ago as 50,000 BP. Many of the data presented come from the Old Crow Basin, Yukon, an area replete with redeposited fossils and artifacts. A background of the collections is provided along with the geological and stratigraphic contexts, a taphonomic analysis, and an examination (supported with many photographs) of artificially modified bones [i.e., human-modified]. Table 9.4 lists 28 radiocarbon dates. Previously unpublished bone dates include: three bison (*Bison*) humeri dated at 11,450 ± 400 BP (Qu-780), 12,220 ± 1500 BP (Qu-783), and 12,660 ± 560 BP (Qu-782); a horse (*Equus*) humerus dated at 26,460 ± 3760 BP (Qu-784); a bison (*Bison*) humerus dated at 26,640 ± 2800 BP (Qu-781); and a mammoth(?) (cf. *Mammuthus*) long-bone fragment dated at 29,300 ± 2400 BP (I-11050). This is the first reported incidence of bone dates from the Old Crow region that fall between 11,000 and 31,000 BP. None of the other reported radiocarbon dates performed on either plant material or organic muck from the same area have provided dates within that gap. This date break seems to correspond to the time during the Wisconsinan glaciation when a meltwater lake covered the whole region. The anomalous bone dates could relate to animals that were either washed into the lake or that fell through the ice [see Harington 1977a, section on steppe bison (*Bison priscus*) from Old Crow Loc. 11(1)].

977. Morlan, R.E. (1980b): Yukon Refugium Project. *In*: Archaeological Survey of Canada Annual Reviews 1977–1979. (R.J.M. Marois, ed.), pp. 35–47, 65–8.
These are reports of the status of research by the author and other members of the Yukon Refugium Project. The discussion focuses mainly on examination at Old Crow Locality 15 (see Figures 15 and 21), and Pleistocene mammal bones and artifacts found along Disconformity A. Members and Associates of the Project are listed on p. 42, and an extensive list of contributions is provided (pp. 42–7). A preliminary list of vertebrate taxa recovered from three stratigraphic contexts (Disconformity A, Mid-section deposit, Detrital sands above the reworked lower lake clay) at Old Crow Locality 15 is provided (pp. 65–7).

978. Morlan, R.E. (1981): Big bones and tiny stones: Early evidence from the northern Yukon Territory. X Congreso, Union Internacional de Ciencas Prehistoricas y Protohistoricas (October 19–24, Mexico, D.F.). pp. 1–26.
Hundreds of green-fractured mammoth (*Mammuthus*) limb bones in Old Crow Basin, Yukon, as well as more than 100 bones and flakes of mammoth bone have been found. Only one of these has been directly radiocarbon dated: 29,300 ± 2400 BP (I-11050, Harington 1980d). The others can be attributed to stratigraphic contexts within Unit 2a or 2b on the basis of states of preservation, and this inference assigns them to ages of >25,000 BP. Further, mammoth limb-bone fragments broken when fresh have been recovered from Disconformity A at three localities in Old Crow Basin, and a mammoth-bone flake was excavated from deeper deposits. These finds show that proboscideans were being butchered and their bones broken and flaked by people (*Homo sapiens*) as early as Early Wisconsinan time. Ground and polished bones from Old Crow Basin may represent tools. Another important category of evidence from bones and antlers consists of cuts made by stone tools either during artifact manufacture or during butchering. Possible 'microdebitage' (stone particles possibly derived from stone-tool manufacture) from Pleistocene deposits at Hungry Creek and Eagle River, Yukon, are also discussed. The author points out that, unfortunately, archaeologists cannot always find ideal evidence (e.g., undisturbed, stratified, dated sites containing artifacts, faunal and human skeletal materials) anymore than paleontologists can do all their work on the basis of articulated skeletons and primary-death assemblages.

979. Morlan, R.E. (1983): Counts and estimates of taxonomic abundance in faunal remains: Microtine rodents from Bluefish Cave I. Canadian Journal of Archaeology 7(1):61–76.

The author discusses different statistical tools for quantifying faunal abundances for Bluefish Cave 1, 54 km SW of Old Crow in the northern Yukon. The methods compared are NISP (number of identified specimens), MNI (minimum number of individuals), and the Peterson Index (Fieller and Turner 1982), which is used by biologists to estimate sizes of living populations. Morlan does not endorse any one method, but lists the advantages and disadvantages of each. Concentration graphs comparing the results of the three methods include the following microtine species: red- or black-backed vole (*Clethrionomys*), brown lemming (*Lemmus*), collared lemming (*Dicrostonyx*), chestnut-cheeked or taiga vole (*Microtus xanthognathus*), singing vole (*Microtus miurus*), and tundra vole (*Microtus oeconomus*). Other remains mentioned include: shrew (*Sorex*), hare (*Lepus*), ground squirrel (*Spermophilus*), groundhog (*Marmota monax*), muskrat(?) (cf. *Ondatra*), mustelid (*Mustela*), canid (*Canis*), fox (*Alopex* and *Vulpes*), horse (*Equus*), caribou (*Rangifer*), bison (*Bison*), sheep (*Ovis*), birds (Aves), and fish (Pisces).

980. Morlan, R.E. (1984a): Biostratigraphy and biogeography of Quaternary microtine rodents from northern Yukon Territory, Eastern Beringia. *In*: Contributions in Quaternary Vertebrate Palaeontology: A Volume in Memorial to John E. Guilday (H.H. Genoways and M.R. Dawson, eds.). Carnegie Museum of Natural History Special Publication No. 8:184–99.

This paper presents new data and reviews previous information on microtine rodent fossils from Late Pleistocene and Holocene sites in northern Yukon Territory. Red-backed vole (*Clethrionomys*), vole (*Microtus*), brown lemming (*Lemmus*), and forms like collared lemming (Dicrostonychini) are present throughout the sequence, which is also punctuated by two appearances of heather vole (*Phenacomys*). Evolutionary changes can be recognized within the collared lemming (*Dicrostonyx*) lineage. Vole fossils bearing similarities to *Microtus paroperarius* are labelled '*Microtus* sp. X' and are separated from tundra voles (*Microtus oeconomus*) by means of a series of measurements. The Beringian microtine record is incomplete, poorly dated, and composed of often very small samples, however some tentative proposals are presented concerning biostratigraphy and biogeography. Gaps in the record are identified. The Yukon sequence is illustrated and arranged in terms of aged intervals, and possible correlations with Alaska and Siberian sediments and faunas are discussed. A map is included illustrating the localities mentioned in the text as well as the limits of glacial advance. Illustrations are also provided of the molar cusp patterns of the specimens.

981. Morlan, R.E. (1984b): Problems in the interpretation of bones from Old Crow Basin, northern Yukon Territory. Typescript of a paper presented at the First International Bone Modification Conference (August, Carson City). 5 pp.

The purpose of this paper is to outline problems encountered in studying evidence of human (*Homo sapiens*) modification of Pleistocene mammal bones collected from Old Crow Basin, Yukon, and to propose solutions. The author mentions several areas of uncertainty: stratigraphic, chronometric, taphonomic, and analogic. All the artifacts have been found in reworked contexts wherein the specimens are older than the surrounding sediments. All or part of nine such artifacts have been sacrificed for radiocarbon dating: tibia flesher, polished antler hide grainer, two green-fractured mammoth (*Mammuthus*) limb bones, and five mammoth bone cores. Seven of the dates range from 30,490 to 24,800 BP. One mammoth bone core was dated to 13,335 BP, and another yielded a date of 41,460 BP (possibly contaminated). The author believes the cluster of dates between 30,000 and 25,000 BP marks a period of occupation by people. The only specimens in his collections that might indicate human presence in the region in earlier times are a cut-marked bison (*Bison*) rib and a few possibly green-fractured mammoth bones from an Early Wisconsinan floodplain perhaps 80,000 to 60,000 years old. The Old Crow fossils could represent an eastern facies of the Siberian proto-Dyuktai and Dyuktai traditions as defined by Mochanov. If so, then it is not surprising that our best evidence from Old Crow Basin begins about 30,000 BP, shortly after the arrival of the proto-Dyuktai people in the Aldan valley of Siberia.

982. Morlan, R.E. (1986): Pleistocene archaeology in Old Crow Basin: A critical reappraisal. *In*: New Evidence for the Pleistocene Peopling of the Americas (A.L. Bryan, ed.). Center for the Study of Early Man, University of Maine at Orono. pp. 27–48.

Some Quaternary vertebrate bones found among the rich

fossil deposits of Old Crow Basin in northern Yukon have been interpreted as artifacts made prior to their permineralization. Previous reports have attempted to show that artificial [human (*Homo sapiens*)] alterations are exhibited by fossils recovered from an Early Wisconsinan floodplain dated to about 80,000 BP. Here, all such specimens, from Disconformity A and deeper deposits, are reexamined with alternate interpretations presented. The alternate interpretations do not prove that people were not present in Early Wisconsinan time, but they show that such early human presence cannot be demonstrated on the basis of evidence gathered so far. None of the definitive artifacts from modern banks and bars of Old Crow River has known stratigraphic context, although several artifacts have now been directly dated by radiocarbon techniques. Directly dated specimens show that people (*Homo sapiens*) were present in Eastern Beringia by at least 25,000 to 30,000 years ago. Table 4 includes dates and comments on selected Quaternary mammal bone and antler from northern Yukon [10 uranium/thorium dates on bison (*Bison*), mammoth (*Mammuthus*) and other proboscidean (Proboscidea) bone; two uranium/protactinium dates on bison and mammoth; three radiocarbon dates on bone apatite from caribou (*Rangifer tarandus*) and proboscidean bone; and seven radiocarbon dates on collagen from proboscidean bone and a caribou antler.

983. Morlan, R.E. (1987): The Pleistocene archaeology of Beringia. *In*: The Evolution of Human Hunting (M.H. Nitecki and D.V. Nitecki, eds.). Plenum Publishing Corporation. pp. 267–307.
Beringian archaeological sites are compared to determine when humans (*Homo sapiens*) first populated the area. The earliest accepted date for Eastern Beringia is about 12,000 BP; however some artifacts much older than this date would seem to indicate human presence. A cutmarked bison (*Bison*) rib found at Disconformity A at the Old Crow River Loc. HH69-21 is believed to represent 105,000–87,000 BP; the condition of the bone cannot necessarily be attributed to human alteration. The oldest archaeological sites in Eastern Beringia are the Bluefish Caves, which are 54 km SW of Old Crow in the northern Yukon. The vertebrate remains in the caves range from Late Pleistocene to Holocene, although most are from the Late Pleistocene. The remains include: mammoth (*Mammuthus*), horse (*Equus*), bison (*Bison*), caribou (*Rangifer tarandus*), moose (*Alces alces*), Dall sheep (*Ovis dalli*), wapiti (*Cervus elaphus*), arctic fox (*Alopex lagopus*), red fox (*Vulpes vulpes*), bear (*Ursus*), wolf (*Canis lupus*), large cat (*Panthera leo atox*), mustelids (*Mustela*), arctic hare (*Lepus arcticus*), snowshoe hare (*Lepus americanus*), marmot (*Marmota*), ground squirrel (*Spermophilus*), vole (*Microtus*), shrews (Soricidae), birds (Aves), and fish (Pisces). Radiocarbon dates for the preceding specimens range from 12,000 to 25,000 BP. Stone and bone (Dall sheep flesher) tools have been found at the lower interval of loess of Cave 1 (dated on birch pollen use from 14,000 to 12,000 BP). In the lower part of the loess within and just outside of Cave 2 were a caribou flesher, a mammoth bone core with flake [dated at about 23,500 BP by accelerator mass spectrometry – see Cinq-Mars and Morlan 1999], and bone tools. The Pleistocene megafauna bones have butcher marks, whereas none of the Holocene specimens do. Radiocarbon dates on some Alaskan bone tools (species not mentioned) were: from Akmak, 9570 ± 150 BP (K-1583, bone collagen); from Trail Creek Cave 2, 9070 ± 150 BP (K-980, bone collagen); and from Chindadn, 11,090 ± 170 BP (GX-1341, bone apatite). The author contends that humans first came to the Americas about 20,000 BP.

984. Morlan, R.E. (1989): Paleoecological implications of Late Pleistocene and Holocene microtine rodents from the Bluefish Caves, northern Yukon Territory. Canadian Journal of Earth Sciences 26:149–56.
The author correlates taxonomic abundances and assemblages of certain rodents from the three Bluefish Caves (54 km SW of Old Crow in the northern Yukon). The results agree with the predicted environments of the different times. Late Pleistocene and Holocene components are clearly distinguishable from each other in each of the three caves, and each component can be subdivided within Cave 1. This paper discusses: (1) variations in taxonomic abundance through time and between sites; (2) contrasts in the microhabitat between N-facing Cave 1 and S-facing cave 2; and (3) decreases in tooth size that may reflect a reduction in the length of the growing season. The fauna include: red-backed vole (*Clethrionomys rutilus*), collared lemming (*Dicrostonyx torquatus*), brown lemming (*Lemmus sibiricus*), singing vole (*Microtus miurus*), tundra vole (*Microtus oeconomus*), meadow vole (*Microtus pennsylvanicus*), yellowcheeked or taiga vole (*Microtus xanthognathus*), muskrat (*Ondatra zibethicus*), and northern bog lemming (*Synap-*

tomys borealis). The presence of certain species indicates a certain paleoenvironment. The Pleistocene components of the caves include an upper and lower layer of loess, radiocarbon-dated, respectively, at 12,900 ± 100 BP from a horse (*Equus*) femur in Cave 1 (GSC-2881), and 15,500 ± 130 BP from a mammoth (*Mammuthus*) scapula in Cave 2 (GSC-3053). The lowermost loess units (6, 7) represent an herb-tundra zone with collared lemming, brown lemming, and singing vole. The dwarf-birch zone units (4, 5) show a decreased brown lemming population, an absence of the other herb-tundra fauna, and an increase of red-backed vole and taiga vole. A northern bog lemming molar was found here. The uppermost loess units (2, 3) have low fossil yields, so abundances are difficult to determine. Some species, such as brown lemming and tundra vole, represent wet habitats, whereas singing vole and collared lemming prefer dry ones. Red-backed vole can be found in both wet and xeric habitats. A general decrease in diversity is shown to involve increased dominance and decreased species richness and evenness. These changes are attributed to postglacial zonation of habitat.

985. Morlan, R.E. (1991): The Yukon Refugium Project: Contributions to vertebrate paleontology. Canadian Museum of Civilization Archives Manuscript No. 2126. 151 pp. [Copy in Quaternary Vertebrates of Northern North America Publication File at Canadian Museum of Nature (Paleobiology).]
The Yukon Refugium Project (YRP) was organized in 1974 as a multidisciplinary research program devoted to the Quaternary prehistory and paleoecology of unglaciated areas of Yukon. From 1975 through 1981, it was one of the principal projects funded in part by the National Museum of Man (now the Canadian Museum of Civilization). From the outset, the search conducted by the YRP was directed toward vertebrate fossils, and the archaeological aspects of the project were concerned primarily with the discovery and interpretation of bone alteration patterns. This report consists of a brief text and several appendices. An annotated list of reports on the vertebrate fossils is given under *Previous Work* (p. 3). Appendices include: (1) Lists of vertebrates by locality and sample. Each list is arranged in taxonomic order so that future investigators can scan the lists for taxa of interest; (2) A numerical list of catalogue numbers assigned to the author by the Paleobiology Division of the Canadian Museum of Nature. This represents, in part, a cross-reference to Appendix 1, but it also includes many surface finds not listed in that appendix; (3) A map of Old Crow River drawn from air photographs, showing the valley walls, oxbow lakes, and river terraces. The distance by river is marked in 1 km intervals with '0' at the river mouth. All localities relevant to this collection (and many others) are shown; (4) A series of 19 reports entitled 'Investigations at [Locality Number].' Each report assembles all observations by year for a given locality. Field sample numbers are referenced in the text, and all photographs and drawings are assembled and cited.

986. Morlan, R.E. (1993): A compilation and evaluation of radiocarbon dates in Saskatchewan. Saskatchewan Archaeology 13:3–84.
This review article conveniently lists all the archaeological sites (*Homo sapiens*) known from the province, with the Borden-site designation and latitude and longitude. All radiocarbon dates on bone or plant material from each locality are included. A pertinent discussion on the reliability of radiocarbon dating outlines some of the problems inherent in the process. The author concludes that in general there is good agreement between stratigraphic sequences and radiocarbon dating in Saskatchewan archaeology. The paper includes an extensive bibliography, and summaries of the radiocarbon dates, by site, in chart form.

987. Morlan, R.E. (1994): Oxbow bison procurement as seen from the Harder Site, Saskatchewan. Journal of Archaeological Science 21:757–77.
The Oxbow complex arose on the northern plains about 6000 BP and persisted there for at least two millennia. Like other archaeologically known cultures of the region, Oxbow subsistence was based on bison (*Bison*). Faunal remains from the Harder Site, the largest known Oxbow campsite, confirm winter seasonality and reveal an attritional-mortality profile dominated by prime-age adults with nearly equal representation of bulls and cows. These findings suggest that Harder was provisioned by communal hunting (*Homo sapiens*) of bison nursery herds and by hunting of solitary bulls and cows or small bull herds. Stable carbon isotope measurements of bison bones suggest a more xeric grassland than occurred in the region during historic time. Faunal assemblages reported for the Oxbow complex are listed in Table 1. The vertebrates include: bison, pronghorn antelope (*Antilocapra*), moose

(*Alces alces*), wapiti (*Cervus elaphus*), deer (*Odocoileus*), wolf (*Canis lupus*), coyote (*Canis latrans*), domestic dog (*Canis familiaris*), fox (*Vulpes*), fisher (*Martes*), rabbit or hare (*Sylvilagus/Lepus*), beaver (*Castor canadensis*), ground squirrel (*Spermophilus*), birds (Aves), and fishes and/or amphibians (Pisces/Amphibia). Vertebrate taxa recovered from the Harder Site itself include: unidentified bird (Aves), snowshoe hare? (?*Lepus americanus*), wolf, dog, coyote, red fox (*Vulpes vulpes*), swift fox (*Vulpes velox*), fisher (*Martes pennanti*), badger (*Taxidea taxus*), deer (Cervidae), moose (*Alces alces*), and bison (*Bison bison*). In contrast to bison from the older Gowen and Norby sites, also located in S-central Saskatchewan, the bison from the Harder site are very similar in size to modern plains bison (*Bison bison bison*). Three bison-bone samples – a mandible and humerus; a humerus and tibia?; and a humerus – yielded radiocarbon dates of 3420 ± 140 BP (S-3453), 4410 ± 150 BP (S-3444), and 4190 ± 90 (S-3452), respectively.

988. Morlan, R.E. (1996): Late and Middle Pleistocene vertebrate fossils from Old Crow Basin, Locality CRH 15, northern Yukon Territory. *In*: Palaeoecology and Palaeoenvironments of Late Cenozoic Mammals: Tributes to the Career of C.S. (Rufus) Churcher (K.M. Stewart and K.L. Seymour, eds.). University of Toronto Press. pp. 483–521.

Vertebrate fossils are described from four stratigraphic levels at Old Crow Basin Locality CRH 15 (67°51′30″N, 139°48′40″W) about 1 km upstream from the juncture of Old Crow River and Johnson Creek in the northern Yukon. Stratigraphy consists of four units: (1) massive clay of unknown thickness extending below the level of Old Crow River; (2) about 20 m of fluvial sediments, incorporating minor lacustrine deposits; (3) about 8 m of massive glaciolacustrine clay; and (4) about 3.5 m of silt and peat forming the modern surface. All vertebrate remains described are from Unit 2 (see Figure 3). The largest assemblage includes four fishes (Pisces), two birds (Aves), and 23 mammals (Mammalia) dating to the Late Pleistocene. The oldest assemblage, tentatively correlated with the Middle Pleistocene Olyorian fauna of Siberia, is relatively poor, with only six mammal taxa and an unidentified fish. Between is an assemblage intermediate in both age and diversity that cannot be dated at present. Vertebrate taxa from the locality are summarized in Table 2, whereas Table 1 lists nine radiocarbon dates (six on bone) and five uranium-thorium dates on bone. The fauna from Disconformity A may include a mix of species that lived during the last (Sangamonian) interglacial, the colder Early Wisconsinan, and the relatively warm Mid-Wisconsinan interstadial. During every field season, workers at the site found mammoth (*Mammuthus*) bones apparently fractured and flaked by people (*Homo sapiens*) soon after the animal's death, but it was rare to find large, interpretable fragments *in situ*. A selection of mammoth bones considered to reflect human activity was radiocarbon dated (Morlan et al., 1990), and the results suggest that human modification of mammoth bones began shortly after 40,000 years ago.

989. Morlan, R.E. (1999): Canadian Archaeological Radiocarbon Database. www.canadianarchaeology.com/radiocarbon/card/card.htm.

The Canadian Archaeological Radiocarbon Database (CARD) is a compilation of radiocarbon dates indicating the ages of archaeological and vertebrate paleontological sites in Canada. During the 50 years since the radiocarbon-dating technique was invented, Canadian archaeologists and paleontologists have invested heavily in this method, acquiring over 7000 radiocarbon dates. The dates are widely scattered in published and unpublished sources, and many have not been reported. The dates are not all of equal value, as they represent the results of analytical methods that have evolved over half a century. This database seeks to improve the understanding and use of radiocarbon dates dealing with Canada by compiling and evaluating the analyses. A related GIS application, Mapping Ancient History, presents most of the data sites on maps depicting the deglaciation and landscape evolution of North America during the last 14,000 years.

990. Morlan, R.E., and Bonnichsen, R. (1975): Early human occupations in Old Crow Flats: The nature of the evidence. Paper presented to the Symposium on 'Correlation of the Ancient Cultures of Siberia and Adjoining Territories of the Pacific Coast.' (October, Novosibirsk). 14 manuscript pp.

Remains of extinct animals found in reworked river deposits in the Old Crow Basin, Yukon, exhibit evidence of having been altered by humans (*Homo sapiens*). Most of the modified bones and artifacts were recognized in the laboratory rather than in the field by using comparative morphological collections showing different types of natural, biological, and cultural alterations, thus providing criteria for separating human-modified bones from those modified by other agents. Furthermore, experiments in bone fracture have clearly demonstrated that green (fresh) bone breaks differently than partially mineralized Pleistocene bone from Old Crow Basin. Accordingly, it would not have been feasible for modern Indians to rework fossil specimens. The authors are confident that the Old Crow specimens are Pleistocene-age artifacts. The idea that the presence of humans can be identified on the basis of bone alteration may have widespread significance. The Old Crow data suggest that most implements were produced from bone in Mid-Wisconsinan times by using flaking techniques commonly associated with stone. It is hoped that future fieldwork will reveal an *in situ* layer with remains that will provide a better understanding of the adaptive patterns of these early North Americans.

991. Morlan, R.E., and Bonnichsen, R. (1979): Pervonachal 'noe zaselenie chelovekom doliny Old Crow, Yukon [Initial peopling of the Old Crow Basin, Yukon]. *In*: Drevnie kultury Sibiri I Tikhookeansko-go Basseina [Ancient Cultures of Siberia and the Pacific Basin (R.C. Vasilevskii, ed.). Akademia Nauk CCCP [USSR Academy of Sciences, Institute of History, Philology and Philosophy]. pp. 48–57.

The authors summarize the history of research in the Old Crow Basin, Yukon, listing some of the Pleistocene mammal remains found at Old Crow Localities 11(1), 14N, and 44, as well as pertinent radiocarbon dates. The paper concludes with new data listed under such headings as 'mineralization' and 'technology' and provides figures of several bones interpreted as artifacts from Old Crow Localities 11, 14N, 22 and 29.

992. Morlan, R.E., and Cinq-Mars, J. (1982): Ancient Beringians: Human occupation in the Late Pleistocene of Alaska and the Yukon Territory. *In*: Paleoecology of Beringia (D.M. Hopkins, J.V. Matthews Jr., C.E. Schweger, and S.B. Young, eds.). Academic Press, New York. pp. 353–81.

The authors make a case for human (*Homo sapiens*) presence in Eastern Beringia tens of thousands of years before the established date of 12,000 BP, although they note that it is difficult to distinguish between some natural and artificial (human) bone alteration. Many artifacts are examined. A caribou (*Rangifer tarandus*) flesher from the Old Crow Basin initiated interest in this field as it was radiocarbon dated at 27,000 +3000 –2000 BP (GX-1640, bone apatite) [this has since been re-dated as a Late Holocene artefact at 1350 ± 150 BP (RIDDL-145, AMS); see Nelson et al. 1986]. Two mammoth (*Mammuthus*) long bones, which were flaked while fresh, yielded similar dates of 29,100 +3000 –2000 BP (GX-1567) and 25,750 +1800 –1500 BP (GX-1568) [since the caribou flesher date was found to be in error, it is likely that the mammoth bone dates are also in error; see Irving 1987 and Nelson et al. 1986]. Another specimen that seems to have been altered by humans is a bison (*Bison*) bone, fractured while fresh, that was found along Old Crow River and radiocarbon dated to >37,000 BP (GSC-2792, enclosing matrix). Fractured bones, including those of mammoth (*Mammuthus*) or mastodon (*Mammut americanum*) long bones, cut caribou-antler fragments, and a cut bison skull were found at placer-mining sites in the Klondike region, Yukon, and are similar to finds from the Fairbanks, Alaska, muck. The Fairbanks muck contains mammal remains ranging from Late Illinoian to the present, with the majority of specimens from the Late Wisconsinan to the present. A placer mine near Lost Chicken in eastern Alaska has produced Late Wisconsinan altered mammal remains: several cut caribou antlers, a steppe bison (*Bison priscus*) tibia fractured while fresh [actually 10,370 ± 160 BP (I-8582)], and a mammoth (*Mammuthus*) limb bone fragment, fractured while fresh. At Jack Wade Creek, near Lost Chicken, Alaska, an assemblage was found dominated by sheep (*Ovis*), and also included altered remains of: bison, caribou, a small Yukon horse (*Equus lambei*), moose (*Alces alces*), and steppe bison (*Bison priscus*), whose horncore was radiocarbon dated (collagen) at 29,700 ± 240 BP (QL-1178). At Trail Creek, Alaska, Cave 9, lowest-level altered mammal remains included: Dall sheep (*Ovis dalli*) bones; a horse (*Equus*) scapula radiocarbon dated (organic fraction) at 15,750 ± 350 BP (K-1210); and broken bison (*Bison*) calcaneum (perhaps by man, for carnivore teeth marks are absent) radiocarbon dated (collagen) at 13,070 ± 280 BP (K-1327). Altered remains from Bluefish Caves in the northern Yukon [see Cinq-Mars 1979, 1982, 1990] include: a horse femur from Cave 1 upper loess radiocarbon dated at 12,900 ± 100 BP (GSC-2881); a mammoth scapula from Cave 2 lower loess radiocarbon dated at 15,500 ± 130 BP (GSC-3053); as well as longitudinally cut sheep and caribou bones. Many photographs of specimens are included.

993. Morlan, R.E., and Matthews, J.V., Jr (1978): New dates for early man. GEOS, Winter:2–5.

This popular paper mentions new finds from the northern Yukon that indicate humans (*Homo sapiens*) were in North America long before some archaeologists thought. Bone tools and other bones broken by people more than 30,000 years ago have been found in areas unglaciated during the Pleistocene. Previous estimates had put the first human inhabitants at 14,000 years ago. During glacial phases, the broad Bering Sea continental shelf was exposed, forming a land connection between unglaciated parts of northeastern Siberia and Alaska/Yukon. Together, these land areas are known as Beringia – a refugium where many species of plants and animals survived periods of glaciation and from which the survivors dispersed following the melting of the ice sheets. Several projects have recently been initiated to explore the paleoenvironmental history of Beringia. The Yukon Refugium Project was begun in the hope that a careful study of the sedimentary sequence (e.g., Figure 1 shows the composite stratigraphy of sediments in Old Crow Basin including the fossil-bearing strata) would narrow the search for early human remains. Bones of fishes (Pisces), birds (Aves), and mammals (Mammalia) are abundantly preserved and are studied by C.R. Harington, National Museum of Natural Sciences [now Canadian Museum of Nature]. Artifacts are analysed by R.E. Morlan, National Museum of Man [now Canadian Museum of Civilization]. Among the bones from the thick interlake sediments in Old Crow Basin are several that seem to have been modified by people during butchering, consumption, and use of the large mammals as sources of new material for tool making. These activities took place before fossilization of the bones, and represent the oldest dated evidence of human occupation yet found in the western hemisphere. More than 50 of the 170 fossil-collecting localities in the Old Crow area have yielded modified bones and artifacts. Mammoth (*Mammuthus*), horse (*Equus*), and caribou (*Rangifer tarandus*) bones were most often altered by humans. Nearly 100 artifacts have been identified, as well as hundreds of spirally fractured bones apparently broken by people when the bones were fresh. Artifacts include: three caribou antler wedges, a caribou tibia fleshing tool (see photographs on p. 5), a caribou antler pestle, and a distinctive tool thought to have been used in the manufacture of stone artifacts. A stratigraphic and chronological framework has been developed to provide specific hypotheses for testing in the field and laboratory. The search continues for the needle in the stratigraphic haystack: an undisturbed archaeological site.

994. Morlan, R.E., Nelson, D.E., Brown, T.A., Vogel, J.S., and Southon, J.R. (1990): Accelerator mass spectrometry dates on bones from Old Crow Basin, northern Yukon Territory. Canadian Journal of Archaeology 14:75–92.

This paper compares traditional radiocarbon dating methods (bone apatite and collagen) with the newer accelerator mass spectrometry (AMS) method. The results show that AMS is generally more reliable. Of the 55 AMS radiocarbon dates listed in the paper, those that are compared with the older methods are the following. Old Crow and Porcupine rivers proboscidean (Proboscidea) bone core radiocarbon dates are (AMS first, original second): from Loc. CRH 74 41,400 ± 1600 BP (RIDDL-234) consistent with 41,460 +5560 –3290 BP (CRNL-1219); from Loc. CRH 14N 27,000 ± 400 BP (RIDDL-232) compared with 25,970 ± 560 BP (CRNL-1234); from Loc. CRH 87 39,900 ± 1300 BP (RIDDL-233) inconsistent with 13,335 ± 390 BP (CRNL-1218); tusk from Cadzow Bluff, Porcupine River, upper unit 24,700 ± 250 BP (RIDDL-229) consistent with 25,170 ± 630 BP (CRNL-1232); and a long bone from Loc. CRH 4 31,120 ± 450 BP (RIDDL-122) and 30,680 ± 600 BP (RIDDL-231) consistent with 30,490 ± 550 BP (CRNL-1235). Other Old Crow River megafauna dates include: bison (*Bison*) radius from Loc. CRH 70 37,300 ± 750 BP (RIDDL-136) consistent with >37,000 BP (GSC-2792); and a caribou (*Rangifer tarandus*) antler pestle from Loc. CRH 29 24,700 ± 300 BP (RIDDL-230) consistent with 24,800 ± 650 BP (CRNL-1233). The authors conclude, based on these figures, that radiocarbon dates approaching 40,000 BP using the traditional methods are less accurate than AMS.

995. Morris, T.F., McAndrews, J.H., and Seymour, K.L. (1993): Glacial Lake Arkona–Whittlesey transition near Leamington, Ontario: Geology, plant, and muskox fossils. Canadian Journal of Earth Sciences 30:2436–47.

A fragmentary metacarpal possibly belonging to a shrubox (cf. *Euceratherium*; ROM 41960) was found in an exposed subaquatic fan deposited in Glacial Lake Maumee exposed in the Bondi Aggregate Pit located 5 km NW of Leamington, Ontario (42°05′17″N, 82°38′15″W; Figure 1). When the lake level dropped, these fans were exposed and developed vegetation mats. AMS radiocarbon dates of the bone [13 410 ± 100 BP (TO-1803)] and organic material [13 225 ± 200 BP (BGS-1404) and 13 150 ± 100 BP (WE-01-89)] indicate deposition during the Glacial Lake Arkona–Glacial Lake Whittlesey transition. Pollen and plant macrofossils indicate a forest-tundra environment with a mean July temperature of 14°C. The discovery of cf. *Euceratherium* is surprising, as its previous range was well S and W of the site. A previously reported muskox [*Ovibos moschatus* = *Ovibos proximus*; ROM 1194, McDonald and Ray 1989)] yielded a date of 11,580 ± 70 BP (TO-3652). Another specimen attributed by Churcher and Karrow (1977) to a metacarpal of *Ovibos* is now identified as a recent [290 ± 50 BP (TO-3651)] right metatarsal of domestic cattle (*Bos taurus*). Sedimentology of the site is described, and photographs of the bone are included (Figure 5).

996. Mott, R.J., and Stalker, A.M. (1972): Palynology of the 'Kansan' carbonaceous clay unit near Medicine Hat, Alberta. Geological Survey of Canada Paper 72-1 (Part B):117, 119.

This report provides a paleoenvironmental context for the Medicine Hat, Alberta, Pleistocene mammal remains from 'preglacial' deposits near the base of high bluffs exposed along the South Saskatchewan River. The fauna, described in Stalker (1969b) and Stalker and Churcher (1970), includes: beaver (*Castor canadensis*), Harlan's ground sloth (*Glossotherium harlani* = *Paramylodon harlani*), a smaller ground sloth (*Nothrotheriops*), wolf (*Canis* cf. *Canis etruscus*), southern mammoth (*Mammuthus meridionalis* = *Mammuthus imperator haroldcooki*), Scott's horse (*Equus scotti*), stilt-legged ass (*Equus calobatus*), Minidoka camel (*Camelops minidokae*), long-legged llama (*Hemiauchenia*? *macrocephala* = *Tanupolama*? *hollomani*), and a prongbuck (Antilocapridae). In an effort to better understand the past landscape, Mott analysed fossil pollen samples from Island and Mitchell bluffs. The carbonaceous unit, with its abundant wood, is the only one near Medicine Hat in which beaver has been found. The site, determined by stratigraphy to be Kansan in age, was open grassland on upland sites, with trees and shrubs occurring along the rivers and on other suitable spots, and possibly pine and spruce forests in the Cypress Hills area. The climate at the time was probably similar to the present or slightly cooler and moister.

997. Muhs, D.R., Ager, T., Stafford, T.W., Jr, Pavich, M., Begét, J.E., and McGeehin, J.P. (1997): The last interglacial-glacial cycle in Late Quaternary loess, central interior Alaska. Beringian Paleoenvironments Workshop (September 20–3, Florissant, Colorado). Program and Abstracts. pp. 109–12.

In their discussion of the Eva Creek, Alaska, section as an example of valley-bottom localities, the authors mention three wood-bearing paleosols with ages of about 8800 BP that can be traced laterally to a beaver (*Castor canadensis*) dam of the same age (p. 110).

998. Müller-Beck, H. (1966): Paleohunters in America: Origins and diffusion. Science 152(3726): 1191–1210.

The author provides a summary of the history of paleohunters (*Homo sapiens*) in the plains of Eurasia since more than 65,000 years ago and their adaptations to changing climate – to subarctic conditions. His ideas are presented explicitly in a series of four maps (polar views showing Eurasia and North America). The expansion of these hunters over northern Eurasia and their crossing of the Bering Isthmus ('land bridge'), as the first known human invaders of America, about 28,000 to 26,000 years ago are reconstructed. Afterward the invaders were isolated by ice advances of the Wisconsinan maximum in southern North America and separated from the continuing technological evolution of the Old World. The contact between Asia and interior America was not feasible again until the melting of the inland ice barrier, when Aurignacoid groups invaded or expanded over North America for the first time. The first invaders must be considered ancestors of the Plains Indians; the second, as those of the Inuit ('Eskimo') and Aleuts. The fact that an animal migration took place during the 'land-bridge periods' (between about 50,000 and 40,000 years ago and 28,000 and 10,000 years ago) is shown by first appearance of woolly mammoths (*Mammuthus primigenius*), reindeer/caribou (*Rangifer tarandus*) [but see Harington 1999b] and other species during the Late Pleistocene ('Neopleistocene').

999. Müller-Beck, H. (1967): On migrations of hunters across the Bering Land Bridge in the Upper Pleistocene. *In*: The Bering Land Bridge (D.M. Hopkins, ed.). Stanford University Press, Stanford, California. pp. 373–408.
This is a review of climatic and geographic history of the Bering Land Bridge, as well as Palaeolithic technological developments in the Old and New worlds during the Late Pleistocene. It allows a partial reconstruction of diffusion and migration events in the Bering region. Evidence from the stone tool cultures of the Old and New worlds indicate that human (*Homo sapiens*) groups must have crossed the Bering Land Bridge in the course of hunting activities about 28,000 years ago.

1000. Munroe, M. (1999): Ancient stone tools chip away date of early humans' arrival. National Post, January 16: B12.
This article summarizes information on finds of stone artifacts from sites on the Bow River near Calgary, in a gravel pit near Grimshaw, and at two localities W of Lethbridge, Alberta, considered by an archaeologist, Dr Jiri Chlachula, to be between 20,000 and 40,000 BP. He states: 'Plenty of animals roamed the land [Prairies] then and there's no reason to think people didn't.' The article includes a map of North America with the Bering Isthmus and the Cordilleran and Laurentide ice sheets delineated. Superimposed on this map are seven archaeological sites possibly more than 12,000 years old, 14 sites 11,000 to 12,000 years old, and 11 sites 10,000 to 11,000 years old. Two possible migration routes for early humans (*Homo sapiens*) are also designated: one from Siberia to Central America E of the Cordillera and a route from Siberia along the Pacific coast of North America to Central America. Three views of one of Chlachula's stone tools are illustrated, at actual size, below the map.

1001. Murray, A. (1852): On the geology of the region between the Ottawa, the St. Lawrence, and the Rideau rivers, with notes on economic minerals. Geological Survey of Canada, Report of Progress 1851-52:57–91.
Capelin (*Mallotus villosus*) and lumpfish (*Cyclopterus lumpus*) are reported in nodules from Green Creek, Ontario.

1002. Nadler, C.F., Hoffmann, R.S., Vorontsov, N.N., and Sukernik, R.I. (1984): Evolutionary relationships of some Beringian mammals. *In*: Beringia in the Cenozoic Era (V.A. Kontrimavichus, ed.). Amerind Publishing Co. Pvt. Ltd., New Delhi. pp. 425–40.
Beringia served as a dispersal route between Asia and North America several times. Some taxa have diverged considerably since their earlier residence in Beringia while other more recent inhabitants are still considered conspecific throughout their Holarctic range. Some arctic ground squirrels (*Spermophilus parryii* = *Citellus parryii*) provide excellent models for the study of Beringian evolutionary relationships. Chromosomes, electrophoretically separable haemoglobins, transferrins, and isozymes, as well as cranial measurements examined by multivariate statistical techniques were compared in populations of Holarctic arctic ground squirrels (2n = 34), Asian ground squirrels (*Spermophilus undulatus*), and American Columbian ground squirrels (*Spermophilus columbianus*), which both display 2n = 32. After isolation by the Bering Strait for 12,800 years, *Spermophilus parryii* has maintained similar chromosomes, and one in 10 proteins has changed in electrophoretic mobility, but skull morphology has diverged. In the *Spermophilus undulatus–Spermophilus columbianus* complex there was no chromosomal divergence and Giemsa-Band patterns substantiate homology after isolation for at least 100,000 years; however, during this time two in 11 proteins differentiated, and cranial divergence has been striking. These data are applied to a zoogeographic reappraisal of the arctic and long-tailed ground squirrel group. Voles (*Microtus*), red-backed voles (*Clethionomys*), and mountain sheep (*Ovis*) are considered similarly. Regarding mountain sheep, the authors propose that: (1) early pachycerine sheep with 2 n = 54 evolved their distinct morphological characteristics while isolated in an Illinoian (?) Beringian refugium; (2) these sheep then migrated southward to western United States, where, after isolation during the Wisconsinan glaciation, they differentiated into modern bighorn sheep (*Ovis canadensis*); (3) isolation of Beringian sheep in eastern Siberia from those in Alaska and Yukon by Bering Strait during the Sangamonian Interglacial led to morphological and cytological differentiation between snow sheep (*Ovis nivicola*) and Dall sheep (*Ovis dalli*).

1003. Nagorsen, D.W. (1987): *Marmota vancouverensis*. Mammalian Species No. 270:1–5.

Among other things, the author discusses the fossil record of the Vancouver Island marmot. The oldest known specimens are remains from an archaeological site at Shoemaker Bay, Alberni Inlet, Vancouver Island, British Columbia (see map, Figure 3), that were dated at about 1500 BP. Hoffman et al. (1979) suggested that the Vancouver Island marmot (*Marmota vancouverensis*) is a Pleistocene isolate derived from mainland populations of the hoary marmot (*Marmota caligata*). Similarly, Heard (1977) and McCabe and Cowan (1945) hypothesized that this species survived the last (Wisconsinan) glaciation in refugia on Vancouver Island. Fossil evidence is needed to confirm the existence of this species on Vancouver Island during the Pleistocene.

1004. Nagorsen, D.W., Keddie, G., and Hebda, R.J. (1995): Early Holocene black bears, *Ursus americanus*, from Vancouver Island. Canadian Field-Naturalist 109(1):11–18.
Skeletal remains of three bears were found in Windy Link Pot Cave (located about 900 m asl on White Ridge near Gold River) on Vancouver Island in 1983. A composite bone sample consisting of two tibias, three ribs, and two vertebrae were submitted for conventional radiocarbon dating. The conventional radiocarbon date was 9760 ± 140 BP, with a corrected date of 9830 ± 140 BP (Beta-10714). Based on extensive measurements, the three specimens (BCPM 17198, 17199, and 19030) were identified as black bear (*Ursus americanus*). Pollen analysis of samples from the site confirms that the Early Holocene environment around the cave consisted of mixed coniferous forest, indicating a climate much warmer and drier than today. Abundant charcoal in the sample is further evidence for warm, dry conditions with high fire frequency. Photographs and measurements of the bear specimens are included.

1005. Nagorsen, D.W., Keddie, G., and Luszcz, T. (1996): Vancouver Island marmot bones from subalpine caves: Archaeological and biological significance. British Columbia Parks, Occasional Paper No. 4:1–56.
Since 1985, remains of the Vancouver Island marmot (*Marmota vancouverensis*) have been discovered in four high-altitude cave sites: Clayoquot Plateau, Mariner Mountain, Limestone Mountain, and Golden Hinde. Cut marks on bones and artifacts recovered in Mariner Mountain Cave indicate that the remains result from human (*Homo sapiens*) hunting. Radiocarbon dates (Table 1 lists 11 dates from cave sites) on bone show that these sites are prehistoric, ranging from 2630 to 830 BP. The faunal remains provide additional evidence for a range decline in this marmot. Although black bear (*Ursus americanus*), black-tailed deer (*Odocoileus hemionus columbianus*), marten (*Martes americana*), and red squirrel (*Tamiasciurus hudsonicus*) are represented in the faunal remains, the predominance of Vancouver Island marmots suggests that aboriginal people travelled to these remote mountain areas to hunt marmots. Remains of Vancouver Island marmots are known from seven prehistoric sites: Shoemaker Bay, the archaeological sites mentioned here, and paleontological sites at Nimpkish Lake and Weymer Creek.

1006. Naldrett, D.L. (1988): Seal (*Phoca* sp.) from Champlain Sea deposits near Ottawa, Canada. Canadian Journal of Earth Sciences 25:787–90.
A left tibia of a juvenile seal (*Phoca*) was found in the Brazeau Sand Pit, on Cedarview Road, Nepean, Ontario, in the interchannel sand and gravel facies of the Twin Elm Ridge. The sand and gravel are a subaqueous outwash deposit, and the tibia showed abrasion probably due to transport. In addition, 23 Champlain Sea seal specimens have been found, of which 10 are from the Ottawa area. These include harp seal (*Phoca groenlandica*), bearded seal (*Erignathus barbatus*), and ringed seal (*Phoca hispida*). [See Harington 1988.]

1007. Nancy, D. (1999): Des collections bien particulières: La laboratoire de paléoanthropologie dévoile ses trésors caches. Forum (Université de Montréal) 34(9):9.
This article mentions paleontological treasures discovered by Michel Chartier at the Saint-Nicolas sandpit, near Quebec City. A beluga (*Delphinapterus leucas*) mandible, a walrus (*Odobenus rosmarus*) pelvis dated to 9790 years BP, two seal (*Phoca*) tibias, and a Thick-Billed Murre (*Uria lomvia*) humerus are among the bones excavated from Champlain Sea sediments. A photo shows Michel Chartier holding the murre humerus – also shown is the beluga mandible, the walrus pelvis, and two seal tibias from the Saint-Nicolas sandpit.

1008. Nelson, D.E., Morlan, R.E., Vogel, J.S., Southon, J.R., and Harington, C.R. (1986): New dates on north-

ern Yukon artifacts: Holocene not Upper Pleistocene. Science 232:749–51.
Several radiocarbon dates cited in Irving and Harington (1973) and in Morlan and Cinq-Mars (1982) were in error by a large factor. In particular, a caribou (*Rangifer tarandus*) tibia flesher found at Loc. 14N, Old Crow Flats, Yukon (69°51′N, 139°46′W) was originally dated at 27,000 +3000 –2000 (GX-1640, apatite). This was the basis for theories of pre-Clovis (>12,000 BP) people (*Homo sapiens*) in North America. The re-dated artifact yielded an accelerator mass spectrometry (AMS) date of only 1350 ± 150 BP (RIDDL-145). Other specimens from the same locality and position yielded dates ranging from 25,000 to >47,000 BP, in particular flaked mammoth (*Mammuthus*) long bones 29,100 +3000 –2000 BP (GX-1567, apatite) and 25,750 +1800 –1500 (GX-1568, apatite). Other caribou (*Rangifer tarandus*) specimens originally dated at Late Wisconsinan were also re-dated: antler billet 2930 ± 140 BP (RIDDL-133); and antler wedges 1730 ± 100 BP (RIDDL-140) and 1880 ± 140 BP (RIDDL-141). Old Crow Loc. 14N is a Holocene terrace deposit with redeposited, transported Late Pleistocene remains; therefore, none of these dates could be determined by stratigraphy alone. Since the four caribou dates were in error, it is likely that the mammoth dates were also in error since they were all dated using the apatite fraction of their bones, which is a less accurate method than AMS radiocarbon dating.

1009. Nelson, E.W. (1887): Report upon natural history collections made in Alaska in the years 1877–1881. Arctic Series, Signal Service, United States Army No. 3:1–337.
Although this report deals almost exclusively with living animals (fishes, birds, mammals, and butterflies) of Alaska, a skull of a muskox (*Ovibos moschatus*) 'very much broken down and worn' procured by Mr Murdoch near the ?mouth of the Colville River, Alaska, 'may be subfossil' (p. 233).

1010. Nelson, R.E., and Carter, L.D. (1985): Pollen analysis of a Late Pliocene and Early Pleistocene section from the Gubik Formation of Arctic Alaska. Quaternary Research 24:295–306.
The authors provide data on a 14-m-thick section of marine and nonmarine sediments of the Gubik Formation of northern Alaska, exposed in bluffs near Ocean Point on the Colville River (70°05′N, 151°24′W; Figure 1). Pollen from the marine sediments, of probable Late Pliocene age, records a boreal forest of spruce and birch with minor amounts of alder in the adjacent terrestrial vegetation. The suggested environment for the Arctic Slope during the time represented by the marine sediments is similar to that of Anchorage, Alaska, today. Vertebrate remains from the marine beds (although slightly beyond the purview of this bibliography), also suggest a climate much milder than today and compatible with the interpretation from fossil pollen. Repenning (1983) has described a sea otter fossil from the marine beds in the same section – collected at the same time as the pollen samples. The historic northern range limit for the modern sea otter (*Enhydra lutris*) was the Aleutian and Kommandorskiy Islands – S of the limit of thick sea ice. The species is nonmigratory, and seldom ranges more than 1 km from the coast. While the Gubik Formation fossil is not assignable to the modern species, it appears to be directly ancestral and likely had similar behaviour patterns. This independently suggests that environmental conditions of the Arctic Slope during G-I time were no more severe than those of the southern coast of Alaska today.

1011. Nero, R.W., and McCorquodale, B.A. (1958): Report of an excavation at the Oxbow Dam site. Blue Jay 16(2):82–90.
A fossiliferous stratigraphic section was discovered just below a dam on the Souris River near Oxbow, Saskatchewan. Bison (*Bison*) remains could be found in all strata. The uppermost layer, a 5-cm silty layer, also contained remains of wapiti(?) (cf. *Cervus elaphus* = cf. *Cervus canadensis*); the next layer, 30 cm of sand, contained remains of kit fox (*Vulpes velox*), wolf(?) (*Canis* cf. *Canis lupus*), coyote(?) (*Canis* cf. *Canis latrans*); and frog (*Rana*) remains were found in the lowermost layer, which contained a hearth with charcoal, indicating the presence of humans (*Homo sapiens*). Radiocarbon dates on the charcoal yielded values of 5100 ± 210 and 5350 ± 250 BP (no lab numbers given). The species identified from the site are indicative of a Plains habitat and Transition Life Zone conditions – not much different from the situation today. [See Morlan 1994.]

1012. Nicolas, F. (1925): Index to paleontology (geological publications 1847–1916). Geological Survey of Canada, Ottawa. 383 pp.

This publication lists all fossils in Geological Survey of Canada reports, by species, to 1916. [It is rather difficult to use.]

1013. Nicolas, F. (1930): Index to paleontology (geological publications 1917–1926). Geological Survey of Canada Miscellaneous Series No. 2:385–482.
Few, if any, Canadian Pleistocene mammals are mentioned.

1014. Nielsen, E., Churcher, C.S., and Lammers, G.E. (1988): A woolly mammoth (Proboscidea, *Mammuthus primigenius*) molar from the Hudson Bay Lowland of Manitoba. Canadian Journal of Earth Sciences 25:933–8.
The first mammoth specimen from the Hudson Bay Lowland, a left upper molar of a woolly mammoth (*Mammuthus primigenius*), was found on a Holocene point bar on the Limestone River near Bird, Manitoba, about 8 km upstream from its junction with the Nelson River. Although the specimen was not found in place, sediments along the banks are of Early Wisconsinan or Sangamonian age. The tooth's relatively low lamellar frequency of 9.0 suggests that it is most likely from an individual that lived near the end of the Sangamonian interglacial or during an Early Wisconsinan interstadial when ice sheets were absent from the Hudson Bay Lowland. Pollen and insect remains from the Nelson River sediments were deposited in steppe or steppe-tundra conditions – in accord with known habitat requirements of the woolly mammoths. Other Manitoba mammoth specimens include: tusk fragments from Transcona, Birds Hill, and Benito; molars from Birds Hill, Dufresne, Rathwell, Souris, and 12 km SE of St Malo; and bone fragments from Snake Creek. Other Hudson Bay Lowland specimens include: a mastodon (*Mammut americanum*) jaw and molar from the bed of the Moose River, Ontario, downstream from the confluence of the Missinaibi and Mattagami rivers, 72 km SW of Moose Factory; and a Columbian mammoth (*Mammuthus columbi*) molar from Long Island, North of Cape Jones (= Pointe Louis XIV) off the southeastern Eastmain shore of Hudson Bay at James Bay, Quebec.

1015. Nielsen, E., Gryba, E.M., and Wilson, M.C. (1984): Bison remains from a Lake Agassiz spit complex in the Swan River Valley, Manitoba: Depositional environment and paleoecological implications. Canadian Journal of Earth Sciences 21(7):829–42.
Fossil bison (*Bison*) remains have been found in gravels on an extensive spit complex between the Upper Campbell (357 m asl, beaches and wave-cut scarps) and Lower Campbell (349 m asl, beaches) levels of Glacial Lake Agassiz in the Swan River Valley in western Manitoba. An air photo included in the paper shows three gravel pits on the spit complex, which is S of Hubble Creek and E of Trout Creek. The remains were not found *in situ* and radiocarbon dates indicate that the spit facies formed during the high-water Emerson Phase of Glacial Lake Agassiz, which began about 10,000 BP. The radiocarbon dates on bison bone in the three pits yielded ages of: 10,300 ± 200 (BGS-617), 9400 ± 125 (BGS-887), and 9500 ± 150 BP (BGS-840). Other Manitoba remains include: (1) a western bison (*Bison bison 'occidentalis'*) cranium from the Assiniboine River floodplain near Russell radiocarbon dated at 6320 ± 140 BP (GSC-280); (2) another western bison-like cranium from the Jansson site on the Winnipeg River associated with Old Copper artifacts, which may date to about 5000 BP, perhaps younger; (3) the remains of two western bison skulls (*Bison bison occidentalis = Bison occidentalis*) reported by Hay (1924) from the Upper Campbell beach near Arden, and one from the bed of the Roaring River, SE of the town of Swan River; and (4) a partial tundra muskox (*Ovibos moschatus*) skull found in gravels near Grandview, 80 km S of the Swan River valley, radiocarbon dated at 8620 ± 190 BP (I-1623). Measurements of bison bones recovered from the spit complex are given in Table 1, and there are photographs of two of the heavily abraded bones (Figure 11).

1016. Nikolskiy, P. (1997): Extinct moose in Beringia. Beringian Paleoenvironments Workshop (September 20–3, Florissant, Colorado). Program and Abstracts. pp. 113–14.
Moose (Alcini) evolution in Eastern Beringia evidently matches that in Western Beringia. The author concludes that all morphotypes described in Eastern Beringia have analogues in Western Beringia. He suggests that remains of giant moose (*Alces latifrons = Cervalces latifrons*) can be divided into two groups: one with extremely long, slender antler beams and poorly molarized dentitions, the other with slightly shorter, more robust antler beams and molarized P_3s. The remains of the first group are about 1.2–0.7 Ma in Western Beringia, so the primitive forms

of giant moose first entered the New World at least 1 Ma. Moose remains intermediate in size between the giant moose and moose (*Alces alces*) are fairly common in Alaska and Yukon, but their age is unclear and they can be confused with Scott's moose (*Alces scotti* = *Cervales scotti*). All remains of Scott's moose are S of 51°N, and it is not known when that species separated from the Beringian population. Since Scott's moose have primitive features (e.g., an archaic skull) and advanced ones (e.g., complicated antler palmations), two hypotheses are suggested: (1) the separation may have occurred before the origin of modern moose (*Alces alces*) – i.e., before 250,000 BP – and then was separated by an ice barrier from the Beringian population; or (2) more advanced giant moose were isolated in Eastern Beringia for a long time – the separation occurring later. Evidently the first *Alces alces* spread into Eastern Beringia in the earliest Holocene and soon after displaced Scott's moose in North America.

1017. Noble, W.C. (1971): Archaeological surveys and sequences in central District of Mackenzie, N.W.T. Arctic Anthropology 8(1):82–90.

At the Acasta Lake archaeological site, Northwest Territories (65°24′N, 115°31′W), the fauna includes fishes (Pisces), birds (Aves) – probably Golden Eagle (*Aquila chrysaëtos*) or possibly Bald Eagle (*Haliaeetus leucocephalus*), and the following mammals (Mammalia): humans (*Homo sapiens* – indirect evidence from artifacts and hearths), hare (*Lepus*), beaver (*Castor canadensis*), black bear (*Ursus americanus*), and caribou (*Rangifer tarandus*). Fossils are from sand below 4 in of leaf mould on the eastern end of an esker. Evidently Plano people periodically hunted caribou in the area about 7000 BP (charcoal dates of 4900 ± 150 B.C. (GaK-3277) and 5020 ± 360 B.C. (I-3957).

1018. Noe-Nygaard, A. (1932): Remarks on *Mytilus edulis* L. in raised beaches in East Greenland. Meddelelser om Grønland 95:1–25.

The author (p. 5) mentions that in the summer of 1932 a botanist, T. Sørenson, found a whale [Cetacea, presumably a bowhead whale (*Balaena mysticetus*)] bone a couple of metres long in an old raised beach 50–60 m asl and now partly covered with vegetation on Ymer Island, East Greenland. Bennike (1997, p. 906) presumes it is of Holocene age.

1019. Nolin, L., and Cinq-Mars, J. (1994): NOGAP Radiocarbon Dates. http://www.civilization.ca/index1e.html.

This database offers a compilation of 54 conventional and 42 AMS radiocarbon dates obtained during the NOGAP (Northern Oil and Gas Archaeology Project) program between 1985 and 1994. Most are uncorrected dates based on analyses of bone or charcoal samples. They are presented along with comments by the researchers who submitted them following the format used in the journal *Radiocarbon*. Areas treated are: (1) Yukon – Herschel Island, Northern Yukon; and (2) Northwest Territories – Mackenzie Delta, Richards Island, Tuktoyaktuk Peninsula, Mackenzie River, Anderson Plain, Cape Bathurst Peninsula.

1020. Nowak, R.M. (1979): North American Quaternary *Canis*. Monograph of the Museum of Natural History of Kansas Number 6:1–154.

This statistical study of skulls of fossil and modern *Canis* of North America is based on an examination of approximately 5000 specimens. A hypothetical phylogenetic diagram for the genus is given in Figure 55. The wolf (*Canis lupus*) is shown entering the New World from the Old World between Irvingtonian and Rancholabrean time – probably during the Illinoian glaciation. Figure 51 includes the following Canadian and Alaskan localities for wolf fossils: Medicine Hat, Alberta; Fort Qu'Appelle, Saskatchewan; Old Crow area, Yukon; Dawson City area, Yukon; Eschscholtz Bay (Historic Bluff), Alaska; Buckland River, Alaska; Fairbanks area, Alaska. Fossil coyotes (*Canis latrans*) have been reported from 109 localities across North America from Florida to Alaska, and from Oaxaca to Pennsylvania. Figure 46 shows fossils from Medicine Hat, Alberta, and Cripple Creek Mine near Fairbanks, Alaska. The extinct dire wolf (*Canis dirus*) did not appear in North America until the Rancholabrean and may have originated in South America or descended from Armbruster's wolf (*Canis armbrusteri*). This large, highly specialized species was not ancestral to modern wolves, and its skull is easily distinguished from that of *Canis lupus*. Apparently it reached southern Alberta (Medicine Hat). [The single lower canine reported by Cowan (1954) from Castleguard icefield in Banff National Park, Alberta, has been reexamined and probably represents a bear (*Ursus*) rather than a dire wolf.]

1021. Obalski, T. (1904): Les grands mammifères fossiles dans le Yukon et l'Alaska. Bulletin du Muséum d'Histoire naturelle (Année 1905) 5:213–17.
Mammal remains (horncores, bone fragments, and ivory tusks) were found by gold placer miners in the Klondike, Yukon, at the contact between muck and placer gravels. The specimens were well preserved and were later identified by the author based on his own photographs (he did not take samples) as: mammoth (*Mammuthus*), mastodon (*Mammut americanum*), muskox (*Ovibos*), bison (*Bison*), moose (*Alces alces*), caribou (*Rangifer tarandus*), wapiti (*Cervus elaphus*), bighorn sheep [probably Dall sheep (*Ovis dalli*) rather than *Ovis canadensis*], and horse (*Equus*). He also found a mammoth (*Mammuthus*) molar in Alaska. The fauna found in the Klondike placer deposits seem to indicate a steppe or grassland environment, whereas the present environment is dominated by mosses.

1022. Occhietti, S., Chartier, M., Hillaire-Marcel, C., Cournoyer, M., Cumbaa, S.L., and Harington, C.R. (2001): Paléoenvironnements de la Mer de Champlain dans la région de Québec, entre 11 300 et 9750 BP: Le site de Saint-Nicolas. Géographie physique et Quaternaire 55(1):23–46.
At Saint-Nicolas, Quebec (Fig. 1), formerly in the Quebec City Strait, tidal current sands of the Pleistocene-Holocene transition preserve an exceptional marine fossil fauna: 33 kinds of invertebrates in addition to microfossils, three species of marine mammals [small seal (*Phoca*), walrus (*Odobenus rosmarus*), white whale (*Delphinapterus leucas*)], three species of seabirds [Oldsquaw (*Clangula hyemalis*), Thick-billed Murre (*Uria lomvia*), and a large bird (Aves)], and five species of fish [sturgeon (*Acipenser*), Atlantic wrymouth (*Cryptacanthodes maculatus*), eelpout (*Lycodes*), capelin (*Mallotus villosus*), and a salmonid (Salmonidae)] (Table 2). Radiocarbon dates on faunal remains lie between 10,060 and 9810 BP (Table 3). The fossil-bearing sands are overlain by tidalites corresponding to the last phase of the Champlain Sea until about 9750 BP. The lithological, geochemical, and paleontological data collected in the area suggest that sedimentation took place in an archipelago which was located at the entrance to the Champlain Sea and was subjected to tides that reached levels of about 10 m. Faunal diversity is associated with high marine productivity, which is tied to the mixing of fresh and salt waters. The food web consisted of molluscs, which were abundant at low tide, and of fishes, walruses, seals, and birds that probably nested on the small, rocky islands (Fig. 2).

1023. Oetelaar, G.A., Siegfried, E., and Jones, K. (1996): The natural and human landscape in the Calgary area 8,500 years ago. Paper presented at the Canadian Association of Geographers Meetings (May 11–16, Saskatoon, Saskatchewan). 12 pp.
The Tuscany site (Eg Pn-377) (*Homo sapiens*) in NW Calgary, Alberta, provides information on mammal remains located below the Mazama ash, which thus predate 6800 BP. The paleosol immediately below this tephra includes only bison (*Bison*, probably the western bison *Bison bison occidentalis* = *Bison occidentalis*), and the paleoenvironment suggested is grassland. The faunal assemblage associated with the lower paleosol includes: bison, wapiti (*Cervus elaphus*), antelope (*Antilocapra americana*), bear (*Ursus*), muskrat (*Ondatra zibethicus*), hare (*Lepus*), and two species of bird (Aves). Of these, the bear and wapiti suggest wooded environments, whereas the bison, antelope, and hare indicate grassland settings. Further, the muskrat suggests a rather moist environment, and the gastropods in the paleosol suggest a moist parkland environment.

1024. Ogle, A. (1985): Ancestors Part 5: Hunting grounds grew to include New World. Edmonton Journal, September 27. p. C12.
The earliest colonizers (*Homo sapiens*) of North America moved from W to E across a broad 1600-km-wide plain – part of Beringia. Probably the spread resulted from a gradual population growth and expansion of hunting ranges. Archaeologists originally believed North America was populated after about 12,000 BP. But a major Canadian research effort in the Old Crow region of northern Yukon produced a caribou (*Rangifer tarandus*) fleshing tool that gave a radiocarbon date of about 27,000 BP. Recent dating, however, shows it is only about 2000 years old. But other much cruder tools shaped from mammoth (*Mammuthus*) bones give dates of 45,000 BP to 25,000 BP. The oldest primary archaeological site in Eastern Beringia with clear evidence of human activity is Bluefish Caves, Yukon. There, a wide range of animal bones dating from 22,000 to 10,000 BP was found with stone flakes left from tool making. So human occupation

continued during a period characterized by harsh periglacial conditions. Most prehistorians regard the overkill scenario, as defined by Paul Martin, as significant for the extinction of the Pleistocene megafauna 'as an extreme and unlikely case.' Perhaps climate and environmental changes at the end of the Pleistocene were more likely reasons for the extinction. The article gives a picture of Eastern Beringia during the Late Pleistocene as being steppe tundra, with zones ranging from polar desert to grasslands, which provided a rich environment for a variety of grazing animals, including: mammoth (*Mammuthus*); bison (*Bison*); more than one kind of horse (*Equus*); caribou (*Rangifer tarandus*); and muskox (*Ovibos moschatus*). Its climate was dry with high summer and low winter temperatures. [The article is based on discussions with archaeologists R. Morlan, W.N. Irving, J. Cinq-Mars, A. Bryan, and P.S. Martin.]

1025. Olivier, R.C.D. (1982): Ecology and behavior of living elephants: Bases for assumptions concerning the extinct mammoths. *In*: Paleoecology of Beringia (D.M. Hopkins, J.V. Matthews, Jr, C.E. Schweger, S.B. Young, and V. Stanley, eds.). Academic Press, Toronto. pp. 291–305.

This review concludes that trends in body size and dental features suggest that elephants (Elephantidae) are highly adapted to cope with abrasive, nutritionally poor diets of a high fibre-to-protein ratio, and that this capacity is responsible for their success. Caecal fermenters such as elephants are therefore primarily limited by the quantity not the quality of food. Riverine environments are generally more lush and would be favoured. In the markedly seasonal environment of the arctic steppe, the high-quality grasses must have been more abundant in the warm months. Some degree of migration is assumed but the extent is still conjecture. It is suggested that perhaps fat reserves and behavioural traits such as snow cratering could have allowed the mammoth (*Mammuthus*) to survive with only limited seasonal migrations (i.e., less than 650 km). The author further concludes that the woolly mammoth (*Mammuthus primigenius*) probably had a similar social organization to the extant elephants (i.e., family groups dominated by a matriarch and adult males either solitary or gathering in bachelor groups). The author argues that there were two major causes for the extinction of the woolly mammoth: (1) climatic factors that changed the habitat from arctic steppe to tundra, and (2) increases in human (*Homo sapiens*) predation.

1026. Olsson, I.U. (1980): Content of ^{14}C in marine mammals from northern Europe. Radiocarbon 22(3):662–75.

The reservoir effect of Scandinavian sea water has been determined by dating bones of seals (Phocidae) and whales (Cetacea) killed well before known impact on the natural ^{14}C concentration became significant. All results were normalized to $\delta^{13}C = -25‰$ vs PDB scale. The determinations yield values of the reservoir effect in agreement with earlier results obtained from marine mollusc shells and mammals. The first results are discussed in light of previous variations of ^{14}C content in the atmosphere. Using a smoother curve, the reservoir effect is slightly smaller than was hitherto believed. Other measurements than those from Scandinavia are dealt with in Table 2, and radiocarbon dates listed from Greenland are: (1) 645 BP (K-347) (expected age = 110 ± 20 BP, reservoir age = 535 ± 50) on walrus (*Odobenus rosmarus*) bone from Thule, northern Greenland; (2) 545 BP (K-346) (expected age = 110 ± 20 BP and the reservoir age = 435 ± 55) on harp seal (*Phoca groenlandica* = *Pagophilus groenlandica*) bone from an unspecified locality in western Greenland; (3) 585 BP (K-348) (expected age = 110 ± 20 BP, reservoir age = 475 ± 55) on polar bear (*Ursus maritimus*) bone from Kap Rink, eastern Greenland; (4) 515 BP (LU-779) (expected age 110 ± 20 BP, reservoir age = 405 ± 50) on polar bear (*Ursus maritimus*) from Kap Stephensen, eastern Greenland; (5) 715 BP (K-350) (expected age = 200 ± 20 BP, reservoir age = 515 ± 75) on human (*Homo sapiens*) bone from Angmagssalik, eastern Greenland; and (6) 705 BP (K-351) (expected age = 200 ± 20 BP, reservoir age = 505 ± 75) on human (*Homo sapiens*) bone from Skjoldungen, eastern Greenland.

1027. O'Neil, L.-C. (1951): Un os de baleine, long de 14 pieds, est découvert dans la rivière de Cookshire. La Tribune, Sherbrooke. 19 avril. pp. 1–2.

This brief article notes the rare find of a 14 ft 2 in long lower jaw of a large whale [resembles a bowhead (*Balaena mysticetus*)] on April 14, 1951. It was found by Victor Paré near Eaton [not Heaton] River, Cookshire (45°25′N, 71°38′W), about 20 km E of Sherbrooke, Quebec. Léon Marcotte, curator of the Musée du Séminaire St Charles [now Le Musée du Séminaire de Sherbrooke] told a reporter that the specimen was thousands of years old and that it was from Champlain Sea deposits that covered the area, mentioning the find at Daveluyville, Quebec [see Laverdière 1950]. The article shows two of Paré's children with the jaw to show its large size. [This specimen, on display at Le Musée du Séminaire du Sherbrooke, yielded a normal collagen date of 370 ± 50 BP (Beta-70093), so was definitely not from Champlain Sea deposits, and was likely transported inland by people (*Homo sapiens*) – see Harington 1988.]

1028. O'Reilly, P., Reimchen, T.E., Beech, R., and Strobeck, C. (1993): Mitochondrial DNA in *Gasterosteus* and Pleistocene glacial refugium on the Queen Charlotte Islands, British Columbia. Evolution 47(2):678–84.

The present three-spined stickleback (*Gasterosteus aculeatus*) population on the Queen Charlotte Islands, British Columbia, exhibits endemism at the subspecies level, and gigantism among other features. This species lives in freshwater and was believed to have populated the islands about 16,000 BP. This date is close to the Wisconsinan glacial maximum, but the Queen Charlotte Islands needed to be at least partially ice-free. Taking into account the highly derived condition for sticklebacks on the Queen Charlotte Islands and its symbiotic relationship with an unusual taxonomy of dinoflagellate, in concert with the predicted divergence time derived from mitochondrial DNA, an extended preglacial history in the Argonaut Plain (see map, Figure 1) is suggested. 'If a refugium was large enough to support freshwater fish throughout [a large part of] the Pleistocene, then other aquatic and terrestrial biota may also have persisted during this period.'

1029. Osborn, H.F. (1910): The age of mammals in Europe, Asia and North America. Macmillan, New York. 635 pp.

The author mentions major features of Pleistocene mammal species and their distribution based on fossil finds.

1030. Osborn, H.F. (1936): Proboscidea: A Monograph of the Discovery, Evolution, Migration and Extinction of the Mastodonts and Elephants of the World. Volume I: Moeritherioidea, Deinotherioidea, Mastodontoidea. American Museum Press, New York. 802 pp.

This classic compendium of information, photographs, diagrams, and maps mentions the following American mastodon (*Mammut americanum* = *Mastodon americanus*) specimens from northern North America: (1) '*Elephas Rupertianus* Richardson' from Swan River, Lake Winnipeg Basin, Manitoba, discovered in 1854 and referred to *Mastodon americanus rupertianus* by Osborn; and (2) '*Mastodon americanus alaskensis* Frick' from near Fairbanks, Alaska, discovered in 1933 (p. 137, Figure 86, with a detailed figure and table of measurements on p. 176). An estimated four American mastodon localities are mapped in southern Ontario on Figure 123b (p. 177).

1031. Osborn, H.F. (1942): Proboscidea: A Monograph of the Discovery, Evolution, Migration and Extinction of the Mastodonts and Elephants of the World. Volume II: Stegodontoidea and Elephantoidea. American Museum Press, New York. pp. 805–1675.

Canadian and Alaskan mammoths (*Mammuthus*) recorded in this classic compendium include: (1) Osborn refers (p. 1069) a lower jaw from Hamilton, Wentworth County, Ontario, Canada, described in 1863 to '*Parelephas jacksoni*' [now considered to represent the Columbian mammoth (*Mammuthus columbi*)]; (2) measurements of a deciduous fourth premolar (Dp^4) from the Yukon River(?), Alaska ('Alaska, Yukon') and a lower second molar (M_2) from Alaska are given on p. 1088. Both have 'seven ridge plates in a 100 mm. line', (3) the photograph (front and right side) of a woolly mammoth (*Mammuthus primigenius* = *Mammonteus primigenius*) lower jaw from Alaska is figured with the cranium of a woolly mammoth from Siberia, Russia (Figure 962c and c_1 on p. 1091); (4) the most progressive form of woolly mammoth (*Mammuthus primigenius compressus* = *Mammonteus primigenius compressus*) is described from Alaska (p. 1099); (5) the 1907 discovery at Historic Bluff on Eschscholtz Bay, Alaska, of a partial woolly mammoth skeleton with some skin and hair (p. 1127); (6) a map (Figure 1004) shows fossil mammals (Mammalia) of Alaska and the Yukon based on Quackenbush (1909), Maddren (1905), Frick (1930), and others, and includes: woolly mammoth (*Mammuthus primigenius* = *Mammonteus primigenius* and *Mammonteus compressus*); Yukon horse (*Equus lambei*); Alaska horse (*Equus alaskae*); caribou (*Rangifer tarandus* = *Rangifer*); giant short-faced bear (*Arctodus simus* = *Arctodus yukonensis*); American mastodon (*Mammut americanum* = *Mastodon americanus*); muskox (*Ovibos moschatus* = *Ovibos yukonensis*); western camel (*Camelops hesternus* = *Camelops*); helmeted muskox (*Bootherium bombifrons* = *Symbos tyrelli*), as well as bison (*Bison*); woolly mammoth; horse (*Equus*); American lion (*Panthera leo atrox* = *Felis atrox alaskensis*); and dire wolf (*Canis dirus* = *Aenocyon dirus alaskensis*) [currently, only the wolf *Canis lupus* is believed to have occupied this region] from the Fairbanks area of Alaska; (7) a list of the Alaska-Yukon fauna on p. 1135 includes, in addition, steppe bison (*Bison priscus* = *Bison crassicornis* and *Bison alleni*); Dall sheep (*Ovis dalli* = *Ovis*); [and, erroneously, the mountain goat (*Oreamnos*)]; (8) Figure 1010 on p. 1145 shows a front-view photograph of a male woolly mammoth cranium with tusks from gravel 42 ft below the surface at Quartz Creek near Dawson City, Yukon, in March 1904 [the same photograph (Figure 1032, p. 1166) is erroneously said to represent a skull with half-grown tusks of a male from Yukon River, Alaska]. Figure 1011 on this page diagrams growth stages in the jaws and teeth of three woolly mammoths from Alaska (the juvenile jaw is from near Anvik and the oldest jaw is from Elephant Point); (9) a detailed description (pp. 1156–9) is given of the paratype of a 'progressive' subspecies of woolly mammoth described by Osborn in 1924 (*Mammuthus primigenius* = *Mammonteus primigenius compressus*); Figure 1024 shows this right upper third molar from Historic Bluff, Eschscholtz Bay, Alaska; (10): a detailed description (pp. 1159–61) of another subspecies (*Mammuthus primigenius* = *Mammonteus primigenius alaskensis*) is based on four crania (cotypes) from the Fairbanks area, Alaska (Figure 1026; Table XV gives cranial and dental measurements); (11) the author notes that woolly mammoths ranged almost exclusively N of the 40th parallel, being well adapted to the borders of retreating glaciers at the close of the ice age, and that, in Alaska, it was extremely abundant and was occasionally found in frozen form along the shores of Eschscholtz Bay (p. 1169); (12) a highly-curved pair of Alaskan adult woolly mammoth tusks is shown in a photograph on p. 1062.

1032. Osborn, S. (1853): Enclosure No. 21 of Sir Edward Belcher's Letter: Commander Sherard Osborn's Sledge Journey, 1853. Great Britain Parliament, House of Commons Sessional Papers [Blue Books]. Accounts and Papers, 1854–55, 35 (1898): 187–235.

On p. 218 (May 29, 1853), the author mentions finding just north of Cape Success [presumably Success Point on northern Cameron Island, Nunavut] fossil bones he thought to be those of fox, wolf, and seal [perhaps arctic fox (*Alopex lagopus*), wolf (*Canis lupus*), and ringed seal (*Phoca hispida*)].

1033. Osborne, R.H. (1950): The Fairbanks Pleistocene. [Prepared while the author was engaged in paleontological research for the American Museum of Natural History and the University of Alaska in 1940 and 1941]. Typescript is in the Quaternary Vertebrates of Northern North America Publication File, at Canadian Museum of Nature]. 53 pp.

The author laments the lack of stratigraphic data provided with most of the earlier Alaskan Pleistocene fossils and reports on some of his own collections, which do have a degree of stratigraphic context: (1) a mummified juvenile mammoth (*Mammuthus*) foot and an associated stone tool (turned over to the University of Alaska and later lost). This foot could possibly be associated with a juvenile mammoth trunk, head, and forelimb recovered downstream eight years later in 1948; (2) an almost complete mummy of a female muskox [since reported in McDonald and Ray (1989) as a helmeted muskox (*Bootherium bombifrons*)] deposited in association with a shed wapiti (*Cervus elaphus*) antler; (3) a largely intact skeleton of a cow moose (*Alces alces*) with the bones in roughly the correct anatomical position and with the bones of a foetus in approximately the correct area within the rib cage; (4) a skull of a large bull moose (*Alces alces*); (5) rodent nests and runways made by ground squirrels (*Spermophilus* = *Citellus*) but more commonly found containing bones of later inhabitants, such as voles (*Microtus*), than of the creators. The author concludes that the silt layer is aeolian in origin while the muck layer is reworked or redeposited silt combined with local weathering products largely from surface zones; hence the high organic component. He further concludes that the silt was deposited during periods of colder climate. He also supplies a possible view of relative age of the stratigraphic layers to each other.

1034. Osgood, W.H. (1905a): Mastodon remains in the

Yukon valley. Proceedings of the Biological Society of Washington 17:254–5.
A well-preserved mastodon (*Mammut americanum*) lower molar was found beneath 8 m of Pleistocene muck and gravel deposits on claim no. 14, Gold Run Creek, near Dawson City, Yukon. The author also mentions that Richardson (1854) reported several mastodon scapulae from Swan River, Manitoba, and that Robert Bell (1879) mentioned a mastodon tooth in the bed of Moose River, 46 mi from Moose Factory, Ontario.

1035. Osgood, W.H. (1905b): *Scaphoceros tyrrelli*, an extinct ruminant from the Klondike gravels. Smithsonian Miscellaneous Collections 48, Part 2:173–85.
Two imperfect skulls from a new species of ruminant, the helmeted muskox, (*Bootherium bombifrons* = *Scaphoceros tyrrelli*) were found 20 m below the surface in gravels of Lovett Gulch, Bonanza Creek, Klondike District, Yukon. The specimens were compared with muskox (*Ovibos moschatus*), helmeted muskox (*Bootherium bombifrons* = *Bootherium sargenti* = *Symbos cavifrons* = *Ovibos cavifrons*), cattle (*Bos*), bison (*Bison*), and sheep (*Ovis*). The specimen's horncores were smaller than *Ovibos*, it was smaller in body size than *Symbos cavifrons* [male of *Bootherium bombifrons*], and it exhibited morphological features found in many ruminants, so the author named a new genus *Scaphoceros*. The author speculates that *Symbos cavifrons* and *Symbos tyrrelli* may be the male and female of one species, as they are found together at many localities, such as horncores at Anvik, Alaska.

1036. Osgood, W.H. (1907): A specimen of *Bison occidentalis* from Northwest Canada. Proceedings of the Biological Society of Washington 20:65–6.
An incomplete western bison (*Bison bison occidentalis* = *Bison occidentalis*) skull was found in the cut banks of a small tributary of the Pelly River, 20 km upstream from Fort Selkirk, Yukon. Bone measurements are provided.

1037. Owen, P.R., Bell, C.J., and Mead, E.M. (2000): Fossils, diet, and conservation of black-footed ferrets (*Mustela nigripes*). Journal of Mammalogy 81(2): 422–33.
A new Middle Pleistocene record of the black-footed ferret (*Mustela nigripes*) is reported from Cathedral Cave, Nevada, which prompted the authors to review the fossil record of this species. Forty-two percent of fossil faunas known to contain this ferret [including those from Sixtymile Loc. 3, Yukon, and Hunker Creek Loc. 12 near Dawson City, Yukon – Figure 1, numbers 9, 10] do not contain the prairie dog (*Cynomys*) – the common prey of black-footed ferrets in historic times. [See Youngman 1993, 1994]. Consideration of dietary information and habitat requirements of *Mustela nigripes* and its sister taxon the steppe ferret (*Mustela eversmanni*) reveals ancestral behavioural repertoires for ferrets. The authors suggest that the historically documented 'obligate' predator–prey relationship between *Mustela nigripes* and *Cynomys* was a secondary effect of colonization by these ferrets of prairie dog–dominated habitats sometime in the past 800,000 years. A phylogenetic perspective on behaviour of ferrets combined with fossil data indicates a broader range of possibilities for conservation of the endangered *Mustela nigripes*.

1038. Padgham, M. (1982): Crown must show artifacts came from here. Whitehorse Star, July 26:3.
In a court case involving a mammoth (*Mammuthus*) tusk seized at the Whitehorse airport in December 1980, Kenneth Shore, the owner, testified that he had purchased the tusk in Vancouver and had brought it to Dawson City, Yukon, in July 1980. Crown counsel Ian McKinnon argued there was enough inconsistency in Shore's and other testimony to raise a doubt that the tusk had originated in Vancouver. Deputy Judge Ralph Hudson said the Crown must show that it is probable that the specimen in question originated in the Yukon. Under the Yukon Act it is illegal to export from the Yukon archaeological artifacts without a permit. Articles so exported may be seized and then ordered forfeit to the Crown by the court.

1039. Paetkau, D., and Strobeck, C. (1996): Mitochondrial DNA and the phylogeography of Newfoundland black bears. Canadian Journal of Zoology 74:192–6.
The degree of genetic differentiation separating Newfoundland black bears (*Ursus americanus hamiltoni*) from continental Canadian black bears (*Ursus americanus americanus* and *Ursus americanus cinnamonum*) was assessed using sequence data from part of the mitochondrial DNA molecule. Given the observation of reduced genetic diversity in Newfoundland black bears, this subspecies likely arose through rapid genetic drift as-

sociated with a founder effect during postglacial colonization of the island, and not through long periods of isolation in a glacial refugium.

1040. Panton, J.H. (1891): The mastodon and mammoth in Ontario, Canada. Geological Magazine, London, Series 3(8):504–5.
Mastodon (*Mammut americanum*) remains were found in a marl bed near Highgate, and mammoth (*Mammuthus*) remains were found 'under similar conditions' near Shelburne, both localities in S-central Ontario. A table of bone measurements is provided. Other proboscidean (Proboscidea) remains have been found in this part of Ontario at or near: St Catharines, Dunnville, Goat Island, Niagara Falls, and Kimbal.

1041. Panton, J.H. (1892): The mastodon and mammoth in Ontario, Canada. Report of the British Association for the Advancement of Science, 61st Meeting (Cardiff, Wales, 1891). pp. 654–5.
[See Panton 1891.]

1042. Paraskevas, J. (2001): Horse skeleton links humans to extinction. Calgary Herald, May 3, pp. A1, A8.
This article tells about the 1996 discovery of a remarkable Late Wisconsinan archaeological-paleontological site in the St Mary Reservoir near Cardston in southern Alberta. The discovery of a horse (*Equus*) skeleton provides researchers with evidence that Paleoindians (*Homo sapiens*) had a part in extinction of that species – at least locally. The skeleton had smashed vertebrae and butcher-marks on it. Further, protein residue testing (matching protein from horse blood and fat on spearheads found at the site to that from the actual horse skeleton) proved that horses had been hunted by Paleoindians [the article includes a colour restoration of Paleoindians hunting horses in the area about 11,000 BP, as well as a photograph of University of Calgary paleontologist Len Hills with an 11,300-year-old horse cranium from the site]. Hunters may have ambushed horses coming up the shallow slope from the proto-St Mary River according to Brian Kooyman, a University of Calgary archaeologist. Also, tracks preserved at the site reveal clues about the habits of the animals and their chronological age distribution. The tracks indicate 'a declining population, a stressed population,' Paul McNeil said. 'So hunting could then lead to extinction.'

1043. Parfit, M. (2000): The dawn of humans: Hunt for the first Americans. National Geographic 198(6):40–67.
This popular article begins with a photograph of Timothy Heaton excavating in SE Alaska [presumably at On Your Knees Cave on Prince of Wales Island], where animal bones covering the past 40,000 years have been recovered. In this extensive, well-illustrated piece, the author stresses that recent finds have opened up the discussion of this topic. Instead of the 'standard' view that people from Siberia entered what is now North America by the Bering Isthmus about 14,000 BP and headed S to the heartland of America along the eastern margins of the Cordillera, some scientists place people in the Americas 15,000, 20,000, or even 30,000 or more years ago, with possible migration routes S along the Pacific Coast, or even across the Atlantic Ocean [fostering the 'thin-point' stone blade technology (e.g., Cactus Hill, Virginia, at 15,000 BP and Clovis at 13,500 BP), supposedly stemming from the Solutrean culture of SW Europe] – see map pp. 46, 47. The Pacific route is discussed in the light of 10,500-year-old human (*Homo sapiens*) bones (casts are shown in a photograph on p. 46) from an Alaskan cave. Testing of the bones revealed that the individual had a marine diet. Also, archaeologist Daryl Fedja found a stone tool at a depth of 175 ft below the ocean surface on what may have been a habitable riverbank 11,500 years ago (pp. 50, 51) in what is now the Queen Charlotte Islands of British Columbia. The inland route involves both the Mesa site in northern Alaska (photograph on p. 49) and Broken Mammoth site in central Alaska. Michael Kunz, director at Mesa, believes that the tools found there are a link between Alaska and the Great Plains because of similar stone tool shapes and flaking techniques, but others argue that the Mesa site represents people of later years moving N instead of S, and yet others think that the tools were left behind by occasional visitors from other parts of Alaska. A bone needle (photograph on p. 48) found at the Broken Mammoth site with artifacts that date to 13,800 BP is crucial because early people there had to be able to make weather-tight clothing to survive the northern winters. An accompanying coloured map (31 in high and 20 in wide, with illustrations of ancient people, prey and stone blades) shows Beringia, North America (with Laurentide ice sheet at maximum extent), Central America, and South America with proposed immigration routes for people: (1) coastal route (20,000–15,000 BP)

down the Pacific coast; (2) overland route along the E side of the western mountain system (14,000 BP); and (3) Atlantic paleomaritime route (24,000–18,000 BP). Archaeological sites are mapped (sites with animal remains and artifacts are designated by squares, and those with human remains and artifacts are designated by circles), the colours being: (1) orange for Pre-Clovis (more than 13,500 BP); (2) blue for Paleoindian (13,500–10,900 BP); and (3) yellow for Early Archaic (10,900–7000 BP). The following North American Pre-Clovis sites mapped are: Trail Creek, Campus (Fairbanks), Broken Mammoth, Swan Point in Alaska; Old Crow, Bluefish Caves, and Dawson City in Yukon; and Vermilion Lakes in Alberta. [It should be noted that Klo-Kut near Old Crow, Yukon, is not Paleoindian as designated, but much later.]

1044. Parks, K.M. (1955): Uplands bones tell of era when whales swam over Ottawa. The Ottawa Journal, 19 November.
White whale (*Delphinapterus leucas*) remains were found in the Foster Sandpit near the Ottawa International Airport (Uplands). The first specimen was found by G.B. Rolland, foreman at the sandpit, who passed on more white whale remains (at least one bone belonging to the earlier skeleton) to Charles M. Sternberg of the National Museum. Sternberg mentioned that before the turn of the century a seal (*Phoca*) skeleton had been found by H.M. Ami and Ruggles Wright in the old Wright Brickyard in Hull West. Other Ontario white whale specimens have been found at Pakenham, Lanark County; Odell's Brickyard in Vanier (= Ottawa East); and Cornwall. Porpoise (*Phocoena*) remains were found at Pontiac, Quebec, adjacent to Ottawa. There was another whale find at Smiths Falls, Ontario. These marine mammals occupied the Champlain Sea, which covered the region about 10,000 years ago. [See Harington 1988.]

1045. Parks, W.A. (1925): Buried Indian workshop with remains of an extinct mammal. Bulletin of the Geological Society of America 36:429–34.
Human (*Homo sapiens*) artifacts (points and tools) and pronghorn antelope (*Antilocapra americana* = *Neomeryx finni* nov. sp.) remains (three molars and an incisor) were found in 1923 at a prehistoric human (*Homo sapiens*) workshop buried in a bog on the Finn property 8 km N of Dundurn, 24 km S of Saskatoon, Saskatchewan. The bog is about 92 m in diameter and is underlain by 10 cm of muck, which is underlain by 15 cm of sand, whose grain size gets coarser with depth. The specimens were found at the bottom of this sand layer at the contact with yellow clay.

1046. Payne, J. (1971): Ancient bones to tell fascinating tales. Houston Today. June 30, p. 4.
The presence of mammoths (*Mammuthus*) or mastodons (*Mammut americanum*) in the Babine Lake area of British Columbia is confirmed by a find of bones at the Noranda Bell copper mine that is being developed on the N side of the lake. The bones were found in peaty material indicative of an ancient swamp. Apparently the swamp deposits built up to a depth of about 40 ft when a change in climate occurred and glacial ice swept in from the E covering the swamp deposits with an additional 60 ft of glacial till. If the swamp had not overlain a deposit of copper ore, the bones may never have been found. Fossil bone was first recognized by supervisory personnel looking at material a curious worker brought them. Bill Allen, manager of the Noranda Bell operation, notified the University of British Columbia, and as a result Dr H.W. Tipper of the federal Department of Energy Mines and Resources, who was doing geological work near Smithers, was sent with his field crew to investigate the find. In the course of excavation, some material (e.g., a tusk) was damaged by heavy equipment. Kurt Rusger, foreman, measured the pieces of tusk at 6 ft long and 8 in diameter. Tipper's crew began excavating on June 23 and recovered leg bones nearly 4 ft long and 6 in across. Use of a backhoe to pare down the overburden was invaluable. The next day many small bones were found (several being ribs over 4 ft long and 3 in across), and vertebrae the size of dinner plates. The big find was what appeared to be a shoulder blade – 3 ft long by at least 2 ft across. Other pieces of tusk were found, including a piece of 8 in diameter and 1 ft long. The bones were taken to the Noranda Bell camp, where they were swathed in paper towels and packed in fibreglass insulation. The article includes seven photographs showing: the excavation of a scapula; a neck vertebra; H. Tipper with a pile of bones; a general view of the site in relation to its surroundings; and one of a backhoe, pick, and geological hammer in use to excavate the bones. [See Harington et al. 1974.]

1047. Perkins, G.H. (1908): Fossil Cetacea of the Pleistocene of the United States and Canada, with

special reference to *Delphinapterus vermontanus*, Thompson. Report of the State Geologist of Vermont 1907–8:76–112.

The author mentions finds of Champlain Sea white whales (*Delphinapterus leucas = Delphinapterus vermontanus*) from: (1) the Mile End quarries, Montreal, Quebec; (2) Cornwall, Ontario; (3) Smiths Falls, Ontario [part of a humpback whale (*Megaptera novaeangliae*) may have been mistaken for a white whale – see J.W. Dawson 1883a]; (4) Leda Clay at Montreal, Quebec; and (5) Pakenham, Ontario.

1048. Perkins, G.H. (1910): On a skeleton of a whale in the Provincial Museum, Halifax, Nova Scotia; with notes on the fossil Cetacea of North America. Transactions of the Nova Scotian Institute of Science 12:139–64.

A whale skeleton was found in a clay bed overlain by 3.7 m of sand and gravel at the intersection of Jacquet River and the Intercolonial Railroad, New Brunswick. After comparison of its periotic bone with those of white whale (*Delphinapterus leucas*) and narwhal (*Monodon monoceras*) by Drs F.W. True, R.C. Andrews, and W.D Matthew of the American Museum of Natural History, it was conclusively identified as narwhal. The paper, which includes many photographs of the bones, gives a detailed description of the morphology of the skeleton, which has elements of both whale species [which is reasonable as both belong to the Monodontidae]. Other cetacean finds mentioned include: several caudal vertebrae of white whale found at the Mile End Quarries in Montreal by W. Logan in 1858; a few bones of white whale found at Rivière-du-Loup, Quebec [not Ontario, as stated on p. 150] collected by J.W. Dawson in 1864; a nearly complete skeleton of (?)white whale (cf. *Delphinapterus leucas*) found at Cornwall by E. Billings in 1870; two vertebrae and part of a rib of a humpback whale (*Megaptera novaeangliae*) found in Welsh's Gravel Pit on the Canadian Pacific Railway line 5 km N of Smiths Falls, Ontario (J.W. Dawson 1883a, b); a nearly complete skeleton of white whale collected by J.W. Dawson in 1895; some other whale bones found at a depth of 6.8 m in brick clay near Papineau Road, Smith's Brickyard, Montreal; several whale bones found at Williamstown, Ontario, by E. Ardley in 1901; and most of a skull and several vertebrae of a white whale found near Pakenham, Lanark County, Ontario [see Whiteaves 1907 – the specimen is now in the Quaternary Zoology Collection, Canadian Museum of Nature].

1049. Pessl, F., Jr (1962): Glacial geology and geomorphology of the Sortehjorne area, East Greenland. Arctic 15(1):73–6.

During 1961 the author noted whalebone and baleen (Balaenidae) occurred on the fourth level of a marine delta on the NW side of Storedal in central East Greenland (Figure 1) at 46 m asl in deltaic sands overlain by coarse gravels. The whale remains and marine shells 'are expected to give dates for these higher sea-levels.' [See Bennike 1997.]

1050. Peterson, R.L. (1965a): A well-preserved grizzly bear skull recovered from a late glacial deposit near Lake Simcoe, Ontario. Nature 208(5016):1233–4.

Well preserved grizzly bear (*Ursus* cf. *Ursus arctos horribilis*) bones were found in a load of gravel 9 m deep from a pit 'close to Orillia,' Ontario. A radiocarbon date on a limb bone yielded 11,700 ± 250 BP [no lab number]. This is the first example of grizzly bear found in the eastern Great Lakes region. Cranial measurements of the specimen (ROM 34226), as well as a side view of the skull and a palatal view of the cranium, are given.

1051. Peterson, R.L. (1965b): The Lake Simcoe grizzly. Meeting Place, Journal of the Royal Ontario Museum 1(4):82–7.

This popular article provides a great deal of background information on the brown bear (*Ursus arctos*) remains from the Orillia area, Ontario, dated at 11,700 ± 250 BP. [See Peterson 1965a.]

1052. Pettipas, L.F. (1971): Roaring River relic. Zoology 12(1):9–10.

A bison (*Bison*) skull, currently the property of F.A. Twilley of Swan River, Manitoba, was collected several years ago from the bed of Roaring River in the Swan River valley of Manitoba. The posterior part of the cranium with horncores is preserved (see photograph p. 9). The author states that the dimensions and morphology of the specimen strongly suggest that it belonged to 'a member of a now-extinct species.' [It looks like a large western bison (*Bison bison occidentalis*), or a young male or adult female of the steppe bison (*Bison priscus*).] In any case, the horncore dimensions of the specimen exceed the maximum size for both plains bison (*Bison bison bison*) and wood bison (*Bison bison athabascae*). Other Manitoba bison remains mentioned include: (1) bison bone occurring in alluvial fill in small gullies tributary to the Assiniboine River near Treesbank [an area that has produced several western bison (*Bison bison occidentalis*) cranial and postcranial specimens between 1915 and 1967 (Harington 1977a, p. 849, Hay 1924, p. 200)]; (2) bison remains thought to be of western bison from near Douglas; (3) bones of an unidentified species of bison collected at Arden from the base of the Upper Campbell beach 30 ft below the surface.

1053. Pettipas, L.F. (1976): Environmental change and cultural dynamics during the Paleo-Indian period, with special reference to Manitoba. Papers in Manitoba Archaeology, Miscellaneous Papers 1:20–40.

A reconstruction of the Paleoindian prehistory of the Glacial Lake Agassiz region is based on paleoecology. Palynology seems to indicate a spruce forest habitat and there are few remains of grazing animals before 10,000 BP. The grazing animals moved into the area around 10,000 BP, when the forest was replaced by grassland, and they include: (1) bison (*Bison*) remains found in southwestern Manitoba, radiocarbon dated at 9110 ± 110 (Y-415) and 8020 ± 100 BP (Y-416); (2) a western bison (*Bison bison occidentalis = Bison preoccidentalis*?) skull found near Russell, radiocarbon dated at 6320 ± 140 BP (GSC-280); (3) extinct bison remains from Swan River, Millwood, Treesbank, and Douglas; (4) a muskox [*Ovibos moschatus*] skull from Grandview radiocarbon dated at 8620 ± 190 BP (I-1623), but the author (Harington 1970b) believes that this date may be too late. Also mentioned is the find of a mammoth (*Mammuthus*) from Kyle, Saskatchewan, radiocarbon dated at 12,000 ± 200 BP (S-246). Clovis projectile points [(*Homo sapiens*), 12,000–11,000 BP] are not expected to be found in Manitoba at this time because much of the province was under Glacial Lake Agassiz; however, a few of these artifacts have been found in the southern part of the province.

1054. Pettipas, L. (1988): Extinct bison of Manitoba. Manitoba Culture, Heritage and Recreation, Historic Resources:1–10.

This popular publication deals with ancient bison (*Bison bison antiquus = Bison antiquus antiquus*) and western bison (*Bison bison occidentalis = Bison antiquus occidentalis*). A map of the Manitoba fossil localities includes: Assiniboine River valley localities, which are by far the most productive, Millwood, Russell, Binscarth, Wigle Springs, Treesbank, Winnipeg. These are radiocarbon dated between 7000 and 4000 BP. Bison remains from the Swan River locality are radiocarbon dated at just over 8000 BP. A bison skull was found 10 m below the Glacial Lake Agassiz beach at Arden, indicating that the specimen is at least older than the lake itself. Human (*Homo sapiens*) artifacts found in association with bison remains, whose estimated ages are about 8000 BP, include: tools made of beaten copper with a bison skull found at Mud Falls; and spear points found with bison remains at Great Falls.

1055. Pettipas, L. (1990): The fossil elephants of Manitoba. Manitoba Culture, Heritage and Recreation, Historic Resources, pp. 1–23.

Fifteen fossil proboscidean (Proboscidea) localities have been found in Manitoba. The localities (map, Figure 5) include: mammoth (*Mammuthus*) teeth found at Birds Hill, Dufresne, Rathwell, Rivers, Souris, St Malo, and Bird (near Gillam); mammoth bone fragments found at Snake Creek (near Erikson); and mammoth tusks found at Transcona, Benito, Boissevain, and Birds Hill. Blumenort and Moosenose each yielded a single mastodon (*Mammut americanum*) tooth. Hind (1860) and Tyrrell (1892) report a find of a complete proboscidean skeleton buried 6 m beneath the surface of an exposed bank of Shell River. Hind also reported a find of skeletal remains of a mammoth from the bank of the Valley River where it cuts the 'Old Lake Ridge.'

1056. Pettipas, L.F., and Buchner, A.P. (1983): Paleo-Indian prehistory of the Glacial Lake Agassiz region in southern Manitoba, 11,500 to 6500 BP. *In*: Glacial Lake Agassiz (J.T. Teller and L. Clayton, eds.). Geological Association of Canada Special Paper 26:422–51.

An attempt is made to fit early people (*Homo sapiens*) into their paleoenvironment, which is partly deduced from the paleontological evidence. At about 10,000 BP Glacial Lake Agassiz covered much of the province, but the area to the SW was not. Before that, the area was boreal forest, but about 10,000 BP it had changed abruptly to a grassland environment. Consequently, the area witnessed an influx of grazing animals. The paper mentions remains of: several bison (*Bison*) specimens radiocarbon dated at 10,300 ± 200 (BGS-617), 9110 ± 110 (Y-415), and 8020 ± 100 BP (Y-416); western bison (*Bison bison occidentalis = Bison preoccidentalis*) found near Russell radiocarbon dated at 6320 ± 140 BP (GSC-280); other unidentified bison found 9 m beneath the surface in gravels at the base of the Upper Campbell Beach near Arden, where the stratigraphic context indicates an Early Holocene age; bison described in Pettipas (1971); the Grandview muskox skull and a moose (cf. *Alces alces*) antler tool and a stone chopper from the Rainy River District of northern Ontario found in deltaic sand and gravelly sand entering Glacial Lake Agassiz radiocarbon dated at 7861 ± 423 BP (SM-696-2). Evidently early humans hunted many large mammals, but preferred mammoth (*Mammuthus*) and mastodon (*Mammut americanum*).

1057. Péwé, T.L. (1957): Permafrost and its effect on life in the North. 18th Annual Biology Colloquium (April, Corvallis). pp. 12–25.
[See Péwé (1966) for revised version.]

1058. Péwé, T.L. (1966): Permafrost and its effect on life in the North. Pamphlet published by Oregon State University Press, Corvallis. 40 pp. Also published in Arctic Biology [Second Edition, 1966 (H.P. Hansen, ed.)]. Oregon State University Press, Corvallis.
The pertinent section is 'Fossil flora and fauna of the frozen ground' (pp. 16–20). The author gives a brief historical perspective on Pleistocene mammals of Alaska. Ice age mammal bones are abundant and widespread in the frozen sediments there – most occurring in the nonglaciated part. It was not until large-scale gold-mining operations began near Fairbanks in 1928 that hopes of early collectors were, in part, realized. Tens of thousands of fossils have been collected in the area during the last 25 years [since about 1941], but until 1947, little attempt was made to relate them to Quaternary stratigraphy. A great variety of animal remains have been preserved in organic silt of the valley bottoms. Most abundant are remains of bison (*Bison*), mammoth (*Mammuthus*), and horse (*Equus*). The majority of bones are isolated, and only rarely are complete skeletons found. At Dome Creek near Fairbanks, a fairly complete skull and tusks of an enormous mammoth were recovered in the early 1950s. The tusks are 13 ft 7 in long (outside curve) and weigh about 350 lbs apiece. Hair from the specimen was radiocarbon dated to 32,700 ± 980 BP (St-1632). A partial carcass of a steppe bison (*Bison priscus = Bison superbison crassicornis*) was found in 1951 (Figure 4). It consists of head, complete with hide, horns, one ear, four legs with hooves, and much torso hide (about 3 mm thick). A date of more than 28,000 years (L-127) was obtained from a piece of carcass and, in 1965, a date of 31,400 +2040 -1815 BP was obtained from the Geological Survey of Sweden. Pieces of fur and hide of a steppe bison from Fairbanks Creek have yielded a date of 11,950 ± 135 BP (St-1633). Table 1, compiled by the author in 1956, lists about 32 species of fossil mammals reported from perennially frozen Quaternary creek-valley silt deposits near Fairbanks.

1059. Péwé, T.L. (1968): Loess deposits of Alaska. Proceedings of the XXIII International Geological Congress 8:297–309.
Loess is probably the most widely distributed sediment of Quaternary age in Alaska. Most is aeolian in origin, although a great deal of water erosion can influence the eventual deposition. Much of the loess falling on slopes and summits has been washed into valley bottoms to form thick deposits of bedded to massive silt that is rich in organic debris and locally known as 'muck.' The periods of greatest loess deposition correspond to glacial maxima. In Illinoian time, the hills were blanketed with loess derived from the floodplain of the Tanana River and glacial outwash plains S of the Fairbanks area. Much of this windblown silt was retransported to creek-valley bottoms, incorporated much organic debris, including vertebrate remains [the most common vertebrate remains in the Wisconsinan muck, in order of abundance are: bison (*Bison*), mammoth (*Mammuthus*), and horse (*Equus*)], and became perennially frozen (pp. 305–6). In Wisconsinan time, additional loess was deposited in the uplands, then followed the same retransportation sequence. This valley-bottom loess is 3 to 46 m thick and contains abundant vertebrate and plant fossils, including partial carcasses of vertebrates that were entombed in the silt and perennially frozen. Frequently the remains of beaver (*Castor canadensis*) dams and beaver-pond sediments are preserved in the frozen loess. A diagrammatic cross-section of Baldwin Peninsula showing Quaternary loess deposits (Figure 6) indicates that Late Wisconsinan

beaver dams and beaver-pond sediments lie below the modern turf and ice wedges in 'Recent' (Holocene) alluvium [see McCulloch and Hopkins 1966]. Recent silt is 0.3 to 8 m thick, and all but the upper 1.5 m is perennially frozen. This silt contains no bones of extinct animals. The section pertinent here is 'Vertebrates' (pp. 91–103). Table 11 lists pre-Wisconsinan mammal remains from Alaska, whereas Table 12 lists mammal remains reported from perennially frozen creek-valley silts in the Fairbanks area, and Table 13 provides a list of radiocarbon dates applying to the latter. The author makes the following points: (1) Pleistocene mammal remains are widespread in Alaska, but the stratigraphic context is unknown for most of the early collections; (2) Most of the remains are of Wisconsinan age, but there is a growing number of significant pre-Wisconsinan specimens; (3) Many Pleistocene mammal taxa were present in North America earlier than generally recognized because of work in Alaska over the past 20 years; (4) All carcasses found so far are Wisconsinan in age – thus destroying the long-held idea that the animals must have lived only in interglacial times [see Matthews 1982]; (5) All specimens show some sign of decay, and the animals were not frozen instantaneously by cataclysmic events; (6) the author thinks that loss of grassy habitat and increased predation by people (*Homo sapiens*) may have caused the extinction of the Late Pleistocene grazers in Alaska 10,000 years ago.

1060. Péwé, T.L. (1975a): Quaternary geology of Alaska. Geological Survey Professional Paper 835:1–145.
The author provides a study of the glacial, periglacial, aeolian, fluvial, lacustrine, marine, and volcanic deposits of Quaternary age in Alaska and paleoclimatic fluctuations in light of formation and disappearance of glaciers and permafrost and changes in the distribution of plants and animals. Chapters include: History of research, the Pliocene-Pleistocene boundary, Glacial geology, Eolian deposits, Permafrost and periglacial deposits, Fluvial deposits, Lacustrine deposits, Marine deposits, Volcanic ash deposits, Flora, Fauna, Climate, and Late Cenozoic history of Alaska. Most pertinent is the section on vertebrates (pp. 91–103), which is summarized on p. 103. Table 12 lists about four dozen mammalian taxa from frozen silt deposits of Pleistocene age near Fairbanks. It is complemented by Table 13 – 28 radiocarbon dates on Pleistocene mammal remains from the Fairbanks area. The text is well supported with illustrations, maps, photographs, lists of specimens, and a very extensive bibliography.

1061. Péwé, T.L. (1975b): Quaternary stratigraphic nomenclature in unglaciated central Alaska. Geological Survey Professional Paper 862:1–32.
Fifteen long-recognized stratigraphic units of Quaternary age in unglaciated central Alaska are here defined and given formal names. Quaternary vertebrates, where found, are commented on under the different formation names from earliest to latest. A systematic classification is proposed under which the Quaternary sand, gravel, volcanic ash, loess, retransported loess, and organic deposits can be readily identified. The loess deposits and retransported loess can be correlated widely in central Alaska and recognized from the Canadian border to the Bering Sea. The stratigraphy of underlying creek-gravel deposits can, however, be correlated only locally. All sediments perhaps originated in a periglacial climate and are in large part perennially frozen today. This comprehensive work features diagrams and photographs of the named stratigraphic units, and more than 150 radiocarbon dates that are applicable to the Late Quaternary stratigraphy arranged by geological formation and locality. Most of the dated Pleistocene mammal specimens are from the Fairbanks area, and are listed in Table 4 (Radiocarbon dates from central Alaska, pp. 26–32).

1062. Péwé, T.L. (1976): Late Cenozoic history of Alaska. *In*: Quaternary Stratigraphy of North America (W.C. Mahaney, ed.). Dowden, Hutchinson and Ross Inc., Stroudsburg. pp. 493–506.
This review, condensed from Péwé's publication *Quaternary Geology of Alaska* (1975), covers the geological, biogeographical, and climatic history of Alaska during the last five million years. During this time there is a record of repeated dispersal of land mammals from Asia to Alaska, the environment being such that most northern mammalian species could migrate across the Bering Isthmus. The climate of the Illinoian glacial interval was rigorous, and permafrost was widespread. Central Alaska and the North Slope were refugia, probably with a tundra environment. Mammals occupying this tundra environment – such as ground squirrels (*Spermophilus*), lemmings (*Dicrostonyx* and *Lemmus*), goats [unsupported by sound fossil evidence] and mountain sheep (*Ovis*) limited

to range above or near tree line today lived then at lower altitudes. Mammoth (*Mammuthus*), horse (*Equus*) and bison (*Bison*) were most abundant. Caribou (*Rangifer tarandus*), Staudinger's muskox (*Praeovibos*), Dall sheep (*Ovis dalli*), moose (*Alces*), tundra muskox (*Ovibos moschatus*), saiga antelope (*Saiga tatarica*), helmeted muskox (*Bootherium bombifrons* = *Bootherium* and *Symbos*), and collared lemming (*Dicrostonyx*), as well as others – were widespread in Alaska in Illinoian time. There is an absence of mammal carcasses for this period. During the Wisconsinan, central and northern Alaska were essentially ice-free, as in the Illinoian. Wisconsinan loess was retransported to valley bottoms to form organic-rich, fetid, perennially frozen deposits. These deposits accumulated up to 60 m thick and incorporated abundant vertebrate and plant fossils, including partial carcasses of mammals that did not completely decay prior to burial. Forest was not entirely absent as indicated by tree remains in beaver dams (including alder, spruce, willow, poplar, and birch wood) of Late Wisconsinan age. The following tree-line mammals existed in lowlands (now forested) near Fairbanks: ground squirrel (*Spermophilus*), beaver (*Castor canadensis*), brown lemming (*Lemmus sibiricus*), collared lemming (*Dicrostonyx torquatus*), and Dall sheep (*Ovis dalli*). Throughout the Wisconsinan the refugia were occupied by an extensive mammal population, many species of which no longer exist. Many more grazers were present then than now. As in Illinoian time, mammoth (*Mammuthus*), horse (*Equus*), and bison (*Bison*) were the most common large mammals. Toward the end of the Wisconsinan, as the forest began to return, the ideal habitat of many grazers began to disappear. This loss, coupled with increased human (*Homo sapiens*) predation, resulted in extinction of many mammals about 10,000 years ago. Sea level rose between 11,000 and 10,000 BP, severing the isthmus between Asia and North America.

1063. Péwé, T.L., and Hopkins, D.M. (1967): Mammal remains of pre-Wisconsin age in Alaska. *In*: The Bering Land Bridge (D.M. Hopkins, ed.). Stanford University Press, Stanford, California. pp. 266–70.

This important paper lists the pre-Wisconsinan mammal (Mammalia) fossils found to 1967 in Alaska, and relates the various taxa to the areas in which they were found and their geological age (based mainly on stratigraphic evidence – unfortunately most Alaskan Pleistocene vertebrate material now in museums has no known stratigraphic context). The authors conclude from a study of mining records that most yak (*Bos grunniens*) and saiga (*Saiga tatarica*) fossils described by Frick (1937) came from Wisconsinan silt. However, pre-Wisconsinan fossils are of special interest because some may represent species that arrived in Alaska long before their earliest appearance in the conterminous United States. The list includes: beaver (*Castor*) – Illinoian, Fairbanks (Sheep Creek); muskrat (*Ondatra zibethicus*) – Late Illinoian or Sangamon, Kotzebue Sound (skeleton from ancient deposits of Kobuk River); collared lemming (*Dicrostonyx torquatus*) – pre-Illinoian interglacial(?), Kotzebue Sound (mandible with teeth from Baldwin Peninsula); dhole (*Cuon* = *Xenocyon*) – Illinoian, Fairbanks (mandible from 'Cripple Creek Sump'); wolf (*Canis lupus*) – Illinoian, Fairbanks (from 'Cripple Creek Sump' – loess deposited on down-warped or down-faulted gravel beneath the present valley of Cripple Creek); red fox (*Vulpes vulpes*) – Illinoian, Fairbanks ('Cripple Creek Sump'); 'large cat' like American lion? (*Panthera leo atrox*? = *Felis atrox*?) – Illinoian, Fairbanks ('Cripple Creek Sump'); American mastodon (*Mammut americanum*) – Illinoian, Fairbanks ('Cripple Creek Sump'); mammoth (*Mammuthus*) – pre-Illinoian interglacial?, Fairbanks (in gravel beneath Illinoian loess) – Illinoian, Fairbanks ('Cripple Creek Sump' and many other exposures) – Illinoian, Nenana Valley (in outwash of Healy Glaciation near Ferry); wapiti (*Cervus elaphus*) – Illinoian, Fairbanks ('Cripple Creek Sump'); giant moose (*Alces latifrons* = *Cervalces alaskensis*) – Illinoian, Fairbanks ('Cripple Creek Sump'); moose (*Alces*) – Illinoian, Fairbanks ('Cripple Creek Sump'); caribou (*Rangifer tarandus*) – pre-Illinoian interglacial?, Fairbanks (beneath Illinoian loess) – Illinoian, Fairbanks ('Cripple Creek Sump' and many other exposures) – pre-Illinoian interglacial?, Kotzebue Sound (pelvis from Baldwin Peninsula); helmeted muskox (*Bootherium bombifrons*) – Illinoian, Fairbanks ('Cripple Creek Sump'); tundra muskox (*Ovibos moschatus*) – Illinoian, Fairbanks ('Cripple Creek Sump') – Illinoian, Nome (cranium from outwash of Third Beach); Dall sheep (*Ovis dalli*) – Illinoian, Fairbanks ('Cripple Creek Sump') – Illinoian, Nenana Valley ('cf. *Ovis*'? in outwash of Healy Glaciation near Ferry); Yukon horse (*Equus lambei*) – Illinoian? Yukon Flats (partial lower jaw with teeth from outwash near Venetie; horse (*Equus*) – Illinoian, Fairbanks ('Cripple Creek Sump' and many other exposures); walrus (*Odobenus rosmarus*) – Sangamon interglacial (tusks from Second Beach = Pelukian transgression). Where known, the institution holding the specimens is recorded.

1064. Péwé, T.L., Hopkins, D.M., and Giddings, J.L. (1965): The Quaternary geology and archaeology of Alaska. *In*: The Quaternary of the United States (H.E. Wright, Jr, and D.G. Frey, eds.). Princeton University Press, Princeton. pp. 355–74.
The earliest geological report dealing with Quaternary deposits of what is now Alaska is Kotzebue's (1821) *A Voyage of Discovery into the South Sea and Bering's Straits, for the purpose of Exploring a Northeast Passage*. It contains an account of the famous bone-bearing (Mammalia) frozen silt and mysterious masses of buried ice at Elephant Point on Eschscholtz Bay, Kotzebue Sound, about 30 km S of the Arctic Circle. The development of most of the present landscape and the entry of people (*Homo sapiens*) to Alaska took place during the Quaternary. It was a time of advance and retreat of glaciers, formation and thawing of permafrost, transgression and regression of seas, formation and erosion of sand dunes and loess deposits, and filling of basins with fluvial sediments. Because loess provides an excellent medium for preservation, more effort should be concentrated on the search for artifacts in Alaskan loess (p. 362). The authors indicate that the Bering Isthmus existed during most of the Wisconsinan glaciation (though it was probably temporarily closed during the Woronzofian transgression about 45,000–33,000 BP) and remained available to migrating plants, animals, and people until it was drowned by the rising sea about 11,000 years ago (p. 365). The oldest Aleut site recognized is at Anangula Island, and was occupied as early as 8400 years ago. The nearby Chaluka site on Umnak Island shows continuity of a single way of life throughout the past 4000 years. The oldest well-dated Eskimo [Paleoeskimo] culture is represented by the Denbigh Flint Complex, characterized by microblades and cores, etc. found in deposits about 5000 years old. A few artifact assemblages from western and northern Alaska may be of Late Pleistocene age, and may represent Indian [Paleoindian] rather than Eskimo [Paleoeskimo] manifestations.

1065. Péwé, T.L., Westgate, J.A., and Stemper, B.A. (1989): Refinement of age interpretation of Quaternary events in Fairbanks area, Alaska. 28th International Geological Congress, Abstracts 2:602.
A detailed geological record of the past 2 million years or more is preserved in the nonglacial sediments of the Fairbanks area in central Alaska (Figure 1). Perennially frozen gravel, loess, and retransported loess with several interbedded tephra layers, preserve a rich floral and faunal history and record climatic changes in a periglacial Quaternary environment only tens of km from the glaciated Alaska Range. The stratigraphy, from oldest to youngest, with vertebrate fossils (where mentioned) and approximate ages follow: (1) the oldest unconsolidated deposit is the gold-rich Cripple Gravel of Pliocene and/or Early Pleistocene age; (2) another gold-bearing bed deposited during rigorous periglacial climate, the Fox gravel, contains large bones of mammoth (*Mammuthus*) and steppe bison (*Bison priscus*); (3) Dawson Cut Formation containing forest beds is believed to represent an Early Pleistocene interglacial, similar to the present; (4) Gold Hill Loess (up to 57 m thick) extends from more than 1 Ma to about 130,000 BP, and may represent two or more glacial advances. The lower part of this deposit is Early Pleistocene. Arctic and alpine mammals occur throughout the deposit – Cripple Sump is rich in fossils, including remains of Staudinger's muskox (*Praeovibos*), and pika (*Ochotona*) is found at the base of Gold Hill. The most primitive known lemming (*Dictrostonyx*) occurs near the middle of this deposit at Gold Hill; (5) Eva Formation, a forest bed 1 m thick, contains logs dated at >57,000 BP [and probably represents the last (Sangamonian) interglacial (Péwé et al. 1997)]; (6) Goldstream Formation, 10–35 mm thick, contains remains of mammoth, horse (*Equus*), bison (*Bison*), and small mammals, as well as rare frozen carcasses. Many radiocarbon ages confirm a Wisconsinan age for this formation; (7) Engineer Loess and Ready Bullion deposits are younger than about 10,000 BP. [See Péwé 1975a, 1976.]

1066. Péwé, T.L., Berger, G.W., Westgate, J.A., Brown, P.M., and Leavitt, S.W. (1997): Eva Interglaciation Forest Bed, unglaciated east-central Alaska: Global warming 125,000 years ago. Geological Society of America, Special Paper 319:1–54.
This forest bed represents a frozen, buried, ancient boreal forest in the Yukon-Tanana Upland of east-central Alaska. It consists of excellently preserved peat lenses, sticks, roots, and logs, as well as rooted and unrooted

stumps of trees, mainly spruce and birch. The underlying Old Crow Tephra dated at 140,000 ± 10,000 BP, and thermoluminescence sediment dating applied to the loess above and below the forest bed indicates its age is probably 125,000 BP with a duration of the Eva Interglaciation of only a few thousand years. Evidence of deep permafrost thawing is supported by the absence of ice wedges, buried pingos, and mammal (Mammalia) carcasses in the presently refrozen loess of pre-Wisconsinan age. Beaver-gnawed wood has not been reported from the Eva Forest Bed. However, beaver-gnawed (Castoridae – probably *Castor canadensis*) wood does exist in at least one (Birch Creek) of the two Eva Forest Bed localities adjacent to the Yukon-Tanana Upland. Apparently all the beaver dams and beaver-gnawed wood cited over the years from mining cuts in the Fairbanks region are from the Giddings Forest Bed and the Ready Bullion Formation – all Holocene in age. Many radiocarbon dates are now available on such wood, and all known dates are 10,000 BP or younger.

1067. Pielou, E.C. (1991): After the Ice Age: The Return of Life to Glaciated North America. University of Chicago Press, Chicago and London. 366 pp.
This is a valuable synthesis of facts and ideas about climate, geography, and life during the past 20,000 years in glaciated North America (Canada, Alaska, and part of the northern conterminous states of the United States). Above all, it documents, based on data from the fields of glacial geology, geomorphology, paleontology, as well as the systematics and biogeography of modern organisms, the return of life to the region. Sections are provided on: large mammals S of the ice sheets about 18,000 years ago (pp. 108–12); vertebrates of Beringia and the Ice-Free Corridor (pp. 126–8 and 147–60); Champlain Sea mammals (p. 217); megafaunal extinctions at the end of the Pleistocene (pp. 251–65), including the Overkill Hypothesis. The author concludes that the great wave of extinctions at the end of the Pleistocene has yet to be convincingly explained.

1068. Piérard, J., and Tremblay, E. (1980): Description d'une dent de mastodonte (*Mammut americanum* Kerr, 1972) [should be 1792] provenant de Chambord, Lac Saint-Jean, Québec. Naturaliste Canadien 107:277–83.
A mastodon (*Mammut americanum*) molar (Figure 2) was found by a child during the summer of 1978 at the NE portion of elevated land on Pointe-de-Chambord, a peninsula jutting in a NW direction into Lac St-Jean, Quebec. It is the first record of the species from Quebec, as well as the most northerly site for eastern North America. The specimen is part of the personal collection of Eve Tremblay. The article includes a detailed description and measurements of the tooth. Although a hypothesis is presented about the age of the tooth, its age will remain unknown until it is radiocarbon dated. The discovery of this tooth in the Lac St-Jean area raises interesting points regarding the extinction of the mastodon and its distribution in the Northeast. [See Dreimanis 1962, 1968.]

1069. Piers, H. (1912): Mastodon remains in Nova Scotia. Transactions of the Nova Scotian Institute of Science 13:163–74.
A mastodon (*Mammut americanum*) thigh bone and molar were found 12 cm below the surface in sandy soil on the McRae farm, 0.8 km W of Lower Middle River, Victoria County, Cape Breton Island, and 2.4 km E of the junction of Leonard MacLeod Brook and the Middle River, Nova Scotia. [See Harington, Grant, and Mott 1993.]

1070. Pigott, P. (1999): Tracking a phantom grizzly. Equinox No. 102:64–74.
The author gives anecdotal and scientific reasons for believing that grizzly bears (*Ursus arctos*) once roamed Labrador. What started the author's quest for more information was seeing a grizzly bear skull on display in the Newfoundland Museum in St John's, that had been unearthed by an anthropologist at an eighteenth-century Inuit settlement on Okak Island off the coast of Northern Labrador. It is unlikely that this skull had been acquired by trade. Furthermore, there are quite convincing Hudson's Bay Company records. John McLean, a Scottish fur trader for the Hudson's Bay Company recorded the acquisition of four 'grey grizzly pelts.' McLean could have been mistaken, but it is unlikely that an experienced fur trader who had handled hundreds of grizzly or brown bear pelts on the western side of the continent could have misidentified the species. By the end of the nineteenth century even anecdotal reports of the Labrador grizzly had declined, and if it existed, it must have disappeared early in the twentieth century. Hudson's Bay Company records show four grizzly skins brought to Fort Chimo in

the winter of 1838–9, and trading posts operated by the Moravian Brothers along the Labrador coast collected 16 between 1835 and 1863, one in 1914, and another in 1926. From a paleontological angle, Dick Harington mentions brown bear fossils from Lake Simcoe in southern Ontario (dated to about 12,000 BP) and another from a cave near La Rédemption on Quebec's Gaspé Peninsula that shared tundra-like terrain with Ungava lemmings (*Dicrostonyx hudsonius*). He conjectures that as the Laurentide ice sheet retreated, the lemmings migrated N to its isolated range in northern Quebec. Perhaps grizzlies were able to follow that route too. Furthermore, black bears (*Ursus americanus*), since the supposed disappearance of the eastern grizzly, seem to be taking on not only the character of the grizzly, but are moving out into their tundra habitat. Alasdair Veitch, who studies these black bears, says they have learned to ambush caribou and to fish in the rivers, as well as evolving smaller body size and a longer denning period. Last, but not least, traditional knowledge bears on this problem. Tshisheniss Pasteen, a Labrador Innu elder, mentions shooting a grizzly near Uptik.

1071. Pilny, J., and Morgan, A.V. (1984): Paleoenvironmental reconstruction of an (?) interstadial organic deposit near Innerkip, Ontario. Geological Association of Canada and Mineralogical Association of Canada Joint Annual Meeting (May 14–16, London, Ontario). Program with Abstracts 9:96.
Vertebrate remains, including several species of rodent (Rodentia), a deer (*Odocoileus*), and a few turtle (*Emydoidea*) shell plates, were found by a stream section in an exposed 1-m-thick layer of organics buried beneath till 3 km SW of Innerkip, Ontario (43°10′30″N, 80°43′00″W). Wood fragments from the organic layer were radiocarbon dated at >50,000 BP (GSC-2010-2). [See Pilny and Morgan (1987) for species identifications.]

1072. Pilny, J.J., and Morgan, A.V. (1987): Paleoentomology and paleoecology of a possible Sangamonian site near Innerkip, Ontario. Quaternary Research 28:157–74.
The species from the Innerkip, Ontario, site (Pilny and Morgan 1984) are: white-tailed deer (*Odocoileus virginianus*), muskrat (*Ondatra zibethicus*), vole (*Microtus*), and Blanding's turtle (*Emydoidea blandingi*). A distribution map of the turtle, whose modern range is mixed and deciduous woodland, is included. The 1-m section of silts and peaty detritus is overlain by two Late Wisconsinan tills, so the site could be Sangamonian in age. Plant, invertebrate, and vertebrate fossils from the site indicate a locally rich and varied flora and fauna – probably deposited in a well-vegetated pond. Modern distributions and habitats of the identified fossil insects suggest temperatures similar to those found in southern Ontario today.

1073. Pilny, J.J., Morgan, A.V., and Morgan, A. (1987): Paleoclimatic implications of a Late Wisconsinan insect assemblage from Rostock, southwestern Ontario. Canadian Journal of Earth Sciences 24:617–30.
Woolly mammoth (*Mammuthus primigenius*) bone fragments (portion of skull with molar, vertebra, scapula, limb, and tusk fragments) were found in 1982 in till overlain by glacial deposits in a peat-covered depression on the Poole farm near Rostock, 15 km N of Stratford, Ontario (43°30′N, 81°00′W; 333 m asl). The peat–marl transition zone (approximately contemporaneous with the spruce–pine transition in southern Ontario) where the mammoth was found occurred about 10,600 BP in this locality, and a radiocarbon date on tusk collagen yielded 10,790 ± 150 BP (WAT-999). Insect fossils from the site suggest a shift from cold, tree-line conditions to a cool, temperate environment between about 13,000 and 11,000 BP.

1074. Pinsof, J.D. (1996): Current status of North American Sangamonian local faunas and vertebrate taxa. *In*: Palaeoecology and Palaeoenvironments of Late Cenozoic Mammals: Tributes to the Career of C.S. (Rufus) Churcher (K.M. Stewart and K.L. Seymour, eds.). University of Toronto Press, Toronto. pp. 156–90.
At present there are at least 62 vertebrate-bearing localities considered to be of Sangamonian age in Canada, the United States, and Mexico. These sites have yielded 43 fish, 51 amphibian, 99 reptile, 180 bird, and 289 mammal taxa. About 22% (107 of 494) of the vertebrate species of this age are extinct. The Sangamonian interglacial, which preceded the Wisconsinan glaciation, is characterized as a stable period ecologically, which in recent years has become synonymous with oxygen-isotope stage 5. It began about 125,000 years ago and lasted until about 75,000

years ago. The paper has an extensive bibliography and two large appendices, one naming the site and its principal reference and the second listing taxonomically the vertebrate species by locality. Alaskan localities include: Amchitka, Second Beach (Seward Peninsula), Canyon Creek, and Tofty Mining District. Yukon localities include: Old Crow River, Loc. 44, and Herschel Island. Northwest Territories localities include: Lower Carp Lake. British Columbia localities include: Quesnel, Quesnel Forks, and Westwold. Alberta localities include: Watino, Bindloss, and Medicine Hat #7. Saskatchewan localities include: Riddell [Saskatoon] and Fort Qu'Appelle. Manitoba localities include: Bird and Minnedosa River. Ontario localities include: Moose River Crossing, Innerkip, and Don Valley Brickyard. The only New Brunswick locality is Hillsborough. Nova Scotia localities include: Miller Creek, East Millford, and Lower Middle River. [See Harington 1990a.]

1075. Pirozynski, K.A., Carter, A., and Day, R.G. (1984): Fungal remains in Pleistocene ground squirrel dung from Yukon Territory, Canada. Quaternary Research 22:375–82.

Vertebrate remains were found at Dominion Creek, Dawson Loc. 28, about 43 km SE of Dawson City, Yukon. Fungi in the dung of an arctic ground squirrel (*Spermophilus parryii*) were studied and most proved to be either widespread saprobes or nonspecific coprophiles. However, both *Chaetomium simile* and *Thecaphora deformans* represent species that may be more characteristic of rodent dung than that of other animals. A radiocarbon date on organic material associated with the rodent remains yielded 12,200 ± 100 BP (GSC-2641). Other vertebrates found in the area include: wolf (*Canis lupus*), American lion (*Panthera leo atrox*), woolly mammoth (*Mammuthus primigenius*), Yukon horse (*Equus lambei = Equus (Asinus) lambei*), caribou (*Rangifer tarandus*), steppe bison (*Bison priscus = Bison crassicornis*), helmeted muskox (*Bootherium bombifrons = Symbos cavifrons*), and Dall-like sheep (*Ovis* cf. *Ovis dalli*). The authors suggest that paleontologists should consider paleomycology as a potential additional source of data.

1076. Pirozynski, K.A., Jarzen, D.M., Carter, A., and Day, R.G. (1988): Palynology and mycology of organic clay balls accompanying mastodon bones – New Brunswick, Canada. Grana 27:123–39.

Mastodon (*Mammut americanum*) remains were found *in situ* in a layer of blue organic clay dating from an interstadial of the Wisconsinan glaciation, at Hillsborough, New Brunswick (Loc. 0206, shown on included map). [But see Harington, Grant, and Mott (1993), who suggest a Sangamonian interglacial age for the mastodon.] Clay balls adhering to the mastodon bones contained wood and other organic detritus radiocarbon dated at 37,200 ± 1310 BP (GSC-2469). The carbonate matrix of the clay balls was radiocarbon dated at 51,500 ± 1270 BP (GSC-2467), and a peat layer adjacent to the mastodon remains at >43,000 BP (GSC-1680). By studying the pollen and fungi spores, this paper tries to determine whether the enigmatic clay balls are faecal formations or some other geological formation. The conclusion reached is that the clay balls (containing about 2.6% organic material) were indeed within the animal at the time of death. Accidental ingestion of clay adhering to plant roots and geophagy are both discounted as the source of the clay content.

1077. Plaskin, R. (1986): Archaeologist trying to prove theory of early Yukon humans. Whitehorse Star. December 19, p. 33.

This article is about University of Toronto archaeologist Max Friesen's study of mammoth (*Mammuthus*) bones (25,000 to 100,000 years old) from ancient deltaic deposits in Old Crow Basin, Yukon. Friesen believed that some specimens (including nearly 10,000 bones) were broken by people (*Homo sapiens*), implying that humans arrived in North America much earlier than most of the scientific community thinks. Evidently the breaks occurred a few months after the animals' deaths, 'and this really reduces the odds of various other forces than man being involved.' Friesen discounted the theory that animal scavengers may have caused the breaks while the carcasses were still fresh. Of 1300 bones in the collection studied, about 5% were broken. However, among the limb bones, which he said were more likely to have served as tools, the breakage rate was 13%. Whereas scavengers evidently gnawed on some of the bones, none of the broken bones showed gnaw marks. 'If these sites stand up, as I think they eventually will, something comparable to *Homo erectus* (the evolutionary precursor to modern humans) may have come across the [Bering] land bridge between Asia and North America, rather than fully evolved humans,' Friesen said. [See Friesen (1989) for the opposite view.]

1078. Pohorecky, Z. (1988): The Saskatoon site – Palaeontological or archaeological? *In*: Out of the Past: Sites, Digs and Artifacts in the Saskatoon Area (U. Linnamae and T.E.H. Jones, eds.). Saskatoon Archaeological Society. pp. 47–64.
Several mammoth (*Mammuthus*) bones were found 10 m beneath the surface in a sandpit near the NW bank of the South Saskatchewan River in the northern part of Saskatoon, Saskatchewan (NW¼ Sec. 14 Twp. 37 Rge. 5 W. 3rd Mer., Figure 6.1). In 1953, bones of a small horse (*Equus*) and a proboscidean (Proboscidea) were found at this site, and some nearby wood was radiocarbon dated at >34,000 BP (S-426). Wood between the Floral and Battleford Formations was radiocarbon dated at '38,000' BP (GSC-104). Evidence of human (*Homo sapiens*) presence was indicated by: the apparently non-natural wearing of the bones; the presence of three left horse dentaries with no right components, and the fact that native North Americans are known to have used dentaries as tools; mammoth scapula marked by parallel cuts possibly used as a cutting table; and flaked stone with the mammoth remains. It is highly improbable that sand could have abraded the bones in such a way. Included are photographs of the specimens.

1079. Pohorecky, Z.S., and Wilson, J. (1968): Preliminary archaeological report on the Saskatoon site. Napao 1(2):35–70.
The Saskatoon site (52°10′N, 106°35′W) was discovered in 1968 by quarrymen working in a sandpit 2 mi N of Saskatoon, Saskatchewan. It proved to contain hundreds of disarticulated fossil bones and possible evidence for the coexistence of mammoths (*Mammuthus*) and humans (*Homo sapiens*) during one of the last interstadial intervals of the Pleistocene. A mammoth limb bone was radiocarbon dated at the Radiocarbon Laboratory of the Saskatchewan Research Council to an age of 20,200 ± 550 BP. Other species discovered include: Niobrara horse (*Equus* cf. *Equus niobrarensis*), left dentary and many teeth; several mammoth (*Mammuthus*) bones; bison (*Bison*) teeth; and some cervid pelvis fragments. With the paper are photographs of the site, fossils (some *in situ*), and stones and bones interpreted as artifacts. [See also Christiansen 1968; Harington 1990a; Lammers 1968; and SkwaraWoolf 1981.]

1080. Porsild, A.E., Harington, C.R., and Mulligan, G.A. (1967): *Lupinus arcticus* Wats. grown from seeds of Pleistocene age. Science 158(3797):113–14.
Rodent (Rodentia) remains and seeds were found buried in a muck layer 8–12 m thick overlying river-deposited gravel. The muck was exposed by placer mining at the mouth of Miller Creek where it enters Sixtymile River, Yukon (64°00′N, 140°49′W). The rodent burrows were 3–6 m below the surface and included a nest with fecal matter and seeds in different chambers. The Miller Creek specimens included collared lemming (*Dicrostonyx torquatus* = *Dicrostonyx groenlandicus*). Collared lemming was also present in Pleistocene deposits 80 km W of the Fairbanks, Alaska, area. This species preferred arctic and high alpine tundra environments, which existed at Miller Creek before 10,000 BP, when postglacial warming took place. The seeds from the burrow were also estimated to be Late Pleistocene in age and were still viable. A similar nest of arctic ground squirrel (*Spermophilus parryii* = *Citellus undulatus*) found in well-preserved burrows from central Alaska was radiocarbon dated at 14,860 ± 840 BP. Before 1932, muck deposits near the mouth of Sixtymile River produced bones representing: woolly mammoth (*Mammuthus* cf. *Mammuthus primigenius*), Yukon horse (*Equus* cf. *Equus lambei*), bison (*Bison*), and caribou (*Rangifer tarandus*). [See Harington 1997a.]

1081. Porter, L. (1986): Jack Wade Creek: An *in situ* Alaskan Late Pleistocene vertebrate assemblage. Arctic 39(4):297–9.
Partial remains of a Late Pleistocene ungulate community are described from Jack Wade Creek (64°09′00″N, 141°27′30″W; Figure 1), Alaska. Sixty fossil bones representing at least 11 Dall sheep (*Ovis dalli*), three steppe bison (*Bison priscus*), two caribou (*Rangifer tarandus*), one small Yukon horse (*Equus lambei* = *Equus* (*Asinus*) *lambei*), and one moose (*Alces alces*) were unearthed by gold miner George Robinson. Bison horncore collagen was radiocarbon dated at 29,700 ± 240 BP (QL-1178) and hare (*Lepus*) coprolites from a depth of 5 m yielded a date of 14,500 ± 3000 BP (QL-1213). The fact that the hindquarter bones of all these large herbivores are missing suggests that there may have been some human (*Homo sapiens*) involvement at this site. This assemblage is one of the few Late Pleistocene *in situ* assemblages known from Eastern Beringia, and is unusual because of the large proportion of Dall sheep in the fauna. The paper

includes a photograph (Figure 3) of a virtually complete cranium of Dall sheep (USNM 262462) and a pollen bar graph of the Jack Wade section (Figure 4).

1082. Porter, L. (1988): Late Pleistocene fauna of Lost Chicken Creek, Alaska. Arctic 41(4):303–13.
The fossil remains of one invertebrate and 16 vertebrate genera have been recovered from Late Quaternary sediments of a large placer gold mine in E-central Alaska. The location is on Lost Chicken Creek (64°03.2′N, 141°52.6′E; Figure 1), Alaska, a stream, less than 2000 m long, that flows into the South Fork of the Fortymile River. Forty-six of the 1055 fossils collected between 1976 and 1983 were recovered *in situ* from nine stratigraphic units recognized at the site. Radiocarbon dating was on organic material associated with fossils or with boundaries of stratigraphic units. Based on these dates, the fossils range in age from approximately 1400 BP on moose (*Alces alces*) to >50,400 BP on wood overlying Sheep Creek Ash. The fauna includes: a snail (*Succinea avara*); ptarmigan (*Lagopus*); collared lemming (*Dicrostonyx torquatus*); wolverine (*Gulo gulo*); American lion (*Panthera leo atrox*); Yukon horse (*Equus lambei = Equus (Asinus) lambei*); kiang-like ass (*Equus (Asinus)* cf. *kiang*); caribou (*Rangifer tarandus*); moose; wapiti (*Cervus elaphus*); saiga antelope (*Saiga tatarica*); Dall sheep (*Ovis ?dalli*); helmeted muskox (*Bootherium bombifrons = Symbos cavifrons*); muskox (*?Ovibos moschatus*); steppe bison (*Bison priscus*); bison (*Bison bison = Bison crassicornis*); woolly mammoth (*Mammuthus primigenius*); and humans (inferred) (*Homo sapiens*). A stratigraphic diagram and explanation compare the Lost Chicken Creek site with two other similar sites: Eva Creek, central Alaska, and Canyon Creek, E central Alaska. Several stratigraphic diagrams, tables of measurements, and photographs of saiga and wolverine skulls are included.

1083. Porter, L., and Hopkins, D.M. (1982): Butchered caribou skulls of the Pleistocene and Recent from Eastern Beringia. XI INQUA Abstracts 1:257.
An ancient style of butchering (*Homo sapiens*) caribou (*Rangifer tarandus*), which seems to have persisted until recently, is seen on seven skulls from Alaska and Yukon. This method involves removing the anterior two-thirds of the cranium in order to obtain the brains for tanning hides. Two of the crania from central Alaska are radiocarbon dated at 29,700 ± 240 BP (QL-1178). [No other date is given and this date is questionable as it evidently refers to a steppe bison (*Bison priscus*) horncore – see Table 1.] Other specimens range from Late Pleistocene in Yukon to Holocene from the North Slope of Alaska.

1084. Powers, W.R., and Hamilton, T.D. (1978): Dry Creek: A Late Pleistocene human occupation in central Alaska. *In*: Early Man in America from a Circum-Pacific Perspective (A.L. Bryan, ed.). Occasional Papers No. 1, Department of Anthropology, University of Alberta, Published by Archaeological Researchers International, Edmonton. pp. 72–7
The Dry Creek site (Figure 1), located about 180 km SW of Fairbanks, Alaska, is a relatively undisturbed site that displays evidence of two periods of human (*Homo sapiens*) occupation: one during the Holocene and one in the Late Pleistocene. The latter includes vertebrate remains along with the lithic tools and toolmaking debris. Based on radiocarbon dates of charcoal, the earliest human occupation of the site was about 11,000 BP. Bison (*Bison*) and horse (*Equus*) bones were identified from the early occupation (p. 75).

1085. Powers, W.R., Guthrie, R.D., and Hoffecker, J.F. (1983): Dry Creek: Archeology and paleoecology of a Late Pleistocene Alaskan hunting camp. Report submitted to the United States National Park Service, Contract CX-9000-7-0047. 461 pp.
Excavations at Dry Creek, which lies in the Nenana River valley of central Alaska about 180 km SW of Fairbanks near the town of Healy (Figure 2.1), have demonstrated human (*Homo sapiens*) occupation (apparently an autumn–winter hunting camp) of the valley about 11,000 to 10,000 BP. These people hunted remnants of the steppe-adapted Late Pleistocene grazing fauna. Chapter 6 (pp. 209–87) by Guthrie is a discussion of the paleoecology of the site and its implications for early hunters. Ungulate remains described include Dall sheep (*Ovis dalli*), steppe bison (*Bison priscus*), and wapiti (*Cervus elaphus = Cervus canadensis*). From about 10,500 to 11,500 BP such large grazers dominated the megafaunal community. Lower levels of Dry Creek were occupied by a local megafauna characterized by Pleistocene body size, i.e., prior to Holocene ‘dwarfing.’ The view from Dry Creek site would have offered an excellent chance to site all

three large grazers. The diversity of large-mammal species suggests an opportunistic hunting strategy that selected for a wide range of age classes.

1086. Preble, E.A. (1923): Mammals of the Pribilof Islands. *In*: A Biological Survey of the Pribilof Islands, Alaska. I. Birds and mammals (E.A. Preble and W.L. McAtee, eds.). North American Fauna 46:102–20.
The known records of polar bear (*Ursus maritimus* = *Thalarctos maritimus*) and woolly mammoth (*Mammuthus primigenius*) are reviewed for this part of Alaska. Polar bear fossils have been found on St Paul Island, including skulls and other bones of two juveniles, suggesting that breeding may have occurred there. In recent times polar bears are usually only seen as stragglers from drifting ice. Remains of woolly mammoth have been found on several occasions. In 1836, a tusk was found on St George Island, two tusks have been reported by natives, from St Paul Island, and two other teeth were collected from St Paul Island. [See also Ray (1971), who casts doubt on the validity of the last two specimens, and who reports on more recent finds.]

1087. Prest, V.K. (1970): Quaternary Geology of Canada. *In*: Geology and Economic Minerals of Canada. Economic Geology Report 1 (R.J.W. Douglas, ed.). Geological Survey of Canada, Queen's Printer, Ottawa. pp. 676–764.
Several Canadian vertebrate finds are mentioned. A beaver (*Castor canadensis*) cut stick radiocarbon dated at >33,800 ± BP (GSC-33) was found in glaciofluvial sediments slumped in a sinkhole in a gypsum quarry (Plate XII-1), 4 km S of Milford Station, Nova Scotia. Water-worn vertebrae of bison (*Bison*), part of a lower jaw of a bear (*Ursus*), a Scott's moose (*Alces scotti* = *Cervalces borealis*) antler, and mammoth (*Mammuthus*) or mastodon (*Mammut americanum*) ivory were found in the Christie St. sandpits, Toronto, Ontario (Coleman 1933). Remains of woolly mammoth (*Mammuthus primigenius*) and horse (*Equus*) have been found in the Early or Late Pleistocene Saskatchewan gravels of Manitoba, Saskatchewan, and Alberta. Arctic ground squirrel (*Spermophilus parryii* = *Citellus undulatus*) and large vole (*Microtus*) remains were found in 12 cm of silt overlain by three till layers separated by silt and sand layers, exposed by a road cut on the N side of the Minnedosa River valley near Riding Mountain, Manitoba. Fossiliferous beds, probably Sangamonian in age, containing the remains of bison (*Bison*) teeth and bones, mammoth (*Mammuthus*), horse (*Equus*), wolf (*Canis lupus*), and bear (Ursidae)[?], are overlain by about 60 m of drift deposits at Fort Qu'Appelle, Saskatchewan.

1088. Pringle, H. (1986): Vision Quest: Recent archaeological findings suggest a bounteous prehistoric life in the Rockies. Equinox 26:72–85.
Evidence has been found for human (*Homo sapiens*) habitation in the Rocky Mountains of British Columbia and Alberta. Projectile points and bison (*Bison*) kill sites dated to around 8000 BP have been found from 208 localities in the Red Rock Canyon at the South Kootenay Pass. Remains of bighorn sheep (*Ovis canadensis*), bison (*Bison*), deer (*Odocoileus*), beaver (*Castor canadensis*), and hare (*Lepus*), dated to about 10,000 BP, were found in the Banff, Alberta, area. Traditionally, archaeologists claimed that the Rocky Mountains were not occupied by humans because the adjacent plains were sparsely populated, but new findings suggest that the intermontane plateau had a more clement climate at that time.

1089. Pringle, H. (1999): Ice-age sleuth. Canadian Geographic 119(2):36–44.
A popular article dealing with a Canadian paleontologist's (Dick Harington) experience in collecting and studying Yukon ice age vertebrate fossils during the past 30 years. The article features a map of Beringia and new restorations of giant beaver (*Castoroides ohioensis*), American lion (*Panthera leo atrox*), steppe bison (*Bison priscus*), and Jefferson's ground sloth (*Megalonyx jeffersonii*).

1090. Provancher, L. (1869): Un morse fossile. Naturaliste Canadien 2:19.
A complete walrus (*Odobenus rosmarus* = *Trichecus rosmarus*) skeleton was found buried 35 cm beneath the surface in clay about 30 m asl near the Intercolonial Railroad at Ste-Cécile-du-Bic, Quebec. Bone measurements are included in the paper. The specimen was placed in the museum at Collège de Rimouski.

1091. Quackenbush, L.S. (1909): Notes on Alaskan mammoth expeditions of 1907 and 1908. Bulletin of the American Museum of Natural History 26:87–130.
This is an account of the 1907 and 1908 expeditions to

Yukon and Alaska launched by the American Museum of Natural History. The principal purpose was to excavate a supposedly entire mammoth found near Nome and to collect other fossil vertebrates. The trip itinerary is provided, along with lists of fossils and localities (see map following p. 130). Fossils from Alaska (pp. 126–7) include: woolly mammoth (*Mammuthus primigenius*), horse (*Equus*), steppe bison (*Bison priscus* = *Bison crassicornis* = *Bison alleni*), western bison (*Bison bison occidentalis* = *Bison occidentalis*), helmeted muskox (*Bootherium bombifrons* = *Symbos tyrrelli*), tundra muskox (*Ovibos moschatus* = *Ovibos yukonensis*), moose (*Alces alces* = *Alce*), caribou (*Rangifer tarandus*), bear (*Ursus*), wolf (*Canis lupus*), beaver (*Castor canadensis*), and walrus (*Odobenus rosmarus*). Fossils from Yukon include: woolly mammoth, American mastodon (*Mammut americanum* = *Mastodon americanus*), horse, steppe bison, western bison, helmeted muskox (*Bootherium bombifrons* = *Symbos tyrrelli*), tundra muskox, sheep (*Ovis*), moose, caribou, wapiti (*Cervus elaphus* = *Cervus canadensis*), bear (*Ursus*), and wolf. One of the Alaskan mammoth specimens retained some muscle tissue and patches of skin with black, brown, or reddish-brown hair still attached. Ten other authentic records of Pleistocene mammal remains found in place in Alaskan and Klondike deposits are listed (pp. 124–5).

1092. Quinn, J.H. (1957): Pleistocene Equidae of Texas. University of Texas, Bureau of Economic Geology Report of Investigations 33:1–51.
Quinn (pp. 14–16) referred the Yukon horse (*Equus lambei*) to *Onager lambei* (Hay 1917), but gives an incorrect specimen number (8226, rather than USNM 8426) for the type specimen, and incorrectly states that it is from Alaska rather than Yukon, Canada.

1093. Quintner, D. (1975): Fossil finding a mammoth family hobby. Toronto Star. June ?
The article mentions the Mississauga recreation room of Paul and Ledy Hoberlin, which contains an 11-ft-long tusk of a woolly mammoth (*Mammuthus primigenius*). The tusk, which Mrs Hoberlin first noticed protruding from a sandbar in the Porcupine River near the settlement of Old Crow, Yukon, was picked up in 1965. Evidently the tusk was radiocarbon dated, yielding an age of about 12,500 years [no laboratory number is given for this date]. A photograph of the complete adult tusk is provided.

1094. Rainey, F.G. (1939): Archaeology in central Alaska. Anthropological papers of the American Museum of Natural History 36(4):351–405.
Pertinent is the section 'Artifacts found through gold dredging operations' (pp. 390–401). The following items are worth noting: (1) About 1938 'almost the entire skin of a super-bison [steppe bison (*Bison priscus*)], the hair remaining, was found in the Fairbanks area. (2) Near Fairbanks, mammoth (*Mammuthus*) and bison (*Bison*) bones are most common, but mastodon (*Mammut americanum*), horse (*Equus*), two species of muskox [presumably the tundra muskox (*Ovibos moschatus*) and the helmeted muskox (*Bootherium bombifrons*)], sabretooth tiger [presumably the scimitar cat (*Homotherium serum*)], lion (*Panthera leo atrox*), camel (*Camelops*), 'gazelle' [?], antelope [presumably the saiga antelope (*Saiga tatarica*)], extinct bear [presumably the giant short-faced bear (*Arctodus simus*)], rodents (Rodentia), and sheep (*Ovis dalli*) were also present. (3) Extinct animal remains appear to be most common in the lower levels of the muck, and often remains are frozen in ice hummocks. (4) In 1933 two long, slender bone points (Figure 9, Nos. 1, 2), hafted as lance or spearheads, were found about 90 m apart in a thin gravel lens containing many animal bones about 1.5 m above the creek gravels or between 14 and 12 m below the original surface on Goldstream Creek. A large 'Folsom-like' chert point was also discovered there in 1933. (5) During 1936–7, Peter Maas found several implements as well as many burned stones associated with mammoth, mastodon, bison, and horse bones at the base of muck deposits (20 m below the original surface) on Ester Creek. Two of the implements – 'Yuma type' flint blades – were found still frozen in at the base of the muck, one (10 cm long and regularly flaked on both surfaces) being associated with the maxillary bone of a young mastodon. A fragment of 'mastodon tusk' (20 cm long and 6 cm in diameter) had been cut to a blunt point at one end and bevelled in the form of a broad arched wedge at the other. In September 1937, Maas found more than 100 young mammoth teeth scattered over an area a few metres in diameter. A horse splint bone, sharpened at one end to form an awl, and a mammoth tusk scraper about 8 cm in diameter were found in this section – both frozen *in situ*. A large collection of Pleistocene faunal (mammoth, mastodon, horse, and bison) remains was found in the same section. (6) An exposure at Little Eldorado Creek produced a large collec-

tion of mammal bones (mammoth, super-bison, horse, etc.) and a crudely flaked flint blade (apparently *in situ* at a depth of 50 cm). (7) In 1938 at Engineer Creek Peter Maas found a rib [perhaps bison, caribou, or moose (*Alces alces*)] with a blunt point and a series of grooves on both sides frozen *in situ* at the base of the muck 40 cm below the original surface.

1095. Rains, R.B. (1987): Holocene alluvial sediments and a radiocarbon dated relict beaver dam, Whitemud Creek, Edmonton, Alberta. Canadian Geographer 31:272–7.
The author provides background information and radiocarbon dates [e.g., 3180 ± 85 BP (S-1796) on beaver (*Castor canadensis*)-dam wood, and 3200 ± 90 BP (S-1794) and 3255 ± 90 BP (S-1795) on bison (*Bison*) bones trapped in the dam] on a relict beaver dam in terrace deposits on Whitemud Creek near Edmonton, Alberta.

1096. Rains, B., and Welch, J. (1988): Out-of-phase Holocene terraces in part of the North Saskatchewan River Basin, Alberta. Canadian Journal of Earth Sciences 25:454–64.
This study focuses on alluvial terraces in part of the North Saskatchewan River valley near Edmonton, Alberta, and in the small tributary valleys of Whitemud and Strawberry creeks. Characteristic terrace morphologies and alluvial sediment sequences are described and deposits at many sites are radiocarbon dated. Twenty-seven new radiocarbon dates, along with previously reported dates, provide a chronological framework for the study. Two unidentified bones (Mammalia) collected from sand a few centimetres above the gravel/sand contact at the Legislature Building excavation site are dated at 11,345 ± 420 BP (S-2385) and 10,740 ± 470 BP (S-1923), respectively – representing part of the T-1n sedimentation phase (Figures 3, 6). [From a photograph of the upper bone in Figure 6C, I can identify it as a distal end of a right radius of a horse – probably the Mexican horse (*Equus conversidens*).] A sample from 'numerous bison [*Bison*] bones' from fine-grained T-4n alluvium near Fort Saskatchewan, Alberta, yielded a date of 6955 ± 80 BP (S-1706). They were associated with a white tephra layer (Figure 3) – assumed to be the Mazama Ash, dated elsewhere at about 6600 BP. [From a photograph (Figure 7C), I can identify some of these bison bones as: phalanx, metapodial, distal end of a femur, proximal part of a rib, part of a thoracic vertebra.] A date on 'bison bone' from about 2 m below a T-4n tread in Edmonton was 2780 ± 85 BP (S-1799). At an exposure of T-1t alluvium in Strawberry Creek valley, a wolf (*Canis lupus*) right mandible was collected and yielded a date of 5640 ± 310 BP (S-1788). A date of 5865 ± 135 BP (S-1789) was obtained on a bone of a juvenile bison buried under landslide debris on another T-1t tread surface in Strawberry Creek valley. A date of 8195 ± 1090 BP (S-1798) was obtained for a bison pelvic fragment found on the surface of a T-2t tread surface in Whitemud Creek valley. A bison skull exposed in a Strawberry Creek valley terrace gave a date of 1135 ± 80 BP (S-1786), but seemed to the authors to be anomalously young. A date of 4685 ± 260 BP (S-1800) was obtained on an unidentified bone fragment buried in T-2t alluvium in Strawberry Creek valley. Dates of 5490 ± 230 BP (S-2387) and 4225 ± 150 BP 9S-1799) were derived from 'bone samples' from Whitemud Creek T-2t alluvium. A date of 3180 ± 85 BP (S-1796) from beaver-dam wood (Figure 9 shows the dam) from T-3t alluvium in Whitemud Creek compares with dates of 3200 ± 90 BP (S-1794) and 3255 ± 90 BP (S-1795) from bison bones trapped in the dam [see Rains 1987]. A bison skull extracted from lag gravels 30 cm above the base of a section of T-3t alluvium in Strawberry Creek yielded a date of 1625 ± 80 BP (S-1785), which is 'questionably young.' A bison vertebra from point-bar sands dated at 760 ± 85 BP (S-1784) 'is also anomalously young.' A bison bone from T-4t alluvium in Strawberry Creek, containing overbank fines gave a date of 1965 ± 75 BP (S 1783) it was probably redeposited from older materials. From another Whitemud Creek T-4t section, a bison scapula in overbank fines was dated at 1220 ± 70 BP (S-1793). [Drs P. Currie and C. Schweger assisted in identifying the bones.]

1097. Rains, R.B., Burns, J.A., and Young, R.R. (1994): Postglacial alluvial terraces and an incorporated bison skeleton, Ghostpine Creek, southern Alberta. Canadian Journal of Earth Sciences 31:1504–9.
Andrew Kopjar chanced upon a semi-articulated bison skeleton on his property near Three Hills, Alberta, in 1986. This largely intact skeleton of a plains bison (*Bison bison bison*; Figure 7) lay in point-bar sediment beside a T-2 terrace paleochannel scar in Ghostpine Creek valley (see map, Figure 3). Measurements of the skull and postcranial elements reveal that the animal seems to represent an adult bull plains bison of moderate to large size (Table

1). A radiocarbon date of 2580 ± 90 BP (AECV-9440) was obtained on collagen from a right metacarpal. Taphonomic analysis suggests that the animal died in the T-2 channel and was soon buried high on a point bar – probably during the waning flow of a single flood. Several small compression fractures, similar to coyote (*Canis latrans*) tooth marks, were present on the head of the right humerus, and spinous processes of the 19th to 15th thoracic vertebrae had been chewed off. The bones showed no signs of disease, although the right 7th and 10th ribs had been broken and healed before the bison died.

1098. Rajnovich, G. (1980): Extinct bison bones dug up near Kenora. Arch Notes 80(3):21–2.
An almost complete skull with horncores, one scapula, two rib fragments, a radius, and a phalanx of a bison identified as *Bison crassicornis* [Manitoba Museum of Man and Nature catalogue # V-780] was dug up from the bottom of a 4.5-m-deep peat bog near Kenora, Ontario. [See McAndrews (1982) for more details.] The office of the Northwestern Regional Archaeologist is planning to radiocarbon date one of the rib fragments. [Further study to confirm the identification seems warranted.]

1099. Rampton, V.N., and Paradis, S. (1981): Quaternary geology of Moncton map area (21I), New Brunswick. Natural Resources New Brunswick Map Report 81-2.
The authors note (p. 21) the presence of a whale like the minke (cf. *Balaenoptera acutorostrata*), based on a vertebra from 1.5 km W of St-Louis-de-Kent, New Brunswick. [The bone has been radiocarbon dated at 12,600 ± 140 BP (GSC-3284) and 11,830 ± 950 BP (S-1969-B) – a date of 18,570 ± 500 BP (S-1969-A) is apparently erroneous.] [See Rampton et al. 1984.]

1100. Rampton, V.N., Gauthier, R.C., Thibault, J., and Seeman, A.A. (1984): Quaternary geology of New Brunswick. Geological Survey of Canada Memoir 416.
The authors note (pp. 32, 34) the find of a Late Pleistocene vertebra of a whale like the minke (cf. *Balaenoptera acutorostrata*) from near St-Louis-de-Kent, New Brunswick (46°43.4′N, 64°59.6′W). It lay at the base of a 2–4-m-thick clay unit under an 0.8 m veneer of marine sand at an elevation of about 8 m. [See Rampton and Paradis 1981.]

1101. Rand, A.L. (1948): Glaciation, an isolating factor in speciation. Evolution 2:314–21.
Presumably before the ice age there was a northern transcontinental belt occupied by plants and animals. But when glaciers came to cover most of northern North America, the plants and animals living there were reduced to existing in refugia south of the ice, and in ice-free refugia within the glaciated area. Later, with the melting and retreat of the ice, the biota again spread to occupy the northern part of the continent. Speed of dispersal as well as point of origin and habitat preference have played a part in present bird (Aves) distribution. Relationships of some northern North American birds [e.g., gulls (*Larus*), loons (*Gavia*), and Canada Geese (*Branta canadensis* and relatives)] seem to be best explained by continental Wisconsinan glaciation, which provided temporary geographical barriers so that fragments of the species had isolation in which to develop. With the retreat of the glaciers, these populations met again, and now show various degrees of relationships between that of subspecies and species. Some may be 'species in the making,' but others may be merging. Refugia mentioned are: one or more SE of the ice; two or more SW of the ice in the Rocky Mountains; one or more in the Yukon–Bering Sea area; and perhaps some in the Arctic Islands.

1102. Rand, A.L. (1954): The ice age and mammal speciation in North America. Arctic 7(1):31–5.
The author extrapolates his discussion of the effects of glaciation on birds (Rand 1948) to mammals in northern North America. He states his hypothesis as follows: 'Geographical isolation by physical barriers seems necessary for a population to evolve further, to the species level. In this isolation the two parts of the population gradually change by random mutation toward a state in which interbreeding with other populations is impossible. Then, when the barriers disappear or are surmounted, the two descendants of the one ancestral form meet.' If the process of speciation has not gone to the point where biological discontinuity is complete, they meet as subspecies and interbreed. If biological discontinuity has been achieved and their ways of life have become different enough, they invade each other's territory and live as sympatric species. His discussion of various animals [shrews (*Sorex*), red squirrels (*Tamiasciurus*), marten (*Martes*), mountain sheep (*Ovis*), ground squirrel (*Sper-*

mophilus = *Citellus*), brown bears (*Ursus*), voles (*Microtus*), red-backed voles (*Clethrionomys*), walrus (*Odobenus*), and caribou (*Rangifer*)] lead him to conclude that these genera have been isolated and changed in the following refugia: one or more SE of the Laurentide ice; two or more SW of the ice in the Rocky Mountains; one or more in the Alaska–Bering Sea area; and one in the Arctic Islands (not necessarily permanent in location). Probably various glacial and interglacial phases each had their effect: some early effects may have been intensified, cancelled, or complicated by later ones. Some well-established species that now live together may have had their origin in an early glacial period; some obvious subspecies may have had their limits influenced by the last (Wisconsinan) glaciation.

1103. Rausch, R.L. (1969): Origin of the terrestrial mammalian fauna of the Kodiak Archipelago. *In*: The Kodiak Island Refugium: Its Geology, Flora, Fauna and History (T.N.V. Karlstrom and G.E. Ball, eds.). Boreal Institute, University of Alberta and Ryerson Press, Edmonton and Toronto. pp. 216–34.
Terrestrial mammals representing six species [red fox (*Vulpes vulpes harrimani*), brown bear (*Ursus arctos middendorffi*), ermine (*Mustela erminea kodiacensis*), otter (*Lutra canadensis kodiacensis* = *Lontra canadensis kodiacensis*), northern vole (*Microtus oeconomus operarius*), and ground squirrel (*Spermophilus parryii kodiacensis* = *Citellus undulatus kodiacensis*)] may be indigenous to the Kodiak Archipelago, Alaska. Considering the extent of these islands (more than 12,500 km^2), their proximity to the mainland, and the diversity of existing habitats, this fauna seems remarkably sparse. Zoogeographic evidence suggests that these mammals became established during postglacial time when the Archipelago was isolated by relatively narrow, but formidable, sea barriers, which seem to account for the impoverished nature of the mammalian fauna. This study produced no evidence that the Kodiak Island Refugium had any importance for the extant mammalian fauna of the Archipelago. However, among the species considered, the brown bear appears to be best suited for survival in the refugium. For a small population the vegetation might have been adequate to provide sustenance during the summers, and the rigorous winters would be sidestepped by denning. No other island population of brown bears is known to exhibit a comparable degree of divergence from presumed ancestral populations on the mainland.

1104. Ray, C.E. (1965): A new chipmunk, *Tamias aristus*, from the Pleistocene of Georgia. Journal of Paleontology 39(5):1016–22.
In describing chipmunk (*Tamias*) specimens from North American Quaternary sites, the author mentions remains from near Ottawa, Ontario [perhaps confused with the skull and front part of a skeleton of an American marten (*Martes americana*): NMC 2041 preserved in a nodule of Champlain Sea age from Green Creek near Ottawa], and two lower jaws (NMC 6824) and a skull, jaws, and partial skeleton (NMC 6825) collected by H.M. Ami in 1890 from marine gravels and sands (Saxicava Sands) of Champlain Sea age at Moose Creek, Stormont County, Ontario (p. 1018). Measurements of alveolar length are given in Table 2.

1105. Ray, C.E. (1971): Polar bear and mammoth on the Pribilof Islands. Arctic 24(1):9–18.
The author reviews all the known records from this part of Alaska. He places them in historical and chronological context and comments on the validity of each one, taking care to find the original source of the incorrect records. The polar bear (*Ursus maritimus*) has not been a regular member of the fauna of the Pribilofs during the period of European observation. It occurred seasonally and sporadically (depending on the extent of the pack ice). The likely source of bears is St Matthew Island, 225 miles N. The population on St Matthew was formerly abundant but was essentially exterminated by the turn of the century. The discovery of remains (some juvenile) from a cave on Bogoslof Hill, St Paul Island, in 1897 by F.A. Lucas (catalogued at USNM as 26108) suggests that the species might have denned on the island sometime in the past. Photographs of USNM 26108 specimens are provided. A cheek tooth (USNM 23455) of a woolly mammoth (*Mammuthus primigenius*), previously unreported, is described. It was excavated in 1965 by Robert Carroll from about 2.5 ft below the surface on the NE Point of St Paul Island. The apparently natural occurrence of the woolly mammoth on St Paul Island, Unalaska Island, and St Lawrence Island demonstrates that the species ranged over much of the Bering Land Bridge. The presence of mammoth in this region is essential to the most compelling theories concerning the first invasion of the New

World by humans (*Homo sapiens*), which call for mammoth hunters, of possible Clovis type, occupying the Bering land mass and the unglaciated areas to the E and W during Late Wisconsinan time.

1106. Rayner, B. (2000): Jurassic beavers: Bear-sized prehistoric rodents dwarfed modern cousins. St Croix Courier, March 14:A1–2.
While searching for native artifacts on Indian Island, New Brunswick, Ralph Welch came up with a rare find – the tooth of a giant beaver (*Castoroides ohioensis*). Welch, a consulting engineer, said he enjoys looking around the local beaches for arrowheads. The tooth, about 10 in long, was not actually found on the beach – he spotted it as he walked along a gravel ridge that separated the beach from a salt pond. It was found about two years ago and, only recently, was taken to the New Brunswick Museum, where it was identified as representing the giant beaver. Dr Randy Miller of the museum said *Castoroides* specimens have only been found at three Canadian sites – the Yukon; Toronto, Ontario; and Indian Island in Passamaquoddy Bay. Miller said, 'I would not be surprised if it had been washed up rather than eroding from the island but we are not sure.' The article mentions the physical appearance and size of the giant beaver as well as its distribution during the ice age – it died out about 10,000 years ago. Welch donated his find to the New Brunswick Museum, so it will become part of the provincial collection.

1107. Reimchen, T.H.F. (1968): Pleistocene mammals from the Saskatchewan gravels in Alberta. MSc thesis, Department of Geology, University of Alberta. 92 pp.
The Saskatchewan gravels are stratified fluvial deposits composed largely of quartzite pebbles that were deposited prior to the advance of the Laurentide ice into central and southern Alberta. In north-central Alberta, three pieces of wood found in nonglacial sediments immediately above the Saskatchewan gravels have been radiocarbon dated at 34,900 ± 3000/2000 BP (I-2616), 35,500 ± 2300/1800 BP (I-2516), and 35,500 ± 3300/2000 BP (I-2615). The Saskatchewan gravels are therefore Middle Wisconsinan and older. Fossils were collected at the following locations: 1a. S bank South Saskatchewan River near Medicine Hat (50°08′00″N, 110°39′42″W); 1b. E bank South Saskatchewan River near Medicine Hat (50°06′08″N, 110°41′20″W); 1c. NE bank South Saskatchewan River near Medicine Hat (50°05′52″N, 110°38′36″W); 1d. NE bank South Saskatchewan River near Medicine Hat (50°04′29″N, 110°38′00″W); 1e. W bank South Saskatchewan River near Medicine Hat (50°05′00″N, 110°39′20″W); 1f. S bank South Saskatchewan River near Medicine Hat (50°05′21″N, 110°41′12″W); 1g. E bank South Saskatchewan River near Medicine Hat (50°08′16″N, 110°38′28″W); 2. E bank Oldman River upstream from the Lethbridge power plant (49°41′42″N, 112°51′30″W); 3. N bank Red Deer River bordering the municipal golf course (52°16′09″N, 113°50′53″W); 4a. N bank Battle River (52°57′32″N, 112°58′56″W); 4b. N bank Battle River (52°57′32″N, 112°58′56″W); 4c. S bank Battle River (52°56′55″N, 112°58′34″W); 5a. N bank North Saskatchewan River near Edmonton (53°29′22″N, 113°36′58″W); 5b. E bank North Saskatchewan River near Edmonton (53°29′15″N, 113°19′24″W); 5c. 0.5 mi NW of Clover Bar Post Office (53°34′28″N, 113°19′24″W); 6. 2.5 mi N of Wabamun townsite (53°35′25″N, 114°28′13″W); 7. E bank Athabasca River (54°17′45″N, 114°50′42″W); 8. N bank Smoky River, near Watino townsite (55°42′59″N, 117°37′51″W). Localities also list the legal description (i.e., section, township, range, and meridian). The following specimens were collected at the above locations: Niobrara horse (*Equus* cf. *Equus niobrarensis*) – 1e, 2, 3, 4a, 4b, 4c, 6, 8; horse (*Equus*) – 1a, 1d, 1e, 1f, 1g, 4a, 4b, 4c, 5b, 7; Columbian mammoth (*Mammuthus* cf. *Mammuthus columbi*) – 5c; mammoth (*Mammuthus*) – 4c; mammoth? (?*Mammuthus*) – 4c, 5c; ground squirrel (*Spermophilus* = *Citellus*) – 8; giant camel (*Titanotylopus*) – 5c; western camel (*Camelops*) – 1a, 1b, 1e; antelope (*Antilocapra*) – 1c; and bison (*Bison*) – 5a. The fossils indicate that the Saskatchewan gravels may be properly assigned to the Rancholabrean and perhaps as young as the Late Wisconsinan in southern Alberta and Illinoian in central Alberta. The author concludes that central and southern Alberta were not glaciated by continental ice during the Nebraskan and Kansan glacial stages. The thesis is accompanied by: a map (Figure 1); tables of measurements of the specimens; detailed descriptions of sections where fossils were found; and photographs of some specimens and localities.

1108. Reimchen, T.H.F., and Rutter, N.W. (1972): Quaternary geology, Dawson Creek, British

Columbia (93P). Geological Survey of Canada Paper 72-1, Part A:176–7.
A mammoth (*Mammuthus*) partial molar [seemingly rather primitive in that the lamellar frequency is an estimated 6/100 mm] was found in Cordilleran gravels overlying Laurentide outwash deposits; and a horse (*Equus*) molar was found in the upper phase I lacustrine deposits overlying the Cordilleran gravels, at Dawson Creek, British Columbia.

1109. Rempel, L.L., and Smith, D.G. (1998): Postglacial fish dispersal from the Mississippi refuge to the Mackenzie River Basin. Canadian Journal of Fisheries and Aquatic Sciences 55(4):893–9.
Wisconsinan glaciation had a profound impact on fish (Pisces) faunas of North America, and deglaciation led to the dispersal of approximately 28 species from the Mississippi glacial refuge into the Mackenzie River Basin. A hypothesized dispersal 11,500 years ago via Glacial Lakes Agassiz and Peace is difficult to verify. Geomorphic evidence indicates that Glacial Lake Agassiz drained into the Mackenzie Basin via the Clearwater River, Saskatchewan, 9900 years ago, creating a second opportunity for fish dispersal northward. Fish distribution data indicate a 96% similarity between Mississippi species in the Mackenzie Basin and species occupying the former Agassiz–Clearwater corridor. Fifteen species dispersed into the headwaters of the Clearwater River during linkage to Glacial Lake Agassiz and are now isolated above an 18.5 m waterfall. Previous genetic data suggest lake trout (*Salvelinus namaycush*) used the Agassiz–Clearwater corridor for two-way dispersal between Beringia and the Great Lakes Basin. Lake whitefish (*Coregonus clupeaformis*) populations of Mississippi origin are distinguished by a marker allele that is absent from modern populations within the former Glacial Lake Peace region, and the distribution of Mississippi whitefish across Canada is best resolved by the Agassiz–Clearwater dispersal hypothesis. Accompanying maps show North American fish refugia and possible routes northwestward toward the Mackenzie River Basin. Table 1 gives the refugial origin and occurrence of fish species in the Clearwater River for three reaches mentioned in the text.

1110. Repenning, C.A. (1967): Palearctic-Nearctic mammalian dispersal in the Late Cenozoic. *In*: The Bering Land Bridge (D.M. Hopkins, ed.). Stanford University Press, Stanford, California. pp. 288–311.
The Late Cenozoic mammal fauna of North America is compared with that of the Old World for two purposes: first, to establish the reliability of temporal correlations based upon the fauna and upon the sparse but impressive data of radiometric geochronology; and second, to determine as clearly as possible what mammals (Mammalia) were involved in these intercontinental exchanges and what significance these may have for the history of the Bering Land Bridge. Extensive lists of species with their earliest-known fossil occurrence are provided. The author also provides a chart of conventional glacial ages with suggested Nearctic–Palearctic correlations.

1111. Repenning, C.A. (1980): Faunal exchanges between Siberia and North America. Canadian Journal of Anthropology 1(1):37–44.
The microtine rodents (Microtinae), because of their rapid rate of evolution in recent geological time, their arctic to temperate preferences, and their rapid rate of dispersal in appropriate environments, provide the most precise record of faunal movement between Siberia and North America. This record has been calibrated by radiocarbon dates, the paleomagnetic polarity scale, oceanic climatic indicators, and the similar history of microtine invasions from Siberia into Europe. The development of provincialism in the North American fossil microtine fauna clearly documents the effective beginning of the ice-free corridor E of the Cordilleran ice sheet. The microtine invasion events that are reliably dated and correlated to the ice-free corridor are: before 4.8 Ma, then 3.7 Ma, 2.5 Ma, 1.8 Ma, 1.2 Ma, and 0.47 Ma. There is at least one more microtine dispersal event that is younger than 0.47 Ma, but it has not yet been accurately dated. Based on oxygen isotope stages, it is estimated at 0.17 Ma. The very oldest known microtine invasion event happened sometime before 5 Ma. The Cape Deceit fossil fauna in Alaska [see Guthrie and Matthews 1971] does not conform to these microtine faunal stages. It is a Siberian fauna found on the 'wrong side' of the Bering Land Bridge. Lack of similarity to other North American faunas indicates that it lived in Alaska during a period when there was no microtine dispersal event in progress, which leads to the conclusion that northern forests were the barrier preventing microtines from dispersing at that time. Many of the data discussed are summarized in Figure 1, a correlation of microtine dispersal events in North

America with mammal ages and their subdivision based on microtine faunal units and to other faunal events.

1112. Repenning, C.A. (1983): New evidence for the age of the Gubik Formation, Alaskan North Slope. Quaternary Research 19:356–72.
The Gubik Formation occurs across the entire Arctic Coastal Plain. It is an overlapping series of thin marine and nonmarine units that blanket a marine unit of supposed Cretaceous age. The age of the Gubik Formation has been cited as Late Pleistocene, and Late Pliocene or Early Pleistocene. This paper discusses fossils collected at Teshekpuk Lake about 139 km ESE of Point Barrow (153°06′W) and near Ocean Point (151°24′W), Alaska (see map, Figure 1). The fossil evidence suggests that the Gubik Formation at Ocean Point is of Late Pliocene age (about 2.2–1.7 Ma), while the Teshekpuk Lake Gubik Formation is much younger (about 131,000–77,000 BP). Clearly this formation was a long time in forming, and both of the earlier age estimates are correct. Mammals collected at Teshekpuk Lake include: grey whale (*Eschrichtius*), white whale (*Delphinapterus*?), harbour seal (*Phoca vitulina*), ribbon seal (*Histriophoca fasciata*), ringed seal (*Phoca hispida* = *Pusa*), and fox (*Vulpes praeglacialis*). Mammals collected at Ocean Point include: sea otter (*Enhydra*) and harp seal (*Phoca groenlandica* = *Pagophilus*). Some of the fossils are illustrated, and there is a correlation chart of Ocean Point and Teshekpuk Lake faunas with time, and a correlation of paleomagnetic events, climate, ages, and faunas of North America and Europe.

1113. Repenning, C.A. (1985): Pleistocene mammal faunas: Climate and evolution. Acta Zoologica Fennica 170:173–6.
Shortly before the Pleistocene, climatic fluctuations throughout the Holarctic began to increase in intensity. This caused the development of strong latitudinal zonation in the mammalian fauna of the world, and accelerated speciation in those more northerly land masses with greatest longitudinal extent. Accelerated speciation, centring primarily in the faunal province of Siberia and North American Beringia, rapidly changed the mammalian fauna, and these newly evolved faunas dispersed southward with each major cold spell into more temperate lands. The dispersal of this Pleistocene fauna was inhibited by a variety of ecological barriers, intensified by the developing latitudinal zonality. Low-latitude N-S environmental avenues were greatly limited and there seems to have been very little exchange between North and South America. North America, S of Beringia, experienced only four immigration waves, and extensive ice barriers possibly prevented other immigrations. These events can be correlated with new precision because of the development of paleomagnetic and paleoclimatic stratigraphies.

1114. Repenning, C.A. (1987): Biochronology of the microtine rodents of the United States. *In*: Cenozoic Mammals of North America, Geochronology and Biostratigraphy (M.O. Woodburne, ed.). University of California Press, Los Angeles. pp. 236–68
This is an outline of the history of the microtine rodents of North America over the last six million years. Although the focus of this article is the United States, the Wellsch Valley fauna, Saskatchewan, is mentioned on p. 260.

1115. Repenning, C.A. (1990): Of mice and ice in the Late Pliocene of North America. Arctic 43(4):314–23.
Between 2.5 and 1.8 Ma, changes in arctic climate and in meadow mouse (*Microtus*) dispersal routes correlate with part of the history of uplift and glacial erosion of the Chugach and Saint Elias mountains in Alaska and adjacent Canada. Pertinent here is the statement that a wave of late-arriving *Microtus* is present to the E of the Rocky Mountains, and that the genus arrived in Canada nearly 2 Ma, where it is present in the Wellsch Valley fauna of southwestern Saskatchewan. Figure 3 shows the modern distribution of *Microtus* in North America compared with the approximate distribution of the genus 2 Ma.

1116. Repenning, C.A. (2001): Beringian climate during intercontinental dispersal: A mouse eye view. Quaternary Science Reviews 20(2001):24–40.
Microtine rodents [meadow mice (= voles), muskrats, and lemmings] have evolved so rapidly over the past 5 million years that they can be correlated with specific climatic events, and their habitats are so specific that they suggest climatic conditions in different areas at the time. Examination of these records not only provides strong evidence of the nature of climate over North America and the Bering Land Bridge, but also suggests some unconventional reasons for the development of continental

glaciation. Of key importance in summarizing the author's view is Figure 1 dealing with microtine rodent dispersal (1) from Asia to western United States and (2) from Asia to eastern United States during the last 5 million years, and showing how the dispersals relate to time and paleomagnetic events. Genera mentioned are: *Mimomys, Pliopotamys, Ogmodontomys, Mictomys, Pliophenacomys, Plioctomys, Phenacomys, Microtus, Allophaiomys, Lasiopodomys, Clethrionomys,* and *Terricola.* It seems that the beginning of major glacial cycles, and the peaks of minor cycles, were the optimum times for dispersal to the United States from Beringia and Asia. Many large mammals, particularly predators, are adaptable to greater variation in habitat, and all seem to have evolved more slowly. About 87% of late Neogene and Pleistocene microtines that crossed between Asia and North America are now temperate or cool temperate and subarctic animals. The habitat restrictions of these mammals, as indicated by their fossil associations, strongly indicate temperate climates in Beringia at times of dispersal to North America.

1117. Repenning, C.A., Hopkins, D.M., and Rubin, M. (1964): Tundra rodents in a Late Pleistocene fauna from the Tofty Placer District, central Alaska. Arctic 17(3):177–97.

The Tofty placer district, about 160 km W of Fairbanks, Alaska (Figures 1, 2), has a rich history of Pleistocene large-mammal fossil finds. In 1956, at the Sullivan Mining Pit on Sullivan Creek, a sequence of gravel, silt, and peat was found to contain abundant small fossils. Along with some molluscs and insects, the following vertebrates have been identified (Table 1): small fish (Pisces); hare (*Lepus*); arctic ground squirrel (*Spermophilus parryii = Spermophilus (Spermophilus) undulatus*); collared lemming (*Dicrostonyx torquatus*); brown lemming (*Lemmus sibiricus*); bog lemming (*Synaptomys*); red-backed vole (*Clethrionomys*); singing vole (*Microtus (Stenocranius) miurus*); large vole (*Microtus*); beaver-chewed wood (*Castor canadensis* – inferred); proboscidean (Proboscidea); probably Yukon horse (*Equus* lambei = *Equus*, small); caribou (cf. *Rangifer*); moose (cf. *Alces*); hypsodont bovid [probably sheep (*Ovis*)]; and bison (*Bison*). Radiocarbon dating of associated wood indicates that the fossils were deposited in their present positions very recently; however, because of the climatic changes they suggest and because they are associated with fossils of larger mammals now extinct, the fossils must have been derived from nearby sediments probably of Late Pleistocene age. The data suggest that bog lemmings, red-backed voles, and beavers existed in a forested environment not unlike the present environment. The fossil tundra rodent fossils are probably derived from an older sediment layer and likely lived in the area during Wisconsinan time, becoming locally extinct when the area was recolonized by forest vegetation more than 6800 years ago.

1118. Repenning, C.A., Brouwers, E.M., Carter, L.D., Marincovich, L., Jr, and Ager, T.A. (1987): The Beringian ancestry of *Phenacomys* (Rodentia: Cricetidae) and the beginning of the modern Arctic Ocean Borderland biota. United States Geological Survey Bulletin 1687:1–31.

The Late Pliocene faunas of northern North America are an important window on vertebrate evolution in the Quaternary, and therefore are mentioned here. A fossil heather vole (*Phenacomys*) from the Late Pliocene part of the Gubik Formation along Fish Creek, northern Alaska, resembles the extinct Asiatic Cromer vole (*Cromeromys*), a progressive offshoot of the extinct Holarctic *Mimomys*, suggesting the probable ancestry. Molluscs and mammals in the Fish Creek fauna with *Phenacomys* indicate that the Arctic Ocean was not perennially frozen during deposition of the upper part of the section. Pollen in the Fish Creek section (about 15 mi S of the Arctic Ocean at Harrison Bay, Alaska at 70°16′N, 152°01′W, Figures 1, 2) records a transition from shrub tundra with nearby conifer forests in the lower part to herbaceous tundra with the molluscs and mammals of the Fish Creek fauna. The first development of widespread, low-elevation tundra in the Arctic Ocean borderland appears about 2.4 Ma – the indicated age of the Fish Creek fauna in the upper part of the section. This fauna, from a 5.0- to 7.6-cm-thick sandy organic deposit believed to be of tidal origin, includes remains of: fish (Pisces), birds (Aves), heather voles (*Phenacomys intermedius* and *Phenacomys gryci*), pika (*Ochotona* cf. *Ochotona alpina*), arctic ground squirrel? (*Spermophilus parryii*? = *Spermophilus undulatus*?), and a crab-eating otter (*Enhydriodon*). The authors also mention a toothless lower jaw of a sea otter (*Enhydra lutris*, USNM 264275) found in a marginal marine unit of the Gubik Formation at Ocean Point, Alaska, about 16 mi SE of the

Fish Creek locality. Perhaps evolution of the North Pacific sea otter took place in the Arctic Ocean. It is noted that the Deering heather vole (*Phenacomys deeringensis* = *Pliomys deeringensis*) has been reported from the Cape Deceit fauna (Guthrie and Matthews 1971). Among the figures provided are: (1) a stratigraphic section at Fish Creek (Figure 3) showing the mammal zones in the grey sand unit between 7 and 9 m from the base; and (2) diagrams of various rodent teeth (Figures 4-7). The age and correlation of the Alaskan Ocean Point and Fish Creek faunas are shown in relation to the Cape Deceit, Alaska, and Kap København, Greenland, mammal faunas (Figure 13).

1119. Rhoads, S.N. (1898): Notes on the fossil walrus of eastern North America. Proceedings of the Philadelphia Academy of Natural Sciences 50: 196–201.
A walrus (*Odobenus rosmarus* = *Trichecus rosmarus* = *Trichecus virginianus*) skull and partially fossilized tusks were found at Sable Island, Nova Scotia. The author believes that the specimens are Pleistocene in age due to their fossilization state, and he compares them with similar specimens from the United States. The bone measurements are provided. [See Harington 1997b, Harington and Occhietti 1988, and Dyke et al. 1999.]

1120. Richard, S.H. (1978): Age of Champlain Sea and 'Lampsilis Lake' episode in the Ottawa-St. Lawrence Lowlands. Geological Survey of Canada Paper 78-1C:23–8.
Whale (Cetacea) bones found in beach sands 168 m asl at White Lake, Renfrew County, Ontario, were radiocarbon dated at 11,500 ± 90 BP (GSC-2269) (Table 5.1). This paper also contains many radiocarbon dates based on bivalves to determine the extent and history of the Champlain Sea.

1121. Richards, H.G. (1962): Studies of the marine Pleistocene. Transactions of the American Philosophical Society 52(3):11.
The fauna of three Ontario arms of the Champlain Sea (Prescott, Smiths Falls, and Green Creek E of Ottawa) is mentioned. Remains of capelin (*Mallotus villosus*), lumpfish (*Cyclopterus lumpus*), and stickleback (*Gasterosteus*) were found at Green Creek near Ottawa, Ontario. The author suggests that this part of the Champlain Sea episode may date from the Two Creeks Interstadial. [See McAllister et al. 1988.]

1122. Richards, R.L., Churcher, C.S., and Turnbull, W.D. (1996): Distribution and size variation in North American short-faced bears, *Arctodus simus*. *In*: Palaeoecology and Palaeoenvironments of Late Cenozoic Mammals: Tributes to the Career of C.S. (Rufus) Churcher (K.M. Stewart and K.L. Seymour, eds.). University of Toronto Press, Toronto. pp. 191–246
Arctodus simus fossils have been recorded from 102 localities in North America. Many of these have been examined to establish the geographic and temporal variation in size of the giant short-faced bear (*Arctodus simus*). The similarity of *Arctodus simus* and the lesser short-faced bear (*Arctodus pristinus*) is examined. Two distinct subspecies (*Arctodus simus yukonensis* and *Arctodus simus simus*) appear to have existed. This subspecific designation is supported using bone dimensions but is not as clear using dental morphology. *Arctodus simus yukonensis* occupied western North America during the Irvingtonian and persisted in Beringia, western Canada, and portions of both western and eastern United States into Rancholabrean times. By Rancholabrean times the smaller *Arctodus simus simus* had differentiated S of the Wisconsinan ice and was widespread throughout much of the United States and Mexico. The authors speculate that the change in size might be related in part to character displacement in hunting sets of opportunistic carnivorous omnivores. Perhaps the lesser short-faced bear was more of a forest animal, with the giant short-faced bear occupying more open environments. North American *Arctodus* localities are mapped (Figure 1). Alaskan localities include: Chatanika River; Chicken Creek; Cleary, Fairbanks Island?; Cripple Creek; Dawson Cut; Engineer Creek; Ester Creek; Gold Hill; Goldstream; No. 2G Strip area (probably the same as Goldstream); Upper Cleary River; Tofty Placer District; and Ikpikpuk River. Canadian localities include: North Saskatchewan River Pits, near Edmonton, Alberta; Lebret, Saskatchewan; Fort Qu'Appelle, Saskatchewan, and the following Yukon specimens: Lower Hunker Creek; Gold Run Creek; Hunker Creek; Sixtymile Loc. 3; Old Crow Basin, Loc. 14; Old Crow Basin, Loc. 11A; Old Crow Basin, between Localities 14 and 14N; Old Crow Basin, Loc. 22; Old Crow Basin, Loc. 29; Old Crow Basin, Loc. 66; Old Crow Basin, Loc. 67; and Old Crow Basin, Loc. 136. A

full list of specimens examined is provided (Appendix 1), along with an extensive bibliography, and numerous statistical measurements (Appendix 2).

1123. Richardson, James (1870): [Capelin from the north shore of the Lower St. Lawrence River]. Geological Survey of Canada Paper 1854:305–11.
Remains of capelin (*Mallotus villosus*) were found in blue clay between Sault au Cochon and Rivière Portneuf, Quebec, on the north shore of the St Lawrence River.

1124. Richardson, John (1854): Observations on the fossil bone deposit in Eschscholtz Bay. *In*: The Zoology of the Voyage of H.M.S. Herald; under the Command of Captain Henry Kellett, R.N., C.B., during the Years 1845–51: Vertebrals Including Fossil Mammals (E. Forbes, ed.). Lovell Reeve, London. 172 pp.
A description of the fossil cliff at Eschscholtz Bay, Alaska, is followed by a list of the fossils found at the bottom. Many of the descriptions include measurements. The species include: woolly mammoth (*Mammuthus primigenius* = *Elephas primigenius*), horse (*Equus*), moose (*Alces alces* = *Cervus alces*), caribou (*Rangifer tarandus* = *Cervus tarandus*), tundra muskox (*Ovibos moschatus*), muskox (*Ovibos maximus*), and steppe bison (*Bison priscus* = *Bison crassicornis*). The author (p. 108) also remarks on a pair of damaged American mastodon (*Mammut americanum* = *Elephas rupertianus*) scapulae from the banks of Swan River, Manitoba, near the western side of Lake Winnipeg.

1125. Richardson, John (1861): The Polar Regions. Adam and Charles Black, Edinburgh.
The core of the information on ice age mammals from NW North America is found on pp. 295–6. Mammoth (= elephant) teeth are abundant in Alaska and have long been articles that Inuit (= Eskimo) have traded to the Chukchi people. The author mentions the account of the fossil-bearing ice cliffs of Eschscholtz Bay in the *Zoology of the Herald* [see Richardson 1854] – a site first discovered by Admiral Kotzebue of the Russian Navy. The fossils lie on top of the ice, embedded in, and more or less completely covered with, boggy or sandy soil. The bones had lost little of their animal matter [collagen?] and those of the mammoth (*Mammuthus*), when the earthy substance was removed by acids, clearly showed the haversian canals. Hair was dug up with the mammoth skulls, and the whole deposit smelled like a charnel house. Species found were: woolly mammoth (*Mammuthus primigenius* = *Elephas primigenius*), moose (*Alces alces* = *Cervus alces*), horse (*Equus* = *Equus fossilis*), caribou (*Rangifer tarandus* = *Tarandus*), muskox (*Ovibos moschatus* = *Ovibos maximus*), and steppe bison (*Bison priscus* = *Bison crassicornis* and *Bison latifrons*). At least 15 individual mammoths must have been represented by bones collected by Admirals Kotzebue and Beechey and Captain Kellett in Eschscholtz Bay. Mammoth teeth are found as far eastward as Point Barrow, Alaska, and an entire skeleton was discovered by Indians inland near the source of the Yukon River [presumably in Yukon]. The author states that no mastodon (*Mammut americanum*) remains have been found N of the valley of Saskatchewan River about 51°N. He thinks it probable that many, or all, of these ice age mammals were migratory, like those living there now.

1126. Riddle, D.K. (1983): Discovering the past. An introduction to the archaeology and culture history of the Neepawa and Area Planning District. Papers in Manitoba Archaeology, Popular Series 5:1–151.
Manitoba was covered with a thick mantle of ice until about 12,000 years ago, denying use of the province by animals and people (*Homo sapiens*). Later, forest developed in ice-free areas. Animals attracted to this forest were probably few and not the preferred prey of the people who were able to inhabit the region. So, only small groups of human hunters may have visited the area – preferring other more suitable environments in which to secure their living (p. 55). Clovis people were nomadic hunters and gatherers. The basis of their diet was mammoth (*Mammuthus*), giant bison (*Bison*), and other members of the now-extinct ice age megafauna. Two of the few known Clovis spear points were found near Neepawa, one near Mentore, and the other NE of Ericson along the Rolling River (p. 54) – from the uplands W of Glacial Lake Agassiz. The next hunters to use the area were of the Lindenmeier culture (9000–8000 BC or 11,000–10,000 BP) – distinguished by Folsom and Midland points. Their main prey was a now-extinct giant bison (*Bison*), an ancient cousin of the modern bison. These hunters used atlatls (spear-throwers) and stampeded bison into ravines to their death. Plano hunters (8000–5000 BC or 10,000–7000 BP) were full-fledged, plains-oriented bison hunters by the time they reached

the Neepawa area. Bison bones were found 10 m deep in the Campbell Beach of Glacial Lake Agassiz at Arden – about 10,000–9500 BP. The region was now rich in the main natural resource sought after by the Plano people.

1127. Rink, H. (1853): Udsigt over Nordgrönlands Geognosie. Det kongelige Danske Videnskabernes Selskabs Skrifter 5:71–98.
According to Bennike (1997, p. 901), the author reports remains of capelin (*Mallotus villosus*) in calcareous concretions from West Greenland.

1128. Ritchie, J.C. (1984): Past and Present Vegetation of the Far Northwest of Canada. University of Toronto Press, Toronto. 251 pp.
This book deals with the modern and Late Quaternary vegetation of northern Yukon and adjacent Mackenzie Delta, Northwest Territories. Factors such as permafrost, geology, climate, Tertiary origins, glacial history, and periglacial features and their impacts on vegetation are discussed at length. Considerable effort was made to reconstruct the Quaternary fossil plant community using palynology. Pertinent here are discussions about whether or not paleobotanical data from the region fit with vertebrate paleontological data (pp. 157–73). The author notes that Guthrie's (1968a) first serious attempt to consider the paleoecological significance of 'these enormous collections of mammalian fossils' was also most influential in establishing a concept of Eastern Beringian Quaternary environments. Guthrie concludes: 'The high percentage of grazers in the fossil community suggests that Interior Alaska was a grassland environment during the late Pleistocene.' Ritchie then proceeds to dissect and criticize this conclusion. He also attempts (pp. 171, 173) to answer the question, 'Does the paleobotanical evidence contribute usefully to testing the various hypotheses preferred to explain the extinction of several of the large mammal species?' [See Bliss and Richards 1982, Guthrie 1985, and Schweger and Habgood 1976.]

1129. Roberts, A. (1948): Whales near Parliament Hill. Forest and Outdoors 44(9):30.
Remains of white whale (*Delphinapterus leucas*) were found 9 m beneath the surface in a sandpit at the Ottawa International Airport at Uplands. The specimen was collected by G.B. Rolland, who said: 'We've come across a lot of bones while loading the trucks with sand, but we try to pick them out to keep the sand clean ... when I found the skull I knew it didn't belong to a horse or cow, so we phoned the National Museum.' To scientists, the bones were a clue to an age when the Ottawa and Rideau valleys were at the bottom of the Champlain Sea. A photograph shows J. Skillen, museum technician, with a white whale skull, nine vertebrae, and four ribs. [However, instead of being '100,000 years old,' the bones are closer to 10,000 years old.]

1130. Robertson, P. (1986): Neighbours: They preserve our heritage. Carlington Summit 3(6):13.
This article features Jerry Fitzgerald and tells of his field and laboratory work in the field of ice age mammals at the National Museum of Natural Sciences in Ottawa, Ontario. A photograph shows Jerry restoring a 60,000-year-old walrus (*Odobenus rosmarus*) skeleton from Qualicum Beach, Vancouver Island, British Columbia.

1131. Roe, F.G. (1970): The North American Buffalo: A Critical Study of the Species in Its Wild State. Second Edition. University of Toronto Press. 991 pp.
The author reviews many observations and opinions bearing on the origin of North American bison (*Bison*) (pp. 12–25). Some are for bison originating in North America, others for a Eurasian origin. One of the earliest, best-considered views is from Canada, Department of the Interior (1926), *Bringing Back the Buffalo*: 'The species is believed to have arrived on this continent in mid-Pleistocene times or during the warm inter-Glacial period. It probably crossed from Asia by the northern land route [i.e., the Bering Isthmus] and gradually spread southwards until it covered more than one-third of North America.' The author adds (p. 16): 'The buffalo pelage points very strongly towards a species developed in a cold winter climate.' He also refers to Lucas (1899) and Skinner and Kaisen (1947), summarizing their views on the subject. In addition, he notes that fossil bison remains have been found in the Don Valley, near Toronto [Sangamonian interglacial – see Coleman 1932, 1941], and he refers to a fossil bison skull 'estimated to be at least 10,000 years old' that was unearthed during sewer excavations in the city of Toronto according to a report in the *Edmonton Journal*, June 11, 1932. Others are recorded in the Klondike, Yukon, for E.T. Seton in *Lives of Game Animals*, III, (1929, p. 604) notes the discovery there, about 1905, of fossil remains of bison and other animals 'in the

low level Creek Gravels, possibly Pleistocene.' Roe also quotes H.F. Osborn (1910) regarding the western bison (*Bison bison occidentalis* = *Bison occidentalis*): 'There lived also in Alaska, probably in late Pleistocene times and ranged down into Kansas, the species *B. occidentalis.*' The author was informed by G.C. Carl, provincial zoologist of Victoria, British Columbia, that a fossilized skeleton was discovered near Kelowna, British Columbia, in 1919 (p. 309). This report is detailed in Appendix L (p. 853): M.H. Ruhmann in his article 'The Bison in the Okanagan Valley' (First Annual Report of the Okanagan Historical and Natural History Society, September, 1926, p. 6, records that in June 1919, Harry Mills, while prospecting on Mission Creek, 12 mi from Kelowna, discovered seven large bones in deposits of the Pleistocene period on bedrock 12 ft below the soil surface. The bones consisted of: two metapodials, two tibias, a humerus, and two ulnas, and many fragments. Five sedimentary units covered the bone (from bottom up): blue clay, coarse gravel, yellow clay, fine gravel, sandy loam. M.Y. Wilkers [Williams] of University of British Columbia confirmed the age as Pleistocene and W.D. Matthew [not D.W. Mathews] of the American Museum of Natural History confirmed the identifications of the limb bones. The bones were deposited with the Okanagan Historical and Natural History Society in autumn 1925. In Appendix LL (pp. 923–4) the author shows a photograph comparing a partial skull with very long horncores from the Bliss Gravel Pit near Fort Qu'Appelle, Saskatchewan, to those of modern bison (presumably the plains *Bison bison bison*) and suggested the fossil may represent either a steppe bison (*Bison priscus* = *Bison crassicornis*) or the giant bison (*Bison latifrons*). [See Khan 1970 – the fossil is identified as giant bison, and its age is considered to be Sangamonian interglacial.] Roe reprints McCorquodale's 1957 paper on the specimen.

1132. Rogers, R.A., Rogers, L.A., Hoffman, R.S., and Martin, L.D. (1991): Native American biological diversity and the biogeographic influence of ice age refugia. Journal of Biogeography 18:623–30.
The Wisconsinan glaciation separated North America into a number of ice-free refugia for flora and fauna. Patterns of variation of some North American mammalian (Mammalia) species have been interpreted as a result of biological differentiation arising from isolation in these refugia. A tripartite biological and linguistic division of human (*Homo sapiens*) populations in North America was proposed by several researchers in the 1980s. The tripartite division in biological and linguistic traits correlates with three ice-free refugia (Beringian, Northwest Coast, and Southern mapped in Figure 1) that existed during the Wisconsinan glaciation. These refugia would have provided the necessary isolation for the development of biologically and linguistically divergent groups. The human data can be seen as part of a broader pattern of glacial influences on biological variation among North American mammals. The section on Glacial Isolation as a General Factor in Biotic Differentiation will be of interest to mammalian paleontologists and mammalogists as it specifies mammals [e.g., different kinds of caribou (*Rangifer tarandus*)] that occupied the three refugia and their possible relationships.

1133. Romer, A.S. (1933): Pleistocene vertebrates and their bearing on the problem of human antiquity in North America. *In*: The American Aborigines: Their Origin and Antiquity (D. Jenness, ed.). University of Toronto Press, Toronto. pp. 49–83
This is a discussion of the value of fossil or subfossil remains in aging the culture of humans (*Homo sapiens*) associated with the fossils. The author begins by presenting a list of Pliocene and Pleistocene mammals listed according to age in the faunal record. Pleistocene mammals are separated into Early, Middle [there is a passing mention (p. 66) under Middle Pleistocene faunas implying that a fauna at Toronto, Ontario, may be of 'approximately Sangamon date'], and Late (pp. 63–74). [It is interesting to note that this prominent paleontologist attests to the sparse data from northwestern North America by stating: 'So little is known about Pleistocene faunas of the north-west that discussion would be unprofitable' (p. 73).] A section then follows describing the succession of various faunas. Here Romer states that certain kinds of Pleistocene mammals have been found 'well up into Ontario, suggesting a shorter time since their extinction.' In his summary (p. 81), the author states that the association of people with 'certain fossil forms is unquestioned' and the evidence is growing; however, there is almost no paleontological evidence suggesting that humans were in North America earlier than the withdrawal of the last Pleistocene ice sheet.

1134. Ronaghan, B., and Newton, B. (1990): The Bow

Corridor First Albertans Project. Alberta Past 6(2):8–9.
Several sites in the Bow River Valley, Alberta, were examined. Fluted points, confirming human (*Homo sapiens*) presence, were found at a site on Lake Minewanka inside the Mountain front, and at another site at Sibbald Creek. From Lac des Arcs, charcoal was found in a windblown silt deposit, radiocarbon dated at 7800 BP [no lab number or error given]; stone tools and remains of bison (*Bison*) and bighorn sheep (*Ovis canadensis*) were found in alluvial fan deposits. The Vermilion Lakes site in the Bow River valley near Banff is currently the oldest-dated site in Alberta (about 10,400 BP).

1135. Roosa, W.B. (1977): Great Lakes Paleoindian: The Parkhill site, Ontario. Annals of the New York Academy of Sciences 288:349–54.
This archaeological site near Parkhill, Ontario, probably dating between 10,570 BP (GSC-1006, 1028) and 9750 ± 135 BP (I-5786), has produced major portions of more than 80 fluted points and knives and many scrapers and gravers. [Perhaps, as at the Udora site (see Storck and Spiess 1994), caribou (*Rangifer tarandus*), hare (*Lepus*), and arctic fox (*Alopex lagopus*) were among the prey.]

1136. Russell, L.S. (1941): Pleistocene horse remains from Saskatchewan. Proceedings and Transactions of the Royal Society of Canada, Series 3, 35:188. (Abstracts of Papers).
Bones of Pleistocene mammals have been obtained from a gravel pit near Sutherland, Saskatchewan, a few miles E of Saskatoon. The exact age of the deposit is not yet determined, but it may date from the withdrawal of the Wisconsinan ice sheet. Teeth and bones of horses are the commonest fossils, making up the first good collection of such remains from the Canadian Prairies. The teeth resemble those of the Niobrara horse (*Equus niobrarensis*) from the Pleistocene of western United States, but differ in some details. Accurate dating of these fossils will throw light on the duration of the native horses of North America.

1137. Russell, L.S. (1943): Pleistocene horse teeth from Saskatchewan. Journal of Paleontology 17(1):110–14.
Glacial gravel near Sutherland, Saskatchewan (the gravel pit lies near the SE corner of sec. 25, twp. 37, rge. 5, W 3rd mer. or about 3 mi N of Sutherland railway station and 6 mi NE of the centre of Saskatoon) has yielded remains of mammals, including several horse teeth (Figures 1–11). Most of these can be referred to the Niobrara horse (*Equus niobrarensis* Hay). Close comparisons are possible with teeth of that species from Hay Springs, Nebraska. The age of the Sutherland material is not established, but the author thinks that it is Late Wisconsinan or possibly 'Peorian (Fourth Interglacial).' [See Skwara-Woolf 1980 and 1981.] A stratigraphic diagram of the gravel pit is included (Figure 13), showing about 15 ft of gravel and sand overlying a till unit. Bones and teeth are from the base of the gravel and sand unit. The author notes that this seems to be the first record of an extinct horse from Saskatchewan, and that teeth and bones from a bog N of Dundurn (Parks 1925) include horse teeth referred by Russell to the modern horse (*Equus caballus*). He also mentions that Hay (1927, p. 273) records an upper molar of the Yukon horse (*Equus lambei*) from Empress, Alberta.

1138. Russell, L.S. (1948): Post-glacial occurrence of mastodon remains in southwestern Ontario. Transactions of the Royal Canadian Institute 27:57–64.
Mastodon (*Mammut americanum*) molars, tusk fragments, and bone fragments and a deer (*Odocoileus*) were found in a black organic muck, underlain by sand and gravel of a Glacial Lake Whittlesey beach, on the Campbell farm in Aldborough Township, Elgin County, about 4 km W of Rodney, Ontario (see map, Figure 1). The author concludes that the Rodney Mastodon died, and its remains were buried, considerably less than 10,000 years ago. Such a dating opens up the possibility that mastodon remains may be found in association with those of humans (*Homo sapiens*) in the Great Lakes region. [See Jackson 1987.]

1139. Russell, L.S. (1956): Additional occurrences of fossil horse remains in western Canada. Annual Report of the National Museum for the Fiscal Year 1954–55, National Museum of Canada Bulletin 142:153–4.
A Niobrara(?) horse (*Equus* cf. *Equus niobrarensis*) incisor with other bone fragments, and a mammoth (*Mammuthus*) tooth, vertebra, and pelvis fragment were found in a gravel pit near Katepwa Beach, in the Qu'Appelle Valley between Regina and Melville, Saskatchewan. A Niobrara horse (*Equus niobrarensis*) molar and

metacarpal were found near Fort Qu'Appelle, Saskatchewan, about 16 km from the Katepwa site. Another Niobrara(?) horse lower-right cheek tooth was found in a gravel pit near Lancer, about 100 km NW of Swift Current, Saskatchewan. A horse (*Equus*) vertebra and tibia and an ancient bison (*Bison bison antiquus* = *Simobison antiquus*) horncore were found in a sand and gravel deposit 6 m beneath the surface near Smoky River, about 43 km E of Grande Prairie, Alberta.

1140. Russell, L.S. (1958): A horse astragalus from the Hand Hills conglomerate of Alberta. National Museum of Canada Natural History Papers 1:1–3.
The Hand Hills conglomerate forms the summit of the Hand Hills about 15 mi E of Drumheller, Alberta. A horse astragalus (CMN 9449, Figures 1, 2), smaller than average for the modern horse (*Equus caballus*) but having features suggesting an advanced type of horse, was found in a large gravel pit located in the NW¼, sec. 32, twp. 29, rge. 17, W of the 4th meridian. The specimen is tentatively identified as 'cf. *Plesihippus*.' [Storer (1976) thinks the specimen probably represents *Equus*, indicating that the deposit is in part Blancan (latest Pliocene to earliest Pleistocene) or younger.] [See Burns and Young 1993.]

1141. Russell, L.S. (1959): Continental zoology of the North American Pleistocene. Problems of the Pleistocene and Arctic, Publications of McGill University Museum 1(1):39–45.
This paper deals in general with Late Pliocene and Pleistocene faunal ranges and migration patterns into and through North America. The northern North American Pleistocene species mentioned include: pronghorn (*Antilocapra americana*), peccary (*Platygonus*), rabbit (*Sylvilagus*), wolf (*Canis lupus*), American lion (*Panthera leo atrox* = *Felis atrox*), sabretooth cat (*Smilodon*), short-faced bear (*Arctodus*) [but not of Eurasian origin as stated], Niobrara horse (*Equus niobrarensis*), mastodon (*Mammut americanum*), woolly mammoth (*Mammuthus primigenius*), wapiti (*Cervus elaphus*), moose (*Alces alces*), caribou (*Rangifer tarandus*), mule deer (*Odocoileus hemionus*), white-tailed deer (*Odocoileus virginianus*), bighorn sheep (*Ovis canadensis*), muskox (Ovibovini), bison (*Bison*), mountain goat (*Oreamnos*), pocket gopher (*Thomomys*), muskrat (*Ondatra zibethicus*), porcupine (*Erethizon dorsatum*), beaver (*Castor canadensis*), giant beaver (*Castoroides ohioensis*), and unspecified genera of lemming, mouse, and squirrel. The author gives some of the salient features of Beringia and notes that camels, horses, and tapirs invaded Eurasia, while bears, deer (e.g., moose and wapiti), bovids, and mammoths entered North America.

1142. Russell, L.S. (1962): Mammalian migrations in the Pleistocene. Problems of the Pleistocene and Arctic, Publications of McGill University Museums 2(2):48–55
The author examines the Pleistocene faunal movements between North America and Eurasia via the Beringian Isthmus. Species that are common to both continents are: wapiti (*Cervus elaphus*), moose (*Alces alces*), caribou (*Rangifer tarandus*), bison (*Bison*), brown or grizzly bear (*Ursus arctos*), wolf (*Canis lupus*), wolverine (*Gulo gulo*), weasel (*Mustela*), and otter (unidentified genera). Species restricted to North America are: pronghorn (*Antilocapra americana*), mountain goat (*Oreamnos*), and deer (*Odocoileus*). These two volumes [see Russell 1959] were compiled by G.R. Lowther in the McGill University Museum series.

1143. Russell, L.S. (1965): The mastodon. Royal Ontario Museum Series What? Why? When? How? Where? Who? pp. 1–16.
This popular publication lists Ontario mastodon (*Mammut americanum*) remains from: Elgin County, near Welland, Rodney, Tupperville, Shelburne, Wellandport, Toronto, Hamilton, St Catharines, and near London. A map shows these localities, whose greatest concentration is on the N shore of Lake Erie. The first mastodon find in Canada (1834) was at Victoria County, Cape Breton Island, Nova Scotia. Finds occur from the maritime provinces to the Yukon Territory. [See Harington 1996c.]

1144. Rutter, N.W. (1977): Multiple glaciation in the area of Williston Lake, British Columbia. Geological Survey of Canada Bulletin 273. p. 21
This paper gives a detailed account of the glacial history of the Williston Lake area, British Columbia. A mammoth (*Mammuthus*) tusk was found in delta formation deposits in the Portage Mountain end moraine and was radiocarbon dated at 11,600 ± 1000 BP [actually >11,600 BP (I-2244a)]. A bighorn sheep (*Ovis canadensis*) skull was found in ice-contact deposits just S of Finlay Forks

(Rutter et al. 1972) and was radiocarbon dated at 9280 ± 200 BP (GSC-1497). These dates help to define the Late Portage Mountain Advance (Table 4).

1145. Rutter, N.W., Geist, V., and Shackleton, D.M. (1972): A bighorn sheep skull 9280 years old from British Columbia. Journal of Mammalogy 53(3): 641–4.

A bighorn sheep (*Ovis canadensis canadensis*) skull was found in ice-contact deposits 3 m beneath the surface in a gravel pit at Mile 62 on the W side of the Finlay-Parsnip Access Road (Figure 1), 20 km SE of Finlay Forks, British Columbia (55°58′00″N, 123°38′30″W; see map, Figure 1). It was radiocarbon dated at 9280 ± 200 BP (GSC-1497). The skull's measurements were compared with those of other bighorn sheep, Dall sheep (*Ovis dalli dalli*), Stone's sheep (*Ovis dalli stonei*), and Late Pleistocene sheep (*Ovis catclawensis*). The authors mention Churcher's (1968a) report of a bighorn (*Ovis canadensis*) metapodial dating to about 11,000 BP from Bow River gravels near Cochrane, Alberta, and a mammoth (*Mammuthus*) tusk found in a recessional moraine spanning the former path of Peace River near Portage Mountain, British Columbia. The Parsnip River skull suggests that bighorn sheep reached their maximum northward distribution between 11,600 and 9280 BP and that they did not come into contact with Dall sheep (*Ovis dalli*) at that time.

1146. Ryder, J.M. (1987): Neoglacial history of the Stikine-Iskut area, northern Coast Mountains, British Columbia. Canadian Journal of Earth Sciences 24:1294–1301.

A bull caribou (*Rangifer tarandus*) antler from a rapidly shrinking snowbank on Iskut-Ningemsow Ridge (56°51′N, 130°18′W, 1920 m asl; see map, Figure 1) yielded a date of 3760 ± 70 BP (S-2279), and suggests that climate has been relatively cool and moist for the past 4000 years. This is one of 'many ancient caribou antlers' uncovered at the site (p. 1300).

1147. Sagi, D. (1977): Diggers hope to find all mammoth bones. Vancouver Sun, June.

Dr Bill Mathews, a University of British Columbia geologist, and a group of graduate students from UBC and Simon Fraser University were in Hudson's Hope, British Columbia, to collect the rest of a mammoth (*Mammuthus*) found two weeks earlier (June 9, by a workman digging up gravel for the concrete of a new BC Hydro dam on the Peace River about 65 mi W of Fort St John). Mathews estimated the bones to be at least 20,000 years old. The field party identified: a jaw bone, shoulder blade, part of a tusk, shoulder joints, and several leg bones, and partly uncovered what appears to be most of the skull. An accompanying photograph shows the bones.

1148. Sattler, R.A. (1997): Large mammals in Lower Rampart Cave 1, Alaska: Interspecific utilization of an Eastern Beringian cave. Geoarchaeology: An International Journal 12(6):657–88.

Lower Rampart Cave 1 (LRC-1) is a small shelter located at the top of a limestone outcrop near the northern bank of Porcupine River in northeastern Alaska (Figures 1, 2). The fill consists of cryogenic breccia with calcic Pleistocene and organic Holocene sediments. The Pleistocene large-mammal fauna consists mainly of limb bones of extinct herbivores. A sparse (in numbers of fossils) Pleistocene fauna includes: woolly mammoth (*Mammuthus primigenius*), horse (*Equus*), bison (*Bison*), mountain sheep (*Ovis dalli*), caribou (*Rangifer tarandus*), red fox (*Vulpes vulpes*), and weasel or mink (*Mustela*). Carnivore gnaw marks occur on most of the bones, and the accumulation evidently results from carnivore scavenging at kill sites and differential transport to the cave. The author infers that wolves (*Canis lupus*) used the cave during the Late Wisconsinan. Holocene faunal remains include: black bear [(*Ursus americanus* – not brown bears (*Ursus arctos*) as reported by Dixon (1984)], caribou, and at least three different kinds of mammalian hair. Six radiocarbon dates on bone collagen are listed in Table I: horse 13,350 ± 120 BP (Beta-37056, ETH-6578) and 21,780 ± 310 BP (DIC-1334); mammoth 13,725 ± 110 (Beta-29166, ETH-5014); bison 21,050 +320/ -340 BP (DIC-1333); Dall sheep 38,260 ± 830 BP (Beta-37057, ETH-6579) and 22,480 ± 200 BP (Beta-29165, ETH-5013). In addition, Table I includes six Holocene charcoal dates ranging from about 6500 to 3500 BP. Culturally modified (*Homo sapiens*) faunal remains associated with stone artifacts are of Holocene age. Photographs show two bones evidently gnawed by carnivores (Figures 8, 10), and rodent (Rodentia) gnawing on the shaft of a small mammal longbone (Figure 12).

1149. Sattler, R.A., and Dixon, E.J. (1997): Timing the

post-glacial arrival of black bears (*Ursus americanus*) in southeast Alaska and Eastern Beringia. Current Research in the Pleistocene 14:155–7.
Radiocarbon ages on black bear (*Ursus americanus*) remains from coastal regions of southeastern Alaska and from unglaciated regions in northern and western Alaska suggest that black bears dispersed northward along the Cordillera during the Pleistocene–Holocene transition and radiated through Eastern Beringia by modern times. The earliest dated remains of a black bear [41,600 ± 1500 BP (AA-16831)] in northwestern North America are from a cave in the southern Alexander Archipelago. Another black bear from the same cave was dated at 28,700 ± 360 BP (AA-21569), so there were black bears on the western side of the Cordillera N of the Queen Charlotte Islands during the Mid-Wisconsinan interstade. Indeed, the Alexander Archipelago may have been a glacial refugium from which terrestrial mammals dispersed to the mainland. Many skeletal remains of black bears in southeastern Alaska date to the Pleistocene–Holocene transition [radiocarbon dates on seven black bear bones range from 11,715 ± 120 BP (AA-15226) to 6415 ± 130 BP (AA-10447)]. Further, rare skeletal elements and series of deciduous teeth of this species are known from Holocene deposits in Porcupine River and Trail Creek caves, and bones from an adjacent cave to one on Porcupine River are probably older than 7900 BP. In northern and western Alaska black bears seem to have survived winter denning throughout the Holocene, and hibernated in small caves in northeastern Alaska about 8000 BP. The geochronology of fossil black bears (*Ursus americanus*) demonstrates a northward migration along the Cordillera after deglaciation as conifer forests recolonized formerly glaciated and unglaciated regions.

1150. Sattler, R.A., and Gillespie, T.E. (1998): Paleoenvironments, cave faunas, and human migration in Late-Pleistocene Beringia: A comparison of calibrated age ranges. Current Research in the Pleistocene 15:124–6.
The timing and environmental circumstances of human colonization of Beringia during the Late Pleistocene remains central to research about the origins of North American natives. The authors have calibrated two radiocarbon-date lists, and illustrated them in relation to the calibrated age ranges of major paleoenvironmental events (Figure 1). One list consists of radiocarbon ages of Beringian cave faunas; the other consists of the earliest-dated human (*Homo sapiens*) occupations of Eastern Beringia. These data support Yesner's (1996) assertion that disintegration of the full-glacial 'Mammoth Steppe' community antedates human colonization of Eastern Beringia. Mammoth (*Mammuthus*), saiga (*Saiga tatarica*), and horse (*Equus*) are common indicator species for the Mammoth Steppe. Calibrated radiocarbon dates from the known cave faunas that include these species virtually all fall within the last glacial maximum (Duvanny Yar Interval). The earliest undoubted human occupations in Eastern Beringia are the Nenana complex, Mesa complex, and earliest phases of the Chindadn complex. These occupations occurred during the initial climatic warming of the Late-Glacial Birch Interval (apparently correlative with the European Bolling and Allerod pollen zones). Faunas associated with these open-air sites (Dry Creek and Broken Mammoth) exclude the mammoth and horse diagnostic of the Duvanny Yar Interval Mammoth Steppe, but are associated with wapiti (*Cervus elaphus*), bison (*Bison*), mountain sheep (*Ovis dalli*), and caribou (*Rangifer tarandus*). In contrast, the association of the presumed Western Beringian antecedents of Nenana, Mesa, and Chindadn complexes with the Mammoth Steppe fauna has been established (e.g., Upper Paleolithic occupation of Dyuktai Cave).

1151. Savage, D.E. (1951): Late Cenozoic vertebrates of the San Francisco Bay region. University of California Publications, Department of Geological Sciences Bulletin 28:215–314.
As part of an attempt to organize sensibly the confused classification of North American Pleistocene horses, Savage (p. 252) referred the Yukon horse (*Equus lambei*) to *Equus* cf. *Equus caballus*.

1152. Savage, H. (1975): A faunal analyst in the northern Yukon. Arch Notes, October 1–3.
The author discusses the value of the fossils of the Old Crow Basin, Yukon, to the faunal analyst and the study of the human (*Homo sapiens*) involvement in Pleistocene extinctions. Faunal remains from Old Crow Locs. 14N and 11A are mentioned, as are Pleistocene mammal (Mammalia) bones evidently modified by humans (e.g., 'a recurring pattern of fractures through heel bones in order to avoid butchering the incredibly tough heel tendons').

1153. Savelle, J.M., Dyke, A.S., and McCartney, A.P. (2000): Holocene bowhead whale (*Balaena mysticetus*) mortality patterns in the Canadian Arctic Archipelago. Arctic 53(4):414–21.

On the basis of emergence curves for 10 localities surveyed in the Canadian Arctic Islands (Figure 1 shows a locality on southern Somerset Island and nine other localities in northern Baffin Island, Nunavut), stranded bowhead whale (*Balaena mysticetus*) remains tend to be relatively common in beaches dating before 8000 BP and between 5500 and 2500 BP, but relatively rare otherwise. Beaches developed during the abundant periods relate to relatively ice-free summers, while the others developed during times of rather pervasive summer sea ice that prevented bowheads from entering the region. Length-based mortality profiles of the stranded whale assemblages suggest that random processes, such as ice entrapment, were responsible for the death of most whales found on these beaches. [See Dyke and Morris 1990.]

1154. Scharff, R.F. (1912): Distribution and origin of life in America. Macmillan Company, New York. 497 pp. [note particularly chapters I to IV dealing with northern North America].

The author discusses various ideas on the origin of caribou (*Rangifer tarandus*) – none with much conviction (pp. 4–6). He mentions that the muskox (*Ovibos moschatus*), according to R. Kowarzik, must have originated in North America, the helmeted muskox (*Bootherium*) being probably its direct ancestor. The author concludes that the muskoxen of Greenland and Arctic America probably survived the ice age where they now live. He states that 'the banded lemming (*Dicrostonyx torquatus*) is not a recent immigrant to Greenland, but has persisted there from pre-Glacial times as indicated by Colonel Fielden, who discovered its remains with those of caribou and muskox in post-Tertiary [Quaternary] deposits from sea-level to an altitude of 1000 ft in northern Greenland' [see Fielden and de Rance 1878]. He seems to think that voles (*Microtus*) originated in North America long before the Pleistocene and that they crossed the Bering land connection in passing from the New World to the Old (p. 29). In discussing moose (*Alces*), he states: 'we are led to assume that Bering Strait was dry land when the ancestors of the existing moose entered the New World.' He mentions the source of some of the North American and Greenland biota as stemming from Europe [see the map (Figure 2) showing the North Atlantic in the Late Pliocene]. That the arctic fauna advanced far S of their present range is demonstrated by the occurrence of a few fossil caribou (*Rangifer tarandus*) antlers S of the area covered by glacial drift (p. 35). The author opines that near the beginning of the Pleistocene the land from the banks [e.g., Grand Banks] of Newfoundland S to Cape Cod was raised high asl forming a refuge for the survival of arctic and Old World fauna and flora. He argues (pp. 55–6) that coastal regions of Labrador and Newfoundland remained ice-free, giving shelter to vertebrate and other survivors from 'pre-Glacial times,' and that this view is supported by the presence of many kinds of mammals characteristic of Newfoundland – e.g., caribou, muskrat (*Ondatra*), vole, hare (*Lepus*), lynx (*Felis lynx*), fox (*Vulpes*), otter (*Lutra*), marten (*Martes*), black bear (*Ursus amerianus*), and wolf (*Canis lupus*) – and the lack of Asian mammals, such as moose (*Alces alces*), wapiti (*Cervus elaphus*), and brown bear (*Ursus arctos*) [see Cameron 1958]. He notes that the presence of ground squirrels (*Spermophilus*) in Eurasia and North America tends to prove that the Bering Strait was dry land. The origin and distribution of Holarctic ptarmigan (*Lagopus*) are also important in this respect, and the presence of flying squirrels (*Glaucomys*) in America seems to be due to migration from Asia, which took place when a land bridge existed between the two continents (p. 64). Muskrats (*Ondatra zibethicus*) are typically North American and probably originated there. In considering fossil bison from Asia and North America, Scharff notes that it seems likely that the ancestor of American bison (*Bison*) came originally from Asia. He mentions that the steppe bison (*Bison priscus* = *Bison crassicornis*) ranged not only in Alaska but throughout Siberia and a large part of Europe (p. 67). Likewise, the wapiti had a similar history, originating in Central Asia. Chapter IV is devoted to Alaska. Regarding mammoths (*Mammuthus*), the author mentions that since the beginning of the nineteenth century 'it was known that fossil ivory might be gathered on the shores of Alaska' (p. 78). Russian authorities had noted its occurrence there as well as on the Pribilof Islands. The author summarizes some of Maddren's (1905) finds, concluding that the most rational explanation for the extinction of the megafauna at the end of the Pleistocene is the gradual alteration of climate from one that was more temperate to the more severe climate of today. Such changes left only the animals capable of adapting to the new conditions [e.g., muskoxen (*Ovibos moschatus*)]. Scharff also summarizes the results of C.W. Gilmore's (1908) Smithsonian expedition to Alaska, giving a list of 11 genera of Pleistocene mammals on p. 80. He mentions Quackenbush's (1909) discovery of a mammoth (*Mammuthus*) with flesh, skin, and hair. He mentions F.A. Lucas's opinion favouring a geologically recent land connection between Asia and North America using the distribution of mountain sheep (*Ovis*) and brown bears (*Ursus arctos*) as evidence, and that the presence of pike (*Esox*) and the Columbian mammoth (*Mammuthus columbi = Elephas columbi*) in North America favoured an older land connection. J.A. Allen (1903) also agrees with a geologically recent land connection between Asia and North America because of the great similarity of the mammals [e.g., weasels (Mustelidae), ground squirrels, and mountain sheep (*Ovis*)], on either side of the Bering Strait. The author devotes much space to showing how the close relationship of mammals on either side of the Bering Strait arose, and under what climatic conditions.

1155. Scheffer, V.B. (1967): Marine mammals and the history of Bering Strait. *In*: The Bering Land Bridge (D.M. Hopkins, ed.). Stanford University Press, Stanford, California. pp. 350–63.

The author discusses the question of whether the Bering Strait (Bering Land Bridge) played a role in marine mammal zoogeography during the Cenozoic. During each glacial stage of the Pleistocene, when a land bridge appeared as a barrier to the passage of marine mammals, it is assumed that the permanent ice over the Arctic Ocean was thick and extensive. The roles of interrupted, disturbed or compelled seasonal migration; colonization by straying individuals; increased ice cover; and taxonomic influence of the 'antitropical' behaviour of most cetaceans and pinnipeds are reviewed by mammalian orders. The orders include Cetacea, Pinnipedia [now Carnivora], Sirenia, and one of the Carnivora (the sea otter). The author concludes that the Pleistocene series of land-ice barriers was effective in shaping the evolution of subspecies of beluga (*Delphinapterus leucas*), ringed seal (*Phoca hispida = Pusa hispida*), bearded seal (*Erignathus barbatus*), walrus (*Odobenus rosmarus*), harbour seal (*Phoca vitulina*), and possibly grey whale (*Eschrichtius robustus*). It may also have split the population of the common ancestor of the ribbon seal (*Histriophoca fasciata*) and the harp seal (*Phoca groenlandica = Pagophilus groenlandicus*). In addition, there are a number of species that are restricted in current distribution to either the North Atlantic or North Pacific. These include the Dall porpoise (*Phocoenoides dalli*), Baird beaked whale (*Berardius bairdii*), Steller sea cow (*Hydrodamalis gigas = Hydrodamalis stelleri*), northern fur seal (*Callorhinus ursinus*), Steller sea lion (*Eumetopias jubatus*), and sea otter (*Enhydra lutris*) in the Pacific, and the North Atlantic bottlenosed whale (*Hyperoodon ampullatus*) and narwhal (*Monodon monoceros*) in the Atlantic. The inference is that cold water, not always coincident with ice, has acted as a barrier.

1156. Schmidt, J. (1962): Wildlife Graveyard? Calgary Herald, December 7, p. 30.

Henry McAreavy ploughed up (using a 20-in breaker plough) thousands of ancient animal bones in a 28-acre field. The field, 17 mi NE of Rimbey, Alberta (NE quarter of sec. 36, twp. 42, rge. 5), is in a green, low-lying valley with three small rivers running through the district (Big Medicine, Little Medicine, and a spring-fed creek). About 75% of the land is peat, and McAreavy discovered the bones of young and old animals while breaking low-lying willow land of heavy black peat. Many of the bones suffered decay, but the farmer identified them as mostly buffalo (*Bison*) and moose (*Alces*). The moose antlers were somewhat different than the specimens seen today, some having a 45-in antler spread. The moose antlers were unlike ordinary trophy ones but were just as big. Instead of wide, platter-like palmations, they were thick at the base, and had a palmation about saucer-size and long, spreading tines. One adult buffalo skull had a horncore spread of 27 in. Other smaller skulls belonged to calves.

1157. Schumacher, C.F.S. (1791): Fortegnelse og Beskrivelse over nogle grønlandske Mineralier. Naturhistorie Selskabets Skrifter 4:206–33.

According to Bennike (1997, p. 901), the author reports remains of capelin (*Mallotus villosus*) in calcareous nodules from West Greenland.

1158. Schweger, C., and Habgood, T. (1976): The Late Pleistocene steppe-tundra of Beringia – A critique. American Quaternary Association, Abstracts of the Fourth Biennial Meeting (October 9–10, Tempe, Arizona):80–1.

The hypothesis of a Pleistocene steppe-tundra environment in unglaciated Alaska and Yukon is based partly on the discovery of fossil collections of large-mammal bones dominated by horse (*Equus*), bison (*Bison*), and mammoth (*Mammuthus*). The authors argue that the faunas recovered during gold-mining operations and from river bars are mixed thanatocoenoses and do not represent a community. Without detailed stratigraphic information, these faunal collections lose their significance. An alternate hypothesis is that two simpler faunas are represented: one for cold, arid glacial maxima, and another dominant during warmer, wetter interstades. Faunal remains from Upper Paleolithic sites may better represent the contemporaneous faunal community. The authors suggest that rather than steppe-tundra, a tundra-steppe mosaic is more realistic, and that local drainage, permafrost, soil, and microclimatic conditions would have played an important role in determining the various regional combinations of steppe, grassland, or tundra vegetation that may have existed.

1159. Schweger, C., and Martin, J. (1976): Grazing strategies of Pleistocene steppe-tundra fauna. American Quaternary Association, Abstracts of the Fourth Biennial Meeting (October 9–10, Tempe, Arizona):157.
The Pleistocene steppe-tundra fauna was dominated by grazers. Ruminants such as bison (*Bison*), sheep (*Ovis*), camel (Camelidae), wapiti (*Cervus*), and caribou (*Rangifer tarandus*) evolved fermentation digestion by microorganisms in modified stomachs, whereas monogastrics such as horse (*Equus*) and mammoth (*Mammuthus*) must ingest essential amino acids and cannot use cellulose. Monogastrics are less efficient than ruminants, but their higher food consumption permits higher nutrient intake. If poor-quality food is abundant, a horse may do better than a large ruminant, but when food is limited, the ruminant may have a survival advantage. Animals that could best use the summer growth period and could reach maturity quickly may have been best adapted to steppe-tundra environments such as those that existed in Eastern Beringia during the Pleistocene. The authors provide an interesting discussion of winter adaptations required by Pleistocene mammals, noting for example that larger animals have an advantage, and that if snow is the water source, up to half of the total caloric intake may be required to convert snow to water.

1160. Schweger, C.E., Matthews, J.V., Jr, Hopkins, D.M., and Young, S.B. (1982): Paleoecology of Beringia – A synthesis. *In*: Paleoecology of Beringia (D.M. Hopkins, J.V. Matthews, Jr, C.E. Schweger, S.B. Young, and V. Stanley, eds.). Academic Press, Toronto. pp. 425–44
This chapter provides a synopsis of the book's contents, including a valuable table (Table 1) that identifies the well-dated Pleistocene Beringian faunas by locality. This table is presented by species (29). Some of the more puzzling Pleistocene paradoxes are discussed, as is the impact of humans (*Homo sapiens*) on the fauna and the impact of the paleoclimate on humans.

1161. Scott, W.B. (1937): A History of Land Mammals in the Western Hemisphere. Macmillan, New York. 786 pp.
This is a comprehensive, lucid, well-illustrated, and perceptive treatment that, in part, juxtaposes Pleistocene and Recent mammals of North America.

1162. Scotter, G.W., and Simmons, N.M. (1976): Mortality of Dall's sheep within a cave. Journal of Mammalogy 57(2):387–9.
A cave at the mouth of First Canyon on the South Nahanni River in Nahanni National Park, Northwest Territories, revealed remains of 93 Dall sheep (*Ovis dalli*) at 12 different locations (Figure 1). Of 92 sheep, 33 were males, 43 were females, and 16 were unsexed. The stratigraphic position was generally on top of silt or other deposits, except for those partly covered by ice. A few skeletons were relatively undisturbed, but most had been scattered by other animals such as bushytail woodrats (*Neotoma cinerea*) and porcupines (*Erethizon dorsatum*), which now live in the cave. Hair, horn, and cartilage from one sheep yielded the following corrected ages: hair 2400 ± 160 BP (GSC-1649); horn 2440 ± 80 BP (GSC-1757); cartilage 2370 ± 150 BP (GSC-1774). Escape from extreme cold and presence of mineral salts (e.g., epsomite and gypsum) are possible reasons for Dall sheep to enter the cave.

1163. Selwyn, A.R.C. (1872): Paleontological Collections. Geological Survey of Canada. Report of Progress for 1870–71. p. 10.
A white whale (*Delphinapterus leucas = Beluga Vermontana*) skeleton was found in a Pleistocene clay pit at

Cornwall, Ontario. The specimen was purchased from Charles Pierce of Cornwall.

1164. Selwyn, A.R.C. (1887): [Harp seal found near Montreal and beluga found near Cornwall]. Geological Survey of Canada Summary Report for 1886. p. 46A.
A harp seal (*Phoca groenlandica*) skeleton was obtained in 1861 by T.C. Weston from the Pleistocene clays of Montreal, Quebec, and that of a white whale (*Delphinapterus leucas = Delphinapterus catadon*) was found in a Pleistocene clay pit at Cornwall, Ontario. They were mounted by Jules Bailly.

1165. Selwyn, A.R.C. (1893): Summary report on the operations of the Geological Survey for the year 1890. Geological Survey of Canada Annual Report (new series), Volume 5, 1890–91, Part A.
A white whale (*Delphinapterus leucas*) is reported from Cornwall, Ontario.

1166. Seymour, K.L. (1983): The Felinae (Mammalia: Felidae) from the Late Pleistocene tar seeps at Talara, Peru, with a critical examination of the fossil and Recent felines of North and South America. MSc thesis, University of Toronto. 240 pp.
The author includes a succinct section on the taxonomic history of fossil lions in North America (pp. 65–70). Since his study showed 'such a wide variation amongst modern species,' he considered fossil subspecies inappropriate and refers to the American lion (*Panthera leo atrox*) as *Panthera leo*. His map (Figure 17) of 39 American lion fossil localities in North America is updated from Harington (1969), and includes the following sites in northern North America: Medicine Hat, Alberta; Bindloss, Alberta; Dawson area, Yukon; Fairbanks, Alaska; Lost Chicken Creek, Alaska; Kaolak River, Alaska.

1167. Seymour, K. (1989): The tail of a whale. Rotunda 22(1):53, 55.
On April 18, 1988, an isolated, fairly complete whale vertebra was discovered at Queen's Quay and Bay Street during the excavation for Toronto, Ontario's new Harbourfront Light Rapid Transit line. The specimen was reported to have been found approximately 1.5 m below landfill that was first deposited in Lake Ontario during the early 1900s. This was apparently the first evidence of marine life in the Toronto area dating after the last glaciation, and no other such signs had been found W of a line joining Pembroke and Brockville, Ontario. The bone is from the lower back of a large-sized whale (Cetacea), although the species is undetermined. A small sample of the whale bone submitted for dating gave a result of AD 1828 ± 125 years (no laboratory number is given). [In the following issue (Rotunda 22(2):63. 1989), Seymour states that the date mentioned applied if the whale had lived in the Arctic Ocean, but that the vertebra would date from AD 1531 ± 120 years if it had lived in the Atlantic Ocean. Could it have been brought inland from the Atlantic coast by people (*Homo sapiens*)? See Harington (1988), p. 238.]

1168. Shackleton, D.M., and Hills, L.V. (1973): New world immigrants. Alberta Conservationist (Spring): 14–16.
This popular article mentions the find of an extinct medium-horned bison (*Bison*) found at the Milan farm about 14 km ESE of Three Hills, Alberta [see Shackleton and Hills (1977) for a more complete description]. A photograph of the specimen, as excavated, is included. The authors reconstruct the Eurasian dispersal of the bison and their spread into North America via the Bering Land Bridge during the Pleistocene.

1169. Shackleton, D.M., and Hills, L.V. (1977): Post-glacial ungulates (*Cervus* and *Bison*) from Three Hills, Alberta. Canadian Journal of Earth Sciences 14(5):963–86.
A western bison (*Bison bison occidentalis = Bison occidentalis*) partial skull with horncore and limb bones was found in medium-coarse laminated sand and partially in clay, at the Milan farm about 14 km ESE of Three Hills, Alberta (NE¼, sec. 20, twp. 31, rge. 22, W4). Disarticulated remains of wapiti (*Cervus elaphus nelsoni*) were found solely in the clay layer at the same locality. The bison ribs were radiocarbon dated (uncorrected) at 9630 ± 300 (GSC-1894) and 9670 ± 160 BP (I-8579). Photographs of bison crania with horncores and a well-preserved wapiti skull with antlers are included. Detailed morphological comparisons are made with modern and extinct species of bison and wapiti.

1170. Sharpe, D.R. (1992): Quaternary geology of Wollaston Peninsula, Victoria Island, Northwest

Territories. Geological Survey of Canada Memoir 434.
Radiocarbon ages of five unidentified whale bone samples from Wollaston Peninsula, Victoria Island, are presented. [Dyke's (Dyke and Savelle 2001) inspection of some of these bones indicates that they came from large whales indistinguishable from those of the bowhead (*Balaena mysticetus*).] Four of the samples, from 28 to 57 m asl, date from the Early Holocene (about 9800 to 8600 BP). The other sample, from 4 m asl, was dated at about 1400 BP. At least two of Sharpe's sites are 7 km NW of Point Caen.

1171. Sher, A.V. (1974): Pleistocene Mammals and Stratigraphy of the Far Northeast U.S.S.R. and North America. [Originally published in 1971 by the USSR Academy of Sciences]. Translation by D. Vitaliano, International Geology Review 16:1–284.
This book reviews the state of knowledge of the stratigraphy and paleontology of Pleistocene deposits in Siberia. Important comparisons are made between fossils from the Würm deposits of northern Siberia and the comparable Wisconsinan deposits in North America. Of 22 species in Siberia, 21 are also known from North America. Only the woolly rhinoceros (*Coelodonta antiquitatis*) does not appear to have migrated across the Bering Land Bridge. Numerous illustrations and photographs of Siberian Pleistocene mammal fossils are provided along with written descriptions. Also included are tables of measurements, distribution maps, and stratigraphic diagrams.

1172. Sher, A.V. (1984): The role of Beringian land in the development of Holarctic mammalian fauna in the Late Cenozoic. In: Beringia in the Cenozoic Era (V.L. Kontrimavichus, ed.). Amerind Publishing Co. Pvt. Ltd., New Delhi. pp. 296–316.
The author proposes the name Beringida (Figure 2) to designate the vast region that played the major role in the history of Beringian faunal exchange. The narrow, active zone of Beringida within which the land connection and strait alternately appear he calls Beringia. Evidently during the Late Cenozoic the fauna of Beringida (including Alaska) developed as one. The early cooling in the high latitudes caused the evolution of a subarctic mammalian fauna in Beringida. Since the Early Pleistocene this fauna can be considered as cryoxerotic. Due to the progressive general cooling of Holarctic climate, the cold-resistant Beringidan species could disperse to the temperate latitudes of Eurasia and North America. Data support the autochthonous evolution of the cold-resistant 'Arctic' fauna within Beringida. In the Late Cenozoic Beringida was not mainly an isthmus for faunal exchange between continents, but an independent (though closely related to the Palearctic) region in which a peculiar fauna evolved that repeatedly dispersed into Old and New worlds.

1173. Sher, A.V. (1993): Global change at the Arctic crossroads: Current problems of the Late Cenozoic of Beringia. 23rd Annual Arctic Workshop of the Byrd Polar Research Center (April 1–3, Columbus, Ohio), BPRC Miscellaneous Series M-322 (Abstracts): 70.
The aim of this contribution is to outline the most important events in the terrestrial environment and biota of Beringia, which were related on a major scale (e.g., global changes in the Northern Hemisphere during the last 3 Ma). The author mentions key sites of Cape Deceit, Alaska, and Old Crow Basin, Yukon, as well as the Olyorian of Siberia corresponding to the 2.0–0.85 Ma cooling. During this period increasing continentality resulted in the development of treeless grassland and open parkland vegetation, permafrost aggradation, and evolution of a rich grazing fauna (Olyorian Fauna). The period coincides with important evolutionary changes in lineages such as collared lemming (*Dicrostonyx*) and vole (*Microtus*). The extinction of many Olyorian mammals almost certainly took place during the Middle Pleistocene. The Mammoth Fauna characterizes the cold-dry environment of Beringia during the late part of the Middle Pleistocene and all of the Late Pleistocene. There were no important changes in faunal composition then. The most radical change in environment and community structures took place near the Pleistocene/Holocene boundary when Pleistocene communities (and some species) became extinct.

1174. Sigalet, H. (1989): Of fish and blondes. Up Here, March/April:33–4.
On p. 34 the author writes of the Hyland River, Yukon: 'At one time, they say, Indians feared to venture to the headwaters because the bones of huge animals (fossils) indicated evil lurked in the area.' [See, for example, Bélanger 1988.]

1175. Símonarson, L.A. (1981): Upper Pleistocene and Holocene marine deposits and faunas of the north

coast of Nûgssuaq, West Greenland. Grønlands Geologiske Undersøgelse Bulletin 140:1–107.
The author describes the field relationships and composition of Quaternary molluscan and barnacle faunas at Pátorfik, Kûtsiaq, and Sarfâgfîk on the N coast of Nûgssuaq, West Greenland. The marine Quaternary deposits at Pátorfik are older than 35,000 years. The fauna indicates that the Pátorfik deposits were formed during the last (Sangamonian) interglacial. Fishes (Pisces), birds (Aves), and seals (Phocidae) occur in the Pátorfik beds (but are not dealt with here) along with foraminifera, serpulids, ostracods, decapods, bryozoans, ophiuroids, and echinoids.

1176. Simpson, G.G. (1945): The principles of classification and a classification of mammals. Bulletin of the American Museum of Natural History 85:1–350.
This is a classic reference on the taxonomy of living and extinct mammals.

1177. Simpson, G.G. (1947): Holarctic mammalian faunas and continental relationships during the Cenozoic. Bulletin of the Geological Society of America 58:613–88.
A series of tables lists land mammals common to North America and Eurasia, their probable time of migration by subepochs throughout the Cenozoic, and important groups that did not migrate. Major faunal interchanges occurred in Early Eocene, Late Eocene, Early Oligocene, Late Miocene, Middle to Late Pliocene, and Pleistocene times (a strong exchange at the Plio-Pleistocene transition and another Mid-Pleistocene surge waning toward the end of the epoch – p. 628). Each interchange involved migration in both directions, but probably there was more movement from Eurasia to North America (direction of migration is outlined on pp. 628–32). The most important selective influence was probably the relatively cold climate of the Beringian land connection. This connection was probably from Siberia to Alaska throughout the Cenozoic, and was in almost continuous existence with important interruptions in parts of the Eocene and Oligocene, and perhaps shorter breaks at later times. The measurement of faunal resemblance is discussed (pp. 671–85) and is charted for the Cenozoic (Figures 3, 4). Important Pleistocene migrants, perhaps from Eurasia, include: pika (*Ochotona*), hare (*Lepus*), ground squirrel (*Spermophilus* = *Citellus*), bear (*Ursus*), wolverine (*Gulo*), cats (*Felis*), mammoths (*Mammuthus*), moose (*Alces* = *Alce*), caribou (*Rangifer*), saiga antelope (*Saiga*), bison (*Bison*), mountain sheep (*Ovis*), and muskoxen (Ovibovini = Ovibovinae). Migrants from North America may include: marmot and woodchuck (*Marmota*), collared lemming (*Dicrostonyx*), brown lemming (*Lemmus*), vole (*Microtus*), fox (*Vulpes*), and camel (Camelidae). Despite the orthodox view, Simpson considers the horse (*Equus*) as a possible migrant from Eurasia to North America.

1178. Skarland, I. (1949): The geography of Alaska in Pleistocene and early post glacial time: A study of the environment from an anthropological viewpoint. PhD thesis, Harvard University.
The author gives a partial list of Pleistocene mammals from the Fairbanks area of Alaska (pp. 132–3).

1179. Skinner, M.F., and Kaisen, O.C. (1947): The fossil *Bison* of Alaska and preliminary revision of the genus. American Museum of Natural History Bulletin 89(3):129–256.
This is a seminal paper in *Bison* taxonomy that also sets the standards for measurement and aging techniques. Sexual differences, growth patterns, tooth-wear characteristics, and ancient and modern distribution are discussed. Fossil species are revised as well as the modern representatives, and measurement statistics for both fossil and modern species are summarized in 24 tables. This study is based mainly on a collection of 180 complete and partial diagnostic male bison skulls from the Fairbanks area of Alaska. Some Yukon specimens are used also. The steppe bison (*Bison priscus* = *Bison* (*Superbison*) *crassicornis*) make up 87% of identifiable skulls. The division of the genus into six named subgenera and one primitive, tentatively unnamed subgenus is based primarily on characters observed in skulls and identifiable patterns of horncore growth. By the close of the Early Pleistocene several species of *Bison* had reached Europe and Siberia, some of which crossed to North America for the first time. Recurrent glaciation isolated and forced these first immigrants southward to the North American plains. The living North American bison are the descendants of a later, second migration from Siberia. Bison species apparently have undergone progressive horncore enlargement followed by a reduction in size [e.g., wood bison (*Bison bison athabascae*) and plains bison (*Bison bison*

bison)]. A problem throughout the study was getting accurate geological evidence, so the geological age of most fossils is questionable.

1180. Skinner, R.G. (1973): Quaternary stratigraphy of the Moose River Basin, Ontario. Geological Survey of Canada Bulletin 73(225):1–77.
Minnow or carp (Cyprinidae) remains and beaver (*Castor canadensis*)-gnawed sticks were recovered in 1972 from silty-clay rhythmites (including forest-bed and peat near the base) in the Missinaibi Formation near Moose River Crossing, Ontario. Evidently fish and beaver lived in an oligotrophic, semi-permanent pond or small lake during the last (Sangamonian) interglacial.

1181. Skwara, T. (1988): The ice age in the Saskatoon area: Setting the stage. *In*: Out of the Past: Sites, Digs and Artifacts in the Saskatoon Area (U. Linnamae and T.E.H. Jones, eds.). Saskatoon Archaeological Society. pp. 25–37
This is a reconstruction of the Pleistocene of the Saskatoon, Saskatchewan, area. The Saskatoon fauna mentioned includes that which was reported in Lammers (1968) and SkwaraWoolf (1980, 1981). The author states (p. 30) that the Riddell Local Fauna represents a time when climate was temperate and equable – lacking the extremes of the present climate. Fossils recovered indicate that summers were cooler and moister and that the winters were milder. 'The fauna derives from a major interglacial,' and may well be as old as 175,000 BP.

1182. Skwara, T., and Walker, E.G. (1989): Extinct muskox and other additions to the Late Pleistocene Riddell Local Fauna, Saskatoon, Canada. Canadian Journal of Earth Sciences 26:881–93.
Additions to the Late Rancholabrean fauna of the Floral Formation fluvial sand at the Riddell site, N of the University of Saskatchewan, in Saskatoon, include: helmeted muskox (*Bootherium bombifrons* = *Symbos cavifrons*), well-preserved incomplete skull; beaver (*Castor canadensis*), incomplete ulna; Mexican horse–Niobrara horse complex (*Equus conversidens–Equus niobrarensis* complex); Niobrara horse (*Equus niobrarensis*); western camel (*Camelops* cf. *Camelops hesternus*) humerus, metapodial fragments, and phalanx; and remains of mammoth (*Mammuthus*). The authors state: 'lithologic and stratigraphic relationships of tills and ecological requirements of the fauna limit the Riddell Member to the Sangamonian.' Photographs of the new specimens are included.

1183. SkwaraWoolf, T. (1980): Mammals of the Riddell Local Fauna (Floral Formation, Pleistocene, Late Rancholabrean) Saskatoon, Canada. Natural History Contributions No. 2. Saskatchewan Culture and Youth, Museum of Natural History, Regina. 129 pp.
The Late Rancholabrean fauna found in Floral Formation fluvial sand at the Riddell site, N of the University of Saskatchewan, in Saskatoon (sec. 13, twp. 37, rge. 5 W 3rd mer.; Figure 1) include: unidentified fish (Pisces), hare (*Lepus*), snowshoe hare (*Lepus* cf. *Lepus americanus*), Richardson's ground squirrel (*Spermophilus richardsonii*), arctic ground squirrel (*Spermophilus* cf. *Spermophilus parryii* = *Spermophilus undulatus*), black-tailed prairie dog (*Cynomys* cf. *Cynomys ludovicianus*), northern pocket gopher (*Thomomys talpoides*), black-backed vole (*Clethrionomys gapperi*), heather vole (*Phenacomys intermedius*), vole (*Microtus*), meadow vole (*Microtus* cf. *Microtus pennsylvanicus*), sagebrush vole (*Lagurus curtatus*), muskrat (*Ondatra zibethicus*), coyote (*Canis* cf. *Canis latrans*), red fox (*Vulpes* cf. *Vulpes vulpes*), proboscidean (Proboscidea), Mexican horse (*Equus conversidens*), western camel (*Camelops* cf. *Camelops hesternus*), deer(?) (Cervidae cf. *Odocoileus*), and pronghorn (*Antilocapra* cf. *Antilocapra americana*). This fauna, which occurs between two hills of the Floral Formation, is considered to be of Late Rancholabrean age (see Figure 20). Habitat preferences suggest that the climate at the time of deposition was probably much like it is today. Evidently the landscape was dominated by open grass, with trees and shrubs in low areas.

1184. SkwaraWoolf, T. (1981): Biostratigraphy and paleoecology of Pleistocene deposits (Riddell Member, Floral Formation, Late Rancholabrean), Saskatoon, Canada. Canadian Journal of Earth Sciences 18(2):311–22.
The author compares three Sangamonian/Late Rancholabrean faunal assemblages: Riddell Member of Saskatoon, Saskatchewan; Fort Qu'Appelle, Saskatchewan; and Medicine Hat, Alberta. The first two have the most similar species. The composite Riddell fauna [comprises all fossils recovered from the Riddell and Saskatoon sites, as well as some isolated specimens

known with reasonable certainty to derive from the Riddell Member in the vicinity of Saskatoon] includes: hare (*Lepus*), snowshoe hare (*Lepus* cf. *Lepus americanus*), Richardson's ground squirrel (*Spermophilus richardsonii*), arctic ground squirrel (*Spermophilus* cf. *Spermophilus parryii* = *Spermophilus undulatus*), black-tailed prairie dog (*Cynomys* cf. *Cynomys ludovicianus*), northern pocket gopher (*Thomomys talpoides*), black-backed vole (*Clethrionomys gapperi*), heather vole (*Phenacomys intermedius*), vole (*Microtus*), meadow vole (*Microtus* cf. *Microtus pennsylvanicus*), sagebrush vole (*Lagurus curtatus*), muskrat (*Ondatra zibethicus*), coyote (*Canis* cf. *Canis latrans*), red fox (*Vulpes* cf. *Vulpes vulpes*), unidentified Proboscidea, Mexican horse (*Equus conversidens*), Niobrara horse (*Equus* cf. *Equus niobrarensis*), western camel (*Camelops* cf. *Camelops hesternus*), deer (*Odocoileus*), pronghorn (*Antilocapra* cf. *Antilocapra americana*), ground sloth (*Megalonyx*), llama (*Hemiauchenia macrocephala*), and giant bison (*Bison* cf. *Bison latifrons*). The fossils are distinctive enough to date the Riddell Member as Late Rancholabrean. Paleoecological analysis indicates that five communities are represented in the composite fauna, the largest and most extensive being dry upland prairie. Temperate, equable climate seems to have prevailed.

1185. SkwaraWoolf, T., and Millar, J.F.V. (1981): Pleistocene muskox (*Ovibos moschatus*) from near Saskatoon, Saskatchewan. Canadian Journal of Earth Sciences 18(5):852–7.
A partial tundra muskox (*Ovibos moschatus*) cranium (Late Rancholabrean) was found in sand and gravel of the Floral Formation in the Muskox Pit on the eastern shore of Patience Lake, near Saskatoon, Saskatchewan (52°07′N, 106°17′W). This is the first known occurrence of tundra muskox remains on the Canadian Prairies. Photographs and bone measurements are included.

1186. Smith, H.C. (1977): A fossil bison skull from western British Columbia. Syesis 10:167–8.
A wood bison (*Bison bison athabascae*) skull with both horncores was found protruding from a gravel bed along the edge of the Kitimat River, 3.2 km E of the Kitimat Bridge, Kitimat, British Columbia. The specimen is important because wood bison are rarely reported from British Columbia and because of its extreme westerly position – this bison had almost reached the Pacific Ocean.

1187. Smith, L.H. (1901): The extinction of the elk in Ontario. Ottawa Naturalist 15(5):96–7.
A set of wapiti (*Cervus elaphus* = *Cervus canadensis*) antlers with part of the skeleton was found on the Wilson farm in the Township of Lobo (lot 15, 12th concession) near Strathroy, Ontario, about AD 1800. Since it was from boggy spring deposits, the animal may have become mired there. The author also mentions his own collection from the vicinity of Adelaide and in adjoining Lambton County.

1188. Smolkin, M.J. (1977): Museum showing Pakenham whale. Arnprior Chronicle, June 15.
In 1906, a Pakenham, Ontario, farmer, Patrick Cannon, dug a well in Champlain Sea clay some 4 mi NW of Pakenham. At about 14 ft he struck bones of a white whale (*Delphinapterus leucas*) that had died there thousands of years earlier. The specimen was first described by paleontologist J.F. Whiteaves (1907). The bones remained in possession of Cannon until 1958, when they were given to J.A. Herrick, Cannon's nephew. A picture of the white whale skeleton was included in a book published in 1963 by Verna Ross McGiffen, *Pakenham, Ottawa Valley Village 1823–1860*. The specimen was rediscovered by two members of the National Museum of Natural Sciences, and in 1973 it was donated to that museum [now Canadian Museum of Nature]. A radiocarbon date on bone of the vertebrae gave an age of about 10,400 years, and the whale died between the ages of two and three years. The Pakenham whale is presently on loan for display in the Arnprior Museum.

1189. Spencer, J.W. (1876): Report on the country between the Upper Assiniboine River and Lakes Winnipegosis and Manitoba. Geological Survey of Canada, Report of Progress for 1874–75. p. 63.
The author, in his report on the Shell River valley, Manitoba, recorded that bones probably belonging to the woolly mammoth (*Mammuthus primigenius* = *Elephas primigenius*) were described to him by a man who had seen the locality where the bones were found. Spencer mentioned that the bones were sent to Fort Ellice before being sent to England. [So perhaps these bones are not the same as those from Swan River sent to Fort Pelly, or his informant was in error. See Tyrrell (1892, p. 129E), and Leith (1949).]

1190. Spencer, J.W. (1883): Surface geology of the region about the western end of Lake Ontario. Canadian Naturalist, Part 3, New Series 10(5):265–312.
Several mammal remains were found during the construction of the Desjardins Canal at Burlington Heights, Hamilton, Ontario, at an elevation of about 20 m (p. 308). These were: mammoth (*Mammuthus*) tusk and vertebra; woolly mammoth (*Mammuthus primigenius* = *Euelephas Jacksoni*) remains; wapiti (*Cervus elaphus* = *Cervus Canadensis*) antlers; and a beaver (*Castor canadensis* = *Castor fiber*) jaw. The author also mentions that several mastodon (*Mammut americanum*) sites were found in swamps north of Lake Erie, Ontario.

1191. Spiess, A.E., and Storck, P.L. (1990): New faunal identifications from the Udora site: A Gainey-Clovis occupation site in southern Ontario. Current Research in the Pleistocene 7:127–9.
A Gainey-Clovis (12,500–10,400 BP) human (*Homo sapiens*) occupation site at Udora, Ontario, contained the following vertebrate remains: deer (Cervidae); caribou (*Rangifer tarandus*); hare (*Lepus*) [either snowshoe hare (*Lepus americanus*) or arctic hare (*Lepus arcticus*)]; fox (*Vulpes* or *Alopex*); and arctic fox (*Alopex lagopus*). The arctic fox is of key interest because it occurs primarily on the tundra, suggesting the Udora site occurred in such surroundings or on the margins of an open forest-tundra environment.

1192. Stalker, A.M. (1968): Geology of the terraces at Cochrane, Alberta. Canadian Journal of Earth Sciences 5(6):1455–66.
The prominent terraces N of the Bow River at Cochrane, Alberta, divide readily into an upper and a lower set. The three upper terraces were constructed between 19,000 and 15,000 years ago, when the Wisconsinan glaciation was near its maximum. Due to inhospitable conditions then, these upper terraces contain few or no fossils. During a warm phase that followed, glaciers shrank and the Bow River deepened and enlarged its valley until ice readvanced about 12,000 years ago. Before glaciers in the Bow and its tributary valleys retreated again, more than 10,000 years ago, Bow River had deposited much fill in its valley. Since then, the river has carved the five lower terraces from that fill. The valley fill (Bighill Creek Formation) is a valuable source of gravel, sand, and vertebrate fossils. Most of the latter came from sands called the Clarke Pit Member, which extends through much of the formation. The vertebrate fossils found in the lower set of terraces were buried during a period of strong ice advance, and represent a fauna that developed during the preceding interstade under climatic conditions resembling those now prevailing [see Churcher 1968a]. A few bones may be found in gravels of the terraces lying below the Clarke Terrace. The number of vertebrate fossils recovered depends directly on the intensity with which the gravel pits are worked, and the care and interest taken in recovering fossils.

1193. Stalker, A.M. (1969a): Geology and age of the early man site at Taber, Alberta. American Antiquity 34(4):425–8.
Human (*Homo sapiens*) bones were found nearly 60 ft below prairie level in sands underlying till at Woodpecker Island Bluff along the E bank of Oldman River about 3 mi N of Taber, Alberta (49°51′15″N, 112°09′20″W). As the till was deposited by Wisconsinan ice that spread over the area more than 22,000 years ago, the bones are at least that old. Correlation with other bluffs along Oldman River indicates the bones are probably more than 37,000 years old.

1194. Stalker, A.M. (1969b): Quaternary stratigraphy in southern Alberta, Report II: Sections near Medicine Hat. Geological Survey of Canada Paper 69(26):1–28.
The author started work in the Medicine Hat area in 1959 with a reconnaissance survey of bluffs along the river, at which time bones (Mammalia) were found in some exposures. Since 1963, the work has continued each summer. During the summers of 1965–7 the author was accompanied by Dr C.S. Churcher, University of Toronto, who supervised the collection and study of the vertebrate fossils. Vertebrate remains are from Surprise, Mitchell, and Island bluffs along the South Saskatchewan River N of Medicine Hat, Alberta (about 50°08′N, 110°38′W). The Basal Gravel and Sand is the oldest surficial deposit of the region, lying directly above bedrock, and contains remains of: camels like the Kansas camel (*Camelops kansanus*); Scott's horse (*Equus scotti*), which Churcher believed confirmed a Kansan date for the unit; mammoth (*Mammuthus*); sloth (Xenarthra); and a possible dog (Canidae). Bones found in the overlying Carbonaceous Deposits have yet to be identified. The Lower Sands were

laid down mainly by a slow-flowing river and contain remains of: prairie dog (*Cynomys* cf. *Cynomys ludovicianus*); Columbian mammoth (*Mammuthus* cf. *Mammuthus columbi* = *Mammuthus* cf. *Jeffersoni*); Mexican horse (*Equus* cf. *Equus conversidens*); and camel (*Camelops*). The overlying Lag Gravel has yielded a few bones, including horse (*Equus*). The Upper Sands and Gravels overlying the Lag Gravel is about 100 ft thick at Mitchell and Island bluffs and 22 ft thick at Surprise Bluff. This unit was deposited over a long time in either shallow lakes or in a broad, interglacial valley by a sluggish, aggrading river. A gravel bed found about two-thirds of the way up the deposit at Mitchell and Island bluffs is called the Artifact Band because of the abundance of chipped chert stones scattered through it. It has been the most prolific of all the deposits in numbers of fossils produced – by the end of 1968 having yielded 1000 bone specimens (mostly from Mitchell Bluff). Wood and mollusc shell dates indicate that the Artifact Band is >30,000 BP and >36,000 BP, respectively. The mammalian fauna (Table II) from the Upper Sands and Gravels includes: Spruce Grouse (*Canachites ?canadensis*); humans (*Homo sapiens*) – indirect evidence; hare (*Lepus*); black-tailed prairie dog (*Cynomys* cf. *Cynomys ludovicianus*); Richardson's ground squirrel (*Spermophilus ?richardsonii* = *Citellus ?richardsonii*); other rodents (Rodentia); dog (*Canis*); fox (*Vulpes*); Columbian mammoth (*Mammuthus columbi* = *Mammuthus jeffersoni*); Mexican horse (*Equus conversidens*); horse (*Equus*); western camel (*Camelops hesternus*); wapiti (*Cervus elaphus* = *Cervus canadensis*), caribou (*Rangifer tarandus*); pronghorn antelope (*Antilocapra americana*); mountain sheep (*Ovis* cf. *Ovis canadensis*); and giant bison (*Bison* cf. *Bison latifrons*). The abundant fauna 'indicates deposition during warm, interglacial conditions, probably in Sangamonian time.' The Lowest Intertill Deposits (between Lowest Till and Contorted Till) have yielded bones only at Island Bluff including: ?horse (*?Equus*); ?mammoth (*?Mammuthus*); possibly artiodactyl (?Artiodactyla); and perhaps elsewhere, woolly mammoth (*Mammuthus primigenius*). The Contorted Till, probably deformed by overriding ice, is Wisconsinan in age. Overlying it and underlying the Brown Till are Intertill Deposits evidently prolific in mammal bones and probably of Mid-Wisconsinan age. Most of the bones have yet to be studied, and only horse (*Equus*), camel (Camelidae), and artiodactyls (Artiodactyla) have been identified. Postglacial deposits overlying the Brown Till were probably laid down 15,000 to 13,000 BP, and have yielded horse and woolly mammoth teeth.

1195. Stalker, A.M. (1977): Indications of Wisconsin and earlier man from the southwest Canadian Prairies. Annals of the New York Academy of Sciences 288:119–36.

In addition to the Taber child (estimated age >37,000 BP) described in Stalker (1969a), other evidence in the form of fractured cherts, found in several Sangamonian sites in southern Alberta and Saskatchewan, may indicate human (*Homo sapiens*) presence in North America as early as 120,000 BP. [The indirect evidence of human remains listed here is questionable as fractured cherts do not necessarily confirm human presence, and the Taber child has since been radiocarbon dated at 3680 BP; see Cohen 1982.] This >37,000 date is much earlier than approximately 12,000 BP, which represents the first confirmed human occupation of this continent. These cherts seem to have been fractured artificially, and are present at several localities. Sangamonian vertebrate remains found in the Surprise, Mitchell, and Island bluffs along the South Saskatchewan River, N of Medicine Hat, Alberta (50°8′N, 110°38′W) include those described in Stalker (1969b), and are updated here: grouse (*Canachites*), hawk (*?Buteo*); human (*Homo sapiens*) – indirect evidence; ground sloth (*Megalonyx*); eastern cottontail (*Sylvilagus floridanus*); hare [white-tailed jackrabbit? (*Lepus* cf. *Lepus townsendii*)]; pocket gopher (*Thomomys* cf. *Thomomys talpoides*); black-tailed(?) prairie dog (*Cynomys* cf. *Cynomys ludovicianus*); Richardson's ground squirrel (*Spermophilus richardsonii* = *Citellus richardsonii*); vole (*Microtus*); muskrat (*Ondatra zibethicus*); porcupine (*Erethizon dorsatum*); American mink (*Mustela vison*); coyote (*Canis* cf. *Canis latrans*); wolf (*Canis lupus*); dire wolf (*Canis* cf. *Canis dirus*); red fox (*Vulpes vulpes*); raccoon (*Procyon lotor*); lynx (*Felis lynx* = *Lynx canadensis*); American lion (*Panthera leo atrox* = *Felis atrox*); Jefferson's mammoth (*Mammuthus columbi jeffersoni*); Niobrara horse (*Equus niobrarensis*); Mexican horse (*Equus conversidens*); neotropical(?) horse (*Amerhippus*); long-legged llama (*Hemiauchenia* = *Tanupolama*); western camel (*Camelops hesternus*); unidentified Cervidae, like moose (cf. *Alces*); deer (*Odocoileus*); wapiti (*Cervus elaphus* = *Cervus canadensis*); caribou (*Rangifer tarandus*); pronghorn

(*Antilocapra* cf. *Antilocapra americana*); bighorn sheep (*Ovis canadensis*); and giant bison (*Bison* cf. *Bison latifrons*). The Reservoir Gully site Unit A, within the city limits of Medicine Hat, Alberta (50°01′55″N, 110°44′05″W) contains remains of: human (*Homo sapiens*) – indirect evidence; dire wolf (*Canis* cf. *Canis dirus*); sabretooth cat (*Smilodon californicus*); mammoth (*Mammuthus*); western camel (*Camelops hesternus*); and Mexican horse (*Equus conversidens*). Bone from Reservoir Gully (Unit A) yielded radiocarbon dates of 15,200 ± 250 (GSC-1399-1, collagen) and 23,200 ± 300 BP (GSC-1399-2, apatite). Wood from Unit A of adjacent Evilsmelling Bluff yielded radiocarbon dates of 24,490 ± 200 (GSC-205), 25,000 ± 800 (GSC-1370), and 28,630 ± 800 (GSC-578). The Bliss Pit site sand and gravel unit, 0.4 km SW of Fort Qu'Appelle, Saskatchewan (50°46′N, 103°48′W), contained mammal remains 90% of which were of horse (*Equus*) bones, as well as chipped cherts. The Saskatoon site, 3 km N of Saskatoon, Saskatchewan, on the W side of the South Saskatchewan River (52°10′N, 106°35′W) contained remains of: human (*Homo sapiens*) – indirect evidence; mammoth (*Mammuthus*); bison (*Bison*); Niobrara(?) horse (*Equus* cf. *Equus niobrarensis*); and unidentified Cervidae.

1196. Stalker, A.M. (1984): Ice age bones – A clue; exotic animals as diverse as Africa's once roamed Canada's great plains. Geos 13(2):11–14.

The author describes the geological history of the Medicine Hat, Alberta, and Wellsch Valley, Saskatchewan, faunas described in the Stalker, Churcher, and Hill (1982) wall chart. An air photograph of the area along the South Saskatchewan River showing the positions of the most important Medicine Hat Quaternary vertebrate localities is included, as is a foldout of the wall chart and a photograph of Stalker excavating a mammoth (*Mammuthus*) vertebra.

1197. Stalker, A.M. (1996): Origin of the vertebrate fossil sites near Medicine Hat, Alberta. *In*: Palaeoecology and Palaeoenvironments of Late Cenozoic Mammals: Tributes to the Career of C.S. (Rufus) Churcher (K.M. Stewart and K.L. Seymour, eds.). University of Toronto Press, Toronto. pp. 247–58.

Quaternary beds exposed along the South Saskatchewan River near Medicine Hat have yielded vertebrate fossils from nine age levels. The oldest predates glaciation in the region and could be up to 1 Ma. The youngest are Late postglacial and bones are still being buried and preserved on some river slip-off slopes. The Medicine Hat fossil beds appear to encompass more than half of the Quaternary period. Most prairie fossil sites yield only one age of fossils. This paper describes the geology of the Medicine Hat region and explains the factors that contributed to the multiplicity of fossil beds.

1198. Stalker, A.M., and Churcher, C.S. (1970): Deposits near Medicine Hat, Alberta, Canada. Geological Survey of Canada Surveys and Mapping Branch (wall chart).

This wall chart contains a stratigraphic cross-section including the Quaternary component (Kansan 1.5 Ma–present) and a composite section of successive faunas for the Medicine Hat, Alberta, area. The faunas shown include: unidentified frog or toad (Anura); grouse (*Canachites*); hawk (*Buteo*); human (*Homo sapiens*) – indirect evidence; ground sloths (*Megalonyx*) and a smaller form (cf. *Nothrotheriops* = cf. *Nothrotherium*); eastern cottontail (*Sylvilagus floridanus*); hare [white-tailed jackrabbit? (*Lepus* cf. *Lepus townsendii*)]; pocket gopher (*Thomomys* cf. *Thomomys talpoides*); black-tailed(?) prairie dog (*Cynomys* cf. *Cynomys ludovicianus*); Richardson's ground squirrel (*Spermophilus richardsonii* = *Citellus richardsonii*); vole (*Microtus*); muskrat (*Ondatra zibethicus*); porcupine (*Erethizon dorsatum*); American mink (*Mustela vison*); coyote (*Canis* cf. *Canis latrans*); wolf (*Canis lupus*); dire wolf (*Canis* cf. *Canis dirus*); red fox (*Vulpes vulpes*); raccoon (*Procyon lotor*); lynx (*Felis lynx* = *Lynx canadensis*); sabretooth cat (*Smilodon californicus*); American lion (*Panthera leo atrox* = *Felis atrox*); mammoth (*Mammuthus*); woolly mammoth (*Mammuthus primigenius*); Columbian mammoth (*Mammuthus columbi*); Jefferson's mammoth (*Mammuthus columbi jeffersoni*); imperial mammoth (*Mammuthus imperator*); southern mammoth (*Mammuthus meridionalis* = *Mammuthus imperator haroldcooki*); Niobrara horse (*Equus niobrarensis*); Mexican horse (*Equus conversidens*); stilt-legged horse (*Equus calobatus*); giant horse (*Equus* cf. *Equus giganteus*); Scott's horse (*Equus scotti*); neotropical(?) horse (*Amerhippus*); unidentified camelid (Camelidae); long-legged llama (*Hemiauchenia Tanupolama*); Steven's long-legged llama (*Hemiauchenia stevensi*); camel (*Camelops*); western camel (*Camelops hesternus*); Minidoka camel (*Camelops *minidokae*); unidentified Cervidae, like moose (cf. *Alces*); deer (*Odocoileus*); wapiti (*Cervus elaphus* = *Cervus canadensis*); caribou (*Rangifer tarandus*); pronghorn (*Antilocapra* cf. *Antilocapra americana*); bighorn sheep (*Ovis canadensis*); bison (*Bison* cf. *Bison bison*); and giant bison (*Bison* cf. *Bison latifrons*). [See also Stalker 1969b, 1996, and Mott and Stalker 1972.]

1199. Stalker, A.M., and Churcher, C.S. (1972): Glacial stratigraphy of the southwestern Canadian Prairies; the Laurentide record. 24th International Geological Congress, Section 12. pp. 110–19.
The authors describe the glacial stratigraphy of the Medicine Hat district, Alberta, and the Wellsch Valley site 40 km N of Swift Current, Saskatchewan. The faunal list for the basal gravel and clay units of the Medicine Hat section (from 50°04′N, 110°52′W to 50°09′N, 110°36′W, Mid-Kansan) include: unidentified ground sloth (Mylodontidae), ground sloth (*Nothrotheriops* = *Nothrotherium*), beaver (*Castor* cf. *Castor canadensis*), extinct large cat (Felidae), Etruscan(?) wolf (*Canis* cf. *Canis etruscus*), southern mammoth (*Mammuthus meridionalis* = *Mammuthus imperator haroldcooki*), Scott's horse (*Equus scotti*), stilt-legged horse (*Equus* cf. *Equus calobatus*), stilt-legged(?) llama (cf. *Hemiauchenia blancoensis* = *Tanupolama blancoensis*), Minidoka camel (*Camelops minidokae*), and extinct pronghorn (Antilocapridae). The faunal list for the basal unit of the Wellsch Valley site (50°39′50″N, 107°52′30″W, Aftonian >1.5 Ma) include: unidentified ground sloth (Xenarthra), Gazin's marsh rabbit(?) (cf. *Hypolagus limnetus*), Meade's ground squirrel (*Spermophilus* cf. *Spermophilus meadensis* = *Citellus* cf. *Citellus meadensis*), unidentified pocket gopher (Geomyidae), unidentified vole (Cricetidae), bone-eating dog (*Borophagus diversidens*), lynx or bobcat (*Felis* cf. *Felis lynx* = *Lynx* cf. *Lynx rufus*), southern mammoth (*Mammuthus meridionalis* = *Mammuthus imperator* cf. *Mammuthus imperator haroldcooki*), Pacific horse (*Equus pacificus*), complex-toothed horse (*Equus complicatus*), peccary (*Platygonus*), camel (*Camelops*), extinct pronghorn (Antilocapridae), and shrubox(?) (Bovidae cf. Ovibovini).

1200. Stalker, A.M., Churcher, C.S., and Hill, R.S. (1982): Ice age deposits and animals from the southwestern part of the Great Plains of Canada. Geological Survey of Canada Surveys and Mapping Branch Miscellaneous Report 31 (wall chart).
This wall chart displays the stratigraphic positions all Pleistocene vertebrate fauna of the Medicine Hat area, Alberta, and Wellsch Valley site, 40 km N of Swift Current, Saskatchewan. The chart, which contains a list of species showing their individual time span and time of greatest abundance, includes: Canadian toad (*Bufo hemiophrys*), grouse (*Canachites*), hawk (cf. *Buteo*), ground sloth (cf. *Megalonyx*), Shasta ground sloth (*Nothrotheriops* cf. *Nothrotheriops shastense* = *Nothrotherium* cf. *Nothrotherium shastense*), Harlan's ground sloth (*Glossotherum harlani* = *Paramylodon harlani*), Gazin's marsh rabbit(?) (cf. *Hypolagus limnetus*), eastern cottontail (*Sylvilagus floridanus*), hare [white-tailed jackrabbit? (*Lepus* cf. *Lepus townsendii*)], pocket gopher (*Thomomys*), northern pocket gopher (*Thomomys talpoides*), white-tailed(?) prairie dog (*Cynomys* cf. *Cynomys leucurus*), Meade prairie dog (*Cynomys* cf. *Cynomys meadensis*), Richardson's ground squirrel (*Spermophilus richardsonii* = *Citellus richardsonii*), Meade ground squirrel (*Spermophilus* cf. *Spermophilus meadensis* = *Citellus* cf. *Citellus meadensis*), beaver (*Castor canadensis*), vole (*Microtus*), extinct voles (*Microtus paroperarius* and *Allophaiomys*), Osborne's extinct tree vole (*Pliophenacomys osborni*), Kansas southern bog lemming (*Synaptomys kansasensis*), muskrat (*Ondatra zibethicus*), porcupine (*Erethizon dorsatum*), black-footed ferret (*Mustela nigripes*), eastern spotted skunk (*Spilogale* cf. *Spilogale putorius*), bone-eating dog (*Borophagus diversidens*), canid (*Canis*), coyote (*Canis* cf. *Canis latrans*), wolf (*Canis lupus*), dire wolf (*Canis dirus*), red fox (*Vulpes vulpes*), raccoon (*Procyon lotor*), bobcat (*Felis* cf. *Felis rufus* = *Lynx* cf. *Lynx rufus*), lynx (*Felis lynx* = *Lynx canadensis*), sabretooth cat (*Smilodon fatalis*), American lion (*Panthera leo atrox* = *Felis leo atrox*), mammoth (*Mammuthus*), woolly mammoth (*Mammuthus primigenius*), Columbian mammoth (*Mammuthus columbi*), imperial mammoth (*Mammuthus imperator*), southern mammoth (*Mammuthus meridionalis* = *Mammuthus imperator haroldcooki*), Niobrara horse (*Equus niobrarensis*), Mexican horse (*Equus conversidens*), stilt-legged horse (*Equus calobatus*), giant horse (*Equus giganteus*), Scott's horse (*Equus scotti*), Pacific horse (*Equus pacificus*), complex-toothed horse (*Equus complicatus*), neotropical(?) horse (*Amerhippus*), Cope's peccary (*Platygonus* cf. *Platygonus bicalcaratus*), unidentified camelid (Camelidae), plains llama (*Hemi-

auchenia), Steven's plains llama (*Hemiauchenia stevensi*), Holloman's plains llama (*Hemiauchenia hollomani*), camel (*Camelops*), western camel (*Camelops hesternus*), Irvingtonian camel (*Camelops minidokae*), unidentified Cervidae, such as moose (cf. *Alces*), Nearctic deer (*Odocoileus*), white-tailed deer (*Odocoileus* cf. *Odocoileus virginianus*), wapiti (*Cervus elaphus* = *Cervus canadensis*), caribou (*Rangifer tarandus*), moose-like deer (cf. *Alces*), four-horned pronghorn (*Tetrameryx*), pronghorn (*Antilocapra* cf. *Antilocapra americana*), bighorn sheep (*Ovis canadensis*), helmeted muskox (*Bootherium bombifrons* = *Symbos cavifrons*), woodland shrubox (cf. *Euceratherium*), bison (*Bison*), bison (*Bison* cf. *Bison bison*), and giant bison (*Bison* cf. *Bison latifrons*).

1201. Stanley-Brown, J. (1892): Geology of the Pribilof Islands. Geological Society of America Bulletin 3:496–500.
The author notes (p. 496) that the Bering Sea is extremely shallow – an elevation of [the land of] 300 or 400 ft 'would convert most of the present sea bottoms into a vast verdure-covered tundra whose gently undulating surface would be dotted with lakes and intersected by sluggish winding streams. Upon such a land surface the four tiny islets to which this brief sketch refers would appear as conspicuous elevations.' He also mentions (p. 499) a mammoth (*Mammuthus*) tusk that was found in the sands of Northeast Point, Saint Paul Island, and a mammoth tooth reported as coming from the shores of Saint George Island. He cautions that the value of this evidence is questionable because there is not a foot of earth upon either island, save that which has resulted from the decomposition of native rock and the decay of vegetation. [See Ray 1971.]

1202. Stansfield, J. (1913): Excursion A8. Mineral deposits of the Ottawa District. *In*: Excursions in the Neighbourhood of Montreal and Ottawa. Geological Survey of Canada Guide Book No. 3:81–115.
Marine shells found in the Leda Clay of Champlain Sea age 'indicate that the climate at that time was similar to that of the present climate of the Labrador coast.' At Green Creek near Ottawa, the clay is rich in calcareous nodules that have yielded many fossils, especially capelin (*Mallotus villosus*) and other fishes (p. 86). [See McAllister et al. 1988.]

1203. Stansfield, J. (1915): Fossils from the Pleistocene and Recent deposits. *In*: The Pleistocene and Recent Deposits of the Island of Montreal. Geological Survey of Canada Memoir 73:65–8.
This is a complete list of all Quaternary fossil species found in Leda Clay on the island of Montreal, and includes: harp seal (*Phoca groenlandica*) and white whale (*Delphinapterus leucas* = *Beluga catodon*).

1204. Stauning, J. (1775): Kort Beskrivelse over Grönland. Viborg.
According to Bennike (1997, p. 901) the author reports capelin (*Mallotus villosus*) remains in calcareous nodules from West Greenland.

1205. Stefansson, V. (1918): The activities of the Canadian Arctic Expedition from October, 1916, to April, 1918. Geographical Review 6:354–69.
The author reports: 'We found a tusk [presumably of a woolly mammoth (*Mammuthus primigenius*)], but no other mammoth remains, in Melville Island.' [See also Blake 1974 and Harington 1990b.]

1206. Stefansson, V. (1921): The Friendly Arctic. MacMillan, New York. 784 pp.
The author reports (pp. 510–11) that on the E coast of Melville Island, Nunavut, they found 'the nearly complete and unfossilized skeleton of a bowhead whale [*Balaena mysticetus*] eight or ten feet above sea level and a hundred and fifty yards inland. This means an elevation, for we know through observation that their skeletons always lodge not at the upper level of wave action, as is the case with driftwood, but at the level of low tide or even lower, where they are commonly buried by sand. This skeleton was undoubtedly originally so buried at or below the level of low tide. The land has since risen and wind and other forces have carried the sand away. Though the skeleton is unfossilized it is thousands of years old, for the same forces which can preserve the flesh of mammoths [*Mammuthus*] so that it may be examined today and is still fresh, can more easily preserve tree trunks and skeletons from decaying.'

1207. Steinbring, J. (1966): A Scottsbluff projectile point from Manitoba. Wisconsin Archaeologist 47(1):2–7.
A Scottsbluff-type projectile point, evidence of human

(*Homo sapiens*) occupation, was found at the surface of an eroded sandy knoll about 32 km W of the Glacial Lake Agassiz shoreline. The site is near Epinette Creek, a tributary of the Assiniboine River, and 5 km N of its confluence with the river. Humans made this artifact from 11,000 to 7000 BP.

1208. Stephenson, R.O., Gerlach, S.C., Guthrie, R.D., Harington, C.R., Mills, R.O., and Hare, G. (2001): Wood bison in Late Holocene Alaska and adjacent Canada: Paleontological, archaeological and historical records. *In*: People and Wildlife in Northern North America: Essays in Honor of R. Dale Guthrie (S.C. Gerlach and M.S. Murray, eds.). British Archaeological Reports International Series 944: 124–58.

Bison have a long history in northwestern North America, with various species and subspecies occupying the area for 400,000 years or more. Bison were one of the most common large mammals in interior Alaska and northwestern Canada during the last 100,000 years. This was also true for northern Europe and northern Asia. The steppe bison (*Bison priscus*) that occupied the North for millennia evolved into a smaller-horned form, the wood bison (*Bison bison athabascae*), shortly after 10,000 years ago. In contrast to some earlier scenarios presuming a Late Pleistocene/Early Holocene decline or demise of bison, radiocarbon dates (Table 1 lists 58 radiocarbon dates on bison between about 12,000 and 200 BP) indicate that bison persisted in Alaska and Yukon until just before European settlement. Prior to this study, there were no published historic accounts describing the presence of bison in Alaska during the Late Holocene. Oral accounts by Athabascan elders and radiocarbon evidence are integrated with zooarchaeological and paleontological data, as well as written records from the region suggesting a pattern of decline and eventual disappearance of bison during the Late Holocene.

1209. Stepp, D. (1995): Mesa tools linked with lower 48. Mammoth Trumpet 10(4):1, 6–11.

This description of the history of the Mesa site, one of the oldest well-documented Paleoindian (*Homo sapiens*) sites in North America, is based on a telephone interview with the site's discoverer. Situated in the periglacial northern flank of Alaska's Brooks Range, it occupies a mesa with an unobstructed 360-degree view of hundreds of square kilometres. Discovered in 1979 by Michael Kunz, the site has been studied and excavated intensively since 1991. Fifteen radiocarbon (AMS) dates from the site range from 11,700 to 9700 BP. Numerous lithic tools, mainly chert, have also been recovered. There is a striking similarity between the artifacts from the Mesa site and those from the Agate Basin culture of eastern Wyoming. Given the ages of the two complexes, the Mesa culture could reasonably be the progenitor of Agate Basin. Kunz even suggests that the Mesa culture could represent a previously unrecognized culture that may predate Clovis hunters, long believed to be the oldest inhabitants of North America. To date, less than 3% of the Mesa site has been excavated.

1210. Sternberg, C.M. (1930): New records of mastodons and mammoths in Canada. Canadian Field-Naturalist 44:59–65.

This is an extensive list of all known mastodon (*Mammut americanum*) sites in southwestern Ontario, and mammoth (*Mammuthus*), woolly mammoth (*Mammuthus primigenius*), imperial mammoth (*Mammuthus imperator*), and Columbian mammoth (*Mammuthus columbi*) sites in Canada. [See Harington and Shackleton 1978.]

1211. Sternberg, C.M. (1951): White whale and other Pleistocene fossils from the Ottawa Valley. Annual Report of the National Museum for the Fiscal Year 1949–1950, National Museum of Canada Bulletin 123:259–61.

White whale (*Delphinapterus leucas*) remains were found: near Pakenham, Ontario; Ottawa International Airport (= Uplands); Vanier, Ontario (= Ottawa East); Jock River, Ontario; and from sandpits 24 km S of Ottawa, Ontario. Harp seal (*Phoca groenlandica*) remains were found at Baie Comeau, Quebec, and at Odell's Brickyard, Ottawa, Ontario. Harbour porpoise (*Phocoena phocoena*) remains were found in sand at Pontiac Point in Pontiac County, Quebec. A partial skeleton of a large whale [considered to be a little piked whale (*Balaenoptera acutorostrata*) by Laverdière (1950) but more likely representing a finback whale (*Balaenoptera physalus*) according to the whale expert R. Kellogg] was found near Daveluyville, Quebec. [See Harington 1988.]

1212. Sternberg, C.M. (1956): A harp seal from the

Leda Clay west of Hull, Quebec. Canadian Field-Naturalist 70:97.
Seven harp seal (*Phoca groenlandica*) dorsal vertebrae were found in a clay nodule at a campsite 37 km W of Hull, Quebec. Harp seal remains have also been found at Wright's Brickyard just W of the centre of Hull and, along with remains of capelin (*Mallotus villosus*) and American marten (*Martes americana*), at Green Creek, E of Ottawa, Ontario. [The author presents no clear evidence that the first two specimens are from harp seals, so the first should be called *Phoca*, and the second represents a ringed seal (*Phoca hispida* – Harington and Sergeant 1972); only the Green Creek specimen is clearly referable to a harp seal.]

1213. Sternberg, C.M. (1963): Additional records of mastodons and mammoths in Canada. National Museum of Canada Natural History Papers No. 19:1–11.
This list updates the one in Sternberg (1930), including: all known mastodon (*Mammut americanum*) sites in southwestern Ontario, and proboscidean (Proboscidea), including woolly mammoth (*Mammuthus primigenius*), and Columbian mammoth (*Mammuthus columbi*) sites in Canada.

1214. Stewart, K. (2000): Summary of paper CARD: Canada's first answer to FAUNMAP. *In*: Transitions in Zooarchaeology: New Methods and New Results (A summary of session papers presented at the Canadian Archaeological Association Annual Conference, May). Canadian Zooarchaeology Number 18:4.
CARD is an acronym for the Canadian Archaeological Radiocarbon Database, a digital database available on the World Wide Web through the home page of the Canadian Archaeological Association <www.canadianarchaeology.com>. It contains nearly 7500 radiocarbon dates, many of which are presented in their paleoenvironmental contexts on a related Web site called Mapping Ancient History, hosted by the Geological Survey of Canada. Although CARD mainly delivers radiocarbon dates, it also records vertebrate faunas that are associated with dated samples. This aspect of the database can be construed as Canada's first answer to FAUNMAP, a digital database of mammalian taxa that have been documented in more than 3200 archaeological and paleontological sites in the continental United States. [See Morlan 1999; FAUNMAP 1994.]

1215. Stewart, T.G., and England, J. (1986): An Early Holocene caribou antler from northern Ellesmere Island, Canadian Arctic Archipelago. Boreas 15:25–31.
A caribou (*Rangifer tarandus pearyi*) antler was found in a 30-km-long structural trough at the N end of Piper Pass, northern Ellesmere Island, Nunavut (82°36.5′N, 68°31′W). It was radiocarbon dated at 8415 ± 135 BP (S-2501, collagen), which agrees with 40 other radiocarbon-dated mollusc shell specimens from the site. The authors conclude that this antler may indicate the presence of caribou in ice-free parts of the Canadian High Arctic during full-glacial time. Other Holocene mammal specimens from the Arctic Archipelago (Northwest Territories and Nunavut) and Greenland include: remains of caribou (*Rangifer tarandus*), muskox (taxon unspecified), and seal (Phocidae) from raised-beach deposits 12 m asl near Alert, Ellesmere Island; remains of caribou (*Rangifer tarandus*) from Baillie Islands, E of the Mackenzie delta; remains of caribou (*Rangifer tarandus*) from Jørgen Brønlund Fjord, Pearyland, northern Greenland, radiocarbon dated at about 8750 BP (K-3865); and a muskox (*Ovibos moschatus*) pelvic bone from Banks Island, radiocarbon dated at 10,660 ± 170 BP (GSC-240, plant detritus). [See Meldgaard 1986, 1991.]

1216. Stewart, T.G., and Hourston-Wright, J. (1990): 6500 BP Oldsquaw duck (*Clangula hyemalis*) from northern Ellesmere Island, Arctic Archipelago, Canada. Arctic 43(3):239–43.
A nearly complete Oldsquaw duck (*Clangula hyemalis*) skeleton, including partially preserved feathers, was found in Holocene marine deposits (prodeltaic, emerged via isostatic rebound, now 16.5 m asl) at Clements Markham Inlet, Ellesmere Island, Nunavut (82°38′N, 68°05′W). The Oldsquaw originally died in the inlet, sank to the ocean bottom, and was buried by prodeltaic sediments in a paleowater depth of 38 m. Isostatic rebound and erosion then exposed the site. Allochthonous terrestrial plant matter radiocarbon dated at 6400 ± 60 BP (SI-4314) was found 2 m lower in the section from the specimen, and the authors estimate that the duck lived about 6500 BP. The other Pleistocene bird fossil from the region was a Dovekie (*Alle alle*) in similar sediments and radiocarbon dated at 41,500 ± 1460 BP (GSC-2786). [See Harington 1990b.]

1217. Stirling, J. (1977): Digging Old Crow … the old-

est evidence of man in North America? North/Nord 24(6):2–5.
A 1976 interview with Dr Bill Irving [University of Toronto and head of the Northern Yukon Research Program (NYRP)] and Jerry Fitzgerald (National Museum of Natural Sciences fieldworker) at Old Crow, Yukon, includes the following data: (1) The Old Crow Flats is a rich, natural laboratory for several scientific disciplines studying the Pleistocene. (2) NYRP is studying the historical biogeography of NE Beringia with special reference to humans (*Homo sapiens*) – its work dovetailing with that of the Beringian Refugium Project (involving representatives of the National Museum of Natural Sciences, Canadian Museum of Man, Geological Survey of Canada, and the University of Alberta). (3) The frozen banks of Old Crow River have yielded man-made (*Homo sapiens*) implements dating between 30,000 and 20,000 BP, as well as remains from about two dozen species of mammals including lions (*Panthera leo atrox*), mammoths (*Mammuthus*), bison (*Bison*), muskoxen (*Ovibos moschatus*), horses (*Equus*), and giant beavers (*Castoroides ohioensis*). (4) About 12,000 bones, implements, and other samples were removed by NYRP from Old Crow Basin in 1975, the figure being somewhat smaller in 1976. (5) Jerry Fitzgerald, with Charlie Thomas of Old Crow, collected Quaternary vertebrate remains from locations that had yielded fossils before, as well as searching for new sites. Some of the items found were as small as a fish (Pisces) operculum or as large as a mammoth (*Mammuthus*) neck vertebra. [Black and white photographs of small specimens, such as a muskrat (*Ondatra zibethicus*) jaw and giant beaver tooth, contrast with pictures of horse leg bones and a mammoth atlas vertebra and teeth.]

1218. Stock, C. (1942): A ground sloth in Alaska. Science 95(2474):552–3.
A ground sloth (*Megalonyx*) phalanx (FAM:30844) from Cripple Creek, Alaska, found by Otto Geist in 1941, is described. It represents the only known ground sloth material from Alaska and places the species considerably farther N than its previously known distribution. [See McDonald et al. 2000.] A list of 10 species of associated fossil mammals from the region is given.

1219. Stock, C., and Richards, H.G. (1949): A *Megalonyx* tooth from the Northwest Territories, Canada. Science 110:709–10.
A ground sloth (*Megalonyx* cf. *Megalonyx jeffersonii*) tooth (Academy of Natural Sciences of Philadelphia 15208; Figure 1) and a fragment of mastodon (*Mammut americanum*) tooth were found at Lower Carp Lake, N of Great Slave Lake, Northwest Territories (63°35′N, 114°10′W). Interest in the specimen arises largely from its occurrence so far N on the North American continent, and so far E of the only other *Megalonyx* specimen (from the Fairbanks area of Alaska) known from northwestern North America. [The remains may be of last (Sangamonian) interglacial age – see Harington 1977a and McDonald et al. 2000.]

1220. Storck, P.L. (1988): Recent excavations at the Udora site: A Gainey/Clovis occupation site in southern Ontario. Current Research in the Pleistocene 5:23–4.
Projectile points confirming human (*Homo sapiens*) presence, dated at 12,500 to 10,400 BP, were found at the Udora site, southern Ontario [see Spiess and Storck 1990]. Fauna associated with the site includes: Cervidae cf. white-tailed deer (*Odocoileus virginianus*) or cf. caribou (*Rangifer tarandus*), hare or rabbit (Leporidae), and fox (Canidae).

1221. Storck, P.L. (1997): Exploring possible ecological factors underlying early-Paleoindian lithic procurement: The 1996 Red Wing Project, southern Ontario, Canada. Current Research in the Pleistocene 14:79–80.
The Red Wing site is an Early Paleoindian lithic-procurement site in the geological source area of Fossil Hill Formation chert on the Blue Mountain/Kolapore uplands in the southern Georgian Bay region of southern Ontario. The site was adjacent to a shallow pond in a spruce parkland environment during the Late Pleistocene from 11,600 ± 100 BP (TO-5785) to 10,170 ± 90 BP (TO-5784). It was hypothesized that the chert for tools was obtained opportunistically while people (*Homo* sapiens) were exploiting fish (Pisces) during the spawning season. The discovery of gravers, scrapers, the tip of a fluted point, and an unfluted preform indicate an occupation by Parkhill-complex people.

1222. Storck, P.L., and Spiess, A.E. (1994): The significance of new faunal identifications attributed to an Early Paleoindian (Gainey Complex) occupation at

the Udora site, Ontario, Canada. American Antiquity 59(1):121–42.
Analysis of calcined bone from the Udora site (adjacent to a strandline of Glacial Lake Algonquin) in S-central Ontario, Canada, indicates that the subsistence of Early Paleoindian (Gainey Complex) peoples (*Homo sapiens*) in the lower Great Lakes region included a mix of both large and small mammals: caribou (*Rangifer tarandus*), hare (*Lepus*), and arctic fox (*Alopex lagopus*). All specimens are catalogued and stored at the Royal Ontario Museum. The bones, along with numerous stone tools, came from Feature 1 of the site, from a depth of approximately 24 to 53 cm below ground surface. Although an effort was made to age the bone, it was found to contain less than 1/1000th of its original protein and was not datable. The Gainey complex is estimated to date between 11,000 and 10,500 years BP in the Great Lakes region. The presence of arctic fox and other paleoecological data indicate that the Paleoindian occupation at Udora occurred in a spruce parkland environment between 10,000 and 10,500 years ago.

1223. Storer, J.E. (1972): Mammals of the Hand Hills Formation of southern Alberta: Preliminary faunal list. Blue Jay 30(2):119–20.
This paper lists all fossils found in the Hand Hills Formation, 24 km E of Drumheller, Alberta, which range in age from Upper Tertiary to Holocene. Those that are Irvingtonian (Pleistocene) or later include: elephants (Elephantidae); Mexican(?) horse (*Equus* cf. *Equus conversidens*); ground squirrel (*Spermophilus* cf. *Spermophilus richardsonii*); vole (*Microtus* cf. *Microtus pennsylvanicus*); pocket gopher (*Geomys*); and rabbit or hare (Leporidae). The author notes that a horse astragalus from the Hand Hills conglomerate, reported by Russell (1958), probably represents *Equus*, and demonstrates that the conglomerate must be, in part, Blancan or younger.

1224. Storer, J.E. (1975): Pleistocene prairie dog in south-central Alberta. Blue Jay 33(4):247.
A black-tailed prairie dog (*Cynomys ludovicianus*) mandible was found in the Hand Hills Formation (Irvingtonian/Late Kansan or younger), 24 km E of Drumheller, Alberta, 410 km NW of its present range. The presence of this species indicates that the Pleistocene fauna of the Hand Hills lived in a climate warmer and drier than prevails there today.

1225. Storer, J.E. (1976): Mammals of the Hand Hills Formation, southern Alberta. *In*: Essays on Palaeontology in Honour of Loris Shano Russell (C.S. Churcher, ed.). Royal Ontario Museum Life Sciences, Miscellaneous Publications, Toronto. pp. 186–209.
This paper gives systematic descriptions of the Hand Hills fauna listed in Storer (1972) including figures and photographs of the specimens. Ranges of the Pleistocene taxa (Table 9) indicate an Irvingtonian (Late Kansan) or younger age.

1226. Storer, J.E. (1989): Quaternary period. *In*: Geological History of Saskatchewan, Saskatchewan Museum of Natural History, Government of Saskatchewan, Regina. pp. 75–82.
This chapter provides a popular perspective on the geological history of Saskatchewan, which includes vertebrate faunal history. The Early Pleistocene (about 1.8 Ma) is represented by the Wellsch Valley local fauna: ground sloth (*Megalonyx*), rabbit (Leporidae), white-tailed prairie dog (*Cynomys leucurus*), pocket gopher (*Thomomys*), voles (*Microtus*), bog lemming (*Synaptomys*), bone-eating dog (*Borophagus*), coyote (*Canis* cf. *Canis latrans*), imperial mammoth (*Mammuthus imperator*), horse (*Equus*), Pacific horse (*Equus pacificus*), peccary (*Platygonus*), camel (*Camelops*), and shrubox (*Euceratherium*). The Early Pleistocene Wellsch Valley fauna indicates a drier-than-usual (and more open than usual) local environment. The Sangamonian (about 120,000 BP) is represented by two sites. The Echo Lake fossils of Fort Qu'Appelle include: badger (*Taxidea taxus*), mammoth (*Mammuthus*), Scott's horse (*Equus scotti*), camel (*Camelops*), Scott's moose (*Alces scotti* = *Cervalces roosevelti*), giant bison (*Bison latifrons*), helmeted muskox (*Bootherium bombifrons*), unidentified fish (Pisces), and grouse (*Canachites*). The fossils indicate a mixed parkland environment. The Saskatoon fauna includes: rabbit (Leporidae), ground squirrel (*Spermophilus*), prairie dog (*Cynomys*), pocket gopher (*Thomomys*), voles (*Microtus*), muskrat (*Ondatra zibethicus*), coyote (*Canis latrans*), fox (Canidae), mammoth (*Mammuthus*), Scott's horse (*Equus scotti*), camel (*Camelops*), deer (*Odocoileus*), and pronghorn (*Antilocapra americana*). Evidently the fauna of this Sangamonian deposit indicates that the landscape was dominated by open grassland with trees and shrubs in lower areas along river valleys. The Late Wisconsinan is represented by several

sites and includes: American lion (*Panthera leo atrox*), sabretooth cat (*Smilodon*), horse (*Equus*), Mexican horse (*Equus conversidens*), mastodon (*Mammut americanum*), and mammoth (*Mammuthus*).

1227. Storer, J.E. (1995): Additions to the mammalian palaeofauna of Saskatchewan, Canada. *In*: Vertebrate Fossils and the Evolution of Scientific Concepts: Writings in Tribute to Beverly Halstead, by Some of His Many Friends (W.A.S. Sarjeant, ed.). Gordon and Breach Publishers, Australia. pp. 555–67
The Pleistocene vertebrate record for Saskatchewan is rather scanty, which is surprising considering that Quaternary deposits cover nearly the entire province. The early part of the record is represented by the Wellsch Valley local fauna, nearly 1.8 Ma. Three important later assemblages occur in Saskatchewan: (1) The Fort Qu'Appelle local fauna, made up almost entirely of large mammals. It has been assigned to the later part of the Sangamonian interglacial, but neither stratigraphic orientation nor faunal content is conclusive: it apparently belongs to the latter part of the Rancholabrean Land Mammal Age. (2) The Saskatoon local fauna near the N end of Saskatoon. (3) The Riddell local fauna at Sutherland near Saskatoon. The last two are in the upper member of the Floral Formation. These assemblages are Rancholabrean, but despite the impressive diversity of the Riddell local fauna, they cannot be correlated precisely with other North American assemblages.

1228. Stuart, A.J. (1991): Mammalian extinctions in the Late Pleistocene of northern Eurasia and North America. Biological Reviews 66:453–62.
Extinctions at the end of the Pleistocene were unique, in that species lost were nearly all large terrestrial mammals. Although global in extent, Late Pleistocene extinctions were most severe in North America, South America, and Australia and moderate in northern Eurasia. In Africa, where nearly all of the Late Pleistocene megafauna survives, losses were slight. In North America, the main reason that losses were severe and sudden is probably the close coincidence of late-glacial climatic/environmental changes with the arrival of Clovis hunters.

1229. Sutton, R. (1982): Farmer plows up a mammoth. Toronto Star, May 8.
A team from the Royal Ontario Museum found the grey, pitted bones ploughed up on the farm of Terry Poole (20 km NW of Stratford, Ontario) to be those of an adolescent mammoth (*Mammuthus*). 'It was obvious we had something' Poole said. 'No cow or horse could have left behind an 8-in molar or neck vertebra 10 in across.'

1230. Szabo, B.J., Stalker, A.M., and Churcher, C.S. (1973): Uranium-series ages of some Quaternary deposits near Medicine Hat, Alberta, Canada. Canadian Journal of Earth Sciences 10(9):1464–9.
^{230}Thorium and ^{231}Palladium dates were obtained on 10 bone samples from Quaternary deposits near Medicine Hat, Alberta. The youngest of these Uranium-series dates on Mexican horse (*Equus conversidens*) long bones (9500 ± 1500 years) agrees reasonably well with a radiocarbon date [11,200 ± 200 BP (GSC-805), chiefly on bison (*Bison*) and horse (*Equus*) bones]. Where samples were collected from more than one bed at the same bluff, the results are in correct chronological order. Probable horse (*Equus*?) bone fragments were dated at 195,000 years (previously estimated to be Yarmouthian, 500,000 BP). Large bones of mammoth? (*Mammuthus*?) and assorted bone fragments estimated to be 25,000–18,000 BP yielded Uranium-series dates of 73,000 ± 6000 years and >64,000 years respectively. Fragments of 'Sangamon Interglacial' Columbian mammoth (*Mammuthus columbi jeffersoni* = *Mammuthus jeffersoni*) bones and a Niobrara horse (*Equus niobrarensis*) ilium and humerus estimated to be 200,000–70,000 BP were dated at 72,000 ± 6000 years and >71,000 years respectively. Imperial mammoth (*Mammuthus imperator*) bone fragments estimated to be 1 Ma–600,000 BP were dated at >200,000 years, whereas Scott's horse (*Equus scotti*) humerus fragments estimated to be 1 Ma–500,000 BP gave a Uranium-series age of >167,000 years. Two samples of bones estimated at 38,000 BP did not yield Uranium-series ages. Except for the late-glacial and 'Sangamon Interglacial' bones, these results reveal large discrepancies with the estimated ages of the beds that yielded the bones.

1231. Taber, S. (1943): Perennially frozen ground in Alaska: Its origin and history. Geological Society of America Bulletin 54(10):1433–1548.
This extensive treatise on perennially frozen ground in Alaska includes remarks on vertebrate remains in creek-valley silts ['muck'] (pp. 1486–90). Fossil bones of Pleis-

tocene mammals (Mammalia) are abundant and widely distributed in the valley silts of the nonglaciated area. They have been found from small headwater streams to the coasts. In valley deposits, remains of bison – mainly steppe bison (*Bison priscus* = *Bison crassicornis* = *Bison alleni*) – were most abundant, with those of horse (*Equus*) and mammoth (*Mammuthus*) next. The mammoth has been reported from more localities in Alaska than any other mammal, probably because of the commercial value of the tusks, ease of identification, and the great popular interest in them. He lists extinct species including: 'dire wolf' – probably wolf (*Canis lupus* = *Aenocyon* or *Canis dirus alaskensis*); short-faced bear (*Arctodus simus yukonensis* = *Arctotherium yukonensis*); steppe bison, western bison (*Bison bison occidentalis* = *Bison occidentalis*); muskox (*Ovibos moschatus* = *Ovibos maximus* = *Ovibos yukonensis*); helmeted muskox (*Bootherium bombifrons* = *Symbos tyrrelli* and *Bootherium sargenti*); western camel (*Camelops hesternus*); American lion (*Panthera leo atrox* = *Felis atrox alaskensis*); horse (*Equus alaskae*); woolly mammoth (*Mammuthus primigenius*); mastodon (*Mammut americanum* = *Mastodon americanus*); and doubtfully, Columbian mammoth (*Mammuthus columbi* = *Parelephas columbi*). The presence of caribou (*Rangifer tarandus*), moose (*Alces alces* = *Alce*), wapiti (*Cervus elaphus* = *Cervus*), and bear (*Ursus*), all late immigrants that have persisted with little or no modification, indicates a land connection with Asia during or shortly before silt and gravel deposition. The woolly mammoth is another Pleistocene immigrant. The author mentions the presence of woolly mammoths on many islands rising from the arctic coastal shelf of Siberia, as well as the record from Banks Island [actually Melville Island, Northwest Territories – see Kindle 1924 and Harington 1990b], King Island (40 mi SW of Seward Peninsula), St George and St Paul islands in the Pribilof group, and Unalaska in the Aleutians. Taber mentions that the range of the beaver (*Castor canadensis*) coincides with the limits of forest growth and notes that beaver-cut wood and even beaver dams are found in frozen silts beyond the present forest line: (1) a beaver dam from Elephant Point; (2) a beaver dam in silts on Old Glory Creek near Deering; (3) beaver-gnawed wood in ice in the Kougarok region; and (4) beaver-gnawed wood in silts on Candle Creek (p. 1488). Nests and tunnels of several species of burrowing rodents, including arctic ground squirrel (*Spermophilus parryii* = *Citellus*), are common in frozen silts near Fairbanks. Other observations include: (1) teeth and coprolites from silts at Dome Creek near Fairbanks believed to be from a shrew (Soricidae); (2) several bones with dried flesh adhering to them from frozen silt near Fairbanks, reported by Wilkerson (1932); (3) part of a mammoth skeleton with attached tendons, flesh, skin, hair, and wood from Elephant Point; (4) mammoth bones and about 300 lbs of fat in frozen clay near the mouth of Naknek River; (5) a few bison bones exposed in a new road cut on the SW side of Dome Creek about 950 ft asl; (6) the scarcity of articulated bones and complete skeletons – except for rodents that died in their burrows (Taber remembers seeing only four bison bones articulated among several tons of bones of larger mammals seen in 1935); (7) some bones and tusks coated with vivianite (part of a biochemical process characteristic of weathering in a wet environment); (8) horncores are common but hornsheaths are rare; (9) generally the proportion of well-preserved to poorly preserved bones tends to increase northwards; (10) soft parts are preserved only when exceptional conditions occur of prompt burial after death, usually in a floodplain or delta deposits as a result of floods, but occasionally under mudflows or in bogs and ponds; and (11) the abundance of plant and animal fossils in Alaskan silts, compared with their occurrence in Pleistocene deposits farther S, must be attributed to better preservation in a colder climate.

1232. Tanton, T.L. (1931): Fort William and Port Arthur, and Thunder Cape map-areas, Thunder Bay District, Ontario. Geological Survey of Canada Memoir 167:83–4.

Relics of prehistoric man were found at places in the Kaministikwia River valley according to Peter McKellar. The relics were found during July 1918 in an excavation made by the Canada Car and Foundry Company about 80 ft N from the turning basin at Westport. About 12 bones of a mammal (Mammalia) and a finely made copper spearhead were found together some 40 ft below the surface. Other relics, including copper tools and a human (*Homo sapiens*) skeleton, were found in the vicinity earlier. According to L.M. Lambe, vertebrate paleontologist with the Geological Survey of Canada, and [C.M.] Sternberg, preparator of geological specimens, a bone from the 1918 collection is from a cloven-hoofed animal – perhaps a buffalo (*Bison*) or domestic cattle (*Bos taurus*). These authorities referred another dozen bones from the collection to a horse (*Equus*) – Sternberg considered that most of them belonged to an individual.

1233. TenBrink, N.W., and Waythomas, C.F. (1978): Late Wisconsin glacial chronology of north-central Alaska Range: A regional synthesis and its implications for early human settlement. National Geographic Research Reports 19:15–33.
The authors provide a radiocarbon date of 13,500 ± 100 BP (QL-1365) on a mammoth (*Mammuthus*) bone from the Teklanika River valley in central Alaska.

1234. Terasmae, J. (1960): A palynological study of the Pleistocene interglacial beds at Toronto, Ontario. Geological Survey of Canada Bulletin 56, Part II: 32–41.
Palynological and paleontological studies show that the interglacial (possibly Sangamon) Don Beds at Don Valley Brickyard, Toronto, were laid down when the mean annual temperature was about 5°F warmer than present, and the overlying Scarborough Beds (tentatively assigned to the St Pierre interval of the Wisconsinan glaciation), in the same vicinity, were deposited when it was about 10°F colder. The author (p. 35) mentions Don Bed mammal fossils based on Coleman (1933): groundhog (*Marmota monax* = *Arctomis monax*), antlers of two species of deer (Cervidae), bison (*Bison*), grizzly? bear (*Ursus*), and giant beaver (*Castoroides ohioensis*).

1235. Terasmae, J. (1967): Notes on the Quaternary palaeoecological problems in the Yukon Territory, and adjacent regions. Geological Survey of Canada Paper 67-46:1–12.
The Yukon Territory is important for North American Quaternary research because parts remained unglaciated throughout the Quaternary. Geological and palynological evidence (see locality map, Figure 1) indicates a long but often broken record. The recycling of deposits by erosion and redeposition, and their frozen conditions tend to complicate the picture further. Evidence also indicates that Quaternary climatic changes in Yukon have been approximately contemporaneous with, and probably of the same magnitude as, those farther S. Paleoecological evidence indicates that human (*Homo sapiens*) migration could have occurred from Asia to Yukon and thence southward in Late Quaternary time, some 14,000 years ago, or possibly even earlier in the Mid-Wisconsinan nonglacial interval.

1236. Thibaudeau, C. (1990): Les frères Daviau ont trouvé le fossile du premier morse de la Mer de Champlain. La Presse, Montréal, Octobre 7:B5.
Brothers Jacques and Raymond Daviau, who farm at Sainte-Julienne, Quebec, near Rawdon, found part of a cranium of a walrus (*Odobenus rosmarus*) in Champlain Sea deposits between 12,700 and 9700 years old. The blade of their tractor hit the skull at a depth of 5 ft in clay according to Jacques. His brother Raymond said it looked like a walrus. Their nephew Marc expressed a scientific interest in the fossil and contacted geologist Michel Bouchard, who sent it to fossil expert Dick Harington, who noted in a 1977 publication, 'It is difficult to understand why a walrus fossil has never been found in Champlain Sea deposits.' The specimen was to be deposited in the Musée de la Civilisation du Québec in Quebec City after it had been carefully studied. [See Bouchard et al. 1993.]

1237. Thomas, D. (1972): Bison skulls span 24,000-year period. Calgary Herald, January 4. p. 32.
This article mentions two fossil bison finds identified by Dr Len Hills, a University of Calgary geologist: (1) a horncore of ancient bison (*Bison bison antiquus* = *Bison antiquus*) from a gravel pit just S of Calgary dating to about 8000 BP; (2) part of a large western bison (*Bison bison occidentalis* = *Bison occidentalis*) skull uncovered by Three Hills, Alberta, farmer Stan Milan while he was digging a waterline to his barn. The age of this skull will be tested by radiocarbon dating, but could be as old as 24,000 BP according to the age of the soil in which it lay. [See Shackleton and Hills 1977.]

1238. Thorson, R.M., and Guthrie, R.D. (1984): River ice as a taphonomic agent: An alternative hypothesis for bone 'artifacts.' Quaternary Research 22:172–88.
The annual freeze-up and violent breakup of temperate and high-latitude rivers produces unique geomorphic features recognizable in ancient sediments. The forces and materials involved in the breakup process would modify contained bones in ways similar to those attributed to human activity [the argument is particularly directed to provide an alternative explanation for controversial bone artifacts from Old Crow Basin, Yukon, mentioned by

Bonnichsen (1979), Irving (1978), and Morlan (1980)]. An intuitive taphonomic model is developed to explain how river ice may affect bones and to predict the modifications expected. Partial testing of the model by experiments [mainly consisting of dragging bones in frozen matrix, weighted from above, over various road surfaces near Fairbanks] was designed to simulate river breakup-produced bone modifications such as fractures, flakes, facets, striations, isolated cuts, and polish. The taphonomic effects of river ice must be discounted before such features can be interpreted as human (*Homo sapiens*) in origin.

1239. Thorson, R.M., and Hamilton, T.D. (1977): Geology of the Dry Creek site; a stratified early man site in interior Alaska. Quaternary Research 7:149–76.
The Dry Creek archaeological site on the Nenana River contains a stratified record of Late Pleistocene human (*Homo sapiens*) occupation in central Alaska. Four archaeological components occur within a sequence of multiple loess and sand layers that together form a 2-m cap above weathered glacial outwash. The two oldest components appear to be of Late Pleistocene age and contain the bones of extinct game animals. Geological mapping, stratigraphic correlations, radiocarbon dating, and sediment analyses indicate that the basal loess units formed part of a widespread blanket that was associated with an arctic-steppe environment and with stream aggradation during waning phases of the last major glaciation of the Alaska Range. These basal loess beds contain artifacts for which radiocarbon dates and typological correlations suggest a time range of perhaps 12,000–9,000 years ago. A long subsequent episode of cultural sterility was associated with waning loess deposition and development of a cryoturbated tundra soil above shallow permafrost. The youngest archaeological component, which is associated with the deepest forest soil, indicates intermittent human occupation of the site between about 4700 and 3400 BP. Stratigraphic columns are illustrated. All of the radiocarbon dates are taken from peat or charcoal. A list of the lithic artifacts and faunal remains (collected in 1974 and 1975) is presented in table form. Mammal fossils collected from the two oldest components include: horse (*Equus*) teeth, bison (*Bison*), and an unidentifiable proboscidean bone. Table 3 shows archaeological material relative to faunal remains on a stratigraphic chart.

1240. Thorson, R.M., Dixon, E.J., Smith, G.S., and Batten, A.R. (1981): Interstadial proboscidean from south-central Alaska: Implications for biogeography, geology, and archeology. Quaternary Research 16:404–17.
In 1980 a large proboscidean femur, probably mammoth (*Mammuthus*), was found *in situ* in a bluff exposure at the mouth of the Tyone River in the northwestern part of the Copper River Basin, Alaska (see map, Figure 1). The regional setting, stratigraphy, radiocarbon chronology, flora, and implications of the fossil locality, which represents the first documented occurrence of Pleistocene terrestrial mammalian fauna in southern Alaska, are described. Radiocarbon dates and stratigraphic relations at the site indicate that the sediments containing the fossil accumulated during the transition from interstadial to glacial conditions during terminal Middle Wisconsinan time. During this interval the immediate vicinity was unforested and large areas of S-central Alaska may have been available for faunal and possibly human (*Homo sapiens*) habitation. This fossil find, dated at 29,450 ± 610 BP (DIC-1819), extends the known range for Pleistocene mammals and possibly for steppe-tundra conditions southward at least 150 km, and suggests that mountain passes through the Alaska Range to the N were ice-free during the last part of the Middle Wisconsinan interstadial. Other radiocarbon dates taken of wood and peat from above and below the femur bracket the age nicely. The date from the femur was the only bone date; all the other dates were taken from wood or peat. A measured stratigraphic column is illustrated, as is the bone *in situ* and some plant macrofossil fragments. Pleistocene terrestrial mammals have been reported from many areas of southern Alaska [about 11 localities are mentioned (pp. 404–5), but none has been adequately documented, making this record significant].

1241. Thorsteinsson, R. (1958): Cornwallis and Little Cornwallis Islands, District of Franklin, Northwest Territories. Geological Survey of Canada Memoir 294:1–134.
This paper mainly deals with geology; however, in a section discussing emerged strand lines (p. 17), the author describes four complete walrus skeletons found several miles from the present shore at between 225 and 300 ft elevation. The exact locations [three on Cornwallis Island and one on Little Cornwallis Island, Nunavut (p. 17)] of

the specimens are shown on Thorsteinsson's Map 1054A. [See Dyke et al. 1999, and Harington 1975c.]

1242. Tobien, H. (1984): Migrations of proboscideans and lagomorphs (Mammalia) across the Bering Land Bridge in the Late Cenozoic. *In*: Beringia in the Cenozoic Era (V.L. Kontrimavichus, ed.). Amerind Publishing Co. Pvt. Ltd., New Delhi. pp. 327–38.
This chapter includes discussions of possible reasons for migration, distribution and evolution of proboscideans and lagomorphs in the Late Cenozoic.

1243. Tokaryk, T.T. (1992): Additions to the Cenozoic avian fauna of Saskatchewan. Saskatchewan Museum of Natural History Occasional Short Notes No. 4:1–4.
In a list of fossil bird taxa from Saskatchewan, the author mentions a Ruffed Grouse (*Bonasa umbellus*) from Rancholabrean deposits (p. 3). [See Weigel 1963.]

1244. Tomenchuk, J. (1988): Engineering use-wear studies of Early Paleoindian tools. Current Research in the Pleistocene 5:35–6.
Studies on the wear and hardness of stone tools and associated wood at archaeological (*Homo sapiens*) sites in southwestern Ontario indicate the many activities that occurred at the Fisher and Udora [see Storck and Spiess 1994] sites (e.g., butchering, hide working, and probably fish (Pisces) processing at the Fisher site). Wood (jack pine and white birch, tentatively identified on their apparent hardnesses) from the Fisher site suggests an occupation of that site between 10,600 and 10,500 BP. Age estimates on some 80 flake tools at the Udora site were unsuccessful, but the author maintains that with further study, these methods will be a useful data source in the future.

1245. Tovell, W.M., and Deane, R.E. (1966): Grizzly bear skull: Site of a find near Lake Simcoe. Science 154:158.
A well-preserved grizzly bear (*Ursus arctos = Ursus arctos-horribilis*) skull was found in strongly cross-bedded gravel 9 m beneath the local grade at the Ennis Gravel Pit, which is located on a Glacial Lake Algonquin Beach – the Ardtrea Beach near Lake Simcoe (lot 12, concession II, Orillia twp., Simcoe County). The specimen was radiocarbon dated at 11,700 ± 250 BP (no lab number). [See Peterson 1965a, b, and Pigott 1999.]

1246. Trylich, C., and Bayrock, L.A. (1966): *Bison occidentalis* Lucas found at Taber, Alberta, Canada. Canadian Journal of Earth Sciences 3(7):987–95.
An articulated, incomplete western bison (*Bison bison occidentalis = Bison occidentalis*) skeleton was found at the top of the southern bank of the Oldman River in an old oxbow lake alluvial deposit of the same river, near Taber, Alberta (49°48.6′N, 112°10.5′W). Wood from a sand layer below the bison remains yielded radiocarbon dates of 11,000 ± 250 (S-68) and 10,500 ± 200 BP (GSC-3). A geological section showing the position of the bison find is included (Figure 1). The specimen's horncore measurements were compared with those of other western bison, plains bison (*Bison bison bison*), wood bison (*Bison bison athabascae*), steppe bison (*Bison priscus = Bison preoccidentalis*), and ancient bison (*Bison bison antiquus = Bison antiquus*), and were closest to those of western bison. Parkland vegetation prevailed at the time, and a pebble chopper indicates the presence of humans (*Homo sapiens*).

1247. Tyrrell, J.B. (1892): Report on north-western Manitoba with portions of the adjacent districts of Assiniboia and Saskatchewan. Geological Survey of Canada Annual Report (New Series) 5, Part I:1E–235E.
The author quotes Richardson's (1854) description of the scapulae of an American mastodon (*Mammut americanum = Mastodon giganteus*) obtained through Dr Rae 'from Swan River (properly Swan River District) near the western side of the basin of Lake Winnipeg' in Manitoba (p. 13E). There is a further note on the mastodon specimens (pp. 129E–130E): 'The Indians allege that at this point [51°43′N, a short distance upstream from the confluence of the E branch of Shell River on a wooded terrace on the W side of the river], huge bones were found at the bottom of a landslide and were brought to the officer in charge [W.J. Christie of Brockville, Ontario], by whom they were forwarded to England.' Christie informed the author that the bones were shoulder blades and that in 1853, some years after the first bones were brought in, he visited the place, and examined the spot carefully where a shoulder blade was taken out of the river at low water. A landslide had occurred and carried the bones into the river at low water. 'I found from questioning my guide that the Indians had collected the bones, and burnt them on the bank, from superstition, and buried

what would not burn. I examined the spot where they had buried the bones, but what remained crumbled to pieces when touched.' The author then quotes Richardson's (1854) description, concluding: 'The probability is that the Swan River (Swan River district) bones belonged to the *Mastodon giganteus* and that the range of that species must be extended northward in Rupert's Land to the fifty-second parallel of latitude.' Near the largest ridge on Wilson River (p. 95E) the author describes a section of 'post-tertiary deposits.' In it, at the top of a fine rounded gravel unit, 'sharp-angled and apparently chipped fragments of quartzite were found. Two of them were roughly in the shape of arrow heads, and one was a very thin flake [*Homo sapiens*?]. The others were very irregular. Associated with them was a broken fragment of one of the toe bones of a moose (*Alces*) or elk (*Cervus*). These irregular fragments were all lying with their longer axes horizontal, and were overlaid by a sandy loam' unit 2 ft 3 in thick at the top of the section.

1248. Tyrrell, J.B. (1912): The gold of the Klondike. Transactions of the Royal Society of Canada 6(3):29–53.
The author lists all Pleistocene mammalian fossils found in the Klondike region, Yukon (between 63° and 64° N, and to a point 80 km E of 141° W). They were found in gold placer deposits, gravels, and muck, and included (p. 35): steppe bison (*Bison priscus* = *Bison crassicornis*), western bison (*Bison bison occidentalis* = *Bison occidentalis*), woolly mammoth (*Mammuthus primigenius* = *Elephas primigenius*), mastodon (*Mammut americanum* = *Mastodon americanus*), helmeted muskox (*Bootherium bombifrons* = *Symbos Tyrrelli*), tundra muskox (*Ovibos moschatus*), wapiti (*Cervus elaphus* = *Cervus Canadensis*), horse (*Equus*), sheep (*Ovis*), moose (*Alces* = *Alce*), caribou (*Rangifer tarandus*), Yukon giant short-faced bear (*Arctodus simus yukonensis* = *Arctotherium yukonense*), bear (*Ursus*), and canid (*Canis*).

1249. Unger, R. (1979): 30,000 years ago. Yukon Indian News 6(5):8.
This newspaper article reports the finding of a caribou (*Rangifer tarandus*) flesher from the Old Crow Basin, Yukon, described in Morlan and Cinq-Mars (1982) [it was radiocarbon dated at 27,000 + 3000 - 2000 BP (GX-1640, bone apatite), but it has since been re-dated as a Late Holocene artifact at 1350 ± 150 BP (RIDDL-145, AMS) in Nelson et al. (1986)]. A general reconstruction of the Yukon Pleistocene fauna includes: American lion (*Panthera leo atrox*), sabretooth cat (*Smilodon*) [probably refers to the scimitar cat (*Homotherium serum*)], horse (*Equus*), woolly mammoth (*Mammuthus primigenius*), steppe bison (*Bison priscus*), helmeted muskox (*Bootherium bombifrons*), sheep (*Ovis*), giant moose (*Alces latifrons*), and spotted skunk (*Spilogale*) [probably refers to the extinct short-faced skunk (*Brachyprotoma obtusata*)].

1250. Uyeno, T., and Miller, R.R. (1963): Summary of Late Cenozoic freshwater fish records for North America. Occasional Papers of the Museum of Zoology, University of Michigan, No. 631:1–34.
Pleistocene fishes (Pisces) of Canada listed on p. 26 are: (1) sticklebacks (Gasterosteidae) from the Ottawa River valley near Ottawa, Ontario; (2) 'Salmonidae' [actually Atlantic cod (*Gadus* cf. *Gadus morhua*) – see Gruchy 1971] from Goose River [now Rivière des Outardes], Quebec; (3) minnows (Cyprinidae) from Lillestrom, 16 mi SW of Moose Jaw, Saskatchewan (age about 10,000 years).

1251. Valdez, R., and Krausman, P.R. (1999): Description, distribution, and abundance of mountain sheep in North America. *In*: Mountain Sheep of North America (R. Valdez and P.R. Krausman, eds.). University of Arizona Press, Tucson. pp. 3–22.
Most pertinent is the section on evolution, in which the fossil record is mentioned as well as genetic evidence of relationships of the different kinds of sheep. Two principal hypotheses are given regarding the evolution and origin of North American wild sheep: (1) the reversed migration hypothesis of Severtzov, which postulates that the first mountain sheep to reach North America across the Bering Isthmus were argaliforms, which, after becoming isolated S of the continental ice sheets, evolved into pachyceriforms. After a postglacial period, these southern pachyceriforms reached Beringia and then gave rise to the Siberian snow sheep (*Ovis nivicola*) and Dall sheep (*Ovis dalli*); and (2) Cowan's hypothesis that pachyceriforms evolved in Beringia rather than in North America S of the continental ice sheets. All fossil North American sheep described so far are pachyceriforms, which supports the latter hypothesis. Based on paleontological, biogeographical, and chromosomal data, Korobitsyna et al.

(1974) conclude: (1) that pachyceriforms evolved in a Beringian refugium; (2) that pachyceriforms migrated southward into the western United States when the ice sheets melted – probably during the Sangamonian interglacial; (3) that isolation of pachyceriforms in eastern Siberia from those in Alaska by waters of Bering Strait probably in Sangamonian time, but possibly as late as the Holocene (10,000 BP), led to differentiation of those two species whose modern descendants are snow sheep and Dall sheep; (4) that isolation of pachyceriforms in the western United States during the Wisconsinan glaciation led to the separation of modern Rocky Mountain bighorn and desert bighorn (*Ovis canadensis*) subspecies.

1252. Van Dyke, S., Hanna, S., Unfreed, W., and Neal, B. (1990): Section 3.0 1989 Oldman Dam prehistoric archaeological mitigation: Final report. Provincial Museum of Alberta, Archaeological Survey, Edmonton. Report by Bison Historical Services Ltd., Permit 89-25. 756 pp.

In this report, along with other archaeological sites, J-Crossing (DjPm-16) is mentioned. It is situated on the S bank of Crowsnest River about 4 km upstream from its confluence with the Oldman River, Alberta. A buried soil at a depth of 265–220 cm yielded an assemblage assigned to the Late Paleoindian Alberta complex. A radiocarbon date on bison (*Bison*) bone collagen yielded a normalized age of 9600 ± 210 BP (AECV-746C). Associated with bison are: ground squirrel (*Spermophilus*), rabbits and hares (Leporidae), canids (Canidae), and a small ungulate – possibly a deer (Cervidae?). [See Morlan 1999.]

1253. Vangengeim, E.A. (1967): The effect of the Bering Land Bridge on the Quaternary mammalian faunas of Siberia and North America. *In*: The Bering Land Bridge (D.M. Hopkins, ed.). Stanford University Press, Stanford, California. pp. 281–7.

A general paleoecological and paleoclimatic reconstruction of Beringia and its mammalian fauna includes species common to both North America and Asia: saiga antelope (*Saiga tatarica*), caribou [= reindeer (*Rangifer tarandus*)], woolly mammoth (*Mammuthus primigenius*), steppe bison [= Siberian (*Bison priscus*)], arctic fox (*Alopex lagopus*), collared lemming (*Dicrostonyx torquatus*), wapiti [= elk (*Cervus*)], tundra muskox (*Ovibos moschatus*), helmeted muskox (*Bootherium*), and grizzly [= brown bear (*Ursus arctos*)]. The author suggests that there were at least three periods of intercontinental faunal exchange. Analysis of fossil faunas indicates a possible connection between the Asiatic and American continents during the Villafranchian. A close similarity between faunas of eastern Siberia and northwestern North America suggests Beringian land connections existed from the beginning of the Samarov (= Illinoian) glaciation until the end of the Sartan (Late Wisconsinan) glaciation – perhaps with interruptions [e.g., during the Sangamonian interglacial] that need not have been lengthy.

1254. Vereshschagin, N.K., and Baryshnikov, G.F. (1982): Paleoecology of the mammoth fauna in the Eurasian Arctic. *In*: Paleoecology of Beringia (D.M. Hopkins, J.V. Matthews, Jr, C.E. Schweger, S.B. Young, and V. Stanley, eds.). Academic Press, Toronto. pp. 267–80.

This review outlines what is known about the ecology and cold-weather adaptations of the Late Pleistocene Mammoth Fauna in the Siberian Arctic. This fauna is broken into two groups of species: (1) woolly mammoth (*Mammuthus primigenius*), horse (*Equus lenensis*), woolly rhinoceros (*Coelodonta antiquitatis*), steppe bison (*Bison priscus*), yak (*Bos grunniens = Poephagus mutus*), tundra muskox (*Ovibos moschatus*), saiga antelope (*Saiga tatarica*), and cave lion (*Panthera leo spelaea = Panthera spelaea*) are discussed as species now extinct in the region; (2) reindeer/caribou (*Rangifer tarandus*), arctic ground squirrel (*Spermophilus parryii = Citellus parryii*), brown lemming (*Lemmus*), collared lemming (*Dicrostonyx*), narrow-skulled vole (*Microtus gregalis*), and arctic fox (*Alopex lagopus*) are species that survived the Pleistocene and still inhabit the region. An analysis of the morphology and ecology of the Late Pleistocene Mammoth Fauna of arctic Eurasia indicates that they lived in a cold, dry climate in steppe and steppe tundra characterized by hard, frozen ground. The decimation of the mammoth fauna came as a result of temperature increases during interstades within the Valdai (i.e., Würmian of Europe, Wisconsinan of North America) cold interval and the establishment of taiga and tundra vegetation at the end of this interval. The animals surviving the extinctions were able to persist in the severe conditions of present-day tundra as a result, in some cases, of their capacity for long migrations and, in others, of physiological adaptations that enabled them to cope with deep snow and occasional winter thaws.

1255. Vibe, C. (1967): Arctic animals in relation to climatic fluctuations. Meddelelser om Grønland 170(5): 1–227.

Two finds of bowhead whales (*Balaena mysticetus*, pp. 83–4) from Greenland prove that this species existed in Davis Strait and Baffin Bay long before humans (*Homo sapiens*) reached these parts from the W: (1) the back of a cranium was excavated at 60 m asl at Holsteinsborg, by H.C. Pedersen in 1963, much earlier than a Sarqaq settlement there at 17 m asl; (2) the back of a cranium was excavated 6 m below the surface in permafrost of a river bank, 25 m asl and 560 m from the coast near Thule. A photograph of the specimen is shown (Figure 47). It is now in the Zoological Museum at Copenhagen. Muskoxen (*Ovibos moschatus*) seem to have been common in NE Greenland during several periods (pp. 158–61): (1) About 1900, A.G. Nathorst found an ancient skull that was so old that the horny substance was gone, and it was overgrown by lichen and so weathered that only the hardest parts remain, 'giving it the appearance of fossil skulls found in the deposits of glacial periods in Europe'. (2) The author noted a very heavy fragment of muskox skull (Figure 87), in front of tent rings at Hurry Fiord, that looked as if it had been fossilized. It is now in the Zoological Museum at Copenhagen. (3) An ancient cranial fragment collected in NE Greenland (Danmarkshavn?) was collected in 1908 by A.L.W. Manniche. [See Weidick 1974.]

1256. Vickers, J.R., and Beaudoin, A.B. (1989): A limiting AMS date for the Cody Complex occupation at the Fletcher site, Alberta, Canada. Plains Anthropologist 34(125):261–4.

Four Alberta and four Scottsbluff projectile points (*Homo sapiens*) were found, respectively, in a bone bed and from dugout spoil, at the Fletcher site in southern Alberta (49°35′30″N, 111°49′20″W). Their age was estimated, from stratigraphy and cultural indices, at between 11,000 and 7000 BP; Alberta points were used between 9650 and 8600 BP (Agenbroad 1978). A later AMS date yielded an age of 9380 ± 110 BP (TO-1097, plant detritus), which agrees well with the other estimates.

1257. Vilà, C., Leonard, J.A., Götherström, A., Marklund, S., Sandberg, K., Lidén, K., Wayne, R.K., and Ellegren, H. (2001): Widespread origins of domestic horse lineages. Science 291:474–7.

The authors analysed mitochondrial DNA control region sequences of 191 domestic horses and found a high diversity of matrilines. Sequence analysis of equids (Equidae) from archaeological sites and Late Pleistocene deposits showed that this diversity was not due to an accelerated mutation rate or an ancient domestication event. Consequently high mt DNA sequence diversity of horses implies an unprecedented and widespread integration of matrilines and extensive use and taming of wild horses. Among the horses sampled were eight from frozen sediments near Fairbanks, Alaska, radiocarbon dated between 28,000 and 12,000 BP, the samples being from metapodials in the American Museum of Natural History (FAM 4695, 4305, 6735, 6206, 133514, 6300, and 6171). Six of the eight sequences cluster in a group ancestral to modern sequences, possibly representing a sister taxon of the domestic horses or a lineage not present in modern domestic horses [perhaps these represent the most common Eastern Beringian Late Pleistocene horse (*Equus lambei*)]. However, the other two sequences cluster with another clade, and differ by as little as 1.2% from modern counterparts.

1258. Vincent, J.-S. (1989): Quaternary geology of the northern Canadian Interior Plains. *In*: Quaternary Geology of Canada and Greenland. Geology of Canada No. 1 (R.J. Fulton, ed.). Geological Survey of Canada, Ottawa. pp. 100–37.

A table (Table 2.8, p. 124) of Quaternary vertebrate remains from northern Yukon and western Northwest Territories, with localities and ages specified (age is Pleistocene if not otherwise indicated), includes: (1) arctic ground squirrel (*Spermophilus parryii*), lemmings (*Dicrostonyx* and *Lemmus*), vole (*Microtus*), and Yukon horse (*Equus lambei*), from Hungry Creek, Yukon. (2) dog (*Canis familiaris*); mammoth (*Mammuthus*); Yukon horse dated at 16,200 ± 150 BP (RIDDL-765); bison (*Bison* cf. *Bison priscus* = *Bison* cf. *Bison crassicornis*); helmeted muskox (*Bootherium bombifrons*); tundra muskox (*Ovibos moschatus*); and whale (Cetacea – pre-Early Wisconsinan), from Herschel Island. (3) woolly mammoth (*Mammuthus primigenius*), horse (*Equus*), caribou (*Rangifer tarandus*), saiga antelope (*Saiga tatarica*), tundra muskox, bowhead whale (*Balaena mysticetus*), ringed seal (*Phoca hispida*), and polar bear (*Ursus maritimus*) from Baillie Islands. (4) Yukon horse from the Beaufort Sea (Immerk), Richards Island, East Channel

(MacKenzie River) at Tununuk, and Summer Island. (5) mammoth from East Channel (MacKenzie River) Tununuk [also *Mammuthus* cf. *Mammuthus primigenius* dated at 19,440 ± 290 BP (I-8578)], Southern Eskimo Lakes (Middle Pleistocene), Garry Island, Kay Point, and Phillips Bay. (6) Proboscidea, like a mammoth from Nicholson Peninsula, Maitland Point, Inlet east of Maitland Point, Bathurst Peninsula, and Ballast Brook, Banks Island. (7) bison (*Bison*) from Richards Island, Tuktoyaktuk [like steppe bison (cf. *Bison priscus*)], Nicholson Peninsula, and Maitland Point. (8) giant moose (*Alces* cf. *Alces latifrons*) from Southern Eskimo Lakes (Middle Pleistocene). (9) caribou from Masik River, Banks Island. (10) collared lemming (*Dicrostonyx torquatus*) and ptarmigan-like bird (cf. *Lagopus*) from Morgan Bluffs, Banks Island (Middle Pleistocene). (11) tundra muskox from Bernard River, Banks Island dated at >34,000 BP (S-288), and Garry Island. (12) bowhead whale from Cape Wollaston, Victoria Island, dated at 9285 ± 140 BP (S-2729). (13) whale (Cetacea) from east of Cape Baring, Victoria Island, radiocarbon dated at 9780 ± 250 (S-2686) and 8565 ± 220 BP (S-2687), and E of Washburn Lake (Early Holocene), Victoria Island.

1259. Vinson, D. (1988): Preliminary report on faunal identifications from Trail Creek Caves. *In*: The Bering Land Bridge National Preserve: An Archaeological Survey (J. Scharf, ed.). National Park Service – Alaska Region, Research/Resources Management Report AR-14, Volume I:410–38.

Five caves at the Trail Creek Caves site (BEN-001; Figure 2), located some 45 km SW of Deering, Alaska, contained remains of arctic mammal (Mammalia) and bird (Aves) populations. Two caves also contained remains of Late Pleistocene large mammals. No cultural materials were recovered from the subsurface deposits [a bone lance socket (*Homo sapiens*) was found on the surface of one cave], and no evidence of butchering has been demonstrated on the faunal remains. Mammoth (*Mammuthus*) bones provided two collagen radiocarbon dates of 14,270 ± 950 BP (Beta-20027) and 11,360 ± 100 BP (Beta-13811). The presence of mammoth, as well as horse (*Equus*) and bison (*Bison*), indicates that at the end of the Wisconsinan glaciation the Trail Creek area could support grazing animals associated with a tundra/steppe environment. Taxa identified from the five Trail Creek caves (Tables I–III) include: bird (Aves), shrew (*Sorex*), tundra vole (*Microtus oeconomus*), narrow-skulled vole (*Microtus gregalis*), meadow vole (*Microtus pennsylvanicus*), collared lemming (*Dicrostonyx torquatus*), brown lemming (*Lemmus sibiricus*), red-backed vole (*Clethrionomys rutilus*), ground squirrel (*Spermophilus*), hare (*Lepus*), brown bear (*Ursus arctos*), fox (*Vulpes*), mammoth, caribou (*Rangifer tarandus*), horse, bison, and sheep – probably Dall sheep (*Ovis dalli*). There is great potential for the faunal record preserved at Trail Creek (over 2581 faunal elements) to add to the general knowledge of Beringian paleontology and paleoenvironment during the Late Pleistocene and Holocene. [See Larsen 1968, and West, C.F. 1996.]

1260. Vinson, D.M. (1993): Taphonomic analysis of faunal remains from Trail Creek Caves, Seward Peninsula, Alaska. MA thesis, Department of Anthropology, University of Alaska Fairbanks.

At Trail Creek Caves site, Alaska, caribou (*Rangifer tarandus*) bones dominate the faunal assemblage, comprising 52% of 450 bones from the site. Steppe bison (*Bison priscus*) remains in contrast, make up only 1.3%; Yukon horse (*Equus lambei* 0.2%); Dall Sheep (*Ovis dalli* 2.6%); woolly mammoth (*Mammuthus primigenius* 2%); bear (*Ursus* 2.4%); red fox (*Vulpes vulpes* 7.1%), arctic fox (*Alopex lagopus* 2.8%); hare (*Lepus* 22.4%); arctic ground squirrel (*Spermophilus parryii* 5.3%); hoary marmot (*Marmota caligata* 0.4%); and ptarmigan (*Lagopus* 0.7%). [See Yesner (2001), Table 1, from which these data are drawn.]

1261. Vorontsev, N.N., and Lyapunova, E.A. (1984): Genetics and problems of trans-Beringian connections of Holarctic mammals. *In*: Beringia in the Cenozoic Era (V.A. Kontrimavichus, ed.). Amerind Publishing Co. Pvt., Ltd., New Delhi. pp. 441–63.

The authors discuss the theoretical background to the use of genetic methods for paleobiogeographical reconstruction. Phylogenetically, variance systems involving genes, genomes, chromosomes, and karyotypes have the advantage of being almost immune to convergence. Convergence in boreal and arctic environments leads to similarity in the morphology of karyologically remote forms: *Microtus agrestis* (2n = 50) and *Microtus pennsylvanicus* (2n = 46); *Stenocranius gregalis* (2n = 36) and *Microtus miurus* (2n = 54); and *Sorex* from North America (2n = 29/20) and Siberia (2n = 32/30-36/37) – the latter once assigned to the single Holarctic species *Sorex arcticus*. Genetic evidence has been obtained for the repeated existence of Beringian associations, exemplified by the distribution of Palearctic genera such as marmot (*Marmota*), ground squirrel (*Spermophilus* = *Citellus*), and pika (*Ochotona*). In addition to pre-Pleistocene Beringian associations, the authors consider Pleistocene pre-Wisconsinan associations: e.g., the origins of mountain sheep [the bighorn (*Ovis canadensis*), Dall sheep (*Ovis dalli*), and snow sheep (*Ovis nivicola*) (p. 450)]; and the divergence of American lemmings (*Dicrostonyx*) into the collared lemming (*Dicrostonyx torquatus*) group, and the Ungava lemming (*Dicrostonyx hudsonius*) group. Among Late Pleistocene Beringian connections, the authors consider relationships among northern ground squirrels (*Spermophilus parryii* = *Citellus parryi*) on either side of Bering Strait (p. 451). The authors also discuss a karyological method of determining the direction of migration (p. 452).

1262. Waddington, J. (1979): An Introduction to Ontario Fossils. Royal Ontario Museum Encounter Series. 28 pp.
Most of this well-illustrated, popular booklet deals with invertebrate fossils. However, last (Sangamonian) interglacial deposits in the Toronto Brick Company Quarry [Don Beds] at Toronto, Ontario, are said to contain the following vertebrate remains: deer (Cervidae), large bear (*Ursus*), woodchuck = groundhog (*Marmota monax*), bison (*Bison*), and giant beaver (*Castoroides ohioensis*). Near Welland remains of several mastodons (*Mammut americanum*) were found. A photograph shows a man collecting a mastodon humerus near Ridgemount (Figure 9). A skeleton of a capelin (*Mallotus villosus*) in a calcareous nodule from Champlain Sea [mistakenly termed 'the last Pleistocene interglacial stage'] deposits near Ottawa is shown (Figure 23).

1263. Wagner, F.J.E. (1967): Published references to Champlain Sea faunas 1837–1966 and list of fossils. Geological Survey of Canada Paper 67-16:1–82.
This report comprises a list of all published references to Champlain Sea fauna that the writer could find. References are arranged in chronological order, followed by lists of species recovered. Vertebrate fossils are mentioned under 37 references, beginning with Charles Lyell's (1845) report of capelin (*Mallotus villosus*) from the Ottawa River near Bytown [Ottawa, Ontario] in concretions obtained by W.E. Logan, and ending with a report of Dyck et al. (1966) of white whale (*Delphinapterus leucas*) from Foster Sandpit near Uplands Airport [now Macdonald-Cartier International Airport), Ottawa, Ontario.

1264. Wagner, F.J.E. (1970): Faunas of the Pleistocene Champlain Sea. Geological Survey of Canada Bulletin 181:1–104.
This report describes the nature, origin, and timing, as well as fossils, of the Champlain Sea that have been recorded since 1837. It gives results of the author's field and laboratory work. The publication deals almost entirely with invertebrates, except for part of p. 48 dealing with capelin (*Mallotus villosus*) in calcareous clay nodules from the Green Creek–Hiawatha Park area along the Ottawa River, Ontario. She collected a specimen at Hiawatha Park in 1954, and an illustration from Dawson (1893) is reproduced on p. 102. In her accompanying faunal list (Table I), she mentions another specimen collected by P.F. Karrow in 1955 from the S shore of the St Lawrence River, 2.25 m NE of St Pierre-les-Becquets, Quebec.

1265. Wagner, F.J.E. (1984): Fossils of Ontario. Part 2: Macroinvertebrates and vertebrates of the Champlain Sea with a listing of nonmarine species. Royal Ontario Museum Life Sciences Miscellaneous Publications 17(2):1–64.
Pages 39–50 list the Champlain Sea (Late Wisconsinan) vertebrate specimens. These are: white whale (*Delphinapterus leucas*) from Ottawa, Pakenham (Lanark County), Cornwall, and Williamstown, Ontario; humpback whale (*Megaptera novaeangliae*) from a gravel pit 128 m asl at Warwick, near Smiths Falls, Ontario; bowhead whale (*Balaena mysticetus*) from White Lake, Ontario; ringed seal (*Phoca hispida*) from Ottawa International Airport (Uplands), Ontario, and Hull (Tétreauville), Quebec; and bearded seal (*Erignathus barbatus*) from Finch, Ontario. Also included is the Green Creek fauna (east of Ottawa, Ontario): capelin (*Mallotus villosus*), smelt (*Osmerus mordax*), tomcod (*Microgadus tomcod*), three-spined stickleback (*Gasterosteus aculeatus*), sculpin (*Artediellus*), deepwater sculpin (*Myoxocephalus thompsoni*), lumpfish (*Cyclopterus lumpus*), lake trout (*Salvelinus namaycush*), cisco (*Coregonus* cf. *Core-*

gonus artedii), American marten (*Martes americana*), harp seal (*Phoca groenlandica*), and feathers and vertebra of unidentified bird (Aves). [See McAllister et al. 1988.]

1266. Waits, L.P., Talbot, S.L., Ward, R.H., and Shields, G.F. (1998): Mitochondrial DNA phylogeography of the North American brown bear and implications for conservation. Conservation Biology 12(2):408–17.
To examine the evolutionary history of the brown bear (*Ursus arctos*) in North America and to assess the genetic divergence between individuals from different geographic regions, the authors obtained 294 nucleotides of mitochondrial DNA sequence data from the control region for 317 free-ranging brown bears. Phylogenetic analyses using maximum parsimony recorded four major mitochondrial DNA phylogeographic groups or clades. These clades do not correlate with taxonomic classifications for *Ursus arctos*, and the authors hypothesize that the clades were formed prior to migration of this species into North America. Discordant results were also obtained for subspecific classifications and mtDNA phylogeny in the grey wolf (*Canis lupus*), suggesting that the morphological differences used to define subspecies in brown bear and grey wolf may represent phenotypic plasticity in differing environments rather than long-term genetic isolation. It seems unreasonable to dramatically alter the current taxonomy based on results from a single mtDNA region, especially with the knowledge that mtDNA gene trees may not always reflect true species trees.

1267. Walde, K. (1994): Archaeological salvage programme at KaVn-2, KbVo-1, and KdVo-3 Alaska Highway, Shakwak Project segments 16A, 17A, and 18B, Beaver Creek area, S.W. Yukon. Permit Report 93-3ASR. Prepared for Heritage Branch, Yukon Government. 140 pp.
KaVn-2 is the oldest of the three sites, with a radiocarbon date of 7810 ± 80 years BP (Beta 68509) on a charcoal sample. A total of 715 bones were examined from KaVn-2. Unfortunately most were either fragments or burned. Only one large-mammal (Mammalia) vertebra remained unburned and recognizable (but was not identified). The radiocarbon-dated charcoal was recovered from a layer between two cultural bearing sediments. The layer below contained leaf-shaped bipoints and tools of siltstone and black basalt, while the layer above contained obsidian artifacts of a more finished appearance. The paper concludes that the upper layer represents a Little Arm Phase occupation, while the lower layer predates that time and represents a population not utilizing a microblade technology. The early occupation of KaVn-2 likely occurred during or shortly after a spruce forest ecosystem replaced that of a shrub tundra. This environmental change affected both temperature and vegetation and likely created changes in the faunal populations as well. KbVo-1 and KdVo-3 were both less than 2000 years old and also contained numerous stone tools. A map locating the sites (p. 67) is included.

1268. Walker, A.E. (1896): Description of the railway cutting. Journal and Proceedings of the Hamilton Association for the Cultivation of Science, Literature, and Art. pp. 147–50.
Several Late Pleistocene mammal specimens were found in a railroad cutting 6 m beneath the surface at the western end of the cutting near Garth Street, between Locke and Pearl streets, in Hamilton, Ontario. The specimens are: moose? (cf. *Alces*), and canid or bear (Canidae or Ursidae).

1269. Walker, D.A., Bockheim, J.G., Chapin, F.S. III, Eugster, W., Nelson, F.E., and Ping, C.L. (2001): Calcium-rich tundra, wildlife, and the 'Mammoth Steppe.' Quaternary Science Reviews 20:149–63.
Moist calcareous tundra is intermediate between steppe tundra and tussock tundra, and provides insights regarding the transition from cold, arid Beringian ecosystems to present-day moist acidic tundra. Modern areas of calcium-rich tundra have many of the ecosystem properties that Guthrie (1990) hypothesized would be important to support the diverse Beringian megafauna [e.g., horse (*Equus*), saiga antelope (*Saiga tatarica*), steppe bison (*Bison priscus*), and woolly mammoth (*Mammuthus primigenius*)], including firm, relatively warm, well-drained, nutrient-rich soils, high diversity of plant species and habitats, and plants low in secondary protective compounds.

1270. Walker, E.G. (1988): The Gowen site: A Mummy Cave occupation within the city limits of Saskatoon. *In*: Out of the Past: Sites, Digs and

Artifacts in the Saskatoon Area (U. Linnamae and T.E.H. Jones, eds.). Saskatoon Archaeological Society. pp. 65–74.
Projectile points (*Homo sapiens*) and mammalian remains were found at the Gowen site, on a major terrace of the South Saskatchewan River, Saskatoon, Saskatchewan (52°05′45″N, 106°42′20″W; map Figure 7.1). The mammalian remains included: bison (*Bison bison*) – 95% of bones; wolf (*Canis lupus*), dog or coyote (*Canis familiaris* or *Canis latrans*), pronghorn (*Antilocapra americana*), and small unspecified mammals. The mean radiocarbon date for the assemblage was 5870 BP (no lab number).

1271. Walker, L. (1984): The 'moon-egg' that contained an arctic ground squirrel. BIOME 4(1):3.
In 1981, in the Sixtymile area W of Dawson City, Yukon, placer miner Manfred Peschke was hosing down the banks at Glacier Creek to reach the precious underlying layer of gold-bearing gravel when he noticed a curious egg-shaped object (about 7.5 × 11 cm) floating in the backwash from the bank. This 'moon-egg' (as it was christened) was given to Dan Drummond of the Yukon Wildlife Branch, who contacted Dick Harington at the National Museum, who had collected fossils in the region for many years. When Harington opened the package from Dan, he saw a dried ball of silt with bits of buffy-brown hair projecting. An X-ray of this strange object revealed the complete skeleton of an arctic ground squirrel (*Spermophilus parryii*) curled up nose to tail, which is the common hibernating position of these animals. Harington speculated that the animal died during hibernation more than 10,000 years ago. Evidently the specimen, which seems to have been preserved by a natural process of freeze-drying, was buried some 6 m below the surface. Clues as to the age of the specimen are provided by several ground squirrel nests that have been collected from frozen organic silt in the Dawson City area. One example from Dominion Creek contained nesting grasses, part of a seed cache, droppings, and the skeleton of a squirrel. The nesting grasses gave a radiocarbon date of about 12,000 BP. [The Glacier Creek specimen eventually yielded an AMS radiocarbon date of about 47,000 BP – see Harington 1989c.]

1272. Wayne, R., Cooper, A., and Leonard, J. (2000): Population genetics of ice age carnivorans. Journal of Vertebrate Paleontology, Abstracts of Papers 20 (Supplement to Number 3): 77A.
The Pleistocene was a dynamic period for Holarctic mammal species, complicated by episodes of glaciation, local extinctions, and intercontinental migration. The genetic effects of these events are difficult to resolve from the study of present-day populations. To provide the first direct view of population genetics in the Late Pleistocene, mitochondrial DNA sequence variation was measured in seven brown bear (*Ursus arctos*) specimens preserved in permafrost dating from 42,000–14,000 BP. About 36,000 BP the Beringian brown bear population had a higher genetic diversity than any extant North American population, but by 15,000 BP genetic diversity appears similar to the modern day. The older, genetically diverse Beringian population contained sequences from three clades now restricted to local regions within North America, indicating that current phylogeographic patterns may provide misleading data for evolutionary studies and conservation management. The Late Pleistocene phylogeographic data also indicate possible colonization routes to areas S of the Cordilleran ice sheet. Finally, the authors provide preliminary data on genetic diversity in wolf-like canids (Canidae) of the northern permafrost.

1273. Weber, F.R., Hamilton, T.D., Hopkins, D.M., Repenning, C.A., and Haas, H. (1981): Canyon Creek: A Late Pleistocene vertebrate locality in interior Alaska. Quaternary Research 16(2):167–80.
A number of specimens, representing an interstadial fauna about 40,000 years old, were found at the Canyon Creek vertebrate-fossil locality in sand and gravel deposits, at a roadcut on the Richardson Highway through a S-facing bluff overlooking the Tanana River near the mouth of Canyon Creek, 85 km SE of Fairbanks, Alaska (64°17′N, 146°29′W). These were: unidentified rodents (Rodentia –indicated by burrows), woolly mammoth (*Mammuthus primigenius*), horse (*Equus*), Yukon horse (*Equus lambei*), western camel (*Camelops hesternus*), steppe bison (*Bison* cf. *Bison priscus* = *Bison* cf. *Bison crassicornis*), sheep (*Ovis* cf. *Ovis dalli*), wolf (*Canis* cf. *Canis lupus*), tundra hare (*Lepus* cf. *Lepus othus* or *Lepus arcticus*), and caribou (*Rangifer*). All these were older than 39,360 ± 1740 BP (SMU-640, bone collagen mainly from *Equus*, but including other unidentifiable bone). The assemblage suggests an open landscape in which trees and tall shrubs were either absent or confined to

sheltered and moist areas. The stratigraphic section of Canyon Creek (Figure 3) appears to demonstrate that the Delta Glaciation of the N-central Alaska Range is at least, in part, of Early Wisconsinan age and was separated from the succeeding Donnelly Glaciation by an interstadial rather than an interglacial period. [See Hamilton and Bischoff (1984) for revised dates.]

1274. Weber, J.N. (1955): Hamilton cave yields Pleistocene fossils. National Speleological Society News 13(4):2.
Systematic investigation of rock-shelter caves in the Hamilton area, Ontario, resulted in the discovery of a deposit of 'perfectly preserved Pleistocene fossils.' The cave, about 20 ft deep, was excavated in an extensive bed of open-work conglomerate showing cross-bedding, which was laid down in a large delta deposit in the waters of Glacial Lake Iroquois. Embedded in the cave roof were remains of fish (Pisces) and mammals such as: fox (*Vulpes vulpes*), woodchuck or 'ground-hog' (*Marmota monax*), and 'rat' (Rodentia). Well-preserved gastropod shells and insect remains were excavated in several places. Further excavations were to be carried out by Drs H.R. Thompson, R.V. Best, D.M. Davies, and D.E. Delzell of McMaster University in Hamilton. [See Karrow and Warner 1988, p. 45, and Wetmore 1958.]

1275. Weidick, A. (1972): Notes on Holocene glacial events in Greenland. *In*: Climatic Change in Arctic Areas during the Last Ten-Thousand Years (Y. Vasari, H. Hyvärinen, and S. Hicks, eds.). Acta Universitas Ouluensis, Series A, Scientiae Rerum Naturalium 3, Geologica 1:177–204.
This review incorporates new information on W Greenland Holocene ice-margin deposits that have been correlated and dated by means of former shorelines. The main deglaciation of this area took place between 10,000 and 6000 BP, and there is evidence that the recession just prior to about 6000 BP had reduced the Inland Ice so that its margin was situated farther E than present. The general recession was interrupted by halts and readvances, of which the youngest, about 6000 BP, marks the end of the recession. Concretions from Inugpait qûat and Manîtsup tunua to Qaleragdlît sermia SW Greenland (map, Figure 1) contain remains of capelin (*Mallotus villosus*) and rosefish (*Sebastes marinus*) respectively (Table 1b). The concretions from Inugpait qûat originated from clayey-silt layers that during historical time were folded up into moraines at the front of the glacier. The silt layers form a terrace about 30 m asl, the age of which is estimated as 6500 BP.

1276. Weidick, A. (1974): C^{14} dating of survey material performed in 1973. Grønlands Geologiske Undersøgelse Rapport 66:42–5.
This paper includes two dates on bone collected in East Greenland by Christian Vibe and A.L.W. Manniche. The material is mentioned in Vibe (1967). The first is on a muskox (*Ovibos moschatus*) skull found in ruins of an Eskimo settlement on a sandy delta surface at the mouth of Gåseelv, Hurry Inlet (70°44′N, 22°37′W). The specimen was collected by Vibe and yielded a date of 400 ± 100 BP (1550AD) (GGU-79302:K-2156). Part of another muskox skull found on the surface at Danmarks Havn (76°46′N, 18°45′W) by Manniche in 1908 gave a date of 770 ± 100 BP (1180AD) (GGU-79303:K-2155).

1277. Weidick, A. (1992): Jakobshavn Isbræ area during the climatic optimum. Grønlands Geologiske Undersøgelse Rapport 155:67–72.
Jakobshavn Isbræ, West Greenland (see map Figure 2) is a tidal outlet of the Inland Ice considered to have the highest calf-ice production of Greenland glaciers. This place and its surroundings were considered to have had about 4000 years ago a retracted ice position comparable to neighbouring areas to N and S. A small walrus (*Odobenus rosmarus*) tusk (Figure 4) was collected at an altitude of about 60 m asl in 1998 and yielded a date of 4290 ± 100 BP (Ua-2350), which fits well with the dating range of neoglacial material from other parts of the ice-margin area. However, due to uncertainties of the subglacial topography of the present Jakobshavn Isbræ frontal area, and to the presumed change in flow direction of the ice cover during the past 100 years, the source of the tusk cannot be definitely determined. It probably comes from somewhere in the 'ice bay.'

1278. Weidick, A., Oerter, H., Reeh, N., Thomsen, H.H., and Thorning, L. (1990): The recession of the Inland Ice margin during the Holocene climatic optimum in the Jakobshavn Isfjord area of West Greenland. Palaeogeography, Palaeoclimatology, Palaeoecology 82:389–99.
Recent subsurface mapping of parts of the Greenland ice

margin in the Jakobshavn Isbræ indicates that the fiord system in the period of at least 2700–4700 calendar years BP was more ice-free than at present, and that the front of the glacier was at least 15 km behind the present position. The ^{14}C-datings of subfossils brought to the present ice margin fit with the climatic records from ice cores and confirm the favourable conditions for Greenland's first settlers, the Sarqaq people, who arrived in the region about 4000 years ago to find hunting grounds 10–20% larger than present. Reindeer/caribou bones (*Rangifer tarandus* – GGU no. 79318) found in the transitional area close to clear glacier ice about 50 m above the present ice surface have been dated to 3040 ± 60 BP (K-4572). Marl concretions, many with well-preserved shells, are also common in the transitional area. An undated fragment of walrus (*Odobenus rosmarus*) tusk is mentioned from the same region.

1279. Weigel, R. (1963): Ruffed Grouse from the Pleistocene of Saskatchewan. Wilson Bulletin 75(2):206.
Ruffed Grouse (*Canachites canadensis* = *Bonasa umbellus*) remains were found from an intercalated sand deposit in a gravel pit in the Qu'Appelle Valley near the village of Fort Qu'Appelle. The fossil was associated with muskox (*Ovibos* or *Bootherium*), horse (*Equus*), mammoth (*Mammuthus*), camel (*Camelops*), and bison (*Bison*). The age of the assemblage is estimated at no older than Sangamonian. [See Khan 1970, and Skwara-Woolf 1980, 1981.]

1280. West, C.F. (1996): Trail Creek Caves, Seward Peninsula. *In*: American Beginnings: The Prehistory and Palaeoecology of Beringia (F.H. West, ed.). University of Chicago Press, Chicago and London. pp. 482–4.
The caves are located on Trail Creek (1 mi above its junction with Cottonwood Creek), Seward Peninsula, Alaska. Many natural cavities are found in the limestone cliffs that rise to a height of about 300 ft. Only in Cave 2 were artifacts older than Mid-Holocene. It is a tunnel about 21.4 m long. Both Cave 2 and Cave 9 yielded a large number of bones – mainly modern forms, with caribou (*Rangifer tarandus*) common at all levels. Caribou bones from layer III in Cave 2 were radiocarbon dated at 9070 ± 250 BP (K-980). Other species represented are: wapiti (*Cervus elaphus*) in the lower layer towards the rear of Cave 2; and bison (*Bison*) and horse (*Equus*) in the lower clay outside the S entrance to Cave 9. Collagen dates were 13,070 ± 280 BP (K-1327) and 15,750 ± 350 BP (K-1210) respectively. The artifacts thought to be the oldest – microblades and associated antler points – were deeply buried near the front of Cave 2. [See Larsen 1968; Vinson 1988.]

1281. Westgate, J.A. (1969): The Quaternary geology of the Edmonton area, Alberta. *In*: Pedology and Quaternary Research (S. Pawluk, ed.). Joint Symposium Sponsored by the National Research Council of Canada and the University of Alberta, Edmonton. pp. 129–51.
An illustrated (photographs, stratigraphic columns, and diagrams) review of the geology of the Edmonton area, including the known and hypothesized ages of the various levels with occasional mentions of the fossil features: (1) horse (*Equus*) and bison (*Bison*) remains show that the valley-fill member within Beverly Valley (Figure 3) is probably Late Pleistocene in age (Reimchen 1968), but the radiocarbon age of associated wood shows that it is >33,000 BP (S-215); (2) the higher and older terrace gravels within Beverly Valley have yielded remains of Columbian mammoth (*Mammuthus* cf. *Mammuthus columbi*) and giant camel (*Titanotylopus*), thereby confirming a Pleistocene age assignment; (3) in a section of a postglacial terrace 60–70 ft above the level of North Saskatchewan River at Emily Murphy Park, a horse metatarsal was found 7 ft below the surface in sand and a bison femur came from 22 ft below the surface (Table 2). Westgate (p. 150) indicates the horse may represent the Mexican horse (*Equus conversidens*) and the bison may represent the western bison (*Bison bison occidentalis* = *Bison occidentalis*); an age of 11,000–10,000 BP is suggested.

1282. Westgate, J.A., Kalas, L., and Evans, M.E. (1976): Geology of the Edmonton Area, Alberta. Geological Association of Canada and Mineralogical Association of Canada Joint Annual Meeting (May 19–21, Edmonton), Field Trip C-8 Guide Book:1–49.
In the Beverly area (approximately 53°33′15″N, 113°23′35″W; Edmonton, Alberta), clean, crossbedded sands at Lsd. 3, sec. 28, twp. 53, rge. 23, W 4th contain remains of bison (*Bison*) and wapiti (*Cervus elaphus*) – antlers of the latter have yielded a collagen date of 9860 ± 140 BP (I-8483). Gravels in the Apex Pit on the oppo-

site side of the North Saskatchewan River at Lsd. 9 and 10, sec. 33, twp. 53 rge. 23, W 4th have yielded remains of the Mexican horse [*Equus* cf. *Equus conversidens* = *Equus* (*Asinus*) cf. *conversidens*] and steppe bison (*Bison priscus* = *Bison crassicornis*) 'and are probably of similar age.' Fossils of western bison (*Bison bison occidentalis* = *Bison occidentalis*) are frequently recovered from fine-gravel overbank deposits covering channel gravels and sands at the Beverly locality (coordinates given above). At the Alberta Concrete Products site (53°33′40″N, 113°21′45″W) near Clover Bar, Saskatchewan Gravels have yielded remains of a medium-sized horse like Scott's horse (*Equus scotti*), a large-horned bison like the giant bison (*Bison* cf. *Bison latifrons* = *Bison* cf. *alleni*), large camel (*Titanotylopus* [perhaps more likely to be *Camelops*]), and Columbian mammoth (*Mammuthus* cf. *Mammuthus columbi*). Thus bison with very robust horns and medium-sized horses lived in the Edmonton region before glaciation and were succeeded by smaller-horned bison and smaller horses after the area had been first affected by continental glaciation (C.R. Harington, personal communication 1975). Evidently molluscan and pollen assemblages from the Saskatchewan Gravels, as well as a dominance of grazers in the vertebrate fossil assemblage 'point to grassland as a significant element in the landscape immediately prior to continental glaciation' (p. 32).

1283. Westgate, J.A., Briggs, N.D., Stalker, A.M., and Churcher, C.S. (1978): Fission-track age of glass from tephra beds associated with Quaternary vertebrate assemblages in the southern Canadian Plains. Abstracts with Programs, Joint Meeting of the Geological Association of Canada, the Mineralogical Association of Canada, and the Geological Society of America (October 23–6, Toronto, Ontario). pp. 514–15.

An early Quaternary age for vertebrate remains from the Wellsch Valley site, near Swift Current, Saskatchewan, is based on the presence of Late Blancan–Irvingtonian forms and the magnetic polarity reversal pattern recorded in the associated sediments. The Mid-Wisconsinan age assigned to the Rancholabrean vertebrate assemblage from the Galt Island site, near Medicine Hat, Alberta, is based on radiocarbon dates: wood 1–2m above the fossiliferous beds has a radiocarbon age of 39,000–38,000 BP. Fission-track dates (regarded as minimum estimates) indicate that the Galt Island vertebrate assemblage is much older than Mid-Wisconsinan and suggest the possibility that the Wellsch Valley fauna is of Middle rather than Early Quaternary age. A table of fission dates (two for Wellsch Valley and one for Galt Island) is provided.

1284. Weston, B. (1975): The late Pleistocene wolverine from Old Crow Basin, Yukon. Term paper for Biology 490. 61 pp, plus Appendices. [Copy in Quaternary Vertebrates of Northern North America Publication File at Canadian Museum of Nature (Paleobiology).]

The purpose of this project was to study statistically a large sample of Late Pleistocene wolverine (*Gulo gulo*) fossils (in the Quaternary Zoology Collection at the Canadian Museum of Nature) from Old Crow Basin, Yukon. The author notes: (1) The size difference between Old Crow Basin fossil and Recent wolverine dental dimensions is not sufficient to preclude the possibility that the fossils were early representatives of *Gulo gulo luscus*; (2) Recent wolverine populations are highly polymorphic in regard to dentition and size; (3) Some lower premolars (P_1 and P_2) and molars (M_2) appear to be vestigial and the variability in their dimensions seem to be increasing in modern populations; (4) The lower carnassial exhibits sexual dimorphism. Mean dimensions are fairly constant between Recent populations but are smaller in the Old Crow fossils. Growth gradients between the length and width of the lower first molar (M_1) suggest evolutionary stability. Allometric comparisons of M_1 in relation to other teeth might reveal a taxonomically significant trend. [See Harington 1977a and Bryant 1986.]

1285. Weston, T.C. (1899): [Skeleton of harp seal]. *In*: Reminiscences among the Rocks in Connection with the Geological Survey of Canada. Warwick Brothers and Rutter, Toronto. pp. 52–3.

The author mentions his recovery of a fossil harp seal (*Phoca groenlandica*) skeleton from post-Tertiary [Quaternary] clays 30 ft below the surface in Peel and Conte's Clay Pit near Montreal, Quebec. In 1887 the skeleton was restored and remounted by Jules Bailly. Weston also mentions white whale (*Delphinapterus leucas*) remains that are sometimes found in Champlain Sea clay as well as capelin (*Mallotus villosus*) in nodules from Green Creek near Ottawa, Ontario.

1286. Wetmore, A. (1958): Miscellaneous notes on fossil birds, Section V. Pleistocene bird records from Ontario. Smithsonian Miscellaneous Collections 135(8):1–11.
Of particular interest are specimens collected in two small caves in the Hamilton, Ontario, area. The bone-bearing deposit consisted of sediments ranging from medium sand to raisin gravel, underlying much coarser deposits 3–4 ft in depth that constituted a part of the former shoreline of Pleistocene Glacial Lake Iroquois during the period of shrinkage of that body of water. The site is presently about 275 ft asl, and is dated as Late Pleistocene [but see Churcher and Karrow (1963), Karrow (1984), and Karrow and Warner (1988, p. 45), who reinterpreted this diverse assemblage as intrusive and of much younger age than Glacial Lake Iroquois]. Four species of birds are represented: Wood Duck (*Aix sponsa*) – the first Pleistocene record of this species; Barred Owl (*Strix varia*), Red-Winged Blackbird (*Agelaius phoeniceus*), and Common Grackle (*Quiscalus quiscula*). Other species found in the same deposits include: short-tailed shrew (*Blarina brevicauda*), chipmunk (*Tamias striatus*), grey squirrel (*Sciurus carolinensis*), red squirrel (*Tamiasciurus hudsonicus*), flying squirrel (*Glaucomys sabrinus*), white-footed mouse (*Peromyscus*), meadow vole (*Microtus pennsylvanicus*), pine mouse (*Microtus pinetorum = Pitymys pinetorum*), muskrat (*Ondatra zibethicus*), red fox (*Vulpes vulpes = Vulpes fulva*), a frog (*Rana*), a colubrid snake, and a pickerel (*Esox*). The cave and vertebrate remains were discovered by Weber (1955). [See Karrow and Warner 1988.]

1287. Whiteaves, J.F. (1889): Palaeontology and zoology. Annual Report of the Geological and Natural History Survey of Canada 3(1):47A–113A.
The author mentions the addition of a lower jaw, several vertebrae, ribs, and limb bones of a mastodon (*Mammut americanum*) to a display of Post Tertiary [Quaternary] and Tertiary vertebrate remains of Canada. The specimen was recovered in 1871 from a peat bog near Woodstock, Ontario (p. 48A). Mention is also made (p. 110A) of the recovery of a juvenile harp seal (*Phoca groenlandica*) skeleton from post-Pleistocene clay at Hull, Quebec [now reidentified as a ringed seal (*Phoca hispida*) – see Harington and Sergeant 1972]; a wapiti (*Cervus elaphus = Cervus Canadensis*) antler ploughed up in a former cedar swamp 3 mi from Sydenham, Ontario (p. 111A); a mammoth (*Mammuthus = Euelephas*) molar from St Catherines, Ontario, dug up in 1887 (p. 113A); and nearly a hundred Green Creek nodules collected by H.M. Ami (p. 109A).

1288. Whiteaves, J.F. (1903): Crania of extinct bisons from the Klondike creek gravels. Ottawa Naturalist 16(12):240–1.
This report describes four skulls or portions of skulls of extinct bison from the Dawson City area, Yukon, received by the Museum of the Geological Survey of Canada in the previous five years. Specimens 1 and 2 are steppe bison (*Bison priscus = Bison crassicornis*) skulls. No. 1 was collected in 1900, 15 ft below the surface by R.G. McConnell at Gold Run Creek, Claim 17. No. 2 is almost complete and was collected by W.G. Luker in 1902, 30 ft below the surface at Dominion Creek, Claim 83 below Lower Discovery. No. 3 is the basal portion of a skull tentatively identified as a juvenile male steppe bison. It was also collected by W.G. Luker in 1902, but from 45 ft below the surface at Bear Creek. No. 4 is a portion of a hornsheath with its core, which can only be identified as bison (*Bison*), collected by W. Ogilvie in 1898, 18 ft below the surface at Bonanza Creek, Claim 39 above Discovery.

1289. Whiteaves, J.F. (1907): Notes on the skeleton of a white whale or beluga, recently discovered in Pleistocene deposits at Pakenham, Ontario. Ottawa Naturalist 20:214–16.
In 1906, near Pakenham, Ontario, a skeleton of a juvenile beluga (*Delphinapterus leucas*) was discovered by P. Cannon embedded in Pleistocene clay, 14 ft below the surface. The skull, atlas, axis, and six other vertebrae were recovered during the well-digging. Presumably the remainder of the skeleton is still *in situ*. [See Harington and Fitzgerald 1973.]

1290. Whitmore, F.C., Jr, and Foster, H.L. (1967): *Panthera atrox* (Mammalia: Felidae) from central Alaska. Journal of Paleontology 41(1):247–51.
A right mandible (USNM 23619) of an American lion (*Panthera leo atrox = Panthera atrox*) was found on an alluvial flat near the mouth of Lost Chicken Creek, Alaska, where it enters the south fork of Forty Mile River. This jaw is the latest in a series of mammal bones discovered at this site, which include: Niobrara horse

(*Equus* cf. *Equus niobrarensis alaskae*; USNM 23620 and 23621); horse (*Equus*; USNM 23622 and 23623); bison (*Bison bison*; USNM 23626); bison (*Bison*; USNM 23627); caribou (*Rangifer tarandus* = *Rangifer arcticus*; USNM 23628); caribou (*Rangifer*; USNM 23629); and wapiti (*Cervus elaphus* = *Cervus*; USNM 23630). Measurements of the lion mandible are included along with a discussion on biogeography of the large cats, and the stratigraphy of the site. [See Harington 1980c, Porter 1988.]

1291. Whitmore, F.C., Jr, Emery, K.O., Cooke, H.B.S., and Swift, D.J.P. (1967): Elephant teeth from the Atlantic Continental Shelf. Science 156(3781): 1477–81.
Teeth of mastodons (*Mammut americanum*) and mammoths (*Mammuthus*) have been recovered by fishermen from at least 40 sites on the continental shelf as deep as 120 m. Also present are submerged shorelines, peat deposits, lagoonal mollusc shells, and relict sands. Evidently elephants and other large mammals ranged this region during the glacial stage of low sea level of the last 25,000 years (the outer part of the shelf was exposed for about 10,000 years; the inner part about 20,000 years). Of interest are several specimens from Georges Bank (three mammoth and four mastodon teeth are mapped in Figure 2, with details listed in Table 1). The size of the teeth indicates that animals were small – perhaps a dwarf variety of woolly mammoth (*Mammuthus primigenius*) or Columbian mammoth (*Mammuthus columbi* = *Mammuthus jeffersonii*). However, the authors state that they can assign no specific name to the specimens. [See Cooke et al. 1993.]

1292. Whyard, F. (1995a): Territory boasts much Beringia history. Whitehorse Star, May 3:21.
The writer briefly describes the nature of Beringia, the land connecting Asia and North America during the ice age. The Yukon part of the Beringian refugium seems to be one of the best-kept secrets of the territory, known only to scientists and the northerners who have worked with them in Old Crow Basin. A Yukon item of note is the giant short-faced bear (*Arctodus simus*) that reached its northernmost range and maximum size in Yukon and Alaska. The most powerful land predator during the ice age, it became extinct about 10,000 years ago. In 1994, placer miner Pete Risby donated a complete cranium of this species from the Indian River area near Dawson City to the Yukon Heritage Branch.

1293. Whyard, F. (1995b): Studying the woolly mammoth. Whitehorse Star, May 10:24.
The woolly mammoth (*Mammuthus primigenius*) is the best-known mammal of the ice age because whole carcasses have been found preserved in frozen ground in Siberia. One of the best-preserved Canadian specimens, most of an entire skeleton from Whitestone River, Yukon, died there about 30,000 years ago. Mammoth limb bones found in Old Crow Basin show signs of breakage by humans (*Homo sapiens*). The woolly mammoth became extinct about 11,000 years ago in most of its range except for Wrangel Island off the Siberian coast, where a dwarf form survived until 4500 years ago.

1294. Whyard, F. (1995c): Studying the scimitar cat. Whitehorse Star, May 17:21.
The scimitar cat (*Homotherium serum*) was related to the sabretooth tiger (*Smilodon fatalis*), and was about the size of a lion (*Panthera leo*) with razor-sharp stabbing teeth. It seems to have specialized in killing young mammoths (*Mammuthus*) and other prey such as mastodons (*Mammut americanum*). In Canada, its remains are known from ice age deposits in western Yukon. [This information has been taken almost directly from the National Museum of Natural Sciences (Ottawa) 'Natural History Notebook' series.]

1295. Whyard, F. (1995d): Muskox is a common fossil. Whitehorse Star, June 7:19.
Helmeted muskoxen (*Bootherium bombifrons* = *Symbos cavifrons*) were the most common ice age muskoxen in North America according to the number of fossils recovered. Their range extended from Yukon and Alaska S to New Mexico. They were taller and longer in the body with deeper skulls and more flaring horns than tundra muskoxen (*Ovibos moschatus*), and died out about 11,000 years ago.

1296. Whyard, F. (1995e): Lion's remains found here. Whitehorse Star, June 14:20.
Slightly larger than the African lion (*Panthera leo*), the American lion (*Panthera leo atrox*) ranged through Yukon, Alaska, and Alberta from more than 70,000 to 20,000 years ago. It became extinct about 10,000 years

ago. The University of Alaska's prize exhibit, 'Blue Babe,' a well-preserved carcass of a 36,000-year-old steppe bison (*Bison priscus*) found at a placer mine near Fairbanks, Alaska, met its end as a result of an attack by American lions.

1297. Whyard, F. (1995f): The Jefferson's ground sloth. Whitehorse Star, June 21:19.

Jefferson's ground sloth (*Megalonyx jeffersonii*), a distant relative of the tree sloths of South America, was bear-sized and may have reached northern Canada (Yukon and Northwest Territories) during a relatively warm period in the ice age. This species foraged on the leaves of trees and shrubs and died out about 9000 years ago.

1298. Whyard, F. (1995g): The giant beaver was huge. Whitehorse Star, June 28:18.

The giant beaver (*Castoroides ohioensis*) reached the size of a black bear (*Ursus americanus*) and weighed as much as 200 kg – the modern beaver (*Castor canadensis*) weighs 30. The Old Crow region of the Yukon has the northernmost fossils. The only other giant beaver site in Canada is interglacial deposits of the Don River valley in Toronto, Ontario. [But see Miller et al. 2000.] Since 1965, the people of Old Crow have worked on summer 'digs' with scientists from the National Museum in Ottawa, and they continue to be involved in the Beringian project.

1299. Whyard, F. (1995h): [American mastodon]. Whitehorse Star, July 5:16.

Compared with the mammoth (*Mammuthus*), the American mastodon (*Mammut americanum*) was squat, between 2 and 3 m in shoulder height, longer in the body, and had straighter tusks. Early hunters killed mastodons from the Pacific coast to the Great Lakes about 12,000 years ago. By 9000 years ago, rapidly changing climate may have contributed to their extinction. Smaller mammals, including mummified arctic ground squirrel (*Spermophilus parryii*) and black-footed ferret (*Mustela nigripes*) specimens, dated to 40,000 years ago or more, were turned over to scientists by placer miners near Sixtymile, Yukon.

1300. Whyard, F. (1995i): Beringia series wraps up. Whitehorse Star, July 19:16.

Klondike miners have helped develop one of the best ice age mammal collections in North America at the Canadian Museum of Nature according to Dr Richard Harington. He thanks them for providing invaluable scientific information through their 'curiosity and astuteness.' Beringia was a giant land of giant animals, and the Yukon had its fair share. This is the final instalment of a series that has examined some of the species appropriate for display at the Yukon's planned Beringia Centre [now the Yukon Beringia Interpretive Centre in Whitehorse].

1301. Wilkerson, A.S. (1932): Some frozen deposits in the gold-fields of interior Alaska: A study of the Pleistocene deposits of Alaska. American Museum Novitates 525:1–22.

A description of the frozen deposits in the vicinity of Fairbanks, Alaska, is given (Figure 1 is an excellent map showing localities mentioned). The deposits consist of muck, sand, gravel, peat, volcanic ash, and ice. The materials, other than ash and ice, are of slope-wash and fluviatile origin. The muck (frozen muds) is prolific in fossil remains, particularly of steppe bison (*Bison priscus*), horse (*Equus*), moose (*Alces*), caribou (*Rangifer tarandus*), muskox (*Ovibos moschatus*), wapiti (*Cervus elaphus*), American lion (*Panthera leo atrox*), bear (Ursidae), wolf (*Canis lupus*), and many rodents (Rodentia) of Pleistocene age. The gravels are also known to contain fossils of the same age. Most fossils from the gravels were recovered by men on dredges. The bones came from all parts of the gravels, and no particular species is found alone in the gravels (p. 4). An illustration (Figure 2) of an almost complete steppe bison skeleton, assembled from bones of different individuals, is included, as well as excellent photographs of a lower jaw and teeth of a mammoth (Figure 10), and an arctic ground squirrel (*Spermophilus parryii*) nest exposed on the muck surface with a pocket watch for scale (Figure 11). During the summer of 1931 the author collected about two tons of bones, which were sent to the American Museum of Natural History in New York. Dried flesh adhered to a small percentage of the bones. A frozen fossil rodent's nest was found containing most of a mummified arctic ground squirrel. Most of the collected bones were obtained from 'bone pits' in several widely separated localities. The mining process allows bones to be collected without the collector having to excavate them (p. 22).

1302. Williams, J.R. (1962): Geologic reconnaissance of the Yukon Flats District, Alaska. United States Geological Survey Bulletin 1111-H:289–331.
In east-central Alaska near Venetie on the Chandalar River (see map, Plate 42), residents have reported the occurrence of mammal remains in the older part of the fan there and points upstream from the village. A horse mandible fragment with teeth found at the base of the river bank just upstream from Venetie was identified by Jean Hough as the Yukon horse (*Equus lambei*). Apparently, she prefers that *Equus lambei* be considered a subspecies of *Equus caballus*. Although the jaw fragment was not collected *in situ*, the author speculates that it is of Early to Middle Pleistocene age.

1303. Williams, M.Y. (1937): Three skulls of *Bison crassicornis* Richardson from the Yukon. Transactions of the Royal Society of Canada, Series 3 (Section 4) 31:105–10.
This paper provides a detailed description of three steppe bison (*Bison priscus = Bison crassicornis*) skulls in the collection of the Geological Museum of the University of British Columbia. The skulls were found in the gold-bearing gravels in the Dawson City area, Yukon. Collection localities are: No. 1, Examiner Pup, Bonanza Creek; No. 2, Eldorado Claim, Bonanza Creek [Eldorado Creek?]; No. 3, no specific locality data. Measurements and photographs are provided. In addition, the paper lists the known steppe bison specimens to date.

1304. Wilmeth, R. (1968): A fossilized bone artifact from southern Saskatchewan. American Antiquity 33(1):100–1.
A fossilized bone point (or dart foreshaft) from near Grenfell, Saskatchewan, is described. It was discovered during waterhole excavations during the early 1900s at a depth of 8 ft below surface level under 2–3 ft of decayed vegetation, 2–3 ft of bluish clay, a layer of small shells, and in a layer of soil in which the remains of bulrushes were visible. The point (Figure 2) is bone, rather than antler, from the shaft of a long bone of a large mammal, probably a mammoth (*Mammuthus*) or mastodon (*Mammut americanum*). Similar points (*Homo sapiens*) have been radiocarbon dated at 8500–9500 BP. The age of this point is estimated at 10,000 BP, about the time of the ice front at that locality (Figure 1).

1305. Wilson, M.C. (1974): Fossil bison and artifacts from the Mona Lisa site, Calgary, Alberta. Part 1: Stratigraphy and artifacts. Plains Anthropologist 19(63):34–45.
In 1968, during basement construction in Calgary (51°02′N, 114°05′W) a Paleoindian (*Homo sapiens*) bison-kill site was discovered. A radiocarbon date of 8080 ± 150 BP (GSC-1209) was obtained from bone high up in the section. The unit could therefore be as old as 10,000 BP. A related site near Calgary (Gallelli Gravel Pit) has yielded an ancient bison (*Bison bison antiquus = Bison antiquus*) skull aged at 8145 ± 320 BP (GX-2104), and an unaged western camel (*Camelops*; very large). The bison kill probably occurred in fall or early winter, since a cranium of a juvenile bison was recovered, but fetal bones were totally lacking. Some lithic material that could have been used as stone tools was found in association with the bones. Additionally, several modified long bones (tibia and femur) were discovered that were used as butchering tools. Photographs of the site, diagrams of the site stratigraphy, a diagram of the bone bed, and illustrations of the stone and bone tools are provided.

1306. Wilson, M.C. (1978a): Late Pleistocene and Holocene bison on the northern Great Plains. American Quaternary Association, Fifth Biennial Meeting, (Abstracts). (September 2–4, Edmonton, Alberta). p. 97.
For the past 12,000 years, northern plains bison have decreased in size at a fairly constant rate. Bison even 1000 to 2000 years ago were notably larger than the modern plains form (*Bison bison bison*). Based on the dwarfism chronocline, the author suggests that ancient bison (*Bison bison antiquus = Bison antiquus*) and western bison (*Bison bison occidentalis = Bison occidentalis*) are best viewed as early subspecies of *Bison bison*.

1307. Wilson, M.C. (1978b): Holocene geology and archaeology of the Bow River floodplain at Calgary, Alberta. American Quaternary Association. Fifth Biennial Meeting (Abstracts). (September 2–4, Edmonton, Alberta). p. 181.
Very large bison (*Bison*) bones were found in Bow River valley gravels in the Calgary, Alberta area. These bones have provided a radiocarbon date of 11,300 ± 290 BP (RL-757). This date implies that the postglacial drainage of the area was earlier than previously supposed, and that

archaeological (*Homo sapiens*) sites between 8400 and 10,000 BP should be sought on the lake-plain rather than in the river valleys.

1308. Wilson, M. (1980): Morphological dating of late Quaternary bison on the northern plains. Canadian Journal of Anthropology 1(1):81–5.
For 12,000 years, plains bison (*Bison bison bison*) have decreased in size at a fairly uniform rate. Dwarfing was under way earlier, but few data bear upon rates through the Wisconsinan glaciation. Cranial data from Wyoming populations, plus data from finds in nearby states and provinces show that dwarfing continued through the Holocene. Even 2000 to 1000 years ago, bison were larger than the modern plains bison. This paper deals mainly with morphological (tip-to-tip horncore measurements) and radiocarbon dates of five Alberta Holocene bison skulls [from Three Hills, W of High River, Okotoks and Calgary (2 sites), Table 1]. The results of morphological dating are encouraging, since the dates in most cases bracketed the radiocarbon dates. An exception was the Milan site – the oldest sample tested against the graph. For specimens other than the Milan sample, a tolerance of 2500 years adequately brackets the radiocarbon date. An increased tolerance is advisable for earliest Holocene specimens. While this is far less accurate than radiocarbon dating, it remains one of the few nonradiometric dating techniques readily available at plains archaeological (*Homo sapiens*) sites. It could prove useful in preliminary (and inexpensive) assessments. Figures 1 and 2 are chronological plots of bison horncore diminution through Holocene time on the plains.

1309. Wilson, M. (1981): Once upon a river: Archaeology and geology of the Bow River valley at Calgary, Alberta. PhD thesis, Department of Archaeology, University of Calgary.
This thesis mentions particularly the mammal remains from Bow River gravels in and near Calgary, Alberta, and places them in their geological context.

1310. Wilson, M.C. (1982): University of Calgary. [Report on Paleontology Research] Geological Association of Canada Paleontology Division Newsletter No. 15:12–13.
In this report of activities during 1981, the author explains that white-tailed deer (*Odocoileus virginianus*) and canid (Canidae) remains have been recovered from the Hitching Post Ranch site (3500 years old) in Alberta. He also indicates that he will soon publish with R.G. Forbis and D.W. Harvey, and that the human (*Homo sapiens*) fossil from the Taber site, Alberta, could be of Holocene age and not necessarily pre-Wisconsinan age as previously purported. [See Cohen 1982, and Stalker 1977.]

1311. Wilson, M.C. (1983): The pit and the conundrum: The Donald site (EePl-218), a Middle Period bone-filled pit feature in Okotoks, Alberta. Alberta Archaeological Review No. 7:3–16.
During basement construction in 1977, a Middle Period Paleoindian (*Homo sapiens*) site was exposed in the town of Okotoks, Alberta (Borden Number – EePl-218; see map, Figure 1). The site consists of a single feature: a bone-filled pit with two stone artifacts and four bone artifacts. All of the bones were identified as plains bison (*Bison bison bison*). Two individual male bison can be isolated. No. 48 is an old male, much older than 10 years, with very worn teeth and measurements well above the plains bison norm. No. 48 was radiocarbon dated at 3800 ± 300 BP. Specimen no. 49, also an old male (although the teeth are not quite as well worn as no. 48), provides measurements near the upper extreme for the modern plains bison. No. 49 was dated at <5500 BP. Other bones from the feature were dated to 3660 ± 150 BP (RL-901). A discussion concerning the function of the bone pit leads the author to conclude that this feature represents a frozen meat cache, surmounted by bison skulls to mark and protect the cache. The text is well supported by measurement tables, illustrations, and photographs.

1312. Wilson, M.C. (1996): Late Quaternary vertebrates and the opening of the ice-free corridor, with special reference to the genus *Bison*. Quaternary International 32:97–105.
One way to gauge the ecological opening of the ice-free corridor is to establish the chronology for the arrival of immigrant animal species. Bison (*Bison*) have excellent potential because of the abundance of their remains. Bison of 'southern' appearance [referable to ancient bison (*Bison bison antiquus*)] were present as far N as the Peace River region until about 10,000 BP. Bison populations in western Canada apparently underwent a rapid change at that time, such that barely 500 years later, bison of 'northern' appearance [referable to western bison

(*Bison bison occidentalis*)] were established. The rapidity and pervasiveness of this change seem to defy an evolutionary explanation rooted in punctuated equilibrium or phenotypic change, and could indicate a sudden population influx through the newly opened corridor.

1313. Wilson, M.C., and Churcher, C.S. (1978): Late Pleistocene *Camelops* from the Gallelli Pit, Calgary, Alberta: Morphology and geologic setting. Canadian Journal of Earth Sciences 15:729–40.
The Gallelli Pit site (50°57′45″N, 114°01′W) in Calgary, Alberta, has yielded bones of ancient bison (*Bison bison antiquus* – two crania and several postcranial elements under study by Hills, Shackleton, and Wilson), western camel (*Camelops* cf. *Camelops hesternus* – left humerus, UCG 6283), and Mexican horse (*Equus conversidens* – ischial radius and parts of acetabulum and ischial blade, UCG 6417). The Gallelli Pit gravels, where the fossils were found, are inferred to have an age of greater than 8000 BP based on radiocarbon dating of a bison cranium at 8145 ± 320 years BP (GX-2104) and a bison vertebra at 11,300 ± 290 (RL-757). Stratigraphic depictions of the site are included. Three radiocarbon dates are also given for the Cochrane, Alberta, late-glacial fauna: 10,760 ± 160 BP (GSC-612); 11,370 BP [no error listed (GSC-613)]; 11,100 ± 160 BP (GSC-989). The ungulate fauna from the Gallelli and Galvin pits suggests a semiforest [parkland?] habitat with extensive shrub and grassland elements.

1314. Wilson, M.C., and Churcher, C.S. (1984): The Late Pleistocene Bighill Creek Formation and its equivalents in Alberta: Correlative potential and vertebrate paleofauna. *In*: Correlation of Quaternary Chronologies (W.C. Mahaney, ed.). Geo Books, Norwich, England. pp. 159–75.
The Late Pleistocene Bighill Creek Formation is a paraglacial gravelly fill in the Bow River valley of Alberta. Similar fossiliferous fills are present in other Alberta drainages that share an origin in mountainous terrain and thus likely share a paraglacial genesis. The Bighill Creek Formation represents a warm interval following deglaciation, and is characterized by a rich large-mammal paleofauna. In the Bow River gravels the following specimens have been found: mammoth (cf. *Mammuthus*), Mexican horse (*Equus conversidens*), western camel (*Camelops hesternus*), caribou (*Rangifer tarandus*), wapiti (*Cervus elaphus*) [but see comment under Churcher 1968a], bighorn sheep (*Ovis canadensis*), and ancient bison (*Bison bison* cf. *Bison bison antiquus*). Elsewhere in the Bighill Creek Formation, tundra muskox (*Ovibos moschatus*) and Niobrara horse (*Equus* cf. *Equus niobrarensis*) have been collected. Age of the gravels has been determined (see Churcher 1968a) at between 11,500 and 10,000 BP. Evidence suggests that the deposits are slightly younger northward.

1315. Wilson, M.C., and Hills, L.V. (1984): Holocene white-tailed deer (*Odocoileus virginianus*) from the foothills northwest of Calgary, Alberta, Canada. Canadian Journal of Earth Sciences 21(3):365–71.
Excavations at the Hitching Post Ranch site (EiPo-51), 40km NW of Calgary (51°24′N, 114°29′W; see map, Figure 1) in 1980 yielded seven antlers or antler fragments, two molars, and four postcranial elements of white-tailed deer (*Odocoileus virginianus*). The deer remains were directly related to a dense bone bed of bison and are the first such subfossils from Alberta. A plains bison (*Bison bison bison*) rib provided a radiocarbon date of 3600 ± 70 BP (Beta-1627) for the site. Specimens are deposited in the collections of the University of Calgary, Department of Archaeology. The bones indicate that the deer were resident (shed antlers were found) and some show signs of human (*Homo sapiens*) modification. Based on the known snow preferences of the species, it is hypothesized that snow depths and winter temperatures 3600 years ago were not significantly more severe than those of the present.

1316. Wilson, M.C., Harvey, D.W., and Forbis, R.G. (1983): Geoarchaeological investigations of the age and context of the Stalker (Taber Child) site, DlPa-4, Alberta. Canadian Journal of Archaeology 7(2):179–207.
The Stalker (Taber Child) site is on the E side of the Oldman River valley, 5.6 km N of Taber, Alberta. This paper details the stratigraphy and ontogeny of the site. Discovery of Holocene mud slumps overlying the Pleistocene orange sand deposits and the presence of clay matrix on the Taber Child (*Homo sapiens*) fossil places its supposed Pleistocene age in doubt. It is more likely that this fossil is of Holocene age. [See Cohen 1982, and Stalker 1977.]

1317. Winchell, A. (1863): Description of elephantine

molars in the Museum of the University. Canadian Naturalist and Geologist 8:398–400.
The author mentions Canadian 'elephantine' [Proboscidea] specimens in the collections of the museum of the University of Michigan in response to a request by E. Billings, a Montreal paleontologist and an editor of this journal: (1) cast of a lower jaw and a tusk of a mastodon (*Mammut americanum*) from near St Thomas, C.W. [Canada West = Ontario]; (2) cast of a left lower molar found near Toronto, C.W. [Ontario] – measurements are provided on p. 400.

1318. Wisner, G. (1995): Pre-glacial site in Alberta suggests early human presence. Mammoth Trumpet 10(2):1, 3–4.
A researcher in Quaternary geology at University of Alberta has produced evidence that people (*Homo sapiens*) were in southwestern Alberta more than 20,000 years ago. Jiri Chlachula says the evidence consists of presumably artifactually flaked quartzite and hard limy siltstone cobbles found deeply buried in Late Pleistocene geological contexts at two locations in upper Bow River valley on the western edge of Calgary, Alberta. The recovery of possible artifacts – some found in place below more than 20 m of sediments of Glacial Lake Calgary – strongly suggests early human activity there. At Varsity Estates (Site 1) investigations carried on since 1990 have yielded approximately 40 artifacts. About 2 km upstream on the N side of the valley, Silver Springs (Site 2) has yielded an assemblage of more numerous flakes. Although no radiocarbon dates are available, Chlachula suggests they date to a time before the 25,000–21,000 BP period – the glacial ice advance. There have been no discoveries of skeletal remains at any of the sites. Comments from Nat Rutter, Jack Ives, and Alwynne Beaudoin – mainly critical of the evidence – are given.

1319. Wisner, G. (1998): North with the thaw? Clues to ice age Albertans prove 'only' 13,000 years old. Mammoth Trumpet 13(3):16–17.
Alan L. Bryan of the University of Alberta indicated that Paleoindians (*Homo sapiens*) were in Alberta's Bow River valley near the end of the Pleistocene, but not as early as researchers had thought. Further examination of evidence of artifacts (flakes, cores, pebble tools) in place on or in glacial till, below more than 20 m of glacial lake deposits at Varsity Estates in NW Calgary, Alberta, are now believed to be about 13,000 years old, rather than more than 20,000 years old [see Chlachula 1994, 1996]. New evidence suggests that the till containing some artifacts was exposed only briefly before Glacial Lake Calgary formed over it. Overlying sediments had insufficient organic material that could help date the artifacts, leaving researchers with a relative date derived from the geological context. Bryan concluded, 'We must look for more sites in Late Pleistocene geological contexts and not restrict the search to surficial and postglacial contexts.'

1320. Woodburne, M.O. (ed.) (1987): Cenozoic Mammals of North America, Geochronology and Biostratigraphy. University of California Press, Los Angeles, 336 pp.
Many of the principal investigators of the time contributed to this volume. Select chapters by a variety of authors are annotated separately within this bibliography. [See, for example, Lundelius et al. 1987, and Repenning 1987.]

1321. Woollett, J.M., Henshaw, A.S., and Wake, C.P. (2000): Palaeoecological implications of archaeological seal bone assemblages: Case studies from Labrador and Baffin Island. Arctic 53(4):395–413.
The usefulness of animals such as ringed (*Phoca hispida*), harp (*Phoca groenlandica*), bearded (*Erignathus barbatus*), and harbour seals (*Phoca vitulina*) as paleoecological indicator species is due to the close relationship of their range, reproductive cycles, and life histories to sea ice. Proportions of seal species represented in archaeological deposits may therefore reflect environmental conditions characterizing past seascapes. This paper examines zooarchaeological data from several Thule and historic Inuit (*Homo sapiens*) archaeological sites occupied over the last 700 years in Baffin Island, Nunavut, and Labrador, Newfoundland, to determine whether regional relationships can be demonstrated between subsistence economies, seal populations, and sea-ice conditions.

1322. Workman, W.B. (1974): First dated traces of Early Holocene man in the southwest Yukon Territory, Canada. Arctic Anthropology 11 (Supplement):94–103.
This article describes a hearth discovered 7 ft under the surface at the Canyon site (JfVg-1) at Mile 998 of the Alaska Highway (see maps, Figures 1, 2). This site bor-

ders the Aishihik River in Yukon Territory. Discovered at the site were: bone artifacts; stone flakes and tools (having closest connections with a series of Plano-derived points); a right ramus fragment and a right radius fragment of bison (*Bison*); and a right mandible fragment and a right innominate bone of beaver (*Castor canadensis*). The beaver was determined to be unrelated to the human activity and could have been deposited when Glacial Lake Champagne covered the area until about 9000 to 10,000 years ago. A radiocarbon date of 7195 ± 100 BP (SI-1117) obtained from hearth charcoal places the site after the drainage of Glacial Lake Champagne. A dry grassland environment quite unlike the spruce parkland prevailing today is inferred. This is the earliest-documented human (*Homo sapiens*) presence in the Yukon as of 1974, and shows that people were present in the Yukon before the advent of modern floral and faunal conditions.

1323. Wright, M.D. (1964): Annotated bibliography of Saskatchewan geology 1961–1963. Saskatchewan Department of Mineral Resources Report No. 9 (2nd Supplement):1–54.
Vertebrate fossil references are on p. 51. [See Kupsch 1959, 1961.]

1324. Wright, M. (1992): Saskatoon Mountain: 9,500 years of human occupation in the Grande Prairie Region. Alberta Past 7(3):8–9.
Saskatoon Mountain, 40 km W of Grande Prairie, Alberta, was excavated starting in 1988. It was a seasonal site occupied mainly in the summer. The oldest-known human (*Homo sapiens*) occupation dates to 9500 years ago, and there is an archaeological record that continues to the present ground surface. There is evidence indicating that bison (*Bison*) and wapiti (*Cervus elaphus*) have lived in this region for the last 10,000 years, 'no doubt attracted to this rich grassland in the midst of the Boreal Forest.' The site is highly stratified with aeolian sediments that allow a reliable chronology to be developed. This is the earliest-known date of human occupation of Alberta as of 1992. [But see Fedje et al. 1995.]

1325. Yesner, D.R. (1996): Human adaptation at the Pleistocene-Holocene boundary (circa. 13,000 to 8,000 BP) in Eastern Beringia. *In*: Humans at the End of the Ice Age: The Archaeology of the Pleistocene-Holocene Transition (L.G. Straus, B.V. Ericksen, J.M. Erlandsen, and D.R. Yesner, eds.). Plenum, New York. pp. 255–72.
The author asserts that disintegration of the full-glacial 'Mammoth Steppe' community antedates human (*Homo sapiens*) colonization of Eastern Beringia.

1326. Yesner, D.R. (1997): Human dispersal into interior Alaska: Antecedent conditions, mode of colonization, and adaptations. Beringian Paleoenvironments Workshop (September 20–3, Florissant, Colorado). Program and Abstracts. pp. 159–61.
Sedimentological and radiocarbon data from the deeply stratified Broken Mammoth, Mead, and Swan Point sites in the central Tanana River valley suggest that human (*Homo sapiens*) occupation began about 12,000 BP, soon after the cessation of Late Pleistocene dune formation and widespread sand deposition. The well-preserved fauna associated with these sites suggests relatively dry, open parkland conditions with Arctic species [arctic fox (*Alopex lagopus*), snowshoe hare (*Lepus americanus*), arctic ground squirrel (*Spermophilus parryii*), pika (*Ochotona*), and lemming (*Dicrostonyx torquatus* and/or *Lemmus sibiricus*)] continuing until 9300 BP. Wetlands were locally important, particularly in the earliest phase of occupation of the Broken Mammoth site (11,800 to 11,000 BP), representing a source of waterfowl (Anseriformes) and fish (Pisces). Most of the biomass was provided by two species: steppe bison (*Bison* cf. *Bison priscus*) and wapiti (*Cervus elaphus*), which were extensively butchered for meat, marrow, and perhaps tool production. Mammoth (*Mammuthus*) utilization was limited to tusk material for tool production – for which there is considerable evidence. Radiocarbon dates on mammoth ivory at these sites range from 18,000 to 12,050 BP, with the last date suggesting that mammoth extinction, if it occurred before human occupation of the region, did not precede it by much.

1327. Yesner, D.R. (1999): Bison, wapiti, caribou, and moose: Changing forest composition, animal abundance, and human subsistence patterns during the Holocene of east-central Alaska. Canadian Archaeological Association, 32nd Annual Conference (April 28–May 2, Whitehorse, Yukon). Programme and Abstracts. pp. 66–7.
Rapidly changing environments during the Late Pleistocene and Early Holocene had profound consequences

for human populations in Eastern Beringia. The reduction of katabatic winds, revegetation of braided stream valleys, and increased moisture regimes toward the close of the Pleistocene about 12,000 BP made interior Alaska more habitable by people (*Homo sapiens*) of the Nenana Complex, as well as their prey. The dwarf birch/grassland association supported bison (*Bison*) herds as well as small game, while local wetlands supported waterfowl (Anatidae) and some aquatic mammals. Around 11,000 BP an aspen parkland developed, with wapiti (*Cervus elaphus*) as a key big-game resource. The Younger Dryas climatic reversal about 11,000 to 10,500 BP may have led to a brief depopulation of the region, but the effects were minimal compared with those of the Younger Younger Dryas (about 9000 to 7500 BP), associated with greater aridity and windiness. The appearance of Denali Complex hunters around 7500 BP seems to be associated with the return of stable environments and an open spruce parkland. It was not until after 4000 BP that a more closed boreal forest developed, associated with podzols, the modern fauna, and proto-Athapaskan peoples.

1328. Yesner, D.R. (2001): Human dispersal into interior Alaska: Antecedent conditions, mode of colonization, and adaptations. Quaternary Science Reviews 20(2001):315–27.

The oldest dates for the human (*Homo sapiens*) occupation of Eastern Beringia lie between 15,000 and 12,000 BP. Possibly older cave sites – Bluefish Caves, Yukon, Lime Hills Caves, and Trail Creek Caves, Alaska, 'have stratigraphic and taphonomic problems that are not easily resolved' [but these are not given, and no mention is made of a mammoth (*Mammuthus*) bone flake and parent core from Bluefish Caves dating to 23,500 BP (Cinq-Mars and Morlan 1999)]. For northern and interior Alaska the earliest colonization appears to have been pushed by the Late Pleistocene drowning of the Bering Isthmus about 12,000 BP, and pulled by ameliorating climate and favourable vegetation changes. By 14,000 BP, the 'Birch Period' in interior Alaska (when dwarf birch and willow colonized the steppe tundra) may have led to the extinction of woolly mammoth (*Mammuthus primigenius*) and horse (*Equus*) – both obligate grazers of the steppe tundra – shortly after 13,000 BP. Mixed open parkland after 11,000 BP was apparently good habitat for both bison (*Bison*) and wapiti (*Cervus elaphus*), whose remains dominate early archaeological sites such as Dry Creek and Broken Mammoth. While bison herds may have diminished over those of the LGM, and caribou (*Rangifer tarandus*) may have retreated to higher altitudes and latitudes, browsers such as wapiti and moose (*Alces alces*) make their first significant Pleistocene appearance. The development of Bering Isthmus may have interrupted both the North Pacific Flyway for migratory waterfowl wintering farther S but nesting in the far N during the short summer. Also, it may have isolated sea bird populations in both Arctic and Pacific oceans leading to 'weak' speciation of taxa such as Murres (*Uria*) and simultaneous subspeciation of harbour seals (*Phoca vitulina*). However, data from Broken Mammoth suggest that the North Pacific Flyway was probably reestablished by 12,000 BP, making waterfowl available to interior Alaska as an additional human attraction. Further, perhaps whitefish (*Coregonus*) and other salmonids [e.g., trout (*Salmo*) and anadromous salmonids (*Onchorhyncus*)] were available to people in interior Alaska as far back as 20,000–14,000 BP. Thus, the region became increasingly attractive for humans before about 12,000 BP. Regarding the paleoeconomy of early archaeological sites in northern and interior Alaska, the author notes that: (1) several caribou teeth and a large caribou scapula with a 7900 BP radiocarbon date were recovered from secondary channel fill associated with the Onion Portage site on the Kobuk River; (2) several caribou bones were found in Late Pleistocene contexts at Lime Hills sites near the Kuskokwim River; (3) caribou dominated the Trail Creek Caves fauna – bison remains were few (1.3%), whereas small game (30%) were important and ptarmigan (*Lagopus*) (1%) were also notable; (4) bison remains have been found in Early Holocene contexts in the Great Kobuk Sand Dunes (where at least eight sites are present) – one bone being broken by people for marrow; (5) in interior Alaska, the Healy Lake site shows a wide diversity of activities – as does the Broken Mammoth blufftop site, which probably reflects the activities of supra-nuclear family groups, but not ones occupying long-term villages; (6) the Dry Creek site in the Nenana Valley has produced remains of bison, wapiti, and Dall sheep (*Ovis dalli*). Faunal data from the Broken Mammoth site with good organic preservation show that many vertebrate species were used: steppe bison (*Bison priscus*) and wapiti were the main large game exploited, but caribou, moose, and Dall sheep were less important in the diet. Only tusk fragments of woolly mammoths (*Mammuthus primigenius*) dated to 15,800

BP and 11,800 BP are represented – much of the ivory being made into finished tools. Although wolf (*Canis*) and bear (*Ursus*) remains have been recovered, most carnivore remains are of arctic fox (*Alopex lagopus*) – red fox (*Vulpes vulpes*) were from Mid-Holocene and later deposits. Among small game, by far the largest numbers are of arctic ground squirrel (*Spermophilus parryii*) remains (some probably dating to 9000–8000 BP), 3% of which are burned – suggesting an occasional food source. Voles (Microtini) and arctic shrews (*Sorex arcticus*) are numerous. Hares (*Lepus*), hoary marmot [*Marmota caligata* – not *'flavescens'* as stated in Tables 1 and 2], collared pikas (*Ochotona collaris*), and the river otter (*Lutra canadensis*) were also food sources. Many bird remains have been collected from early deposits at Broken Mammoth: (1) 65% are of Tundra Swan (*Cygnus columbianus*); (2) Canada Goose (*Branta canadensis*), Snow Goose (*Chen caerulescens* [not *hyperborea*, as listed]), and White-fronted Goose (*Anser albifrons*) make up about 15% of the identifiable bird remains; (3) another 10% are of five species of dabbling ducks [mallards (*Anas platyrhyncos*), Pintails (*Anas acuta*), Gadwalls (*Anas strepera*), American Widgeons (*Anas americana*), and Green-winged Teals (*Anas crecca*)]; (4) the final 10%, as at Trail Creek Caves, are of Willow Ptarmigan (*Lagopus lagopus*) – probably year-round residents. Fish vertebrae and scales from Culture Zone 3 appear to represent grayling (*Thymallus signifer*) or other salmonids (Salmonidae). These fauna indicate hunting in several environments near the Broken Mammoth site: open park land (bison, wapiti, and moose); wetlands (waterfowl, beaver [not mentioned in Table 2], and otter); and uplands (Dall sheep, hoary marmot, and possible caribou). Interior Alaska may have been largely abandoned during the Mid-Holocene (about 9000–7500 BP) dry, windy conditions.

1329. Yesner, D.R., Pearson, G.A., and Stone, D.E. (2000): Additional organic artifacts from the Broken Mammoth site, Big Delta, Alaska. Current Research in the Pleistocene 17:87–9.

Broken Mammoth is a stratified pre-Clovis site located about 13 km NW of Big Delta in east-central Alaska. Archaeological and paleontological materials are entrained in a 2-m-thick loess cap above 50 cm of late-glacial sands, resting on an 80 m schistose bedrock bluff overlooking the central Tanana River valley. Late Pleistocene and Early Holocene materials are embedded in two distinct paleosol complexes. Radiocarbon dates on charcoal, bone, and mammoth (*Mammuthus*) ivory associated with these paleosols are internally consistent and range from 11,800 to 9300 BP. The site is known for well-preserved bones derived mainly from: steppe bison (*Bison priscus*), wapiti (*Cervus elaphus*), arctic hare (*Lepus othus*), arctic fox (*Alopex lagopus*), and several birds, including tundra swan (*Cygnus columbianus*), geese (Anserinae), dabbling ducks (Anatidae), and ptarmigan (*Lagopus*). Considering the diversity of small animal species taken, particularly birds, nets and traps were undoubtedly used to get maximum returns. Excavation in 1998 demonstrated that moose (*Alces alces*) were more important to Late Pleistocene inhabitants than suspected. Mammoth remains, limited to ivory tusk fragments, probably were used for tool manufacture, as indicated by a microchip in a tusk groove. Most predate human (*Homo sapiens*) occupation at the site, but a new date of about 11,800 BP on ivory fragments from the lowest paleosol complex raises the possibility that people and mammoths coexisted there. Butcher marks are frequently noted on bird bones and mammal skull fragments. Animal bones outnumber organic artifacts by a 100:1 ratio. This site has provided the only organic tools known from the Nenana Complex, including bi-bevelled mammoth ivory rods, an ivory point fragment, a possibly ivory atlatl handle, two other bone points, an eyed needle, and a bone toggle. In 1998, a bone rod (Figure 1) was uncovered from the basal paleosol complex associated with a stone-lined hearth dated to about 10,500 BP, as well as scattered bison teeth, small mammal bones, bird bones, and mammoth ivory fragments. The rod, although broken in three, is nearly complete (about 160 mm long). Hatch marks suggestive of attached cordage are present on both extremities and the midsection and probably served as a foreshaft for a projectile point, such as a triangular, basally thinned 'Chindadn' point.

1330. Young, D. (2000): One bi-i-ig beaver. Saint John Times-Globe. January 25, p. A1.

Ralph Welch, a consulting engineer who has been scanning beaches for artifacts for decades, found a tooth a few years ago while hunting for arrowheads on Indian Island in Passamaquoddy Bay, New Brunswick. Puzzled by its appearance, he picked it out of the grass on a gravel ridge that separated the beach from a salt marsh. Finally, he decided to bring the specimen to the New Brunswick Museum, for checking. Randy Miller of the museum thought that it might be a prehistoric beaver tooth and sent photographs to Dr Richard Harington at the Canadian Museum of Nature. He confirmed that it was a cutting tooth (incisor) of a giant beaver (*Castoroides ohioensis*), and stated that this was only the third locality in Canada where remains of this black-bear–sized beaver had been found – the others being Toronto and the northern Yukon [Old Crow Basin]. Dr Miller believes this tooth was washed up during a storm. Parts of the Continental Shelf were once land, so it is possible that the beaver bones had been underwater for a long time. This tooth expands the known range of the giant beaver. A photograph shows Randy Miller with the giant beaver tooth in one hand and a modern beaver (*Castor canadensis*) skull in the other. [See Miller et al. 2000.]

1331. Young, H.R. (1966): Proboscidean molars from Manitoba. Canadian Field-Naturalist 80(2):95–8.
A summary of mammoth (*Mammuthus*) remains in Manitoba (see map, Figure 1) is provided and includes: (1) a tusk fragment from the S bank of Swan River, 5.6 km N of Benito; (2) a tusk fragment from a gravel pit near Transcona (sec. 12, twp.11, rge. 4, E); (3) a milk tooth from Birds Hill; (4) a molar tooth from a gravel pit 8 km E of Dufresne; (5) a small lower molar of a woolly mammoth? (*Mammuthus primigenius*? = *Mammonteus primigenius*?) from a gravel pit on the S bank of the Assiniboine River, 14 km NE of Rathwell (specimen at Manitoba Museum of Man and Nature in Winnipeg); and (6) a large upper third molar of a Columbian mammoth (*Mammuthus columbi* = *Parelephas jeffersoni*) from a gravel pit 15 km S of Rivers. The Rathwell and Rivers teeth are illustrated (Figures 2-5). [For more recent reviews of Manitoba specimens see Nielsen et al. 1988.]

1332. Young, R.R. (1992): Late Pleistocene geomorphology of the Hand Hills area, southern Alberta. *In*: Isotrace Laboratory 1991 Annual Report (A.E. Litherland and W.E. Kieser eds.). p. 17.
Several bone-collagen radiocarbon dates were obtained from specimens of Middle Wisconsinan Churcher's prairie dog (*Cynomys niobrarius churcherii* = *Cynomys churcherii*) found in the Hand Hills area of south-central Alberta. The dates established that the prairie dogs occupied the area from at least 33,000 to about 22,000 BP. These dates show that during the Middle Wisconsinan, ice-free conditions persisted in Alberta several thousand years longer than previously supposed. Using the fossils and the commonly occurring burrow casts of the prairie dog as a marker, it was possible to reconstruct the Late Pleistocene stratigraphy of the Hand Hills, showing that the Late Wisconsinan Laurentide ice sheet submerged these high-elevation sites. Glaciotectonic deformation of the prairie dog's burrow casts indicates that the ice-flow direction during maximum stages of the Late Wisconsinan glaciation was from NW to SE.

1333. Young, R.R., Rains, R.B., and Osborn, G. (1998): Comment on 'Geology and Quaternary environments of the first preglacial palaeolithic sites found in Alberta, Canada' by Jiri Chlachula. Quaternary Science Reviews 17:449–53.
The authors do not assert that the rock pieces described by Chlachula are not human (*Homo sapiens*) artifacts. However, they think that his unwavering confidence in his interpretation of the impossibility of a natural origin for the rock pieces may be misleading to those not familiar with the sites. They state that the Calgary sites 'seem to lack that combination of evidence that would lead to general acceptance.' They mention (p. 451) evidence for an extensive vertebrate fauna and related stratigraphy from 30 sites around Alberta that exceeds 10,000 specimens from 25 species. This record shows that grazing herbivores reflecting steppe-like habitats dominated for at least the 15,000 years preceding Late Wisconsinan glaciation. Although wood was present before about 35,000 BP, the occurrence of bone for dating after that was probably because of the onset of more steppe-like conditions. This interpretation differs greatly from Chlachula's, which is partly based on 'fresh-looking and well-preserved' pollen from the Varsity Estates site. [See Chlachula 1998.]

1334. Young, R.R., Burns, J.A., Smith, D.G., Arnold, L.D. and Rains, R.B. (1994): A single, Late Wisconsin, Laurentide glaciation, Edmonton area and southwestern Alberta. Geology 22:683–6.
The absence of igneous and metamorphic clasts of Canadian Shield origin in the Saskatchewan gravels and sands (preglacial) valley fills in seven active gravel pits studied over a five-year period, coupled with 29 radiocarbon dates (42,910–21,300 BP), suggest that Laurentide

glaciation prior to the Late Wisconsinan did not reach the Edmonton area, Alberta. It is likely that large areas of higher elevation in Alberta, S and W of Edmonton, only underwent the Late Wisconsinan event, not multiple events. Dates of some of the fossils include: proboscidean bone 36,150 ± 2960 (AECV-940c); mammoth (*Mammuthus*) bone 27,860 ± 880 (AECV-721c); tusk 28,890 ± 960 (AECV-612c); pelvis 22,820 ± 520 (AECV-538c); humerus 26,750 ± 790 (AECV-1102c); bone 37,120 ± 2370 (AECV-934c); bone 38,960 ± 3520 (AECV-935c); bone 22,750 ± 1650 (S-1964); pelvis 26,050 ± 880 (I-10650); tusk 39,960 ± 3950 (AECV-718c); horse (*Equus*) tibia 29,380 ± 4970 (AECV-720c); humerus 21,330 ± 340 (AECV-1664c); metatarsal 31,220 ± 1260 (AECV-1202c); radio-ulna 27,520 ± 850 (AECV-937c); metapodial 35,840 ± 2370 (AECV-938c); metatarsal >41,090 (AECV-1493c); bison (*Bison*) humerus 25,210 ± 760 (AECV-1201c); radio-ulna 31,750 ± 850 (AECV-936c); horncore >41,260 (AECV-942c), ground sloth (*Megalonyx*) vertebra >41,610 (AECV-1315c); deer (Cervidae) antler 31,520 ± 450 (TO-1828); caribou (*Rangifer*) antler 27,730 ± 1060 (AECV-599c). So, for the late Middle Wisconsinan in the Edmonton and Hand Hills area, only bone was available for dating, probably because of the onset of more steppe-like conditions (note the species present) then. The authors conclude it is likely that large areas of higher elevation in Alberta, S and W of Edmonton, only underwent the Late Wisconsinan event, not multiple glaciations.

1335. Youngman, P.M. (1986): The extinct short-faced skunk *Brachyprotoma obtusata* (Mammalia, Carnivora): First records for Canada and Beringia. Canadian Journal of Earth Sciences 23(3):419–24.
A mandible (CMN 25529) of the extinct short-faced skunk (*Brachyprotoma obtusata*) was collected at Old Crow Loc. 44, Yukon, from sandy sediments overlying the basal clay unit. Other specimens from the same fossil-bearing sediments gave dates from >39,000 BP (I-3572) to >54,000 BP (GSC-2066); hence, a Sangamonian interglacial age is postulated for the specimen. Another record of this species, a lower canine (MgVo-1-D8-4-10) from Bluefish Cave I, Yukon, in collections of the Archaeological Survey of Canada is identified tentatively and because of possible cryoturbation can only be aged at 10,000 BP or older. Both specimens are illustrated photographically in three views. The short-faced skunk specimen from the Old Crow area represents a paleorange extension of more than 4300 km to the NW of other known sites. This species is now known to have occurred both in Beringia and in the refugium to the S of the main ice sheets. The author argues that the extinction of some small mammals, including the short-faced skunk, and the retreat of the ranges of others concurrent with the extinction or change in distribution of the megafauna emphasize environmental change as the overall regulatory factor rather than human (*Homo sapiens*) impact.

1336. Youngman, P.M. (1987): Freeze-dried mummies are hot items with paleontologists. BIOME 7(2):4.
The author mentions an earlier report (Walker 1984) of a mummified arctic ground squirrel (*Spermophilus parryii*) about 47,000 years old found in 1981 by a placer miner at Sixtymile, Yukon. During the summer of 1984, Lynn and Chuck McDougall took a small carcass from the family dog, and passed it on to Dick Harington for study. It turned out to be a black-footed ferret (*Mustela nigripes*) that was AMS radiocarbon dated at about 40,000 years old. The mummy, an adult male, is exceptionally well preserved – the colour of the fur differs little from pelts of this species trapped 30 years ago. An X-ray showed that the ferret had remains of its last meal [probably a vole (*Microtus*) according to hair analysis] in its intestine. The author sent bits of muscle to a laboratory in California in the hope that some DNA might be recovered and compared with that from the living black-footed ferret (that survives in small numbers in western United States) and near relatives in Asia. The proximity of the two mummies may indicate that the deaths and preservation of both animals may have occurred during a relatively dry climatic event in the middle of the last (Wisconsinan) glaciation in northwestern North America. The author also mentions collecting a 1300-year-old jaw of this species from northern Yukon.

1337. Youngman, P.M. (1991): Freeze-dried mummies are hot items with paleontologists. Klondike Sun, June 13, p. 34.
This article is a reprint of one published (Youngman 1987). It differs from the original in that Palma Berger of Dawson City has included the following note at the end detailing the discovery of the black-footed-ferret (*Mustela nigripes*) carcass from Sixtymile, Yukon: 'The 40,000 year old "weasel" was unearthed as the Mc-

Dougalls washed down a hillside with their monitor. The resulting muck was diverted into ditches to be channeled into the sluice box. It was in one of these ditches that Molly found the animal … Molly is a Border Collie into whose ancestry a Dachshund sneaked, hence the short legs that kept her low to the ground. But her "fossilling" days are over, as the washed ground has now become overgrown with willows and rabbits abound. Thus Molly's interests have been diverted elsewhere, and science suffers.'

1338. Youngman, P.M. (1993): The Pleistocene small carnivores of Eastern Beringia. Canadian Field-Naturalist 107(2):139–63.
Twenty-six species of fissiped carnivore (four canids, 13 mustelids, one hyaenid, four felids, and four ursids) lived in Eastern Beringia during the Pleistocene. Of these species, six are extinct: the dhole (*Cuon*); giant short-faced bear (*Arctodus simus*); short-faced skunk (*Brachyprotoma obtusata*); an unidentified mephitine (skunk); one hyaena (Hyaenidae); scimitar cat (*Homotherium serum*); sabretooth cat (*Smilodon fatalis*) [there appears to be no good evidence of this species from Eastern Beringia]; and four species [black-footed ferret (*Mustela nigripes*), Beringian ferret (*Mustela eversmanni*), American badger (*Taxidea taxus*), and American lion (*Panthera leo*)] that no longer live in the region. The status of 568 specimens, representing 17 species of small carnivores, is examined. Those not mentioned previously include: arctic fox (*Alopex lagopus* = *Vulpes lagopus*), red fox (*Vulpes vulpes*), lynx (*Felis lynx*), sea otter (*Enhydra*), wolverine (*Gulo gulo*), river otter (*Lutra canadensis*), American marten (*Martes americana*), fisher (*Martes pennanti*), least weasel (*Mustela nivalis*), and mink (*Mustela vison*). Radiocarbon dates mentioned include: a dentary of a domestic dog (*Canis familiaris*) from Old Crow Loc. 11A, Yukon AMS dated at 2110 ± 40 BP (TO-276); a phalanx of cougar (*Felis concolor*) from Bluefish Cave III AMS dated at 18,970 ± 1490 BP (TO-1266); a large jaw of a Beringian ferret (*Mustela eversmanni*) from Bluefish Cave III AMS dated at 33,500 ± 350 BP (TO-1196); and partial carcasses of black-footed ferret from Sixtymile Loc. 3, Yukon, and Dawson Loc. 12, Yukon, AMS dated at 39,560 ± 490 BP (TO-214) and 30,370 ± 560 BP (Beta 23,347, ETH 3537), respectively.

1339. Youngman, P.M. (1994): Beringian ferrets: Mummies, biogeography, and systematics. Journal of Mammalogy 75(2):454–61.
Two mummified black-footed ferrets (*Mustela nigripes*) are reported from Pleistocene deposits in the Yukon (see map, Figure 1). One, CMN 43786, from Sixtymile River Loc. 3, Yukon (64°00′N, 140°44′W), was AMS dated to 39,560 ± 490 BP (TO-214). The other, a partial mummy, CMN 44305, from Hunker Creek Loc. 12 (63°58′N, 138°59′W), was AMS dated to 30,370 ± 560 BP (Beta-23347). A dentary (ASC 85-139a) [should be ASC M-9-66], collected at Bluefish Cave III (67°08′N, 140°47′35″W), one of six dentaries and a partial skull identified as a Beringian ferret (*Mustela eversmanni*) was AMS dated to 33,550 ± 350 BP (TO-1196). The presence of both species in the western Yukon during the Mid-Wisconsinan (between 39,560 BP and 30,370 BP) suggests that they lived together in a grassland environment in the Beringian refugium.

1340. Youngman, P.M., and Schueler, F.W. (1991): *Martes nobilis* is a synonym of *Martes americana*, not an extinct Pleistocene-Holocene species. Journal of Mammalogy 72(3):567–77.
Large fossil martens from Pleistocene and Early Holocene cave deposits in western North America have been assigned to the supposedly extinct noble marten (*Martes nobilis*) (e.g., Old Crow, Yukon specimens). After examining quantitative differences between these fossils and American marten (*Martes americana*), comparing two distinctive subspecies, the authors found the fossils to be indistinguishable from the modern *caurina* subspecies except for size. Accordingly, the noble marten and American marten are synonymized. [See Anderson 1970, 1984; Harington 1977a.]

1341. Yurtsev, B.A. (2001): The Pleistocene 'Tundra-Steppe' and the productivity paradox: The landscape approach. Quaternary Science Reviews 20:165–74.
'Tundra-steppe' means either a certain type of plant community with co-dominance of both steppe and tundra species (including prostrate shrubs), or a type of landscape co-dominated by both steppe and tundra. A discrepancy between Pleistocene glacial climates that were much colder and drier than at present in Beringia and the highly diverse herbivorous fossil fauna (the 'productivity paradox') is explained in terms of much greater diversity of herbaceous vegetation (grasses, sedges, and forbs) in

the mosaic of Beringian 'tundra-steppe' landscapes. Analysis of the relic distribution of some predominantly herbaceous plant communities throughout Beringia provides the phytogeographic and landscape-ecological grounds for the reconstruction of plant cover of these landscapes.

1342. Zazhigin, V.S. (1984): Early evolutionary stages of collared lemmings (Dicrostonychini, Microtinae, Rodentia) as characteristic representatives of Beringian subarctic fauna. *In*: Beringia in the Cenozoic Era (V.L. Kontrimavichus, ed.). Amerind Publishing Co. Pvt. Ltd., New Delhi. pp. 367–78.
The succession of morphological changes in the molars (especially M^{1-3}) suggests that the evolution of the Tribe up to the Middle Pleistocene may be represented by a single phyletic line *Predicrostonyx hopkinsi* – *Predicrostonyx*(?) *compitalis* – *Dicrostonyx renidens* – *Dicrostonyx simplicior*. The continuous evolutionary history of collared lemmings in the climatic conditions of the Beringian subarctic, from the Pliocene to the present, gives good grounds to consider the subarctic as the area of origin.

1343. Zazhigin, V.S. (1997): Late Pliocene and Pleistocene rodent faunas in the Kolyma Lowland: Possible correlations with North America. *In*: Terrestrial Paleoenvironmental Studies in Beringia (M.E. Edwards, A.V. Sher, and R.D. Guthrie, eds.). Alaska Quaternary Center, University of Alaska, Fairbanks. pp. 25–9.
The Kolyma Lowland in NE Siberia has produced an abundance of mammalian (Mammalia) fossils from Early–Late Pleistocene and even as early as Late Pliocene. Most sites have provided smaller bones as well as the larger ones. The fossils demonstrate a clear chronological trend of morphological development through the evolutionary history of collared lemmings and suggests that the group may be presented as a single phyletic lineage: *Predicrostonyx hopkinsi* – *Predicrostonyx compitalis* – *Dicrostonyx renidens* – *Dicrostonyx simplicior* – *Dicrostonyx gulielmi* – *Dicrostonyx torquatus* [see Zazhigin 1984]. Practically all of these evolutionary stages have been recorded in Eastern Beringia, primarily from sites in the Old Crow Basin. This shows that collared lemmings had an amphiberingian distribution during the last 1 to 1.5 Ma and evolved synchronously in Western and Eastern Beringia. Specimens identified as *Dicrostonyx* cf. *Dicrostonyx simplicior* by Morlan (1984) are referred to by the author as *Dicrostonyx renidens*. Faunal exchange between Western and Eastern Beringia was intensive during the Late Pliocene and Quaternary. At least two genera – *Synaptomys* and *Pliophenacomys* – were involved in that exchange in the Late Pliocene. During the Quaternary the exchange involved *Ochotona*, *Mimomys*, *Cromeromys*, *Phenacomys*, and *Allophaiomys*. [See Repenning 1980, 1985; Repenning et al. 1987.]

1344. Zimmerman, M.R., and Tedford, R.H. (1976): Histologic structures preserved for 21,300 years. Science 194(4261):183–4.
Preserved tissues of Pleistocene mammals from the Fairbanks area, Alaska (from the Childs Frick Collection of the American Museum of Natural History) were sampled and histological preparations were made. The following were investigated: woolly mammoth (*Mammuthus primigenius*) forefoot and face F:AM 99927; hare (*Lepus*) F:AM 99926; lynx (*Felis lynx* = *Lynx*) F:AM 99925; lemming or vole (Rodentia) F:AM 99928; and horse (*Equus*) F:AM 99929. The mammoth had been radiocarbon dated to an age of 21,300 ± 1300 years BP (L-601). On the basis of stratigraphic evidence, the rabbit, lynx, and rodent are aged at between 15,000 to 25,000 BP. The horse, from Gold Hill, is known to be pre-Wisconsinan (?Illinoian) in age. Histological structure was found to be preserved in several of the specimens, although most showed clearly that a considerable degree of decay had taken place before entombment and preservation, contrary to the popular notion that mummified remains indicate rapid freezing under conditions of catastrophic climatic change. Photomicrographs of a mammoth eye (Figure 1) and hare liver (Figure 2) are provided.

1345. Zwarun, S. (1979): The mystery of an old child. Maclean's, September 3. pp. 42–3.
This article provides background information on the Taber child (*Homo sapiens*) remains at Woodpecker Island bluff, N of Taber, Alberta. The specimen (a photograph of the parts are included) was found on a sandbank above the Oldman River by Archie Stalker's field assistant nearly two decades earlier [see Langston and Oschinsky 1963] and was considered to be between 58,000 and 22,000 years old. The specimen itself was too

small and too valuable to be destroyed in order to provide a radiocarbon date. [See Stalker 1977; Cohen 1982.]

1346. Zymela, S., Schwarcz, H.P., Grün, R., Stalker, A.M., and Churcher, C.S. (1988): ESR dating of Pleistocene fossil teeth from Alberta and Saskatchewan. Canadian Journal of Earth Sciences 25:235–45.
Teeth collected from Pleistocene surficial sediments along the South Saskatchewan River in Alberta and Saskatchewan have been dated by electron spin resonance (ESR). The age of tooth enamel, as determined by ESR analyses, generally increases with increasing absolute age, as inferred from independent geological or paleontological criteria. There are, however, major discrepancies between the results and previous age estimates and also within the ESR results themselves. The authors provide ESR ages for the sites and also supply a table with comparative dates of the geologic ages using both ESR and conventional dating. The paper is mainly a discussion of the validity of ESR dating; however, one specimen is identified to species and locality: a steppe bison (*Bison priscus*) from Lindoe Bluff (50°04′50″N, 110°39′35″W), Alberta. ESR data for this site, obtained on three bison teeth, yielded dates in the ranges of 12,000–8000 BP and 20,000 –14,000 BP. Wood from this gravel bed was radiocarbon dated at 11,200 ± 200 BP (GSC-805) (Lowdon and Blake 1968).

ADDENDUM

Seton, E.T. (1929): Life LVIII – The muskox. *In*: Lives of Game Animals 3(2). Doubleday, Doran and Company, Inc. pp. 595–637.
Pleistocene records of the tundra muskox (*Ovibos moschatus*) listed (p. 604) are: (1) R.G. McConnell's report of remains from low-level creek gravels in the Klondike District, Yukon; (2) part of a cranium with horncores of a small individual (*Praeovibos* = *Ovibos*) from Upper Porcupine River, Yukon, in the British Museum (Lydekker 1885); (3) a tooth from Pleistocene gravel near Midway, British Columbia, reported by Lambe (1907) [the specimen looks like a modern sheep (*Ovis*) tooth]; and (4) a muskox skull from 'old gravel' near Edmonton, Alberta, reported by G.M. Dawson (1901a).

INTRODUCTION TO THE RADIOCARBON-DATE TABLE

This radiocarbon-date table (Table 1) attempts to list systematically [arbitrarily following Simpson (1945) because it corresponds to the organization of CMN's Quaternary Zoology Collection and my Quaternary Mammal Species File] direct dates on bone (mammalian soft tissues, beaver-cut sticks, ground squirrel nesting grasses, and droppings, etc., being exceptions) of vertebrates (fishes, amphibians, reptiles, birds, and mammals) from northern North America (Canada, Alaska, and Greenland). Unless otherwise mentioned, the dates are on bone collagen.

To be most useful to paleontologists, zoologists, and zooarchaeologists, the taxa are grouped as follows: Class, Order, Family, Genus, and Species. Following this information are columns giving Measured Date and/or Normalized Date (the latter in parentheses) and laboratory numbers (Lab No) for each specimen. A complementary table provides a key to designation and other information on the various laboratories mentioned. Where not given by the laboratory, often normalized dates have been derived from information provided by A.S. Dyke and R.E. Morlan. Also, the dates have been listed in approximate order from oldest to youngest so that the known span and incidence of dates for a given genus or species may be more easily recognized. Where known, the specimen sampled has been designated under Material, and in some cases, field or specimen numbers are given. If there is no entry in this column, it is assumed that the material dated is bone.

Where possible, location is designated by a place name or distance and direction from a place as well as latitude, longitude, and elevation, supplemented by Province/Territory/State and Country (e.g., AB – Alberta; AK – Alaska; BC – British Columbia, etc.; regions of Greenland have not always been designated). The final column, References/Remarks, gives names and dates of references in which the specimen dated and/or its radiocarbon date are mentioned. Significant comments on the specimen or date may also be included.

Although an effort has been made to note controversial dates (see the References/Remarks column), dates listed have not been selected, 'refined,' or 'sanitized' [e.g., using the rating system developed and applied by Meade, J.I., and Meltzer, D.J. (1984): North American Late Quaternary extinctions and the radiocarbon record. *In:* Quaternary Extinctions: A Prehistoric Revolution (P.S. Martin and R.G. Klein, eds.). University of Arizona Press, Tucson. pp. 440–50]. So users must be cautious.

Data are computerized so that they may be added to, changed, or removed, and so that specific information can be gained from the table (e.g., 'List of all radiocarbon-dated bison from Alaska'). This is only an initial attempt to put such radiocarbon data in readily usable order, so, if you see ways of correcting and augmenting the information, please advise me. I am particularly grateful to Dr R.E. Morlan (Archaeological Survey of Canada, Canadian Museum of Civilization), as well as Drs Roger McNeely and A.S. Dyke (Terrain Sciences Division, Geological Survey of Canada) for sharing their databases and providing particular information such as normalized dates. Additional radiocarbon dates from SE Alaska and Alberta were supplied by Drs T.H. Heaton (University of South Dakota, Vermillion) and Alwynne Beaudoin (Provincial Museum of Alberta, Edmonton) respectively. Jennifer Leonard (University of California at Los Angeles) kindly supplied eight unpublished dates on Alaskan Pleistocene horses, and Rick Farnell (Yukon Fish and Wildlife Branch, Whitehorse) contributed two Late Holocene bison dates from southern Yukon. Dr C. Carlson (University College of the Cariboo, Kamloops) provided important information on dates for two fishes from the Kamloops area, BC.

The following publications have been useful in checking data in Table 1: Geological Survey of Canada Radiocarbon Dates I and II, Geological Survey of Canada Paper 63-21 (1963) to Geological Survey of Canada Radiocarbon Dates XXXII, Geological Survey of Canada Current Research 1995-G (1996); and for British Columbia, Alberta, and Saskatchewan – Late Quaternary Geology and Geochronology of British Columbia by J.J. Clague (Part 1: Radiocarbon Dates, Geological Survey of Canada Paper 80-13, 1980, and Part 2: Summary and Discussion of Radiocarbon-Dated Quaternary History, Geological Survey of Canada Paper 80-35, 1981); Alberta Radiocarbon Dates by L.E. Jackson, Jr, and M. Pawson (Geological Survey of Canada Paper 83-25, 1984), and A Compilation and Evaluation of Radiocarbon Dates in Saskatchewan by R.E. Morlan (Saskatchewan Archaeology 13:3–84, 1993).

For database information, please consult the following annotations: Danish/Greenlandic Network (1999): Archaeological Research in Greenland; McNeely, R. (2000). The Geological Survey of Canada 'Date Locator File' (includes hundreds of dates from Arctic marine mammals collected by Dr A.S. Dyke and his field parties under 'Bone Dates'); Morlan, R.E. (1999): Canadian Archaeological Radiocarbon Database; and Nolin, L., and Cinq-Mars, J. (1994): NOGAP Radiocarbon Dates.

RADIOCARBON-DATE LABORATORIES

Laboratory Acronym	*Laboratory Name And Address*
A	University of Arizona (USA)
AA	University of Arizona Accelerator, Tucson, Arizona (USA)
AAR	University of Aarhus (DENMARK)
AECV*	Alberta Environmental Centre at Vegreville, Alberta (CAN)
Beta	Beta Analytic Inc., Miami, Florida (USA)
BGS	Brock Geological Sciences, Brock University, St Catherines, Ontario (CAN)
BM	British Museum, London (UK)
CAMS	Center for Accelerator Mass Spectrometry, Lawrence Livermore Laboratories, California (USA)
CRNL*	Chalk River Nuclear Laboratories, Chalk River, Ontario (CAN)
DIC*	Dicarb Radioisotope Company (USA)
GaK	Gakushuin University, Tokyo (JAPAN)
Gif	Gif sur Yvette (FRANCE)
GSC	Geological Survey of Canada, Ottawa, Ontario (CAN)
GX	Geochron Laboratories Inc., Cambridge, Massachusetts (USA)
I	Isotopes – a Teledyne Company, Westwood, New Jersey (USA)
K	Danish National Museum, Copenhagen (DENMARK)
L	Lamont-Doherty Geological Observatory, Columbia University, Palisades, New York (USA)
LU	Leningrad University, St Petersburg State University (RUSSIA)
M*	University of Michigan, Ann Arbor, Michigan (USA)
NSRL	Nuclear Structure Research Laboratory, Institute of Arctic and Alpine Research, University of Colorado, Boulder, Colorado (USA)
NUTA	Nagoya University Tandem Accelerator Centre, Nagoya (JAPAN)
OxA	Oxford Radiocarbon Accelerator Unit (UK)
P*	University of Pennsylvania, Philadelphia, Pennsylvania (USA)
QC*	Queens College, City University of New York, Flushing, New York (USA)
QL	Quaternary Isotope Laboratory, University of Washington, Seattle, Washington (USA)
QU*	Centre de Recherches Minérales, Ministère des Richesses Naturelles, Gouvernement du Québec (CAN)
RIDDL*	Radioisotope Direct Detection Laboratory, McMaster University, Hamilton, Ontario (CAN)
RL*	Radiocarbon Ltd., New York (USA)
S*	Saskatchewan Research Council, Saskatoon, Saskatchewan (CAN)
SFU*	Simon Fraser University, Burnaby, British Columbia (CAN)
SI*	Smithsonian Institution, Washington, DC (USA)
SM*	Socony & Mobil Oil (USA)
St	Stockholm (SWEDEN)
TO	IsoTrace Laboratory, University of Toronto, Toronto, Ontario (CAN)
U	Fysiska institutionen, Uppsala Universitet, Uppsala (SWEDEN)
Ua	Uppsala Tandem Accelerator Laboratory (SWEDEN)
Uga	University of Georgia, Athens, Georgia (USA)
UQ	University of Quebec at Montreal, Montreal, Quebec (CAN)
USGS	United States Geological Survey, Menlo Park, California (USA)
W	United States Geological Survey, Reston, Virginia (USA)
WAT	University of Waterloo, Waterloo, Ontario (CAN)

*Laboratories no longer operating.

LOCATION ABBREVIATIONS

Province/Territory/ State/Country	*Abbreviation*
Alaska	AK
Alberta	AB
British Columbia	BC
Canada	CAN
Greenland	GRE
Manitoba	MB
New Brunswick	NB
Newfoundland and Labrador	NF
Nova Scotia	NS
Northwest Territories	NT
Nunavut	NU
Ontario	ON
Prince Edward Island	PE
Quebec	QC
Saskatchewan	SK
United States of America	USA
Yukon	YT

RADIOCARBON-DATE TABLE

Table 3. Radiocarbon Dates on Quaternary Vertebrates of Northern North America

Pisces (Fish)

Salmoniformes, Salmonidae

Onchorhynchus sp. (salmon)

Measured Date (Normalized)	Lab No	Material	Location	References/Remarks
18110 ± 90 (18160 ± 90)	Beta-76622 CAMS-16369	skeleton	SE shore of Kamloops L., Cherry Bluff (50°43'N, 120°36'N), BC, CAN	[Carlson and Klein (1996) - specimen recovered from a buried nodule]
15480 ± 60 (15500 ± 60)	Beta-70750 CAMS-11912	skeleton	SE shore of Kamloops L., Cherry Bluff (50°43'N, 120°36'N), BC, CAN	[Carlson and Klein (1996) - specimen recovered from a buried nodule]
(1210 ± 60)	CAMS-15363		Lyell Bay, N Midden, Gwaii Haanas, Queen Charlotte I. (52°39'59"N, 131°39'50"W), BC, CAN	[Southon and Fedje (2000) In: R.E. Morlan, CMC database (1999)]
(1030 ± 70)	CAMS-15358		Echo Point Town, Gwaii Haanas, Queen Charlotte I. (52°41'39"N, 131°44'57"W), BC, CAN	[Southon and Fedje (2000) In: R.E. Morlan, CMC database (1999)]

Stenodus leucichthys (inconnu)

Measured Date (Normalized)	Lab No	Material	Location	References/Remarks
4420 ± 50 (4440 ± 50)	Beta-115210	left dentary (CR-97-65)	Old Crow Basin (approx. 68°05'N, 139°45'W), YT, CAN	C.R. Harington [unpublished]

Gadiformes, Gadidae

Gadus morhua (Atlantic cod)

Measured Date (Normalized)	Lab No	Material	Location	References/Remarks
640 ± 50 (700 ± 50)	GSC-3283	vertebrae	Red Bay, NF (Labrador), CAN	R.E. Morlan, CMC database (1999); S.L. Cumbaa, pers. comm. (2000)

Perciformes, Xiphiidae

Xiphias gladius (swordfish)

Measured Date (Normalized)	Lab No	Material	Location	References/Remarks
340 ± 80 (585 ± 85)	GSC-1855	rostrum (CMN 12456)	Reservoir 2.25 km WNW of St-Louis-de-Champlain (46.5°25.7'N, 72.5°37.5'W), QC, CAN	R. McNeely, GSC cardbox database (1999); R.E. Morlan, CMC database (1999). [Not Champlain Sea, probably imported by humans].

Perciformes, Scorpaenidae

Sebastes sp. (rosefish)

Measured Date (Normalized)	Lab No	Material	Location	References/Remarks
(950 ± 90)	CAMS-15359		Echo Point Town, Gwaii Haanas, Queen Charlotte I. (52°41'39"N, 131°44'57"W), BC, CAN	[Southon and Fedje (2000) In: R.E. Morlan, CMC database (1999)]

Aves (Birds)

Gaviiformes, Gaviidae

Gavia sp. (loon)

Measured Date (Normalized)	Lab No	Material	Location	References/Remarks
(3920 ± 60)	TO-6032	ulna	Threepoint L., 110 km W of Thompson drainage, MB, CAN	[E.L. Syms, pers. comm. (1998) In: R.E. Morlan, CMC database (1999)]

Anseriformes, Anatidae

Chen caerulescens (Snow Goose - Blue Goose)

Measured Date (Normalized)	Lab No	Material	Location	References/Remarks
(39020 ± 690)	TO-2698	left ulna shaft (CMN 38229)	Sixtymile area (Sixtymile Loc. 3), 701 m asl., (64°1.00'N, 140°44'W), YT, CAN	C.R. Harington [unpublished]; Harington (1997a)

Olor columbianus (Whistling Swan)

Measured Date (Normalized)	Lab No	Material	Location	References/Remarks
(11500 ± 80)	CAMS-8261		Broken Mammoth site, near confluence of Shaw Ck. and Tanana R, about 22 km NW of Delta Junction (64°16'N, 146°07'W), AK, USA	Holmes (1996)

Charadriiformes, Alcidae

Alle alle (Dovekie)

Measured Date (Normalized)	Lab No	Material	Location	References/Remarks
35560 ± 1550 (35500 ± 1500)	Ua-2212	sternum (CMN 43813)	Cape Storm, Ellesmere I., NU, CAN	Blake (1980, 1996); Harington (1990b); R.E. Morlan, CMC database (1999)

Strigiformes, Strigidae

Nyctea scandiaca (Snowy Owl)

Measured Date (Normalized)	Lab No	Material	Location	References/Remarks
13350 ± 100 (13630 ± 100)	Beta-129151		Bluefish Cave II, 600 m asl. (67°08'N, 140°47'W), YT, CAN	C.R. Harington [unpublished]; [See also McCuaig-Balkwill and Cinq-Mars (1998)]

Passeriformes, Corvidae

Corvus corax (Common Raven)

Measured Date (Normalized)	Lab No	Material	Location	References/Remarks
(10290 ± 100)	CAMS-2317	scapula	Charlie L. Cave, NW of Fort St. John (56°16'35"N, 120°56'15"W), BC, CAN	Driver, et al. (1996, 1998b); R.E. Morlan, CMC database (1999)
(9490 ± 140)	CAMS-2138		Charlie L. Cave, NW of Fort St. John (56.275°N, 120.930°W), BC, CAN	[Driver, et al. (1996) - incorrectly cited as CAMS-2318]; [Fladmark, et al. (1988, 1996); Driver and Hobson (1992); In: R.E. Morlan, CMC database (1999)]

Mammalia (Mammals)

Primates, Hominidae

Homo sapiens (humans)

Measured Date (Normalized)	Lab No	Material	Location	References/Remarks
(9880 ± 50)	CAMS-3238	pelvis	49-PET-408 Cave, northern Prince of Wales I., AK, USA	Lysek (1997); [See Dixon et al. (1997) - oldest reliably dated human in Alaska]
(9730 ± 60)	CAMS-29873	mandible	49-PET-408 Cave, northern Prince of Wales I., AK, USA	Lysek (1997)
(8480 ± 390)	CRNL-1216	calcined bone	Lake Superior, Thunder Bay District (48°24'23"N, 89°21'19"W), ON, CAN	[Dawson (1983); Julig (1994) In: R.E. Morlan, CMC database (1999)]
8250 ± 115 (8340 ± 115)	S-1737	humerus (CMC-1085)	Gore Ck., S Thompson R. (50°41'50"N, 119°49'33"W), BC, CAN	[Cybulski, et al. (1981); Chisholm and Nelson (1983); Stryd and Rousseau (1996) In: R.E. Morlan, CMC database (1999)]
(7080 ± 90)	TO-1942		E end of Big Trout L., Kenora District, ON, CAN	[Hamilton (1991); LeBlanc (1992) In: R.E. Morlan, CMC database (1999)]

(6800 ± 90)	TO-1941		E end of Big Trout L., Kenora District, ON, CAN	[Hamilton (1991); LeBlanc (1992) In: R.E. Morlan, CMC database (1999)]
(6630 ± 90)	TO-1943		E end of Big Trout L., Kenora District, ON, CAN	[Hamilton (1991); LeBlanc (1992) In: R.E. Morlan, CMC database (1999)]
6350 ± 110 (6450 ± 110)	GaK-6494		Whitemouth Falls, S of Lac du Bonnet, 267 m asl (50.117°N, 96.017°W), MB, CAN	[Buchner and Pujo (1977); Buchner (1979); Ens (1998) In: R.E. Morlan, CMC database (1999)]
6230 ± 80 (6330 ± 80)	S-1596	innominate, femoral head (CMC-1004)	Garden I., Prince Rupert Harbour, BC, CAN	[Wilmeth (1978); MacDonald and Inglis (1981); Ames (1998) In: R.E. Morlan, CMC database (1999)]
(6090 ± 90)	NSRL-3128		Whitemouth Falls, Whitemouth drainage (50°07'10"N, 96°02'00"W), MB, CAN	[Ens (1998) In: R.E. Morlan, CMC database (1999)]
6070 ± 110 (6170 ± 110)	GaK-6493		Whitemouth Falls, S of Lac du Bonnet, 267 m asl (50.117°N, 96.017°W), MB, CAN	[Buchner and Pujo (1977); Buchner (1979); Ens (1998) In: R.E. Morlan, CMC database (1999)]
5910 ± 165 (6010 ± 165)	GX-5193G		Albion twp., 328-335 m asl., Peel County, ON, CAN	[Katzenberg and Sullivan (1979) In: R.E. Morlan, CMC database (1999)]
5620 ± 320 (5720 ± 320)	SFU-297		About 8 km NW of Swift Current, South Saskatchewan valley, 747 m asl (50.350°N, 107.900°W), SK, CAN	[Longin (1971); So and Wade (1975); Millar (1978, 1981); Gibson (1981); Chisholm, et al. (1983); Walker (1984); Callaghan (1986); Lovell, et al. (1986) In: R.E. Morlan, CMC database (1999)]; Morlan (1993); [Morlan and McNeely (1996) In: A.S. Dyke (GSC database)]
(5590 ± 100)	SFU-344	long-bones	Namu R. (51.860°N, 127.860°W), BC, CAN	[Cannon (1991) In: A.S. Dyke (GSC database)]; [Chisholm (1986); Carlson (1991, 1996) In: R.E. Morlan, CMC database (1999)]
5320 ± 160 (5420 ± 160)	SFU-296		About 8 km NW of Swift Current, South Saskatchewan valley, 747 m asl (50.350°N, 107.900°W), SK, CAN	[Longin (1971); So and Wade (1975); Millar (1978, 1981); Gibson (1981); Chisholm, et al. (1983); Walker (1984); Callaghan (1986); Lovell, et al. (1986) In: R.E. Morlan, CMC database (1999)]; Morlan (1993); [Morlan and McNeely (1996) In: A.S. Dyke (GSC database)]
5240 ± 80 (5340 ± 80)	S-509	long-bones (CMC-360)	Outaouais region, QC, CAN	[Wilmeth (1978); Taillon et Barre (1987) In: R.E. Morlan, CMC database (1999)]
(5170 ± 220)	RIDDL-100		Pender Canal, Pender I., Gulf of Georgia (48°45'50"N, 123°15'20"W), BC, CAN	[Chisholm (1986); Carlson and Hobler (1993) In: R.E. Morlan, CMC database (1999)]
5150 ± 160 (5250 ± 160)	SFU-294		About 8 km NW of Swift Current, South Saskatchewan valley, 747 m asl (50.350°N, 107.900°W), SK, CAN	[Longin (1971); So and Wade (1975); Millar (1978, 1981); Gibson (1981); Chisholm, et al. (1983); Walker (1984); Callaghan (1986); Lovell, et al. (1986) In: R.E. Morlan, CMC database (1999)]; Morlan (1993); [Morlan and McNeely (1996) In: A.S. Dyke (GSC database)]
5100 ± 390 (5200 ± 390)	S-647		About 8 km NW of Swift Current, S Saskatchewan valley, 747 m asl (50.350°N, 107.900°W), SK, CAN	[Longin (1971); So and Wade (1975); Millar (1978, 1981); Gibson (1981); Chisholm, et al. 1983); Walker (1984); Callaghan (1986); Lovell, et al. (1986) In: R.E. Morlan, CMC database (1999)]; Morlan (1993); [Morlan and McNeely (1996) In: A.S. Dyke (GSC database)]
4985 ± 75 (5085 ± 80)	S-520	femur	St. Brieux, Lenore L. (52°34'00"N, 104°53'00"W), SK, CAN	[Walker (1984) In: R.E. Morlan, CMC database (1999)]; Morlan (1993)

4955 ± 165 (5055 ± 165)	S-619		About 8 km NW of Swift Current, S Saskatchewan valley, 747 m asl (50.350°N, 107.900°W), SK, CAN	[Longin (1971); So and Wade (1975); Millar (1978, 1981); Gibson (1981); Chisholm, et al. 1983); Walker (1984); Callaghan (1986); Lovell, et al. (1986) In: R.E. Morlan, CMC database (1999)]; Morlan (1993); [Morlan and McNeely (1996) In: A.S. Dyke (GSC database)]
4950 ± 170 (5075 ± 170)	SFU-298		Clinton, Fraser drainage (51°25'N, 122°05'W), BC, CAN	[Chisholm (1986); Kuijt (1989) In: R.E. Morlan, CMC database (1999)]
4900 ± 80 (5000 ± 80)	S-1263	femur (CMC-889)	Lac des Deux-Montagnes, Montreal area, QC, CAN	[Cybulski (1978), Marois (1987), Taillon et Barre (1987) In: R.E. Morlan, CMC database (1999)]
4900 ± 155 (5055 ± 155)	S-2352	ribs (CMC-1297)	Graham I., Queen Charlotte I., BC, CAN	[Severs (1974); Wilmeth (1978) In: R.E. Morlan, CMC database (1999)]
4890 ± 150 (5005 ± 150)	S-2351	ribs (CMC-1296)	Graham I., Queen Charlotte I., BC, CAN	[Severs (1974); Wilmeth (1978) In: R.E. Morlan, CMC database (1999)]
(4860 ± 50)	Beta-88851		SE of Morrison's I., Outaouais region, QC, CAN	[Wilmeth (1978); Taillon et Barre (1987); Clermont and Chapdelaine (1998) In: R.E. Morlan, CMC database (1999)]
4775 ± 130 (4970 ± 130)	S-2326	rib and innominate fragments (CMC-1282)	Namu R. (51.860°N, 127.860°W), BC, CAN	[Cannon (1991) In: A.S. Dyke (GSC database)]; [Chisholm (1986); Carlson (1991, 1996) In: R.E. Morlan, CMC database (1999)]
4750 ± 160 (4850 ± 160)	SFU-295		About 8 km NW of Swift Current, South Saskatchewan valley, 747 m asl (50.350°N, 107.900°W), SK, CAN	[Longin (1971); So and Wade (1975); Millar (1978, 1981); Gibson (1981); Chisholm, et al. (1983); Walker (1984); Callaghan (1986); Lovell, et al. (1986) In: R.E. Morlan, CMC database (1999)]; Morlan (1993); [Morlan and McNeely (1996) In: A.S. Dyke (GSC database)]
4700 ± 125 (4880 ± 125)	S-2328	ribs (CMC-1284)	Namu R. (51.860°N, 127.860°W), BC, CAN	[Cannon (1991) In: A.S. Dyke (GSC database)]; [Chisholm (1986); Carlson (1991) In: R.E. Morlan, CMC database (1999)]
(4680 ± 160)	SFU-342	rib and miscellaneous fragments	Namu R. (51.860°N, 127.860°W), BC, CAN	[Cannon (1991) In: A.S. Dyke (GSC database)]; [Carlson (1991, 1996) In: R.E. Morlan, CMC database (1999)]
(4630 ± 40)	Beta-88852		SE of Morrison's I., Outaouais region, QC, CAN	[Wilmeth (1978); Taillon et Barre (1987); Clermont and Chapdelaine (1998) In: R.E. Morlan, CMC database (1999)]
4625 ± 80 (4725 ± 80)	S-2227		Qu'appelle valley (50°51'48"N, 103°47'18"W), SK, CAN	[Dyck (1983) In: R.E. Morlan, CMC database (1999)]; Morlan (1993)
(4620 ± 40)	Beta-88725		SE of Morrison's I., Outaouais region, QC, CAN	[Wilmeth (1978); Taillon et Barre (1987); Clermont and Chapdelaine (1998) In: R.E. Morlan, CMC database (1999)]
(4600 ± 130)	RIDDL-514		About 8 km NW of Swift Current, South Saskatchewan valley, 747 m asl (50.350°N, 107.900°W), SK, CAN	[Longin (1971); So and Wade (1975); Millar (1978, 1981); Gibson (1981); Chisholm, et al. 1983); Walker (1984); Callaghan (1986); Lovell, et al. (1986) In: R.E. Morlan, CMC database (1999)]; Morlan (1993); [Morlan and McNeely (1996) In: A.S. Dyke (GSC database)]
(4600 ± 170)	RIDDL-513		About 8 km NW of Swift Current, South Saskatchewan valley, 747 m asl (50.350°N, 107.900°W), SK, CAN	[Longin (1971); So and Wade (1975); Millar (1978, 1981); Gibson (1981); Chisholm, et al. 1983); Walker (1984); Callaghan (1986); Lovell, et al. (1986) In: R.E. Morlan, CMC database (1999)]; Morlan (1993); [Morlan and McNeely (1996) In: A.S. Dyke (GSC database)]

(4580 ± 550)	RIDDL-218		Pender I., Gulf of Georgia, BC, CAN	[Borden (1954, 1962); Duff (1955, 1956); Wilmeth (1978); Chisholm (1986); Carlson and Hobler (1993); Dale (1994) In: R.E. Morlan, CMC database (1999)]
4570 ± 120 (4815 ± 125)	Gak-3794a	(CMC-460)	Thames valley, Middlesex County, ON, CAN	[Wilmeth (1978) In: R.E. Morlan, CMC database (1999) - bone apatite date]
(4510 ± 140)	RIDDL-512		About 8 km NW of Swift Current, South Saskatchewan valley, 747 m asl (50.350°N, 107.900°W), SK, CAN	[Longin (1971); So and Wade (1975); Millar (1978, 1981); Gibson (1981); Chisholm, et al. 1983); Walker (1984); Callaghan (1986); Lovell, et al. (1986) In: R.E. Morlan, CMC database (1999)]; Morlan (1993)
4500 ± 145 (4675 ± 145)	S-2349	ribs (CMC-1295)	Graham I., Queen Charlotte I., BC, CAN	[Severs (1974); Wilmeth (1978) In: R.E. Morlan, CMC database (1999)]
(4430 ± 170)	RIDDL-104		Pender Canal, Pender I., Gulf of Georgia (48°45'50"N, 123°15'20"W), BC, CAN	[Chisholm (1986); Carlson and Hobbler (1993) In: R.E. Morlan, CMC database (1999)]
(4420 ± 190)	RIDDL-515		About 8 km NW of Swift Current, South Saskatchewan valley, 747 m asl (50.350°N, 107.900°W), SK, CAN	[Longin (1971); So and Wade (1975); Millar (1978, 1981); Gibson (1981); Chisholm, et al. 1983); Walker (1984); Callaghan (1986); Lovell, et al. (1986) In: R.E. Morlan, CMC database (1999)]; Morlan (1993)
(4390 ± 160)	SFU-343	rib fragments	Namu R. (51.860°N, 127.860°W), BC, CAN	[Cannon (1991) In: A.S. Dyke (GSC database)]; [Carlson (1991, 1996) In: R.E. Morlan, CMC database (1999)]
4390 ± 105 (4490 ± 105)	S-1447	ribs	Greenwater L., near Greenwater L. Provincial Pk., Red Deer drainage, SK, CAN	Morlan (1993); [Walker (1981) In: R.E. Morlan, CMC database (1999)]
4340 ± 250 (4440 ± 250)	GX-3373		S Saskatchewan valley (50.350°N, 107.900°W), SK, CAN	[Longin (1971); So and Wade (1975); Millar (1978, 1981); Gibson (1981); Chisholm, et al. 1983); Walker (1984); Callaghan (1986); Lovell, et al. (1986) In: R.E. Morlan, CMC database (1999)]; Morlan (1993); [Morlan and McNeely (1996) In: A.S. Dyke (GSC database)]
(4320 ± 150)	RIDDL-105		Pender I., Gulf of Georgia, BC, CAN	[Borden (1954, 1962); Duff (1955, 1956); Wilmeth (1978); Chisholm (1986); Carlson and Hobler (1993); Dale (1994) In: R.E. Morlan, CMC database (1999)]
(4320 ± 220)	RIDDL-96		Pender I., Gulf of Georgia, BC, CAN	[Borden (1954, 1962); Duff (1955, 1956); Wilmeth (1978); Chisholm (1986); Carlson and Hobler (1993); Dale (1994) In: R.E. Morlan, CMC database (1999)]
4300 ± 125 (4480 ± 125)	S-2327	ribs (CMC-1283)	Namu R. (51.860°N, 127.860°W), BC, CAN	[Cannon (1991) In: A.S. Dyke (GSC database)]; [Chisholm (1986); Carlson (1991, 1996) In: R.E. Morlan, CMC database (1999)]
4235 ± 100 (4335 ± 100)	S-1220	(CMC-554)	Swift Current valley, S Saskatchewan drainage (50°09'00"N, 107°40'00"W), SK, CAN	[Bryan and Steele (1976) In: R.E. Morlan, CMC database (1999)]; Morlan (1993)
4120 ± 120 (4220 ± 120)	Beta-38357		N of Pt. Roberts, about 2 m asl., Strait of Georgia, BC, CAN	[Bernick and Howe (1990); Hanson (1991); Arcas Consulting (1994) In: R.E. Morlan, CMC database (1999)]
(4070 ± 150)	RIDDL-268		Pender I., Gulf of Georgia, BC, CAN	[Borden (1954, 1962); Duff (1955, 1956); Wilmeth (1978); Chisholm (1986); Carlson and Hobler (1993); Dale (1994) In: R.E. Morlan, CMC database (1999)]
4030 ± 110 (4120 ± 120)	Beta-11902		Belly R., near Cardston (49°16'N, 113°34'W), AB, CAN	[Ball (1986); Beaudoin (1987) In: R.E. Morlan, CMC database (1999)]
(4010 ± 70)	NSRL-3125		Slave Falls, Winnipeg drainage (50°15'N, 95°35'W), MB, CAN	[Wyman (1993); Ens (1998) In: R.E. Morlan, CMC database (1999)]

3970 ± 60 (4165 ± 60)	SFU-541	Pender I., Gulf of Georgia, BC, CAN	[Borden (1954, 1962); Duff (1955, 1956); Wilmeth (1978); Chisholm (1986); Carlson and Hobler (1993); Dale (1994) In: R.E. Morlan, CMC database (1999)]
(3940 ± 140)	RIDDL-107	Pender I., Gulf of Georgia, BC, CAN	[Borden (1954, 1962); Duff (1955, 1956); Wilmeth (1978); Chisholm (1986); Carlson and Hobler (1993); Dale (1994) In: R.E. Morlan, CMC database (1999)]
3930 ± 130 (4030 ± 130)	I-4678	Port au Choix, NF, CAN	[Tuck (1970, 1971, 1976, 1978); Wilmeth (1978); Kennedy (1981, 1987) In: R.E. Morlan, CMC database (1999)]
3880 ± 50 (3980 ± 50)	Beta-38353	N of Pt. Roberts, about 2 m asl., Strait of Georgia, BC, CAN	[Bernick and Howe (1990); Hanson (1991); Arcas Consulting (1994) In: R.E. Morlan, CMC database (1999)]
3800 ± 60 (3900 ± 60)	Beta-38354	N of Pt. Roberts, about 2 m asl., Strait of Georgia, BC, CAN	[Bernick and Howe (1990); Hanson (1991); Arcas Consulting (1994) In: R.E. Morlan, CMC database (1999)]
(3780 ± 180)	RIDDL-103	Pender I., Gulf of Georgia, BC, CAN	[Borden (1954, 1962); Duff (1955, 1956); Wilmeth (1978); Chisholm (1986); Carlson and Hobler (1993); Dale (1994) In: R.E. Morlan, CMC database (1999)]
3755 ± 100 (3855 ± 100)	S-646	About 8 km NW of Swift Current, S Saskatchewan valley, 747 m asl (50.350°N, 107.900°W), SK, CAN	[Longin (1971); So and Wade (1975); Millar (1978, 1981); Gibson (1981); Chisholm, et al. 1983); Walker (1984); Callaghan (1986); Lovell, et al. (1986) In: R.E. Morlan, CMC database (1999)]; Morlan (1993); [Morlan and McNeely (1996) In: A.S. Dyke (GSC database)]
3750 ± 180 (3850 ± 180)	S-707	About 8 km NW of Swift Current, S Saskatchewan valley, 747 m asl (50.350°N, 107.900°W), SK, CAN	[Longin (1971); So and Wade (1975); Millar (1978, 1981); Gibson (1981); Chisholm, et al. 1983); Walker (1984); Callaghan (1986); Lovell, et al. (1986) In: R.E. Morlan, CMC database (1999)]; Morlan (1993); [Morlan and McNeely (1996) In: A.S. Dyke (GSC database)]
(3750 ± 160)	RIDDL-271	Pender I., Gulf of Georgia, BC, CAN	[Borden (1954, 1962); Duff (1955, 1956); Wilmeth (1978); Chisholm (1986); Carlson and Hobler (1993); Dale (1994) In: R.E. Morlan, CMC database (1999)]
3680 ± 80 (3780 ± 80)	Beta-38356	N of Pt. Roberts, about 2 m asl., Strait of Georgia, BC, CAN	[Bernick and Howe (1990); Hanson (1991); Arcas Consulting (1994) In: R.E. Morlan, CMC database (1999)]
(3630 ± 140)	RIDDL-274	Pender I., Gulf of Georgia, BC, CAN	[Borden (1954, 1962); Duff (1955, 1956); Wilmeth (1978); Chisholm (1986); Carlson and Hobler (1993); Dale (1994) In: R.E. Morlan, CMC database (1999)]
(3600 ± 160)	RIDDL-272	Pender I., Gulf of Georgia, BC, CAN	[Borden (1954, 1962); Duff (1955, 1956); Wilmeth (1978); Chisholm (1986); Carlson and Hobler (1993); Dale (1994) In: R.E. Morlan, CMC database (1999)]
3550 ± 295 (3650 ± 295)	S-693	About 8 km NW of Swift Current, S Saskatchewan valley, 747 m asl (50.350°N, 107.900°W), SK, CAN	[Longin (1971); So and Wade (1975); Millar (1978, 1981); Gibson (1981); Chisholm, et al. 1983); Walker (1984); Callaghan (1986); Lovell, et al. (1986) In: R.E. Morlan, CMC database (1999)]; Morlan (1993); [Morlan and McNeely (1996) In: A.S. Dyke (GSC database)]

(3550 ± 500)	CRNL-1229	part of the skull and the right tibia (CMC-1229)	Woodpecker I. Bluff, Oldman R. (49°51'15"N, 112°09'20"W), AB, CAN	Cohen (1982); [Brown, et al. (1983); Wilson, et al. (1983); Moffatt and Wainwright (1983); Stalker (1983); Wilson (1984) In: R.E. Morlan, CMC database (1999)]; [Taber child bones]
(3520 ± 170)	RIDDL-275		Pender I., Gulf of Georgia, BC, CAN	[Borden (1954, 1962); Duff (1955, 1956); Wilmeth (1978); Chisholm (1986); Carlson and Hobler (1993); Dale (1994) In: R.E. Morlan, CMC database (1999)]
3500 ± 60 (3600 ± 60)	Beta-39228		N of Pt. Roberts, about 2 m asl., Strait of Georgia, BC, CAN	[Bernick and Howe (1990); Hanson (1991); Arcas Consulting (1994) In: R.E. Morlan, CMC database (1999)]
3490 ± 125 (3590 ± 125)	S-2350	ribs (CMC-1298)	Duke Pt., Vancouver I. (49°09'15"N, 123°53'10"W), BC, CAN	R.E. Morlan, CMC database (1999)
3485 ± 195 (3585 ± 195)	S-706		About 8 km NW of Swift Current, S Saskatchewan valley, 747 m asl (50.350°N, 107.900°W), SK, CAN	[Longin (1971); So and Wade (1975); Millar (1978, 1981); Gibson (1981); Chisholm, et al. 1983); Walker (1984); Callaghan (1986); Lovell, et al. (1986) In: R.E. Morlan, CMC database (1999)]; Morlan (1993); [Morlan and McNeely (1996) In: A.S. Dyke (GSC database)]
3415 ± 105 (3515 ± 105)	S-1450		About 8 km NW of Swift Current, S Saskatchewan valley, 747 m asl (50.350°N, 107.900°W), SK, CAN	Morlan (1993); [Longin (1971); So and Wade (1975); Millar (1978); Gibson (1981); Wade (1981); Chisholm, et al. (1983); Walker (1984); Callaghan (1986); Lovell, et al. (1986) In: R.E. Morlan, CMC database (1999)]
(3390 ± 90)	OxA-773	part of the skull and the right tibia	Woodpecker I. Bluff, Oldman R. (49°51'15"N, 112°09'20"W), AB, CAN	[R.E. Morlan, CMC database (1999)]; [Taber child bones]
(3380 ± 150)	RIDDL-273		Pender I., Gulf of Georgia, BC, CAN	[Borden (1954, 1962); Duff (1955, 1956); Wilmeth (1978); Chisholm (1986); Carlson and Hobler (1993); Dale (1994) In: R.E. Morlan, CMC database (1999)]
(3370 ± 280)	RIDDL-102		Pender I., Gulf of Georgia, BC, CAN	[Borden (1954, 1962); Duff (1955, 1956); Wilmeth (1978); Chisholm (1986); Carlson and Hobler (1993); Dale (1994) In: R.E. Morlan, CMC database (1999)]
3360 ± 100 (3460 ± 100)	S-651	tibia	Near Eriksdale, Lake Manitoba drainage, MB, CAN	[Syms (1983) In: R.E. Morlan, CMC database (1999)]
3310 ± 70 (3410 ± 75)	Beta-47925		Vancouver I., BC, CAN	[Stryd, et al. (1991); Matson and Coupland (1995); Dahlstrom (1997) In: R.E. Morlan, CMC database (1999)]
3300 ± 90 (3400 ± 90)	S-2945	ribs (CMC-1382)	Willowbunch, 732 m asl., Big Muddy drainage, SK, CAN	Morlan (1993); R.E. Morlan, CMC database (1999)
(3270 ± 220)	RIDDL-108		Pender I., Gulf of Georgia, BC, CAN	[Borden (1954, 1962); Duff (1955, 1956); Wilmeth (1978); Chisholm (1986); Carlson and Hobler (1993); Dale (1994) In: R.E. Morlan, CMC database (1999)]
(3260 ± 200)	RIDDL-99		Pender I., Gulf of Georgia, BC, CAN	[Borden (1954, 1962); Duff (1955, 1956); Wilmeth (1978); Chisholm (1986); Carlson and Hobler (1993); Dale (1994) In: R.E. Morlan, CMC database (1999)]
3250 ± 50 (3350 ± 55)	S-1574		S Saskatchewan valley, 3.5 km S of Saskatoon (52.079°N, 106.754°W), SK, CAN	Morlan (1993); [Morlan and McNeely (1996) In: A.S. Dyke (GSC database)]; [Walker (1984) In: R.E. Morlan, CMC database (1999)]
3190 ± 80 (3290 ± 80)	S-3361	ribs (CMC-1399)	W side of Kaien I., Prince Rupert Harbour, BC, CAN	[Inglis (1974); MacDonald and Inglis (1981); Chisholm (1986); In: R.E. Morlan, CMC database (1999)]

3145 ± 80 (3245 ± 80)	S-2251	ribs	Moose Mountain Ck., Souris drainage (49.638°N, 102.756°W), SK, CAN	Morlan (1993); [Walker (1982, 1984) In: R.E. Morlan, CMC database (1999)]; [Morlan and McNeely (1996) In: A.S. Dyke (GSC database)]
(3140 ± 200)	RIDDL-97		Pender I., Gulf of Georgia, BC, CAN	[Borden (1954, 1962); Duff (1955, 1956); Wilmeth (1978); Chisholm (1986); Carlson and Hobler (1993); Dale (1994) In: R.E. Morlan, CMC database (1999)]
3120 ± 50 (3240 ± 50)	Beta-74076	fibula	Gabriola I., Strait of Georgia (49°08'25"N, 123°46'10"W), BC, CAN	[Curtin (1998) In: R.E. Morlan, CMC database (1999)]
3100 ± 70 (3200 ± 75)	S-1741	vertebrae (CMC-1089)	Fairview Pt., Prince Rupert Harbour, BC, CAN	[Inglis (1974, 1976); MacDonald and Inglis (1981); Chisholm (1986); Ames (1998) In: R.E. Morlan, CMC database (1999)]
(3050 ± 150)	RIDDL-106		Pender I., Gulf of Georgia, BC, CAN	[Borden (1954, 1962); Duff (1955, 1956); Wilmeth (1978); Chisholm (1986); Carlson and Hobler (1993); Dale (1994) In: R.E. Morlan, CMC database (1999)]
3040 ± 60 (3140 ± 65)	SFU-545		Pender I., Gulf of Georgia, BC, CAN	[Borden (1954, 1962); Duff (1955, 1956); Wilmeth (1978); Chisholm (1986); Carlson and Hobler (1993); Dale (1994) In: R.E. Morlan, CMC database (1999)]
3030 ± 80 (3130 ± 80)	S-1433	ribs and sacrum (CMC-949)	Fairview Pt., Prince Rupert Harbour, BC, CAN	[Inglis (1974, 1976); MacDonald and Inglis (1981); Chisholm (1986); Ames (1998) In: R.E. Morlan, CMC database (1999)]
2920 ± 50 (3020 ± 55)	S-1598	innominate (CMC-1006)	W side of Kaien I., Prince Rupert Harbour, BC, CAN	[Inglis (1974); MacDonald and Inglis (1981); Chisholm (1986); In: R.E. Morlan, CMC database (1999)]
2920 ± 170 (3165 ± 175)	Gak-3794b	(CMC-460)	Thames valley, Middlesex County, ON, CAN	[Wilmeth (1978) In: R.E. Morlan, CMC database (1999) - bone apatite date]
2915 ± 85 (3015 ± 85)	S-1449		About 8 km NW of Swift Current, S Saskatchewan valley, 747 m asl (50.350°N, 107.900°W), SK, CAN	Morlan (1993); [Longin (1971); So and Wade (1975); Millar (1978); Gibson (1981); Wade (1981); Chisholm, et al. (1983); Walker (1984); Callaghan (1986); Lovell, et al. (1986) In: R.E. Morlan, CMC database (1999)]
2875 ± 75 (2975 ± 80)	GaK-3794C	11 ribs, 3 vertebrae, sternum (CMC-813)	Thames valley, Middlesex County, ON, CAN	[Wilmeth (1978) In: R.E. Morlan, CMC database (1999)]
2860 ± 120 (2960 ± 120)	S-1431	ribs (CMC-947)	Elizabeth Pt., Digby I., Prince Rupert Harbour, BC, CAN	[MacDonald and Inglis (1981); Chisholm (1986); Stewart and Stewart (1996); Ames (1998) In: R.E. Morlan, CMC database (1999)]
2830 ± 75 (2930 ± 80)	S-1426	ribs and scapulae (CMC-942)	Dodge I., Prince Rupert Harbour, BC, CAN	[MacDonald (1969); Wilmeth (1978); MacDonald and Inglis (1981); Chisholm (1986) In: R.E. Morlan, CMC database (1999)]
(2810 ± 70)	WAT-561		Beach Grove, Fraser Delta (49°01'55"N, 123°03'30"W), BC, CAN	[Ball (1979, 1981); Ham (1981); Chisholm (1986) In: R.E. Morlan, CMC database (1999)]
2800 ± 75 (2900 ± 80)	S-441		Bradwell, S Saskatchewan drainage (51°48'00"N, 106°09'00"W), SK, CAN	[Edmunds, et al. (1938); Kupsch, et al. (1970); Walker (1984) In: R.E. Morlan, CMC database (1999)]; Morlan (1993)
2800 ± 50 (2900 ± 55)	S-1595	ribs (CMC-1003)	Garden I., Prince Rupert Harbour, BC, CAN	[Wilmeth (1978); MacDonald and Inglis (1981); Ames (1998) In: R.E. Morlan, CMC database (1999)]
2800 ± 70 (2900 ± 75)	S-1284	femora (CMC-900)	Elizabeth Pt., Digby I., Prince Rupert Harbour, BC, CAN	[MacDonald and Inglis (1981); Chisholm (1986); Stewart and Stewart (1996); Ames (1998) In: R.E. Morlan, CMC database (1999)]
(2760 ± 60)	TO-1145	zygoma	Gabriola I., Strait of Georgia (49°08'25"N, 123°46'10"W), BC, CAN	[Skinner (1991); Curtin (1998) In: R.E. Morlan, CMC database (1999)]

2725 ± 95 (2825 ± 95)	S-1962		Highwood R., 1006 m asl., 20 km S and E of Calgary, Bow drainage, AB, CAN	Brumley and Rushworth (1983); [Brink and Baldwin (1988) In: R.E. Morlan, CMC database (1999)]
(2720 ± 80)	SFU-26		Beach Grove, Fraser Delta (49°01'55"N, 123°03'30"W), BC, CAN	[Ball (1981); Chisholm (1986) In: R.E. Morlan, CMC database (1999)]
2680 ± 70 (2780 ± 75)	SFU-581	femur	Gabriola I., Strait of Georgia (49°08'25"N, 123°46'10"W), BC, CAN	[Skinner (1991); Curtin (1998) In: R.E. Morlan, CMC database (1999)]
2670 ± 60 (2830 ± 60)	Beta-37845	rib and vertebra fragments	Gabriola I., Strait of Georgia (49°08'25"N, 123°46'10"W), BC, CAN	[Curtin (1998) In: R.E. Morlan, CMC database (1999)]
2620 ± 50 (2720 ± 55)	SFU-537		Pender I., Gulf of Georgia, BC, CAN	[Borden (1954, 1962); Duff (1955, 1956); Wilmeth (1978); Chisholm (1986); Carlson and Hobler (1993); Dale (1994) In: R.E. Morlan, CMC database (1999)]
2610 ± 150 (2710 ± 150)	S-2723	ribs (CMC-1346)	Ridley I., Prince Rupert Harbour, BC, CAN	[MacDonald and Inglis (1981); Ames (1998) In: R.E. Morlan, CMC database (1999)]
2610 ± 65 (2710 ± 70)	S-1285	femur (CMC-901)	Elizabeth Pt., Digby I., Prince Rupert Harbour, BC, CAN	[MacDonald and Inglis (1981); Chisholm (1986); Stewart and Stewart (1996); Ames (1998) In: R.E. Morlan, CMC database (1999)]
2600 ± 75 (2700 ± 80)	S-1425	ribs (CMC-941)	Dodge I., Prince Rupert Harbour, BC, CAN	[MacDonald (1969); Wilmeth (1978); MacDonald and Inglis (1981); Chisholm (1986) In: R.E. Morlan, CMC database (1999)]
(2580 ± 180)	RIDDL-101		Pender I., Gulf of Georgia, BC, CAN	[Borden (1954, 1962); Duff (1955, 1956); Wilmeth (1978); Chisholm (1986); Carlson and Hobler (1993); Dale (1994) In: R.E. Morlan, CMC database (1999)]
2575 ± 75 (2675 ± 80)	S-1432	ribs and scapulae (CMC-948)	Elizabeth Pt., Digby I., Prince Rupert Harbour, BC, CAN	[MacDonald and Inglis (1981); Chisholm (1986); Stewart and Stewart (1996); Ames (1998) In: R.E. Morlan, CMC database (1999)]
2565 ± 60 (2665 ± 65)	S-1666	innominates (CMC-1048)	Elizabeth Pt., Digby I., Prince Rupert Harbour, BC, CAN	[MacDonald and Inglis (1981); Chisholm (1986); Stewart and Stewart (1996); Ames (1998) In: R.E. Morlan, CMC database (1999)]
(2530 ± 160)	SFU-341	long-bone and rib fragments	Namu R. (51.860°N, 127.860°W), BC, CAN	[Cannon (1991) In: A.S. Dyke (GSC database)]; [Carlson (1991, 1996) In: R.E. Morlan, CMC database (1999)]
2530 ± 260 (2660 ± 260)	GSC-2886	femur shaft (CMC-1056)	Garden I., Prince Rupert Harbour, BC, CAN	[Wilmeth (1978); MacDonald and Inglis (1981); Ames (1998) In: R.E. Morlan, CMC database (1999)]
2490 ± 70 (2590 ± 75)	S-1429	ribs (CMC-945)	Garden I., Prince Rupert Harbour, BC, CAN	[Wilmeth (1978); MacDonald and Inglis (1981); Ames (1998) In: R.E. Morlan, CMC database (1999)]
2470 ± 90 (2570 ± 90)	S-912		S of the Frenchman R., near Bracken (49.333°N, 108.300°W), SK, CAN	[Morlan and McNeely (1996) In: A.S. Dyke (GSC database)]; [King (1961); Walker (1982) In: R.E. Morlan, CMC database (1999)]; Morlan (1993)
2465 ±75 (2565 ± 80)	S-1427	ribs and scapula (CMC-943)	Dodge I., Prince Rupert Harbour, BC, CAN	[MacDonald (1969); Wilmeth (1978); MacDonald and Inglis (1981); Chisholm (1986) In: R.E. Morlan, CMC database (1999)]
2440 ± 70 (2670 ± 70)	GSC-2888	humerus (CMC-1055)	Garden I., Prince Rupert Harbour, BC, CAN	[Wilmeth (1978); MacDonald and Inglis (1981); Ames (1998) In: R.E. Morlan, CMC database (1999)]
2425 ± 85 (2525 ± 85)	S-1739	ribs and scapulae (CMC-1087)	Fairview Pt., Prince Rupert Harbour, BC, CAN	[Inglis (1974, 1976); MacDonald and Inglis (1981); Chisholm (1986); Ames (1998) In: R.E. Morlan, CMC database (1999)]
2420 ± 70 (2520 ± 75)	SFU-546	26 ribs from two adults	Gabriola I., Strait of Georgia (49°08'25"N, 123°46'10"W), BC, CAN	[Skinner (1991); Curtin (1998) In: R.E. Morlan, CMC database (1999)]

2410 ± 85 (2510 ± 85)	I-8878		Lake St. Clair, Kent County, ON, CAN	[Fox (1978); Spence, et al. (1990) In: R.E. Morlan, CMC database (1999)]
2345 ± 50 (2445 ± 55)	S-1664	innominates (CMC-1046)	Elizabeth Pt., Digby I., Prince Rupert Harbour, BC, CAN	[MacDonald and Inglis (1981); Chisholm (1986); Stewart and Stewart (1996); Ames (1998) In: R.E. Morlan, CMC database (1999)]
2335 ± 105 (2435 ± 105)	S-2943	ribs (CMC-1380)	S Saskatchewan valley (50°36'10"N, 108°08'45"W), SK, CAN	Morlan (1993)
2330 ± 50 (2430 ± 55)	S-1665	ribs (CMC-1047)	Elizabeth Pt., Digby I., Prince Rupert Harbour, BC, CAN	[MacDonald and Inglis (1981); Chisholm (1986); Stewart and Stewart (1996); Ames (1998) In: R.E. Morlan, CMC database (1999)]
2325 ± 90 (2425 ± 90)	S-1735	ribs and scapulae (CMC-1083)	Elizabeth Pt., Digby I., Prince Rupert Harbour, BC, CAN	[MacDonald and Inglis (1981); Chisholm (1986); Stewart and Stewart (1996); Ames (1998) In: R.E. Morlan, CMC database (1999)]
2300 ± 600 (2400 ± 600)	UQ-780	rib fragments	5 km NW of Three Rivers, QC, CAN	[Clermont, et al. (1986); Taillon and Barre (1987) In: R.E. Morlan, CMC database (1999)]
2290 ± 60 (2450 ± 60)	Beta-37843	rib and vertebra fragments	Gabriola I., Strait of Georgia (49°08'25"N, 123°46'10"W), BC, CAN	[Curtin (1998) In: R.E. Morlan, CMC database (1999)]
2250 ± 75 (2350 ± 80)	S-2549	ribs (CMC-1337)	Parizeau Pt., Prince Rupert (54°17'18"N, 130°22'25"W), BC, CAN	[Ames (1998) In: R.E. Morlan, CMC database (1999)]
2250 ± 75 (2350 ± 80)	S-2548	ribs and scapulae (CMC-1336)	Parizeau Pt., Prince Rupert (54°17'18"N, 130°22'25"W), BC, CAN	[Ames (1998) In: R.E. Morlan, CMC database (1999)]
(2240 ± 50)	CAMS-54729		Buckley Bay, Vancouver I., BC, CAN	[A.R. Mason, pers. comm. (1999) In: R.E. Morlan, CMC database (1999)]
2220 ± 60 (2320 ± 60)	SFU-639	femur	Saltspring I., Georgia Strait, 4 m asl, BC, CAN	[Johnstone (1987, 1988); Hutchinson (1992) In: R.E. Morlan, CMC database (1999)]
2210 ± 65 (2310 ± 70)	S-1282	humerus (CMC-898)	Fairview Pt., Prince Rupert Harbour, BC, CAN	[Inglis (1974, 1976); MacDonald and Inglis (1981); Chisholm (1986); Ames (1998) In: R.E. Morlan, CMC database (1999)]
2170 ± 70 (2300 ± 70)	Beta-37846	rib and vertebra fragments	Gabriola I., Strait of Georgia (49°08'25"N, 123°46'10"W), BC, CAN	[Curtin (1998) In: R.E. Morlan, CMC database (1999)]
2170 ± 70 (2270 ± 75)	SFU-610	unburned long bone, vertebrae, and rib fragments	Gabriola I., Strait of Georgia (49°08'25"N, 123°46'10"W), BC, CAN	[Skinner (1991); Curtin (1998) In: R.E. Morlan, CMC database (1999)]
2150 ± 70 (2320 ± 70)	Beta-37844	rib and vertebra fragments	Gabriola I., Strait of Georgia (49°08'25"N, 123°46'10"W), BC, CAN	[Curtin (1998) In: R.E. Morlan, CMC database (1999)]
2140 ± 50 (2240 ± 55)	S-1730	ribs and scapulae (CMC-1078)	Fairview Pt., Prince Rupert Harbour, BC, CAN	[Inglis (1974, 1976); MacDonald and Inglis (1981); Chisholm (1986); Ames (1998) In: R.E. Morlan, CMC database (1999)]
2135 ± 65 (2235 ± 70)	S-1283	scapulae and clavicles (CMC-899)	W side of Kaien I., Prince Rupert Harbour, BC, CAN	[Inglis (1974); MacDonald and Inglis (1981); Chisholm (1986); In: R.E. Morlan, CMC database (1999)]
2135 ± 45 (2235 ± 50)	S-1728	ribs and scapulae (CMC-1076)	Fairview Pt., Prince Rupert Harbour, BC, CAN	[Inglis (1974, 1976); MacDonald and Inglis (1981); Chisholm (1986); Ames (1998) In: R.E. Morlan, CMC database (1999)]
2105 ± 50 (2295 ± 50)	S-1729	ribs and scapulae (CMC-1077)	Fairview Pt., Prince Rupert Harbour, BC, CAN	[Inglis (1974, 1976); MacDonald and Inglis (1981); Chisholm (1986); Ames (1998) In: R.E. Morlan, CMC database (1999)]
2090 ± 60 (2190 ± 65)	S-1734	ribs and scapulae (CMC-1082)	Elizabeth Pt., Digby I., Prince Rupert Harbour, BC, CAN	[MacDonald and Inglis (1981); Chisholm (1986); Stewart and Stewart (1996); Ames (1998) In: R.E. Morlan, CMC database (1999)]

2080 ± 65 (2180 ± 70)	S-1736	ribs and scapulae (CMC-1084)	Elizabeth Pt., Digby I., Prince Rupert Harbour, BC, CAN	[MacDonald and Inglis (1981); Chisholm (1986); Stewart and Stewart (1996); Ames (1998) In: R.E. Morlan, CMC database (1999)]
2060 ± 90 (2160 ± 90)	Beta-40985		N of Pt. Roberts, about 2 m asl., Strait of Georgia, BC, CAN	[Bernick and Howe (1990); Hanson (1991); Arcas Consulting (1994) In: R.E. Morlan, CMC database (1999)]
2015 ± 95 (2115 ± 95)	S-1740	ribs and scapulae (CMC-1088)	Fairview Pt., Prince Rupert Harbour, BC, CAN	[Inglis (1974, 1976); MacDonald and Inglis (1981); Chisholm (1986); Ames (1998) In: R.E. Morlan, CMC database (1999)]
1990 ± 100 (2090 ± 100)	S-896	(CMC-607)	Morrison's I., Outaouais region, QC, CAN	[Taillon and Barre (1987); Spence, et al. (1990) In: R.E. Morlan, CMC database (1999)]
1945 ± 75 (2045 ± 80)	S-776	ribs (CMC-555)	N bank of Saugeen R., 3.5 km upstream from Lake Huron, Annabel twp., Bruce County (44°35'N, 81°35W), ON, CAN	[McAllister (1962); Wright and Anderson (1963); Finlayson (1977); Spence, et al. (1990) In: R.E. Morlan, CMC database (1999)]
1905 ± 145 (2005 ± 145)	S-1773	ribs and scapulae (CMC-1081)	Elizabeth Pt., Digby I., Prince Rupert Harbour, BC, CAN	[MacDonald and Inglis (1981); Chisholm (1986); Stewart and Stewart (1996); Ames (1998) In: R.E. Morlan, CMC database (1999)]
1895 ± 60 (1995 ± 65)	S-1732	femur (CMC-1080)	Fairview Pt., Prince Rupert Harbour, BC, CAN	[Inglis (1974, 1976); MacDonald and Inglis (1981); Chisholm (1986); Ames (1998) In: R.E. Morlan, CMC database (1999)]
1865 ± 50 (1965 ± 55)	S-1667	scapula, clavicles, humerus, radius, ulna (CMC-1049)	Elizabeth Pt., Digby I., Prince Rupert Harbour, BC, CAN	[MacDonald and Inglis (1981); Chisholm (1986); Stewart and Stewart (1996); Ames (1998) In: R.E. Morlan, CMC database (1999)]
1860 ± 60 (1960 ± 65)	S-1738	ribs and scapulae (CMC-1086)	W side of Kaien I., Prince Rupert Harbour, BC, CAN	[Inglis (1974); MacDonald and Inglis (1981); Chisholm (1986); In: R.E. Morlan, CMC database (1999)]
(1840 ± 60)	CAMS-13182		Whaley Cairn, Winnipeg drainage (50°15'N, 95°55'W), MB, CAN	[Wyman (1993); Ens (1998); Syms, pers. comm. (1998) In: R.E. Morlan, CMC database (1999)]
(1840 ± 60)	CAMS-13183		Whaley Cairn, Winnipeg drainage (50°15'N, 95°55'W), MB, CAN	[Wyman (1993); Ens (1998); Syms, pers. comm. (1998) In: R.E. Morlan, CMC database (1999)]
1830 ± 50 (1930 ± 55)	DIC-107		Trent R., Seymour twp., Northumberland County, ON, CAN	[Kenyon (1986); Spence, et al. (1990) In: R.E. Morlan, CMC database (1999)]
(1810 ± 60)	CAMS-13184		Pas Reserve, N bank of Saskatchewan R., 267 m asl, MB, CAN	[Tamplin (1977); Long and Tamplin (1977); Tamplin, et al. (1983); Syms (1992) pers. comm. (1998) In: R.E. Morlan, CMC database (1999)]
1790 ± 100 (2115 ± 105)	GaK-3794C	(CMC-460)	Thames valley, Middlesex County, ON, CAN	[Wilmeth (1978) In: R.E. Morlan, CMC database (1999)]
1760 ± 60 (1860 ± 65)	S-1599	femur (CMC-1007)	W side of Kaien I., Prince Rupert Harbour, BC, CAN	[Inglis (1974); MacDonald and Inglis (1981); Chisholm (1986); In: R.E. Morlan, CMC database (1999)]
(1760 ± 60)	NSRL-3126		Whaley Cairn, Winnipeg drainage (50°15'N, 95°55'W), MB, CAN	[Wyman (1993); Ens (1998) In: R.E. Morlan, CMC database (1999)]
(1750 ± 60)	CAMS-13186		Pas Reserve, N bank of Saskatchewan R., 267 m asl, MB, CAN	[Tamplin (1977); Long and Tamplin (1977); Tamplin, et al. (1983); Syms (1992) pers. comm. (1998) In: R.E. Morlan, CMC database (1999)]
1750 ± 70 (1850 ± 75)	S-1428	ribs (CMC-944)	Garden I., Prince Rupert Harbour, BC, CAN	[Wilmeth (1978); MacDonald and Inglis (1981); Ames (1998) In: R.E. Morlan, CMC database (1999)]
1750 ± 45 (1850 ± 50)	S-1731	ribs and scapulae (CMC-1079)	Fairview Pt., Prince Rupert Harbour, BC, CAN	[Inglis (1974, 1976); MacDonald and Inglis (1981); Chisholm (1986); Ames (1998) In: R.E. Morlan, CMC database (1999)]
1750 ± 40 (1850 ± 45)	S-1663	tibia (CMC-1045)	Fairview Pt., Prince Rupert Harbour, BC, CAN	[Inglis (1974, 1976); MacDonald and Inglis (1981); Chisholm (1986); Ames (1998) In: R.E. Morlan, CMC database (1999)]

(1740 ± 60)	CAMS-13185		Pas Reserve, N bank of Saskatchewan R., 267 m asl, MB, CAN	[Tamplin (1977); Long and Tamplin (1977); Tamplin, et al. (1983); Syms (1992) pers. comm. (1998) In: R.E. Morlan, CMC database (1999)]
(1720 ± 60)	CAMS-13190	limb-bone	Wapisu L., Churchill drainage, MB, CAN	[E.L. Syms, pers. comm. (1998) In: R.E. Morlan, CMC database (1999)]
1720 ± 55 (1820 ± 60)	DIC-732		Trent R., Seymour twp., Northumberland County (44°15'N, 77°25'W), ON, CAN	[Kenyon (1986); Spence, et al. (1990) In: R.E. Morlan, CMC database (1999)]
(1710 ± 190)	RIDDL-267		Pender I., 0-3 m asl, Gulf of Georgia, BC, CAN	[Chisholm (1986); Hanson (1991, 1995); Carlson and Hobler (1993); Dale (1994) In: R.E. Morlan, CMC database (1999)]
(1700 ± 70)	CAMS-13191	tooth	Wapisu L., Churchill drainage, MB, CAN	[E.L. Syms, pers. comm. (1998) In: R.E. Morlan, CMC database (1999)]
1670 ± 100 (1770 ± 100)	Beta-38348		N of Pt. Roberts, about 2 m asl., Strait of Georgia, BC, CAN	[Bernick and Howe (1990); Hanson (1991); Arcas Consulting (1994) In: R.E. Morlan, CMC database (1999)]
1635 ± 55 (1735 ± 60)	S-1430	ribs and scapulae (CMC-946)	Elizabeth Pt., Digby I., Prince Rupert Harbour, BC, CAN	[MacDonald and Inglis (1981); Chisholm (1986); Stewart and Stewart (1996); Ames (1998) In: R.E. Morlan, CMC database (1999)]
1590 ± 120 (1690 ± 120)	S-946		Ottawa R., McNab twp., Renfrew County, ON, CAN	[Kennedy (1972) In: R.E. Morlan, CMC database (1999)]
1585 ± 110 (1685 ± 110)	S-2220	femur (CMC-1261)	Nass R. (55°01'45"N, 129°34'27"W), BC, CAN	[Cybulski (1992) In: R.E. Morlan, CMC database (1999)]
(1570 ± 60)	NSRL-3127		N shore of Whitemouth R., 259 m asl (50.117°N, 96.035°W), MB, CAN	[Buchner (1979, 1982); Meiklejohn, et al. (1994); Ens (1998) In: R.E. Morlan, CMC database (1999)]
1570 ± 100 (1670 ± 100)	S-2226	ribs (CMC-1266)	Carroll, S Saskatchewan R. (51°40'30"N, 106°58'40"W), SK, CAN	[Walker (1984) In: R.E. Morlan, CMC database (1999)]; Morlan (1993)
1550 ± 60 (1650 ± 60)	Beta-39231		N of Pt. Roberts, about 2 m asl., Strait of Georgia, BC, CAN	[Bernick and Howe (1990); Hanson (1991); Arcas Consulting (1994) In: R.E. Morlan, CMC database (1999)]
1540 ± 70 (1690 ± 70)	Beta-58221		Somenos Ck., Vancouver I., BC, CAN	[Warner (1993) In: R.E. Morlan, CMC database (1999)]
1520 ± 70 (1620 ± 70)	Beta-38349		N of Pt. Roberts, about 2 m asl., Strait of Georgia, BC, CAN	[Bernick and Howe (1990); Hanson (1991); Arcas Consulting (1994) In: R.E. Morlan, CMC database (1999)]
1520 ± 50 (1620 ± 50)	Beta-38352		N of Pt. Roberts, about 2 m asl., Strait of Georgia, BC, CAN	[Bernick and Howe (1990); Hanson (1991); Arcas Consulting (1994) In: R.E. Morlan, CMC database (1999)]
1500 ± 60 (1600 ± 60)	Beta-40986		N of Pt. Roberts, about 2 m asl., Strait of Georgia, BC, CAN	[Bernick and Howe (1990); Hanson (1991); Arcas Consulting (1994) In: R.E. Morlan, CMC database (1999)]
1500 ± 50 (1600 ± 50)	Beta-39226		N of Pt. Roberts, about 2 m asl., Strait of Georgia, BC, CAN	[Bernick and Howe (1990); Hanson (1991); Arcas Consulting (1994) In: R.E. Morlan, CMC database (1999)]
(1460 ± 130)	RIDDL-270		Pender I., Gulf of Georgia, BC, CAN	[Borden (1954, 1962); Duff (1955, 1956); Wilmeth (1978); Chisholm (1986); Carlson and Hobler (1993); Dale (1994) In: R.E. Morlan, CMC database (1999)]
1430 ± 60 (1530 ± 60)	Beta-39225		N of Pt. Roberts, about 2 m asl., Strait of Georgia, BC, CAN	[Bernick and Howe (1990); Hanson (1991); Arcas Consulting (1994) In: R.E. Morlan, CMC database (1999)]
(1420 ± 90)	RIDDL-98		Pender I., Gulf of Georgia, BC, CAN	[Borden (1954, 1962); Duff (1955, 1956); Wilmeth (1978); Chisholm (1986); Carlson and Hobler (1993); Dale (1994) In: R.E. Morlan, CMC database (1999)]
1410 ± 60 (1510 ± 60)	Beta-39229		N of Pt. Roberts, about 2 m asl., Strait of Georgia, BC, CAN	[Bernick and Howe (1990); Hanson (1991); Arcas Consulting (1994) In: R.E. Morlan, CMC database (1999)]
1410 ± 60 (1510 ± 60)	SFU-583		N of Pt Roberts, about 2 m asl., Strait of Georgia, BC, CAN	[Bernick and Howe (1990); Hanson (1991); Arcas Consulting (1994) In: R.E. Morlan, CMC database (1999)]

1400 ± 80 (1500 ± 80)	GaK-3800	ribs (CMC-501)	N bank of Saugeen R., 3.5 km upstream from Lake Huron, Annabel twp., Bruce County, ON, CAN	[McAllister (1962); Wright and Anderson (1963); Finlayson (1977); Spence, et al. (1990) In: R.E. Morlan, CMC database (1999)]
1400 ± 50 (1500 ± 50)	Beta-39230		N of Pt. Roberts, about 2 m asl., Strait of Georgia, BC, CAN	[Bernick and Howe (1990); Hanson (1991); Arcas Consulting (1994) In: R.E. Morlan, CMC database (1999)]
(1370 ± 60)	Beta-56200 CAMS-3830		Harrison R., Fraser drainage (49°15'N, 121°55'W), BC, CAN	[Blake, et al. (1983) In: R.E. Morlan, CMC database (1999)]
1360 ± 60 (1460 ± 65)	S-1303	(CMC-910)	12 km W of Brandon along the N side of Assiniboine R., MB, CAN	[Tisdale (1978); Badertscher, et al. (1987) In: R.E. Morlan, CMC database (1999)]
1350 ± 110 (1450 ± 110)	S-2333	ribs (CMC-1289)	Nass R. (55.292°N, 129.574°W), BC, CAN	[Cybulski (1992) In: A.S. Dyke (GSC database)]; [D. Balkwill, pers. comm. (1996) In: R.E. Morlan, CMC database (1999)]
1350 ± 60 (1450 ± 60)	Beta-38351		N of Pt. Roberts, about 2 m asl., Strait of Georgia, BC, CAN	[Bernick and Howe (1990); Hanson (1991); Arcas Consulting (1994) In: R.E. Morlan, CMC database (1999)]
(1340 ± 150)	RIDDL-269		Pender I., Gulf of Georgia, BC, CAN	[Borden (1954, 1962); Duff (1955, 1956); Wilmeth (1978); Chisholm (1986); Carlson and Hobler (1993); Dale (1994) In: R.E. Morlan, CMC database (1999)]
1280 ± 70 (1380 ± 70)	Beta-40987		N of Pt. Roberts, about 2 m asl., Strait of Georgia, BC, CAN	[Bernick and Howe (1990); Hanson (1991); Arcas Consulting (1994) In: R.E. Morlan, CMC database (1999)]
1270 ± 40 (1450 ± 40)	Beta-126882	rib (CMC-1523)	Mungo Cannery, Fraser R. (49°09'25"N, 122°56'30"W), BC, CAN	[A. Bowman, pers. comm. (1999) In: R.E. Morlan, CMC database (1999)]
1260 ± 80 (1360 ± 80)	GaK-5447		N shore of Whitemouth R., 259 m asl (50.117°N, 96.035°W), MB, CAN	[Buchner (1979, 1982); Meiklejohn et al. (1994); Ens (1998) In: R.E. Morlan, CMC database (1999)]
1260 ± 60 (1360 ± 60)	Beta-38350		N of Pt. Roberts, about 2 m asl., Strait of Georgia, BC, CAN	[Bernick and Howe (1990); Hanson (1991); Arcas Consulting (1994) In: R.E. Morlan, CMC database (1999)]
1260 ± 40 (1400 ± 40)	Beta-90639	mandible	Gabriola I., Strait of Georgia (49°08'25"N, 123°46'10"W), BC, CAN	[Curtin (1998) In: R.E. Morlan, CMC database (1999)]
(1230 ± 60)	Beta-11056		Graham I., Queen Charlotte I. (54°05'N, 132°05'W), BC, CAN	[Skinner (1983) In: R.E. Morlan, CMC database (1999)]
(1230 ± 50)	TO-1385	ribs	Lake Superior, Thunder Bay District (48°55'N, 88°15'W), ON, CAN	[Arthurs (1995) In: R.E. Morlan, CMC database (1999)]
1190 ± 80 (1290 ± 80)	S-2487	ribs (CMC-1316)	Nass R. (55.292°N, 129.574°W), BC, CAN	[Cybulski (1992) In: A.S. Dyke (GSC database)]; [D. Balkwill, pers. comm. (1996) In: R.E. Morlan, CMC database (1999)]
(1190 ± 60)	TO-1925	4th metatarsal	Niagara Escarpment, near Winona, Niagara County, ON, CAN	[Spence and Fox (1992) In: R.E. Morlan, CMC database (1999)]
1160 ± 80 (1260 ± 80)	Beta-40984		N of Pt. Roberts, about 2 m asl., Strait of Georgia, BC, CAN	[Bernick and Howe (1990); Hanson (1991); Arcas Consulting (1994) In: R.E. Morlan, CMC database (1999)]
1160 ± 50 (1260 ± 50)	Beta-39277		N of Pt. Roberts, about 2 m asl., Strait of Georgia, BC, CAN	[Bernick and Howe (1990); Hanson (1991); Arcas Consulting (1994) In: R.E. Morlan, CMC database (1999)]
1150 ± 60 (1250 ± 60)	Beta-38355		N of Pt. Roberts, about 2 m asl., Strait of Georgia, BC, CAN	[Bernick and Howe (1990); Hanson (1991); Arcas Consulting (1994) In: R.E. Morlan, CMC database (1999)]
(1110 ± 60)	TO-1812	femur	Rogers Ossuary, 640 m asl., S of Brantford twp., Brant County, ON, CAN	[Mullen and Hoppa (1992) In: R.E. Morlan, CMC database (1999)]
(1110 ± 60)	TO-1926	ribs	Long Pt., Lake Erie, Norfolk County, ON, CAN	[Fox and Molto (1994) In: R.E. Morlan, CMC database (1999)]
(1110 ± 60)	TO-1811	femur	Rogers Ossuary, 640 m asl., S of Brantford twp., Brant County, ON, CAN	[Mullen and Hoppa (1992) In: R.E. Morlan, CMC database (1999)]

1100 ± 70 (1200 ± 70)	Beta-38347	skull	N of Pt. Roberts, about 2 m asl., Strait of Georgia, BC, CAN	[Bernick and Howe (1990); Hanson (1991); Arcas Consulting (1994) In: R.E. Morlan, CMC database (1999)]
(1090 ± 130)	RIDDL-95		Pender I., Gulf of Georgia, BC, CAN	[Borden (1954, 1962); Duff (1955, 1956); Wilmeth (1978); Chisholm (1986); Carlson and Hobler (1993); Dale (1994) In: R.E. Morlan, CMC database (1999)]
1060 ± 80 (1260 ± 80)	Beta-11055	ribs and vertebra fragments	Saltspring I., Georgia Strait, BC, CAN	[Skinner (1983); Chisholm (1986) In: R.E. Morlan, CMC database (1999)]
1045 ± 105 (1145 ± 105)	S-2219	ribs (CMC-1260)	Nass R. (55.292°N, 129.574°W), BC, CAN	[Cybulski (1992) In: A.S. Dyke (GSC database)]; [D. Balkwill, pers. comm. (1996) In: R.E. Morlan, CMC database (1999)]
(970 ± 100)	RIDDL-556	rib	Saunaktuk, Eskimo L. (69.170°N, 132.500°W), NT, CAN	[Cinq Mars (1991); Morrison and Arnold (1994); C. Arnold, pers. comm. (1998) In: R.E. Morlan, CMC database (1999)]
960 ± 50 (1150 ± 50)	Beta-126881	rib (CMC-1522)	Mungo Cannery, Fraser R. (49°09'25"N, 122°56'30"W), BC, CAN	[A. Bowman, pers. comm. (1999) In: R.E. Morlan, CMC database (1999)]
(930 ± 65)	Beta-26467 ETH-4270		Place Royale, region de Quebec, QC, CAN	[Chapdelaine (1990); Allard and Seguin (1992) In: R.E. Morlan, CMC database (1999)]
920 ± 90 (1020 ± 90)	S-2944	ribs (CMC-1381)	Buffalo Pound, Qu'Appelle valley (50°36'00"N, 105°24'00"W), SK, CAN	Morlan (1993)
905 ± 60 (1005 ± 65)	UGa-2488		Rice L., Otonabee twp., Peterborough County, ON, CAN	[Johnston (1968, 1979); Wilmeth (1978); Fox (1980); Kenyon (1986); Spence, et al. (1990); Williamson (1990) In: R.E. Morlan, CMC database (1999)]
900 ± 60 (1000 ± 65)	S-1840	2 femora (CMC-1113)	L. Erie, S Wallingham twp., Norfolk County, ON, CAN	[Spence, et al. (1978, 1981, 1990); Fox (1980); Williamson (1990) In: R.E. Morlan, CMC database (1999)]
875 ± 70 (975 ± 75)	S-1329	ribs	Souris R. (49.117°N, 102.979°W), SK, CAN	Morlan (1993); [Syms (1979); Walker (1983) In: R.E. Morlan, CMC database (1999)]; [Morlan and McNeely (1996) In: A.S. Dyke (GSC database)]
860 ± 60 (960 ± 60)	Beta-40983		N of Pt. Roberts, about 2 m asl., Strait of Georgia, BC, CAN	[Bernick and Howe (1990); Hanson (1991); Arcas Consulting (1994) In: R.E. Morlan, CMC database (1999)]
800 ± 70 (990 ± 70)	BGS-1556	proximal femur	Grand R., Brant County, ON, CAN	[Woodley (1994) In: R.E. Morlan, CMC database (1999)]
790 ± 60 (890 ± 65)	Beta-39588	ribs (CMC-1398)	Nass R. (55.029°N, 129.575°W), BC, CAN	[Cybulski (1992) In: A.S. Dyke (GSC database)]; [D. Balkwill, pers. comm. (1996) In: R.E. Morlan, CMC database (1999)]
790 ± 100 (890 ± 100)	RL-1831		Place Royale, region de Quebec, QC, CAN	[Chapdelaine (1990); Allard and Seguin (1992) In: R.E. Morlan, CMC database (1999)]
770 ± 60 (870 ± 60)	Beta-65249	ribs	Harbour House, Saltspring I., Gulf of Georgia (48°55'N, 123°25'W), BC, CAN	[Brolly, et al. (1993) In: R.E. Morlan, CMC database (1999)]
735 ± 110 (835 ± 110)	S-2332	tibia (CMC-1288)	Nass R. (55.292°N, 129.574°W), BC, CAN	[Cybulski (1992) In: A.S. Dyke (GSC database)]; [D. Balkwill, pers. comm. (1996) In: R.E. Morlan, CMC database (1999)]
725 ± 90 (825 ± 90)	S-2946	ribs (CMC-1383) ribs (CMC-1383)	Boundary Dam, Souris drainage (49°02'55"N, 102°59'40"W), SK, CAN	Morlan (1993)
(715 ± 70)	Beta-26468 ETH-4271		Place Royale, region de Quebec, QC, CAN	[Chapdelaine (1990); Allard and Seguin (1992) In: R.E. Morlan, CMC database (1999)]
(710 ± 50 (1275AD), 1265-	K-5889		Abels Gard, Qaqortoq Municipality, GRE	[Vebaek (1992) In: Danish/Greenlandic Network (1999)]
685 ± 60 (785 ± 65)	S-2092		Writing-On-Stone, Milk R. (49°05'N, 111°35'W), AB, CAN	[Getty (1971) In: R.E. Morlan, CMC database (1999)]; Brumley and Rushworth (1983)

670 ± 180 (770 ± 180)	S-743	long-bone fragments	Glacial outwash ridge about 200 m S of Pipestone Ck., near Reston, MB, CAN	[Braddell, et al. (1970); Syms (1979) In: R.E. Morlan, CMC database (1999)]
660 ± 60 (760 ± 65)	UGa-2489		Rice L., Otonabee twp., Peterborough County, ON, CAN	[Johnston (1968, 1979); Wilmeth (1978); Fox (1980); Kenyon (1986); Spence, et al. (1990); Williamson (1990) In: R.E. Morlan, CMC database (1999)]
(630 ± 80)	Beta-16257		Belly R., near Cardston (49°16'N, 113°34'W), AB, CAN	[Beaudoin (1987) In: R.E. Morlan, CMC database (1999)]
(590 ± 40)	TO-5228	radius	Southern Indian L., Churchill drainage, MB, CAN	[Brownlee and Syms (1999) In: R.E. Morlan, CMC database (1999)]
530 ± 70 (630 ± 75)	S-2236		Moose Jaw valley, Qu'appelle drainage (50°00'08"N, 105°43'37"W), SK, CAN	[Dyck (1983) In: R.E. Morlan, CMC database (1999)]; Morlan (1993)
510 ± 210 (610 ± 210)	S-3001	ribs (CMC-1389)	Nass R. (55.292°N, 129.574°W), BC, CAN	[Cybulski (1992) In: A.S. Dyke (GSC database)]; [D. Balkwill, pers. comm. (1996) In: R.E. Morlan, CMC database (1999)]
510 ± 60 (610 ± 65)	UGa-2487		Rice L., Otonabee twp., Peterborough County, ON, CAN	[Johnston (1968, 1979); Wilmeth (1978); Fox (1980); Kenyon (1986); Spence, et al. (1990); Williamson (1990) In: R.E. Morlan, CMC database (1999)]
480 ± 50 (580 ± 55)	S-1822		Pelee I., Essex County, ON, CAN	R.E. Morlan, CMC database (1999)
475 ± 70 (630 ± 70)	S-2610		Charles Arm (49°21'N, 55°17'W), NF, CAN	[E.G. Walker, pers. comm. (1999) In: R.E. Morlan, CMC database (1999)]
450 ± 55 (550 ± 60)	S-972		Pelee I., Essex County, ON, CAN	R.E. Morlan, CMC database (1999)
(450 ± 60)	NSRL-3130		Fort Alexander, Winnipeg drainage (50°37'N, 96°18'W), MB, CAN	[Ens (1998) In: R.E. Morlan, CMC database (1999)]
440 ± 140 (470 ± 140)	GSC-1394	femur and 2 tibiae (CMC-391)	Ottawa R. McNab twp., Renfrew County, ON, CAN	[Kennedy (1972) In: R.E. Morlan, CMC database (1999)]
430 ± 80 (530 ± 80)	S-2010	(CMC-1199)	S Nation R., 3.2 km NE of Spencerville, Grenville County, (44°51'30"N, 75°30'45"W), ON, CAN	[Pendergast (1966); Wright (1985); Jamieson (1990) In: R.E. Morlan, CMC database (1999)]
420 ± 30 (440 ± 30)	Beta-107745		Southern Indian L., Churchill drainage, MB, CAN	[Brownlee and Syms (1999) In: R.E. Morlan, CMC database (1999)]
400 ± 50 (500 ± 55)	Beta-39029		Northumberland County, ON, CAN	[McKillop and Jackson (1991) In: R.E. Morlan, CMC database (1999)]
(370 ± 75)	OxA-384	rib	Belly R., near Cardston (49°16'N, 113°34'W), AB, CAN	[Ball (1986); Beaudoin (1987) In: R.E. Morlan, CMC database (1999)]
(340 ± 70)	OxA-682	femur	Belly R., near Cardston (49°16'N, 113°34'W), AB, CAN	[Ball (1986); Beaudoin (1987) In: R.E. Morlan, CMC database (1999)]
310 ± 140 (410 ± 140)	AECV-195C		Belly R. (49°16'N, 113°34'W), AB, CAN	[Beaudoin (1987) In: R.E. Morlan, CMC database (1999)]
(250 ± 60)	NSRL-3129		Fort Alexander, Winnipeg drainage (50°37'N, 96°18'W), MB, CAN	[Ens (1998) In: R.E. Morlan, CMC database (1999)]
100 ± 100 (200 ± 100)	M-1712		Sandwich W twp., Essex County, ON, CAN	R.E. Morlan, CMC database (1999)

Edentata, Megalonychidae

Megalonyx sp. (groundsloth)

Measured Date (Normalized)	Lab No	Material	Location	References/Remarks
(>41610)	AECV-1315C	vertebrae	Consolidated Concrete pit 48, 670 m asl., Edmonton area (53°38.5'N, 113°17'W), AB, CAN	Young, et al. (1994) Burns and Young (1994); Burns (1996); R.E. Morlan, CMC database (1999)

Lagomorpha, Ochotonidae

Ochotona whartoni (giant pika)

Measured Date (Normalized)	Lab No	Material	Location	References/Remarks
(8670 ± 220)	TO-2566		Elba Cave, Niagara Escarpment, ON, CAN	[Savage (1992), Mead and Grady (1996) In: R.E. Morlan (CMC database (1999)]

Lagomorpha, Leporidae

Lepus arcticus (arctic hare)

Measured Date (Normalized)	Lab No	Material	Location	References/Remarks
(770 ± 40)	AAR-3219	jawbone	Qeqrtaaraq, Hatherton Bay, Thule District, Inglefield Land, GRE	[Heinemeier and Rasmussen (1998) In: Danish/Greenlandic Network (1999)]; [3,280 (1280BC), 1240-1285BC]
(155 ± 110)	GX-7122		Knud Pen., 7-8 m asl., Ellesmere I. (79.120°N, 76.190°W), NU, CAN	[Schledermann (1980, pers. comm. (1998); McCullough (1989); Morrison (1989); Bowman, et al. (1990) In: R.E. Morlan, CMC database (1999)]

Rodentia, Sciuridae

Marmota caligata (hoary marmot)

Measured Date (Normalized)	Lab No	Material	Location	References/Remarks
(34860 ± 470)	TO-6350		Eagle Cave: N side of Crowsnest L. 1430 m asl., Crowsnest Pass (49°38'N, 114°38'W), AB, CAN	[Wilmeth (1978), Brumley and Rushworth (1983), Burns (1991, 1996), R. McNeely, pers. comm. (1998) In: R.E. Morlan, CMC database (1999)]
(32900 ± 2400)	AA-21565	incisor	On Your Knees Cave, Prince of Wales I., 135 m asl. (56.250°N, 133.500°W), AK, USA	Heaton and Grady (1996, 1997); Dixon, et al. (1997)
(29180 ± 300)	TO-6351	skull fragments	Eagle Cave: N side of Crowsnest L. 1430 m asl., Crowsnest Pass (49°38'N, 114°38'W), AB, CAN	[Wilmeth (1978), Brumley and Rushworth (1983), Burns (1991, 1996), R. McNeely, pers. comm. (1998) In: R.E. Morlan, CMC database (1999)]
(23560 ± 770)	AA-21566	incisor	On Your Knees Cave, Prince of Wales I., 135 m asl. (56.250°N, 133.500°W), AK, USA	Heaton and Grady (1996, 1997); Sattler and Dixon (1997)

Marmota vancouverensis (Vancouver Island marmot)

Measured Date (Normalized)	Lab No	Material	Location	References/Remarks
(2630 ± 50)	TO-1224	femur	Clayoquot Plateau, Vancouver I., BC, CAN	Nagorsen et al. (1996); [corrected date = ± 1 standard deviation]
(2490 ± 50)	TO-692	tibia	Clayoquot Plateau, Vancouver I., BC, CAN	Nagorsen, et al. (1996)
(1070 ± 60)	TO-4267	femur	Limestone Mountain, Vancouver I., BC, CAN	Nagorsen, et al. (1996)
(1030 ± 60)	TO-4266	femur	Limestone Mountain, Vancouver I., BC, CAN	Nagorsen, et al. (1996)
(990 ± 50)	TO-3564	femur	Mariner Mountain, Vancouver I., BC, CAN	Nagorsen, et al. (1996)
(970 ± 90)	TO-3567	femur	Mariner Mountain, Vancouver I., BC, CAN	Nagorsen, et al. (1996)
(920 ± 50)	TO-3563	femur	Mariner Mountain, Vancouver I., BC, CAN	Nagorsen, et al. (1996)
(890 ± 50)	TO-3562	femur	Mariner Mountain, Vancouver I., BC, CAN	Nagorsen, et al. (1996)
(890 ± 50)	TO-3566	femur	Mariner Mountain, Vancouver I., BC, CAN	Nagorsen, et al. (1996)
(850 ± 50)	TO-3565	femur	Mariner Mountain, Vancouver I., BC, CAN	Nagorsen, et al. (1996)

(830 ± 60)	TO-4265	mandible	Golden Hinde, Vancouver I., BC, CAN	Nagorsen, et al. (1996)

Marmota sp. (marmot)

Measured Date (Normalized)	Lab No	Material	Location	References/Remarks
(>44500)	AA-8871A	incisor	Devil's Canopy Cave, Prince of Wales I. (56.250°N, 133.500°W), AK, USA	Heaton (1995c); Heaton and Grady (1997)

Cynomys niobrarius churcherii (Churcher's white-tailed prairie dog)

Measured Date (Normalized)	Lab No	Material	Location	References/Remarks
(33650 ± 340)	TO-1142		Winter: Hand Hills, 1090 m asl, Red Deer drainage, 28 km NE of Drumheller (51.570°N, 112.330°W), AB, CAN	Young (1991); [Storer (1972, 1975), Graham et al. (1987), Burns (1996) In: R.E. Morlan, CMC database (1999)]
(33650 ± 340)	TO-871		Courtenay West, 26.5 km NE of Drumheller, Hand Hills, Red Deer drainage (51.570°N, 112.330°W), AB, CAN	[Graham et al. (1987), Young (1991); Burns (1996) In: R.E. Morlan, CMC database (1999)]
(29610 ± 220)	TO-1304		Winter: Hand Hills, 1090 m asl, Red Deer drainage, 28 km NE of Drumheller (51.570°N, 112.330°W), AB, CAN	Young (1991); [Storer (1972, 1975), Graham et al. (1987), Burns (1996) In: R.E. Morlan, CMC database (1999)]
(28000 ± 250)	TO-872		Schowalter: 30 km SE of Drumheller, Wintering Hills, Red Deer drainage (51.570°N, 112.330°W), AB, CAN	[Graham et al. (1987), Young (1991), Burns (1996) In: R.E. Morlan, CMC database (1999)]
(25980 ± 180)	TO-1307		Courtenay West, 26.5 km NE of Drumheller, Hand Hills, Red Deer drainage (51.570°N, 112.330°W), AB, CAN	Young (1991); [Graham et al. (1987), Burns (1996) In: R.E. Morlan, CMC database (1999)]
(23000 ± 150)	TO-1305		Heaton: 30 km SE of Drumheller, Wintering Hills, Red Deer drainage (51.570°N, 112.330°W), AB, CAN	[Graham, et al (1987), Young (1991), Burns (1996) In: R.E. Morlan, CMC database (1999)]
(22200 ± 320)	RIDDL-681		Winter: Hand Hills, 1090 m asl., Red Deer drainage 28 km NE of Drumheller (51.570°N, 112.330°W), AB, CAN	[Storer (1972, 1975), Graham et al. (1987), Young (1991), Burns (1996) In: R.E. Morlan, CMC database (1999)]
(17060 ± 180)	TO-1143		Seward: Hand Hills, Red Deer drainage, AB, CAN	Young, 1991; [Graham (1987) In: R.E. Morlan, CMC database (1999)]; [Anomalously young; very low collagen yield (A.S. Dyke, pers. comm. (2001))].

Spermophilus cf. Spermophilus parryii (ground squirrel)

Measured Date (Normalized)	Lab No	Material	Location	References/Remarks
(45160 ± 1200)	TO-2704	droppings (CR-83-0)	Sixtymile area (Sixtymile Loc. 3), 701 m asl., (64°1.00'N, 140°44'W), YT, CAN	C.R. Harington [unpublished]; Harington (1997a)

Spermophilus parryii (arctic ground squirrel)

Measured Date (Normalized)	Lab No	Material	Location	References/Remarks
(47500 ± 1900)	Beta-16157	skin and hair from dried carcass	Glacier Ck., Sixtymile area (Sixtymile Loc. 6) (64°2.00'N, 140°53'W), YT, CAN	Harington (1984, 1989, 1997)
14760 ± 850	GX-250	nesting grasses	Chatanika R. (65.080°N, 147.750°W), AK, USA	Krueger and Weeks (1966); Péwé (1975)
14510 ± 300 (14510 ± 500)	W-2703	faeces	Chatanika R. (65.080°N, 147.750°W), AK, USA	Péwé (1975); Harington (1977a); R.E. Morlan, CMC database (1999)
13910 ± 70 (13910 ± 70)	Beta-111606	twigs and grass from a ground squirrel nest	Hunker Ck. (Dawson Loc. 76), (64°N, 139°W), YT, CAN	[E. Kotler, pers. comm. (1999) In: R.E. Morlan, CMC database (1999)]
12200 ± 100 (12200 ± 100)	GSC-2641	organic components of a nest (CMN 21094)	Dominion Ck. (63.800°N, 138.680°W), YT, CAN	Harington (1977, 1997); Blake (1984); R.E. Morlan, CMC database (1999); The nest contained a nearly complete ground squirrel skeleton.

(11460 ± 80)	TO-4875		Stirling Bend Section, Stewart R., YT, CAN	[R. McNeely, pers. comm. (1998) In: R.E. Morlan, CMC database (1999)]

Spermophilus parryii? (arctic ground squirrel)

Measured Date (Normalized)	Lab No	Material	Location	References/Remarks
26950 +6030 / -3405	QC-672	nesting grasses	Cripple Ck. (64.817°N, 148.017°W), AK, USA	Guthrie (1990a)
24525 +1680 / -1390	QC-675	nesting grasses	Ester Ck. (65.000°N, 147.000°W), AK, USA	Guthrie (1990a)
24000 +1500 / -1200	QC-666	nesting grasses	Goldstream Ck. (64.950°N, 147.583°W), AK, USA	Guthrie (1990a)
23380 +450 / -2680	QC-667	nesting grasses	Cripple Ck. (64.817°N, 148.017°W), AK, USA	Guthrie (1990a)
23130 +1765 / -1445	QC-670	nesting grasses	Goldstream Ck. (64.950°N, 147.583°W), AK, USA	Guthrie (1990a)
22280 +870 / -770	QC-669	nesting grasses	Ester Ck. (65.000°N, 147.000°W), AK, USA	Guthrie (1990a)
21750 +1400 / -1200	QC-660	nesting grasses	Cripple Ck. (64.817°N, 148.017°W), AK, USA	Guthrie (1990a)
19660 ± 30	QC-673	nesting grasses	Little Eldorado Ck. (65.000°N, 148.000°W), AK, USA	Guthrie (1990a)
18230 ± 410	QC-668	nesting grasses	Engineer Ck. (64.950°N, 147.167°W), AK, USA	Guthrie (1990a)
17980 ± 575	QC-661	nesting grasses	Cripple Ck. (64.817°N, 148.017°W), AK, USA	Guthrie (1990a)
13350 ± 265 (13350 ± 265)	QC-664	nesting grasses	Lower Sulphur Ck. (63°39'N, 138°40'W), YT, CAN	Guthrie (1990a); R.E. Morlan, CMC database (1999)
11170 ± 205 (11170 ± 205)	QC-662	nesting grasses	Glacier Ck. (64°02'N, 140°45'W), YT, CAN	Guthrie (1990a); R.E. Morlan, CMC database (1999)

Spermophilus sp. (ground squirrel)

Measured Date (Normalized)	Lab No	Material	Location	References/Remarks
14860 ± 860	GX-251	nesting grasses	Chatanika R. (65.080°N, 147.750°W), AK, USA	Guthrie (1985, 1990)

Rodentia, Castoridae

Castor canadensis (North American beaver)

Measured Date (Normalized)	Lab No	Material	Location	References/Remarks
(>38600)	GSC-120	beaver-gnawed wood	Rat R., W of Mackenzie R., (67°39.5'N, 135°28'W), NT, CAN	R.E. Morlan, CMC database (1999)
9940 ± 90 (9940 ± 90)	GSC-2022	beaver-gnawed wood (Populus sp.)	5 km SE of Sabine Pt., 32 m asl. (69°02'N, 137°38'W), YT, CAN	Lowdon and Blake (1976); R.E. Morlan, CMC database (1999)
9590 ± 160	I-2477	beaver-gnawed wood	Near Donkin, Cape Breton County (46.000°N, 61.000°W), NS, CAN	[Harington (1978b) - Beaver-gnawed wood in organic silt and peat]
(9560 ± 120)	GSC-2298	beaver-gnawed wood (Populus sp.)	Grandview Hills, 275 m asl. (67.100°N, 133.240°W), NT, CAN	[Lowdon and Blake (1979) - Assoc. fauna: beaver probably C. canadensis]; R.E. Morlan, CMC database (1999)
9500 ± 90 (9490 ± 90)	GSC-1814	beaver-cut sticks (CR-72-42A)	Roadcut on Dempster Hwy, 32 km W of Arctic Red R., and W of Frog Ck., Ft. Macpherson area (67.420°N, 134.470°W), NT, CAN	[Lowdon and Blake (1979) - Associated fauna: Castor (gnawed sticks) probably C. canadensis]; Harington (1978); R.E. Morlan, CMC database (1999)
9330 ± 300	W-2160	wood from beaver dam	Washington Ck. (65°44'N, 164°52'W), AK, USA	C.R. Harington (card file)
(9180 ± 90)	GSC-2023	beaver-gnawed willow wood (Salix sp.)	Eskimo L, 3 m asl., (69.417°N, 131.975°W), NT, CAN	Blake (1987); R.E. Morlan, CMC database (1999)
8480 ± 300	W-2596	birch or aspen log from ancient beaver dam	Mouth of Old Wound N SP (66°29'55"N, 164°46'30"W, AK, USA	Hopkins, et al. (1981); [Associated fauna: log in fossil beaver dam (CRH card file)]

8080 ± 300	W-2808	birch wood	Mud Ck., near Candle, NE SP (65°56'40"N, 161°59'00"W), AK, USA	Hopkins, et al. (1981); [Associated fauna: beaver-chewed log (CRH card file)]
(8000 ± 80)	BGS-137	beaver-gnawed wood	30 km SW of Fort Macpherson, 300 m asl., (67.27°N, 135.23°W), NT, CAN	[Delorme, et al. (1977) In: R.E. Morlan, CMC database (1999)]
(6920 ± 105)	Beta-28763 ETH-4760	skull (CMN 45290)	Hunker Ck. (Dawson Loc. 37) (64.020°N, 139.150°W), YT, CAN	C.R. Harington [unpublished]
6820 ± 200	W-733	large beaver-gnawed birch or aspen log	Sullivan Ck., Hot Springs District (65°05'N, 150°54'W), AK, USA	Hopkins, et al. (1981)
(5490 ± 190)	RIDDL-249	ilium	Highland valley, 1208-1211 m asl., Thompson drainage, BC, CAN	[Lawhead, et al. (1986); Richards and Rousseau (1987); Stryd and Rousseau (1996) In: R.E. Morlan, CMC database (1999)]
3700 ± 150	L-434	beaver-gnawed wood	Cook Inlet (60.780°N, 151.200°W), AK, USA	A.S. Dyke (GSC database)
3220 ± 60 (3220 ± 60)	GSC-6228	beaver-gnawed wood	Riviere des Ha Ha (approx. 48°19'N, 70°51'W), QC, CAN	C.R. Harington (card file); [C.R. Harington, pers. comm. (1999) In: R.E. Morlan, CMC database (1999)]
900 ± 40 (980 ± 45)	S-1714	(CMC-1064)	Lake Huron, Bruce County, (44°26'20"N, 81°23'30"W), ON, CAN	[Stewart (1974); Wright (1974, 1985) In: R.E. Morlan, CMC database (1999)]
530 ± 80 (610 ± 80)	S-2012	(CMC-1201)	S Nation R., 3.2 km NE of Spencerville, Grenville County, (44°51'30"N, 75°30'45"W), ON, CAN	[Pendergast (1966); Wright (1985); Jamieson (1990) In: R.E. Morlan, CMC database (1999)]
<200	W-1111	wood (beaver-gnawed)	Sullivan Ck. (65°05'N, 150°53'W), AK, USA	A.S. Dyke (GSC database)
	GX-5495-G		SE shoreline of Wanipigow L., 320 m asl, MB, CAN	[S. Saylor (1976, 1977, 1989); Buchner (1979); B. Saylor (1989); Zoltai (1989); Hambly (1994) In: R.E. Morlan, CMC database (1999)]

Castor sp. (beaver)

Measured Date (Normalized)	Lab No	Material	Location	References/Remarks
(36900 ± 300)	GSC-2422	beaver-gnawed wood (Picea sp. or Salix sp.)	Hungry Ck., Bonnet Plume Basin (65.5°34.5'N, 135.5°30'W), YT, CAN	Hughes, et al. (1981); McNeely (1989); R.E. Morlan, CMC database (1999)
(9730 ± 110)	AECV-1620C	beaver-gnawed wood	Wood Bog: E of Grande Prairie, N of Wapiti R., in the Peace drainage (55.150°N, 118.780°W), AB, CAN	Beaudoin et al. (1996); R.E.Morlan, CMC database (1999)
9350 ± 80	Beta-58408	beaver-gnawed wood sticks	Giddings Forest Bed, Fairbanks, AK, USA	Péwé, et al. (1997)
8940 ± 80	Beta-46215	beaver-gnawed sticks	Giddings Forest Bed, Fairbanks, AK, USA	Péwé, et al. (1997)
8800 ± 70	Beta-50685	beaver-gnawed log	Giddings Forest Bed, Fairbanks, AK, USA	Péwé, et al. (1997)
8780 ± 100	Beta-48788	beaver-gnawed sticks	Giddings Forest Bed, Fairbanks, AK, USA	Péwé, et al. (1997)
7280 ± 80	Beta-50683	beaver dam log	Giddings Forest Bed, Fairbanks, AK, USA	Péwé, et al. (1997)
6100 ± 80	Beta-52700	beaver-gnawed wood	Giddings Forest Bed, Fairbanks, AK, USA	Péwé, et al. (1997)
3180 ± 85	S-1796	beaver-dam wood	Whitemud Ck., N Saskatchewan R. valley, near Edmonton (53.500°N, 113.670°W), AB, CAN	Rains (1987) Rains (1987)

Castor? sp. (beaver)

Measured Date (Normalized)	Lab No	Material	Location	References/Remarks
10150 ± 140	AECV-1596C		(52.080°N, 111.450°W), AB, CAN	[Beaudoin (1992) In: A.S. Dyke (GSC database)]

Rodentia, Cricetidae

Neotoma cinerea (bushy-tailed woodrat)

Measured Date (Normalized)	Lab No	Material	Location	References/Remarks
140 ± 60 (110 ± 60)	GSC-2035	faeces	Grotte Valerie Cave, 760 m asl., South Nahanni R. (61.280°N, 124.080°W), NT, CAN	Blake (1987)

Dicrostonyx sp. (collared lemming)

Measured Date (Normalized)	Lab No	Material	Location	References/Remarks
14000 ± 500	I-1171	pellets	Barrow (71.300°N, 156.710°W), AK, USA	[Brown (1966) In: A.S. Dyke (GSC database)]

Lemmus sibiricus (brown lemming)

Measured Date (Normalized)	Lab No	Material	Location	References/Remarks
23100 ± 860 (23180 ± 860)	GaK-5438		January Cave, Plateau Mountain, 80 km SW of Calgary (50.180°N, 114.520°W), AB, CAN	Burns (1980, 1991); Jackson and Pawson (1984)

Cetacea, Monodontidae

Delphinapterus leucas (white whale, beluga)

Measured Date (Normalized)	Lab No	Material	Location	References/Remarks
10420 ± 150 (10580 ± 150)	GSC-454		Foster Sand Pit, Ottawa (45.330°N, 75.700°W), ON, CAN	Harington (1977b, 1978b); Harington and Occhietti (1988); R.E. Morlan, CMC database (1999)
(10400 ± 80)	GSC-2418	cervical vertebrae (CMN 21336)	6 km NW of Pakenham (45.370°N, 76.340°W), ON, CAN	Whiteaves (1907); Harington (1977, 1978); Harington and Fitzgerald (1973); Harington and Occhietti (1988); R.E. Morlan, CMC database (1999)
9470 ± 170 (9630 ± 170)	Beta-27511	vertebra	Sandpit on S side of Mont-Saint-Hilaire (45.570°N, 73.180°W), QC, CAN	Harington (1988); R.E. Morlan, CMC database (1999)
(4320 ± 50)	TO-3713	mandibular teeth (CMN 30501)	Great Whale R. (55.258°N, 77.708°W), QC, CAN	C.R. Harington [unpublished]
(2980 ± 60)	CAMS-33465	teeth	Head of Koluktoo Bay, Milne Inlet, 10 m asl., Baffin I. (72.058°N, 81.092°W), NU, CAN	[A.S. Dyke, pers. comm. (1998) In: R.E. Morlan, CMC database (1999)]; [Dyke (2000b) GSC Map 1962A]
500 ± 80 (660 ± 80)	Gif-1819		Hudson Bay, Nouveau-Quebec region (60.03°N, 77.2°W), QC, CAN	R.E. Morlan, CMC database (1999)

Monodon monoceras (narwhal)

Measured Date (Normalized)	Lab No	Material	Location	References/Remarks
(8540 ± 60)	TO-5013	tusk	14 km W of Stanley Pt., 23 m asl., Baffin I. (73.820°N, 85.520°W), NU, CAN	[Dyke (2000d) GSC Map 1964A]
(8340 ± 60)	TO-5012	tusk	Vista R. valley, NW of Kakiak Pt., 22 m asl., Baffin I. (72°43'00"N, 86°48.5'00"W), NU, CAN	[Dyke (2000c) GSC Map 1963A]
7970 ± 90 (8380 ± 90)	TO-1966	tusk	About 8 km W of Cape Lockwood on the S coast of Greely Fiord, N Fosheim Pen., 34 m asl., Ellesmere I. (80.280°N, 84.300°W), NU, CAN	[Bell (1992) In: Morlan, CMC database (1999)]; [Bell (1996) - date normalized using delta 13C = -25o/oo and corrected for a reservoir effect of 410 years]
(7270 ± 50)	CAMS-33464	tusk	Cape York, 7.5 m asl., Baffin I. (73.791°N, 87.092°W), NU, CAN	[A.S. Dyke, pers. comm. (1998) In: R.E. Morlan, CMC database (1999)]; [Dyke (2000d) GSC Map 1964A]
7075 ± 120 (7230 ± 120)	S-2911	tusk	2.1 km NE of the head of Smith Bay, 7 m asl., Prince of Wales I. (73.230°N, 99.750°W), NU, CAN	Dyke and Morris (1990); Dyke, et al. (1991); R.E. Morlan, CMC database (1999)

6830 ± 50 (7240 ± 50)	TO-476	tusk	Wootton Pen., 34 m asl., Ellesmere I. (82.270°N, 85.330°W), NU, CAN	Evans (1989); [R.E. Morlan, pers. comm. (1998) indicates a corrected date using delta 13C correction = -25o/oo]; Bednarski (1990)
(6500 ± 50)	TO-5011	tusk	Vista R. valley, 7 km NW of Kakiak Pt., 43 m asl., Baffin I. (72.720°N, 86.600°W), NU, CAN	[Dyke (2000c) GSC Map 1963A]
(6210 ± 50)	CAMS-39649	tusk	E bank of Taggart R., Strathcona Fiord area, 12.5 m asl., Ellesmere I. (78.567°N, 82.250°W), NU, CAN	[A.S. Dyke, pers. comm. (1998), In: R.E. Morlan, CMC database (1999)]
(5800 ± 60)	TO-5015	tusk	Kuuruluk R., 25 m asl., Baffin I. (72.390°N, 86.280°W), NU, CAN	Dyke (2000c) GSC Map 1963A]
(5530 ± 50)	TO-5007	tusk	Near McBean Bay, Prince Regent Inlet coast of Brodeur Pen., 27.5m asl., Baffin I. (72.630°N, 89.700°W), NU, CAN	[Dyke (2000c) GSC Map 1963A]
5120 ± 140 (5300 ± 140)	S-3608	tusk	Kuuruluk R., 21.5 m asl., Baffin I. (72.410°N, 86.300°W), NU, CAN	[A.S. Dyke, pers. comm. (1998) In: R.E. Morlan, CMC database (1999)]; [Dyke (2000c) GSC Map 1963A]
(5090 ± 50)	CAMS-33463	tusk	Vista R., 5.4 km NW of Kakiak Pt. 14.5-15 m asl., Baffin I. (72.720°N, 86.760°W), NU, CAN	[A.S. Dyke, pers. comm. (1998) In: R.E. Morlan, CMC database (1999)]; [Dyke (2000c) GSC Map 1963A]
(4500 ± 60)	TO-5010	tusk	3.8 km NW of the mouth of Tikiraq R., 20 m asl., Baffin I. (71.780°N, 86.120°W), NU, CAN	[Dyke and Hooper (2000) GSC Map 1960A]
(4400 ± 60)	TO-5062	tusk	18.3 km NNW of Owen Pt., 4.4 km SE of Pt. Hogarth, 15 m asl., Devon I. (76.140°N, 92.660°W), NU, CAN	[A.S. Dyke, pers. comm. (1999)]
4180 ± 80 (4360 ± 80)	S-2914	tusk	S shore of NW arm of Smith Bay, 12 m asl., Prince of Wales I. (73.370°N, 100.370°W), NU, CAN	Dyke and Morris (1990); Dyke, et al. (1991); R.E. Morlan, CMC database (1999)
(2880 ± 60)	TO-5017	tusk	7 km SE of Van Koenig Pt., 20 m asl., Baffin I. (71.030°N, 89.400°W), NU, CAN	[Hooper and Dyke (2000) GSC Map 1959A]
2075 ± 140 (2235 ± 135)	S-2672	tusk fragment	7 km E of Cape Grant, 9 m asl., NW Russell I. (73.290°N, 99.290°W), NU, CAN	Dyke and Morris (1990); Dyke et al. (1991); R.E. Morlan, CMC database (1999)
(2030 ± 50)	TO-5014	tusk	Kuuruluk R., 8-8.5 m asl., Baffin I. (72.440°N, 86.270°W), NU, CAN	[Dyke (2000c) GSC Map 1963A]
(1180 ± 50)	GSC-1918	tusk	E side of Erik Harbour, 3 m asl., Ellesmere I. (78.625°N, 74.730°W), NU, CAN	Blake (1980, 1987, 1992)
990 ± 130 (1190 ± 130)	S-3609	tusk	Tiriganialaag R., 21 m asl., Baffin I. (72.220°N, 86.470°W), NU, CAN	[A.S. Dyke, pers. comm. (1998) In: R.E. Morlan, CMC database (1999)]; [Dyke (2000c) GSC Map 1963A]

Monodon sp. (narwhal)

Measured Date (Normalized)	Lab No	Material	Location	References/Remarks
(700 ± 50)	TO-3706	solid tusk (CR-87-200)	Raised beach overlooking Strand Fiord, Axel Heiberg I. (79°14.5'N, 90°20.7'W), NU, CAN	C.R. Harington [unpublished]

Cetacea, Eschrictiidae

Eschrichtius robustus (grey whale)

Measured Date (Normalized)	Lab No	Material	Location	References/Remarks
(>38000)	AA-312C	gelatin	Near Beaufort Sea coast (70.745°N, 153.110°W), AK, USA	[Stafford, et al. (1987) In: A.S. Dyke (GSC database)]

Cetacea, Balaenopteridae

Balaenoptera physalus (common finback whale)

Measured Date (Normalized)	Lab No	Material	Location	References/Remarks
11300 ± 90 (11400 ± 90)	GSC-2871	vertebra (CR-78-21)	3.6 km SW of Daveluyville, 85 m asl. (46.175°N, 72.170°W), QC, CAN	Lowdon and Blake (1981); Harington (1988); Harington and Occhietti (1988); R.E. Morlan, CMC database (1999)
(3480 ± 90)	K-6096		Storstrommen (77°N, 22°W), GRE	Bennike (1997); [Add 400 yr. for conventional normalization (A.S. Dyke, pers. Comm. (2001))]

Balaenoptera acutorostrata (minke whale or little piked whale)

Measured Date (Normalized)	Lab No	Material	Location	References/Remarks
18570 ± 500	S-1969-A	vertebra (CMN 36170)	Kouchibouguacis R., St-Louis-de-Kent (46°43.4'N, 64°59.6'W), NB, CAN	Rampton and Paradis (1981); Blake (1984) [incorrect age determination - see GSC-3284; S-1969-B]; [Anomalously old (A.S. Dyke, pers. comm. (2001))]
12600 ± 140 (12620 ± 140)	GSC-3284	vertebra (CMN 36170)	S bank of Kouchibouguacis R., 1.5 km W of St-Louis-de-Kent, 8 m asl. (46°43.4'N, 64°59.6'W), NB, CAN	Rampton and Paradis (1981); Blake (1984); Harington and Occhietti (1988); Miller (1990); [Rampton, et al. (1984) In: R.E. Morlan, CMC database (1999)]
11830 ± 950 (11850 ± 950)	S-1969-B	vertebra (CMN 36170)	Kouchibouguacis R., St-Louis-de-Kent (46°43.4'N, 64°59.6'W), NB, CAN	Rampton and Paradis (1981); Rampton, et al. (1984); Blake (1984); Harington and Occhietti (1988); Miller (1990); R.E. Morlan, CMC database (1999)

Cetacea, Balaenidae

Balaena mysticetus (bowhead whale)

Measured Date (Normalized)	Lab No	Material	Location	References/Remarks
(39130 ± 550)	TO-5241	rib	Cape Storm, Ellesmere I., NU, CAN	Blake (1996)
>37000	GSC-2229-2	rib	Cape Storm, 55 m asl., Ellesmere I. (76.400°N, 87.570°W), NU, CAN	A.S. Dyke (GSC database)
>33000	S-3565	periotic	Sheills Pen., 24 m asl., Devon I. (76.250°N, 94.990°W), NU, CAN	Dyke (1999)
(28700 ± 1110)	GSC-2229	rib	Cape Storm, 55 m asl., Ellesmere I. (76.400°N, 87.570°W), NU, CAN	Lowdon and Blake (1980); R.E. Morlan, CMC database (1999); [Superseded by GSC-2229-2, A.S. Dyke, pers. comm. (2001)]
13345 ± 230 (13505 ± 230)	S-3047	14th or 15th lumbar vertebra (CMN 45042)	Romaines R., 8 m asl., (48.555°N, 58.680°W), NF, CAN	[Grant (1991) GSC Map 1737A]; McNeely and Jorgensen (1993); GSC palaeontology Collection No. 104326; R.E. Morlan, CMC database (1999)
(13240 ± 90)	TO-316		22 km E of Thackeray Pt., 56 m asl., Prince of Wales I. (71°41'00"N, 98°56'00"W), NU, CAN	[R. McNeely, pers. comm. (1998) In: R.E. Morlan, CMC database (1999)]; [Anomalously old, A.S. Dyke, pers. comm. (2001)]
11720 ± 205 (11880 ± 200)	S-2149	lumbar vertebra	Trout R., 45 m asl., (49°28.4'N, 58°7.1'W), NF, CAN	[I. Brookes, pers. comm. (1986) In: A.S. Dyke (GSC database)]; R.E. Morlan, CMC database (1999); Two other dates on same specimen - see Beta-33189 and TO-2509
11400 ± 90 (11500 ± 90)	GSC-2269	humerus (CMN CR-75-55)	2.1 km W of the town of White L., 170 m asl., (45.370°N, 76.530°W), ON, CAN	Harington (1977, 1978); Richard (1978); Lowdon and Blake (1979); Harington and Occhietti (1988); R.E. Morlan, CMC database (1999)
10530 ± 145 (10690 ± 140)	S-2591	skull	2 km S of salt water lake, S side of Reliance Bay, 54 m asl., Prince of Wales I. (73.720°N, 100.000°W), NU, CAN	Dyke and Morris (1990); Dyke, et al. (1991); R.E. Morlan, CMC database (1999)

(10210 ± 70)	TO-7755	periotic	10.7 km SSE of Cape Baring, Wollaston Pen., 56 m asl., Victoria I. (69°57'33"N, 117°16'15"W), NT, CAN	Dyke and Savelle (2001)
10005 ± 120 (10165 ± 120)	S-2916	skull	Richard Collinson, 62 m asl., Prince of Wales I. (72.700°N, 102.370°W), NU, CAN	Dyke and Morris (1990); Dyke, et al. (1991); R.E. Morlan, CMC database (1999)
10000 ± 145 (10160 ± 140)	S-2922	skull	20.5 km NNE of Cape Richard Collinson, 66.5 m asl., Prince of Wales I. (72.930°N, 102.470°W), NU, CAN	Dyke and Morris (1990); Dyke, et al. (1991); R.E. Morlan, CMC database (1999)
9935 ± 190 (10095 ± 190)	S-2954	rib	W Storkerson Pen., 68 m asl., Victoria I. (72.850°N, 108.020°W), NU, CAN	[Hodgson (1993) In: R.E. Morlan, CMC database (1999)]
(9930 ± 80)	TO-8012	rib	11.1 km SSE of Cape Baring, Wollaston Pen., 52 m asl., Victoria I. (69°56'45"N, 117°16'30"W), NT, CAN	Dyke and Savelle (2001)
9920 ± 180 (10055 ± 180)	S-3595	periotic	6.1 km NNE of Cape Home, Croker Bay, 13 m asl., Devon I. (74.590°N, 83.530°W), NU, CAN	Dyke (1998); R.E. Morlan, CMC database (1999)
(9840 ± 80)	TO-2509	lumbar vertebra	Trout R., 45 m asl., (49°28.4'N, 58°7.1'W), NF, CAN	[I.A. Brookes, pers. comm. (1991)]; Two other dates on same specimen - see S-2149 and Beta-33189
9770 ± 150 (9930 ± 150)	S-3066	periotic	4 km SSW of the mouth of McBean Bay, 81.4 m asl., Baffin I. (72.610°N, 89.670°W), NU, CAN	R.E. Morlan, CMC database (1999); [Dyke (2000c) GSC Map 1963A]
9740 ± 200 (9885 ± 200)	S-3591	otic capsule	3.5 km NW of Cape Home, Croker Bay, 29 m asl., Devon I. (74.570°N, 83.650°W), NU, CAN	Dyke (1998); R.E. Morlan, CMC database (1999)
9605 ± 140 (9765 ± 140)	S-2590	rib fragment	20 km SW of Hollist Pt., 66 m asl., Prince of Wales I. (72.825°N, 101.970°W), NU, CAN	Dyke and Morris (1990); Dyke, et al. (1991); R.E. Morlan, CMC database (1999)
9600 ± 145 (9760 ± 145)	S-2994		7 km NW of Point Caen, Wollaston Pen., 57 m asl., Victoria I. (69.350°N, 116.120°W), NU, CAN	[A.S. Dyke, pers. comm. (1998) In: R.E. Morlan, CMC database (1999)]; Dyke and Savelle (2001)
9590 ± 115 (9750 ± 115)	S-1381	vertebra (DCA-77-B1)	7.8 km SSW of Cape Anne, 69 m asl., Somerset I. (74.030°N, 94.800°W), NU, CAN	[Dyke (1980) These dates supersede earlier incorrect dates]; Dyke and Morris (1990); R.E. Morlan, CMC database (1999)
9540 ± 200 (9680 ± 200)	S-3580	otic capsule	3.9 km NW of Cape Home, Croker Bay, 32.5 m asl., Devon I. (74.570°N, 83.660°W), NU, CAN	Dyke (1998); R.E. Morlan, CMC database (1999)
9505 ± 120 (9665 ± 120)	S-2918	rib	5 km NW of Smith Bay, 74.5 m asl., Prince of Wales I. (73.070°N, 99.470°W), NU, CAN	Dyke and Morris (1990); Dyke, et al. (1991); R.E. Morlan, CMC database (1999)
(9500 ± 80)	TO-7759	periotic	3 km inland from the coast just N of Penny Bay, N of Kugaluk R., Wollaston Pen., 43 m asl., Victoria I. (69°45'15"N, 117°06'00"W), NT, CAN	Dyke and Savelle (2001)
9470 ± 120 (9600 ± 120)	U-2510		Cape Storm, 118 m asl., Ellesmere I. (76°25'N, 87°33'W), NU, CAN	Olsson et al. (1974)
9450 ± 190 (9580 ± 190)	S-3592	periotic	3.7 km NW of Cape Home, Croker Bay, 21 m asl., Devon I. (74.570°N, 83.630°W), NU, CAN	Dyke (1998); R.E. Morlan, CMC database (1999)
9450 ± 120 (9590 ± 120)	GSC-1496-4	mandible	Cape Storm, 118 m asl., Ellesmere I. (76°25'N, 87°33'W), NU, CAN	Blake (1979, 1980)
9440 ± 135 (9600 ± 135)	S-2912	periotic	11.5 km E of the head of Smith Bay, 70 m asl., Prince of Wales I. (73.220°N, 99.430°W), NU, CAN	Dyke and Morris (1990); Dyke, et al. (1991); R.E. Morlan, CMC database (1999)
9420 ± 200 (9555 ± 200)	S-3572	periotic	6.5 km SW of Cape Sparbo-Hardy Lowland, 64.5 m asl., Devon I. (75.770°N, 83.850°W), NU, CAN	Dyke (1999); R.E. Morlan, CMC database (1999)

9375 ± 140 (9535 ± 140)	S-2706	skull	18 km SSW of Cape Hardy, 69-71 m asl., Prince of Wales I. (73.740°N, 97.680°W), NU, CAN	Dyke and Morris (1990); Dyke, et al. (1991); R.E. Morlan, CMC database (1999)
(9340 ± 70)	TO-7756	periotic	5.2 km inland from Penny Bay, N of Kugaluk R., Wollaston Pen., 41 m asl., Victoria I. (69°39'50"N, 116°43'45"W), NT, CAN	Dyke and Savelle (2001)
9335 ± 145 (9495 ± 145)	S-2913	otic capsule	1.2 km inland from the SW side of Prescott I., 57.5 m asl. (73.020°N, 97.050°W), NU, CAN	Dyke and Morris (1990); Dyke, et al. (1991); R.E. Morlan, CMC database (1999)
(9300 ± 80)	TO-7758	periotic	2.5 km inland from the coast just N of Penny Bay, N of Kugaluk R., Wollaston Pen., 41.5 m asl., Victoria I. (69°43'25"N, 116°59'55"W), NT, CAN	Dyke and Savelle (2001)
9285 ± 140 (9445 ± 140)	S-2729	ear bone	6.5 km SW of Cape Wollaston, 21 m asl., Victoria I. (71.075°N, 118.180°W), NT, CAN	Vincent (1989); R.E. Morlan, CMC database (1999)
9285 ± 135 (9445 ± 135)	S-2593	skull	3 km SW of the W end of Allen L., 71 m asl., Prince of Wales I. (73.620°N, 98.630°W), NU, CAN	Dyke and Morris (1990); Dyke, et al. (1991); R.E. Morlan, CMC database (1999)
(9270 ± 80)	TO-7752	squamosal	14 km SSE of Cape Baring, Wollaston, Pen., 39 m asl., Victoria I. (69°55'04"N, 117°16'45"W), NT, CAN	Dyke and Savelle (2001)
9260 ± 150 (9410 ± 150)	GSC-1496-C		Cape Storm, 118 m asl., Ellesmere I. (76°25'N, 87°33'W), NU, CAN	Blake (1975); [same sample as GSC-1496-4, A.S. Dyke, pers. comm. (2001)]
9225 ± 215 (9385 ± 215)	S-2597	skull	Guillemard Bay, 2 km E of Fisher L., in beach gravel 99 m asl., Prince of Wales I. (72.170°N, 97.830°W), NU, CAN	Dyke and Morris (1990); Dyke, et al. (1991); R.E. Morlan, CMC database (1999)
9210 ± 120 (9370 ± 120)	S-1390	humerus	10 km W of Rodd Bay, 76 m asl., Somerset I. (73.925°N, 90.620°W), NU, CAN	Dyke and Morris (1990); R.E. Morlan, CMC database (1999)
9200 ± 80 (9340 ± 80)	GSC-1496-3		Cape Storm, 118 m asl., Ellesmere I. (76°25'N, 87°33'W), NU, CAN	Blake (1979, 1980) [This determination has the smallest error term, and is identical to that of high level marine mollusc shells nearby (Blake, 1979)]
9200 ± 170 (9340 ± 170)	S-3612	periotic	2.8 km NW of Cape Home, Croker Bay, 19 m asl., Devon I. (74.560°N, 83.660°W), NU, CAN	Dyke (1998); R.E. Morlan, CMC database (1999)
(9200 ± 90)	TO-7749	otic capsule	6 km SSE of Cape Baring, Wollaston Pen., 36 m asl., Victoria I. (69°59'21"N, 117°19'00"W), NT, CAN	Dyke and Savelle (2001)
9190 ± 135 (9350 ± 135)	S-2993		7 km NW of Point Caen, Wollaston Pen., 51 m asl., Victoria I. (69.350°N, 116.120°W), NU, CAN	[A.S. Dyke, pers. comm. (1998) In: R.E. Morlan, CMC database (1999)]
9170 ± 200 (9290 ± 200)	S-3585	periotic	3.2 km NW of Cape Home, Croker Bay, 16 m asl., Devon I. (74.560°N, 83.690°W), NU, CAN	Dyke (1998); R.E. Morlan, CMC database (1999)
9150 ± 190 (9280 ± 190)	S-3590	periotic	4.2 km NW of Cape Home, Croker Bay, 20.2 m asl., Devon I. (74.570°N, 83.710°W), NU, CAN	Dyke (1998); R.E. Morlan, CMC database (1999)
9140 ± 200 (9275 ± 200)	S-3573	periotic	5 km S of Cape Hardy, 57 m asl., Devon I. (75.780°N, 83.750°W), NU, CAN	Dyke (1998); R.E. Morlan, CMC database (1999)
9120 ± 200 (9265 ± 200)	S-3597	periotic	3.5 km NE of Lovell Pt., 76.5 m asl., Devon I. (74.960°N, 91.990°W), NU, CAN	Dyke (1998); R.E. Morlan, CMC database (1999)
(9120 ± 80)	TO-7751	periotic	10.5 km SSE of Cape Baring, Wollaston Pen., 37 m asl., Victoria I. (69°56'55"N, 117°18'00"W), NT, CAN	Dyke and Savelle (2001)

9110 ± 110 (9245 ± 110)	S-3310	periotic	NW coast of Griffith I., 98.5 m asl. (74.595°N, 95.775°W), NU, CAN	Dyke (1993); R.E. Morlan, CMC database (1999)
9110 ± 190 (9245 ± 190)	S-3593	periotic	2.6 km NNW of Cape Home, Croker Bay, 16 m asl., Devon I. (74.570°N, 83.610°W), NU, CAN	Dyke (1998); R.E. Morlan, CMC database (1999)
9080 ± 140 (9230 ± 140)	GSC-1496-B		Cape Storm, 118 m asl., Ellesmere I. (76°25'N, 87°33'W), NU, CAN	Blake (1975)
(9070 ± 100)	TO-7753	periotic	14.2 km SSE of Cape Baring, Wollaston, Pen., 40 m asl., Victoria I. (69°54'55"N, 117°16'45"W), NT, CAN	Dyke and Savelle (2001)
9060 ± 210 (9190 ± 210)	S-3581	periotic	1.6 km NW of Cape Home, Croker Bay, 14.2 m asl., Devon I. (74.550°N, 83.640°W), NU, CAN	Dyke (1998); R.E. Morlan, CMC database (1999)
(9060 ± 50)	CAMS-60325	periotic	9.8 km N of Cape Larsen and 20 km S of Cape Baring, Wollaston Pen., 31.5 m asl., Victoria I. (69°52'45"N, 117°16'00"W), NT, CAN	Dyke and Savelle (2001)
9050 ± 180 (9180 ± 180)	S-3575	periotic	7.7 km SW of Cape Hardy, 43 m asl., Devon I. (75.770°N, 83.910°W), NU, CAN	Dyke (1998); R.E. Morlan, CMC database (1999)
9050 ± 100 (9210 ± 100)	S-3069	periotic	10.5 km W of Stanley Pt., 30.5-31.5 m asl., Baffin I. (73.820°N, 85.440°W), NU, CAN	R.E. Morlan, CMC database (1999); [Dyke (2000d) GSC Map 1964A]
9040 ± 130 (9200 ± 130)	S-2836	rib	2 km N of Crooked L., 79.5 m asl., Prince of Wales I. (72.670°N, 98.280°W), NU, CAN	Dyke and Morris (1990); Dyke, et al. (1991); R.E. Morlan, CMC database (1999)
9020 ± 100 (9180 ± 100)	S-3070	periotic	8.4 km W of Stanley Pt., 30-30.5 m asl., Baffin I. (73.820°N, 85.370°W), NU, CAN	R.E. Morlan, CMC database (1999); [Dyke (2000d) GSC Map 1964A]
(9020 ± 80)	TO-7757	periotic	2.8 km inland from Penny Bay, N of Kugaluk R., Wollaston Pen., 29 m asl., Victoria I. (69°41'00"N, 116°50'00"W), NT, CAN	Dyke and Savelle (2001)
9010 ± 100 (9170 ± 100)	S-3083	periotic	3.5 km S of the mouth of McBean Bay, 66.4 m asl., Baffin I. (72.610°N, 89.630°W), NU, CAN	R.E. Morlan, CMC database (1999); [Dyke (2000c) GSC Map 1963A]
9000 ± 130 (9160 ± 130)	S-2910	periotic	W side of mouth of Smith Bay, 59 m asl., Prince of Wales I. (73.320°N, 100.600°W), NU, CAN	Dyke and Morris (1990); Dyke, et al. (1991); R.E. Morlan, CMC database (1999)
8990 ± 140 (9150 ± 140)	GSC-450	vertebra	Cunningham Inlet., 67 m asl., Somerset I. (73.980°N, 93.670°W), NU, CAN	[Dyke and Morris (1990) In: R.E. Morlan, CMC database (1999)]
8990 ± 130 (9150 ± 130)	S-2923	skull	10.8 km E of the head of Smith Bay, 61.5 m asl., Prince of Wales I. (73.230°N, 99.450°W), NU, CAN	Dyke and Morris (1990); Dyke, et al. (1991); R.E. Morlan, CMC database (1999)
8970 ± 190 (9105 ± 190)	S-3594	periotic	5.8 km NNE of Cape Home, Croker Bay, 17.5 m asl., Devon I. (74.590°N, 83.550°W), NU, CAN	Dyke (1998); R.E. Morlan, CMC database (1999)
8950 ± 180 (9085 ± 180)	S-3576	periotic	7 km SW of Cape Hardy, 46 m asl., Devon I. (75.770°N, 83.880°W), NU, CAN	Dyke (1998); R.E. Morlan, CMC database (1999)
(8940 ± 90)	GSC-964-2	rib	Cape Storm, 82 m asl., Ellesmere I. (76.680°N, 86.320°W), NU, CAN	Blake (1975)
8940 ± 90	GSC-1748-2	rib	S Cape Fiord, Ellesmere I. (76.425°N, 85.050°W), NU, CAN	Lowdon, et al. (1977)
8920 ± 140 (9080 ± 140)	Beta-14819		Truelove Lowland, 62 m asl., Devon I. (75.650°N, 84.580°W), NU, CAN	King (1991); [A.S. Dyke (GSC database) incorrectly cites lab number as Beta-14891]; Dyke (1998); R.E. Morlan, CMC database (1999)
8905 ± 405 (9065 ± 405)	S-2716	rib	1.6 km NE of Muskox Hill, .8 km W of Browne Bay, 100 m asl., Prince of Wales I. (72.820°N, 97.300°W), NU, CAN	Dyke and Morris (1990); Dyke, et al. (1991); R.E. Morlan, CMC database (1999)

8900 ± 170 (9040 ± 170)	GSC-1496-A		Cape Storm, 118 m asl., Ellesmere I. (76°25'N, 87°33'W), NU, CAN	Blake (1975); This sample was contaminated by air leakage.
8875 ± 120 (9035 ± 120)	S-2917	rib fragments and mandible	N shore of mouth of SE arm of Smith Bay, 65 m asl., Prince of Wales I. (73.300°N, 100.130°W), NU, CAN	Dyke and Morris (1990); Dyke, et al. (1991); R.E. Morlan, CMC database (1999)
8875 ± 135 (9035 ± 135)	S-2588	rib fragment	4 km S of Pt. Gell, 58 m asl., Prince of Wales I. (73.030°N, 101.900°W), NU, CAN	Dyke and Morris (1990); Dyke, et al. (1991); R.E. Morlan, CMC database (1999)
8860 ± 180 (9040 ± 180)	GSC-1496-2		Cape Storm, 118 m asl., Ellesmere I. (76°25'N, 87°33'W), NU, CAN	Blake (1975)
(8850 ± 130)	TO-7750	periotic	10.1 km SSE of Cape Baring, Wollaston Pen., 32.5 m asl., Victoria I. (69°57'08"N, 117°18'00"W), NT, CAN	Dyke and Savelle (2001)
(8830 ± 80)	TO-7760	periotic	2.5 km inland from the coast just N of Penny Bay, N of Kugaluk R., Wollaston Pen., 28 m asl., Victoria I. (69°43'07"N, 116°58'22"W), NT, CAN	Dyke and Savelle (2001)
8820 ± 130 (8980 ± 130)	S-3138	periotic	4.2 km SW of Cape York, 54 m asl., Baffin I. (73.790°N, 87.130°W), NU, CAN	McNeely and Jorgensen (1993); R.E. Morlan, CMC database (1999); [Dyke (2000d) GSC Map 1964A]
8820 ± 200 (8940 ± 200)	S-3589	periotic	4.2 km WNW of Cape Home, Croker Bay, 9 m asl., Devon I. (74.560°N, 83.730°W), NU, CAN	Dyke (1998); R.E. Morlan, CMC database (1999)
8820 ± 100 (8930 ± 100)	S-3637	mandible	7 km inland from the N coast of Stefansson I., 53-54 m asl. (73.720°N, 105.370°W), NU, CAN	[A.S. Dyke, pers. comm. (1998) In: R.E. Morlan, CMC database (1999)]; Dyke and Savelle (2001)
8805 ± 70 (8965 ± 75)	S-1388	maxilla	9 km NW of mouth of Creswell R., 73 m asl., Somerset I. (72.860°N, 93.570°W), NU, CAN	[Dyke (1980). This date supersede earlier (5205 ± 70 BP) incorrect date]; R.E. Morlan, CMC database (1999)
8790 ± 110 (8925 ± 110)	S-3017	periotic	4.5 km SSW of the mouth of McBean Bay, 86 m asl., Baffin I. (72.600°N, 89.675°W), NU, CAN	R.E. Morlan, CMC database (1999); [Dyke (2000c) GSC Map 1963A]
8770 ± 170 (8895 ± 170)	S-3596	periotic	18.5 km NNE of Cape Home, Croker Bay, 11.8 m asl., Devon I. (74.700°N, 83.380°W), NU, CAN	Dyke (1998); R.E. Morlan, CMC database (1999)
8770 ± 145 (8930 ± 145)	S-3018	periotic	8 km SSW of the mouth of McBean Bay, 70 m asl., Baffin I. (72.575°N, 89.700°W), NU, CAN	R.E. Morlan, CMC database (1999); [Dyke (2000c) GSC Map 1963A]
8765 ± 160 (8925 ± 160)	S-2965	periotic	NW coast of Prescott I., 50.5 m asl. (73.130°N, 97.050°W), NU, CAN	Dyke and Morris (1990); Dyke, et al. (1991); R.E. Morlan, CMC database (1999)
8750 ± 200 (8890 ± 200)	S-3582	periotic	1.5 km NW of Cape Home, Croker Bay, 11.7 m asl., Devon I. (74.550°N, 83.640°W), NU, CAN	Dyke (1998); R.E. Morlan, CMC database (1999)
8740 ± 100 (8900 ± 100)	S-3051	periotic	9.2 km W of Stanley Pt., 30 m asl., Baffin I. (73.820°N, 85.380°W), NU, CAN	R.E. Morlan, CMC database (1999); [Dyke (2000d) GSC Map 1964A]
8730 ± 190 (8855 ± 190)	S-3584	periotic	2.7 km NW of Cape Home, Croker Bay, 11.7 m asl., Devon I. (74.560°N, 83.670°W), NU, CAN	Dyke (1998); R.E. Morlan, CMC database (1999)
8720 ± 190 (8860 ± 190)	S-3600	otic capsule	6.5 km NNW of Lovell Pt., 64.5 m asl., Devon I. (74.997°N, 92.140°W), NU, CAN	Dyke (1998); R.E. Morlan, CMC database (1999)
8675 ± 135 (8835 ± 135)	S-2603	rib	332 km W of Coningham Bay, 75.5 m asl., 1.5 km N of the S shore of Prince of Wales I. (71.590°N, 97.350°W), NU, CAN	Dyke and Morris (1990); Dyke, et al. (1991); R.E. Morlan, CMC database (1999)
8660 ± 90 (8820 ± 90)	S-3068	periotic	10 km W of Stanley Pt., 28 m asl., Baffin I. (73.820°N, 85.420°W), NU, CAN	R.E. Morlan, CMC database (1999); [Dyke (2000d) GSC Map 1964A]

8655 ± 130 (8815 ± 130)	S-2864		N coast of a small unnamed inlet, 7 km N of Muskox Hill, 68 m asl., Prince of Wales I. (72.880°N, 97.500°W), NU, CAN	Dyke and Morris (1990); Dyke, et al. (1991); R.E. Morlan, CMC database (1999)
8650 ± 110 (8810 ± 110)	S-3065	periotic	12.7 km S of the mouth of Fitzgerald Bay, 70.4 m asl., Baffin I. (71.950°N, 89.920°W), NU, CAN	R.E. Morlan, CMC database (1999); [Dyke and Hooper (2000) GSC Map 1960A]
8645 ± 135 (8805 ± 135)	S-2707	mandible	11 km SE of Cape Hardy, 51.5 m asl., Prince of Wales I. (73.840°N, 97.280°W), NU, CAN	Dyke and Morris (1990); Dyke, et al. (1991); R.E. Morlan, CMC database (1999)
8645 ± 205 (8805 ± 205)	S-2599		N side of Dalgety Pt., 87 m asl., Prince of Wales I. (72.010°N, 96.510°W), NU, CAN	Dyke and Morris (1990); Dyke, et al. (1991); R.E. Morlan, CMC database (1999)
8640 ± 140 (8800 ± 140)	GSC-266	rib	29 km N of Ferguson L., 113 m asl., Victoria I., NU, CAN	R.E. Morlan, CMC database (1999)
8630 ± 135 (8790 ± 135)	S-2992		70 km E of Washburn L., 132 m asl., Victoria I. (70.100°N, 104.930°W), NU, CAN	[A.S. Dyke, pers. comm. (1998) In: R.E. Morlan, CMC database (1999)]
8630 ± 195 (8790 ± 195)	S-2598	skull fragment	Guillemard Bay, 2 km E of Fisher L., in beach gravel 81.5 m asl., Prince of Wales I. (72.170°N, 97.830°W), NU, CAN	Dyke and Morris (1990); Dyke, et al. (1991); R.E. Morlan, CMC database (1999)
8620 ± 135 (8740 ± 135)	S-3048	periotic	17.6 km W of Stanley Pt., 16.5 m asl., Baffin I. (73.825°N, 85.825°W), NU, CAN	R.E. Morlan, CMC database (1999); [Dyke (2000d) GSC Map 1964A]
8620 ± 170 (8740 ± 170)	S-3601	otic capsule	8.1 km SSE of Lovell Pt., 57 m asl., Devon I. (74.870°N, 92.002°W), NU, CAN	Dyke (1998); R.E. Morlan, CMC database (1999)
8590 ± 260 (8770 ± 260)	GSC-1496-1		Cape Storm, 118 m asl., Ellesmere I. (76°25'N, 87°33'W), NU, CAN	Blake (1975)
8570 ± 200 (8705 ± 200)	S-3587	otic capsule	7.9 km WNW of Cape Home, Croker Bay, 8.2 m asl., Devon I. (74.550°N, 83.860°W), NU, CAN	Dyke (1998); R.E. Morlan, CMC database (1999)
8570 ± 110 (8730 ± 110)	S-3052	periotic	10 km W of Stanley Pt., 33.5-34.5 m asl., Baffin I. (73.820°N, 85.420°W), NU, CAN	R.E. Morlan, CMC database (1999); [Dyke (2000d) GSC Map 1964A]
8565 ± 125 (8725 ± 125)	S-2835	skull	NW coast of Pandora I., 66 m asl., Prince of Wales I. (72.880°N, 96.780°W), NU, CAN	Dyke and Morris (1990); Dyke, et al. (1991); R.E. Morlan, CMC database (1999)
8555 ± 165 (8715 ± 165)	S-2964	periotic	11.8 km NE of Cape Richard Collinson, 57 m asl., Prince of Wales I. (72.850°N, 102.580°W), NU, CAN	Dyke and Morris (1990); Dyke, et al. (1991); R.E. Morlan, CMC database (1999)
8520 ± 190 (8680 ± 190)	S-2604	rib	5 km W of Coningham Bay, 84 m asl., S shore of Prince of Wales I. (71.600°N, 97.350°W), NU, CAN	Dyke and Morris (1990); Dyke, et al. (1991); R.E. Morlan, CMC database (1999)
8520 ± 90 (8680 ± 90)	S-3078	periotic	4.5 km SW of the mouth of McBean Bay, 61 m asl., Baffin I. (72.600°N, 89.680°W), NU, CAN	R.E. Morlan, CMC database (1999); [Dyke (2000c) GSC Map 1963A]
8500 ± 200	W-48	baleen	Thule (76.330°N, 69.000°W), GRE	Davies, et al. (1963); Bennike (1997)
8490 ± 200 (8630 ± 200)	S-3588	periotic	8.8 km NNE of Cape Home, Croker Bay, 5.5 m asl., Devon I. (74.610°N, 83.470°W), NU, CAN	Dyke (1998); R.E. Morlan, CMC database (1999)
8470 ± 190 (8600 ± 190)	S-3579	periotic	.85 km NW of Cape Home, Croker Bay, 10.2 m asl., Devon I. (74.550°N, 83.620°W), NU, CAN	Dyke (1998); R.E. Morlan, CMC database (1999)
8420 ± 190 (8540 ± 190)	S-3583	periotic	.7 km N of Cape Home, Croker Bay, 5.5-7 m asl., Devon I. (74.550°N, 83.600°W), NU, CAN	Dyke (1998); R.E. Morlan, CMC database (1999)
8400 ± 135 (8560 ± 135)	S-3024	otic capsule	5.5 km SSW of the mouth of McBean Bay, 63 m asl., Baffin I. (72.600°N, 89.690°W), NU, CAN	R.E. Morlan, CMC database (1999); [Dyke (2000c) GSC Map 1963A]
8390 ± 200 (8520 ± 200)	S-3586	periotic	3.8 km WNW of Cape Home, Croker Bay, 3.5 m asl., Devon I. (74.550°N, 83.720°W), NU, CAN	Dyke (1998); R.E. Morlan, CMC database (1999)

8380 ± 110 (8540 ± 110)	S-3064	periotic	S side of Fitzgerald Bay, 3.5 km inland, 72.5 m asl., Baffin I. (72.050°N, 89.880°W), NU, CAN	R.E. Morlan, CMC database (1999); [Dyke and Hooper (2000) GSC Map 1960A]
8375 ± 105 (8535 ± 105)	Beta-33189 ETH-5897	lumbar vertebra (CMN 38465)	Trout R., 45 m asl., (49°28.4'N, 58°7.1'W), NF, CAN	C.R. Harington [unpublished]; R.E. Morlan, CMC database (1999); Two other dates on same specimen - see S-2149 and TO-2509
8345 ± 135 (8475 ± 135)	S-3023	periotic	13.5 km S of the mouth of McBean Bay, 50-51 m asl., Baffin I. (72.525°N, 89.630°W), NU, CAN	R.E. Morlan, CMC database (1999); [Dyke (2000c) GSC Map 1963A]
8330 ± 120 (8490 ± 120)	S-3137	periotic	4.1 km SW of Cape York, 35 m asl., Baffin I. (73.790°N, 87.130°W), NU, CAN	R.E. Morlan, CMC database (1999); [Dyke (2000d) GSC Map 1964A]
8310 ± 90 (8470 ± 90)	S-3049	periotic	11.2 km W of Stanley Pt., 21.5-23 m asl., Baffin I. (73.820°N, 85.470°W), NU, CAN	R.E. Morlan, CMC database (1999); [Dyke (2000d) GSC Map 1964A]
8290 ± 100 (8140 ± 100)	Lu-3661		Thule (76.720°N, 69.430°W), GRE	Bennike (1997)
8270 ± 120 (8430 ± 120)	S-3135	periotic	2.8 km S of Cape York, 27.5 m asl., Baffin I. (73.790°N, 87.000°W), NU, CAN	R.E. Morlan, CMC database (1999); [Dyke (2000d) GSC Map 1964A]
8270 ± 150 (8430 ± 150)	GSC-991	skull	Truelove Inlet, 42.4 m asl., Devon I. (75.670°N, 84.380°W), NU, CAN	[Barr (1971); Dyke (1988); King (1991); R.E. Morlan, CMC database (1999)
8250 ± 100 (8375 ± 100)	S-3638	left periotic	Cape Martyr, Resolute Bay, 49 m asl., Cornwallis I. (74.680°N, 95.000°W), NU, CAN	[A.S. Dyke, pers. comm. (1998) In: R.E. Morlan, CMC database (1999)]
8230 ± 100 (8350 ± 100)	S-3291	periotic	Cape Charles York, Borden Pen., 9-9.5 m asl., Baffin I. (73.650°N, 83.000°W), NU, CAN	R.E. Morlan, CMC database (1999); [Dyke (2000e) GSC Map 1965A]
8230 ± 100 (8390 ± 100)	S-3082	periotic	15 km S of the mouth of McBean Bay, 52.4 m asl., Baffin I. (72.510°N, 89.650°W), NU, CAN	Dyke (1998); R.E. Morlan, CMC database (1999)
8210 ± 170 (8345 ± 170)	S-3611	periotic	9. km SW of Cape Hardy, 29 m asl., Devon I. (75.770°N, 83.960°W), NU, CAN	Dyke (1998); R.E. Morlan, CMC database (1999)
8210 ± 100 (8370 ± 100)	S-3062	periotic	S side of Fitzgerald Bay, 2 km inland, 58.2 m asl., Baffin I. (72.070°N, 89.900°W), NU, CAN	[A.S. Dyke, pers. comm. (1998) In: R.E. Morlan, CMC database (1999)]
8170 ± 170 (8290 ± 170)	S-3578	periotic	8.6 km SW of Cape Hardy, 31.5 m asl., Devon I. (75.770°N, 83.950°W), NU, CAN	Dyke (1998); R.E. Morlan, CMC database (1999)
8150 ± 190 (8290 ± 190)	S-3524	otic capsule	4.9 km SSW of Firkin Pt., 32 m asl., Devon I. (75.560°N, 85.610°W), NU, CAN	Dyke (1998); R.E. Morlan, CMC database (1999)
8150 ± 100 (8265 ± 100)	S-3292	periotic	Cape Charles York, Borden Pen., 7.5 m asl., Baffin I. (73.650°N, 83.000°W), NU, CAN	R.E. Morlan, CMC database (1999); [Dyke (2000e) GSC Map 1965A]
8110 ± 190 (8240 ± 190)	S-3570	periotic	5.7 km SW of Cape Hardy, 26 m asl., Devon I. (75.790°N, 83.890°W), NU, CAN	Dyke (1998); R.E. Morlan, CMC database (1999)
8080 ± 100 (8215 ± 100)	S-3290	periotic	Cape Charles York, Borden Pen., 2.5-3 m asl., Baffin I. (73.650°N, 83.000°W), NU, CAN	R.E. Morlan, CMC database (1999); [Dyke (2000e) GSC Map 1965A]
8070 ± 180 (8205 ± 180)	S-3525	otic capsule	8.9 km SW of Firkin Pt., 30-30.5 m asl., Devon I. (75.540°N, 85.770°W), NU, CAN	Dyke (1998); R.E. Morlan, CMC database (1999)
8040 ± 200 (8160 ± 200)	S-3569	periotic	5.7 km SW of Cape Hardy, 26 m asl., Devon I. (75.790°N, 83.890°W), NU, CAN	Dyke (1998); R.E. Morlan, CMC database (1999)
8005 ± 155 (8165 ± 155)	S-1384		9 km SW of Cape Anne, 50 m asl., Somerset I. (74.025°N, 94.880°W), NU, CAN	Dyke and Morris (1990); R.E. Morlan, CMC database (1999)
7940 ± 190 (8075 ± 190)	S-3558	periotic	5 km SSW of Firkin Pt., 32 m asl., Devon I. (75.560°N, 85.620°W), NU, CAN	Dyke (1998); R.E. Morlan, CMC database (1999)

7920 ± 180 (8055 ± 180)	S-3523	otic capsule	5.9 km SW of Firkin Pt., 32.5 m asl., Devon I. (75.560°N, 85.670°W), NU, CAN	Dyke (1998); R.E. Morlan, CMC database (1999)
7890 ± 170 (8025 ± 170)	S-3421	periotic	Bay NW of Lavoie Pt., Borden Pen., 31 m asl., Baffin I. (72.830°N, 80.540°W), NU, CAN	R.E. Morlan, CMC database (1999); [Dyke (2000b) GSC Map 1962A]
7880 ± 110 (8040 ± 110)	S-3136	periotic	3.3 km SW of Cape York, Baffin I. (73.790°N, 87.120°W), NU, CAN	R.E. Morlan, CMC database (1999); [Dyke (2000d) GSC Map 1964A]
7865 ± 155 (8025 ± 155)	S-2955	skull fragment	E Stefansson I., 29 m asl. (73.220°N, 104.750°W), NU, CAN	[Hodgson (1993) In: R.E. Morlan, CMC database (1999)]
7790 ± 100 (7950 ± 100)	S-3063	periotic	S side of Fitzgerald Bay, 2.5 km inland, 64.6 m asl., Baffin I. (72.070°N, 89.880°W), NU, CAN	R.E. Morlan, CMC database (1999); [Dyke and Hooper (2000) GSC Map 1960A]
7750 ± 180 (7885 ± 180)	S-3538	periotic	5.2 km SW of Firkin Pt., 24.5-25 m asl., Devon I. (75.560°N, 85.660°W), NU, CAN	Dyke (1998); R.E. Morlan, CMC database (1999)
7710 ± 180 (7850 ± 180)	S-3598	periotic	4 km N of Lovell Pt., 40 m asl., Devon I. (74.980°N, 92.090°W), NU, CAN	Dyke (1998); R.E. Morlan, CMC database (1999)
7650 ± 120 (7810 ± 120)	S-2602		25 km W of Coningham Bay, 98 m asl., near Prince of Whales I. (71°35'00"N, 97°20'00"W), NU, CAN	Dyke, et al. (1991); [A.S. Dyke, pers. comm. (1999); age is anomalously young. Mimimum age is 9200 BP]; R.E. Morlan, CMC database (1999)
7630 ± 160 (7770 ± 160)	S-3599	periotic	9.7 km S of Lovell Pt., 35 m asl., Devon I. (74.850°N, 92.030°W), NU, CAN	Dyke (1998); R.E. Morlan, CMC database (1999)
7475 ± 220 (7635 ± 220)	S-3035	otic capsule	Nansen Sound coast, 40 m asl., Ellesmere I., NU, CAN	Bednarski (1990); Dyke, et al. (1996); R.E. Morlan, CMC database (1999)
(7380 ± 140)	GSC-1193	vertebra	Resolute, 50 m asl., Cornwallis I. (74.700°N, 94.980°W), NU, CAN	Harington (1975); Dyke, et al. (1991)
7280 ± 100 (7440 ± 100)	S-3056	periotic	Vista R., 4.2 km NW of Kakiak Pt., 28.5-29 m asl., Baffin I. (72.710°N, 86.770°W), NU, CAN	R.E. Morlan, CMC database (1999); [Dyke (2000c) GSC Map 1963A]
7180 ± 120 (7340 ± 120)	S-3175	periotic	16 km SW of Cape Crawford, 6 m asl., Baffin I. (73.625°N, 85.120°W), NU, CAN	R.E. Morlan, CMC database (1999); [Dyke (2000d) GSC Map 1964A]
7160 ± 100 (7320 ± 100)	S-3057	periotic	Vista R., 7.4 km NW of Kakiak Pt., 37 m asl., Baffin I. (72.720°N, 86.830°W), NU, CAN	R.E. Morlan, CMC database (1999); [Dyke (2000c) GSC Map 1963A]
7105 ± 90 (7265 ± 90)	S-1386	rib	5 km SW of Cape Anne, 28 m asl., Somerset I. (74.060°N, 94.800°W), NU, CAN	[Dyke (1980) These dates supersede earlier incorrect dates taken]; Dyke and Morris (1990); R.E. Morlan, CMC database (1999)
7070 ± 100 (7230 ± 100)	S-3139	periotic	1.9 km SW of Cape York, 12 m asl., Baffin I. (73.800°N, 87.050°W), NU, CAN	R.E. Morlan, CMC database (1999); [Dyke (2000d) GSC Map 1964A]
6940 ± 155 (7100 ± 155)	S-2601	ribs	20 km W of Coningham Bay, 48 m asl., 1 km N of the S shore of Prince of Wales I. (71°41'00"N, 97°15'00"W), NU, CAN	Dyke and Morris (1990); Dyke, et al. (1991); Dyke (1998); R.E. Morlan, CMC database (1999)
6920 ± 100 (7080 ± 100)	S-3143	periotic	7.1 km NW of the mouth of Tikiraq R., 53.5 m asl., Baffin I. (71.800°N, 86.180°W), NU, CAN	R.E. Morlan, CMC database (1999); [Dyke and Hooper (2000) GSC Map 1960A]
6850 ± 80 (6995 ± 80)	S-3340	periotic	.5 km N of Eqalulik R., Borden Pen., 27 m asl., Baffin I. (72.650°N, 85.680°W), NU, CAN	R.E. Morlan, CMC database (1999); [Dyke (2000c) GSC Map 1963A]
6820 ± 220 (6940 ± 220)	S-3510	periotic	31.4 km NNE of Fort Ross, 40 m asl., Somerset I. (72.270°N, 93.900°W), NU, CAN	[A.S. Dyke, pers. comm. (1998) In: R.E. Morlan, CMC database (1999)]
6790 ± 90	Beta-19201		Truelove Lowland, Devon I., 21.8 m asl. (75°30'00"N, 84°35'00"W), NU, CAN	King (1991); Dyke (1998)

6660 ± 160 (6820 ± 160)	S-2594		Head of Guillemard Bay, 3 km SE from the mouth of the Fisher R., 81 m asl., Prince of Wales I. (71°50'00"N, 98°10'00"W), NU, CAN	Dyke, et al. (1991); Dyke (1998); [A.S. Dyke, pers. comm. (1999) - Age is anomalously young for elevation. Minimum age is 8700 BP]; R.E Morlan, CMC database (1999)
6610 ± 110 (6770 ± 110)	S-3170	periotic	3.8 km S of the mouth of Kuuruluk R., 23.5 m asl., Baffin I. (72.410°N, 86.280°W), NU, CAN	R.E. Morlan, CMC database (1999); [Dyke (2000c) GSC Map 1963A]
6610 ± 140 (6740 ± 140)	S-3409	periotic	Bay NW of Lavoie Pt., Borden Pen., 22.5 m asl., Baffin I. (72.820°N, 80.470°W), NU, CAN	R.E. Morlan, CMC database (1999); [Dyke (2000b) GSC Map 1962A]
6585 ± 110 (6690 ± 105)	S-3013	periotic	27 km WSW of the head of Bernier Bay, 49 m asl., Baffin I. (70.930°N, 87.700°W), NU, CAN	R.E. Morlan, CMC database (1999); [Dyke and Hooper (2000) GSC Map 1960A]
6450 ± 150 (6605 ± 150)	S-3422	periotic	Bay NW of Lavoie Pt., Borden Pen., 20.5 m asl., Baffin I. (72.830°N, 80.540°W), NU, CAN	R.E. Morlan, CMC database (1999); [Dyke (2000b) GSC Map 1962A]
6420 ± 90 (6580 ± 90)	S-3145	periotic	9 km NW of the mouth of Tikiraq R., 40.5 m asl., Baffin I. (71.825°N, 86.180°W), NU, CAN	R.E. Morlan, CMC database (1999); [Dyke and Hooper (2000) GSC Map 1960A]
6360 ± 100 (6520 ± 100)	S-3159	periotic	1.8 km NW of the mouth of Tiriganialaag R., 35.5 m asl., Baffin I. (72.190°N, 86.470°W), NU, CAN	R.E. Morlan, CMC database (1999); [Dyke (2000c) GSC Map 1963A]
6350 ± 130 (6490 ± 130)	S-3433	periotic	10 km NW of Pt. Kendall, 72 m asl., Crown Prince Frederick I., Baffin I. (70.070°N, 86.670°W), NU, CAN	R.E. Morlan, CMC database (1999); [Hooper and Dyke (2000) GSC Map 1959A]
6275 ± 175 (6435 ± 175)	S-2595	vertebra	1 km E of Fisher R., 1.5 km upstream from the river's mouth, 103.9 m asl., Prince of Wales I. (71°50'00"N, 98°5'00"W), NU, CAN	Dyke, et al. (1991); Dyke (1998); [A.S. Dyke, pers. comm. (1999) - Age anomalously young. Minimum age is 9200 BP]; R.E. Morlan, CMC database (1999)
6240 ± 100 (6400 ± 100)	S-3165	periotic	1.3 km WNW of the mouth of Tiriganialaag R., 37 m asl., Baffin I. (72.180°N, 86.460°W), NU, CAN	R.E. Morlan, CMC database (1999); [Dyke (2000c) GSC Map 1963A]
6200 ± 110 (6360 ± 110)	S-3126	otic capsule	1 km W of the mouth of Tiriganialaag R., 39 m asl., Baffin I. (72.180°N, 86.470°W), NU, CAN	R.E. Morlan, CMC database (1999); [Dyke (2000c) GSC Map 1963A]
6150 ± 90 (6290 ± 90)	S-3359	periotic	8 km NW of head of Bernier Bay, 47 m asl., Baffin I. (71.050°N, 87.280°W), NU, CAN	R.E. Morlan, CMC database (1999); [Dyke and Hooper (2000) GSC Map 1960A]
6110 ± 125 (6260 ± 125)	S-432	ear bone	Truelove Lowland, 11 m asl., Devon I. (75.650°N, 84.580°W), NU, CAN	Barr 1971; King (1991); Dyke (1998); R.E. Morlan, CMC database (1999)
6070 ± 90 (6230 ± 90)	S-3180	periotic	9 km NW of the mouth of Saputing R., Bell Bay, 44 m asl., NW Baffin I. (70.950°N, 85.080°W), NU, CAN	R.E. Morlan, CMC database (1999); [Hooper and Dyke (2000) GSC Map 1959A]
6060 ± 160 (6205 ± 160)	S-3539	periotic	7.5 km SW of Firkin Pt., 12.5 m asl., Devon I. (75.550°N, 85.740°W), NU, CAN	Dyke (1998); R.E. Morlan, CMC database (1999)
(5920 ± 60)	TO-315	rib	17 km S of the head of Young Bay, 104.5 m asl., Prince of Wales I. (72°27'00"N, 97°19'00"W), NU, CAN	[A.S. Dyke, pers. comm. (1999); indicates a corrected date of 5510 ± 60 BP - anomalously young]; [R. McNeely, pers. comm. (1998) In: R.E. Morlan, CMC database (1999)]
5890 ± 80 (6050 ± 80)	S-3067	periotic	6 km SW of the mouth of McBean Bay, 20.1 m asl., Baffin I. (72.600°N, 89.720°W), NU, CAN	R.E. Morlan, CMC database (1999); [Dyke (2000c) GSC Map 1963A]
5870 ± 80 (6030 ± 80)	S-3232	periotic	1.9 km W of the mouth of Tikiraq R., 28 m asl., Baffin I. (71.760°N, 86.075°W), NU, CAN	R.E. Morlan, CMC database (1999); [Dyke and Hooper (2000) GSC Map 1960A]
5870 ± 70 (6030 ± 75)	S-3230	periotic	2.1 km NW of the mouth of Tikiraq R., 19.5 m asl., Baffin I. (71.770°N, 86.070°W), NU, CAN	R.E. Morlan, CMC database (1999); [Dyke and Hooper (2000) GSC Map 1960A]

5850 ± 120 (6010 ± 120)	S-3152	periotic	2 km SW of the mouth of Tikiraq R., 38 m asl., Baffin I. (71.760°N, 86.070°W), NU, CAN	R.E. Morlan, CMC database (1999); [Dyke and Hooper (2000) GSC Map 1960A]
5780 ± 160 (5910 ± 160)	S-3540	periotic	7.5 km SW of Firkin Pt., 12.5 m asl., Devon I. (75.550°N, 85.740°W), NU, CAN	Dyke (1998); R.E. Morlan, CMC database (1999)
5770 ± 140 (5915 ± 140)	S-3413	periotic	Bay NW of Lavoie Pt., Borden Pen., 17 m asl., Baffin I. (72.810°N, 80.440°W), NU, CAN	R.E. Morlan, CMC database (1999); [Dyke (2000b) GSC Map 1962A]
5760 ± 90 (5920 ± 90)	S-3163	periotic	1.4 km NW of the mouth of Tiriganialaag R., 30 m asl., Baffin I. (72.180°N, 86.470°W), NU, CAN	R.E. Morlan, CMC database (1999); [Dyke (2000c) GSC Map 1963A]
5700 ± 120 (5835 ± 120)	S-3432	otic capsule	13 km NW of Pt. Kendall, 50 m asl., Crown Prince Frederick I., Baffin I. (70.080°N, 86.720°W), NU, CAN	R.E. Morlan, CMC database (1999); [Hooper and Dyke (2000) GSC Map 1959A]
5690 ± 80 (5850 ± 80)	S-3147	periotic	2.1 km WNW of the mouth of Tikiraq R., 29 m asl., Baffin I. (71.770°N, 86.070°W), NU, CAN	R.E. Morlan, CMC database (1999); [Dyke and Hooper (2000) GSC Map 1960A]
5690 ± 100 (5850 ± 100)	S-3252	periotic	2.8 km S of the mouth of Kuuruluk R., 23 m asl., Baffin I. (72.420°N, 86.270°W), NU, CAN	R.E. Morlan, CMC database (1999); [Dyke (2000c) GSC Map 1963A]
5670 ± 100 (5805 ± 100)	S-3014	periotic	30 km WSW of the head of Bernier Bay, 37 m asl., Baffin I. (70.930°N, 87.780°W), NU, CAN	R.E. Morlan, CMC database (1999); [Hooper and Dyke (2000) GSC Map 1959A]
5645 ± 115	I-6235		Holsteinsborg (66.930°N, 53.630°W), GRE	Bennike (1997)
5640 ± 70 (5800 ± 75)	S-3233	periotic	1.9 km W of the mouth of Tikiraq R., 31 m asl., Baffin I. (71.760°N, 86.075°W), NU, CAN	R.E. Morlan, CMC database (1999); [Dyke and Hooper (2000) GSC Map 1960A]
5620 ± 95 (5780 ± 95)	S-2681	humerus	4 km N of Collinson Inlet, 37 m asl., King William I., NU, CAN	Dyke and Morris (1990); Dyke et al. (1991); R.E. Morlan, CMC database (1999)
5610 ± 90 (5770 ± 90)	S-3238	periotic	8.6 km NW of the mouth of Tikiraq R., 29.5 m asl., Baffin I. (71.825°N, 86.180°W), NU, CAN	R.E. Morlan, CMC database (1999); [Dyke and Hooper (2000) GSC Map 1960A]
5600 ± 90 (5760 ± 95)	S-3042	periotic	14 km N of Morin Pt., mouth of Bernier Bay, 34 m asl., Baffin I. (71.470°N, 89.870°W), NU, CAN	R.E. Morlan, CMC database (1999); [Dyke and Hooper (2000) GSC Map 1960A]
(5600 ± 60)	GSC-979	rib	Cape Storm, 23 m asl., Ellesmere I. (76.400°N, 87.500°W), NU, CAN	Lowdon and Blake (1980); R.E. Morlan, CMC database (1999)
5560 ± 680	S-3516	otic capsule	34.5 km NNE of Fort Ross, Somerset I., NU, CAN	A.S. Dyke (pers. comm. 1999)
5535 ± 475 (5695 ± 475)	S-2668	skull fragment	4 km S of Cape Walker, 36.6 m asl., E Russell I. (74°1.3'00"N, 97°37.5'00"W), NU, CAN	Dyke, et al. (1991); Dyke (1998); R.E. Morlan, CMC database (1999)
5520 ± 70 (5680 ± 75)	S-3098	periotic	5 km SE of Van Koenig Pt., mouth of Bernier Bay, 27 m asl., Baffin I. (71.050°N, 89.370°W), NU, CAN	R.E. Morlan, CMC database (1999); [Dyke and Hooper (2000) GSC Map 1960A]
5485 ± 100 (5645 ± 100)	S-3021	periotic	12 km SSW of the mouth of McBean Bay, 18.6-18.9 m asl., Baffin I. (72.540°N, 89.770°W), NU, CAN	R.E. Morlan, CMC database (1999); [Dyke (2000c) GSC Map 1963A]
(5420 ± 70)	GSC-979-2		Cape Storm, Ellesmere I., NU, CAN	Lowdon and Blake (1980)
5410 ± 70 (5570 ± 75)	S-3103	otic capsule	6 km NE of the mouth of Fitzgerald Bay, 16 m asl., Baffin I. (72.090°N, 89.870°W), NU, CAN	R.E. Morlan, CMC database (1999); [Dyke and Hooper (2000) GSC Map 1960A]
5395 ± 95 (554 ± 95)	S-3026	periotic	Vista R., 5.4 NW of Kakiak Pt., 14-14.5 m asl., Baffin I. (72.720°N, 86.770°W), NU, CAN	R.E. Morlan, CMC database (1999); [Dyke (2000c) GSC Map 1963A]
5390 ± 90 (5530 ± 90)	S-3307	otic capsule	Jungersen Bay, Borden Pen., 38.5 m asl., Baffin I. (71.528°N, 84.430°W), NU, CAN	R.E. Morlan, CMC database (1999); [Dyke and Hooper (2000) GSC Map 1960A]

5390 ± 80 (5550 ± 80)	S-3153	otic capsule	6 km NW of the mouth of Tikiraq R., 20 m asl., Baffin I. (71.800°N, 86.120°W), NU, CAN	R.E. Morlan, CMC database (1999); [Dyke and Hooper (2000) GSC Map 1960A]
5310 ± 120 (5440 ± 120)	S-3431	periotic	8 km WSW of Pt. Kendall, 50 m asl., Crown Prince Frederick I., Baffin I. (70.000°N, 86.670°W), NU, CAN	R.E. Morlan, CMC database (1999); [Hooper and Dyke (2000) GSC Map 1959A]
5220 ± 80 (5380 ± 80)	S-3100	periotic	5.8 km NE of Morin Pt., mouth of Bernier Bay, 30 m asl., Baffin I. (71.380°N, 89.700°W), NU, CAN	R.E. Morlan, CMC database (1999); [Dyke and Hooper (2000) GSC Map 1960A]
5200 ± 90 (5340 ± 90)	S-3360	periotic	8 km NW of head of Bernier Bay, 37 m asl., Baffin I. (71.050°N, 87.280°W), NU, CAN	R.E. Morlan, CMC database (1999); [Dyke and Hooper (2000) GSC Map 1960A]
5180 ± 70 (5340 ± 75)	S-3150	periotic	2.3 km SSW of the mouth of Tikiraq R., 31 m asl., Baffin I. (71.750°N, 86.050°W), NU, CAN	R.E. Morlan, CMC database (1999); [Dyke and Hooper (2000) GSC Map 1960A]
5180 ± 200 (5310 ± 200)	S-3515	otic capsule	13 km NNE of Fort Ross, 25 m asl., Somerset I., NU, CAN	[A.S. Dyke, pers. comm. (1998) In: R.E. Morlan, CMC database (1999)]
5170 ± 80 (5325 ± 80)	S-3288	periotic	NW coast of Griffith I., 23 m asl. (74.565°N, 95.777°W), NU, CAN	Dyke (1993); R.E. Morlan, CMC database (1999)
5130 ± 260 (5265 ± 260)	S-3541	periotic	4. km SW of Firkin Pt., 11-11.5 m asl., Devon I. (75.570°N, 85.640°W), NU, CAN	Dyke (1998); R.E. Morlan, CMC database (1999)
5120 ± 120 (5265 ± 120)	S-3434	periotic	8 km WSW of Pt. Kendall, 47 m asl., Crown Prince Frederick I., Baffin I. (70.000°N, 86.670°W), NU, CAN	R.E. Morlan, CMC database (1999); [Hooper and Dyke (2000) GSC Map 1959A]
5110 ± 90 (5260 ± 90)	S-3345	periotic	7 km SW of S end of Navarana L., Foss Fiord, 37.5 m asl., Baffin I. (70.400°N, 87.250°W), NU, CAN	R.E. Morlan, CMC database (1999); [Hooper and Dyke (2000) GSC Map 1959A]
5080 ± 80 (5240 ± 80)	S-3128	otic capsule	8 km SE of the head of Berlinguet Inlet, 4 km S of the S coast, 29.5 m asl., Baffin I. (70.950°N, 86.650°W), NU, CAN	R.E. Morlan, CMC database (1999); [Dyke and Hooper (2000) GSC Map 1960A]
5080 ± 130 (5220 ± 130)	S-3415	periotic	Bay NW of Lavoie Pt., Borden Pen., 15 m asl., Baffin I. (72.820°N, 80.440°W), NU, CAN	R.E. Morlan, CMC database (1999); [Dyke (2000b) GSC Map 1962A]
5070 ± 90 (5230 ± 90)	S-3226	periotic	2.7 km NW of the mouth of Tikiraq R., 25 m asl., Baffin I. (71.770°N, 86.100°W), NU, CAN	R.E. Morlan, CMC database (1999); [Dyke and Hooper (2000) GSC Map 1960A]
5050 ± 165 (5190 ± 165)	S-3040	periotic	36 km WSW of the head of Bernier Bay, 28 m asl., Baffin I. (70.930°N, 87.930°W), NU, CAN	R.E. Morlan, CMC database (1999); [Dyke and Hooper (2000) GSC Map 1960A]
5050 ± 170 (5195 ± 179)	S-3563	otic capsule	34.5 km NNE of Fort Ross, 23 m asl., Somerset I. (72.290°N, 93.850°W), NU, CAN	[A.S. Dyke, pers. comm. (1998) In: R.E. Morlan, CMC database (1999)]
5050 ± 160 (5205 ± 160)	S-3526	periotic	4.9 km SW of Firkin Pt., 8.5 m asl., Devon I. (75.570°N, 85.670°W), NU, CAN	Dyke (1998); R.E. Morlan, CMC database (1999)
(5050 ± 180)	GSC-979-3	rib	Cape Storm, 23 m asl., Ellesmere I. (76.400°N, 87.500°W), NU, CAN	Lowdon and Blake (1980)
5020 ± 70 (5180 ± 75)	S-3077	periotic	S side of the mouth of McBean Bay, 16.5 m asl., Baffin I. (72.625°N, 88.630°W), NU, CAN	R.E. Morlan, CMC database (1999); [Dyke (2000c) GSC Map 1963A]
5000 ± 80 (5160 ± 80)	S-3146	otic capsule	2.6 km WNW of the mouth of Tikiraq R., 25 m asl., Baffin I. (71.770°N, 86.080°W), NU, CAN	R.E. Morlan, CMC database (1999); [Dyke and Hooper (2000) GSC Map 1960A]
4990 ± 90 (5150 ± 90)	S-3241	periotic	8.4 km NW of the mouth of Tikirraq R., 27 m asl., Baffin I. (71.825°N, 86.190°W), NU, CAN	R.E. Morlan, CMC database (1999); [Dyke and Hooper (2000) GSC Map 1960A]
4960 ± 160 (5110 ± 160)	S-3557	periotic	5.3 km SW of Firkin Pt., 8.5 m asl., Devon I. (75.570°N, 85.680°W), NU, CAN	Dyke (1998); R.E. Morlan, CMC database (1999)
4930 ± 80 (5090 ± 80)	S-3178	periotic	2 km W of Bell Bay, 31 m asl., Baffin I. (70.950°N, 85.000°W), NU, CAN	R.E. Morlan, CMC database (1999); [Dyke and Hooper (2000) GSC Map 1960A]

4920 ± 80 (5055 ± 80)	S-3299	periotic	Jungersen Bay, Borden Pen., 31 m asl., Baffin I. (71.490°N, 84.490°W), NU, CAN	R.E. Morlan, CMC database (1999); [Dyke and Hooper (2000) GSC Map 1960A]
4900 ± 170 (5040 ± 170)	S-3574	periotic	S of Cape Hardy, 4.5 m asl., Devon I. (75.790°N, 83.680°W), NU, CAN	Dyke (1998); R.E. Morlan, CMC database (1999)
4900 ± 140 (5045 ± 140)	S-3420	periotic	Bay NW of Lavoie Pt., Borden Pen., 13 m asl., Baffin I. (72.830°N, 80.510°W), NU, CAN	R.E. Morlan, CMC database (1999); [Dyke (2000b) GSC Map 1962A]
4870 ± 95 (5030 ± 95)	S-2600	rib	20 km W of Coningham Bay, 17 m asl., .5 km N of the S shore of Prince of Wales I. (71.670°N, 98.250°W), NU, CAN	Dyke and Morris (1990); Dyke, et al. (1991); R.E. Morlan, CMC database (1999)
4860 ± 80 (5000 ± 80)	S-3358	periotic	28 km NW of head of Bernier Bay, 30 m asl., Baffin I. (71.130°N, 87.620°W), NU, CAN	R.E. Morlan, CMC database (1999); [Dyke and Hooper (2000) GSC Map 1960A]
4850 ± 100 (4990 ± 100)	S-3313	periotic	Jungersen Bay, Borden Pen., 30.5 m asl., Baffin I. (71.483°N, 84.480°W), NU, CAN	R.E. Morlan, CMC database (1999); [Dyke and Hooper (2000) GSC Map 1960A]
4810 ± 95 (4955 ± 95)	S-3019	periotic	9.5 km SSW of the mouth of McBean Bay, 14-14.3 m asl., Baffin I. (72.570°N, 89.770°W), NU, CAN	R.E. Morlan, CMC database (1999); [Dyke (2000c) GSC Map 1963A]
4780 ± 80 (4940 ± 80)	S-3088	periotic	14 km NE of the mouth of Fitzgerald Bay, 10.5 m asl., Baffin I. (72.125°N, 89.670°W), NU, CAN	R.E. Morlan, CMC database (1999); [Dyke (2000c) GSC Map 1963A]
4760 ± 260 (4900 ± 260)	S-3521	periotic	4.6 km NE of Owen Pt., 26 m asl., Devon I. (76.010°N, 92.490°W), NU, CAN	Dyke (1998); R.E. Morlan, CMC database (1999)
4750 ± 140 (4900 ± 140)	S-3419	periotic	Bay NW of Lavoie Pt., Borden Pen., 13.5-14 m asl., Baffin I. (72.830°N, 80.510°W), NU, CAN	R.E. Morlan, CMC database (1999); [Dyke (2000b) GSC Map 1962A]
4740 ± 150 (4885 ± 150)	S-3577	periotic	8.9 km SW of Cape Hardy, 7 m asl., Devon I. (75.780°N, 83.998°W), NU, CAN	Dyke (1998); R.E. Morlan, CMC database (1999)
4720 ± 100 (4880 ± 100)	S-3156	periotic	2.2 km NW of the mouth of Tiriganialaag R., 20 m asl., Baffin I. (72.200°N, 86.470°W), NU, CAN	R.E. Morlan, CMC database (1999); [Dyke (2000c) GSC Map 1963A]
4705 ± 100 (4865 ± 100)	S-3171	periotic	5 km S of the mouth of Kuuruluk R., 21 m asl., Baffin I. (72.400°N, 86.260°W), NU, CAN	R.E. Morlan, CMC database (1999); [Dyke (2000c) GSC Map 1963A]
4670 ± 80 (4830 ± 80)	S-3061	periotic	7 km S of the mouth of Fitzgerald Bay, 24.1 m asl., Baffin I. (72.015°N, 89.950°W), NU, CAN	R.E. Morlan, CMC database (1999); [Dyke (2000c) GSC Map 1963A]
4670 ± 80 (4830 ± 80)	S-3151	periotic	2.2 km SSW of the mouth of Tikiraq R., 28 m asl., Baffin I. (71.740°N, 86.030°W), NU, CAN	R.E. Morlan, CMC database (1999); [Dyke and Hooper (2000) GSC Map 1960A]
4635 ± 80 (4795 ± 80)	S-2915	periotic	NE coast of Prescott I., 14 m asl. (73.160°N, 96.720°W), NU, CAN	Dyke and Morris (1990); Dyke, et al. (1991); R.E. Morlan, CMC database (1999)
4620 ± 140 (4765 ± 140)	S-3425	periotic	Bay NW of Lavoie Pt., Borden Pen., 11 m asl., Baffin I. (72.825°N, 80.480°W), NU, CAN	R.E. Morlan, CMC database (1999); [Dyke (2000b) GSC Map 1962A]
4610 ± 60 (4755 ± 60)	S-3346	periotic	5.5 km S of S end of Navarana L., Foss Fiord, 31 m asl., Baffin I. (70.420°N, 87.170°W), NU, CAN	R.E. Morlan, CMC database (1999); [Hooper and Dyke (2000) GSC Map 1959A]
4600 ± 100 (4760 ± 100)	S-3168	periotic	1 km NW of the mouth of Kuuruluk R., 19 m asl., Baffin I. (72.450°N, 86.270°W), NU, CAN	R.E. Morlan, CMC database (1999); [Dyke and Hooper (2000) GSC Map 1960A]
(4580 ± 60)	GSC-1021-2	rib	Cape Storm, 16.5 m asl., Ellesmere I. (76.400°N, 87.500°W), NU, CAN	Lowdon and Blake (1980)
4570 ± 85 (4730 ± 85)	S-1387		11 km W of Rodd Bay, 14 m asl., Somerset I. (73.955°N, 90.630°W), NU, CAN	Dyke and Morris (1990); R.E. Morlan, CMC database (1990)

4570 ± 90 (4730 ± 90)	S-3084	periotic	3.5 km E of the mouth of Fitzgerald Bay, 21 m asl., Baffin I. (72.070°N, 89.920°W), NU, CAN	R.E. Morlan, CMC database (1999); [Dyke (2000c) GSC Map 1963A]
4560 ± 140 (4705 ± 140)	S-3418	otic capsule	Bay NW of Lavoie Pt., Borden Pen., 12.5 m asl., Baffin I. (72.830°N, 80.510°W), NU, CAN	R.E. Morlan, CMC database (1999); [Dyke (2000b) GSC Map 1962A]
4550 ± 90 (4710 ± 90)	S-2919	periotic	7.5 km NE of Cape Richard Collinson, 16.5 m asl., Prince of Wales I. (72.830°N, 102.730°W), NU, CAN	Dyke and Morris (1990); Dyke, et al. (1991); R.E. Morlan, CMC database (1999)
4530 ± 80 (4675 ± 80)	S-3355	periotic	32 km NW of head of Bernier Bay, 20.5 m asl., Baffin I. (71.150°N, 87.700°W), NU, CAN	R.E. Morlan, CMC database (1999); [Dyke and Hooper (2000) GSC Map 1960A]
4520 ± 100 (4680 ± 100)	S-3125	otic capsule	32 km SW of Imek Pt., 1.2 km S of Berlinguet Inlet, 33 m asl., Baffin I. (71.080°N, 85.620°W), NU, CAN	R.E. Morlan, CMC database (1999); [Dyke and Hooper (2000) GSC Map 1960A]
4510 ± 165 (4670 ± 165)	S-2664		Near Palmerston Pt., 15.6 m asl., Russell I. (74.125°N, 97.730°W), NU, CAN	Dyke and Morris (1990); Dyke, et al. (1991); R.E. Morlan, CMC database (1999)
4505 ± 85 (4665 ± 85)	S-2861	rib	About 17 km NW of Charles Dickens Pt., .5-.25 km W of Guillemard Bay, 18 m asl., Prince of Wales I. (71.670°N, 98.400°W), NU, CAN	Dyke and Morris (1990); Dyke, et al. (1991); R.E. Morlan, CMC database (1999)
4500 ± 100 (4660 ± 100)	S-3176	periotic	3 km SW of the head of Bell Bay, 28 m asl., Baffin I. (70.920°N, 85.020°W), NU, CAN	R.E. Morlan, CMC database (1999); [Dyke and Hooper (2000) GSC Map 1960A]
4500 ± 80 (4645 ± 80)	S-3305	periotic	Jungersen Bay, Borden Pen., 29 m asl., Baffin I. (71.422°N, 84.480°W), NU, CAN	R.E. Morlan, CMC database (1999); [Dyke and Hooper (2000) GSC Map 1960A]
4500 ± 80 (4660 ± 80)	S-3045	periotic	2 km ENE of the head of Thibault Bay, mouth of Bernier Bay, 28 m asl., Baffin I. (70.970°N, 89.120°W), NU, CAN	R.E. Morlan, CMC database (1999); [Hooper and Dyke (2000) GSC Map 1959A]
4500 ± 110	K-1628		NW region (76.630°N, 70.600°W), AK, USA	A.S. Dyke (GSC database)
(4490 ± 60)	GSC-1021	rib	Cape Storm, 16.5 m asl., Ellesmere I. (76.400°N, 87.500°W), NU, CAN	Blake (1975b); Lowdon and Blake (1980); R.E. Morlan, CMC database (1999)
4480 ± 90 (4640 ± 90)	S-3247	periotic	6.6 km N of the mouth of Tiriganialaag R., 16.5 m asl., Baffin I. (72.240°N, 86.450°W), NU, CAN	R.E. Morlan, CMC database (1999); [Dyke (2000c) GSC Map 1963A]
4470 ± 80 (4630 ± 80)	S-3060	periotic	10 km S of the mouth of Fitzgerald Bay, 18.9-20.4 m asl., Baffin I. (71°58'00"N, 89°57'00"W), NU, CAN	[Dyke and Hooper (2000) GSC Map 1960A]
4465 ± 80 (4625 ± 80)	S-3227	periotic	3 km NW of the mouth of Tikiraq R., Baffin I. (71.770°N, 86.110°W), NU, CAN	R.E. Morlan, CMC database (1999); [Dyke and Hooper (2000) GSC Map 1960A]
4460 ± 150 (4640 ± 150)	S-3555	periotic	6.3 km ENE of Owen Pt., 23 m asl., Devon I. (76.030°N, 92.560°W), NU, CAN	Dyke (1998); R.E. Morlan, CMC database (1999)
(4450 ± 70)	GSC-1021-3	rib	Cape Storm, 16.5 m asl., Ellesmere I. (76.400°N, 87.500°W), NU, CAN	Lowdon and Blake (1980)
4450 ± 140 (4595 ± 140)	S-3417	periotic	Bay NW of Lavoie Pt., Borden Pen., 11-12 m asl., Baffin I. (72.830°N, 80.510°W), NU, CAN	R.E. Morlan, CMC database (1999); [Dyke (2000b) GSC Map 1962A]
4440 ± 80 (4600 ± 80)	S-3229	periotic	3.8 km NW of the mouth of Tikiraq R., 20 m asl., Baffin I. (71.775°N, 86.120°W), NU, CAN	R.E. Morlan, CMC database (1999); [Dyke and Hooper (2000) GSC Map 1960A]
4430 ± 65 (4590 ± 70)	S-1385	rib	5 km NW of mouth of Creswell R., 11 m asl., Somerset I. (72.830°N, 93.550°W), NU, CAN	Dyke and Morris (1990); R.E. Morlan, CMC database (1999)

4415 ± 90 (4575 ± 90)	S-3025	periotic	7 km S of the mouth of Fitzgerald Bay, 13.7-14.9 m asl., Baffin I. (72.010°N, 89.960°W), NU, CAN	R.E. Morlan, CMC database (1999); [Dyke (2000c) GSC Map 1963A]
4390 ± 90 (4550 ± 90)	S-3085	periotic	3.3 km E of the mouth of Fitzgerald Bay, 14 m asl., Baffin I. (72.075°N, 89.920°W), NU, CAN	R.E. Morlan, CMC database (1999); [Dyke (2000c) GSC Map 1963A]
4380 ± 70 (4540 ± 75)	S-3046	periotic	10 km N of Morin Pt., mouth of Bernier Bay, 23 m asl., Baffin I. (71.420°N, 89.870°W), NU, CAN	R.E. Morlan, CMC database (1999); [Dyke and Hooper (2000) GSC Map 1960A]
4380 ± 60 (4530 ± 60)	S-3337	periotic	Sunday Bay, Borden Pen., 21 m asl., Baffin I. (71.790°N, 85.410°W), NU, CAN	R.E. Morlan, CMC database (1999); [Dyke and Hooper (2000) GSC Map 1960A]
4370 ± 80 (4530 ± 80)	S-3149	periotic	2.2 km S of the mouth of Tikiraq R., 21.5 m asl., Baffin I. (71.740°N, 86.020°W), NU, CAN	R.E. Morlan, CMC database (1999); [Dyke and Hooper (2000) GSC Map 1960A]
4370 ± 60 (4510 ± 60)	S-3302	periotic	Jungersen Bay, Borden Pen., 17.5 m asl., Baffin I. (71.525°N, 84.470°W), NU, CAN	R.E. Morlan, CMC database (1999); [Dyke and Hooper (2000) GSC Map 1960A]
4350 ± 70 (4530 ± 70)	S-3074	periotic	7 km NE of Morin Pt., mouth of Bernier Bay, 24 m asl., Baffin I. (71.370°N, 89.680°W), NU, CAN	R.E. Morlan, CMC database (1999); [Dyke and Hooper (2000) GSC Map 1960A]
4310 ± 60 (4465 ± 60)	S-3335	otic capsule	Sunday Bay, Borden Pen., 18.5 m asl., Baffin I. (71.775°N, 85.470°W), NU, CAN	R.E. Morlan, CMC database (1999); [Dyke and Hooper (2000) GSC Map 1960A]
4310 ± 90 (4470 ± 90)	S-1382	skull	10 km NNW of mouth of Creswell R., 22 m asl., Somerset I. (72.870°N, 93.530°W), NU, CAN	Dyke and Morris (1990); R.E. Morlan, CMC database (1999)
4310 ± 60 (4470 ± 65)	S-3071	periotic	Vista R., 5.5 km NW of Kakiak Pt., 12-12.5 m asl., Baffin I. (73.720°N, 86.750°W), NU, CAN	R.E. Morlan, CMC database (1999); [Dyke (2000d) GSC Map 1964A]
4300 ± 110 (4460 ± 110)	S-3124	otic capsule	33 km SW of Imek Pt., S side of Berlinguet Inlet, 26.5 m asl., Baffin I. (71.130°N, 85.670°W), NU, CAN	R.E. Morlan, CMC database (1999); [Dyke and Hooper (2000) GSC Map 1960A]
4265 ± 65 (4425 ± 70)	S-1389	skull	5.5 km SW of Cape Anne, 18 m asl., Somerset I. (74.055°N, 94.810°W), NU, CAN	Dyke and Morris (1990); R.E. Morlan, CMC database (1999)
4250 ± 80 (4395 ± 80)	S-3301	periotic	Jungersen Bay, Borden Pen., 26 m asl., Baffin I. (71.483°N, 84.480°W), NU, CAN	R.E. Morlan, CMC database (1999); [Dyke and Hooper (2000) GSC Map 1960A]
4250 ± 70 (4410 ± 75)	S-3096	periotic	14 km E of Easter Cape, mouth of Bernier Bay, 23 m asl., Baffin I. (70.920°N, 89.050°W), NU, CAN	R.E. Morlan, CMC database (1999); [Dyke and Hooper (2000) GSC Map 1960A]
4220 ± 80 (4380 ± 80)	S-3181	otic capsule	33 km SW of Imek Pt., .2 km inland, S side of Berlinguet Inlet, 19 m asl., Baffin I. (71.130°N, 85.670°W), NU, CAN	R.E. Morlan, CMC database (1999); [Dyke and Hooper (2000) GSC Map 1960A]
4190 ± 90 (4345 ± 90)	S-3022	periotic	11.5 km SSW of the mouth of McBean Bay, 11-11.3 m asl., Baffin I. (72.550°N, 89.770°W), NU, CAN	R.E. Morlan, CMC database (1999); [Dyke (2000c) GSC Map 1963A]
4190 ± 100 (4350 ± 100)	S-3161	periotic	1.5 km NW of the mouth of Tiriganialaag R., 10 m asl., Baffin I. (72.190°N, 86.470°W), NU, CAN	R.E. Morlan, CMC database (1999); [Dyke (2000c) GSC Map 1963A]
4155 ± 100 (4315 ± 100)	S-2589	vertebra	9 km S of Hollist Pt., 12-12.3 m asl., Prince of Wales I. (72.830°N, 101.530°W), NU, CAN	Dyke and Morris (1990); Dyke, et al. (1991); R.E. Morlan, CMC database (1999)
4115 ± 85 (4265 ± 85)	S-3015	periotic	28 km WSW of the head of Bernier Bay, 20 m asl., Baffin I. (70.930°N, 87.750°W), NU, CAN	R.E. Morlan, CMC database (1999); [Dyke and Hooper (2000) GSC Map 1960A]
4100 ± 80 (4245 ± 80)	S-3336	otic capsule	Sunday Bay, Borden Pen., 18.5 m asl., Baffin I. (71.775°N, 85.470°W), NU, CAN	R.E. Morlan, CMC database (1999); [Dyke and Hooper (2000) GSC Map 1960A]
4090 ± 80 (4240 ± 80)	S-3349	periotic	19 km W of Berlinguet R., 20.5 m asl., Baffin I. (71.030°N, 86.680°W), NU, CAN	R.E. Morlan, CMC database (1999); [Dyke and Hooper (2000) GSC Map 1960A]

4060 ± 70 (4220 ± 75)	S-3072	periotic	Vista R., 2.8 km NW of Kakiak Pt., 9.5-10 m asl., Baffin I. (72.700°N, 86.720°W), NU, CAN	R.E. Morlan, CMC database (1999); [Dyke (2000c) GSC Map 1963A]
4050 ± 150 (4190 ± 150)	S-3542	periotic	4.2 km SW of Firkin Pt., 5.5-6 m asl., Devon I. (75.570°N, 85.660°W), NU, CAN	Dyke (1998); R.E. Morlan, CMC database (1999)
4050 ± 80 (4200 ± 80)	S-3296	periotic	Jungersen Bay, Borden Pen., 21.5 m asl., Baffin I. (71.480°N, 84.490°W), NU, CAN	R.E. Morlan, CMC database (1999); [Dyke and Hooper (2000) GSC Map 1960A]
4040 ± 100 (4200 ± 100)	S-3122	otic capsule	4.2 km S of the mouth of Kuuruluk R., 15 m asl., Baffin I. (72.410°N, 86.230°W), NU, CAN	R.E. Morlan, CMC database (1999); [Dyke (2000c) GSC Map 1963A]
4030 ± 110 (4130 ± 110)	S-3429	periotic	Bay NW of Lavoie Pt., Borden Pen., 6 m asl., Baffin I. (72.825°N, 80.470°W), NU, CAN	R.E. Morlan, CMC database (1999); [Dyke (2000b) GSC Map 1962A]
4010 ± 80 (4155 ± 80)	S-3351	periotic	22 km W of Berlinguet R., 23 m asl., Baffin I. (71.020°N, 86.770°W), NU, CAN	R.E. Morlan, CMC database (1999); [Dyke and Hooper (2000) GSC Map 1960A]
4010 ± 70 (4170 ± 75)	S-3097	periotic	13.5 km SW of the head of Bernier Bay, 22 m asl., Baffin I. (70.950°N, 87.320°W), NU, CAN	R.E. Morlan, CMC database (1999); [Dyke and Hooper (2000) GSC Map 1960A]
3990 ± 80 (4145 ± 80)	S-3326	periotic	Sunday Bay, Borden Pen., 19.5 m asl., Baffin I. (71.791°N, 85.430°W), NU, CAN	R.E. Morlan, CMC database (1999); [Dyke and Hooper (2000) GSC Map 1960A]
3990 ± 80 (4150 ± 80)	S-3183	periotic	40 km SW of Imek Pt., .4 km inland, S side of Berlinguet Inlet, 16 m asl., Baffin I. (71.120°N, 85.850°W), NU, CAN	R.E. Morlan, CMC database (1999); [Dyke and Hooper (2000) GSC Map 1960A]
3975 ± 260 (4120 ± 260)	S-3545	periotic	22 km NNW of Owen Pt., 0.2 km ESE Pt Hogarth, 19 m asl., Devon I. (76.160°N, 92.760°W), NU, CAN	Dyke (1998); R.E. Morlan, CMC database (1999)
3950 ± 70 (4110 ± 75)	S-3242	periotic	6 km NW of the mouth of Tikiraq R., 20 m asl., Baffin I. (71.800°N, 86.120°W), NU, CAN	[A.S. Dyke, pers. comm. (1998) In: R.E. Morlan, CMC database (1999)]
3950 ± 70 (4110 ± 75)	S-3043	periotic	6 km ENE of Morin Pt., mouth of Bernier Bay, 21 m asl., Baffin I. (71.350°N, 89.620°W), NU, CAN	R.E. Morlan, CMC database (1999); [Dyke and Hooper (2000) GSC Map 1960A]
3940 ± 90 (4100 ± 90)	S-3133	otic capsule	6 km SE of the head of Berlinguet Inlet, 1.5 km S of the S coast, 19.5 m asl., Baffin I. (70.970°N, 86.680°W), NU, CAN	R.E. Morlan, CMC database (1999); [Dyke and Hooper (2000) GSC Map 1960A]
3940 ± 110 (4100 ± 110)	S-3430	otic capsule	Bay NW of Lavoie Pt., Borden Pen., 7 m asl., Baffin I. (72.825°N, 80.470°W), NU, CAN	R.E. Morlan, CMC database (1999); [Dyke (2000b) GSC Map 1962A]
3930 ± 130 (4065 ± 130)	S-3414	periotic	Bay NW of Lavoie Pt., Borden Pen., 9-9.5 m asl., Baffin I. (72.820°N, 80.440°W), NU, CAN	R.E. Morlan, CMC database (1999); [Dyke (2000b) GSC Map 1962A]
3930 ± 80 (4090 ± 80)	S-3075	cranium	8 km SW of the head of Bernier Bay, 25 m asl., Baffin I. (70.930°N, 87.230°W), NU, CAN	R.E. Morlan, CMC database (1999); [Dyke and Hooper (2000) GSC Map 1960A]
3930 ± 70 (4095 ± 70)	S-3333	periotic	Sunday Bay, Borden Pen., 17.5 m asl., Baffin I. (71.772°N, 85.480°W), NU, CAN	R.E. Morlan, CMC database (1999); [Dyke and Hooper (2000) GSC Map 1960A]
3920 ± 80 (4080 ± 80)	S-3240	periotic	8.1 km NW of the mouth of Tikiraq R., 18 m asl., Baffin I. (71.825°N, 86.170°W), NU, CAN	R.E. Morlan, CMC database (1999); [Dyke and Hooper (2000) GSC Map 1960A]
3910 ± 140 (4070 ± 140)	S-3427	otic capsule	Bay NW of Lavoie Pt., Borden Pen., 8.5 m asl., Baffin I. (72.825°N, 80.480°W), NU, CAN	R.E. Morlan, CMC database (1999); [Dyke (2000b) GSC Map 1962A]
3880 ± 80 (4040 ± 80)	S-3228	otic capsule	3.8 km NW of the mouth of Tikiraq R., 20 m asl., Baffin I. (71.775°N, 86.120°W), NU, CAN	R.E. Morlan, CMC database (1999); [Dyke and Hooper (2000) GSC Map 1960A]
3880 ± 80 (4030 ± 80)	S-3353	periotic	28 km NW of head of Bernier Bay, 18.5 m asl., Baffin I. (71.120°N, 87.670°W), NU, CAN	R.E. Morlan, CMC database (1999); [Dyke and Hooper (2000) GSC Map 1960A]

3870 ± 80 (4030 ± 80)	S-3182	otic capsule	33 km SW of Imek Pt., .3 km inland, S side of Berlinguet Inlet, 20.5 m asl., Baffin I. (71.130°N, 85.670°W), NU, CAN	R.E. Morlan, CMC database (1999); [Dyke and Hooper (2000) GSC Map 1960A]
3870 ± 80 (4020 ± 80)	S-3297	periotic	Jungersen Bay, Borden Pen., 18 m asl., Baffin I. (71.498°N, 84.500°W), NU, CAN	R.E. Morlan, CMC database (1999); [Dyke and Hooper (2000) GSC Map 1960A]
3860 ± 80 (4020 ± 80)	S-3328	periotic	Sunday Bay, Borden Pen., 17 m asl., Baffin I. (71.828°N, 85.460°W), NU, CAN	R.E. Morlan, CMC database (1999); [Dyke and Hooper (2000) GSC Map 1960A]
3860 ± 50 (4025 ± 50)	S-3338	periotic	Sunday Bay, Borden Pen., 17.5 m asl., Baffin I. (71.772°N, 85.480°W), NU, CAN	R.E. Morlan, CMC database (1999); [Dyke and Hooper (2000) GSC Map 1960A]
3850 ± 90 (4010 ± 90)	S-3158	periotic	1.4 km NW of the mouth of Tiriganialaag R., 15 m asl., Baffin I. (72.180°N, 86.470°W), NU, CAN	R.E. Morlan, CMC database (1999); [Dyke (2000c) GSC Map 1963A]
3850 ± 100 (4010 ± 100)	S-3160	periotic	2.3 km NW of the mouth of Tiriganialaag R., 16 m asl., Baffin I. (72.200°N, 86.470°W), NU, CAN	R.E. Morlan, CMC database (1999); [Dyke (2000c) GSC Map 1963A]
3840 ± 130 (4000 ± 130)	S-3412	periotic	Bay NW of Lavoie Pt., Borden Pen., 8 m asl., Baffin I. (72.820°N, 80.440°W), NU, CAN	R.E. Morlan, CMC database (1999); [Dyke (2000b) GSC Map 1962A]
3825 ± 75 (3985 ± 80)	S-1393		11 km W of Rodd Bay, 10 m asl., Somerset I. (73.960°N, 90.630°W), NU, CAN	Dyke and Morris (1990); R.E. Morlan, CMC database (1999)
3820 ± 60 (3980 ± 65)	S-3234	periotic	1 km W of the mouth of Tikiraq R., 18 m asl., Baffin I. (71.760°N, 86.050°W), NU, CAN	R.E. Morlan, CMC database (1999); [Dyke and Hooper (2000) GSC Map 1960A]
3820 ± 80 (4005 ± 80)	S-3329	periotic	Sunday Bay, Borden Pen., 12 m asl., Baffin I. (71.848°N, 85.475°W), NU, CAN	R.E. Morlan, CMC database (1999); [Dyke and Hooper (2000) GSC Map 1960A]
3820 ± 100 (3980 ± 100)	S-3173	otic capsule	6.1 km S of the mouth of Kuuruluk R., 13 m asl., Baffin I. (72.390°N, 86.300°W), NU, CAN	R.E. Morlan, CMC database (1999); [Dyke (2000c) GSC Map 1963A]
3800 ± 150 (3955 ± 150)	S-3532	otic capsule	7.9 km NNW of Porden Pt., Prince Alfred Bay, 18 m asl., Devon I. (76.330°N, 93.720°W), NU, CAN	Dyke (1998); R.E. Morlan, CMC database (1999)
3790 ± 80 (3950 ± 80)	S-3179	otic capsule	2 km W of the head of Bell Bay, 21 m asl., Baffin I. (70.930°N, 85.000°W), NU, CAN	R.E. Morlan, CMC database (1999); [Dyke and Hooper (2000) GSC Map 1960A]
3780 ± 80 (3935 ± 80)	S-3357	otic capsule	26 km NW of head of Bernier Bay, 19 m asl., Baffin I. (71.120°N, 87.600°W), NU, CAN	R.E. Morlan, CMC database (1999); [Dyke and Hooper (2000) GSC Map 1960A]
3770 ± 140 (3900 ± 140)	S-3514	otic capsule	4.5 km E of Fort Ross, 17 m asl., Somerset I. (72.020°N, 94.110°W), NU, CAN	R.E. Morlan, CMC database (1999); [Hooper and Dyke (2000) GSC Map 1959A]
3760 ± 40 (3920 ± 40)	S-3342	periotic	3.3 km N of Eqalulik R., Borden Pen., 1-1.5 m asl., Baffin I. (72.677°N, 85.670°W), NU, CAN	R.E. Morlan, CMC database (1999); [Dyke (2000c) GSC Map 1963A]
3745 ± 80 (3905 ± 80)	S-3254	periotic	6.8 km S of the mouth of Kuuruluk R., 13-13.5 m asl., Baffin I. (72.380°N, 86.275°W), NU, CAN	R.E. Morlan, CMC database (1999); [Dyke (2000c) GSC Map 1963A]
3740 ± 130 (3915 ± 130)	S-3426	otic capsule	Bay NW of Lavoie Pt., Borden Pen., 6.5 m asl., Baffin I. (72.825°N, 80.480°W), NU, CAN	R.E. Morlan, CMC database (1999); [Dyke (2000b) GSC Map 1962A]
3740 ± 70	S-3311	otic capsule	Jungersen Bay, Borden Pen., Baffin I. (71.478°N, 84.490°W), NU, CAN	Dyke, et al. (1996); [Dyke and Hooper (2000) GSC Map 1960A]
3740 ± 80 (3900 ± 80)	S-3354	periotic	32 km NW of head of Bernier Bay, 20.5 m asl., Baffin I. (71.150°N, 87.750°W), NU, CAN	R.E. Morlan, CMC database (1999); [Dyke and Hooper (2000) GSC Map 1960A]
3740 ± 80 (3900 ± 80)	S-3142	otic capsule	5.1 km NW of the mouth of Tikiraq R., 18 m asl., Baffin I. (71.790°N, 86.130°W), NU, CAN	R.E. Morlan, CMC database (1999); [Dyke and Hooper (2000) GSC Map 1960A]

3720 ± 80 (3865 ± 80)	S-3314	periotic	Jungersen Bay, Borden Pen., 19.5 m asl., Baffin I. (71.475°N, 84.490°W), NU, CAN	R.E. Morlan, CMC database (1999); [Dyke and Hooper (2000) GSC Map 1960A]
3700 ± 70 (3870 ± 70)	S-3332	periotic	Sunday Bay, Borden Pen., 15.5-16 m asl., Baffin I. (71.772°N, 85.480°W), NU, CAN	R.E. Morlan, CMC database (1999); [Dyke and Hooper (2000) GSC Map 1960A]
3700 ± 80 (3865 ± 80)	S-3350	periotic	5 km W of Berlinguet R., 20.5 m asl., Baffin I. (71.070°N, 86.430°W), NU, CAN	R.E. Morlan, CMC database (1999); [Dyke and Hooper (2000) GSC Map 1960A]
3690 ± 70 (3850 ± 75)	S-3055	periotic	Vista R., 6.7 km NNW of Kakiak Pt., 11-11.5 m asl., Baffin I. (72.730°N, 86.770°W), NU, CAN	R.E. Morlan, CMC database (1999); [Dyke (2000c) GSC Map 1963A]
3685 ± 100 (3845 ± 100)	S-3172	periotic	6.5 km S of the mouth of Kuuruluk R., 12 m asl., Baffin I. (72.380°N, 86.270°W), NU, CAN	R.E. Morlan, CMC database (1999); [Dyke (2000c) GSC Map 1963A]
3685 ± 155 (3845 ± 155)	S-2671		8 km SW of Cape Walker, 10.8 m asl., Russell I. (74.010°N, 97.650°W), NU, CAN	Dyke and Morris (1990); Dyke, et al. (1991); Dyke (1998); R.E. Morlan, CMC database (1999)
3680 ± 90 (3840 ± 90)	S-3144	otic capsule	7.4 km NW of the mouth of Tikiraq R., 16 m asl., Baffin I. (71.810°N, 86.150°W), NU, CAN	R.E. Morlan, CMC database (1999); [Dyke and Hooper (2000) GSC Map 1960A]
3675 ± 110 (3835 ± 110)	S-2662	rib	5 km E of Krabbe Pt., 14.6 m asl., N coast of Russell I. (74.040°N, 98.600°W), NU, CAN	Dyke and Morris (1990); Dyke et al. (1991); R.E. Morlan, CMC database (1999)
3670 ± 70 (3815 ± 70)	S-3321	periotic	Sunday Bay, Borden Pen., 15 m asl., Baffin I. (71.790°N, 85.440°W), NU, CAN	R.E. Morlan, CMC database (1999); [Dyke and Hooper (2000) GSC Map 1960A]
3670 ± 130 (3820 ± 130)	S-3312	periotic	Jungersen Bay, Borden Pen., 20.5 m asl., Baffin I. (71.482°N, 84.500°W), NU, CAN	Dyke, et al. (1996); [Dyke and Hooper (2000) GSC Map 1960A]
3660 ± 80 (3820 ± 80)	S-3239	periotic	5.6 km NW of the mouth of Tikiraq R., 16.5-17 m asl., Baffin I. (71.800°N, 86.130°W), NU, CAN	R.E. Morlan, CMC database (1999); [Dyke and Hooper (2000) GSC Map 1960A]
3650 ± 60 (3800 ± 60)	S-3298	periotic	Jungersen Bay, Borden Pen., 18.5-19.5 m asl., Baffin I. (71.482°N, 84.500°W), NU, CAN	R.E. Morlan, CMC database (1999); [Dyke and Hooper (2000) GSC Map 1960A]
3650 ± 70 (3820 ± 70)	S-3325	periotic	Sunday Bay, Borden Pen., 16.5 m asl., Baffin I. (71.791°N, 85.430°W), NU, CAN	R.E. Morlan, CMC database (1999); [Dyke and Hooper (2000) GSC Map 1960A]
3640 ± 140 (3720 ± 140)	S-3543	periotic	2.2 km SW of Firkin Pt., 4.5 m asl., Devon I. (75.590°N, 85.610°W), NU, CAN	Dyke (1998); R.E. Morlan, CMC database (1999)
3640 ± 60 (3795 ± 60)	S-3295	otic capsule	Jungersen Bay, Borden Pen., 19-20 m asl., Baffin I. (71.478°N, 84.490°W), NU, CAN	R.E. Morlan, CMC database (1999); [Dyke and Hooper (2000) GSC Map 1960A]
3610 ± 70 (3770 ± 75)	S-3079	periotic	8 km SSW of the mouth of McBean Bay, 9.1 m asl., Baffin I. (72.575°N, 89.750°W), NU, CAN	R.E. Morlan, CMC database (1999); [Dyke (2000c) GSC Map 1963A]
3600 ± 80 (3760 ± 80)	S-3317	otic capsule	Jungersen Bay, Borden Pen., 19.5 m asl., Baffin I. (71.460°N, 84.500°W), NU, CAN	R.E. Morlan, CMC database (1999); [Dyke and Hooper (2000) GSC Map 1960A]
3580 ± 140 (3725 ± 140)	S-3513	otic capsule	5.8 km NNE of Fort Ross, 13 m asl., Somerset I. (72.030°N, 94.080°W), NU, CAN	R.E. Morlan, CMC database (1999); [Hooper and Dyke (2000) GSC Map 1959A]
3570 ± 100 (3730 ± 100)	S-3162	periotic	1.2 km NW of the mouth of Tiriganialaag R., 12 m asl., Baffin I. (72.180°N, 86.470°W), NU, CAN	R.E. Morlan, CMC database (1999); [Dyke (2000c) GSC Map 1963A]
3550 ± 150 (3695 ± 150)	S-3571	periotic	5 km E of Cape Skogn, 2.3 m asl., Devon I. (75.770°N, 84.030°W), NU, CAN	Dyke (1998); R.E. Morlan, CMC database (1999)
3540 ± 80 (3700 ± 80)	S-3250	periotic	5 km S of the mouth of Kuuruluk R., 14 m asl., Baffin I. (72.400°N, 86.250°W), NU, CAN	R.E. Morlan, CMC database (1999); [Dyke (2000c) GSC Map 1963A]
3520 ± 130 (3660 ± 130)	S-3511	otic capsule	19 km NNE of Fort Ross, 8 m asl., Somerset I. (72.170°N, 94.010°W), NU, CAN	R.E. Morlan, CMC database (1999); [Hooper and Dyke (2000) GSC Map 1959A]

3500 ± 80 (3660 ± 80)	S-2711	vertebra	5 km SE of stream entering Birthday Bay, near Cape Hardy, 11.5 m asl., Prince of Wales I. (73.780°N, 97.080°W), NU, CAN	Dyke and Morris (1990); Dyke, et al. (1991); R.E. Morlan, CMC database (1999)
3490 ± 80 (3650 ± 80)	S-3253	periotic	7.1 km S of the mouth of Kuuruluk R., 11 m asl., Baffin I. (72.375°N, 86.275°W), NU, CAN	R.E. Morlan, CMC database (1999); [Dyke (2000c) GSC Map 1963A]
3480 ± 80 (3640 ± 80)	S-3102	periotic	4 km SW of the mouth of McBean Bay, 10.5 m asl., Baffin I. (72.620°N, 89.730°W), NU, CAN	R.E. Morlan, CMC database (1999); [Dyke (2000c) GSC Map 1963A]
3450 ± 70 (3605 ± 70)	S-3334	periotic	Sunday Bay, Borden Pen., 14.5 m asl., Baffin I. (71.775°N, 85.470°W), NU, CAN	R.E. Morlan, CMC database (1999); [Dyke and Hooper (2000) GSC Map 1960A]
3390 ± 80 (3550 ± 80)	S-3246	periotic	3.3 km N of the mouth of Tiriganialaag R., 15 m asl., Baffin I. (72.210°N, 86.470°W), NU, CAN	R.E. Morlan, CMC database (1999); [Dyke (2000c) GSC Map 1963A].
3360 ± 150 (3510 ± 150)	S-3528	otic capsule	18.3 km NNW of Owen Pt., 4.4 km SE of Pt. Hogarth, 15 m asl., Devon I. (76.140°N, 92.660°W), NU, CAN	Dyke (1998); R.E. Morlan, CMC database (1999)
3350 ± 100 (3510 ± 100)	S-3174	periotic	6.2 km S of the mouth of Kuuruluk R., 10 m asl., Baffin I. (72.390°N, 86.270°W), NU, CAN	[A.S. Dyke, pers. comm. (1998) In: R.E. Morlan, CMC database (1999)]
3340 ± 80 (3490 ± 80)	S-3347	periotic	1.5 km S of the mouth of Berlinguet R., Baffin I. (71.080°N, 86.230°W), NU, CAN	Dyke, et al. (1996); [Dyke and Hooper (2000) GSC Map 1960A]
3330 ± 100 (3485 ± 100)	S-3356	periotic	26 km NW of head of Bernier Bay, 17 m asl., Baffin I. (71.120°N, 87.600°W), NU, CAN	R.E. Morlan, CMC database (1999); [Dyke and Hooper (2000) GSC Map 1960A]
3320 ± 90 (3480 ± 90)	S-3166	periotic	3.1 km NNW of the mouth of Tiriganialaag R., 13 m asl., Baffin I. (72.210°N, 86.470°W), NU, CAN	R.E. Morlan, CMC database (1999); [Dyke (2000c) GSC Map 1963A]
3320 ± 80 (3480 ± 80)	S-3258	otic capsule	35 km SW of Imek Pt., 1 km inland, S side of Berlinguet Inlet, 18 m asl., Baffin I. (71.130°N, 85.730°W), NU, CAN	R.E. Morlan, CMC database (1999); [Dyke and Hooper (2000) GSC Map 1960A]
3315 ± 75 (3475 ± 80)	S-2921	periotic	W coast of Prescott I., 9 m asl. (73.100°N, 97.120°W), NU, CAN	Dyke and Morris (1990); Dyke, et al. (1991); R.E. Morlan, CMC database (1999)
3310 ± 90 (3470 ± 90)	S-3244	periotic	1.3 km N of the mouth of Tiriganialaag R., 10 m asl., Baffin I. (72.190°N, 86.460°W), NU, CAN	R.E. Morlan, CMC database (1999); [Dyke (2000c) GSC Map 1963A]
3300 ± 130 (3460 ± 130)	S-3423	otic capsule	Bay NW of Lavoie Pt., Borden Pen., 6.8 m asl., Baffin I. (72.830°N, 80.530°W), NU, CAN	R.E. Morlan, CMC database (1999); [Dyke (2000b) GSC Map 1962A]
3280 ± 80 (3425 ± 80)	S-3348	periotic	6 km W of Berlinguet R., 18 m asl., Baffin I. (71.050°N, 86.350°W), NU, CAN	Dyke, et al. (1996); [Dyke and Hooper (2000) GSC Map 1960A]
3270 ± 60 (3430 ± 65)	S-3059	periotic	S side of Fitzgerald Bay, 6 km NE of mouth, 11.3-11.6 m asl., Baffin I. (72.090°N, 89.870°W), NU, CAN	R.E. Morlan, CMC database (1999); [Dyke (2000c) GSC Map 1963A]
3255 ± 85 (3400 ± 85)	S-3020	periotic	10 km SSW of the mouth of McBean Bay, 4-4.6 m asl., Baffin I. (72.570°N, 89.770°W), NU, CAN	R.E. Morlan, CMC database (1999); [Dyke (2000c) GSC Map 1963A]
3250 ± 80 (3410 ± 80)	S-3263	periotic	6 km ESE of the head of Berlinguet Inlet, 1 km S of the S coast, 19.5 m asl., Baffin I. (70.970°N, 86.680°W), NU, CAN	R.E. Morlan, CMC database (1999); [Dyke and Hooper (2000) GSC Map 1960A]
3240 ± 90 (3380 ± 90)	S-3400	periotic	Head of Koluktoo Bay, 14 m asl., Baffin I. (72.050°N, 81.100°W), NU, CAN	R.E. Morlan, CMC database (1999); [Dyke (2000b) GSC Map 1962A]
3230 ± 80 (3390 ± 80)	S-3256	periotic	5 km S of the head of Bell Bay, 1 km W of Saputing R., 22.5 m asl., Baffin I. (70.880°N, 84.900°W), NU, CAN	R.E. Morlan, CMC database (1999); [Dyke and Hooper (2000) GSC Map 1960A]

3220 ± 80 (3380 ± 80)	S-3294	periotic	Jungersen Bay, Borden Pen., 11.5-13 m asl., Baffin I. (71.475°N, 84.500°W), NU, CAN	R.E. Morlan, CMC database (1999); [Dyke and Hooper (2000) GSC Map 1960A]
3210 ± 140 (3355 ± 140)	S-3509	periotic	28 km NNE of Fort Ross, 11 m asl., Somerset I. (72.240°N, 93.920°W), NU, CAN	[A.S. Dyke, pers. comm. (1998) In: R.E. Morlan, CMC database (1999)]
3190 ± 90 (3350 ± 90)	S-3257	periotic	5 km S of the head of Bell Bay, 1 km W of Saputing R., 22.5 m asl., Baffin I. (70.880°N, 84.900°W), NU, CAN	R.E. Morlan, CMC database (1999); [Dyke and Hooper (2000) GSC Map 1960A]
3180 ± 150 (3320 ± 150)	S-3564	periotic	Sheills Pen., 16.5 m asl., Devon I. (76.280°N, 94.820°W), NU, CAN	Dyke (1998); R.E. Morlan, CMC database (1999)
3140 ± 80 (3300 ± 80)	S-3248	periotic	2.2 km S of the mouth of Kuuruluk R., 10 m asl., Baffin I. (72.425°N, 86.240°W), NU, CAN	R.E. Morlan, CMC database (1999); [Dyke (2000c) GSC Map 1963A]
3140 ± 60 (3300 ± 60)	S-3341	periotic	4.7 km N of mouth of Eqalulik R., Borden Pen., 6 m asl., Baffin I. (72.690°N, 85.670°W), NU, CAN	R.E. Morlan, CMC database (1999); [Dyke (2000c) GSC Map 1963A]
3120 ± 90 (3290 ± 90)	S-3403	periotic	Head of Koluktoo Bay, 12-14 m asl., Baffin I. (72.060°N, 81.100°W), NU, CAN	R.E. Morlan, CMC database (1999); [Dyke (2000b) GSC Map 1962A]
3090 ± 110 (3240 ± 110)	S-3428	otic capsule	Bay NW of Lavoie Pt., Borden Pen., 5 m asl., Baffin I. (72.825°N, 80.470°W), NU, CAN	R.E. Morlan, CMC database (1999); [Dyke (2000b) GSC Map 1962A]
3080 ± 150 (3250 ± 150)	S-3556	periotic	18.8 km NE of Porden Pt., 2.0 km ESE of Reid Ck., Prince Alfred Bay, 14-14.5 m asl., Devon I. (76.360°N, 93.070°W), NU, CAN	Dyke (1998); R.E. Morlan, CMC database (1999)
3080 ± 110 (3220 ± 110)	S-3404	periotic	Head of Koluktoo Bay, 16.5 m asl., Baffin I. (72.075°N, 81.080°W), NU, CAN	R.E. Morlan, CMC database (1999); [Dyke (2000b) GSC Map 1962A]
3050 ± 130 (3210 ± 130)	GSC-361		16 km NE of Discovery Pt., 3 m asl., Prince Patrick I. (77.170°N, 119.920°W), NT, CAN	McNeely (1989); R.E. Morlan, CMC database (1999)
3040 ± 60 (3180 ± 60)	S-3300	periotic	Jungersen Bay, Borden Pen., 15 m asl., Baffin I. (71.495°N, 84.506°W), NU, CAN	R.E. Morlan, CMC database (1999); [Dyke and Hooper (2000) GSC Map 1960A]
3040 ± 70 (3195 ± 70)	S-3315	periotic	Jungersen Bay, Borden Pen., 15 m asl., Baffin I. (71.540°N, 84.520°W), NU, CAN	R.E. Morlan, CMC database (1999); [Dyke and Hooper (2000) GSC Map 1960A]
3025 ± 90 (3185 ± 90)	S-2592	rib fragment	E side of outlet of salt water lake, S side of Reliance Bay, 7 m asl., Prince of Wales I. (73.770°N, 100.030°W), NU, CAN	Dyke and Morris (1990); Dyke, et al. (1991); R.E. Morlan, CMC database (1999)
2980 ± 90 (3140 ± 90)	S-3054	periotic	Vista R., 2.8 km NW of Kakiak Pt., 6.5 m asl., Baffin I. (72.700°N, 86.720°W), NU, CAN	R.E. Morlan, CMC database (1999); [Dyke (2000c) GSC Map 1963A]
2960 ± 80 (3120 ± 80)	S-3132	otic capsule	4.3 km S of the mouth of Kuuruluk R., 8 m asl., Baffin I. (72.410°N, 86.230°W), NU, CAN	R.E. Morlan, CMC database (1999); [Dyke (2000c) GSC Map 1963A]
2950 ± 70 (3110 ± 75)	S-3076	otic capsule	5 km SW of the mouth of McBean Bay, 7 m asl., Baffin I. (72.610°N, 89.730°W), NU, CAN	R.E. Morlan, CMC database (1999); [Dyke (2000c) GSC Map 1963A]
2940 ± 70 (3100 ± 75)	S-3089	otic capsule	9 km S of the mouth of Fitzgerald Bay, 10.1 m asl., Baffin I. (71.980°N, 89.970°W), NU, CAN	R.E. Morlan, CMC database (1999); [Dyke and Hooper (2000) GSC Map 1960A]
2900 ± 85 (3060 ± 85)	S-433	skull	Truelove Lowland, 3 m asl., Devon I. (75.670°N, 84.620°W), NU, CAN	Barr 1971; King (1991); Dyke (1998); R.E. Morlan, CMC database (1999)
2900 ± 80 (3060 ± 80)	S-3127	otic capsule	8.5 km NW of the mouth of Tikiraq R., 12 m asl., Baffin I. (71.825°N, 86.150°W), NU, CAN	R.E. Morlan, CMC database (1999); [Dyke and Hooper (2000) GSC Map 1960A]
2890 ± 110 (3050 ± 110)	S-3407	otic capsule	Bay NW of Lavoie Pt., Borden Pen., 4.5 m asl., Baffin I. (72.825°N, 80.470°W), NU, CAN	R.E. Morlan, CMC database (1999); [Dyke (2000b) GSC Map 1962A]

2890 ± 130 (3045 ± 130)	S-3416	otic capsule	Bay NW of Lavoie Pt., Borden Pen., 4 m asl., Baffin I. (72.820°N, 80.440°W), NU, CAN	R.E. Morlan, CMC database (1999); [Dyke (2000b) GSC Map 1962A]
2870 ± 60 (3030 ± 65)	S-3148	periotic	1.7 km S of the mouth of Tikiraq R., 9 m asl., Baffin I. (71.750°N, 86.030°W), NU, CAN	R.E. Morlan, CMC database (1999); [Dyke and Hooper (2000) GSC Map 1960A]
2860 ± 70 (3020 ± 75)	S-3243	periotic	1.4 km N of the mouth of Tiriganialaag R., 8 m asl., Baffin I. (72.190°N, 86.460°W), NU, CAN	R.E. Morlan, CMC database (1999); [Dyke (2000c) GSC Map 1963A]
2850 ± 60 (3005 ± 60)	S-3319	periotic	Sunday Bay, Borden Pen., 9.5 m asl., Baffin I. (71.782°N, 85.580°W), NU, CAN	R.E. Morlan, CMC database (1999); [Dyke and Hooper (2000) GSC Map 1960A]
2820 ± 70 (2970 ± 70)	S-3322	periotic	Sunday Bay, Borden Pen., 10.5 m asl., Baffin I. (71.792°N, 85.560°W), NU, CAN	R.E. Morlan, CMC database (1999); [Dyke and Hooper (2000) GSC Map 1960A]
2820 ± 170 (2965 ± 170)	S-3566	periotic	6.5 km NW of mouth of Lyall R., Grinnell Pen., 12 m asl., Devon I. (77.050°N, 95.400°W), NU, CAN	Dyke (1998); R.E. Morlan, CMC database (1999)
2820 ± 130 (2230 ± 130)	GSC-362		Cape Andreas, 4.5 m asl., Prince Patrick I. (77.350°N, 118.830°W), NT, CAN	R.E. Morlan, CMC database (1999)
2820 ± 70 (2975 ± 70)	S-3316	periotic	Jungersen Bay, Borden Pen., 11 m asl., Baffin I. (71.465°N, 84.505°W), NU, CAN	R.E. Morlan, CMC database (1999); [Dyke and Hooper (2000) GSC Map 1960A]
2800 ± 70 (2960 ± 75)	S-3259	otic capsule	33 km SW of Imek Pt., .4 km inland, S side of Berlinguet Inlet, 17 m asl., Baffin I. (71.130°N, 85.670°W), NU, CAN	R.E. Morlan, CMC database (1999); [Dyke and Hooper (2000) GSC Map 1960A]
2720 ± 80 (2875 ± 80)	S-3399	otic capsule	E of the mouth of Phillips Ck., 6-6.5 m asl., Baffin I. (72.050°N, 81.100°W), NU, CAN	R.E. Morlan, CMC database (1999); [Dyke (2000a) GSC Map 1961A]
2720 ± 80 (2880 ± 80)	S-3104	otic capsule	8.2 km S of the mouth of Fitzgerald Bay, 11.5 m asl., Baffin I. (71.990°N, 89.970°W), NU, CAN	R.E. Morlan, CMC database (1999); [Dyke and Hooper (2000) GSC Map 1960A]
2720 ± 40 (2880 ± 40)	S-3339	otic capsule	Sunday Bay, Borden Pen., 6-6.5 m asl., Baffin I. (71.772°N, 85.360°W), NU, CAN	R.E. Morlan, CMC database (1999); [Dyke and Hooper (2000) GSC Map 1960A]
2710 ± 120 (2860 ± 120)	S-3411	otic capsule	Bay NW of Lavoie Pt., Borden Pen., 5.5 m asl., Baffin I. (72.820°N, 80.470°W), NU, CAN	R.E. Morlan, CMC database (1999); [Dyke (2000b) GSC Map 1962A]
2700 ± 90 (2860 ± 90)	S-3262	periotic	6 km ESE of the head of Berlinguet Inlet, 1 km S of the S coast, 15 m asl., Baffin I. (70.970°N, 86.670°W), NU, CAN	R.E. Morlan, CMC database (1999); [Dyke and Hooper (2000) GSC Map 1960A]
2660 ± 150 (2820 ± 150)	S-3533	periotic	7.5 km NW of Porden Pt., Prince Alfred Bay, 13 m asl., Devon I. (76.320°N, 93.760°W), NU, CAN	Dyke (1998); R.E. Morlan, CMC database (1999)
2650 ± 80 (2810 ± 80)	S-3261	periotic	14 km E of the head of Berlinguet Inlet, 16 m asl., Baffin I. (70.980°N, 86.430°W), NU, CAN	R.E. Morlan, CMC database (1999); [Dyke and Hooper (2000) GSC Map 1960A]
2590 ± 140 (2750 ± 140)	S-2670	skull fragment	10 km SW of Cape Walker, 16.1 m asl., SE Russell I. (74.000°N, 97.700°W), NU, CAN	Dyke and Morris (1990); Dyke, et al. (1991); Dyke (1998); R.E. Morlan, CMC database (1999)
2590 ± 120 (2705 ± 120)	S-3410	periotic	Bay NW of Lavoie Pt., Borden Pen., 3 m asl., Baffin I. (72.820°N, 80.470°W), NU, CAN	R.E. Morlan, CMC database (1999); [Dyke (2000b) GSC Map 1962A]
2580 ± 70 (2745 ± 70)	S-3330	otic capsule	Sunday Bay, Borden Pen., 9 m asl., Baffin I. (71.774°N, 85.490°W), NU, CAN	R.E. Morlan, CMC database (1999); [Dyke and Hooper (2000) GSC Map 1960A]
2580 ± 160 (2740 ± 160)	S-3522	otic capsule	4.5 km NW of Providence Mtn., 9 m asl., Devon I. (75.670°N, 92.010°W), NU, CAN	Dyke (1998); R.E. Morlan, CMC database (1999)
2580 ± 200 (2725 ± 200)	S-3537	periotic	9 km NW of Bere Bay, Grinnell Pen., 10.5 m asl., Devon I. (76.960°N, 94.550°W), NU, CAN	Dyke (1998); R.E. Morlan, CMC database (1999)

2570 ± 60 (2730 ± 65)	S-3044	periotic	12 km SE of Von Koenig Pt., mouth of Bernier Bay, 12 m asl., Baffin I. (71.020°N, 89.330°W), NU, CAN	R.E. Morlan, CMC database (1999); [Dyke and Hooper (2000) GSC Map 1960A]
(2570 ± 220)	GSC-3055-2	rib	SW side of Rosse Bay, Cape Herschel, Ellesmere I. (78.630°N, 74.800°W), NU, CAN	Blake (1987)
2550 ± 120 (2700 ± 120)	S-3408	otic capsule	Bay NW of Lavoie Pt., Borden Pen., 4.5 m asl., Baffin I. (72.825°N, 80.480°W), NU, CAN	R.E. Morlan, CMC database (1999); [Dyke (2000b) GSC Map 1962A]
2540 ± 90 (2695 ± 90)	S-3398	periotic	E of the mouth of Phillips Ck., 11.5-12 m asl., Baffin I. (71.880°N, 80.860°W), NU, CAN	R.E. Morlan, CMC database (1999); [Dyke (2000a) GSC Map 1961A]
2530 ± 70 (2680 ± 70)	S-3289	periotic	NW coast of Griffith I., 7.5-8 m asl. (74.610°N, 95.802°W), NU, CAN	Dyke (1993); R.E. Morlan, CMC database (1999)
2410 ± 70 (2570 ± 70)	S-3331	periotic	Sunday Bay, Borden Pen., 11 m asl., Baffin I. (71.774°N, 85.490°W), NU, CAN	R.E. Morlan, CMC database (1999); [Dyke and Hooper (2000) GSC Map 1960A]
2350 ± 110 (2510 ± 110)	S-3406	periotic	Head of Koluktoo Bay, 10.5-11 m asl., Baffin I. (72.060°N, 81.100°W), NU, CAN	R.E. Morlan, CMC database (1999); [Dyke (2000b) GSC Map 1962A]
2350 ± 60	S-3344	periotic	6.7 km N of Eqalulik R., Borden Pen., Baffin I. (72.710°N, 85.680°W), NU, CAN	Dyke, et al. (1996); [Dyke (2000c) GSC Map 1963A]
2320 ± 70 (2480 ± 75)	S-3058	periotic	S side of the mouth of Fitzgerald Bay, 2 km inland, W coast of Brodeur Bay, 6.7 m asl., Baffin I. (72.070°N, 89.970°W), NU, CAN	R.E. Morlan, CMC database (1999); [Dyke (2000c) GSC Map 1963A]
2310 ± 80 (2470 ± 80)	S-3260	periotic	40 km SW of Imek Pt., .3 km inland, S side of Berlinguet Inlet, 12 m asl., Baffin I. (71.120°N, 85.850°W), NU, CAN	R.E. Morlan, CMC database (1999); [Dyke and Hooper (2000) GSC Map 1960A]
2310 ± 70 (2465 ± 70)	S-3327	periotic	Sunday Bay, Borden Pen., 7.5-8 m asl., Baffin I. (71.825°N, 85.450°W), NU, CAN	R.E. Morlan, CMC database (1999); [Dyke and Hooper (2000) GSC Map 1960A]
2295 ± 80 (2455 ± 80)	S-3255	periotic	6.6 km S of the mouth of Kuuruluk R., 7.5 m asl., Baffin I. (72.380°N, 86.260°W), NU, CAN	R.E. Morlan, CMC database (1999); [Dyke (2000c) GSC Map 1963A]
2290 ± 70 (2455 ± 70)	S-3352	otic capsule	12 km WNW of head of Bernier Bay, 9 m asl., Baffin I. (71.030°N, 87.280°W), NU, CAN	R.E. Morlan, CMC database (1999); [Dyke and Hooper (2000) GSC Map 1960A]
2270 ± 230 (2430 ± 230)	S-2715		1.6 km NE of Muskox Hill, 10 m asl., Prince of Wales I. (72.870°N, 97.280°W), NU, CAN	Dyke and Morris (1990); Dyke, et al. (1991); R.E. Morlan, CMC database (1999)
2230 ± 110 (2380 ± 110)	S-3405	periotic	Head of Koluktoo Bay, 8.5 m asl., Baffin I. (72.075°N, 81.050°W), NU, CAN	R.E. Morlan, CMC database (1999); [Dyke (2000b) GSC Map 1962A]
2230 ± 80 (2390 ± 80)	S-3157	periotic	1.7 km NW of the mouth of Tiriganialaag R., 6 m asl., Baffin I. (72.190°N, 86.450°W), NU, CAN	R.E. Morlan, CMC database (1999); [Dyke (2000c) GSC Map 1963A]
2220 ± 75 (2361 ± 75)	S-3016	periotic	20 km SW of the head of Bernier Bay, 7 m asl., Baffin I. (70.930°N, 87.500°W), NU, CAN	R.E. Morlan, CMC database (1999); [Dyke and Hooper (2000) GSC Map 1960A]
(2210 ± 60)	TO-5716	earbone	Dundee Bight, 7 m asl., Bathurst I., NU, CAN	[R. McNeely, pers. comm. (1998) In: R.E. Morlan, CMC database (1999)]
2180 ± 70 (2330 ± 70)	S-3304	periotic	Jungersen Bay, Borden Pen., 10 m asl., Baffin I. (71.470°N, 84.510°W), NU, CAN	R.E. Morlan, CMC database (1999); [Dyke and Hooper (2000) GSC Map 1960A]
2175 ± 130 (2340 ± 130)	S-3041	vertebra	10 km ESE of Morin Pt., mouth of Bernier Bay, 5 m asl., Baffin I. (71.320°N, 89.520°W), NU, CAN	R.E. Morlan, CMC database (1999); [Dyke and Hooper (2000) GSC Map 1960A]
2160 ± 70 (2320 ± 75)	S-3245	periotic	1 km N of the mouth of Tiriganialaag R., 8-8.5 m asl., Baffin I. (72.180°N, 86.440°W), NU, CAN	R.E. Morlan, CMC database (1999); [Dyke (2000c) GSC Map 1963A]

2110 ± 200 (2260 ± 200)	S-3568	periotic	4 km NW of the head of Triton Bay, 7 m asl., Devon I. (76.620°N, 92.540°W), NU, CAN	Dyke (1998); R.E. Morlan, CMC database (1999)
2090 ± 70 (2250 ± 75)	S-3141	otic capsule	4.6 km NW of the mouth of Tikiraq R., 5 m asl., Baffin I. (71.800°N, 86.120°W), NU, CAN	R.E. Morlan, CMC database (1999); [Dyke and Hooper (2000) GSC Map 1960A]
2060 ± 100 (2220 ± 100)	S-3177	periotic	.5 km W of the head of Bell Bay, 9 m asl., Baffin I. (70.930°N, 84.970°W), NU, CAN	R.E. Morlan, CMC database (1999); [Dyke and Hooper (2000) GSC Map 1960A]
2030 ± 80 (2190 ± 80)	S-3164	periotic	1 km NW of the mouth of Tiriganialaag R., 7 m asl., Baffin I. (72.180°N, 86.450°W), NU, CAN	R.E. Morlan, CMC database (1999); [Dyke (2000c) GSC Map 1963A]
2020 ± 140 (2175 ± 140)	S-3529	periotic	8.2 km SE of Owen Pt., 9.5 m asl., Devon I. (75.930°N, 92.350°W), NU, CAN	Dyke (1998); R.E. Morlan, CMC database (1999)
2010 ± 90 (2170 ± 90)	S-3167	periotic	2.2 km S of the mouth of Kuuruluk R., 5.5 m asl., Baffin I. (72.425°N, 86.230°W), NU, CAN	R.E. Morlan, CMC database (1999); [Dyke (2000c) GSC Map 1963A]
1980 ± 130 (2125 ± 130)	S-3508	periotic	29 km NNE of Fort Ross, 6 m asl., Somerset I. (72.250°N, 93.900°W), NU, CAN	R.E. Morlan, CMC database (1999); [Hooper and Dyke (2000) GSC Map 1959A]
1950 ± 90 (2110 ± 90)	S-3154	periotic	6.8 km NW of the mouth of Tikiraq R., 7-8 m asl., Baffin I. (71.800°N, 86.130°W), NU, CAN	R.E. Morlan, CMC database (1999); [Dyke and Hooper (2000) GSC Map 1960A]
1930 ± 170 (2110 ± 170)	S-3534	otic capsule	7.5 km NNW of Porden Pt., Prince Alfred Bay, 8 m asl., Devon I. (76.320°N, 93.730°W), NU, CAN	Dyke (1998); R.E. Morlan, CMC database (1999)
(1870 ± 90)	GSC-1227-2	cranium	Cape Storm, 3 m asl., Ellesmere I. (76.400°N, 87.500°W), NU, CAN	Lowdon and Blake (1980)
1860 ± 80 (2020 ± 80)	S-1405		12 km W of Rodd Bay, 4.5 m asl., Somerset I. (73.965°N, 90.630°W), NU, CAN	Dyke and Morris (1990); R.E. Morlan, CMC database (1999)
1830 ± 90 (1960 ± 90)	S-3053	periotic	Vista R., 7.3 km NNW of Kakiak Pt., 1.5-2.5 m asl., Baffin I. (72.730°N, 86.750°W), NU, CAN	R.E. Morlan, CMC database (1999); [Dyke (2000c) GSC Map 1963A]
1810 ± 180 (1965 ± 180)	S-3530	otic capsule	18.3 km NNW of Owen Pt., 4.4 km SE of Pt. Hogarth, 6-7 m asl., Devon I. (76.130°N, 92.670°W), NU, CAN	Dyke (1998); R.E. Morlan, CMC database (1999)
(1660 ± 60)	TO-8011	rib	2.9 km NNE of Cape Larsen and S of Cape Baring, Wollaston Pen., 11 m asl., Victoria I. (69°48'58"N, 117°15'00"W), NT, CAN	Dyke and Savelle (2001)
1640 ± 70 (1800 ± 75)	S-3225	periotic	8.5 km NW of Tikiraq R., 4.5-5 m asl., Baffin I. (71.780°N, 86.100°W), NU, CAN	R.E. Morlan, CMC database (1999); [Dyke and Hooper (2000) GSC Map 1960A]
(1610 ± 50)	GSC-1227	cranium	Cape Storm, 3 m asl., Ellesmere I. (76.400°N, 87.500°W), NU, CAN	Blake (1975b); Lowdon and Blake (1980); R.E. Morlan, CMC database (1999)
1585 ± 70 (1745 ± 75)	S-2920	otic capsule	W coast of Prescott I., 3.5 m asl. (73.050°N, 97.120°W), NU, CAN	Dyke and Morris (1990); Dyke, et al. (1991); R.E. Morlan, CMC database (1999)
1570 ± 60 (1730 ± 65)	S-3237	periotic	4.6 km NW of the mouth of Tikiraq R., 5 m asl., Baffin I. (71.800°N, 86.120°W), NU, CAN	R.E. Morlan, CMC database (1999); [Dyke and Hooper (2000) GSC Map 1960A]
1565 ± 125 (1725 ± 125)	S-2666		S of Addington Pt., 61.3 m asl., NE Russell I. (74° 5'00"N, 98°3.5'00"W), NU, CAN	Dyke, et al. (1991); Dyke (1998); R.E. Morlan, CMC database (1999); [Anomalously young, A.S. Dyke, pers. comm. (2001)]
(1560 ± 120)	QU-240		Millut Bay, 1.5 m asl., Baffin I. (66°38'N, 67°34'W), NU, CAN	Dyke (1979b)
1510 ± 80 (1670 ± 80)	S-3134	periotic (89-DCA-10)	Lowther I., 4.5 m asl. (74.596°N, 97.502°W), NU, CAN	McNeely and Jorgensen (1993); Dyke (1993); R.E. Morlan, CMC database (1999)
1500 ± 60 (1660 ± 65)	S-3231	periotic	1.7 km NW of the mouth of Tikiraq R., 4.5 m asl., Baffin I. (71.770°N, 86.060°W), NU, CAN	R.E. Morlan, CMC database (1999); [Dyke and Hooper (2000) GSC Map 1960A]

1455 ± 80 (1615 ± 80)	S-1383	vertebra	4 km SW of Cape Anne, 7.5 m asl., Somerset I. (74.070°N, 94.810°W), NU, CAN	Dyke and Morris (1990); R.E. Morlan, CMC database (1999)
1440 ± 170 (1590 ± 130)	S-3507	otic capsule	17.8 km NNE of Fort Ross, 4 m asl., Somerset I. (72.150°N, 94.020°W), NU, CAN	Dyke, et al. (1996); R.E. Morlan, CMC database (1999)
1410 ± 120 (1565 ± 120)	S-3424	periotic	Bay NW of Lavoie Pt., Borden Pen., 1.5 m asl., Baffin I. (72.840°N, 80.500°W), NU, CAN	R.E. Morlan, CMC database (1999); [Dyke (2000b) GSC Map 1962A]
1400 ± 60 (1560 ± 60)	S-3320	periotic	Sunday Bay, Borden Pen., 2.5 m asl., Baffin I. (71.774°N, 85.520°W), NU, CAN	Dyke, et al. (1996); R.E. Morlan, CMC database (1999); [Dyke and Hooper (2000) GSC Map 1960A]
1390 ± 70 (1540 ± 70)	S-3293	periotic	Cape Charles York, Borden Pen., 4 m asl., Baffin I. (71.475°N, 84.500°W), NU, CAN	Dyke, et al. (1996); R.E. Morlan, CMC database (1999); [Dyke (2000e) GSC Map 1965A]
1390 ± 70 (1550 ± 75)	S-2995		7 km NW of Point Caen, Wollaston Pen., 4 m asl., Victoria I. (69.350°N, 116.120°W), NU, CAN	R.E. Morlan, CMC database (1999); Dyke and Savelle (2001)
1380 ± 130 (1540 ± 130)	GSC-452	periotic	Holder Hills, 2 m asl., Ellesmere I. (78°54'00"N, 85°10'00"W), NU, CAN	A.S. Dyke, pers. comm. (1999); R.E. Morlan, CMC database (1999)
1340 ± 70 (1500 ± 75)	S-1423	rib (CMC-939)	Grinnell Pen., 4 m asl., Devon I. (76.250°N, 92.670°W), NU, CAN	[Park (1989) In: A.S. Dyke (GSC database)]; Dyke (1998); [Morrison (1989) In: R.E. Morlan, CMC database (1999)]
(1340 ± 80)	TO-8363	vertebra	6 km SSW of Cape Baring, Wollaston Pen., 3 m asl., Victoria I. (69°59'12"N, 117°23'05"W), NT, CAN	Dyke and Savelle (2001)
1320 ± 60 (1465 ± 60)	S-3323	periotic	Sunday Bay, Borden Pen., 1.5-2 m asl., Baffin I. (71.790°N, 85.460°W), NU, CAN	Dyke, et al. (1996); R.E. Morlan, CMC database (1999); [Dyke and Hooper (2000) GSC Map 1960A]
1300 ± 50 (1460 ± 55)	S-3235	periotic	1 km N of the mouth of Tikiraq R., 4 m asl., Baffin I. (71.760°N, 86.050°W), NU, CAN	R.E. Morlan, CMC database (1999); [Dyke and Hooper (2000) GSC Map 1960A]
1290 ± 50 (1450 ± 55)	S-3236	periotic	1 km NW of the mouth of Tikiraq R., 3 m asl., Baffin I. (71.770°N, 86.040°W), NU, CAN	Dyke, et al. (1996); R.E. Morlan, CMC database (1999); [Dyke and Hooper (2000) GSC Map 1960A]
1290 ± 140 (1425 ± 140)	S-3559	otic capsule	9.2 km NNE of Porden Pt., 9 km WSW of Reid Ck., Prince Alfred Bay, 3 m asl., Devon I. (76.330°N, 93.450°W), NU, CAN	Dyke, et al. (1996); Dyke (1998); R.E. Morlan, CMC database (1999)
1280 ± 70 (1430 ± 70)	S-3303	periotic	Jungersen Bay, Borden Pen., 3 m asl., Baffin I. (71.470°N, 84.510°W), NU, CAN	Dyke, et al. (1996); R.E. Morlan, CMC database (1999); [Dyke and Hooper (2000) GSC Map 1960A]
1240 ± 170 (1390 ± 170)	S-3567	periotic	9.7 km NW of Lyall R., Grinnell Pen., 3.5 m asl., Devon I. (77.060°N, 95.550°W), NU, CAN	Dyke (1998); R.E. Morlan, CMC database (1999)
1170 ± 90 (1330 ± 90)	S-3123	otic capsule	.7 km S of the mouth of Kuuruluk R., 3 m asl., Baffin I. (72.440°N, 86.250°W), NU, CAN	R.E. Morlan, CMC database (1999); [Dyke (2000c) GSC Map 1963A]
1140 ± 70 (1300 ± 75)	S-3073	periotic	Vista R., 6.8 km NNW of Kakiak Pt., 1.5-2 m asl., Baffin I. (72.730°N, 86.730°W), NU, CAN	R.E. Morlan, CMC database (1999); [Dyke (2000c) GSC Map 1963A]
1130 ± 60 (1285 ± 60)	S-3318	periotic	Sunday Bay, Borden Pen., 4 m asl., Baffin I. (71.784°N, 85.580°W), NU, CAN	R.E. Morlan, CMC database (1999); [Dyke and Hooper (2000) GSC Map 1960A]
(1110 ± 80)	GSC-1899	rib?	Muskox Fiord, Ellesmere I., 3.5 m asl (76.410°N, 87.130°W), NU, CAN	[Wilmeth (1978); Morrison (1989) In: R.E. Morlan, CMC database (1999)]
1110 ± 130 (1270 ± 130)	GSC-355	baleen	Tullett Pt., 0 m asl., Prince Patrick I. (76.730°N, 121.170°W), NT, CAN	Dyck, et al. (1966); R.E. Morlan, CMC database (1999)

1110 ± 60 (1270 ± 65)	S-3099	periotic	10 km SE of Van Koenig Pt., mouth of Bernier Bay, 27 m asl., Baffin I. (71.030°N, 89.370°W), NU, CAN	Dyke, et al. (1996); R.E. Morlan, CMC database (1999); [Dyke and Hooper (2000) GSC Map 1960A]
1100 ± 100 (1260 ± 100)	S-3155	periotic	2 km NW of the mouth of Tiriganialaag R., 2.5 m asl., Baffin I. (72.200°N, 86.460°W), NU, CAN	Dyke, et al. (1996); R.E. Morlan, CMC database (1999); [Dyke (2000c) GSC Map 1963A]
1095 ± 80	SI-758-B	baleen	Chesterfield Inlet (63.680°N, 90.080°W), NU, CAN	[Wilmeth (1978); Staab (1979); McCartney (1970, 1977, pers. comm. 1998); Morrison (1989) In: R.E. Morlan, CMC database (1999)]
1090 ± 60 (1120 ± 60)	GSC-3802	whale fat	Hall Beach Hamlet, Melville Pen., frozen beach gravel 8 m asl., NU, CAN	R.E. Morlan, CMC database (1999)
1070 ± 80 (1230 ± 80)	S-3397	periotic	E of the mouth of Phillips Ck., 1.5 m asl., Baffin I. (71.880°N, 80.900°W), NU, CAN	Dyke, et al. (1996); R.E. Morlan, CMC database (1999); [Dyke (2000a) GSC Map 1961A]
1060 ± 140 (1225 ± 140)	S-3535	otic capsule	18 km WNW of Porden Pt., Prince Alfred Bay, 4.3 m asl., Devon I. (76.280°N, 94.300°W), NU, CAN	Dyke, et al. (1996); Dyke (1998); R.E. Morlan, CMC database (1999)
1060 ± 70 (1220 ± 75)	S-3027		Hall Beach Hamlet, Melville Pen., beach gravel 8 m asl (68.5°46.5'N, 81.5°13.0'W), NU, CAN	[R. McNeely, pers. comm. (1998) In: R.E. Morlan, CMC database (1999)]
(1040 ± 60)	TO-8010	rib	6.1 km ENE of Cape Baring, Wollaston Pen., 22 m asl., Victoria I. (70°04'08"N, 117°13'35"W), NT, CAN	Dyke and Savelle (2001)
990 ± 80 (1155 ± 80)	S-3395	periotic	E of the mouth of Phillips Ck., 1 m asl., Baffin I. (71.880°N, 80.890°W), NU, CAN	Dyke, et al. (1996); R.E. Morlan, CMC database (1999); [Dyke (2000a) GSC Map 1961A]
980 ± 140 (1140 ± 140)	GSC-488	ear bone	Tullett Pt., 0 m asl., Prince Patrick I. (76.730°N, 121.170°W), NT, CAN	Dyck, et al. (1966); R.E. Morlan, CMC database (1999)
970 ± 60 (1130 ± 60)	GSC-3850		Hall Beach Hamlet, Melville Pen., frozen beach gravel 8 m asl., NU, CAN	R.E. Morlan, CMC database (1999)
930 ± 130 (1090 ± 130)	GSC-489	long-bone from flipper	Tullett Pt., 0 m asl., Prince Patrick I. (76.730°N, 121.170°W), NT, CAN	Dyck, et al. (1966); R.E. Morlan, CMC database (1999)
930 ± 150 (1080 ± 150)	S-3536	otic capsule	5.6 km NW of Porden Pt., Prince Alfred Bay, 2.7-3 m asl., Devon I. (76.290°N, 93.790°W), NU, CAN	Dyke, et al. (1996); Dyke (1998); R.E. Morlan, CMC database (1999)
930 ± 50	SI-758-C	baleen	Chesterfield Inlet (63.680°N, 90.080°W), NU, CAN	[Wilmeth (1978); Staab (1979); McCartney (1970, 1977, pers. comm. 1998); Morrison (1989) In: R.E. Morlan, CMC database (1999)]
920 ± 130 (1075 ± 130)	S-3517	periotic	E of the mouth of Phillips Ck., 6.5 m asl., Baffin I., NU, CAN	R.E. Morlan, CMC database (1999); [Dyke (2000a) GSC Map 1961A]
900 ± 50 (1060 ± 50)	S-3343	otic capsule	Lavasseur Inlet, Borden Pen., 1 m asl., Baffin I. (72.590°N, 85.525°W), NU, CAN	Dyke, et al. (1996); R.E. Morlan, CMC database (1999); [Dyke (2000c) GSC Map 1963A]
820 ± 60	GSC-5840	baleen	3 km SW of Black Pt., NE Baffin I., NU, CAN	A.S. Dyke (pers. comm., 1999)
780 ± 80 (935 ± 80)	S-3396	periotic	E of the mouth of Phillips Ck., 6.5 m asl., Baffin I. (71.880°N, 80.890°W), NU, CAN	R.E. Morlan, CMC database (1999); [Dyke (2000a) GSC Map 1961A]
770 ± 80 (930 ± 80)	QU-241		Head of Shark Fiord, Cumberland Sound, 0 m asl., Baffin I. (66°36'N, 66°41'W), NU, CAN	Dyke (1979b); R.E. Morlan, CMC database (1999)
720 ± 70 (880 ± 75)	S-3249	periotic	1.7 km S of the mouth of Kuuruluk R., 2 m asl., Baffin I. (72.430°N, 86.240°W), NU, CAN	Dyke, et al. (1996); R.E. Morlan, CMC database (1999); [Dyke (2000c) GSC Map 1963A]
680 ± 70 (840 ± 75)	S-3251	periotic	.6 km N of the mouth of Kuuruluk R., 1 m asl., Baffin I. (72.440°N, 86.260°W), NU, CAN	Dyke, et al. (1996); R.E. Morlan, CMC database (1999); [Dyke (2000c) GSC Map 1963A]

660 ± 95 (820 ± 95)	S-2596	vertebra	E side of Guillemard Bay, 2 km SE from the mouth of Fisher R., 4 m asl., Prince of Wales I. (71°50'00"N, 98°10'00"W), NU, CAN	Dyke and Morris (1990); Dyke, et al. (1991); R.E. Morlan, CMC database (1999)
635 ± 50 (795 ± 55)	S-1391	vertebra	2 km SW of Cape Anne, 4 m asl., Somerset I., (74.080°N, 94.780°W), NU, CAN	Dyke and Morris (1990); Dyke, et al. (1996); R.E. Morlan, CMC database (1999)
605 ± 80 (765 ± 80)	S-3169	otic capsule	3.5 km S of the mouth of Kuuruluk R., 1.5 m asl., Baffin I. (72.420°N, 86.230°W), NU, CAN	Dyke, et al. (1996); R.E. Morlan, CMC database (1999); [Dyke (2000c) GSC Map 1963A]
580 ± 80 (750 ± 80)	S-3402	periotic	Head of Koluktoo Bay, 1.5 m asl., Baffin I. (72.060°N, 81.070°W), NU, CAN	Dyke, et al. (1996); R.E. Morlan, CMC database (1999); [Dyke (2000b) GSC Map 1962A]
540 ± 130 (695 ± 130)	S-3531	periotic	5.5 km WNW of Owen Pt., 2.7-3 m asl., Devon I. (76.020°N, 92.670°W), NU, CAN	Dyke, et al. (1996); Dyke (1998); R.E. Morlan, CMC database (1999)
480 ± 60 (640 ± 60)	S-3324	otic capsule	Sunday Bay, Borden Pen., 1 m asl., Baffin I. (71.800°N, 85.440°W), NU, CAN	Dyke, et al. (1996); R.E. Morlan, CMC database (1999); [Dyke and Hooper (2000) GSC Map 1960A]
440 ± 80 (590 ± 80)	S-3401	otic capsule	Head of Koluktoo Bay, 0.3 m asl., Baffin I. (72.070°N, 81.020°W), NU, CAN	Dyke, et al. (1996); R.E. Morlan, CMC database (1999); [Dyke (2000b) GSC Map 1962A]
410 ± 100	K-567	baleen	Snenaes (76.820°N, 19.350°W), GRE	[Knuth (1951) In: A.S. Dyke (GSC database)]
320 ± 60 (480 ± 65)	S-3140	otic capsule	10 km NW of the mouth of Tikiraq R., 3.5-4 m asl., Baffin I. (71.790°N, 86.120°W), NU, CAN	Dyke, et al. (1996); R.E. Morlan, CMC database (1999); [Dyke and Hooper (2000) GSC Map 1960A]
<100 (<100)	S-2667	vertebra	NE Russell I., 3.6 m asl. (74°6.5'00"N, 97°50.2'00"W), NU, CAN	Dyke, et al. (1991); Dyke (1998); R.E. Morlan, CMC database (1999); [Anomalously young, A.S. Dyke, pers. comm. (2001)]

Balaena sp. (bowhead whale or right whale)

Measured Date (Normalized)	Lab No	Material	Location	References/Remarks
8900 ± 40 (9010 ± 40)	Beta-83411	rib (CRH-95-15)	Cook's Harbour, NF, CAN	Harington (1998a)
(7260 ± 80)	GSC-1498	skull fragment?	Cape Storm, 38 m asl., Ellesmere I. (76.410°N, 87.500°W), NU, CAN	Lowdon and Blake (1980); R.E. Morlan, CMC database (1999)
(7240 ± 80)	GSC-1498-3	skull fragment?	Cape Storm, 38 m asl., Ellesmere I. (76.410°N, 87.500°W), NU, CAN	Lowdon and Blake (1980)
(6560 ± 170)	GSC-1498-2	skull fragment?	Cape Storm, 38 m asl., Ellesmere I. (76.410°N, 87.500°W), NU, CAN	Lowdon and Blake (1980)
(6500 ± 260)	GSC-3600	rib	N side of Baird Inlet, 80 m asl., Ellesmere I., NU, CAN	Blake (1987)
1440 ± 140 (1490 ± 140)	GSC-3086	humerus	Red Bay, NF (Labrador), CAN	R.E. Morlan, CMC database (1999); S.L. Cumbaa, pers. comm. (2000)
(980 ± 60)	GSC-980	vertebra	Cape Storm, 1.5 m asl., Ellesmere I. (76.375°N, 87.500°W), NU, CAN	Lowdon and Blake (1980); R.E. Morlan, CMC database (1999)
(940 ± 70)	GSC-980-2	vertebra	Cape Storm, 1.5 m asl., Ellesmere I. (76.375°N, 87.500°W), NU, CAN	Lowdon and Blake (1980)

cf. Balaena mysticetus (bowhead whale)

Measured Date (Normalized)	Lab No	Material	Location	References/Remarks
(>40000)	Ua-2451	large whale rib	Andersrag Beach, 55 m asl., Cape Storm, Ellesmere I., NU, CAN	[GSC Paper (1988) 87-7:64 (for other dates on same specimen)]
(9700 ± 70)	CAMS-62937	periotic	Near Clinton Pt. (69°25'N, 120°17'W), NT, CAN	A.S. Dyke, pers. comm. (2000)
(9560 ± 70)	CAMS-62455	vertebra	About 1 km inland, S of Holman airport, Victoria I. (70°45'22"N, 117°49'50"W), NT, CAN	A.S. Dyke, pers. comm. (2000)
(9070 ± 90)	GSC-1748	rib	S Cape Fiord, 80.5 m asl., Ellesmere I. (76.425°N, 85.050°W), NU, CAN	Blake (1975b); Lowdon, et al. (1977); R.E. Morlan, CMC database (1999)
(8580 ± 85)	Lu-1095		Scoresby Land (72.130°N, 23.630°W), GRE	Bennike (1997)

(7750 ± 120)	I-16356		Disko (69.860°N, 54.760°W), GRE	Bennike, et al. (1994)
(7590 ± 150)	I-16366	rib	Disko (69.900°N, 54.850°W), GRE	Bennike, et al. (1994)
7120 ± 25 (7530 ± 25)	TO-33	rib	SW side of Rosse Bay, 54 m asl., Ellesmere I. (78.630°N, 74.800°W), NU, CAN	Blake (1980, 1992); [R.E. Morlan, pers. comm. (1998) indicates a corrected date of 7530 ± 25 using delta13C correction -25o/oo]
(6910 ± 80)	Ua-1785		Disko (69.950°N, 54.220°W), GRE	Bennike, et al. (1994)
6820 ± 90 (6920 ± 90)	GSC-3055	rib	SW side of Rosse Bay, Cape Herschel, 54 m asl., Ellesmere I. (78.630°N, 74.800°W), NU, CAN	Blake (1987); [Blake (1992) two determinations were made - see GSC-3055-2]; R.E. Morlan, CMC database (1999)
6050 ± 105 (5900 ± 105)	K-5977		Kronprins Christian Land (81.580°N, 16,620°W), GRE	Bennike (1997)
(780 ± 80)	I-16414		Nuussuaq Pen., 5 m asl. (70°43'N, 54°36'W), GRE	Bennike, et al. (1994)
(750 ± 60)	Beta-70094	lumbar vertebra (CR-94-3)	Sherbrooke (45°25'N, 71°54'W), QC, CAN	C.R. Harington [unpublished]; R.E. Morlan, CMC database (1999)
370 ± 50 (370 ± 50)	Beta-70093	lower jaw (CR-94-2)	Eaton R., Cookshire (45°25'N, 71°38'W), QC, CAN	C.R. Harington [unpublished]; R.E. Morlan, CMC database (1999)

Balaena? mysticetus (bowhead whale)

Measured Date (Normalized)	Lab No	Material	Location	References/Remarks
(6040 ± 60)	P-2141		Griffith I., 22 km S of Cornwallis I. (74°33'N, 95°30'W), NU, CAN	[A.S. Dyke (1993) - Anomalously young]; R.E. Morlan, CMC database (1999)
(4740 ± 70)	Lu-3658	2 vertebrae and a rib (2MKS1/1980)	Hvalterrasserne at Frigg Fjord (83°6.7'N, 32°58'W), GRE	Knuth (1983)

Carnivora, Canidae

Canis lupus (wolf)

Measured Date (Normalized)	Lab No	Material	Location	References/Remarks
>47170 (>47240)	Beta 89988	rib (CMN 51772)	Dominion Ck. (Dawson Loc. 78) (63°49'15"N, 138°40'30"W), YT, CAN	C.R. Harington [unpublished]; R.E. Morlan, CMC database (1999)
(27920 ± 650)	Beta-33191 ETH-5899	skull and atlas vertebra (CMN 45574)	Eldorado Ck. (Dawson Loc. 63, Nugget Gulch) (63°53'N, 139°17'W), YT, CAN	C.R. Harington [unpublished]
(18610 ± 165)	AA-3912	(F:AM-295 5219)	Fairbanks Ck., AK, USA	Guthrie (1990c)
(6785 ± 100)	Ua-1346	left humerus (MGUH VP 3332)	Volvedal, Peary Land (83.000°N, 35.000°W), GRE	Bennike, et al. (1994); Bennike (1996)
5640 ± 310	S-1788	right mandible	Strawberry Ck., N Saskatchewan R. valley, near Edmonton (53.500°N, 113.830°W), AB, CAN	Rains and Welch (1988)
(1570 ± 100)	AAR-830	left mandible (ZMK 112a11950)	Midsommersoen, Peary Land (82.250°N, 34.000°W), GRE	Bennike, et al. (1994)
(790 ± 100)	Ua-950	limb	Blaso (79.670°N, 22.250°W), GRE	Bennike, et al. (1994); limb bones from same adult - (ZMK 112a /1950)
(520 ± 70)	AAR-827	left mandible (CN 3961)	Heilprin Land (82.170°N, 31.500°W), GRE	Bennike, et al. (1994)
(215 ± 90)	AAR-829		Jameson Land, E region (71.500°N, 23.500°W), GRE	[Bennike (1996) In: A.S. Dyke (GSC database)]
(100 ± 85)	AAR-831		Jameson Land, E region (70.920°N, 22.750°W), GRE	[Bennike (1996) In: A.S. Dyke (GSC database)]

Canis familiaris (domestic dog)

Measured Date (Normalized)	Lab No	Material	Location	References/Remarks
(5400 ± 50)	Beta-87709	radius	Namu R. (51.860°N, 127.860°W), BC, CAN	[Chisholm (1986); Carlson (1991, 1996) In: R.E. Morlan, CMC database (1999)]
(2110 ± 40)	TO-276	tooth from mandible	Old Crow R. (Old Crow Loc. 11(A)), 259 m asl. (67.480°N, 139.920°W), YT, CAN	Beebe (1980); Harington (1989c); R.E. Morlan, CMC database (1999)

Measured Date (Normalized)	Lab No	Material	Location	References/Remarks
990 ± 170 (1070 ± 170)	S-1950		Frank Bay, Lake Nipissing, ON, CAN	[Ridley (1954); Brizinski and Savage (1983) In: R.E. Morlan, CMC database (1999)]
(890 ± 60)	Beta-39224	skull	N of Pt. Roberts, about 2 m asl., Strait of Georgia (49.033°N, 123.10°W), BC, CAN	[Bernick and Howe (1990); Hanson (1991); Arcas Consulting (1994) In: R.E. Morlan, CMC database (1999)]
690 ± 170 (770 ± 170)	S-1949		Frank Bay, Lake Nipissing, ON, CAN	[Ridley (1954); Brizinski and Savage (1983) In: R.E. Morlan, CMC database (1999)]
(600 ± 70)	AA-3905	(F:AM-30434)	Cleary Ck. (65.080°N, 147.33°W), AK, USA	Guthrie (1990c)
(490 ± 50)	AA-3904	(F:AM-199 8698)	Ester Ck. (65.000°N, 147.000°W), AK, USA	Guthrie (1990c)
(370 ± 105)	AA-3911	(F:AM-482 2548)	Cripple Ck., AK, USA	Guthrie (1990c)
(355 ± 60)	AA-3901	(F:AM-672 8171)	Fairbanks region, AK, USA	Guthrie (1990c)
(330 ± 70)	AA-3907	(F:AM-30473)	Fairbanks area (64°50'N, 147°30'W), AK, USA	Guthrie (1990c)
(230 ± 55)	AA-3909	(F:AM-197 8697)	Ester Ck. (New Cut) (65.000°N, 147.000°W), AK, USA	Guthrie (1990c)
(140 ± 80)	AA-3910	(F:AM-340 1038)	Little El Dorado Ck. (65.100°N, 147.680°W), AK, USA	Guthrie (1990c)
(140 ± 80)	AA-3903	(F:AM-82 1006)	Ester Pit, AK, USA	Guthrie (1990c)
(130 ± 50)	AA-3908	(F:AM-200 4345)	Little El Dorado Ck. (65.100°N, 147.680°W), AK, USA	Guthrie (1990c)
(105 ± 50)	AA-3900	(F:AM-688 70932)	Fairbanks Ck., AK, USA	Guthrie (1990c)

Canis sp. (wolf or domestic dog)

Measured Date (Normalized)	Lab No	Material	Location	References/Remarks
3825 ± 90 (3905 ± 90)	S-2369		S bank of Saskatchewan R., 10.5 km upstream from the Francois-Finlay Dam (53.170°N, 104.000°W), SK, CAN	Morlan (1993); [Quigg (1986); Brandzin-Low (1997) In: R.E. Morlan, CMC database (1999)]; [Morlan and McNeely (1996) In: A.S. Dyke (GSC database)]

Alopex lagopus (arctic fox)

Measured Date (Normalized)	Lab No	Material	Location	References/Remarks
(10800 ± 90)	TO-1197	mandible (CMN 45286)	Lafleche Cavern (45.650°N, 75.780°W), QC, CAN	P.M. Youngman [unpublished];
3520 ± 90 (3600 ± 90)	I-12137		Bettison Pt., Back Bay, 18 m asl., Prince of Wales I. (73.930°N, 99.290°W), NU, CAN	Dyke, et al. (1991); [Ramsden and Murray (1995) In: R.E. Morlan, CMC database (1999)]
2470 ± 120 (2530 ± 120)	SFU-82		Karluk I., 10-11 m asl., (75.500°N, 97.270°W), NU, CAN	[Helmer (1980, 1981) In: Dyke, et al. (1999)]; R.E. Morlan, CMC database (1999)
(1175 ± 40)	AAR-3222	jawbone	Smith Sound area (78.333°N, 72.75°W), GRE	Appelt, et al. (1998); [Heinemeier and Rasmussen (1998) In: Danish/Greenlandic Network (1999)]; [Reservoir corrected 14C age = 930 ± 40 BP]; [950 AD1050-1155, AD1030-1170]
890 ± 110 (970 ± 110)	RL-1666		Nelson R., Banks I. (71.080°N, 123.000°W), NT, CAN	[Arnold and Stimmell (1983); Morrison (1989); Arnold (1994) In: R.E. Morlan, CMC database (1999)]
575 ± 75 (655 ± 85)	I-11769		Cape Walker, 4 m asl., Russell I. (73.930°N, 99.290°W), NU, CAN	Dyke, et al. (1991); [Morrison (1989) In: Dyke, et al., (1999)]; [Ramsden (1980) In: R.E. Morlan, CMC database (1999)]

Vulpes vulpes (red fox)

Measured Date (Normalized)	Lab No	Material	Location	References/Remarks
(11275 ± 90)	AA-21567	pelvis	On Your Knees Cave, Prince of Wales I., 135 m asl. (56.250°N, 133.500°W), AK, USA	Heaton and Grady (1997)

900 ± 105 (1000 ± 105)	GX-4143		N shore of Whitemouth R., 259 m asl (50.122°N, 96.035°W), MB, CAN	[Meiklejohn et al. (1994) In: A.S. Dyke (GSC database)]; [Buchner (1979, 1982); Ens (1998) In: R.E. Morlan, CMC database (1999)]

Carnivora, Ursidae

Arctodus simus (giant short-faced bear)

Measured Date (Normalized)	Lab No	Material	Location	References/Remarks
(44240 ± 930)	TO-2699	left metacarpal (CMN 42388)	Sixtymile area (Sixtymile Loc. 3), 701 m asl., (64°1.00'N, 140°44'W), YT, CAN	Harington (1997a)
(39565 ± 1126)	AA-17514	(AMNH 99209)	Ester Ck., Fairbanks area (65.000°N, 147.000°W), AK, USA	P. Matheus, pers. comm. (1996)
(34974 ± 652)	AA-17515		Birch Ck., N of Fairbanks, AK, USA	P. Matheus, pers. comm. (1996)
29600 ± 1200 (29695 ± 1200)	I-11037	humerus (CMN 37577)	Lower Hunker Ck. (Dawson Loc. 10) (64.000°N, 139.000°W), YT, CAN	Harington (1977, 1980, 1989); Matheus (1995); R.E. Morlan, CMC database (1999)
(27511 ± 289)	AA-17513	(A-37-10)	Upper Cleary Ck., Fairbanks area (65.080°N, 147.333°W), AK, USA	P. Matheus, pers. comm. (1996)
(27190 ± 280)	TO-2539	left humerus (ROM:VP 43646)	Ikpikpuk R. (70°N, 154°45'W), AK, USA	Churcher, et al. (1993); P. Matheus, pers. comm. (1996)
(26040 ± 270)	TO-2696	cranium (CMN 7438)	Gold Run Ck. (Dawson Loc. 31) (63°43.5'N, 138°48'W), YT, CAN	Matheus (1995); [C.R. Harington, pers. comm. (1998) In: R.E. Morlan, CMC database (1999)]
(25496 ± 224)	AA-17512	(F:AM 30494)	Ester Ck., Fairbanks area (65.000°N, 147.000°W), AK, USA	P. Matheus, pers. comm. (1996)
(24850 ± 150)	TO-3707	cranial bone (CR-91-5) (CMN 50367)	Hunker Ck. (Dawson Loc. 37) (64.017°N, 139.150°W), YT, CAN	C.R. Harington [unpublished]
(20524 ± 180)	AA-17511	(F:AM 30492)	Cleary Ck., Fairbanks area (65.000°N, 147.000°W), AK, USA	P. Matheus, pers. comm. (1996)
20210 ± 110 (20250 ± 110)	Beta-79852 CAMS-18415	cranium (CRH-95-3)	Ophir Ck. (Dawson Loc. 77) (64.133°N, 139.500°W), YT, CAN	C.R. Harington [unpublished]; R.E. Morlan, CMC database (1999)

Arctodus? simus (giant short-faced bear)

Measured Date (Normalized)	Lab No	Material	Location	References/Remarks
(26720 ± 290)	OxA-9259	right ulna (CMN-49874)	Ester Ck. (65.000°N, 147.000°W), AK, USA	A. Cooper, pers. comm. (2000)

Ursus americanus (black bear)

Measured Date (Normalized)	Lab No	Material	Location	References/Remarks
(41600 ± 1500)	AA-16831	tibia (small tibia)	On Your Knees Cave, Prince of Wales I., 135 m asl. (56.250°N, 133.500°W), AK, USA	Heaton and Grady (1997); Lysek (1997); Dixon, et al. (1997); Sattler and Dixon (1997)
(29820 ± 400)	AA-21570	vertebra	On Your Knees Cave, Prince of Wales I., 135 m asl. (56.250°N, 133.500°W), AK, USA	Heaton and Grady (1996)
(28695 ± 360)	AA-21569	calcaneum	On Your Knees Cave, Prince of Wales I., 135 m asl. (56.250°N, 133.500°W), AK, USA	Heaton and Grady (1996, [1997) incorrectly cited as 28,700 ± 360 BP]; Sattler and Dixon (1997)
(11565 ± 115)	AA-10448	cranium (juvenile)	El Capitan Cave, Prince of Wales I. (56.250°N, 133.500°W), AK, USA	Heaton and Grady (1993); Dixon, et al. (1997)
(11540 ± 110)	AA-10446	skull (complete)	El Capitan Cave, Prince of Wales I. (56.250°N, 133.500°W), AK, USA	Heaton, et al. (1996)
(10745 ± 75)	AA-7793	humerus (complete skeleton)	El Capitan Cave, Prince of Wales I. (56.000°N, 133.500°W), AK, USA	Heaton and Grady (1992b, 1993)
9760 ± 140 (9830 ± 140)	Beta-10714	composite sample (tibiae, ribs and vertebrae)	Windy Link Pot Cave: White Ridge near Gold R., Vancouver I. (49°46'55"N, 125°59'10"W), BC, CAN	Nagorsen, et al. (1995); R.E. Morlan, CMC database (1999)

9380 ± 140	(no lab # given)		NW coast of Vancouver I., BC, CAN	Lehman (1984)
(9330 ± 155)	AA-17451R	femur	Kushtaka Cave, 45 m asl., Prince of Wales I. (56.210°N, 133.520°W), AK, USA	Heaton, et al. (1996)
(8725 ± 70)	AA-17451	femur	Kushtaka Cave, 45 m asl., Prince of Wales I. (56.210°N, 133.520°W), AK, USA	Heaton, et al. (1996); Dixon, et al. (1997)
(8630 ± 60)	CAMS-24967	rib	Kushtaka Cave, 45 m asl., Prince of Wales I. (56.210°N, 133.520°W), AK, USA	Heaton, et al. (1996)
(8600 ± 60)	CAMS-24378	rib	Kushtaka Cave, 45 m asl., Prince of Wales I. (56.210°N, 133.520°W), NT, CAN	Dixon, et al. (1997)
(6415 ± 130)	AA-10447	cranium (fused)	El Capitan Cave, Prince of Wales I. (56.250°N, 133.500°W), AK, USA	Heaton, et al. (1996); Sattler and Dixon (1997)
(3960 ± 50)	CAMS-31068	dentary	On Your Knees Cave, Prince of Wales I., 135 m asl. (56.250°N, 133.500°W), AK, USA	Heaton, et al. (1996)
920 ± 70 (1000 ± 75)	S-1711	(CMC-1061)	Lake Huron, Bruce County, (44°26'20"N, 81°23'30"W), ON, CAN	[Stewart (1974); Wright (1974, 1985) In: R.E. Morlan, CMC database (1999)]
710 ± 40 (790 ± 45)	S-1710	(CMC-1060)	Lake Huron, Bruce County, (44°26'20"N, 81°23'30"W), ON, CAN	[Stewart (1974); Wright (1974, 1985) In: R.E. Morlan, CMC database (1999)]
(298 ± 39)	OxA-9258	right femur (CMN-47036)	Caribou Ck., YT, CAN	A. Cooper, pers. comm. (2000)
(222 ± 38)	OxA-9257	(FAM:95683)	Grouse Ck., AK, USA	A. Cooper, pers. comm. (2000)
(199 ± 33)	OxA-9256	(FAM:95686)	Cripple Ck., (64°50'N, 148°W), AK, USA	A. Cooper, pers. comm. (2000)

Ursus arctos (brown bear)

Measured Date (Normalized)	Lab No	Material	Location	References/Remarks
(48164 ± 3224)	AA-17510		N slope, Colville R. (70.000°N, 151.500°W), AK, USA	P. Matheus, pers. comm. (1996)
(47100 ± 3100)	OxA-9260	ulna (FAM:95666)	Cripple Ck., (64°50'N, 148°W), AK, USA	A. Cooper, pers. comm. (2000)
(>43400)	Beta-68923 CAMS-10527	cranium (CMN 42105)	Boliden Ck. (62°21'N, 137°15'W), YT, CAN	C.R. Harington [unpublished]
41000 ± 1050 (41085 ± 1050)	Beta-16159	mandible (CMN 35965)	Hunker Ck. (Dawson Loc. 16) (63.930°N, 138.900°W), YT, CAN	Harington (1977, 1978, 1980, 1989, 1997); Harington and Morlan (1992); Matheus (1995); R.E. Morlan, CMC database (1999)
(36500 ± 1150)	Beta-16162	right humerus (CMN 38279)	Sixtymile area (Sixtymile Loc. 3), 701 m asl., (64°1.00'N, 140°44'W), YT, CAN	Harington (1989, 1997); Matheus (1995)
(36400 ± 1700)	AA-1889		Fairbanks area (65.000°N, 147.000°W), AK, USA	P. Matheus, pers. comm. (1996)
(36137 ± 783)	AA-17509	(F:AM 95601)	Cripple Ck., Fairbanks area (64.817°N, 148.017°W), AK, USA	P. Matheus, pers. comm. (1996)
(35365 ± 800)	AA-15227	femur (large)	On Your Knees Cave, Prince of Wales I., 135 m asl. (56.250°N, 133.500°W), AK, USA	Heaton, et al. (1995, 1996, 1997); Lysek (1997); Dixon, et al. (1997); Heaton and Grady (1998)
(33700 ± 1100)	AA-1887		Fairbanks area (65.000°N, 147.000°W), AK, USA	P. Matheus, pers. comm. (1996)
(20080 ± 160)	OxA-9261	phalanx (FAM:30771)	Lower Goldstream (64°55'N, 147°36'W), AK, USA	A. Cooper, pers. comm. (2000)
(19027 ± 132)	AA-17507	(AMNH 30422)	Probably Fairbanks area (65.000°N, 147.000°W), AK, USA	P. Matheus, pers. comm. (1996)
(15830 ± 100)	OxA-9263	ulna (FAM:95670)	Gold Hill, AK, USA	A. Cooper, pers. comm. (2000)
(14150 ± 90)	OxA-9262	femur (FAM:95642)	Ester Ck. (65.000°N, 147.000°W), AK, USA	A. Cooper, pers. comm. (2000)

Measured Date (Normalized)	Lab No	Material	Location	References/Remarks
(12441 ± 101)	AA-17508	(F:AM 95595)	Goldstream Ck., Fairbanks (64.920°N, 147.600°W), AK, USA	P. Matheus, pers. comm. (1996)
(12295 ± 120)	AA-10445	humerus (small)	El Capitan Cave, Prince of Wales I. (56.250°N, 133.500°W), AK, USA	Heaton, et al. (1993, 1996, 1997); Dixon, et al. (1997)
(11715 ± 120)	AA-15226	humerus (complete skeleton)	Enigma Cave, Dall I. (54.500°N, 133.000°W), AK, USA	Heaton, et al. (1996); Sattler and Dixon (1997)
(11640 ± 80)	AA-15222	humerus (complete skeleton)	Bumper Cave, Prince of Wales I. (56.170°N, 133.500°W), AK, USA	Heaton, et al. (1996)
(11225 ± 110)	AA-15223	humerus (large)	Bumper Cave, Prince of Wales I. (56.170°N, 133.500°W), AK, USA	Heaton, et al. (1996)
(10970 ± 85)	AA-15225	molar (large)	Bumper Cave, Prince of Wales I. (56.170°N, 133.500°W), AK, USA	Heaton, et al. (1996)
(10015 ± 62)	AA-17506	(F:AM 95612)	Ester Ck., Fairbanks area (65.000°N, 147.000°W), AK, USA	P. Matheus, pers. comm. (1996)
(9995 ± 95)	AA-10451	skull (juvenile skeleton)	Blowing in the Wind Cave, Prince of Wales I. (56.200°N, 133.580°W), AK, USA	Heaton, et al. (1993, 1996)
(9760 ± 75)	AA-7794	humerus (large)	El Capitan Cave, Prince of Wales I. (56.000°N, 133.500°W), AK, USA	Heaton, et al. (1992b)
(7205 ± 65)	AA-15224	lower jaw	Bumper Cave, Prince of Wales I. (56.170°N, 133.500°W), AK, USA	Heaton, et al. (1996, 1997); Dixon, et al. (1997)
1290 ± 60 (1390 ± 65)	S-3364	humerus (CMC-1402)	Overlooking S shore of Amundsen Gulf, NE corner of Pearce Pt. Harbour, 1 m asl, NT, CAN	[D.A. Morrison, pers. comm. (1999) In: R.E. Morlan, CMC database (1999)]

Ursus cf. Ursus arctos horribilis (brown bear, grizzly bear)

Measured Date (Normalized)	Lab No	Material	Location	References/Remarks
11700 ± 250 (11800 ± 250)	(no lab # given)	limb	Orillia twp., Simcoe County (44.58°N, 79.42°W), ON, CAN	Peterson (1965a, b); Tovell and Deane (1966); R.E. Morlan, CMC database (1999). This is the first example of grizzly bear found in the eastern Great Lakes region.

Ursus maritimus (polar bear)

Measured Date (Normalized)	Lab No	Material	Location	References/Remarks
(2070 ± 110)	AAR-1357	ulna	Vandfeldsnaes, 1.9 m asl., (82.160°N, 130.270°W), GRE	[Bennike (1996) In: A.S. Dyke (GSC database)]
1975 ± 120 (2135 ± 120)	Beta-18129	canine tooth (CMN 43803)	19 km NNE of Cape Richard Collinson, 64 m asl., Prince of Wales I. (72.925°N, 102.580°W), NU, CAN	C.R. Harington [unpublished]; R.E. Morlan, CMC database (1999)
1440 ± 45	K-352	right mandible	Kolnaes site (82.590°N, 19.670°W), GRE	[Knuth (1981) In: A.S. Dyke (GSC database), incorrectly cited as Ovibos moschatus]; Bennike (1991)
1390 ± 60 (1560 ± 65)	Gif-7512		W end of Naujat Pen., Victoria I., NU, CAN	[Le Mouel (1999) In: R.E. Morlan, CMC database (1999)]
(1350 ± 40)	Gif-8434		W end of Naujat Pen., Victoria I., NU, CAN	[Le Mouel (1999) In: R.E. Morlan, CMC database (1999)]
(1310 ± 40)	Gif-8178		W end of Naujat Pen., Victoria I., NU, CAN	[Le Mouel (1999) In: R.E. Morlan, CMC database (1999)]

Ursus sp. (bear)

Measured Date (Normalized)	Lab No	Material	Location	References/Remarks
480 ± 70 (560 ± 75)	S-2006	(CMC-1195)	S Nation R., 3.2 km NE of Spencerville, Grenville County, (44°51'30"N, 75°30'45"W), ON, CAN	[Pendergast (1966); Wright (1985); Jamieson (1990) In: R.E. Morlan, CMC database (1999)]

Carnivora, Mustelidae

Mustela eversmanni (Beringian ferret)

Measured Date (Normalized)	Lab No	Material	Location	References/Remarks
(33550 ± 350)	TO-1196	mandible	Bluefish Cave III, 600 m asl. (67°08'N, 140°47'W), YT, CAN	Youngman (1993)

Mustela nigripes (black-footed ferret)

Measured Date (Normalized)	Lab No	Material	Location	References/Remarks
(39560 ± 490)	TO-214	fibula (CMN 43786)	Sixtymile area (Sixtymile Loc. 3), 701 m asl., (64°1.00'N, 140°44'W), YT, CAN	Harington (1989, 1997); Youngman (1993)
(30370 ± 560)	Beta-23347 ETH-3537	mummy (CMN 44305)	Hunker Ck. (Dawson Loc. 12) (63.970°N, 138.980°W), YT, CAN	Youngman (1993)

Martes americana (American marten)

Measured Date (Normalized)	Lab No	Material	Location	References/Remarks
(510 ± 60)	(no lab # given)	adult female skull	Elba Cave, Niagara Escarpment, ON, CAN	[Savage (1992), Mead and Grady (1996) In: R.E. Morlan, CMC database (1999)]

Gulo gulo (wolverine)

Measured Date (Normalized)	Lab No	Material	Location	References/Remarks
(41420 ± 1100)	TO-2701	left mandible (CMN 42492)	Sixtymile area (Sixtymile Loc. 3), 701 m asl., (64°1.00'N, 140°44'W), YT, CAN	Youngman (1993); Harington (1997a)

Taxidea taxus (American badger)

Measured Date (Normalized)	Lab No	Material	Location	References/Remarks
37930 ± 750 (37990 ± 750)	Beta-83413	maxilla (CRH-95-17)	Hunker Ck. (Dawson Loc. 10) (64°N, 139°5.00'W), YT, CAN	Harington (1980, 1997); R.E. Morlan, CMC database (1999) - Specimen in University of Calgary collection.
15190 ± 130 (15240 ± 130)	Beta-81133	skull (CMN 17260)	Dominion Ck. (Dawson Loc. 28) (63°49;N, 138°39'W), YT, CAN	Harington (1977, 1997); R.E. Morlan, CMC database (1999)

Lutra canadensis (northern river otter)

Measured Date (Normalized)	Lab No	Material	Location	References/Remarks
(3290 ± 60)	AA-10450	limb	El Capitan Cave, Prince of Wales I. (56.250°N, 133.500°W), AK, USA	Heaton, et al. (1996)
(1990 ± 95)	AA-21568	skull (juvenile)	On Your Knees Cave, Prince of Wales I., 135 m asl. (56.250°N, 133.500°W), AK, USA	Heaton and Grady (1996)

Enhydra lutris (sea otter)

Measured Date (Normalized)	Lab No	Material	Location	References/Remarks
(9270 ± 100)	CAMS-14438		Echo Bay, Gwaii Haanas, SE Moresby I., Queen Charlotte I., 1 m asl. (1 m below high tide), (52°41'52"N, 131°45'58"W), BC, CAN	Josenhans, et al. (1995); [Fedje and Christensen (1999); Southon and Fedje (2000) In: R.E. Morlan, CMC database (1999)]

Carnivora, Felidae

Felis concolor (cougar, mountain lion)

Measured Date (Normalized)	Lab No	Material	Location	References/Remarks
(18970 ± 1490)	TO-1266	phalanx	Bluefish Cave III, 600 m asl. (67°08'N, 140°47'W), YT, CAN	Cinq-Mars (1979, 1990); Morlan and Cinq-Mars (1982);Youngman (1993); Harington and Cinq-Mars (1995); Burke and Cinq-Mars (1996)

Panthera leo atrox (American lion)

Measured Date (Normalized)	Lab No	Material	Location	References/Remarks
39300 ± 1000 (39380 ± 1000)	Beta-79858 CAMS-18421	left mandible (CMN 45988)	Porcupine R. (67°34'N, 138°25'W), YT, CAN	C.R. Harington [unpublished]; R.E. Morlan, CMC database (1999)
22680 ± 300	SI-456	tendon from left tibia	Upper Ester Ck. (65.000°N, 147.500°W), AK, USA	Péwé (1975b)

Homotherium serum (American scimitar cat)

Measured Date (Normalized)	Lab No	Material	Location	References/Remarks
>52200 (>52300)	Beta-68926 CAMS-10530	humerus (CMN 46442)	Sixtymile area (Sixtymile Loc. 3), 701 m asl., (64°1.00'N, 140°44'W), YT, CAN	Harington (1997); R.E. Morlan, CMC database (1999)
43380 ± 1100 (43430 ± 1100)	Beta-89989	left humerus (CMN 46442)	Sixtymile area (Sixtymile Loc. 3), 701 m asl., (64°1.00'N, 140°44'W), YT, CAN	Harington (1997); R.E. Morlan, CMC database (1999)

Carnivora, Otariidae

Eumetopias jubatus (Steller's sea lion)

Measured Date (Normalized)	Lab No	Material	Location	References/Remarks
13020 ± 90 (13180 ± 90)	Beta-79860 CAMS-18423	left humerus (CRH-95-11)	From the side of a 1.8 m well at Bowen I. (49°23.8'N, 123°21.8'W), BC, CAN	C.R. Harington [unpublished]; R.E. Morlan, CMC database (1999). Check date on Beta-16164.
(12700 ± 270)	Beta-16164	left humerus (CRH-8)	Bowen I., Strait of Georgia (49°23.8'N, 123°21.8'W), BC, CAN	C.R. Harington [unpublished]; [R. McNeely, pers. comm. (1998) In: R.E. Morlan, CMC database (1999)]
12570 ± 70 (12720 ± 70)	Beta-115202	neck vertebra (Courtenay and District Museum 996.20.13b)	Lerwick Road site (49°43.06'N, 124°58.44'W), Courtenay, BC, CAN	C.R. Harington [unpublished]

Carnivora, Odobenidae

Odobenus rosmarus (walrus)

Measured Date (Normalized)	Lab No	Material	Location	References/Remarks
45630 ± 1580 (47720 ± 1580)	Beta-70839 CAMS-11917	mandible	Pauline Cove, Herschel I., YT, CAN	R.E. Morlan, CMC database (1999)
>40000 (>40000)	I-11017	vertebra and rib fragment (CMN 38490)	Qualicum Beach, Vancouver I. (49°23'00"N, 124°34'50"W), BC, CAN	[Harington and Beard (1992). Chronological age of the specimen has been determined through cementum counts to be 12 years.]; Dyke, et al. (1999); R.E. Morlan, CMC database (1999)
(12760 ± 90)	TO-1927	cranium	Bay of Fundy, 18 km off Black R. (45.000°N, 65.500°W), NB, CAN	Miller (1990); Bouchard, et al. (1993); Miller (1997); [Dyke, et al. (1999)
12390 ± 50 (12550 ± 55)	Beta-83414	cranium (CRH-95-18)	Tail of Grand Banks Pen. (44.552°N, 49.151°W), NF, CAN	Harington (1998a); Dyke, et al. (1999)
11490 ± 160 (11650 ± 160)	Beta-16518	rib fragment	Forteau (51.480°N, 56.975°W), NF (Labrador), CAN	Harington, et al. (1993); Dyke, et al. (1999)
(10270 ± 70)	Beta-89281	mandible (powdered bone) (NBMG 9105)	Bay of Fundy S of St. Martins (45.250°N, 65.500°W), NB, CAN	Miller (1997); Dyke, et al. (1999)
10090 ± 60 (10500 ± 60)	TO-2224	skull (MCQ 91-117)	Sainte-Julienne-de-Montcalm, 60 km N of Montreal (45.970°N, 73.700°W), QC, CAN	Bouchard, et al. (1993); Harington, et al. (1993); Dyke, et al. (1999)
(9980 ± 60)	Beta-69386	mandible (powdered bone) (NBMG 8621)	Bay of Fundy off Quaco Head (45.250°N, 65.500°W), NB, CAN	Miller (1997); Dyke, et al. (1999)
9960 ± 50 (10130 ± 50)	CAMS-43273	cranium	Les Capucins, Matane/Gaspesie region (49.042°N, 66.850°W), QC, CAN	Harington (1975); Dyke, et al. (1999)
9790 ± 60 (9940 ± 60)	Beta-115199	ilium shaft (CR-97-61)	Saint-Nicolas (46°42'N, 71°23'W), QC, CAN	C.R. Harington [unpublished]
9700 ± 130 (9860 ± 130)	Beta-16161	humerus (skeleton)	Moncton (46.100°N, 64.780°W), NB, CAN	Miller (1990); Bouchard, et al. (1993); Harington, et al. (1993); Dyke, et al. (1998, 1999) "Moncton Walrus"

9470 ± 110 (9630 ± 110)	S-3081	tusk	11.5 km SSW of the mouth of McBean Bay, 49.5 m asl., Baffin I. (72.540°N, 89.730°W), NU, CAN	Dyke, et al. (1999); [Dyke (2000c) GSC Map 1963A]
9380 ± 100 (9540 ± 100)	S-3086	tusk	7.5 km SE of the mouth of Fitzgerald Bay, 64 m asl., Baffin I. (72.020°N, 89.880°W), NU, CAN	Dyke, et al. (1999); [Dyke (2000c) GSC Map 1963A]
(9360 ± 90)	TO-1554	fragments from skull	Bay of Fundy (45.080°N, 66.750°W), NB, CAN	Miller (1990); Bouchard, et al. (1993); Harington, et al. (1993); R.E. Morlan, CMC database (1999)
9150 ± 60 (9300 ± 60)	CAMS-38289	tusk (CMN 34514)	9 km NW of mouth of Creswell R., 83 m asl., Somerset I. (72.861°N, 93.567°W), NU, CAN	Dyke, et al. (1999); R.E. Morlan, CMC database (1999)
(8850 ± 80)	TO-5960	tusk	Cape York: 27.5 m asl., Baffin I. (73.790°N, 87.000°W), NU, CAN	Dyke, et al. (1999); R.E. Morlan, CMC database (1999); [Dyke (2000d) GSC Map 1964A]
(8850 ± 80)	TO-5962	tusk	S of Cape Hardy, 44 m asl., Devon I. (75.770°N, 83.920°W), NU, CAN	Dyke (1998); Dyke, et al. (1999); R.E. Morlan, CMC database (1999)
8570 ± 100 (8690 ± 100)	GSC-2899	tusk and cranium	Coburg I., 5.1 m asl., W side of the isthmus connecting Marina Pen. with Coburg I. (75.875°N, 79.030°W), NU, CAN	Harington (1975); Blake (1987); Dyke, et al. (1999)
8570 ± 110 (8730 ± 110)	S-3087	tusk	7 km SSE of the mouth of Fitzgerald Bay, 17 m asl., Baffin I. (72.010°N, 89.950°W), NU, CAN	Dyke, et al. (1999); [Dyke (2000c) GSC Map 1963A]
8510 ± 50 (8670 ± 55)	Beta-83412	vertebra (CRH-95-16)	Cook's Harbour (51.542°N, 55.950°W), NF, CAN	Harington (1998) [given as 8670 ± 50 apply a delta 13 correction of -15]; Dyke, et al. (1999)
8480 ± 110 (8640 ± 110)	S-3092	tusk	15.5 km W of Stanley Pt., 24.5 m asl., Baffin I., (73.825°N, 85.580°W), NU, CAN	Dyke, et al. (1999); [Dyke (2000d) GSC Map 1964A]
8350 ± 80 (8510 ± 80)	S-3090	tusk	12.4 km W of Stanley Pt., 25 m asl., Baffin I. (73.820°N, 85.500°W), NU, CAN	Dyke, et al. (1999); [Dyke (2000d) GSC Map 1964A]
(7980 ± 40)	CAMS-38025	tusk	NW coast of Prescott I., 6.5 m asl., Prince of Wales I. (73.150°N, 97.000°W), NU, CAN	Dyke, et al. (1999)
7320 ± 120 (7480 ± 120)	I-7795	maxilliary (CMN 13747)	Bathurst I., 53.3 m asl. (75.717°N, 98.417°W), NU, CAN	Harington (1975); Harington, et al. (1993); Dyke, et al. (1999); R.E. Morlan, CMC database (1999)
7150 ± 160 (7290 ± 160)	S-3512	tusk	20 km NNE of Fort Ross, 41 m asl., Somerset I. (72.175°N, 94.030°W), NU, CAN	Dyke, et al. (1999)
7090 ± 130 (7300 ± 130)	CAMS-43271	tusk	Approx. 5 km NE of Freeport (44.333°N, 66.233°W), NS, CAN	Dyke, et al. (1998, 1999)
7070 ± 100 (7250 ± 100)	S-3641	tusk	Kuuruluk R., 34 m asl, Baffin I. (72.400°N, 86.300°W), NU, CAN	Dyke, et al. (1999); [Dyke (2000c) GSC Map 1963A]
6880 ± 50 (7050 ± 50)	CAMS-43274	tusk	Magdelen I. (47.267°N, 61.717°W), QC, CAN	Dyke, et al. (1999)
6870 ± 180 (7050 ± 180)	S-3527	right tusk	9.3 km SW of Firkin Pt., 19 m asl., Devon I. (75.540°N, 85.790°W), NU, CAN	Dyke (1998); Dyke, et al. (1999); R.E. Morlan, CMC database (1999)
6790 ± 100 (6980 ± 100)	S-3643	tusk	4.9 km SW of Firkin Pt., 13.5-14.5 m asl., Devon I. (75.570°N, 85.660°W), NU, CAN	Dyke (1998); Dyke, et al. (1999); R.E. Morlan, CMC database (1999)
6410 ± 50 (6590 ± 50)	CAMS-43269	tusk	Sable I. (43.933°N, 60.000°W), NS, CAN	Dyke, et al. (1998, 1999)
6200 ± 80 (6360 ± 89)	S-3093	tusk	3 km NE of the head of Thibault Bay, 38 m asl., Baffin I. (70.970°N, 89.120°W), NU, CAN	Dyke, et al. (1999); [Hooper and Dyke (2000) GSC Map 1959A]
5390 ± 380 (5550 ± 380)	RL-834	tusk	Skraeling I. Alexandria Fiord, 29 m asl., Ellesmere I. (78.900°N, 75.620°W), NU, CAN	[Schlederman (1978, 1990) In: R.E. Morlan, CMC database (1999)]; Dyke, et al. (1999)
5370 ± 50 (5550 ± 50)	CAMS-43268	tusk	Sable I. (43.767°N, 60.000°W), NS, CAN	Dyke, et al. (1999)

5000 ± 90 (5190 ± 90)	S-3642	tusk	14 km E of the head of Berlinguet Inlet, S shore, 200 m inland, 1 m asl., Baffin I. (70.980°N, 86.420°W), NU, CAN	Dyke, et al. (1999); R.E. Morlan, CMC database (1999); [Dyke and Hooper (2000) GSC Map 1960A]
(4940 ± 70)	TO-5016	tusk	5 km NW of Pt. Kendall, 41 m asl., Crown Prince Frederick I., Baffin I. (70.030°N, 86.600°W), NU, CAN	[Dyke, et al. (1998) In: R.E. Morlan, CMC database (1999)]; [Hooper and Dyke (2000) GSC Map 1959A]
4810 ± 90 (4970 ± 90)	S-3095	tusk	23 km SE of Easter Cape, 30 m asl., Baffin I. (70.730°N, 89.100°W), NU, CAN	Dyke, et al. (1999); [Hooper and Dyke (2000) GSC Map 1959A]
(4750 ± 60)	TO-5961	tusk	9 km NW of the mouth of Lyall R., 25 m asl., Grinnell Pen., Devon I. (77.055°N, 95.510°W), NU, CAN	[Dyke, et al. (1998) In: R.E. Morlan, CMC database (1999)]; Dyke, et al. (1999)
(4730 ± 50)	TO-4985	upper first molar	Bay NW of Lavoie Pt., Borden Pen., 10 m asl., Baffin I. (72.820°N, 80.440°W), NU, CAN	[Dyke (2000b) GSC Map 1962A]
(4670 ± 70)	CAMS-38434	tusk	3 km NW of Point Kendall, 26 m asl., Crown Prince Frederick I., Baffin I. (70.033°N, 86.550°W), NU, CAN	[Dyke, et al. (1998) In: R.E. Morlan, CMC database (1999)]; [Hooper and Dyke (2000) GSC Map 1959A]
(4530 ± 50)	TO-5008	tusk	McBean Bay, in raised beach gravel 45 m asl., Baffin I. (72.630°N, 89.720°W), NU, CAN	Dyke, et al. (1999); [Dyke (2000c) GSC Map 1963A]
(4500 ± 60)	CAMS-38433	tusk	5.5 km W of the E end of Kimakto Pen., 31.5 m asl., Baffin I. (70.283°N, 87.767°W), NU, CAN	Dyke, et al. (1999); [Hooper and Dyke (2000) GSC Map 1959A]
4440 ± 70 (4600 ± 75)	S-3094	tusk	5 km SE of Van Koenig Pt., 26 m asl., Baffin I. (71.050°N, 89.370°W), NU, CAN	Dyke, et al. (1999); [Dyke and Hooper (2000) GSC Map 1960A]
4320 ± 90 (4480 ± 90)	S-3101	tusk	Near Easter Cape, 23 m asl., Baffin I. (71.050°N, 89.030°W), NU, CAN	Dyke, et al. (1999); [Dyke and Hooper (2000) GSC Map 1960A]
3920 ± 130 (4080 ± 130)	K-1041	tusk	Igloolik I., 42-54 m asl (69.375°N, 81.800°W), NU, CAN	Dyke, et al. (1998, 1999); [Rainey and Ralph (1959); Meldgaard (1960, 1962); Wilmeth (1978); Maxwell (1985) In: R.E. Morlan, CMC database (1999)]
3890 ± 100 (4290 ± 100)	Ua-2350	tusk	Jakobshaven Isbrae, 60 m asl (68.950°N, 50.170°W), GRE	Weidick (1992); Bennike (1997), Dyke, et al. (1999)
3880 ± 130 (4040 ± 130)	K-1040	tusk	Igloolik I., 42-54 m asl (69.375°N, 81.800°W), NU, CAN	Dyke, et al. (1998, 1999); [Rainey and Ralph (1959); Meldgaard (1960, 1962); Wilmeth (1978); Maxwell (1985) In: R.E. Morlan, CMC database (1999)]
3870 ± 90 (4270 ± 90)	K-5978	tusk	Holsteinsborg (66.950°N, 53.630°W), GRE	Bennike (1997); Dyke, et al. (1999)
3700 ± 120 (3860 ± 120)	K-1077	tusk	Igloolik I. (Lyon Hill), 41 m asl (69.333°N, 81.800°W), NU, CAN	Dyke, et al. (1998, 1999); [O. Bennike, pers. comm. (1997) In: R.E. Morlan, CMC database (1999)]
3660 ± 95 (3820 ± 95)	I-9995	humerus shaft (CMN 26080)	Sable I. (43.420°N, 65.400°W), NS, CAN	Harington and Occhietti (1988); Bouchard, et al. (1993); Dyke, et al. (1999)
3480 ± 70 (3640 ± 75)	S-3080	tusk	11 km SSW of the mouth of McBean Bay, 10.2 m asl., Baffin I. (72.550°N, 89.750°W), NU, CAN	Dyke, et al. (1999); [Dyke (2000c) GSC Map 1963A]
(3480 ± 60)	TO-5063	tusk	6.1 km NW of Porden Pt., Prince Alfred Bay, 22-23 m asl., Devon I. (76.290°N, 93.810°W), NU, CAN	Dyke (1998)
3410 ± 50 (3510 ± 50)	GSC-2951	tusk	2.6 km SSE of Read Bay, 170 m asl., Cornwallis I. (75.020°N, 93.570°W), NU, CAN	Blake (1987); Dyke, et al. (1999); R.E. Morlan, CMC database (1999)
3250 ± 120 (3410 ± 120)	K-1042	tusk	Jens Munk I., Igloolik area 41 m asl (69.667°N, 80.000°W), NU, CAN	Dyke, et al. (1998, 1999); R.E. Morlan, CMC database (1999)
3130 ± 140 (3290 ± 140)	CAMS-43272	tusk	Sable I. (43.933°N, 60.000°W), NS, CAN	Dyke, et al. (1999)

2910 ± 110 (3070 ± 110)	K-1074	tusk	Igloolik I. (Tikilik) 23 m asl (69.333°N, 81.800°W), NU, CAN	Dyke, et al. (1998, 1999); [O. Bennike, pers. comm. (1997) In: R.E. Morlan, CMC database (1999)]
(2890 ± 40)	Beta-71157	cranium (powdered bone) (NBMG 4584)	Portage I., Miramichi Bay (47.170°N, 65.000°W), NB, CAN	Miller (1990, 1997); Dyke, et al. (1993)
2880 ± 120 (3040 ± 120)	K-1043	tusk	Jens Munk I., Igloolik area 24 m asl (69.667°N, 80.000°W), NU, CAN	Dyke, et al. (1998, 1999); R.E. Morlan, CMC database (1999)
2760 ± 110 (2920 ± 110)	K-1073	tusk	Igloolik I. (Freuchen) 22 m asl (69.333°N, 81.800°W), NU, CAN	Dyke, et al. (1998, 1999); [O. Bennike, pers. comm. (1997) In: R.E. Morlan, CMC database (1999)]
2670 ± 130 (2830 ± 130)	GSC-701	(CMC-117)	Sugluk I., Hudson Strait, Nouveau-Quebec region (62.260°N, 75.560°W), NT, CAN	[Lowdon, et al. (1969), Jackson and Pawson (1984) In: Dyke, et al. (1999)]; R.E. Morlan, CMC database (1999)
2660 ± 100 (2820 ± 100)	K-1075	tusk	Igloolik I. (Freuchen) 25 m asl (69.367°N, 81.800°W), NU, CAN	Dyke, et al. (1998, 1999); R.E. Morlan, CMC database (1999)
2630 ± 60 (3030 ± 60)	K-5979	tusk	Frederikshab Isblink (62.500°N, 50.170°W), GRE	Bennike (1997); Dyke, et al. (1999)
2630 ± 130 (2790 ± 130)	GSC-703	(CMC-116)	Sugluk I., Hudson Strait, Nouveau-Quebec region (62.260°N, 75.560°W), NT, CAN	[Lowdon, et al. (1969), Jackson and Pawson (1984) In: Dyke, et al. (1999)]; R.E. Morlan, CMC database (1999)
2620 ± 110 (2780 ± 110)	K-1076	tusk	Igloolik I. (Tikilik) 22 m asl (69.367°N, 81.800°W), NU, CAN	Dyke, et al. (1998, 1999); [O. Bennike, pers. comm. (1997) In: R.E. Morlan, CMC database (1999)]
2570 ± 110 (2730 ± 110)	K-1072	tusk	Igloolik I. (Tikilik) 19 m asl (69.367°N, 81.800°W), NU, CAN	Dyke, et al. (1998, 1999); R.E. Morlan, CMC database (1999)
2460 ± 110 (2620 ± 110)	K-1070	tusk	Igloolik I. (Tikilik) 18 m asl (69.367°N, 81.800°W), NU, CAN	Dyke, et al. (1998, 1999); R.E. Morlan, CMC database (1999)
2440 ± 180 (2450 ± 180)	GSC-3081	mandible (CMN 34510-B)	10 km SSW of Cape Anne, 103 m asl., Somerset I. (74.030°N, 94.850°W), NU, CAN	Blake (1987); Dyke, et al. (1999); R.E. Morlan, CMC database (1999)
2420 ± 65 (2580 ± 70)	S-1392	tusk	SW of Cape Anne, 103 m asl., Somerset I. (74°02'00"N, 94°51'00"W), NU, CAN	Dyke (1979); Dyke, et al. (1999); R.E. Morlan, CMC database (1999) .
2380 ± 110 (2540 ± 110)	K-1071	tusk	Igloolik I. (Freuchen) 18.5 m asl (69.367°N, 81.800°W), NU, CAN	Dyke, et al. (1998, 1999); R.E. Morlan, CMC database (1999)
2370 ± 110 (2530 ± 110)	K-1069	tusk	Igloolik I. (Tikilik) 17 m asl (69.367°N, 81.800°W), NU, CAN	Dyke, et al. (1998, 1999); R.E. Morlan, CMC database (1999)
2110 ± 120 (2270 ± 100)	K-1048	tusk	Igloolik (Alarnerk) 19 m asl (69.200°N, 81.333°W), NU, CAN	Dyke, et al. (1998, 1999); R.E. Morlan, CMC database (1999)
2010 ± 70 (2170 ± 70)	SI-3892		Avayalik I. 2, (60.500°N, 64.670°W), NF (Labrador), CAN	[Jordan (1980) In: Dyke, et al. (1999)]; R.E. Morlan, CMC database (1999)
1970 ± 80 (2180 ± 80)	CAMS-43270	cranium	Sable I. (43.933°N, 60.000°W), NS, CAN	Dyke, et al. (1999)
(1870 ± 60)	Lu-3660	tusk	Peary Land (82.130°N, 29.830°W), GRE	[Bennike (1997), corrected date = 1320 ± 60 BP]; Dyke, et al. (1999)
(1870 ± 40)	CAMS-38024	calcaneum	E of Claxton Pt., 60 m asl., SW Cornwallis I. (74.883°N, 95.683°W), NU, CAN	Dyke, et al. (1999)
(1870 ± 50)	CAMS-38290	two cheek teeth	Cape Charles York, Borden Pen., 6 m asl., Baffin I. (73.667°N, 82.967°W), NU, CAN	Dyke, et al. (1999); [Dyke (2000e) GSC Map 1965A]
1810 ± 110 (1970 ± 110)	K-1047	tusk	Igloolik (Alarnerk) 14 m asl (69.200°N, 81.333°W), NU, CAN	Dyke, et al. (1998, 1999); R.E. Morlan, CMC database (1999)
(1630 ± 50)	TO-5009	typanic bulla	Cape York, 10 m asl., Baffin I. (73.800°N, 87.050°W), NU, CAN	Dyke, et al. (1999); [Dyke (2000d) GSC Map 1964A]
1580 ± 50 (1740 ± 50)	S-3091	tusk	12.5 km W of Stanley Pt., 35 m asl., Baffin I. (73.820°N, 85.500°W), NU, CAN	Dyke, et al. (1999); [Dyke (2000d) GSC Map 1964A]
1520 ± 80 (1680 ± 80)	Beta-13866	rib	Cornwallis I., 76 m asl. (74°52.5'N, 95°42'W), NU, CAN	[R. McNeely (pers. comm. 1998 to R.E. Morlan)]; R.E. Morlan, CMC database (1999)
1490 ± 100 (1650 ± 100)	K-1068	tusk	Fury and Hecla Strait (Abverdjar) 8-10.5 m asl (69.750°N, 82.500°W), NU, CAN	Dyke, et al. (1998, 1999); R.E. Morlan, CMC database (1999)

1490 ± 170 (1680 ± 170)	S-3544	tusk	12 km ESE of Owen Pt., 37 m asl., Devon I. (75.960°N, 92.140°W), NU, CAN	Dyke (1998); Dyke, et al. (1999); R.E. Morlan, CMC database (1999)
1490 ± 75 (1890 ± 75)	K-5980	tusk	Peary Land (82.170°N, 30.180°W), GRE	[Bennike (1997), corrected date = 1340 ± 75 BP]; Dyke, et al. (1999)
1240 ± 100 (1400 ± 100)	K-1067	tusk	Fury and Hecla Strait (Abverdjar) 8 m asl (69.750°N, 82.500°W), NU, CAN	Dyke, et al. (1998, 1999); R.E. Morlan, CMC database (1999)
1240 ± 100 (1400 ± 100)	K-1046	tusk	Igloolik (Alarnerk) 11 m asl (69.200°N, 81.333°W), NU, CAN	Dyke, et al. (1998, 1999); R.E. Morlan, CMC database (1999)
1170 ± 100 (1330 ± 100)	K-1045	tusk	Igloolik (Alarnerk) 6.5 m asl (69.200°N, 81.333°W), NU, CAN	Dyke, et al. (1998, 1999); R.E. Morlan, CMC database (1999)
1150 ± 70 (1200 ± 70)	GSC-3108	femur	Knud Pen., Ellesmere I. (79.120°N, 76.150°W), NU, CAN	[Blake (1988) In: Dyke, et al. (1999)]; [Schledermann (1980, pers. comm. (1998); Morrison (1989); McCullough (1989); Bowman, et al. (1990) In: R.E. Morlan, CMC database (1999)]
(1150 ± 100)	BM-1754R		Knud Pen., 8-9 m asl., Ellesmere I. (79.120°N, 76.190°W), NU, CAN	Dyke, et al. (1999); [Bowman, et al. (1990), Schledermann (199) In: R.E. Morlan, CMC database (1999), revision of BM-1754]
(1150 ± 100)	K-1488	tusk	Nares Strait area (78.830°N, 69.250°W), AK, USA	[Tauber (1973) In: Dyke, et al. (1999)]
(1135 ± 40)	BM-1754		Knud Pen., Ellesmere I. (79.120°N, 76.190°W), NU, CAN	[Burleigh et al. (1982) In: Dyke, et al. (1999)]
(1120 ± 100)	K-1487	tusk	Nares Strait area (78.830°N, 69.250°W), AK, USA	[Tauber (1973) In: Dyke, et al. (1999)]
1100 ± 100 (1260 ± 100)	K-1044	tusk	K'aersut I., 6.5 m asl., Igloolik area (69.483°N, 80.317°W), NU, CAN	Dyke, et al. (1998, 1999); R.E. Morlan, CMC database (1999)
(1040 ± 100)	K-1078	tusk	NW region (76.780°N, 70.330°W), GRE	[Tauber (1973) In: Dyke, et al. (1999)]
(1010 ± 100)	K-1080	tusk	NW region (76.780°N, 70.330°W), GRE	[Tauber (1973) In: Dyke, et al. (1999)]
(990 ± 60)	CAMS-38435	tusk	Little Cornwallis I., 70 m asl. (75.417°N, 96.667°W), NU, CAN	Thorsteinsson (1958); Dyke, et al. (1999)
(910 ± 50)	CAMS-39511	tusk	15.5 km S of Bere Bay, Belcher Ch. coast of Grinnell Pen., 250 m asl., Devon I. (76.748°N, 94.207°W), NU, CAN	Dyke, et al. (1999)
(720 ± 50)	CAMS-42768	cranium	Cow Head Harbour, Gros Morne National Pk. (49.917°N, 57.800°W), NF, CAN	Dyke, et al. (1998, 1999)
710 ± 40 (870 ± 40)	CAMS-43266	cranium	La Chaloupe, Cote-Nord region (50.283°N, 65.133°W), QC, CAN	Harington (1975); Dyke, et al. (1999)
(560 ± 100)	BM-1753R	vertebra	Knud Pen., Ellesmere I. (79.120°N, 76.190°W), NU, CAN	[Schledermann (1980), McCullough (1989), Morrison (1989), Bowman et al. (1990) In: R.E. Morlan, CMC database (1999), revision of BM-1753]; Dyke, et al., (1999)
470 ± 60 (630 ± 60)	Beta-61073	mandible	Willows I., Baffin I. (62.767°N, 65.467°W), NU, CAN	[Fitzhugh (1993) In: R.E. Morlan, CMC database (1999)]; [Manley and Jennings (1996) In: Dyke, et al. (1999)]
430 ± 110 (590 ± 110)	Beta-8972	(BCPM 467)	Victoria (48.420°N, 123.280°W), BC, CAN	C.R. Harington [unpublished]; R.E. Morlan, CMC database (1999)
(360 ± 25)	BM-1753	vertebra	Knud Pen., Ellesmere I. (79.120°N, 76.190°W), NU, CAN	[Burleigh et al. (1982) In: Dyke, et al. (1999)]
(210 ± 100)	BM 1751R	radius	RCMP station, Alexandra Fiord, Bache Pen. area, Ellesmere I. (79.120°N, 76.190°W), NU, CAN	[Bowman et al. (1990) In: Dyke, et al. (1999), revision of BM-1751]
180 ± 90 (340 ± 90)	Beta-13865		Little Cornwallis I., 84 m asl. (75°27.4'N, 96°39'W), NU, CAN	[R. McNeely (pers. comm. 1998 to R.E. Morlan), date probably correct on relatively recent inland "crawler"]; R.E. Morlan, CMC database (1999)
(155 ± 40)	BM-1751	radius	RCMP station, Alexandra Fiord, Bache Pen. area, Ellesmere I. (79.120°N, 76.190°W), NU, CAN	[Burleigh et al. (1982) In: Dyke, et al. (1999)]

Carnivora, Phocidae

Phoca hispida (ringed seal)

Measured Date (Normalized)	Lab No	Material	Location	References/Remarks
41910 ± 800 (42320 ± 800)	TO-2294	innominate and scapula	10 km S of Worth Pt., 2 m asl., Banks I. (72.175°N, 125.660°W), NT, CAN	R.E. Morlan, CMC database (1999)
(20660 ± 80)	CAMS-33980	ulna	On Your Knees Cave, Prince of Wales I., 135 m asl. (56.250°N, 133.500°W), AK, USA	Heaton, et al. (1996), [(1997 incorrectly cited as 20,670 ± 80 BP); see AA-18450 for original date on same specimen]; Dixon, et al. (1997)
(20060 ± 500)	AA-22884	humerus	On Your Knees Cave, Prince of Wales I., 135 m asl. (56.250°N, 133.500°W), AK, USA	Heaton and Grady (1997)
(19060 ± 275)	AA-18450R	ulna	On Your Knees Cave, Prince of Wales I., 135 m asl. (56.250°N, 133.500°W), AK, USA	Heaton, et al. (1996)
(17565 ± 160)	AA-18450	ulna	On Your Knees Cave, Prince of Wales I., 135 m asl. (56.250°N, 133.500°W), AK, USA	[Heaton, et al. (1996, 1997), same specimen redated under CAMS-33980 at 20,660 ± 80 BP]; Dixon, et al. (1997)
(13690 ± 240)	AA-21564	humerus	On Your Knees Cave, Prince of Wales I., 135 m asl. (56.250°N, 133.500°W), AK, USA	Heaton and Grady (1996, 1997); Dixon, et al. (1997)
10420 ± 50 (10520 ± 50)	Beta-92375	left humerus (CMN 41849)	NW corner of Natural Heritage Bldg., Aylmer (45.44412°N, 75.80947°W), QC, CAN	C.R. Harington [unpublished]; R.E. Morlan, CMC database (1999)
8680 ± 120	K-4687	skeleton (GGU 313897)	Eastern Hall Land (81.500°N, 60.000°W), GRE	Bennike, et al. (1989)
(4470 ± 150)	QL-4129	fibula	Resolute Bay, 16 m asl., Cornwallis I. (74.735°N, 94.735°W), NU, CAN	[Washburn (1997) In: R.E. Morlan, CMC database (1999)]
4450 ± 60 (4610 ± 65)	S-1660	long-bones (CMC-1025)	Grinnell Pen., 24 m asl., Devon I. (76.320°N, 94.630°W), NU, CAN	Dyke (1998); [Wilmeth (1978); McGhee (1979, 1980) In: R.E. Morlan, CMC database (1999)]
4160 ± 50 (4260 ± 50)	Beta-115208	rib (CMN 43817)	Clements Markham Inlet, Ellesmere I. (82°36.5'N, 68°31'W), NU, CAN	C.R. Harington [unpublished]
3790 ± 60 (3950 ± 65)	S-1662	long-bones (CMC-1027)	Grinnell Pen., 24 m asl., Devon I. (76.320°N, 94.630°W), NU, CAN	Dyke (1998); [Wilmeth (1978); McGhee (1979, 1980) In: R.E. Morlan, CMC database (1999)]
3430 ± 60 (3590 ± 65)	S-1661	long-bones (CMC-1026)	Grinnell Pen., 24 m asl., Devon I. (76.320°N, 94.630°W), NU, CAN	Dyke (1998); [Wilmeth (1978); McGhee (1979, 1980) In: R.E. Morlan, CMC database (1999)]
2895 ± 100 (3055 ± 100)	S-521		E shore of Button Bay, 29-35 m asl., Churchill R., MB, CAN	[Nash (1969); Meyer (1977); Wilmeth (1978) In: R.E. Morlan, CMC database (1999)]
2070 ± 50 (2230 ± 55)	S-1689	long-bones (CMC-1024)	Grinnell Pen., 24 m asl., Devon I. (76.320°N, 94.630°W), NU, CAN	Dyke (1998); [Wilmeth (1978); McGhee (1979, 1980) In: R.E. Morlan, CMC database (1999)]
1765 ± 90	SI-2160		Walakpa, NW coast (71.130°N, 157.100°W), AK, USA	[Savelle and McCartney (1988) In: Dyke, et al. (1999)]
1695 ± 95	SI-2159		Walakpa, NW coast (71.130°N, 157.100°W), AK, USA	[Savelle and McCartney (1988) In: Dyke, et al. (1999)]
1555 ± 95	SI-2162		Walakpa, NW coast (71.130°N, 157.100°W), AK, USA	[Savelle and McCartney (1988) In: Dyke, et al. (1999)]
1310 ± 70 (1470 ± 75)	S-1424	ribs (CMC-940)	Grinnell Pen., 4 m asl., Devon I. (76.250°N, 92.670°W), NU, CAN	[Park (1989) In: A.S. Dyke (GSC database)]; Dyke (1998); [Morrison (1989) In: R.E. Morlan, CMC database (1999)]
1070 ± 90	SI-2161		Walakpa, NW coast (71.130°N, 157.100°W), AK, USA	[Savelle and McCartney (1988) In: Dyke, et al. (1999)]
780 ± 90 (940 ± 90)	RL-288		On an island in the upper part of Cumberland Sound, 8 km W of Immigen, Baffin I., 11.5 m asl, NU, CAN	[Schledermann (1975, 1979); Morrison (1989) In: R.E. Morlan, CMC database (1999)]

Phoca sp. (seal)

Measured Date (Normalized)	Lab No	Material	Location	References/Remarks
5910 ± 195 (6070 ± 195)	S-2665	humerus	N-central Russell I., 39.9 m asl. (74°4.8'00"N, 98°09'00"W), NU, CAN	Dyke, et al. (1991); Dyke (1998); R.E. Morlan, CMC database (1999)
(5260 ± 60)	TO-3710	head of left humerus (CR-92-200)	Near Cape Weynton on Committee Bay (67°44'N, 88°08'W), NU, CAN	C.R. Harington [unpublished]
(3570 ± 80)	K-5127		Disko area, 5.5 m asl. (68.583°N, 51.083°W), GRE	[Gronnow (1994) In: Dyke, et al. (1999)]
(1670 ± 40)	Gif-8179		W end of Naujat Pen., Victoria I., NU, CAN	[Le Mouel (1999) In: R.E. Morlan, CMC database (1999)]
(1480 ± 70)	Gif-8373		W end of Naujat Pen., Victoria I., NU, CAN	[Le Mouel (1999) In: R.E. Morlan, CMC database (1999)]
(1470 ± 80)	Gif-8375		W end of Naujat Pen., Victoria I., NU, CAN	[Le Mouel (1999) In: R.E. Morlan, CMC database (1999)]
(1430 ± 90)	Gif-8374		W end of Naujat Pen., Victoria I., NU, CAN	[Le Mouel (1999) In: R.E. Morlan, CMC database (1999)]
(1270 ± 40)	Gif-8806		W end of Naujat Pen., Victoria I., NU, CAN	[Le Mouel (1999) In: R.E. Morlan, CMC database (1999)]
1240 ± 90 (1360 ± 90)	SI-2158		Walakpa, NW coast (71.130°N, 157.100°W), AK, USA	[Savelle and McCartney (1988) In: Dyke, et al. (1999)]
(980 ± 40 (1030AD), 1015-	K-3200		Niaquusat, Norse farm V48, Nuuk Municipality, GRE	[Sorensen (1982); Arneborg (1991) In: Danish/Greenlandic Network (1999)]
(920 ± 50 (1065-1160AD), 1020-	K-3198		Niaquusat, Norse farm V48, Nuuk Municipality, GRE	[Sorensen (1982); Arneborg (1991) In: Danish/Greenlandic Network (1999)]
(870 ± 40 (1195AD), 1160-	K-3202		Niaquusat, Norse farm V48, Nuuk Municipality, GRE	[Sorensen (1982); Arneborg (1991) In: Danish/Greenlandic Network (1999)]
(820 ± 65 (1230AD), 1170-	K-4739		Niaquusat, Norse farm V48, Nuuk Municipality, GRE	[Arneborg (1991) In: Danish/Greenlandic Network (1999)]
(800 ± 65 (1255AD), 1210-	K-4737		Niaquusat, Norse farm V48, Nuuk Municipality, GRE	[Arneborg (1991) In: Danish/Greenlandic Network (1999)]
(790 ± 65 (1265AD), 1215-	K-4732		Niaquusat, Norse farm V48, Nuuk Municipality, GRE	[Arneborg (1991) In: Danish/Greenlandic Network (1999)]
(780 ± 65 (1275AD), 1220-	K-4736		Niaquusat, Norse farm V48, Nuuk Municipality, GRE	[Arneborg (1991) In: Danish/Greenlandic Network (1999)]
(780 ± 65 (1275AD), 1220-	K-4730		Niaquusat, Norse farm V48, Nuuk Municipality, GRE	[Arneborg (1991) In: Danish/Greenlandic Network (1999)]
(670 ± 50 (1300AD), 1290-	K-4741		Niaquusat, Norse farm V48, Nuuk Municipality, GRE	[Arneborg (1991) In: Danish/Greenlandic Network (1999)]
620 ± 235 (780 ± 235)	S-2663		NE Russell I., 25.4 m asl. (74°06'00"N, 97°59.5'00"W), NU, CAN	Dyke, et al. (1991); R.E. Morlan, CMC database (1999)
(560 ± 50 (1405AD), 1320-	K-3204		Niaquusat, Norse farm V48, Nuuk Municipality, GRE	[Arneborg (1991) In: Danish/Greenlandic Network (1999)]
355 ± 115 (515 ± 115)	S-2669	hip bones	3 km S of Cape Walker, 81.5 m asl., E Russell I. (74°2.5'00"N, 97°38'00"W), NU, CAN	Dyke, et al. (1991); Dyke (1998); R.E. Morlan, CMC database (1999)

Erignathus barbatus (bearded seal)

Measured Date (Normalized)	Lab No	Material	Location	References/Remarks
10710 ± 80 (11120 ± 80)	TO-99	atlas vertebra fragment (ROM 985)	McWilliams' gravel pit, near Finch, Stormont County (45.130°N, 75.080°W), ON, CAN	Harington and Occhietti (1988); [R.E. Morlan, pers. comm. (1998) indicates a corrected date of 11,120 ± 80 BP using delta 13C correction = 25o/oo)]; R.E. Morlan, CMC database (1999)
10200 ± 40 (10410 ± 40)	Beta-153816	vertebra (CR-00-10)	Near La Durantaye (46°50'N, 70°51'W), QC, CAN	Harington and Locat (in prep.)

Proboscidea, Mammutidae

Mammut americanum (American mastodon)

Measured Date (Normalized)	Lab No	Material	Location	References/Remarks
(32000 ± 630)	GSC-1220	femur	Lower Middle R., Cape Breton I. (46.130°N, 60.922°W), NS, CAN	Blake (1984); R.E. Morlan, CMC database (1999)
(31300 ± 500)	GSC-1220-2	femur	Lower Middle R., Cape Breton I. (46.130°N, 60.922°W), NS, CAN	Blake (1984); R.E. Morlan, CMC database (1999)
24980 ± 1300 (25060 ± 1300)	Beta-16163	molar	Miller Ck., Sixtymile area (Sixtymile Loc. 5) (63°59'N, 140°47'W), YT, CAN	Harington (1989, 1997); R.E. Morlan, CMC database (1999)
(13600 ± 220)	GSC-1222		Hillsborough area, 30 m asl., 22 km S of Moncton, NB, CAN	Blake (1983); Harington (1990); Harington, et al. (1993); R.E. Morlan, CMC database (1999) "Hillsborough Mastodon"
11380 ± 170 (11460 ± 170)	GSC-611		2.5 km NE of Thamesville (42.570°N, 81.960°W), ON, CAN	Dreimanis (1967, 1968); Lowdon and Blake (1968); McAndrews and Jackson (1988)
8910 ± 150 (8990 ± 150)	GSC-614		3.2 km SSW of Tupperville (42.590°N, 82.285°W), ON, CAN	Dreimanis (1967, 1968); Lowdon and Blake (1968); [McAndrews and Jackson (1988) questionable dates]; [Jackson (1988) In: R.E. Morlan, CMC database (1999)

Mammut sp. (American mastodon)

Measured Date (Normalized)	Lab No	Material	Location	References/Remarks
(51500 ± 1270)	GSC-2467	faeces	Hillsborough area, 22 km S of Moncton, 30 m asl., NB, CAN	Harington (1990); Harington, et al. (1993) [Wood in dung balls of "Hillsborough Mastodon"]
>49980	Beta-173288	root of lower molar (LM3) (MAM 1996.53)	Pointe de Chambord (48°27.8'N, 72°05.4'W), QC, CAN	C.R. Harington [unpublished]
>47410	Beta-167189	lower molar (RM3) dentine (uncatalogued)	SW margin Great Bear Lake (65°25.60'N, 122°47.90'W), NT, CAN	C.R. Harington [unpublished]

Proboscidea, Elephantidae

Mammuthus imperator (imperial mammoth)

Measured Date (Normalized)	Lab No	Material	Location	References/Remarks
10600 ± 140 (10680 ± 145)	S-232	femur	S Saskatchewan drainage, near Wiseton (51.320°N, 107.650°W), SK, CAN	Agenbroad (1984); R.E. Morlan, CMC database (1999)

Mammuthus columbi (Columbian mammoth)

Measured Date (Normalized)	Lab No	Material	Location	References/Remarks
20190 ± 400 (29270 ± 400)	I-10971	mandible (CMN 29270)	Quartz Ck. (Dawson Loc. 8) (64.000°N, 139.000°W), YT, CAN	Harington (1977, 1980); R.E. Morlan, CMC database (1999)
12000 ± 200 (12035 ± 200)	S-246		S Saskatchewan drainage, near Kyle (50.830°N, 108.120°W), SK, CAN	Harington (1978); Agenbroad (1984); Morlan (1993); R.E. Morlan, CMC database (1999) "Kyle Mammoth"
(10930 ± 100)	TO-8514	lower molar (M3)	Bindloss gravel pit (50°57'N, 110°08'W), AB, CAN	Hills and Harington (in press)
8650 ± 400 (8685 ± 400)	A-619	apatite extraction	S Saskatchewan drainage, near Kyle, SK, CAN	[Agenbroad (1984) date incorrectly cited at 8680 ± 400 BP]; Morlan (1993); R.E. Morlan, CMC database (1999) "Kyle Mammoth" [See S-246]

Mammuthus cf. Mammuthus columbi (Columbian mammoth)

Measured Date (Normalized)	Lab No	Material	Location	References/Remarks
33900 ± 690 (34000 ± 690)	GSC-1754	anterior rib fragment (CMN 17915)	Noranda open pit mine on Newman Pen., Babine L. (55.000°N, 126.230°W), BC, CAN	Harington, et al. (1974); Harington (1978); Lowdon and Blake (1979); Clague (1980); Agenbroad (1984); R.E. Morlan, CMC database (1999)

Mammuthus primigenius (woolly mammoth)

Measured Date (Normalized)	Lab No	Material	Location	References/Remarks
(41400 ± 1500)	CAMS-66371	tooth	North Head, Richards I (69°34'N, 134°30'W), NT, CAN	A.S. Dyke, pers. comm. (2000)
(39800 ± 1200)	CAMS-66370	tooth	Near Inuvik (68°00'N, 134°00'W), NT, CAN	A.S. Dyke, pers. comm. (2000)
37220 ± 830 (37300 ± 830)	Lu-3010	right humerus (CMN 49952c)	Quartz Ck., about 595 m asl. (Dawson Loc. 60), YT, CAN	C.R. Harington [unpublished]; R.E. Morlan, CMC database (1999)
32700 ± 980	St-1632	hair from skull	Dome Ck., near Fairbanks (64.830°N, 147.500°W), AK, USA	Gillespie (1970); Péwé (1966, 1975b); Harington (1978); Agenbroad (1984); [Guthrie (1985, 1990) specimen incorrectly cited within Alaskan Pleistocene Bison Table]
32250 ± 1750 (32330 ± 1750)	I-4226	thoracic vertebra	Gold Run Ck. (Dawson Loc. 32) (63.725°N, 138.680°W), YT, CAN	Harington and Clulow (1973); Harington (1977), [1978, 1989 incorrectly cited as 32,350 ± 1750 BP]; Agenbroad (1984); R.E. Morlan, CMC database (1999)
30300 ± 2000 (30380 ± 2000)	I-3576	rib (most of skeleton)	Whitestone R. (Loc. 43) 400 m asl. (66.000°N, 138.000°W), YT, CAN	Harington (1978, 1980); Agenbroad (1984); Guthrie (1985, 1990); R.E. Morlan, CMC database (1999) "Whitestone Mammoth"
(27400 ± 580)	GSC-2034	upper molar (CMN 17679)	Ostero Gravel Pit: near Taylor (56°06.8'N, 121°47.9'W), BC, CAN	Harington (1978); Harington and Shackleton (1978); Mathews (1978b); Churcher and Wilson (1979); Lowdon and Blake (1979); Clague (1980); Agenbroad (1984); Bobrowsky, et al. (1991); R.E. Morlan, CMC database (1999)
(26750 ± 790)	AECV-1102C	humerus	Consolidated Concrete pit 48, 670 m asl., Edmonton area (53°38.5'N, 113°17'W), AB, CAN	Burns and Young (1994); Young et al. (1994); Burns (1996); R.E. Morlan, CMC database (1999)
(23910 ± 200)	RIDDL-224	limb	Bluefish Cave II, 600 m asl. (67°08'N, 140°47'W), YT, CAN	Cinq-Mars (1979, 1990); Morlan and Cinq-Mars (1982); Cinq-Mars, et al. (1991); Burke and Cinq-Mars (1996)
23620 ± 110	USGS-1438	jaw	Epiguruk (67.070°N, 158.170°W), AK, USA	Hamilton, et al. (1993)
(23200 ± 250)	RIDDL-225	limb	Bluefish Cave II, 600 m asl. (67°08'N, 140°47'W), YT, CAN	Cinq-Mars (1979, 1990); Morlan and Cinq-Mars (1982); Cinq-Mars, et al. (1991); Burke and Cinq-Mars (1996)
(22740 ± 90)	CAMS-23470	limb-fragment	Bluefish Cave II, 600 m asl. (67°08'N, 140°47'W), YT, CAN	Cinq-Mars (1979, 1990); Morlan and Cinq-Mars (1982); Cinq-Mars, et al. (1991); Burke and Cinq-Mars (1996)
(22430 ± 260)	RIDDL-558	rib	Bluefish Cave III, 600 m asl. (67°08'N, 140°47'W), YT, CAN	Cinq-Mars (1979, 1990); Morlan and Cinq-Mars (1982); Harington and Cinq-Mars (1995); Burke and Cinq-Mars (1996)
21300 ± 1300	L-601	dessicated flesh of a partial carcass of baby mammoth (F:AM 99927)	Fairbanks area (65.050°N, 147.170°W), AK, USA	Farrand (1961); Péwé (1975); Zimmerman and Tedford (1976); Harington (1980); Agenbroad 1984; [Guthrie (1985, 1990) specimen incorrectly cited as Alaskan Pleistocene Bison within Table]
(20230 ± 180)	RIDDL-223	scapula	Bluefish Cave II, 600 m asl. (67°08'N, 140°47'W), YT, CAN	Cinq-Mars (1979, 1990); Morlan and Cinq-Mars (1982); Cinq-Mars, et al. (1991); Burke and Cinq-Mars (1996)
(19640 ± 170)	RIDDL-330	scapula	Bluefish Cave II, 600 m asl. (67°08'N, 140°47'W), YT, CAN	Cinq-Mars (1979, 1990); Morlan and Cinq-Mars (1982); Cinq-Mars, et al. (1991); Burke and Cinq-Mars (1996)
17950 ± 120 (18030 ± 120)	Beta-70099	(GR 1)	Gold Run Ck. (approx. 63°42'N, 138°36'W), YT, CAN	C.R. Harington [unpublished]; R.E. Morlan, CMC database (1999)
17900 ± 500	CRNL-1221-a	scapula	Bluefish Cave II, 600 m asl. (67°08'N, 140°47'W), YT, CAN	Cinq-Mars (1979, 1990); Morlan and Cinq-Mars (1982); Cinq-Mars, et al. (1991); (Burke and Cinq-Mars (1996)

17880 ± 470 (17880 ± 330)	CRNL-1221	scapula (CMC-1221)	Bluefish Cave II, 600 m asl. (67°08'N, 140°47'W), YT, CAN	Cinq-Mars (1979, 1990); Morlan and Cinq-Mars (1982); Cinq-Mars, et al. (1991); Burke and Cinq-Mars (1996); [(Normalized age is based on an average of two runs: see next two entries) R.E. Morlan, CMC database (1999)]
17860 ± 440	CRNL-1221-b	scapula	Bluefish Cave II, 600 m asl. (67°08'N, 140°47'W), YT, CAN	Cinq-Mars (1979, 1990); Morlan and Cinq-Mars (1982); Cinq-Mars, et al. (1991); (Burke and Cinq-Mars (1996)
15380 ± 300	SI-453	flesh from lower leg	Fairbanks Ck. (65.067°N, 147.167°W), AK, USA	Péwé (1975); Agenbroad (1984); [Guthrie (1985, 1990) specimen cited incorrectly as Alaskan Pleistocene Bison within Table]; Dixon (1993)
15090 ± 170	Beta-5691	lumbar vertebra	Colorado Ck., McGrath (63°38'N, 156°01'W), AK, USA	[Guthrie (1985, 1990) - referred to as B-5691]; Dixon (1993)
(13940 ± 160)	RIDDL-559	humerus	Bluefish Cave I, 600 m asl. (67°08'N, 140°47'W), YT, CAN	Cinq-Mars (1979, 1990); Morlan and Cinq-Mars (1982); Burke and Cinq-Mars (1996)
13820 ± 340 (13900 ± 340)	Beta-13867	long-bone shaft (CMN 43522)	Old Crow R. (Old Crow Loc. 92) (67.850°N, 139.970°W), YT, CAN	Harington (1989); R.E. Morlan, CMC database (1999)
10790 ± 150 (10870 ± 150)	WAT-999	tusk	Rostock, 15 km N of Stratford (43.500°N, 81.000°W), ON, CAN	Pilny, et al. (1987); [McAndrews and Jackson (1988) portion of skull with molar, vertebra, scapula, limb and tusk fragments]; R.E. Morlan, CMC database (1999)

Mammuthus cf. Mammuthus primigenius (woolly mammoth)

Measured Date (Normalized)	Lab No	Material	Location	References/Remarks
29100 +3000 /-2000 (29245 +	GX-1567	radius (CMC 330)	Old Crow (Loc. 14N) (67°51'N, 139°46'W), YT, CAN	Irving and Harington (1973); Morlan and Bonnichsen (1975) Harington (1977); Agenbroad (1984); Morlan (1986); [R.E. Morlan, pers. comm. (1998) indicates a corrected date of 29,000 +3000 / -2000 BP]; R.E. Morlan, CMC database (1999)
25680 ± 580 (25760 ± 580)	I-8583	left metatarsal IV (CMN 25996)	Hunker Ck. (Dawson Loc. 10) (64.070°N, 139.370°W), YT, CAN	Harington (1980, 1997); R.E. Morlan, CMC database (1999)
19440 ± 290 (19520 ± 290)	I-8578		Tununuk (69.000°N, 134.670°W), NT, CAN	Harington (1978); Agenbroad (1984); Vincent (1989); Guthrie (1985, 1990); Dallimore, et al. (1997); R.E. Morlan, CMC database (1999); [Specimen in Northern Life Museum, Fort Smith, NT].
15500 ± 130 (15540 ± 130)	GSC-3053	scapula	Bluefish Cave II, 600 m asl. (67°08'N, 140°47'W), YT, CAN	Cinq-Mars (1979, 1990); Morlan and Cinq-Mars (1982); Harington (1989); Morlan (1989); Cinq-Mars (1991); [Guthrie (1985, 1990) & Dixon (1993) incorrectly cite the date as 15,550 ± 130 BP]; R.E. Morlan, CMC database (1999)
8280 ± 60	Beta-115204	molar (CR-97-64)	Kendall I. (approx. 69°29'N, 135°15'W), NT, CAN	C.R. Harington [unpublished]

Mammuthus sp. (mammoth)

Measured Date (Normalized)	Lab No	Material	Location	References/Remarks
>50800	Beta-79859 CAMS-18422	tusk (CRH-95-10)	4 km E of North Star Harbour, Harrowby Bay (70°14'N, 127°34'W), NT, CAN	C.R. Harington [unpublished]
49600 ± 3600 (49620 ± 3600)	Beta-79851 CAMS-18414	rib shaft (CRH-95-2) (CMN 14195)	Old Crow R. (Old Crow Loc. 12) (67.750°N, 139.750°W), YT, CAN	C.R. Harington [unpublished]; Morlan et al. (1990); R.E. Morlan, CMC database (1999)
(>42000)	SI-2823 CO	patella	Old Crow R. (Old Crow Loc. 11(A)), 259 m asl. (67.850°N, 139.970°W), YT, CAN	Harington (1977); [Irving, et al. (1977); Beebe (1980); Morlan (1980) In: R.E. Morlan, CMC database (1999)]
(>42000)	SI-2814 CO	left patella	Old Crow R. (Old Crow Loc. 11(A)), 259 m asl. (67.850°N, 139.970°W), YT, CAN	Harington (1977); [Irving, et al. (1977); Beebe (1980); Morlan (1980) In: R.E. Morlan, CMC database (1999)]

(40100 ± 680)	TO-2697	left magnum (CMN 12079)	S side of Liverpool Bay (69°36'40"N, 130°00'26"W), NT, CAN	C.R. Harington [unpublished]
(39500 ± 1600)	RIDDL-127	scapula	Johnson Ck. (Old Crow Loc. 71) (67.917°N, 139.750°W), YT, CAN	Morlan et al. (1990); R.E. Morlan, CMC database (1999)
(39360 ± 3950)	AECV-718C	tusk	Consolidated Pit 46: Villeneuve, N Saskatchewan R., NW of Edmonton (53.660°N, 113.820°W), AB, CAN	Young, et al. (1994); Burns (1996b); R.E. Morlan, CMC database (1999)
(38960 ± 3520)	AECV-935C		Consolidated Pit 48, N Saskatchewan R. Edmonton (53°38.5'N, 113°17'W), AB, CAN	Young, et al. (1994); Burns and Young (1994); Burns (1996); R.E. Morlan, CMC database (1999)
>38800	W-1132		Sullivan Ck. (65°05'N, 150°53'W), AK, USA	Agenbroad (1984)
(38450 ± 1400)	RIDDL-195	limb (collagen extraction method 1)	Old Crow R. (Old Crow Loc. 12) (67.750°N, 139.750°W), YT, CAN	Morlan et al. (1990); R.E. Morlan, CMC database (1999). [Suggested age for this bone is 38,325 ± 910 BP (average of RIDDL-195 and 307)]
(38200 ± 1200)	RIDDL-307	limb (collagen extraction method 2)	Old Crow R. (Old Crow Loc. 12) (67.750°N, 139.750°W), YT, CAN	Morlan et al. (1990); R.E. Morlan, CMC database (1999). [Suggested age for this bone is 38,325 ± 910 (average of RIDDL-195 and 307)]
(37800 ± 800)	RIDDL-733	mandible	Old Crow R. (Loc. 47) (68.250°N, 140.500°W), YT, CAN	Morlan et al. (1990); [Westgate, et al. (1995) In: R.E. Morlan, CMC database (1999)]
(37700 ± 1300)	RIDDL-124	limb	Old Crow R. (Old Crow Loc. 85) (68.250°N, 140.167°W), YT, CAN	Morlan et al. (1990); R.E. Morlan, CMC database (1999)
(37500 ± 1200)	RIDDL-129	limb	Old Crow R. (Old Crow Loc. 22) (68.083°N, 139.500°W), YT, CAN	Harington (1977); Morlan et al. (1990); R.E. Morlan, CMC database (1999)
(37120 ± 2370)	AECV-934C		Consolidated Pit 48, N Saskatchewan R. Edmonton (53°38.5'N, 113°17'W), AB, CAN	Young, et al. (1994); Burns and Young (1994); Burns (1996); R.E. Morlan, CMC database (1999)
36650 ± 1300	SI-2816 CO	humerus	Old Crow R. (Old Crow Loc. 11(A)), 259 m asl. (67.850°N, 139.970°W), YT, CAN	C.R. Harington (card file)
(36500 ± 1000)	RIDDL-194	limb	Johnson Ck. (Old Crow Loc. 71) (67.917°N, 139.750°W), YT, CAN	Morlan et al. (1990); R.E. Morlan, CMC database (1999)
(35800 ± 1000)	RIDDL-123	limb	Old Crow R. (Old Crow Loc. 13) (67.917°N, 139.750°W), YT, CAN	Morlan et al. (1990); R.E. Morlan, CMC database (1999)
35500 ± ?	I-9373		Ikpikpuk R. (70°N, 154°45'W), AK, USA	Guthrie (1984)
(35400 ± 900)	RIDDL-189	limb (collagen extraction method 1)	Old Crow R. (Old Crow Loc. 69) (67.855°N, 139.800°W), YT, CAN	Harington (1977); Morlan et al. (1990); R.E. Morlan, CMC database (1999). [Suggested age for this bone is 34,400 ± (average of RIDDL-189 and 304)]
(>35000)	S-224	tusk	Clover Bar Pit, E Edmonton, N Saskatchewan R. (53°35'N, 113°19.5'W), AB, CAN	Young et al. (1994); Burns and Young (1994); Burns (1996); R.E. Morlan, CMC database (1999)
>35000	I-9342		Ikpikpuk R. (70°N, 154°45'W), AK, USA	Guthrie (1985, 1990)
34620 ± 560	USGS-1440		Epiguruk (67.070°N, 158.170°W), AK, USA	Hamilton, et al. (1993)
(34400 ± 850)	RIDDL-126	limb	Old Crow R. (Old Crow Loc. 69) (67.855°N, 139.800°W), YT, CAN	Harington (1977); Morlan et al. (1990); R.E. Morlan, CMC database (1999)
34250 ± 580 (34290 ± 580)	Beta-79855 CAMS-18418	molar tooth (CMN 9927)	Hunker Ck. (Dawson Loc. 12) (63°58'N, 138°59'W), YT, CAN	C.R. Harington [unpublished]; R.E. Morlan, CMC database (1999)
>34200	S-483		Saskatoon (52.170°N, 106.580°W), SK, CAN	Morlan (1993); [Christiansen (1968); Lammers (1968); Pohorecky and Wilson (1968); Pohorecky (1988) In: R.E. Morlan, CMC database (1999)];
(33860 ± 330)	TO-4639	tusk	Turtle Mountain, 22 km SW of Boissevain and 19 km SE of Deloraine, MB, CAN	Fulton (1995); R.E. Morlan, CMC database (1999)
(33700 ± 800)	RIDDL-135	limb	Old Crow R. (Old Crow Loc. 60) (68.250°N, 140.167°W), YT, CAN	Morlan et al. (1990); R.E. Morlan, CMC database (1999)

(33400 ± 650)	RIDDL-304	limb (collagen extraction method 2)	Old Crow R. (Old Crow Loc. 69) (67.855°N, 139.800°W), YT, CAN	Harington (1977); Morlan et al. (1990); R.E. Morlan, CMC database (1999). [Suggested age for this bone is 34,400 ± 525 (average of RIDDL-189 and 304)]
32700 ± 980	SI-1632	hair from a skull	Dome Ck. (64.830°N, 147.500°W), AK, USA	Harington (1980)
32340 +1070 / -1250	DIC-2123		Ikpikpuk R. (70°N, 154°45'W), AK, USA	[Guthrie (1985, 1990) specimen also cited within Alaskan Pleistocene BisonTable]
(32300 ± 1050)	BGS-1022	tusk and bone	Johnson Ck., Mayo region (63.670°N, 135.850°W), YT, CAN	Harington (1995b); [C.R. Harington, pers. comm. (1998), In: R.E. Morlan, CMC database (1999)]
(32000 ± 600)	RIDDL-132	limb	Old Crow R. (Old Crow Loc. 60) (68.250°N, 140.167°W), YT, CAN	Morlan et al. (1990); R.E. Morlan, CMC database (1999)
(31900 ± 650)	RIDDL-188	limb	Old Crow R. (Old Crow Loc. 69) (67.855°N, 139.800°W), YT, CAN	Harington (1977); Morlan et al. (1990); R.E. Morlan, CMC database (1999)
(31300 ± 1400)	RIDDL-186	limb	Old Crow R. (Old Crow Loc. 69) (67.855°N, 139.800°W), YT, CAN	Morlan et al. (1990); R.E. Morlan, CMC database (1999)
30490 ± 550 (30570 ± 550)	CRNL-1235 = ("CMN-1235")	limb	Old Crow R. (Old Crow Loc. 4) (67.750°N, 139.890°W), YT, CAN	Morlan (1980, 1986); Morlan, et al. (1990); R.E. Morlan, CMC database (1999)
(30100 ± 600)	RIDDL-190	limb	Old Crow R. (Old Crow Loc. 66) (68.083°N, 139.500°W), YT, CAN	Morlan et al. (1990); R.E. Morlan, CMC database (1999)
(30100 ± 600)	RIDDL-125	limb	Johnson Ck. (Old Crow Loc. 71) (67.917°N, 139.750°W), YT, CAN	Morlan et al. (1990); R.E. Morlan, CMC database (1999)
29450 ± 610	DIC-1819	femur	Tyone R., AK, USA	Thorson, et al. (1981); Agenbroad (1984)
29420 +670/-740	DIC-1570	tusk fragment	Canyon Village, Porcupine R., AK, USA	Dixon, et al. (1979, 1985)
29100 ± 3000	GX-1567-Du	radius (CMC 330) apatite extraction	Old Crow (Loc. 14N) (67°51'N, 139°46'W), YT, CAN	Irving and Harington (1973)
28920 ± 2250	GX-5740-A		Bering Sea coast area (67.780°N, 165.470°W), AK, USA	A.S. Dyke (GSC database)
(28890 ± 960)	AECV-612C	tusk	Apex Pit, E Edmonton, North Saskatchewan R. (53°36.5'N, 113°20'W), AB, CAN	Young, et al. (1994); [Burns (1996) In: R.E. Morlan, CMC database (1999)]
(28780 ± 350)	RIDDL-301	limb (collagen extraction method 2)	Old Crow R. (Old Crow Loc. 42) (67.750°N, 139.750°W), YT, CAN	Morlan et al. (1990). [Suggested age for this bone is 28,790 ± 275 (average of RIDDL-130 and 301)]
(28600 ± 350)	RIDDL-305	limb (collagen extraction method 2)	Old Crow R. (Old Crow Loc. 22) (67.850°N, 139.970°W), YT, CAN	Harington (1977); Morlan et al. (1990); R.E. Morlan, CMC database (1999). [Suggested age for this bone is 27,850 ± 320 (average of RIDDL-192 and 305)]
(27860 ± 880)	AECV-721C		Apex Pit, E Edmonton, North Saskatchewan R. (53°36.5'N, 113°20'W), AB, CAN	Young, et al. (1994); Burns (1996); R.E. Morlan, CMC database (1999)
27700 ± 460 (27780 ± 460)	SI-2812 CO	axis	Old Crow R. (Old Crow Loc. 11(A)), 259 m asl. (67.850°N, 139.970°W), YT, CAN	Harington (1977); Morlan (1980); R.E. Morlan, CMC database (1999)
27500 ± 1800	GX-1568-Du	long-bone shaft (CMC 331) apatite extraction	Old Crow (Loc. 14N) (67°51'N, 139°46'W), YT, CAN	Irving and Harington (1973)
>27200	GX-5741	apatite extraction	Bering Sea coast area (67.780°N, 165.470°W), AK, USA	[Dixon (1983) In: A.S. Dyke (GSC database)
(27100 ± 800)	RIDDL-192	limb (collagen extraction method 1)	Old Crow R. (Old Crow Loc. 22) (68.167°N, 139.500°W), YT, CAN	Morlan et al. (1990); R.E. Morlan, CMC database (1999). [Suggested age for this bone is 27,850 ± 329 (average of RIDDL-192 and 305)]
26770 +490 / -520	DIC-2124		Tanana R., Richardson Hwy. (64°18'N, 146°15'W), AK, USA	Guthrie (1985, 1990)
26050 ± 880 (26130 ± 880)	I-10650	pelvis (CMN 37578)	Horse Hills Pit: E Edmonton, N Saskatchewan R. (53.700°N, 113.230°W), AB, CAN	Young, et al. (1994); Burns (1996); R.E. Morlan, CMC database (1999)

25910 ± 680 (25990 ± 680)	SI-2818 CO	calcaneum	Old Crow R. (Old Crow Loc. 11(A)), 259 m asl. (67.850°N, 139.970°W), YT, CAN	Harington (1977); Morlan (1980); R.E. Morlan, CMC database (1999)
(25800 ± 320)	GSC-2859	tusk	Rocky Mountain Portage, 13 km W of Hudson Hope (56°01'N, 122°07'W), BC, CAN	Lowdon and Blake (1979); Clague (1980); Mathews (1978a) incorrectly cited in Bobrowsky and Rutter (1992); [Fladmark (1981) In: R.E. Morlan, CMC database (1999)]
(25450 ± 450)	RIDDL-193	limb (collagen extraction method 1)	Old Crow R. (Old Crow Loc. 22) (68.167°N, 139.500°W), YT, CAN	Morlan et al. (1990); R.E. Morlan, CMC database (1999). [Suggested age for this bone is 25,350 ± 250 (average of RIDDL-193 and 306)]
(25250 ± 300)	RIDDL-306	limb (collagen extraction method 2)	Old Crow R. (Old Crow Loc. 20) (68.083°N, 139.500°W), YT, CAN	Morlan et al. (1990); R.E. Morlan, CMC database (1999). [Suggested age for this bone is 25,350 ± 250 (average of RIDDL-193 and 306)]
(25200 ± 400)	RIDDL-191	limb	Old Crow R. (Old Crow Loc. 20) (68.083°N, 139.500°W), YT, CAN	Morlan et al. (1990); R.E. Morlan, CMC database (1999)
(24640 ± 940)	AECV-1185C	tusk	Beaverlodge, Red Willow R., Peace drainage, AB, CAN	Burns (1996); R.E. Morlan, CMC database (1999)
(22820 ± 520)	AECV-538C	pelvis	Clover Bar Sand & Gravel Pit(B), Edmonton area (53°35'N, 113°19.5'W), AB, CAN	Burns and Young (1994); Young et al. (1994); Burns (1996); R.E. Morlan, CMC database (1999)
22750 ± 1650 (22830 ± 1650)	S-1964		Horse Hills Pit: E Edmonton, N Saskatchewan R. (53.700°N, 113.230°W), AB, CAN	Young, et al. (1994); Burns (1996); R.E. Morlan, CMC database (1999)
22600 ± 600 (22680 ± 600)	I-3573	femur	Old Crow R. (Old Crow Loc. 14N), 267 m asl. (67.850°N, 139.768°W), YT, CAN	Irving and Harington (1973); Morlan and Bonnichsen (1975); Morlan (1980); Morlan et al. (1990); R.E. Morlan, CMC database (1999)
22600 ± 320 (22700 ± 320)	GSC-2232	tusk	3.2 km ENE of Vedder Crossing Bridge, Chilliwack R. (49.100°N, 121.920°W), BC, CAN	Harington (1975, 1977); Lowdon and Blake (1979); Clague (1980); Hicock, et al. (1982)
(22020 ± 450)	AECV-719C	tusk	Bushe R., Peace drainage (58.520°N, 117.130°W), AB, CAN	Young et al. (1994); Burns and Young (1994); Burns (1996); R.E. Morlan, CMC database (1999)
(21600 ± 240)	SFU-66	tusk	Chilliwack, gravel pit adjacent to Chilliwack R., BC, CAN	R.E. Morlan, CMC database (1999)
(21400 ± 240)	SFU-65	tusk	Chilliwack, gravel pit adjacent to Chilliwack R., BC, CAN	R.E. Morlan, CMC database (1999)
20200 ± 500	S-499	skull	Saskatoon (52°10'N, 106°35'W), SK, CAN	Morlan (1993); [Anomalously young, A.S. Dyke, pers. comm. (2001)]
20190 ± 190 (20270 ± 190)	Beta-50901	molar	NW end of Little L., 11.2 km W of the bridge over the Quesnel R. at Likely (52°37'N, 121°44'W), BC, CAN	Carlson and Carlson (1998); R.E. Morlan, CMC database (1999)
19520 ± 470	SI-2814 AP	left patella	Old Crow R. (Old Crow Loc. 11(A)), 259 m asl. (67.850°N, 139.970°W), YT, CAN	R. McNeely, GSC cardbox database (1999)
19080 ± 280	SI-2812 AP	axis	Old Crow R. (Old Crow Loc. 11(A)), 259 m asl. (67.850°N, 139.970°W), YT, CAN	R. McNeely, GSC cardbox database (1999)
19060 ± 90	USGS-1439		Epiguruk (67.070°N, 158.170°W), AK, USA	Hamilton, et al. (1993)
18560 ± 70	USGS-1485		Epiguruk (67.070°N, 158.170°W), AK, USA	Hamilton, et al. (1993)
(17000 ± 240)	GSC-2829	humerus	Saanich Pen., Vancouver I. (48.530°N, 123.380°W), BC, CAN	Harington (1975, 1976, 1977, 1979); Keddie (1979); Clague (1980); Blake (1982); GSC Paper 82-7, 1982; R.E. Morlan, CMC database (1999)
16880 ± 250	SI-2823 AP	right patella	Old Crow R. (Old Crow Loc. 11(A)), 259 m asl. (67.850°N, 139.970°W), YT, CAN	R. McNeely, GSC cardbox database (1999)
(16150 ± 230)	AA-683	hair, tissue	Colorado Ck. (63.640°N, 156.020°W), W central, AK, USA	[Thorson and Guthrie (1992) In: A.S. Dyke (GSC database)]

16100 ± 130 (16200 ± 130)	GSC-1893	tusk fragment (CMN 21301)	Scroggie Ck., 620 m asl. (63.130°N, 138.620°W), YT, CAN	Harington (1977); R.E. Morlan, CMC database (1999); [Collector H.S. Bostock]
15830 ± 70	Beta-67690 CAMS-9898	tusk	Broken Mammoth site, near confluence of Shaw Ck. and Tanana R, about 22 km NW of Delta Junction (64°16'N, 146°07'W), AK, USA	Holmes (1996)
15280 ± 120	Beta-16996		Colorado Ck., AK, USA	[Dixon (1993) unpublished data, on file, UA Museum, same bone as Beta-9906]
14650 ± 360 (14730 ± 360)	S-498	tusk fragments	Saskatoon (52°10'N, 106°35'W), SK, CAN	Lammers (1968); Pohorecky and Wilson (1968); Pohorecky (1988); Morlan (1993); [Christiansen (1968) In: R.E. Morlan, CMC database (1999)]; [Anomalously young, A.S. Dyke, pers. comm. (2001)]
14270 ± 950	Beta-20027	vertebra	Trail Ck. Caves site, (Cave 3), 45 km SW of Deering (65°48'N, 163°13'W), AK, USA	Dixon (1993); Vinson (1988)
13725 ± 110	Beta-29166 ETH-5014	collagen	Lower Rampart Cave 1, Porcupine R. (67°30'N, 142°W), AK, USA	Sattler and Dixon (1997)
13500 ± 100	QL-1365		Teklanika valley, Central region (63.580°N, 149.500°W), AK, USA	TenBrink and Waythomas (1978)
13340 ± 115	DIC-2130		Teklanika R. (63.642°N, 149.500°W), AK, USA	Guthrie (1985, 1990); Dixon (1993)
(13280 ± 390)	CRNL-1220-c	tibia	Bluefish Cave I, 600 m asl. (67°08'N, 140°47'W), YT, CAN	Cinq-Mars (1979, 1982); Morlan and Cinq-Mars (1982); Burke and Cinq-Mars (1996); R.E. Morlan, CMC database (1999). [Suggested normalized age for this bone is 12,845 ± 250 (average of three runs CRNL-1220a, b, and c)]
(13070 ± 400)	CRNL-1220-a	tibia	Bluefish Cave I, 600 m asl. (67°08'N, 140°47'W), YT, CAN	Cinq-Mars (1979, 1982); Morlan and Cinq-Mars (1982); Burke and Cinq-Mars (1996); R.E. Morlan, CMC database (1999). [Suggested normalized age for this bone is 12,845 ± 250 (average of three runs CRNL-1220a, b, and c)]
12980 ± 250	Beta-9906	V-48-178	Colorado Ck. (63.647°N, 156.017°W), AK, USA	Guthrie (1985); [Dixon (1993) incorrect reference cited, i.e. Guthrie (1985) for this specimen - (same bone as Beta-16996)]
12845 ± 430 (A. Dyke cardbox)	CRNL-1220	tibia	Bluefish Cave I, 600 m asl. (67°08'N, 140°47'W), YT, CAN	Cinq-Mars (1979, 1982); Morlan and Cinq-Mars (1982); Burke and Cinq-Mars (1996); R.E. Morlan, CMC databasse (1999); [Normalized age is an average of three runs]
12622 ± 750	W-401		Fairbanks (65°N, 147°W), AK, USA	Agenbroad (1984); Dixon (1993)
12230 ± 60 (12270 ± 60)	Beta-115203	molar (NS 92-6)	George's Bank (41°55'N, 67°30'W), NS, CAN	Cooke, et al. (1993)
(12190 ± 500)	CRNL-1220-b	tibia	Bluefish Cave I, 600 m asl. (67°08'N, 140°47'W), YT, CAN	Cinq-Mars (1979, 1982); Morlan and Cinq-Mars (1982); Burke and Cinq-Mars (1996); R.E. Morlan, CMC database (1999). [Suggested normalized age for this bone is 12,845 ± 250 (average of three runs CRNL-1220a, b, and c)]
12130 ± 80	GrA-22177	cuboid	Near Muirkirk, Kent County (approx. 42°31'N, 81°46'W), ON, CAN	Harington and Mol (in prep.)
(12060 ± 70)	CAMS-17045	tusk (NSRL-2001)	Swan Pt. site, Tanana R. valley, 322 m asl. (63.300°N, 146.033°W), AK, USA	Holmes, et al. (1996)

12000 ± 320 (12080 ± 320)	S-482	tusk fragments	Saskatoon (52°10'N, 106°35'W), SK, CAN	Morlan (1993); [Christiansen (1968); Lammers (1968); Pohorecky and Wilson (1968); Pohorecky (1988) In: R.E. Morlan, CMC database (1999)];
>11600	I-2244-A	tusk	Rocky Mountain Portage (13 km W of Hudson Hope) (56°01'N, 122°07'W), BC, CAN	Clague (1980) - see Table 1, footnote e for detailed explanation; Mathews (1978b); Harington and Shackleton (1978); Rutter (1977). The latest and presumably most reliable date on this specimen is 25,800 ± 320 BP (see GSC-2859).
11360 ± 100	Beta-13811		Trail Ck. Caves site, 45 km SW of Deering (65°48'N, 163°13'W), AK, USA	Dixon (1993); Vinson (1988)
11170 ± 60 (11120 ± 60)	Beta-79915	molar (CMN 50369)	1 km SE of the confluence of Barton Ck, with Callum Ck, 1400 asl, on the E side of Hwy 22, Oldman drainage, Pincher Ck. (47°57.23'N, 114°08.53'W), AB, CAN	C.R. Harington [unpublished]; [R. McNeely, pers. comm. In: R.E. Morlan, CMC database (1999)]
10240 ± 325 (10320 ± 320)	GX-2918-D	tibia	James R. (51.900°N, 114.570°W), AB, CAN	Graham, et al. (1987); [Burns (1996) In: R.E. Morlan, CMC database (1999)]
8310 ± 200 (8390 ± 200)	Beta-17869	right radius (CMN 6747)	Near Muirkirk, Kent County (approx. 42°31'N, 81°46'W), ON, CAN	C.R. Harington [unpublished]; [McAndrews and Jackson (1988) In: R.E. Morlan, CMC database (1999)]; [Anomalously young; found in Spruce pollen zone (>10ka), A.S. Dyke, pers. comm. (2001)]
7670 ± 170 (7915 ± 175)	I-2244	tusk	Rocky Mountain Portage (13 km W of Hudson Hope) (56°01'N, 122°07'W), BC, CAN	Harington (1978); [Harington and Shackleton (1978). This date was based on carbonate sample and therefore considered unreliable.]; Agenbroad (1984); R.E. Morlan, CMC database (1999)
6440 ± 60 (6510 ± 60)	Beta-115209	interior tusk (CMN 6747)	Near Muirkirk, Kent County (approx. 42°31'N, 81°46'W), ON, CAN	C.R. Harington [unpublished]; [McAndrews and Jackson (1988) In: R.E. Morlan, CMC database (1999)]; [Anomalously young, see Beta-17869 and GrA-22177]
4290 ± 120 (4370 ± 120)	WAT-945	scapula	Rostock, 15 km N of Stratford (43.500°N, 81.000°W), ON, CAN	Pilny, et al. (1987); [McAndrews and Jackson (1988) - questionable date]; R.E. Morlan, CMC database (1999)
4000 ± 130 (4080 ± 130)	RL-1058	tusk	Cambrian Plaza: 14th St., and Northmount Dr., Calgary (51°05'00"N, 114°05'00"W), AB, CAN	Jackson and Pawson (1984); [Wilson (1983) In: R.E. Morlan, CMC database (1999)]
(940 ± 70)	TO-313	tusk	Birds Hill: NE of Winnipeg, MB, CAN	[G. Matile, pers. comm. to R. McNeely (1998) In: R.E. Morlan, CMC database (1999)]

cf. Mammuthus sp. (mammoth)

Measured Date (Normalized)	Lab No	Material	Location	References/Remarks
(>40000)	I-10824	tusk fragment	Old Crow R. (Old Crow Loc. 11(1)), 260 m asl. (67.850°N, 139.970°W), YT, CAN	C.R. Harington [unpublished]. [Specimen probably eroded from older sediments]
(>39900)	I-4228	thoracic vertebra	Old Crow R. (Old Crow Loc. 44) (68.220°N, 140.000°W), YT, CAN	Harington (1977, 1978); [Bonnichsen (1978) incorrectly cites lab number as I-4428 in Table 2, and Agenbroad (1984) incorrectly cited as I-4428 in Table 3.4]; R.E. Morlan, CMC database (1999)
(37300 ± 1000)	RIDDL-131	limb	Old Crow R. (Old Crow Loc. 24) (67.750°N, 139.750°W), YT, CAN	Morlan et al. (1990); R.E. Morlan, CMC database (1999)
(31120 ± 450)	RIDDL-122	limb	Old Crow R. (Old Crow Loc. 4) (67.755°N, 139.892°W), YT, CAN	Morlan (1980); Morlan et al. (1990)
(30680 ± 600)	RIDDL-231	limb	Old Crow R. (Old Crow Loc. 4) (67.755°N, 139.892°W), YT, CAN	Morlan et al. (1990)

29300 ± 1200 (29380 ± 1200)	I-11050	long-bone fragments	Old Crow R. (Old Crow Loc. 3) (67.740°N, 139.875°W), YT, CAN	[Harington (1980) - specimen incorrectly cited as I-11051]; Morlan (1980, 1986); R.E. Morlan, CMC database (1999)
(28800 ± 450)	RIDDL-130	limb (collagen extraction method 1)	Old Crow R. (Old Crow Loc. 42) (67.750°N, 139.750°W), YT, CAN	Morlan et al. (1990). [Suggested age for this bone is 28,790 ± 275 (average of RIDDL-130 and 301)]
25750 +1800 / -1500 (25995 +	GX-1568	long-bone shaft (CMC 331)	Old Crow (Loc. 14N) (67°51'N, 139°46'W), YT, CAN	Irving and Harington (1973); Morlan and Bonnichsen (1975); Harington (1977); Agenbroad (1984); Morlan (1986); R.E. Morlan, CMC database (1999)
(21900 ± 320)	GSC-1760	tusk (CMN 11833)	Near Cape James Ross, Dundas Pen., Melville I. (75.700°N, 114.420°W), NU, CAN	Blake (1974); Harington (1978b, 1990b); Agenbroad (1984); Guthrie (1985, 1990). [Tusk collected by Northern Party of Stefansson's Canadian Arctic Expedition. The most northern mammoth specimen recorded for North America].
(21600 ± 230)	GSC-1760-2	tusk (CMN 11833)	Near Cape James Ross, Dundas Pen., Melville I. (75.700°N, 114.420°W), NU, CAN	Harington (1978b, 1990b); Blake (1987)
(20270 ± 270)	TO-2355	left tibia shaft (CMN 38655)	Ballast Brook area, Banks I. (74.300°N, 123.080°W), NT, CAN	C.R. Harington [unpublished]; [Collector L.V. Hills. Second most northern record of mammoth in North America].

Mammuthus? sp. (mammoth)

Measured Date (Normalized)	Lab No	Material	Location	References/Remarks
(44600 +1600/ -1900)	RIDDL-724		Old Crow R., YT, CAN	R.E. Morlan, CMC database (1999)
(42100 +1200/ -1400)	RIDDL-732	femur	Old Crow R. (Loc. 14N) (67.916°N, 139.750°W), YT, CAN	Morlan et al. (1990); R.E. Morlan, CMC database (1999)
41460 +5670/ -3290 (41540 +	CRNL-1219 = ("CMN 1219")		Old Crow R. (67.800°N, 139.921°W), YT, CAN	Morlan (I980, 1986); Morlan, et al. (1990); R.E. Morlan, CMC database (1999)
(41400 ± 1600)	RIDDL-234		Old Crow R. (Old Crow Loc. 74), 265 m asl. (67.800°N, 139.917°W), YT, CAN	Morlan et al. (1990); R.E. Morlan, CMC database (1999)
(39900 ± 1300)	RIDDL-233		Old Crow R. (Old Crow Loc. 87), 275 m asl. (68.130°N, 139.950°W), YT, CAN	Morlan et al. (1990); R.E. Morlan CMC database (1999)
(39800 ± 1100)	RIDDL-731	limb-bone fragment	Old Crow R. (Old Crow Loc. 15) (67.916°N, 139.750°W), YT, CAN	Morlan (1980, 1986, 1996); Morlan, et al. (1990); R.E. Morlan, CMC database (1999)
(39700 ± 1000)	RIDDL-134	limb	Old Crow R. (Old Crow Loc. 14N), 267 m asl. (67.850°N, 139.768°W), YT, CAN	Harington (1977); Morlan et al. (1990); R.E. Morlan, CMC database (1999)
(38700 ± 900)	RIDDL-128		Old Crow R. (Old Crow Loc. 74), 265 m asl. (67.800°N, 139.917°W), YT, CAN	Morlan et al. (1990); R.E. Morlan, CMC database (1999)
(37200 ± 700)	RIDDL-725		Old Crow R. (Old Crow Loc. 74), 265 m asl. (67.800°N, 139.917°W), YT, CAN	R.E. Morlan, CMC database (1999)
(36600 ± 1000)	RIDDL-187		Old Crow R. (Old Crow Loc. 87), 275 m asl. (68.130°N, 139.950°W), YT, CAN	Morlan et al. (1990); R.E. Morlan, CMC database (1999)
(34600 ± 900)	RIDDL-185		Old Crow R. (Old Crow Loc. 14N), 267 m asl. (67.850°N, 139.768°W), YT, CAN	Harington (1977); Morlan et al. (1990); R.E. Morlan, CMC database (1999)
34460 +2190 / -1720 (34540 +	CRNL-1231	tusk (CMC-1231)	Kittigazuit, Mackenzie R. (69.350°N, 132.670°W), NT, CAN	[McGhee (1974); Wilmeth (1978); Morrison (1989); Cinq-Mars (1991); Vogel, et al. (1991) In: R.E. Morlan, CMC database (1999)]
(34200 ± 500)	RIDDL-726	limb	Old Crow R. (Old Crow Loc. 14N), 267 m asl. (67.850°N, 139.768°W), YT, CAN	Harington (1977); Morlan et al. (1990); R.E. Morlan, CMC database (1999)

Measured Date (Normalized)	Lab No	Material	Location	References/Remarks
(32600 ± 600)	RIDDL-302	limb	Old Crow R. (Old Crow Loc. 14N), 267 m asl. (67.850°N, 139.768°W), YT, CAN	Harington (1977); Morlan et al. (1990); R.E. Morlan, CMC database (1999)
(31200 ± 500)	RIDDL-727		Old Crow R. (Old Crow Loc. 87), 275 m asl. (68.130°N, 139.950°W), YT, CAN	Morlan et al. (1990); R.E. Morlan, CMC database (1999)
(27000 ± 400)	RIDDL-232	limb	Old Crow R. (Old Crow Loc. 14N), 267 m asl. (67.850°N, 139.768°W), YT, CAN	Harington (1977); Morlan et al. (1990); R.E. Morlan, CMC database (1999)
25970 ± 560 (26050 ± 560)	CRNL-1234 = ("CMN 1234")	limb (CMC-1234)	Old Crow R. (Old Crow Loc. 14N), 267 m asl. (67.850°N, 139.768°W), YT, CAN	Harington (1977); Morlan (1986); Morlan et al. (1990); R.E. Morlan, CMC database (1999)
(25620 ± 300)	RIDDL-303		Old Crow R. (Old Crow Loc. 87), 275 m asl. (68.130°N, 139.950°W), YT, CAN	Morlan et al. (1990); R.E. Morlan, CMC database (1999)
(25600 ± 300)	RIDDL-300		Old Crow R. (Old Crow Loc. 74), 265 m asl. (67.800°N, 139.917°W), YT, CAN	R.E. Morlan, CMC database (1999)
25170 ± 630 (25250 ± 630)	CRNL-1232 = ("CMN-1232")	tusk (CMC-1232)	Cadzow Bluff, left bank of Porcupine R., 305 m asl. (67.550°N, 138.920°W), YT, CAN	Morlan (1986); Morlan, et al. (1990); R.E. Morlan, CMC database (1999)
(24700 ± 250)	RIDDL-229	tusk	Cadzow Bluff, left bank of Porcupine R., 305 m asl. (67.583°N, 138.750°W), YT, CAN	Morlan (1986); Morlan et al. (1990); R.E. Morlan, CMC database (1999)
13335 ± 390 (13415 ± 390)	CRNL-1218 = ("CMN-1218")		Old Crow R. (Old Crow Loc. 87), 275 m asl. (68.130°N, 139.950°W), YT, CAN	Morlan (l980, 1986); Morlan, et al. (1990); R.E. Morlan, CMC database (1999)

Perissodactyla, Equidae

Equus scotti (Scott's horse)

Measured Date (Normalized)	Lab No	Material	Location	References/Remarks
2760 ± 90 (2840 ± 90)	I-8581	right metatarsal (CMN 21326)	Riddell site, 4.8 km N of Sutherland, S Saskatchewan R., SK, CAN	C.R. Harington [unpublished (questionable date)]; Skwara Woolf (1981); Skwara (1988); R.E. Morlan, CMC database (1999)

Equus cf. Equus verae (Vera's horse)

Measured Date (Normalized)	Lab No	Material	Location	References/Remarks
(46660 ± 840)	TO-2702	left metatarsal (CMN 46507)	Sixtymile area (Sixtymile Loc. 3), 701 m asl., (64°1.00'N, 140°44'W), YT, CAN	Harington (1997a)

Equus conversidens (Mexican horse)

Measured Date (Normalized)	Lab No	Material	Location	References/Remarks
11330 ± 80 (11330 ± 70)	TO-7696		St. Mary's Reservoir (49°07'N, 113°12'W), AB, CAN	Hall (1999b); McNeil, et al. (1999); Hills, et al. (1999)
10870 ± 45	CAMS-82411	astragalus (CMN 12444)	Pashley gravel pit (49°59'00"N, 110°29'40"W), AB, CAN	Hills and Harington (in press)

Equus lambei (Yukon horse)

Measured Date (Normalized)	Lab No	Material	Location	References/Remarks
31450 ± 1300 (31530 ± 1300)	I-10935	metatarsal (CMN 34964)	Dublin Gulch (64.000°N, 139.000°W), YT, CAN	Harington (1977, 1989, 1995b); R.E. Morlan, CMC database (1999)
(28700 ± 460)	USGS-804	metacarpal and toes	Titaluk R. (69.680°N, 155.370°W), N Slope, AK, USA	A.S. Dyke (GSC database)
(26280 ± 210)	Beta-67407 CAMS-9769	humerus (CR-93-80)	Last Chance Ck. (15 Pup) near Dawson City (63.900°N, 139.367°W), YT, CAN	Harington and Eggleston-Stott (1996); [From partial carcass of Yukon horse].
22680 ± 530 (22760 ± 530)	CRNL-1237	limb (CMC 1237)	Bluefish Cave II, 600 m asl. (67°08'N, 140°47'W), YT, CAN	Cinq-Mars (1979, 1990); Morlan and Cinq-Mars (1982); Cinq-Mars, et al. (1991); Burke and Cinq-Mars (1996, 1998); R.E. Morlan, CMC database (1999)
20660 ± 100	Beta-70102	mandible (SC 1)	Scottie Ck. (approx. 62°30'N, 141°00'W), YT, CAN	MacIntosh (1997); Hare (1994)

(17440 ± 220)	RIDDL-278	metatarsal	Bluefish Cave I, 600 m asl. (67°08'N, 140°47'W), YT, CAN	Cinq-Mars (1979, 1990); Morlan and Cinq-Mars (1982); Burke and Cinq-Mars (1996, 1998)
(16700 ± 200)	RIDDL-766	cranium (CMN 43815)	Herschel I., 35 m asl. (69°37'N, 138°56'W), YT, CAN	Harington (1990b); [Dallimore (unpublished), R. McNeely, pers. comm. (1998) In: R.E. Morlan, CMC database (1999)]
(16200 ± 150)	RIDDL-765	cranium (CMN 43815)	Herschel I., 35 m asl. (69°37'N, 138°56'W), NT, CAN	Harington (1989); Klassen (1989); R.E. Morlan, CMC database (1999)
14870 ± 260 (14950 ± 260)	I-3569	right metacarpal	Dominion Ck. Dawson (Loc. 28) (63.820°N, 138.650°W), YT, CAN	Harington (1969, 1977, 1980, 1997); R.E. Morlan, CMC database (1999)
(12900 ± 100)	GSC-2881	femur	Bluefish Cave I, 600 m asl. (67°08'N, 140°47'W), YT, CAN	Cinq-Mars (1979, 1990); Morlan and Cinq-Mars (1982); Harington (1989); Morlan (1989); Guthrie (1985, 1990); Burke and Cinq-Mars (1996, 1998)
12290 ± 440 (12370 ± 440)	CRNL-1236	limb (CMC 1236)	Bluefish Cave III, 600 m asl. (67°08'N, 140°47'W), YT, CAN	Cinq-Mars (1979, 1990); Morlan and Cinq-Mars (1982); Harington and Cinq-Mars (1995); Burke and Cinq-Mars (1996, 1998); R.E. Morlan, CMC database (1999)

Equus sp. (horse)

Measured Date (Normalized)	Lab No	Material	Location	References/Remarks
(>41090)	AECV-1493C	metatarsal	Riverview Pit, E Edmonton, N Saskatchewan R. (53°36'N, 113°19.5'W), AB, CAN	Burns (1996); [Young, et al. (1994); Burns and Young (1994) In: R.E. Morlan, CMC database (1999)]
>40000	I-9320		Ikpikpuk R. (70°N, 154°45'W), AK, USA	Guthrie and Stoker (1990); [Guthrie (1985, 1990) duplicate date of 24,070 ± 680 BP, and locality data = Birch Ck. cited for this specimen incorrectly within Table]
>40000	I-9319		Ikpikpuk R. (70°N, 154°45'W), AK, USA	Guthrie (1985, 1990); Guthrie and Stoker (1990)
>39900 (>39900)	I-4223	pelvis (innominate)	Old Crow R. (Old Crow Loc. 44) (68.220°N, 140.000°W), YT, CAN	Harington (1977, 1978); R.E. Morlan, CMC database (1999)
37320 +1780 / -2300	DIC-3095		Lost Chicken Ck. (64.050°N, 141.877°W), AK, USA	Guthrie (1985, 1990)
36850 ± 750	USGS-1514		Epiguruk (67.070°N, 158.170°W), AK, USA	Hamilton, et al. (1993)
36700 ± 800	SI-2820-B	tibia (apatite extraction) (11A75-3742)	Old Crow R. (Old Crow Loc. 11(A)), 259 m asl. (67.850°N, 139.970°W), YT, CAN	B. Beebe [unpublished]. This date is considered less reliable than SI-2820-A on the same specimen (see below).
36160 ± 530 (36140 ± 530)	Beta-70841 CAMS-11919	phalanx	Pauline Cove, Herschel I., YT, CAN	R.E. Morlan, CMC database (1999)
(35840 ± 2370)	AECV-938C	metapodial	Consolidated Pit 48, N Saskatchewan R. Edmonton (53°38.5'N, 113°17'W), AB, CAN	Young, et al. (1994); Burns and Young (1994); Burns (1996); R.E. Morlan, CMC database (1999)
35620 +1530 / -1900	DIC-3100		Fairbanks Ck. (65°03'N, 147°10'W), AK, USA	Guthrie (1985, 1990)
34000 ± 2600 (34080 ± 2600)	I-4222	metatarsal	Old Crow R. (Old Crow Loc. 28) (68.170°N, 138.930°W), YT, CAN	Harington (1977); Morlan and Bonnichsen (1975); Harington (1977a); R.E. Morlan, CMC database (1999)
32800 ± 800 (32880 ± 800)	SI-2820-A	tibia (collagen extraction) (11A75-3742)	Old Crow R. (Old Crow Loc. 11(A)), 259 m asl. (67.850°N, 139.970°W), YT, CAN	Harington (1977); Morlan (1980); R.E. Morlan, CMC database (1999). This date is considered more reliable than SI-2820-B on the same specimen (see above).
32270 ± 1500	I-9275		Ikpikpuk R. (70°N, 154°45'W), AK, USA	Guthrie (1985, 1990); Guthrie and Stoker (1990)
(31220 ± 1260)	AECV-1202C	metatarsal	Clover Bar Pit, E Edmonton, N Saskatchewan R. (53°35'N, 113°19.5'W), AB, CAN	Young, et al. (1994); Burns and Young et al. (1994); Burns (1996); R.E. Morlan, CMC database (1999)
30250 +890 / -990	DIC-3092		Cleary Ck. (65.080°N, 147.33°W), AK, USA	Guthrie (1985, 1990)

(29380 ± 4970)	AECV-720C	tibia	Apex Pit, E Edmonton, North Saskatchewan R. (53°36.5'N, 113°20'W), AB, CAN	Young, et al. (1994); Burns (1996); R.E. Morlan, CMC database (1999)
28600 +690 / -760	DIC-3099		Goldstream Ck. (64.950°N, 147.580°W), AK, USA	Guthrie (1985, 1990)
(28340 ± 850)	AA-37607	metapodial (F:AM-4695)	Gold Hill, Fairbanks, AK, USA	[J. Leonard, pers. comm. (2001)]
28200 ± 500 (28280 ± 500)	SI-2824-A	metatarsal (11A75-5645)	Old Crow R. (Old Crow Loc. 11(A)), 259 m asl. (67.850°N, 139.970°W), YT, CAN	Harington (1977); Morlan (1980); R.E. Morlan, CMC database (1999); collagen fraction.
27680 ± 650 (27760 ± 650)	Beta-8865	right metatarsal (CR-79-10)	Dominion Ck. (Dawson Loc. 29) (63.800°N, 138.600°W), YT, CAN	C.R. Harington [unpublished]; R.E. Morlan, CMC database (1999)
(27520 ± 850)	AECV-937C	radio-ulna	Consolidated Pit 48, N Saskatchewan R. Edmonton (53°38.5'N, 113°17'W), AB, CAN	Young, et al. (1994); Burns and Young (1994); Burns (1996); R.E. Morlan, CMC database (1999)
26830 +1230 / -1450	DIC-3097		Lost Chicken Ck. (64.050°N, 141.877°W), AK, USA	Guthrie (1985, 1990)
26760 ± 300	SI-355		Lost Chicken Ck. (64.050°N, 141.875°W), AK, USA	Péwé (1975); Harington (1978); [Agenbroad (1984) - specimen incorrectly cited as Mammuthus primigenius within Table, and Guthrie (1985, 1990) - specimen incorrectly cited within Alaskan Pleistocene Bison Table]
(26710 ± 800)	AA-37610	metapodial (F:AM-6206)	Ester Ck., Fairbanks, AK, USA	[J. Leonard, pers. comm. (2001)]
26460 ± 3760 (26540 ± 1880)	QU-784	humerus	Old Crow R. (Old Crow Loc. 11(A)), 259 m asl. (67.480°N, 139.920°W), YT, CAN	Harington (1977); Morlan (1980); [R.E. Morlan, pers. comm. (1998) indicates corrected age of 26,460 ± 1880 BP]; R.E. Morlan, CMC database (1999)
25750 +910 / -1040	DIC-2125		Tanana R., Richardson Hwy. (64°18'N, 146°15'W), AK, USA	[Guthrie (1985, 1988, 1990) specimen incorrectly cited within Alaskan Pleistocene Bison Table]
(24400 ± 500)	AA-37612	metapodial (F:AM-6300)	Ester Ck., Fairbanks, AK, USA	[J. Leonard, pers. comm. (2001)]
24070 +380 / -410	DIC-3098		Lost Chicken Ck. (64.050°N, 141.877°W), AK, USA	Guthrie (1985, 1990)
23920 ± 620	I-9321		Fairbanks area (65.000°N, 148.000°W), AK, USA	Guthrie (1985, 1990)
23910 ± 470	I-9318		Ikpikpuk R. (70°N, 154°45'W), AK, USA	Guthrie (1985, 1990); Guthrie and Stoker (1990)
23340 ± 388	DIC-3094		Lillian Ck., AK, USA	Guthrie (1985, 1990)
22370 +300 / -310	DIC-3093		Lillian Ck., AK, USA	Guthrie (1985, 1990)
21780 ± 310	DIC-1334	ulna	Porcupine R. Cave (67.000°N, 143.000°W), AK, USA	Dixon (1984); [Guthrie (1985, 1990) specimen incorrectly cited within Alaskan Pleistocene Bison Table]; Sattler and Dixon (1997)
(21330 ± 340)	AECV-1664C	humerus	Clover Bar Pit, E Edmonton, N Saskatchewan R. (53°35'N, 113°19.5'W), AB, CAN	Young, et al. (1994); Burns (1996); [Burns and Young (1994) In: R.E. Morlan, CMC database (1999)]
20810 ± 410	I-9274		Ikpikpuk R. (70°N, 154°45'W), AK, USA	Guthrie (1985, 1990); Guthrie and Stoker (1990)
(20670 ± 350)	AA-37613	metapodial (F:AM-6171)	Goldstream, Fairbanks, AK, USA	[J. Leonard, pers. comm. (2001)]
(20220 ± 310)	AA-37608	metapodial (F:AM-4305)	Fairbanks Ck., Fairbanks, AK, USA	[J. Leonard, pers. comm. (2001)]
(19800 ± 200)	AA-37611	metapodial (F:AM-133514)	Ester Ck., Fairbanks, AK, USA	[J. Leonard, pers. comm. (2001)]
19250 ± 360	I-9371		Ikpikpuk R. (70°N, 154°45'W), AK, USA	Guthrie (1985, 1990); Guthrie and Stoker (1990)
19200 ± 225	SI-2824-B	metatarsal (apatite extraction) (11A75-5645)	Old Crow R. (Old Crow Loc. 11(A)), 259 m asl. (67.850°N, 139.970°W), YT, CAN	B. Beebe [unpublished]
18640 ± 205	DIC-3096		Lost Chicken Ck. (64.050°N, 141.877°W), AK, USA	Guthrie (1985, 1990)

18450 +200 / -210	DIC-3091		Cleary Ck. (65.080°N, 147.33°W), AK, USA	Guthrie (1985, 1990)
17190 ± 240	DIC-2418	hoof keratin	Titaluk R. (69.683°N, 155.367°W), AK, USA	Guthrie (1985, 1990); Guthrie and Stoker (1990)
16270 ± 230 (16350 ± 230)	I-9271		Dominion Ck. (64.067°N, 139.367°W), YT, CAN	Guthrie (1985, 1990); R.E. Morlan, CMC database (1999)
15750 ± 350	K-1210	proximal part of right scapula	Trail Ck. Cave 9, Seward Pen. (65.800°N, 163.220°W), AK, USA	Morlan and Cinq-Mars (1982); [Guthrie (1985, 1990) This specimen and date incorrectly cited as Alaskan Pleistocene Bison (at 14,410 ± 315 BP) within the Table]; West (1996)
15340 ± 500 (15420 ± 500)	S-1305	metapodial	Riddell site, 4.8 km N of Sutherland, S Saskatchewan R. (52°09'N, 106°36'W), SK, CAN	Skwara Woolf (1981); Skwara (1988); Morlan (1993); R.E. Morlan, CMC database (1999)
15090 ± 70	Beta-109267	radius	Gerstle R. quarry site, about 1 km SE of Gerstle R. Bridge on the Alaska Hwy, AK, USA	Holmes (1998)
14990 ± 220 (15070 ± 220)	I-9316		Dominion Ck. (64.067°N, 139.367°W), YT, CAN	Guthrie (1985, 1990); R.E. Morlan, CMC database (1999)
13640 ± 410	I-9422		Fairbanks area (65.000°N, 148.000°W), AK, USA	Guthrie (1985, 1990)
13350 ± 120	Beta-37056 ETH-6578	collagen	Lower Rampart Cave 1, Porcupine R (67°30'N, 142°W), AK, USA	Sattler and Dixon (1997)
(12840 ± 140)	AA-37614	metapodial (F:AM-6282)	Ester Ck., Fairbanks, AK, USA	[J. Leonard, pers. comm. (2001)]
(12380 ± 120)	AA-37609	metapodial (F:AM-6735)	Ester Ck., Fairbanks, AK, USA	[J. Leonard, pers. comm. (2001)]
(11180 ± 150)	RIDDL-662		Milk R. Ridge, Milk R. (49.050°N, 111.750°W), AB, CAN	Burns (1996); R.E. Morlan, CMC database (1999)
815 ± 110 (895 ± 110)	GX-201		Hemlock Pk Farm, Frontenac County (44°18'40"N, 76°28'20"W), ON, CAN	Krueger and Weeks (1966); R.E. Morlan, CMC database (1999); [See corrected date (Beta-151996) showing the horse is modern, C.R. Harington (unpublished)]
195 ± 90 (275 ± 90)	I-5403	left mandible fragment (CMN 9924) from complete skull	Hunker Ck. (Dawson Loc. 12) (63°58'N, 138°59'W), YT, CAN	C.R. Harington [unpublished]; R.E. Morlan, CMC database (1999); [Probably recent horse].
110 ± 40 (170 ± 40)	Beta-151996	right third carpal from skeleton	Hemlock Pk Farm, Frontenac County (44°18'40"N, 76°28'20"W), ON, CAN	[Correction to GX-201date shows horse is modern, C.R. Harington (unpublished)]
106 ± 0.7%	Beta-116673	left scapula (CR-97-62A)	Spencerville (approx. 44°51'N, 75°33'W), ON, CAN	C.R. Harington [unpublished]; [Probably recent horse]

Artiodactyla, Camelidae

Camelops hesternus (western camel)

Measured Date (Normalized)	Lab No	Material	Location	References/Remarks
43580 ± 1100 (43620 ± 1100)	Beta-89985	left astragalus (CMN 29194)	Sixtymile area (Sixtymile Loc. 3), 701 m asl., (64°1.00'N, 140°44'W), YT, CAN	Harington (1997a); R.E. Morlan, CMC database (1999)
(23320 ± 640)	Beta-8864	humerus (CMN 38227)	Sixtymile area (Sixtymile Loc. 3), 701 m asl., (64°1.00'N, 140°44'W), YT, CAN	Harington (1989, 1997)

Camelops sp. (camel)

Measured Date (Normalized)	Lab No	Material	Location	References/Remarks
43230 ± 510 (43270 ± 510)	Beta-115205	radius (CMN 46728)	Sixtymile area (Sixtymile Loc. 3), 701 m asl., (64°1.00'N, 140°44'W), YT, CAN	Harington (1997a); R.E. Morlan, CMC database (1999)
38980 ± 350 (39030 ± 350)	Beta-115207	left mandible of juvenile (CMN 42549)	Sixtymile area (Sixtymile Loc. 3), 701 m asl., (64°1.00'N, 140°44'W), YT, CAN	Harington (1997a); R.E. Morlan, CMC database (1999)
27900 +1,000/ -1,100	I-2117	metapodial	Fairbanks area (64°50'N, 148°W), AK, USA	Harington (1977a), p. 684

Artiodactyla, Cervidae

Cervus elaphus (wapiti, elk)

Measured Date (Normalized)	Lab No	Material	Location	References/Remarks
(11060 ± 90)	CAMS-7204		Broken Mammoth site, near confluence of Shaw Ck. and Tanana R, about 22 km NW of Delta Junction (64°16'N, 146°07'W), AK, USA	Holmes (1996)
(10820 ± 60)	CAMS-23467	radius	Bluefish Cave III, 600 m asl. (67°08'N, 140°47'W), YT, CAN	Cinq-Mars (1979, 1990); Morlan and Cinq-Mars (1982); Harington and Cinq-Mars (1995); Burke and Cinq-Mars (1996)
10050 ± 150	I-9998	left antler (CMN 25873)	Lost Chicken Ck. (64.050°N, 141.875°W), AK, USA	Harington (1978, 1980); [Agenbroad (1984) incorrectly cites specimen as Mammoth within Table 3.4]; Dixon (1993)
9920 ± 220 (10000 ± 220)	AECV-272C	rib	Smoky R., 4.5 km S of Watino (55.720°N, 117.625°W), AB, CAN	[Churcher (1984) In: A.S. Dyke (GSC database)]; Burns (1986); R.E. Morlan, CMC database (1999)
9075 ± 305 (9155 ± 305)	S-2614	antler from nearly complete skeleton	Smoky R., 4.5 S of Watino (55.720°N, 117.625°W), AB, CAN	Burns (1986); R.E. Morlan, CMC database (1999)
4570 ± 100 (4650 ± 100)	I-7796	humerus shaft (CMN 17048)	Old Crow R. (Old Crow Loc. 1) (67.850°N, 139.97°W), YT, CAN	Harington (1977); Bonnichsen (1978); R.E. Morlan, CMC database (1999)
(4250 ± 100)	AECV-586C	antler	Pulp Mill, Peace R. (56°22'N, 117°11'W), AB, CAN	[Bobrowsky permit 88-5, in Luther and Purdon (1991) In: R.E. Morlan, CMC database (1999)]; Beaudoin (1991)
3625 ± 75 (3705 ± 80)	S-538	antler	Near Dog Ck., Williams L. area (51.580°N, 122.250°W), BC, CAN	Clague (1980); R.E. Morlan, CMC database (1999)
3385 ± 100 (3465 ± 100)	I-10917	antler (CMN 34967)	Fort Saskatchewan, N Saskatchewan R. (53.700°N, 113.230°W), AB, CAN	C.R. Harington [unpublished]; R.E. Morlan, CMC database (1999)
(3010 ± 60)	TO-5681	upper M3	Firth R., Engigstciak (69.360°N, 139.530°W), YT, CAN	Fox (1998); R.E. Morlan, CMC database (1999)
2225 ± 90 (2305 ± 90)	I-9993	antler	Okanagan L., Kelowna (49.870°N, 119.500°W), BC, CAN	Clague (1980); R.E. Morlan, CMC database (1999)
2010 ± 100 (2090 ± 100)	S-2557	antler	S bank of the Saskatchewan R., 10.5 km upstream from the Francois-Finlay Dam (53.170°N, 104.000°W), SK, CAN	Morlan (1993); [Quigg (1986); Brandzin-Low (1997) In: R.E. Morlan, CMC database (1999)]
1740 ± 75 (1820 ± 80)	S-454	antler	McGlashan L., Kamloops area (50.500°N, 120.150°W), BC, CAN	Clague (1980); R.E. Morlan, CMC database (1999)
1430 ± 85 (1510 ± 85)	S-495	antler	Dominic L., Kamloops area (50.580°N, 120.680°W), BC, CAN	Clague (1980); R.E. Morlan, CMC database (1999)
1200 ± 65 (1280 ± 70)	S-540	antler	Near Squawk L., Williams L. area (52.050°N, 121.570°W), BC, CAN	Clague (1980); R.E. Morlan, CMC database (1999)
1105 ± 65 (1185 ± 70)	S-541	antler	Near Chezacut, Williams L. area (52.330°N, 124.020°W), BC, CAN	Clague (1980); R.E. Morlan, CMC database (1999)
615 ± 70 (695 ± 75)	S-539	antler	Moose Meadow, Williams L. area (51.820°N, 121.800°W), BC, CAN	Clague (1980); R.E. Morlan, CMC database (1999)
390 ± 75 (470 ± 80)	S-496	antler	3.22 km S of the W end of Upper Loon L., 40.23 km ENE of Clinton, Thompson drainage (51.170°N, 121.080°W), BC, CAN	Clague (1980); R.E. Morlan, CMC database (1999)
367 ± 75 (447 ± 80)	S-497	antler	3.22 km S of the W end of Upper Loon L., 40.23 km ENE of Clinton, Thompson drainage (51.170°N, 121.080°W), BC, CAN	Clague (1980); R.E. Morlan, CMC database (1999)
270 ± 40 (290 ± 40)	GSC-2402	antler (APM #1)	Lanark twp., Mississippi R. (45.080°N, 76.380°W), ON, CAN	Lowdon and Blake (1979); R.E. Morlan, CMC database (1999)
184 ± 75 (264 ± 80)	S-493	antler	Pennask L., Kamloops area (50.000°N, 120.150°W), BC, CAN	Clague (1980); R.E. Morlan, CMC database (1999)
110 ± 55 (190 ± 60)	S-537	antler	Near Maze L., Williams L. area (51.950°N, 121.700°W), BC, CAN	Clague (1980); R.E. Morlan, CMC database (1999)

Cervus sp. (wapiti)

Measured Date (Normalized)	Lab No	Material	Location	References/Remarks
4850 ± 100 (4930 ± 100)	Beta-11453		Oregon Jack Ck., 1 km W of the Thompson R. (50.500°N, 121.333°W), BC, CAN	[Rousseau and Richards (1988) In: A.S. Dyke (GSC database)]; [Stryd and Rousseau (1996) In: R.E. Morlan, CMC database (1999)]
4490 ± 130	Beta-14402		Kusawa Bluffs site (JdVa-2) (60°35'N, 136°05'W), YT, CAN	[Greer (1986) In: Stephenson, et al. (in press), located below bison bones in archaeological site]

Odocoileus hemionus (mule deer)

Measured Date (Normalized)	Lab No	Material	Location	References/Remarks
(8180 ± 70)	AA-10574	humerus	Nautilus Cave, Heceta I. (55.750°N, 134.000°W), AK, USA	Heaton, et al. (1995, 1996); Heaton and Grady (1998)
(5210 ± 60)	CAMS-31069	pelvis	On Your Knees Cave, Prince of Wales I., 135 m asl. (56.250°N, 133.500°W), AK, USA	Heaton, et al. (1996); Heaton and Grady (1998)

Odocoileus sp. (deer)

Measured Date (Normalized)	Lab No	Material	Location	References/Remarks
7670 ± 80 (7750 ± 80)	SFU-866	humerus	Landels, Thompson R. (50°35'N, 121°25'W), BC, CAN	[Rousseau (1989, 1991); Stryd and Roussseau (1996); Pokotylo and Mitchell (1998) In: R.E. Morlan, CMC database (1999)]
6000 ± 80 (6080 ± 80)	SFU-886		About 13 km SW of Ashcroft, Thompson R., BC, CAN	[Rousseau (1989, 1991); Stryd and Rousseau (1996); Pokotylo and Mitchell (1998) In: R.E. Morlan, CMC database (1999)]
(5510 ± 90)	Beta-14483	metapodial	E side of Big Divide L., Thompson drainage, BC, CAN	[Lawhead, et al. (1986); Richards and Rousseau (1987); Stryd and Rousseau (1996) In: R.E. Morlan, CMC database (1999)]
5480 ± 70 (5560 ± 75)	Beta-37977		About 13 km SW of Ashcroft, Thompson R., BC, CAN	[Rousseau (1989, 1991); Stryd and Rousseau (1996); Pokotylo and Mitchell (1998) In: R.E. Morlan, CMC database (1999)]
4100 ± 120 (4180 ± 120)	I-14490	antler	Van der Windt, Rice L. region, ON, CAN	[Jackson and McKillop (1991) In: R.E. Morlan, CMC database (1999)]
3520 ± 70 (3600 ± 75)	SFU-626		About 13 km SW of Ashcroft, Thompson R., BC, CAN	[Rousseau (1989, 1991); Stryd and Rousseau (1996); Pokotylo and Mitchell (1998) In: R.E. Morlan, CMC database (1999)]
2470 ± 65 (2550 ± 70)	S-1288		N shore of L. Erie, 171 m asl. (42°36'N, 80°28'W), ON, CAN	[Spence, et al. (1978, 1981, 1990); Fox (1980); Williamson (1990) In: R.E. Morlan, CMC database (1999)]
1300 ± 90 (1380 ± 90)	Beta-10073	antler	Twinkie Bog: Rice L. region, ON, CAN	[Jackson and McKillop (1991) In: R.E. Morlan, CMC database (1999)]
1050 ± 330 (1130 ± 330)	S-2064	(CMC-1200)	S Nation R., 3.2 km NE of Spencerville, Grenville County (44°51.5'N, 75°30.75'W), ON, CAN	[Pendergast (1966); Wright (1985); Jamieson (1990) In: R.E. Morlan, CMC database (1999)]
(980 ± 100)	SFU-57	antler	Shuswap R., 6.43 km E of township of Enderby, Thompson drainage (50.503°N, 119.068°W), BC, CAN	[Mohs (1980b); Richards and Rousseau (1987) In: R.E. Morlan, CMC database (1999)]
800 ± 40 (880 ± 45)	S-1712	(CMC-1062)	Lake Huron, Bruce County, (44°26'20"N, 81°23'30"W), ON, CAN	[Stewart (1974); Wright (1974, 1985) In: R.E. Morlan, CMC database (1999)]
780 ± 120 (860 ± 120)	S-1717	(CMC-1067)	Lake Huron, Bruce County, (44°26'20"N, 81°23'30"W), ON, CAN	[Stewart (1974); Wright (1974, 1985) In: R.E. Morlan, CMC database (1999)]
430 ± 70 (510 ± 75)	S-2011	(CMC-1200)	S Nation R., 3.2 km NE of Spencerville, Grenville County, (44°51'30"N, 75°30'45"W), ON, CAN	[Pendergast (1966); Wright (1985); Jamieson (1990) In: R.E. Morlan, CMC database (1999)]

Measured Date (Normalized)	Lab No	Material	Location	References/Remarks
410 ± 100 (490 ± 100)	S-2009	(CMC-1198)	S Nation R., 3.2 km NE of Spencerville, Grenville County, (44°51'30"N, 75°30'45"W), ON, CAN	[Pendergast (1966); Wright (1985); Jamieson (1990) In: R.E. Morlan, CMC database (1999)]
(400 ± 40)	Beta-94421	vertebra	Near Courtenay, Vancouver I., BC, CAN	[Dallstrom, et al. (1996) In: R.E. Morlan, CMC database (1999)]

Odocoileus? sp. (deer)

Measured Date (Normalized)	Lab No	Material	Location	References/Remarks
8400 ± 90 (8480 ± 90)	SFU-867		Landels, Thompson R. (50°35'N, 121°25'W), BC, CAN	[Rousseau (1989, 1991); Stryd and Roussseau (1996); Pokotylo and Mitchell (1998) In: R.E. Morlan, CMC database (1999)]

Alces latifrons (giant moose)

Measured Date (Normalized)	Lab No	Material	Location	References/Remarks
40200 ± 1100 (40230 ± 1100)	Beta-79862 CAMS-18425	cranium (frontal) (CRH-95-13)	Friday Gulch (Dawson Loc. 18) (63°46'N, 138°54'W), YT, CAN	C.R. Harington [unpublished]; R.E. Morlan, CMC database (1999)
(35610 ± 340)	TO-3712	antler beam (CMN 25969)	Dominion Ck. (Dawson Loc. 29) (63°48'N, 138°36'W), YT, CAN	C.R. Harington [unpublished]; R.E. Morlan, CMC database (1999)
33800 ± 2000 (33880 ± 2000)	I-4229	antler beam	Old Crow R. (Old Crow Loc. 22) (67.850°N, 139.970°W), YT, CAN	Morlan and Bonnichsen (1975); Harington (1977); [Bonnichsen (1978) incorrectly cites lab number as I-4429); Morlan et al. (1990); R.E. Morlan, CMC database (1999)
(33000 ± 750)	AA-3897	skin (F:AM-274 4002)	Little El Dorado Ck. (65.100°N, 147.680°W), AK, USA	Guthrie (1990d)
32040 +870/ -980	DIC-3090	skin (UA V64)	Livengood (65.000°N, 150.000°W), AK, USA	Guthrie (1990d)
(25330 ± 200)	AA-3896	skin (F:AM-274 4001)	Little El Dorado Ck. (65.100°N, 147.680°W), AK, USA	Guthrie (1990d)

Alces alces (moose)

Measured Date (Normalized)	Lab No	Material	Location	References/Remarks
(11570 ± 60)	CAMS-23472	metacarpal	Bluefish Cave I, 600 m asl. (67°08'N, 140°47'W), YT, CAN	Cinq-Mars (1979, 1990); Morlan and Cinq-Mars (1982); Burke and Cinq-Mars (1996)
8740 ± 70	DIC-2414	antler (F:AM-2039)	Fairbanks region, AK, USA	Guthrie (1990d)
7861 ± 423 (7041 ± 423)	SM-696-1	antler tool	Morson, Rainy R. District (49.030°N, 94.300°W), ON, CAN	Pettipas and Buchner (1983); Kenyon and Churcher (1965); R.E. Morlan, CMC database
7484 ± 423	SM-696-2	antler tool	Morson, Rainy R. District (49.030°N, 94.300°W), ON, CAN	Pettipas and Buchner (1983); Kenyon and Churcher (1965); R.E. Morlan, CMC database
5380 ± 55	DIC-2416	antler (uncatalogued)	Pearl Ck. (65.983°N, 147.317°W), AK, USA	Guthrie (1990d)
4400 ± 130	QU-717		St-Elzear Cave, Gaspe (48.235°N, 65.360°W), QC, CAN	[LaSalle (1984) In: A.S. Dyke (GSC database)]
4390 ± 120	QU-745		St-Elzear Cave, Gaspe (48.235°N, 65.360°W), QC, CAN	[LaSalle (1984) In: A.S. Dyke (GSC database)]
(4360 ± 100)	RIDDL-545	humerus	Anderson R. (69.000°N, 132.000°W), NT, CAN	Cinq-Mars (1991); R.E. Morlan, CMC database (1999)
(4040 ± 60)	TO-2356	antler (CMN 47691)	Miller Ck., Sixtymile area (Sixtymile Loc. 5) (63°59'N, 140°47'W), YT, CAN	Harington (1997a)
2330 ± 50	Beta-135071	metatarsal	Killik R. (Site KIR 275), AK, USA	[C. Martin and T. Birkedal, pers. comm. In: Stephenson et al. (2001). Reidentification based on DNA analysis (A. Cooper, pers. comm.)]

(985 ± 70)	Beta-33190 ETH-5898	antler (CMN 44668)	Sulphur Ck. (Dawson Loc. 55, Green Gulch) (67.800°N, 139.920°W), YT, CAN	Harington (1997a) [Incorrect laboratory numbers are given in Harington (1997) for this specimen. It is Beta-33190 (ETH-5898) rather than Beta-33189 (ETH-5897). [This specimen did not provide good quality collagen and the date was done on the insoluble fraction].
(920 ± 100)	OxA-385		Lac du Bonnet, Winnipeg R. (50.500°N, 97.000°W), MB, CAN	[Buchner and Roberts (1990) In: R.E. Morlan, CMC database (1999)]
740 ± 40	DIC-2415	antler (F:AM-539)	Goldstream Ck. (64.950°N, 147.583°W), AK, USA	Guthrie (1990d)
(650 ± 40)	RIDDL-550		SE shore of Richards I., Mackenzie Delta (68.830°N, 135.250°W), NT, CAN	[Friesen (1987); Morrison (1989); Cinq-Mars (1991); Arnold (1991, 1994); Balkwill and Rick (1994); Nolin (1994); Friesen and Arnold (1995) In: R.E. Morlan, CMC database (1999)]
(640 ± 90)	AECV-1014C		SE shore of Richards I. Mackenzie Delta (69.417°N, 134.167°W), NT, CAN	[Arnold (1991, 1994); Nolin (1994) In: R.E. Morlan, CMC database (1999)]
620 ± 75 (700 ± 85)	GaK-4712		N shore of Whitemouth R., 259 m asl (50.122°N, 96.035°W), MB, CAN	[Buchner (1979, 1982); Meiklejohn, et al. (1994); Ens (1998) In: R.E. Morlan, CMC database (1999)]
(610 ± 80)	AECV-1016C		SE shore of Richards I. Mackenzie Delta (69.417°N, 134.167°W), NT, CAN	[Arnold (1991, 1994); Nolin (1994) In: R.E. Morlan, CMC database (1999)]
(550 ± 120)	RIDDL-544	axis	Trail R. (69.330°N, 138.000°W), YT, CAN	[Morrison (1989); Nagy (1989, 1991); Cinq-Mars (1991); Vogel, et al. (1991) In: R.E. Morlan, CMC database (1999)]
410 ± 120	QU-714		St-Elzear Cave, Gaspe (48.235°N, 65.360°W), QC, CAN	[LaSalle (1984) In: A.S. Dyke (GSC database)]
180 ± 55 (260 ± 60)	S-973	(CMC-745)	McDonald Farm: on a shoreline at the W end of Olding I., Pictou County, NS, CAN	[Smith and Wintemberg (1929) In: R.E. Morlan, CMC database (1999)]

Alces sp. (moose)

Measured Date (Normalized)	Lab No	Material	Location	References/Remarks
(200 ± 100)	TO-3711	antler beam (CMN 12231)	Near Rimbey (approx. 52°38'N, 114°14'W), AB, CAN	C.R. Harington [unpublished]; [C.R. Harington, pers. comm. (1999) In: R.E. Morlan, CMC database (1999)]

Rangifer tarandus (caribou, reindeer)

Measured Date (Normalized)	Lab No	Material	Location	References/Remarks
52500 +2700/ -2000	USGS-1843	antler	Noatak Basin, AK, USA	Elias, et al. (1999)
(45470 ± 1150)	TO-2700	cranial fragment with base of right antler (CMN 42431)	Sixtymile area (Sixtymile Loc. 3), 701 m asl., (64°1.00'N, 140°44'W), YT, CAN	Harington (1997a)
(40640 ± 420)	TO-3714	antler fragment (CMN 43810)	S of Riviere Verte - from a gravel pit opened into the St-Antonin moraine, near Riviere-du-Loup (47°46'N, 69°29'W), QC, CAN	C.R. Harington [unpublished]
>40000	Ua-1112	antler	Kap Kobenhavn (Peary Land) 53 m asl (82°29'N, 21°01'W), GRE	Meldgaard and Bennike (1989); Bennike (1997)
31680 +1600 / -1380	K-4918	antler	Kap Kobenhavn (Peary Land) 53 m asl (82°29'N, 21°01'W), GRE	[Meldgaard and Bennike (1989) - see AMS date (Ua-1112) of >40,000 BP which is considered to be more reliable]
(27730 ± 1060)	AECV-599C	antler	Consolidated Pit 46, Villeneuve, N Saskatchewan R., NW of Edmonton (53.660°N, 113.820°W), AB, CAN	Young, et al. (1994); Burns (1996); R.E.Morlan, CMC database (1999)
27000 +3000 / -2000 (27245 +	GX-1640	tibia flesher (CMC 342) - bone apatite date	Old Crow (Loc. 14N) (67°51'N, 139°46'W), YT, CAN	Irving and Harington (1973); Morlan and Bonnichsen (1975); Harington (1977); Morlan (1986); Irving (1987); R.E. Morlan, CMC database (1999); [See AMS collagen re-dating of 1350 ± 150 (RIDDL-145)]

(24820 ± 110)	RIDDL-226	tibia	Bluefish Cave II, 600 m asl. (67°08'N, 140°47'W), YT, CAN	Cinq-Mars (1979, 1990); Morlan and Cinq-Mars (1982); Cinq-Mars, et al. (1991); Burke and Cinq-Mars (1996)
24800 ± 650 (24880 ± 650)	CRNL-1233 = ("CMN-1233")	antler (CMC-1233)	Old Crow R. (Old Crow Loc. 29) (68.180°N, 139.960°W), YT, CAN	Harington et al. (1975); Morlan (1986); Morlan et al. (1990); R.E. Morlan, CMC database (1999)
(24700 ± 300)	RIDDL-230	antler	Old Crow R. (Old Crow Loc. 29) (68.180°N, 139.958°W), YT, CAN	Harington et al. 1975; Morlan et al. (1990); R.E. Morlan, CMC database (1999)
23900 ± 470 (23980 ± 470)	I-8580	antler (CMN 25926)	Hunker Ck. (Dawson Loc. 16) (63.930°N, 138.980°W), YT, CAN	Harington (1977, 1978, 1980, 1989, 1997); Bonnichsen (1979); Harington and Morlan (1992); Matheus (1995); R.E. Morlan, CMC database (1999)
15690 ± 140	Beta-67669	humerus	Lime Hills, Cave 1, 527 m asl. (61.750°N, 155.500°W), AK, USA	Ackerman (1996b)
13130 ± 180	Beta-67671	metapodial	Lime Hills, Cave 1, 527 m asl. (61.750°N, 155.500°W), AK, USA	Ackerman (1996b)
(12830 ± 60)	CAMS-23468	radius	Bluefish Cave I, 600 m asl. (67°08'N, 140°47'W), YT, CAN	Cinq-Mars (1979, 1990); Morlan and Cinq-Mars (1982); Burk and Cinq-Mars (1996)
(12210 ± 210)	RIDDL-277	metatarsal	Bluefish Cave I, 600 m asl. (67°08'N, 140°47'W), YT, CAN	Cinq-Mars (1979, 1990); Morlan and Cinq-Mars (1982); Burke and Cinq-Mars (1996)
(11350 ± 110)	Beta-27512 ETH-4582	antler (KlVi-1:1)	Hunker Ck. (Dawson Loc. 16) (63.920°N, 137.830°W), YT, CAN	[Harington and Morlan (1992) - caribou antler tool. Earliest evidence of people in the Dawson area.]; R.E. Morlan, CMC database (1999)
(10555 ± 110)	AA-18449R	metacarpal	Bumper Cave, Prince of Wales I., (56.170°N, 133.500°W), AK, USA	Heaton, et al. (1996); Heaton and Grady (1998)
(10515 ± 90)	AA-18449	metacarpal	Bumper Cave, Prince of Wales I. (56.170°N, 133.500°W), AK, USA	Heaton, et al. (1996); Heaton and Grady (1998)
9570 ± 150	K-1583		Nares Strait area (67.100°N, 158.250°W), AK, USA	A.S. Dyke (GSC database)
9070 ± 150	K-980	marrow-cracked bones	Trail Ck. Cave 2, Seward Pen. (65.800°N, 163.220°W), AK, USA	A.S. Dyke (GSC database)
8435 ± 85	Beta-49163		Tingmeakpuk site (KIP-273), Brooks Range, AK, USA	Mason, et al. (2001, Table 2)
7980 ± 115	K-3865	antler	Jorgen Bronlund Fjord, Peary land (82.000°N, 32.000°W), GRE	Stewart and England (1986); Meldgaard (1986); Bennike (1997)
7060 ± 70	Beta-49166		Tingmeakpuk site (KIP-273), Brooks Range, AK, USA	Mason, et al. (2001, Table 2)
6650 ± 180 (6730 ± 180)	S-3486	(CMC-1428)	Hutchison Bay, 15 m asl., Beaufort Sea coast, NT, CAN	[Nolin (1994) In: R.E. Morlan, CMC database (1999)]
6450 ± 135 (6530 ± 135)	I-4221	antler	Old Crow R. (Old Crow Loc. 69) (67.855°N, 139.800°W), YT, CAN	Harington (1977, 1980); Morlan et al. (1990); R.E. Morlan, CMC database (1999)
6240 ± 100	K-4348	antler	Jorgen Bronlund Fjord (82.000°N, 32.000°W), GRE	Meldgaard (1986)
6200 ± 70	Lu-1096		Skeldal (73.000°N, 24.000°W), GRE	Bennike (1997)
5800 ± 80	Beta-27514 ETH-4583	mandible in clay nodule (Sondrestrom 1)	Sondrestrom Fiord (67.300°N, 52.000°W), GRE	Bennike (1997)
(5690 ± 60)	TO-1564		Old Horton R. channel, Cape Bathurst Pen. (70.080°N, 128.580°W), NU, CAN	[Nolin (1994) In: R.E. Morlan, CMC database (1999)]
(5610 ± 65)	Beta-62584 ETH-10483	(CMC-1451)	Old Horton R. channel, Cape Bathurst Pen. (70.080°N, 128.580°W), NT, CAN	[Nolin (1994) In: R.E. Morlan, CMC database (1999)]
5470 ± 95	K-3866	antler	Jorgen Bronlund Fjord, Peary land (82.000°N, 32.000°W), GRE	Meldgaard (1986)
5360 ± 140 (5440 ± 140)	S-3485	(CMC-1427)	Hutchison Bay, 15 m asl., Beaufort Sea coast, NT, CAN	[Nolin (1994) In: R.E. Morlan, CMC database (1999)]
5010 ± 100 (5090 ± 100)	I-8642	radio-ulna (CMN 26011)	Quartz Ck. (Dawson Loc. 8) (63.800°N, 139.070°W), YT, CAN	Harington (1977, 1980); R.E. Morlan, CMC database (1999)

4430 ± 110	I-1946		S region (59.000°N, 155.500°W), AK, USA	Dumond (1984)
(4430 ± 240)	RIDDL-322	metatarsal	Great Bear R. (65.000°N, 124.000°W), NT, CAN	Cinq-Mars (1991); R.E. Morlan, CMC database (1999) - specimen with butchering marks
4360 ± 1160 (4440 ± 1160)	CRNL-1230	antler (CMC-1230)	Grinnell Pen., 24 m asl., Devon I. (76°17'N, 94°48'W), NU, CAN	[Wilmeth (1978); McGhee (1979, 1980) In: R.E. Morlan, CMC database (1999)]
(4280 ± 90)	AECV-1713C		Harrowby Bay, Cape Bathurst Pen., NT, CAN	Cinq-Mars (1991); LeBlanc (1994); [Nolin (1994) In: R.E. Morlan, CMC database (1999)]
(4100 ± 90)	AECV-1711C		Harrowby Bay, Cape Bathurst Pen., NT, CAN	Cinq-Mars (1991); LeBlanc (1994); [Nolin (1994) In: R.E. Morlan, CMC database (1999)]
(3980 ± 80)	AECV-1710C		Harrowby Bay, Cape Bathurst Pen., NT, CAN	R.E. Morlan, CMC database (1999)
(3950 ± 80)	AECV-1712C		Harrowby Bay, Cape Bathurst Pen., NT, CAN	Cinq-Mars (1991); LeBlanc (1994); [Nolin (1994) In: R.E. Morlan, CMC database (1999)]
(3890 ± 180)	RIDDL-323	tibia	Great Bear R. (65.000°N, 124.000°W), NT, CAN	Cinq-Mars (1991); R.E. Morlan, CMC database (1999) - specimen with butchering marks
(3880 ± 80)	AECV-1709C		Harrowby Bay, Cape Bathurst Pen., NT, CAN	R.E. Morlan, CMC database (1999)
3760 ± 70 (3840 ± 70)	S-2279	antler	Near confluence of Ningunsaw and Iskut rivers (56°56.3'N, 130°21'W), BC, CAN	Ryder (1987); R.E. Morlan, CMC database (1999)
3710 ± 80	K-4059	antler	Jorgen Bronlund Fjord, Peary Land (82.000°N, 32.000°W), GRE	Meldgaard (1986)
3700 ± 300 (3780 ± 300)	K-505	antler	Igloolik I., 42-54 m asl (69.380°N, 81.800°W), NU, CAN	[Rainey and Ralph (1959); Meldgaard (1960, 1962); Wilmeth (1978); Maxwell (1985) In: R.E. Morlan, CMC database (1999)]
3700 ± 70 (3780 ± 75)	Beta-25032		Truelove Lowland, Devon I. (75.670°N, 83.670°W), NU, CAN	[McCartney (1989); McCartney and Helmer (1989) In: R.E. Morlan, CMC database (1999)]; [Helmer (1991) In: A.S. Dyke (GSC database)]; Dyke (1998)
3670 ± 70 (3750 ± 75)	Beta-62585	(CMC-1452)	Old Horton R. channel, Cape Bathurst Pen. (70.080°N, 128.580°W), NT, CAN	R.E. Morlan, CMC database (1999)
3670 ± 90 (4030 (2030-	K-6192		Nipisat, Sisimiut, Sisimiut Kommune, Sisimiut District, GRE	[Mobjerg (1997) In: Danish/Greenlandic Network (1999)]; Mobjerg (1998)
3620 ± 125	S-1261		Near Thelon R. (62.750°N, 105.417°W), NT, CAN	Gordon (1996)
3560 ± 123 (3640 ± 123)	P-208	antler	Igloolik I., 42-54 m asl (69.380°N, 81.800°W), NU, CAN	[Rainey and Ralph (1959); Meldgaard (1960, 1962); Wilmeth (1978); Maxwell (1985); In: R.E. Morlan, CMC database (1999)]
3555 ± 140 (3635 ± 140)	S-2883	(CMC-1368)	Navy Board Inlet, 6.5 m asl., northern Baffin I., NU, CAN	R.E. Morlan, CMC database (1999)
(3525 ± 50)	AAR-3576		For locality, see Mobjerg (1998) Figure 1, GRE	Mobjerg (1998); [3,880 (1880-1790BC), 1910-1750BC]
(3515 ± 65)	AAR-3577		For locality, see Mobjerg (1998) Figure 1, GRE	Mobjerg (1998); [3,870 (1870-1705BC), 1910-1740BC]
3490 ± 80 (3860 (1860-	K-6031		Nipisat, Sisimiut, Sisimiut Kommune, Sisimiut District, GRE	[Mobjerg (1997) In: Danish/Greenlandic Network (1999)]; Mobjerg (1998)
3490 ± 80 (3860 (1860-	K-5192		Angujaartorfik, Sondre Stromfjord, Manitsoq Kommune, Sisimiut District (66.833°N, 51.083°W), GRE	[Kapel (1996) In: Danish/Greenlandic Network (1999)]
(3480 ± 50)	AAR-3578		For locality, see Mobjerg (1998) Figure 1, GRE	Mobjerg (1998); [3,760 (1760BC), 1880-1700BC]

3440 ± 80 (3740 (1740BC),	K-5194		Angujaartorfik, Sondre Stromfjord, Manitsoq Kommune, Sisimiut District (66.833°N, 51.083°W), GRE	[Kapel (1996) In: Danish/Greenlandic Network (1999)]
3370 ± 80 (3740 (1740BC),	K-5191		Angujaartorfik, Sondre Stromfjord, Manitsoq Kommune, Sisimiut District (66.833°N, 51.083°W), GRE	[Kapel (1996) In: Danish/Greenlandic Network (1999)]
3310 ± 85 (3600 (1600-	K-6195		Nipisat, Sisimiut, Sisimiut Kommune, Sisimiut District, GRE	[Mobjerg (1997) In: Danish/Greenlandic Network (1999)]; Mobjerg (1998)
3300 ± 85 (3530 (1530BC),	K-6196		Nipisat, Sisimiut, Sisimiut Kommune, Sisimiut District, GRE	[Mobjerg (1997) In: Danish/Greenlandic Network (1999)]; Mobjerg (1998)
3208 ± 156 (3288 ± 156)	P-228	antler	Firth R., 152 m asl., Engigstciak (69.360°N, 139.525°W), YT, CAN	Mackay, et al. (1961); [Wilmeth (1978) In: A.S. Dyke (GSC database)]; R.E. Morlan, CMC database (1999)
3190 ± 75 (3440 (1440BC),	K-5193		Angujaartorfik, Sondre Stromfjord, Manitsoq Kommune, Sisimiut District (66.833°N, 51.083°W), GRE	[Kapel (1996) In: Danish/Greenlandic Network (1999)]
(3160 ± 800)	RIDDL-546	humerus	Harrowby Bay, Cape Bathurst Pen. (69.000°N, 132.000°W), NT, CAN	Cinq-Mars (1991); R.E. Morlan, CMC database (1999)
3140 ± 120 (3220 ± 120)	GX-1702	long-bone fragments (CMC-359)	Lapointe, Coppermine drainage (67.750°N, 115.370°W), NT, CAN	[McGhee (1969); Wilmeth (1978); Morrison (1979); Gordon (1996) In: R.E. Morlan, CMC database (1999)]
(3110 ± 180)	RIDDL-364	long-bone fragments	Harrowby Bay, Cape Bathurst Pen. (69.000°N, 132.000°W), NT, CAN	Cinq-Mars (1991); R.E. Morlan, CMC database (1999)
(3085 ± 45)	AAR-3570		For locality, see Mobjerg (1998) Figure 1 (66.75°N, 53.58°W), GRE	Mobjerg (1998); [3,385 (1385-1325BC), 1405-1265BC]
(3065 ± 40)	AAR-3572		For locality, see Mobjerg (1998) Figure 1 (66.75°N, 53.58°W), GRE	Mobjerg (1998); [3,370 (1370-1315BC), 1395-1265BC]
3040 ± 60	K-4572	(GGU 79318)	Jakobshavn Isfjord (69°25'N, 50°15'W), GRE	Weidick, et al. (1990)
(3040 ± 80)	AAR-3571		For locality, see Mobjerg (1998) Figure 1 (66.75°N, 53.58°W), GRE	Mobjerg (1998); [3,290 (1290-1270BC), 1300 1210BC]
3010 ± 95 (3260 (1260-	K-6459		Nipisat, Sisimiut, Sisimiut Kommune, Sisimiut District, GRE	[Mobjerg (1997) In: Danish/Greenlandic Network (1999)]; Mobjerg (1998)
(2950 ± 50)	AAR-3573		For locality, see Mobjerg (1998) Figure 1, GRE	Mobjerg (1998); [3,150 (1150-1130BC), 1260-1040BC]
2940 ± 80 (2860 (2860± 80),	K-5584		Nipisat, Sisimiut, Sisimiut Kommune, Sisimiut District, GRE	[Mobjerg (1997) In: Danish/Greenlandic Network (1999)]; Mobjerg (1998); [Mobjerg (1998) In: Dyke, et al. (1999)]
(2930 ± 140)	RIDDL-133	antler	Old Crow R. (Loc. 29) (68.180°N, 139.958°W), YT, CAN	Nelson et al. (1986); R.E. Morlan, CMC database (1999)
2920 ± 85 (3120 (1120BC),	K-6193		Nipisat, Sisimiut, Sisimiut Kommune, Sisimiut District, GRE	[Mobjerg (1997) In: Danish/Greenlandic Network (1999)]; Mobjerg (1998)]
2910 ± 80 (3110 (1110-	K-6460		Nipisat, Sisimiut, Sisimiut Kommune, Sisimiut District, GRE	[Mobjerg (1997) In: Danish/Greenlandic Network (1999)]; Mobjerg (1998)]
2898 ± 136 (2978 ± 136)	P-210	antler	Jens Munk I., Igloolik area 44 m asl (69.170°N, 83.980°W), NU, CAN	[Rainey and Ralph (1959); Meldgaard (1960, 1962); Wilmeth (1978); Maxwell (1985); In: R.E. Morlan, CMC database (1999)]
2860 ± 80 (3000 (1000BC),	K-5864		Nipisat, Sisimiut, Sisimiut Kommune, Sisimiut District, GRE	[Mobjerg (1997) In: Danish/Greenlandic Network (1999); Mobjerg (1998)]
(2815 ± 45)	AAR-3575		For locality, see Mobjerg (1998) Figure 1, GRE	Mobjerg (1998); [2,970 (970-930BC), 1005-905BC]
2810 ± 110	K-979	marrow-cracked bones	Trail Ck. Cave 2, Seward Pen. (65.800°N, 163.220°W), AK, USA	A.S. Dyke (GSC database)

2805 ± 140 (2885 ± 140)	S-2999	(CMC-1393)	Old Horton R. channel, Cape Bathurst Pen. 60 m asl (70.080°N, 128.580°W), NT, CAN	LeBlanc (1994); R.E. Morlan, CMC database (1999)
2770 ± 110	K-1290		Trail Ck. Cave 9, Seward Pen. (65.800°N, 163.220°W), AK, USA	A.S. Dyke (GSC database)
2730 ± 90 (2810 ± 90)	S-467	(CMC-303)	400 m W of Coppermine River (67.750°N, 115.370°W), NT, CAN	[McGhee (1969); Wilmeth (1978); Morrison (1979); Gordon (1996) In: R.E. Morlan, CMC database (1999)]
2706 ± 118	P-139		Central region (59.450°N, 151.730°W), AK, USA	[Workman et al. (1980) In: A.S. Dyke (GSC database)]
2700 ± 110	K-983	antler	Trail Ck. Cave 2, Seward Pen. (65.800°N, 163.220°W), AK, USA	A.S. Dyke (GSC database)
2670 ± 85 (2810 (810BC), 900-	K-6194		Nipisat, Sisimiut, Sisimiut Kommune, Sisimiut District, GRE	[Mobjerg (1997) In: Danish/Greenlandic Network (1999)]; Mobjerg (1998)]
2620 ± 120	K-147	antler	Trail Ck. Cave 2, Seward Pen. (65.800°N, 163.220°W), AK, USA	[Tauber (1960) In: A.S. Dyke (GSC database)]
2605 ± 370 (2685 ± 370)	CRNL-1259	antler (CMC-1259)	Grinnell Pen., 24 m asl., Devon I. (76°17'N, 94°48'W), NU, CAN	[Wilmeth (1978); McGhee (1979, 1980) In: R.E. Morlan, CMC database (1999)]
(2540 ± 100)	AECV-1108C		Old Horton R. channel, Cape Bathurst Pen. 60 m asl (70.080°N, 128.500°W), NT, CAN	LeBlanc (1994); R.E. Morlan, CMC database (1999)
(2540 ± 100)	AECV-1107C		Old Horton R. channel, Cape Bathurst Pen. 60 m asl (70.080°N, 128.500°W), NT, CAN	LeBlanc (1994); R.E. Morlan, CMC database (1999)
2535 ± 140 (2615 ± 140)	S-3036	(CMC-1394)	Old Horton R. channel, Cape Bathurst Pen. 60 m asl (70.080°N, 128.580°W), NU, CAN	LeBlanc (1994); R.E. Morlan, CMC database (1999)
(2530 ± 110)	AECV-1104C		Old Horton R. channel, Cape Bathurst Pen. 60 m asl. (70.080°N, 128.580°W), NT, CAN	LeBlanc (1994); R.E. Morlan, CMC database (1999)
(2520 ± 80)	AECV-1795C		NW of Ivujivik, 23 m asl., Hudson Strait, Nouveau-Quebec region, QC, CAN	[Nagy (1997) In: R.E. Morlan, CMC database (1999)]
(2455 ± 50)	AAR-3574		For locality, see Mobjerg (1998) Figure 1, GRE	Mobjerg (1998); [2,520 (520BC), 760-410BC]
(2450 ± 50)	TO-6871	faeces	Thandlat Mountain, W of Kusawa L., YT, CAN	Kuzyk, et al. (1999); [Gotthardt (1998) In: R.E. Morlan, CMC database (1999)]
2440 ± 110 (2520 ± 110)	RL-797	antler	S of Longview on Pekisko Ck., a tributary of Highwood R., in the Bow drainage (50°25'N, 114°15'W), AB, CAN	Brumley and Rushworth (1983); [Wilson (1977, 1983) In: R.E. Morlan, CMC database (1999)]
(2410 ± 90)	AECV-1105C		Old Horton R. channel, Cape Bathurst Pen. 60 m asl (70.080°N, 128.580°W), NT, CAN	LeBlanc (1994); R.E. Morlan, CMC database (1999)
(2410 ± 100)	AECV-1106C		Old Horton R. channel, Cape Bathurst Pen. 60 m asl (70.080°N, 128.500°W), NT, CAN	LeBlanc (1994); R.E. Morlan, CMC database (1999)
2404 ± 137 (2484 ± 137)	P-212	antler	Igloolik (69.380°N, 81.800°W), NU, CAN	[Maxwell (1985) In: A.S. Dyke (GSC database)]; R.E. Morlan, CMC database (1999)
2354 ± 135 (2434 ± 140)	P-211	antler	Jens Munk I., Igloolik area 24 m asl (69.170°N, 83.980°W), NU, CAN	[Rainey and Ralph (1959); Meldgaard (1960, 1962); Wilmeth (1978); Maxwell (1985); In: R.E. Morlan, CMC database (1999)]
2320 ± 50 (2400 ± 55)	S-1613	(CMC-1019)	Navy Board Inlet, Borden Pen., Baffin I. (73.025°N, 80.630°W), NU, CAN	R.E. Morlan, CMC database (1999)
2250 ± 130	P-175		W region (66.270°N, 161.970°W), AK, USA	Mason and Jordan (1993)
2160 ± 110	K-1289		Trail Ck. Cave 9, Seward Pen. (65.800°N, 163.220°W), AK, USA	A.S. Dyke (GSC database)
2080 ± 75	K-3867	antler	Jorgen Bronlund Fjord, Peary land (82.000°N, 32.000°W), GRE	Meldgaard (1986)

1925 ± 70 (2005 ± 75)	S-756	(CMC-595)	Eclipse Sound, 5 m asl., Baffin I., NU, CAN	R.E. Morlan, CMC database (1999)
(1900 ± 60)	TO-4531	tibia	Little Cornwallis I. (75.583°N, 95.833°W), NU, CAN	[Helmer, et al. (1995) In: R.E. Morlan, CMC database (1999)]; Dyke, et al. (1999)
(1880 ± 140)	RIDDL-141	antler	Porcupine R. (67.483°N, 139.917°W), YT, CAN	Nelson et al. (1986); R.E. Morlan, CMC database (1999)
1880 ± 90 (1960 ± 90)	S-880	(CMC-656)	Navy Board Inlet, Borden Pen., Baffin I. (73.025°N, 80.630°W), NU, CAN	[Mary-Rousselière (1976) In: Dyke, et al. (1999)]; R.E. Morlan, CMC database (1999)
1870 ± 70 (1920 ± 70)	Beta-65519		Satkualuk, Richards I., NT, CAN	R.E. Morlan, CMC database (1999)
1860 ± 120 (1940 ± 120)	S-1444	antler (CMC-959)	Navy Board Inlet, Borden Pen., Baffin I. (73.025°N, 80.630°W), NU, CAN	R.E. Morlan, CMC database (1999)
1830 ± 70	K-3868	antler (ZMK Collection)	Kap Morris Jesup (83.000°N, 34.000°W), GRE	Meldgaard (1986)
1830 ± 80 (1910 ± 80)	S-1208	(CMC-851)	Navy Board Inlet, N Baffin I., NU, CAN	R.E. Morlan, CMC database (1999)
1790 ± 70 (1870 ± 75)	S-465	(CMC-301)	400 m W of Coppermine River (67.750°N, 115.370°W), NT, CAN	[McGhee (1969); Wilmeth (1978); Gordon (1996) In: R.E. Morlan, CMC database (1999)]
1740 ± 110 (1820 ± 110)	K-1066	antler	Igloolik I. (Freuchen) 18.5 m asl (69.333°N, 81.800°W), NU, CAN	R.E. Morlan, CMC database (1999)
1730 ± 160 (1810 ± 160)	S-3368	antler (CMC-1406)	N end of Cape Krusenstern, 3.5 m asl, NU, CAN	[Morrison (1983, 1989) In: R.E. Morlan, CMC database (1999)]
(1730 ± 100)	RIDDL-140	antler	Porcupine R. (67.583°N, 139.750°W), YT, CAN	Nelson et al. (1986); R.E. Morlan, CMC database (1999)
1690 ± 150 (1770 ± 150)	S-847	(CMC-651)	Navy Board Inlet, Borden Pen., Baffin I. (73.025°N, 80.630°W), NU, CAN	R.E. Morlan, CMC database (1999)
1680 ± 70	K-4138	antler (CN 3694)	Vahl Fjord (66.250°N, 34.420°W), GRE	Meldgaard (1986)
1619 ± 210	P-98		NW region (68.320°N, 166.700°W), AK, USA	[Ralph and Ackerman (1961) In: A.S. Dyke (GSC database)]
1590 ± 100 (1670 ± 100)	S-848	(CMC-652)	Navy Board Inlet, Borden Pen., Baffin I. (73.025°N, 80.630°W), NU, CAN	[Mary-Rousseliere (1976) In: Dyke, et al. (1999)]; R.E. Morlan, CMC database (1999)
(1550 ± 50)	TO-3747	long-bone	Little Cornwallis I. (75.583°N, 95.833°W), NU, CAN	[Helmer, et al. (1995) In: Dyke, et al. (1999)]; R.E. Morlan, CMC database (1999)
1545 ± 155 (1625 ± 155)	S-2189	(CMC-1227)	Eclipse Sound, 2.7 m asl., Baffin I., NU, CAN	R.E. Morlan, CMC database (1999)
1530 ± 80 (1610 ± 80)	S-755	(CMC-594)	Eclipse Sound, 2.7 m asl., Baffin I., NU, CAN	R.E. Morlan, CMC database (1999)
1520 ± 70 (1600 ± 75)	S-849	(CMC-653)	Navy Board Inlet, Borden Pen., Baffin I. (73.025°N, 80.630°W), NU, CAN	R.E. Morlan, CMC database (1999)
1510 ± 30 (1470 ± 30)	Beta-88718	(CMC-1473)	N side of Wood R., Ellesmere I., NU, CAN	[Hattersley-Smith (1973) In: R.E. Morlan, CMC database (1999)]
(1470 ± 60)	Beta-65520 CAMS-9005	antler	Satkualuk, Richards I., NT, CAN	R.E. Morlan, CMC database (1999)
1450 ± 80 (1530 ± 80)	S-468	(CMC-306)	300 m N of Lapointe (67.750°N, 115.370°W), NT, CAN	R.E. Morlan, CMC database (1999)
(1450 ± 60)	TO-4532	antler	Little Cornwallis I. (75.583°N, 95.833°W), NU, CAN	[Helmer, et al. (1995) In: Dyke, et al. (1999)]; R.E. Morlan, CMC database (1999)
(1420 ± 70)	Gif-8181		W end of Naujat Pen., Victoria I., NU, CAN	[Le Mouel (1999) In: R.E. Morlan, CMC database (1999)]
1369 ± 102	P-138		(59.450°N, 151.730°W), AK, USA	[Workman et al. (1980) In: A.S. Dyke (GSC database)]
(1350 ± 150)	RIDDL-145	tibia	Old Crow Flats (Old Crow Loc. 14N) (67.850°N, 139.768°W), YT, CAN	[Irving (1987) - redated]; R.E. Morlan, CMC database (1999); [Reliable AMS date on the fleshing tool].
1325 ± 90 (1405 ± 90)	S-881	(CMC-657)	Eclipse Sound, 2.7 m asl., Baffin I., NU, CAN	R.E. Morlan, CMC database (1999)

1320 ± 90 (1400 ± 90)	S-879	(CMC-655)	Navy Board Inlet, Borden Pen., Baffin I. (73.025°N, 80.630°W), NU, CAN	R.E. Morlan, CMC database (1999)
1315 ± 80 (1395 ± 80)	S-2190	(CMC-1228)	Eclipse Sound, 5 m asl., Baffin I., NU, CAN	R.E. Morlan, CMC database (1999)
1280 ± 60 (1360 ± 65)	S-1207	(CMC-850)	Navy Board Inlet, Borden Pen., Baffin I. (73.025°N, 80.630°W), NU, CAN	R.E. Morlan, CMC database (1999)
1260 ± 70	K-4350	antler (CN 3390)	Tingmiarmiut Fjord (62.670°N, 43.170°W), GRE	Meldgaard (1986)
1250 ± 70	K-4343	antler (CN 1007)	Angmagssalik Fjord (65.570°N, 36.830°W), GRE	Meldgaard (1986)
1180 ± 70 (1260 ± 75)	S-3370	(CMC-1408)	Overlooking S shore of Amundsen Gulf, NE corner of Pearce Pt. Harbour, 1 m asl, NT, CAN	[D.A. Morrison, pers. comm. (1999) In: R.E. Morlan, CMC database (1999)]
1180 ± 90 (1260 ± 90)	S-850	(CMC-654)	Eclipse Sound, 5 m asl., Baffin I., NU, CAN	R.E. Morlan, CMC database (1999)
(1160 ± 100)	BM-1803R	antler	Brooman Pt., Bathurst I. (75.390°N, 97.290°W), NU, CAN	[Bowman et al. (1990) In: Dyke, et al. (1999), revision of BM-1803]; [McGhee (1984); Morrison (1989) In: R.E. Morlan, CMC database (1999)]
1145 ± 125 (1225 ± 125)	S-3037	(CMC-1395)	SW corner of Ferguson L., Victoria I., NU, CAN	R.E. Morlan, CMC database (1999)
1110 ± 70	K-4139	antler (CN 3789)	Sermilik Fjord (65.980°N, 37.250°W), GRE	Meldgaard (1986)
(1100 ± 100)	BM-1804R	antler	Brooman Pt., Bathurst I. (75.390°N, 97.290°W), NU, CAN	[Bowman et al. (1990) In: Dyke, et al. (1999), revision of BM-1804]; [McGhee (1984); Morrison (1989) In: R.E. Morlan, CMC database (1999)]
1100 ± 100	K-982	marrow-cracked bones	Trail Ck. Cave 2, Seward Pen. (65.800°N, 163.220°W), AK, USA	A.S. Dyke (GSC database)
1090 ± 90 (1170 ± 90)	S-1205	(CMC-848)	Navy Board Inlet, Borden Pen., Baffin I. (73.025°N, 80.630°W), NU, CAN	R.E. Morlan, CMC database (1999)
1080 ± 125 (1160 ± 125)	S-3039	(CMC-1397)	Ekalluk R., 5 m asl., Victoria I., NU, CAN	R.E. Morlan, CMC database (1999)
(1030 ± 50)	TO-3749	femur	Little Cornwallis I. (75.583°N, 95.833°W), NU, CAN	[Helmer, et al. (1995) In: Dyke, et al. (1999)]; R.E. Morlan, CMC database (1999)
(1020 ± 50)	TO-2518	antler	Little Cornwallis I. (75.583°N, 95.833°W), NU, CAN	[Helmer, et al. (1993) In: Dyke, et al. (1999)]
1010 ± 160 (1090 ± 160)	GaK-1258	burned bone (CMC-166)	Great Slave L. (62.780°N, 108.930°W), NT, CAN	[Noble (1971); Wilmeth (1978); Gordon (1996) In: R.E. Morlan, CMC database (1999)]
1010 ± 50 (1090 ± 55)	S-2429	long-bone fragments (CMC-1312)	S bank of Netsilik R., 8 m asl (69.080°N, 93.750°W), NT, CAN	[Savelle (1987); Morrison (1989) In: R.E. Morlan, CMC database (1999)]
1005 ± 50 (1135 ± 55)	S-2428	antler (CMC-1311)	Sherman Inlet, Crescent Harbour (67.750°N, 98.500°W), NT, CAN	[Savelle (1987); Morrison (1989) In: R.E. Morlan, CMC database (1999)]
1000 ± 50 (1070 ± 50)	Beta-79850 CAMS-18413	left pelvis fragment (CRH-95-1)	Simpson Pt., Pauline Cove, Herschel I. (69°34'N, 138°55'W), YT, CAN	C.R. Harington [unpublished]
990 ± 70	K-4140	antler (CN 2267)	Kulusuk (65.420°N, 36.830°W), GRE	Meldgaard (1986)
970 ± 70	K-4345	antler (ZMK 11/1939)	Kulusuk (65.420°N, 36.830°W), GRE	Meldgaard (1986)
955 ± 125 (1035 ± 125)	S-3038	(CMC-1396)	SW corner of Ferguson L., Victoria I., NU, CAN	R.E. Morlan, CMC database (1999)
(950 ± 70 (1040AD), 1010-	K-3514		Aasivisuit, Sisimiut District, GRE	[Gullov (1996) In: Danish/Greenlandic Network (1999)]
940 ± 70	K-4344	antler (CN 1006)	Angmagssalik O (65.520°N, 38.000°W), GRE	Meldgaard (1986)
940 ± 70 (1020 ± 75)	GaK-2344	(CMC-329)	Sandgirt Lake, 455-456 m asl., (53.900°N, 65.315°W), NF (Labrador), CAN	[Wilmeth (1978) In: R.E. Morlan, CMC database (1999)]

(930 ± 120)	TO-4535	antler	Little Cornwallis I. (75.583°N, 95.833°W), NU, CAN	[Helmer, et al. (1995) In: Dyke, et al. (1999)]; R.E. Morlan, CMC database (1999)
900 ± 80 (980 ± 80)	I-12341		Buchanan Lake, 7 m asl., Axel Heiberg I. (79.500°N, 87.500°W), NU, CAN	[Morrison (1989); Kalkreuth and Sutherland (1998) In: R.E. Morlan, CMC database (1999)]
895 ± 150 (975 ± 150)	S-512	(CMC-378)	Bluefish Basin (67.580°N, 139.680°W), YT, CAN	Morlan (1973); Bonnichsen (1979); R.E. Morlan, CMC database (1999)
(890 ± 55)	AAR-3395	(GUS x3702)	"Farm beneath the Sand" near Nuuk (SW region) (64°06'N, 50°05'W), GRE	Arneborg, et al. (1998); Danish/Greenlandic Network (1999); [830 (AD1170, AD1040-1220)]
(880 ± 50)	AECV-1708C		Frobisher Bay, 10 m asl., Baffin I. (63.762°N, 68.683°W), NU, CAN	[Manley and Jenning (1996) In: A.S. Dyke (GSC database)]; [Stenton and Rigby (1995) In: R.E. Morlan, CMC database (1999)]
875 ± 95 (955 ± 95)	S-587	charred bone (CMC-383)	E shore of Artillery L. (62.930°N, 108.220°W), NT, CAN	[Noble (1971); Wilmeth (1978); Gordon (1996) In: R.E. Morlan, CMC database (1999)]
870 ± 65	K-4346	antler (ZMK 14/1932)	Angmagssalik O (65.470°N, 37.420°W), GRE	Meldgaard (1986)
(870 ± 30)	BM-1803		Brooman Pt., Bathurst I. (75.390°N, 97.290°W), NU, CAN	[McGhee (1984) In: Dyke, et al. (1999)]
855 ± 120 (935 ± 120)	S-515	(CMC-381)	Bluefish Basin (67.580°N, 139.680°W), YT, CAN	Morlan (1973); Bonnichsen (1979); R.E. Morlan, CMC database (1999)
850 ± 100 (930 ± 100)	S-882	(CMC-658)	Navy Board Inlet, Borden Pen., 12 m asl., Baffin I. (73.025°N, 80.630°W), NU, CAN	[Mary-Rousseliere (1976) In: Dyke, et al. (1999)]; [Wilmeth (1978); Mary-Rousseliere (1979); Morrison (1989) In: R.E. Morlan, CMC database (1999)]
840 ± 105 (920 ± 105)	S-2200	mandible, scapula, ribs, long bone fragments (CMC-1248)	1 km W of Thom Bay Mission, 5-8 m asl., Boothia Pen. (70.000°N, 95.250°W), NU, CAN	[Savelle (1987); Morrison (1989) In: R.E. Morlan, CMC database (1999)]
830 ± 150 (910 ± 150)	S-514	(CMC-380)	Bluefish Basin (67.580°N, 139.680°W), YT, CAN	Morlan (1973); Bonnichsen (1979); R.E. Morlan, CMC database (1999)
825 ± 100 (905 ± 100)	S-1313		W coast of Hudson Bay (62.050°N, 92.250°W), NU, CAN	[Clark (1979) In: A.S. Dyke (GSC database)]; Dyke (1998); [Clark (1979); Morrison (1989) In: R.E. Morlan, CMC database (1999)]
810 ± 100 (890 ± 100)	S-513	(CMC-379)	Bluefish Basin (67.580°N, 139.680°W), YT, CAN	Morlan (1973); Bonnichsen (1979); R.E. Morlan, CMC database (1999)
(800 ± 30)	BM-1804	antler	Brooman Pt., Bathurst I. (75.390°N, 97.290°W), NU, CAN	[McGhee (1984) In: Dyke, et al. (1999)]
785 ± 75 (865 ± 80)	S-2971	antler (CMN 47731)	E Prince of Wales I. (72.720°N, 97.600°W), NU, CAN	C.R. Harington [unpublished]; R.E. Morlan, CMC database (1999)
(780 ± 50)	TO-3748	rib	Little Cornwallis I. (75.583°N, 95.833°W), NU, CAN	[Helmer, et al. (1995) In: Dyke, et al. (1999)]; R.E. Morlan, CMC database (1999)
775 ± 80 (855 ± 100)	S-511	(CMC-377)	Porcupine R. (67.580°N, 139.680°W), YT, CAN	Morlan (1973); Bonnichsen (1979); R.E. Morlan, CMC database (1999)
770 ± 70 (850 ± 75)	S-3369	antler (CMC-1407)	N end of Cape Krusenstern, 3.5 m asl, NU, CAN	[Morrison (1983, 1989) In: R.E. Morlan, CMC database (1999)]
(770 ± 120)	AECV-728C		Nettilling L., Baffin I., NU, CAN	[Stenton (1991) In: R.E. Morlan, CMC database (1999)
(750 ± 60)	Gif-8180		W end of Naujat Pen., Victoria I., NU, CAN	[Le Mouel (1999) In: R.E. Morlan, CMC database (1999)]
745 ± 90 (825 ± 90)	S-1314		W coast of Hudson Bay (62.050°N, 92.250°W), NU, CAN	[Clark (1979) In: A.S. Dyke (GSC database)]; [Clark (1979); Morrison (1989) In: R.E. Morlan, CMC database (1999)]
745 ± 40 (825 ± 45)	SI-768-B	antler	Chesterfield Inlet, 38 m asl (63.680°N, 90.080°W), NU, CAN	[Wilmeth (1978); Staab (1979); McCartney (1970, 1977, pers. comm. 1998); Morrison (1989) In: R.E. Morlan, CMC database (1999)]

(740 ± 120)	RIDDL-548	antler	SE shore of Richards I., Mackenzie Delta (68.830°N, 135.250°W), NT, CAN	[Friesen (1987); Morrison (1989); Cinq-Mars (1991); Arnold (1991, 1994); Balkwill and Rick (1994); Nolin (1994); Friesen and Arnold (1995) In: R.E. Morlan, CMC database (1999)]
740 ± 65	K-4349	antler (CN 2791)	Independence Fjord (81.750°N, 32.000°W), GRE	Meldgaard (1986)
(740 ± 70)	AECV-1349C		Frobisher Bay, 16 m asl., Baffin I. (63.762°N, 68.683°W), NU, CAN	[Manley and Jenning (1996) In: A.S. Dyke (GSC database)]; [Stenton and Rigby (1995) In: R.E. Morlan, CMC database (1999)]
(720 ± 50 (1290AD), 1275-	K-4733		Niaquusat, Norse farm V48, Nuuk Municipality, GRE	[Arneborg (1991) In: Danish/Greenlandic Network (1999)]
709 ± 94 (790 ± 100)	P-177	skull	Arnaud R., Nouveau-Quebec region (59.280°N, 73.330°W), QC, CAN	R.E. Morlan, CMC database (1999)
(700 ± 80)	AECV-537C		Nettilling L., Baffin I., NU, CAN	[Stenton (1991) In: R.E. Morlan, CMC database (1999)]
(700 ± 90)	AECV-1012C		SE shore of Richards I. Mackenzie Delta (69.417°N, 134.167°W), NT, CAN	[Friesen (1987); Morrison (1989); Arnold (1991, 1994); Cinq-Mars (1991); Nolin (1994); Balkwill and Rick (1994); Friesen and Arnold (1995) In: R.E. Morlan, CMC database (1999)]
690 ± 60 (740 ± 60)	Beta-111668	(CMC-1503)	Pond Inlet village, Baffin I., NU, CAN	[Mary-Rousseliere (1976); Wilmeth (1978); Maxwell (1985) In: R.E. Morlan, CMC database (1999)]
(690 ± 100)	Gif-8182		W end of Naujat Pen., Victoria I., NU, CAN	[Le Mouel (1999) In: R.E. Morlan, CMC database (1999)]
(680 ± 50 (1300AD), 1285-	K-4742		Niaquusat, Norse farm V48, Nuuk Municipality, GRE	[Arneborg (1991) In: Danish/Greenlandic Network (1999)]
675 ± 65 (755 ± 70)	S-1317		W coast of Hudson Bay (62.880°N, 92.150°W), NU, CAN	[Clark (1979) In: A.S. Dyke (GSC database)]; [Clark (1977, 1979); Morrison (1989) In: R.E. Morlan, CMC database (1999)]
660 ± 100 (740 ± 100)	K-1079	antler	Jens Munk I., Igloolik area (69.667°N, 80.000°W), NU, CAN	[Rainey and Ralph (1959), Meldgaard (1960, 1962), Wilmeth (1978), Maxwell (1985), O. Bennike, pers. comm. to A.S. Dyke (1997) In: R.E. Morlan, CMC database (1999)]
(660 ± 65 (1300AD), 1285-	K-4734		Niaquusat, Norse farm V48, Nuuk Municipality, GRE	[Arneborg (1991) In: Danish/Greenlandic Network (1999)]
649 ± 100 (730 ± 105)	P-176	antler	Arnaud R., Nouveau-Quebec region (59.280°N, 73.330°W), QC, CAN	R.E. Morlan, CMC database (1999)
(630 ± 70)	Gif-8433		W end of Naujat Pen., Victoria I., NU, CAN	[Le Mouel (1999) In: R.E. Morlan, CMC database (1999)]
600 ± 150 (680 ± 150)	K-504	antler	K'aersut I., 8 m asl., Igloolik area (69.480°N, 80.320°W), NU, CAN	R.E. Morlan, CMC database (1999)
(600 ± 70)	AECV-536CR		Nettilling L., Baffin I., NU, CAN	[Stenton (1991) In: R.E. Morlan, CMC database (1999)]
(600 ± 130)	RIDDL-363	antler fragment	Whitefish Station (west) (68.916°N, 136.417°W), YT, CAN	Cinq-Mars et al. (1991); [Morrison (1989); Vogel, et al. (1991) In: R.E. Morlan, CMC database (1999)]
(600 ± 50 (1330-1395AD), 1305-	K-4735		Niaquusat, Norse farm V48, Nuuk Municipality, GRE	[Arneborg (1991) In: Danish/Greenlandic Network (1999)]
570 ± 55 (650 ± 60)	S-1310		W coast of Hudson Bay (62.880°N, 92.150°W), NU, CAN	[Clark (1979) In: A.S. Dyke (GSC database)]; [Clark (1977, 1979); Morrison (1989) In: R.E. Morlan, CMC database (1999)]
570 ± 80 (650 ± 80)	I-18243	femur	Firth drainage, 220 m asl, YT, CAN	[G. Adams, pers. comm. (1998) In: R.E. Morlan, CMC database (1999)]
(565 ± 65)	AAR-3223	rib	Smith Sound area (78.333°N, 72.75°W), GRE	Appelt, et al. (1998); [Heinemeier and Rasmussen (1998) In: Danish/Greenlandic Network (1999)]; [600 (AD1400, AD1310-1430)]

(560 ± 160)	RIDDL-324	antler	Maxwell Bay site, SW coast of Devon I. (75.000°N, 89.000°W), NU, CAN	Cinq-Mars (1991); [Morrison (1989) In: R.E. Morlan, CMC database (1999)]
(550 ± 50)	AECV-1348C		Frobisher Bay, 6 m asl., Baffin I. (63.762°N, 68.683°W), NU, CAN	[Manley and Jenning (1996) In: A.S. Dyke (GSC database)]; [Stenton and Rigby (1995) In: R.E. Morlan, CMC database (1999)]
550 ± 40 (630 ± 40)	Beta-126879	metatarsal (CMC-1520)	Overlooking S shore of Amundsen Gulf, NE corner of Pearce Pt. Harbour, 1 m asl, NT, CAN	[D.A. Morrison, pers. comm. (1999) In: R.E. Morlan, CMC database (1999)]
(550 ± 90)	AECV-1006C		SE shore of Richards I. Mackenzie Delta (69.417°N, 134.167°W), NT, CAN	[Friesen (1987); Morrison (1989); Arnold (1991, 1994); Cinq-Mars (1991); Nolin (1994); Balkwill and Rick (1994); Friesen and Arnold (1995) In: R.E. Morlan, CMC database (1999)]
540 ± 70 (620 ± 75)	S-2948	phalanx (CMC-1373)	Iglulualuit, 2 m asl., Franklin Bay (70.000°N, 127.000°W), NT, CAN	[Morrison (1991) In: A.S. Dyke (GSC database)]; [LeBlanc (1987); Morrison (1989, 1990); Cinq-Mars (1991); Vogel, et al. (1991) In: R.E. Morlan, CMC database (1999)]
540 ± 95 (620 ± 95)	I-4550	charred bone	E shore of Artillery L. (62.930°N, 108.220°W), NT, CAN	Wright (1975); [Noble (1971); Wilmeth (1978); Gordon (1996) In: R.E. Morlan, CMC database (1999)]
(540 ± 90)	AECV-1011C		SE shore of Richards I. Mackenzie Delta (69.417°N, 134.167°W), NT, CAN	[Friesen (1987); Morrison (1989); Arnold (1991, 1994); Cinq-Mars (1991); Nolin (1994); Balkwill and Rick (1994); Friesen and Arnold (1995) In: R.E. Morlan, CMC database (1999)]
530 ± 60 (610 ± 65)	Gif-7550		W end of Naujat Pen., Victoria I., NU, CAN	[Le Mouel (1999) In: R.E. Morlan, CMC database (1999)]
(530 ± 90)	AECV-1008C		SE shore of Richards I. Mackenzie Delta (69.417°N, 134.167°W), NT, CAN	[Friesen (1987); Morrison (1989); Arnold (1991, 1994); Cinq-Mars (1991), Nolin (1994); Balkwill and Rick (1994); Friesen and Arnold (1995) In: R.E. Morlan, CMC database (1999)]
510 ± 100	K-1291	dark and light bones	Trail Ck. Cave 9, Seward Pen. (65.800°N, 163.220°W), AK, USA	A.S. Dyke (GSC database)
(500 ± 90)	RIDDL-547	antler	Saunaktuk, Eskimo L. (69.170°N, 132.500°W), NT, CAN	[Cinq Mars (1991); Morrison and Arnold (1994); C. Arnold, pers. comm (1998) In: R.E. Morlan, CMC database (1999)]
(490 ± 70)	AECV-1351C		Frobisher Bay, 8 m asl., Baffin I. (63.762°N, 68.683°W), NU, CAN	[Manley and Jenning (1996) In: A.S. Dyke (GSC database)]; [Stenton and Rigby (1995) In: R.E. Morlan, CMC database (1999)]
490 ± 105 (570 ± 105)	S-2196	skull, distal femur (CMC-1244)	King William I., 13-19 m asl (68.330°N, 97.420°W), NU, CAN	[McCartney and Savelle (1985) In: A.S. Dyke (GSC database)]; [Savelle (1987); Morrison (1989) In: R.E. Morlan, CMC database (1999)]
(490 ± 170)	RIDDL-344	antler	Kittigazuit, Mackenzie R. (69.350°N, 132.670°W), NT, CAN	[McGhee (1974); Wilmeth (1978); Morrison (1989); Vogel, et al. (1991); Cinq-Mars (1991) In: R.E. Morlan, CMC database (1999)]
480 ± 70 (560 ± 75)	S-2947	phalanx (CMC-1371)	Iglulualuit, 2 m asl., Franklin Bay (70.000°N, 127.000°W), NT, CAN	[Morrison (1991) In: A.S. Dyke (GSC database)]; [LeBlanc (1987); Morrison (1989, 1990); Cinq-Mars (1991); Vogel, et al. (1991) In: R.E. Morlan, CMC database (1999)]
480 ± 95 (560 ± 95)	S-2845	(CMC-1363)	Burnside R., 375 m asl (67.000°N, 109.000°W), NU, CAN	[Morrison (1989); Gordon (1988, 1994) In: R.E. Morlan, CMC database (1999)]
480 ± 60 (560 ± 65)	S-2199	2 long-bone fragments (CMC-1247)	King William I., 7-10 m asl (68.330°N, 97.000°W), NU, CAN	[McCartney and Savelle (1985) In: A.S. Dyke (GSC database)]; [Savelle (1987); Morrison (1989) In: R.E. Morlan, CMC database (1999)]
470 ± 60 (550 ± 65)	GX-23599		Backhouse R., YT, CAN	[G. Adams, pers. comm. (1998); Gotthardt (1998) In: R.E. Morlan, CMC database (1999)]

460 ± 90 (540 ± 90)	Beta-101514	(CMC-1494)	Hazard Inlet, Somerset I., NU, CAN	[McCartney and Savelle (1985) In: R.E. Morlan, CMC database (1999)]
(460 ± 90)	AECV-1017C		SE shore of Richards I. Mackenzie Delta (69.417°N, 134.167°W), NT, CAN	[Arnold (1991, 1994); Nolin (1994) In: R.E. Morlan, CMC database (1999)]
430 ± 210 (510 ± 210)	S-3002	(CMC-1390)	Iglulualuit, Franklin Bay (70.000°N, 127.000°W), NT, CAN	Morrison (1989, 1990, 1991); [LeBlanc (1987); Cinq-Mars (1991); Vogel, et al. (1991) In: R.E. Morlan, CMC database (1999)]
(430 ± 50)	TO-3708	left metacarpal (CR-92-36)	Kuujjua R., Victoria I. (71°07'N, 115°43'W), NT, CAN	C.R. Harington [unpublished]
425 ± 225 (505 ± 225)	S-1312		W coast of Hudson Bay (62.050°N, 92.250°W), NU, CAN	[Clark (1979) In: A.S. Dyke (GSC database)]; [Clark (1979); Morrison (1989) In: R.E. Morlan, CMC database (1999)]
420 ± 60 (500 ± 65)	Beta-101515	(CMC-1495)	Hazard Inlet, Somerset I., NU, CAN	[McCartney and Savelle (1985) In: R.E. Morlan, CMC database (1999)]
420 ± 75 (500 ± 80)	I-11755		Buchanan L., 7 m asl., Axel Heiberg I. (79.500°N, 87.500°W), NU, CAN	[Morrison (1989); Kalkreuth and Sutherland (1998) In: R.E. Morlan, CMC database (1999)]
410 ± 55 (490 ± 60)	S-1309		W coast of Hudson Bay (62.880°N, 92.150°W), NU, CAN	[Clark (1979) In: A.S. Dyke (GSC database)]; [Clark (1977, 1979); Morrison (1989) In: R.E. Morlan, CMC database (1999)]
410 ± 30 (510 ± 30)	Beta-126880	thoracic vertebra (CMC-1521)	Overlooking S shore of Amundsen Gulf, NE corner of Pearce Pt. Harbour, 1 m asl, NT, CAN	[D.A. Morrison, pers. comm. (1999) In: R.E. Morlan, CMC database (1999)]
(380 ± 50)	Beta-51302 ETH-9418	tibia (CMC-1422)	SW Anderson Plain, NT, CAN	[Nolin (1992, 1994); Pilon (1992, 1994) In: R.E. Morlan, CMC database (1999)]; L. Nolin and J. Cinq-Mars, NOGAP database (1994)
(360 ± 80)	AECV-1002C		SE shore of Richards I. Mackenzie Delta (69.417°N, 134.167°W), NT, CAN	[Friesen (1987); Morrison (1989); Arnold (1991, 1994); Cinq-Mars (1991); Nolin (1994); Balkwill and Rick (1994); Friesen and Arnold (1995) In: R.E. Morlan, CMC database (1999)]
(360 ± 90)	AECV-1010C		SE shore of Richards I. Mackenzie Delta (69.417°N, 134.167°W), NT, CAN	[Friesen (1987); Morrison (1989); Arnold (1991, 1994); Cinq-Mars (1991); Nolin (1994); Balkwill and Rick (1994); Friesen and Arnold (1995) In: R.E. Morlan, CMC database (1999)]
(350 ± 90)	AECV-1009C		SE shore of Richards I. Mackenzie Delta (69.417°N, 134.167°W), NT, CAN	[Friesen (1987); Morrison (1989); Arnold (1991, 1994); Cinq-Mars (1991); Nolin (1994); Balkwill and Rick (1994); Friesen and Arnold (1995) In: R.E. Morlan, CMC database (1999)]
345 ± 85 (425 ± 85)	S-1316		W coast of Hudson Bay (62.050°N, 92.250°W), NU, CAN	[Clark (1979) In: A.S. Dyke (GSC database)]; [Clark (1979); Morrison (1989) In: R.E. Morlan, CMC database (1999)]
345 ± 85 (425 ± 85)	S-1311		W coast of Hudson Bay (62.050°N, 92.250°W), NU, CAN	[Clark (1979) In: A.S. Dyke (GSC database)]; [Clark (1979); Morrison (1989) In: R.E. Morlan, CMC database (1999)]
(330 ± 50)	AECV-1707C		Mallikjuak I., 23 m asl., SW Baffin I. (64.233°N, 76.583°W), NU, CAN	[Manley and Jenning (1996) In: A.S. Dyke (GSC database)]; R.E. Morlan, CMC database (1999)
330 ± 70 (410 ± 75)	Beta-64368	antler	Pauline Cove, Herschel I., YT, CAN	[Morrison (1989); Yorga (1980); Friesen and Hunston (1994) In: R.E. Morlan, CMC database (1999)]
(290 ± 90)	RIDDL-553	radio-ulna fragment	Anderson Plain, NT, CAN	R.E. Morlan, CMC database (1999)
(<290)	RIDDL-342	scapula	Trail R. (69.330°N, 138.000°W), YT, CAN	[Morrison (1989); Nagy (1989, 1991); Cinq-Mars (1991); Vogel, et al. (1991) In: R.E. Morlan, CMC database (1999) - specimen exhibited butchering marks]

285 ± 100 (365 ± 100)	S-1315		W coast of Hudson Bay (62.050°N, 92.250°W), NU, CAN	[Clark (1979) In: A.S. Dyke (GSC database)]; [Clark (1979); Morrison (1989) In: R.E. Morlan, CMC database (1999)]
(280 ± 110)	AECV-1003C	bone and antler collagen	SE shore of Richards I. Mackenzie Delta (69.417°N, 134.167°W), NT, CAN	[Friesen (1987); Morrison (1989); Arnold (1991, 1994); Cinq-Mars (1991); Nolin (1994); Balkwill and Rick (1994); Friesen and Arnold (1995) In: R.E. Morlan, CMC database (1999)]
(270 ± 80)	AECV-1005C		SE shore of Richards I. Mackenzie Delta (69.417°N, 134.167°W), NT, CAN	[Friesen (1987); Morrison (1989); Arnold (1991, 1994); Cinq-Mars (1991); Nolin (1994); Balkwill and Rick (1994); Friesen and Arnold (1995) In: R.E. Morlan, CMC database (1999)]
260 ± 50 (350 ± 50)	Beta-70226	antler	Herschel I., YT, CAN	[Nolin (1994) In: R.E. Morlan, CMC database (1999)]
(260 ± 100)	RIDDL-543	antler tool fragmemt	Iglulualuit, Franklin Bay (70.000°N, 127.000°W), NT, CAN	R.E. Morlan, CMC database (1999)
(260 ± 120)	RIDDL-365	antler fragment	Trail R. (69.330°N, 138.000°W), YT, CAN	[Morrison (1989); Nagy (1989, 1991); Cinq-Mars (1991); Vogel, et al. (1991) In: R.E. Morlan, CMC database (1999)]
250 ± 205 (330 ± 205)	S-3003	(CMC-1391)	Iglulualuit, Franklin Bay (70.000°N, 127.000°W), NT, CAN	Morrison (1989, 1990, 1991); [LeBlanc (1987); Cinq-Mars (1991); Vogel, et al. (1991) In: R.E. Morlan, CMC database (1999)]
(230 ± 100)	RIDDL-551	long-bone splinter	Anderson Plain, Jiggle L. 3, NT, CAN	R.E. Morlan, CMC database (1999)
(<230)	RIDDL-549	astralagus	Saunaktuk, Eskimo L. (69.170°N, 132.500°W), NT, CAN	[Cinq Mars (1991); Morrison and Arnold (1994); C. Arnold, pers. comm. (1998) In: R.E. Morlan, CMC database (1999)]
220 ± 70 (320 ± 70)	Beta-94898	(CMC-1492)	Old Horton R. channel, Cape Bathurst Pen. (70.080°N, 127.500°W), NU, CAN	[Morrison (1997) In: R.E. Morlan, CMC database (1999)]
(<210)	RIDDL-554	long-bone splinter	Anderson Plain, NT, CAN	[Pilon (1985, 1987); Cinq-Mars (1991) In: R.E. Morlan, CMC database (1999)]
210 ± 90 (290 ± 90)	I-3956		NE end of Kamut Lake, NT, CAN	[Noble (1971); Wilmeth (1978) In: R.E. Morlan, CMC database (1999)]
200 ± 105 (280 ± 105)	S-2197	long-bone fragments (CMC 1245)	King William I., 13-19 m asl (68.330°N, 97.420°W), NU, CAN	[McCartney and Savelle (1985) In: A.S. Dyke (GSC database)], [Savelle (1987); Morrison (1989) In: R.E. Morlan, CMC database (1999)]
200 ± 100 (300 ± 100)	Beta-94897	(CMC-1491)	Old Horton R. channel, Cape Bathurst Pen. (70.080°N, 127.500°W), NU, CAN	[Morrison (1997); C.R. Harington, pers. comm. (1998) In: R.E. Morlan, CMC database (1999)]
(200 ± 90)	RIDDL-542	antler	Trail R. (69.330°N, 138.000°W), YT, CAN	[Cinq-Mars (1991); Vogel, et al. (1991) In: R.E. Morlan, CMC database (1999)]
190 ± 90 (270 ± 90)	SI-1688	antler	Pangnirtung Pass, Cumberland Pen., Baffin I. (66.520°N, 65.470°W), NU, CAN	INSTAAR List; [pers. comm. R. McNeely to R.E. Morlan (1998)]; R.E. Morlan, CMC database (1999)
190 ± 70 (270 ± 75)	S-3004	(CMC-1392)	Iglulualuit, Franklin Bay (70.000°N, 127.000°W), NT, CAN	Morrison (1989, 1990, 1991); [LeBlanc (1987); Cinq-Mars (1991); Vogel, et al. (1991) In: R.E. Morlan, CMC database (1999)]
190 ± 50 (280 ± 50)	Beta-94896	(CMC-1490)	Old Horton R. channel, Cape Bathurst Pen. (70.080°N, 127.500°W), NU, CAN	[Morrison (1997); C.R. Harington, pers. comm. (1998) In: R.E. Morlan, CMC database (1999)]
(<180)	RIDDL-541	antler	Old Crow Basin, YT, CAN	[Cinq-Mars (1991); Vogel, et al. (1991); Fafard (1999) In: R.E. Morlan, CMC database (1999)]
160 ± 105 (240 ± 105)	S-2773	(CMC-1353)	Burnside R., 375 m asl., (67.000°N, 109.000°W), NU, CAN	Morrison (1989); [Gordon (1988, 1994) In: R.E. Morlan, CMC database (1999)]
(160 ± 90)	RIDDL-552	radio-ulna fragment	Anderson Plain, Sandy L. 3, NT, CAN	R.E. Morlan, CMC database (1999)
(<150)	RIDDL-555	long-bone splinter	Anderson Plain, NT, CAN	[Pilon (1985, 1987); Cinq-Mars (1991) In: R.E. Morlan, CMC database (1999)]

Measured Date (Normalized)	Lab No	Material	Location	References/Remarks
100 ± 0.7% modern (70 ± 50)	Beta-94895	(CMC-1489)	Old Horton R. channel, Cape Bathurst Pen. (70.080°N, 127.500°W), NU, CAN	[Morrison (1997); C.R. Harington, pers. comm. (1998) In: R.E. Morlan, CMC database (1999)]
<100 (<100)	S-3549	(CMC-1456)	Old Horton R. channel, Cape Bathurst Pen. 1.5 m asl (70.080°N, 127.500°W), NT, CAN	[Morrison (1997) In: R.E. Morlan, CMC database (1999)]
(<100)	S-3439	(CMC-1414)	Gutchiak, Eskimo L., NT, CAN	[Morrison and Swayze (1991); Morrison (1994, 1998); Nolin (1994) In: R.E. Morlan, CMC database (1999)]
<100 (<100)	S-3438	(CMC-1413)	Gutchiak, Eskimo L., NT, CAN	[Morrison and Swayze (1991); Morrison (1994, 1998); Nolin (1994) In: R.E. Morlan, CMC database (1999)]
<100 (<100)	S-3437	(CMC-1412)	Gutchiak, Eskimo L., NT, CAN	[Morrison and Swayze (1991); Morrison (1994, 1998); Nolin (1994) In: R.E. Morlan, CMC database (1999)]
70 ± 50 (140 ± 50)	Beta-70225	antler	Herschel I., YT, CAN	[Nolin (1994) In: R.E. Morlan, CMC database (1999)]
(40 ± 220)	GSC-1395	metapodial	1.6 km SE of Ingonish Centre, 23 m asl., Cape Breton I., NS, CAN	Lowdon, et al. (1971); R.E. Morlan, CMC database (1999)

Rangifer tarandus pearyi (Peary caribou)

Measured Date (Normalized)	Lab No	Material	Location	References/Remarks
8415 ± 135 (8495 ± 135)	S-2501	antler	Piper Pass, Clements Markham Inlet, Ellesmere I. (82.610°N, 68.520°W), NU, CAN	Stewart and England (1986); R.E. Morlan, CMC database (1999)

Rangifer tarandus? (caribou, reindeer)

Measured Date (Normalized)	Lab No	Material	Location	References/Remarks
2170 ± 60 (2230 ± 60)	Beta-77809 CAMS-17243	CMC-1464	Satkualuk, Richards I., NT, CAN	[Nolin (1994); Sutherland (1994, 1995) In: R.E. Morlan, CMC database (1999)]
101 ± 0.7% modern	Beta-51301 ETH-9417	scapula fragment (CMC-1421)	Lure L., Anderson Plain, NT, CAN	[Nolin (1992, 1994), Pilon (1992) In: R.E. Morlan, CMC database (1999)]

Rangifer sp. (caribou, reindeer)

Measured Date (Normalized)	Lab No	Material	Location	References/Remarks
(9940 ± 120)	AA-3285	antler	Steep Rock L, Rainy R. District (48.5°N, 90.5°W), ON, CAN	[Jackson (1988, 1989), Jackson and McKillop (1989) In: R.E. Morlan (CMC database (1999)]
(4950 ± 80)	TO-1107	calcaneum	Webb Bay, Rice L. region, ON, CAN	[Pavlish and Alcock (1984), Jackson and McKillop (1991) In: R.E.Morlan, CMC database (1999)]
(3880 ± 80)	AECV-109C		Qugyuk site, Cape Bathurst Pen., NU, CAN	L. Nolin and J. Cinq-Mars, NOGAP database (1994)
(3415 ± 85)	Beta-26660	tibia	Paynter Marsh: Rice L. region, ON, CAN	[Jackson and McKillop (1991) In: R.E. Morlan, CMC database (1999)]
(2530 ± 70)	TO-1092		Paynter site: Rice L. region, ON, CAN	[L.J. Jackson, pers. comm. (1998) In: R.E. Morlan, CMC database (1999)]

Rangifer? tarandus (caribou, reindeer)

Measured Date (Normalized)	Lab No	Material	Location	References/Remarks
2185 ± 120 (2265 ± 120)	M-1504	antler, teeth, bone (CMC-2)	Ferguson R., Victoria I. (69.400°N, 106.250°W), NU, CAN	[Taylor (1967, 1972); Wilmeth (1978) In: R.E. Morlan, CMC database (1999)]

Torontoceros hypogaeus (Toronto deer)

Measured Date (Normalized)	Lab No	Material	Location	References/Remarks
11315 ± 325 (11395 ± 320)	GX-4914	antler (ROM 75974)	Glacial L. Iroquois, Toronto (43.640°N, 79.525°W), ON, CAN	Churcher and Peterson (1982); [Spiess, et al. (1985), Jackson (1988) In: R.E. Morlan, CMC database (1999)]

Artiodactyla, Antilocapridae

Antilocapra americana (pronghorn antelope)

Measured Date (Normalized)	Lab No	Material	Location	References/Remarks
6930 ± 260 (7010 ± 260)	RL-903		Metke site near High River (50.670°N, 113.780°W), AB, CAN	Brumley and Rushworth (1983); Graham, et al. (1987); R.E. Morlan, CMC database (1999)

Artiodactyla, Bovidae

Bos taurus (cattle)

Measured Date (Normalized)	Lab No	Material	Location	References/Remarks
(290 ± 50)	TO-3651	metatarsal	Scarborough Bluffs, E of Toronto, ON, CAN	Churcher and Karrow (1977); Harington (1990); Morris, et al. (1993); R.E. Morlan, CMC database (1999)

Bison priscus (steppe bison)

Measured Date (Normalized)	Lab No	Material	Location	References/Remarks
36425 ± 2575	DIC-891	skin (Blue Babe)	Pearl Ck. (65.983°N, 147.317°W), AK, USA	[Guthrie (1990) In: A.S. Dyke (GSC database)]
33800 ± 2000 (33880 ± 2000)	I-4227	humerus	Old Crow R. (Old Crow Loc. 14N), 267 m asl. (67.850°N, 139.768°W), YT, CAN	Harington (1977); [Bonnichsen (1978) incorrectly cites lab number as I-4427 and date of 33,800 ± 200 BP within Table 2]; Morlan et al. (1990); R.E. Morlan, CMC database (1999)
31980 ± 4490	SI-843	hornsheath (AMNH 30530)	Fairbanks Ck. (65.070°N, 147.170°W), AK, USA	Péwé (1975); Bonnichsen (1979); Guthrie (1985, 1990)
31400 +2,040 /-1815	St-1721	hide and hair from carcass [See Figure 1, this paper]	Dome Ck. (64.830°N, 147.500°W), AK, USA	[Péwé (1975) In frozen silt near contact with underlying gravel. A great concentration of mammal bones of various species occur in this part of the formation]; Harington (1977); [Guthrie (1985, 1990) specimen incorrectly cited within Mammoth Tables]
30300 ± 1850 (30380 ± 1850)	I-3571	horncore	Quartz Ck. (Dawson Loc. 7) (63.780°N, 139.100°W), YT, CAN	Harington (1969, 1977); R.E. Morlan, CMC database (1999)
29700 ± 240	QL-1178	horncore	Jack Wade Ck., near Lost Chicken, AK, USA	[Morlan and Cinq-Mars (1982) - altered remains]; Porter and Hopkins (1982)
>28000	L-127	tissue	Dome Ck., near Fairbanks (64.830°N, 147.500°W), AK, USA	Péwé (1966, 1975); Harington (1977a); [from same carcass as ST-1721]
24900 ± 1000 (24980 ± 1000)	I-3575	part of left horncore and frontal	Flat Ck. (Dawson Loc. 25) (63.940°N, 138.625°W), YT, CAN	Harington (1977); R.E. Morlan, CMC database (1999)
(23710 ± 100)	CAMS-23471	radio-ulna	Bluefish Cave III, 600 m asl. (67°08'N, 140°47'W), YT, CAN	Cinq-Mars (1979, 1990); Morlan and Cinq-Mars (1982); Cinq-Mars, et al. (1991); Harington and Cinq-Mars (1995); Burke and Cinq-Mars (1996);
22200 ± 1400 (22280 ± 1400)	I-3570	horncore	Gold Run Ck. (Dawson Loc. 32) (63.725°N, 138.680°W), YT, CAN	Harington and Clulow (1973); Harington (1977, 1978, 1989); R.E. Morlan, CMC database (1999)
(20370 ± 230)	Beta-28762 ETH-4759	foot muscle tissue (CMN 45289)	Hunker Ck. (Dawson Loc. 12) (63.970°N, 138.980°W), YT, CAN	C.R. Harington [unpublished]
16400 ± 2000	M-38	hornsheath	Fairbanks area (64°50'N, 147°30'W), AK, USA	Farrand (1961); [Péwé (1975) black carbon method]; Guthrie (1985, 1990)
(12610 ± 70)	TO-34	metacarpal	Old Crow R. (Old Crow Loc. 11(1)), 260 m asl. (67.820°N, 139.840°W), YT, CAN	Harington (1977); R.E. Morlan, CMC database (1999)
12460 ± 320	SI-290	hornsheath	Upper Cleary Ck. (65.000°N, 148.000°W), AK, USA	Péwé (1975); Guthrie (1985, 1990); Dixon (1993)
12460 ± 220 (12540 ± 220)	I-3574	posterior lumbar vertebra	Old Crow R. (Old Crow Loc. 11(1)), 260 m asl. (67.820°N, 139.840°W), YT, CAN	Harington (1977, 1978); Bonnichsen (1978); Morlan (1079, 1980); Dixon (1993); R.E. Morlan, CMC database (1999)
12275 ± 180 (12335 ± 180)	I-7764	horncore	Old Crow R. (Old Crow Loc. 11(A)), 259 m asl. (67.480°N, 139.920°W), YT, CAN	Harington (1977, 1978); Morlan (1979); [Bonnichsen (1978) incorrectly cites date of 12,460 ± 180 BP on one of the samples for iron content analyses on bone]; Morlan (1980); Dixon (1993); R.E. Morlan, CMC database (1999)
11950 ± 135	St-1633	hide	Fairbanks Ck. (65.067°N, 147.167°W), AK, USA	Péwé (1966, 1975); Harington (1977, 1978); Guthrie (1985, 1990); Dixon (1993); Stephenson, et al. (in press).

11910 ± 180 (11990 ± 180)	I-7765	scapula fragment	Old Crow R. (Old Crow Loc. 11(1)), 260 m asl. (67.820°N, 139.840°W), YT, CAN	Harington (1977, 1978); Morlan (1979, 1980); Dixon (1993); R.E. Morlan, CMC database (1999); Stephenson, et al. (in press).
11735 ± 130	St-1631	hornsheath	Cleary Ck. (65.000°N, 147.000°W), AK, USA	Péwé (1975); [Harington (1977) internal piece of hornsheath from type specimen]; Guthrie (1985, 1990); Dixon (1993); Stephenson, et al. (in press).
10370 ± 160	I-8582	tibia, distal portion (CMN 25845)	Lost Chicken Ck. (64.050°N, 141.875°W), AK, USA	Bonnichsen (1979); Harington (1978, 1980); Porter (1988); [Morlan and Cinq-Mars (1982) distinct impact mark and concentric and radial cracks indicating breakage by humans); [Agenbroad (1984) incorrectly cites specimen as Mammoth within Table 3.4]

Bison cf. Bison priscus (steppe bison)

Measured Date (Normalized)	Lab No	Material	Location	References/Remarks
(39290 ± 540)	TO-5650	right metacarpal	Confluence of Thomas Gulch and Klondike R., 340 m asl., 3 km E of Dawson, YT, CAN	Harington and Clulow (1973)
(10200 ± 100)	GSC-2902	left tibia	Smoky R., S of Watino (55.720°N, 117.720°W), AB, CAN	Churcher and Wilson (1979); Lowdon and Blake (1979); Jackson and Pawson (1984); Driver (1988); R.E. Morlan, CMC database (1999)
(10200 ± 100)	GSC-2895	cervical vertebra	Smoky R., Watino (55.720°N, 117.625°W), AB, CAN	Churcher and Wilson (1979); Lowdon and Blake (1979); Jackson and Pawson (1984); Driver (1988); R.E. Morlan, CMC database (1999)
(9880 ± 130)	GSC-2865	2 tibiae	Peace R. gravel pit, Peace R. (56.225°N, 117.300°W), AB, CAN	Churcher and Wilson (1979); Lowdon and Blake (1979); Jackson and Pawson (1984); Driver (1988); R.E. Morlan, CMC database (1999)

Bison alaskensis (Alaskan bison)

Measured Date (Normalized)	Lab No	Material	Location	References/Remarks
>39900 (>39900)	I-5405	most of left horncore (CMN 13506)	Gold Run Ck. (Dawson Loc. 33) (63.720°N, 138.680°W), YT, CAN	Harington and Clulow (1973); Harington (1977, 1978, 1989), [Bonnichsen (1979) incorrectly cites date as >39,000 BP]; R.E. Morlan, CMC database (1999)

Bison cf. Bison latifrons (giant bison)

Measured Date (Normalized)	Lab No	Material	Location	References/Remarks
35430 ± 1080 (35480 ± 1080)	Beta-78574	humerus	Approx. 1 km downstream from Chuchi L. (55.190°N, 124.380°W), BC, CAN	Harington, et al. (1996); R.E. Morlan, CMC database (1999)
34740 ± 420 (34800 ± 420)	Beta-78573 CAMS-17566	humerus	Approx. 1 km downstream from Chuchi L. (55.190°N, 124.380°W), BC, CAN	Harington, et al. (1996); R.E. Morlan, CMC database (1999)
(30740 ± 220)	TO-3653	horncore	Approx. 1 km downstream from Chuchi L. (55.190°N, 124.380°W), BC, CAN	Harington, et al. (1996); R.E. Morlan, CMC database (1999)

Bison bison occidentalis (western bison)

Measured Date (Normalized)	Lab No	Material	Location	References/Remarks
11370 ± 170 (11450 ± 170)	GSC-613		Clarke Pit: Bow R., Cochrane (51.180°N, 114.450°W), AB, CAN	Lowdon, et al. (1967); Churcher (1968); Jackson and Pawson (1984); [Stalker (1968), Wilson (1983), Wilson and Churcher (1984) In: R.E. Morlan, CMC database (1999)]

11100 ± 160 (11180 ± 160)	GSC-989		Clarke Pit: Bow R., Cochrane (51°10'40"N, 114°27'30"W), AB, CAN	Lowdon, et al. (1967); Wilson and Churcher (1978); Jackson and Pawson (1984); [Churcher (1968, 1975), Stalker (1968), Wilson (1983), Wilson and Churcher (1984) In: R.E. Morlan, CMC database (1999)]
10760 ± 160 (10840 ± 160)	GSC-612		Griffin Gravel Pit: Bow R., Cochrane (51.180°N, 114.450°W), AB, CAN	Lowdon, et al. (1967); [Churcher (1968, 1975), Stalker (1968), Wilson (1983), Wilson and Churcher (1984) In: R.E. Morlan, CMC database (1999)]
9670 ± 160 (9770 ± 160)	I-8579	ribs	Milan, 11.3 km E and 3.2 km S of Three Hills (51.690°N, 113.075°W), AB, CAN	Shackleton and Hills (1977)
9630 ± 300 (9720 ± 300)	GSC-1894	ribs	Milan: 11.3 km E and 3.2 km S of Three Hills (51.690°N, 113.075°W), AB, CAN	Shackleton and Hills (1977); Lowdon and Blake (1979)
7220 ± 480 (7300 ± 480)	GSC-612-2		Griffin Gravel Pit: Bow R., Cochrane (51.180°N, 114.450°W), AB, CAN	Jackson and Pawson (1984); [Churcher (1968, 1975), Stalker (1968), Wilson (1983), Wilson and Churcher (1984) In: R.E. Morlan, CMC database (1999)
6880 ± 280 (6960 ± 280)	GX-1769		Little Bow R., near Travers, AB, CAN	[Reeves (1973) In: R.E. Morlan, CMC database (1999)]
6320 ± 140 (6320 ± 140)	GSC-280	skull	Assiniboine R., floodplain 10 km W of Russell (50.770 °N, 101.430°W), MB, CAN	Pettipas (1976); Nielsen, et al. (1984); [Reeves (1973) In: R.E. Morlan, CMC database (1999)]
5670 ± 150 (5750 ± 150)	GSC-988		Griffin Pit: Bow R., Cochrane (51°10'40"N, 114°27'10"W), AB, CAN	Jackson and Pawson (1984); [Churcher (1968, 1975), Stalker (1968), Wilson (1983), Wilson and Churcher (1984) In: R.E. Morlan, CMC database (1999)]
5340 ± 110	SI-845	hornsheath	Goldstream Ck. (64.950°N, 147.583°W), AK, USA	[Péwé (1975) spurious date]; Bonnichsen (1979); Holmes and Bacon (1982); Guthrie (1985, 1990); Dixon (1993); Stephenson, et al. (in press)
3240 ± 90 (3320 ± 90)	I-11638	cranial fragment	Banff, Bow drainage (51.280°N, 115.500°W), AB, CAN	Harington (1984c); R.E. Morlan, CMC database (1999)
470 ± 90	SI-852	hornsheath	Chester Ck., Anchorage, AK, USA	Holmes and Bacon (1982); Guthrie (1985, 1990); Dixon (1993); [Probably wood bison (Bison bison athabascae (Stephenson, et al. (in press)].

Bison bison cf. Bison bison occidentalis (western bison)

Measured Date (Normalized)	Lab No	Material	Location	References/Remarks
(10750 ± 180)	RIDDL-220	left tibia	Clayhurst Gravel Pit: Peace R. bank S of Clayhurst (56°07'40"N, 120°05'00"W), BC, CAN	[Apland and Harington (1994) four AMS dates were processed on this specimen - see AA-1219 (10,600 ± 160 BP); CAMS-398 (10,580 ± 210 BP); and CAMS 150 (10,340 ± 150 BP): the weighted average age = 10,500 ± 90 BP]; R.E. Morlan, CMC database (1999)
(10600 ± 160)	AA-1219	left tibia	Clayhurst Gravel Pit: Peace R. bank S of Clayhurst (56°07'40"N, 120°05'00"W), BC, CAN	[Apland and Harington (1994) four AMS dates were processed on this specimen - see also RIDDL-220, CAMS-398, and CAMS 150: the weighted average age = 10,500 ± 90 BP]; R.E. Morlan, CMC database (1999)

Bison bison antiquus (ancient bison)

Measured Date (Normalized)	Lab No	Material	Location	References/Remarks
11300 ± 290 (11380 ± 290)	RL-757	vertebra	Gallelli Pit, E bank of Bow R., 1010 m asl, Calgary (51.170°N, 114.420°W), AB, CAN	Brumley and Rushworth (1983); Jackson and Pawson (1984); Wilson and Churcher (1978, 1984); Wilson (1978); R.E. Morlan, CMC database (1999)

10730 ± 160	BGS-1016	tibia	S region, 420 m asl (53.500°N, 113.750°W), AB, CAN	[Wilson (1996) In: A.S. Dyke (GSC database)]

Bison bison athabascae (wood bison)

Measured Date (Normalized)	Lab No	Material	Location	References/Remarks
11130 ± 70 (11130 ± 90)	TO-7693		St. Mary's Reservoir (49°07'N, 113°12'W), AB, CAN	Hall (1999b); McNeil, et al. (1999); Hills, et al. (1999)
8145 ± 320 (8225 ± 320)	GX-2104	skull	Gallelli Pit: on the left (E) bank of Bow R., 1010 m asl, Calgary (51.030°N, 114.080°W), AB, CAN	Brumley and Rushworth (1983); Jackson and Pawson (1984); Wilson and Churcher (1978, 1984); R.E. Morlan, CMC database (1999)
(8080 ± 60)	TO-3709	radius (CR-92-37C)	Kuujjua R., Victoria I. (71°07'N, 115°43'W), NT, CAN	C.R. Harington [unpublished]; Stephenson, et al. (in press). [Earliest-known, and northernmost record for wood bison].
(4880 ± 80)	Beta-25120 ETH-3899	nasal bone fragment (CMN 44760)	Carmacks (62.080°N, 136.250°W), YT, CAN	C.R. Harington [unpublished]; Stephenson, et al. (in press)
2130 ± 60 (2200 ± 60)	Beta-79854 CAMS-18417	cranium (ear capsule) (CRH-95-5)	Finlayson R., YT, CAN	C.R. Harington [unpublished]
2130 ± 30 (2180 ± 30)	Beta-91755	cranium	Congdon Ck., Kluane L., YT, CAN	[R.M. Gotthardt, pers. comm. (1998) In: R.E. Morlan, CMC database (1999)]; [M. Hoefs, pers. comm. In: Stephenson, et al. (in press)]
2110 ± 40 (2150 ± 40)	Beta-91756	horncore	Takhini R., YT, CAN	[R.M. Gotthardt, pers. comm. (1998) In: R.E. Morlan, CMC database (1999)]; [M. Hoefs, pers. comm. In: Stephenson, et al. (in press)]
1350 ± 95 (1430 ± 95)	I-5404	frontal (CMN 17519)	Quartz Ck. (Dawson Loc. 6) (63°47'N, 139°6.00'W), YT, CAN	Harington (1977, 1980); Dixon (1993); R.E. Morlan, CMC database (1999); Stephenson, et al. (in press)
930 ± 90 (940 ± 90)	Beta-69762 CAMS-11018	posterior cranial fragment with horncores (CR-94-1) female specimen	Cowley L. (61°37'N, 130°25'W), YT, CAN	C.R. Harington [unpublished]; Stephenson, et al. (in press)

Bison bison bison (plains bison)

Measured Date (Normalized)	Lab No	Material	Location	References/Remarks
<5500	#49		Okotoks (50.750°N, 113.920°W), AB, CAN	Wilson (1983)
3800 ± 300	#48		Okotoks (50.750°N, 113.920°W), AB, CAN	Wilson (1983)
3600 ± 70 (3680 ± 75)	Beta-1627	ribs	Hitching Post Ranch, 40 km NW of Calgary (51.500°N, 114.500°W), AB, CAN	Brumley and Rushworth (1983); [Wilson (1983); Graham, et al. (1987) In: R.E. Morlan, CMC database (1999)]; Wilson and Hills (1984)

Bison bison (bison)

Measured Date (Normalized)	Lab No	Material	Location	References/Remarks
(5045 ± 45)	AA-4379 (VP-4157)	skull (male)	Fort Yukon (66.5°N, 145°W), AK, USA	[R.D. Guthrie In: Stephenson, et al. (in press)]. [Probably wood bison].
4660 ± 40	Beta-152446	jaw (JcVa-14)	Sandpiper Ice Patch site, 72 km SW of Whitehorse (60°18'N, 136°00'W), YT, CAN	[R. Farnell, pers. comm. (2001)]; [Probably wood bison]
4495 ± 60	Beta-65662	horncore and cranium (male)	Black R., Yukon Flats (66.5°N, 144°W), AK, USA	Stephenson, et al. (in press). [Probably wood bison].
4390 ± 70	Beta-136731	skull with hornsheaths (male)	Black R., Yukon Flats (66.5°N, 144°W), AK, USA	Stephenson, et al. (in press). [Probably wood bison].
3520 ± 40	Beta-104823	skull with both hornsheaths	Fort Yukon (66.5°N, 145°W), AK, USA	Stephenson, et al. (in press). [Probably wood bison].
3425 ± 105 (3505 ± 105)	S-668	ribs (CMC-547)	Harder site near Saskatoon (52.220°N, 107.050°W), SK, CAN	[Dyck (1970, 1972, 1977); Gibson (1981); Morlan, 1993, 1994) In: R.E. Morlan, CMC database (1999); Graham, et al. (1987)
(2900 ± 80)	AA-3220	skull [AMNH (A-508-5331)]	Fairbanks railroad terminal (66°50'N, 147°48'W), AK, USA	[R.D. Guthrie In: Stephenson, et al. (in press)]. [Probably wood bison].

Measured Date (Normalized)	Lab No	Material	Location	References/Remarks
(2545 ± 80)	AA-3217	skull [AMNH (A-479-4783)]	Yukon Flats, downstream from Circle (65°49'N, 144°22'W), AK, USA	[R.D. Guthrie In: Stephenson, et al. (in press)]. [Probably wood bison].
2460 ± 40	Beta-137731	skeleton	Braeburn, YT, CAN	[M. Hoefs, pers. comm. In: Stephenson, et al. (in press)]. [Probably wood bison].
1730 ± 60	Beta-62999		Black R., Yukon Flats (66.5°N, 144°W), AK, USA	Stephenson, et al. (in press). [Probably wood bison].
(1270 ± 55)	AA-3218	skull [AMNH (A-393-1013)]	Tetlin-Tanaross area (63°N, 143°W), AK, USA	[R.D. Guthrie In: Stephenson, et al. (in press)]. [Probably wood bison].
1100 ± 150 (1180 ± 150)	SFU-224		Assiniboine drainage, 12 km W of Brandon (49.820°N, 100.090°W), MB, CAN	[Tisdale (1978); Badertscher, et al. (1987) In: R.E. Morlan, CMC database (1999)]
370 ± 40	Beta-152441	tooth (JeUx-16)	Annie Ned Ck. site, 48 km W of Whitehorse (60°46'N, 135°52'W), YT, CAN	[R. Farnell, pers. comm. (2001)]; [Probably wood bison]
170 ± 30	Beta-136732	skull with hornsheaths	Anchorage (61°12'N, 149°48'W), AK, USA	Stephenson, et al. (in press)

Bison sp. (bison)

Measured Date (Normalized)	Lab No	Material	Location	References/Remarks
>47500	Beta-70840 CAMS-11918	calcaneum	Pauline Cove, Herschel I., YT, CAN	R.E. Morlan, CMC database (1999)
45200 ± 2100 (45200 ± 2100)	Beta-79857 CAMS-18420	right metatarsal (CMN 33876)	Upper Dominion Ck. (Dawson Loc. 36) (63°51'N, 138°54'W), YT, CAN	C.R. Harington [unpublished]; R.E. Morlan, CMC database (1999)
44600 ± 1900 (44660 ± 1900)	Beta-79856 CAMS-18419	thoracic vertebra	Hunker Ck. (Dawson Loc. 10) (64°N, 139°5.00'W), YT, CAN	Harington (1980, 1997); R.E. Morlan, CMC database (1999)
(42000 ± 1200)	RIDDL-142	rib	Old Crow R. (Old Crow Loc. HH69-21) (67.917°N, 139.750°W), YT, CAN	Morlan et al. (1980, 1986); Morlan, et al. (1990); R.E. Morlan, CMC database (1999)
(>41260)	AECV-942C	horncore	Riverview Pit (C), E Edmonton, N Saskatchewan R. (53°36'N, 113°19.5'W), AB, CAN	Young, et al. (1994); Burns and Young (1994); Burns (1996); R.E. Morlan, CMC database (1999)
>41000	SI-2816-B	humerus (apatite extraction) (11A75-2310)	Old Crow R. (Old Crow Loc. 11(A)), 259 m asl. (67.850°N, 139.970°W), YT, CAN	B. Beebe [unpublished]
(39500 ± 900)	RIDDL-730	innominate	Old Crow R. (Old Crow Loc. 15) (67.916°N, 139.750°W), YT, CAN	Morlan et al. (1980, 1986, 1996); Morlan, et al. (1990); R.E. Morlan, CMC database (1999)
>39000	SI-840	hornsheath (AMNH 4038)	Cripple Ck. (64.820°N, 148.020°W), AK, USA	Péwé (1975); Bonnichsen (1979); Guthrie (1985, 1990)
>38000	SI-2817-A	metacarpal (11A75-2387)	Old Crow R. (Old Crow Loc. 11(A)), 259 m asl. (67.850°N, 139.970°W), YT, CAN	Harington (1977); Morlan (1980); R.E. Morlan, CMC database (1999)
(37850 ± 3850)	AECV-1559C		Ostero Gravel Pit: NW side of Peace R. near Taylor (57°07'40"N, 120°05'W), BC, CAN	[Mathews (1978); Clague (1980); Bobrowski et al. (1991); C.R. Harington, pers. comm. (1998) In: R.E. Morlan, CMC database (1999)]
(37300 ± 750)	RIDDL-136	radius	Johnson Ck. (Old Crow Loc. 70) (67.917°N, 139.750°W), YT, CAN	Morlan et al. (1979, 1980); Morlan, et al. (1990); R.E. Morlan, CMC database (1999)
(37000 ± 950)	RIDDL-196	humerus	Old Crow R. (Old Crow Loc. 20) (68.083°N, 139.500°W), YT, CAN	Morlan et al. (1990); R.E. Morlan, CMC database (1999)
>37000	I-9273		Tanana Bluffs, AK, USA	Guthrie (1985, 1990)
36850 ± 1300 (36730 ± 1300)	SI-2816-A	humerus (11A75-2310)	Old Crow R. (Old Crow Loc. 11(A)), 259 m asl. (67.850°N, 139.970°W), YT, CAN	Harington (1977); Morlan (1980); [Morlan and McNeely (1996) In: A.S. Dyke (GSC database)]; R.E. Morlan, CMC database (1999)
(36500 ± 650)	RIDDL-144	rib	Old Crow R. (Old Crow Loc. 15) (67.916°N, 139.750°W), YT, CAN	Morlan et al. (1980, 1986, 1996); Morlan, et al. (1990); R.E. Morlan, CMC database (1999)
(36500 ± 1000)	RIDDL-138	tibia	Old Crow R. (Old Crow Loc. REM78-1) (67.917°N, 139.750°W), YT, CAN	Morlan et al. (1990); R.E. Morlan, CMC database (1999)

36425 +2575 / -1974	QC-891		Pearl Ck., AK, USA	Guthrie (1985, 1990) "Blue Babe" carcass.
(35200 ± 750)	RIDDL-310	rib	Old Crow R. (Old Crow Loc. 15) (67.916°N, 139.750°W), YT, CAN	Morlan et al. (1980, 1986, 1996); Morlan, et al. (1990); R.E. Morlan, CMC database (1999)
(35200 ± 750)	RIDDL-309	tibia	Old Crow R. (Old Crow Loc. REM78-1) (67.917°N, 139.750°W), YT, CAN	Morlan et al. (1990); R.E. Morlan, CMC database (1999)
>35000	SI-844	hornsheath (AMNH 46890)	Little El Dorado Ck. (65.100°N, 147.680°W), AK, USA	Péwé (1975); Bonnichsen (1979); Guthrie (1985, 1990)
>35000	RL-402		Ester Ck. (65.000°N, 147.000°W), AK, USA	Guthrie (1985, 1990)
34900 ± 500	SI-2825-B	metacarpal (apatite extraction) (11A75-7909)	Old Crow R. (Old Crow Loc. 11(A)), 259 m asl. (67.850°N, 139.970°W), YT, CAN	B. Beebe [unpublished]
(34700 ± 600)	RIDDL-139	humerus	Old Crow R. (Old Crow Loc. 15) (67.916°N, 139.750°W), YT, CAN	Morlan et al. (1980, 1986, 1996); R.E. Morlan, CMC database (1999)
34000 ± 1000 (34080 ± 1000)	SI-2822-A	distal tibia (11A75-4588)	Old Crow R. (Old Crow Loc. 11(A)), 259 m asl. (67.850°N, 139.970°W), YT, CAN	Harington (1977); Morlan (1980); R.E. Morlan, CMC database (1999)
(31750 ± 850)	AECV-936C	radio-ulna	Consolidated Pit 48, N Saskatchewan R. Edmonton (53°38.5'N, 113°17'W), AB, CAN	Young, et al. (1994); Burns and Young (1994); Burns (1996); R.E. Morlan, CMC database (1999)
(31730 ± 230)	CAMS-23469	tibia	Bluefish Cave II, 600 m asl. (67°08'N, 140°47'W), YT, CAN	Cinq-Mars (1979, 1990); Morlan and Cinq-Mars (1982); Cinq-Mars, et al. (1991); Burk and Cinq-Mars (1996);
(30810 ± 970)	Beta-33192 ETH-5900	right radio-ulna (CMN 46320)	Eldorado Ck. (Dawson Loc. 63, Nugget Gulch) (63°53'N, 139°17'W), YT, CAN	C.R. Harington [unpublished]
30200 ± 830	SI-2822-B	distal tibia (apatite extraction) (11A75-4588)	Old Crow R. (Old Crow Loc. 11(A)), 259 m asl. (67.850°N, 139.970°W), YT, CAN	B. Beebe [unpublished]
29295 ± 2440	SI-842	hornsheath (AMNH 46928)	Cripple Ck. (64.820°N, 148.020°W), AK, USA	Péwé (1975); Bonnichsen (1979); Guthrie (1985, 1990)
28050 ± 500 (28130 ± 500)	SI-2825-A	metacarpal (11A75-7909)	Old Crow R. (Old Crow Loc. 11(A)), 259 m asl. (67.850°N, 139.970°W), YT, CAN	Harington (1977); Morlan (1980); R.E. Morlan, CMC database (1999)
27950 ± 560	Beta-67670	astragalus	Lime Hills, Cave 1, 527 m asl. (61.750°N, 155.500°W), AK, USA	Ackerman (1996b)
27400 ± 1050	SI-2817-B	metacarpal (apatite extraction) (11A75-2387)	Old Crow R. (Old Crow Loc. 11(A)), 259 m asl. (67.850°N, 139.970°W), YT, CAN	B. Beebe [unpublished]
26640 ± 2800 (26720 ± 1400)	QU-781	humerus	Porcupine R. (Porcupine R. Loc. 100), 297 m asl. (67.480°N, 139.920°W), YT, CAN	Morlan (1980) [R.E. Morlan, pers. comm. (1998) indicates corrected age of 26,640 ± 1400 BP]; R.E. Morlan, CMC database (1999)
(26350 ± 280)	TO-393	left radius fragment (CMN 43794)	Ketza R. locality, Ross R. area, YT, CAN	Jackson and Harington (1991); R.E. Morlan, CMC database (1999)
(25210 ± 760)	AECV-1201C	humerus	E of Edmonton along N Saskatchewan R., AB, CAN	Young, et al. (1994) Burns and Young (1994); Burns (1996); R.E. Morlan, CMC database (1999)
24140 ± 2200	SI-445		Fairbanks Ck. (65.067°N, 147.167°W), AK, USA	Guthrie (1985, 1990)
22870 ± 540 (22950 ± 540)	I-10825	horncore (CMN 34962)	Ostero Gravel Pit: NW side of Peace R., near Taylor (56.140°N, 120.675°W), BC, CAN	C.R. Harington [unpublished]; Mathews (1978); Clague (1980); Bobrowsky et al. (1991); R.E. Morlan, CMC database (1999)
21065 ± 1365	SI-839	hornsheath (AMNH 4037)	Cripple Ck. (64.820°N, 148.020°W), AK, USA	Péwé (1975); Bonnichsen (1979)
21050 +320/-340	DIC-1333	metacarpal	Porcupine R. Lower Rampart Cave 1 (67.000°N, 143.000°W), AK, USA	Dixon (1984); [Guthrie (1985,1990) specimen incorrectly cited within Horse Table]; Sattler and Dixon (1997)
20445 ± 885	SI-837	hornsheath (AMNH 1042)	Fairbanks Ck. (65.070°N, 147.170°W), AK, USA	Bonnichsen (1979); Guthrie (1985, 1990)
18000 ± 200	SI-841	hornsheath (AMNH 4002)	Manley Hot Springs (65.000°N, 149.000°W), AK, USA	Péwé (1975); Bonnichsen (1979); Guthrie (1985, 1990)

17170 ± 840	SI-838	hornsheath (AMNH 1014)	Fairbanks Ck. (65.070°N, 147.170°W), AK, USA	Péwé (1975); Bonnichsen (1979); Guthrie (1985, 1990)
13070 ± 280	K-1327	calcaneum	Trail Ck. Cave 9, Seward Pen. (65.800°N, 163.220°W), AK, USA	Morlan and Cinq-Mars (1982) [broken perhaps by man, for carnivore teeth marks are absent]; (Dixon 1993); West (1996)
12660 ± 560 (12740 ± 280)	QU-782	humerus	Old Crow R. (Old Crow Loc. 11(A)) (67.480°N, 139.920°W), YT, CAN	Harington (1977); Morlan (1980); [R.E. Morlan, pers. comm. (1998) indicates corrected age of 12,660 ± 280 BP]; R.E. Morlan, CMC database (1999). [Probable source Loc. 11(1)].
12220 ± 1500 (12300 ± 750)	QU-783	humerus	Old Crow R. (Old Crow Loc. 11), 260 m asl. (67.480°N, 139.870°W), YT, CAN	Morlan (1980); [R.E. Morlan, pers. comm. (1998) indicates corrected age of 12,220 ± 750 BP]; R.E. Morlan, CMC database (1999). [Probable source Loc. 11(1)].
11900 ± 70	Beta-67494	skull with horn sheaths (female)	Birch Ck., Yukon Flats (66.5°N, 146.5°W), AK, USA	Stephenson, et al. (in press)
11750 ± 110 (11830 ± 110)	SFU-(unknown)	skull	32 km N of Victoria at North Saanich, Vancouver I. (48.5°N, 123.5°W), BC, CAN	[Mackie (1987), Hebda (1988) In: R.E. Morlan, CMC database (1999)]
(11620 ± 170)	AECV-1203C	metacarpal	Clover Bar Sand & Gravel Pit(B), Edmonton area (53°35'N, 113°19.5'W), AB, CAN	Burns and Young (1994); Young et al. (1994); Burns (1996); R.E. Morlan, CMC database (1999)
11450 ± 400 (11530 ± 200)	QU-780	humerus	Old Crow R. (Old Crow Loc. 11(A)), 259 m asl. (67.480°N, 139.920°W), YT, CAN	Morlan (1980); [R.E. Morlan, pers. comm. (1998) indicates corrected age of 11,450 ± 200 BP]; R.E. Morlan, CMC database (1999). [Probable source Loc. 11(1)].
10770 ± 120 (10850 ± 120)	SFU-454		Charlie L. Cave, NW of Fort St. John (56.275°N, 120.930°W), BC, CAN	[Driver (1988b) In: Bobrowsky and Rutter (1992)]; Driver, et al. (1996); R.E. Morlan, CMC database (1999)
(10730 ± 160)	AECV-1111C	horncore	Clover Bar Pit, E Edmonton, N Saskatchewan R. (53°35'N, 113°19.5'W), AB, CAN	Burns (1996); [Burns and Young (1994); Young, et al. (1994) In: R.E. Morlan, CMC database (1999)]
10715 ± 225	St-1561		Dry Ck. (63°53'N, 149°02'W), AK, USA	[Guthrie (1985) In: Stephenson, et al. (in press)]
(10580 ± 210)	CAMS-398	left tibia	Clayhurst Gravel Pit: Peace R. bank S of Clayhurst (56°07'40"N, 120°05'00"W), BC, CAN	Apland and Harington (1994) four AMS dates were processed on this specimen - see also RIDDL-220, AA-1219, and CAMS 150: the weighted average age = 10,500 ± 90 BP]; R.E. Morlan, CMC database (1999)
(10560 ± 80)	CAMS-2134		Charlie L. Cave, NW of Fort St John (56.275°N, 120.930°W), BC, CAN	Driver, et al. (1996); Driver, (1998b); R.E. Morlan, CMC database (1999)
10450 ± 150 (10530 ± 150)	SFU-300		Charlie L. Cave, NW of Fort St. John (56.275°N, 120.930°W), BC, CAN	[Driver (1988b) In: Bobrowsky and Rutter (1992)]; Driver, et al. (1996); R.E. Morlan, CMC database (1999)
10380 ± 160 (10460 ± 160)	SFU-378		Charlie L. Cave, NW of Fort St. John (56.275°N, 120.930°W), BC, CAN	[Driver (1988b) In: Bobrowsky and Rutter (1992)]; Driver, et al. (1996); R.E. Morlan, CMC database (1999)
(10340 ± 150)	CAMS-150	left tibia	Clayhurst Gravel Pit: Peace R. bank S of Clayhurst (56°07'40"N, 120°05'00"W), BC, CAN	Apland and Harington (1994) four AMS dates were processed on this specimen - see also RIDDL-220, AA-1219, and CAMS 398: the weighted average age = 10,500 ± 90 BP]; R.E. Morlan, CMC database (1999)
10300 ± 200 (10380 ± 200)	BGS-617		Hubbell Ck. 13 km NW of town of Swan R. (52.180°N, 101.430°W), MB, CAN	Pettipas and Buchner (1983); Nielsen, et al. (1984); [Nielsen (1988) In: R.E. Morlan, CMC database (1999)]
(10240 ± 160)	AECV-1206C	tibia	Ostero Gravel Pit: NW side of Peace R., near Taylor (56.145°N, 120.690°W), BC, CAN	Clague (1980); Bobrowsky, et al. (1991); Bobrowsky and Rutter (1992); R.E. Morlan, CMC database (1999)
(10230 ± 140)	AECV-1558C	scapula	Clayhurst Gravel Pit: Peace R. bank S of Clayhurst (56°07'40"N, 120°05'00"W), BC, CAN	P. Bobrowsky (pers. comm. 1992); Apland and Harington (1994); R.E. Morlan, CMC database (1999)

10210 ± 100 (10290 ± 100)	S-3118		Great Sand Hills, S Saskatchewan R. valley, E of Prelate (50.670°N, 109.330°W), SK, CAN	Morlan (1993); [Linnamae (1990); Linnamae and Cazakoff (1990); Corbeil (1995) In: R.E. Morlan, CMC database (1999)]
(10200 ± 280)	GSC-3065	skull	Aquitaine Pit, 1045 m asl., Calgary (51.050°N, 114.072°W), AB, CAN	Jackson and Pawson (1984); [Blake (1986) - on a terrace 5 m above the Bow R.]; [Jackson, et al. (1982); Wilson (1983); Wilson and Churcher (1984); In: R.E. Morlan, CMC database (1999)]
(10080 ± 150)	AECV-439C		Bushe R. (58.500°N, 117.000°W), AB, CAN	[Beaudoin (1988) In: A.S. Dyke (GSC database)]; [Beaudoin (1988), PMA database]
10000 ± 160 (10200 ± 160)	GSC-1205	2 tibiae	Upper terrace of Athabasca R., Athabasca (54.720°N, 113.280°W), AB, CAN	Lowdon and Blake (1979); R.E. Morlan, CMC database (1999)
(9990 ± 150)	RIDDL-393		Charlie L. Cave, BC (56.275°N. 120.930°W), BC, CAN	Driver et al. (1996, 1988); [Fladmark, et al. (1988); Fladmark (1996) In: R.E. Morlan, CMC database (1999)]
(9990 ± 80)	CAMS-13192		Twin Pines, James Pass (51°41'N, 114°29'W), AB, CAN	[A. Beaudoin (PMA database)]
(9930 ± 60)	CAMS-12913		Twin Pines, James Pass (51°41'N, 114°29'W), AB, CAN	[A. Beaudoin (PMA database)]
(9870 ± 180)	RIDDL-362	distal end of right tibia, metacarpal, and metatarsal	Firth R., Engigstciak (69.360°N, 139.530°W), YT, CAN	Cinq-Mars et al. (1991); R.E. Morlan, CMC database (1999); Stephenson, et al. (in press)
9860 ± 140	I-8483		(53.500°N, 113.500°W), AB, CAN	[Westgate, et al. (1976) In: A.S. Dyke (GSC database)]
(9770 ± 180)	RIDDL-281	distal end of right metacarpal	Firth R., Engigstciak (69.360°N, 139.530°W), YT, CAN	Cinq-Mars et al. (1991); R.E. Morlan, CMC database (1999); Stephenson, et al. (in press)
9760 ± 160 (9840 ± 160)	SFU-355		Charlie L. Cave, NW of Fort St. John (56.275°N, 120.930°W), BC, CAN	Driver, et al. (1996); [Fladmark, et al. (1988); Fladmark (1996) In: R.E. Morlan, CMC database (1999)]
(9750 ± 80)	TO-2999	distal metacarpal	James Pass Meadow (51.750°N, 116.500°W), AB, CAN	Beaudoin et al. (1996); [Ronaghan (1993) In: R.E. Morlan, CMC database (1999)]
(9670 ± 150)	CAMS-2136		Charlie L. Cave, NW of Fort St. John (56.275°N, 120.930°W), BC, CAN	Driver, et al. (1996); R.E. Morlan, CMC database (1999)
(9600 ± 210)	AECV-746C		Oldman River Dam area (49.580°N, 114.050°W), AB, CAN	[Beaudoin In: A.S. Dyke (GSC database)]; Van Dyke, et al. (1990); [Beaudoin (1991), PMA database]; [Van Dyke, et al. (1991) In: R.E. Morlan, CMC database (1999); Beaudoin (1991, 1998)]
9565 ± 190 (9645 ± 190)	I-9997	cranial fragment	Muskeg R. (60.320°N, 123.350°W), NT, CAN	C.R. Harington [unpublished]; R.E. Morlan, CMC database (1999); Stephenson, et al. (in press)
9540 ± 60 (9560 ± 60)	Beta-79861 CAMS-18424	left scapula (CRH-95-12)	Old Horton R. channel, Cape Bathurst Pen. (70.080°N, 127.500°W), NT, CAN	C.R. Harington [unpublished]; [Cinq-Mars (1991), LeBlanc (1994), Nolin (1994) In: R.E. Morlan, CMC database (1999)]; Stephenson, et al. (in press)
9520 ± 240 (9600 ± 240)	GX-956	cervical vertebra	Oldman R. 1400 m asl (49.860°N, 114.375°W), AB, CAN	Lowdon, et al. (1972); R.E. Morlan, CMC database (1999)
9500 ± 150 (5980 ± 120)	BGS-840		Hubbell Ck. 13 km NW of town of Swan R. (52.180°N, 101.430°W), MB, CAN	Nielsen, et al. (1984); [Nielsen (1988) In: R.E. Morlan, CMC database (1999)]
9400 ± 125 (9480 ± 125)	BGS-887		Hubbell Ck. 13 km NW of town of Swan R. (52.180°N, 101.430°W), MB, CAN	Nielsen, et al. (1984); [Nielsen (1988) In: R.E. Morlan, CMC database (1999)]
(9400 ± 230)	RIDDL-319	distal end of left metatarsal	Firth R., Engigstciak (69.360°N, 139.530°W), YT, CAN	Cinq-Mars et al. (1991); R.E. Morlan, CMC database (1999); Stephenson, et al. (in press)
9210 ± 110 (9290 ± 110)	S-3308		Great Sand Hills, S Saskatchewan valley, E of Prelate (50.744°N, 109.374°W), SK, CAN	[Morlan and McNeely (1996) In: A.S. Dyke (GSC database)]; R.E. Morlan, CMC database (1999)

9000 ± 250	Beta-18552		Arch Cave, Porcupine R. (67°N, 143°W), AK, USA	Dixon (1993); Stephenson, et al. (in press)
8930 ± 120 (9010 ± 120)	S-3114		Great Sand Hills, S Saskatchewan R. valley, E of Prelate (50.670°N, 109.330°W), SK, CAN	Morlan (1993); Corbeil (1995); R.E. Morlan, CMC database (1999)
8920 ± 130 (9000 ± 130)	S-3309		Great Sand Hills, S Saskatchewan valley, E of Prelate (50.744°N, 109.374°W), SK, CAN	[Morlan and McNeely (1996) In: A.S. Dyke (GSC database)]; R.E. Morlan, CMC database (1999)
8670 ± 230	AECV-303C	scapula, glenoid fossa	Callahan Sand Pit, Porcupine Hills (50°33'N, 113°48'W), AB, CAN	[Beaudoin (1987), PMA database]
(8580 ± 160)	AECV-1198C		Oldman River Dam area (49.580°N, 114.050°W), AB, CAN	[Beaudoin In: A.S. Dyke (GSC database)]; [A. Beaudoin (PMA database)]; [Van Dyke, et al. (1990, 1991) In: R.E. Morlan, CMC database (1999)]; Beaudoin (1991, 1998)
8550 ± 270 (8630 ± 270)	RL-873		Crowsnest Pass (49.630°N, 114.600°W), AB, CAN	Driver (1982); Landals (1986); R.E. Morlan, CMC database (1999)
(8390 ± 120)	AECV-1380C		Oldman River Dam area (49.580°N, 114.050°W), AB, CAN	[Beaudoin In: A.S. Dyke (GSC database)]; [A. Beaudoin (PMA database)]; [Van Dyke, et al. (1990); Beaudoin (1998) In: R.E. Morlan, CMC database (1999)]
(8370 ± 110)	AECV-1375C		Oldman River Dam area (49.580°N, 114.050°W), AB, CAN	[Beaudoin In: A.S. Dyke (GSC database)]; [A. Beaudoin (PMA database)]; [Van Dyke, et al. (1990); Beaudoin (1998) In: R.E. Morlan, CMC database (1999)]
8300 ± 280 (8545 ± 280)	GX-6397-A		Bow R., Calgary (51.030°N, 114.080°W), AB, CAN	Brumley and Rushworth (1983); Jackson and Pawson (1984); [Wilson (1974, 1983) In: R.E. Morlan, CMC database (1999)]
(8210 ± 110)	AECV-1382C		Oldman River Dam area (49.580°N, 114.050°W), AB, CAN	[Beaudoin In: A.S. Dyke (GSC database)]; [A. Beaudoin (PMA database)]
8195 ± 1090	S-1798	pelvic fragment	Whitemud Ck., N Saskatchewan R. valley, near Edmonton (53.500°N, 113.670°W), AB, CAN	Rains and Welch (1988)
(8190 ± 120)	AECV-1369C		Oldman River Dam area (49.580°N, 114.050°W), AB, CAN	[Beaudoin In: A.S. Dyke (GSC database)]; [A. Beaudoin (PMA database)]; [Van Dyke, et al. (1990); Beaudoin (1998) In: R.E. Morlan, CMC database (1999)]
(8180 ± 110)	AECV-1379C		Oldman River Dam area (49.580°N, 114.050°W), AB, CAN	[Beaudoin In: A.S. Dyke (GSC database)]; [A. Beaudoin (PMA database)]; [Van Dyke, et al. (1990); Beaudoin (1998) In: R.E. Morlan, CMC database (1999)]
(8170 ± 110)	AECV-1381C		Oldman River Dam area (49.580°N, 114.050°W), AB, CAN	[Beaudoin In: A.S. Dyke (GSC database)]; [A. Beaudoin (PMA database)]; [Van Dyke, et al. (1990); Beaudoin (1998) In: R.E. Morlan, CMC database (1999)]
(8160 ± 110)	AECV-1374C		Oldman River Dam area (49.580°N, 114.050°W), AB, CAN	[Beaudoin In: A.S. Dyke (GSC database)]; [A. Beaudoin (PMA database)]; [Van Dyke, et al. (1990); Beaudoin (1998) In: R.E. Morlan, CMC database (1999)]
8160 ± 200 (8240 ± 200)	S-3208		Great Sand Hills, S Saskatchewan valley, E of Prelate (50.670°N, 109.330°W), SK, CAN	Morlan (1993); [Linnamae (1990); Linnamae and Cazakoff (1990); Corbeil (1995) In: R.E. Morlan, CMC database (1999)]
(8130 ± 110)	AECV-1368C		Oldman River Dam area (49.580°N, 114.050°W), AB, CAN	[Beaudoin In: A.S. Dyke (GSC database)]; [A. Beaudoin (PMA database)]; [Van Dyke and Unfreed (1992); Beaudoin (1998) In: R.E. Morlan, CMC database (1999)]

(8110 ± 110)	AECV-1370C		Oldman River Dam area (49.580°N, 114.050°W), AB, CAN	[Beaudoin In: A.S. Dyke (GSC database)]; [A. Beaudoin (PMA database)]; [Van Dyke, et al. (1990); Beaudoin (1998) In: R.E. Morlan, CMC database (1999)]
(8090 ± 110)	AECV-1372C		Oldman River Dam area (49.580°N, 114.050°W), AB, CAN	[Beaudoin In: A.S. Dyke (GSC database)]; [A. Beaudoin (PMA database)]; Van Dyke, et al. (1990); Beaudoin (1998) In: R.E. Morlan, CMC database (1999)]
(8080 ± 150)	GSC-1209		Mona Lisa Site, Bow R., Calgary (51.030°N, 114.080°W), AB, CAN	Brumley and Rushworth (1983); Jackson and Pawson (1984); Lowdon, et al. (1971); Wilson, et al. (1974); [Wilson (1983) In: R.E. Morlan, CMC database (1999)]
8020 ± 200 (8100 ± 200)	RL-448		Crowsnest Pass (49°35'N, 114°25'W), AB, CAN	[Reeves (1975); Driver (1978) In: R.E. Morlan, CMC database (1999)]; Brumley and Rushworth (1983)
(8000 ± 120)	AECV-1378C		Oldman River Dam area (49.580°N, 114.050°W), AB, CAN	[Beaudoin In: A.S. Dyke (GSC database)]; [A. Beaudoin (PMA database)]; [Van Dyke, et al. (1990); Beaudoin (1998) In: R.E. Morlan, CMC database (1999)]
(7980 ± 180)	AECV-760C		Oldman River Dam area (49.580°N, 114.050°W), AB, CAN	[Beaudoin (1989) In: A.S. Dyke (GSC database)]; [Beaudoin (1991), PMA database]; [Van Dyke, et al. (1991); Beaudoin (1991) In: R.E. Morlan, CMC database (1999)]
(7980 ± 110)	AECV-1376C		Oldman River Dam area (49.580°N, 114.050°W), AB, CAN	[Beaudoin In: A.S. Dyke (GSC database)]; [A. Beaudoin (PMA database)]; [Van Dyke, et al. (1990); Beaudoin (1998) In: R.E. Morlan, CMC database (1999)]
(7920 ± 100)	AECV-1373C		Oldman River Dam area (49.580°N, 114.050°W), AB, CAN	[Beaudoin In: A.S. Dyke (GSC database)]; [A. Beaudoin (PMA database)]; [Van Dyke, et al. (1990); Beaudoin (1998) In: R.E. Morlan, CMC database (1999)]
(7890 ± 130)	AECV-762C		Oldman River Dam area (49.580°N, 114.050°W), AB, CAN	[Beaudoin (1989) In: A.S. Dyke (GSC database)]; [Beaudoin (1991), PMA database]; [Van Dyke, et al. (1991); Beaudoin (1991) In: R.E. Morlan, CMC database (1999)]
(7890 ± 110)	AECV-1371C		Oldman River Dam area (49°35'N, 114°03'W), AB, CAN	[Beaudoin (1992), PMA database]; [Van Dyke, et al. (1990); Beaudoin (1998) In: R.E. Morlan, CMC database (1999)]
(7880 ± 110)	AECV-1367C		Oldman River Dam area (49.580°N, 114.050°W), AB, CAN	[Beaudoin In: A.S. Dyke (GSC database)]; [A. Beaudoin (PMA database)]; [Van Dyke and Unfreed (1992); Beaudoin (1998) In: R.E. Morlan, CMC database (1999)]
(7830 ± 160)	AECV-1207C		Oldman River Dam area (49.580°N, 114.050°W), AB, CAN	[Beaudoin In: A.S. Dyke (GSC database); [A. Beaudoin (PMA database)]; [Van Dyke, et al. (1990); Beaudoin (1998) In: R.E. Morlan, CMC database (1999)]
(7810 ± 120)	AECV-1050C		Bow valley (51°04'N, 115°11'W), AB, CAN	[Beaudoin (1991), PMA database]
(7810 ± 120)	AECV-624-C		Lac des Arcs, Bow valley (51.050°N, 115.200°W), AB, CAN	Beaudoin (1991); [Magne and Ives (1991); Newton (1991) In: R.E. Morlan, CMC database (1999)]
7790 ± 120	AECV-336C	radius	Oldman River Dam area (49°36'N, 114°04'W), AB, CAN	[Beaudoin (1988), PMA database]
(7740 ± 190)	AECV-743C		Oldman River Dam area (49.580°N, 114.050°W), AB, CAN	[Beaudoin (1989) In: A.S. Dyke (GSC database)]; [Beaudoin (1991), PMA database]; [Van Dyke, et al. (1991); Beaudoin (1991) In: R.E. Morlan, CMC database (1999)]

(7690 ± 210)	AECV-742C		Oldman River Dam area (49.580°N, 114.050°W), AB, CAN	[Beaudoin (1989) In: A.S. Dyke (GSC database)]; [Beaudoin (1991), PMA database]; [Van Dyke, et al. (1991); Beaudoin (1991) In: R.E. Morlan, CMC database (1999)]
(7680 ± 140)	AECV-1383C		Oldman River Dam area (49.580°N, 114.050°W), AB, CAN	[Beaudoin In: A.S. Dyke (GSC database)]; [A. Beaudoin (PMA database)]; [Van Dyke and Unfreed (1992); Beaudoin (1998) In: R.E. Morlan, CMC database (1999)]
7655 ± 110 (7735 ± 110)	S-1084		Chin Coulee, 50 km SSE of Taber (49.6°N, 111.817°W), AB, CAN	Forbis (1968); Brumley and Rushworth (1983); [Quigg (1976); Vickers and Beaudoin (1989); Wilson, et al. (1991) In: R.E. Morlan, CMC database (1999)]
(7450 ± 100)	AECV-1377C		Oldman River Dam area (49.580°N, 114.050°W), AB, CAN	[Beaudoin In: A.S. Dyke (GSC database)]; [A. Beaudoin (PMA database)]; [Van Dyke, et al. (1990); Beaudoin (1998) In: R.E. Morlan, CMC database (1999)]
7200 ± 230 (7280 ± 230)	RL-876		Crowsnest Pass (49.630°N, 114.600°W), AB, CAN	Driver (1982); Brumley and Rushworth (1983); [Driver (1985, 1987); Landals (1986) In: R.E. Morlan, CMC database (1999)]
7165 ± 320 (7245 ± 320)	S-2453		Notukeu Ck., Wood R. drainage (49.720°N, 107.400°W), SK, CAN	[Morlan and McNeely (1996) In: A.S. Dyke (GSC database)]; R.E. Morlan CMC database (1999)
7150 ± 150	AECV-605C		S Saskatchewan valley (50.670°N, 111.670°W), AB, CAN	[Evans and Campbell (1992) In: A.S. Dyke (GSC database)]
(7110 ± 160)	AECV-1048C		Bow valley (51°04'N, 115°11'W), AB, CAN	[Beaudoin (1991), PMA database]
7065 ± 175 (7145 ± 175)	S-2500		Head-Smashed-in Buffalo Jump: in the Porcupine Hills, 1060 m asl, 16 km W of Fort MacLeod (49.717°N, 113.650°W), AB, CAN	[Beaudoin (1987, 1988, 1991, 1998), Brink and Dawe (1989), Brink et al. (1985, 1986), Brumley and Rushworth (1983), Morlan (1985), Reeves (1970, 1978), Wilmeth (1978) In: R.E. Morlan, CMC database (1999)]
6960 ± 100 (7040 ± 100)	BGS-1100	metacarpal and ungual phalanx	Swan R., 2.5 km SW of Harlington (52.000°N, 101.470°W), MB, CAN	[Nielsen (1988) In: R.E. Morlan, CMC database (1999)]
6955 ± 80	S-1706		N Saskatchewan R., near Fort Saskatchewan (53.700°N, 113.230°W), AB, CAN	[Rains and Welch (1988) In A.S. Dyke (GSC database)]
6810 ± 90 (6985 ± 90)	S-3644	tibia	Oxbow Dam, Souris R. (49°13'N, 102°10'W), SK, CAN	[Green (1998) In: R.E. Morlan, CMC database (1999)]
(6730 ± 110)	AECV-875C		Oldman River Dam area (49.600°N, 114.100°W), AB, CAN	[Beaudoin (1989) In: A.S. Dyke (GSC database)]; [Beaudoin (1991), PMA database]; [Landals permit 88-38; Beaudoin (1991) In: R.E. Morlan, CMC database (1999)]
6730 ± 260	W-1108		Sullivan Pit, Sullivan Ck., Tofty area, AK, USA	Harington (1977a); [Repenning, et al. (1964) In: Stephenson, et al. (in press)]
(6720 ± 80)	Beta-37840	fragments of humerus, femur, and ribs	Lac des Arcs, Bow valley (51.050°N, 115.200°W), AB, CAN	[Beaudoin In: A.S. Dyke (GSC database)]; [A. Beaudoin (1992), PMA database]; [Newton (1991) In: R.E. Morlan, CMC database (1999)]; Beaudoin (1998)
(6690 ± 310)	AECV-770C		Oldman River Dam area (49.600°N, 114.050°W), AB, CAN	[Beaudoin (1989) In: A.S. Dyke (GSC database)]; [Beaudoin (1991), PMA database]; [Van Dyke, et al. (1991); Beaudoin (1991) In: R.E. Morlan, CMC database (1999)]
(6630 ± 100)	AECV-1657C	skull fragments	Bow Corridor (51°04'N, 115°08'W), AB, CAN	[A. Beaudoin (PMA database)]
6610 ± 160 (6690 ± 160)	RL-510		Village of Bellevue, Crowsnest Pass, AB, CAN	[Reeves (1975) In: R.E. Morlan, CMC database (1999)]; Driver (1976, 1978); Brumley and Rushworth (1983)

6580 ± 205 (6825 ± 205)	GX-6396-A		Bow R., Calgary (51.030°N, 114.080°W), AB, CAN	Brumley and Rushworth (1983); Jackson and Pawson (1984); [Wilson (1974, 1983) In: R.E. Morlan, CMC database (1999)]
6570 ± 95 (6650 ± 95)	S-2892		Battle valley: on a terrace in the Battle Ck. valley, 1080 m asl, near Fort Walsh (49.520°N, 109.520°W), SK, CAN	[Klassen (1992) In: R.E. Morlan, CMC database (1999)]
(6460 ± 60)	TO-4984		Todd Creek: E/W road allowance across Todd Ck, 1160 msl, 7 km NE of Lundbreck, Oldman drainage, AB, CAN	[Jackson and Leboe (1998) In: R.E. Morlan, CMC database (1999)]
6340 ± 140 (6420 ± 140)	GSC-705		Castle R., near Pincher Ck. (49.480°N, 114.060°W), AB, CAN	Lowdon and Blake (1968); R.E. Morlan, CMC database (1999)
6300 ± 100 (6380 ± 100)	BGS-1019	right horncore	Waggle Springs: Assiniboine R. (49.780°N, 99.720°W), MB, CAN	[E. Nielsen, pers. comm. (1998), M.C. Wilson, unpublished report to E. Nielsen (1984) In: R.E. Morlan, CMC database (1999)]
6230 ± 180 (6310 ± 180)	RL-447		Crowsnest Pass (49°35'N, 114°25'W), AB, CAN	[Reeves (1975); Driver (1978) In: R.E. Morlan, CMC database (1999)]; Brumley and Rushworth (1983)
(6170 ± 700)	TO-934		Notukeu Ck., Wood R. drainage (49.720°N, 107.400°W), SK, CAN	[Morlan and McNeely (1996) In: A.S. Dyke (GSC database)]; [Hanna, et al. (1983); Meyer (1985, 1986); Pettipas (1986); Meyer and Liboiron (1990) In: R.E. Morlan, CMC database (1999)]
6150 ± 140 (6230 ± 140)	GSC-447	maxilla	Castle R., 6 mi. W of Pincher Ck. (49.480°N, 114.060°W), AB, CAN	[Lowdon, et al. (1967); Jackson and Pawson (1984) In: R.E. Morlan, CMC database (1999)]
6100 ± 180 (6180 ± 180)	GSC-490	tooth dentine	Castle R., 6 mi. W of Pincher Ck. (49.480°N, 114.060°W), AB, CAN	[Lowdon, et al. (1967); Jackson and Pawson (1984) In: R.E. Morlan, CMC database (1999)]
5960 ± 170 (6040 ± 170)	RL-560		Chin Coulee, 50 km SSE of Taber (49.6°N, 111.817°W), AB, CAN	Forbis (1968); Brumley and Rushworth (1983); [Quigg (1976); Vickers and Beaudoin (1989); Wilson, et al. (1991) In: R.E. Morlan, CMC database (1999)]
5930 ± 230 (6020 ± 230)	AECV-1148C		Bow Corridor (51.070°N, 115.180°W), AB, CAN	[Beaudoin In: A.S. Dyke (GSC database)]; [A. Beaudoin (PMA database)]; [Newton (1991); Beaudoin (1998) In: R.E. Morlan, CMC database (1999)]
(5920 ± 170)	AECV-1200C		Oldman River Dam area (49.580°N, 114.050°W), AB, CAN	[Beaudoin In: A.S. Dyke (GSC database)]; [A. Beaudoin (PMA database)]; [Van Dyke, et al. (1990, 1991); Van Dyke and Unfreed (1992) In: R.E. Morlan, CMC database (1999)]; Beaudoin (1991, 1998)
5900 ± 140 (6040 ± 140)	GSC-1207	tibia and rib	Pakan, N Saskatchewan R. (54.030°N, 112.330°W), AB, CAN	Lowdon and Blake (1979); R.E. Morlan, CMC database (1999)
5865 ± 135	S-1789		Strawberry Ck., N Saskatchewan R. valley, near Edmonton (53.500°N, 113.830°W), AB, CAN	Rains and Welch (1988)
5800 ± 70 (5840 ± 70)	Beta-70100 CAMS-11243	(MC 1)	McIntyre Ck., Whitehorse (approx. 60°46'N,135°06'W), YT, CAN	G. Hare [unpublished]; R.E. Morlan, CMC database (1999); Stephenson, et al. (in press)
5470 ± 150 (5715 ± 150)	GX-6395-A		Bow R., Calgary (51.030°N, 114.080°W), AB, CAN	Brumley and Rushworth (1983); Jackson and Pawson (1984); [Wilson (1974, 1983) In: R.E. Morlan, CMC database (1999)]
5460 ± 160 (5540 ± 160)	GX-6130-G	talus, phalanx, 4 long-bone fragments	2.5 km W of Battle R., 648 m asl, AB, CAN	Brumley and Rushworth (1983); Beaudoin (1987); [Quigg (1984) In: R.E. Morlan, CMC database (1999) - Note: GX-6130-A, apatite fraction, 4370 ± 210; normalized, 4675 ± 215 BP]
(5430 ± 70)	TO-6733	molar	Ste. Agathe, MB, CAN	[R. McNeely, pers. comm. In: R.E. Morlan, CMC database (1999)]

5410 ± 300 (5490 ± 300)	GSC-803		Head-Smashed-In Buffalo Jump (Ft. MacLeod) (49.710°N, 113.645°W), AB, CAN	[Jackson and Pawson (1984) In: A.S. Dyke (GSC database)]; [Reeves (1970, 1978); Wilmeth (1978); Brink, et al. (1985, 1986); Brink and Dawe (1989) In: R.E. Morlan, CMC database (1999)]; Brumley and Rushworth (1983); Morlan (1985); Beaudoin (1987, 1988, 1991, 1998)
5350 ± 120 (5430 ± 120)	BGS-728	atlas and calcaneum	Bowsman R., 6.4 km W of Bowsman, MB, CAN	[Nielsen, et al. (1993) In: R.E. Morlan, CMC database (1999)]
(5330 ± 90)	AECV-1910C		Spring Kill site (51°04'N, 115°08'W), AB, CAN	[A. Beaudoin (PMA database)]
5280 ± 90 (5380 ± 90)	S-3662	tibia	7 km S of Morris, about 68 km S of Winnipeg, about 228 m asl, in the Red River Vallley, MB, CAN	[R. McNeely, pers. comm. In: R.E. Morlan, CMC database (1999)]
(5230 ± 200)	RIDDL-321	distal metacarpal	Harrowby Bay, Cape Bathurst Pen. (69.000°N, 132.000°W), NT, CAN	Harington (1990); Cinq-Mars (1991); LeBlanc (1994); R.E. Morlan, CMC database (1999); Stephenson, et al. (in press)
5145 ± 170 (5390 ± 170)	GX-6394-A		Bow R., Calgary (51.030°N, 114.080°W), AB, CAN	Brumley and Rushworth (1983); Jackson and Pawson (1984); [Wilson (1974, 1983) In: R.E. Morlan, CMC database (1999)]
(5120 ± 100)	AECV-843C		Head-Smashed-In Buffalo Jump (49.720°N, 113.650°W), AB, CAN	[Beaudoin (1989) In: A.S. Dyke (GSC database)]; [Beaudoin (1991), PMA database]; [Beaudoin (1991) In: R.E. Morlan, CMC database (1999)]
5070 ± 160 (5150 ± 160)	RL-1584		Elbow R. (50°55'N, 114°45'W), AB, CAN	[Quigg (1982) In: R.E. Morlan, CMC database (1999)]; Brumley and Rushworth (1983)
5010 ± 90 (5090 ± 90)	S-3007		Wanuskewin Heritage Pk., S Saskatchewan R. (52.170°N, 106.330°W), SK, CAN	Morlan (1993); [Ramsay (1993) In: R.E. Morlan, CMC database (1999)]; [Morlan and McNeely (1996) In: A.S. Dyke (GSC database)]
4940 ± 155 (4940 ± 155)	S-2438		Beaver Ck., S Saskatchewan R. (51°55'N, 106°45'W), SK, CAN	[Walker (1988, 1992) In: R.E. Morlan, CMC database (1999)]; Morlan (1993)
4860 ± 150 (4940 ± 150)	GaK-4248		Whitemouth Falls, S of Lac du Bonnet, 267 m asl (50.117°N, 96.017°W), MB, CAN	[Buchner and Pujo (1977); Buchner (1979); Ens (1998) In: R.E. Morlan, CMC database (1999)]
4780 ± 195 (4860 ± 195)	S-1067		Katepwa Beach, Qu'Appelle valley (50.670°N, 103.500°W), SK, CAN	Morlan (1993); [Morlan and McNeely (1996) In: A.S. Dyke (GSC database)]; R.E.Morlan, CMC database (1999)
(4750 ± 90)	AECV-1911C		Head-Smashed-In Buffalo Jump (49°43'N, 113°39'W), AB, CAN	[A. Beaudoin (PMA database)]
4725 ± 150 (4805 ± 150)	GX-6129-G	distal metacarpal	2.5 km W of Battle R., 648 m asl, AB, CAN	Brumley and Rushworth (1983); Beaudoin (1987); [Quigg (1984) In: R.E. Morlan, CMC database (1999) - Note: GX-6129-A, apatite fraction, 4345 ± 160; normalized, 4590 ± 165 BP]
4720 ± 160 (4800 ± 160)	S-906		Fisherman L., Liard drainage (60.370°N, 123.870°W), NT, CAN	[Millar (1968); Fedirchuk (1970, 1975) In: R.E. Morlan, CMC database (1999)]
4705 ± 155 (4705 ± 150)	GX-4416		Whitemouth Falls, S of Lac du Bonnet, 267 m asl (50.117°N, 96.017°W), MB, CAN	[Buchner (1979) In: A.S. Dyke (GSC database)]; [Buchner and Pujo (1997); Ens (1998) In: R.E. Morlan, CMC database (1999)]
4630 ± 160 (4710 ± 160)	RL-902	scapula	Calgary (51.080°N, 114.080°W), AB, CAN	Brumley and Rushworth (1983); [Wilson (1983) In: R.E. Morlan, CMC database (1999)]
(4620 ± 70)	TO-1993	long-bone fragments	Lac des Arcs, Bow valley (51.050°N, 115.200°W), AB, CAN	[Beaudoin In: A.S. Dyke (GSC database)]; [A. Beaudoin (1992), PMA database]; [Newton (1991) In: R.E. Morlan, CMC database (1999)]; Beaudoin (1998)
(4610 ± 90)	AECV-1924C	metatarsal shaft and P2	Pine Coulee Reservoir (50°13'N, 113°46'W), AB, CAN	[A. Beaudoin (PMA database)]

4560 ± 115 (4640 ± 115)	S-1306	fragments	Riddell site, 4.8 km N of Sutherland, S Saskatchewan R. (52°09'N, 106°36'W), SK, CAN	Skwara (1988); Morlan (1993); R.E. Morlan, CMC database (1999)
4540 ± 100 (4620 ± 100)	BGS-924		Bottomly Ck. E bank of the Red R., NE of Winnipeg, MB, CAN	[Nielsen, et al. (1993) In: R.E. Morlan, CMC database (1999)]
4490 ± 60 (4580 ± 60)	Beta-97670	radius (CMC-1493)	Wanuskewin Heritage Pk., S Saskatchewan R. (52°13'12"N, 106°35'35"W), SK, CAN	[Amundson (1986); Amundson and Kelly (1987); E.G. Walker, pers. comm. (1998) In: R.E. Morlan, CMC database (1999)]; Morlan (1993)
4470 ± 120 (4550 ± 120)	S-1082		Chin Coulee, 50 km SSE of Taber (49.6°N, 111.817°W), AB, CAN	Brumley and Rushworth (1983); [Forbis (1968); Quigg (1976); Vickers and Beaudoin (1989); Wilson, et al. (1991) In: R.E. Morlan, CMC database (1999)]
4410 ± 150 (4515 ± 150)	S-3444	humerus and tibia? (CMC-1419)	Harder site near Sakatoon (52.220°N, 107.050°W), SK, CAN	[Dyck (1970, 1972, 1977); Gibson (1981); Morlan, 1993, 1994) In: R.E. Morlan, CMC database (1999)]
(4360 ± 90)	AECV-757C		Oldman River Dam area (49.100°N, 114.000°W), AB, CAN	[Beaudoin In: A.S. Dyke (GSC database)]; [Beaudoin (1991), PMA database]; [Van Dyke, et al. (1991); Beaudoin (1991) In: R.E. Morlan, CMC database (1999)]
(4340 ± 40)	AECV-1925C	tarsals and phalanges	Pine Coulee Reservoir (50°13'N, 113°46'W), AB, CAN	[A. Beaudoin (PMA database)]
4295 ± 85 (4375 ± 85)	S-2525	metacarpal (CMC-1344)	10.5 km upstream from the Francois-Finlay Dam(53.170°N, 104.000°W), SK, CAN	Morlan (1993); [Quigg (1986); Brandzin-Low (1997) In: R.E. Morlan, CMC database (1999)]; [Morlan and McNeely (1996) In: A.S. Dyke (GSC database)]
4280 ± 80 (4360 ± 80)	S-3009	humerus	Wanuskewin Heritage Pk., S Saskatchewan R. (52.170°N, 106.330°W), SK, CAN	Morlan (1993); [Ramsay (1993) In: R.E. Morlan, CMC database (1999)]; [Morlan and McNeely (1996) In: A.S. Dyke (GSC database)]
4270 ± 160 (4350 ± 160)	SFU-382		Charlie L. Cave, NW of Fort St. John (56.275°N, 120.930°W), BC, CAN	[Fladmark, et al. (1988); Driver (1988, 1996); Driver and Hobson (1992); Driver, et al. (1996) In: R.E. Morlan, CMC database (1999)]
4270 ± 130 (4350 ± 130)	GX-1770	horncore	Del Bonita, Milk R. drainage, AB, CAN	[Reeves (1973), Wilson (1978), Brumley and Rushworth (1983) In: R.E. Morlan, CMC database (1999)]
4225 ± 150	S-1797		Stawberry Ck., N. Saskatchewan R. near Edmonton (53.500°N, 113.670°W), AB, CAN	Rains and Welch (1988)
(4210 ± 90)	AECV-751C		Oldman River Dam area (49.100°N, 114.000°W), AB, CAN	[Beaudoin In: A.S. Dyke (GSC database)]; [Beaudoin (1991), PMA database]; [Van Dyke, et al. (1991); Beaudoin (1991) In: R.E. Morlan, CMC database (1999)]
4190 ± 90 (4335 ± 90)	S-3452	humerus (CMC-1423)	Harder site near Sakatoon (52.220°N, 107.050°W), SK, CAN	[Dyck (1970, 1972, 1977); Gibson (1981); Morlan, 1993, 1994) In: R.E. Morlan, CMC database (1999)]
4150 ± 105 (4230 ± 105)	GX-1023		Belly R. (49°35'N, 113°35'W), AB, CAN	[Forbis and Reeves (1966); Byrne (1973) In: R.E. Morlan, CMC database (1999)]; Brumley and Rushworth (1983)
4130 ± 115 (4210 ± 115)	S-1083		Chin Coulee, 50 km SSE of Taber (49.6°N, 111.817°W), AB, CAN	Brumley and Rushworth (1983); [Forbis (1968); Quigg (1976); Vickers and Beaudoin (1989); Wilson, et al. (1991) In: R.E. Morlan, CMC database (1999)]
(4100 ± 90)	AECV-1686C	partial pelvis	Fort Saskatchewan Urban Park (53.700°N, 113.230°W), AB, CAN	[A. Beaudoin (PMA database)]
4040 ± 90 (4145 ± 90)	S-3645		Wanuskewin Heritage Pk., S Saskatchewan R., SK, CAN	[Mack (1999); Webster (1999) In: R.E. Morlan, CMC database (1999)]
3995 ± 80 (4075 ± 80)	S-2526	radius (CMC-1345)	10.5 km upstream from the Francois-Finlay Dam (53.170°N, 104.000°W), SK, CAN	Morlan (1993); [Quigg (1986); Brandzin-Low (1997) In: R.E. Morlan, CMC database (1999)]; [Morlan and McNeely (1996) In: A.S. Dyke (GSC database)]

(3980 ± 80)	AECV-1575C	femur	Western Heritage Centre, Bow R. (51.200°N, 114.470°W), AB, CAN	[Beaudoin In: A.S. Dyke (GSC database)]; [A. Beaudoin (PMA database)]; [Beaudoin (1998) In: R.E. Morlan, CMC database (1999)]
(3970 ± 110)	AECV-1225C	skull	Kirriemuir Bison: Sounding Ck. drainage (51.820°N, 110.320°W), AB, CAN	[A. Beaudoin (PMA database)]; [Beaudoin (1998) In: R.E. Morlan CMC database (1999)]
3970 ± 80 (4050 ± 80)	S-2893		Conglomerate valley: on a bench near the confluence with Frenchman R., 950 m asl (49.500°N, 109.050°W), SK, CAN	[Klassen (1992) In: R.E. Morlan, CMC database (1999)]
3880 ± 70 (3980 ± 70)	S-3375	radius	Wanuskewin Heritage Pk., S Saskatchewan R. (52.170°N, 106.500°W), SK, CAN	Morlan (1993); [Ramsay (1993) In: R.E. Morlan, CMC database (1999)]; [Morlan and McNeely (1996) In: A.S. Dyke (GSC database)]
3860 ± 70 (3965 ± 70)	S-3374		Wanuskewin Heritage Pk., S Saskatchewan R. (52.170°N, 106.500°W), SK, CAN	Morlan (1993); [Ramsay (1993) In: R.E. Morlan, CMC database (1999)]; [Morlan and McNeely (1996) In: A.S. Dyke (GSC database)]
3760 ± 80 (3870 ± 80)	S-3648	tibia	Lowest terrace of the Souris R. valley, 520 m asl, near Oxbow Dam, SK, CAN	[Nero and McCorquodale (1958); Walker (1980); Spurling and Ball (1981); Dyck (1983); Morlan (1993); Green (1995, 1998) In: R.E. Morlan, CMC database (1999)]
3675 ± 100 (3755 ± 100)	I-10852	cranium (CMN 34966)	Fort Saskatchewan, N Saskatchewan R. (53.700°N, 113.230°W), AB, CAN	C.R. Harington [unpublished]; R.E. Morlan, CMC database (1999)
(3670 ± 130)	AECV-1190C		Oldman River Dam area (49.580°N, 114.050°W), AB, CAN	[Beaudoin In: A.S. Dyke (GSC database)]; [A. Beaudoin (PMA database)]; [Van Dyke, et al. (1990, 1991); Van Dyke and Unfreed (1992) In: R.E. Morlan, CMC database (1999)]; Beaudoin (1991, 1998)
3660 ± 150 (3740 ± 150)	RL-901		Okotoks (50.750°N, 113.920°W), AB, CAN	[Wilson (1979) In: R.E. Morlan, CMC database (1999)]; Brumley and Rushworth (1983); Wilson (1983)
3630 ± 100 (3710 ± 100)	BGS-927	ribs and vertebrae	St. Andrews, E bank of the Red R., MB, CAN	[Nielsen, et al. (1993) In: R.E. Morlan, CMC database (1999)]
3610 ± 105 (3690 ± 105)	S-2524	radius	S bank of the Saskatchewan R., 10.5 km upstream from the Francois-Finlay Dam (53.170°N, 104.000°W), SK, CAN	Morlan (1993); [Quigg (1986); Brandzin-Low (1997) In: R.E. Morlan, CMC database (1999)]
(3600 ± 90)	AECV-857C		Oldman River Dam area (49.600°N, 114.050°W), AB, CAN	[Beaudoin (1989) In: A.S. Dyke (GSC database)]; [Beaudoin (1991), PMA database]; [Landals permit 88-38; Beaudoin (1991) In: R.E. Morlan, CMC database (1999)]
(3600 ± 100)	AECV-740C		Oldman River Dam area (49°35'N, 114°01'W), AB, CAN	[Beaudoin (1991), PMA database]
3595 ± 150 (3675 ± 150)	S-2061	scapula	S Saskatchewan R., 500 m asl (51.438°N, 107.091°W), SK, CAN	Morlan (1993); [Morlan and McNeely (1996) In: A.S. Dyke (GSC database)]; [Dyck (1983); Dyck and Morlan (1995) In: R.E. Morlan, CMC database (1999)]
(3570 ± 220)	AECV-763C		Oldman River Dam area (49.600°N, 114.050°W), AB, CAN	[Beaudoin (1989) In: A.S. Dyke (GSC database)]; [Beaudoin (1991), PMA database]; [Van Dyke, et al. (1991); Beaudoin (1991) In: R.E. Morlan, CMC database (1999)]
3560 ± 100 (3640 ± 100)	BGS-926		Red R. (50°01'30"N, 97°02'15"W), MB, CAN	[Nielsen, et al. (1993, 1996) In: R.E. Morlan, CMC database (1999)]
3530 ± 115 (3610 ± 115)	S-2062	femur	S Saskatchewan R., 500 m asl (51.438°N, 107.091°W), SK, CAN	[Dyck (1983); Dyck and Morlan (1995) In: R.E. Morlan, CMC database (1999)]; Morlan (1993); [Morlan and McNeely (1996) In: A.S. Dyke (GSC database)]

(3520 ± 210)	AECV-764C		Oldman River Dam area (49.580°N, 114.020°W), AB, CAN	[Beaudoin (1989) In: A.S. Dyke (GSC database)]; Van Dyke, et al. (1990); [Beaudoin (1991), PMA database]; [Van Dyke, et al. (1991) In: R.E. Morlan, CMC database (1999)]; Beaudoin (1991, 1998)]
3510 ± 90 (3590 ± 90)	S-1077		Kame Hills, NW corner of Southern Indian L., 259 m asl (57.500°N, 98.550°W), MB, CAN	[Dickson (1976, 1980); Gordon (1996) In: R.E. Morlan, CMC database (1999)]
3510 ± 120 (3590 ± 120)	RL-1583		Elbow R. (50°55'N, 114°45'W), AB, CAN	[Quigg (1982) In: R.E. Morlan, CMC database (1999)]; Brumley and Rushworth (1983)
3485 ± 100 (3565 ± 100)	I-10851	humerus (CMN 34965)	Fort Saskatchewan, N Saskatchewan R. (53.700°N, 113.230°W), AB, CAN	C.R. Harington [unpublished]; R.E. Morlan, CMC database (1999)
3480 ± 150 (3560 ± 150)	RL-866		N bank of Crowsnest R., Crowsnest Pass (49.630°N, 114.600°W), AB, CAN	[Driver (1993) In: A.S. Dyke (GSC database)]; [Reeves (1974); Driver (1978, 1985, 1993) In: R.E. Morlan, CMC database (1999)]
3480 ± 80 (3580 ± 80)	S-3372	femur	Wanuskewin Heritage Pk., S Saskatchewan R. (52.170°N, 106.500°W), SK, CAN	Morlan (1993); [Ramsay (1993) In: R.E. Morlan, CMC database (1999)]; [Morlan and McNeely (1996) In: A.S. Dyke (GSC database)]
3470 ± 80 (3570 ± 80)	S-3373	scapula, vertebra, rib and long-bone fragments	Wanuskewin Heritage Pk., S Saskatchewan R. (52.170°N, 106.500°W), SK, CAN	Morlan (1993); [Ramsay (1993) In: R.E. Morlan, CMC database (1999)]; [Morlan and McNeely (1996) In: A.S. Dyke (GSC database)]
3470 ± 70	Beta-136362	tibia	Ruby Range, Kluane District, YT, CAN	[M. Hoefs, pers. comm. In: Stephenson, et al. (in press)]
3420 ± 140 (3570 ± 140)	S-3453	mandible and humerus (CMC-1424)	Harder site near Sakatoon (52.220°N, 107.050°W), SK, CAN	[Dyck (1970, 1972, 1977); Gibson (1981); Morlan, 1993, 1994) In: R.E. Morlan, CMC database (1999)]
(3420 ± 100)	AECV-847C		Mitigation (51.970°N, 112.220°W), AB, CAN	[Beaudoin (1989) In: A.S. Dyke (GSC database)]; [Beaudoin (1991), PMA database]; [Light permit 88-76 in Luther and Purdon (1991), Beaudoin (1991) In: R.E. Morlan, CMC database (1999)]
3405 ± 175 (3485 ± 175)	GX-4415		Whitemouth Falls, S of Lac du Bonnet, 267 m asl (50.117°N, 96.017°W), MB, CAN	[Buchner (1979) In: A.S. Dyke (GSC database)]; [Buchner and Pujo (1997); Ens (1998) In: R.E. Morlan, CMC database (1999)]
3400 ± 200 (3480 ± 200)	S-2	charred bone	Besant valley, Qu'Appelle drainage, SK, CAN	[Wettlaufer (1955); Dyck (1983); Reeves (1983) In: R.E. Morlan, CMC database (1999)]; Morlan (1993)
3255 ± 90	S-1795		Whitemud Ck., N Saskatchewan R. valley, near Edmonton (53.500°N, 113.670°W), AB, CAN	Rains (1987) Rains (1987)
3250 ± 95 (3330 ± 95)	S-2969	radius	N bank of Souris R., 542 m asl (49.330°N, 103.500°W), SK, CAN	[Finnigan, et al. (1990) In: R.E. Morlan, CMC database (1999)]; Morlan (1993)
(3250 ± 110)	AECV-865C		Oldman River Dam area (49.600°N, 114.100°W), AB, CAN	[Beaudoin (1989) In: A.S. Dyke (GSC database)]; [Beaudoin (1991), PMA database]; [Landals permit 88-38; Beaudoin (1991) In: R.E. Morlan, CMC database (1999)]
3200 ± 90	S-1794		Whitemud Ck., N Saskatchewan R. valley, near Edmonton (53.500°N, 113.670°W), AB, CAN	Rains (1987) Rains (1987)
(3160 ± 100)	AECV-1196C		Oldman River Dam area (49.580°N, 114.050°W), AB, CAN	[Beaudoin In: A.S. Dyke (GSC database)]; [A. Beaudoin (PMA database)]; [Van Dyke, et al. (1990, 1991) In: R.E. Morlan, CMC database (1999)]; Beaudoin (1991, 1998)
3150 ± 550 (3230 ± 550)	BGS-1717		9.6 km W of Elm Ck., 4 km Highway 2, Assiniboine drainage, MB, CAN	[Nero (1997) In: R.E. Morlan, CMC database (1999)]

3125 ± 135 (3205 ± 135)	GX-4145		N shore of Whitemouth R., 259 m asl (50.122°N, 96.035°W), MB, CAN	[Meiklejohn et al. (1994) In: A.S. Dyke (GSC database)]; [Buchner (1979, 1982); Ens (1998) In: R.E. Morlan, CMC database (1999)]
(3110 ± 90)	AECV-1195C		Oldman River Dam area (49.580°N, 114.050°W), AB, CAN	[Beaudoin In: A.S. Dyke (GSC database)]; [A. Beaudoin (PMA database)]; [Van Dyke, et al. (1990, 1991) In: R.E. Morlan, CMC database (1999)]; Beaudoin (1991, 1998)
3105 ± 105 (3185 ± 105)	GX-4147		N shore of Whitemouth R., 259 m asl (50.122°N, 96.035°W), MB, CAN	[Meiklejohn et al. (1994) In: A.S. Dyke (GSC database)]; [Buchner (1979, 1982); Ens (1998) In: R.E. Morlan, CMC database (1999)]
(3100 ± 80)	TO-362		Notukeu Ck., Wood R. drainage (49.717°N, 107.383°W), SK, CAN	[Morlan and McNeely (1996) In: A.S. Dyke (GSC database)]; [Hanna, et al. (1983); Meyer (1985, 1986); Pettipas (1986); Meyer and Liboiron (1990) In: R.E. Morlan, CMC database (1999)]
(3000 ± 80)	AECV-1569C	humerus	Aldon Plant, S Saskatchewan drainage (50.030°N, 110.720°W), AB, CAN	[Beaudoin In: A.S. Dyke (GSC database)]; [A. Beaudoin (PMA database)]; [Beaudoin (1998) In: R.E. Morlan, CMC database (1999)]
(3000 ± 170)	AECV-741C		Oldman River Dam area (49°35'N, 114°03'W), AB, CAN	[Beaudoin (1991), PMA database]
2990 ± 115 (3070 ± 115)	S-2793		Qu'Appelle valley (50.670°N, 103.670°W), SK, CAN	Morlan (1993); [Dyck (1983); Smith (1986); Smith and Walker (1988) In: R.E. Morlan, CMC database (1999)]
2980 ± 130	GSC-1195	vertebrae, sacrum, and ribs	Bellis, N Saskatchewan R. (54.050°N, 112.230°W), AB, CAN	Lowdon and Blake (1979)
2980 ± 105 (3060 ± 105)	S-2791		Qu'Appelle valley (50.670°N, 103.670°W), SK, CAN	Morlan (1993); [Morlan and McNeely (1996) In: A.S. Dyke (GSC database)]; [Dyck (1983); Smith (1986); Smith and Walker (1988) In: R.E. Morlan, CMC database (1999)]
2980 ± 110	AECV-333C	innominate, rib and radius	Oldman River Dam area (49°33'N, 114°01'W), AB, CAN	[Beaudoin (1988), PMA database]
2925 ± 110 (3005 ± 110)	S-2264	(CMC-1270)	Finn Bog, SE of Strehlow Pond, 517 m asl., 8.5 km N of Dundurn (51.830°N, 106.330°W), SK, CAN	[McCann (1983) In: R.E. Morlan, CMC database (1999)]; Morlan (1993); [Morlan and McNeely (1996) In: A.S. Dyke (GSC database)]
2915 ± 125 (2995 ± 125)	BGS-1753		9.6 km W of Elm Ck., 4 km Highway 2, Assiniboine drainage, MB, CAN	[Nero (1997) In: R.E. Morlan, CMC database (1999)]
(2880 ± 100)	AECV-1233C	humerus	Smyth site (Oldman River Dam area) (49.600°N, 114.070°W), AB, CAN	[Beaudoin In: A.S. Dyke (GSC database)]; [A. Beaudoin (PMA database)]; [Landals (1991); Beaudoin [1988, 1991, 1998) In: R.E. Morlan, CMC database (1999)]
(2850 ± 135)	GX-8400-G		Forty Mile Coulee, AB, CAN	[Wilson (1983); Beaudoin (1987); Brumley and Dau (1988) In: R.E. Morlan, CMC database (1999)
2840 ± 130 (2920 ± 130)	GX-4149		N shore of Whitemouth R., 259 m asl (50.122°N, 96.035°W), MB, CAN	[Buchner (1979) In: A.S. Dyke (GSC database)]; [Buchner and Pujo (1997); Ens (1998) In: R.E. Morlan, CMC database (1999)]
2830 ± 80 (2910 ± 80)	S-3666	humerus	Eagle Ck. (51°35'N, 108°25'W), SK, CAN	[Rollins (1999) In: R.E. Morlan, CMC database (1999)]
(2820 ± 80)	SFU-119		E side of the N Saskatchewan R., eastern limits of Edmonton (53.550°N, 113.370°W), AB, CAN	[Ives and Newton (1980); Pollock (1981); Kooyman (1987, 1988) In: A.S. Dyke (GSC database)]; [Brumley and Rushworth (1983)
2780 ± 85	S-1799		Strawberry Ck., N Saskatchewan R. valley, near Edmonton (53.500°N, 113.670°W), AB, CAN	Rains and Welch (1988)
(2770 ± 100)	AECV-1232C	right radius	Smyth site (Oldman River Dam area) (49.600°N, 114.070°W), AB, CAN	[Beaudoin In: A.S. Dyke (GSC database)]; [A. Beaudoin (PMA database)]; [Landals (1991); Beaudoin [1988, 1991, 1998) In: R.E. Morlan, CMC database (1999)]

(2770 ± 120)	AECV-773C		Oldman River Dam area (49.580°N, 114.020°W), AB, CAN	[Beaudoin (1989) In: A.S. Dyke (GSC database)]; [Beaudoin (1991), PMA database]; [Van Dyke, et al. (1991); Beaudoin (1991) In: R.E. Morlan, CMC database (1999)]
2765 ± 90 (2845 ± 90)	S-804		Weed Ck., near Thorsby (53.420°N, 114.000°W), AB, CAN	[Shelford (1975) In: R.E. Morlan, CMC database (1999)]
2750 ± 95 (2830 ± 95)	GaK-4713		N shore of Whitemouth R., 259 m asl (50.122°N, 96.035°W), MB, CAN	[Buchner (1979, 1982); Meiklejohn, et al. (1994); Ens (1998) In: R.E. Morlan, CMC database (1999)]
(2750 ± 160)	AECV-1228C	scapula	Smyth site (Oldman River Dam area) (49.600°N, 114.070°W), AB, CAN	[Beaudoin In: A.S. Dyke (GSC database)]; [A. Beaudoin (PMA database)]; [Landals (1991); Beaudoin [1988, 1991, 1998) In: R.E. Morlan, CMC database (1999)]
(2730 ± 90)	AECV-1570C	femur	Aldon Plant, S Saskatchewan drainage (50.030°N, 110.720°W), AB, CAN	[Beaudoin In: A.S. Dyke (GSC database)]; [A. Beaudoin (PMA database)]; [Beaudoin (1998) In: R.E. Morlan, CMC database (1999)]
(2720 ± 110)	AECV-869C		Oldman River Dam area (49.600°N, 114.050°W), AB, CAN	[Beaudoin (1989) In: A.S. Dyke (GSC database)]; [Beaudoin (1991), PMA database]; [Landals (1991); Beaudoin (1991) In: R.E. Morlan, CMC database (1999)]
2720 ± 60 (2800 ± 60)	Beta-70101	rib (P 6)	Gravel Quarry, Montague House, YT, CAN	[R.M. Gotthardt and C.R. Harington, pers. comm. (1998) In: R.E. Morlan, CMC database (1999)]; Stephenson, et al. (in press). [Probably wood bison].
2705 ± 120 (2785 ± 120)	GX-4144		N shore of Whitemouth R., 259 m asl (50.122°N, 96.035°W), MB, CAN	[Meiklejohn et al. (1994) In: A.S. Dyke (GSC database)]; [Buchner (1979, 1982); Ens (1998) In: R.E. Morlan, CMC database (1999)]
2700 ± 140 (2800 ± 140)	GSC-1195-2	vertebrae, sacrum, and ribs	Bellis, N Saskatchewan R. (54.050°N, 112.230°W), AB, CAN	Lowdon and Blake (1979); R.E. Morlan, CMC database (1999)
(2690 ± 140)	AECV-759C		Oldman River Dam area (49.580°N, 114.020°W), AB, CAN	[Beaudoin (1989) In: A.S. Dyke (GSC database)]; [Beaudoin (1991), PMA database]; [Van Dyke, et al. (1991); Beaudoin (1991) In: R.E. Morlan, CMC database (1999)]
(2650 ± 140)	AECV-868C		Oldman River Dam area (49.600°N, 114.050°W), AB, CAN	[Beaudoin (1989) In: A.S. Dyke (GSC database)]; [Beaudoin (1991), PMA database]; [Landals (1991); Beaudoin [1988, 1991, 1998) In: R.E. Morlan, CMC database (1999)]
2650 ± 80 (2770 ± 80)	S-3665		Red River (49°31'40"N, 97°13'05"W), MB, CAN	[R. McNeely, pers. comm. (1998) In: R.E. Morlan, CMC database (1999)]
2650 ± 100	AECV-339C	calcaneum	Oldman River Dam area (49°37'N, 114°03'W), AB, CAN	[Beaudoin (1988), PMA database]
(2630 ± 120)	AECV-1229C	pelvis fragments	Smyth site (Oldman River Dam area) (49.600°N, 114.070°W), AB, CAN	[Beaudoin In: A.S. Dyke (GSC database)]; [A. Beaudoin (PMA database)]; [Landals (1991); Beaudoin [1988, 1991, 1998) In: R.E. Morlan, CMC database (1999)]
(2590 ± 90)	AECV-1230C		Smyth site (Oldman River Dam area) (49.600°N, 114.070°W), AB, CAN	[Beaudoin In: A.S. Dyke (GSC database)]; [A. Beaudoin (PMA database)]; [Landals (1991); Beaudoin [1988, 1991, 1998) In: R.E. Morlan, CMC database (1999)]
(2590 ± 90)	AECV-767C		Oldman River Dam area (49.570°N, 114.020°W), AB, CAN	[Beaudoin In: A.S. Dyke (GSC database)]; [Beaudoin (1991), PMA database]; [Van Dyke, et al. (1991); Beaudoin (1991) In: R.E. Morlan, CMC database (1999)]
2585 ± 290 (2665 ± 290)	S-2798		Qu'Appelle valley (50.670°N, 103.670°W), SK, CAN	Morlan (1993); [Morlan and McNeely (1996) In: A.S. Dyke (GSC database)]; [Dyck (1983); Smith (1986); Smith and Walker (1988) In: R.E. Morlan, CMC database (1999)]

2580 ± 90	AECV-9440	right metacarpal	Ghostpine Ck., near Three Hills, AB, CAN	Rains, et al. (1994)
(2560 ± 110)	AECV-1236C	vertebra	Smyth site (Oldman River Dam area) (49.600°N, 114.070°W), AB, CAN	[Beaudoin In: A.S. Dyke (GSC database)]; [A. Beaudoin (PMA database)]; [Landals (1991); Beaudoin (1988, 1991, 1998) In: R.E. Morlan, CMC database (1999)]
(2540 ± 100)	AECV-1234C	thoracic vertebra and metatarsal	Smyth site (Oldman River Dam area) (49.600°N, 114.070°W), AB, CAN	[Beaudoin In: A.S. Dyke (GSC database)]; [A. Beaudoin (PMA database)]; [Landals (1991); Beaudoin [1988, 1991, 1998) In: R.E. Morlan, CMC database (1999)]
(2530 ± 140)	GSC-1744		Windle site, Kananaskis R. (50.940°N, 115.130°W), AB, CAN	[Wilmeth (1978), Wilson (1983), Brumley and Rushworth (1983) In: R.E. Morlan, CMC database (1999)]
(2530 ± 90)	AECV-768C		Oldman River Dam area (49.580°N, 114.000°W), AB, CAN	[Beaudoin In: A.S. Dyke (GSC database)]; [Beaudoin (1991), PMA database]; [Van Dyke, et al. (1991); Beaudoin (1991) In: R.E. Morlan, CMC database (1999)]
(2530 ± 70)	AECV-1923C	humerus	Pine Coulee Reservoir (50°13'N, 113°46'W), AB, CAN	[A. Beaudoin (PMA database)]
(2520 ± 110)	AECV-766C		Oldman River Dam area (49.600°N, 114.120°W), AB, CAN	[Beaudoin (1989) In: A.S. Dyke (GSC database)]; [Beaudoin (1991), PMA database]; [Van Dyke, et al. (1991); Beaudoin (1991) In: R.E. Morlan, CMC database (1999)]
2515 ± 140 (2595 ± 140)	GX-10863	vertebra	E bank of the Red R., N of Winnipeg (50.083°N, 96.917°W), MB, CAN	[Buchner (1988); Deck and Shay (1992); Nielsen, et al. (1993) In: R.E. Morlan, CMC database (1999)]; Roberts (1992)
(2510 ± 90)	AECV-755C		Oldman River Dam area (49.600°N, 114.120°W), AB, CAN	[Beaudoin (1989) In: A.S. Dyke (GSC database)]; [Beaudoin (1991), PMA database]; [Van Dyke, et al. (1991); Beaudoin (1991) In: R.E. Morlan, CMC database (1999)]
(2510 ± 90)	AECV-1231C	cervical vertebrae	Smyth site (Oldman River Dam area) (49.600°N, 114.070°W), AB, CAN	[Beaudoin In: A.S. Dyke (GSC database)]; [A. Beaudoin (PMA database)]; [Landals (1991); Beaudoin [1988, 1991, 1998) In: R.E. Morlan, CMC database (1999)]
2505 ± 90 (2585 ± 90)	S-2058		S Saskatchewan R., 500 m asl (51.438°N, 107.091°W), SK, CAN	Morlan (1993); [Morlan and McNeely (1996) In: A.S. Dyke (GSC database)]; [Dyck (1983); Dyck and Morlan (1995) In: R.E. Morlan, CMC database (1999)]
2500 ± 85 (2580 ± 85)	S-2060		S Saskatchewan R., 500 m asl (51.438°N, 107.091°W), SK, CAN	Morlan (1993); [Morlan and McNeely (1996) In: A.S. Dyke (GSC database)]; [Dyck (1983); Dyck and Morlan (1995) In: R.E. Morlan, CMC database (1999)]
(2480 ± 100)	AECV-772C		Oldman River Dam area (49.580°N, 114.000°W), AB, CAN	[Beaudoin (1989) In: A.S. Dyke (GSC database)]; [Beaudoin (1991), PMA database]; [Van Dyke, et al. (1991); Beaudoin (1991) In: R.E. Morlan, CMC database (1999)]
(2460 ± 110)	AECV-1237C	ribs and radius	Smyth site (Oldman River Dam area) (49.600°N, 114.070°W), AB, CAN	[Beaudoin In: A.S. Dyke (GSC database)]; [A. Beaudoin (PMA database)]; [Landals (1991); Beaudoin (1988, 1991, 1998) In: R.E. Morlan, CMC database (1999)]
(2450 ± 100)	AECV-1356C		Oldman River Dam area (49.600°N, 114.000°W), AB, CAN	[Beaudoin In: A.S. Dyke (GSC database)]; [A. Beaudoin (PMA database)]; [Van Dyke, et al. (1990, 1991) In: R.E. Morlan, CMC database (1999)]; Beaudoin (1991, 1998)
(2450 ± 90)	AECV-863C		Oldman River Dam area (49.600°N, 114.100°W), AB, CAN	[Beaudoin (1989) In: A.S. Dyke (GSC database)]; [Beaudoin (1991), PMA database]; [Landals permit 88-38; Beaudoin (1991) In: R.E. Morlan, CMC database (1999)]

2450 ± 120 (2530 ± 120)	RL-1658		Happy valley Kill site, Bow drainage, AB, CAN	[Shortt (1993); Novecosky (1999) In: R.E. Morlan, CMC database (1999)]
2440 ± 120 (2520 ± 120)	RL-1657		Happy valley Kill site, Bow drainage, AB, CAN	[Shortt (1993); Novecosky (1999) In: R.E. Morlan, CMC database (1999)]
2380 ± 130 (2460 ± 130)	RL-863		N bank of Crowsnest R., Crowsnest Pass (49.630°N, 114.600°W), AB, CAN	[Driver (1993) In: A.S. Dyke (GSC database)]; [Reeves (1974); Driver (1978, 1985, 1993) In: R.E. Morlan, CMC database (1999)]
2375 ± 140 (2455 ± 140)	S-2365		Opamihaw, Wanuskewin Heritage Pk., S Saskatchewan R. (52°10'N, 106°30'W), SK, CAN	[Walker (1988) In: R.E. Morlan, CMC database (1999)]; Morlan (1993)
(2370 ± 90)	AECV-1235C	pelvis	Smyth site (Oldman River Dam area) (49.600°N, 114.070°W), AB, CAN	[Beaudoin In: A.S. Dyke (GSC database)]; [A. Beaudoin (PMA database)]; [Landals (1991); Beaudoin (1998) In: R.E. Morlan, CMC database (1999)]
2370 ± 100 (2450 ± 100)	BGS-925		Bottomly Ck. E bank of the Red R., NE of Winnipeg, MB, CAN	[Nielsen, et al. (1993) In: R.E. Morlan, CMC database (1999)]
2365 ± 70 (2445 ± 75)	S-1074	fragments	Eagle Ck., N Saskatchewan drainage (51.780°N, 107.500°W), SK, CAN	Morlan (1993); R.E. Morlan, CMC database (1999)
2355 ± 105 (2435 ± 105)	S-2059		S Saskatchewan R., 500 m asl (51.438°N, 107.091°W), SK, CAN	Ryder (1987); Morlan (1993); [Morlan and McNeely (1996) In: A.S. Dyke (GSC database)]; [Dyck (1983); Dyck and Morlan (1995) In: R.E. Morlan, CMC database (1999)]
2350 ± 80 (2430 ± 80)	Beta-51285	thoracic vertebra	Happy valley Kill site, Bow drainage, AB, CAN	[Shortt (1993); Novecosky (1999) In: R.E. Morlan, CMC database (1999)]
(2320 ± 140)	AECV-996C		Strathcona (53.550°N, 113.380°W), AB, CAN	[Beaudoin (1989) In: A.S. Dyke (GSC database)]; [Beaudoin (1991), PMA database]; [Oetelaar permit 89-10, in Luther and Purdon (1991); Beaudoin (1991) In: R.E. Morlan, CMC database (1999)]
2310 ± 50 (2400 ± 50)	Beta-70707	sacrum (CMC-1487)	Wanuskewin Heritage Pk., S Saskatchewan R. (52.220°N, 106.593°W), SK, CAN	Morlan (1988, 1993); [Walker, et al. (1986, 1987); Walker (1988) In: R.E. Morlan, CMC database (1999)]; [Morlan and McNeely (1996) In: A.S. Dyke (GSC database)]
2300 ± 140 (2380 ± 140)	M-1724		Waldron Ranch: 800 m above junction of Callum Ck. and Oldman R., 30 m upstream from Callum Ck. (49.760°N, 114.090°W), AB, CAN	[Wagner (1966) In: R.E. Morlan, CMC database (1999)]
(2290 ± 110)	AECV-867C		Oldman River Dam area (49.600°N, 114.050°W), AB, CAN	[Beaudoin (1989) In: A.S. Dyke (GSC database)]; [Beaudoin (1991), PMA database]; [Landals (1991); Beaudoin [1988, 1991, 1998) In: R.E. Morlan, CMC database (1999)]
2270 ± 100	AECV-335C	humerus	Oldman River Dam area (49°33'N, 114°01'W), AB, CAN	[Beaudoin (1988), PMA database]
2260 ± 310 (2340 ± 310)	RL-865		N bank of Crowsnest R., Crowsnest Pass (49.630°N, 114.600°W), AB, CAN	[Driver (1993) In: A.S. Dyke (GSC database)]; [Reeves (1974); Driver (1978, 1985, 1993) In: R.E. Morlan, CMC database (1999)]
(2260 ± 180)	AECV-1909C		Spring Kill site (51°04'N, 115°08'W), AB, CAN	[A. Beaudoin (PMA database)]
(2240 ± 90)	AECV-1192C		Oldman River Dam area (49.580°N, 114.050°W), AB, CAN	[Beaudoin In: A.S. Dyke (GSC database)]; [A. Beaudoin (PMA database)]; [Van Dyke, et al. (1990, 1991); Van Dyke and Unfreed (1992) In: R.E. Morlan, CMC database (1999)]; Beaudoin (1991, 1998)
2220 ± 110 (2300 ± 110)	AECV-344C	axis	Oldman River Dam area (49°36'N, 114°04'W), AB, CAN	[Beaudoin (1988), PMA database]; [Landals (1991); Beaudoin (1988, 1991) In: R.E. Morlan, CMC database (1999)]

2170 ± 70 (2250 ± 75)	Beta-6410		Forty Mile Coulee (49°35'N, 111°25'W), AB, CAN	[Wilson (1983); Beaudoin (1987); Brumley and Dau (1988) In: R.E. Morlan, CMC database (1999)
(2160 ± 90)	AECV-1189C		Oldman River Dam area (49.580°N, 114.050°W), AB, CAN	[Beaudoin In: A.S. Dyke (GSC database)]; [A. Beaudoin (PMA database)]; [Van Dyke, et al. (1990, 1991); Van Dyke and Unfreed (1992) In: R.E. Morlan, CMC database (1999)]; Beaudoin (1991, 1998)
2160 ± 60 (2250 ± 60)	Beta-70706	scapula (CMC-1486)	Wanuskewin Heritage Pk., S Saskatchewan R. (52.220°N, 106.593°W), SK, CAN	Morlan (1988, 1993); [Walker, et al. (1986, 1987); Walker (1988) In: R.E. Morlan, CMC database (1999)]; [Morlan and McNeely (1996) In: A.S. Dyke (GSC database)]
(2150 ± 100)	AECV-1360C		Oldman River Dam area (49.600°N, 114.000°W), AB, CAN	[Beaudoin In: A.S. Dyke (GSC database)]; [A. Beaudoin (PMA database)]; [Van Dyke, et al. (1990, 1991) In: R.E. Morlan, CMC database (1999)]; Beaudoin (1991, 1998)
2140 ± 100 (2220 ± 100)	RL-1408		St. Mary R. (49°07'N, 113°12'W), AB, CAN	Brumley and Rushworth (1983); [Ronaghan and Reeves (1980) In: R.E. Morlan, CMC database (1999)]
2130 ± 125 (2210 ± 125)	S-2366		Meewasin Ck., Wanuskewin Heritage Pk., S Saskatchewan R. (52°13'12"N, 106°35'00"W), SK, CAN	[Walker (1988) In: R.E. Morlan, CMC database (1999)]; Morlan (1993)
(2120 ± 90)	AECV-756C		Oldman River Dam area (49.600°N, 114.120°W), AB, CAN	[Beaudoin In: A.S. Dyke (GSC database)]; Van Dyke, et al. (1990); [Beaudoin (1991), PMA database]; [Van Dyke, et al. (1991) In: R.E. Morlan, CMC database (1999)]; Beaudoin (1991, 1998)]
2090 ± 165 (2170 ± 165)	S-1767		S Saskatchewan R., 500 m asl (51.420°N, 107.090°W), SK, CAN	Morlan (1993); [Morlan and McNeely (1996) In: A.S. Dyke (GSC database)]; [Dyck (1983); Dyck and Morlan (1995) In: R.E. Morlan, CMC database (1999)]
2065 ± 105 (2145 ± 105)	S-2827		S bank of Willow Ck., 358 m asl., which enters Assiniboine R., 1 km S of Chater (49.858°N, 99.750°W), MB, CAN	[Nicholson (1985b, 1994a, 1994b) In: R.E. Morlan, CMC database (1999)]
(2060 ± 90)	AECV-1359C		Oldman River Dam area (49.600°N, 114.000°W), AB, CAN	[Beaudoin In: A.S. Dyke (GSC database)]; [A. Beaudoin (PMA database)]; [Van Dyke, et al. (1990, 1991) In: R.E. Morlan, CMC database (1999)]; Beaudoin (1991, 1998)
(2060 ± 120)	AECV-753C		Oldman River Dam area (49.100°N, 114.000°W), AB, CAN	[Beaudoin In: A.S. Dyke (GSC database)]; Van Dyke, et al. (1990); [Beaudoin (1991), PMA database]; [Van Dyke, et al. (1991) In: R.E. Morlan, CMC database (1999)]; Beaudoin (1991, 1998)]
2060 ± 80 (2155 ± 80)	S-3647	horncore	Eagle Ck. (51°35'N, 108°25'W), SK, CAN	[Rollins (1999) In: R.E. Morlan, CMC database (1999)]
2045 ± 140 (2125 ± 140)	S-2368		Tipperary Ck., N of Saskatoon (52.220°N, 106.220°W), SK, CAN	[Morlan (1988, 1993); Walker 1988); Walker, et al. (1986, 1987) In: R.E. Morlan, CMC database (1999)]; [Morlan and McNeely (1996) In: A.S. Dyke (GSC database)]
(2010 ± 80)	AECV-1568C	tibia	Old Bindloss Ferry, S Saskatchewan drainage (50.650°N, 110.170°W), AB, CAN	[Beaudoin In: A.S. Dyke (GSC database)]; [A. Beaudoin (PMA database)]; [Brumley permit 92-14; Beaudoin (1998) In: R.E. Morlan, CMC database (1999)]
(2000 ± 90)	AECV-1243C	pelvis fragments	Oldman River Dam area (49.620°N, 114.070°W), AB, CAN	[Beaudoin In: A.S. Dyke (GSC database)]; [A. Beaudoin (PMA database)]; [Landals (1991); Beaudoin (1998) In: R.E. Morlan, CMC database (1999)]

(1980 ± 80)	AECV-1567C	humerus	Old Bindloss Ferry, S Saskatchewan drainage (50.650°N, 110.170°W), AB, CAN	[Beaudoin In: A.S. Dyke (GSC database)]; [A. Beaudoin (PMA database)]; [Brumley permit 92-14; Beaudoin (1998) In: R.E. Morlan, CMC database (1999)]
1965 ± 75	S-1783		Strawberry Ck., N Saskatchewan R. valley, near Edmonton (53.500°N, 113.670°W), AB, CAN	Rains and Welch (1988)
1960 ± 90 (2040 ± 90)	S-491		Qu'Appelle drainage, about 19 km E of the elbow of the S Saskatchewan R. (51.070°N, 106.350°W), SK, CAN	Morlan (1993); [Phenix (1969); Ramsay (1991) In: R.E. Morlan, CMC database (1999)]; [Morlan and McNeely (1996) In: A.S. Dyke (GSC database)]
(1940 ± 120)	AECV-455C	radius, pelvis, rib and tarsal fragments	Lyndon Ck. Bison Kill (50.050°N, 113.900°W), AB, CAN	[Beaudoin (1988) In: A.S. Dyke (GSC database)]; [Beaudoin (1988), PMA database]
(1940 ± 90)	AECV-1365C		Oldman River Dam area (49.580°N, 113.980°W), AB, CAN	[Beaudoin In: A.S. Dyke (GSC database)]; [A. Beaudoin (PMA database)]; [Van Dyke, et al. (1990) In: R.E. Morlan, CMC database (1999)]; Beaudoin (1998)
1940 ± 160	AECV-343C	humerus, radius, and innominate	Oldman River Dam area (49°36'N, 114°04'W), AB, CAN	[Beaudoin (1988), PMA database]
(1930 ± 80)	AECV-1763C	bone fragments	Dersch (49°42'N, 113°39'W), AB, CAN	[A. Beaudoin (PMA database)]
1910 ± 70 (1990 ± 75)	S-1640	(CMC-1036)	Qu'Appelle drainage, about 19 km E of the elbow of the S Saskatchewan R. (51.070°N, 106.330°W), SK, CAN	Morlan (1993); [Phenix (1969); Ramsay (1991) In: R.E. Morlan, CMC database (1999)]; [Morlan and McNeely (1996) In: A.S. Dyke (GSC database)]
1905 ± 110 (1985 ± 110)	S-2855		Qu'Appelle drainage, about 19 km E of the elbow of the S Saskatchewan R. (51.070°N, 106.350°W), SK, CAN	Morlan (1993); [Morlan and McNeely (1996) In: A.S. Dyke (GSC database)]; [Phenix (1969); Ramsay (1991) In: R.E. Morlan, CMC database (1999)]
1890 ± 80	AECV-318C	scapula fragments	Cranford Gravel Pit (49°50'N, 112°21'W), AB, CAN	[Beaudoin (1987), PMA database]
(1880 ± 80)	AECV-1479C	radius	Head-Smashed-In Buffalo Jump (49.720°N, 113.650°W), AB, CAN	[Beaudoin In: A.S. Dyke (GSC database)]; [A. Beaudoin (PMA database)]; [Brink permit 88-29; Beaudoin (1998) In: R.E. Morlan, CMC database (1999)]
(1880 ± 100)	AECV-749C		Oldman River Dam area (49.580°N, 114.050°W), AB, CAN	[Beaudoin (1989) In: A.S. Dyke (GSC database)]; Van Dyke, et al. (1990); [Beaudoin (1991), PMA database]; [Van Dyke, et al. (1991) In: R.E. Morlan, CMC database (1999)]; Beaudoin (1991, 1998)]
(1880 ± 100)	AECV-1199C		Oldman River Dam area (49.580°N, 114.050°W), AB, CAN	[Beaudoin In: A.S. Dyke (GSC database)]; [A. Beaudoin (PMA database)]; [Van Dyke, et al. (1990, 1991) In: R.E. Morlan, CMC database (1999)]; Beaudoin (1991, 1998)
(1860 ± 90)	AECV-856C		Oldman River Dam area (49.600°N, 114.050°W), AB, CAN	[Beaudoin (1989) In: A.S. Dyke (GSC database)]; [Beaudoin (1991), PMA database]; [Landals permit 88-38; Beaudoin (1991) In: R.E. Morlan, CMC database (1999)]
1860 ± 200 (1940 ± 200)	S-1765	(CMC-1110)	S Saskatchewan R., 500 m asl (51.420°N, 107.090°W), SK, CAN	[Dyck (1983); Dyck and Morlan (1995) In: R.E. Morlan, CMC database (1999); Morlan (1993)]; [Morlan and McNeely (1996) In: A.S. Dyke (GSC database)]
(1850 ± 100)	AECV-1244C	calcaneum	Oldman River Dam area (49.620°N, 114.070°W), AB, CAN	[Beaudoin In: A.S. Dyke (GSC database)]; [A. Beaudoin (PMA database)]; [Landals (1991); Beaudoin (1998) In: R.E. Morlan, CMC database (1999)]
(1850 ± 110)	AECV-454C	metatarsal	Lyndon Ck. Bison Kill (50.050°N, 113.900°W), AB, CAN	[Beaudoin (1988) In: A.S. Dyke (GSC database)]; [Beaudoin (1988), PMA database]

1840 ± 70 (1920 ± 75)	S-91		Old Women's Buffalo Jump, S of Calgary (50°28'N, 113°53'W), AB, CAN	Brumley and Rushworth (1983); [Forbis (1962); Wilmeth (1978) In: R.E. Morlan, CMC database (1999)]
(1820 ± 90)	AECV-761C		Oldman River Dam area (49.580°N, 114.020°W), AB, CAN	[Beaudoin (1989) In: A.S. Dyke (GSC database)]; Van Dyke, et al. (1990); [Beaudoin (1991), PMA database]; [Van Dyke, et al. (1991) In: R.E. Morlan, CMC database (1999)]; Beaudoin (1991, 1998)]
(1820 ± 70)	AECV-1915C		Head-Smashed-In Buffalo Jump (49°43'N, 113°39'W), AB, CAN	[A. Beaudoin (PMA database)]
1810 ± 190 (1890 ± 190)	S-3028		13 km SW of Leader, Great Sand Hills (50.830°N, 109.630°W), SK, CAN	Morlan (1993); R.E. Morlan, CMC database (1999)
1810 ± 90 (1890 ± 90)	I-5407	right hornsheath (CMN 17505)	S.E. Baillie I. (70.580°N, 128.070°W), NT, CAN	Harington (1980); Harington and Cinq-Mars (1995); R.E. Morlan, CMC database (1999); Stephenson, et al. (in press)
(1800 ± 70)	AECV-1764C	metatarsal, tibia, astragalus, naviculo-culboid	Dersch (49°42'N, 113°39'W), AB, CAN	[A. Beaudoin (PMA database)]
1790 ± 75 (1870 ± 80)	S-2816	metatarsal	Tipperary Ck., N of Saskatoon (52.170°N, 106.500°W), SK, CAN	Morlan (1988, 1993); [Walker, et al. (1986, 1987); Walker (1988) In: R.E. Morlan, CMC database (1999)]; [Morlan and McNeely (1996) In: A.S. Dyke (GSC database)]
1780 ± 150	AECV-341C	lumbar vertebrae	Oldman River Dam area (49°36'N, 114°03'W), AB, CAN	[Beaudoin (1988), PMA database]
1770 ± 80	AECV-312C	humerus fragments	Cranford Gravel Pit (49°50'N, 112°21'W), AB, CAN	[Beaudoin (1987), PMA database]
(1760 ± 90)	AECV-862C		Oldman River Dam area (49.600°N, 114.100°W), AB, CAN	[Beaudoin (1989) In: A.S. Dyke (GSC database)]; [Beaudoin (1991), PMA database]; [Landals permit 88-38; Beaudoin (1991) In: R.E. Morlan, CMC database (1999)]
1710 ± 50 (1790 ± 55)	S-1641	(CMC-1037)	Qu'Appelle drainage, about 19 km E of the elbow of the S Saskatchewan R. (51.070°N, 106.330°W), SK, CAN	Morlan (1993); [Phenix (1969); Ramsay (1991) In: R.E. Morlan, CMC database (1999)]; [Morlan and McNeely (1996) In: A.S. Dyke (GSC database)]
(1690 ± 80)	AECV-1480C		Head-Smashed-In Buffalo Jump (49.720°N, 113.650°W), AB, CAN	[Beaudoin In: A.S. Dyke (GSC database)]; [A. Beaudoin (PMA database)]; [Brink permit 89-29; Beaudoin (1998) In: R.E. Morlan, CMC database (1999)]
(1690 ± 90)	AECV-1191C		Oldman River Dam area (49.580°N, 114.050°W), AB, CAN	[Beaudoin In: A.S. Dyke (GSC database)]; [A. Beaudoin (PMA database)]; [Van Dyke, et al. (1990); Beaudoin (1998) In: R.E. Morlan, CMC database (1999)]
(1680 ± 90)	AECV-1385C	tibia, metatarsal, astragalus, and fragments	NOVA Donalda (52.370°N, 113.070°W), AB, CAN	[Beaudoin In: A.S. Dyke (GSC database)]; [A. Beaudoin (PMA database)]; [Poole permit 90-33; Beaudoin (1998) In: R.E. Morlan, CMC database (1999)]
1680 ± 80 (1840 ± 80)	S-3664	fragments	About 4 km NE of Morris, and 58 km S of Winnipeg, and 228 m asl, in the Red River valley, MB, CAN	[R. McNeely, pers. comm. In: R.E. Morlan, CMC database (1999)]
1680 ± 100	AECV-337C	metacarpal	Oldman River Dam area (49°37'N, 114°03'W), AB, CAN	[Beaudoin (1988), PMA database]
(1680 ± 70)	Beta-19794		Forty Mile Coulee, AB, CAN	[Brumley and Dau (1988) In: R.E. Morlan, CMC database (1999)]
1675 ± 145 (1755 ± 145)	S-2202	tibia (CMC-1255)	S Saskatchewan drainage, SW of Flaxcombe (51.330°N, 109.500°W), SK, CAN	Morlan (1993); [Finnigan and Johnson (1984) In: R.E. Morlan, CMC database (1999)]

1675 ± 145 (1755 ± 145)	S-1081		Chin Coulee, 50 km SSE of Taber (49.6°N, 111.817°W), AB, CAN	Brumley and Rushworth (1983); [Forbis (1968); Quigg (1976); Vickers and Beaudoin (1989); Wilson, et al. (1991) In: R.E. Morlan, CMC database (1999)]
(1670 ± 80)	AECV-1387C	tibia	NOVA Donalda (52.370°N, 113.070°W), AB, CAN	[Beaudoin In: A.S. Dyke (GSC database)]; [A. Beaudoin (PMA database)]; [Poole permit 90-33; Beaudoin (1998) In: R.E. Morlan, CMC database (1999)]
1660 ± 130 (1740 ± 130)	RL-1461	scapula	Highwood R. (50°25'N, 114°35'W), AB, CAN	[Quigg (1982) In: R.E. Morlan, CMC database (1999)]; Brumley and Rushworth (1983)
1650 ± 60 (1730 ± 65)	S-90	charred bone	Old Woman's Buffalo Jump, S of Calgary (50°28'N, 113°53'W), AB, CAN	Brumley and Rushworth (1983); [Forbis (1962); Wilmeth (1978) In: R.E. Morlan, CMC database (1999)]
(1640 ± 90)	AECV-1386C	postcranial elements	NOVA Donalda (52.370°N, 113.070°W), AB, CAN	[Beaudoin In: A.S. Dyke (GSC database)]; [A. Beaudoin (PMA database)]; [Poole permit 90-33; Beaudoin (1998) In: R.E. Morlan, CMC database (1999)]
1635 ± 105 (1715 ± 105)	S-2797		Qu'Appelle valley (50.670°N, 103.670°W), SK, CAN	Morlan (1993); [Morlan and McNeely (1996) In: A.S. Dyke (GSC database)]; [Dyck (1983); Smith (1986); Smith and Walker (1988) In: R.E. Morlan, CMC database (1999)]
1630 ± 200 (1710 ± 200)	S-1764	(CMC-1109)	S Saskatchewan R., 500 m asl (51.420°N, 107.090°W), SK, CAN	[Dyck (1983); Dyck and Morlan (1995) In: R.E. Morlan, CMC database (1999); Morlan (1993)]; [Morlan and McNeely (1996) In: A.S. Dyke (GSC database)]
1625 ± 80	S-1785	skull	Strawberry Ck., N Saskatchewan R. valley, near Edmonton (53.500°N, 113.670°W), AB, CAN	[Rains and Welch (1988) - age seems to be anomalously young]
(1610 ± 90)	AECV-758C		Oldman River Dam area (49.100°N, 114.000°W), AB, CAN	[Beaudoin In: A.S. Dyke (GSC database)]; Van Dyke, et al. (1990); [Beaudoin (1991), PMA database]; [Van Dyke, et al. (1991) In: R.E. Morlan, CMC database (1999)]; Beaudoin (1991, 1998)]
(1600 ± 90)	AECV-855C		Oldman River Dam area (49.567°N, 114.017°W), AB, CAN	[Beaudoin (1989) In: A.S. Dyke (GSC database)]; [Beaudoin (1991), PMA database]; [Landals permit 88-38; Beaudoin (1991) In: R.E. Morlan, CMC database (1999)]
1600 ± 110 (1680 ± 110)	S-1318		L. Diefenbaker shore near Birsay, S Saskatchewan valley (51.040°N, 107.830°W), SK, CAN	Morlan (1993); R.E. Morlan, CMC database (1999)
(1590 ± 80)	AECV-1361C		Oldman River Dam area (49.600°N, 114.000°W), AB, CAN	[Beaudoin In: A.S. Dyke (GSC database)]; [A. Beaudoin (PMA database)]; [Van Dyke, et al. (1990, 1991) In: R.E. Morlan, CMC database (1999)]; Beaudoin (1991, 1998)
1580 ± 159 (1660 ± 159)	S-22	charred bone	Besant valley, Qu'Appelle drainage, SK, CAN	Morlan (1993); [Wettlaufer (1955); Dyck (1983); Reeves (1983) In: R.E. Morlan, CMC database (1999)]
1575 ± 115 (1655 ± 115)	S-2856		Qu'Appelle drainage, about 19 km E of the elbow of the S Saskatchewan R. (51.070°N, 106.350°W), SK, CAN	Morlan (1993); [Morlan and McNeely (1996) In: A.S. Dyke (GSC database)]; [Phenix (1969); Ramsay (1991) In: R.E. Morlan, CMC database (1999)]
1565 ± 205 (1645 ± 205)	S-2623		5 km E of the town of Avonlea, Moose Jaw valley, Qu'Appelle drainage (50.020°N, 105.000°W), SK, CAN	[Kehoe and McCorquodale (1961); Klimko (1985); Klimko and Hanna (1988); Kehoe, et al. (1988) In: R.E. Morlan, CMC database (1999)]; Morlan (1993); [Morlan and McNeely (1996) In: A.S. Dyke (GSC database)]
1560 ± 120 (1640 ± 120)	RL-904	skull	Edworthy Pk., NW Calgary on S side of Bow valley (51.080°N, 114.080°W), AB, CAN	Brumley and Rushworth (1983); [Wilson (1983, 1986) In: R.E. Morlan, CMC database (1999)]

(1550 ± 90)	AECV-1177C	left metatarsal, 2nd phalanx	Dersch Locality area (49.700°N, 113.650°W), AB, CAN	[Beaudoin In: A.S. Dyke (GSC database)]; [A. Beaudoin (PMA database)]; [Fedirchuk (1991); Tiffany (1993) In: R.E. Morlan, CMC database (1999)]; Beaudoin (1998)
(1540 ± 90)	AECV-1178C	partial left scapula	Dersch Locality area (49.700°N, 113.650°W), AB, CAN	[Beaudoin In: A.S. Dyke (GSC database)]; [A. Beaudoin (PMA database)]; [Fedirchuk (1991); Tiffany (1993) In: R.E. Morlan, CMC database (1999)]; Beaudoin (1998)
(1540 ± 70)	AECV-1705C	long-bone fragments	Head-Smashed-In Buffalo Jump (49°42'N, 113°39'W), AB, CAN	[A. Beaudoin (PMA database)]
1535 ± 75 (1615 ± 80)	S-2885	femur (CMC-1370)	Wanuskewin Heritage Pk., S Saskathcewan R., N of Saskatoon (52.170°N, 106.500°W), SK, CAN	Morlan (1993); [Walker, et al. (1986, 1987); Morlan (1988) In: R.E. Morlan, CMC database (1999)]; [Morlan and McNeely (1996) In: A.S. Dyke (GSC database)]
1530 ± 105 (1610 ± 105)	S-2799		Qu'Appelle valley (50.670°N, 103.670°W), SK, CAN	Morlan (1993); [Morlan and McNeely (1996) In: A.S. Dyke (GSC database)]; [Dyck (1983); Smith (1986); Smith and Walker (1988) In: R.E. Morlan, CMC database (1999)]
(1530 ± 90)	AECV-1366C		Oldman River Dam area (49.580°N, 113.980°W), AB, CAN	[Beaudoin In: A.S. Dyke (GSC database)]; [A. Beaudoin (PMA database)]; [Van Dyke, et al. (1990) In: R.E. Morlan, CMC database (1999)]; Beaudoin (1998)
(1530 ± 90)	AECV-866C		Oldman River Dam area (49.600°N, 114.050°W), AB, CAN	[Beaudoin (1989) In: A.S. Dyke (GSC database)]; [Beaudoin (1991), PMA database]; [Landals permit 88-38; Beaudoin (1991) In: R.E. Morlan, CMC database (1999)]
1500 ± 100 (1580 ± 100)	S-45	charred bone	5 km E of Avonlea, Moose Jaw valley, Qu'Appelle drainage (50.020°N, 105.000°W), SK, CAN	[Kehoe and McCorquodale (1961); Klimko (1985); Klimko and Hanna (1988); Kehoe, et al. (1988) In: R.E. Morlan, CMC database (1999)]; Morlan (1993)
1490 ± 90 (1570 ± 90)	Beta-69005	4th cervical vertebra (CMC-1426)	Moose Woods Sand Hills, 548 m asl., 15 km S of Saskatoon (51.978°N, 106.518°W), SK, CAN	[Long and Rippeteau (1974); Hjermstad (1996) In: R.E. Morlan, CMC database (1999)]; [Morlan and McNeely (1996) In: A.S. Dyke (GSC database)]
(1470 ± 70)	AECV-1550C	radius	Head-Smashed-In Buffalo Jump (49.720°N, 113.650°W), AB, CAN	[Beaudoin In: A.S. Dyke (GSC database)]; [A. Beaudoin (PMA database)]; [Reeves (1970, 1978); Wilmeth (1978); Brink, et al. (1985, 1986); Brink and Dawe (1989) In: R.E. Morlan, CMC database (1999)]; Brumley and Rushworth (1983); Morlan (1985); Beaudoin (1987b, 1988, 1991, 1998)
1470 ± 90	AECV-338C	tibia	Oldman River Dam area (49°37'N, 114°03'W), AB, CAN	[Beaudoin (1988), PMA database]
(1450 ± 90)	Beta-19806		Forty Mile Coulee, AB, CAN	[Brumley and Dau (1988) In: R.E. Morlan, CMC database (1999)]
(1450 ± 80)	AECV-1706C	long-bone fragments	Head-Smashed-In Buffalo Jump (49°42'N, 113°39'W), AB, CAN	[A. Beaudoin (PMA database)]
1445 ± 85 (1565 ± 85)	I-11051	tibia	Hunker Ck. (Dawson Loc. 16) (64.070°N, 139.370°W), YT, CAN	Harington (1977, 1978, 1980, 1997); Harington and Morlan (1992), Matheus (1995); R.E. Morlan, CMC database (1999)
(1440 ± 90)	AECV-1363C		Oldman River Dam area (49.600°N, 114.120°W), AB, CAN	[Beaudoin In: A.S. Dyke (GSC database)]; [A. Beaudoin (PMA database)]; [Van Dyke, et al. (1990); Beaudoin (1998) In: R.E. Morlan, CMC database (1999)]
1440 ± 60 (1520 ± 65)	S-3130		W of Souris, 427 m asl (49.500°N, 100.518°W), MB, CAN	[Wowchuk (1990) In: A.S. Dyke (GSC database)]; [Peach and Burton-Coe (1994) In: R.E. Morlan, CMC database (1999)]

1430 ± 80 (1510 ± 80)	S-688		Souris R., S of Melita, 435 m asl (49.200°N, 101.030°W), MB, CAN	[Long and Tamplin (1977); Wilmeth (1978) In: R.E. Morlan, CMC database (1999)]
(1400 ± 80)	AECV-1482C	long-bone and rib fragments	Head-Smashed-In Buffalo Jump (49.720°N, 113.650°W), AB, CAN	[Beaudoin In: A.S. Dyke (GSC database)]; [A. Beaudoin (PMA database)]; Van Dyke, et al. (1990); [Brink permit 89-29; Beaudoin (1998) In: R.E. Morlan, CMC database (1999)]
1390 ± 40 (1470 ± 45)	S-1575	(CMC-1012)	Arm R., NE of Bethune, Qu'Appelle drainage (50.780°N, 105.120°W), SK, CAN	[Dawson and Walker (1988) In: R.E. Morlan, CMC database (1999)]; Morlan (1993)
1390 ± 90 (1470 ± 90)	S-1017	vertebra, scapula, metapodial (CMC-790)	Shaw Burial site, Red Deer drainage (50.600°N, 110.190°W), AB, CAN	[Brumley (1976); Wilmeth (1978) In: R.E. Morlan, CMC database (1999)]; [A. Beaudoin (PMA database)]; Brumley and Rushworth (1983)
1380 ± 200 (1460 ± 200)	S-1762	(CMC-1107)	S Saskatchewan R., 500 m asl (51.420°N, 107.090°W), SK, CAN	[Dyck (1983); Dyck and Morlan (1995) In: R.E. Morlan, CMC database (1999); Morlan (1993)]; [Morlan and McNeely (1996) In: A.S. Dyke (GSC database)]
1380 ± 70	S-722		Manyfingers (49.280°N, 113.550°W), AB, CAN	A.S. Dyke (GSC database)
1380 ± 190 (1460 ± 190)	S-1763	(CMC-1108)	S Saskatchewan R., 500 m asl (51.420°N, 107.090°W), SK, CAN	[Dyck (1983); Dyck and Morlan (1995) In: R.E. Morlan, CMC database (1999); Morlan (1993)]; [Morlan and McNeely (1996) In: A.S. Dyke (GSC database)]
1370 ± 110 (1450 ± 110)	RL-1563	femur, tibia, 2 long-bone fragments and a rib	Highwood R. (50°25'N, 114°35'W), AB, CAN	[Quigg (1982) In: R.E. Morlan, CMC database (1999)]; Brumley and Rushworth (1983)
1365 ± 110 (1445 ± 110)	S-2324	femur	N bank of Saskatchewan R., about 6.5 km upstream from Francois-Finlay Dam (53.170°N, 104.000°W), SK, CAN	[Finnigan, et al. (1983) In: R.E. Morlan, CMC database (1999)]; Morlan (1993)
(1360 ± 80)	AECV-1770C	fragments - astragalus, rib	Head-Smashed-In Buffalo Jump (49°42'N, 113°39'W), AB, CAN	[A. Beaudoin (PMA database)]
1350 ± 140 (1465 ± 140)	S-3554	ulna and horncore	Souris R. (49°24'N, 103°35'W), SK, CAN	[Finnigan (1988) In: R.E. Morlan, CMC database (1999)]
1340 ± 190 (1420 ± 190)	S-1761	(CMC-1106)	S Saskatchewan R., 500 m asl (51.420°N, 107.090°W), SK, CAN	Morlan (1993); [Dyck (1983); Dyck and Morlan (1995) In: R.E. Morlan, CMC database (1999)]; [Morlan and McNeely (1996) In: A.S. Dyke (GSC database)]
1340 ± 60 (1420 ± 65)	Beta-69004	humerus (CMC-1425)	Moose Woods Sand Hills, 548 m asl., 15 km S of Saskatoon (51.978°N, 106.518°W), SK, CAN	[Long and Rippeteau (1974); Hjermstad (1996) In: R.E. Morlan, CMC database (1999)]; [Morlan and McNeely (1996) In: A.S. Dyke (GSC database)]
1320 ± 190 (1400 ± 190)	S-1760	(CMC-1105)	S Saskatchewan R., 500 m asl (51.420°N, 107.090°W), SK, CAN	Morlan (1993); [Dyck (1983); Dyck and Morlan (1995) In: R.E. Morlan, CMC database (1999)]; [Morlan and McNeely (1996) In: A.S. Dyke (GSC database)]
1320 ± 100 (1400 ± 100)	S-687		Souris R., S of Melita, 435 m asl (49.200°N, 101.030°W), MB, CAN	[Long and Tamplin (1977); Wilmeth (1978) In: R.E. Morlan, CMC database (1999)]
(1320 ± 80)	AECV-1548C	metatarsal	Head-Smashed-In Buffalo Jump (49.720°N, 113.650°W), AB, CAN	[Beaudoin In: A.S. Dyke (GSC database)]; [A. Beaudoin (PMA database)]; [Reeves (1970, 1978); Wilmeth (1978); Brink, et al. (1985, 1986); Brink and Dawe (1989) In: R.E. Morlan, CMC database (1999)]; Brumley and Rushworth (1983); Morlan (1985); Beaudoin (1987, 1988, 1991, 1998)
(1320 ± 80)	AECV-1774C	astragalus, calcaneum, part of articulated bone unit	Head-Smashed-In Buffalo Jump (49°43'N, 113°39'W), AB, CAN	[A. Beaudoin (PMA database)]

1310 ± 100 (1390 ± 100)	S-2826		N bank of the Souris R., 427 m asl (49.467°N, 100.583°W), MB, CAN	[Nicholson (1988, 1994b) In: R.E. Morlan, CMC database (1999)]
(1290 ± 120)	AECV-739C		Oldman River Dam area (49.620°N, 114.120°W), AB, CAN	[Beaudoin (1989) In: A.S. Dyke (GSC database)]; [Beaudoin (1991), PMA database]; [Van Dyke, et al. (1991); Beaudoin (1991) In: R.E. Morlan, CMC database (1999)]
(1290 ± 80)	AECV-1772C	radius, part of an articulated group of bones	Head-Smashed-In Buffalo Jump (49°42'N, 113°39'W), AB, CAN	[A. Beaudoin (PMA database)]
(1290 ± 80)	AECV-1765C	long-bone fragments	Head-Smashed-In Buffalo Jump (49°42'N, 113°39'W), AB, CAN	[A. Beaudoin (PMA database)]
1280 ± 190 (1360 ± 190)	S-2700		Lovstrom, S of Brandon, 373.5 m asl (49.500°N, 99.942°W), MB, CAN	[Nicholson (1986, 1990, 1994b, 1996); Nicholson and Malainey (1991) In: R.E. Morlan, CMC database (1999)]
1270 ± 140 (1350 ± 140)	S-3546	6th cervical vertebra (CMC-1453)	Moose Woods Sand Hills, 548 m asl., 15 km S of Saskatoon (51.978°N, 106.518°W), SK, CAN	[Long and Rippeteau (1974); Hjermstad (1996) In: R.E. Morlan, CMC database (1999)]; [Morlan and McNeely (1996) In: A.S. Dyke (GSC database)]
(1260 ± 160)	AECV-765C		Oldman River Dam area (49.620°N, 114.120°W), AB, CAN	[Beaudoin (1989) In: A.S. Dyke (GSC database)]; [Beaudoin (1991), PMA database]; [Van Dyke, et al. (1991); Beaudoin (1991) In: R.E. Morlan, CMC database (1999)]
1260 ± 115 (1340 ± 115)	S-2691		Qu'Appelle valley (50.670°N, 103.670°W), SK, CAN	[Dyck (1983); Smith (1986); Smith and Walker (1988) In: R.E. Morlan, CMC database (1999)]; [Morlan and McNeely (1996) In: A.S. Dyke (GSC database)]; Morlan (1993)
(1260 ± 80)	AECV-1771C	2 astragali	Head-Smashed-In Buffalo Jump (49°42'N, 113°39'W), AB, CAN	[A. Beaudoin (PMA database)]
(1260 ± 80)	AECV-1769C	fragments - tibia, radius, proximal phalanx, ribs	Head-Smashed-In Buffalo Jump (49°42'N, 113°39'W), AB, CAN	[A. Beaudoin (PMA database)]
(1240 ± 80)	AECV-1481C	femur	Head-Smashed-In Buffalo Jump (49.720°N, 113.650°W), AB, CAN	[Beaudoin In: A.S. Dyke (GSC database)]; [A. Beaudoin (PMA database)]; [Reeves (1970, 1978); Wilmeth (1978); Brink, et al. (1985, 1986); Brink and Dawe (1989) In: R.E. Morlan, CMC database (1999)]; Brumley and Rushworth (1983); Morlan (1985); Beaudoin (1987, 1988, 1991, 1998)
1235 ± 75 (1315 ± 80)	S-2815		Wanuskewin Heritage Pk., S Saskatchewan R. (52.170°N, 106.500°W), SK, CAN	Morlan (1988, 1993); [Walker, et al. (1986, 1987); Walker (1988) In: R.E. Morlan, CMC database (1999)]; [Morlan and McNeely (1996) In: A.S. Dyke (GSC database)]
1230 ± 100 (1310 ± 100)	S-2825		N bank of the Souris R., 427 m asl (49.467°N, 100.583°W), MB, CAN	[Nicholson (1988, 1994b) In: R.E. Morlan, CMC database (1999)]
1225 ± 85 (1305 ± 85)	S-2968	rib	N bank of Souris R., 546 m asl (49.330°N, 103.330°W), SK, CAN	[Finnigan, et al. (1990); Shay, et al. (1990, 1993); Rollans and McKean (1992) In: R.E. Morlan, CMC database (1999)]; Morlan (1993)
1220 ± 70	S-1793	scapula	Whitemud Ck., N Saskatchewan R. valley, near Edmonton (53.500°N, 113.670°W), AB, CAN	Rains and Welch (1988)
1215 ± 320 (1295 ± 320)	S-2699		Lovstrom, S of Brandon, 373.5 m asl (49.500°N, 99.942°W), MB, CAN	[Nicholson (1986, 1990, 1994b, 1996); Nicholson and Malainey (1991) In: R.E. Morlan, CMC database (1999)]
(1210 ± 70)	AECV-1773C	atlas vertebra	Head-Smashed-In Buffalo Jump (49°43'N, 113°39'W), AB, CAN	[A. Beaudoin (PMA database)]
1205 ± 100 (1285 ± 100)	S-2792		Qu'Appelle valley (50.670°N, 103.670°W), SK, CAN	Morlan (1993); [Dyck (1983); Smith (1986); Smith and Walker (1988) In: R.E. Morlan, CMC database (1999)]

(1200 ± 80)	AECV-1551C	humerus	Head-Smashed-In Buffalo Jump (49.720°N, 113.650°W), AB, CAN	[Beaudoin In: A.S. Dyke (GSC database)]; [A. Beaudoin (PMA database)]; [Reeves (1970, 1978); Wilmeth (1978); Brink, et al. (1985, 1986); Brink and Dawe (1989) In: R.E. Morlan, CMC database (1999)]; Brumley and Rushworth (1983); Morlan (1985); Beaudoin (1987, 1988, 1991, 1998)
(1200 ± 80)	AECV-1766C	long-bone fragments	Head-Smashed-In Buffalo Jump (49°42'N, 113°39'W), AB, CAN	[A. Beaudoin (PMA database)]
1190 ± 60 (1260 ± 60)	Beta-70705	atlas (CMC-1485)	Wanuskewin Heritage Pk., S Saskatchewan R. (52.220°N, 106.593°W), SK, CAN	Morlan (1988, 1993); [Walker, et al. (1986, 1987); Walker (1988) In: R.E. Morlan, CMC database (1999)]; [Morlan and McNeely (1996) In: A.S. Dyke (GSC database)]
(1190 ± 90)	AECV-752C		Oldman River Dam area (49.600°N, 114.120°W), AB, CAN	[Beaudoin (1989) In: A.S. Dyke (GSC database)]; [Beaudoin (1991), PMA database]; [Van Dyke, et al. (1991); Beaudoin (1991) In: R.E. Morlan, CMC database (1999)]
(1190 ± 70)	AECV-1767C	tibia, femur-shaft fragments	Head-Smashed-In Buffalo Jump (49°42'N, 113°39'W), AB, CAN	[A. Beaudoin (PMA database)]
1180 ± 80 (1260 ± 80)	S-2966	cranium	N bank of Souris R., 565 m asl (49.000°N, 103.000°W), SK, CAN	[Finnigan and McKeand (1992) In: R.E. Morlan, CMC database (1999)]; Morlan (1993); [Morlan and McNeely (1996) In: A.S. Dyke (GSC database)]
1170 ± 130 (1260 ± 130)	GSC-990	3 vertebrae	NE of Carberry (49.900°N, 99.260°W), MB, CAN	Lowdon, et al. (1971); R.E. Morlan, CMC database (1999)
(1170 ± 90)	AECV-998C		Strathcona (53.550°N, 113.380°W), AB, CAN	[Beaudoin (1989) In: A.S. Dyke (GSC database)]; [Beaudoin (1991), PMA database]; [Oetelaar permit 89-10, in Luther and Purdon (1991); Beaudoin (1991) In: R.E. Morlan, CMC database (1999)]
(1170 ± 70)	AECV-1914C		Head-Smashed-In Buffalo Jump (49°43'N, 113°39'W), AB, CAN	[A. Beaudoin (PMA database)]
1160 ± 170 (1240 ± 170)	S-3547	metacarpal (CMC-1454)	Moose Woods Sand Hills, 548 m asl., 15 km S of Saskatoon (51.978°N, 106.518°W), SK, CAN	[Long and Rippeteau (1974); Hjermstad (1996) In: R.E. Morlan, CMC database (1999)]; [Morlan and McNeely (1996) In: A.S. Dyke (GSC database)]
1155 ± 75 (1235 ± 80)	S-2814		Wanuskewin Heritage Pk., S Saskatchewan R. (52.170°N, 106.500°W), SK, CAN	Morlan (1988, 1993); [Walker, et al. (1986, 1987); Walker (1988) In: R.E. Morlan, CMC database (1999)]; [Morlan and McNeely (1996) In: A.S. Dyke (GSC database)]
1150 ± 80	AECV-314C	radius	Cranford Gravel Pit (49°50'N, 112°21'W), AB, CAN	[Beaudoin (1987), PMA database]
1140 ± 70 (1220 ± 75)	S-1367		Assiniboine drainage (49.810°N, 100.550°W), MB, CAN	[Syms (1983) In: R.E. Morlan, CMC database (1999)]
(1140 ± 90)	AECV-461C		Kettles Buffalo Jump (49.533°N, 113.833°W), AB, CAN	[Beaudoin (1988) In: A.S. Dyke (GSC database)]; [Beaudoin (1988), PMA database]
1135 ± 80	S-1786	skull	Strawberry Ck., N Saskatchewan R. valley, near Edmonton (53.500°N, 113.670°W), AB, CAN	[Rains and Welch (1988) - age seems to be anomalously young]
1120 ± 60 (1200 ± 65)	S-3382	tibia	S of Saskatoon, S Saskatchewan R. (52.000°N, 106.500°W), SK, CAN	Morlan (1993); [Clarke and Meyer (1992); Clarke (1995) In: R.E. Morlan, CMC database (1999)]; [Morlan and McNeely (1996) In: A.S. Dyke (GSC database)]
(1120 ± 80)	AECV-1768C	fragments - humerus, rib, shaft	Head-Smashed-In Buffalo Jump (49°42'N, 113°39'W), AB, CAN	[A. Beaudoin (PMA database)]
1110 ± 70 (1170 ± 70)	Beta-111672	rib (CMC-1510)	Pembina R. plain, 485 m asl., Pembina drainage (49.117°N, 98.217°W), MB, CAN	[Dyck (1998) In: A.S. Dyke (GSC database)]; [Capes (1963); Dyck (1999) In: R.E. Morlan, CMC database (1999)]

1100 ± 80	S-87	decomposed matted hair, hide, and buffalo bones	Old Woman's Buffalo Jump, S of Calgary (50°28'N, 113°53'W), AB, CAN	Brumley and Rushworth (1983); [Forbis (1962); Wilmeth (1978) In: R.E. Morlan, CMC database (1999)]
1100 150 (1180 ± 150)	SFU-229		Assiniboine drainage, 12 km W of Brandon (49.820°N, 100.090°W), MB, CAN	[Tisdale (1978); Badertscher, et al. (1987) In: R.E. Morlan, CMC database (1999)]
(1100 ± 80)	AECV-1685C	complete femur	Lyle Marcyk, Grande Prairie (55°09'N, 118°34'W), AB, CAN	[A. Beaudoin (PMA database)]
(1070 ± 90)	AECV-1188C		Oldman River Dam area (49.580°N, 114.050°W), AB, CAN	[Beaudoin In: A.S. Dyke (GSC database)]; [A. Beaudoin (PMA database)]; [Van Dyke, et al. (1990, 1991) In: R.E. Morlan, CMC database (1999)]; Beaudoin (1991, 1998)
(1060 ± 60)	AECV-1552C	humerus	Head-Smashed-In Buffalo Jump (49.720°N, 113.650°W), AB, CAN	[Beaudoin In: A.S. Dyke (GSC database)]; [A. Beaudoin (PMA database)]; [Reeves (1970, 1978); Wilmeth (1978); Brink, et al. (1985, 1986); Brink and Dawe (1989) In: R.E. Morlan, CMC database (1999)]; Brumley and Rushworth (1983); Morlan (1985); Beaudoin (1987, 1988, 1991, 1998)
1050 ± 80 (1130 ± 80)	S-3219	scapula	Confluence of Jewel Ck., and Roughbark Ck., Souris drainage (49.330°N, 103.500°W), SK, CAN	[Gibson, pers. comm. (1992) In: R.E. Morlan, CMC database (1999)]; Morlan (1993); [Morlan and McNeely (1996) In: A.S. Dyke (GSC database)]
(1040 ± 70)	AECV-1775C	humerus shaft, calcaneum	Highway 2/3 Junction Impact Mitigation (49°42'N, 113°27'W)), AB, CAN	[A. Beaudoin (PMA database)]
1035 ± 60 (1115 ± 65)	S-266		Frenchman's Flat, floodplain of the S Saskatchewan R, S of Saskatoon (51.900°N, 106.750°W), SK, CAN	Morlan (1993); R.E. Morlan, CMC database (1999)
1020 ± 60 (1100 ± 65)	S-1366		Assiniboine drainage (49.810°N, 100.550°W), MB, CAN	[Syms (1983) In: R.E. Morlan, CMC database (1999)]
1020 ± 50 (1100 ± 55)	Beta-6715		Forty Mile Coulee (49°35'N, 111°25'W), AB, CAN	[Brumley and Dau (1988) In: R.E. Morlan, CMC database (1999)]
(1010 ± 90)	AECV-1357C		Oldman River Dam area (49.600°N, 114.000°W), AB, CAN	[Beaudoin In: A.S. Dyke (GSC database)]; [A. Beaudoin (PMA database)]; [Van Dyke, et al. (1990, 1991) In: R.E. Morlan, CMC database (1999)]; Beaudoin (1991, 1998)
(1000 ± 120)	AECV-1362C		Oldman River Dam area (49.600°N, 114.120°W), AB, CAN	[Beaudoin In: A.S. Dyke (GSC database)]; [A. Beaudoin (PMA database)]; [Van Dyke, et al. (1990, 1991) In: R.E. Morlan, CMC database (1999)]; Beaudoin (1991, 1998)
1000 ± 90	AECV-313C	humerus fragments	Cranford Gravel Pit (49°50'N, 112°21'W), AB, CAN	[Beaudoin (1987), PMA database]
1000 ± 100	AECV-315C	phalanges, metatarsal fragments	Cranford Gravel Pit (49°50'N, 112°21'W), AB, CAN	[Beaudoin (1987), PMA database]
990 ± 70 (1070 ± 75)	S-1368		Assiniboine drainage (49.808°N, 100.550°W), MB, CAN	[Syms (1983) In: R.E. Morlan, CMC database (1999)]
(980 ± 80)	AECV-1918C	3 calcanea	Pine Coulee Reservoir (50°13'N, 113°46'W), AB, CAN	[A. Beaudoin (PMA database)]
965 ± 65 (1045 ± 70)	S-1015	(CMC-788)	Ramilies Kill site, Suffield Military Reserve, S Saskatchewan drainage, 777 m asl (50.360°N, 110.900°W), AB, CAN	[Brumley (1976); Wilmeth (1978) In: R.E. Morlan, CMC database (1999)]; [A. Beaudoin (PMA database)]; Brumley and Rushworth (1983)
(960 ± 80)	AECV-860C		Oldman River Dam area (49.600°N, 114.000°W), AB, CAN	[Beaudoin (1989) In: A.S. Dyke (GSC database)]; [Beaudoin (1991), PMA database]; [Landals permit 88-38; Beaudoin (1991) In: R.E. Morlan, CMC database (1999)]

950 ± 190 (1030 ± 190)	S-1759	(CMC-1104)	S Saskatchewan R., 500 m asl (51.420°N, 107.090°W), SK, CAN	Morlan (1993); [Dyck (1983); Dyck and Morlan (1995) In: R.E. Morlan, CMC database (1999)]; [Morlan and McNeely (1996) In: A.S. Dyke (GSC database)]
(950 ± 90)	AECV-861C		Oldman River Dam area (49.600°N, 114.100°W), AB, CAN	[Beaudoin (1989) In: A.S. Dyke (GSC database)]; [Beaudoin (1991), PMA database]; [Landals permit 88-38; Beaudoin (1991) In: R.E. Morlan, CMC database (1999)]
945 ± 135 (1025 ± 135)	S-2813		Wanuskewin Heritage Pk., S Saskatchewan R. (52.170°N, 106.500°W), SK, CAN	Morlan (1988, 1993); [Walker, et al. (1986, 1987); Walker (1988) In: R.E. Morlan, CMC database (1999)]; [Morlan and McNeely (1996) In: A.S. Dyke (GSC database)]
(940 ± 90)	AECV-750C		Oldman River Dam area (49.580°N, 114.050°W), AB, CAN	[Beaudoin (1989) In: A.S. Dyke (GSC database)]; [Beaudoin (1991), PMA database]; [Van Dyke, et al. (1991); Beaudoin (1991) In: R.E. Morlan, CMC database (1999)]
(940 ± 70)	AECV-1920C	femur	Pine Coulee Reservoir (50°13'N, 113°46'W), AB, CAN	[A. Beaudoin (PMA database)]
920 ± 45 (1000 ± 50)	S-1631	(CMC-1035)	Dunfermline Sand Hills, 506 m asl., about 16 km NW of Saskatoon (52.220°N, 106.930°W), SK, CAN	[Dyck (1972); Walker (1979); Prentice (1983); Linnamae (1988) In: R.E. Morlan, CMC database (1999)]; Morlan (1993); [Morlan and McNeely (1996) In: A.S. Dyke (GSC database)]
(910 ± 80)	AECV-1524C	radius and metatarsal fragments	Pincher Ck., Oldman drainage (49.480°N, 114.030°W), AB, CAN	[Beaudoin In: A.S. Dyke (GSC database)]; [A. Beaudoin (PMA database)]; [Dau (1991); Beaudoin (1998) In: R.E. Morlan, CMC database (1999)]
(910 ± 70)	AECV-1635C	distal right humerus	Highway 2/3 Junction (49°42'N, 113°27'W), AB, CAN	[A. Beaudoin (PMA database)]
880 ± 80 (960 ± 80)	Beta-62705	humerus	NE of Lauder, Souris drainage (49.400°N, 100.813°W), MB, CAN	[Nicholson (1994); pers. comm. (1998) In: R.E. Morlan, CMC database (1999)]; [Nicholson (1996) In: A.S. Dyke (GSC database)]
880 ± 70 (960 ± 75)	S-2812		Wanuskewin Heritage Pk., S Saskatchewan R. (52.170°N, 106.500°W), SK, CAN	Morlan (1988, 1993); [Walker, et al. (1986, 1987); Walker (1988) In: R.E. Morlan, CMC database (1999)]; [Morlan and McNeely (1996) In: A.S. Dyke (GSC database)]
880 ± 100	AECV-334C	metacarpal, 1st and 3rd phalanges	Oldman River Dam area (49°33'N, 114°01'W), AB, CAN	[Beaudoin (1988), PMA database]
(880 ± 70)	AECV-1636C	distal right humerus	Highway 2/3 Junction (49°42'N, 113°27'W), AB, CAN	[A. Beaudoin (PMA database)]
(870 ± 110)	AECV-854C		Oldman River Dam area (49.570°N, 114.020°W), AB, CAN	[Beaudoin In: A.S. Dyke (GSC database)]; [Beaudoin (1991), PMA database]; [Landals permit 88-38; Beaudoin (1991) In: R.E. Morlan, CMC database (1999)]
(870 ± 70)	AECV-1921C	tibia	Pine Coulee Reservoir (50°13'N, 113°46'W), AB, CAN	[A. Beaudoin (PMA database)]
(860 ± 70)	AECV-1640C	right metatarsal	Highway 2/3 Junction (49°42'N, 113°27'W), AB, CAN	[A. Beaudoin (PMA database)]
(860 ± 70)	AECV-1637C	assorted postcranial elements	Highway 2/3 Junction (49°42'N, 113°27'W), AB, CAN	[A. Beaudoin (PMA database)]
(850 ± 90)	AECV-845C		Mitigation (51.970°N, 112.220°W), AB, CAN	[Beaudoin (1989) In: A.S. Dyke (GSC database)]; [Beaudoin (1991), PMA database]; [Light permit 88-76 in Luther and Purdon (1991), Beaudoin (1991) In: R.E. Morlan, CMC database (1999)]

850 ± 115 (930 ± 115)	S-3029		Lovstrom, S of Brandon, 373.5 m asl (49.500°N, 99.942°W), MB, CAN	[Peach and Burton-Coe (1994) In: A.S. Dyke (GSC database)]; [Nicholson (1986, 1990, 1994, 1996); Nicholson and Malainey (1991); D.E. Nelson, pers. comm. (1998) In: R.E. Morlan, CMC database (1999)]
850 ± 80 (930 ± 80)	S-690		W of Virden, along Pipestone Ck. (49.800°N, 101.330°W), MB, CAN	[Rushowick (1975); Wilmeth (1978) In: R.E. Morlan, CMC database (1999)]
850 ± 60 (930 ± 65)	S-1073	fragments	Eagle Ck., N Saskatchewan drainage (51.780°N, 107.500°W), SK, CAN	Morlan (1993); R.E. Morlan, CMC database (1999)
850 ± 110 (930 ± 110)	RL-1475		17 km SW of Bragg Ck, on the N side of Elbow R., AB, CAN	Brumley and Rushworth (1983); [Quigg (1982) In: R.E. Morlan, CMC database (1999)]
(840 ± 70)	AECV-1641C	partial proximal left femur	Highway 2/3 Junction (49°42'N, 113°27'W), AB, CAN	[A. Beaudoin (PMA database)]
820 ± 110 (900 ± 110)	S-2884	metacarpal (CMC-1369)	S Saskatchewan R., SK, CAN	Morlan (1993); [J. Finnigan, pers. comm. (1993) In: R.E. Morlan, CMC database (1999)]
(820 ± 70)	AECV-1634C	distal left femur and cervical vertebra	Highway 2/3 Junction (49°42'N, 113°27'W), AB, CAN	[A. Beaudoin (PMA database)]
810 ± 205 (890 ± 205)	S-2857		Qu'Appelle drainage, about 19 km E of the elbow of the S Saskatchewan R. (51.070°N, 106.350°W), SK, CAN	Morlan (1993); [Morlan and McNeely (1996) In: A.S. Dyke (GSC database)]; [Phenix (1969); Ramsay (1991) In: R.E. Morlan, CMC database (1999)]
(810 ± 90)	AECV-1364C		Oldman River Dam area (49.600°N, 114.120°W), AB, CAN	[Beaudoin In: A.S. Dyke (GSC database)]; [A. Beaudoin (PMA database)]; [Van Dyke, et al. (1990); Beaudoin (1998) In: R.E. Morlan, CMC database (1999)]
810 ± 75	S-1790		Whitemud Ck., N Saskatchewan R. valley, near Edmonton (53.500°N, 113.670°W), AB, CAN	Rains and Welch (1988)
(800 ± 90)	AECV-997C		Strathcona (53.550°N, 113.380°W), AB, CAN	[Beaudoin (1989) In: A.S. Dyke (GSC database)]; [Beaudoin (1991), PMA database] [Oetelaar permit 89-10, in Luther and Purdon (1991); Beaudoin (1991) In: R.E. Morlan, CMC database (1999)]
(800 ± 90)	AECV-1193C		Oldman River Dam area (49.580°N, 114.050°W), AB, CAN	[Beaudoin In: A.S. Dyke (GSC database)]; [A. Beaudoin (PMA database)]; [Van Dyke, et al. (1990); Beaudoin (1998) In: R.E. Morlan, CMC database (1999)]
(800 ± 80)	AECV-1642	postcranial elements	Highway 2/3 Junction (49°42'N, 113°27'W), AB, CAN	[A. Beaudoin (PMA database)]
790 ± 135 (870 ± 135)	S-2810		Wanuskewin Heritage Pk., S Saskatchewan R. (52.170°N, 106.500°W), SK, CAN	Morlan (1988, 1993); [Walker, et al. (1986, 1987); Walker (1988) In: R.E. Morlan, CMC database (1999)]; [Morlan and McNeely (1996) In: A.S. Dyke (GSC database)]
(780 ± 90)	AECV-744C		Oldman River Dam area (49.580°N, 114.050°W), AB, CAN	[Beaudoin (1989) In: A.S. Dyke (GSC database)]; [Beaudoin (1991), PMA database]; [Van Dyke, et al. (1991); Beaudoin (1991) In: R.E. Morlan, CMC database (1999)]
780 ± 110 (860 ± 110)	S-3031		Lovstrom, S of Brandon, 373.5 m asl (49.500°N, 99.942°W), MB, CAN	[Peach and Burton-Coe (1994) In: A.S. Dyke (GSC database)]; [Nicholson (1986, 1990, 1994, 1996); Nicholson and Malainey (1991); D.E. Nelson, pers. comm. (1998) In: R.E. Morlan, CMC database (1999)]
(780 ± 80)	AECV-1633C	left tibia and left astragalus	Highway 2/3 Junction (49°42'N, 113°27'W), AB, CAN	[A. Beaudoin (PMA database)]
775 ± 90 (855 ± 90)	S-2824		Lovstrom, S of Brandon, 373.5 m asl (49.500°N, 99.942°W), MB, CAN	[Nicholson (1986, 1990, 1994b, 1996); Nicholson and Malainey (1991) In: R.E. Morlan, CMC database (1999)]

(760 ± 110)	AECV-873C		Oldman River Dam area (49.600°N, 114.100°W), AB, CAN	[Beaudoin (1989) In: A.S. Dyke (GSC database)]; [Beaudoin (1991), PMA database]; [Landals permit 88-38; Beaudoin (1991) In: R.E. Morlan, CMC database (1999)]
760 ± 85	S-1784	vertebra	Pt. bar sands, N Saskatchewan R. valley, near Edmonton, AB, CAN	[Rains and Welch (1988) - age seems to be anomalously young]
(760 ± 70)	AECV-1842C	radius	Highway 2/3 Junction Impact Mitigation (49°42'N, 113°27'W), AB, CAN	[A. Beaudoin (PMA database)]
(750 ± 150)	AECV-771C		Oldman River Dam area (49.600°N, 114.120°W), AB, CAN	[Beaudoin (1989) In: A.S. Dyke (GSC database)]; [Beaudoin (1991), PMA database]; [Van Dyke, et al. (1991); Beaudoin (1991) In: R.E. Morlan, CMC database (1999)]
(740 ± 90)	AECV-874C		Oldman River Dam area (49.600°N, 114.100°W), AB, CAN	[Beaudoin (1989) In: A.S. Dyke (GSC database)]; [Beaudoin (1991), PMA database]; [Landals permit 88-38; Beaudoin (1991) In: R.E. Morlan, CMC database (1999)]
(740 ± 120)	AECV-458C	humerus	Point Beazer, above W bank of St. Mary R. (49°07'N, 113°12'W), AB, CAN	Brumley and Rushworth (1983); [Beaudoin (1988), PMA database]
(740 ± 70)	AECV-1639C	partial cervical vertebra	Highway 2/3 Junction (49°42'N, 113°27'W), AB, CAN	[A. Beaudoin (PMA database)]
740 ± 100 (820 ± 100)	BGS-1377		The Forks (Red and Assiniboine Rivers), MB, CAN	[L.D. Arnold, letter to P. Priess, 89.05.04, forwarded to E. Nielsen, R. McNeely, and R.E. Morlan; Ebell (1988); Kroker (1989); Kroker and Goundry (1990, 1993); Nicholson (1996); Quaternary Consultants Ltd., (1990) In: R.E. Morlan, CMC database (1999)]
730 ± 190 (810 ± 190)	S-1758	(CMC-1103)	S Saskatchewan R., 500 m asl (51.420°N, 107.090°W), SK, CAN	Morlan (1993); [Dyck (1983); Dyck and Morlan (1995) In: R.E. Morlan, CMC database (1999)]; [Morlan and McNeely (1996) In: A.S. Dyke (GSC database)]
720 ± 110 (800 ± 110)	RL-1564	tibia, metatarsal, long-bone shaft	Highwood R. (50°25'N, 114°35'W), AB, CAN	[Quigg (1982) In: R.E. Morlan, CMC database (1999)]; Brumley and Rushworth (1983)
715 ± 110 (795 ± 110)	S-3034		Lovstrom, S of Brandon, 373.5 m asl (49.500°N, 99.942°W), MB, CAN	[Peach and Burton-Coe (1994) In: A.S. Dyke (GSC database)]; [Nicholson (1986, 1990, 1994, 1996); Nicholson and Malainey (1991); D.E. Nelson, pers. comm. (1998) In: R.E. Morlan, CMC database (1999)]
(710 ± 100)	AECV-1146C	calcaneum	Bow Corridor (51.070°N, 115.180°W), AB, CAN	[Beaudoin In: A.S. Dyke (GSC database)]; [A. Beaudoin (PMA database)]; [Quigg (1978); Newton (1991) In: R.E. Morlan, CMC database (1999)]; Brumley and Rushworth (1983); Beaudoin (1998)
(710 ± 130)	AECV-1239C	ribs	Castle Forks Buffalo Jump (Oldman River Dam area) (49.600°N, 114.000°W), AB, CAN	[Beaudoin In: A.S. Dyke (GSC database)]; [A. Beaudoin (PMA database)]; [Landals (1990); Beaudoin (1991, 1998) In: R.E. Morlan, CMC database (1999)]
(710 ± 90)	AECV-870C		Oldman River Dam area (49.600°N, 114.030°W), AB, CAN	[Beaudoin (1989) In: A.S. Dyke (GSC database)]; [Beaudoin (1991), PMA database]; [Landals permit 88-38; Beaudoin (1991) In: R.E. Morlan, CMC database (1999)]
705 ± 75 (785 ± 80)	S-2951		Iglulualuit, Franklin Bay (70.000°N, 127.000°W), NT, CAN	[Nicholson (1996) In: A.S. Dyke (GSC database)]; [Nicholson (1986, 1990, 1994); Nicholson and Malainey (1991); D.E. Nelson, pers. comm. (1998) In: R.E. Morlan, CMC database (1999)]

705 ± 70	S-1791		Whitemud Ck., N Saskatchewan R. valley, near Edmonton (53.500°N, 113.670°W), AB, CAN	Rains and Welch (1988)
(690 ± 70)	AECV-1841C	radius	Highway 2/3 Junction Impact Mitigation (49°42'N, 113°27'W), AB, CAN	[A. Beaudoin (PMA database)]
675 ± 80 (755 ± 80)	S-2953		Lovstrom, S of Brandon, 373.5 m asl (49.500°N, 99.942°W), MB, CAN	[Nicholson (1996) In: A.S. Dyke (GSC database)]; [Nicholson (1986, 1990, 1994); Nicholson and Malainey (1991); D.E. Nelson, pers. comm. (1998) In: R.E. Morlan, CMC database (1999)]
675 ± 70 (755 ± 75)	S-2952		Lovstrom, S of Brandon, 373.5 m asl (49.500°N, 99.942°W), MB, CAN	[Nicholson (1996) In: A.S. Dyke (GSC database)]; [Nicholson (1986, 1990, 1994); Nicholson and Malainey (1991); D.E. Nelson, pers. comm. (1998) In: R.E. Morlan, CMC database (1999)]
(670 ± 90)	AECV-1271C	5th cervical vertebra	Deadlodge Canyon: Dinosaur Provincial Pk, Red Deer (50.770°N, 111.500°W), AB, CAN	[Seemann (1992), Beaudoin (1998) In: R.E. Morlan, CMC database (1999)]
660 ± 115 (740 ± 115)	S-1016	(CMC-789)	Ramilies Kill site, Suffield Military Reserve, S Saskatchewan drainage, 777 m asl (50.360°N, 110.900°W), AB, CAN	[Brumley (1976); Wilmeth (1978) In: R.E. Morlan, CMC database (1999)]; [A. Beaudoin (PMA database)]; Brumley and Rushworth (1983)
(660 ± 80)	AECV-1631C	left scapula	Highway 2/3 Junction (49°42'N, 113°27'W), AB, CAN	[A. Beaudoin (PMA database)]
(650 ± 50)	GSC-2256		Kipp, Ponomar Gravel Pit on the NE side of Oldman R. 860 m asl, about 13 km WNW of Lethbridge (49.770°N, 113.022°W), AB, CAN	R.E. Morlan, CMC database (1999)
645 ± 70 (725 ± 75)	S-2527	metacarpal	S bank of the Saskatchewan R., 10.5 km upstream from the Francois-Finley Dam (53.170°N, 104.000°W), SK, CAN	Morlan (1993); [Quigg (1986); Brandzin-Low (1997) In: R.E. Morlan, CMC database (1999)]; [Morlan and McNeely (1996) In: A.S. Dyke (GSC database)]
(640 ± 70)	AECV-1839C	scapula	Highway 2/3 Junction Impact Mitigation (49°42'N, 113°27'W), AB, CAN	[A. Beaudoin (PMA database)]
640 ± 105 (720 ± 105)	S-864	(CMC-646)	Manyfingers, right bank of Belly R. (49.280°N, 113.550°W), AB, CAN	Brumley and Rushworth (1983); [Quigg (1974, 1975, 1988) In: R.E. Morlan, CMC database (1999)]; [A. Beaudoin (PMA database)]
(630 ± 140)	AECV-457C	skull fragments	Point Beazer, above W bank of St. Mary R. (49°07'N, 113°12'W), AB, CAN	Brumley and Rushworth (1983); [Beaudoin (1988), PMA database]
(620 ± 140)	AECV-871C		Oldman River Dam area (49.600°N, 114.100°W), AB, CAN	[Beaudoin (1989) In: A.S. Dyke (GSC database)]; [Beaudoin (1991), PMA database]; [Landals permit 88-38; Beaudoin (1991) In: R.E. Morlan, CMC database (1999)]
620 ± 100	AECV-317C	tarsals	Cranford Gravel Pit (49°50'N, 112°21'W), AB, CAN	[Beaudoin (1987), PMA database]
(610 ± 70)	AECV-1825C	cervical vertebrae	Highway 2/3 Junction Impact Mitigation (49°42'N, 113°27'W), AB, CAN	[A. Beaudoin (PMA database)]
600 ± 60 (680 ± 65)	S-3131		Lovstrom, S of Brandon, 373.5 m asl (49.500°N, 99.942°W), MB, CAN	[Nicholson (1986, 1990, 1994, 1996); Nicholson and Malainey (1991) In: R.E. Morlan, CMC database (1999)]; [Wowchuk (1990) In: A.S. Dyke (GSC database)]
580 ± 190 (660 ± 190)	S-1757	thoracic vertebra (CMC-1102)	S Saskatchewan R., 500 m asl (51.420°N, 107.090°W), SK, CAN	Morlan (1993); [Dyck (1983); Dyck and Morlan (1995) In: R.E. Morlan, CMC database (1999)]; [Morlan and McNeely (1996) In: A.S. Dyke (GSC database)]
580 ± 100 (660 ± 100)	S-3460	long-bone fragments	Tiger Hills, N of Ninette, Pembina drainage (49.500°N, 99.617°W), MB, CAN	[Nicholson (1994, 1996) In: R.E. Morlan, CMC database (1999)]

(580 ± 70)	AECV-1834C	humerus	Highway 2/3 Junction Impact Mitigation (49°42'N, 113°27'W), AB, CAN	[A. Beaudoin (PMA database)]
(580 ± 70)	AECV-1837C	lumbar vertebrae	Highway 2/3 Junction Impact Mitigation (49°42'N, 113°27'W), AB, CAN	[A. Beaudoin (PMA database)]
575 ± 70 (655 ± 75)	BGS-1436		Assiniboine R., 26 km W of junction with Red R. (49.840°N, 97.500°W), MB, CAN	[Rannie et al. (1989), Nielsen, pers. comm. (1998) In: R.E. Morlan, CMC database (1999)]
(570 ± 60)	Beta-19807		Forty Mile Coulee (49°45'N, 111°25'W), AB, CAN	[Brumley and Dau (1988) In: R.E. Morlan, CMC database (1999)]
(560 ± 70)	AECV-1831C	humerus	Highway 2/3 Junction Impact Mitigation (49°42'N, 113°27'W), AB, CAN	[A. Beaudoin (PMA database)]
(560 ± 60)	AECV-1833C	humerus	Highway 2/3 Junction Impact Mitigation (49°42'N, 113°27'W), AB, CAN	[A. Beaudoin (PMA database)]
(560 ± 80)	AECV-1632C	metatarsal and metacarpal	Highway 2/3 Junction (49°42'N, 113°27'W), AB, CAN	[A. Beaudoin (PMA database)]
(550 ± 80)	AECV-846C		Mitigation (51.970°N, 112.220°W), AB, CAN	[Beaudoin (1989) In: A.S. Dyke (GSC database)]; [Beaudoin (1991), PMA database - C13 value highly suspect]; [Light permit 88-76 in Luther and Purdon (1991); Beaudoin (1991) In: R.E. Morlan, CMC database (1999)]
(550 ± 80)	AECV-1630C	distal 3/4 right tibia	Highway 2/3 Junction (49°42'N, 113°27'W), AB, CAN	[A. Beaudoin (PMA database)]
(550 ± 70)	AECV-1828C	humerus	Highway 2/3 Junction Impact Mitigation (49°42'N, 113°27'W), AB, CAN	[A. Beaudoin (PMA database)]
540 ± 60 (620 ± 65)	Beta-65952		Souris drainage, NE of Lauder (49.414°N, 100.808°W), MB, CAN	[Nicholson (1996) In: R.E. Morlan, CMC database (1999)]
535 ± 100 (615 ± 100)	S-2275	scapula	N bank of Saskatchewan R., 325 m asl (53.170°N, 104.000°W), SK, CAN	[Quigg (1983); Meyer and Russell (1987) In: R.E. Morlan, CMC database (1999)]; Morlan (1993); [Morlan and McNeely (1996) In: A.S. Dyke (GSC database)]
(530 ± 80)	AECV-769C		Oldman River Dam area (49.620°N, 114.120°W), AB, CAN	[Beaudoin (1989) In: A.S. Dyke (GSC database)]; Van Dyke, et al. (1990); [Beaudoin (1991), PMA database]; [Van Dyke, et al. (1991) In: R.E. Morlan, CMC database (1999)]; Beaudoin (1991, 1998)]
(530 ± 90)	AECV-1274C	rib	Burn Coulee: Dinosaur Provincial Pk, Red Deer (50.770°N, 111.500°W), AB, CAN	[Seemann (1992), Beaudoin (1998) In: R.E. Morlan, CMC database (1999)]
(530 ± 70)	Beta-19804		Forty Mile Coulee, AB, CAN	[Brumley and Dau (1988) In: R.E. Morlan, CMC database (1999)]
(520 ± 90)	AECV-1130C		Oldman River Dam area (49.670°N, 113.150°W), AB, CAN	[Beaudoin In: A.S. Dyke (GSC database)]; [Beaudoin (1992), PMA database]; [Loveseth and Dau (1991); Beaudoin (1998) In: R.E. Morlan, CMC database (1999)]
(520 ± 90)	AECV-858C		Oldman River Dam area (49.600°N, 114.000°W), AB, CAN	[Beaudoin (1989) In: A.S. Dyke (GSC database)]; [Beaudoin (1991), PMA database]; [Landals permit 88-38; Beaudoin (1991) In: R.E. Morlan, CMC database (1999)]
(520 ± 80)	AECV-1829C	humerus	Highway 2/3 Junction Impact Mitigation (49°42'N, 113°27'W), AB, CAN	[A. Beaudoin (PMA database)]
(520 ± 70)	AECV-1840C	humerus	Highway 2/3 Junction Impact Mitigation (49°42'N, 113°27'W), AB, CAN	[A. Beaudoin (PMA database)]
(520 ± 100)	AECV-456C	ilium, metatarsal	Point Beazer, above W bank of St. Mary R. (49°07'N, 113°12'W), AB, CAN	Brumley and Rushworth (1983); [Beaudoin (1988), PMA database]

510 ± 70 (590 ± 75)	S-2809	pelvis (innominate)	Wanuskewin Heritage Pk., S Saskatchewan R. (52.170°N, 106.500°W), SK, CAN	Morlan (1988, 1993); [Walker, et al. (1986, 1987); Walker (1988) In: R.E. Morlan, CMC database (1999)]; [Morlan and McNeely (1996) In: A.S. Dyke (GSC database)]
(510 ± 90)	AECV-1240C	ribs	Castle Forks Buffalo Jump (Oldman River Dam area) (49.600°N, 114.000°W), AB, CAN	[Beaudoin In: A.S. Dyke (GSC database)]; [A. Beaudoin (PMA database)]; [Landals (1990); Beaudoin (1991, 1998) In: R.E. Morlan, CMC database (1999)]
(500 ± 70)	AECV-1638C	occipital and cervical vertebra	Highway 2/3 Junction (49°42'N, 113°27'W), AB, CAN	[A. Beaudoin (PMA database)]
(490 ± 80)	AECV-1525C	humerus and tibia fragments	Pincher Ck., Oldman drainage (49.480°N, 114.030°W), AB, CAN	[Beaudoin In: A.S. Dyke (GSC database)]; [A. Beaudoin (PMA database)]; [Dau (1991); Beaudoin (1998) In: R.E. Morlan, CMC database (1999)]
480 ± 100 (545 ± 100)	S-3033		Lovstrom, S of Brandon, 373.5 m asl (49.500°N, 99.942°W), MB, CAN	[Peach and Burton-Coe (1994) In: A.S. Dyke (GSC database)]; [Nicholson (1986, 1990, 1994, 1996); Nicholson and Malainey (1991); D.E. Nelson, pers. comm. (1998) In: R.E. Morlan, CMC database (1999)]
(480 ± 90)	AECV-1358C		Oldman River Dam area (49.600°N, 114.000°W), AB, CAN	[Beaudoin In: A.S. Dyke (GSC database)]; [A. Beaudoin (PMA database)]; [Van Dyke, et al. (1990, 1991) In: R.E. Morlan, CMC database (1999)]; Beaudoin (1991, 1998)
(470 ± 70)	Beta-19802		Forty Mile Coulee, AB, CAN	[Brumley and Dau (1988) In: R.E. Morlan, CMC database (1999)]
(470 ±150)	Beta-19796		Forty Mile Coulee, AB, CAN	[Brumley and Dau (1988) In: R.E. Morlan, CMC database (1999)]
(460 ± 80)	AECV-859C		Oldman River Dam area (49.600°N, 114.000°W), AB, CAN	[Beaudoin (1989) In: A.S. Dyke (GSC database)]; [Beaudoin (1991), PMA database]; [Landals permit 88-38; Beaudoin (1991) In: R.E. Morlan, CMC database (1999)]
(450 ± 90)	AECV-1272C	partial right mandible	Dunbar Coulee: Dinosaur Provincial Pk., Red Deer (50.780°N, 111.470°W), AB, CAN	[Seemann (1992), Beaudoin (1998) In: R.E. Morlan, CMC database (1999)]
<450 (<450)	RL-418	vertebrae	N bank of Crowsnest R. 2 km E of Burmis, AB, CAN	Brumley and Rushworth (1983); [Quigg and Reeves (1975); Driver (1978, 1985) In: R.E. Morlan, CMC database (1999)]
(440 ± 60)	AECV-1908C		Spring Kill site (51°04'N, 115°08'W), AB, CAN	[A. Beaudoin (PMA database)]
440 ± 50 (520 ± 55)	Beta-6710		Forty Mile Coulee (49°35'N, 111°25'W), AB, CAN	[Brumley and Dau (1988) In: R.E. Morlan, CMC database (1999)]
(440 ± 70)	AECV-1836C	radius	Highway 2/3 Junction Impact Mitigation (49°42'N, 113°27'W), AB, CAN	[A. Beaudoin (PMA database)]
(430 ± 70)	AECV-1838C	metacarpal	Highway 2/3 Junction Impact Mitigation (49°42'N, 113°27'W), AB, CAN	[A. Beaudoin (PMA database)]
(430 ± 90)	Beta-19799		Forty Mile Coulee (49°35'N, 111°25'W), AB, CAN	[Brumley and Dau (1988) In: R.E. Morlan, CMC database (1999)]
(420 ± 65)	Beta-28765 ETH-4761	skull (Oa Rw-2:1) small adult male	Old Horton R. channel, Cape Bathrust Pen. (70.080°N, 127.500°W), NT, CAN	C.R. Harington [unpublished]; [Morrison (1997); C.R. Harington, pers. comm. (1998) In: R.E. Morlan, CMC database (1999); Stephenson, et al. (in press)
(420 ± 70)	AECV-1835C	tibia	Highway 2/3 Junction Impact Mitigation (49°42'N, 113°27'W), AB, CAN	[A. Beaudoin (PMA database)]
(410 ± 60)	Beta-82792	metapodial	NE of Lauder, Souris drainage (49.414°N, 100.808°W), MB, CAN	[Nicholson (1996) In: A.S. Dyke (GSC database)]; [Nicholson, pers. comm. (1998) In: R.E. Morlan, CMC database (1999)]

(410 ± 90)	AECV-1186C		Oldman River Dam area (49.580°N, 114.050°W), AB, CAN	[Beaudoin In: A.S. Dyke (GSC database)]; [A. Beaudoin (PMA database)]; [Van Dyke, et al. (1990); Beaudoin (1998) In: R.E. Morlan, CMC database (1999)]
410 ± 100 (490 ± 100)	RL-1474		Red Deer valley, about 10 km S of Drumheller, AB, CAN	Brumley and Rushworth (1983); [Quigg (1981) In: R.E. Morlan, CMC database (1999)]
405 ± 110 (485 ± 110)	S-3032		Lovstrom, S of Brandon, 373.5 m asl (49.500°N, 99.942°W), MB, CAN	[Peach and Burton-Coe (1994) In: A.S. Dyke (GSC database)]; [Nicholson (1986, 1990, 1994, 1996); Nicholson and Malainey (1991); D.E. Nelson, pers. comm. (1998) In: R.E. Morlan, CMC database (1999)]
400 ± 255 (480 ± 255)	S-2801		Qu'Appelle valley (50.670°N, 103.670°W), SK, CAN	Morlan (1993); [Dyck (1983) Smith (1986); Smith and Walker (1988) In: R.E. Morlan, CMC database (1999)]
400 ± 100 (480 ± 100)	RL-1059		Ormman, Bow R., along Memorial Dr., in Hillhurst, Calgary, AB, CAN	[Wilson (1983) In: R.E. Morlan, CMC database (1999)]
(390 ± 70)	AECV-1927C	tibia	Pine Coulee Reservoir, Diversion Weir Kill site (50°08'N, 113°46'W), AB, CAN	[A. Beaudoin (PMA database)]
385 ± 155 (465 ± 155)	S-2274	femur	N bank of Saskatchewan R., 325 m asl (53.170°N, 104.000°W), SK, CAN	[Quigg (1983); Meyer and Russell (1987) In: R.E. Morlan, CMC database (1999)]; Morlan (1993); [Morlan and McNeely (1996) In: A.S. Dyke (GSC database)]
380 ± 70 (460 ± 75)	S-2806	radius	Wanuskewin Heritage Pk., S Saskatchewan R. (52.170°N, 106.500°W), SK, CAN	Morlan (1988, 1993); [Walker, et al. (1986, 1987); Walker (1988) In: R.E. Morlan, CMC database (1999)]; [Morlan and McNeely (1996) In: A.S. Dyke (GSC database)]
(380 ± 70)	AECV-1643C	distal right metatarsal and distal left radius	Highway 2/3 Junction (49°42'N, 113°27'W), AB, CAN	[A. Beaudoin (PMA database)]
(370 ± 100)	AECV-1241C		Castle Forks Buffalo Jump (Oldman River Dam area) (49.600°N, 114.000°W), AB, CAN	[Beaudoin In: A.S. Dyke (GSC database)]; [A. Beaudoin (PMA database)]; [Landals (1990); Beaudoin (1991, 1998) In: R.E. Morlan, CMC database (1999)]
370 ± 80 (450 ± 80)	S-1372	femur	Bow R. (51°04'N, 115°11'W), AB, CAN	[Quigg (1978); Newton (1991) In: R.E. Morlan, CMC database (1999)]; Brumley and Rushworth (1983)
(360 ± 90)	AECV-844C		Mitigation (51.970°N, 112.220°W), AB, CAN	[Beaudoin (1989) In: A.S. Dyke (GSC database)]; [Beaudoin (1991), PMA database]; [Light permit 88-76 in Luther and Purdon (1991), Beaudoin (1991) In: R.E. Morlan, CMC database (1999)]
350 ± 130 (430 ± 130)	S-689		Souris R., S of Melita, 435 m asl (49.200°N, 101.030°W), MB, CAN	[Long and Tamplin (1977); Wilmeth (1978) In: R.E. Morlan, CMC database (1999)]
350 ± 60 (430 ± 65)	S-3129		W of Souris, 427 m asl (49.500°N, 100.518°W), MB, CAN	[Peach and Burton-Coe (1994) In: R.E. Morlan, CMC database (1999)]
350 ± 60	AECV-288C	ulna	Miller site (52°15'N, 113°22'W), AB, CAN	[Beaudoin (1987), PMA database]
(350 ± 80)	Beta-19808		Forty Mile Coulee (49°45'N, 111°25'W), AB, CAN	[Brumley and Dau (1988) In: R.E. Morlan, CMC database (1999)]
(340 ± 80)	AECV-872C		Oldman River Dam area (49.600°N, 114.100°W), AB, CAN	[Beaudoin (1989) In: A.S. Dyke (GSC database)]; [Beaudoin (1991), PMA database]; [Landals permit 88-38; Beaudoin (1991) In: R.E. Morlan, CMC database (1999)]
(340 ± 80)	AECV-1827C	femur	Highway 2/3 Junction Impact Mitigation (49°42'N, 113°27'W), AB, CAN	[A. Beaudoin (PMA database)]
330 ± 140 (410 ± 140)	S-3499		Souris R. floodplain, E of Souris (49.606°N, 100.183°W), MB, CAN	[Nicholson (1996) In: A.S. Dyke (GSC database)]; [Nicholson (1994, pers. comm. (1998) In: R.E. Morlan, CMC database (1999)]

(330 ± 90)	AECV-864C		Oldman River Dam area (49.600°N, 114.100°W), AB, CAN	[Beaudoin (1989) In: A.S. Dyke (GSC database)]; [Beaudoin (1991), PMA database]; [Landals permit 88-38; Beaudoin (1991) In: R.E. Morlan, CMC database (1999)]
(330 ± 70)	AECV-1826C	cervical vertebrae	Highway 2/3 Junction Impact Mitigation (49°42'N, 113°27'W), AB, CAN	[A. Beaudoin (PMA database)]
325 ± 100 (405 ± 100)	BGS-1410		W end of Winnipeg (49.880°N, 97.230°W), MB, CAN	[Rannie, et al. (1989), Nielsen, et al. (1993) In: R.E. Morlan, CMC database (1999)]
(320 ± 90)	AECV-1273C	8th thoracic vertebra	Burn Coulee: Dinosaur Provincial Pk, Red Deer (50.770°N, 111.500°W), AB, CAN	[Seemann (1992), Beaudoin (1998) In: R.E. Morlan, CMC database (1999)]
320 ± 60 (400 ± 65)	Beta-6712		Forty Mile Coulee (49°35'N, 111°25'W), AB, CAN	[Brumley and Dau (1988) In: R.E. Morlan, CMC database (1999)]
310 ± 70 (390 ± 75)	S-1951	distal humerus (CMC-1171)	Swan R. 350-358 m asl., Swan River valley (52.180°N, 101.470°W), MB, CAN	[Ehrlich, et al. (1962); Elson (1967); Gryba (1968); Klassen (1969); Colwill (1973); Badertscher (1980) In: R.E. Morlan, CMC database (1999)]
(310 ± 120)	AECV-1194C		Oldman River Dam area (49.580°N, 114.050°W), AB, CAN	[Beaudoin In: A.S. Dyke (GSC database)]; [A. Beaudoin (PMA database)]; [Van Dyke, et al. (1990); Beaudoin (1998) In: R.E. Morlan, CMC database (1999)]
(310 ± 70)	AECV-1832C	femur	Highway 2/3 Junction Impact Mitigation (49°42'N, 113°27'W), AB, CAN	[A. Beaudoin (PMA database)]
(300 ± 70)	Beta-83864	cervical vertebra	NE of Lauder, Souris drainage (49.414°N, 100.808°W), MB, CAN	[Nicholson (1996) In: A.S. Dyke (GSC database)]; [Nicholson, pers. comm. (1998) In: R.E. Morlan, CMC database (1999)]
300 ± 50 (380 ± 55)	Beta-6717		Forty Mile Coulee (49°35'N, 111°25'W), AB, CAN	[Brumley and Dau (1988) In: R.E. Morlan, CMC database (1999)]
300 ± 90	AECV-340C	metatarsal, navicular cuboid, P2 and lumbar vertebrae	Oldman River Dam area (49°36'N, 114°03'W), AB, CAN	[Beaudoin (1988), PMA database]
(300 ± 80)	AECV-1916C	tibia	Pine Coulee Reservoir (50°13'N, 113°46'W), AB, CAN	[A. Beaudoin (PMA database)]
(290 ± 50)	Beta-83865	radius	NE of Lauder, Souris drainage (49.414°N, 100.808°W), MB, CAN	[Nicholson (1996) In: A.S. Dyke (GSC database)]; [Nicholson, pers. comm. (1998) In: R.E. Morlan, CMC database (1999)]
(280 ± 80)	AECV-1830C	2 cervical vertebrae	Highway 2/3 Junction Impact Mitigation (49°42'N, 113°27'W), AB, CAN	[A. Beaudoin (PMA database)]
(279 ± 70)	AECV-1898C	hair	Old Woman's Buffalo Jump, S of Calgary (50°28'N, 113°53'W), AB, CAN	[A. Beaudoin (PMA database)]
(270 ± 90)	AECV-1131C		Oldman River Dam area (49.670°N, 113.150°W), AB, CAN	[Beaudoin In: A.S. Dyke (GSC database)]; [Beaudoin (1992), PMA database]; [Loveseth and Dau (1991); Beaudoin (1998) In: R.E. Morlan, CMC database (1999)]
(260 ± 130)	AECV-459C	bone fragments	Calderwood Buffalo Jump (49.717°N, 113.650°W), AB, CAN	[Beaudoin (1988) In: A.S. Dyke (GSC database)]; [Beaudoin (1988), PMA database]
(260 ± 90)	AECV-1238C	thoracic vertebra	Castle Forks Buffalo Jump (Oldman River Dam area) (49.600°N, 114.000°W), AB, CAN	[Beaudoin In: A.S. Dyke (GSC database)]; [A. Beaudoin (PMA database)]; [Landals (1990); Beaudoin (1998) In: R.E. Morlan, CMC database (1999)]
(260 ± 80)	SFU-1144		Head-Smashed-In Buffalo Jump (49°42'40"N, 113°38'45"W), AB, CAN	Beaudoin (1991); R.E, Morlan, CMC database (1999)
250 ± 80 (330 ± 80)	Beta-6716		Forty Mile Coulee (49°35'N, 111°25'W), AB, CAN	[Brumley and Dau (1988) In: R.E. Morlan, CMC database (1999)]

230 ± 90 (310 ± 90)	S-2823		Lovstrom, S of Brandon, 373.5 m asl (49.500°N, 99.942°W), MB, CAN	[Nicholson (1986, 1990, 1994b, 1996); Nicholson and Malainey (1991) In: R.E. Morlan, CMC database (1999)]
(230 ± 90)	Beta-19801		Forty Mile Coulee (49°35'N, 111°25'W), AB, CAN	[Brumley and Dau (1988) In: R.E. Morlan, CMC database (1999)]
210 ± 60	AECV-293C	femur, tibia	Miller site (52°15'N, 113°22'W), AB, CAN	[Beaudoin (1987), PMA database]
(200 ± 120)	AECV-460C	bone fragments	Calderwood Buffalo Jump (49.717°N, 113.650°W), AB, CAN	[Beaudoin (1988) In: A.S. Dyke (GSC database)]; [Beaudoin (1988), PMA database]
200 ± 70 (350 ± 70)	S-3663	tibia	About 3 km NE of Morris, and 60 km S of Winnipeg, and 228 m asl, in the Red River valley, MB, CAN	[R. McNeely, pers. comm. In: R.E. Morlan, CMC database (1999)]
>190 (>190)	I-11636	horncore (ROM 1198)	Lake Ontario, Toronto (43°40'N, 79°23'W), ON, CAN	C.R. Harington [unpublished]; R.E. Morlan, CMC database (1999)
(190 ± 100)	AECV-1242C	partial pelvis, vertebra, patella, calcaneum	Oldman River Dam area (49.580°N, 114.000°W), AB, CAN	[Beaudoin In: A.S. Dyke (GSC database)]; [A. Beaudoin (PMA database)]; [Landals (1990); Beaudoin (1998) In: R.E. Morlan, CMC database (1999)]
(180 ± 90)	AECV-1270C	1st thoracic vertebra	Fryberger Coulee: Dinosaur Provincial Pk, Red Deer (50.770°N, 111.550°W), AB, CAN	[Seemann (1992), Beaudoin (1998) In: R.E. Morlan, CMC database (1999)]
180 ± 90 (300 ± 90)	AECV-1147C		Bow Corridor (51.070°N, 115.180°W), AB, CAN	[Beaudoin In: A.S. Dyke (GSC database)]; [A. Beaudoin (PMA database)]; Beaudoin (1998); R.E. Morlan, CMC database (1999)
(180 ± 70)	AECV-1523C		Oldman River Dam area (49.600°N, 114.120°W), AB, CAN	[Beaudoin In: A.S. Dyke (GSC database)]; [A. Beaudoin (PMA database)]; [Dau (1991); Beaudoin (1998) In: R.E. Morlan, CMC database (1999)]
(180 ± 70)	AECV-1899C	hair	Old Woman's Buffalo Jump, S of Calgary (50°28'N, 113°53'W), AB, CAN	[A. Beaudoin (PMA database)]
170 ± 105 (250 ± 105)	S-2323		N bank of Saskatchewan R., 325 m asl (53.170°N, 104.330°W), SK, CAN	[Finnigan, et al. (1983) In: R.E. Morlan, CMC database (1999)]; Morlan (1993)
(170 ± 70)	AECV-1917C	metatarsal	Pine Coulee Reservoir (50°13'N, 113°46'W), AB, CAN	[A. Beaudoin (PMA database)]
160 ± 60 (240 ± 65)	S-1014	3 vertebrae (CMC-787)	See-everywhere, Suffield Military Reserve, S Saskatchewan drainage (50.360°N, 110.900°W), AB, CAN	[Wilmeth (1978); Brumley and Rushworth (1983) In: R.E. Morlan, CMC database (1999)]; [A. Beaudoin (PMA database)]
140 ± 140 (220 ± 140)	S-3500	cervical vertebra	Tiger Hills, N of Ninette, Pembina drainage (49.500°N, 99.617°W), MB, CAN	[Nicholson (1994, 1996, pers. comm. (1998) In: R.E. Morlan, CMC database (1999)]
130 ± 85 (210 ± 85)	S-844	(CMC-627)	St. Mary R. (49°07'N, 113°12'W), AB, CAN	Brumley and Rushworth (1983); [Quigg (1975) In: R.E. Morlan, CMC database (1999)]
<100 (<100)	S-3030		Lovstrom, S of Brandon, 373.5 m asl (49.500°N, 99.942°W), MB, CAN	[Peach and Burton-Coe (1994) In: A.S. Dyke (GSC database)]; [Nicholson (1986, 1990, 1994, 1996); Nicholson and Malainey (1991); D.E. Nelson, pers. comm. (1998) In: R.E. Morlan, CMC database (1999)]
(100 ± 90)	AECV-1197C		Oldman River Dam area (49.580°N, 114.050°W), AB, CAN	[Beaudoin In: A.S. Dyke (GSC database)]; [A. Beaudoin (PMA database)]; [Van Dyke, et al. (1990); Beaudoin (1998) In: R.E. Morlan, CMC database (1999)]
<100 (<100)	S-2805	radius and femur	Wanuskewin Heritage Pk., S Saskatchewan R. (52.170°N, 106.500°W), SK, CAN	Morlan (1988, 1993); [Walker, et al. (1986, 1987); Walker (1988) In: R.E. Morlan, CMC database (1999)]; [Morlan and McNeely (1996) In: A.S. Dyke (GSC database)]

<100 (<100)	S-2367		Meewasin Ck., Wanuskewin Heritage Pk., S Saskatchewan R. (52°13'12"N, 106°35'00"W), SK, CAN	[Walker, pers. comm. (1999) In: R.E. Morlan, CMC database (1999)]
60 ± 90	AECV-342C	metacarpal, P1 and P2, thoracic vertebrae, rib	Oldman River Dam area (49°36'N, 114°04'W), AB, CAN	[Beaudoin (1988), PMA database]
0 ± 140 (0 ± 140)	GSC-704		Island Bluff: W side of gully tributary to the NE bank of the S Saskatchewan R, 685 m asl, about 11 km N of Medicine Hat, AB, CAN	Lowdon and Blake (1968); R.E. Morlan, CMC database (1999)
	Beta-19800		Forty Mile Coulee (49°35'N, 111°25'W), AB, CAN	[Brumley and Dau (1988) In: R.E. Morlan, CMC database (1999)]
	Beta-19798		Forty Mile Coulee (49°35'N, 111°25'W), AB, CAN	[Brumley and Dau (1988) In: R.E. Morlan, CMC database (1999)]
	Beta-6719		Forty Mile Coulee (49°35'N, 111°25'W), AB, CAN	[Brumley and Dau (1988) In: R.E. Morlan, CMC database (1999)]
	Beta-6713		Forty Mile Coulee (49°35'N, 111°25'W), AB, CAN	[Brumley and Dau (1988) In: R.E. Morlan, CMC database (1999)]
	Beta-6720		Forty Mile Coulee (49°35'N, 111°25'W), AB, CAN	[Brumley and Dau (1988) In: R.E. Morlan, CMC database (1999)]
	Beta-6412		Forty Mile Coulee (49°35'N, 111°25'W), AB, CAN	[Wilson (1983); Brumley and Dau (1988) In: R.E. Morlan, CMC database (1999)]
	Beta-6718		Forty Mile Coulee (49°35'N, 111°25'W), AB, CAN	[Brumley and Dau (1988) In: R.E. Morlan, CMC database (1999)]
	Beta-6711		Forty Mile Coulee (49°35'N, 111°25'W), AB, CAN	[Brumley and Dau (1988) In: R.E. Morlan, CMC database (1999)]
	AECV-359C		Sounding Ck. Basin near Red Deer R. (50°55'N, 110°08'W), AB, CAN	[Beaudoin (1998) In: R.E. Morlan, CMC database (1999)]; [Beaudoin (1988), PMA database]

Bison? sp. (bison)

Measured Date (Normalized)	Lab No	Material	Location	References/Remarks
(10230 ± 140)	RIDDL-561	metacarpal	Bluefish Cave II, 600 m asl. (67°08'N, 140°47'W), YT, CAN	Cinq-Mars (1979, 1990); Morlan and Cinq-Mars (1982); Cinq-Mars et al. (1991); Burke and Cinq-Mars (1996);
8250 ± 330 (8330 ± 330)	RL-1554		Bow valley, SW margin of the Nose Hill uplands along Bow R., AB, CAN	R.E. Morlan, CMC database (1999)
(5550 ± 70)	TO-7052	rib	Upper terrace along the shore at Red River, 234 m asl, 3 km NE of Ste. Agathe, MB, CAN	[R. McNeely and E. Nielsen, pers. comm. (1998) In: R.E. Morlan, CMC database (1999)]
(3000 ± 90)	AECV-418C		Pincher Ck. Buffalo Jump (49.567°N, 113.933°W), AB, CAN	[Beaudoin (1987) In: A.S. Dyke (GSC database)]; [Beaudoin (1988), PMA database]
(2910 ± 90)	AECV-373C	metatarsal, tarsal, metacarpal fragments	Calderwood Buffalo Jump (49.717°N, 113.650°W), AB, CAN	[Beaudoin (1988) In: A.S. Dyke (GSC database)]; [Beaudoin (1988), PMA database]
(2860 ± 170)	AECV-372C	bone fragments	Calderwood Buffalo Jump (49.717°N, 113.650°W), AB, CAN	[Beaudoin (1988) In: A.S. Dyke (GSC database)]; [Beaudoin (1988), PMA database]
(2810 ± 90)	AECV-422C		Pincher Ck. Buffalo Jump (49.567°N, 113.933°W), AB, CAN	[Beaudoin (1987) In: A.S. Dyke (GSC database)]; [Beaudoin (1988), PMA database]
(2310 ± 100)	AECV-370C	metapodial, ribs, tarsal, phalanx	Calderwood Buffalo Jump (49.717°N, 113.650°W), AB, CAN	[Beaudoin (1988) In: A.S. Dyke (GSC database)]; [Beaudoin (1988), PMA database]
(2170 ± 90)	AECV-367C	humerus and ulna	Calderwood Buffalo Jump (49.717°N, 113.650°W), AB, CAN	[Beaudoin (1988) In: A.S. Dyke (GSC database)]; [Beaudoin (1988), PMA database]
(1910 ± 90)	AECV-369C	bone fragments	Calderwood Buffalo Jump (49.717°N, 113.650°W), AB, CAN	[Beaudoin (1988) In: A.S. Dyke (GSC database)]; [Beaudoin (1988), PMA database]

1790 ± 80 (1870 ± 80)	AECV-234C	scapula, bone fragments	Head-Smashed-In Buffalo Jump (49.717°N, 113.650°W), AB, CAN	[Beaudoin (1987) In: A.S. Dyke (GSC database)]; [Beaudoin (1987), PMA database]
1660 ± 80 (1740 ± 80)	Beta-7718	humerus	Head-Smashed-In Buffalo Jump (49.717°N, 113.650°W), AB, CAN	[Beaudoin (1987) In: A.S. Dyke (GSC database); [Beaudoin (1987), PMA database]; [Brink, et al. (1985); Morlan (1985); Beaudoin (1987) In: R.E. Morlan, CMC database (1999)]
1620 ± 180	AECV-240C	bone fragments	Head-Smashed-In Buffalo Jump (49.717°N, 113.650°W), AB, CAN	[Beaudoin (1987) In: A.S. Dyke (GSC database)]; [Beaudoin (1987), PMA database]
1620 ± 80 (1700 ± 80)	Beta-7792		Head-Smashed-In Buffalo Jump (49.717°N, 113.650°W), AB, CAN	[Beaudoin (1987) In: A.S. Dyke (GSC database); [Beaudoin (1987), PMA database]; [Brink, et al. (1985); Beaudoin (1987) In: R.E. Morlan, CMC database (1999)]
(1530 ± 90)	AECV-368C	scapula, femur, radius, carpals, long-bone fragments	Calderwood Buffalo Jump (49.717°N, 113.650°W), AB, CAN	[Beaudoin (1988) In: A.S. Dyke (GSC database)]; [Beaudoin (1988), PMA database]
(1460 ± 120)	AECV-361C	bone fragments	Calderwood Buffalo Jump (49.717°N, 113.650°W), AB, CAN	[Beaudoin (1988) In: A.S. Dyke (GSC database)]; [Beaudoin (1988), PMA database]
(1420 ± 150)	AECV-417C		Pincher Ck. Buffalo Jump (49.567°N, 113.933°W), AB, CAN	[Beaudoin (1987) In: A.S. Dyke (GSC database)]; [Beaudoin (1988), PMA database]
(1380 ± 90)	AECV-362C	bone fragments	Calderwood Buffalo Jump (49.717°N, 113.650°W), AB, CAN	[Beaudoin (1988) In: A.S. Dyke (GSC database)]; [Beaudoin (1988), PMA database]
(1360 ± 140)	AECV-375C		Head-Smashed-In Buffalo Jump (49.717°N, 113.650°W), AB, CAN	[Beaudoin (1988) In: A.S. Dyke (GSC database)]; [Beaudoin (1988), PMA database]; [Reeves (1970, 1978); Wilmeth (1978); Brink, et al. (1985, 1986); Brink and Dawe, 1989) In: R.E. Morlan, CMC database (1999)]; Brumley and Rushworth (1983); Morlan (1985); Beaudoin (1987, 1988, 1991, 1998)
1320 ± 90	AECV-82C		Blood Coulee Buffalo Jump (49.570°N, 112.920°W), AB, CAN	[Beaudoin (1987) In: A.S. Dyke (GSC database)]; [Beaudoin (1987), PMA database]
1310 ± 110	AECV-320C	bone fragments	Calderwood Buffalo Jump (49.717°N, 113.650°W), AB, CAN	[Beaudoin (1987) In: A.S. Dyke (GSC database)]; [Beaudoin (1987), PMA database]
1300 ± 70	AECV-237C	bone fragments	Head-Smashed-In Buffalo Jump (49.717°N, 113.650°W), AB, CAN	[Beaudoin (1987) In: A.S. Dyke (GSC database)]; [Beaudoin (1987), PMA database]
(1290 ± 90)	AECV-363C	bone fragments	Calderwood Buffalo Jump (49.717°N, 113.650°W), AB, CAN	[Beaudoin (1988) In: A.S. Dyke (GSC database)]; [Beaudoin (1988), PMA database]
1280 ± 100	AECV-231C	axis and tibia	Head-Smashed-In Buffalo Jump (49.717°N, 113.650°W), AB, CAN	[Beaudoin (1987) In: A.S. Dyke (GSC database)]; [Beaudoin (1987), PMA database]
1260 ± 90	AECV-236C	bone fragments	Head-Smashed-In Buffalo Jump (49.717°N, 113.650°W), AB, CAN	[Beaudoin (1987) In: A.S. Dyke (GSC database)]; [Beaudoin (1987), PMA database]
1250 ± 180	AECV-232C	calcaneum and radius	Head-Smashed-In Buffalo Jump (49.717°N, 113.650°W), AB, CAN	[Beaudoin (1987) In: A.S. Dyke (GSC database)]; [Beaudoin (1987), PMA database]
(1200 ± 90)	AECV-423C		Pincher Ck. Buffalo Jump (49.567°N, 113.933°W), AB, CAN	[Beaudoin (1987) In: A.S. Dyke (GSC database)]; [Beaudoin (1988), PMA database]
(1170 ± 100)	AECV-425C		Pincher Ck. Buffalo Jump (49.567°N, 113.933°W), AB, CAN	[Beaudoin (1987) In: A.S. Dyke (GSC database)]; [Beaudoin (1988), PMA database]
1140 ± 90 (1220 ± 90)	AECV-63C		Head-Smashed-In Buffalo Jump (49.717°N, 113.650°W), AB, CAN	[Beaudoin (1987) In: A.S. Dyke (GSC database)]; [Beaudoin (1987), PMA database]; [Brink, et al. (1986); Beaudoin (1987) In: R.E. Morlan, CMC database (1999)]

1080 ± 90	AECV-241C	bone fragments	Head-Smashed-In Buffalo Jump (49.717°N, 113.650°W), AB, CAN	[Beaudoin (1987) In: A.S. Dyke (GSC database)]; [Beaudoin (1987), PMA database]
1030 ± 150	AECV-233C	scapula, humerus, cervical vertebrae	Head-Smashed-In Buffalo Jump (49.717°N, 113.650°W), AB, CAN	[Beaudoin (1987) In: A.S. Dyke (GSC database)]; [Beaudoin (1987), PMA database]
965 ± 70 (1045 ± 75)	S-786	(CMC-605)	W of Virden, along Pipestone Ck. (49.800°N, 101.330°W), MB, CAN	[Rushowick (1975); Wilmeth (1978) In: R.E. Morlan, CMC database (1999)]
(940 ± 110)	AECV-419C		Pincher Ck. Buffalo Jump (49.567°N, 113.933°W), AB, CAN	[Beaudoin (1987) In: A.S. Dyke (GSC database)]; [Beaudoin (1988), PMA database]
890 ± 110	AECV-321C	bone fragments and teeth	Calderwood Buffalo Jump (49.717°N, 113.650°W), AB, CAN	[Beaudoin (1987) In: A.S. Dyke (GSC database)]; [Beaudoin (1987), PMA database]
870 ± 90	AECV-238C	bone fragments	Head-Smashed-In Buffalo Jump (49.717°N, 113.650°W), AB, CAN	[Beaudoin (1987) In: A.S. Dyke (GSC database)]; [Beaudoin (1987), PMA database]
860 ± 380 (940 ± 380)	AECV-67C		Head-Smashed-In Buffalo Jump (49.717°N, 113.650°W), AB, CAN	[Beaudoin (1987) In: A.S. Dyke (GSC database)]; [Brink, et al. (1986); Beaudoin (1987) In: R.E. Morlan, CMC database (1999)]
830 ± 80	AECV-235C	vertebral column	Head-Smashed-In Buffalo Jump (49.717°N, 113.650°W), AB, CAN	[Beaudoin (1987) In: A.S. Dyke (GSC database)]; [Beaudoin (1987), PMA database]
(800 ± 80)	AECV-421C		Pincher Ck. Buffalo Jump (49.567°N, 113.933°W), AB, CAN	[Beaudoin (1987) In: A.S. Dyke (GSC database)]
755 ± 60 (835 ± 65)	S-785	(CMC-604)	W of Virden, along Pipestone Ck. (49.800°N, 101.330°W), MB, CAN	[Rushowick (1975); Wilmeth (1978) In: R.E. Morlan, CMC database (1999)]
(690 ± 90)	AECV-420C		Pincher Ck. Buffalo Jump (49.567°N, 113.933°W), AB, CAN	[Beaudoin (1987) In: A.S. Dyke (GSC database)]; [Beaudoin (1988), PMA database]
(560 ± 90)	AECV-424C		Pincher Ck. Buffalo Jump (49.567°N, 113.933°W), AB, CAN	[Beaudoin (1987) In: A.S. Dyke (GSC database)]; [Beaudoin (1988), PMA database]
470 ± 150 (550 ± 150)	AECV-62C		Head-Smashed-In Buffalo Jump (49.717°N, 113.650°W), AB, CAN	[Beaudoin (1987) In: A.S. Dyke (GSC database)]; [Beaudoin (1987), PMA database]; [Brink, et al. (1986); Beaudoin (1987) In: R.E. Morlan, CMC database (1999)]
470 ± 90 (550 ± 90)	AECV-65C		Head-Smashed-In Buffalo Jump (49.717°N, 113.650°W), AB, CAN	[Beaudoin (1987) In: A.S. Dyke (GSC database)]; [Brink, et al. (1986); Beaudoin (1987) In: R.E. Morlan, CMC database (1999)]
(400 ± 80)	AECV-630C		Marchand Kill site (51.233°N, 114.650°W), AB, CAN	[Beaudoin (1989) In: A.S. Dyke (GSC database)]; [A. Beaudoin (1991), PMA database]
390 ± 100	AECV-319C	bone fragments	Calderwood Buffalo Jump (49.717°N, 113.650°W), AB, CAN	[Beaudoin (1987) In: A.S. Dyke (GSC database)]; [Beaudoin (1987), PMA database]
360 ± 180	AECV-238C	bone fragments	Head-Smashed-In Buffalo Jump (49.717°N, 113.650°W), AB, CAN	[Beaudoin (1987) In: A.S. Dyke (GSC database)]; [Beaudoin (1987), PMA database]
310 ± 80 (390 ± 80)	Beta-7793		Head-Smashed-In Buffalo Jump (49.717°N, 113.650°W), AB, CAN	[Beaudoin (1987) In: A.S. Dyke (GSC database); [Beaudoin (1987), PMA database]; [Brink, et al. (1985); Beaudoin (1987) In: R.E. Morlan, CMC database (1999)]
190 ± 80 (270 ± 80)	AECV-64C		Head-Smashed-In Buffalo Jump (49.717°N, 113.650°W), AB, CAN	[Beaudoin (1987) In: A.S. Dyke (GSC database)]; [Beaudoin (1987), PMA database]; [Brink, et al. (1986); Beaudoin (1987) In: R.E. Morlan, CMC database (1999)]

Bison sp.? (bison)

Measured Date (Normalized)	Lab No	Material	Location	References/Remarks
6340 ± 160 (6420 ± 160)	RL-508		Crowsnest Pass (49.630°N, 114.600°W), AB, CAN	Brumley and Rushworth (1983); [Driver (1985, 1987); Landals (1986) In: R.E. Morlan, CMC database (1999)]

4160 ± 150 (4405 ± 150)	GX-6384-A		Crowsnest Pass (49.630°N, 114.600°W), AB, CAN	Brumley and Rushworth (1983); [Driver (1985, 1987); Landals (1986) In: R.E. Morlan, CMC database (1999)]
4140 ± 150 (4220 ± 150)	RL-877		Crowsnest Pass (49.630°N, 114.600°W), AB, CAN	Brumley and Rushworth (1983); [Driver (1985, 1987); Landals (1986) In: R.E. Morlan, CMC database (1999); Driver (1994)
2800 ± 140 (2880 ± 140)	RL-872		Crowsnest Pass (49.630°N, 114.600°W), AB, CAN	Brumley and Rushworth (1983); [Driver (1985, 1987); Landals (1986) In: R.E. Morlan, CMC database (1999)]
2680 ± 80 (2760 ± 80)	GX-1196		Waterton Lakes National Park, AB, CAN	Brumley and Rushworth (1983); [Reeves (1972) In: R.E. Morlan, CMC database (1999)]
2530 ± 120 (2610 ± 120)	RL-864		N bank of Crowsnest R., Crowsnest Pass (49.630°N, 114.600°W), AB, CAN	[Driver (1993) In: A.S. Dyke (GSC database)]; [Reeves (1974); Driver (1978, 1985, 1993) In: R.E. Morlan, CMC database (1999)]
2410 ± 130 (2490 ± 130)	RL-868		Jim Kerr Ranch, 2.5 km E of Crowsnest L., Crowsnest Pass, AB, CAN	Brumley and Rushworth (1983); [Reeves (1974); Driver (1978, 1985) In: R.E. Morlan, CMC database (1999)]
1900 ± 90 (1980 ± 90)	GX-2050		Waterton Lakes National Park, AB, CAN	Brumley and Rushworth (1983); [Reeves (1972) In: R.E. Morlan, CMC database (1999)]
1470 ± 120 (1550 ± 120)	GX-2049		Waterton Lakes National Park, AB, CAN	Brumley and Rushworth (1983); [Reeves (1972) In: R.E. Morlan, CMC database (1999)]
1345 ± 160 (1425 ± 160)	GX-1272		Waterton Lakes National Park, AB, CAN	Brumley and Rushworth (1983); [Reeves (1972) In: R.E. Morlan, CMC database (1999)]
1270 ± 150 (1350 ± 150)	GSC-696	charred bone (CMC-99)	Muhlbach Site near Stettler (52.250°N, 112.830°W), AB, CAN	Brumley and Rushworth (1983); [Gruhn (1969); Wilmeth (1978) In: R.E. Morlan, CMC database (1999)]
1020 ± 80 (1100 ± 85)	S-89	decomposed matted hair, hide, and buffalo bones	Old Woman's Buffalo Jump, S of Calgary (50°28'N, 113°53'W), AB, CAN	Brumley and Rushworth (1983); [Forbis (1962); Wilmeth (1978) In: R.E. Morlan, CMC database (1999)]
1010 ± 140 (1040 ± 140)	GSC-983	charred bone (CMC-169)	Head-Smashed-In Buffalo Jump (Ft. MacLeod) (49.720°N, 113.670°W), AB, CAN	[Reeves (1970, 1978); Wilmeth (1978); Brink, et al. (1985, 1986); Brink and Dawe (1989) In: R.E. Morlan, CMC database (1999)]; Brumley and Rushworth (1983); Morlan (1985); Beaudoin (1987, 1988, 1991, 1998)
680 ± 170 (700 ± 170)	GSC-992	charred bone (CMC-167)	Head-Smashed-In Buffalo Jump (Ft. MacLeod) (49.720°N, 113.670°W), AB, CAN	[Reeves (1970, 1978); Wilmeth (1978); Brink, et al. (1985, 1986); Brink and Dawe (1989) In: R.E. Morlan, CMC database (1999)]; Brumley and Rushworth (1983); Morlan (1985); Beaudoin (1987, 1988, 1991, 1998)
400 ± 110 (480 ± 110)	GX-2048		Waterton Lakes National Park, AB, CAN	Brumley and Rushworth (1983); [Reeves (1972) In: R.E. Morlan, CMC database (1999)]

Bison? (bison)

Measured Date (Normalized)	Lab No	Material	Location	References/Remarks
1675 ± 115 (1755 ± 115)	S-2348	(CMC-1299)	Great Sand Hills, (50.500°N, 109.170°W), SK, CAN	Morlan (1993); [Johnson (1983) In: R.E. Morlan, CMC database (1999)]

Saiga tatarica (saiga antelope)

Measured Date (Normalized)	Lab No	Material	Location	References/Remarks
(37000 ± 990)	GSC-3050	right horncore (USGS M1422)	Near northern limit of Lookout Ridge, Usuktuk (also spelled Esuktuk) R. (69.975°N, 156.550°W), AK, USA	Harington (1980); Blake (1984); Guthrie, et al. (in press)
(36650 ± 850)	AA-3216	skull (UA/FAM:30408)	Cripple Ck. (64°49'N, 148°01'W), AK, USA	Guthrie, et al. (in press)
(31390 ± 780)	AA-3076	skull (USNM-374407)	Lost Chicken Ck. (64.050°N, 141.875°W), AK, USA	Guthrie, et al. (in press)

(28930 ± 560)	AA-3073	skull (private collection)	Braux Ck., AK, USA	Guthrie, et al. (in press)
(25750 ± 450)	AA-3214	skull (UA/FAM:30407)	Gold Hill (64°55'N, 147°36'W), AK, USA	Guthrie, et al. (in press)
(14920 ± 160)	Beta-25119 ETH-3898	interior of horncore (CMN 12090)	S.E. Baillie I. (70.580°N, 128.250°W), NT, CAN	Harington (1978); Harington and Cinq-Mars (1995); Dallimore, et al. (1997); Guthrie, et al. (in press)
(13390 ± 180)	RIDDL-279	cranial fragment	Bluefish Cave III, 600 m asl. (67°08'N, 140°47'W), YT, CAN	Cinq-Mars (1979, 1990); Morlan and Cinq-Mars (1982); Harington and Cinq-Mars (1995); Burke and Cinq-Mars (1996); Guthrie, et al. (in press)
13220 ± 70 (13250 ± 70)	Beta-79853 CAMS-18416	right radius (CMN 43360)	Old Crow (Loc. 149) (68°2.00'N, 139°35'W), YT, CAN	C.R. Harington [unpublished]; R.E. Morlan, CMC database (1999); Guthrie, et al. (in press)
(12780 ± 140)	AA-3215	skull (FAM:30495)	Fairbanks (65°04'N, 147°10'W), AK, USA	[Guthrie, et al. (in press) - type of Saiga ricei]
(12620 ± 110)	AA-3074	skull (UA-V-54-970)	Lost Chicken Ck. (64.050°N, 141.875°W), AK, USA	Guthrie, et al. (in press)
(12390 ± 120)	AA-3075	skull (UA-V-54-196)	Lost Chicken Ck. (64.050°N, 141.875°W), AK, USA	Guthrie, et al. (in press)
(12220 ± 130)	AA-3077	skull (private collection)	Specific locality unknown, AK, USA	Guthrie, et al. (in press)

cf. Saiga tatarica (saiga antelope)

Measured Date (Normalized)	Lab No	Material	Location	References/Remarks
(32000 ± 2200)	AA-22883	horncore tip	On Your Knees Cave, Prince of Wales I., 135 m asl. (56.250°N, 133.500°W), AK, USA	Heaton and Grady (1996, 1997); Heaton and Grady (1998)

Bootherium bombifrons (helmeted muskox)

Measured Date (Normalized)	Lab No	Material	Location	References/Remarks
>43100 (>43200)	Beta-68924 CAMS-10528	mandible (CMN 42517)	Sixtymile area (Sixtymile Loc. 3), 701 m asl., (64°1.00'N, 140°44'W), YT, CAN	Harington (1997); R.E. Morlan, CMC database (1999)
>40000	SI-291	winter fecal pellets (A-204-4254)	Little Eldorado Ck., Fairbanks (65.100°N, 147.680°W), AK, USA	[Mielke and Long (1969) In: A.S. Dyke (GSC database)]; Péwé (1975); Harington (1980); McDonald and Ray (1989); [Guthrie (1985, 1990) specimen incorrectly cited within Alaskan Pleistocene Bison Table]
>40000	PITT-0218	left hornsheath (keratin) of a female	Lilian Ck. (65°30'N, 148°30'W), AK, USA	Guthrie (1991)
25090 ± 1070	SI-850	hornsheaths (Am-Frick Coll. 33124)	Upper Cleary Ck. (65.080°N, 147.330°W), AK, USA	Péwé (1975); Bonnichsen (1979); Harington (1980); McDonald and Ray (1989); [Guthrie (1985,1990) specimen incorrectly cited as Alaskan Pleistocene Bison within Table]
24140 ± 2200	SI-455	muscle from scalp (FAM-293-5268)	Fairbanks Ck. (65.067°N, 147.167°W), AK, USA	[Péwé (1975) incorrectly cited as Ovibos]; McDonald (1981); McDonald and Ray (1989); [Guthrie (1985, 1990) INCORRECTLY cited as SI-445]; [same animal as SI-454]
22540 ± 900	SI-292	hornsheath (Am-Frick Coll. 30508)	Fairbanks Ck. (65.000°N, 148.000°W), AK, USA	Péwé (1975); McDonald and Ray (1989); [Guthrie (1985, 1990) specimen incorrectly cited as Alaskan Pleistocene Bison within Table]
20500 ± 390	I-10649	horncore (CMN 25892)	Lost Chicken Ck. (64.050°N, 141.875°W), AK, USA	Harington (1978, 1980); [Agenbroad (1984) incorrectly cites specimen as Mammoth within Table 3.4]; McDonald and Ray (1989)
17695 ± 445	SI-851	hornsheath (Am-Frick Coll. A-651-3006	Dome Ck. (64.830°N, 147.500°W), AK, USA	Bonnichsen (1979); Harington (1980); McDonald and Ray (1989); [Guthrie (1985, 1990) specimen incorrectly cited as Alaskan Pleistocene Bison within Table]

17210 ± 500	SI-454	hair from hind limb (FAM-293-5268)	Fairbanks Ck. (65.067°N, 147.167°W), AK, USA	[Péwé (1975) incorrectly cited as Ovibos]; McDonald (1981); McDonald and Ray (1989); [Guthrie (1985, 1990) specimen incorrectly cited as Alaskan Pleistocene Bison within Table]
10980 ± 90 (10980 ± 80)	TO-7691		St. Mary's Reservoir (49°07'N, 113°12'W), AB, CAN	Hall (1999b); McNeil, et al. (1999)
6410 ± 120	GSC-2130	organic detritus from skull	Taylor Hwy. W of Alaska-Yukon border, AK, USA	Lowdon and Blake (1979)

cf. Euceratherium sp. (shrubox)

Measured Date (Normalized)	Lab No	Material	Location	References/Remarks
(13410 ± 100)	TO-1803	metacarpal (ROM 41960)	Bondi, 5 km NW of Leamington (42.080°N, 82.630°W), ON, CAN	Morris, et al. (1993); R.E. Morlan, CMC database (1999)

Ovibos moschatus (muskox, tundra muskox)

Measured Date (Normalized)	Lab No	Material	Location	References/Remarks
43100 ± 490 (43180 ± 490)	Beta-115201	cranial fragment (CMN 6765)	N Saskatchewan R., Edmonton (approx. 53°36'N, 113°20'W), AB, CAN	Dawson (1901a); C.R. Harington [unpublished]
40220 ± 670 (40310 ± 670)	Beta-173287	metacarpal (CR-02-2)	artificial island, McKinley Bay (69°54'N, 131°10'W), NT, CAN	C.R. Harington [unpublished]
>34000	S-288	metapodial	Bernard R. and Thesiger Bay, Banks I. (73.380°N, 121.900°W), NT, CAN	Maher (1968); Harington (1978, 1980); Vincent (1989); R.E. Morlan, CMC database (1999)
23720 ± 80 (23800 ± 80)	Beta-116674	cranial fragment (frontal) (CMN 9816)	Gee Pits at Ponoka (52°39'12"N, 113°38'47"W), AB, CAN	C.R. Harington [unpublished]; [C.R. Harington, pers. comm. (1999) In: R.E. Morlan, CMC database (1999)]
21160 ± 280 (21240 ± 280)	Beta-13869	cranium (CMN 42047)	Miller Ck., Sixtymile area (Sixtymile Loc. 3) part way up Miller Ck., 701 m asl., (63°59'N, 140°47'W), YT, CAN	Harington (1989, 1997); R.E. Morlan, CMC database (1999)
(14370 ± 130)	RIDDL-557	metatarsal	Bluefish Cave III, 600 m asl. (67°08'N, 140°47'W), YT, CAN	Cinq-Mars (1979, 1990); Morlan and Cinq-Mars (1982); Harington and Cinq-Mars (1995); Burke and Cinq-Mars (1996)
(13300 ± 60)	Beta-71232 CAMS-12070	right metacarpal (CR-94-221)	Simpson Pt., Pauline Cove, Herschel I. (69°34'N, 138°55'W), YT, CAN	C.R. Harington [unpublished]
(11580 ± 70)	TO-3652	skull (ROM 1194)	Toronto (43°41'N, 79°16'W), ON, CAN	Harington (1970); McDonald and Ray (1989); Morris, et al. (1993); formerly designated as Ovibos proximus (ROM 1194); R.E. Morlan, CMC database (1999)
8620 ± 190 (8700 ± 190)	I-1623	skull	4.8 km E of Grandview, MB, CAN	Harington (1970, 1978); Nielsen, et al. (1984); R.E. Morlan, CMC database (1999) [In private collection].
6725 ± 130 (6805 ± 130)	I-10919	horncore (CMN 34511)	About 16 km W of Goodsir Inlet, Bathurst I. (75.500°N, 99.000°W), NU, CAN	Harington (1980); R.E. Morlan, CMC database (1999)
(6240 ± 60)	CAMS-34960	femur	N mouth of Stuart R. valley, 24 m asl., Bathurst I., NU, CAN	R.E. Morlan, CMC database (1999)
3870 ± 85	K-3366	horncore	Solbakken, HLV (81.570°N, 61.550°W), GRE	[Knuth (1981) In: A.S. Dyke (GSC database)]
(3850 ± 80)	AAR-1023		Sdr. Mellemland, Dove Bay, GRE	[Andreasen (1996) In: Danish/Greenlandic Network (1999)]; [4,290 (2290BC), 2455-2145BC]
(3840 ± 70)	TO-1556		Knud Pen., Ellesmere I. (79.100°N, 76.150°W), NU, CAN	[Schledermann (1990) In: R.E. Morlan, CMC database (1999)]
3830 ± 85	K-3364		Midternaes, JBF (82.150°N, 30.150°W), GRE	[Knuth (1981) In: A.S. Dyke (GSC database)]
3800 ± 85	K-3362		Kap Peter Henrik, IPS (81.980°N, 26.120°W), GRE	[Knuth (1981) In: A.S. Dyke (GSC database)]
3670 ± 80	K-3531		Adam C. Knuth Site, Hvalterrasserne, Frigg Fjord (83.120°N, 33.800°W), GRE	[Knuth (1981) In: A.S. Dyke (GSC database)]; Knuth (1983)

3630 ± 80	K-3532		Adam C. Knuth Site, Hvalterrasserne, Frigg Fjord (83.120°N, 33.800°W), GRE	[Knuth (1981) In: A.S. Dyke (GSC database)]; Knuth (1983)
(3610 ± 60)	TO-1555		Knud Pen., Ellesmere I. (79.100°N, 76.150°W), NU, CAN	[Schledermann (1990) In: R.E. Morlan, CMC database (1999)]
3590 ± 60	K-3365		Martins Site, NMS (82.220°N, 33.370°W), GRE	[Knuth (1981) In: A.S. Dyke (GSC database)]
(3330 ± 50)	TO-5679		Firth R., Engigstciak (69.360°N, 139.530°W), YT, CAN	Fox (1998); R.E. Morlan, CMC database (1999)
3280 ± 90 (3360 ± 90)	I-10985	cranium (CMN 36137)	Miller Ck., Sixtymile area (Sixtymile Loc. 7) (64.020°N, 140.870°W), YT, CAN	Harington (1980); R.E. Morlan, CMC database (1999)
(3270 ± 50)	TO-5678		Firth R., Engigstciak (69.360°N, 139.530°W), YT, CAN	Fox (1998); R.E. Morlan, CMC database (1999)
2830 ± 100 (2910 ± 100)	I-3568	right horncore	Stewart R. (Brewer Ck.) (63.180°N, 139.000°W), YT, CAN	Harington (1977, 1980); R.E. Morlan, CMC database (1999)
(2550 ± 90)	AAR-1779		Magefjeldet, GRE	[Andreasen (1996) In: Danish/Greenlandic Network (1999)]; [2,780 (780BC), 800-530BC]
2520 ± 100 (2600 ± 100)	I-3577	left horncore	Banks I. (72°50'N, 120°00'W), NT, CAN	C.R. Harington [unpublished]; R.E. Morlan, CMC database (1999)
(2520 ± 50)	AAR-3941		Egil Knuth Site, GRE	[Andreasen (1996) In: Danish/Greenlandic Network (1999)]; [2,770 (770BC), 790-530BC]
(2510 ± 90)	AAR-1770		Egil Knuth Site, GRE	[Andreasen (1996) In: Danish/Greenlandic Network (1999)]; [2,760 (760-600BC), 800-410BC]
(2455 ± 45)	AAR-3939		Egil Knuth Site, GRE	[Andreasen (1996) In: Danish/Greenlandic Network (1999)]; [2,520 (520BC), 760-410BC]
2430 ± 75	K-3363		Kap Mylius-Erichsen (82.110°N, 30.000°W), GRE	[Knuth (1981) In: A.S. Dyke (GSC database)]
2390 ± 110 (2470 ± 110)	RL-767	scapula	3 km SE of Masik R., Banks I., NT, CAN	[Arnold (1980, 1981); LeBlanc (1994) In: R.E. Morlan, CMC database (1999)]
2370 ± 75	K-3361		Kap Ludovika, I. IFS (81.980°N, 23.180°W), GRE	[Knuth (1981) In: A.S. Dyke (GSC database)]
2370 ± 75	K-3360		Juniskaeret, DMFV (82.100°N, 22.000°W), GRE	[Knuth (1981) In: A.S. Dyke (GSC database)]
2320 ± 120 (2300 ± 120)	RL-765	scapula	3 km SE of Masik R., Banks I., NT, CAN	[Arnold (1980, 1981); LeBlanc (1994) In: R.E. Morlan, CMC database (1999)]
2290 ± 120 (2370 ± 120)	RL-766	scapula	3 km SE of Masik R., Banks I., NT, CAN	[Arnold (1980, 1981); LeBlanc (1994) In: R.E. Morlan, CMC database (1999)]
(2285 ± 60)	AAR-1015		Magefjeldet, GRE	[Andreasen (1996) In: Danish/Greenlandic Network (1999)]; [2,370 (370BC), 390-205BC]
(2275 ± 75)	AAR-1024		Magefjeldet, GRE	[Andreasen (1996) In: Danish/Greenlandic Network (1999)]; [2,370 (370BC), 400-200BC]
2240 ± 120 (2330 ± 120)	SFU-81		Markham Pt, Bathurst I. (75.500°N, 97.500°W), NU, CAN	[Helmer (1980, 1981) In: Dyke, et al. (1999)]; R.E. Morlan, CMC database (1999)
2210 ± 120 (2290 ± 120)	S-1673	(CMC-1022)	Karluk I. (75.500°N, 97.270°W), NU, CAN	[Helmer (1980) In: Dyke, et al. (1999)]; R.E. Morlan, CMC database (1999)
(1280 ± 50)	TO-5474		Franklin Pierce Bay, Ellesmere I., NU, CAN	R.E. Morlan, CMC database (1999)
(1250 ± 40 (780AD), 710-	K-6871		Qeqrtaaraq, Hatherton Bay, Thule District, Inglefield Land, GRE	Danish/Greenlandic Network (1999)
1130 ± 110 (1210 ± 110)	RL-1665		Nelson R., Banks I. (71.080°N, 123.000°W), NT, CAN	[Arnold and Stimmell (1983); Morrison (1989); Arnold (1994) In: R.E. Morlan, CMC database (1999)]
(1060 ± 80)	TO-4533	horncore	Little Cornwallis I. (75.583°N, 95.833°W), NU, CAN	[Helmer, et al. (1995) In: Dyke, et al. (1999)]
985 ± 45 (3015 (1015BC), 1015-	K-6703		Qeqrtaaraq, Hatherton Bay, Thule District, Inglefield Land (78.333°N, 72.75°W), GRE	Appelt, et al. (1998); Heinemeier and Rasmussen (1998); [Appelt and Gullov (1999) In: Danish/Greenlandic Network (1999)

950 ± 45 (3040 (1040-1150BC),	K-6702		Qeqrtaaraq, Hatherton Bay, Thule District, Inglefield Land (78.333°N, 72.75°W), GRE	[Appelt and Gullov (1999) In: Danish/Greenlandic Network (1999)]
(930 ± 60 (1050-1155AD), 1025-	Ka-6970	tooth	Qallunatalik Site, Hatherton Bay, GRE	Danish/Greenlandic Network (1999)
925 ± 38 (3060 (1060-1155BC),	K-6706	jawbone	Smith Sound area (78.333°N, 72.75°W), GRE	Heinemeier and Rasmussen (1998); Danish/Greenlandic Network (1999)
(920 ± 50)	TO-4534	mandible	Little Cornwallis I. (75.583°N, 95.833°W), NU, CAN	[Helmer, et al. (1995) In: Dyke, et al. (1999)]; R.E. Morlan, CMC database (1999)
920 ± 45 (3065 (1065-1165BC),	K-6704		Qeqrtaaraq, Hatherton Bay, Thule District, Inglefield Land (78.333°N, 72.75°W), GRE	[Appelt and Gullov (1999) In: Danish/Greenlandic Network (1999)]
(905 ± 40 (1165AD), 1045-	K-6874		Qeqrtaaraq, Hatherton Bay, Thule District, Inglefield Land, GRE	Danish/Greenlandic Network (1999)
(870 ± 50)	TO-5680		Firth R., Engigstciak (69.360°N, 139.530°W), YT, CAN	Fox (1998); R.E. Morlan, CMC database (1999)
(850 ± 40)	TO-3716	skull (OaRw-2)	Old Horton R. channel, Cape Bathurst Pen. (70.080°N, 127.500°W), NT, CAN	C.R. Harington [unpublished]; [Morrison (1997); C.R. Harington, pers. comm. (1998) In: R.E. Morlan, CMC database (1999)]
(770 ± 100 (1180AD))	K-2155	skull (GGU-79303)	Danmarks Havn (76°46'N, 18°45'W), GRE	Weidick (1974)
740 ± 70 (820 ± 75)	I-12340		Gilman Bay, Ellesmere I. (81.920°N, 70.000°W), NU, CAN	[Morrison (1989); Sutherland (1989) In: R.E. Morlan, CMC database (1999)]
(700 ± 70)	TO-2703	left humerus (CMN 49941)	Thompson Glacier, Axel Heiberg I. (79°30'N, 90°35'W), NU, CAN	C.R. Harington [unpublished]
670 ± 70 (720 ± 70)	Beta-10702		Somerset I., NU, CAN	R.E. Morlan, CMC database (1999)
610 ± 70 (650 ± 70)	Beta-11683		Somerset I., NU, CAN	R.E. Morlan, CMC database (1999)
560 ± 75 (640 ± 85)	I-11756		Slidre Fiord, 8 m asl., Ellesmere I. (80.020°N, 86.000°W), NU, CAN	[Morrison (1989); Sutherland (1993) In: R.E. Morlan, CMC database (1999)]
540 ± 70	K-2017		Mellembygden, NMS, WD (82.240°N, 33.750°W), GRE	[Knuth (1981) In: A.S. Dyke (GSC database)]
540 ± 70	K-2834		Uranienborg, AS, WD (82.240°N, 37.380°W), GRE	[Knuth (1981) In: A.S. Dyke (GSC database)]
440 ± 100	K-2016		Blanknaes, BG, OMS, WD (82.260°N, 35.850°W), GRE	[Knuth (1981) In: A.S. Dyke (GSC database)]
(400 ± 100 (1550AD))	K-2156	skull (GGU-79302)	Hurry Inlet (70°44'N, 22°37'W), GRE	Weidick (1974)
(380 ± 110)	BM-1766R	horncore	CMN station on E side of Goodsir R., Bathurst I., NU, CAN	R.E. Morlan, CMC database (1999)
(310 ± 110)	BM-1767R	horncore	CMN station on E side of Goodsir R., Bathurst I., NU, CAN	R.E. Morlan, CMC database (1999)
300 ± 60 (380 ± 65)	S-2198	distal humerus, proximal humerus, vertebra (CMC-1246)	King William I., 10 m asl (68.670°N, 98.000°W), NU, CAN	[McCartney and Savelle (1985) In: A.S. Dyke (GSC database)]; [Savelle (1987); Morrison (1989) In: R.E. Morlan, CMC database (1999)]
280 ± 95 (360 ± 95)	S-2213	(CMC-1252)	Gilman Bay, Ellesmere I. (81.920°N, 70.000°W), NU, CAN	[Morrison (1989); Sutherland (1989) In: R.E. Morlan, CMC database (1999)]
(280 ± 115)	GX-7123-G		Knud Pen., 7-8 m asl., Ellesmere I. (79.120°N, 76.190°W), NU, CAN	[Schledermann (1980, pers. comm. (1998); McCullough (1989); Morrison (1989); Bowman, et al. (1990) In: R.E. Morlan, CMC database (1999)]

Ovibos sp. (muskox)

Measured Date (Normalized)	Lab No	Material	Location	References/Remarks
2950 ± 90 (3030 ± 90)	I-9996	mandible	Approx. 16 km W of Goodsir Inlet on Bathurst I. (75.500°N, 99.000°W), NU, CAN	Harington (1980); R.E. Morlan, CMC database (1999)

Ovis dalli (Dall sheep)

Measured Date (Normalized)	Lab No	Material	Location	References/Remarks
>49800 (>49900)	Beta-68925 CAMS-10529	interior horncore (CMN 44711)	Sixtymile area (Sixtymile Loc. 3), 701 m asl., (64°1.00'N, 140°44'W), YT, CAN	Harington (1997); R.E. Morlan, CMC database (1999)
38260 ± 830	Beta-37057 ETH-6579	collagen	Lower Rampart Cave 1, Porcupine R (67°30'N, 142°W), AK, USA	Sattler and Dixon (1997)
23000 ± 600 (23080 ± 600)	I-4225	horncore	Bonanza Ck. (Dawson Loc. 2) (67.720°N, 139.800°W), YT, CAN	Harington (1977, 1980); R.E. Morlan, CMC database (1999)
22480 ± 200	Beta-29165 ETH-5013	collagen	Lower Rampart Cave 1, Porcupine R. (67°30'N, 142°W), AK, USA	Sattler and Dixon (1997)
3470 ± 60 (3510 ± 60)	Beta-89986	right mandible	Firth R., Engigstciak (69.360°N, 139.530°W), YT, CAN	C.R. Harington [unpublished]
2390 ± 80 (2440 ± 80)	GSC-1757	horn	Grotte Valerie: about 700 m asl., Nahanni National Park (61.280°N, 124.100°W), NT, CAN	Lowdon and Blake (1979); Scotter and Simmons (1976); R.E. Morlan, CMC database (1999)
2330 ± 150 (2370 ± 150)	GSC-1774	cartilage	Grotte Valerie: at the mouth of First Canyon, about 700 m asl., S Nahanni R., in Nahanni National Park (61.280°124.100°W), NT, CAN	Scotter and Simmons (1976); R.E. Morlan, CMC database (1999)
2330 ± 160 (2400 ± 160)	GSC-1649	hair	Grotte Valerie: about 700 m asl., Nahanni National Park (61.280°N, 124.100°W), NT, CAN	Lowdon and Blake (1979); Scotter and Simmons (1976); R.E. Morlan, CMC database (1999)
2270 ± 50 (2310 ± 50)	Beta-89987	right mandible	Firth R., Engigstciak (69.360°N, 139.530°W), YT, CAN	C.R. Harington [unpublished]

Ovis canadensis (bighorn sheep)

Measured Date (Normalized)	Lab No	Material	Location	References/Remarks
9200 ± 200 (9280 ± 200)	GSC-1497	horn	Finlay Forks: gravel pit W side of Finlay-Parsnip Access Rd., Williston L. Basin, Peace R. drainage (55.800°N, 123.630°W), BC, CAN	Rutter et al. (1972); Rutter (1977); Lowdon and Blake (1979); Clague (1980); R.E. Morlan, CMC database (1999)
(1810 ± 90)	NUTA-4687		Stirling Ck. Bridge, near the village of Hedley, Similkameen R., Okanagan drainage, BC, CAN	[Copp (1996); pers. comm. (1998) In: R.E. Morlan, CMC database (1999)]

cf. Ovis sp. (sheep)

Measured Date (Normalized)	Lab No	Material	Location	References/Remarks
(13580 ± 80)	CAMS-23473	tibia	Bluefish Cave I, 600 m asl. (67°08'N, 140°47'W), YT, CAN	Cinq-Mars (1979, 1990); Morlan and Cinq-Mars (1982); Burke and Cinq-Mars (1996)

INDEX OF SCIENTIFIC NAMES

INDEX OF COMMON NAMES

INDEX OF LOCALITIES AND STRATIGRAPHIC TERMS

INDEX OF PERSONAL NAMES AND INSTITUTIONS

INDEX OF SUBJECTS